Advances in Intelligent Systems and Computing

Volume 437

Series editor

Janusz Kacprzyk, Polish Academy of Sciences, Warsaw, Poland
e-mail: kacprzyk@ibspan.waw.pl

About this Series

The series "Advances in Intelligent Systems and Computing" contains publications on theory, applications, and design methods of Intelligent Systems and Intelligent Computing. Virtually all disciplines such as engineering, natural sciences, computer and information science, ICT, economics, business, e-commerce, environment, healthcare, life science are covered. The list of topics spans all the areas of modern intelligent systems and computing.

The publications within "Advances in Intelligent Systems and Computing" are primarily textbooks and proceedings of important conferences, symposia and congresses. They cover significant recent developments in the field, both of a foundational and applicable character. An important characteristic feature of the series is the short publication time and world-wide distribution. This permits a rapid and broad dissemination of research results.

More information about this series at http://www.springer.com/series/11156

Millie Pant · Kusum Deep
Jagdish Chand Bansal · Atulya Nagar
Kedar Nath Das
Editors

Proceedings of Fifth International Conference on Soft Computing for Problem Solving

SocProS 2015, Volume 2

Editors
Millie Pant
Department of Applied Science and Engineering
Saharanpur Campus of IIT Roorkee
Saharanpur
India

Atulya Nagar
Department of Mathematics and Computer Science
Liverpool Hope University
Liverpool
UK

Kusum Deep
Department of Mathematics
Indian Institute of Technology Roorkee
Roorkee
India

Kedar Nath Das
Department of Mathematics
National Institute of Technology Silchar
Silchar, Assam
India

Jagdish Chand Bansal
Akbar Bhawan Campus
South Asian University
Chankyapuri, New Delhi
India

ISSN 2194-5357 ISSN 2194-5365 (electronic)
Advances in Intelligent Systems and Computing
ISBN 978-981-10-0450-6 ISBN 978-981-10-0451-3 (eBook)
DOI 10.1007/978-981-10-0451-3

Library of Congress Control Number: 2016930058

Printed on acid-free paper

This Springer imprint is published by SpringerNature
The registered company is Springer Science+Business Media Singapore Pte Ltd.

Preface

It is a matter of pride that, the Annual Series of International Conference, called 'Soft Computing for Problem Solving' is entering its fifth edition as an established and flagship international conference. This annual event is a joint collaboration between a group of faculty members from institutes of repute like Indian Institute of Technology Roorkee; South Asian University, Delhi; NIT Silchar and Liverpool Hope University, UK. The first in the series of SocProS started in 2011 and was held from 20 to 22 December at the IIT Roorkee Campus with Prof. Deep (IITR) and Prof. Nagar (Liverpool Hope University) as the General Chairs. JKLU Jaipur hosted the second SocProS from 28 to 30 December 2012. Coinciding with the Golden Jubilee of the IIT Roorkee's Saharanpur Campus, the third edition of this international conference, which has by now become a brand name, took place at the Greater Noida Extension Centre of IIT Roorkee during 26–28 December 2013. The fourth conference took place at NIT Silchar during 27–29 December 2014. Like earlier SocProS conferences, the focus of SocProS 2015 is on Soft Computing and its applications to real-life problems arising in diverse areas of image processing, medical and healthcare, supply chain management, signal processing and multimedia, industrial optimisation, cryptanalysis, etc. SocProS 2015 attracted a wide spectrum of thought-provoking articles. A total of 175 high-quality research papers were selected for publication in the form of this two-volume proceedings.

We hope that the papers contained in this proceeding will prove helpful towards improving the understanding of Soft Computing at the teaching and research levels and will inspire more and more researchers to work in the field of Soft Computing. The editors express their sincere gratitude to SocProS 2015 Patron, Plenary Speakers, Invited Speakers, Reviewers, Programme Committee Members, International Advisory Committee, and Local Organizing Committee; without whose support the quality and standards of the conference could not be maintained. We express special thanks to Springer and its team for this valuable support in the publication of this proceedings.

Over and above, we express our deepest sense of gratitude to the 'Indian Institute of Technology Roorkee' to facilitate the hosting of this conference. Our sincere thanks to all the sponsors of SocProS 2015.

Saharanpur, India	Millie Pant
Roorkee, India	Kusum Deep
New Delhi, India	Jagdish Chand Bansal
Liverpool, UK	Atulya Nagar
Silchar, India	Kedar Nath Das

Conference Organising Committee

Patron
Prof. Pradipta Banerji, Director, Indian Institute of Technology Roorkee

Conference Chair
Dr. Y.S. Negi, Indian Institute of Technology Roorkee

Honaray Chair
Prof. Dr. Chander Mohan, Retired, Indian Institute of Technology Roorkee

Conveners
Prof. Kusum Deep, Indian Institute of Technology Roorkee
Dr. Millie Pant, Indian Institute of Technology Roorkee
Prof Atulya Nagar, Liverpool Hope University, Liverpool, United Kingdom

Organising Secretary
Dr. J.C. Bansal, South Asian University, New Delhi
Dr. Kedar Nath Das, NIT Silchar
Dr. Tarun Kumar Sharma, Amity University Rajasthan

Joint Organising Secretary
Dr. Dipti Singh, GBU, Greater Noida
Dr. Sushil Kumar, Amity University, Noida

Treasurer
Prof. Kusum Deep, Indian Institute of Technology Roorkee
Dr. Millie Pant, Indian Institute of Technology Roorkee

Best Paper and Best Ph.D. Thesis Chair
Prof V.K. Katiyar, IIT Roorkee
Prof S.G. Deshmukh, ABV-IITM Gwalior
Prof Atulya Nagar, Liverpool Hope University, UK

Publicity Chair
Dr. Musrrat Ali, Sungkyunkyan University, Korea
Dr. Radha Thangaraj, Indian Institute of Technology Roorkee
Dr. Rani Chinnappa Naidu, Northumbria University, UK

Social Media Chairs
Dr. Kedar Nath Das, NIT Silchar
Dr. Anupam Yadav, NIT Uttarakhand

Conference Proceedings and Printing and Publication Chair
Prof. Kusum Deep, IIT Roorkee
Dr. J.C. Bansal, South Asian University, New Delhi

Technical Sessions Chair
Dr. Kedar Nath Das, NIT Silchar, Assam
Dr. Manoj Thakur, IIT Mandi
Dr. Anupam Yadav, NIT Uttarakhand

Special Session Chair
Dr. Musrrat Ali, Sungkyunkyan University, Korea
Dr. Sushil Kumar, Amity University, Noida

Hospitality Chair
Dr. Millie Pant, Indian Institute of Technology Roorkee
Dr. Tarun Kumar Sharma, Amity University Rajasthan
Dr. Sushil Kumar, Amity University, Noida

Cultural Program and Registration Chair
Dr. Divya Prakash, AUR
Dr. Abha Mittal, CSIR-CBRI, Roorkee

Local Organising Committees

Amarjeet Singh, Indian Institute of Technology Roorkee
Asif Assad, Indian Institute of Technology Roorkee
Bilal Mirza, Indian Institute of Technology Roorkee
Garima Singh, Indian Institute of Technology Roorkee
Hira Zaheer, Indian Institute of Technology Roorkee
Kavita Gupta, Indian Institute of Technology Roorkee
Neetu Kushwaha, Indian Institute of Technology Roorkee
Renu Tyagi, Indian Institute of Technology Roorkee
Sunil Kumar Jauhar, Indian Institute of Technology Roorkee
Tushar Bharadwaj, Indian Institute of Technology Roorkee
Vanita Garg, Indian Institute of Technology Roorkee
Vidushi Gupta, Indian Institute of Technology Roorkee

Contents

About the Editors

Dr. Millie Pant is an Associate Professor with the Department of Paper Technology, Indian Institute of Technology Roorkee, Roorkee, India. She has to her credit several research papers in journals of national and international repute and is a well-known figure in the field of Swarm Intelligence and Evolutionary Algorithms.

Prof. Kusum Deep is working as a full time professor in the Department of Mathematics, Indian Institute of Technology Roorkee, Roorkee, India. Over the last 25 years, her research is increasingly well cited making her a central International figure in the area of Nature Inspired Optimization Techniques, Genetic Algorithms and Particle Swarm Optimization.

Dr. Jagdish Chand Bansal is an Assistant Professor with the South Asian University, New Delhi, India. Holding an excellent academic record, he is an excellent researcher in the field of Swarm Intelligence at the National and International Level, having several research papers in journals of national and international repute.

Prof. Atulya Nagar holds the Foundation Chair as Professor of Mathematical Sciences and is the Dean of Faculty of Science, at Liverpool Hope University, Liverpool, UK. Prof. Nagar is an internationally recognised scholar working at the cutting edge of theoretical computer science, applied mathematical analysis, operations research, and systems engineering and his work is underpinned by strong complexity-theoretic foundations.

Dr. Kedar Nath Das is now working as Assistant Professor in the Department of Mathematics, National Institute of Technology Silchar, Assam, India. Over the last 10 years, he has a good contribution towards to research in 'soft computing'. He has many papers to his credit in many journal in national and international level of repute. His area of interest is on Evolutionary and Bio-inspired algorithms for optimization.

About the Book

The proceedings of SocProS 2015 will serve as an academic bonanza for scientists and researchers working in the field of Soft Computing. This book contains theoretical as well as practical aspects using fuzzy logic, neural networks, evolutionary algorithms, swarm intelligence algorithms, etc., with many applications under the umbrella of 'Soft Computing'. The book will be beneficial for young as well as experienced researchers dealing across complex and intricate real-world problems for which finding a solution by traditional methods is a difficult task.

The different application areas covered in the proceedings are: Image Processing, Cryptanalysis, Industrial Optimisation, Supply Chain Management, Newly Proposed Nature-Inspired Algorithms, Signal Processing, Problems related to Medical and Health Care, Networking Optimisation Problems, etc.

Traffic Accident Prediction Model Using Support Vector Machines with Gaussian Kernel

Bharti Sharma, Vinod Kumar Katiyar and Kranti Kumar

Abstract Road traffic accident prediction models play a critical role to the improvement of traffic safety planning. The focus of this study is to extract key factors from the collected data sets which are responsible for majority of accidents. In this paper urban traffic accident analysis has been done using support vector machines (SVM) with Gaussian kernel. Multilayer perceptron (MLP) and SVM models were trained, tested, and compared using collected data. The results of the study reveal that proposed model has significantly higher predication accuracy as compared with traditional MLP approach. There is a good relationship between the simulated and the experimental values. Simulations were carried out using LIBSVM (library for support vector machines) integrated with octave.

Keywords Artificial neural networks · Accident characteristics · Data mining · Road traffic safety

1 Introduction

In today's scenario transportation is both a boom and curse. It has reduced human effort by providing comfort but with the same pace it became the major source of loss of human lives through accidents. According to the World Road Statistics

Bharti Sharma (✉)
College of Engineering Roorkee, Roorkee, India
e-mail: mbharti2000@gmail.com

V.K. Katiyar
Department of Mathematics, Indian Institute of Technology Roorkee, Roorkee, India
e-mail: vktmafma@iitr.ac.in

Kranti Kumar
School of Liberal Studies, Ambedkar University Delhi, Lothian Road, Kashmere Gate, Delhi, India
e-mail: kranti31lu@gmail.com

M. Pant et al. (eds.), *Proceedings of Fifth International Conference on Soft Computing for Problem Solving*, Advances in Intelligent Systems and Computing 437, DOI 10.1007/978-981-10-0451-3_1

(WRS) 2010 [1] compiled by the International Road Federation (IRF), Geneva, USA, maximum number of injuries/accidents stands at 16,30,000 in the world for the year 2008. The highest mortality by road accidents in the world in 2008 was reported in India (1,19,860) followed by China (73,484). The investigation of road accidents data received from States/Union Territories (UT) in India is presented in Table 1.

Information about those traffic parameters is essential which are basically responsible for accidents in order to identify more accident prone places/spots on a particular road/highway. Because of various factors, it is tough to process large quantities of traffic accident data efficiently. Hence, previous traffic studies are reviewed, and SVM which has got the attention as a data mining technique, is preferred. The SVM is used in the investigation of traffic accident data. The performance of SVM is examined in this study.

The paper is organized as follows: In Sect. 2, we present survey of the previous studies reported in literature. Sections 3 and 4 describe briefly the theories related to MLP and SVM, respectively. Model development is presented in Sect. 5. Section 6 consists of analysis of results obtained through MLP and SVM. In the last section conclusion has been given for the current study.

2 Literature Review

Traffic safety and accident studies have been given a great importance in research for past many years. Extensive investigations have been done caused by frightening rate of increase of accidents across the world. However, many researches have been done in this area, but mostly investigation was on the crash severity studies which do not emphasis on the factors and causes of the accidents. La Scala et al. [2] presented in their studied that pedestrian injury rates are correlated to certain factors like gender, age, population density, composition of the local population, unemployment, traffic flow, education, etc. Additionally author noticed that males are involved in pedestrian crashes more than female [2]. McCarthy investigated the

Table 1 Statistics of factors responsible for road accident [1]

Factors responsible for road accident	% of the factor
Drivers negligence	78
Cyclists flaw	1.2
Pedestrians lapse	2.7
Blot in road conditions	1.2
Defect in motor vehicle	1.7
Weather condition	1
All other causes (Includes fault of other drivers and light conditions and other issues)	14.2

impact of speed limits of motor vehicle on the speed and safety of highway and noticed that small changes in the speed limit on non-limited-access roads will have a little impact on speed distribution and highway safety unless accompanied with speed-reducing activities [3]. Mungnimit and Bener gave the form of road traffic accidents and their reasons in developing countries. Thailand and State of Qatar was taken as study areas for the study and investigated that the key reason for traffic accidents was careless driving and the majority of accident victims were in the age between 10 and 40 years [4, 5]. Yau studied five sets of variables in Hongkong that caused fatality of motor cycle accidents, namely region, traffic condition, human, vehicle, and environment. It was found that the day of the week and the time of the accident find the severity of crashes. In a connected study, Yau et al. gave that lighting condition at night was related to severe vehicle crashes [6].

Accidents are increasing in India, but very few researches had investigated on its reasons. Some factors have been investigated about accidents in India. Cropper and Kopits have predicted that fatalities in India would extent a total of about 198,000 beginning of 2042 [7]. Dinesh Mohan [8] investigated that road traffic fatalities have been growing 8 % annually and showing no signs of decreasing.

Adolescents are the focal sufferers of accidents; therefore reasons of their mishaps are desired to be investigated. Many approaches are to be implemented to stop accidents of teens. Stutts et al. observed that while younger drivers may not be more likely to use entertainment devices while driving than older drivers, younger drivers are less skilled at multitasking while driving and are therefore more easily become unfocused than their older counterparts [9]. Lin and Fearn analyzed the factors that teenage drivers at greater risk of accident while driving late at night or driving in the company of teenage passengers [10]. Shults and Compton showed that teenage drivers generally drive too fast for usual conditions [11].

There are numerous models dealing with accident prediction, crash severity, and traffic safety, but among them most common are the regression and logit models. Lajunen et al. [12] presented the literature on the use of the driver behavior questionnaire by documenting the occurrence of damages instead of errors in the occurrence of crashes. Chang and Yeh [13] showed that negligence of potential risk and damage are related with improved accident rate. Relation between age and driving behavior was also examined because younger drivers incline to show dangerous activities in developing countries.

3 Multilayer Perceptron

Multilayer perceptron (MLP) network is one of the frequently used architectures of artificial neural networks (ANN). MLP is a type of feed forward network. The information flows from input layer and goes through the hidden layers to the output layer and finally produces the outputs. Many researchers have shown that MLP with one hidden layer is capable of approximating nonlinear functions with very high accuracy [14].

Each layer of network comprises neurons that are processing elements (PEs). Three-layer network architecture is shown in Fig. 1. Each neuron of any layer connected with other neurons of next/previous layers through links called "weight coefficients." Any modification in coefficients changes the function of the network.

In fact, the main aim of the network training is to find the best weight coefficients to get the desired output. Levenberg Marquardt is learning technique used in the current study. In training, the inputs of the first layer multiplied by weight coefficients (that can be any randomly chosen number generally very small) are transferred to the second layer neurons. Neuron works in two ways:

1. Computing the sum of the inputs, defined as net_i

$$\mathrm{net}_i = \sum w_{ij} x_i \tag{1}$$

2. Introducing the sum in a function called "activation functions."

$$\mathrm{out}_i = F(\mathrm{net}_i) \tag{2}$$

This process repeats in the rest neurons of the hidden layer and lastly the outputs are produced in the last layer. Sigmoid activation functions have been used in the current study.

Figure 2 shows the working of a neuron in a network.

In the current study, the output is accident prediction. During learning, the error between network output and desired output is computed and sent from the output layer to the prior one and therefore the weight coefficients modified using Eqs. 3, 4.

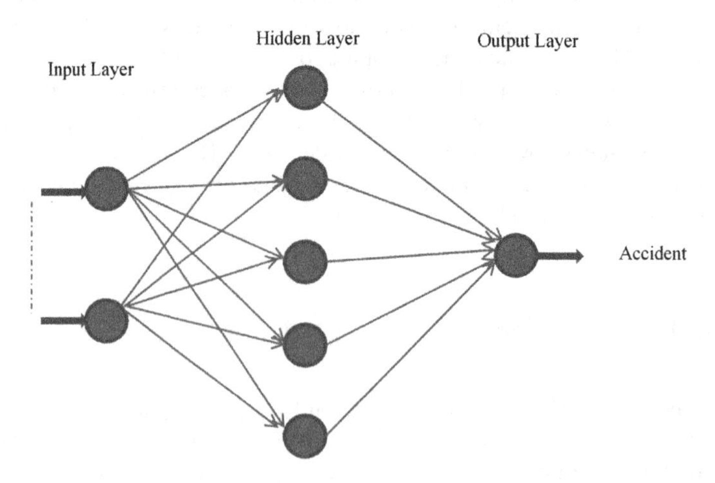

Fig. 1 Three-layer networks

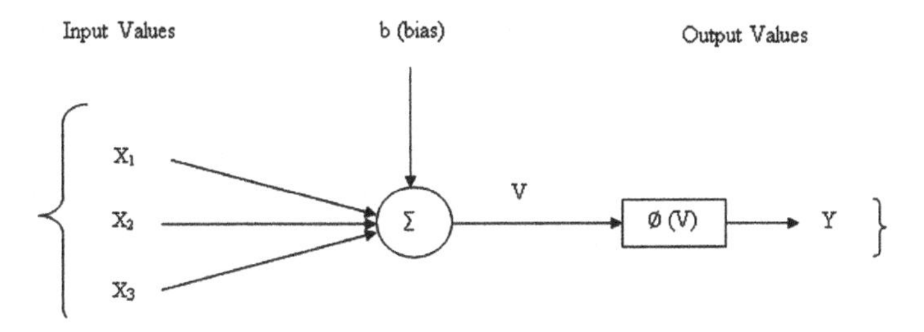

Fig. 2 The neuron cell. *Y* Neuron output, *b* bias, ϕ activation function, *X1*, *X2*, *X3* neuron input, *V* weighted sum of input variables

This method is known as error backpropagation. Once again the network produces output, using the new weight coefficients and also computes the decreasing error and backpropagates it into the network until the error gets to its least value, that is, the desired value.

$$w_{ij}(t+1) = w_{ij}(t) + \eta\delta_{pi}o_{pj} \tag{3}$$

$$\delta_{pi}w_{ij}(t+1) = w_{ij} + \eta\delta_{pi}o_{pj} + \alpha[w_{ij}(t) - w_{ij}(t-1)] \tag{4}$$

where

$w_{ij}(t+1)$	Weight coefficient at time $t + 1$, from neuron$_i$ to neuron$_j$
$w_{ij}(t)$	Weight coefficient at time t from neuron$_i$ to neuron$_j$
η	Learning coefficient
δ_{pi}	Difference between desired output and network output in neuron p of layer i
o_{pi}	Output of neuron p of layer j
p_i	Output of neuron p of layer i
α	Momentum coefficient
$w_{ij}(t-1)$	Weight coefficient at time $t - 1$ from neuron i to neuron j

4 Support Vector Machine

Support vector machines (SVMs) are supervised learning models with related learning methods that investigate data and identify data samples used for classification. A set of training data, each labeled for belonging to one of two classes, an SVM training algorithm constructs a model that labels new data set into one class or the other, constructing it a non-probabilistic binary linear classifier. An SVM model is a form of the samples as points in space, plotted so that the samples of the distinct classes are separated by a clear gap that is as widespread as possible. New samples are then plotted into that same space and projected into a class created on whose side of the gap they fall. Moreover to perform linear classification, SVMs can powerfully do a nonlinear classification with the kernel trick, by mapping inputs data into high-dimensional feature spaces [15].

A classification function includes separating data into training and testing sets. Each sample in the training set consists of one target value and several features. The objective of SVM is to create a model using the training data and predicts the target values of the test data.

Assume a training sets (x^i, y^i); $i = 1 \ldots p$, where $x_i \in R^n$ and y $\in (1, -1)^p$, we need to solve the following optimization problem using support vector machines [15, 16]:

$$\min_{w,b,\varepsilon} \frac{1}{2} w^T w + C \sum_{i=1}^{n} \xi_i \quad (5)$$

subject to

$$y_i(w^T \phi(x_i)) + b) \geq 1 - \xi_i, \; \xi_i \geq 0 \quad (6)$$

Here training sets x_i are mapped into a higher dimensional space by applying function Ø. SVM discovers a linear separating hyperplane with the greatest margin in this higher dimensional space. The error term is defined by $C > 0$.

Additionally, $A(x_i, x_j) = \phi(x_i)^T \phi(x_j)$ is known as the kernel function.

But new kernels has been introduced by researchers. The following four fundamental kernels are:

Linear:

$$A(x_i, x_j) = x_i^T x_j \quad (7)$$

Polynomial:

$$A(x_i, x_j) = (\gamma x_i^T x_j + r)^d, \gamma > 0 \quad (8)$$

Gaussian (Radial basis function (RBF))

$$A(x_i, x_j) = \exp\left(-\gamma \frac{||x_i - x_j||^2}{2\sigma^2}\right), \gamma > 0 \quad (9)$$

Sigmoid:

$$A(x_i, x_j) = \tanh(\gamma x_i^T x_j + r) \quad (10)$$

Here, γ, r, and d are kernel parameters [17].

The Gaussian kernel has been used in the current study.

5 Model Development

Simulations were carried out to predict the traffic accidents using the most critical factors responsible for the accident using SVM and MLP. Comparative analysis was done to find which technique gives better results. Initially various factors such as age, alcohol/drugs, sleepiness, speed, vehicle condition, and weather condition were taken into consideration but after sensitivity analysis it was observed that there are two key factors, i.e., speed and alcohol which play major role in accidents. Both models considered these two factors as input parameters. We also analyzed their effect to accurately predict the accidents.

5.1 Accident Prediction Model Based on Neural Network

Figure 3 shows the architecture of the proposed model. Following units have been used in the proposed model:

1. three units in input layer ($x(0)$ is biased term)
2. two units in hidden layer
3. one unit in output layer

   x/X = this is the set of features

a. $X(1)$ = alcohol content
b. $X(2)$ = driving speed

Initially weights were chosen at random and were learned though backpropagation.

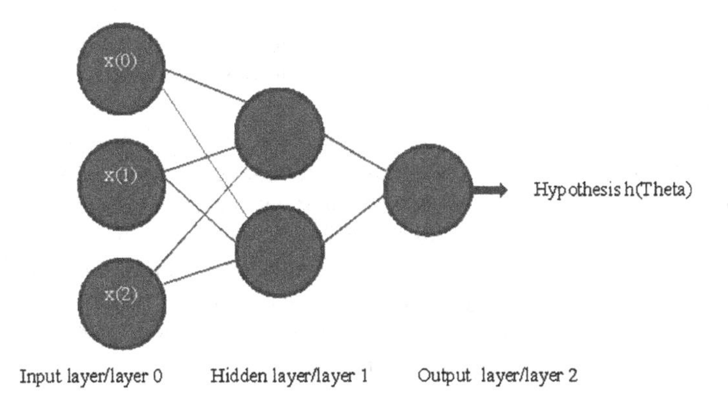

Fig. 3 Multi layer perceptron model for accident prediction

5.2 Support Vector Machine with Gaussian Kernel

LIBSVM (library for support vector machines) has been used in this study. This package has been integrated in octave and the data set has been used in this package. Data set consists of two features: alcohol content and vehicle speed. All the data sets have been generated from the questionnaire filled by different kinds of people like drivers, pedestrians, etc. In this way, a set of 300 data was considered to train MLP and SVM. Whole data were divided in three parts: for training (70 %), cross-validation (20 %), and testing (10 %).

6 Results and Discussion

6.1 Results with MLP

Result shows that MLP was not able to perform effectively. The underfitting or high bias of data has made the learning approach of algorithm poor. The major reason of underfitting is that the number of features is less. The accuracy measured by MLP is approximately 60 %. It shows that MLP model is not able to predict data close to the actual value prediction.

6.2 Results with SVM

SVM has been used to analyze the effect of these two factors (alcohol and speed) in the prediction of accidents. The networks were trained using the training set. Then the cross-validation set (CV) is used to generalize the parameters by plotting error curves called the learning curves and this shows that how well the algorithm is generalizing over the new data sets. Finally, the algorithm is tested over the test set to check for correctness and the plot is obtained to observe the effectiveness of the algorithm.

The accuracy of the SVM model over the test set is 94 %. The SVM with Gaussian kernel is clearly able to divide a large margin boundary between the data set which gives the best fit of the data set and conclusions can be drawn from the Fig. 4. In Fig. 4, speed has been normalized as corresponding to the level of alcohol content. The graph shows that a person with high alcohol consumption driving at a high speed has more probability of meeting an accident. The upper right corner of the graph shows the regions of high alcohol content and high speed. The accident-occurring region is clearly distinguished by the SVM, which gives better result than logistic regression.

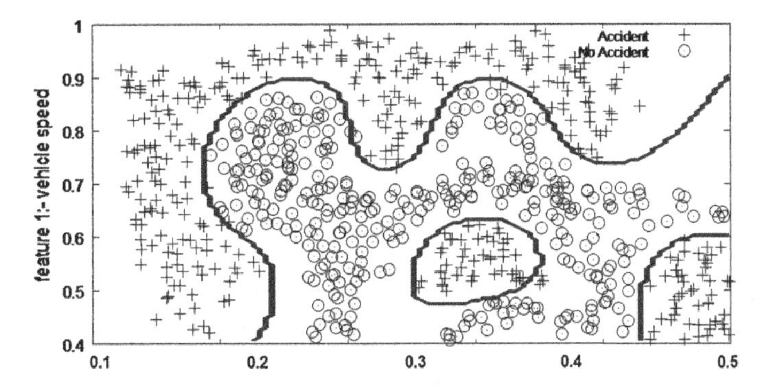

Fig. 4 Graph support vector machine using Gaussian kernel

Table 2 SVMs with different kernels and MLP

MLP	Poly kernel	Gaussian (RBF) kernel
62 %	78 %	94 %

6.3 *Comparative Analysis*

To check the performance of the SVM with different kernels, some tests with same data were done. The performance evaluation of SVM using different kernel and the multilayer neural networks was done by applying the following equation:

$$\text{performance} = \frac{a}{A} \times 100 \quad (11)$$

where a is correct classification of data and A is total number of elements in data set.

In the current study, MLP network is trained using Levenberg–Marquardt algorithm and SVM using POLY and RBF. The training time of MLP was higher than SVM. Table 2 shows the classification accuracy of SVM with different kernel versus the MLP.

7 Conclusion

The current study shows SVM with Gaussian kernel has a higher computation advantage over other kernels and traditional MLP. The study shows that SVM gives much better results than traditional approaches. The driving after getting drunk and at high speed is found to be serious factors in accidents after sensitivity analysis. Less number of features is found to be the main reason for the poor performance of neural network, because as the number of features is less that MLP is not able to fit the parameters well and hence underfitting or high bias occurs. Authors believe that present study will help to reduce number of accidents by strengthening the crackdown on the drunk driver as well those caught moving with high speed.

References

1. Ministry of Road Transport & Highways India 15 Dec 2011 16:26 IST
2. LaScala, E., Gerber, D., Gruenwald, P.J.: Demographic and environmental correlates of pedestrian injury collisions: a spatial analysis. Accid. Anal. Prev. **32**, 651–658 (2000)
3. McCarthy, P.: Effect of speed limits on speed distribution and highway safety: a survey of recent literature. Transp. Rev. **21**(1), 31–50 (2001)
4. Mungnimit, S.: Road Traffic Accident Losses. Transport and Communications Policy and Planning Bureau, Ministry of Transport and Communications, Thailand (2001)
5. Bener, A.: The neglected epidemic: road traffic accidents in a developing country, State of Qatar. Int. J. Inj. Control Saf. Promot. **12**(1), 45–47 (2005)
6. Yau, K.K.W.: Risk factors affecting the severity of single vehicle traffic accidents in Hong Kong. Accid. Anal. Prev. **36**(3), 333–340 (2004)
7. Cropper, M.L., Kopits, E.: Traffic fatalities and economic growth. World Bank Policy Research Working Paper, vol. 3035, pp. 1–42 (2003)
8. Mohan, D.: Road accidents in India. Transportation Research and Injury Prevention Programme, Indian Institute of Technology Delhi, New Delhi, India (2009)
9. Stutts, J.C., Reinfurt, D.W., et al.: The role of driver distraction in traffic crashes. AAA Foundation for Traffic Safety, Washington, DC (2001)
10. Lin, M.-L., Fearn, K.T.: The provisional license: nighttime and passenger restrictions—a literature review. J. Saf. Res. **34**, 51–61 (2003)
11. Shultz, R., Compton, R.: What we know, what we dont know, and what we need to know about graduated driver licensing. J. Saf. Res. **34**, 107–115 (2003)
12. Lajunen, T., Parker, D., et al.: The manchester driver behaviour questionnaire: a cross-cultural study. Accid. Anal. Prev. **36**(2), 231–238 (2004)
13. Chang, H.L., Yeh, T.H.: Motorcyclist accident involvement by age, gender, and risky behaviors in Taipei, Taiwan. Transp. Res. Part F: Psychol. Behav. **10**(2), 109–122 (2007)
14. Schalkoff, R.J.: Artificial neural networks, pp. 146–188. Mcgrawhill, New York (1997)
15. Baby Deepa, V.: Classification of EEG data using FHT and SVM based on bayesian network. Int. J. Comput. Sci. Issues (IJCSI)
16. Boser, B.E.: The support vector machines (SVM) (1992)
17. Hsu, C.W., Chang, C.C., et.al.: A practical guide to support vector classification. Technical report, Department of Computer Science, National Taiwan University (2003)

A Brief Overview on Least Square Spectral Analysis and Power Spectrum

Ingudam Gomita, Sandeep Chauhan and Raj Kumar Sagar

Abstract LSSA, i.e., least square spectral analysis method can be used to calculate the power density function of a fully populated covariance matrix. It fulfills the limitation cause by the fast Fourier transform (FFT), i.e., equally spaced and equally weighted time series. This time of experimental time series used LSSA (which used projection theorem) so that it can also analyze the time series, which is unequally spaced and unequally weighted. Signal-to noise ratio (SNR) is used to find out the probability density function.

Keywords LSSA · Power density function · Signal-to-noise ratio · Projection theorem

1 Introduction

An engineer or scientist cannot be guaranteed that they have an experimental time series with no gaps. Time series can be rendered due to machine failure or may be due to the weather or other natural factors. So the main problem is how to handle a time series when a certain time series has gaps or datum shifts, etc. The obtained time series from an experiment say weather observation is unequally spaced or to avoid aliasing, a variable sampling rate has been introduced. Besides, the instrument development advances and physical phenomenon is understood better so that data accuracy is improved.

Ingudam Gomita (✉) · Sandeep Chauhan · R.K. Sagar
Amity University, Noida, India
e-mail: Ingudamgomita007@gmail.com

Sandeep Chauhan
e-mail: chauhansandee@gmail.com

R.K. Sagar
e-mail: rksagar@amity.edu

M. Pant et al. (eds.), *Proceedings of Fifth International Conference on Soft Computing for Problem Solving*, Advances in Intelligent Systems and Computing 437, DOI 10.1007/978-981-10-0451-3_2

Well a fast Fourier transform (FFT) is normally used to determine power spectrum, but ironically FFT can handle only the time series, which are equally spaced, and does not any datum shift. Because of this limitation FFT cannot be used in astronomical time series as astronomical observations are implicitly unequally spaced and have datum shifts. Additionally, disturbances can originate from instrument failure or replacement or repair may cause datum shift in the time series. And so, instead of using FFT we use least space spectral analysis (LSSA) which was invented by Vanicek [1, 2]. And it is used as an alternative to FFT.

LSSA removes all the limitations of FFT, i.e., equally spaced, datum shift, etc. and so LSSA is more flexible than FFT. One of the important factors in LSSA is the covariance matrix, which should be fully populated, which means values may or may not be fully independent. The covariance matrix should be absolute, meaning a priori variance should be known.

2 Least Square Spectral Analysis

A time series comprises of a signal and a noise. A noise can be colored (periodic noise) [3] or nonstationary. LSSA can handle this kind of time series, which is impossible for FFT. FFT always needs a stationary time series with equal space and no datum shifts.

A noise can always distort a signal and cause datum shifts.

Advantages of using LSSA are as follows: LSSA can suppress the noise without showing any significant shift in the spectral peaks.

1. Unequally spaced time series can be used and do not need to be preprocessed.
2. Time series, which has covariance matrix, can be used.
3. Significant peaks of a time series can be found out.

3 Principles of LSSA

The principle that guides LSSA is the projection theorem in n-dimensional space. By considering the plane to be in a sub-space and then by applying projection theorem we have the shortest distance of a plane and a point in the perpendicular distance between the two, i.e., angle should be 90°.

3.1 *Projection Theorem*

Considering the space to be in Hilbert space (H) [4]. And point f to be a point in the time series $f(t)$, where $f \in H$ and $S \in H$, where S is the sub-space. Minimum or

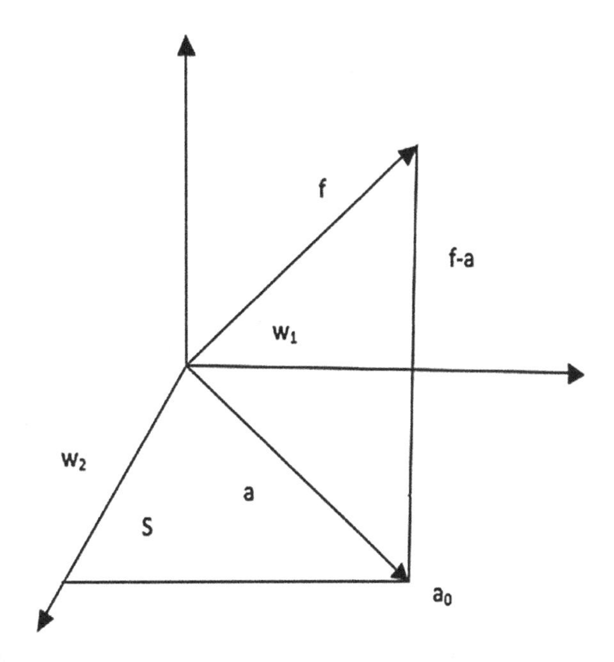

Fig. 1 Projection theorem

shortest distance(s) is achieved by using orthogonal projection as shown in Fig. 1 and a_o is the projection point on S such that $d(f, a_o) \leq d(f, a)$ and $(f - a_o) \perp S$ [4]. By assuming A consists of W_i where W_i is the basic function, then

$$a_o = \sum_i C_i W_i$$

3.2 *Second Projection Theorem*

In second projection as shown in Fig. 2 corresponding with Fig. 1 we can project a back onto f, such that length is given by

$$\frac{\langle f, a \rangle}{\|f\|}$$

3.3 *Mathematical Calculation*

Let $f(t)$ be a time series and belong to Hilbert space H [4]. The observed time series are equally spaced and its value is observed at t_i where $i = 1, 2, 3, 4, \ldots, m$. Let C_f be fully populated covariance matrix and C_f matrices H.

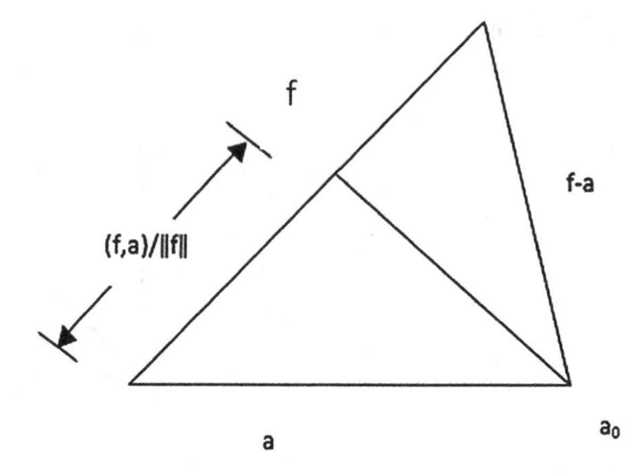

Fig. 2 Second projection theorem

The time series $f(t)$ contains both signal and noise and so the time series f can be represented as

$$g = Wx$$

W is Vander mode matrix [4] and x is the vector of unknown parameters where $W = W_s/W_n$, where W_s and W_n are functional forms and W and x are arbitrary.

And so we need to find the difference between g and f by using least square method. This difference is the residual.

$$\hat{r} = f - \hat{g}$$

where g is the model time series. Figure 2 represents the relationship between f, a, S and $\{W_i, \forall i\}$. any point $s \in S$ [4] can be represented as

$$s = \sum_i C_i W_i \tag{1}$$

And so far n tuple $\{C_i, \forall_i\}$

$$a = \sum_i \hat{C}_i W_i \tag{2}$$

Now we have the condition [4],

$$(f - a) \perp s \equiv (f - a) \perp W_j \forall_j \tag{3}$$

In terms of $\{\hat{C}_i, \forall I\}$ corresponding to a,

$$\forall_j : \langle (x - \sum_i \hat{C}_i W_i), W_j \rangle = 0$$

This can be written as

$$\sum\nolimits_{i} \hat{C}_i \langle W_i,\, W_j \rangle = \langle f,\, W_j \rangle \qquad j = 1, 2, 3 \ldots n \tag{4}$$

We can define

$$N = \begin{bmatrix} \langle W_1,\, W_1 \rangle \langle W_2,\, W_1 \rangle \ldots \langle W_n,\, W_1 \rangle \\ \langle W_1,\, W_2 \rangle \langle W_2,\, W_2 \rangle \ldots \langle W_n,\, W_1 \rangle \\ \ldots \\ \ldots \\ \langle W_1,\, W_n \rangle \langle W_2,\, W_n \rangle \ldots \langle W_n,\, W_n \rangle \end{bmatrix}$$

$$U = \begin{bmatrix} \langle f,\, W_1 \rangle \\ \langle f,\, W_2 \rangle \\ \vdots \\ \langle f,\, W_n \rangle \end{bmatrix}$$

$$\hat{C} = \begin{bmatrix} \hat{C}_1 \\ \hat{C}_2 \\ \vdots \\ \vdots \\ \vdots \\ \hat{C}_n \end{bmatrix}$$

Then Eq. 4 becomes

$$N\hat{C} = \mathrm{u} \tag{5}$$

Since $f - a \perp S$, then normal equation

$$\begin{aligned} N &= W^T W \text{ and} \\ U &= W^T f \end{aligned}$$

Equation 5 becomes

$$\begin{aligned} \hat{C} &= N^{-1} u \\ &= \left(W^T W\right)^{-1} W^T f \end{aligned} \tag{6}$$

And from Eq. 2

$$\mathrm{A} = W\hat{C} = \sum_i W_i \hat{C}_i$$

Therefore, residual

$$\begin{aligned} r &= f - a \\ &= f - W\hat{C} \\ &= f - W\left(W^T W\right)^{-1} W^T f \end{aligned} \tag{7}$$

So by using the projection theorem,

$$\begin{aligned} &r = f - a \\ &\text{so } \hat{r} \perp a \text{ and } r = f - a \perp S \end{aligned}$$

so we can say that $r \perp g$ [4], i.e., they are orthogonal (from Fig. 1). This means the signal x is divided into noise, i.e., residual r and signal g.

Consider Fig. 2, we project a_o to x to know that how much of a_o is present in x [3].

$$\begin{aligned} S &= \langle f, \hat{g} \rangle / |f| / |f| \\ &= \langle f, \hat{g} \rangle / |f|^2 \\ &= f^T C_f^{-1} \hat{g} / f^T C_f^{-1} f \quad \epsilon\,(0, 1) \end{aligned} \tag{8}$$

Signal can be expressed in the form of sine and cosine in spectral analysis, i.e., W can be expressed as

$$W = [\cos \theta_i t,\ \sin\theta_i t], I = 1, 2, 3 \ldots k \tag{9}$$

For each angle there is different spectral value and so LSSA can be represented as

$$S(\theta_i) = f^T C_f^{-1} \hat{g}(\theta_i) / f^T C_f^{-1} f, \quad I = 1, 2, 3, \ldots k \tag{10}$$

This above equation shows it depends on the trigonometric function θ.

Reexamining Eq. (1) and matrix W. W is partitioned into W_S and W_N where W_S is for signal and W_N is for noise. From Eq. (4) we know that this θ is used to describe the components of the series, which is periodic. In least square solution which is used in least square spectral method depends on the parameters. These parameters are driven by noise and can change their phase and amplitude due to the noise. And so, we need to find out the pdf (partial distribution function) of the least square spectrum.

4 Revisiting Least Square Spectrums

$$\hat{u} = (I - K) f \tag{11}$$

where I is the identity matrix and

$$K = W\left(W^{T}C_{f}^{-1}W\right)^{-1}W^{T}\,C_{f}^{-1}$$

Substituting Eq. (11) in (10)

$$S = f^{T}\,C_{f}^{-1}K_{f}/f^{T}\,C_{f}^{-1}f \tag{12}$$

To derive the pdf equation, the two equation should be statistically independent but in the Eq. (12) the two equations are not statistically independent as $C_{f}^{-1}JC_{f}C_{f}^{-1} \neq 0$ (Lemma 3) [3].

By rearranging the denominator of the equation we have

$$f^{T}C_{f}^{-1}f = f^{T}C_{f}^{-1}Jf + f^{T}C_{f}^{-1}(I-K)f \tag{13}$$

The denominator is decomposed into two terms. The first term is the signal and the second term is the noise. Now Eq. 13 is statistically independent.

Using Eq. (13) in (12) and then rearranging it we get

$$\begin{aligned} S &= \left[1 + f^{T}\,C_{f}^{-1}(I-K)f/f^{T}\,C_{f}^{-1}Jf\right]^{-1} \\ &= [1 = q_{n}/q_{s}] \end{aligned} \tag{14}$$

Equation (14) is a function of two quadratic equations q_s and q_n which are statistically independent and the ratio gives the least square spectrum. It is inverse of signal to noise ratio (SNR). The above Eq. (14) can be used to calculate the PSD (power spectral density) by applying logarithm function. The PSD is given by the equation

$$\mathrm{PSD} = 10\log[S/1 - S] \tag{15}$$

This PSD that is given by Eq. (15) is equivalent to the PSD of the FFT method only when the time series is equally weighted and equally spaced. So FFT, PSD is not useful for time series with datum shift.

But the above PSD given by Eq. (15) can be used to calculate any time series even if it is not equally spaced.

5 Conclusion

So by using least square method we can eliminate the limitation of FFT. So for analysis time series with datum biased and unequally spaced value least square spectral method is used. The spectral density is same as the spectral density of the FFT but the only difference is that it can be used for time series with datum shifts and unequally spaced.

References

1. Vanicek, P.: Approximate spectral analysis by least-square fit. Astrophys. Space Sci. **4**, 387–391 (1969)
2. Vanicek, P.: Further development and properties of the spectral analysis by least-squares. Astrophys. Space Sci. **12**, 10–33 (1971)
3. Pagiatakis, S.D.: Application of the least-squares spectral analysis to superconducting gravimeter data treatment and analysis. Cahiers du Centre Europeen de Geodynamique et Seismologie **17**, 103–113 (2000)
4. Wells, D., Vanicek, P., Pagiatakis, S.: least square spectral analysis revisited (1985)

ASCCS: Architecture for Secure Communication Using Cloud Services

Vaibhav Sharma and Kamal Kumar Gola

Abstract Security is an important aspect in an open environment. When two or more entities want to communicate, an entity must be aware about the other entity to whom he/she is communicating. Therefore there has to be a mechanism which will ensure the identities of different entities. As cryptanalysts left no stone unturned to decipher the code, therefore there has to be a mechanism to ensure that only authorized entity will be able to get the encryption−decryption code and it should be available only for limited time or on temporary basis so that no one should have time to analyze the code. As code will only be available to authorized entities, they should be able to get it on any machine. The proposed architecture will cover all the given aspects and it will provide authentication, confidentiality, availability as software as a service.

Keywords Authentication · Confidentiality · Availability · Time bound communication · SAAS

1 Introduction

Authentication, confidentiality, time bound information, and availability are the major issues in secure communication. If A wants to send some confidential information to B and B does not pay attention to the message of A because he is not sure about the identity of A, there may arise a problem because that information may be very important and B is not paying any attention to it. In the second scenario, A wants to send some time bound information to B. B is well aware of A but he does

Vaibhav Sharma (✉) · K.K. Gola
College of Engineering, Teerthanker Mahaveer University, Moradabad, India
e-mail: vaibhavaatrey@gmail.com

K.K. Gola
e-mail: kkgolaa1503@gmail.com

M. Pant et al. (eds.), *Proceedings of Fifth International Conference on Soft Computing for Problem Solving*, Advances in Intelligent Systems and Computing 437, DOI 10.1007/978-981-10-0451-3_3

not know that the information has to be decrypted before specific time, otherwise after specific time that information will be of no use. Due to some reason if B fails to decrypt the information on time, that communication will be pointless. In the third scenario, if A wants to communicate with B and B is aware of A and B knows that the information has to be decrypted before specific time but B is at a place where he is not able to use his personal system on which encryption/decryption algorithm is implemented. The communication will be of no use. This paper proposes an architecture where all the three scenario problems will be entertained. It will provide a software as a service. As only authorized entities will get access to the service, authentication will be there, entities will be able to get the encryption and decryption code for a limited time, so cryptanalysts are left with minimum window size to crack the code. Encryption and decryption keys will be shared on an open network using "Key sharing technique to preserve confidentiality and Integrity," [1] which will ensure the secure sharing of key.

2 Literature Review

In [1] the authors propose a technique to exchange key securely on an unsecured network,where key size is 512 bits, key is shared in two parts, and each part has 256 bit key code and 256 bit hash code of the other part to ensure integrity. As each 512 bit part is encrypted using proper steps key sharing is secure over public network.

In [2] the author proposes a service for secure communication. However, it has many disadvantages such as keys should be as per public key infrastructure, users can not use keys on mutual consensus, communication is restricted within the intranet, and receiver has to be attentive all the time because there is no prior information regarding communication. There is no concept of deadline regarding message validity.

In [3] the authors discuss the various services of cloud computing and the different security concerns of cloud computing. The proposed approach ensures secure communication in cloud environment. There should not be total dependency on the cloud services to provide security.

In [4] the authors propose a technique for secure communication over unsecured network. But they do not ensure the authenticity of users. Security of communication depends on a single thing, i.e., key, which can be hazardous when key is being compromised.

In [5] the authors discuss the Kerberos authentication protocol and its operation in wireless communication networks. The author analyzes the authentication methods for wireless communication. The main issue was the use of public key infrastructure for authentication process. In the case of shared key schemes, compromising any host or key may compromises the entire system.

In [6] the authors search for structural limitations of Kerberos. The limitations are: its acquaintance with the dictionary attack, and the time synchronization assumption. If there is a problem of time synchronization, ticket replay attack is possible. The structural limitations of Kerberos cannot be surmounted without designing a new authentication system. Kerberos also miss distributed services.

The following points emerge that need to be solved:

1. Sender must inform the receiver in advance that there will be a communication (to avoid the expiry of message validity).
2. Common repository should be there to authenticate the users.
3. Encryption and decryption algorithms should be secret and available only to valid users for a limited period of time.
4. Total dependency should not be there on cloud server.
5. Key should only be known to sender and receiver.

3 Proposed Architecture

In this architecture the following entities are involved:

1. Software server
2. Sender
3. Receiver
4. Mailing server
5. Sender's mobile phone
6. Sender's system
7. Receiver's mobile phone
8. Receiver's system

Software server has the following information about its users.

1. User ID
2. Password
3. User mobile number

It also has the encryption and decryption code which will be available to the users on demand and only for a limited period of time.

The user has the following information:

1. His own ID
2. ID of the user he wants to communicate
3. Shared key
4. Email ID of the user he wants to communicate

Users can use any mailing server for communication.

3.1 Algorithm

1. Sender first logs on to the server using provided credentials.
2. After that sender will provide receiver's ID and the deadline before which the message should be decrypted.
3. After checking all the details, i.e., receiver's ID is valid and time is greater than current time, the server will provide the encryption code to the sender. Meanwhile, the server will send an SMS to the receiver which consists of sender ID and deadline.
4. After getting the delivery report from the receiver the server will send an SMS to the sender regarding acknowledgment.
5. After getting the encryption code the sender will encrypt the plain text to cipher text using the key. Key is shared among users using "Key Sharing Technique to Preserve Integrity and Confidentiality."
6. After getting the encrypted code the sender will take that code and send to the receiver using any mailing server.
7. At receiver's end, the receiver will log onto the server using provided credentials.
8. After logon the receiver will provide the sender ID and deadline to the server. After confirming the provided details, the server will provide the decryption code. If current time exceeds the deadline the server will deny to provide the decryption code. Contrarily, if the sender allows for flexible deadline then the server will provide the decryption code.
9. After getting the decryption code, the receiver will provide the shared key to decrypt the message.
10. Now the receiver is with decrypted code.

3.2 Key Sharing Process

1. Here sender will opt for a key of 512 bits and divide the key into two parts and follow the steps of algorithm proposed in "Key Sharing Technique to Preserve Integrity and Confidentiality" [1].
   Now the sender will send the two parts of the key via unsecured network (Fig. 1).
2. Here the receiver will receive the two parts of the key and apply the decryption steps to get the original key [1].

Now both the sender and receiver will have the same secret key. Sender and receiver have to take care of the key and use it when it is required [1].

Fig. 1 Key sharing [1]

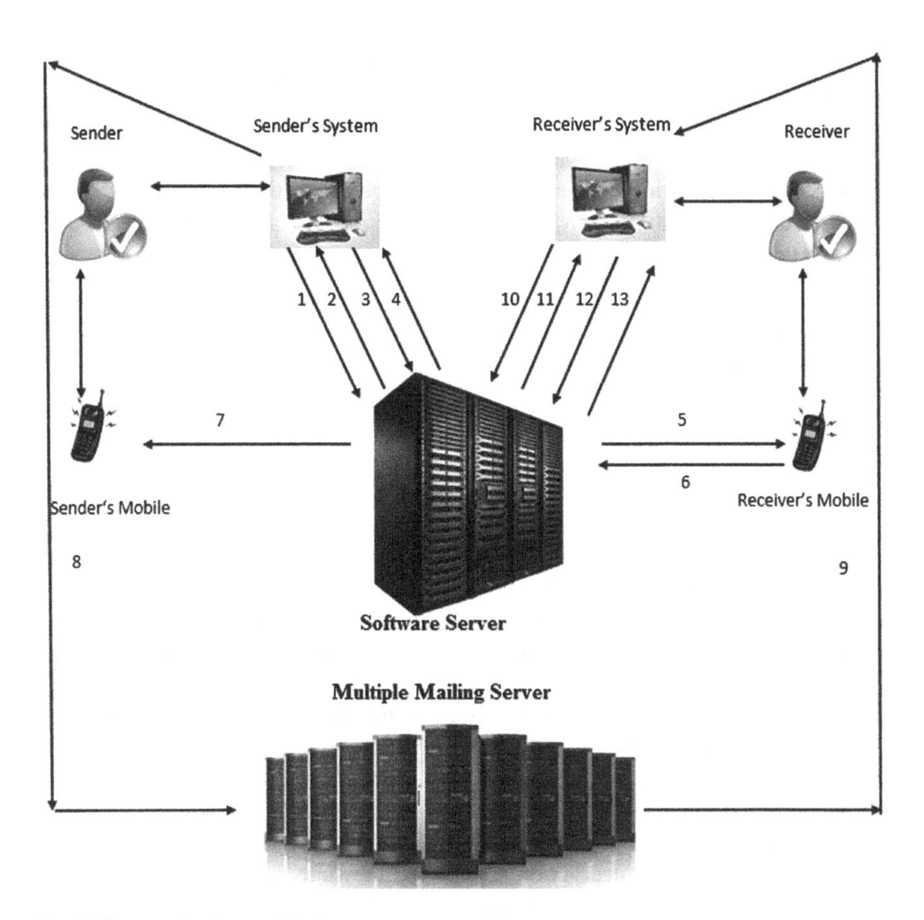

Fig. 2 Communication architecture

3.3 Communication Process

Step-by-step explanation of communication in the architecture (Fig. 2).

1. Sender will provide credentials to logon to the software server.
2. After checking, authenticated user logs on to the software server.
3. Sender will provide the receiver's ID and deadline before which the message must be decrypted.
4. Encryption code will be as SAAS, which will be provided to the sender.
5. Software server will send an SMS to the receiver that contains the sender's ID and deadline.
6. Delivery report will be sent to the software server as acknowledgment.
7. Software server will send an SMS to sender that receiver has a knowledge that you are sending him a message.
8. After getting encryption code the sender will encrypt the message using a pre-shared key and send it to the receiver through any mailing server.
9. The receiver will receive the encrypted message through the mailing server.
10. The receiver will logon to the software server using his credentials.
11. After checking the credentials the server allows to access the account.
12. Receiver will provide the sender's ID and deadline to the server.
13. After confirming the details, the software server will provide the decryption code to the receiver.

After receiving the encrypted message, decryption code, and the key the receiver can easily decrypt the code.

4 Conclusion

In the proposed architecture users exchange keys using key sharing technique to preserve integrity and confidentiality. Their key is shared in an open environment in a safe manner. Afterwards, sender communicates with the software server and authenticates himself. As software server informs the receiver about future communication and about the message validity deadline, the receiver is sure about the identity of sender. Even software server informs the sender about the acknowledgment that receiver receives the information that the sender will communicate him. Encryption and decryption code is available for a limited time with some logic bomb, therefore cryptanalyst is left with very limited time to crack the encryption and decryption code. As sender and receiver can use any mailing system, it is hard to track the encrypted message. The flexibilities of the system include: users can access the encryption and decryption code from anywhere, software server can use any encryption and decryption algorithm, but for a communication it should be the same; users can use any mailing server, users can use any set of key depend on mutual consensus. The only limitation of the system is that only one-to-one communication is possible.

References

1. Sharma, V., Gola, K.K., Khan, G., Rathore, R.: Key sharing technique to preserve integrity and confidentiality. In: Second International Conference on Advances in Computing and Communication Engineering, IEEE Computer Society (2015) doi: 10.1109/ICACCE.2015.48
2. Kohl, J.T., Neuman, B.C.: The kerberos network authentication service, version 5 revision 5, Project Athena, Massachusetts Institute of Technology (1992)
3. Sudha, M., Rao, B.R.K., Monica, M.: A comprehensive approach to ensure secure data communication in cloud environment. Int. J. Comput. Appl. (0975–8887) Volume 12– No. 8, December 2010
4. Graham, S.L., Rivest, R.L.: Secure communications over insecure channels. Commun. ACM **21** (4) (1978)
5. Shinde, K.M:, Shahabade, R.V.: Security in WLAN using Kerberos nternational. J. Sci. Res. Publ. **4**(6) (2014)
6. Harb, H.M., Mahdy, Y.B.: Overcoming Kerberos structural limtions, Nishidha laboratory. http://www.ii.ist.i.kyoto-u.ac.jp/~yasser/Publications/pdf OvercomingKerberosLimitationsOne-Column.pdf pp. 1–12

Multi-objective Optimization-Based Design of Robust Fractional-Order $PI^{\lambda}D^{\mu}$ Controller for Coupled Tank Systems

Nitish Katal and Shiv Narayan

Abstract In this paper, optimal control of coupled tank systems has been proposed using H_{∞} fractional-order controllers. Controller tuning has been posed as multi-objective mixed sensitivity minimization problem for tuning the fractional-order PID (FOPID) controllers and multi-objective variants of bat algorithm (MOBA) and differential evolution (MODE) has been used for optimization. Use of fractional-order controllers provides better characterization of dynamics of the process and their tuning using multi-objective optimization helps in attaining the robust trade-offs between sensitivity and complementary sensitivity. Both the FOPID controllers tuned with MOBA and MODE present robust behavior to external disturbance and the compared results show that MOBA-tuned controller presents efficient tracking of the reference.

Keywords Fractional-order control · Differential evolution · Bat algorithm · Coupled tank systems · Multi-objective optimization

1 Introduction

Most real-world control engineering problems are multi-objective in nature and it is challenging to meet the design constraints. In control systems, the objectives range from time domain specifications like rise time, percentage overshoot, ISE, ITSE, etc., to frequency domain requirements such as gain and phase margin, sensitivity, etc. Most of these objectives are usually conflictive in nature. So, a trade-off among several objectives has to be achieved for optimal performance. Thus, the controller tuning can be posed either as an aggregate objective function (AOF) or as generate-first choose-later (GFCL) multi-objective optimization problem [1].

N. Katal (✉) · S. Narayan
Electrical Engineering Department, PEC University of Technology, Chandigarh, India
e-mail: nitishkatal@gmail.com

M. Pant et al. (eds.), *Proceedings of Fifth International Conference on Soft Computing for Problem Solving*, Advances in Intelligent Systems and Computing 437, DOI 10.1007/978-981-10-0451-3_4

In this paper, a fractional-order $PI^{\lambda}D^{\mu}$ controller has been optimized considering multiple robustness objectives. PID controllers because of their simplicity are most popular and effective controllers and account for the 90 % of the total controllers used in industry today [2]. Fractional-orders systems provide a better characterization of the real dynamic processes [3]. Podlubny [4] introduced the concept and demonstrated the effectiveness of fractional-order $PI^{\lambda}D^{\mu}$ controllers (FOPID) over standard integer-order PID controllers. The extra degrees of freedom introduced by the fractional-order integrator λ and fractional-order differentiator μ increase the difficulty in finding their optimal gains. Obtaining the optimal gains for FOPID such that it satisfies the various time domain and frequency domain characteristics is of great significance both in theoretical and practical terms.

The optimized FOPID controller has been implemented on a coupled tank liquid level control system. In coupled tank systems, liquid level control in multiple connected tanks is a nonlinear control problem and is center to various diverse industrial establishments ranging from petrochemical, wastewater treatment to nuclear power generation. Several researchers globally have applied different control methodologies for the same, such as, second-order SMC [5], FOPID [6], bat algorithm optimized PID [7], multi-objective GA-tuned PID [8], and fuzzy PID controllers [9].

The work focuses on the optimization-based design of FOPID controllers using multi-objective bat algorithm (MOBA) and multi-objective differential evolution (MODE). The optimization problem has been posed as a mixed sensitivity problem, in which the H_{∞} norm of the sensitivity $S(s)$ and complementary sensitivity $T(s)$ has been used to define the multi-objective problem. Thus facilitating obtaining optimal FOPID parameters such that the system offers robust behavior to model uncertainties and external disturbances.

The paper has been organized into the following sections; Sect. 2 provides the mathematical modeling of the coupled tank liquid level control system. In Sect. 3, the prerequisites of fractional-order calculus and FOPID controllers have been discussed. Section 4 deals with defining the mixed sensitivity optimization problem. Sections 5 and 6 deal with basics of multi-objective bat algorithm and multi-objective differential evolution. In Sect. 7 several results obtained have been discussed followed by conclusions and references.

2 Mathematical Modeling of Coupled Tank System

The schematic representation of the coupled tank liquid level control system is shown in Fig. 1. Considering the flow balance equations, the nonlinear mathematical model of the system has been derived as under [7, 8].

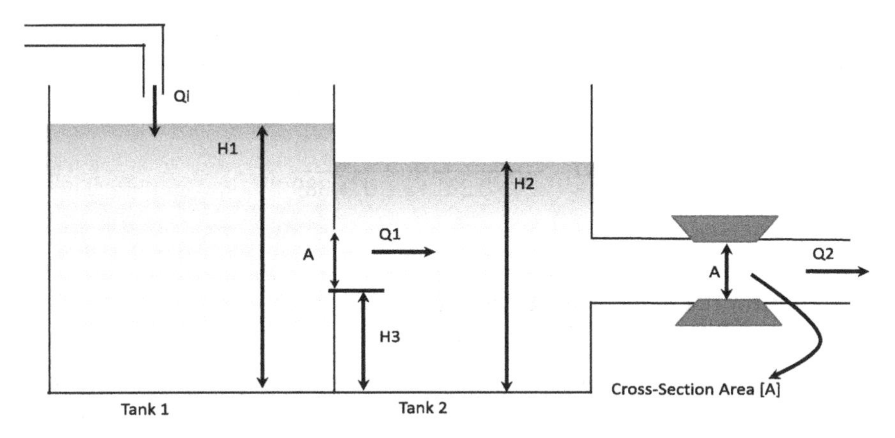

Fig. 1 Schematic representation of coupled tank liquid level system

For tanks 1 and 2:

$$Q_i - Q_1 = A\frac{\mathrm{d}H_1}{\mathrm{d}t} \quad \text{and} \quad Q_1 - Q_2 = A\frac{\mathrm{d}H_2}{\mathrm{d}t} \tag{1}$$

where, H_1, H_2 are the heights of tank 1 and 2; A is the cross-sectional area; Q_1 and Q_2 are the rates of flow of liquid.

Equation 2 shows the steady state representation of the system:

$$\begin{bmatrix} \dot{h}_1 \\ \dot{h}_2 \end{bmatrix} = \begin{pmatrix} -k_1/A & k_1/A \\ k_1/A & -(k_1+k_2)/A \end{pmatrix} \begin{bmatrix} h_1 \\ h_2 \end{bmatrix} + \begin{bmatrix} 1/A \\ 0 \end{bmatrix} q_i \tag{2}$$

Transfer function of the system given by Eq. 3 has been obtained by taking the Laplace transformation of Eq. 2.

$$G(s) = \frac{1/k_2}{\left(\frac{A^2}{k_1 \cdot k_2}\right) \cdot s^2 + \left(\frac{A(2 \cdot k_1 + k_2)}{k_1 \cdot k_2}\right) \cdot s + 1} = \frac{1/k_2}{(T_1 \cdot s + 1)(T_2 \cdot s + 1)} \tag{3}$$

where,

$$T_1 + T_2 = A\frac{2 \cdot k_1 + k_2}{k_1 \cdot k_2}, \quad T_1 \cdot T_2 = \frac{A^2}{k_1 \cdot k_2}$$

$$k_1 = \frac{\alpha}{2\sqrt{H_1 + H_2}} \quad \text{and} \quad k_1 = \frac{\alpha}{2\sqrt{H_2 + H_3}}$$

Following parameters have been considered as constants for obtaining Eq. 4; $H_1 = 18$ cm, $H_2 = 14$ cm, $H_3 = 6$ cm, $\alpha = 9.5$ (constant for coefficient of discharge), $A = 32$.

$$G(s) = \frac{0.002318}{s^2 + 0.201 \cdot s + 0.00389} \quad (4)$$

3 Fractional-Order PID Controllers

Non-integer integrals and derivatives provide better characterization of the dynamic systems and the use of fractional-order calculus in control systems has an ample potential to change the way we model systems and controllers [3]. Fractional-order PID controllers (FOPID) introduced by Podlubny [4] are more flexible and offer better performance in achieving the control objectives. The transfer function of the FOPID controllers is given by Eq. 5.

$$K_{\text{frac}} = k_{\text{P}} + \frac{k_I}{s^{\lambda}} + k_{\text{D}} \cdot s^{\mu} \quad (5)$$

Optimal design of the FOPID controllers involves the design of three parameters k_{P}, k_{I}, and k_{D} and two orders λ and μ, the values of which can be non-integer [10].

4 Mixed Sensitivity Multi-objective Problem Formulation

Mixed sensitivity optimization allows achieving simultaneous trade-offs between performance and robustness. Reduction of the sensitivity $S(j\omega)$ ensures disturbance rejection and complementary sensitivity $T(j\omega)$ reduction handles the robustness issues and the minimization of control effort [11]. The mixed sensitivity optimization problem is given by Eq. 6.

$$\min_{K \text{ stabalizing}} \left\| \begin{matrix} w_1 \cdot S(j\omega) \\ w_2 \cdot T(j\omega) \end{matrix} \right\|_{\infty} \quad (6)$$

In the work presented in this paper, mixed sensitivity reduction has been posed as aggregate objective function (AOF) given by Eq. 7:

$$\|w_1 \cdot S(j\omega)\|_{\infty} + \|w_2 \cdot T(j\omega)\|_{\infty} \leq 1 \quad (7)$$

5 Multi-objective Optimization Using Bat Algorithm

Bat algorithm (BA) was introduced by X.S. Yang in 2010 and extended it for solving multiple objectives in 2012 [12]. Bat algorithm mimics the echolocation behavior of bats, which they use to locate their prey and differentiate between different insects even in complete darkness. In the initial population, each bat updates it position using echolocation in which echoes are created by loud ultrasound waves which are received back with delay and specific preys are characterized by specific sound levels. Following equations characterize the bat motion, i.e., x_i is the position, velocity v_i and the new updated position x_i^t, and velocities v_i^t at time t.

$$f_i = f_{\min} + (f_{\max} - f_{\min}) \cdot \beta$$

$$v_i^{t+1} = v_i^t + \left(x_i^t - x_*\right) \cdot f_i$$

$$x_i^{t+1} = x_i^t + v_i^t$$

where, $\beta \in [0, 1]$ is a uniformly distributed random vector, x^* is the current global best location of all n bats in population. Initially, uniformly derived frequency from $[f_{\min}, f_{\max}]$ is randomly assigned to each bat.

When a bat finds its prey, the loudness A_i usually decreases and the rate of pulse emission r_i increases. Initial loudness A_0 can be set to any value; here A_0 is taken as 1 and considering that bat has bound its prey $A_{\min}$ is taken as 0.

$$A_i^{t+1} = \alpha \cdot A_i^t, \quad r_i^t = r_i^0\left(1 - e^{-\lambda \cdot t}\right)$$

α and γ are constants and for $0 < \alpha < 1$ and $\gamma > 0$,

$$A_i^t \to 0, \quad r_i^t \to r_i^0, \quad \text{as } t \to \infty$$

For defining the objective function as mixed sensitivity, the problem has been expressed as aggregate objective function (AOF), i.e., expressing all objectives J_k as weighted sum given by Eq. 8. As mixed sensitivity problem has been considered, so K is taken as 2.

$$J = \sum_{k=1}^{K} w_k \cdot J_k, \quad \sum_{k=1}^{K} w_k = 1 \tag{8}$$

Since, to introduce diversity in population of bats, loudness and pulse emission rates have been generated randomly and while searching for the solution, these values are updated, only if the new solution shows scope of improvement while converging toward the global minima.

6 Multi-objective Optimization Using Differential Evolution

Differential evolution (DE) is a population-based stochastic direct search optimization algorithm and is based on the use of similar operators like those of genetic algorithm; crossover, mutation, and selection. DE is advantageous in several aspects such as; it can handle nonlinear and multimodal cost functions, less computationally intense, has few control variables so it is easy to use and has very good convergence properties [13].

DE pivots on mutation operator for generating better solutions, whereas GA is dependent on crossover. In DE for searching and selection of global solution in the prospective search space, mutation operation is used. In pursuit of better solutions, scattered crossover among the parents efficiently intermix the knowledge about successful combinations [14].

The optimization index for mixed sensitivity optimization problem has also been formulated as weighted sum of objective functions given by Eq. 9.

$$J = \min_{K \text{ stabalizing}} \left\{ \|S(j\omega)\|_\infty + \|T(j\omega)\|_\infty \right\} \tag{9}$$

In this work, a variant of DE with jitter [15] has been used for optimization.

6.1 *Mutation*

A mutation vector produced for each target vector $x_{i,G}$ is given as in below equation.

$$v_{i,G+1} = x_{i,G} + K \cdot \left(x_{r1,G} - x_{i,G}\right) + F \cdot \left(x_{r2,G} - x_{r3,G}\right)$$

where r_1, r_2, $r_3 \in \{1, 2\ldots \text{NP}\}$ are different from each other and are randomly generated, F is the scaling factor and K is combination factor.

6.2 *Crossover*

Crossover generates the trial vector $u_{ji,G+1}$ by mixing the parent with mutated vector and is given by equation as

$$u_{ji,G+1} = \begin{cases} v_{ji,G+1} & \text{if} \quad (rnd_j \leq \text{CR}) \quad \text{or} \quad j = rn_i \\ q_{ji,G} & \text{if} \quad (rnd_j > \text{CR}) \quad \text{or} \quad j \neq rn_i \end{cases}$$

where, $j = 1, 2, \ldots, D$; random vector $r_j \in [1, 0]$; crossover constant CR $\in [1, 0]$; and randomly chosen index $rn_i \in (1, 2, \ldots, D)$.

6.3 Selection

From the search space, parents can be selected form all the individuals in the population irrespective of their fitness value. After mutation and crossover, the fitness of the child is evaluated and compared with that of parent and the one with better fitness value is selected.

7 Results and Discussion

The optimization and simulation of the closed loop coupled tank system has been carried out in MATLAB and the implementation of the fractional-order PID controller is done using FOMCON toolbox [10]. Table 1 shows the multi-objective bat algorithm parameters and Table 2 shows the multi-objective differential evolution parameters considered for the optimization carried out in the work.

In Fig. 2, closed loop response of the system is shown with varying inputs and it is clear from the figure that the MOBA-tuned FOPID controller tracks the reference efficiently and too with negligible overshoot. Figure 3 shows the control effort employed by the FOPID controllers to ensure the steady and smooth tracking of the reference signal. From Figs. 2 and 3, it can be surmised that in MODE-tuned FOPID controllers, tracking is not proper and a steady state error exists throughout the response.

Table 1 Multi-objective bat algorithm parameters

Parameter	Value
Population size	40
Loudness	0.5
Pulse rate	0.5
Minimum frequency	0
Maximum frequency	2

Table 2 Multi-objective differential evolution parameters

Parameter	Value
DE statergy	DE with Jitter [15]
Population size	20
Step size	0.85
Crossover probability	1

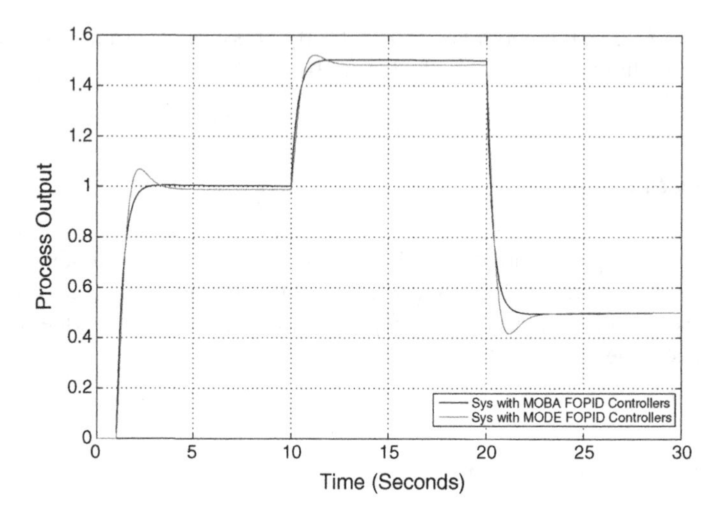

Fig. 2 Response of the closed loop system with MOBA- and MODE-tuned FOPID controllers

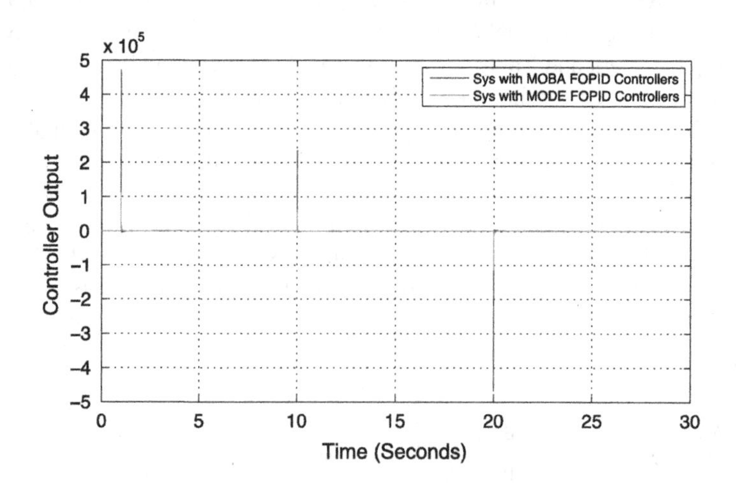

Fig. 3 Controller output with MOBA- and MODE-tuned FOPID controllers

In order to check the robustness of the system to external disturbances, a pulse has been introduced. The magnitude of the pulse is 10 % of the reference signal having a duration of 0.15 s and has been introduced at 4 s. Figures 4 and 5 show the step response of the closed loop system and the controller output, respectively. The Figs. 4 and 5 implies that the closed loop system with MOBA- and MODE-tuned FOPID controllers both efficiently tackle the noise and restore the system to original tracking of reference immediately.

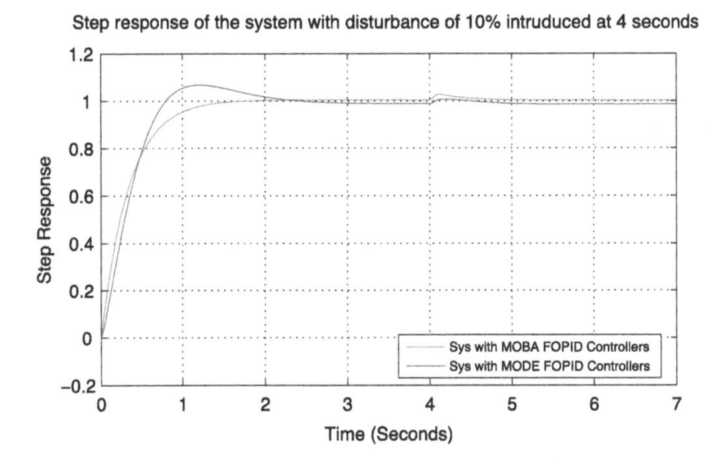

Fig. 4 Step response of the closed loop system with MOBA- and MODE-tuned FOPID controllers with 10 % disturbance introduced at 4 s

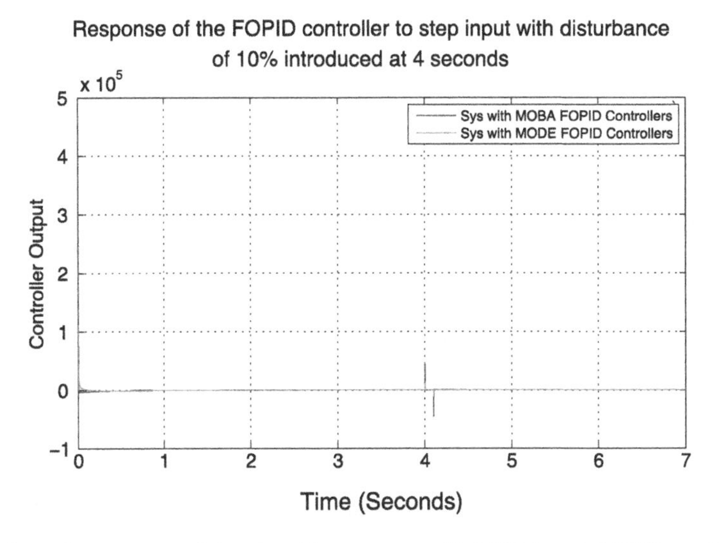

Fig. 5 Controller response of the closed loop system with MOBA- and MODE-tuned FOPID controllers with 10 % disturbance introduced at 4 s

Table 3 shows the obtained optimal FOPID parameters obtained using MOBA and MODE. Table 4 shows the H_∞ norm of the sensitivity function and complementary sensitivity function is shown and are <1 and even their sum is ≤1 which shows better robustness properties and satisfies the performance index defied in Eq. 7. In Table 5, the time domain performance indexes has been compared of both the systems and it can be seen the MOBA-tuned FOPID controllers provides better indexes like reduced overshoot percentage and reduced settling times.

Table 3 Optimal FOPID parameters obtained after optimization

Optimization method	K_p	K_i	K_d	λ	μ
Multi-objective bat algorithm	282.9	6.9	1276.5	0.9	1
Multi-objective DE	100	6.65	1500	0.6905	0.744

Table 4 Compared H_∞ norm of the $S(j\omega)$ and $T(j\omega)$

Optimization method	$\|S(j\omega)\|_\infty$	$\|T(j\omega)\|_\infty$
Multi-objective bat algorithm	0.0036	0.9964
Multi-objective DE	0.0068	0.9932

Table 5 Compared time domain performances

Optimization method	Overshoot (%)	Rise time	Settling time
Multi-objective bat algorithm	0.36	0.7983	1.36
Multi-objective DE	6.09	0.5385	1.98

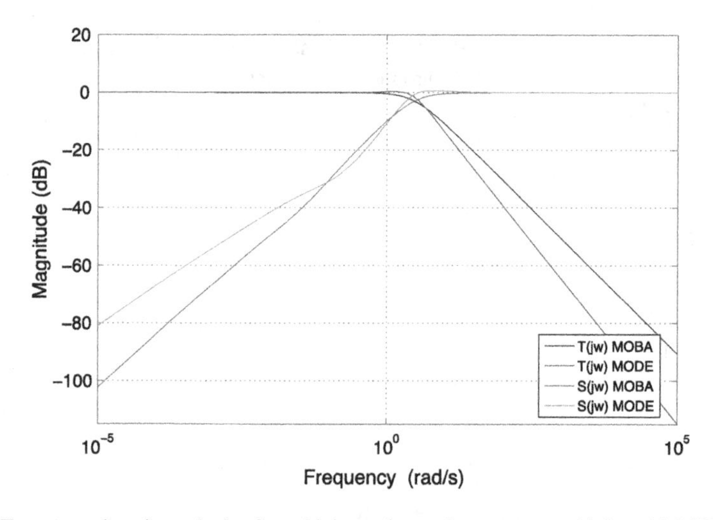

Fig. 6 Frequency domain analysis of sensitivity and complementary sensitivity with MOBA- and MODE-tuned FOPID controllers

Figure 6 shows the frequency domain representation of the sensitivity and complementary sensitivity functions using both the controllers. Gain curve of the Bode plot of sensitivity and complementary sensitivity (Fig. 6) shows how feedback influences the disturbances. Figure 6 shows that maximum sensitivity $S(j\omega)$ for MODE-tuned FOPID controllers is 0.66 dB at 6.11 rad/s, whereas for MOBA tuned FOPID controllers maximum sensitivity is ≈0 dB at that too till 10^{15} rad/s. Thus following inferences can be obtained that feedback will reduce the disturbances less than the gain crossover frequency (ω_{gc}) and for higher frequencies (beyond ω_{gc}) the disturbances will be amplified and the largest amplification factor of 0.66 (MODE) and ≈0 (MOBA). From Fig. 6 for the complementary sensitivity function $T(j\omega)$, maximum peak gains of 0.0184 dB at 0.192 rad/s for MOBA-tuned controllers and 0.351 dB at 1.27 rad/s for MODE-tuned controllers has been obtained. This provides insights about the stability margins and allowable process variations.

8 Conclusions

In real-world engineering applications, most design requirements are multi-objective in nature and for optimal performance several constraints have to be satisfied. Robust behavior of controllers in presence of uncertainties is one of the most complicated control design objectives and use of fractional calculus in controller/system designing offers flexibility and extra degree-of-freedom in achieving such goals. In this paper, mixed sensitivity minimization has been posed as a multi-objective problem. MOBA and MODE have been used for tuning the FOPID controllers implemented for the liquid level control in coupled tank systems. Results obtained show that efficient disturbance rejection and satisfaction of the performance indexes. Such optimal robust behavior ensures the process safety and the quality of products.

References

1. Reynoso-Meza, G., et al.: Controller tuning using evolutionary multi-objective optimisation: current trends and applications. Control Eng. Pract. **28**, 58–73 (2014). doi:10.1016/j.conengprac.2014.03.003
2. A°ström, K.J., Kumar, P.R.: Control: a perspective. Automatica **50**(1), 3–43 (2014). ISSN 0005-1098, doi:10.1016/j.automatica.2013.10.012
3. Chen, Y., Petráš, I., Xue, D.: Fractional order control-a tutorial. In: American Control Conference, 2009. ACC'09. IEEE (2009). doi:10.1109/ACC.2009.5160719
4. Podlubny, I.: Fractional-order systems and PID controllers. IEEE Trans. Autom. Control **44** (1), 208–213 (1999). doi:10.1109/9.739144
5. Abu Khadra, F., Abu Qudeiri, J.: Second order sliding mode control of the coupled tanks system. Math. Probl. Eng. (2015)
6. Kumar, A., Vashishth, M., Rai, L.: Liquid level control of coupled tank system using Fractional PID controller. Int. J. Emerg. Trends Electr. Electron. **3**(1) (2013). 2320-9569
7. Katal, N., Kumar, P., Narayan, S.: Optimal PID controller for coupled-tank liquid-level control system using bat algorithm. In: International Conference on Power, Control and Embedded Systems (ICPCES), IEEE (2014). doi:10.1109/ICPCES.2014.7062818
8. Singh, S.K., Katal, N., Modani, S.G.: Multi-objective optimization of PID controller for coupled-tank liquid-level control system using genetic algorithm. In: Proceedings of the Second International Conference on Soft Computing for Problem Solving (SocProS 2012), December 28–30, 2012. Springer India (2014). doi:10.1007/978-81-322-1602-5_7
9. Xu, J.-X., Hang, C.-C., Liu, C.: Parallel structure and tuning of a fuzzy PID controller. Automatica **36**(5), 673–684 (2000). doi:10.1016/S0005-1098(99)00192-2

10. Tepljakov, A., Petlenkov, E., Belikov, J.: FOMCON: fractional-order modeling and control toolbox for MATLAB. In: Proceedings of 18th International Mixed Design of Integrated Circuits and Systems (MIXDES) Conference (2011)
11. Maciejowski, J.M.: Multivariate feedback design (1989)
12. Yang, X.-S.: Bat algorithm for multi-objective optimisation. International Journal of Bio-Inspired Computation **3**(5), 267–274 (2011). doi:10.1504/IJBIC.2011.042259
13. Storn, R., Price, K.: Differential evolution–a simple and efficient heuristic for global optimization over continuous spaces. J. Global Optim. **11**(4), 341–359 (1997). doi:10.1023/A:1008202821328
14. Okdem, S.: A simple and global optimization algorithm for engineering problems: differential evolution algorithm. Turk J. Elec. Eng. **12**(1) (2004)
15. Storn, R.: Differential Evolution Research–Trends and Open Questions. Advances in Differential Evolution. Springer, Berlin, pp. 1–31 (2008). doi:10.1007/978-3-540-68830-3_1

Using Least-Square Spectral Analysis Method for Time Series with Datum Shifts

Ingudam Gomita, Sandeep Chauhan and Raj Kumar Sagar

Abstract Advancement in the field of technology such as processing power in computers and servers has provided us with an opportunity to improve the performance of legacy system programs. This will improve the performance of these systems along with increase in maintainability and life expectancy. But it is not a simple task to simply pick a legacy system and convert it. In this project the algorithm to compute least-squares spectrum LSSA, which is used for analyzing huge data and plot the spectrum chart will be converted from its original FORTRAN program to C program and implement parallel processing in order to improve it using various available methodologies.

Keywords LSSA · f2c · Par4all · FFT · Spectrum

1 Introduction

Researchers and engineers are often expected to work with experimental time series which have many missing data points. At times, there are time series that are unequally spaced, like that of weather observations, or sometimes to avoid aliasing variable sampling rate may be introduced intentionally. Further, advancements in development of instruments have led us to understand the physical phenomena better and hence resulting in more accurate data. So over long periods the time series collected will be unequally weighted and hence will be nonstationary. There may also be disturbances due to repair and replacement of instruments, which may cause datum shifts [1] in the series (Figs. 1, 2, 3 and 4).

Ingudam Gomita (✉) · Sandeep Chauhan · R.K. Sagar
Amity University, Noida, India
e-mail: Ingudamgomita007@gmail.com

Sandeep Chauhan
e-mail: chauhansande@gmail.com

R.K. Sagar
e-mail: rksagar@amity.edu

M. Pant et al. (eds.), *Proceedings of Fifth International Conference on Soft Computing for Problem Solving*, Advances in Intelligent Systems and Computing 437, DOI 10.1007/978-981-10-0451-3_5

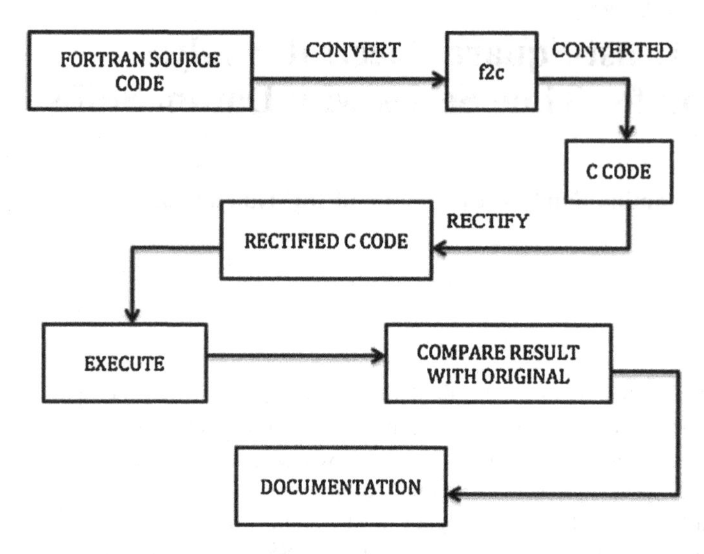

Fig. 1 f2c conversion process

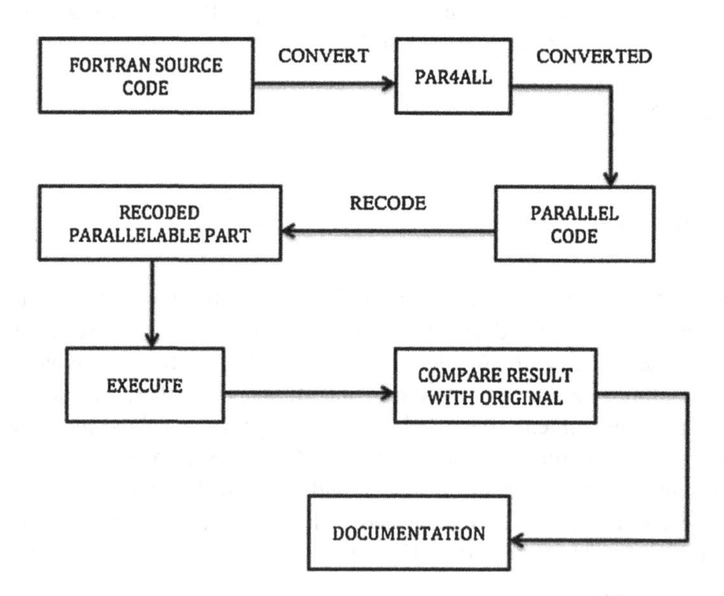

Fig. 2 Conversion using par4all

Entirety of the techniques of fashionable spectral techniques lies with the fast fourier transformation algorithms. These algorithms are used for determining the power spectrum. The FFT is mainly used for two reasons—(i) it is computationally efficient and (ii) for a huge class of signal processes it gives reasonable results (Figs. 5, 6 and 7).

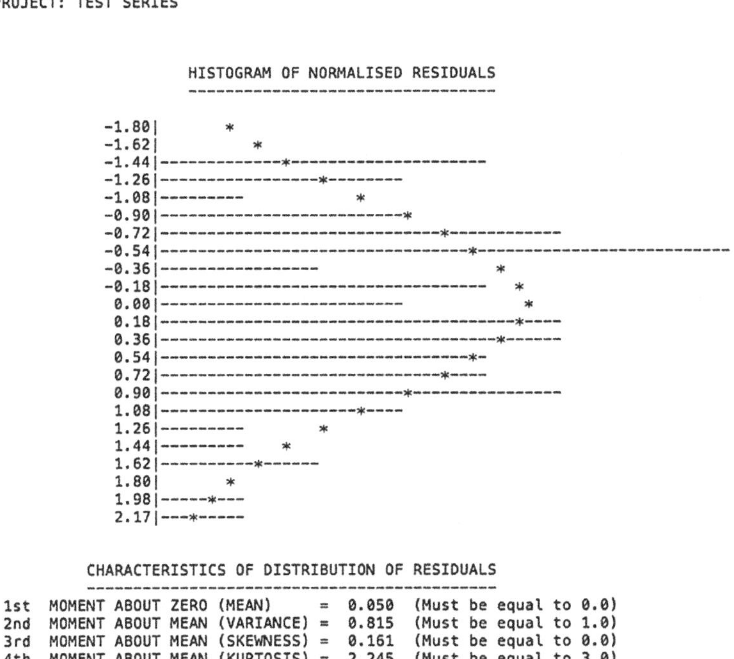

```
      L E A S T   S Q U A R E S   S P E C T R A L   A N A L Y S I S
                              Version 5.02

PROJECT: TEST SERIES


                  HISTOGRAM OF NORMALISED RESIDUALS
                  ---------------------------------

          -1.80|       *
          -1.62|          *
          -1.44|-------------*--------------------
          -1.26|-----------------*--------
          -1.08|--------            *
          -0.90|--------------------------*
          -0.72|------------------------------*------------
          -0.54|---------------------------------*---------------------------
          -0.36|----------------                    *
          -0.18|----------------------------------    *
           0.00|-------------------------              *
           0.18|--------------------------------------*----
           0.36|------------------------------------*------
           0.54|---------------------------------*-
           0.72|------------------------------*----
           0.90|--------------------------*----------------
           1.08|---------------------*----
           1.26|--------         *
           1.44|--------     *
           1.62|----------*------
           1.80|       *
           1.98|-----*---
           2.17|---*-----


           CHARACTERISTICS OF DISTRIBUTION OF RESIDUALS
           --------------------------------------------
 1st  MOMENT ABOUT ZERO (MEAN)      =  0.050  (Must be equal to 0.0)
 2nd  MOMENT ABOUT MEAN (VARIANCE) =  0.815  (Must be equal to 1.0)
 3rd  MOMENT ABOUT MEAN (SKEWNESS) =  0.161  (Must be equal to 0.0)
 4th  MOMENT ABOUT MEAN (KURTOSIS) =  2.245  (Must be equal to 3.0)
```

Fig. 3 LSSA.OUT output

However, it is not a perfect method for spectral analysis. It has certain limitations—the most salient being the fact that the data need to be equally weighted and equally spaced. So in these cases it becomes inevitable to process the data and as a result it performs poorly and unsatisfactorily.

An alternative to the FFT was developed by Vanicek (1969, 1971) [2]. It is known as least-square spectral analysis (LSSA). It overcomes the methods of the fourier methods. The data need not be equally spaced or weighted. Gaps and datum shifts do not make a difference here. The LSSA has been revisited [3] highlighting its significant properties and concentrating on the covariance matrix of input series. This is done to derive the probability density function emphasizing the least-square spectrum. We design the criteria to determine a threshold value. Any least-square spectral peak whose value is above this threshold value is statistically significant. It is imperative to note that we assume the covariance matrix should be fully populated for all our derivations. This means that contrary to previous studies the time series need not be statistically significant.

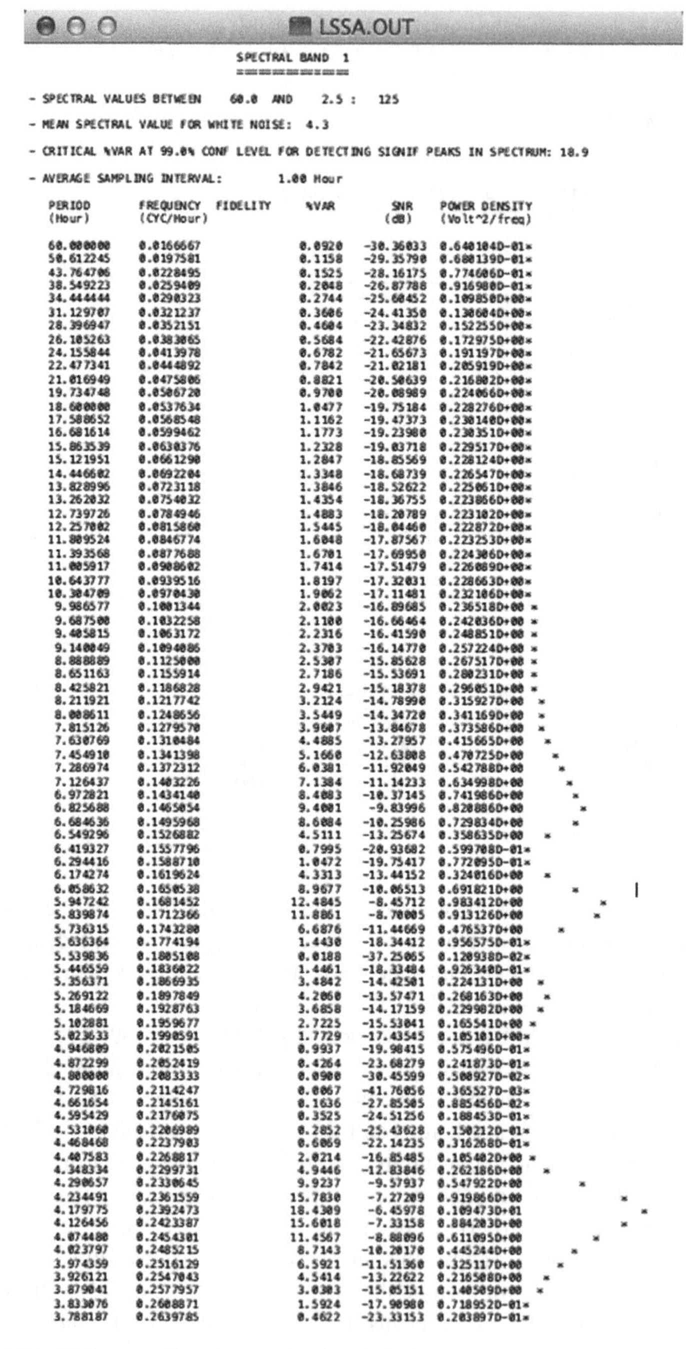

LSSA.OUT

SPECTRAL BAND 1

- SPECTRAL VALUES BETWEEN 60.0 AND 2.5 : 125
- MEAN SPECTRAL VALUE FOR WHITE NOISE: 4.3
- CRITICAL %VAR AT 99.0% CONF LEVEL FOR DETECTING SIGNIF PEAKS IN SPECTRUM: 18.9
- AVERAGE SAMPLING INTERVAL: 1.00 Hour

PERIOD (Hour)	FREQUENCY (CYC/Hour)	FIDELITY	%VAR	SNR (dB)	POWER DENSITY (Volt^2/freq)
60.000000	0.0166667		0.0920	-30.36033	0.640104D-01
50.612245	0.0197581		0.1158	-29.35790	0.680139D-01
43.764706	0.0228495		0.1525	-28.16175	0.774606D-01
38.549223	0.0259409		0.2048	-26.87788	0.916980D-01
34.444444	0.0290323		0.2744	-25.60452	0.109850D+00
31.129707	0.0321237		0.3606	-24.41350	0.130604D+00
28.396947	0.0352151		0.4604	-23.34832	0.152255D+00
26.105263	0.0383065		0.5684	-22.42876	0.172975D+00
24.155844	0.0413978		0.6782	-21.65673	0.191197D+00
22.477341	0.0444892		0.7842	-21.02181	0.205919D+00
21.016949	0.0475806		0.8821	-20.50639	0.216802D+00
19.734748	0.0506720		0.9700	-20.08989	0.224066D+00
18.600000	0.0537634		1.0477	-19.75184	0.228276D+00
17.588652	0.0568548		1.1162	-19.47373	0.230140D+00
16.681614	0.0599462		1.1773	-19.23980	0.230351D+00
15.863539	0.0630376		1.2328	-19.03718	0.229517D+00
15.121951	0.0661290		1.2847	-18.85569	0.228124D+00
14.446602	0.0692204		1.3348	-18.68739	0.226547D+00
13.828996	0.0723118		1.3846	-18.52622	0.225061D+00
13.262032	0.0754032		1.4354	-18.36755	0.223866D+00
12.739726	0.0784946		1.4883	-18.20789	0.223102D+00
12.257002	0.0815860		1.5445	-18.04460	0.222872D+00
11.809524	0.0846774		1.6048	-17.87567	0.223253D+00
11.393568	0.0877688		1.6701	-17.69950	0.224306D+00
11.005917	0.0908602		1.7414	-17.51479	0.226089D+00
10.643777	0.0939516		1.8197	-17.32031	0.228663D+00
10.304709	0.0970430		1.9062	-17.11481	0.232106D+00
9.986577	0.1001344		2.0023	-16.89685	0.236518D+00
9.687500	0.1032258		2.1100	-16.66464	0.242036D+00
9.405815	0.1063172		2.2316	-16.41590	0.248851D+00
9.140049	0.1094086		2.3703	-16.14770	0.257224D+00
8.888889	0.1125000		2.5307	-15.85628	0.267517D+00
8.651163	0.1155914		2.7186	-15.53691	0.280231D+00
8.425821	0.1186828		2.9421	-15.18378	0.296051D+00
8.211921	0.1217742		3.2124	-14.78990	0.315927D+00
8.008611	0.1248656		3.5449	-14.34720	0.341169D+00
7.815126	0.1279570		3.9607	-13.84678	0.373586D+00
7.630769	0.1310484		4.4885	-13.27957	0.415665D+00
7.454910	0.1341398		5.1660	-12.63808	0.470725D+00
7.286974	0.1372312		6.0381	-11.92049	0.542788D+00
7.126437	0.1403226		7.1384	-11.14233	0.634998D+00
6.972821	0.1434140		8.4083	-10.37145	0.741986D+00
6.825688	0.1465054		9.4001	-9.83996	0.820886D+00
6.684636	0.1495968		8.6084	-10.25986	0.729834D+00
6.549296	0.1526882		4.5111	-13.25674	0.358635D+00
6.419327	0.1557796		0.7995	-20.93682	0.599708D-01
6.294416	0.1588710		1.0472	-19.75417	0.772095D-01
6.174274	0.1619624		4.3313	-13.44152	0.324016D+00
6.058632	0.1650538		8.9677	-10.06513	0.691821D+00
5.947242	0.1681452		12.4845	-8.45712	0.983412D+00
5.839874	0.1712366		11.8861	-8.70005	0.913126D+00
5.736315	0.1743280		6.6876	-11.44669	0.476537D+00
5.636364	0.1774194		1.4430	-18.34412	0.956575D-01
5.539836	0.1805108		0.0188	-37.25065	0.120938D-02
5.446559	0.1836022		1.4461	-18.33484	0.926340D-01
5.356371	0.1866935		3.4842	-14.42501	0.224131D+00
5.269122	0.1897849		4.2060	-13.57471	0.268163D+00
5.184669	0.1928763		3.6858	-14.17159	0.229982D+00
5.102881	0.1959677		2.7225	-15.53041	0.165541D+00
5.023633	0.1990591		1.7729	-17.43545	0.105101D+00
4.946809	0.2021505		0.9937	-19.98415	0.575496D-01
4.872299	0.2052419		0.4264	-23.68279	0.241873D-01
4.800000	0.2083333		0.0900	-30.45599	0.500927D-02
4.729816	0.2114247		0.0067	-41.76056	0.365527D-03
4.661654	0.2145161		0.1636	-27.85505	0.885456D-02
4.595429	0.2176075		0.3525	-24.51256	0.188453D-01
4.531060	0.2206989		0.2852	-25.43628	0.150212D-01
4.468468	0.2237903		0.6069	-22.14235	0.316268D-01
4.407583	0.2268817		2.0214	-16.85485	0.105402D+00
4.348334	0.2299731		4.9446	-12.83846	0.262186D+00
4.290657	0.2330645		9.9237	-9.57937	0.547922D+00
4.234491	0.2361559		15.7830	-7.27209	0.919866D+00
4.179775	0.2392473		18.4309	-6.45978	0.109473D+01
4.126456	0.2423387		15.6018	-7.33158	0.884203D+00
4.074480	0.2454301		11.4567	-8.88096	0.611095D+00
4.023797	0.2485215		8.7143	-10.20170	0.445244D+00
3.974359	0.2516129		6.5921	-11.51360	0.325117D+00
3.926121	0.2547043		4.5414	-13.22622	0.216508D+00
3.879041	0.2577957		3.0303	-15.05151	0.140509D+00
3.833076	0.2608871		1.5924	-17.90980	0.718952D-01
3.788187	0.2639785		0.4622	-23.33153	0.203897D-01

Fig. 4 LSSA.OUT output (the *dots* are the time series)

spectrum.dat

```
60.00000000     0.01667     0.00000     0.0920    -30.36033    0.64010D-01
50.61224490     0.01976     0.00000     0.1158    -29.35790    0.68014D-01
43.76470588     0.02285     0.00000     0.1525    -28.16175    0.77461D-01
38.54922280     0.02594     0.00000     0.2048    -26.87788    0.91698D-01
34.44444444     0.02903     0.00000     0.2744    -25.60452    0.10985D+00
31.12970711     0.03212     0.00000     0.3606    -24.41350    0.13060D+00
28.39694656     0.03522     0.00000     0.4604    -23.34832    0.15226D+00
26.10526316     0.03831     0.00000     0.5684    -22.42876    0.17298D+00
24.15584416     0.04140     0.00000     0.6782    -21.65673    0.19120D+00
22.47734139     0.04449     0.00000     0.7842    -21.02181    0.20592D+00
21.01694915     0.04758     0.00000     0.8821    -20.50639    0.21680D+00
19.73474801     0.05067     0.00000     0.9700    -20.08989    0.22407D+00
18.60000000     0.05376     0.00000     1.0477    -19.75184    0.22828D+00
17.58865248     0.05685     0.00000     1.1162    -19.47373    0.23014D+00
16.68161435     0.05995     0.00000     1.1773    -19.23980    0.23035D+00
15.86353945     0.06304     0.00000     1.2328    -19.03718    0.22952D+00
15.12195122     0.06613     0.00000     1.2847    -18.85569    0.22812D+00
14.44660194     0.06922     0.00000     1.3348    -18.68739    0.22655D+00
13.82899628     0.07231     0.00000     1.3846    -18.52622    0.22506D+00
13.26203209     0.07540     0.00000     1.4354    -18.36755    0.22387D+00
12.73972603     0.07849     0.00000     1.4883    -18.20789    0.22310D+00
12.25700165     0.08159     0.00000     1.5445    -18.04460    0.22287D+00
11.80952381     0.08468     0.00000     1.6048    -17.87567    0.22325D+00
11.39356815     0.08777     0.00000     1.6701    -17.69950    0.22431D+00
11.00591716     0.09086     0.00000     1.7414    -17.51479    0.22609D+00
10.64377682     0.09395     0.00000     1.8197    -17.32031    0.22866D+00
10.30470914     0.09704     0.00000     1.9062    -17.11481    0.23211D+00
 9.98657718     0.10013     0.00000     2.0023    -16.89685    0.23652D+00
 9.68750000     0.10323     0.00000     2.1100    -16.66464    0.24204D+00
 9.40581542     0.10632     0.00000     2.2316    -16.41590    0.24885D+00
 9.14004914     0.10941     0.00000     2.3703    -16.14770    0.25722D+00
 8.88888889     0.11250     0.00000     2.5307    -15.85628    0.26752D+00
 8.65116279     0.11559     0.00000     2.7186    -15.53691    0.28023D+00
 8.42582106     0.11868     0.00000     2.9421    -15.18378    0.29605D+00
 8.21192053     0.12177     0.00000     3.2124    -14.78990    0.31593D+00
 8.00861141     0.12487     0.00000     3.5449    -14.34720    0.34117D+00
 7.81512605     0.12796     0.00000     3.9607    -13.84678    0.37359D+00
 7.63076923     0.13105     0.00000     4.4885    -13.27957    0.41566D+00
 7.45490982     0.13414     0.00000     5.1660    -12.63808    0.47072D+00
 7.28697356     0.13723     0.00000     6.0381    -11.92049    0.54279D+00
 7.12643678     0.14032     0.00000     7.1384    -11.14233    0.63500D+00
 6.97282099     0.14341     0.00000     8.4083    -10.37145    0.74199D+00
 6.82568807     0.14651     0.00000     9.4001     -9.83996    0.82089D+00
 6.68463612     0.14960     0.00000     8.6084    -10.25986    0.72983D+00
 6.54929577     0.15269     0.00000     4.5111    -13.25674    0.35864D+00
 6.41932701     0.15578     0.00000     0.7995    -20.93682    0.59971D-01
 6.29441624     0.15887     0.00000     1.0472    -19.75417    0.77209D-01
 6.17427386     0.16196     0.00000     4.3313    -13.44152    0.32402D+00
 6.05863192     0.16505     0.00000     8.9677    -10.06513    0.69182D+00
 5.94724221     0.16815     0.00000    12.4845     -8.45712    0.98341D+00
 5.83987441     0.17124     0.00000    11.8861     -8.70005    0.91313D+00
 5.73631457     0.17433     0.00000     6.6876    -11.44669    0.47654D+00
 5.63636364     0.17742     0.00000     1.4430    -18.34412    0.95657D-01
 5.53983619     0.18051     0.00000     0.0188    -37.25065    0.12094D-02
 5.44655930     0.18360     0.00000     1.4461    -18.33484    0.92634D-01
 5.35637149     0.18669     0.00000     3.4842    -14.42501    0.22413D+00
 5.26912181     0.18978     0.00000     4.2060    -13.57471    0.26816D+00
 5.18466899     0.19288     0.00000     3.6858    -14.17159    0.22998D+00
 5.10288066     0.19597     0.00000     2.7225    -15.53041    0.16554D+00
 5.02363268     0.19906     0.00000     1.7729    -17.43545    0.10510D+00
 4.94680851     0.20215     0.00000     0.9937    -19.98415    0.57550D-01
 4.87229862     0.20524     0.00000     0.4264    -23.68279    0.24187D-01
 4.80000000     0.20833     0.00000     0.0900    -30.45599    0.50093D-02
 4.72981564     0.21142     0.00000     0.0067    -41.76056    0.36553D-03
 4.66165414     0.21452     0.00000     0.1636    -27.85505    0.88546D-02
```

Fig. 5 Spectrum.dat

spectrum.dat

```
4.87229862   0.20524   0.00000    0.4264   -23.68279   0.24187D-01
4.80000000   0.20833   0.00000    0.0900   -30.45599   0.50093D-02
4.72981564   0.21142   0.00000    0.0067   -41.76056   0.36553D-03
4.66165414   0.21452   0.00000    0.1636   -27.85505   0.88546D-02
4.59542928   0.21761   0.00000    0.3525   -24.51256   0.18845D-01
4.53105968   0.22070   0.00000    0.2852   -25.43628   0.15021D-01
4.46846847   0.22379   0.00000    0.6069   -22.14235   0.31627D-01
4.40758294   0.22688   0.00000    2.0214   -16.85485   0.10540D+00
4.34833431   0.22997   0.00000    4.9446   -12.83846   0.26219D+00
4.29065744   0.23306   0.00000    9.9237    -9.57937   0.54792D+00
4.23449061   0.23616   0.00000   15.7830    -7.27209   0.91987D+00
4.17977528   0.23925   0.00000   18.4309    -6.45978   0.10947D+01
4.12645591   0.24234   0.00000   15.6018    -7.33158   0.88420D+00
4.07447974   0.24543   0.00000   11.4567    -8.88096   0.61109D+00
4.02379665   0.24852   0.00000    8.7143   -10.20170   0.44524D+00
3.97435897   0.25161   0.00000    6.5921   -11.51360   0.32512D+00
3.92612137   0.25470   0.00000    4.5414   -13.22622   0.21651D+00
3.87904067   0.25780   0.00000    3.0303   -15.05151   0.14051D+00
3.83307573   0.26089   0.00000    1.5924   -17.90980   0.71895D-01
3.78818737   0.26398   0.00000    0.4622   -23.33153   0.20390D-01
3.74433820   0.26707   0.00000    0.2496   -26.01755   0.10858D-01
3.70149254   0.27016   0.00000    0.9313   -20.26836   0.40334D-01
3.65961633   0.27325   0.00000    1.7275   -17.55008   0.74569D-01
3.61867704   0.27634   0.00000    1.7214   -17.56575   0.73469D-01
3.57864358   0.27944   0.00000    1.4051   -18.46163   0.59114D-01
3.53948620   0.28253   0.00000    1.3446   -18.65531   0.55917D-01
3.50117647   0.28562   0.00000    1.4876   -18.21019   0.61281D-01
3.46368715   0.28871   0.00000    1.4991   -18.17621   0.61101D-01
3.42699217   0.29180   0.00000    1.8798   -17.17648   0.76102D-01
3.39106655   0.29489   0.00000    3.3736   -14.57000   0.13724D+00
3.35588633   0.29798   0.00000    4.9288   -12.85306   0.20167D+00
3.32142857   0.30108   0.00000    4.9803   -12.80557   0.20179D+00
3.28767123   0.30417   0.00000    3.9699   -13.83631   0.15754D+00
3.25459318   0.30726   0.00000    4.4702   -13.29809   0.17653D+00
3.22217410   0.31035   0.00000    7.1514   -11.13383   0.28767D+00
3.19039451   0.31344   0.00000    8.1009   -10.54776   0.32599D+00
3.15923567   0.31653   0.00000    5.0138   -12.77491   0.19330D+00
3.12867956   0.31962   0.00000    2.2112   -16.45671   0.82002D-01
3.09870887   0.32272   0.00000    1.0293   -19.82967   0.37355D-01
3.06930693   0.32581   0.00000    0.1481   -28.28877   0.52759D-02
3.04045770   0.32890   0.00000    0.1852   -27.31623   0.65380D-02
3.01214575   0.33199   0.00000    1.0189   -19.87434   0.35940D-01
2.98435620   0.33508   0.00000    1.3587   -18.60930   0.47649D-01
2.95707472   0.33817   0.00000    1.9057   -17.11587   0.66590D-01
2.93028751   0.34126   0.00000    2.5600   -15.80493   0.89238D-01
2.90398126   0.34435   0.00000    1.4304   -18.38282   0.48848D-01
2.87814313   0.34745   0.00000    0.5335   -22.70502   0.17895D-01
2.85276074   0.35054   0.00000    1.2855   -18.85321   0.43060D-01
2.82782212   0.35363   0.00000    1.5733   -17.96307   0.52393D-01
2.80331575   0.35672   0.00000    1.8395   -17.27237   0.60893D-01
2.77923048   0.35981   0.00000    3.1410   -14.89073   0.10447D+00
2.75555556   0.36290   0.00000    3.5971   -14.28138   0.11918D+00
2.73228057   0.36599   0.00000    3.1212   -14.91904   0.10204D+00
2.70939548   0.36909   0.00000    2.7439   -15.49541   0.88606D-01
2.68689057   0.37218   0.00000    3.4357   -14.48807   0.11081D+00
2.66475645   0.37527   0.00000    4.2531   -13.52418   0.13721D+00
2.64298401   0.37836   0.00000    3.6666   -14.19518   0.11660D+00
2.62156448   0.38145   0.00000    2.0308   -16.83432   0.62988D-01
2.60048934   0.38454   0.00000    0.9156   -20.34299   0.27854D-01
2.57975035   0.38763   0.00000    0.3588   -24.43542   0.10769D-01
2.55933953   0.39073   0.00000    0.0912   -30.39579   0.27082D-02
2.53924915   0.39382   0.00000    0.0898   -30.46445   0.26448D-02
2.51947172   0.39691   0.00000    0.3070   -25.11568   0.89924D-02
2.50000000   0.40000   0.00000    0.8961   -20.43731   0.26202D-01
```

Fig. 5 (continued)

Fig. 6 Residual.dat

residual.dat

```
 4.00000000     0.136902     0.020659     0.099295
 5.00000000     0.390090     0.195939     0.216906
 6.00000000     0.141527    -0.010035     0.028077
 7.00000000    -0.027663     0.006112     0.014774
 8.00000000    -0.307199    -0.112845     0.169611
 9.00000000    -0.155525     0.008707     0.016813
10.00000000     0.018286    -0.006338     0.014675
11.00000000     0.258915     0.071470     0.135608
12.00000000     0.185465    -0.013855     0.025962
13.00000000     0.146284     0.051516     0.047003
14.00000000    -0.289092    -0.242544     0.300716
15.00000000    -0.184411     0.043785     0.071505
16.00000000    -0.574695    -0.092414     0.181769
17.00000000    -0.762129    -0.036807     0.061586
18.00000000    -0.774982    -0.003915     0.018518
19.00000000    -0.375381     0.156270     0.166114
20.00000000    -0.020643     0.133265     0.138337
21.00000000     0.255015     0.148112     0.178431
22.00000000     0.136920    -0.000202     0.013822
23.00000000    -0.033641    -0.071170     0.082602
24.00000000    -0.088730    -0.041569     0.052496
25.00000000    -0.037063     0.060317     0.055394
26.00000000    -0.052219     0.121397     0.101179
27.00000000    -0.272747    -0.013245     0.025685
28.00000000    -0.171839     0.074215     0.097565
29.00000000    -0.179159    -0.088177     0.105701
30.00000000     0.088596     0.006806     0.037650
31.00000000    -0.039406    -0.146157     0.158515
32.00000000    -0.250330    -0.224556     0.169559
33.00000000    -0.138639     0.026616     0.082834
34.00000000     0.024098     0.223322     0.132295
35.00000000    -0.166605     0.021326     0.104316
36.00000000    -0.421756    -0.163866     0.111038
37.00000000    -0.420016    -0.011206     0.109846
38.00000000    -0.393538     0.110551     0.069802
39.00000000    -0.579378    -0.123229     0.238984
40.00000000    -0.353845    -0.006742     0.011877
41.00000000    -0.220510     0.098102     0.047171
42.00000000    -0.377541     0.010176     0.073609
43.00000000    -0.443573     0.000857     0.017898
44.00000000    -0.577567    -0.151378     0.108172
45.00000000    -0.210087     0.203764     0.289928
46.00000000    -0.483433     0.024244     0.045840
47.00000000    -0.717095    -0.058084     0.049688
48.00000000    -0.687379     0.017306     0.047067
49.00000000    -0.572158    -0.000956     0.011938
50.00000000    -0.340500     0.033840     0.033040
51.00000000    -0.432759    -0.155742     0.126587
52.00000000    -0.225004     0.071701     0.170177
53.00000000    -0.327111    -0.009010     0.012806
54.00000000    -0.239819     0.039630     0.057173
55.00000000    -0.250350     0.013172     0.018205
56.00000000    -0.444400    -0.078369     0.051262
57.00000000    -0.332266     0.200468     0.202154
58.00000000    -0.632851    -0.029347     0.172729
59.00000000    -0.567436    -0.055986     0.099174
60.00000000    -0.393928    -0.028075     0.041169
61.00000000    -0.185856     0.121355     0.083462
62.00000000    -0.317383     0.019828     0.060852
63.00000000    -0.353419     0.001687     0.009986
64.00000000    -0.398449    -0.062312     0.048484
65.00000000    -0.515497    -0.129946     0.090929
66.00000000    -0.583248    -0.003118     0.014153
67.00000000    -0.617788     0.200791     0.099294
```

CHARACTERISTICS OF RESIDUAL SERIES

RES. SERIES QUADRATIC NORM: 0.510017D+02 Volt^2
WEIGHTED RMS OF FIT: 0.026745 Volt
REDUCED CHI-SQUARE: 1.108733

SOLUTION FOR KNOWN CONSTITUENTS

Significance of parameters based on: SIGMA*EXPANSION FACTOR
Expansion factor = 3.453

DESCRIPTION	NAME	No	VALUE (Hour)	AMPLITUDE (Volt)	SIGMA (Volt)	PHASE (DEG)	SIGMA (DEG)	SIGNIF 99.0%
Datum shift		1	0.40000000D+01	0.57213082D-01	0.13735621D-01			YES
linear trend				-.93436397D-02	0.33694875D-03			YES
Periodic constituent	NAME 1	1	0.28178300D+02	0.15418611D+00	0.10862991D-01	28.0739	0.3762	YES
Periodic constituent	NAME 2	2	0.19679000D+02	0.45139795D-01	0.10175255D-01	143.9299	0.4839	YES
Periodic constituent	NAME 3	3	0.14055700D+02	0.65152391D-01	0.94792908D-02	297.0741	0.5952	YES
Periodic constituent	NAME 4	4	0.12262000D+02	0.57354734D-01	0.91017295D-02	300.3566	0.5843	YES
Periodic constituent	NAME 5	5	0.10076000D+02	0.18129704D+00	0.11150106D-01	72.6110	0.5777	YES
Periodic constituent	NAME 6	6	0.81954000D+01	0.98219412D-01	0.94454427D-02	254.2986	0.7372	YES
Periodic constituent	NAME 7	7	0.75816000D+01	0.71004406D-01	0.98089119D-02	254.0131	0.5470	YES
Periodic constituent	NAME 7	8	0.49587000D+01	0.83885502D-01	0.85238017D-02	58.3804	0.5501	YES

Fig. 7 Characteristics of residuals

2 Related Work

Pagiatakis used least-square spectrum to find the significant peaks in a time series [3]. A noise is a part of a signal it can white noise or nonwhite random noise and Gaussian rules may or may not be followed to find the spectral density. Periodic and non-periodic noise might also be included in a systematic noise. This non-periodic noise is actually the factor of datum shifts in a time series. So he used LSSA for time series with datum shift and unequally spaced. It is used to find the significant peaks.

3 Motivation

This project is concerned with the conversion of legacy system program for least-square spectral analysis (LSSA), which is in "FORTRAN 77" code into "C" code, so that it is easily understood by those who are not familiar with "FORTRAN" language. Even with current technological advancements we still have to use the old program code, as it is a legacy system and cannot be changed easily. These codes are hard to understand and debug by a person who is new to this language. This program takes a lot of time for processing and plotting the spectrum and histogram chart for a huge set of data e.g., big data-set from satellites. So, there is a need to make this program be able to run in parallel mode so that its performance is increased and produces results faster. It may be even possible to achieve optimal performance without using optimization of the systems.

4 Tools and Technique Applied

4.1 f2c

f2c [4] known as Fortran77-to-c and it is a source code translator. Because f2c is its input and its output is c, and can be compiled natively on Unix operating systems, f2c offers a very transportable compiler solution and offers a means of converting large Fortran libraries (e.g., LAPACK) into c. The converted program needs to be rectified and modified manually to make sure that the program runs as the original program in its native language.

4.2 Par4All

Par4All is an open-source project and it merges many open source developments. This helps in finding which part of the program can be converted into parallel code and accordingly we can convert the sequential code into parallel programming code.

4.3 OpenMP

OpenMP stands for Open Multi-Processing [5]; it is an API which supports multiple platforms and provides developers and programmers with a means to parallel programming granting them with freedom to choose the number of nodes that can be used to execute the program being developed and thus improving the performance.

5 Proposed Methodology

5.1 Conversion to C

Since this code has been written in Fortran, which is a legacy language, it is hard for us to understand, and so we convert the code to C so that we can understand the code properly. We used f2c and then after conversion checked the output.

5.2 Conversion Using Par4all and OpenMP

Since we need to deal with large dataset we need to make the execution process faster. To do this we used Par4all and OpenMP. The flowchart shows the process.

6 Results

The result is shown as snapshots. The converted C programs also give the same output.

6.1 *The Input Time Series File*

The format of the input series is FREE FORMAT. Fields must be separated by space. The time series can be equally or unequally spaced. No two equal times are accepted. If you violate this condition, LSSA will abort and will give you error message in file **lssa.out.**

6.2 *The Output Files*

Lssa.out: This is the output file with the analysis results along with all statistical tests. The spectrum is described in three different forms: percentage variance, power spectral density in dB, and power spectral density in unit2/frequency, where the unit used is the time series value unit and frequency is in cycles per time unit of the series (e.g., cycles/day, cycles/min, etc.). The spectrum is given in six columns as follows:

- Column 1: Period of spectrum in units of time of the series.
- Column 2: Frequency in cycles/unit time.
- Column 3: Fidelity: This is in units of time, and when present, it indicates the location of a significant peak.
- Column 4: Least-squares spectrum in percentage variance (%).
- Column 5: Power spectral density in dB. Please note that this spectral density is equivalent to FFT when the series is equally spaced.
- Column 6: Power spectral density in unit2/frequency. Again this spectral density (normalized by frequency) is equivalent to that of FFT when the series is equally spaced. Obviously LSSA does not require equally spaced series, contrary to FFT.
- Finally, the spectrum is printed with asterisks for quick identification of peaks.

6.3 *Residuals.dat*

This is an ASCII file that lists the input and residuals series in different columns for easy plotting with any plotting package. There are four columns as follows:

- Column 1: Time
- Column 2: Input time series values
- Column 3: Normalized residual series (after the removal of trends, periods, processes, etc.)
- Column 4: Standard deviation of residual.

6.4 *Spectrum.dat*

It is an ASCII file that contains the output spectrum (six columns) as described in **lssa.out**. Simply, this file can be used for plotting purposes. Columns are in the same order as in **lssa.out**.

7 Conclusion and Future Work

Post conversion of FORTRAN 77 code into C program code improved the maintainability of the legacy system program. It is now easier to understand the program and its purpose by looking at the documentation of the C program code. Those who do not know FORTRAN programming language can utilize this and work to maintain and improve the program. The results produced by FORTRAN 77 code and C program code was identical and so it has to be successfully converted to a working C code which can be used alongside the legacy system to help in calculation and plotting of the spectrum. Executing the parallel code of FORTRAN 77 was much faster compared to the sequential program execution, thus saving a lot of resources such as CPU time and memory, etc. The future work is to develop an application of optimization levels inbuilt with the server to further improve the performance and execution time. Increase the intake size or lines of time series in **TEST-SER.DAT** from 1,000,000 lines to 100,000,000 lines.

References

1. Wells, D., Vanicek, P., Pagiatakis, S.: Least square spectral analysis revisited (1985)
2. Vanicek, P.: Approximate spectral analysis by least-square fit. Astrophys. Space Sci. **4**, 387–391 (1969)
3. Pagiatakis, S.D.: Application of the least-squares spectral analysis to superconducting gravimeter data treatment and analysis. Cahiers du Centre Europeen de Geodynamique et Seismologie **17**, 103–113 (2000)
4. f2c (Fortran to C convertor) support site. http://www.webmo.net/support/f2c_linux.html
5. Chan, A., Gropp, W., Lusk, E.: Users guide for MPE: extensions for MPI programs, pp. 7–30. Chicago (1998)

Application of Differential Evolution for Maximizing the Loadability Limit of Transmission System During Contingency

P. Malathy, A. Shunmugalatha and T. Marimuthu

Abstract This paper deals with the improvement of transmission system loadability during single contingency by optimal location and settings of multitype flexible AC transmission system (FACTS) devices such as thyristor-controlled series capacitor (TCSC) and static var compensator (SVC). Contingency severity index (CSI) and fast voltage sensitivity index (FVSI) are used for identifying the optimal location of FACTS devices. To enhance the socio-economical benefits, the solution to the problem is optimized using differential evolution (DE) technique thereby minimizing the installation cost (IC) of FACTS devices and severity of overloading (SOL).

Keywords Maximum loadability limit · FACTS · Contingency · Ranking indices · DE

1 Introduction

Power system stability is a major issue under contingency condition in a restructured environment. Methods to improve voltage stability and to prevent the voltage collapse are discussed in [1]. Voltage stability can be maintained by enhancing the maximum loadability [2]. Continuation power flow (CPF) is the conventional

P. Malathy (✉) · T. Marimuthu
PSNA College of Engineering and Technology, Dindigul, Tamil Nadu, India
e-mail: malathypaulpandi@gmail.com

T. Marimuthu
e-mail: marimuthu1016@gmail.com

A. Shunmugalatha
Velammal College of Engineering and Technology, Madurai, Tamil Nadu, India
e-mail: shunmugalatha@gmail.com

M. Pant et al. (eds.), *Proceedings of Fifth International Conference on Soft Computing for Problem Solving*, Advances in Intelligent Systems and Computing 437, DOI 10.1007/978-981-10-0451-3_6

method for computing the maximum loadability [3]. FACTS devices assist in maximizing the power flow by increasing the power transfer capability of a long transmission line to reduce the real power losses and to improve voltage stability [4] of the system.

TCSC is one among the FACTS devices which offers faster and smooth response during contingency condition [5]. SVC is capable of maintaining voltage within the operating limits under normal and network outage conditions [6]. Contingency screening and ranking [7, 8] are used as tools for online analysis of power system security. Back propagation neural network (BPNN) is used to determine maximum loadability [9]. Genetic algorithm [10] (GA), evolutionary programming (EP), and particle swarm optimization (PSO) [11] have been used to solve the problem of complex stability. PSO exhibits the issue of premature convergence technique resulting from incorporation of PSO with breeding and sub-population process of genetic algorithm and is known as hybrid particle swarm optimization (HPSO) [12]. Differential evolution [13–16] is a heuristic optimization technique which has been widely used to tackle various power system problems. MATPOWER [17] is a MATLAB-based software tool which is used to analyze various power system problems.

The proposed work is to improve voltage stability during line and generator contingency by maximizing the loadability. The maximum loadability is achieved through reactive power compensation which deploys placement of multi-type FACTS devices namely SVC and TCSC. Optimal location and setting of TCSCs and SVCs are obtained based on CSI and FVSI, respectively, so as to reduce the installation cost. In order to validate the proposed approach, the problem is simulated using WSCC-9 bus system and the solution is optimized using DE.

2 Methodology

2.1 Problem Formulation

The objective function to maximize the loading factor λ and to minimize the cost of installation of FACTS devices is given by

$$\begin{aligned} F &= \text{Max}\, Z + \text{Min}\, \text{IC} \\ Z &= \lambda \\ \text{IC} &= S_{\text{ab}} * C_{\text{d}} * 1000 \end{aligned} \tag{1}$$

where

λ Loading factor;

IC Optimal cost of installation of multi-type FACTS devices;

C_{d} Cost of installation of FACTS device(s);

d FACTS device(s) SVC and /or TCSC;
S_{ab} $Q_a - Q_b$, VAr compensation provided by SVC and TCSC in KVAr;
Q_b MVAr flow in the branch before placing the FACTS devices;
Q_a MVAr flow in the branch after placing the FACTS devices.

This objective function is subjected to the equality and inequality constraints. The equality constraints include voltage stability and power balance constraints given by Eqs. (2)–(5). The constraints for FACTS devices contribute for inequality given by Eqs. (6) and (7):

$$\begin{aligned} V_s &= 1 \text{ if } V_{\min i} \le V_i \le V_{\max i} \\ &= 0 \text{ otherwise} \end{aligned} \tag{2}$$

$$\sum_{i=1}^{n_b} P_{gi} = \sum_{i=1}^{n_b} P_{di} + P_{loss} \tag{3}$$

$$P_i = \sum_{i=1}^{n_b} |E_i|\,|E_j|\,\left[G_{ij}\cos\left(\theta_i - \theta_j\right) + B_{ij}\sin\left(\theta_i - \theta_j\right)\right] \tag{4}$$

$$Q_i = \sum_{i=1}^{n_b} |E_i|\,|E_j|\,\left[G_{ij}\sin\left(\theta_i - \theta_j\right) + B_{ij}\cos\left(\theta_i - \theta_j\right)\right] \tag{5}$$

$$-0.5X_{Line} \le X_{TCSC} \le 0.5X_{Line} \tag{6}$$

$$-200\,\text{MVAr} \le Q_{SVC} \le 200\,\text{MVAr} \tag{7}$$

where
P_{gi}, P_{di} Real power generation and real power demand, respectively;
P_{loss} Transmission lines losses;
θ_i, θ_j Phase angles at buses i and j, respectively;
E_i, E_j Voltage magnitudes at buses i and j, respectively;
G_{ij}, B_{ij} Conductance and susceptance of the Y bus matrix;
V_i Voltage at bus i;
X_{Line} Actual line reactance;
X_{TCSC} Reactance offered by TCSC;
Q_{SVC} Var compensation offered by SVC;
P_i, Q_i Real power and reactive power at bus i;
V_{min}, V_{max} Minimum and maximum voltage limits of bus i;
n_b Number of buses.

2.2 Optimal Location of FACTS Devices

The step-by-step procedure for optimal placement of TCSC and SVC is given below.

Step 1: Find the participation matrix. The participation matrix (U) is used for identifying the branches which are overloaded. The size of the binary matrix ($x * y$) consists of 1's or 0's based on whether or not the corresponding branch is overloaded, where q is the total number of branches of interest and p is the total number of single contingency.

Step 2: Find the ratio matrix. The elements of ratio matrix (W) are normalized overflow of all the branches. The size of the matrix is ($x * y$) in which each element denotes the excess power flow with respect to the base case flow through branch y during contingency x.

$$W_{xy} = \frac{P_{xy,\text{Conti}}}{P_{oy,\text{Base}}} - 1 \tag{8}$$

where

$P_{xy,\text{Conti}}$ Power flow through branch y during contingency x;
$P_{oy,\text{Base}}$ Base case power flow through branch y.

Step 3: Find the contingency probability array (P). This array is used to identify the probability of branch outages. It is a ($p * 1$) array. The probability of branch outage is calculated based on the equation given below.

Step 4: Find the CSI for all the branches. The CSI is evaluated for each and every branch considering all possible single contingency using Eq. (10) and the best location for the placement of FACTS devices is identified.

$$P_{p\times 1} = \left[P_1 P_2 \cdots P_p\right] \tag{9}$$

where

$P_{xy,\text{Conti}}$ Power flow through branch y during contingency x;
$P_{oy,\text{Base}}$ Base case power flow through branch y.

$$\text{CSI}_y = \sum_{x=1}^{p} P_x U_{xy} W_{xy} \tag{10}$$

where

CSI_y Contingency severity for the branch y;
P_x Elements of contingency probability array;
U_{xy} Elements of participation matrix U;
W_{xy} Elements of ratio matrix W.

2.3 Optimal Settings of FACTS Devices

In order to minimize the severity of overloading given by Eq. (11), the best possible setting of FACTS devices is determined for all possible single contingency, and the optimization problem will have to be solved using DE technique:

$$\mathrm{SOL} = \sum_{c=1}^{p}\sum_{y=1}^{q}[a_y]\left[\frac{P_y}{P_{y\max}}\right]^4 \tag{11}$$

where

a_y Weight factor;
P_y Real power flow on branch k;
$P_{y\mathrm{max}}$ Maximum real power flow;
P Number of contingencies considered;
Q Number of branches;
C Contingency index.

2.4 Installation Cost

The costs of TCSC and SVC are given by Eqs. (12) and (13):

$$C_{\mathrm{TCSC}} = 0.0001S_{\mathrm{ab}}^2 - 0.71S_{\mathrm{ab}} + 153.75\,(\mathrm{US\$/KVAr}) \tag{12}$$

$$C_{\mathrm{SVC}} = 0.0003S_{\mathrm{ab}}^2 - 0.305S_{\mathrm{ab}} + 127.38(\mathrm{US\$/KVAr}) \tag{13}$$

where $S_{\mathrm{ab}} = Q_{\mathrm{a}} - Q_{\mathrm{b}}$, the reactive power compensation provided by SVC and TCSC in KVAr. Q_{a} and Q_{b} are the KVAr flow through the branch after and before placing them, respectively.

3 Differential Evolution

Differential evolution [13–16] is a population-based heuristic optimization algorithm. To begin with, NP number of populations, D number of variables, scaling factor ($F = 0.8$), and crossover ratio ($\mathrm{CR} = 0.8$) are initialized. The initial population is generated between upper and lower bounds of the variables using Eq. (14). A new population is then reproduced using mutation, crossover, and selection [16] given by Eqs. (15–18), respectively. The search space consists of randomly generated initial population of NP with dimension D given by

$$X_{i,G} = \left\{x_{i,G}^{1}, \ldots, x_{i,G}^{D}\right\}, \quad i = 1, \ldots, \mathrm{NP} \tag{14}$$

$$V_{i,G} = X_{\mathrm{best},G} + F(X_{r1,G} - X_{r2,G}) \quad r_1, r_2 \in \{1, \ldots., \mathrm{NP}\} \tag{15}$$

$X_{\mathrm{best},G}$ is the best individual of the population at Gth generation. Recombination is used to increase the search diversity which can be expressed as

$$ui, G = \begin{cases} v_{i,G,}^{j} \text{ if } (\mathrm{rand}_j\,[0,1] \leq \mathrm{CR}) \text{ or } (j = j_{\mathrm{rand}}) \\ x_{i,G}^{j} \text{ otherwise} \end{cases} \quad j = 1, \ldots, D \tag{16}$$

$$U_{i,G} = (u_{i,G}^{1}, u_{i,G}^{2}, \ldots, u_{i,G}^{D}) \tag{17}$$

The members of next generation can be found out by

$$X_{i,G+1} = \begin{cases} U_{i,G} \text{ if } f(U_{i,G}) \leq f(X_{i,G}) \\ X_{i,G} \quad \text{otherwise} \end{cases} \tag{18}$$

4 Results and Discussion

This section presents a case study based on WSCC 9-bus test system. The aim of this case study is to show that the proposed technique is feasible for maximum loadability problem by considering single line and generator contingency with optimal placement and setting of SVC and TCSC. The devices are located in the system based on the CSI and FVSI values calculated for all the branches and buses by considering all possible single contingencies. The program is done using MATLAB version 7.11.0.584 and simulated using Intel (R) core (TM) i3-2330 M processor with a speed of 2.2 GHz. The ML obtained from base case without any FACTS devices was 35.12 %. The problem is optimized using DE with 50 different initial solutions. The numerical results obtained are tabulated in Tables 1, 2, 3, and 4.

Table 1 Ranking of branches and buses using CSI and FVSI

Rank	Branch no.	CSI	Bus no.	FVSI
1	5	0.5039	3#	0.8232
2	2	0.4124	8	0.7699
3	6	0.4045	5	0.4701
4	8	0.3936	2#	0.4376
5	7*	0.3398	9	0.3964
6	9	0.2708	6	0.3358
7	4*	0.2505	7	0.2875
8	3	0.2505	4	0.2696
9	1*	0.1458	1#	0.2650

Table 2 ML, location, settings, and cost of FACTS devices using DE without contingency

No. of FACTS devices	ML (%)	Location		Settings		Installation cost			Run no.	Iteration no.	Time of convergence
		SVC	TCSC	X (p.u.)	Q (MVAr)	SVC ($/KVAr)	TCSC ($/KVAr)	Total cost ($/KVAr)			
1	35.17	8	–	–	−99.9338	37.5974	–	37.5974	14	29	16.5038
2	36.00	8	–	–	−53.0911	80.0656	–	165.2922	37	2	16.7849
		5	–	–	−74.9030	85.2267	–				
3	**36.50**	**8**	–	–	**−93.3881**	**43.4527**	–	**240.5727**	**26**	**9**	**16.9763**
		5	–	–	**−83.7033**	**77.1733**	–				
		–	**5**	**−0.0816**	–	–	**119.9467**				
4	36.50	8	–	–	−86.2840	49.8366	–	351.3837	49	12	17.3106
		5		−0.6157	−83.7033	76.9349	–				
		–	5	0.1672	–	–	119.8043				
		–	2	–	–	–	104.8079				
5	36.50	8	–	–	−78.0764	57.2497	–	487.6810	19	30	16.9763
		5	–	–	−90.8167	70.6529	–				
		9	–	–	−77.7401	134.9861	–				
		–	5	−0.0670	–	–	116.1181				
		–	2	−0.5968	–	–	108.6741				

Table 3 ML, location, settings, and cost of FACTS devices using DE with line contingency

No. of FACTS devices	ML (%)	Location		Settings		Installation cost		Total cost ($/KVAr)	Run no.	Iteration no.	Time of computation
		SVC	TCSC	X (p.u.)	Q (MVAr)	SVC ($/KVAr)	TCSC ($/KVAr)				
1	34.99	8	–	–	−96.7754	40.1494	–	40.1494	17	32	32.4190
2	34.99	8	–	–	−99.6170	37.8763	–	108.0060	20	21	29.9114
		5	–	–	−91.2916	70.1207	–				
3	34.99	8	–	–	−44.8962	87.6304	–	301.4718	4	28	30.2446
		5	–	–	−72.4651	87.5347	–				
		–	5	−0.1879	–	–	126.3047				
4	**35.00**	**8**	–	–	**−97.5812**	**39.6989**	–	**479.4529**	**2**	**43**	**28.9649**
		–	**5**	**−0.1900**	–	–	**176.4000**				
		–	**2**	**0.6952**	–	–	**161.4439**				
		5	–	–	**−57.0125**	**101.9100**	–				
5	34.99	8	–	–	−91.0516	45.5489	–	609.9992	50	25	32.3722
		–	5	−0.5162	–	–	176.4000				
		5	–	–	−88.8640	72.4398	–				
		–	2	−0.3470	–	–	161.4439				
		–	6	0.0335	–	–	154.1665				

Table 4 ML, location, settings, and cost of FACTS devices using DE with generator contingency

No. of FACTS devices	ML (%)	Location		Settings		Installation Cost		Total cost ($/KVAr)	Run no.	Iteration no.	Time of convergence
		SVC	TCSC	X (p.u.)	Q (MVAr)	SVC ($/KVAr)	TCSC ($/KVAr)				
1	32.99	8	–	–	−99.8209	37.6981	–	37.6981	20	4	23.0472
2	35.98	8	–	–	−98.9246	38.4985	–	120.7227	39	27	23.4055
		5	–	–	−78.2158	82.2242	–				
3	35.99	8	–	–	−86.6230	49.5312	–	236.9167	16	2	24.2106
		5	–	–	−90.3909	71.0424	–				
		–	5	−0.4691	–	–	116.3431				
4	**36.00**	**8**	–	–	**−67.8783**	**66.5171**	–	**367.9040**	**42**	**4**	**23.8779**
		5	–	–	**−90.2234**	**71.1956**	–				
		–	**5**	**−0.4374**	–	–	**116.4317**				
		–	**2**	**−0.0007**	–	–	**113.7596**				
5	35.98	8	–	–	−73.7572	61.1671	–	550.9001	8	7	24.4912
		5	–	–	−98.6438	63.5133	–				
		9	–	–	37.9785	249.6056	–				
		6	–	–	−78.1224	64.5333	–				
		–	5	−0.2793	–	–	112.0809				

The placement of FACTS devices based on the CSI and FVSI values is shown in Table 1. The values are calculated by considering all possible single contingencies. '*' represents the branches, associated with transformers, and is not preferred for the placement of TCSC, as the capacitive reactance of TCSC along with the inductive transformer reactance may create resonance problem. '#' represents the buses associated with generators and are not preferred for the placement of SVC, because all generators are the source of reactive power. It is inferred from Table 1 that the branch 5, which has a high severity index, is given first priority for the placing TCSC. Then preferences are given to other branches such as 2, 6, 8, 9, and 3, as branches 1, 4, 7 are associated with transformers. For the placement of SVC bus 8 is given first priority followed by buses 5, 9, 6, 7, and 4, as buses 1, 2, and 3 are generator buses.

4.1 Maximum Loadability Limit Without Contingency

Table 2 shows the performance of DE for ML problem without contingency. A maximum loadability of 36.50 % is obtained at twenty sixth run with computation time as 16.9763 s. For obtaining this, two SVCs are placed at buses 8 and 5 with reactive power setting of −93.3881 and −83.7033 MVAr respectively and one TCSC with reactance setting of 0.0816 p.u at line 5. The total IC for optimal setting of all these devices is 240.5727 $/KVAr. It is evident that DE yields an ML with minimal IC with moderate time of computation.

4.2 Loadability Limit with Line Contingency

The eigth line with maximum real power loss of 14.154 MW and reactive power loss of 112.57 MVAr is made out of service. Then the locations of FACTS devices are decided based on the CSI and FVSI values by considering all single contingencies for the enhancement of loadability and to maintain voltage stability.

Table 3 shows the performance of DE for ML problem with line contingency. A maximum loadability of 35 % is obtained at second run with computation time as 28.9649 s. For obtaining this, two SVCs are placed at buses 8 and 5 with reactive power setting of −97.5812 and −57.0125 MVAr respectively and two TCSCs at lines 5 and 2 with reactance setting of −0.1900 p.u. and 0.6952 respectively. The total IC for optimal setting of all these devices are 479.4529 $/KVAr. It is evident that, DE yields an ML with a minimal IC with moderate time of computation.

4.3 Loadability Limit with Generator Contingency

For this case, second generator with maximum real power generation of 163 MW is made out of service. Then the locations of FACTS devices are decided based on the

CSI and FVSI values by considering all single contingencies for the enhancement of loadability and to maintain voltage stability. Table 4 shows the performance of DE for ML problem with generator contingency. A maximum loadability of 36 % is obtained at forty second run with computation time as 23.8779 s. For obtaining this, two SVCs are placed at buses 8 and 5 with reactive power setting of −67.8783 and −90.2234 MVAr respectively and two TCSCs at lines 5, 2 with reactance setting of −0.4374, −0.0007 p.u., respectively. The total IC for optimal setting of all these devices is 367.9040 $/KVAr.

4.4 Graphical Analysis

The graphical analyses of maximum loadability and total installation cost, by increasing the number of FACTS devices for the three different categories, are presented in Figs. 1 and 2. The convergence characteristics of differential evolution for maximum loadability for 50 iterations are given in Fig. 3. From Figs. 1 and 2 it is inferred that a maximum loadability of 36.50 % is obtained for base case without contingency by placing three FACTS devices at a total installation cost of 240.5727 $/KVAr. By simulating the system with single line outage (eigth line), a maximum loadability of 35 % is obtained by placing four FACTS devices at a total installation cost of 479.4529 $/KVAr. By simulating the system with single generator outage (second generator), a maximum loadability of 36 % is obtained by placing four FACTS devices at a total installation cost of 367.9040 $/KVAr.

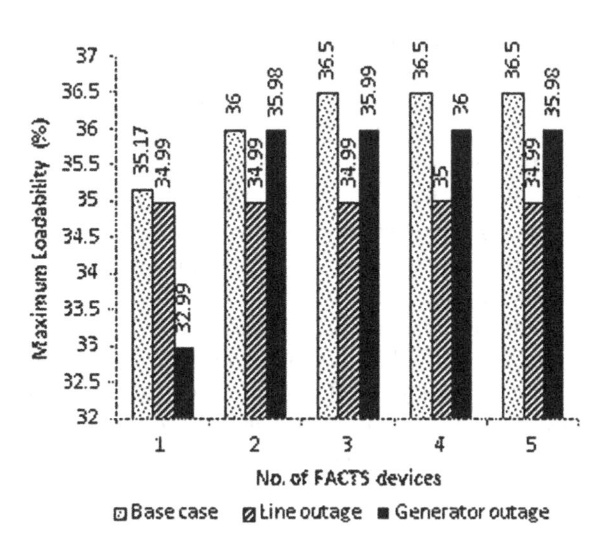

Fig. 1 ML obtained using DE for base case, line, and generator outages

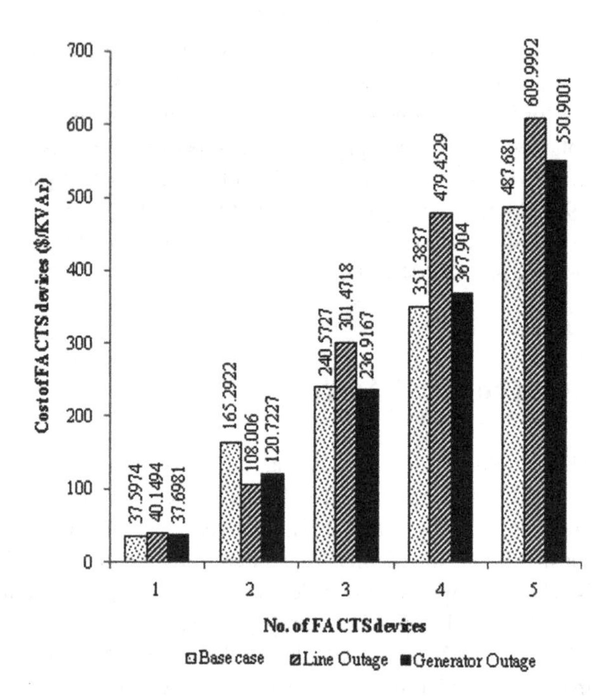

Fig. 2 Cost of FACTS devices using DE for base case, line, and generator outages

The bar chart drawn between maximum loadability and the number of FACTS devices for base case, line, and generator contingencies is shown in Fig. 1. It is inferred that the maximum loadability is increased by increasing the number of devices which in turn increases the total installation cost of FACTS devices.

Figure 2 shows the bar chart representing total IC of multi-type FACTS devices for obtaining maximum loadability using DE for the base case, line, and generator outages. For the base case an ML of 36.50 % is obtained by installing two SVCs and one TCSCs with total IC of 240.5727 $/KVAr. For line contingency an ML of 35 % is obtained by installing two SVCs and two TCSCs with total IC of 479.4529 $/KVAr. For generator contingency an ML of 36 % is obtained by installing two SVCs and two TCSCs with total IC of 367.9040 $/KVAr.

The graph plotted between maximum loadability and number of iterations for the base case without contingency by optimal placement and setting of FACTS devices is shown in Fig. 3. For base case an ML of 36.50 % is obtained at ninth iteration with time of computation as 16.9763 s. For line contingency an ML of 35 % is obtained at forty third iteration with time of computation as 28.9649 s. For generator contingency an ML of 36 % is obtained at fourth iteration with computation time as 23.8779 s. It is evident that DE yields an ML with minimal IC with moderate time of computation.

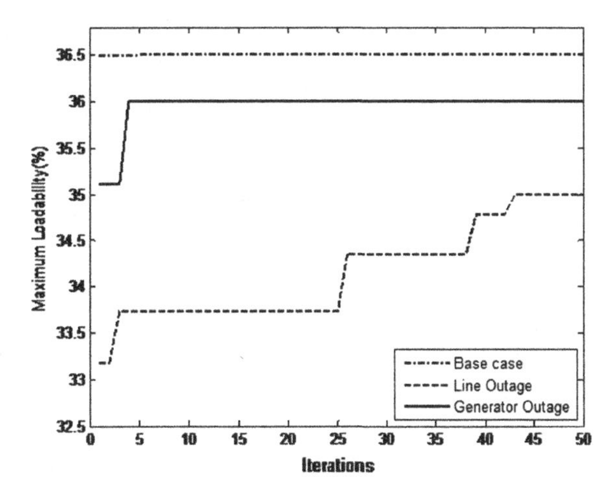

Fig. 3 Convergence characteristics of DE for base case, line, and generator outages

5 Conclusion

This work presents the differential evolution (DE) technique to enhance the voltage stability by maximizing the loadability of a transmission system through the optimal placement of TCSC and SVC by considering all possible single contingencies including generators and branches. The proposed algorithm has proved its efficiency in obtaining optimal or near-optimal solutions. The case study analyzed proves the same. To demonstrate the effectiveness and robustness of the proposed technique, Newton–Raphson power flow method is used for analyzing the voltage stability improvement by increasing the maximum loadability through the placement of series- and shunt-type FACTS devices namely TCSC and SVC. This method shows the optimal location for placement of TCSC and SVC devices and their optimal settings for the improvement of maximum loadability by installing devices economically. The solutions obtained are superior to those obtained using the conventional power flow method. This work can be extended for the enhancement of maximum loadability using other FACTS devices such as unified power flow controller (UPFC), thyristor-controlled phase angle regulator (TCPAR), etc. In future, this work will be proceeded by considering multiple contingencies and can be optimized using new efficient optimization algorithms.

References

1. Nagalakshmi, S., Kamaraj, N.: Comparison of computational intelligence algorithms for loadability enhancement of restructured power system with FACTS devices. Elsevier, Swarm Evol. Comput. **5**, 17–27 (2012)
2. Virk, P.S., Garg, V.K.: Stability enhancement of long transmission line system by using static var compensator (SVC). Int. J. Adv. Res. Electr. Electr. Instrum. Eng. **2**(9), 4361–4365 (2013)

3. Vijayasanthi, N., Arpitha, C.N.: Transmission system loadability enhancement using FACTS devices. Int. J. Electr. Electr. Eng. Telecommun. **2**(1), 137–149 (2013)
4. Mathad, V.G., Ronad, B.F., Jangamshetti, S.H.: Review on comparison of FACTS controllers for power system stability enhancement. Int. J. Scient. Res. Public. **3**(3), 1–4 (2013)
5. Amroune, M., Bourzami, A., Bouktir, T.: Voltage stability limit enhancement using thyristor controlled series capacitor (TCSC). Int. J. Comput. Appl. **72**(20), 46–50 (2013)
6. Manayarkarasi, S.P., SreeRenga Raja, T.: Optimal location and sizing of multiple SVC for voltage risk assessment using hybrid-PSO GSA algorithm. Arabian J. Sci. Eng. **39**(11), 7967–7980 (2014)
7. Vadivelu, K.R., Marutheswar, G.V.: Fast voltage stabilty index based optimal reactive power planning using differential evolution. Electr. Electron. Eng. Int. J. **3**(1), 51–60 (2014)
8. Manikandan, R., Bhoopathi, M.: Contingency analysis in deregulated power market. Int. J. Adv. Res. Electr. Electr. Instrum. Eng. **2**(12), 6310–6317 (2013)
9. Nagalakshmi, S., Kamaraj, N.: On-line evaluation of loadability limit for pool model with TCSC using back propagation neural network. Int. J. Electr. Power Energy Syst. **47**, 52–60 (2013)
10. Medeswaran, R., Kamaraj, N.: Power system loadability improvement by optimal allocation of FACTS devices using real coded genetic algorithm. J. Theoret. Appl. Inf. Technol. **65**(3), 824–830 (2014)
11. Fukuyama, H.Y.K.K.Y., Nakanishi, Y.: A particle swarm optimization for reactive power and voltage control considering voltage stability. In: IEEE International Conference on Intelligent System Applications to Power Systems, Rio de Janeiro, pp. 1–4, April 4–8 (2012)
12. EL-Dib, A.A., Youssef, H.K.M., EL-Metwally, M.M., Osman, Z.: Maximum loadability of power systems using hybrid particle swarm optimization. Elsevier, Electr. Power Syst. Res. **76**, 485–492 (2009)
13. Jithendranath, J., Hemachandra Reddy, K.: Differential evolution approach to optimal reactive power dispatch with voltage stability enhancement by modal analysis. Int. J. Eng. Res. Appl. **3** (4), 66–70 (2013)
14. Gnanambal, K., Babulal, C.K.: Maximum loadability limit of power system using hybrid differential evolution with particle swarm optimization. Elsevier, Electric. Power Energy Syst. **43**, 150–155 (2012)
15. Lenin, K., Ravindhranath Reddy, B.: Reduction of real power loss by Improved differential evolution algorithm. Int. J. Darshan Institute Eng. Res. Emerg. Technol. **3**(1), 18–23 (2014)
16. Das, S., Suganthan, P.N.: Differential evolution: a survey of the state-of-the-art. IEEE Trans. Evol. Comput. Syst. **15**(1), 4–31 (2011)
17. Zimmerman, R.D., Murillo-Sánchez, C.E., Thomas, R.J.: MATPOWER: steady-state operations, planning, and analysis tools for power systems research and education. IEEE Trans. Power Syst. **26**(1) (2011)

Common Fixed-Point Theorem for Set-Valued Occasionally Weakly Compatible Mappings in Fuzzy Metric Spaces

Vishal Gupta, R.K. Saini and Manu Verma

Abstract In our paper, we prove common fixed-point theorem on the basis of control functions in FM-spaces. Using the fuzzy edition of the Banach contraction principle, the present theorem generalized the Banach contraction principle using the continuous control function. Condition of continuity for mappings P, Q, S and T is not required. Our result generalizes the accumulation of compact subsets of Y, where the assemblage H_M itself is an FM-space. Here, we prove the result for two pairs of single-valued and set-valued owc mappings. Also, we give example to justify our result.

Keywords Fuzzy metric space (FM-space) · Occasionally weakly compatible mappings (owc)

1 Introduction

The FM-space acquainted by Kramosil and Michalek [1] is a generalization of metric space using the abstract of fuzzy sets. The theory of fuzzy sets was brought by Zadeh [2]. George and Veeramani [3] modified the belief of FM-spaces using continuous t-norm and generalized the abstract of a probabilistic metric space to a fuzzy state of affairs. Fixed-point theory has wide application areas in FM-space. Common fixed-point theorems required some basic conditions to obtain fixed point. One of the important conditions is the condition of commutativity. Jungck [4] produced the idea

V. Gupta (✉) · M. Verma
M. M. University Mullana, Ambala, Haryana, India
e-mail: vishal.gmn@gmail.com

M. Verma
e-mail: ammanu7@gmail.com

R.K. Saini
Bundelkhand University, Jhansi, UP, India
e-mail: rksaini.bu@gmail.com

M. Pant et al. (eds.), *Proceedings of Fifth International Conference on Soft Computing for Problem Solving*, Advances in Intelligent Systems and Computing 437, DOI 10.1007/978-981-10-0451-3_7

of weak compatibility by introducing the abstract of compatible maps. Al-Thagafi and Shahzad [5] weakened the concept of weakly compatible maps by giving the new idea of occasionally weakly compatible maps (owc). Doric et al. [6] have cleared that weak compatibility and occasionally weakly compatibility do not give different results with respect to single-valued mappings and in presence of unique point of coincidence. Compatibility of mappings is utilized by Chen and Chang [7] to extend a fixed-point theorem for four single-valued and set-valued mappings in complete Menger spaces. Pant et al. [8] valued the results of Chen and Chang [7] using the owc mappings. Chauhan and Kumam [9] verified fixed point for owc mappings in probabilistic metric spaces. Several interesting results for multivalued mappings and owc mappings in FM-spaces also appeared in [10].

Definition 1.1 ([11]) A binary operation $* : [0,1] \times [0,1] \to [0,1]$ is continuous t-norm if $*$ satisfies the following conditions:

(i) $*$ is commutative and associative;
(ii) $*$ is continuous;
(iii) $\alpha * 1 = \alpha$ for all $\alpha \in [0,1]$;
(iv) $\alpha * \beta \leq \gamma * \delta$ whenever $\alpha \leq \gamma$ and $\beta \leq \gamma$ for all $\alpha, \beta, \gamma, \delta \in [0,1]$.

Definition 1.2 ([3]) A triplet $(Y, M, *)$ is said to be FM-space if Y is an arbitrary set, $*$ is a continuous t-norm and M is fuzzy set on $Y^2 \times [0,\infty)$ satisfying the following conditions for all $a, b, c \in Y$ and $s, t > 0$

(F1) $M(a,b,t) > 0$;
(F2) $M(a,b,t) = 1$ for all $t > 0$ iff $a = b$;
(F3) $M(a,b,t) = M(b,a,t)$;
(F4) $M(a,b,s) * M(b,c,t) \leq M(a,c,t+s)$;
(F5) $M(a,b,.) : [0,\infty) \to [0,1]$ is left continuous.

The function $M(a,b,t)$ denotes the degree of nearness between a and b w.r.t t.

Lemma 1.3 ([3]) *In FM-space* $(Y, M, *)$, $M(a,b,.)$ *is nondecreasing for all* $a, b \in Y$.

Definition 1.4 ([3]) Let $(Y, M, *)$ be a FM-space. Then

a. a sequence $\{a_n\}$ in Y is said to be convergent to a point $a \in Y$ if for all $t > 0$, $\lim_{n\to\infty} M(a_n, a, t) = 1$.
b. a sequence $\{a_n\}$ in Y is said to be Cauchy sequence if for all $t > 0$ and $p > 0$, $\lim_{n\to\infty} M(a_{n+p}, a_n, t) = 1$.
c. A FM-space $(Y, M, *)$ is said to be complete if and only if every Cauchy sequence in Y is convergent.

Let CB(Y) denotes the accumulation of all nonempty closed subset of Y and function on CB$(Y) \times$ CB(Y) is given by

$$H_M(P,Q,t) = \min\left\{\inf_{a\in A} M(a,Q,t), \inf_{b\in B} M(P,b,t)\right\} \text{ for any } P,Q \in \mathrm{CB}(Y) \text{ and } t > 0, \tag{1}$$
where $M(c,D,t) = M(D,c,t) = \sup_{d\in D} M(c,d,t)$.

In [12] the authors have proved that H_M is FM-space.

Lemma 1.5 ([12]) *If* $P \subset CB(Y)$, *then* $a \in P$ *if and only if* $M(a,P,t) = 1$ *for* $t > 0$.

Definition 1.6 ([4]) Two self-mappings h and k of an FM-space $(Y,M,*)$ are called compatible if $\lim_{n\to\infty} M(hka_n, kha_n, t) = 1$, whenever $\{a_n\}$ is a sequence in X such that $\lim_{n\to\infty} ha_n = \lim_{n\to\infty} ka_n = z$ for some $z \in Y$.

Therefore, we interpret that h and k will be noncompatible if there exists at least one sequence $\{a_n\}$ in Y such that $\lim_{n\to\infty} ha_n = \lim_{n\to\infty} ka_n = z$ for some $z \in Y$, but $\lim_{n\to\infty} M(hka_n, kha_n, t) \neq 1$ or does not exist.

Definition 1.7 ([13]) Let $(Y,M,*)$ be an FM-space. Mappings $g : Y \to Y$ and $A : Y \to$ CB(Y) are said to be weakly compatible if $gAa = Aga$, whenever $ga \in Aa$.

Definition 1.8 ([5]) Let $(Y,M,*)$ be an FM-space. Maps H, $K : Y \to$ CB(Y) are said to be occasionally weakly compatible if there is a point a in Y such that $Ha \in Ka$ and $HKa \subseteq KHa$

The objective of the paper is to prove a fixed-point theorem for single-valued and set-valued owc mappings in FM-space.

2 Main Result

Chauhan and Kumam [9] prove the following result for two pairs of single-valued and set-valued owc mappings in Menger spaces.

Theorem 2.1 ([9]) *Let* (X,F,Δ) *be a Menger space with continuous t-norm. Further, let* $S,T : X \to X$ *be single-valued and* $A,B : X \to B(X)$ *be two set-valued mappings such that the pairs* $(S,A),(T,B)$ *are each owc satisfying*

$$_{\delta}F_{Ax,By}(t) \geq \phi(F_{Ax,By}(t)) \tag{2}$$

for all $x,y \in X$, *where* $\phi : [0,1] \to [0,1]$ *is a continuous function such that* $\phi(t) > t$ *for every* $0<t<1$, $\phi(0) = 0$ *and* $\phi(1) = 1$. *Then* $A,B,S,$ *and* T *have a unique common fixed point.*

In the present paper, we prove the same result in FM-space as follows:

Theorem 2.2 *Let* $(Y,M,*)$ *be an FM-space with continuous t-norm. Further, let* $P,Q : Y \to CB(Y)$ *be two set-valued mappings and* $S,T : Y \to Y$ *be single-valued mappings such that the pairs* $(S,P),(T,Q)$ *are each owc satisfying*

$$H_M(Pa, Qb, t) \geq \eta(M(Sa, Tb, t)) \tag{3}$$

for all $a, b \in Y$, *where* $\eta : [0, 1] \to [0, 1]$ *is a continuous function such that* $\eta(t) > t$ *for each* $0<t<1, \eta(0) = 0$ *and* $\eta(1) = 1$. *Then* $P, Q, S,$ *and* T *have a unique common fixed point.*

Proof Since the pairs (S, P) and (T, Q) are each owc, there exist points $a, b \in Y$ such that $Sa = Pa$, $SPa = PSa$ and $Tb = Qb$, $TQb = QTb$. By Lemma 1.5 if $P \subseteq CB(Y)$, then $a \in P$ iff $M(a, P, t) = 1$ for $t > 0$. This implies $Sa \in Pa$ and $Tb \in Qb$.

Next, we claim $Sa = Tb$. For if $Sa \neq Tb$, then there exists a real number $t > 0$ such that $M(Sa, Tb, t) < 1$. Applying inequality (3) and condition (1), we have

$$\begin{aligned} M(Sa, Tb, t) &\geq H(Pa, Qb, t) \\ &\geq \eta(M(Sa, Tb, t)) \\ &= \eta(M(Sa, Tb, t)) \\ &> M(Sa, Tb, t), \end{aligned} \tag{4}$$

which is a contradiction. Hence $Sa = Tb$. Since $Sa \in Pa$, therefore $SSa \in SPa = PSa$. Also, from condition (1) we get $M(SSa, Sa, t) \geq H(PSa, Qb, t)$. Further, we claim that $Sa = SSa$. For if $Sa \neq SSa$, then existence of a real number $t > 0$ such that $M(Sa, SSa, t) < 1$. Using inequality (3) and condition (1), we have

$$\begin{aligned} M(SSa, Sa, t) &\geq H(PSa, Qb, t) \\ &\geq \eta(M(SSa, Tb, t)) \\ &= \eta(M(SSa, Tb, t)) \\ &> M(SSa, Sa, t), \end{aligned} \tag{5}$$

this gives a contradiction. Hence $Sa = SSa$. Likewise, it can be proved $Tb = TTb$, which proves that Sa is a common fixed point of P, Q, S and T. The uniqueness of common fixed point can be easily shown by inequality (3).

Example 2.3 Let $Y = [0, \infty)$ with the usual metric and $t \in [0, 1]$.

$$\text{Define} \quad M(a, b, t) = \begin{cases} \frac{t}{t + |a - b|}, & \text{if } t > 0 \\ 0, & \text{if } t = 0 \end{cases} \tag{6}$$

for all $a, b \in Y$ with t-norm defined by $T(c_1, c_2) = \min\{c_1, c_2\}$ for all $c_1, c_2 \in [0, 1]$.

Define mappings $S, T : Y \to Y$ and $P, Q : Y \to \mathrm{CB}(Y)$ by

$$P(a) = \begin{cases} \{a\} & \text{if } 0 \leq a < 1 \\ [1, a + 4] & \text{if } 1 \leq a < \infty \end{cases} \tag{7}$$

$$B(a) = \begin{cases} \{0\} & \text{if } 0 \leq a < 1 \\ [1, a+2] & \text{if } 1 \leq a < \infty \end{cases} \quad (8)$$

$$S(a) = \begin{cases} 0 & \text{if } 0 \leq a < 1 \\ a+2 & \text{if } 1 \leq a < \infty \end{cases} \quad (9)$$

$$T(a) = \begin{cases} a/4 & \text{if } 0 \leq a < 1 \\ 4a+5 & \text{if } 1 \leq a < \infty. \end{cases} \quad (10)$$

Let $\eta : [0, 1] \rightarrow [0, 1]$ be defined by $\eta(t) = t^2$ for $0 \leq t < 1$. Then P, Q, S, and T satisfy all the conditions of Theorem 2.2 and '0' is the unique common fixed point.

Corollary 2.4 *Let $(Y, M, *)$ be an FM-space with continuous t-norm. Further, let $S : Y \rightarrow Y$ be single valued and $P : Y \rightarrow CB(Y)$ be set-valued mapping such that the pair (S, P) is owc satisfying*

$$H_M(Pa, Pb, t) \geq \phi(M(Sa, Sb, t)) \quad (11)$$

for all $a, b \in Y$, where $\eta : [0, 1] \rightarrow [0, 1]$ is a continuous function such that $\eta(t) > t$ for each $0 < t < 1, \eta(0) = 0$ and $\eta(1) = 1$. Then P and S have a unique common fixed point.

3 Conclusion

In this paper, we used the accumulation of all nonempty closed subsets of Y and H_M satisfying the condition of FM-space. The owc mapping is used to find fixed point without considering the continuity of the mappings.

References

1. Kramosil, I., Michalek, J.: Fuzzy metric and statistical metric spaces. Kybernetica **11**(5), 336–344 (1975)
2. Zadeh, L.A.: Fuzzy sets. Inf. Control **8**, 338–353 (1965)
3. George, A., Veeramani, P.: On some results in FM-spaces. Fuzzy Sets Syst. **64**, 395–399 (1994)
4. Jungck, G.: Compatible mappings and common fixed points. Int. J. Math. Math. Sci. **9**(4), 771–779 (1986)
5. Al-Thagafi, M.A., Shahzad, N.: Generalized I-nonexpansive selfmaps and invariant approximations. Acta Math. Sinica, English Series **24**, 867–876 (2008)
6. Doric, D., Kadelburg, Z., Radnovic, S.: A note on owc mappings and common fixed point. Fixed Point Theory. **13**(2), 475–480 (2012)
7. Chen, C.M., Chang, T.H.: Common fixed point theorems in menger spaces. Int. J. Math. Math. Sci. Article ID 75931, 1–15 (2006)

8. Pant, B.D., Chauhan, S., Kumar, S.: A common fixed point theorem for set-valued contraction mappings using implicit relation. J. Adv. Res. Appl. Math. **4**(2), 51–62 (2012)
9. Chauhan, S., Kumam, P.: Common fixed point theorem for owc mappings in probabilistic metric spaces. Thai J. Math. **11**(2), 285–292 (2013)
10. Jungck, G., Rhoades, B.E.: Fixed points theorems for owc mappings. Fixed Point Theor. **7**, 286–296 (2006)
11. Schweizer, B., Sklar, A.: Statistical metric spaces. Pac. J. Math. **10**, 314–334 (1960)
12. Rodriguez-Lopez, J., Romaguera, S.: The Hausdorff fuzzy metric on compact sets. Fuzzy Sets Syst. **147**, 273–283 (2004)
13. Jungck, G., Rhoades, B.E.: Fixed points for set-valued functions without continuity. Indian J. Pure Appl. Math. **29**(3), 227–238 (1998)

Privacy Preservation in Utility Mining Based on Genetic Algorithm: A New Approach

Sugandha Rathi and Rishi Soni

Abstract Privacy preservation in data mining tends to protect the sensitive information from getting exploited by the nefarious users in a huge database. This paper explains the concepts of utility-based privacy mining approach using genetic algorithm for optimized computing and search to enhance security and confidentiality. A brief classification and comparison of PPDM (Privacy Preservation in Data Mining) techniques are also listed along with the techniques and the pros and cons of utility mining techniques. To hide the sensitive information, many approaches have been proposed. In this study, we propose an efficient method, for protecting high utility itemsets using genetic approach to achieve the privacy with balance between privacy and disclosure of information. The basic idea behind is to use the proposed work to enhance effectiveness measurements with certain parameters.

Keywords Data mining · Privacy preservation · Utility mining · Frequent patterns · High utility itemsets · Genetic algorithm

1 Introduction

In the past few years, data mining acted as the most influential data analyzing tool. With the expansion of database technology [1], a huge amount of beneficial data, that comprises of more of individual private information, has been collected in different fields such as company's confidential information, personal information, etc. If the original data are given directly to the diggers, it certainly produces private information disclosure [2]. Few issues which are related to these fields are as given below [3]:

Sugandha Rathi (✉) · Rishi Soni
Department of Computer Science and Engineering, ITM, Gwalior, India
e-mail: sugandharathi@gmail.com

Rishi Soni
e-mail: rishisoni17@gmail.com

M. Pant et al. (eds.), *Proceedings of Fifth International Conference on Soft Computing for Problem Solving*, Advances in Intelligent Systems and Computing 437, DOI 10.1007/978-981-10-0451-3_8

1. **Confidentiality issue:** There are many situations where data need to be shared for mutual gain but that data should not get leaked to anybody else.
2. **Very large databases:** With the technological advancement, the data are increasing exponentially. It is now becoming problem to distinguish between the useful and unnecessary data.
3. **Fraudulent adversaries:** In a multiparty system, a fraud adversary may always create problems by altering the input. This results in the loss of data as the altered input may lead to any random output.

Privacy preservation is now becoming an important and serious issue for further data mining techniques and their development with simple access to data having confidential and secure information [4]. Because of this issue, it became a topic of concern to protect the data and thus a new concept of privacy preservation in data mining was introduced. There are conditions when users do not want to disclose their information until it is assured to them that their privacy and confidentiality is guaranteed [5, 6].

Privacy Preservation in Data Mining (PPDM) was first introduced by Agrawal and Srikant in 2000 [7]. The algorithms based on PPDM are based on the fusion of data mining of the useful data and maintaining its privacy and applying protection mechanism. When a person or extractor mines the information, he/she should not be able access the data easily in the form in which user has given the input. The sensitive information that is contained in the database thus needs to be saved. For this purpose, we need more vigorous techniques in privacy preserving data mining that may change or modify the data and preserve the sensitive information [8].

1.1 Utility Mining

The method of categorizing the itemsets with high utility values is called as *utility mining* [9]. It defines all the itemsets whose utility values are equal to or above the given threshold in a database.

The utility of an itemset V, i.e., $u(V)$, gives the sum of the utilities of itemset V in all the transactions containing V. An itemset V is called a *high utility itemset* if $u(V)$ more than user defined as minimum utility, i.e., threshold [10, 11]. Basically, the main purpose of utility mining is to find regardless of whether the chosen item is profitable. Some key points about utility mining are as follows:

1. The purpose of utility-based privacy preserving techniques is to improve the query answering accuracy on anonymized table.
2. There are two aspects of utility measures, which are given below:
   - First, the less generalized attribute value gives more accurate answers in query answering.
   - Second, different attributes may have different utility in data analysis.

3. The local and external utility is used to calculate the utility of an item or itemset.
4. The local transaction utility gives the total data stored in the database such as the total number of items sold.
5. The external utility of an item is taken as information from items other than transactions such as a profit table.

1.2 Genetic Algorithm

A genetic algorithm [12] is a search technique to fairly find the accurate solutions to optimization and search issues. Genetic algorithms are an important part of evolutionary algorithms inspired by evolution. Few points related to this one are explained in brief below:

1. **Inheritance**—It is the transfer of traits from one generation to the next generation.
2. **Crossover**—In this method, two parent's crossover and the two child chromosomes generated have qualities of both the parents.
3. **Mutation**—For diversity in population this process came into existence. In mutation, one or more genes get changed or we may say change their bits that's by the new generation may have better and stronger features.
4. **Selection**—It is the selection of the fittest individuals from the population.
5. **Fitness function**—It defines a limit for the population to be fit enough to be selected. It acts as a threshold in evolution process.

A flowchart showing the genetic algorithm is shown to clarify the whole functioning of the genetic algorithm (Fig. 1).

The rest of this paper is organized as follows: Sect. 2 describes the various literature reviews for this article and in this section, various authors have proposed their work by which our work had been carried out to the next level. Section 3 explains about the effectiveness measures. This section is more important part of the article because without this the proposed work cannot be defined as per needs. Section 4 gives the idea of the proposed work. In this, the proposed methodology is used to enhance the utility mining work that we will use further. Section 5 contains the conclusion.

2 Literature Review

An additive perturbation method was proposed by Agrawal et al., which adds Gaussian noise to the data and they constructed decision tree on the perturbed data to demonstrate its utility [7].

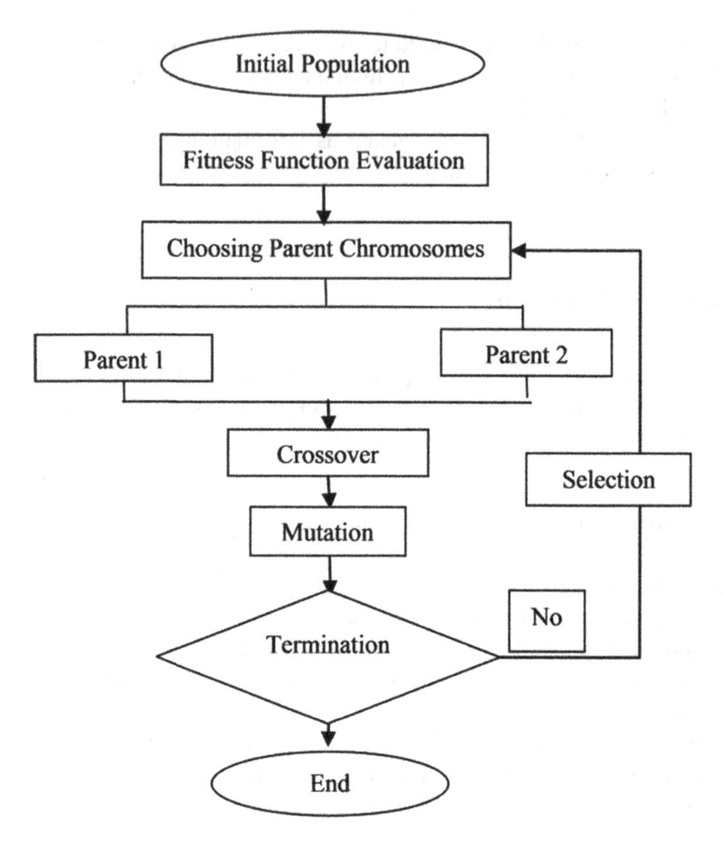

Fig. 1 Flowchart of genetic algorithm

A random rotation method which multiplies the original data matrix with a random orthogonal matrix was proposed by Chen et al. This method can preserve the distances of the original data points [13]. A random projection-based multiplicative perturbation method in which the set of data points from high-dimensional space are projected to a randomly chosen low-dimensional subspace was proposed by Liu and Kargupta [13].

While releasing data for research, one needs to think about the risk of disclosing the data up to a level that is acceptable. To limit the disclosure risk, introduction of the k-anonymity privacy requirement, which requires every record in an anonymized table to be indistinguishable to differentiate with at least k other entries within the dataset, was proposed by Samarati et al. and Sweeney [14]. k-anonymity does not provide sufficient protection against attribute disclosure. l-diversity is used to overcome the attribute level privacy. l-diversity was proposed to solve the homogeneity attack of the k-anonymity. It mainly focuses on the sensitive attribute's range of every group. In this, for each sensitive attribute, anonymized group must hold at least l well-represented values.

A new field was developed for preserving the privacy of the data using cryptographic techniques. This branch became popular [7] for two main reasons:

- Cryptography provides great level of privacy by encrypting the data with various techniques that consists of methods that quantify it.
- Various tools have been developed which provide cryptography in better ways and thus, execute privacy preservation efficiently.

Yeh et al. [15] proposed algorithms for privacy preservation in utility mining. The purpose of HHUIF (Hiding High Utility Item First Algorithm) is to decrease the quantity of the high utility itemsets so that they can have their utility value below the threshold and thus can be preserved from any kind of fraudulent adversaries. It modifies the quantity of the sensitive itemset whose value is highest in the database such that the value modified can not be accessed by any third party. The MSICF (Maximum Sensitive Itemsets Conflict First algorithm) reduces the total number of modified items from the original database. The MSICF picks up an item which has the maximum conflict count among all the sensitive itemsets present in the database.

A GA-based approach to Hide Sensitive High Utility Itemsets [16] is a genetic algorithm-based privacy preservation method to find actual and appropriate transactions to be inserted into the database for hiding sensitive high utility itemsets. The contributions can be shown as below:

- Adoption of a GA-based algorithm to sanitize the original database so as to hide sensitive high utility itemsets through transaction insertion.
- Also a new concept of pre-large concepts is introduced so as to reduce the execution time for re-scanning the original database in chromosome evaluation.
- An effectiveness evaluation function is considered to check the side effects in PPUM such as hiding failures, missing costs, and artificial costs.

The work in [3] discusses about the decision tree method called ID3, which indicates the approximations of the best splitting attributes. A privacy preserving EM clustering over horizontally partitioned datasets was proposed by Lin et al. [16] Their method chooses to disclose the means and the covariance matrix [17]. A new method for horizontally partitioned data was proposed by Cheung et al. Distributed classification has also been proposed [18]. A new approach known as meta-learning approach has been developed that mainly use classifiers trained at various sites to develop a global classifier [18–20].

A secure multi-party computation and secure matrix inverse protocols were proposed by Du et al. [23] and hence designed privacy preserving linear regression protocols for vertically partitioned datasets. Their method protects the data matrix by multiplying with random matrices, which cannot provide theoretical guarantee about privacy [21] (Fig. 2).

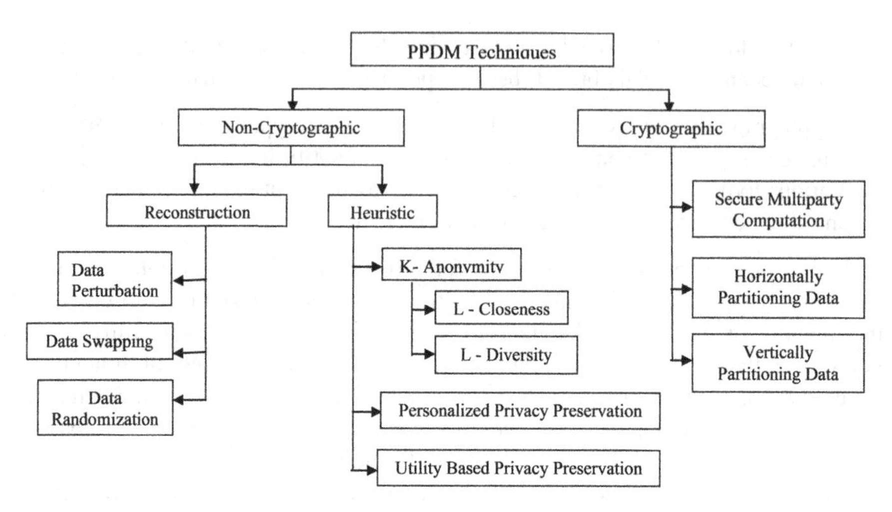

Fig. 2 PPDM techniques

3 Effectiveness Measurements

To measure the effectiveness of the general algorithms, this adopts the performance measures introduced by Oliveira and Zaine [22]. It was observed that there is no PPDM algorithm which satisfies all the possible evaluation criteria. It is important for development and evaluation of PPDM algorithms to determine the suitable evaluation framework that covers all possible criteria related to problems.

The quality of privacy preserving algorithms can be measured with the following parameters that we will apply in our proposed work:

(a) *Hiding failure* (*HF*): The ratio of sensitive itemsets that are disclosed before and after the sanitizing process. It can also be defined as the sensitive information that is not hidden during the process or by applying the application. The hiding failure is calculated as follows:

$$\text{HF} = \frac{|U(D')|}{|U(D)|'} \quad (1)$$

where $U(D)$ and $U(D')$ denote the sensitive itemsets discovered from the original database D and the sanitized database D', respectively. Most of the algorithms developed try to make the hiding failure factor zero. The cardinality of a set S is denoted as $|S|$.

(b) *Miss cost* (*MC*): The difference ratio of legitimate itemsets found in the original and the sanitized databases. The misses cost is measured as follows:

$$\text{MC} = \frac{|\sim U(D) - \sim U(D')|}{|\sim U(D)|} \tag{2}$$

where $\sim U(D)$ and $\sim U(D')$ denote the nonsensitive itemsets discovered from the original database D and the sanitized database D', respectively.

(c) *Time taken*: It defines the total time taken by the algorithm to perform its task of preserving the itemsets and saving it into the new sanitized database DB. This function performs a comparison of the total time taken by the previous HHUIF algorithm and the new hybrid algorithm based on genetic algorithm.

$$\begin{aligned}\text{TT} = &\text{Total time taken from the initialization of} \\ &\text{algorithm till all the itemsets are sanitized.}\end{aligned} \tag{3}$$

(d) *Performance*: It defines the performance of the system related to the size of the database and the time taken to scan the database. Mathematically it can be defined for n transactions

$$\text{Per} = \frac{\text{total size of the database}}{\text{Time}} \tag{4}$$

4 Proposed Work

In the proposed work, we are working on a hybrid algorithm that makes a modification of the HHUIF algorithm using genetic approach. A genetic algorithm is applied to make an algorithm more efficient and more secure. The whole process consists of the following steps:

1. *Collection* Collecting the database and preprocessing the data and sending it as input to find frequent itemsets.
2. *Mining Algorithm* Applying the high utility mining algorithm for generating the high utility itemsets.
3. *Condition to check high utility itemsets* Check the high utility itemsets generated by algorithm whether they are high utility itemsets or not. If they are not high utility itemsets then directly send them to the new sanitized database.
4. *Genetic Algorithm* Else perform crossover and mutation in a repeated manner till the value of high utility itemset reaches below the threshold. Here, in our process fitness function is considered as the threshold.
5. *New Sanitized Database* Save the generated outcomes in a new database which is called sanitized database. The main purpose of the sanitizing algorithm is to sanitize the database, i.e., cleaning the database and preserving the high utility values.

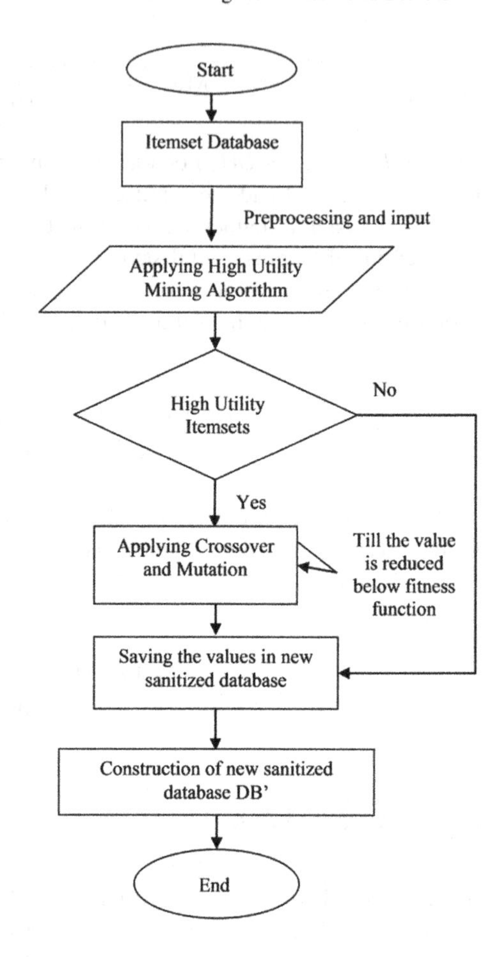

Fig. 3 Flowchart for whole process to find out the sanitized data

The proposed concept is shown with the help of a flowchart. It explains the whole process to find the sanitized data using the new genetic approach by Fig. 3. Further we will develop an algorithm and implement this concept with all the proposed effective measurements.

5 Conclusions

In this paper, we discussed about various existing techniques that focused on privacy preservation data mining so far. The utility mining concept has also been described as an emerging component for data mining to enhance its advantageous features. Two most important algorithms have been used to determine the utility values for the proposed work as we have discussed. Also we have proposed the new concept using genetic algorithm. The purpose of using genetic algorithm's concept is to implement the HHUIF for utility mining as a proposed useful solution. Future work can be carried out to design an algorithm for the proposed solution and compare its aspects with existing algorithms.

References

1. Fayyad, U.M., Shapiro, G.P., Smyth, P.: From data mining to knowledge discovery in databases. AI Mag. 37–53 (1996). ISSN 0738-4602
2. Han, J., Kamber, M.: Data Mining: Concepts and Techniques, 2nd edn. Morgan Kaufmann Publishers
3. Lindell, Y., Pinkas, B.: Privacy-Preserving Data Mining. CRYPTO (2000)
4. Bertino, E., Fovino, I.N., Provenza, L.P.: A framework for evaluating privacy preserving data mining algorithms. Data Min. Knowl. Disc. **11**(2), 121–154 (2005)
5. Mukherjee, S., Chen, Z., Gangopadhyay, A.: A privacy-preserving technique for Euclidean distance-based mining algorithms using Fourier-related transforms. VLDB J. **15**(4), 293–315 (2006)
6. Kantarcioglu, M., Clifton, C.: Privacy-preserving distributed mining of association rules on horizontally partitioned data. IEEE Trans. Knowl. Data Eng. **16**(9), 1026–1037 (2004)
7. Agrawal, R., Srikant, R.: Privacy-preserving data mining. In: Proceedings of ACM SIGMOD'00, pp: 439–450, Dallas, Texas, USA (2000)
8. Boora, R.K., Shukla, R., Misra, A.K.: An improved approach to high level privacy preserving itemset mining. (IJCSIS) Int. J. Comput. Sci. Inf. Secur. **6**(3) (2009)
9. Shankar, S., Purusothoman, T.P., Jayanthi, S., Babu, N.: A fast algorithm for mining high utility itemsets. In: Proceedings of IEEE International Advance Computing Conference (IACC 2009), Patiala, India, pp. 1459–1464 (2009)
10. Yao, H., Hamilton, H., Geng, L.: A unified framework for utilty-based measures for mining itemsets. In: Proceedings of the ACM International Conference on Utility-Based Data Mining Workshop (UBDM), pp. 28–37 (2006)
11. Verykios, V.S., Bertino, E., Fovino, I.N., Provenza, L.P., Saygin, Y., Theodoridis, Y.: State-of-the-art in privacy preserving data mining, SIGMOD Rec. **33**(1), 50–57 (2004)
12. Berkhin, P.: A survey of clustering data mining techniques. In: Grouping Multidimensional Data, pp. 25–71 (2006)
13. Liu, K., Kargupta, H.: Random perturbation-based multiplicative data perturbation for privacy preserving distributed data mining. IEEE Trans. Knowl. Data Eng. **18**(1) (2006)
14. Samarati, P.: Protecting respondent's identities in microdata release. IEEE Trans. Knowl. Data Eng. **13**(6), 1010–1027 (2001)
15. Yeh, J.S., Hsu, P.C. Hhuif and msicf: Novel algorithms for privacy preserving utility mining. Expert Syst. Appl. **37**(7), 4779–4786 (2010)
16. Lin, C.W., Hong, T.P., Wong, J.W., Lan, G.C., Lin, W.Y.: A GA-based approach to hide sensitive high utility itemsets, Hindawi Publishing Corporation. Sci. World J. **2014** (Article ID 804629), 12 (2014)
17. Yao, A.C.: Protocol for secure computations. In: Proceedings of the 23rd Annual IEEE Symposium on Foundation of Computer Science, pp. 160–164, (1982)
18. Cheung, D.W.L., et al.: Efficient mining of association rules in distributed databases. IEEE Trans. Knowl. Data Eng. 8(6), 911–922 (1996)
19. Chan, P.: An extensible meta-learning approach for scalable and accurate inductive learning. PhD Thesis, Department of Computer Science, Columbia University, New York, NY (1996)
20. Chan, P.: On the accuracy of meta-learning for scalable data mining. J. Intell. Inf. Syst. **8**, 5–28 (1997)

21. Polat, H., Du, W.: Privacy-preserving collaborative filtering using randomized perturbation techniques. In: Proceedings of the Third IEEE International Conference on Data Mining (ICDM'03), pp. 625–628, Melbourne, FL, 19–22 Nov 2003
22. Oliveira, S.R.M., Zaïne, O.R.: A framework for enforcing privacy in mining frequent patterns, Technical Report, TR02-13, Computer Science Department, University of Alberta, Canada, June (2000)
23. Du, W., Han, Y., Chen, S.: Privacy preserving multivariate statistical analysis Linear regression and classification. In: Proceedings of the 4th SIAM International Conference on Data Mining, pp. 222–233, Florida, 22–24 Apr 2004

Reliability Evaluation of Software System Using AHP and Fuzzy TOPSIS Approach

Chahat Sharma and Sanjay Kumar Dubey

Abstract The paper evaluates the reliability of software systems using the multi-criteria decision-making (MCDM) approaches. In this paper, object-oriented software systems are evaluated using the analytic hierarchy process (AHP) and fuzzy technique for order preference by similarity to ideal solution (FTOPSIS). The selection criteria are determined on the basis of ISO/IEC 25010 quality model. The approaches evaluate and select the most reliable object-oriented software system considering the fuzzy nature of decision-making process. The work is different in nature and easy in comparison to the other reliability evaluation approaches which can be explored according to the needs of an individual from various paradigms of software industry.

Keywords Reliability · Object-oriented system · AHP · FTOPSIS · Model

1 Introduction

Software reliability serves as a vital characteristic of software quality. Software reliability in the standard form can be defined as the probability of software execution without failure in a specified environment for a specified period of time [1]. IEEE defines software reliability as the ability of a system or component to perform its required functions under stated conditions for a specified period of time [2]. Software becomes unreliable due to software failures which occur as a result of software errors.

The measurement of reliability is a difficult task due to the highly complex nature of software. To achieve reliable software, the gap between software development

C. Sharma (✉) · S.K. Dubey
Amity School of Engineering and Technology, Amity University,
Sec.-125, Noida, UP, India
e-mail: chahat.s.03@gmail.com

S.K. Dubey
e-mail: skdubey1@amity.edu

M. Pant et al. (eds.), *Proceedings of Fifth International Conference on Soft Computing for Problem Solving*, Advances in Intelligent Systems and Computing 437, DOI 10.1007/978-981-10-0451-3_9

process and software reliability is bridged through object-oriented technology which provides broad description of software's nature and internal structure [3].

ISO/IEC 25010, extended model of 9126 quality model is used for the selection of reliability criteria on the basis of which reliability is evaluated [4]. The reliability features are mapped with certain object-oriented features establishing a relationship between software reliability and object-oriented technology as given below:

1. *Maturity*—It refers to the standard to which a particular software or component meet the reliability needs to work under normal condition. The factor is mapped with multi-threaded feature of objectoriented programming [5].
2. *Recoverability*—It refers to the standard to which while an event of failure or interruption of system or component recover the affected part of data and re-establish the original state of system. The mapping of it is done with high performance feature of object-oriented programming [6].
3. *Availability*—It is the standard to which software or a component is available for use, access or operation. The subattribute is mapped with the portability feature of object-oriented programming [7].
4. *Reliability Compliance*—It is the ability of a particular system or a component to conform to the certain set of rules, regulations, and standards of software reliability. The subattribute is mapped to the robust feature of object-oriented programming [8].

As reliability depends on various factors it is a multi-criteria decision-making (MCDM) problem. MCDM administers the problems which require selection of an option from a set of alternatives having various characteristics [9]. There are various MCDM approaches such as analytic hierarchy process (AHP), compromise programming (CP), and preference ranking organization method for enrichment evaluation (PROMETHEE), fuzzy analytic hierarchy process (FAHP), and fuzzy technique for order preference by similarity to ideal solution (FTOPSIS), analytic network process (ANP), etc. [10–16].

In present work, three object-oriented software systems Eclipse (P_1), NetBeans (P_2), and Android (P_3) are taken which are written in object-oriented language. These projects are evaluated with respect to four software reliability selection criteria taken from ISO 25010 quality model which are Maturity (C_1), recoverability (C_2), availability (C_3), andreliability compliance (C_4). These software systems are evaluated using MCDM approaches, AHP and Fuzzy TOPSIS. Fuzzy TOPSIS is used to tackle the imprecision and uncertainty in decision maker's judgment which incorporates fuzzy set theory to select a reliable software system.

2 Materials and Approaches

The relationship between object-oriented software systems and the selection criteria is shown through the software reliability relationship model in Fig. 1. The evaluation is done by applying the analytic hierarchy process (AHP) and fuzzy technique for order preference by imilarity to ideal solution (FTOPSIS).

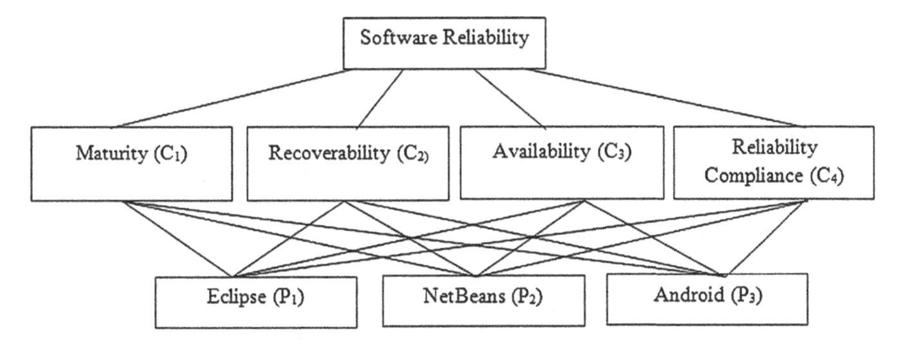

Fig. 1 Software reliability relationship model

2.1 *The Analytic Hierarchy Process Using MakeItRational Tool*

The MakeItRational tool is used for decision-making process which implements the analytic hierarchy process (AHP). It ranks finite number of alternatives with respect to finite number of criteria. The alternative with the highest rank is found to be comparatively more suitable option. The tool deals with the quantitative values and produces results in the form of numeric values or graph visuals. The inputs fed are criteria, alternatives, criteria weights, alternatives evaluation with respect to the criteria. The output is generated in the form of visualization graph.

The procedure of MakeItRational tool is given as [17]:

1. *Defining alternatives* The list of all alternatives which will be evaluated is defined.
2. *Defining criteria* The finite set of criteria for the decision problem is defined.
3. *Prioritizing criteria* The pairwise comparison of criteria on the basis of Saaty rating scale [10].
4. *Scoring alternatives* The pairwise comparison of alternatives in context to criteria based on Saaty rating scale is done.
5. *Final ranking of alternatives* The ranking show the most suitable alternative.

2.2 *The Fuzzy Technique for Order Preference by Similarity to Ideal Solution (FTOPSIS)*

The Technique for order preference by similarity to ideal solution (TOPSIS) was developed by Hwang and Yoon [18] to solve multi-criteria decision-making problem. The extension of this approach is given by Chen [15] which uses linguistic variables instead of numerical values in order to obtain the rating of every alternative with respect to each criterion.

The steps of the TOPSIS approach by Chen are as follows:

1. Establish committee of decision makers. Let there be K number of decision makers and D_K represents the Kth number of decision maker.
2. Obtain the evaluation criterion.
3. Select the linguistic variables for the importance weight of criteria and linguistic variables for the ratings of alternative with respect to criteria [15].
4. Sum the importance weight of criteria of Kth decision maker $\tilde{w}_j$ of criterion C_j and aggregate the fuzzy rating $\tilde{x}_{ij}$ of alternative A_i with respect to criteria C_j from Eqs. (1) and (2) given below

$$\tilde{x}_{ij} = \frac{1}{K}\left[\tilde{x}_{ij}^{1}(+)\ \tilde{x}_{ij}^{2}\ (+)\cdots(+)\ \tilde{x}_{ij}^{K}\right] \tag{1}$$

$$\tilde{w}_j = \frac{1}{K}\left[\tilde{w}_j^{1}(+)\ \tilde{w}_j^{2}\ (+)\cdots(+)\ \tilde{w}_j^{K}\right] \tag{2}$$

5. Obtain the fuzzy decision matrix $\tilde{D}$ and fuzzy weights of alternatives $\tilde{W}$ from Eqs. (3) and (4) using results obtained from step 4,

$$\tilde{D} = \begin{bmatrix} \tilde{x}_{11} & \tilde{x}_{12} & \ldots & \tilde{x}_{1n} \\ \tilde{x}_{21} & \tilde{x}_{22} & \ldots & \tilde{x}_{2n} \\ \vdots & \vdots & \ldots & \vdots \\ \tilde{x}_{m1} & \tilde{x}_{m2} & \ldots & \tilde{x}_{mn} \end{bmatrix} \tag{3}$$

$$\tilde{W} = [\,\tilde{w}_1 \quad \tilde{w}_2 \quad \ldots \quad \tilde{w}_n\,] \tag{4}$$

where, $\tilde{x}_{ij} = \left(a_{ij},\ b_{ij},\ c_{ij}\right)$ and $\tilde{w}_j = \left(w_{j1},\ w_{j2},\ w_{j3}\right); i = 1, 2,\ldots, m$ and $j = 1, 2\ldots n$.

6. Construct the normalized fuzzy decision matrix $\tilde{R}$ using Eqs. (5) and (6), where B is the set of benefit criteria

$$\tilde{R} = \left[\tilde{r}_{ij}\right]_{m*n}$$

$$\tilde{r}_{ij} = \left(\frac{a_{ij}}{c_j^*}, \frac{b_{ij}}{c_j^*}, \frac{c_{ij}}{c_j^*}\right), \quad j \in \mathrm{B} \tag{5}$$

$$c_j^* = \max_i c_{ij}, \quad \text{if } j \in \mathrm{B} \tag{6}$$

7. Now, formulate the weighted normalized fuzzy decision matrix $\tilde{V}$ by multiplying normalized fuzzy decision matrix $\tilde{r}_{ij}$ with importance weight criteria $\tilde{w}_j$ using Eq. (7)

$$\tilde{V} = \left[\tilde{v}_{ij}\right]_{m*n}, \; i = 1, 2, \ldots, m \text{ and } j = 1, 2, \ldots, n$$

$$\tilde{v}_{ij} = \tilde{r}_{ij}(\cdot)\,\tilde{w}_j \tag{7}$$

8. Obtain fuzzy positive ideal solution (FPIS) A^* and fuzzy negative ideal solution (FNIS) A^- using Eqs. (8) and (9)

$$A^* = \left(\tilde{v}_1^*, \tilde{v}_2^*, \ldots, \tilde{v}_n^*\right) \tag{8}$$

$$A^- = \left(\tilde{v}_1^-, \tilde{v}_2^-, \ldots, \tilde{v}_n^-\right) \tag{9}$$

   where, $\tilde{v}_{ij}^* = \max\{\tilde{v}_{ij}\}$ and $\tilde{v}_{ij}^- = \min\{\tilde{v}_{ij}\}$, j = 1, 2,…, n

9. Compute the distance between each alternative from FPIS (d_j^*) and FNIS (d_j^-) using Eqs. (10) and (11) respectively.

$$d_j^* = \sum_{j=1}^{n} d\left(\tilde{v}_{ij}, \tilde{v}_j^*\right) \tag{10}$$

$$d_j^- = \sum_{j=1}^{n} d\left(\tilde{v}_{ij}, \tilde{v}_j^-\right) \tag{11}$$

   where, d(.,.) denotes the distance between two fuzzy numbers.

10. Compute the closeness coefficient of each alternative (CC_i) using Eq. (12)

$$CC_i = \frac{d_i^-}{d_i^* + d_i^-} \tag{12}$$

11. Rank the alternatives in the decreasing order of closeness coefficient.

3 Implementation and Results

In the work, three object-oriented software systems are taken which are Eclipse (P_1), NetBeans (P_2), and Android (P_3). The projects are evaluated on the basis of four selection criteria maturity (C_1), recoverability (C_2), availability (C_3), and reliability compliance (C_4).

3.1 *Evaluation Using AHP Through MakeItRational Tool*

The reliability of the projects is evaluated using AHP approach through MakeItRational tool which is applied on the software reliability relationship model shown in Fig. 1.

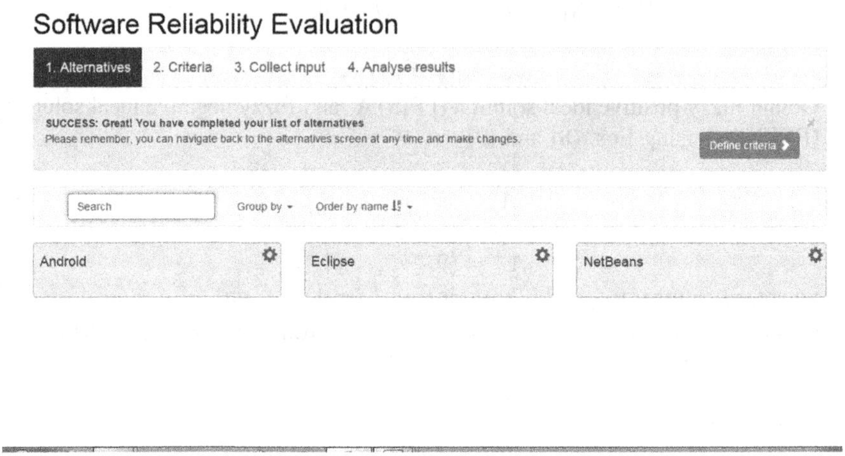

Software Reliability Evaluation

1. Alternatives 2. Criteria 3. Collect input 4. Analyse results

SUCCESS: Great! You have completed your list of alternatives
Please remember, you can navigate back to the alternatives screen at any time and make changes.
Define criteria

Search Group by Order by name

Android Eclipse NetBeans

Fig. 2 Defining alternatives of software system

Step 1: Defining alternatives

The inputs for the alternatives Eclipse (P_1), NetBeans (P_2), and Android (P_3) are defined as shown in Fig. 2.

Step 2: Defining criteria

The inputs for the criteria maturity (C_1), recoverability (C_2), availability (C_3), and reliability compliance (C_4) are defined as shown in Fig. 3.

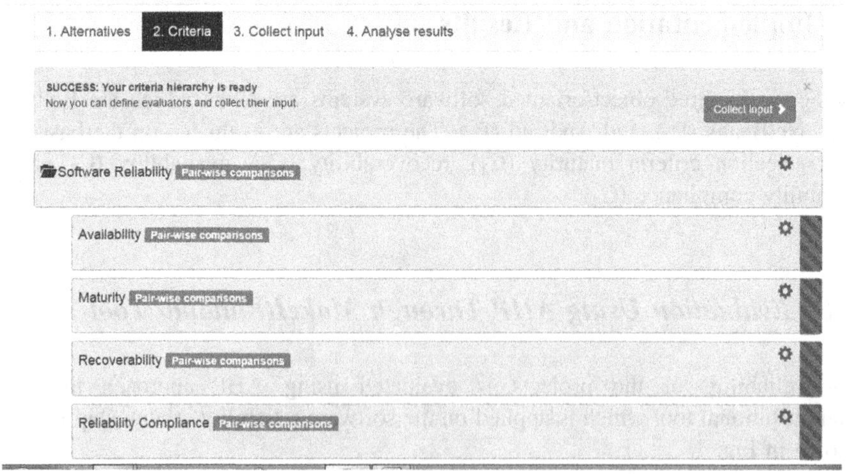

1. Alternatives 2. Criteria 3. Collect input 4. Analyse results

SUCCESS: Your criteria hierarchy is ready
Now you can define evaluators and collect their input.
Collect input

Software Reliability Pair-wise comparisons

Availability Pair-wise comparisons

Maturity Pair-wise comparisons

Recoverability Pair-wise comparisons

Reliability Compliance Pair-wise comparisons

Fig. 3 Defining criteria for software system

Table 1 Pairwise comparison of software reliability criteria

	C_1	C_2	C_3	C_4
C_1 (Maturity)	1	1/3	1/2	1/2
C_2 (Recoverability)	3	1	2	1/2
C_3 (Availability)	2	1/2	1	1
C_4 (Reliability Compliance)	2	2	1	1

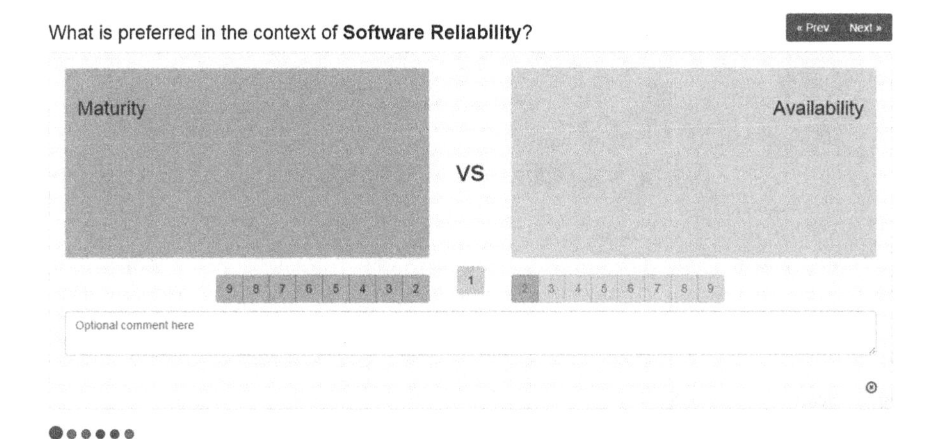

Fig. 4 Prioritizing criteria

Step 3: Prioritizing criteria

The criteria are prioritized through pairwise comparison of one-to-one criteria. Example: Availability is preferred to maturity by scale of 2 as shown in Table 1 and Fig. 4 on the basis of Saaty rating scale. The consistency ratio of software reliability criteria is $0.075 < 0.1$ which means that the judgments taken are consistent [10]. $\lambda_{max} = 4.204$, C.I. = 0.068, C.R. = $0.075 < 0.1$.

Step 4: Scoring alternatives

The alternatives are compared through pairwise comparison with respect to criteria. Example: Android is compared with Eclipse in context of maturity by scale of 3 as shown in Table 2 and Fig. 5. The consistency ratio for maturity is $0.031 < 0.1$. $\lambda_{max} = 3.036$, C.I. = 0.018, C.R. = $0.031 < 0.1$.

Similarly, we execute with respect to other criteria and obtain consistency ratio as given below:

Table 2 Pairwise comparison of object-oriented software systems for maturity

	P_1	P_2	P_3
P_1 (Eclipse)	1	3	1/3
P_2 (NetBeans)	1/3	1	1/5
P_3 (Android)	3	5	1

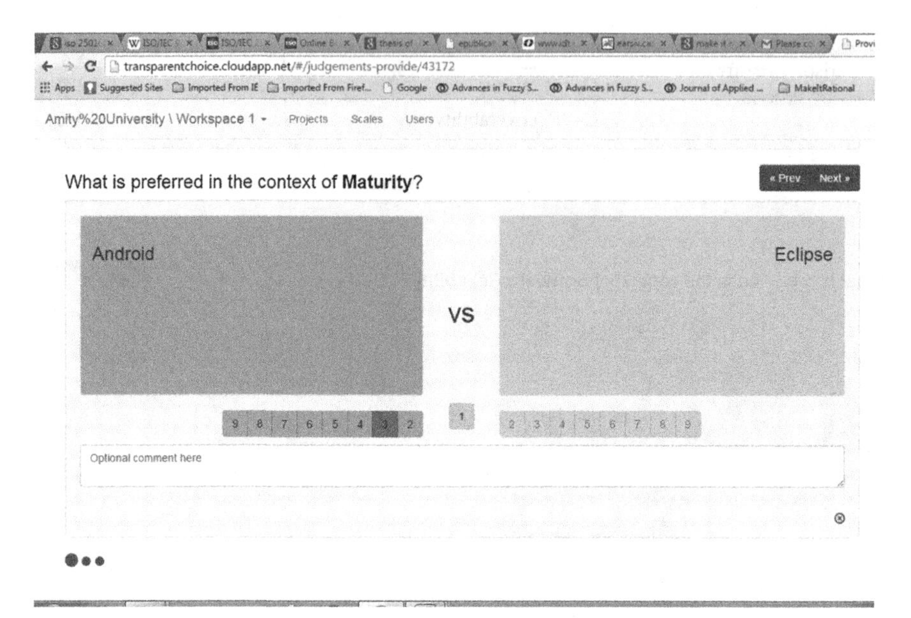

Fig. 5 Scoring of alternatives

Recoverability: λ_{max} = 3.036, C.I. = 0.0315, C.R. = 0.054 < 0.1
Availability: λ_{max} = 3.065, C.I. = 0.018, C.R. = 0.031 < 0.1
Reliability Compliance: λ_{max} = 3.029, C.I. = 0.0145, C.R. = 0.025 < 0.1
This shows that the judgments taken are consistent in nature.

Step 5: Final ranking of alternatives

After scoring of alternatives, the final ranking of alternatives is achieved through graph visualization shown in Fig. 6. The final score and ranking of object-oriented software systems is shown in Table 3.

The software reliability order calculated from AHP is

$$\text{Eclipse} > \text{Android} > \text{NetBeans}$$

3.2 Evaluation Using FTOPSIS

The fuzzy TOPSIS approach is applied on software reliability relationship model represented in Fig. 1.

Step 1: The committee of decision makers is established. Three decision makers (D_1, D_2 and D_3) who are distinguished industry experts are chosen having good knowledge of object-oriented programming

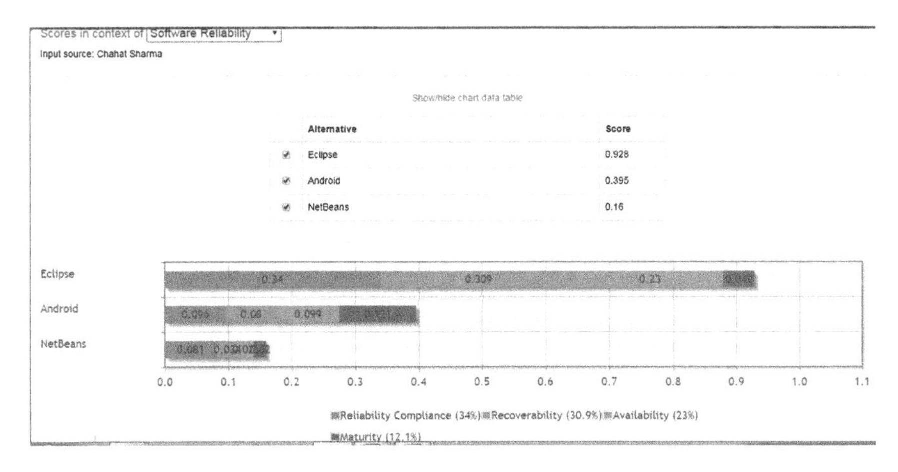

Fig. 6 Final ranking of alternatives

Table 3 Overall score and ranking of object-oriented software systems

Object-oriented software systems	Score	Rank
P_1 (Eclipse)	0.928	1
P_2 (NetBeans)	0.16	3
P_3 (Android)	0.395	2

Step 2: The software reliability criteria C_1 (Maturity), C_2 (Recoverability), C_3 (Availability), and C_4 (Reliability Compliance) are taken as the evaluation criteria

Step 3: The importance weight of software reliability criteria C_1 to C_4 (Table 4) and rating of alternatives (Table 5) is obtained from the linguistic variables filled by the decision makers

Step 4: Aggregate the fuzzy weights of each criteria and the fuzzy decision matrix of alternatives for each criteria using Eqs. (1) and (2) shown in Table 6

Step 5: After formulating the fuzzy normalized decision matrix, compute fuzzy weighed normalized decision matrix using equations in Sect. 2.2 and then obtain FPIS and FNIS from Eqs. (8) and (9)

Table 4 Linguistic variables for the importance weight of each criteria

	D_1	D_2	D_3
C_1 (Maturity)	ML	M	ML
C_2 (Recoverability)	MH	MH	M
C_3 (Availability)	ML	M	M
C_4 (Reliability Compliance)	H	M	M

Table 5 Linguistic variables for the performance ratings

Criteria	Object-oriented software systems	Decision makers		
		D_1	D_2	D_3
C_1	P_1	MG	MP	F
	P_2	MP	MP	MP
	P_3	G	G	MG
C_2	P_1	G	MG	G
	P_2	MP	MP	MP
	P_3	MG	MG	MG
C_3	P_1	MP	MG	MG
	P_2	MP	MP	MP
	P_3	MP	MP	MP
C_4	P_1	MP	MG	MG
	P_2	MP	MP	MP
	P_3	MP	MP	MP

Table 6 The fuzzy decision matrix and the fuzzy weights of three alternatives

	C_1	C_2	C_3	C_4
P_1	3, 5, 7	6.33, 8.33, 10	7, 9, 10	3.67, 5.67, 7.67
P_2	1, 3, 5	1, 3, 5	1, 3, 5	1, 3, 5
P_3	7, 9, 10	5, 7, 9	7, 9, 10	1, 3, 5
Weights	0.17, 0.37, 0.57	0.43, 0.63, 0.83	0.23, 0.43, 0.63	0.43, 0.63, 0.8

Table 7 Closeness coefficient and rank of object-oriented software systems

Object-oriented software systems	Closeness coefficient (CC_i)	Rank
P_1 (Eclipse)	0.420	1
P_2 (NetBeans)	0.231	3
P_3 (Android)	0.388	2

Step 6: Thereafter, calculate the distance between each alternative from FPIS and FNIS in using Eqs. (10) and (11)

Step 7: Obtain the closeness coefficient of alternative using Eqs. (12) and rank the alternatives in decreasing order of closeness coefficient as shown in Table 7

The software reliability order calculated from FTOPSIS is

$$\text{Eclipse} > \text{Android} > \text{NetBeans}$$

4 Conclusion

In this paper, the reliability of object-oriented software systems are evaluated using MCDM approaches; AHP and FTOPSIS. AHP is widely used and popular approach for decision making. The approach suffers from imprecision of judgments and uncertainty. In order to overcome such shortcomings FTOPSIS approach is applied which is based on fuzzy set theory and converts linguistic variables of decision maker to a crisp value for effective assessment of results.

The work evaluates the object-oriented software systems Eclipse (P_1), NetBeans (P_2), and Android (P_3) in context of reliability selection criteria. The evaluation through AHP is done using MakeItRational tool. The figures depict the steps of procedure followed by the graphical visualization in Fig. 6 which shows that Eclipse is more reliable as compared to other software systems. In order to validate the efficacy of results obtained through AHP, FTOPSIS approach is used on the similar set of data to deal with the fuzziness of decision-making process. Thereafter, it is found that Eclipse is the most suitable alternative having highest rank.

References

1. Chatterjee, S., Singh, J.B., Roy, A.: A structure-based software reliability allocation using fuzzy analytic hierarchy process. Int. J. Syst. Sci. (2013). doi:10.1080/00207721.2013.791001
2. Antony, J., Dev, H.: Estimating reliability of software system using object oriented metrics. Int. J. Comput. Sci. Eng. 283–294 (2013)
3. Mishra, A., Dubey, S.K.: Evaluation of reliability of object oriented software system using fuzzy approach. In: 5th International Conference-Confluence, The Next Generation Information Technology Summit, 806–809 (2014)
4. ISO/IEC 25010:2011. https://www.iso.org/obp/ui/#iso:std:iso-iec:25010:ed-1:v1:en. Accessed 1 May 2015
5. Olsina, L., Covella, G., Rossi, G.: Web Quality. In: Mendes, E., Mosley, N. (eds.) Web Engineering, pp. 109–142. Springer, Heidelberg (2006)
6. Losavio, F., Chirinos, L., Lévy, N., Cherif, A.R.: Quality characteristics for software architecture. J. Object Technol. 133–150 (2003)
7. Malaiya, Y.K.: Software Reliability and Security. Taylor & Francis (2005)
8. Java: http://web.cs.wpi.edu/~kal/elecdoc/java/features.html. Accessed 9 May 2015
9. Hsu, H.M., Chen, C.T.: Fuzzy hierarchical weight analysis model for multicriteria decision problem. J. Chinese Inst. Industrial Eng., 129–136 (1994)
10. Coyle, G.: The Analytic Hierarchy Process. Practical Strategy, Pearson Education Limited (2004)
11. Saaty, T.L.: The Analytic Hierarchy Process. Mc-Graw Hill, New York (1980)
12. Charnes, A., Cooper, W.W.: Management Models and Industrial Applications of Linear Programming. Wiley, New York (1961)
13. Brans, J.P., Vincke, P., Marschal, B.: How to select and how to rank projects: The PROMETHEE method. Eur. J. Oper. Res. 228–238 (1986)
14. Deng, H.: Multicriteria analysis with fuzzy pairwise comparison. Int. J. Approximate Reasoning, 215–231 (1999)
15. Chen, C.-T.: Extensions of the TOPSIS for group decision-making under fuzzy environment. Fuzzy Sets Syst. 1–9 (2000)

16. Saaty, T.L.: Decision making with dependence and feedback: the analytic hierarchy process. RWS Publications, Pittsburgh (1996)
17. MakeItRational: http://makeitrational.com/tutorials. Accessed 11 May 2015
18. Hwang, C.L., Yoon, K.: Multiple Attributes Decision Making. Springer, Berlin Heidelberg (1981)

Information Security and Its Insurance in the World of High Rise of Cybercrime Through a Model

Sudha Singh, S.C. Dutta and D.K. Singh

Abstract With the advent of Internet technology and World Wide Web (WWW), the whole world has become a global village. In this globally connected world, information system becomes the unique empire without tangible territory. Cybercrime is any illegal action which performs knowingly or unknowingly to not disturb or disturb the system to get information using technology. Cybercrime is one of the negative results of globalization. No system is fully secured in this global village. It is unrealistic to divide information systems into segments bordered by state boundaries. So it is required to build a legal system for this global village. But till now this type of law does not come into its existence. As we know, India has third largest number of Internet users in the world. This paper is a study of cyber safety in India through modeling and analysis. We also studied cyber-insurance companies in India.

Keywords Cyber security · Cybercriminal · Insurance survey · Cyber fraud

---

Sudha Singh (✉)
Department of Computer Engineering, MGM College of Engineering
and Technology, Kamothe, Navi Mumbai 410209, India
e-mail: sudha_2k6@yahoo.com

S.C. Dutta
Department of Computer Science and Engineering, BIT Sindri,
Dhanbad 828123, India
e-mail: dutta_subhash@yahoo.com

D.K. Singh
Department of Electronics and Communication Engineering,
National Institute of Technology, Patna 800005, India
e-mail: dksingh_bit@yahoo.com

M. Pant et al. (eds.), *Proceedings of Fifth International Conference on Soft Computing for Problem Solving*, Advances in Intelligent Systems and Computing 437, DOI 10.1007/978-981-10-0451-3_10

1 Introduction

Today, the cyber world is at a huge risk of growing cybercrime. Cybercrime is a collection of all illegal activities using computers, the Internet, cyber space, and the worldwide web. After the advent of Internet technology and World Wide Web (WWW), the whole world has become a cyber village where every interest needs space, availability, dependability, sharing, trust, and security.

In the absence of governing authority and laws for this village, cybercrime is on the rise. Of the many cyber attacks, criminal insider (30–40 %), phishing (20–30 %), SQL injection (25–35 %), social engineering (15–25 %), theft of data-bearing devices (25–35 %) and virus, malware, worms, and Trojan (50–65 %) are the most common [1].

Cybercrime, cyber warfare, cyberespionage and hacktivism are the most common motivations behind cyber attack [2]. The rising rate of cybercrime is at an alarming level, e.g., India has registered 107 % of common annual growth rate in the number of cybercrimes registered in the past few years. Since many cases remain unregistered due to defamation, etc., the actual rate of increase in cybercrime will be much more than the given rate. If we study the age of offenders, then more than 70 % belong to 18–30 years of age group [1].

Use of ad hoc and mobile technology gives rise to the vulnerabilities to a great extent [3–5]. Increasing online fraud concerns banking and other sectors to think on security of resources, data, and its insurance. Due to hacking, virus attacks, denial of service attacks, and web content liability there is a need for cyber-insurance policy (Figs. 1, 2).

Cyber security insurance is designed for the protection of intangible data. It can be easily explained as the coverage for professional errors and the risks of doing business on the Internet or working with a network system.

Cyber-insurance is a risk management technique via which network user risks are transferred to an insurance company, in return for a fee, i.e., the insurance premium [6]. Till a few years back, we did not think about cyber-insurance companies, its policies, and coverage. Nowadays when cybercrime is on a rise and intrusions are affecting day-to-day life of netizens, people are forced to search for insurers. Since this concept is new, very few insurance companies are coming to provide cyber security. Also, their cover areas are very limited. According to Klynveld Peat Marwick Goerdeler (KPMG)'s report, cyber-insurance is a type of new business with no profit till date, despite the rising risk. More than 75 % of senior security officers in international business do not know about cyber-insurance of their company. In 2014, cyber attacker from inside the organization are 8 %, external fraudsters are 37 %, both internal and external perpetrators are 47 % and others are 8 % [1].

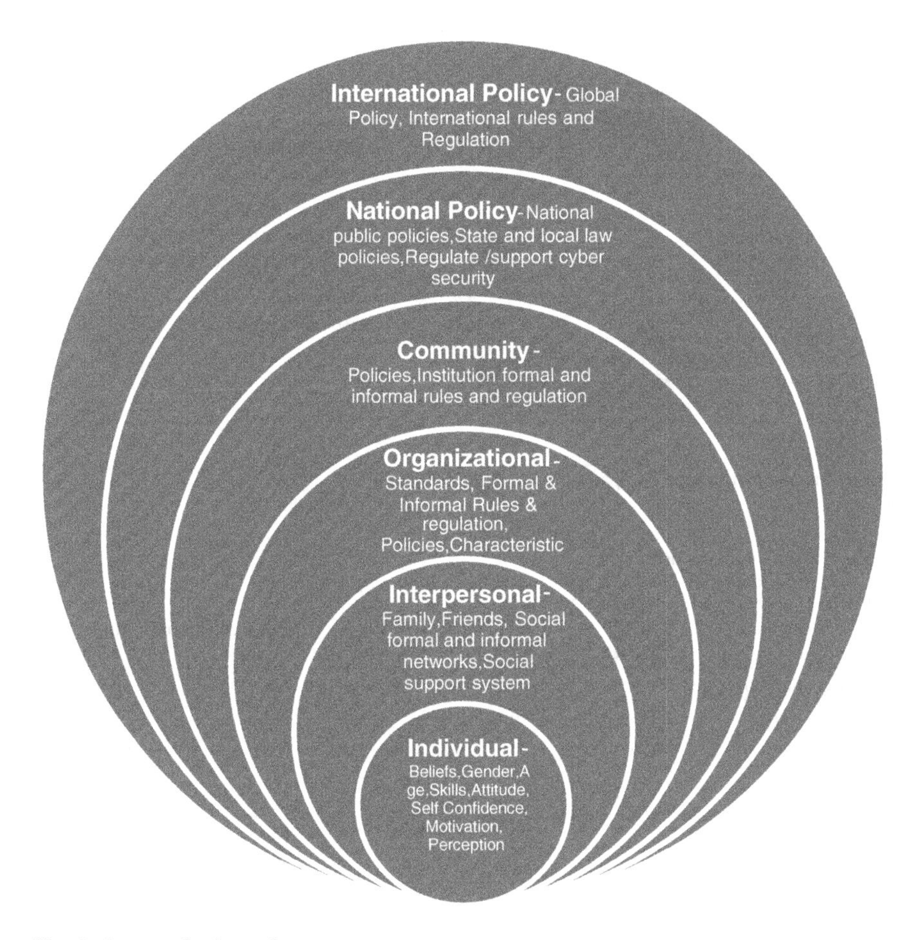

Fig. 1 Layers of cyber safety

2 A Socio-environmental Model

To achieve the goal of cyber safety, we have studied different layers of sharing through the following model.

An Individual can get a cyber victim by e-mail spoofing, phishing, spamming, cyber defamation, cyber stalking, password sniffing, etc. The one on whom an individual believes can get the related information easily and simply attack. Technically sound person can attack by motivating someone or by self confidence of doing such a miscellaneous thing.

In prevention, an individual can take care of information sharing on social website and use proper privacy. Most of the offenders are from the age group 18–30. In 2014, 17.2 % of people became the victims [2] and more than 51 % of nation got attacked.

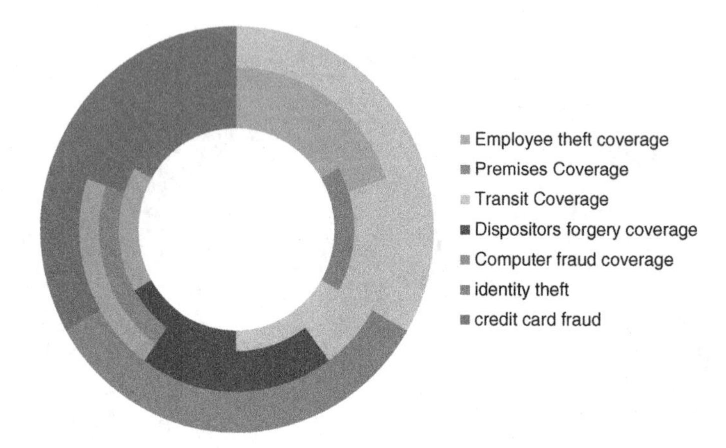

Fig. 2 Study of insurance company based on their policies

Table 1 Comparative study

Policies	Companies				
	HDFC ERGO	TATA AIG	ICICI Lombard	Reliance General Insurnce	Bajaj Alianze
Employee theft coverage	Yes	Yes	Yes	No	No
Premises coverage	Yes	No	No	No	No
Transit coverage	Yes	Yes	Yes	Yes	Yes
Dispositors forgery coverage	Yes	Yes	Yes	No	No
Computer fraud coverage	Yes	No	Yes	No	No
Identity theft	No	Yes	No	Yes	Yes
Credit card fraud	Yes	Yes	Yes	Yes	Yes

Motivations, revenge, etc., are some of the many reasons for cyber attack. The cybercriminal who most probably are not social commit cybercrime. In 2014, 2.3 % of people were under such attack.

Organizations are on a high risk of cybercrime originating from either inside or outside. By study, it is clear that cybercriminals are mostly technically-sound people in the case of well-organized cybercrime [7]. To prevent vulnerability examine cyber security practices, identify important electronic assets, do content reviews, conduct inspections and validation testing. Assess possible cyber threats through vulnerabilities using cyber tools and techniques. Aim should be to do complete risk analysis, its continuous updating, monitoring and strictly not to compromise on the risk factors.

Indian Government passed cybercrime law against cybercrime in the year 2000 known as Information Technology ACT 2000. Further, new amendments were made in it in year 2006, 2008, 2011, and 2012.

There are various ongoing activities and programs by the government to address the cyber security challenges which have significantly contributed to the creation of a platform that is now capable of supporting and sustaining the effort in cyberspace. The cyber security policy is an evolving task and it caters to the whole spectrum of Information and Communication Technology (ICT) users and providers including small users and small, medium and large enterprises, government and non-government entities.

Due to the lack of sufficient international law and its enforcement policies, international cybercrimes often challenge it and becomes a major concern to the cyber police.

Study of the statistical reports on cybercrime in India and solved cyber cases clearly indicate that existing cyber law in India is not sufficient. Many organizations and people do not bother on its serious impacts due to the lack of clear knowledge. Cyber-insurance can be a solution to tackle cybercrime at its alarming level.

3 Cyber-Insurance Companies in India

Companies which provide cyber fraud insurance are (1) HDFC Ergo, (2) ICICI Lombard, (3) Reliance General Insurance, (4) Tata AIG, and (5) Bajaj Allianz.

If we study the documents of HDFC Ergo [1], plan coverage includes (a) employee theft coverage, (b) premises coverage, (c) transit coverage, (d) depositors forgery coverage, and (e) computer fraud coverage.

If we study the documents of ICICI Lombard, it states that their policy covers "Privacy Breach Liability", "Cyber Extortion", "Business Interruption Losses", "Liability from multimedia and public relations costs," "Legal expenses," and "data Theft liability." The premiums are said to be around 0.5–1.5 % of the amount insured.

Similarly after studying the available documents of various companies, we have done a comparative study based on points like employee theft coverage, premises coverage, transit coverage, forgery depository coverage, computer fraud coverage, identity theft, and credit card fraud.

Results are summarized below (Table 1).

4 Conclusion Future Scope

We have studied on cyber safety through socio-environmental model. We have also studied and analyzed on different companies providing cyber-insurance in India. Through this study we can say that the security of our information is at a vulnerable position. Till now we cannot say that our data is 100 % safe. In future, we will work on how to provide safety to our data in the era of mobile technology and internet.

References

1. Downloaded material from https://infosecuritylive.com/indian-perspective-cyber-liability-insurance/, https://www.naavi.org/wp/?p=1556, https://www.tataaiginsurance.in/lifestyle-insurance/identity-n-card-protection.html, https://www.tataaiginsurance.in/taig/taig/tata_aig/about_us/Media_Centre/pdf/policy_wordings/personal-identity-protection-policy-wording.pdf https://www.hackmageddon.com
2. Chaoo, K.K.R.: The cyber threat landscape: challenges and future research directions. Elsevier's Science Direct J. Comput. Secur. **30**(8), 719–731 (2011)
3. Walker, J.J., Jones, T., Bhount, R.: Visualization modeling and predictive analysis of cyber infrastructure oriented system. In: Publish in IEEE conference Technology for Homeland Security (HIS), pp. 81–85 (2011)
4. Sandhu, R., Krishnan, R., White, G.B.: Towards secure information sharing models for community cyber security. In: Publish in IEEE Conference Collaboration Computing: Networking Application and Worksharing International Conference, vol. 6, pp. 1–6 (2010)
5. Lagazio, M., Sherif, N., Cushman, M.: A multi level approach to understanding the impact of cyber crime on the financial sector. Elsevier's Science Direct J. Comput. Secur. **45**(8), 58–74 (2014)
6. Pal, R., Golubchik, L., Konstantinos, P., Hui, P.: Will cyber—insurance improve network security? a market analysis. In: Proceeding of IEEE Conference on Computer Communications (IEEE INFOCOM), pp. 235–243 (2014)
7. Tehrani, P.M., Manap, N.A., Taji, H.: Cyber terrorism challenges: the need for a global response to a multi-jurisdictional crime. Elsevier's Science Direct J. Comput. Law Secur. Rev. **29**, 207–215 (2013)

A Systematic Study on Mitigating Parameter Variation in MOSFETs

Prashant Kumar, Neeraj Gupta, Rashmi Gupta, Janak Kumar B. Patel and Ashutosh Gupta

Abstract The VLSI design was horse-powered mainly by the elegant and simple MOSFET (Metal Oxide-Semiconductor Field Effect Transistor) which amazingly yielded to the designer's scaling wishes; Though its operation is governed by umpteen kinds of parameters which depends on the physical dimensions, many process-related aspects as well as the ambient features of the device. The dependence of these parameters on the above-mentioned factors make them vulnerable to variation. Thus, the effect of parameter variations can be summed up as the reduction in wafer yield and hence profit. Over the years, designers were diligently researching and coming up with solutions to combat the menace of parameter variation according to the extent of variation and stringency of requirement. The solution approaches range from not-so-sophisticated marketing techniques such as frequency-binning to the employment of sophisticated techniques such as body biasing to automatically compensate the parameter variations. This paper aims to probe various reasons and impacts of the parameter variations in MOSFETs and to describe some of the techniques to mitigate the parameter variations in a systematic manner.

Keywords Adaptive body bias · Body bias generator · Process control monitoring · Voltage controlled oscillator

---

Prashant Kumar (✉)
YMCA University of Science & Technology, Faridabad, India
e-mail: pk.vlsi@gmail.com

Neeraj Gupta · Rashmi Gupta · J.K.B. Patel
ASET, Amity University Haryana, Gurgaon, India
e-mail: neerajsingla007@gmail.com

Ashutosh Gupta
ASET, Amity University Uttar Pradesh, Noida, India

M. Pant et al. (eds.), *Proceedings of Fifth International Conference on Soft Computing for Problem Solving*, Advances in Intelligent Systems and Computing 437, DOI 10.1007/978-981-10-0451-3_11

1 Introduction

The main building blocks in an Integrated Circuit (IC) are MOSFETs. MOSFETs with their elegance and simplicity became the natural choice as the workhorse of IC revolution. They amazingly yielded to the scaling wishes of VLSI engineers. Though its operation can be summed up in simple terms, there are a lot of parameters associated with a MOSFET which are used to model the device. These parameters owe their dependence to geometrical, electrical, environmental, and process-related aspects. So, any deviation of the MOSFET from its predictability and uniformity can be expressed as variations in these parameters. For instance, any variation in the channel length can be seen reflected in the threshold voltage (V_T) and the drain current (I_{DS}). The alarming thing about parameter variations is that they show a melancholic trend with scaling thus making the matters worse. Consequently, with scaling the designers have to put more effort to make the designs reliable. Any variations in the device *parameters* will be reflected in the circuit in which it is and the impact on the circuit will be reflected in the system behavior.

1.1 Parameter Variations in MOSFETs

The major agent which brings about parameter variations are process, voltage, temperature, and aging. The roles of the first three agents were traditionally recognized right from the initial days of MOSFET. In fact, their influence was so felt that the variations caused by these agents were often called as PVT-variations [1]. However, the aging-related effects began to create ripples during the recent technology nodes. This is due to the fact that with scaling various wearout effects such as Hot Carrier Injection (HCI) and Bias Temperature Instability (BTI) began to surmount [2]. The negative impact of these agents is aggravated by their ability to form certain feedback loops, thereby worsening the situation [3]. For instance, if the devices in a part of the die have their V_T smaller than elsewhere, then there more current will be flowing and thus more heat will be generated. This creates a gradient in temperature across the chip and therefore temperature-induced variations will be prominent in that particular region. The following sections deal with the modus operandi of these agents causing parameter variations.

1.2 Process-Induced Parameter Variations

Process variations are caused by the inability to precisely control various fabrication processes, especially when we are fabricating smaller and smaller devices. It includes both systematic and random effects. An example of a systematic effect is

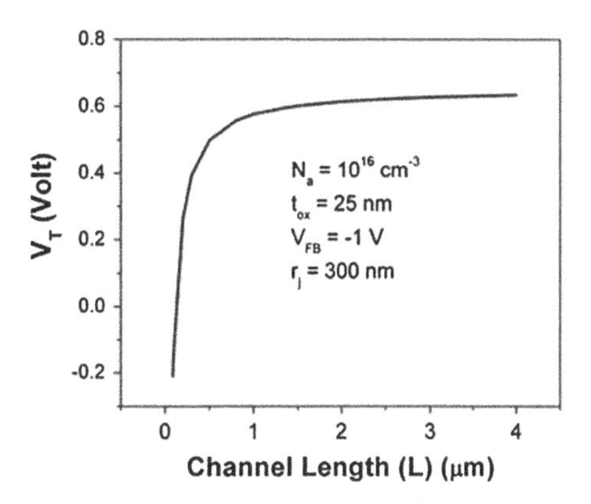

Fig. 1 A typical V_T roll-off curve [5]

the presence of aberrations in the lithographic lens. Whereas, Random Dopant Fluctuation (RDF) can be considered as a random effect which results in parameter variations [4]. With scaling, these imperfections are becoming more severe. One of the major sources of geometrical variation during the fabrication is the diffraction related imperfections in the lithographic process. Though the chip manufacturers try to reduce this effect by using Phase Shift Masks (PSM) and Optical Proximity Correction (OPC), a complete removal of diffraction-related effects is not achievable [2]. Moreover, any variation in the channel length is reflected in the threshold voltage which is widely known as the V_T roll-off effect. A typical VTL curve is shown in Fig. 1. Such a variation results in a shift in the threshold voltage and consequently a change in the MOSFETs drive strength.

1.3 Voltage Variations

Voltage variations seen in a design can be due to manufacture-time or run-time phenomena. Basically, they stem from the nonidealities present in the voltage distribution line. For instance, if the supply-voltage distribution lines are not uniformly laid, there can be variations in the supply-voltage available across different parts of the die. This can be exacerbated by varying activity levels, thus causing significant difference in the IR drops and thereby a variation in the supply voltage at different parts of the circuit [6]. Along with parasitic resistance, inductance can also play a spoilsport which is commonly termed as d*i*/d*t* noise. As a result, there can be a possibility of temporal voltage drop and voltage over-shoots. In contrast to the Process variations, Voltage variations are dynamic in nature and have a time constant in the range of nanoseconds (ns) to microseconds (μs) [2]. Figure 2 shows an example of voltage variations caused due to changes in the activity level.

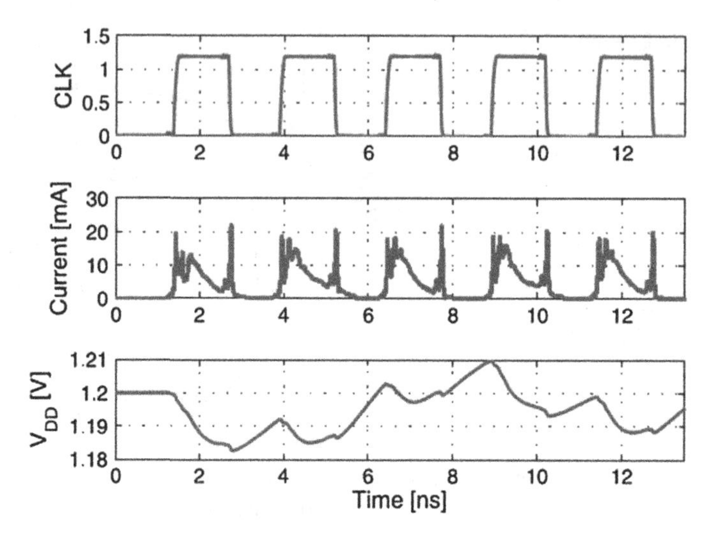

Fig. 2 Voltage variations caused due to changes in the switching activity [2]

The major impact of the Voltage-variations can be seen as variations in the delay of the circuit. The delay of a CMOS gate can be approximated by the following equation.

$$t_{\text{gate}} \propto V_{\text{DD}}(V_{\text{DD}} - V_{\text{T}})$$

The dependence of the delay with the supply voltage is quite obvious with the above equation [1, 7].

1.4 Temperature Variations

Most of the MOSFET parameters have their dependence on the ambient temperature. Temperature variations are dynamic in nature and they have a timeconstant in the range of milliseconds to seconds. A temperature variation across the die happens mainly due to nonuniformities in the Thermal Interface Material (TIM) and variations in the switching activity. Moreover, temperature variations have a direct correlation with the activity factor. If the circuit architecture is such that the switching activity is more in certain parts of the die, then there may occur hotspots that result in WID parameter variations. An increase in temperature typically can have two implications [2]. (i) An increase in the delay due to reduced carrier mobility and increased interconnect resistance. (ii) A decrease in the threshold voltage and thus a reduction in the delay. For, high-VDD circuits the first effect dominates and for the low-VDD case the second one is prominent.

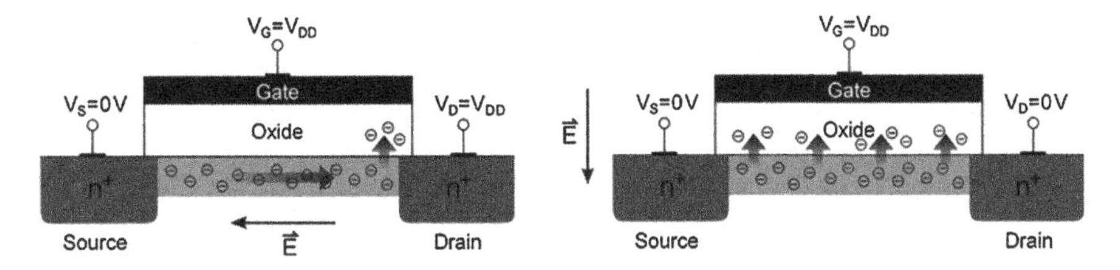

Fig. 3 Occurrence of HCI and BTI in an n-MOSFET [2]

1.5 Aging-Related Variations

With the scaling of the devices, an increase in the fields can be observed which surmounts the aging-related reliability issues. Two major wearout effects of concern are the Hot Carrier Injection (HCI) and Bias Temperature Instability (BTI). HCI mainly occurs during switching of logic gates. Carriers are accelerated in the lateral field under the oxide and gain sufficient kinetic energy to be injected into the gate dielectric, whereas BTI is caused due to high vertical fields which generally happens when the MOSFET is operated in triode region. BTI is also characterized by the trapping of the charges in the oxide. These trapped charges culminate in a shift in the threshold voltage [2]. Occurrence of HCI and BTI in an n-MOSFET is depicted in Fig. 3. The aging-related parameter variations are also dynamic in nature. However, considerable V_T shifts can be observed only after days, weeks, or even years of operation [2].

2 Implication of Parameter Variations

The extent of parameter variations is not limited within the device boundary. Its impact can obviously be seen in the circuit which is comprised of the affected devices. Moreover, it will have system level implications too. The effect of almost all these variations can be considered to be affecting the speed and power consumption of the system. It is the duty of the manufacturer to ensure that these variations do not affect the quality of the system and the expectation of the customer. Such parameter variations may render some of the chips too slow to meet the functional requirements or intolerably power-hungry.

3 Solution Approaches to Parameter Variations

Here, a survey on the different solution approaches adopted by designers, researchers, and engineers is carried out. Some of the popular ways of handling parameter variations are Guard-banding, Design/Process Engineering Techniques, Adaptive Supply

Voltage Scaling, Body-Bias-Based Techniques, and Frequency-Binning. The latter three techniques are often termed as Post-Silicon Tuning. It is due to the fact that these techniques do the actual tackling after the manufacturing of the chip.

3.1 Guard-Banding

Parameter variations can transform a uniform, static (static parameter value) circuit into a nonuniform and dynamic one. In Guard-banding, the circuit will not be considered as a nonuniform and dynamic one but as a uniform and static network with a constant guard band. This allows the optimized static circuit to continue functioning even if the devices have arbitrarily distributed parameters, or become faster or slower over time [8]. In spite of its regular use, the methodology is ineffective and involves significant penalty in the area–performance–power–reliability budget. According to 'Moore's law' two increase in transistor count in every technology generation. As technology scales down to submicron regime for single crystalline Si and grain size becomes comparable to channel length of transistor; such conventional design approaches cannot be continued because the area–performance–power budget is shrinking fast, and the process-induced variation and time-dependent shifts in transistor parameters are increasing rapidly [8]. In short, Guard-banding results in increased design cost and time to market.

3.2 Design Techniques and Process Engineering Techniques

The constant search for methods to combat parameter variations had led the designers and engineers to come up with some design practices and process engineering techniques which proved useful in tackling the parameter variations. Some of such techniques are mentioned here. Oxide thickness (tox)—Implantation energy Adjustment. One of the reason for shifts in threshold voltage as observed by the process engineers was the variation in the gate oxide thickness (tox). A closer study revealed that the tox variation can have a double-fold impact on V_T. First as V_T α $1C$, a decrement in tox results in a decrement in V_T and another impact is related to the threshold adjust implant. Here, smaller tox means that a larger number of dopant atoms can reach the channel region thus increasing the V_T. These two opposing tendencies can be purposefully utilized to cancel the parameter variation in V_T due to oxide thickness variation [7].

3.3 Adaptive Supply Voltage Scaling

Utilization of multiple supply voltages or dynamic and continuous adjustment of supply voltage had been one of the popular low power design techniques. Later on, designers began to seriously consider the prospects of using Adaptive Supply Voltage (ASV) schemes for post-silicon tuning and thereby to achieve yield improvement [9]. Here, the slower dies will be supplied with a higher supply voltage, whereas the faster dies will be given a smaller supply voltage. By the proper use of ASV, it is possible to achieve a controlled tuning of the power dissipation and circuit speed in a dynamic manner. ASV delivers the designer a freedom to trade-off frequency for power consumption. The relation between switching power dissipation (Pswitching) and operating frequency (*f*) is given as

$$\text{Pswitching} = \alpha \times CV_{DD}^2 f$$

Here, α is the activity factor or the fraction of the capacitance (C) which switches on average in a given clock cycle [10]. The switching power having a square dependency on the supply voltage, it is a handy knob in adjusting the power consumption. If a mechanism such as a Voltage Controlled Oscillator (VCO) is employed to alter the chip clock frequency according to the adaptive supply voltage, there can be achieved even a cubic dependency between the power consumption and supply voltage [9].

3.4 Body-Bias-Based Techniques

Body-Bias-based techniques where employed right from the early days of memory design [11]. Here, the post-silicon tuning is achieved by the modulation of body-bias voltage. Application of body bias can alter the threshold voltage (V_T) of the MOSFET and a change in V_T is seen reflected in the delay and leakage of the circuit. Traditionally, a static reverse body bias was used aimed at reducing the subthreshold leakage of MOSFETs. The important requirement for body-bias-based techniques is the need for dedicated power distribution networks for supplying the body voltage. Such distribution networks can be expensive in terms of silicon area even though they do not carry much current [9].

3.5 Adaptive Body Bias (ABB)

Adaptive Body Bias is a dynamic control technique in which the body-bias voltage is dynamically dependent on the delay and power dissipation of the die. The aim of ABB is to bring maximum number of dies within the acceptable limits of power

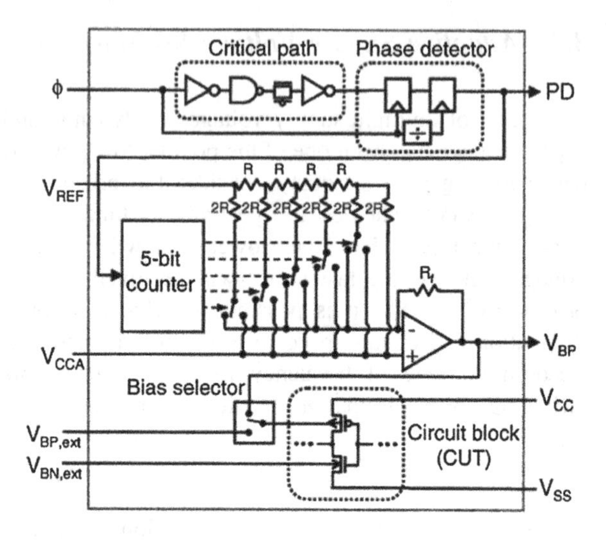

Fig. 4 A typical body-bias generator of an ABB [9]

dissipation and delay metrics. As the application of body bias in ABB is dynamic in nature, there needs to be some sort of sensing mechanism to get a feel of the present delay/leakage condition in the die. This sensing mechanism is the heart of the ABB block. A typical body-bias generator of an ABB is shown in Fig. 4.

The ABB generator contains a chain of circuit elements which models the critical path of the die. It is with this replica of the critical path, the measurements regarding the delay/leakage of the die are found. This critical path replica is fed with a clock signal of nominal frequency. The response of the critical path replica to the nominal clock and the nominal clock itself are given to a phase detector. The output from the phase detector is used to run a digital counter. The digital output from the counter is appropriately converted into analog format to obtain the body-bias voltage. The body-bias voltage will be a function of the value at the counter output, VCCA, as well as the VREF [9].

3.6 *Frequency-Binning/Harvesting*

The parameter variations can cause deterioration in the performance of some of the dies. This leads to reduction in yield as many of the dies cannot be sold. Thus, instead of simply throwing them out a harvesting is done on the low-performing dies. These are often called as 'Harvested' version of the actual product. However, the fact is that the yields will eventually improve with time and then there will be financial hurt as the manufacturer would have created a market for the harvested version of the product. This will force the manufacturer to sell the higher performing product for the cost of the low-performance harvested version.

4 Conclusion

The problems associated with parameter variations are a growing concern for the designers and manufacturers as it affects the yield and the profit. The alarming thing about the parameter variations is that they are getting aggravated with scaling. Though there are some techniques to reduce the source of parameter variations, they have only limited capability in solving the problem of parameter variations. It may not be possible to thoroughly wipe out the parameter variations, however there needs to be found ways to cohabit with the parameter variations while inhibiting their impact as far as possible. The strategy of combat should be to model the variations as far as possible. The actual problem with the variations is that they are not being modeled and in that case these variations remain as an uncertainty. Variation modeling can be done by studying the parameter variations using Process Control Monitoring (PCM) structures. PCM structures are devices fabricated mostly on the scribe lines of the wafer which can be used to extract process related statistics. Once the parameter variations are properly modeled, they are no longer an uncertainty. The solution may not come from the adoption of any single solution, but an amalgam of solutions spanning from circuit design to process engineering.

References

1. Maudslay, H.: Wikipedia. http://en.wikipedia.org/wiki/Maudslay (2014)
2. Wirnshofer, M.: Variation-aware adaptive voltage scaling for digital CMOS circuits. Springer Series in Advanced Microelectronics, vol. 41
3. Humenay, E., Tarjan, D., Skadron, K.: Impact of parameter variations on multi-core chips. In: Workshop on Architectural Support for Gigascale Integration. 33rd International Symposium on Computer Architecture
4. Teodorescu, R., Nakano, J., Tiwari, A., Torrellas, J.: Mitigating parameter variation with dynamic fine-grain body biasing. In: 40th Annual IEEE/ACM International Symposium, pp. 27–42 (2007)
5. Komaragiri, R.S.: A simulation study on the performance improvement of CMOS devices using alternative gate electrode structures. PhD Dissertation, TU Darmstadt (2007)
6. Wilson, R.: The dirty little secret: engineers at design forum vexed by rise in process variations at the die level EE Times, vol. 1 (2002)
7. Borkar, S., Karnik, T., Narendra, S., Tschanz, J., Keshavarzi, A., De, V.: Parameter variations and impact on circuits and microarchitecture. In: Proceedings of the 40th Annual Design Automation Conference, pp. 338–342 (2003)
8. Alam, M.: Reliability-and process-variation aware design of integrated circuits. Microelectron. Reliab. **48**(8–9), 1114–1122 (2008)
9. Tschanz, J., Kao, J.T., Narendra, S.G., Nair, R., Antoniadis, D.A., Chandrakasan, A.P., De, V.: Adaptive body bias for reducing impacts of die to die and within-die parameter variations on microprocessor frequency and leakage. IEEE J. Solid-State Circuits **37**(11), 1396–1402 (2002)

10. Yadav, A.K., Vaishali, Yadav, M.: Implementation of CMOS current mirror for low voltage and low power. Int. J. Electron. Commun. Comput. Eng. **3**(3), 620–624 (2012)
11. Kumar, S., Kim, C.H., Sapatnekar, S.S.: Body bias voltage computations for process and temperature compensation. IEEE Trans. Very Large Scale Integration (VLSI) Syst. **16**(3), 249–262 (2008)

Vendor–Supplier Cooperative Inventory Model with Two Warehouse and Variable Demand Rate Under the Environment of Volume Agility

Isha Sangal and Vandana Gupta

Abstract This theory develops an integrated production model for the vendor and supplier. A supply chain model with volume agile manufacturing for the deteriorating inventory is developed. The demand rate is assumed to be parabolic for the vendor and supplier both. In this paper, supplier has two warehouses one is rented warehouse and second is owned warehouse for the excess inventory. The deterioration rate for the own warehouse is time dependent and for rented warehouse is constant. This integrated theory is examined in the environment of inflation. The purpose of this study is to minimize the integrated cost for the whole system. A numerical solution with the sensitivity analysis is given to illustrate the complete model.

Keywords Supply chain model · Two warehouse · Volume agility · Time dependent deterioration rate · Inflation · Parabolic demand rate

1 Introduction

Supply chain management is the oversight of raw materials, information, and finances as they move in a process from manufacturer to supplier to wholesaler to retailer to consumer. This model is an integrated model of vendor and supplier. In this model, vendor uses the own warehouse (OW) and rented warehouse (RW) to store the extra stock. Many researchers described the theory of two warehouse but very few researchers have discussed the concept of two storage with the supply chain model. The concept of two warehouse was first modified by Hartley [1]. Goswami and Chaudhuri [2] considered an EOQ model with a linear demand rate for the deteriorating items with two levels of storage. Bhunia and Maiti [3] extend

I. Sangal (✉)
Department of Mathematics and Statistics, Banasthali University, Jaipur, India
e-mail: isha.sangal@gmail.com

Vandana Gupta
Banasthali University, Jaipur, India
e-mail: vandana.vandana1983@gmail.com

M. Pant et al. (eds.), *Proceedings of Fifth International Conference on Soft Computing for Problem Solving*, Advances in Intelligent Systems and Computing 437, DOI 10.1007/978-981-10-0451-3_12

the Goswami and Chaudhuri [2] model to consider the concept of shortages. Yang [4] developed a two warehouse inventory model with partial backlogging and inflation for deteriorated items. Das et al. [5] established two warehouse integrated models in the possibility measures. Singh et al. [6] described a two storage inventory model for the deteriorated inventory with the demand depends on stock and the conditions of shortages and permissible delay. Yadav et al. [7] extend the model in fuzzy environment with two warehouse and variable demand rate with the help of genetic algorithm. Singh et al. [8] discussed a three stage integrated model under the inflationary environment with two warehouse, imperfect production, and variable demand rate.

In the existing literature, maximum number of inventory models used the concept of constant production rate. With a constant production rate, average cost of producing items is minimized but during the slumps in the demand, the already produced items needs to be stored and this results the increase in the holding cost. But as the production rate varies then the production cost decreases with time and it will decrease the holding cost. Therefore, the best production scheme always depends on the holding cost and on the agility of the production system. Yet there is a need of volume agility in the inventory models to make model more realistic due to present business situation. To improve production efficiency, manufacturing companies have used the concept of flexible manufacturing systems (FMS). Khouja and Mehrez [9], Khouja [10], Khouja [11] extended the economic lot size manufacturing model to an imperfect production process under the environment of volume agility. Sana and Chaudhuri [12] considered volume flexibility for deteriorated items with the stock dependent demand rate. Sana and Chaudhuri [13] developed inventory model with agile manufacturing for deteriorating goods with shortage which are completely backlogged and time dependent demand. Singh and Urvashi [14] considered integrated models with imperfect production process and agile manufacturing under the inflationary environment. The best production scheme depends on the agility of the production system. Singh et al. [15] discussed an inventory model with volume flexibility for defective items with partial backlogging and multi variate demand. Singh et al. [16] established an EOQ model with agile manufacturing, variable demand rate, Weibull deterioration rate and inflation. Mehrotra et al. [17] developed an inventory model with trapezoidal demand and inflation under the environment of volume flexibility.

In the inventory models, the demand rate is assumed to be cost dependent, time dependent, and stock dependent. In the years, many researchers have recognized that the demand for items should be parabolic. Mandal1 and Kumar [18] developed an Economic order quantity model with parabolic demand rate and time-varying selling price. Singh et al. [19] developed a manufacturing model with effects of learning when demand rate is fluctuating with production rate. In this research, an effort is done to formulate a supply chain production model for deteriorated inventory with two warehouse, volume agility, and parabolic demand rate under the effect of inflation. The rest of this paper is organized as follows. Section 2 describes the assumptions and notations used throughout this paper. The mathematical model and the minimum total integrated cost are given in Sect. 3. Illustrative example, which explain the applications of the theoretical results and their numerical verifications, are given in Sect. 4. Sensitivity analyses are carried out with respect to the different parameters of the system in the Sect. 5, while concluding suggestions for further research are given in Sect. 6.

2 Assumptions and Notation

The inventory model is formulated which is based on the assumptions given below:

(1) Lead time is taken as zero.
(2) The demand rate is parabolic function of time for vendor and supplier both. where $D(t) = d + bt - at^2$, where $a, b > 0$ are positive constants, $a > b$ and d is the demand rate at the time zero.
(3) Demand is less than the production.
(4) The replenishment rate is taken as infinite.
(5) Deterioration rate is taken as constant for the vendor. Deterioration rate is different for the supplier's own warehouse and for rented warehouse. In the OW deterioration rate is time dependent and in the RW deterioration rate is constant.
(6) Inflation rate is also considered.
(7) Capacity of OW remains constant, i.e., W units.
(8) Capacity of RW is unlimited.
(9) The inventory cost in the RW is more than the inventory cost in the OW.
(10) The production cost per unit item is a function of the production rate.
(11) Per order multiple deliveries are considered.

The following notations are assumed throughout the paper:

T Length of the complete cycle of vendor.
T_1 The production period of vendor.
T_2 The nonproduction time period of vendor.
P Production rate per unit.
$C_{1\text{v}}$ Per unit vendor's holding cost.
$C_{2\text{v}}$ Vendor's deterioration cost per unit.
$C_{3\text{v}}$ Vendor's set up cost per production cycle.
$C_{1\text{so}}$ Supplier's holding cost for the OW.
$C_{1\text{sr}}$ Supplier's holding cost for the RW.
$C_{2\text{so}}$ Supplier's deterioration cost in OW.
$C_{2\text{sr}}$ Supplier's deterioration cost in RW.
$C_{3\text{s}}$ Per order supplier's set up cost.
θ Rate of deteriorated inventory for vendor, where $0 < \theta < 1$.
$\zeta(t)$ Deterioration rate in OW of supplier is $\zeta(t) = \zeta$ t, where o $< \zeta < 1$.
η Deterioration rate in RW of the supplier, where $0 < \eta < 1$.

W Capacity of the OW.
n Number of deliveries for the supplier.
r Inflation rate.
T/n One cycle length for supplier which is equal to $T_3 + T_4$.

3 Model Development

In this integrated model a strategy is adopted for a two-level vendor–buyer inventory model. The vendor's go by an EPQ policy to manufacture the goods under the effect of volume agility. The coordination strategy is such that the vendor receives the supplier's demand and manufactures the items and delivers to the supplier. For the surplus stock supplier uses the RW and OW. For the OW rate of deterioration is parabolic function of time and for the RW deterioration rate is constant. Here, demand rate is parabolic for both vendor and supplier. Complete model is studied under the inflationary environment.

3.1 *Vendor's Inventory Model*

The inventory system for vendor is depicted in Fig. 1, can be separated into two phases depicted by T_1 and T_2. During the time period $[0, T_1]$, there is an increment in the inventory because of the production and decrement because of the deterioration and demand both. At the time T_1 if the production breaks then the stock level increases to its maximum level MI_v. During $T_1 < t \leq T_2$ time period there is no production and inventory level decreases because of the demand and deterioration. At time T_2 inventory level will be zero.

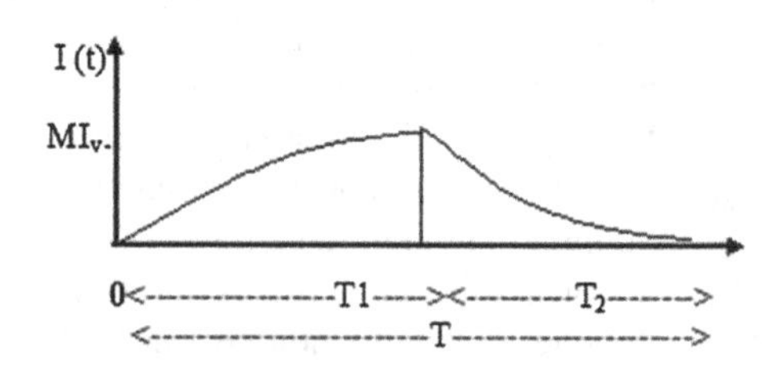

Fig. 1 Vendor's inventory system

Differential equations governing the model are as follows:

$$I'_{v1}(t) + \theta I_{v1}(t) = P - D(t), \quad 0 \le t \le T_1 \tag{1}$$

$$I'_{v2}(t) + \theta I_{v2}(t) = -D(t), \quad 0 \le t \le T_2 \tag{2}$$

Boundary conditions are $I_{v1}(0) = 0$ and $I_{v2}(T_2) = 0$
The solution of the differential Eqs. (1) and (2) are

$$I_{v1}(t) = \left(\frac{P-d}{\theta} + \frac{b}{\theta^2} + \frac{2a}{\theta^3}\right)(1 - e^{-\theta t}) - \left(\frac{\theta(bt - at^2) + 2at}{\theta^2}\right) \tag{3}$$

$$\begin{aligned} I_{v2}(t) = {} & \left(\frac{p-d}{\theta} + \frac{b}{\theta^2} + \frac{2a}{\theta^3}\right)\left(1 - e^{\theta(T_2 - t)}\right) \\ & - \left(\frac{bt - at^2}{\theta} + \frac{2at}{\theta^2}\right) + e^{\theta(T_2 - t)}\left(\frac{bT_2 - aT_2^2}{\theta} + \frac{2aT_2}{\theta^2}\right) \end{aligned} \tag{4}$$

Maximum inventory

$$MI_v = \left(\frac{P-d}{\theta} + \frac{b}{\theta^2} + \frac{2a}{\theta^3}\right)(1 - e^{-\theta T_1}) - \left(\frac{\theta(bT_1 - aT_1^2) + 2aT_1}{\theta^2}\right) \tag{5}$$

(1) Present worth holding cost is

$$\mathrm{HC}_v = C_{1v}\left[\int_0^{T_1} I_{v1}(t)e^{-rt}\mathrm{d}t + e^{-rT_1}\int_0^{T_2} I_{v2}(t)e^{-rt}\mathrm{d}t\right]$$

$$\mathrm{HC}_v = C_{1v}\left[\begin{array}{l} \left(\frac{p-d}{\theta} + \frac{b}{\theta^2}\right)\left\{\frac{-\theta - (\theta + r)e^{-rT_1} + re^{-(\theta+r)T_1}}{r(\theta + r)}\right\} - \frac{1}{\theta^2}\left\{\begin{array}{l} e^{-rT_1}\left(\frac{2a\theta T_1 - r(b\theta T_1 - a\theta T_1^2 + 2aT_1)}{r^2}\right) \\ + \frac{(1 - e^{-rT_1})(r\theta b + 2ar + 2a\theta)}{r^3} \end{array}\right\} \\ + e^{-rT_1}\left\{\begin{array}{l} \left(\frac{p-d}{\theta} + \frac{b}{\theta^2} + \frac{2a}{\theta^3}\right)\left(\frac{1 - e^{-rT_2}}{r} + \frac{e^{-rT_2} - e^{\theta T_2}}{\theta + r}\right) - \left\{\begin{array}{l} e^{-rT_2}\left\{\begin{array}{l} \left(\frac{-bT_2 + aT_2^2}{r\theta} - \frac{2aT_2}{r\theta^2}\right) \\ -\left(\frac{b - 2aT_2}{r^2\theta} + \frac{2a}{r^2\theta^2}\right) + \frac{2aT_2}{\theta r^3} \end{array}\right\} \\ + \left(\frac{b}{\theta} + \frac{2a}{\theta^2}\right)\left(\frac{1}{r^2} - \frac{2a}{\theta r^3}\right) \end{array}\right\} \\ + \left(\frac{bT_2 - aT_2^2}{\theta} - \frac{2aT_2}{\theta^2}\right)\left(\frac{-e^{-rT_2} + e^{\theta T_2}}{\theta + r}\right) \end{array}\right\} \end{array}\right] \tag{6}$$

(2) Present worth deterioration cost is

$$\mathrm{DC}_v = C_{2v}\left[\mathrm{MI}_v - \int_0^{T_1} D(t)e^{-rt}\mathrm{d}t + e^{-rT_1}\int_0^{T_2} D(t)e^{-rt}\mathrm{d}t\right]$$

$$= C_{2v}\left[\begin{array}{l} \left(\dfrac{P-d}{\theta}+\dfrac{b}{\theta^2}+\dfrac{2a}{\theta^3}\right)(1-e^{-\theta T_1}) - \left(\dfrac{\theta(bT_1-aT_1^2)+2aT_1}{\theta^2}\right) \\ -e^{-rT_1}\left\{\begin{array}{l} \left(\dfrac{2a}{r^3}-\dfrac{b}{r^2}\right)(1+e^{-rT_2}) + \left(\dfrac{d}{r}+\dfrac{b}{r^2}-\dfrac{2a}{r^3}\right)(1+e^{rT_1}) \\ -\dfrac{bT_1-aT_1^2}{r}+\dfrac{2aT_1}{r^2}+e^{-rT_2}\left(\dfrac{2aT_2}{r^2}-\dfrac{bT_2-aT_2^2}{r}\right) \end{array}\right\} \end{array}\right] \tag{7}$$

(3) Present worth set up cost is

$$\mathrm{SC}_v = C_{3v} \tag{8}$$

Since we assume a self-manufacturing system in which goods produced by a machine. The cost for setting up the apparatus and the whole system is follows as

Production cost per unit is

$$\psi(P) = \mu + \frac{g}{P} + sP + \beta(p-p_c)H(p-p_c)$$

where $H(p-p_c) = \begin{cases} 1 & P > P_c \\ 0 & P \leq P_c \end{cases}$

The production cost is based on the following factors:

1. The material cost μ per unit item is fixed.
2. As the production rate buildup, some costs like labor and energy costs are uniformly distributed over a huge number of units. Hence, the per unit production cost $\left(\frac{g}{P}\right)$ decreases as the production (P) increases.
3. The third term (sP), connected with the tool/die costs, is proportional to the production rate.
4. The fourth term is associated to a critical design production rate (P_c) for the machine. The produced items are quite likely to be imperfect for a high production rate ($P > P_c$). The excess labor and energy costs along with rework costs will be needed to get perfect stock.

(4) The production cost is

$$\begin{aligned}\text{PC} &= \psi(P)\text{MI}_v \\ &= \left(\mu + \frac{g}{P} + sP + \beta(p - p_c)H(p - p_c)\right)\left[\left(\frac{P-d}{\theta} + \frac{b}{\theta^2} + \frac{2a}{\theta^3}\right)(1 - e^{-\theta T_1}) \right. \\ &\qquad \left. - \left(\frac{\theta(bT_1 - aT_1^2) + 2aT_1}{\theta^2}\right)\right] \end{aligned} \tag{9}$$

Present worth average total cost of the vendor is the sum of holding cost, deterioration cost, set up cost, and production cost.

$$\text{TC}_v(P, T_1) = \frac{\text{HC}_v + \text{DC}_v + \text{SC}_v + \text{PC}}{T} \tag{10}$$

3.2 *Supplier's Inventory Model*

The supplier's inventory level is depicted in Fig. 2. Supplier uses two warehouses. One is own warehouse which has a fixed capacity and other is rented warehouse for any extra stock, which is available with abundant space. The stock of own storage are consumed only after consuming the stock kept in RW.

3.2.1 Supplier's Inventory Model for the OW

In OW, the stock level decreases during the time period [0, T_3], due to deterioration only, but during [0, T_4] the stock level is decreased due to both demand and deterioration. Here deterioration rate is a function of time.

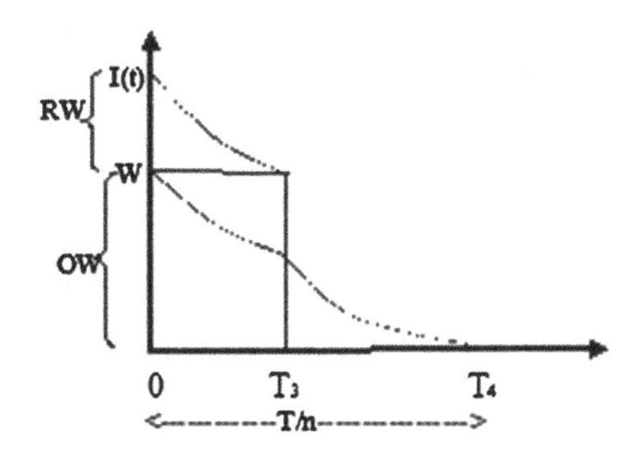

Fig. 2 Supplier's inventory system

$$I'_{1so}(t) = -\zeta(t)I_{1so}(t) \quad 0 \le t \le T_3 \tag{11}$$

$$I'_{2so}(t) = -\zeta I_{2so}(t) - D(t) \quad 0 \le t \le T_4 \tag{12}$$

Boundary conditions are $I_{1so}(0) = \text{W}$ and $I_{2so}(\text{T}_4) = 0$
The solution of the differential Eqs. (11) and (12) are

$$I_{1so}(t) = We^{g(t)-g(0)}, \text{where} \quad g(t) = -\int \varsigma(t)\text{d}t \tag{13}$$

$$I_{2so}(t) = e^{g(t)}\left[\begin{array}{l}(d+bT_4-aT_4^2)g_1(T_4)-(b-2aT_4)g_2(T_4)-2ag_3(T_4)\\ -(d+bt-at^2)g_1(t)+(b-2at)g_2(t)+2ag_3(t)\end{array}\right. \tag{14}$$

where, $g_1(t) = -\int e^{-g(t)}\text{d}t, g_2(t) = \int g_1(t)\text{d}t, g_3(t) = \int g_2(t)\text{d}t,$

(1) Present worth holding cost in the own warehouse

$$\text{HC}_{so} = C_{1so}\left[\int_0^{T_3} I_{1so}(t)e^{-rt}\text{d}t + e^{-rT_3}\int_0^{T_4} I_{2so}(t)e^{-rt}\text{d}t\right]$$

$$\text{HC}_{so} = C_{1so}\left[W\left(T_3+\frac{\varsigma T_3^3}{6}-\frac{rT_3^2}{2}\right)+e^{-rT_3}\left\{\begin{array}{l}\left(1-\frac{\varsigma T_4^2}{2}-rT_4\right)\left(\frac{dT_4^2}{2}+\frac{bT_4^3}{3}+\frac{\varsigma dT_4^4}{8}-\frac{aT_4^4}{2}\right)\\ +(r+\varsigma T_4)\left(\frac{dT_4^2}{2}+\frac{(3b-d)T_4^3}{6}-\frac{7bT_4^4}{24}\right)\\ -\varsigma\left(\frac{dT_4^4}{8}+\frac{3bT_4^5}{4}\right)\end{array}\right\}\right] \tag{15}$$

(2) Present worth deterioration cost in the own warehouse

$$\begin{aligned}\text{DC}_{so} &= C_{2so}\left[W - e^{-rT_3}\int_0^{T_4} D(t)e^{-rt}\text{d}t\right]\\ &= C_{2so}\left[W - e^{-rT_3}\left\{e^{-rT_4}\left(\frac{2a}{r^3}-\frac{(d+bT_4-aT_4^2)}{r}-\frac{(b-2aT_4)}{r^2}\right)\right.\right.\\ &\qquad\left.\left.+\left(\frac{d}{r}+\frac{b}{r^2}-\frac{2a}{r^3}\right)\right\}\right]\end{aligned} \tag{16}$$

3.2.2 Supplier's Inventory Model for Rented Warehouse (RW)

In the interval $[0, T_3]$, the stock level in the RW gradually decreases due to deterioration and demand and it vanishes at $t = T_3$.

$$I'_{\text{sr}}(t) = -\eta I_{\text{sr}}(t) - D(t) \quad 0 \le t \le T_3 \tag{17}$$

With the boundary condition $I_{\text{sr}}(T_3) = 0$, we can solve the above differential equation, we get

$$I_{\text{sr}}(t) = \frac{e^{\eta(T_3-t)}\left(d + bT_3 - aT_3^2\right) - \left(d + bt - at^2\right)}{\eta} + \frac{e^{\eta(T_3-t)}(-b + 2aT_3) - (-b + 2at)}{\eta^2} + \frac{2a\left(1 - e^{\eta(T_3-t)}\right)}{\eta^3} \tag{18}$$

With the condition $I_{\text{sr}}(0) = I_r$, we get

$$I_r = \frac{e^{\eta T_3}\left(d + bT_3 - aT_3^2\right) - d}{\eta} + \frac{e^{\eta T_3}(-b + 2aT_3) + b}{\eta^2} + \frac{2a(1 - e^{\eta T_3})}{\eta^3} \tag{19}$$

(1) Present worth holding cost in the rented warehouse

$$\begin{aligned}\text{HC}_{\text{sr}} &= C_{1\text{sr}} \int_0^{T_3} I_{sr}(t) e^{-rt} \text{d}t \\ &= C_{1\text{sr}} \left[\begin{array}{l} \frac{e^{-rT_3}}{r} \left\{ \frac{2a}{\eta^3}\left(1 + \frac{1}{(\eta + r)}\right) + \frac{1}{r\eta}\left(b - 2a + \frac{2a}{\eta} + \frac{2a}{r}\right) \right\} \\ -\frac{1}{\eta r}\left(\frac{2a}{\eta r} + d - \frac{b}{\eta} - \frac{2a}{\eta^2} + \frac{2a}{r^3}\right) + \left(\frac{\left(d + bT_3 - aT_3^2\right)}{\eta} + \frac{(-b + 2aT_3)}{\eta^2}\right) \\ \left\{ \frac{e^{-rT_3}}{r}\left(1 - \frac{1}{(\eta + r)} - \frac{e^{\eta T_3}}{(\eta + r)}\right) \right\} + \frac{2ae^{\eta T_3}}{\eta^3(\eta + r)} \end{array} \right]\end{aligned} \tag{20}$$

(2) Present worth deterioration cost in the rented warehouse

$$\begin{aligned}\text{DC}_{\text{sr}} &= C_{2\text{sr}} \left[I_r - \int_0^{T_3} D(t) e^{-rt} \text{d}t \right] \\ &= C_{2\text{sr}} \left[I_r - \left\{ e^{-rT_3}\left(\frac{2a}{r^3} - \frac{\left(d + bT_3 - aT_3^2\right)}{r} - \frac{(b - 2aT_3)}{r^2}\right) + \left(\frac{d}{r} + \frac{b}{r^2} - \frac{2a}{r^3}\right) \right\} \right]\end{aligned} \tag{21}$$

(3) Present worth set up cost of the supplier is

$$\mathrm{SC}_s = C_{3s} \tag{22}$$

Present worth average integrated cost of the supplier is the sum of holding cost, set up cost, deterioration cost.

$$\mathrm{TC}_s(T_3) = \frac{\mathrm{HC}_{\mathrm{so}} + \mathrm{HC}_{\mathrm{sr}} + \mathrm{DC}_{\mathrm{so}} + \mathrm{DC}_{\mathrm{sr}} + \mathrm{SC}_s}{T/n} \tag{23}$$

Per cycle there are n deliveries. Per delivery time is $T_3 + T_4 = T/n$.
Integrated total cost of vendor and supplier is the function of three variables as given below
TC $(P,\ T_1,\ T_3) = \mathrm{TC}_v + \mathrm{TC}_s$
To find out the optimal solution we apply the following procedure

$$\partial \mathrm{TC}(P, T_1, T_3)/\partial P = 0, \partial \mathrm{TC}(P, T_1, T_3)/\partial T_1 = 0$$
$$\partial \mathrm{TC}(P, T_1, T_3)/\partial T_3 = 0$$

4 Numerical Illustration for the Model

In this section, a numerical example is formulated to illustrate the complete model. The following values of parameters are used in the example.

$C_{1v} = 0.5\$,\ C_{2v} = 0.2\$,\ C_{3v} = 0.59\$,\ C_{1so} = 3.9\$,$
$C_{1\mathrm{sr}} = 0.26\$,\ C_{2\mathrm{so}} = 0.54\$,\ C_{2\mathrm{sr}} = 0.9\$,\ C_{3s} = 0.31\$,$
$n = 2,\ a = 5,\ b = 4,\ d = 20,\ \theta = 0.001,\ \zeta = 0.08,\ \eta = 0.05,\ W = 400$ unit,
$r = 0.11,\ \mu = 90,\ g = 3500,\ s = 0.01,\ \beta = 0.04,\ P_c = 220,\ T = 30$ days.

We have $\psi(P) = \mu + \frac{g}{P} + sP + \beta(p - p_c), \quad P > P_c$
$= \mu + \frac{g}{P} + sP, \quad P \leq P_c$

Then $\frac{\mathrm{d}\psi}{\mathrm{d}P} = -\frac{g}{P^2} + s + \beta, \quad P > P_c$
$= -\frac{g}{P^2} + s, \quad P \leq P_c$

$\frac{\mathrm{d}^2\psi}{\mathrm{d}P^2} = \frac{2g}{P^3} > 0$, for all $P > 0$
We thus have:

(i) $\psi_{\min}$ at $P = \sqrt{\frac{g}{s+\beta}} = 223.6$, when $P > P_c$

(ii) $\psi_{\min}$ at $P = \sqrt{\frac{g}{s}}$, when $P \leq P_c$

We may, therefore, take $P_c = 220$.

Table 1 When $P > P_c$

n	P	Time period (T_1) days	Time period (T_3)	Total cost (TC)
1	260.71	13.5873	11.9952	1.13164×10^7
2	260.71	13.5873	9.6777	1.14285×10^7
3	260.71	13.5873	8.6673	1.14191×10^7
4	260.71	13.5873	5.6469	1.14097×10^7

Table 2 When $P \leq P_c$ i.e. $\beta = 0$

n	P	Time period (T_1) days	Time period (T_3)	Total cost (TC)
1	554.002	13.5748	11.9952	1.16759×10^7
2	554.002	13.5748	9.6777	1.15646×10^7
3	554.002	13.5748	8.6673	1.16664×10^7
4	554.002	13.5748	5.6469	1.1657×10^7

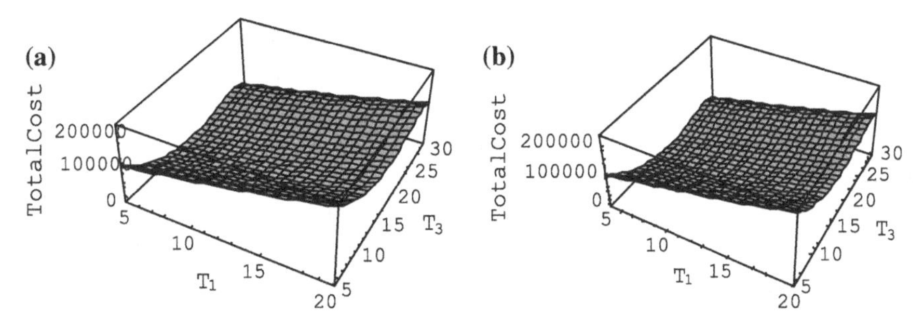

Fig. 3 Graphical representation of convexity of integrated cost. **a** When $P > P_c$. **b** When $P \leq P_c$

From the above graph, we can examine the convexity of the total integrated cost, according that total cost is minimum for the case when $P > P_c$. If start with the condition $P < P_c$ with the same parameter values as in Table 1, it is found (Table 2) that there is no possible solution. Therefore, we have no option but to continue operation of the production system with the production rates $P > P_c$ (Fig. 3).

5 Sensitivity Analysis

The sensitivity of the optimal solution has been analyzed for various system parameters when $P > P_c$ through Tables 3, 4 and 5 as shown below.

We can examine the relative effects of the cost parameters, demand rate, deterioration rate, and inflation rate on the integrated cost of the model from the above sensitivity analysis and these results give us previous warning that in future, which

Table 3 Sensitivity analysis w.r.t. vendor's cost parameters when $P > P_c$

	% of change	T_1	T_3	P	(TC)
C_{1v}	−50	13.8359	11.6777	260.666	7.28908×10^6
	−25	13.6684	11.6777	260.675	9.29301×10^6
	+25	13.5394	11.6777	260.754	1.33495×10^7
	+50	13.5078	11.6777	260.802	1.53856×10^7
C_{2v}	−50	13.5871	11.6777	260.713	1.13143×10^7
	−25	13.5872	11.6777	260.712	1.13158×10^7
	+25	13.5874	11.6777	260.708	1.13188×10^7
	+50	13.5875	11.6777	260.706	1.13204×10^7
C_{3v}	−50	13.5873	11.6777	260.71	1.14285×10^7
	−25	13.5873	11.6777	260.71	1.14285×10^7
	+25	13.5873	11.6777	260.71	1.14285×10^7
	+50	13.5873	11.6777	260.71	1.14285×10^7
Θ	−50	13.458	11.6777	262.601	3.92451×10^7
	−25	13.522	11.6777	261.645	1.8873×10^7
	+25	13.6538	11.6777	259.794	5.8392×10^7
	+50	13.7217	11.6777	258.898	7.81203×10^7

Table 4 Sensitivity analysis w.r.t. supplier's cost parameters

	% of change	T_1	T_3	P	TC
C_{1so}	−50	13.5873	11.8919	260.71	1.14272×10^7
	−25	13.5873	11.7826	260.71	1.14278×10^7
	+25	13.5873	11.5767	260.71	1.14292×10^7
	+50	13.5873	11.4793	260.71	1.14298×10^7
C_{1sr}	−50	13.5873	11.1924	260.71	1.14397×10^7
	−25	13.5873	11.5058	260.71	1.14342×10^7
	+25	13.5873	11.7868	260.71	1.14228×10^7
	+50	13.5873	11.8622	260.71	1.14371×10^7
C_{2so}	−50	13.5873	11.66	260.71	1.14286×10^7
	−25	13.5873	11.669	260.71	1.14285×10^7
	+25	13.5873	11.6865	260.71	1.14285×10^7
	+50	13.5873	11.6953	260.71	1.14289×10^7
C_{2sr}	−50	13.5873	11.7571	260.71	1.14279×10^7
	−25	13.5873	11.7172	260.71	1.14282×10^7
	+25	13.5873	11.6386	260.71	1.14288×10^7
	+50	13.5873	11.5998	260.71	1.14291×10^7
C_{3s}	−50	13.5873	11.6777	260.71	1.14285×10^7
	−25	13.5873	11.6777	260.71	1.14286×10^7
	+25	13.5873	11.6777	260.71	1.14289×10^7
	+50	13.5873	11.6777	260.71	1.14290×10^7

(continued)

Table 4 (continued)

	% of change	T_1	T_3	P	TC
η	−50	13.5873	17.8353	260.71	1.12172×10^7
	−25	13.5873	13.7521	260.71	1.14087×10^7
	+25	13.5873	10.3788	260.71	1.14329×10^7
	+50	13.5873	9.5057	260.71	1.1435×10^7
ζ	−50	13.5873	11.9076	260.71	1.14274×10^7
	−25	13.5873	11.79	260.71	1.14279×10^7
	+25	13.5873	11.5703	260.71	1.1429×10^7
	+50	13.5873	11.4673	260.71	1.14296×10^7

Table 5 Sensitivity analysis w.r.t. the parameters of both vendor and supplier

Parameters	% of change	T_1	T_3	P	Total cost
a	−50	13.5868	11.3314	257.033	5.70943×10^6
	−25	13.5871	11.5539	259.469	8.56899×10^6
	+25	13.5874	11.7568	261.467	1.4288×10^7
	+50	13.5874	11.8117	261.976	1.71475×10^7
b	−50	13.5877	11.6215	260.71	1.14402×10^7
	−25	13.5875	11.6498	260.71	1.14344×10^7
	+25	13.5871	11.7053	260.71	1.14344×10^7
	+50	13.5868	11.7324	260.71	1.14344×10^7
d	−50	13.5873	11.7189	260.71	1.14289×10^7
	−25	13.5873	11.6983	260.71	1.14287×10^7
	+25	13.5873	11.6573	260.709	1.14283×10^7
	+50	13.5873	11.637	260.709	1.14281×10^7
r	−50	16.3038	16.0266	261.774	–
	−25	14.5952	13.1078	261.106	–
	+25	12.9348	10.8807	260.443	4.14732×10^7
	+50	12.4808	10.3622	260.252	4.12048×10^7

parameter is more or less affected on the total cost if there is any variation in parameters.

- If we increase the no of cycles then production time is unchanged and total cost increases.
- With the small increment in the deterioration rate there is a large variation in the total cost of the model.
- If we raise the cost parameter then the total cost increases very extremely.
- The total cost increases with the increment in the inflation rate.

6 Conclusion

This paper is the collaboration of the vendor's strategy and supplier's strategy. In this paper, we have considered a two storage integrated inventory model with some realistic assumptions for the supplier and vendor. Since demand rate effects on the production rate and production rate is a decision variable therefore this theory motivates us for the investigation on variable demand rate because we consider the parabolic demand rate in this paper. The complete model is considered in inflationary environment. Concept of agile manufacturing and two warehouse is very practical. The concept of volume agility is explained on the vendor's part and supplier uses the concept of warehouses. A numerical estimation of the model has been done to point up the theory. In this study, sensitivity analysis has been done by changing the parameters of the model. We can examine with the help of the sensitivity analysis that which parameter is more valuable for the total cost. The study can be extended with fuzzy environment in the further research.

References

1. Hartley, V.R.: Operations Research—a Managerial Emphasis, pp. 315–317. Good Year Publishing Company, California (1976)
2. Goswami, A., Chaudhuri, K.S.: An economic order quantity model for items with two levels of storage for a linear trend in demand. J. Oper. Res. **43**(2), 157–167 (1992)
3. Bhunia, A.K., Maiti, M.: A two warehouses inventory model for deteriorating items with a linear trend in demand and shortages. J. Oper. Res. **49**(1), 287–292 (1998)
4. Yang, H.L.: Two warehouse partial backlogging inventory models for deteriorating items under inflation. Int. J. Prod. Econ. **103**(2), 362–370 (2006)
5. Das, B., Maity, K., Maiti, M.: A two warehouse supply-chain model under possibility necessity credibility measures. Math. Comput. Model. **46**(3–4), 398–409 (2007)
6. Singh, S.R., Kumari, R., Kumar, N.: A deterministic two warehouse inventory model for deteriorating items with stock dependent demand and shortages under the conditions of permissible delay. Int. J. Math. Model. Numer. Optim. **2**(4), 357–375 (2011)
7. Yadav, D., Singh, S.R., Kumari, R.: Inventory model of deteriorating items with two warehouse and stock dependent demand using genetic algorithm in fuzzy environment. YUJOR. ISSN: 0354-0243 (2012)
8. Singh, S.R., Gupta, V., Gupta, P.: Three stage supply chain model with two warehouse, imperfect production, variable demand rate and inflation. Int. J. Ind. Eng. Comput. **4**(1), 81–92 (2013)
9. Khouja, M., Mehrez, A.: An economic production lot size model with variable production rate and imperfect quality. J. Oper. Res. Soc. **45**(12), 1405–1417 (1994)
10. Khouja, M.: The economic production lot size model under volume flexibility. Comput. Oper. Res. **22**(5), 515–523 (1995)
11. Khouja, M.: The scheduling of economic lot size on volume flexibility production system. Int. J. Prod. Econ. **48**(1), 73–86 (1997)
12. Sana, S., Chaudhuri, K.S.: On a volume flexible stock dependent inventory model. Adv. Model. Optim. **5**(3), 197–210 (2003)
13. Sana, S., Chaudhuri, K.S.: On a volume flexible production policy for a deteriorating item with time dependent demand and shortage. Adv. Model. Optim. **6**(1), 57–73 (2004)

14. Singh, S.R., Urvashi.: Supply chain models with imperfect production process and volume flexibility under inflation. IUP J. Supply Chain Manage. **7**(1&2), 61–76 (2010)
15. Singh, S.R., Singhal, S., Gupta, P.K.: A volume flexible inventory model for defective items with multi variate demand and partial backlogging. Int. J. Oper. Res. Optim. **1**(1), 55–69 (2010)
16. Singh, S.R., Gupta, V., Gupta, P.: EOQ model with volume agility, variable demand rate, Weibull deterioration rate and inflation. Int. J. Comput. Appl. **72**(23), 1–6 (2013)
17. Mehrotra, K., Dem, H., Singh, S.R., Sharma, R.: A volume flexible inventory model with trapezoidal demand under inflation. Pak. J. Stat. Oper. Res. **9**(4), 429–442 (2014)
18. Mandal1, M., Kumar, S.: An EOQ model with parabolic demand rate and time varying selling price. Pure Appl. Math. **1**(1), 32–43 (2012)
19. Singh, S.R., Gupta, V., Kumar, N.: A production-inventory model incorporating the effect of learning when demand is fluctuating with production rate. In: Proceeding of International Conference on Recent Trends in Engineering & Technology (ICRTET'2014)

Invariant Feature Extraction for Finger Vein Matching Using Fuzzy Logic Inference

Devarasan Ezhilmaran and Manickam Adhiyaman

Abstract Finger vein is a new member in the biometrics family. In the recent years, as a biometric trait finger vein has been achieved success and attracted lots of attention from the researchers. Current finger vein recognition algorithms provide high accuracy in good quality images. Most of the finger vein recognition systems utilize based on minutiae points. In this paper, we propose a method for finger vein matching using scale-invariant feature transformation (SIFT) key points which are determining the rules using fuzzy inference system and the matching scores are delivered by Euclidean distance. The fuzzy logic system provides the more adequate description of the proposed algorithm. The algorithm has been formulated based on SIFT key points which examine n number of images. MATLAB 7:14:0 is used to implement this technique.

Keywords Biometrics · Finger vein images · SIFT key points · Fuzzy inference system · Matching

1 Introduction

Physical characteristics and behavior are the parameters for identifying a person in biometrics security. The finger vein is a physical characteristic based biometric, which is captured from the near-infrared light. As a biometric characteristic, a finger vein is very suitable for human recognition due to universality, distinctiveness, permanence, collectability, and acceptability. Finger vein is a better recognition trait compared to other traits (such as face, iris, fingerprint, palmprint, etc.) because it is

Devarasan Ezhilmaran · Manickam Adhiyaman (✉)
School of Advanced Sciences, VIT University, Vellore 632014, India
e-mail: adhimsc2013@gmail.com

Devarasan Ezhilmaran
e-mail: ezhil.devarasan@yahoo.com

M. Pant et al. (eds.), *Proceedings of Fifth International Conference on Soft Computing for Problem Solving*, Advances in Intelligent Systems and Computing 437, DOI 10.1007/978-981-10-0451-3_13

a living body identification and an internal characteristic. Typically for all the biometric traits including finger vein, the identification system mainly consists of image acquisition, preprocessing, feature extraction, and matching. The finger veins are not clearly visible. In fact, nothing has been known about numbers, locations, and lengths of finger veins. Today's technologies have made it possible to get sharp vein images.

Kono et al. [5] proposed the finger vein pattern matching as a new method for identification of individuals. Miura et al. [10] extracted the finger vein pattern from unclear image by using repeated line tracing method. Yanagawa et al. [16] described the diversity of finger vein images and its usefulness in personal identification using 2024 finger vein images from 506 persons. Liukui et al. [9] have demonstrated a tri-value template fuzzy matching for finger vein recognition. A novel method of exploiting finger vein features for personal identification is proposed by Yang et al. [17]. Song et al. [14] have derived the mean curvature method, which views the vein image as a geometric shape and finds the valley-like structures with negative mean curvatures. Yang et al. [18] used a method based on multi-instance matching score fusion. Peng et al. [12] have explained a method for finger vein enhancement using Gabor filter and matching using SIFT features. Pang et al. [11] have described a method to solve the common rotation problem of finger vein images. Aziz et al. [1] extracted the minutiae points based on maximum curvature method. Here, only bifurcation will be marked as minutiae points since vein has no ridge ending or termination. Cao et al. [2] have discussed finger vein recognition based on structure feature extraction. Liu et al. [8] have described a singular value decomposition based minutiae matching for finger vein recognition which has involved three stages: (i) minutiae pairing, (ii) false removing, and (iii) score calculating. Gupta et al. [3] have proposed a method for accurate personal authentication system and to reduce the noise, lower local contrast, hairs, and texture. The impact of sensor ageing related pixel defects on the performance of finger vein based recognition systems in terms of the Equal Error Rate (EER) is investigated by Kauba and Uhl [4]. Lee et al. [6] introduced an advanced authentication system for personal identification and proposed an algorithm for finger vein authentication. Xian et al. [15] described organization of the competition and Recognition Algorithm Evaluation Platform (RATE), and then described data sets and test protocols, and finally presented results of the competition.

In this paper, a new method is presented for finger vein matching using SIFT key points and determining the rules using fuzzy inference engine for match scores. The rest of the paper is organized as follows: Sect. 2 explains the concept of fuzzy set theory. Section 3 describes the proposed methodology including SIFT key points based matching. Section 4 discusses about fuzzy inference engine module. Experimental results are presented in Sect. 5 and finally the conclusion is drawn in Sect. 6 (Fig. 1).

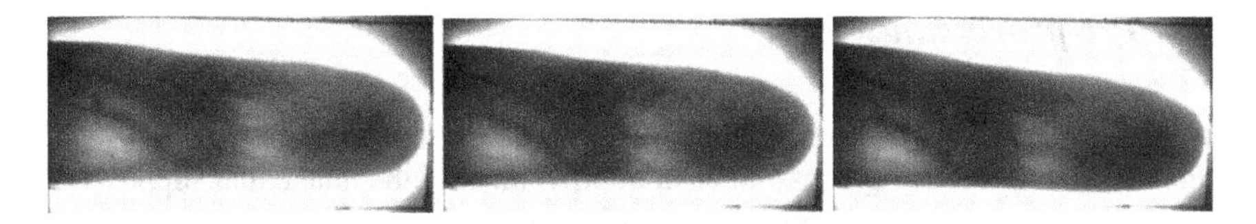

Fig. 1 Example of finger vein images (database of Hong Kong Polytechnic University, Hong Kong)

2 Preliminaries of Fuzzy Set Theory

The concept of fuzzy set theory is defined by Zadeh in 1965. It has a vital role in various fields of image processing. Images are considered to be fuzzy for their gray levels or information lost while mapping and so fuzzy set theory is used, which considers uncertainty in the form of the membership function. Fuzzy logic is an effective tool for handling the ambiguity and uncertainty of real-world systems. Fuzzy Inference Systems (FIS) are also recognized as fuzzy rule-based systems, fuzzy model, fuzzy expert system, and fuzzy associative memory. This is the primary content of a fuzzy logic system. The decision-making is a significant part in the entire system. The FIS formulates suitable rules and based upon the rules the decision is made. This is mainly based on the concepts of the fuzzy set theory, fuzzy IF-THEN rules, and fuzzy reasoning [13]. The database of a rule-based system may contain imperfect information which is inherent in the description of the rules given by the expert. This information may be incomplete, imprecise, fragmentary, not fully reliable, vague, contradictory, or deficient in some other ways. Nowadays, we can handle much of this imperfection of information using fuzzy logic in many active areas such as remote sensing, medical imaging, video surveillance, clustering, pattern recognition, neural network and so on.

Definition 1 The most commonly used range of values of membership functions is the unit interval [0, 1]. We shall denote the membership function of a fuzzy set A by μ_A, which is defined as $\mu_A(x) : X \rightarrow [0, 1]$, such that for each $x \in X$, $\mu_A(x) = \alpha,\ 0 \leq \alpha \leq 1$ [19].

3 Proposed Methodology

In this section, we investigate some techniques for matching such as pre-processing, SIFT key points extraction, and matching.

3.1 Pre-processing

The pre-processing is categorized into two processes which are binarization and thinning. The aim of pre-processing is an improvement of the image that suppresses or enhances some features important for further processing. The gray scale image is reformed into binary image which is labeled as binarization. This means each pixel is stored as a single bit, i.e., 0 or 1. Where 0 refers to black and 1 refers to white. The usability of image thinning is reducing the darkness of ridge lines. It is used in matching process because image quality is required.

3.2 SIFT Feature Extraction

SIFT was originally developed for general purpose of object recognition. SIFT method extracts distinctive invariant features from images, which perform reliable matching between different views of an object or scene. Image content is transformed into local feature coordinates that are invariant to translation, rotation, scale, and other imaging parameters (Fig. 2). There are four major stages of computation used for generating the SIFT features such as scale-space extrema detection, key point localization, orientation assignment, and the key point descriptor. These major stages are performed as follows: pick the key locations at local maximum and minimum of a Difference of Gaussian (DOG) function applied in scale space, which is constructed by successively downsampling the input image. Maximum and minimum of this scale space function are decided by comparing each pixel to its neighbors.

3.3 Matching

The proposed algorithm (Fig. 3) is performed as follows:

Step 1: Introduce the input fingerprint image
Step 2: Acquire the input image

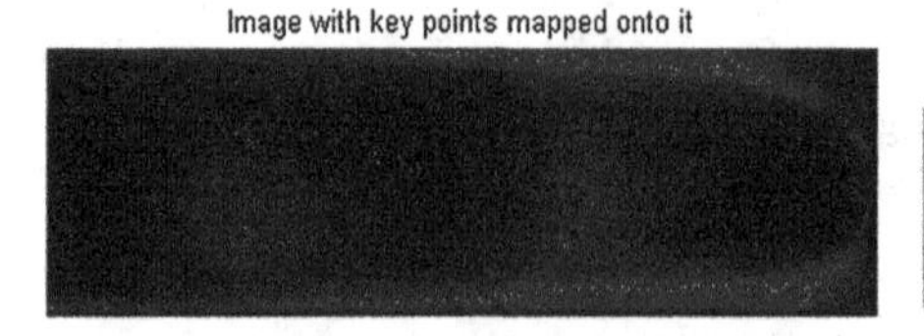

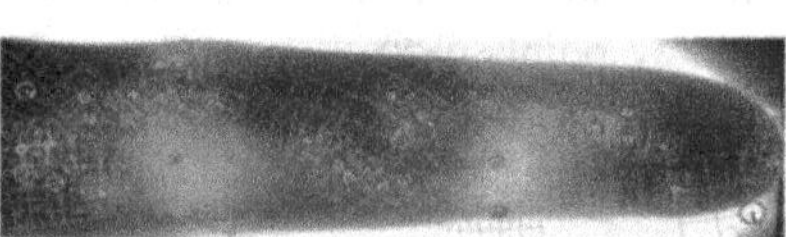

Fig. 2 SIFT key points with threshold parameter of finger vein image

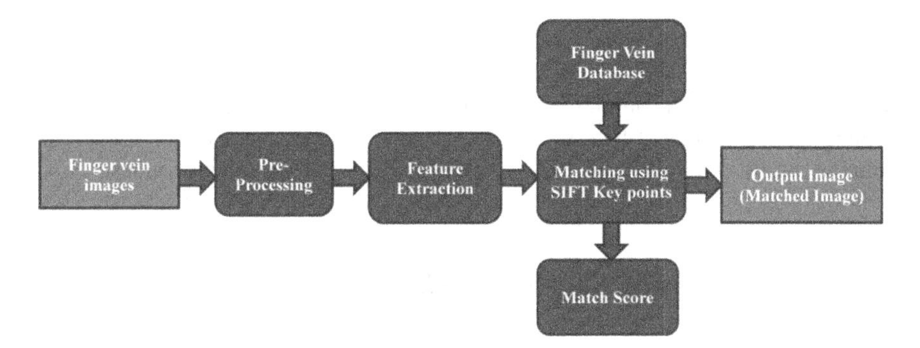

Fig. 3 Block diagram of proposed method

Step 3: Convert the gray scale image into the binary image and apply the thinning process to the image
Step 4: Extract feature from the binary image and count the SIFT key points of the input image
Step 5: Pick up the query image from the database and count the SIFT key points for query image
Step 6: Calculate the Euclidean distance between input image and database image
Step 7: Sort the output image and find out which one is a perfect match to the given query image
Step 8: Using fuzzy rules, find out the similar image for the query image

Mathematically, the Euclidean distance is the ordinary distance between two points in Euclidean space.

$$D(p,q) = \sqrt{\sum_{i=1}^{n} (p_i - q_i)^2} \quad (1)$$

After the completion of matching process, the matched image will be displayed on the screen.

4 Similarity Checking Using Fuzzy Inference System

Typically, fuzzy inference engine consist the following three processes: (i) fuzzification, (ii) processing, and (iii) defuzzification. Those processes are discussed in this section.

4.1 Fuzzification

In the process of fuzzification, membership functions defined on input variables are applied to their actual values so that the degree of fuzzy number for each rule premise can be determined. The crisp values are changed into fuzzy values.

4.2 Fuzzy Inference Engine Module

The rule-based form uses linguistic variables as its antecedents and consequents. The antecedents express an inference or the inequality, which should be satisfied. This is mainly based on the concepts of the fuzzy set theory, fuzzy IF-THEN rules, and fuzzy reasoning.

The following numbers are assigned based on fuzzy logic numeric range [7]:

A = 0; B = 1 to 50; C = 51 to 60; D = 61 to 100; E ≥ 100

These rules are developed using fuzzy inference engine module (Fig. 5) in order to the match points (M) and similar value (S) (Fig. 4):

Rules:

Rule 1: If M is A level then S is exact match

Rule 2: If M is B level then S is high

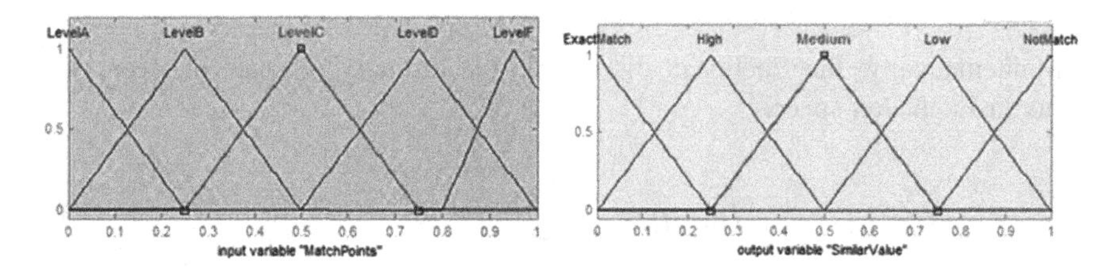

Fig. 4 The input and output membership function of match points and similar value

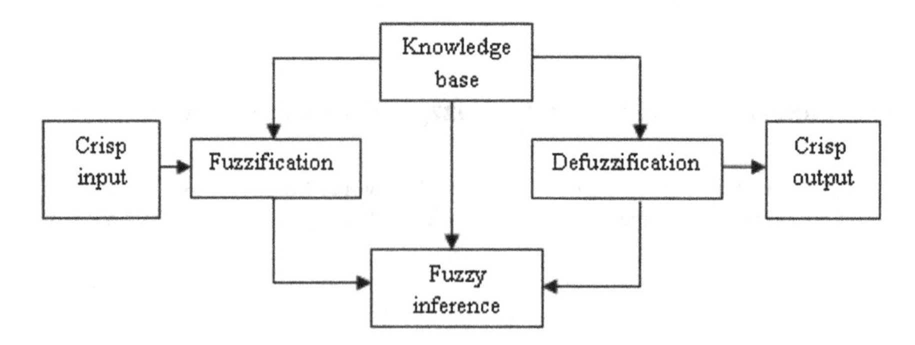

Fig. 5 Fuzzy inference engine

Rule 3: If M is C level then S is medium
Rule 4: If M is D level then S is low
Rule 5: If M is E level then S is not match

Using the above rules, we can find out the similar image to the given query image (Fig. 5).

4.3 Defuzzification

Converting the fuzzy grade into crisp output is called defuzzification. The following defuzzification methods are commonly used in many applications:

1. Centroid method.
2. Max-membership principle.
3. Weighted average method.
4. Middle of maxima method.

Centroid method is often called as center of gravity. It is the most prevalent and physically appealing of all the defuzzification methods.

5 Experiment Results and Discussions

These finger vein images (database) are taken from the Hong Kong Polytechnic University, Hong Kong. There were 156 subjects who offered vein images of the right index finger, the right middle finger, and each finger provides six images. The experiment is designed to evaluate the proposed method. The SIFT key points are used to finger vein matching and it is implemented by MATLAB. This algorithm brings off the SIFT points extraction and counts the key points from the image simultaneously. Subsequently, we can use the Euclidean distance for matching. It is only applicable for point-wise matching and otherwise it is not suitable, as the results of these SIFT features are different from image to image. SIFT feature extraction is only suitable for pair of images but in this algorithm, we have developed n number of image matching using SIFT points. Experimentally, 10 images are taken from the finger vein database and tested with a query image. Then the SIFT key points for both images (database and query) are counted. Furthermore, if a query image matches with any one of the template images, which matches score, then the value is equal to 0. Otherwise similarity value will be shown. Using fuzzy inference engine, the rules are defined. According to similarity value, the ranges are shown. For comparison, we are using the minutiae point in our algorithm from Aziz et al. [1].

6 Conclusions and Future Work

This paper has addressed the problem of finger vein matching. The finger vein matching result is shown based on SIFT key points using fuzzy inference system. Towards the matching this algorithm checks each SIFT key points for all images. It is observed that the proposed method is more effective because using fuzzy inference rule the matching score is high (Fig. 6). The proposed results are found much better than other features (Tables 1 and 2). The finger vein application is widely used for more security such as crime investigation, ATM machine, and credit/debit and VISA cards. This algorithm always not suitable for *n* number of images, because of while matching the similar number of key points are occurring

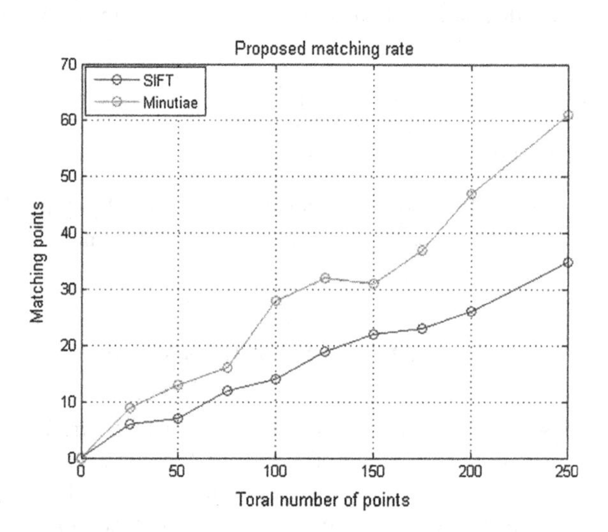

Fig. 6 Finger vein matching for SIFT and minutiae points

Table 1 Matching results

Images	Size	NSDBI	NSQI	MS	FIR
Test-1	225 × 225 (unit 8)	108	108	00.00	Match
Test-2	225 × 225 (unit 8)	102	108	06.00	High
Test-3	225 × 225 (unit 8)	115	108	07.00	High
Test-4	225 × 225 (unit 8)	096	108	12.00	High
Test-5	225 × 225 (unit 8)	094	108	14.00	High
Test-6	225 × 225 (unit 8)	089	108	19.00	High
Test-7	225 × 225 (unit 8)	130	108	22.00	High
Test-8	225 × 225 (unit 8)	131	108	23.00	High
Test-9	225 × 225 (unit 8)	134	108	26.00	High
Test-10	225 × 225 (unit 8)	143	108	35.00	High

NSDBI Number of SIFT points in database images; *NSQI* number of SIFT points in query images; *MS* match scores; *FIR* fuzzy inference rules

Table 2 Comparative results with Minutiae points

Images	Size	NMDBI	NMQI	MS	FIR
Test-1	225 × 225 (unit 8)	051	051	00.00	Match
Test-2	225 × 225 (unit 8)	042	051	09.00	High
Test-3	225 × 225 (unit 8)	064	051	13.00	High
Test-4	225 × 225 (unit 8)	067	051	16.00	High
Test-5	225 × 225 (unit 8)	079	051	28.00	High
Test-6	225 × 225 (unit 8)	083	051	32.00	High
Test-7	225 × 225 (unit 8)	020	051	31.00	High
Test-8	225 × 225 (unit 8)	014	051	37.00	High
Test-9	225 × 225 (unit 8)	004	051	47.00	High
Test-10	225 × 225 (unit 8)	112	051	61.00	Medium

NSDBI Number of minutiae points in database images; *NSQI* number of minutiae points in query images; *MS* match scores; *FIR* fuzzy inference rules

in the same image. So, it is lead to complexity for matching accuracy. Thus, the reader of this paper is encouraged to try it. We believe that the above research problem is open for discussion for future direction of this work.

References

1. Aziz, W.N., Kamaruzzaman Seman, I.A., Mohd, M.N.S.: Finger vein minutiae points extraction based on maximum curvature points in image profile and finger print application methods. Aust. J. Basic Appl. Sci. **7**, 751–756 (2013)
2. Cao, D., Yang, J., Shi, Y., Xu, C.: Structure feature extraction for finger-vein recognition. In: 2nd IAPR Asian Conference on Pattern Recognition, pp. 567–571, (2013)
3. Gupta, P., Gupta, P.: An accurate finger vein based verification system. Digital Signal Process. 384–352 (2015)
4. Kauba, C., Uhl, A.: Sensor ageing impact on finger-vein recognition. In: IEEE International Conference on Biometrics, pp. 113–120 (2015)
5. Kono, M., Ueki, H., Umemura, S.: A new method for the identification of individuals by using of vein pattern matching of a finger. In: 5th Symposium on Pattern Measurement, pp. 9–12 (2000)
6. Lee, J., Ahmadi, M., Ko, K.W., Lee, S.: Development of identification system using the image of finger vein. Adv. Sci. Technol. Lett. **93**:, 44–48 (2015)
7. Lee, K.H.: First course on fuzzy theory and applications. Springer Science and Business Media (2006)
8. Liu, F., Yang, G., Yin, Y., Wang, S.: Singular value decomposition based minutiae matching method for finger vein recognition. Neurocomputing **145**, 75–89 (2014)
9. Liukui, C., Hong, Z.: Finger vein image recognition based on tri-value template fuzzy matching. In: 9th WSEAS International Conference on Multimedia Systems and Signal Processing, pp. 206–211 (2009)
10. Miura, N., Nagasaka, A., Miyatake, T.: Feature extraction of finger-vein patterns based on repeated line tracking and its application to personal identification. Mach. Vis. Appl. **15**, 194–203 (2004)

11. Pang, S., Yin, Y., Yang, G., Li, Y.: Rotation invariant finger vein recognition. Biometric Recognition, pp. 151–156. Springer Berlin, Heidelberg (2012)
12. Peng, J., Wang, N., El-Latif, A.A.A., Li, Q., Niu, X.: Finger-vein verification using Gabor filter and sift feature matching. In: Eighth International Conference on Intelligent Information Hiding and Multimedia Signal Processing, pp. 45–48 (2012)
13. Sivanandam, S.N., Sumathi, S., Deepa, S.N.: Introduction to fuzzy logic using MATLAB, p. 1. Springer, Berlin (2007)
14. Song, W., Kim, T., Kim, H. C., Choi, J.H., Kong, H.J., Lee, S.R.: A finger-vein verification system using mean curvature. Pattern Recogn. Lett. **32**, 1541–1547 (2011)
15. Xian, R., Ni, L., Li, W.: The ICB-2015 competition on finger vein recognition. IEEE Int. Conf. Biometrics 85–89 (2015)
16. Yanagawa, T., Aoki, S., Ohyama, T.: Human finger vein images are diverse and its patterns are useful for personal identification. MHF Preprint Series, Kyushu University, pp. 1–7 (2007)
17. Yang, J., Shi, Y., Yang, J., Jiang, L.: A novel finger-vein recognition method with feature combination. In: 16th IEEE International Conference on Image Processing, pp. 2709–2712 (2009)
18. Yang, Y., Yang, G., Wang, S.: Finger vein recognition based on multi-instance. Int. J. Digital Content Technol. Appl. **6** (2012)
19. Zadeh, L.A.: Fuzzy sets. Inf. Control **8**, 338–353 (1965)

Fuzzy Reasoning Approach (FRA) for Assessment of Workers Safety in Manganese Mines

Shikha Verma and Sharad Chaudhari

Abstract Safety management today is an important element in the mining industry with advancement in sociotechnical complexity in mining systems/operations. One of the most important components of safety management is to maintain safety of workers in the workplace, as this in turn affects the safety of the whole work system. The safety of workers depends on various factors and these factors affect workplace safety simultaneously. So to evaluate the same, a robust holistic approach with potential to exploit uncertain safety-related data and give close to perfect outcome is desired. This paper presents the fuzzy reasoning approach (FRA) to perform assessment of risk associated with workers' safety specifically in mining industry. Initially, factors affecting worker safety are identified, weighted with fuzzy triangular numbers, since the factors have nonphysical structures and fuzzy numbers aid computational ease. Thereafter, fuzzy inference system with expert decisions is developed. As a result, the level of risk associated with the combination of factors comprising the working environment for the worker can be identified and reduced to considerable levels.

Keywords FRA · Fuzzy · Risk assessment · Safety

1 Introduction

The mining industry since ages is known to be one of the most hazard-prone industries. Recent advancements in the techniques and operations performed for ore extraction have shown considerable reduction in the accident rates, yet still there lies huge scope to work further in the enhancement of safety levels in mining sites.

Shikha Verma (✉) · Sharad Chaudhari
Department of Mechanical Engineering, YCCE, R.T.M.N.U, Nagpur
Maharashtra, India
e-mail: shikhaverma2108@gmail.com

Sharad Chaudhari
e-mail: sschaudharipatil@rediffmail.com

M. Pant et al. (eds.), *Proceedings of Fifth International Conference on Soft Computing for Problem Solving*, Advances in Intelligent Systems and Computing 437, DOI 10.1007/978-981-10-0451-3_14

Currently, requirement for risk assessment has been made mandatory in the mining industry (Occupational health and safety act 2000 sections 7 and 8). Risk assessment is a systematic process comprising hazard identification, risk analysis of the hazards, and risk estimation. Risk estimation is the most crucial step in the entire assessment process, and no standard procedure is made mandatory to be followed in specific industries. Industries are free to adopt any technique for risk assessment; as a result the easiest of all techniques is mostly adopted irrespective of whether a considerable outcome is obtained; this indicates the lacuna in risk assessment process adopted in various industries these days [1]. Safety in mining sites depends upon various factors like geo-mechanical, mechanical, electrical, geochemical, chemical, environmental, personal, social, cultural, and managerial [2–4]. The approach to be developed to evaluate mining site safety should be robust enough to consider all the factors mentioned above and give considerable outcome. The present work is restricted to only the safety of workers who are exposed to a hazardous working environment in mining sites. When a worker enters this field, he is at some age, with some or no experience of working based upon which he is assigned the task and shift in which he will be working. Human errors constitute a major contribution in any mishap, which is calculated to be about 88 %; 10 % of actual mechanical/machine failures leads to mishaps, 2 % accidents, it is believed, takes place due to act of God [5]. It can be noticed that human errors are the top priority in creating unfavorable consequences like accidents. But, human errors are found to be the result of a combination of many factors like the age of the worker, his experience, and the shift in which he is working [6–14]. It is found from accident records of mining companies that workers of different ages have different levels of exposure to accidents [15–18]. A fresher who is young works in night shifts; accident statistics have confirmed the night shift to have the maximum chances of meeting with mishaps compared to general and morning shifts. Similarly, if the worker is aged although seasoned with experience, has considerable chance of meeting with mishap in any of the three shifts with maximum risk in night shift, since with age the normal tendency/capacity of human body to work gets reduced and workers often face issues of lack of concentration with increase in age. This work highlights the identified factors that are found to be influencing worker performance and safety conditions along with the best combination of conditions based on identified factors, to be maintained to attain maximum safety levels for the workers while working in sites.

Till date, safety analysis in mining industry specifically in the Indian scenario is performed considering factors, namely consequence of occurrence, frequency of occurrence, and exposure level (RRM technique) [19–21]. Many statistical approaches [22], various types of distributions such as beta [23], Weibull distribution, Poisson distribution [24], time between failures (TBF), time between occurrences (TBO) [25], Monte Carlo simulation [26]; analysis of questionnaire [27–29]; surveys [30], are conducted worldwide to understand and interpret safety conditions in mining sites; incidences evaluation based upon nonfatal days lost (NFDL), no days lost (NTL) [31]; loss time injury rates (LTI), incident rate ratios (IRR) [32] etc., is done and various ranking method, assessment models, frameworks are developed namely, The Scoring and Ranking Assessment Model

(SCRAM), Chemical Hazard Evaluation for Management Strategies (CHEMS-1), EU Risk Ranking method (EURAM), Safety Health and Environment (SHE) Assessment, Human factor analysis and classification systems (HFACS), Intrinsic safety approach [33–35].

The above-mentioned approaches adopted for risk assessment are highly dependent on availability of safety-related data. The data available in such industry are highly uncertain, imprecise, and vague. Conventional techniques cannot be adopted in such cases. The approaches that can exploit uncertainty issues can give robust outcome are most suitable for dynamic industries like mining. In this paper fuzzy reasoning approach is proposed for assessment of risk associated with safety of workers.

2 Methodology

A generic method for evaluation of risk associated with worker safety is explained below.

Step 1: Define locations under study.
Step 2: Refer the accident statistics (accidents occurred in the location during the span under study).
Step 3: Identify potential factors having contribution in accidents.
Step 4: Define and describe factors.
Step 5: Develop conceptual fuzzy model for risk assessment.
Step 6: Identify the input parameters.
Step 7: Develop fuzzy inference engine.
Step 8: Identify output parameters.

FRA model is developed using MATLAB R2009a. Fuzzy logic toolbox is used to analyze the data with "mamdani" inference system. After the identification of input parameters and output parameters, a yardstick comprising of qualitative descriptors with description and associated membership function is constructed, for all the inputs and outputs. These inputs are then referred to develop fuzzy rule base. Fuzzy rules consist of if-then statements with two segments, antecedent and consequent, combined together with logical operators like OR/AND. The fuzzy inference engine consists of the above developed rules. When the engine is given input which is fuzzified with membership function, the rule strength is established combining the fuzzified inputs according to fuzzy rules. The consequence is obtained by combining the rule strength and the output membership function. Then consequences are combined to give output distribution which is then defuzzified and crisp output in terms of risk level scores is obtained. In the present case, centroid method is used for defuzzification and triangular and trapezoidal membership functions are used for input and output variables for computational ease. For rule base AND operator is used.

Conceptual framework is shown in Fig. 1.

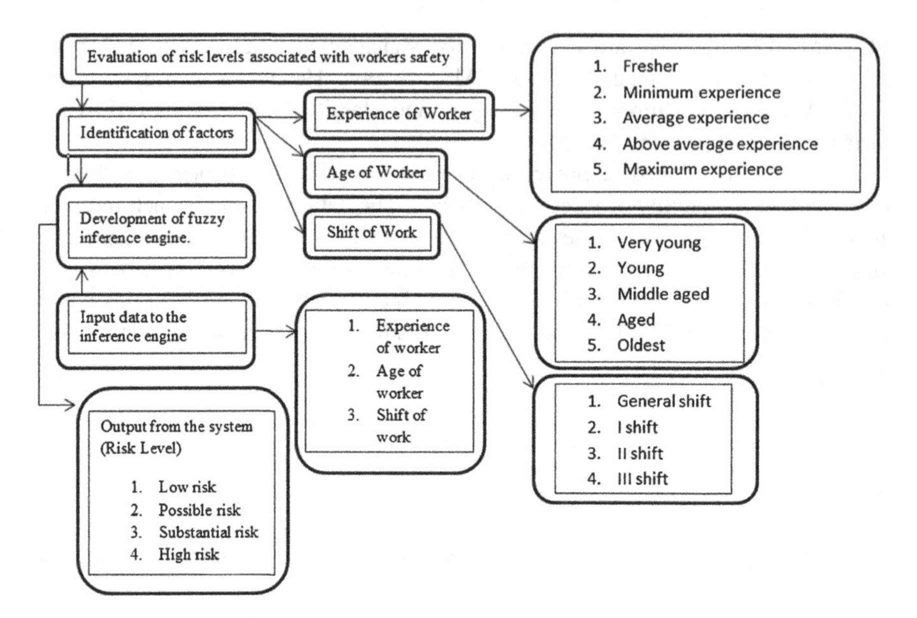

Fig. 1 Conceptual framework

3 Application

As explained above, the first step in application is a description of the location where the study is conducted. The data for the study are collected from 11 manganese mining sites located in Maharashtra and Madhya Pradesh. All the sites are owned by the central government. The sites include both opencast mines and underground mines. To conduct the analysis, accident data for 7 years (2006–2012) is gathered. Factors such as age of the worker, experience of the worker, and shift in which that worker who faced accident was working is analyzed and yardsticks to compute risk levels are developed. Conventionally, most of the work of risk estimation is performed based upon the consequence of the accident, exposure level, and frequency of occurrence. The factors explained above that are considered in the present work have not been considered so far as solely responsible for any mishap to occur. By the end of the analysis, a clear indication regarding the most sensitive combination of these factors that has the highest potential to cause any injury/accident is obtained. The yardsticks developed with qualitative descriptor and their description are given in Tables 1, 2, 3, and 4.

The factors identified will be the input parameter for inference engine and output in terms of risk level will be obtained after defuzzification. Suppose, a worker is hired, he is a fresher and very young in age, and is assigned to work in shifts. Based on the mentioned details the level of risk associated with him while he gets exposed to the site area can be evaluated once these details are fed into the inference engine. Accordingly, it can be decided whether he should continue in the designated shift or can be allotted any other shift that is less risky for him to work in. Similarly, if an older worker is to be assigned a shift to work based upon his experience and the age group he belongs to, the least risky shift can be allotted to him. The inference engine is developed with expert opinion, the experts identified have been working with this industry since 40 years. The inference engine comprises 80 rules. The total number of rules depend upon the input variables; in this case the total number of inputs are three with a total of 5 × 5 × 4 descriptors, but the number of rules are less than the multiplicative outcome of the descriptors, since a few combinations are not possible, like a worker above 58 years of age can not be a fresher with experience between 1 month and 1 year. Similarly, a worker with above average experience, that is, experience between 11 and 20 years can not be very young with age between 18 and 27 years. This way, all the combinations that were logically not possible were discarded by the experts with appropriate justification. With their contribution, the least and most risk combinations of identified factors are generated and fed as rule in the inference engine that finally gave the desired outcome.

Input parameters for fuzzy inference system

Table 1 Yardstick for experience of worker

Qualitative descriptor	Description	Membership function
Fresher	1 month–1 year	0, 0, 0.5, 1.5 (trapezoidal)
Minimum experience	1–5 years	0.5, 1.5, 2.5 (triangle)
Average experience	6 –10 years	1.5, 2.5, 3.5 (triangle)
Above average experience	11–20 years	2.5, 3.5, 4.5 (triangle)
Maximum experience	Above 20 years	3.5, 4.5, 5.5 (trapezoidal)

Table 2 Yardstick for age of worker

Qualitative descriptor	Description	Parameter
Very young	18–27 years	0, 0, 0.5, 1.5 (trapezoidal)
Young	28–37 years	0.5, 1.5, 2.5 (triangle)
Middle aged	38–47 years	1.5, 2.5, 3.5 (triangle)
Aged	48–57 years	2.5, 3.5, 4.5 (triangle)
Oldest	58 years and above	3.5, 4.5, 5.5 (trapezoidal)

Table 3 Shift of work

Qualitative descriptor	Description	Parameter
General	8:00 am–4:00 pm	0, 0, 1, 2 (trapezoidal)
I	6:00 am–2:00 pm	1, 2, 3 (triangle)
II	2:00 pm–1:00 pm	2, 3, 4 (triangle)
III	10:00 pm–6:00 am	3, 4, 5, 5 (trapezoidal)

Table 4 Risk level

Qualitative descriptor	Description	Parameter
Low	Risk is acceptable	0, 0, 3, 4 (trapezoidal)
Possible	Risk is tolerable but should be further reduced if cost-effective to do so	3, 4, 6, 7 (trapezoidal)
Substantial	Risk must be reduced if it is reasonably practical to do so	6, 7, 9, 10 (trapezoidal)
High	Risk must be reduced safe in exceptional circumstances	9, 10, 12, 12 (trapezoidal)

Fuzzy inference system with rule base is shown in Figs. 2 **and** 3

Fig. 2 Fuzzy inference system

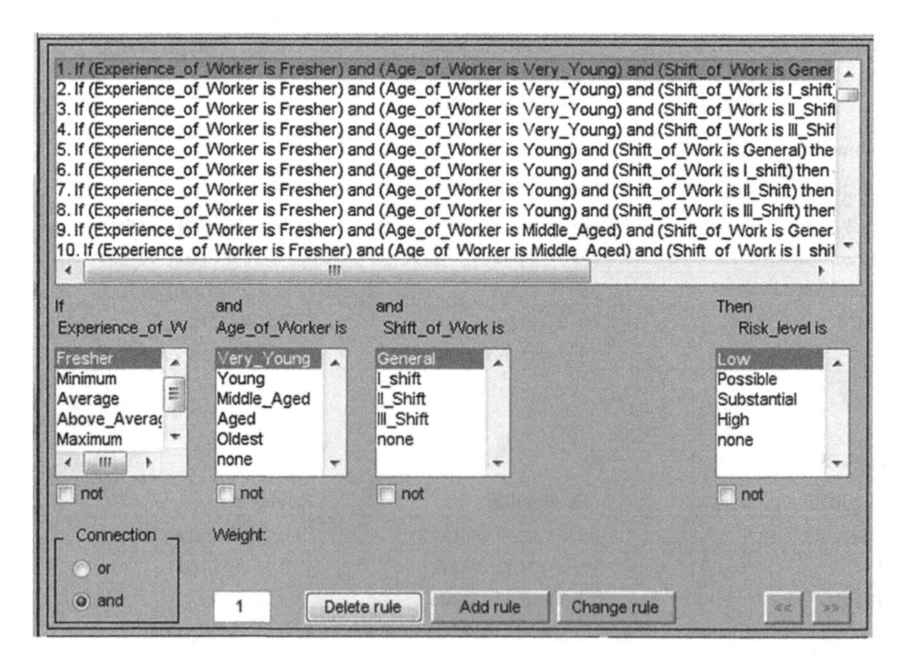

Fig. 3 Fuzzy rule base

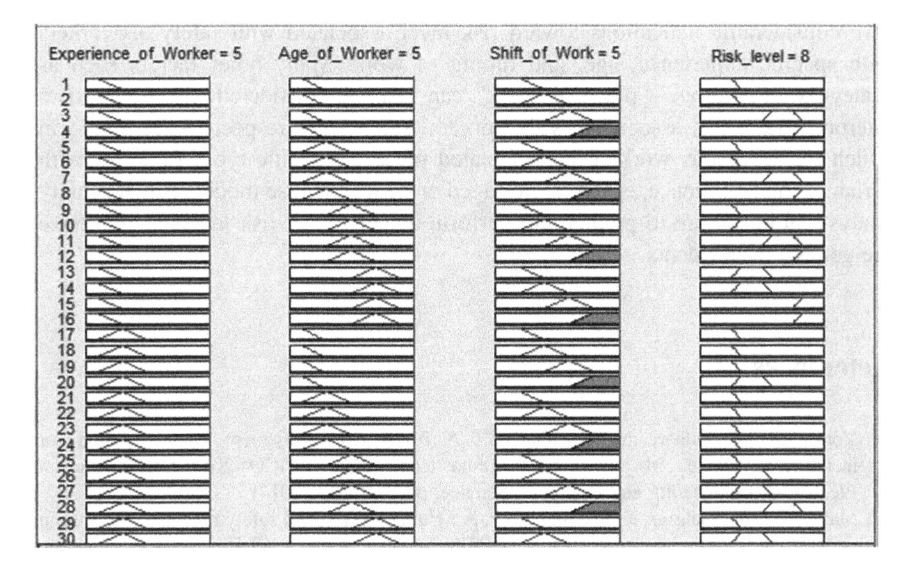

Fig. 4 Risk level interpretation from rule viewer

Once the input is given to the system as the inference engine shown above, the output in terms of risk level can be interpreted as shown in Fig. 4. It can be seen in the figure that with a combination of the inputs given to the engine an output with a certain risk level is obtained. If the worker is above 58 years with experience above 20 years, he is usually given lighter work because of his age, since he is working in night shift. As discussed above, there will also be a substantial level of risk associated with his working. Similar indications are getting confirmed in the rule viewer given below concluding the suitability of the model developed for the scenario under study. Similarly with different combinations of age, experience, and timing of work, the safest working conditions can be identified for every worker and safety levels can be enhanced to a considerable level in mining sites.

4 Conclusion

Mining is a dynamic industry in which there is no set pattern and trend for accident occurrence. Conventional and pro-data-based techniques are not sufficient to predict unsafe conditions that might lead to an accident in mining sites. An expert technique/approach that can provide more structured analysis of the available information (partial/vague/incomplete/uncertain) and as an outcome can give robust and acceptable results that are required to be developed today. This paper highlights three unconventional factors, namely "experience of the worker," "age of the worker," "shift of work" for the development of fuzzy model to evaluate risk levels. These factors are not considered for risk evaluation till date and have potential to give considerable indications toward risk level associated with safety of workers with specific experience, age, and timing of work. Many other factors such as "category of mining," "place of work" can also be considered in the future to interpret risk levels associated with workers' safety with respect to the timing in which the worker is working at designated place in specific types of mine, with certain or no experience as a young or aged employee. These models can aid safety analysts in the future to predict and perform abatement of risk levels to minimize the genesis of accidents.

References

1. Verma, S., Chaudhari, S., Khedkar, S.: A fuzzy risk assessment model applied for metalliferous mines in India. In: Procedia technology, IEMCON2014 Conference on Electronics Engineering and Computer Science, pp. 202–213 (2014)
2. Mahdevari, S., Shahriar, K., Esfahanipour, A.: Human health and safety risks management in underground coal mines using fuzzy TOPSIS. Science of the Total Environment 488–489, pp. 85–99 (2014)
3. Badri, A., Nadeau, S., Gbodossou, A.: A new practical approach to risk management for underground mining project in Qubec. J. Loss Prev. Process Ind. **26**, 1145–1158 (2013)
4. Khanzode, V.V., Maiti, J., Ray, P.K.: A methodology for evaluation and monitoring of recurring hazards in underground coal mining. Saf. Sci. **49**, 1172–1179 (2011)
5. Dhillon, B.S., Raouf, A.: Safety assessment: a quantitative approach. Florida, Boca Raton (1994)
6. Laflamme, L., Menckel, E., Lundholm, L.: The age-related risk of occupational accidents: the case of Swedish iron-ore miners. Accid. Anal. Prev. **28**(3), 349–357 (1996)
7. Jensen, Q.S., Kyed, M., Christensen, A., Bloksgaard, L., Hansen, D.C., Nielsen, J.K.: A gender perspective on work-related accidents. Saf. Sci. **64**, 190–198 (2014)
8. Blank, V., Laflamme, L., Diderichsen, F.: The impact of major transformations of a production process on age-related accident risks: a study of an iron-ore mine. Accid. Anal. Prev. **28**(5), 627–636 (1996)
9. Nenonen, N.: Analysing factors related to slipping, stumbling, and falling accident at work: application of data mining methods to Finnish occupational accidents and diseases statistics database. Applied Ergonomics **44**(2), 215–224 (2013)
10. Silva, J., Jacinto, C.: Finding occupational accident patterns in the extractive industry using a systematic data mining approach. Reliab. Eng. Syst. Saf. **108**, 108–122 (2012)
11. Picard, M., Girard, S., Simard, M., Larocque, R., Leroux, T., Turcotte, F.: Association of work-related accidents with noise exposure in the workplace and noise-induced hearing loss based on the experience of some 240,000 person-years of observation. Accid. Anal. Prev. **40** (5), 1644–1652 (2008)
12. Hajakbari, M., Minaei-Bidgoli, B.: A new scoring system for assessing the risk of occupational accidents: a case study using data mining techniques with Iran's Ministry of Labor data. J. Loss Prev. Process Ind. **32**, 443–453 (2014)
13. Amponsah-Tawiah, K., Jain, A., Leka, S., Hollis, D., Cox, T.: Examining psychosocial and physical hazards in the Ghanaian mining industry and their implications for employees' safety experience. J. Saf. Res. **45**, 75–84 (2013)
14. Amponsah-Tawiah, K., Jain, A., Leka, S., Hollis, D., Cox, T.: The impact of physical and psychosocial risks on employee well-being anquality of life: the case of the mining industry in Ghana. J. Saf. Res. **65**, 28–35 (2014)
15. Directorate general of mines safety annual report (2012)

16. Directorate general of mines safety annual report (2010)
17. Directorate general of mines safety annual report (2009)
18. Directorate general of mines safety annual report (2005)
19. Pathak, K., Sen, P.: Risk assessment in mining industry. Indian Min. Eng. J. **53**(3), 347–350 (2001)
20. Paliwal, R., Jain, M.L.: Assessment & management of risk—an overview. Indian Min. Eng. J. **32**, 99–103 (2001)
21. Tripathy, D., Patra, A.K.: Safety risk assessment in mines. Indian Min. Eng. J. **37**(1), 120–122 (1998)
22. Poplin, S.G., Miller, B.H., Moore, R.J., Bofinger, M.C., Spencer, K.M., Harris, B.R., Burgess, L.J.: International evaluation of injury rates in coal mining: a comparison of risk and compliance-based regulatory approaches. J. Saf. Sci. **46**, 1196–1204 (2008)
23. Coleman, J., Kerkering, C.J.: Measuring mining safety with injury statistics: lost workdays as indicators of risk. J. Saf. Res. **38**, 523–533 (2007)
24. Khanzode, V., Maiti, J., Ray, P.K.: Occupational injury and accident research: a comprehensive review. J. Saf. Sci. **50**, 1355–1367 (2012)
25. Ghasemi, E., Ataei, M., Shahriar, K., Sereshki, F., Jalali, S., Ramazanzadeh, A.: Assessment of roof fall risk during retreat mining in room and pillar coal mines. Int. J. Rock Mech. Min. Sci. **54**, 80–89 (2012)
26. Sari, M., Seluck, S., Karpuz, C., Duzgun, H.: Stochastic modelling of accidents risks associated with an underground coal mine in Turkey. J. Saf. Sci. **47**, 78–87 (2009)
27. Alteren, B.: Implementation and evaluation of the safety element method at four mining sites. J. Saf. Sci. **32**, 231–264 (1999)
28. Hickman, J., Geller, S.: A safety self-management intervention for mining operations. J. Saf. Res. 299–308 (2003)
29. Evans, R., Brereton, D.: Risk assessment as a tool to explore sustainable development issues: lessons from the Australian coal industry. Int. J. Risk Assess. Manag. **7**(5), 607–619 (2007)
30. Komljenovic, D., Groves, A., W., Kecojecvic, J.V.: Analysis of fatalities and injuries involving mining equipment. J. Saf. Sci. **46**, 792–801 (2007)
31. Komljenovic, D., Groves, A.W., Kecojecvic, J.V.: Injuries in U.S. mining operations—a preliminary risk analysis. J. Saf. Sci. **46**, 792–801 (2008)
32. Poplin, S.G., Miller, B.H., Moore, R.J., Bofinger, M.C., Spencer, K.M., Harris, B.R., Burgess, L.J.: International evaluation of injury rates in coal mining: a comparison of risk and compliance-based regulatory approaches. J. Saf. Sci. **46**, 1196–1204 (2008)
33. Singh, K., Ihlenfeld, C., Oates, C., Plant, J., Voulvoulis, N.: Developing a screening method for the evaluation of environmental and human health risks of synthetic chemicals in the mining industry. Int. J. Min. Process. **101**, 1–20 (2011)
34. Lenne, G.M., Salmon, M.P., Liu, C.C., Trotter, M.A.: Systems approach to accident causation in mining: an application of the HFACS method. J. Accid. Anal. Prev. **48**, 111–117 (2012)
35. Kaihuan, Z., Fuchuan, J.: Research on intrinsic safety method for open-pit mining, international symposium on safety science and engineering in China. Proceedia Eng. **43**, 453–458 (2012)

Multiple-Input Multiple-Output Paper Machine System Control Using Fuzzy-PID Tuned Controller

Rajesh Kumar and Minisha Moudgil

Abstract Artificial Intelligence (AI) techniques have grown rapidly in recent year due to good ability to control the nonlinear model. Fuzzy logic is an important artificial intelligence technique. The ability of fuzzy logic to handle imprecise and inconsistent real-world problems has made it suitable for different applications. In this paper, a method for auto tuning MIMO fuzzy-PID (proportional–integral–derivative) conventional methods is proposed. Modeling and control of basis weight and ash content using fuzzy-PID, adjusts online parameters of a conventional PID (proportional–integral–derivative). The performance and stability objectives are assessed by simulated results.

Keywords Wet end system · Fuzzy-PID controller · Membership function

1 Introduction

The fuzzy logic provides a means of converting a linguistic control strategy based on expert knowledge into an automatic control strategy [1]. Fuzzy logic controller (FLC) appears very useful when the two processes are too complex for analysis conventional quantitative techniques or when the information about the input-output of the system is uncertain [2]. Generally, fuzzy control shows good performance for controlling nonlinear and uncertain system that could not be controlled satisfactorily by using PID conventional controller. The fuzzy auto tune procedure adjusts online parameters of a conventional PID (proportional–integral–derivative). The performance and stability objectives are assessed by simulated results. The proposed control scheme offers some advantages over the conventional

Rajesh Kumar (✉) · Minisha Moudgil
Department of Instrumentation & Control Engineering, Graphic Era University, Dehradun, Uttrakhand, India
e-mail: rajtisotra@gmail.com

Minisha Moudgil
e-mail: minishamoudgil20@gmail.com

M. Pant et al. (eds.), *Proceedings of Fifth International Conference on Soft Computing for Problem Solving*, Advances in Intelligent Systems and Computing 437, DOI 10.1007/978-981-10-0451-3_15

fuzzy controller such as tuning becomes simple and control operators can easily understand it working.

The present trends in most of the process industries like paper industry, chemical, and fertilizers is to measure and control the process variables with the help of advance control techniques. Various methods have been applied to improve the wet end process in paper making mill. In present work, basis weight of paper is controlled and analyzed by Fuzzy-PID advance control technique. Basis weight is the weight of the paper per unit area, expressed in g/m^2. The entire process of paper making right from handling the raw material up to the production of paper, has been so designed that the paper of uniform basis weight could be produced on the paper machine.

If both the consistency of the stock and supply to the paper machine are controlled. Then all other flows at various places in the short loop will be smooth and automatically be controlled. Thus, uniform basis weight of paper as desired could be produced with no problem. Dynamic model is an important tool for the process control in the wet end section of paper making machine.

2 Wet End System of Paper Making Machine

The work starts with introducing the operations and processes of a paper industry. From instrumentation view point, these may be termed as subsystems like, raw material preparation, pulping, washing of brown stock, bleaching, stock preparation, approach flow system, wet end operation, drying, calendaring, and chemical recovery operation. The wet end is the first section of paper making machine. The wet end process mainly consists of three subsystems: the approach flow, fiber recovery, and broke handling. At the beginning of the approach flow, the various furnish components are mixed in the right proportions. Before the headbox the approach flow is an extremely sensitive area where it is essential to keep all parameters stable.

First, the diluted stock comes into contact with paper machine and which then passes through the refiner. The refiner flattens the pulp to improve its bonding characteristics. Chemical additive is introduced at the mixing chest, which allows proper blending of the chemical additives with the pulp. The next tank is referred to as machine chest. The machine chest sets the stock consistency prior to the paper machine. In the stock preparation, the consistency range is from the 0.5 % or up to 15 % for high density pulp storage. The finished stock is pumped through the stuff box. The stuff box maintains head pressure and allows air to escape prior to reaching to the headbox. The pulp is screened and cleaned before it reaches the headbox. The screen moves the large contaminants, while cleaners remove mid-sized contaminants. The headbox of the paper machine is final element in the

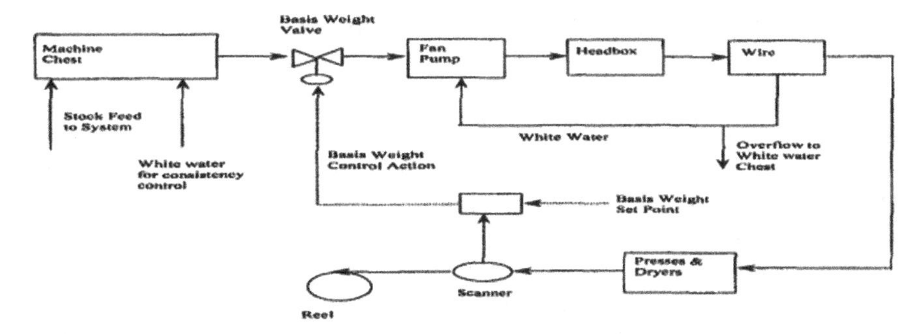

Fig. 1 Typical closed loop of basis weight control

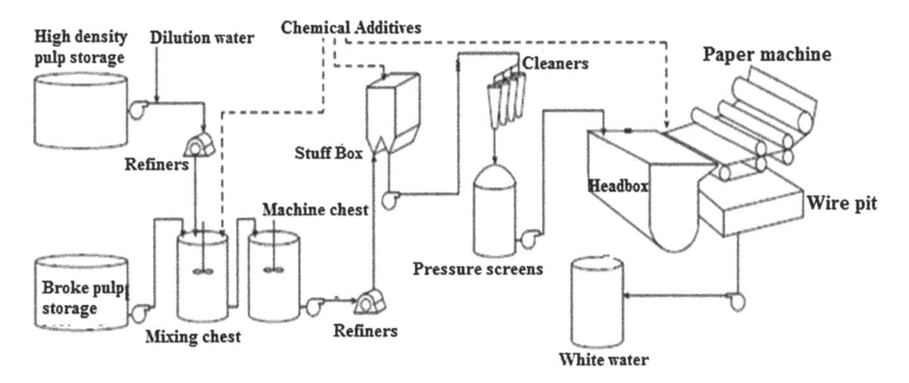

Fig. 2 Wet end of paper making machine

process before the pulp is spread out on the wire mesh. The headbox prevent the flocculation of fiber which cause nonuniformities in the final product, i.e., a paper sheet. The water drains from the wire is called white water and is collected on save all trays installed just below the wire and lead to pit called wire pit. Broke is the completely manufactured paper that is discarded from finishing process. From the broke storage tank, a controlled flow is reintroduced through the blending system into the machine furnish. The fibers and fillers in paper machine white water overflow are reclaimed into the machine furnish [3]. A typical control loop of basis weight control is shown in Fig. 1. The wet end of paper machine flow diagram is shown in Fig. 2. The operator sets the target basis weight set point on the basis weight controller, the output of which becomes the cascaded set point for the basis weight volumetric flow. It sets the set point for the calculated basis weight sub-loop. The basis weight volumetric flow signal is multiplied by the heavy stock consistency and divided by the paper machine wire speed to calculate the basis weight.

3 Design and Tuning of Fuzzy-PID Controller

FLC consists of fuzzifier, rule base, and defuzzifier. Inference machine or rule base depends on the input variables and output variables. There are two inputs (ash content and basis weight) and three outputs, such as K_p, K_i, K_d (PID controller parameters). The rules set of fuzzy controller can be found using the available knowledge in the area of designing the system. These rules are defined by using the linguistic variables. The proposed controller uses a nonlinear fuzzification algorithm and odd number membership functions. The controller is to be considered as a nonlinear PID where parameters are tuned online based on error between controlled variables, i.e., basis weight, ash content, set point, and error derivative, as shown in Fig. 3. On the basis of linguistic variables, one is governing 49 rules and mechanism for each manipulated variable, i.e., filler flow rate and stock flow rate as explained in Tables 1, 2 and 3. The basis weight and ash content dynamic models have been proposed by Eqs. 1 and 2 from Vijay [4].

$$Y_1 = [(0.214 \exp - 68s/125s + 1)] \text{ filler flow rate} \\ + [(-0.192 \exp - 68s/17s + 1)] \text{ stock flow rate} \tag{1}$$

$$Y_2 = [(0.153 \exp - 68s/125s + 1)] \text{ filler flow rate} \\ + [(0.93 \exp - 68s/17s + 1)] \text{ stock flow rate} \tag{2}$$

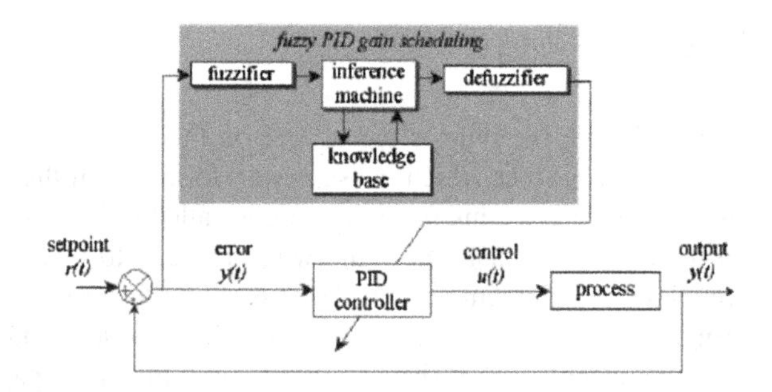

Fig. 3 Fuzzy PID controller

Table 1 Fuzzy rule of K_p

K_p		Ash content						
Basis weight		NB	NM	PS	PM	Z	NS	PB
	NB	B	S	S	S	S	S	B
	NM	B	B	S	B	S	S	B
	PS	B	B	B	B	S	B	B
	PM	B	B	S	B	S	S	B
	NS	B	B	B	B	S	B	B
	PB	B	B	S	B	S	S	B
	Z	B	B	B	B	B	B	B

Table 2 Fuzzy rule of K_i

K_i		Ash content						
Basis weight		NB	NM	PS	PM	Z	NS	PB
	NB	S	B	B	B	B	B	S
	NM	S	S	B	B	B	B	S
	PS	S	S	S	S	B	S	S
	PM	S	B	B	S	B	B	S
	NS	S	S	S	S	B	S	S
	PB	S	S	B	S	B	B	S
	Z	S	S	S	S	S	S	S

Table 3 Fuzzy rule of K_d

K_d		Ash content						
Basis weight		NB	NM	PS	PM	Z	NS	PB
	NB	S	B	B	B	B	B	S
	NM	S	B	B	B	B	B	S
	PS	S	S	B	S	B	B	S
	PM	S	B	B	B	B	B	S
	NS	S	S	B	S	B	B	S
	PB	S	B	B	B	B	B	S
	Z	S	S	S	S	B	S	S

3.1 Steps to Design Fuzzy Logic Controller

- Define the input and output variables. In this paper, a fuzzy logic controller is designed with two input variables, i.e., ash content, basis weight, and three output variables, i.e., K_p, K_i, K_d.
- Define the linguistic variables which provide a language for the expert to express ideas about the control decision making process of fuzzy controller inputs and outputs. Here seven linguistic variables for inputs: negative big (NB), negative medium (NM), negative small (NS), zero (z), positive small (PS), positive medium (PM), positive big (PB) are used. For output, the two linguistic variables such as small (s) and big (B) are considered.
- Construct the membership function for both input and output. Two input variables of fuzzy control have seven fuzzy sets ranging from negative big (NB) to positive big (PB). For the first input, i.e., ash content, Membership ranges from [−60 to 60] as shown in Fig. 4. For the second input basis weight, Membership range from [−80 to 80] as shown in Fig. 5.

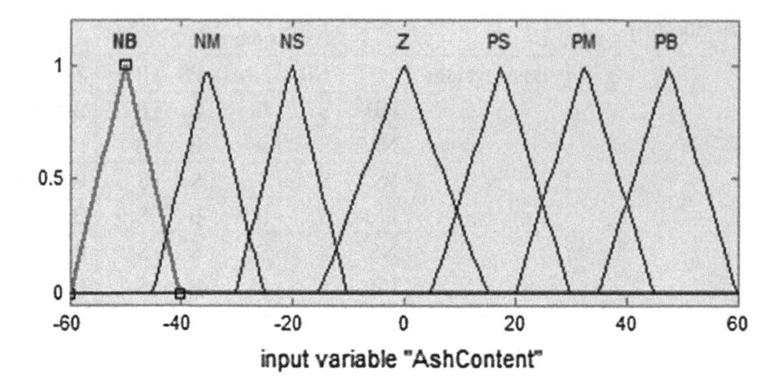

Fig. 4 Membership function of input 1, i.e., ash content

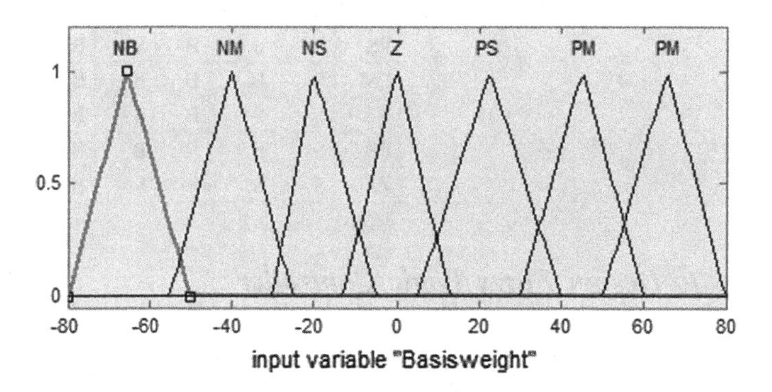

Fig. 5 Membership function of input 2, i.e., basis weight

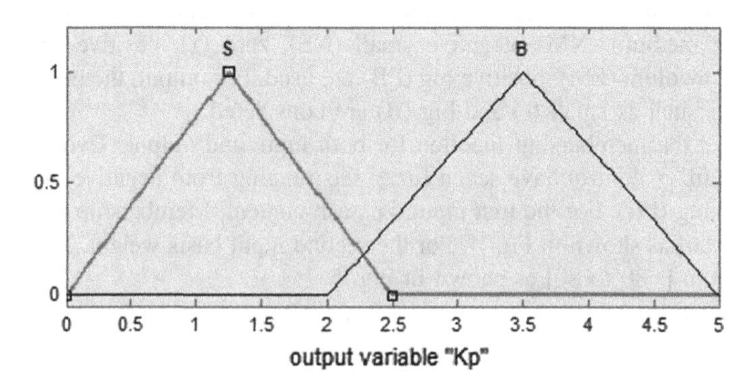

Fig. 6 Membership function of output 1, i.e., 'K_p'

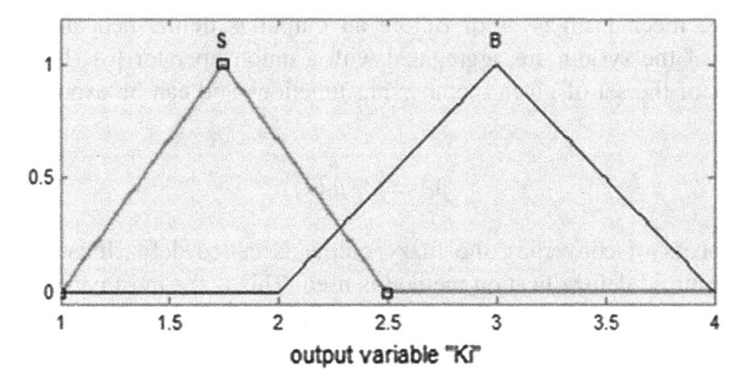

Fig. 7 Membership function of output 2, i.e., 'K_i'

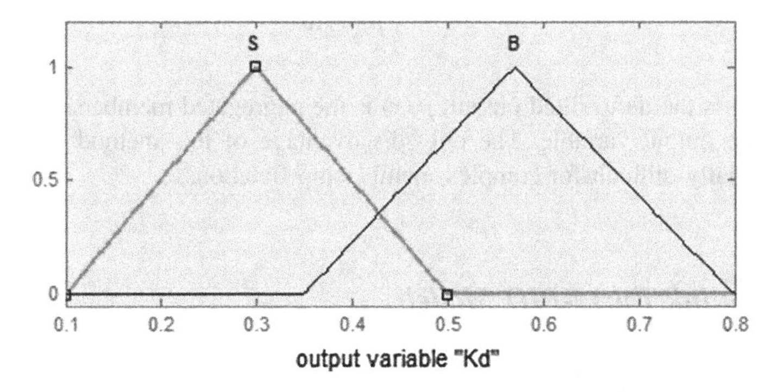

Fig. 8 Membership function of output 2, i.e., 'K_d'

- The three output variables, i.e., K_p, K_i, K_d of fuzzy control has two sets ranging from small (s) to big (B). For the first output 'K_p' range is from [0 5] as shown in Fig. 6. The range for second output 'K_i' is from [1 4] and for 'K_d' the range is from [0.1 0.8] shown in Figs. 7 and 8.
- Define the fuzzy rules base which depends on the human experience or knowledge [5]. The rules for K_p, K_i and K_d are written according to a system with a fast settling time, small overshoot, no steady-state error. For fast rise time, The PID controller has large value of $K_{p,}$ i.e., proportional gain. To reduce the overshoot, the value of proportional gain and integral gain must be small, and derivative gain should be large. The rules of K_p, K_i, K_d are shown in Tables 1, 2 and 3.
- Define the inference mechanism rule to find the input-output relation. Fuzzy logic is a rule-based system written in the form if-then rules. These rules are stored in the knowledge base of the system. In this paper, Mamdani (max–min)

inference mechanism is used. Before an output is defuzzified, all the fuzzy outputs of the system are aggregated with a union operator [6]. The union is the max of the set of given membership functions and can be expressed as

$$\mu A = \cup_i(\mu_i(x)) \tag{3}$$

- The process of converting the fuzzy output is called defuzzification. In this work, centroid defuzzification method is used. This is the most commonly used technique and is very accurate. The centroid defuzzification technique can be expressed as

$$x^* = \frac{\int \mu_i(x)\, x\, \mathrm{d}x}{\int \mu_i(x)\, \mathrm{d}x} \tag{4}$$

where x^* is the defuzzified output, $\mu_i(x)$ is the aggregated membership function and x is the output variable. The only disadvantage of this method is that it is computationally difficult for complex membership functions.

3.2 *Simulink Fuzzy-PID Model*

The closed loop MIMO (Multivariable input multivariable output) model for paper machine basis weight and ash content has been developed on MATLAB Platform. Fuzzy logic controller is designed to tune the PID controller parameters to improve the response of MIMO wet end system, shown in Fig. 9.

4 Result and Analysis

Fuzzy-PID controller and conventional controller have been designed to analyze the response of ash content (Y_1) and basis weight (Y_2) of wet end system. For conventional controller, Output response for basis weight and ash content (MIMO) using PID controller shown in Fig. 10. The step response for basis weight and ash

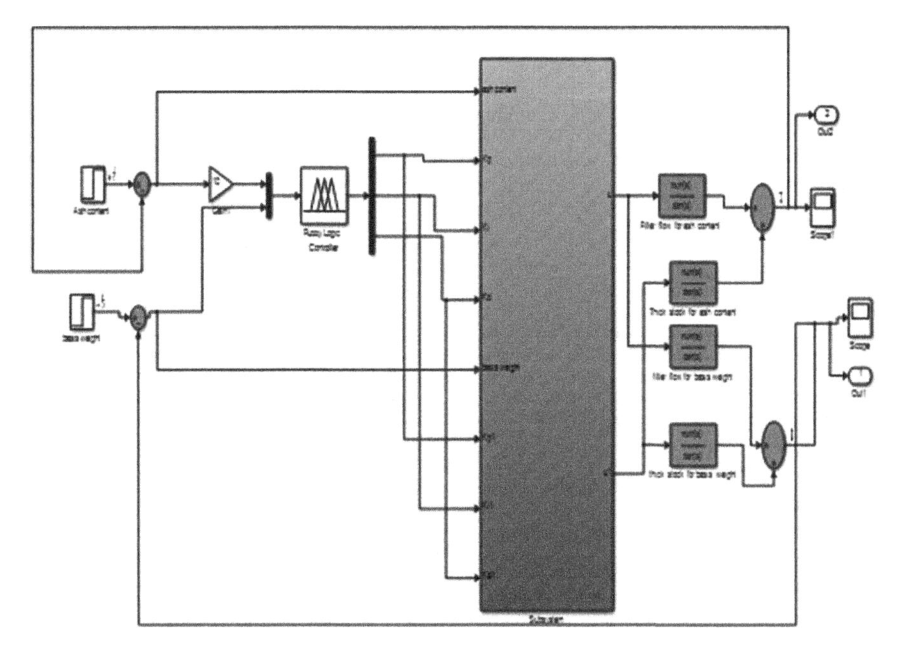

Fig. 9 Fuzzy-PID control model

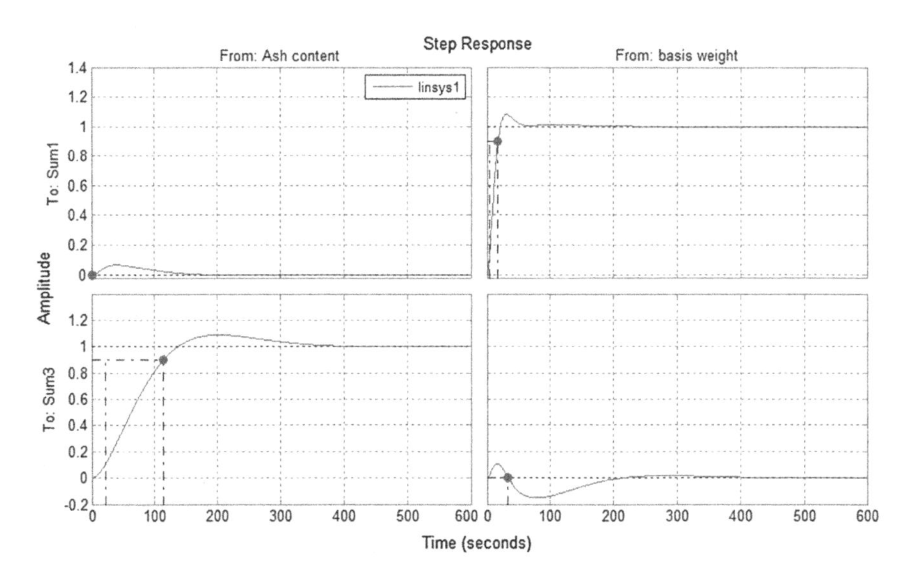

Fig. 10 Output response for basis weight and Ash content (MIMO) using PID controller

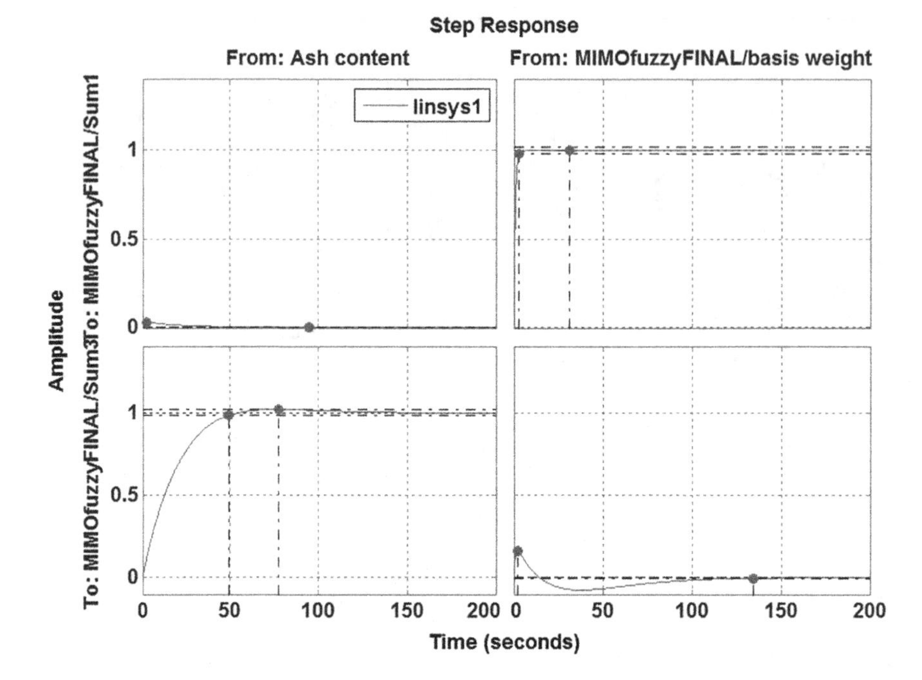

Fig. 11 Step response for ash content and basis weight using fuzzy-PID controller

content using Fuzzy-PID controller is shown in Fig. 11, respectively. The dynamic performance parameters like rise time, peak time, and overshoot and settling time for the wet end system are shown in Table 4. It shows that the parameters of Fuzzy-PID controllers improve the performance of the MIMO system.

Table 4 Fuzzy PID and PID controller performance parameters

Performances	PID controller	Fuzzy-PID controller	PID controller	Fuzzy-PID controller
	Basis weight		Ash content	
Rise time	14 s	1.65 s	93.7 s	33.9 s
Settling time	51.1 s	2.98 s	334 s	49.2 s
Peak time	30.8 s	31 s	202 s	77.6 s
Overshoot	8.35 %	0.279 %	8.89 %	1.83 %

5 Conclusion

In this paper, PID and FLC-PID controllers have been presented for a MIMO system. The Simulation results conclude that the performance of Fuzzy PID controller is superior to the PID controller in terms of peak time, settling time, rise time, and overshoot for wet end system of a paper making machine.

References

1. Zadeh, L.A.: Fuzzy sets. Inf. Control **8**, 338–353 (1965)
2. Mamdani, E.H.: Application of fuzzy algorithms for control of simple dynamic plant. IEEE Proc. **121**(12), 1585–1588 (1974)
3. Smook, G.A.: Handbook for pulp and paper technologists, 2nd edn., pp. 229. Angus wide publication Inc., Vancouver (2001)
4. Vijay, S.M.: Fuzzy PID controller for multi input multi output model. Int. J. Adv. Eng. Technol. **II**, 344–348 (2011)
5. Elaydi, H., Hardouss, I.A., Alassar, A.: Supervisory fuzzy control for 5 DOF robot arm. Int. J. Sci. Adv. Technol. **2**(7) (2012)
6. Shivanandam, S.N., Deepa, S.N.: Principles of Soft Computing, 2nd edn. Wiley India Pvt, Ltd (2011)

Common Fixed Point Results in G_b Metric Space and Generalized Fuzzy Metric Space Using E.A Property and Altering Distance Function

Vishal Gupta, Raman Deep and Naveen Gulati

Abstract Fuzzy logic is a new scientific field which is employed in mathematics, computer science and engineering. Fuzzy logic calculates the extent to which a proposition is correct and allows computer to manipulate the information. In this paper, our objective is to prove fixed point results in the setting of G_b metric space and generalized fuzzy metric space. The concept of G_b metric space is given by Aghajani in 2013, which is generalization of G-metric space and b metric space. Sedghi et al., proved results on G_b metric space using continuity and commutativity. Also, the notion of generalized Fuzzy metric space is given by Sung and Yang. In 2011, Rao et al., proved results in generalized fuzzy metric space using w-compatible mappings. Here, to prove the results notion of E.A property, weakly compatible property and contractive type condition have been utilized. Our results extend and generalize the results of many authors existing in the literature. An example has also been given to justify the result.

Keywords Common fixed point · E.A property · G_b metric space · Generalized fuzzy metric space · Altering distance function · Weakly compatible mapping

Vishal Gupta (✉) · Raman Deep
Department of Mathematics, Maharishi Markandeshwar University, Mullana, Ambala 133001, Haryana, India
e-mail: vishal.gmn@gmail.com

Raman Deep
e-mail: ramandeepvirk02@gmail.com

Naveen Gulati
Department of Mathematics, S.D(P.G) College, Ambala Cantt 133001, Haryana, India
e-mail: dr.naveengulati31@gmail.com

M. Pant et al. (eds.), *Proceedings of Fifth International Conference on Soft Computing for Problem Solving*, Advances in Intelligent Systems and Computing 437, DOI 10.1007/978-981-10-0451-3_16

1 Introduction and Preliminaries

The concept of G-metric space was given by Mustafa and Sims [1] in 2006.

Definition 1 [1] Let X be a non-empty set and let $G: X \times X \times X \to R^+$ be a function satisfying the following properties:

1. $G(x, y, z) = 0$, if $x = y = z$,
2. $G(x, x, y) > 0$ for all $x, y \in X$ with $x \neq y$,
3. $G(x, x, y) \leq G(x, y, z)$ for all $x, y, z \in X$ with $y \neq z$,
4. $G(x, y, z) = G(x, z, y) = G(y, z, x) = \ldots$,
5. $G(x, y, z) \leq G(x, a, a) + G(a, y, z)$ for all $x, y, z, a \in X$.

Then the function G is called a generalized metric or a G-metric on X, and the pair (X, G) is called G-metric space [2].

On the basis of these considerations, many authors like Abbas [3] introduced the notion of generalized probabilistic metric space and obtained fixed point result without using continuity and commutativity. Aydi [4] has proved fixed point results for two self mappings f and g, where f is generalized weakly g-contraction mapping with respect to g. Gu [5] proved a common fixed point theorem for a class of twice power type weakly compatible mapping in generalized metric space. Vahid [6] gives common fixed point results for six self mappings satisfying generalized weakly (ϕ, ψ)-contractive condition in the setting of G-metric space. In 2010, Obiedat and Mustafa [7] presented several fixed point theorems on a nonsymmetric complete G-metric space. Chugh [8] satisfies the property P for the results on G-metric space. After that in 2011, Altun [9] define weakly increasing maps in metric space. On this basis, Shatanawi [10] gave weakly increasing maps in G-metric space and proved fixed point results along an integral application. Also in 2011, Abbas [11] used R-weakly commuting maps. Pant [12] introduced the concept of reciprocal continuity that was extended by him in [13] to weak reciprocal continuity. Further in 2013, Mustafa [14] et al. proved fixed point result in G-metric space using control functions.

The concept of control function was due to Khan, Swalech, and Sessa [15] in 1984.

Definition 2 [15] A mapping $\phi : [0, \infty) \to [0, \infty)$ is called an altering distance function if the following properties are satisfied:

1. ϕ is continuous and non-decreasing,
2. $\phi(t) = 0 \Leftrightarrow t = 0$.

Sessa [16] defined the concept of weakly commutative maps to obtain common fixed point for a pair of maps. Jungck generalized this idea to compatible mappings [17] and then to weakly compatible mappings [18].

Definition 3 [18] Two maps f and g are said to be weakly compatible if they commute at their coincidence points.

In 2002, Amari and Moutawakil [19] introduced a new concept of the property E.A. in metric spaces and proved some common fixed point theorems.

Definition 4 [19] Two self mappings A and S of a metric space (X, d) are said to satisfy the property E.A if there exist a sequence $\{x_n\}$ in X such that $\lim_{n\to\infty} Ax_n = \lim_{n\to\infty} Sx_n = z$ for some $z \in X$.

In 2012, Abbas et al. [20] obtained some unique common fixed-point results for a pair of mappings satisfying E.A property under certain generalized strict contractive conditions, which was further generalized by Long [21] with only one pair satisfies E.A property in the setup of generalized metric spaces. In 2013, Gu [22] proved results employing E.A property but without any commuting and continuity conditions of mappings. In [23] Mustafa introduced new concepts like G-weakly commuting of type G_f and G-R-weakly commuting of type G_f along with E.A property to prove the results in the setting of G-metric space.

The concept of b-metric was suggested by Czerwik in [24].

Definition 5 [24] Let X be a set and $s \geq 1$ be a given number. A function $d : X \times X \to R^+$ is said to be a b-metric if and only if for all $x, y, z \in X$, the following conditions are satisfied:

1. $d(x, y) = 0$ if and only if $x = y$;
2. $d(x, y) = d(y, x)$;
3. $d(x, z) \leq s[d(x, y) + d(y, z)]$.

Then the pair (X, d) is said b-metric space.

Many eminent results have been proved in this space; for more detail reader can see [25, 26].

The theory of G_b-metric space proposed by Aghajani [27] in 2013, which is abstraction of G-metric space and b-metric space.

Definition 6 [27] Let X be a nonempty set and $s \geq 1$ be a given real number. Suppose that a mapping $G : X \times X \times X \to R^+$ satisfies:

1. $G(x, y, z) = 0$, if $x = y = z$;
2. $0 < G(x, x, y)$ for all $x, y \in X$ with $x \neq y$;
3. $G(x, x, y) \leq G(x, y, z)$ for all $x, y, z \in X$ with $y \neq z$;
4. $G(x, y, z) = G(p\{x, y, z\})$, where p is a permutation of x, y, z;
5. $G(x, y, z) \leq s(G(x, a, a) + G(a, y, z))$ for all $x, y, z, a \in X$.

Then G is called a generalized b-metric space and the pair (X, G) is called a generalized b-metric space or G_b metric space.

Example 1 [27] Let (X, G) be a G-metric space, and $G_*(x, y, z) = G(x, y, z)^p$, where $p > 1$ is a real number. Then G_* is a G_b metric with $s = 2^{p-1}$.

Definition 7 [27] A G_b-metric G is said to be symmetric if $G(x, y, y) = G(y, x, x)$ for all $x, y \in X$.

Definition 8 [27] Let (X, G) be a G_b-metric space then for $x_0 \in X, r > 0$, the G_b-ball with centre x_0 and radius r is—$B_G(x_0, r) = \{y \in X : G(x_0, y, y) < r\}$.

For example, let $X = \Re$ and consider the G_b-metric G defined by $G(x, y, z) = \frac{1}{9}(|x - y| + |y - z| + |x - z|)^2$ for all $x, y, z \in \Re$. Then

$$\begin{aligned} B_G(3,4) &= \{y \in X : G(3,3,y) < 4\} \\ &= \left\{y \in X : \frac{1}{9}\left(|y - 3| + |y - 3|^2\right) < 4\right\} \\ &= \left\{y \in X : |y - 3|^2 < 4\right\} \\ &= (0, 6). \end{aligned}$$

Proposition 1 [27] *Let (X, G) is a G_b-metric space, then for each $x, y, z, a \in X$ it follows that:*

1. *if $G(x, y, z) = 0$ then $x = y = z$,*
2. $G(x, y, z) \leq s(G(x, x, y) + G(x, x, z))$,
3. $G(x, y, y) \leq G(y, x, x)$,
4. $G(x, y, z) \leq s(G(x, a, z) + G(a, y, z))$.

Definition 9 [27] Let (X, G) is a G_b-metric space, we define $d_G(x, y) = G(x, y, y) + G(x, x, y)$, it is easy to see that d_G defines a b-metric on X, which we call b-metric associated with G.

Definition 10 [27] Let (X, G) is a G_b-metric space. A sequence $\{x_n\}$ in X is said to be

1. G_b-Cauchy sequence if, for each $\varepsilon > 0$, there exists a positive integer n_0 such that, for all $m, n, l \geq n_0, G(x_n, x_m, x_l) < \varepsilon$;
2. G_b-convergent to a point $x \in X$ if, for each $\varepsilon > 0$, there exists a positive integer n_0 such that, for all $m, n \geq n_0, G(x_n, x_m, x) < \varepsilon$.

Proposition 2 [27] *Let (X, G) is a G_b-metric space, then the following are equivalent:*

1. *the sequence $\{x_n\}$ is G_b-Cauchy,*
2. *for any $\varepsilon > 0$, there exists n_0 such that $G(x_n, x_m, x_m) < \varepsilon$.*

Proposition 3 [27] *Let (X, G) is a G_b-metric space, then the following are equivalent:*

1. *$\{x_n\}$ is G_b-convergent to x,*
2. *$G(x_n, x_n, x) \to 0$ as $n \to +\infty$,*
3. *$G(x_n, x, x) \to 0$ as $n \to +\infty$.*

Definition 11 [27] A G_b-metric space X is called G_b-complete if every G_b-Cauchy sequence is G_b-convergent in X.

Mustafa [14, 28] presented some coupled fixed point results and tripled coincidence point results in G_b metric space by applying control function. In addition, Shaban Sedghi [29] in 2014 also demonstrates coupled fixed point theorem in the same space.

Sung and Yang [30] introduced the notion of generalized fuzzy metric spaces. Generalized fuzzy metric space is a combination of the concept of fuzzy set introduced by Zadeh [31] in 1965 and concept of fuzzy metric space introduced by Kramosil and Michalek [32]. Gupta et al. in [33–37], proved various results in Fuzzy metric space. Saini in [38, 39] gave results using Expansion maps and R-weakly commuting maps in Fuzzy metric space.

Definition 12 [30] A binary operation $* : [0,1] \to [0,1] \times [0,1]$ is a continuous t-norm if $*$ satisfies the following conditions:

1. $*$ is commutative and associative,
2. $*$ is continuous,
3. $a * 1 = a$ for all $a \in [0,1]$,
4. $a * b \leq c * d$ whenever $a \leq c$ and $b \leq d$ for all $a, b, c, d \in [0,1]$.

Definition 13 [30] A 3-tuple $(X, G, *)$ is said to be a G-fuzzy metric space (denoted by GF space) if X is an arbitrary nonempty set, $*$ is a continuous t-norm and G is a fuzzy set on $X^3 \times (0, \infty)$ satisfying the following conditions for each $t, s > 0$:

1. $G(x, x, y, t) > 0$ for all $x, y \in X$ with $x \neq y$,
2. $G(x, x, y, t) > G(x, x, z, t)$ for all $x, y, z \in X$ with $y \neq z$,
3. $G(x, x, z, t) = 1$ if and only if $x = y = z$,
4. $G(x, x, z, t) = G(p(x, y, z), t)$; where p is a permutation function,
5. $G(x, a, a, t) * G(a, y, z, t) \leq G(x, y, z, t + s)$,
6. $G(x, y, z, .) : (0, \infty) \to [0,1]$ is continuous.

Definition 14 [30] Let $(X, G, *)$ be a G-fuzzy metric space, then the following conditions are satisfied:

1. a sequence $\{x_n\}$ in X is said to be convergent to x if, $\lim_{n\to\infty} G(x_n, x_n, x, t) = 1$ for all $t > 0$,
2. a sequence $\{x_n\}$ in X is said to be a Cauchy sequence if, $\lim_{n\to\infty} G(x_n, x_n, x_m, t) = 1$,
3. a GF-metric space is said to be complete if every Cauchy sequence in X is convergent.

Here, we intent to prove common fixed point results working on E.A property and weakly compatible property in the framework of G_b-metric space and generalized fuzzy metric space.

2 Main Results

Theorem 1 *Consider* (X, G) *is a* G_b*-metric space, and* $f, g, S, T : X \to X$ *are mappings such that*

1. (f, S) *satisfy E.A property;*
2. $f(X) \subseteq T(X)$;
3. $S(X)$ *is closed subspace of* X;
4. (f, S) *and* (g, T) *are weakly compatible pair of mappings;*
5. $$G(fx, gy, gz) \leq \frac{1}{s^2}\phi(M(x, y, z)) \tag{1}$$

$$M(x,y,z) = \max\left\{G(Sx, Ty, Tz), G(fx, Sx, Sx), G(Ty, gy, gy), \frac{G(fx, Ty, Tz) + G(Sx, gy, gy)}{2}\right\},$$

where ϕ *is an altering distance function such that* $\phi(t) < t$*. Then* f, g, S, T *have a unique common fixed point in* X*.*

Proof In consideration of the pair of mapping (f, S) satisfy E.A property there exists a sequence $\{x_n\}$ in X such that $\lim_{n\to\infty} fx_n = \lim_{n\to\infty} Sx_n = t$ for some $t \in X$. Further $f(X) \subset T(X)$, accordingly $fx_n = Ty_n = t$, where $\{y_n\}$ is a sequence in X. Hence $\lim_{n\to\infty} Ty_n = t$. Following from (1), we get

$$G(fx_n, gy_n, gy_n) \leq \frac{1}{s^2}\phi(M(x_n, y_n, y_n)). \tag{2}$$

Taking as $n \to \infty$, we get

$$G\left(t, \lim_{n\to\infty} gy_n, \lim_{n\to\infty} gy_n\right) \leq \frac{1}{s^2}\phi\left[\lim_{n\to\infty} M(x_n, y_n, y_n)\right], \tag{3}$$

$$\begin{aligned}\lim_{n\to\infty} M(x_n, y_n, y_n) &= \lim_{n\to\infty} \max\{G(Sx_n, Ty_n, Ty_n), G(fx_n, Sx_n, Sx_n), \\ &\quad G(Ty_n, gy_n, gy_n), \frac{G(fx_n, Ty_n, Ty_n) + G(Sx_n, gy_n, gy_n)}{2}\Bigg\} \\ &= \max\left\{G(t,t,t), G(t,t,t), G\left(t, \lim_{n\to\infty} gy_n, \lim_{n\to\infty} gy_n\right)\right. \\ &\quad \left.\frac{G(t,t,t) + G(t, \lim_{n\to\infty} gy_n, \lim_{n\to\infty} gy_n)}{2}\right\} \\ &= G\left(t, \lim_{n\to\infty} gy_n, \lim_{n\to\infty} gy_n\right).\end{aligned}$$

From (2),

$$G\left(t, \lim_{n\to\infty} gy_n, \lim_{n\to\infty} gy_n\right) \le \frac{1}{s^2}\phi\left(G\left(t, \lim_{n\to\infty} gy_n, \lim_{n\to\infty} gy_n\right)\right)$$
$$<\phi\left(G\left(t, \lim_{n\to\infty} gy_n, \lim_{n\to\infty} gy_n\right)\right)$$
$$<G\left(t, \lim_{n\to\infty} gy_n, \lim_{n\to\infty} gy_n\right),$$

which implies that $G(t, \lim_{n\to\infty} gy_n, \lim_{n\to\infty} gy_n) = 0$, and hence $\lim_{n\to\infty} gy_n = t$. Again, $S(X)$ is a closed subset of X, therefore $t = Su$ for some $u \in X$.

We claim that $fu = t$. In fact, from (1) and using $\lim_{n\to\infty} fx_n = \lim_{n\to\infty} Sx_n = \lim_{n\to\infty} Ty_n = \lim_{n\to\infty} gy_n = t$, we get

$$G(fu, gy_n, gy_n) \le \frac{1}{s^2}\phi(M(u, y_n, y_n)).$$

Letting $n \to \infty$, we obtain

$$G(fu, t, t) \le \frac{1}{s^2}\phi\left(\lim_{n\to\infty} M(u, y_n, y_n)\right), \tag{4}$$

where

$$\lim_{n\to\infty} M(u, y_n, y_n) = \lim_{n\to\infty} \max\{G(Su, Ty_n, Ty_n), G(fu, Su, Su),$$
$$G(Ty_n, gy_n, gy_n), \frac{G(fu, Ty_n, Ty_n) + G(Su, gx_n, gx_n)}{2}\Big\}$$
$$= \max\{G(t, t, t), G(fu, t, t), G(t, t, t),$$
$$\frac{G(fu, t, t) + G(t, t, t)}{2}\Big\}$$
$$= G(fu, t, t).$$

Consequently from (4), $G(fu, t, t) = 0$. Thus $fu = Su = t$. As $f(X) \subset T(X)$, there exists $v \in X$, such that $fu = Tv = t$.

Next, we show that $Tv = gv$. From condition (1),

$$G(fu, gv, gv) \le \frac{1}{s^2}\phi(M(u, v, v)), \tag{5}$$

and

$$\lim_{n\to\infty} M(u,v,v) = \lim_{n\to\infty} \max\{G(Su,Tv,Tv), G(fu,Su,Su),$$
$$G(Tv,gv,gv), \frac{G(fu,Tv,Tv)+G(Su,gv,gv)}{2}\Bigg\}$$
$$= \max\{G(t,t,t), G(t,t,t), G(t,gv,gv),$$
$$\frac{G(t,t,t)+G(t,gv,gv)}{2}\Bigg\}$$
$$= G(t,gv,gv).$$

Using (5), we have $gv = t$. Thus $fu = Su = Tv = gv = t$.

Next, we employ the property of compatibility of (f,S). As u is the point of coincidence of mappings f and S, therefore $f(fu) = ft = f(Su) = S(fu) = St$.

Now we prove that $f(t) = t$. Again from (1), we obtain,

$$G(ft,gv,gv) \leq \frac{1}{s^2}\phi(M(t,v,v)), \tag{6}$$

where

$$\lim_{n\to\infty} M(t,v,v) = \lim_{n\to\infty} \max\{G(St,Tv,Tv), G(ft,St,St),$$
$$G(Tv,gv,gv), \frac{G(ft,Tv,Tv)+G(St,gv,gv)}{2}\Bigg\}$$
$$= \max\{G(ft,t,t), G(ft,ft,ft), G(t,t,t),$$
$$\frac{G(t,t,t)+G(ft,t,t)}{2}\Bigg\}$$
$$= G(ft,t,t).$$

Using (6), we conclude that $ft = t$, this gives $ft = St = t$. By following the same argument one can get $Tt = gt = t$. Hence t is the fixed point of f,S,T,g.

Next, we show that the common fixed point of the mappings is unique. To prove this let w be another fixed point of the mappings in X. Then

$$G(ft,gw,gw) \leq \frac{1}{s^2}\phi(M(t,w,w)), \tag{7}$$

where

$$\lim_{n\to\infty} M(t,w,w) = \lim_{n\to\infty} \max\{G(St,Tw,Tw), G(ft,St,St), G(Tw,gw,gw), \frac{G(ft,Tw,Tw)+G(St,gw,gw)}{2}\}$$
$$= \max\{G(t,w,w), G(t,t,t), G(w,w,w), \frac{G(t,w,w)+G(t,w,w)}{2}\}$$
$$= G(t,w,w).$$

This implies that $t = w$. Hence the result.

On Taking $\phi(t) = t$ in Theorem 1, we have the following corollary:

Corollary 1 *Consider* (X, G) *is a* G_b*-metric space. And* $f, g, S, T : X \to X$ *are mappings such that*

1. (f, S) *satisfy E.A property;*
2. $f(X) \subseteq T(X)$;
3. $S(X)$ *is closed subspace of* X;
4. (f, S) *and* (g, T) *are weakly compatible pair of mappings;*
5. $G(fx, gy, gz) \leq \frac{1}{s^2} M(x, y, z)$

$$M(x,y,z) = \max\left\{G(Sx,Ty,Tz), G(fx,Sx,Sx), G(Ty,gy,gy), \frac{G(fx,Ty,Tz)+G(Sx,gy,gy)}{2}\right\}.$$

Here ϕ *is an altering distance function such that* $\phi(t) < t$. *Then* f, g, S, T *have a unique common fixed point in* X.

Theorem 2 *Consider* (X, G) *is a Generalized Fuzzy metric space and* $f, g, S, T : X \to X$ *be mappings such that*

1. (f, S), (g, T) *satisfy E.A property;*
2. $f(X) \subseteq T(X)$ *and* $g(X) \subseteq S(X)$;
3. $T(X), S(X)$ *is closed subspace of* X;
4. (f, S) *and* (g, T) *are weakly compatible pair of mappings;*
5. $$\mathrm{G(fx, gy, gz, t)} \geq \phi[\mathrm{M(x, y, z)}]. \quad (8)$$

where

$$M(x,y,z) = \min\left\{G(Sx,Ty,Tz,kt), G(gy,Sx,Sx,kt), G(Ty,gy,gy,kt), \left[\frac{G(gy,Ty,Tz,kt)+G(Sx,gy,gy,kt)}{2}\right]\right\}. \quad (9)$$

Here ϕ be an altering distance function. Then f, g, S, T have a unique common fixed point in X.

Proof As the pair of mappings (g, T) satisfies E.A property, therefore there exists a sequence $\{x_n\}$ such that $\lim_{n\to\infty} G(gx_n, u, u, t) = \lim_{n\to\infty} G(Tx_n, u, u, t) = 1$.

Also, $gX \subseteq SX$, $\exists$ a sequence $\{y_n\}$ such that $gx_n = Sy_n$.

We claim that $\lim_{n\to\infty} G(fy_n, u, u, t) = 1$.

Now, $G(fy_n, gx_n, gx_{n+1}, t) \geq \phi[M(y_n, x_n, x_{n+1})]$, here

$$M(y_n, x_n, x_{n+1}) = \min\left\{ G(Sy_n, Tx_n, Tx_{n+1}, kt), G(gx_n, Sy_n, Sy_n, kt), G(Tx_n, gx_n, gx_n, kt), \left[\frac{G(gx_n, Tx_n, Tx_{n+1}, kt) + G(Sy_n, gx_n, gx_n, kt)}{2}\right] \right\}. \tag{10}$$

Therefore, $\lim_{n\to\infty} G(fy_n, l, l, t) = 1$.

Hence, $\lim_{n\to\infty} fy_n = \lim_{n\to\infty} Sy_n = \lim_{n\to\infty} gx_n = \lim_{n\to\infty} Tx_n = l$. Since $S(X)$ is complete F-fuzzy metric space, therefore there exists $x \in X$ such that $Sx = l$. Now if $fx \neq l$, then

$G(fx, gx_n, gx_{n+1}, t) \geq \phi[M(x, x_n, x_{n+1})]$, where

$$M(x, x_n, x_{n+1}) = \phi \min\left\{ G(Sx, Tx_n, Tx_{n+1}, kt), G(gx_n, Sx, Sx, kt), G(Tx_n, gx_n, gx_n, kt), \left[\frac{G(gx_n, Tx_n, Tx_{n+1}, kt) + G(Sx, gx_n, gx_n, kt)}{2}\right] \right\}. \tag{11}$$

Taking $n \to \infty$, we get $\lim_{n\to\infty} G(fx, l, l, t) = 1$ and hence $fx = l = Sx$.

By the condition that (f, S) be weakly compatible, we have $fSx = Sfx$, so $ffx = fSx = Sfx = SSx$.

As $fX \subset TX$, there exists $m \in X$ such that $fx = Tm$.

Next we prove that $Tm = gm$. If not then, $G(fx, gm, gm, t) \geq \phi[M(x, m, m)]$, where

$$M(x, m, m) = \phi \min\left\{ G(Sx, Tm, Tm, kt), G(gm, Sx, Sx, kt), G(Tm, gm, gm, kt), \left[\frac{G(gm, Tm, Tm, kt) + G(Sx, gm, gm, kt)}{2}\right] \right\}. \tag{12}$$

If $gm \neq l$ then $G(fx, gm, gm, t) > G(fx, gm, gm, t)$, it is a contradiction. Thus $Tm = gm = l$.

As mappings g and T are weakly compatible, we have $TTm = Tgm = gTm = ggm$ and $Tl = gl$.

Now we aim to prove that $fl = l$,
$G(fl, gm, gm, t) \geq \phi[M(l, m, m)]$, where

$$M(l,m,n) = \phi \min\left\{G(Sl,Tm,Tm,kt), G(gm,Sl,Sl,kt), G(Tm,gm,gm,kt), \left[\frac{G(gm,Tm,Tm,kt) + G(Sl,gm,gm,kt)}{2}\right]\right\}. \tag{13}$$

If $fl \neq l$ then $G(fl,l,l,t) > G(fl,l,l,t)$, it is a contradiction. Thus $fl = Sl = l$.
Next we show $gl = l$.
$G(fl, gl, gl, t) \geq \phi[M(l,l,l)]$,

$$M(l,l,l) = \min\left\{G(Sl,gl,Tl,kt), G(gl,Sl,Sl,kt), G(Tl,gl,gl,kt), \left[\frac{G(gl,Tl,Tl,kt) + G(Sl,gl,gl,kt)}{2}\right]\right\}. \tag{14}$$

If $gl \neq l$ then $G(l,gl,gl,t) > G(l,gl,gl,t)$, it is a contradiction. Thus $fl = gl = Sl = Tl = l$.

Hence f, g, S and T have common fixed point. Uniqueness follows by taking $x = s, y = l$ and $z = l$ in (8) condition.

Example 2 Let $X = [0,1]$. Define $G : X \times X \times X \to R^+$ by $G(x,y,z) = (|x-y| + |y-z| + |z-x|)^2$ for all $x,y,z \in X$. Then (X,G) is a complete G_b-metric space with $s = 2$.

Also consider pairs of self mappings (f,S) and (g,T) be defined as

$$f(x) = \begin{cases} 1, & \text{if } x \in \left[0, \frac{1}{2}\right] \\ \frac{3}{4}, & \text{if } x \in \left(\frac{1}{2}, 1\right], \end{cases} \qquad S(x) = \begin{cases} 0, & \text{if } x \in \left[0, \frac{1}{2}\right] \\ \frac{3}{4}, & \text{if } x \in \left(\frac{1}{2}, 1\right) \\ \frac{5}{6}, & \text{if } x = 1, \end{cases}$$

$$g(x) = \begin{cases} \frac{4}{5}, & \text{if } x \in \left[0, \frac{1}{2}\right] \\ \frac{3}{4}, & \text{if } x \in \left(\frac{1}{2}, 1\right], \end{cases} \qquad T(x) = \begin{cases} 1, & \text{if } x \in \left[0, \frac{1}{2}\right] \\ \frac{3}{4}, & \text{if } x \in \left(\frac{1}{2}, 1\right) \\ 0, & \text{if } x = 1. \end{cases}$$

The pair (f,S) satisfy E.A property and the pairs $(f,S)(g,T)$ are weakly compatible pairs. Take $\phi(t) = \frac{t}{4}$. Now we prove the mappings f, g, S, T satisfies the contractive condition of Theorem 1.

Case (1): If $x,y,z \in \left[0, \frac{1}{2}\right]$ then we have $G(fx, gy, gz) = G\left(1, \frac{4}{5}, \frac{4}{5}\right) = \frac{4}{25}$.
$M(x,y,z) = \max\left\{G(0,1,1), G(1,0,0), G\left(1, \frac{4}{5}, \frac{4}{5}\right), \frac{G(1,1,1) + G\left(0, \frac{4}{5}, \frac{4}{5}\right)}{2}\right\}$ and $\frac{1}{s^2}\phi(M(x,y,z)) = \frac{1}{4}$.

Case (2): If $x, y \in \left[0, \frac{1}{2}\right]$ and $z \in \left(\frac{1}{2}, 1\right]$ then we have $G(fx, gy, gz) = G\left(1, \frac{4}{5}, \frac{3}{4}\right) = \frac{1}{4}$.

Subcase (1): for $z \in \left(\frac{1}{2}, 1\right)$, $M(x,y,z) = \max\left\{G\left(0, 1, \frac{3}{4}\right), G(1,0,0), G\left(1, \frac{4}{5}, \frac{4}{5}\right), \frac{G\left(1,1,\frac{3}{4}\right) + G\left(0,\frac{4}{5},\frac{4}{5}\right)}{2}\right\}$.

Subcase (2): for $z = 1$, $M(x,y,z) = \max\left\{G(0,1,0), G(1,0,0), G\left(1, \frac{4}{5}, \frac{4}{5}\right), \frac{G(1,1,0) + G\left(0,\frac{4}{5},\frac{4}{5}\right)}{2}\right\}$

and $\frac{1}{s^2}\phi(M(x,y,z)) = \frac{1}{4}$.

Case (3): If $x, z \in \left[0, \frac{1}{2}\right]$ and $y \in \left(\frac{1}{2}, 1\right]$ then we have $G(fx, gy, gz) = G\left(1, \frac{3}{4}, \frac{4}{5}\right) = \frac{1}{4}$.

Subcase (1): for $y \in \left(\frac{1}{2}, 1\right)$, $M(x,y,z) = \max\left\{G\left(0, \frac{3}{4}, 1\right), G(1,0,0), G\left(\frac{3}{4}, \frac{3}{4}, \frac{3}{4}\right), \frac{G\left(1,\frac{3}{4},1\right) + G\left(0,\frac{3}{4},\frac{3}{4}\right)}{2}\right\}$.

Subcase (2): for $y = 1$, $M(x,y,z) = \max\left\{G\left(0, \frac{3}{4}, 1\right), G(1,0,0), G\left(0, \frac{3}{4}, \frac{3}{4}\right), \frac{G(1,0,1) + G\left(0,\frac{3}{4},\frac{3}{4}\right)}{2}\right\}$

and $\frac{1}{s^2}\phi(M(x,y,z)) = \frac{1}{4}$.

Case (4): If $y, z \in \left[0, \frac{1}{2}\right]$ and $x \in \left(\frac{1}{2}, 1\right]$ then we have $G(fx, gy, gz) = G\left(\frac{3}{4}, \frac{4}{5}, \frac{4}{5}\right) = \frac{1}{100}$.

Subcase (1): for $x \in \left(\frac{1}{2}, 1\right)$, $M(x,y,z) = \max\left\{G\left(\frac{3}{4}, 1, 1\right), G\left(\frac{3}{4}, \frac{3}{4}, \frac{3}{4}\right), G\left(1, \frac{4}{5}, \frac{4}{5},\right), \frac{G\left(\frac{3}{4},1,1\right) + G\left(\frac{3}{4},\frac{4}{5},\frac{4}{5}\right)}{2}\right\}$

and $\frac{1}{s^2}\phi(M(x,y,z)) = \frac{1}{64}$.

Subcase (2): for $x = 1$, $M(x,y,z) = \max\left\{G\left(\frac{5}{6}, 1, 1\right), G\left(\frac{3}{4}, \frac{5}{6}, \frac{5}{6}\right), G\left(1, \frac{4}{5}, \frac{4}{5},\right), \frac{G\left(\frac{3}{4},1,1\right) + G\left(\frac{5}{6},\frac{4}{5},\frac{4}{5}\right)}{2}\right\}$

and $\frac{1}{s^2}\phi(M(x,y,z)) = \frac{1}{100}$.

Case (5): If $x \in \left[0, \frac{1}{2}\right]$ and $y, z \in \left(\frac{1}{2}, 1\right]$ then we have $G(fx, gy, gz) = G\left(1, \frac{3}{4}, \frac{3}{4}\right) = \frac{1}{4}$.

Subcase (1): for $y, z \in \left(\frac{1}{2}, 1\right)$, $M(x,y,z) = \max\left\{G\left(1, 1, \frac{3}{4}\right), G(1,0,0), G\left(\frac{3}{4}, \frac{3}{4}, \frac{3}{4},\right), \frac{G\left(1,\frac{3}{4},\frac{3}{4}\right) + G\left(0,\frac{3}{4},\frac{3}{4}\right)}{2}\right\}$.

Subcase (2): for $y = 1$, $M(x,y,z) = \max\left\{G\left(0,0,\frac{3}{4}\right), G(1,0,0), G\left(0,\frac{3}{4},\frac{3}{4},\right), \frac{G\left(1,0,\frac{3}{4}\right)+G\left(0,\frac{3}{4},\frac{3}{4}\right)}{2}\right\}$.

Subcase (3): for $z = 1$, $M(x,y,z) = \max\left\{G\left(0,\frac{3}{4},0\right), G(1,0,0), G\left(\frac{3}{4},\frac{3}{4},\frac{3}{4},\right), \frac{G\left(1,\frac{3}{4},0\right)+G\left(0,\frac{3}{4},\frac{3}{4}\right)}{2}\right\}$

and $\frac{1}{s^2}\phi(M(x,y,z)) = \frac{1}{4}$.

Case (6): If $y \in \left[0,\frac{1}{2}\right]$ and $x,z \in \left(\frac{1}{2},1\right]$ then we have $G(fx,gy,gz) = G\left(\frac{3}{4},\frac{4}{5},\frac{3}{4}\right) = \frac{1}{100}$.

Subcase (1): for $x,z \in \left(\frac{1}{2},1\right)$, $M(x,y,z) = \max\left\{G\left(\frac{3}{4},1,1\right), G\left(\frac{3}{4},\frac{3}{4},\frac{3}{4},\right), G\left(1,\frac{4}{5},\frac{4}{5},\right), \frac{G\left(\frac{3}{4},1,\frac{3}{4}\right)+G\left(\frac{3}{4},\frac{4}{5},\frac{4}{5}\right)}{2}\right\}$

then $\frac{1}{s^2}\phi(M(x,y,z)) = \frac{1}{64}$.

Subcase (2): for $x = 1$, $M(x,y,z) = \max\left\{G\left(\frac{5}{6},1,1\right), G\left(\frac{3}{4},\frac{5}{6},\frac{5}{6},\right), G\left(1,\frac{4}{5},\frac{4}{5},\right), \frac{G\left(\frac{3}{4},1,\frac{3}{4}\right)+G\left(\frac{5}{6},\frac{4}{5},\frac{4}{5}\right)}{2}\right\}$

then $\frac{1}{s^2}\phi(M(x,y,z)) = \frac{1}{100}$.

Subcase (3): for $z = 1$, $M(x,y,z) = \max\left\{G\left(\frac{5}{6},1,0\right), G\left(\frac{3}{4},\frac{5}{6},\frac{5}{6},\right), G\left(1,\frac{4}{5},\frac{4}{5},\right), \frac{G\left(\frac{3}{4},1,0\right)+G\left(\frac{5}{6},\frac{4}{5},\frac{4}{5}\right)}{2}\right\}$

then $\frac{1}{s^2}\phi(M(x,y,z)) = \frac{1}{64}$.

Case (7): If $z \in \left[0,\frac{1}{2}\right]$ and $x,y \in \left(\frac{1}{2},1\right]$ then we have $G(fx,gy,gz) = G\left(\frac{3}{4},\frac{3}{4},\frac{4}{5}\right) = \frac{1}{100}$.

Subcase (1): for $x,y \in \left(\frac{1}{2},1\right)$, $M(x,y,z) = \max\left\{G\left(\frac{3}{4},\frac{3}{4},1\right), G\left(\frac{3}{4},\frac{3}{4},\frac{3}{4},\right), G\left(\frac{3}{4},\frac{3}{4},\frac{3}{4},\right), \frac{G\left(\frac{3}{4},\frac{3}{4},1\right)+G\left(\frac{3}{4},\frac{3}{4},\frac{3}{4}\right)}{2}\right\}$

and $\frac{1}{s^2}\phi(M(x,y,z)) = \frac{1}{64}$.

Subcase (2): for $x = 1$, $M(x,y,z) = \max\left\{G\left(\frac{5}{6},\frac{3}{4},1\right), G\left(\frac{3}{4},\frac{5}{6},\frac{5}{6},\right), G\left(\frac{3}{4},\frac{3}{4},\frac{3}{4},\right), \frac{G\left(\frac{3}{4},1,\frac{3}{4}\right)+G\left(\frac{5}{6},\frac{3}{4},\frac{3}{4}\right)}{2}\right\}$

and $\frac{1}{s^2}\phi(M(x,y,z)) = \frac{1}{64}$.

Subcase (3): for $y = 1$, $M(x, y, z) = \max\left\{G\left(\frac{3}{4}, 0, 1\right), G\left(\frac{3}{4}, \frac{3}{4}, \frac{3}{4},\right), G\left(0, \frac{3}{4}, \frac{3}{4},\right), \frac{G\left(\frac{3}{4}0, 1\right) + G\left(\frac{3}{4}, \frac{3}{4}, \frac{3}{4}\right)}{2}\right\}$

and $\frac{1}{s^2}\phi(M(x, y, z)) = \frac{1}{4}$.

Case (8): If $x, y, z \in \left(\frac{1}{2}, 1\right]$ then we have $G(fx, gy, gz) = G\left(\frac{3}{4}, \frac{3}{4}, \frac{3}{4}\right) = 0$.

Subcase (1): for $x, y, z \in \left(\frac{1}{2}, 1\right)$, $M(x, y, z) = \max\left\{G\left(\frac{3}{4}, \frac{3}{4}, \frac{3}{4}\right), G\left(\frac{3}{4}, \frac{3}{4}, \frac{3}{4},\right), G\left(\frac{3}{4}, \frac{3}{4}, \frac{3}{4},\right), \frac{G\left(\frac{3}{4}, \frac{3}{4}, \frac{3}{4}\right) + G\left(\frac{3}{4}, \frac{3}{4}, \frac{3}{4}\right)}{2}\right\}$

and $\frac{1}{s^2}\phi(M(x, y, z)) = 0$.

Subcase (2): for $x = 1$, $M(x, y, z) = \max\left\{G\left(\frac{5}{6}, \frac{5}{6}, \frac{3}{4}\right), G\left(\frac{3}{4}, \frac{5}{6}, \frac{5}{6},\right), G\left(\frac{3}{4}, \frac{3}{4}, \frac{3}{4},\right), \frac{G\left(\frac{3}{4}, \frac{3}{4}, \frac{3}{4}\right) + G\left(\frac{5}{6}, \frac{3}{4}, \frac{3}{4}\right)}{2}\right\}$

and $\frac{1}{s^2}\phi(M(x, y, z)) = \frac{1}{576}$.

Subcase (3): for $y = 1$, $M(x, y, z) = \max\left\{G\left(\frac{3}{4}, 0, \frac{3}{4}\right), G\left(\frac{3}{4}, \frac{3}{4}, \frac{3}{4},\right), G\left(0, \frac{3}{4}, \frac{3}{4},\right), \frac{G\left(\frac{3}{4}, 0, \frac{3}{4}\right) + G\left(\frac{3}{4}, \frac{3}{4}, \frac{3}{4}\right)}{2}\right\}$

and $\frac{1}{s^2}\phi(M(x, y, z)) = \frac{9}{64}$.

Subcase (4): for $z = 1$, $M(x, y, z) = \max\left\{G\left(\frac{3}{4}, 0, \frac{3}{4}\right), G\left(\frac{3}{4}, \frac{3}{4}, \frac{3}{4},\right), G\left(\frac{3}{4}, \frac{3}{4}, \frac{3}{4},\right), \frac{G\left(\frac{3}{4}, \frac{3}{4}, 1\right) + G\left(\frac{3}{4}, \frac{3}{4}, \frac{3}{4}\right)}{2}\right\}$

and $\frac{1}{s^2}\phi(M(x, y, z)) = \frac{9}{64}$.

In all cases, the mappings satisfy the contractive condition of Theorem 1, and $\frac{3}{4}$ is the common fixed point.

3 Conclusion

The G_b metric space is more general than G-metric space. Results on G-metric space and G_b metric space based on continuity and commutativity can be proved in the setting of G_b metric space using E.A property without using continuity. By taking different contractive conditions we can generalize many results in that framework.

Acknowledgments The authors would like to express their sincere appreciation to the referees for their helpful suggestions and many kind comments.

References

1. Mustafa, Z., Sims, B.: A new approach to generalized metric spaces. J. Nonlinear Convex Anal. **7**, 289–297 (2006)
2. Rao, K.P.R., Altun, I., Bindu, S.H.: Common coupled fixed point theorems in Generalized fuzzy metric spaces. Adv. Fuzzy Syst. Article I.D 986748, 6 pp. (2011)
3. Abbas, M., Rhoades, B.: Common fixed point results for noncommuting mappings without continuity in generalized metric spaces. Appl. Math. Comp. **215**, 262–269 (2009)
4. Aydi, H., Shatanawi, W., Vetro, C.: On generalized weakly G-contraction mapping in G-metric spaces. Comput. Math Appl. **62**, 4222–4229 (2011)
5. Gu, F., Shen, Y., Wang, L.: Common fixed points results under a new contractive condition without using continuity. J. Inequalities Appl. **464** (2014)
6. Parvaneh, V., Abdolrahman, R., Jamal, R.: Common fixed points of six mappings in partially ordered G-metric spaces. Math. Sci. **7**(18) (2013)
7. Obiedat, Hamed, Mustafa, Zead: Fixed point results on a nonsymmetric G-metric spaces. Jordan J. Math. Statis. **3**(2), 65–79 (2010)
8. Chugh, R., Kadian, T., Rani, A., Rhoades, B.E.: Property P in G-metric space. Fixed Point Theor. Appl. Article ID 401684, 12pp (2010)
9. Altun, I., Simsek, H.: Some fixed point theorems on ordered metric spaces and application. Fixed Point Theor. Appl. Article ID 621469, 17pp. (2010)
10. Shatanawi, W.: Some Fixed point theorems in ordered G-metric space and applications. Abstract Appl. Anal. Article ID 126205, 11pp. (2011)
11. Abbas, M., Khan, S.H., Nazir, T.: Common fixed points of R-weakly computing maps in generalized metric spaces. Fixed Point Theor. Appl. **1**(41) (2011)
12. Pant, R.P., Swales, M.: Common fixed points of four mappings. Bull. Calcutta Math. Soc. **90**, 281–286 (1998)
13. Pant, R.P., Arora, R.K.: Weak reciprocal continuity and fixed point theorems. Ann. Uni. Ferra **57**, 181–190 (2011)
14. Mustafa, Z., Roshan, J. R., Parvaneh, V.: Coupled coincidence point results for (ψ, ϕ)-weakly contractive mappings in partially ordered G_b-metric spaces. Fixed Point Theor. Appl. 206 (2013)
15. Khan, M.S., Swales, M.: Fixed point theorems by altering distances between the points. Bull. Aust. Math. Soc. **30**, 1–9 (1984)
16. Sessa, S.: On a weak commutativity condition of mappings in fixed point consideration. Publ. Inst. Math. **32**, 149–153 (1982)
17. Jungck, G.: Compatible mappings and common fixed point. Int. J. Math. Math. Sci. **9**, 771–779 (1986)
18. Jungck, G.: Common fixed points for noncontinuous nonself maps on nonmetric spaces. Far East J. Math. Sci. **4**, 199–215 (1996)
19. Aamri, M., Moutawakil, D.El.: Some new common fixed point theorems under strict contractive conditions. J. Math. Anal. Appl. **270**, 181–188 (2002)
20. Abbas, M., Nazir, T., Doric, D.: Common fixed point of mappings satisfying E.A property in generalized metric spaces. Appl. Math. Comput. **218**(14), 7665–7670 (2012)
21. Long, W., Abbas, M., Nazir, T., Radenovi´c, S.: Common fixed point for two pairs of mappings satisfying (E.A) property in generalized metric spaces. Abstract and Applied Analysis, Article ID 394830, 15 pp. (2012)
22. Gu, F., Shatanawi, W.: Common fixed point of generalized weakly G-contraction mappings satiusfying common (E.A) property in G-metric spaces. Fixed Point Theor. Appl. **309** (2013)
23. Mustafa, Z., Aydi, H., Karapinar, E.: On common fixed points in G-metric spaces using (E.A) property. Comput. Math. Appl. Article in Press
24. Czerwik, S.: Nonlinear set-valued contration mappings in b-metric spaces. Int. J. Mod. Math. **4** (3), 285–301 (2009)

25. Boriceanu, M., Petrusel, A., Rus, I.A.: Fixed point theorems for some multivalued generalized contractions in b-metric spaces. Int. J. Math. Statis. **6**, 65–76 (2010)
26. Yingtaweesittikul, Hatairat: Suzuki type fixed point theorems for generalized multi-valued mappings in b-metric spaces. Fixed Point Theor. Appl. (2013). doi:10.1186/1687-1812-2013-215
27. Aghajani, A., Abbas, M., Roshan, J.R.: Common fixed point of generalized weak contractive mappings in partially ordered Gb-metric spaces. Filomat **28**(6), 1087–1101 (2014)
28. Mustafa, Z., Roshan, J. R., Parvaneh, V.: Existence of a tripled coincidence point in ordered G_b-metric spaces and applications to a system of integral equations. J. Inequ. Appl. **453** (2013)
29. Sedghi, S., Shobkolaei, N., Roshan, J.R., Shatanawi, W.: Coupled fixed point theorems in G_b-metric space. Matematiqki Vesnik **66**(2), 190–201 (2014)
30. Sun, G., Yang, K.: Generalized fuzzy metric spaces with properties. Res. J. Appl. Sci. **2**, 673–678 (2010)
31. Zadeh, L.A.: Fuzzy sets. Inf. Control **8**, 338–353 (1965)
32. Kramosil, O., Michalek, J.: Fuzzy metric and statistical metric spaces. Kybernetica **11**, 326–334 (1975)
33. Gupta, V., Kanwar, A.: Fixed Point Theorem in Fuzzy metric spaces satisfying E.A property. Indian J. Sci. Technol. **5**(12), 3767–3769 (2012)
34. Gupta, V., Kanwar, A., Singh, S.B.: Common fixed point theorem on fuzzy metric spaces using biased maps of type R_m. In: Jimson, M., Singh, H., Rakesh, K.B., Mahesh, P. (eds.) Advances in Energy Aware Computing and Communication Systems, pp. 227–233. McGraw Hill Education (2013)
35. Gupta, V., Mani, N.: Existence and uniqueness of fixed point in fuzzy metric spaces and its applications. In: Janusz, K.W. (Series editor), Babu, B.V., Nagar, A., Deep, K., Pant, M., Bansal, J.C., Ray, K., Gupta, U. (Volume Editors) Advances in Intelligent Systems and Computing, vol. 236, pp. 217–224. Springer (2014)
36. Gupta, V., Mani, N.: Common fixed points using E.A property in fuzzy metric spaces. advances in intelligent systems and computing. In: Janusz, K.W. (Series Editor), Pant, M., Deep, K., Nagar, A., Bansal, J.C. (Volume Editors), vol. 259, pp. 45–54. Springer (2014)
37. Gupta, V., Saini, R.K., Mani, N., Tripathi, A. K.: Fixed point theorems by using control function in fuzzy metric spaces. Cognet Mathematics (Taylor and Francis), vol. 2, no. 1, 7pp. (2015)
38. Saini, R.K., Gupta, V., Singh, S.B.: Fuzzy version of some fixed point theorems on Expansion type maps in Fuzzy metric Space. Thai J. Math. Math. Assoc. Thailand **5**(2), 245–252 (2007)
39. Saini, R.K., Kumar, M., Gupta, V., Singh, S.B.: Common Coincidence points of R-Weakly Commuting Fuzzy Maps. Thai J. Math. Math. Assoc. Thailand **6**(1), 109–115 (2008)

Design Optimization of Sewer System Using Particle Swarm Optimization

Praveen K. Navin and Y.P. Mathur

Abstract Particle swarm optimization (PSO) technique with new modification is applied in this paper for optimally determine the sewer network component sizes of a predetermined layout. This PSO technique is used for dealing with both discrete and continuous variables as requisite by this problem. A live example of a sewer network is considered to show the algorithm performance, and the results are presented. The results show the capability of the proposed technique for optimally solving the problems of sewer networks.

Keywords Sewer network · Particle swarm optimization · Optimal sewer design

1 Introduction

Sewer networks are an essential part of human society, which collect wastewater from residential, commercial and industrial areas and transports to wastewater treatment plant. Construction and maintenance of this large-scale sewer networks required a huge investment. A relatively small change in the component and construction cost of these networks, therefore leads to a substantial reduction in project cost. The design of a sewer network problem includes two sequential sub problems: (1) generation of the network layout and (2) optimal sizing of sewer network components. The component size optimization of sewer network problem consists of many hydraulic and technical constraints which are generally nonlinear, discrete and sequential. Satisfying such constraints to give an optimal design is often challenging even to the modern heuristic search methods. Many optimization techniques have been applied and developed for the optimal design of sewer networks, such as linear programming [1, 2], nonlinear programming [3, 4] and

P.K. Navin (✉) · Y.P. Mathur
Department of Civil Engineering, MNIT, Jaipur 302017, India
e-mail: navin.nitj@gmail.com

Y.P. Mathur
e-mail: ypmathur.ce@mnit.ac.in

M. Pant et al. (eds.), *Proceedings of Fifth International Conference on Soft Computing for Problem Solving*, Advances in Intelligent Systems and Computing 437, DOI 10.1007/978-981-10-0451-3_17

dynamic programming [5–7]. Evolutionary strategies, such as genetic algorithms [8, 9], ant colony optimization algorithms [10, 11], cellular automata [12] and particle swarm optimization algorithms [13], have received significant consideration in sewer network design problems. Recently, Ostadrahimi et al. [14] used multi-swarm particle swarm optimization (MSPSO) approach to present a set of operation rules for a multi-reservoir system. Haghighi and Bakhshipour [15] developed an adaptive genetic algorithm. Therefore, every chromosome, consisting of sewer slopes, diameters, and pump indicators, is a feasible design. The adaptive decoding scheme is set up based on the sewer design criteria and open channel hydraulics. Using the adaptive GA, all the sewer system's constraints are systematically satisfied, and there is no need to discard or repair infeasible chromosomes or even apply penalty factors to the cost function. Moeini and Afshar [16] used tree growing algorithm (TGA) for efficiently solving the sewer network layouts out of the base network while the ACOA is used for optimally determining the cover depths of the constructed layout.

In this paper, PSO algorithm with new modification is applied to get optimal sewer network component sizes of a predetermined layout.

2 Formulation of Sewer System Design

2.1 Sewer Hydraulics

In circular sewer steady-state flow is described by the continuity principle and Manning's equation which is

$$Q = VA \tag{1}$$

$$V = \frac{1}{n} R^{2/3} S^{1/2} \tag{2}$$

where Q = sewage flow rate, V = velocity of sewage flow, A = cross-sectional flow area, R = hydraulic mean depth, n = Manning's coefficient and S = slope of the sewer. Common, partially full specifications for circular sewer sections are also determined from the following equations:

$$K = QnD^{-8/3} S^{-1/2} \tag{3}$$

$$\theta = \frac{3\pi}{2}\sqrt{1 - \sqrt{1 - \sqrt{\pi K}}} \tag{4}$$

$$\left(\frac{d}{D}\right) = \frac{1}{2} \times \left(1 - \cos\frac{\theta}{2}\right) \tag{5}$$

$$R = \frac{D}{4}\left(\frac{\theta - \sin\theta}{\theta}\right) \quad (6)$$

where K = constant, D = sewer diameter, θ = the central angle in radian and (d/D) = proportional water depth. Equation (4) is applicable for K values less than $(1/\pi) = 0.318$ Saatci [17].

2.2 Sewer Design Constraints

For a given network, the optimal sewer design is defined as a set of pipe diameters, slopes and excavation depths which satisfies all the constraints. Typical constraints of sewer networks design are:

1. Pipe flow velocity: each pipe flow velocity must be greater than minimum permissible velocity to prevent the deposit of solids in the sewers and less than maximum permissible velocity to prevent sewer scouring. The minimum permissible velocity of 0.6 m/s and maximum velocity of 3.0 m/s have been adopted in the present paper.
2. Flow depth ratio: wastewater depth ratio of the pipe should be less than 0.8.
3. Choosing pipe diameters from the commercial list.
4. Pipe cover depths: maintaining the minimum cover depth to avoid damage to the sewer line and adequate fall for house connections. The minimum cover depth of 0.9 m and maximum cover depth of 5.0 m have been adopted.
5. Progressive pipe diameters: The diameter of ith sewer should not be less than the diameter of immediately preceding sewer.

The optimal design of a sewer system for a given layout is to determine the sewer diameters, cover depths and sewer slopes of the network in order to minimize the total cost of the sewer system. The objective function can be stated as

$$\text{Minimize}(C) = \sum_{i=1}^{N} (\text{TCOST}_i + \text{PC}_i) \quad (7)$$

where I = 1,…, N (total number of sewers), TCOST_i (total cost) = (Cost of sewer$_i$ + Cost of manhole$_i$ + Cost of earth work$_i$) and PC_i = penalty cost (it is assigned if the design constraint is not satisfied).

3 Particle Swarm Optimization (PSO)

Kennedy and Eberhart [18] were first to introduce particle swarm optimization technique in 1995. In PSO techniques, every problem solution is a flock of birds and denoted to the particle. In this technique, birds develop personal and social behaviour and reciprocally manage their movement towards a destination [13, 19].

Each particle is affected by these components: (i) its own velocity, (ii) the best location or position it has attained so far called particle best position and (iii) the overall best position attained by all particles called global best position. Initially, the group of particles starts their movement in the first iteration randomly, and then they try to search the optimum solution. The procedure can be described mathematically, as below [14, 20–22].

The current location of the ith particle with D-dimensions at tth iteration is indicated as

$$X_i(t) = \{x_{i1}, x_{i2}, x_{i3}, \ldots, x_{id}\}^t \quad (8)$$

Earlier best position or location,

$$P_i(t) = \{p_{i1}, p_{i2}, p_{i3}, \ldots, p_{id}\}^t \quad (9)$$

and velocity

$$V_i(t) = \{v_{i1}, v_{i2}, v_{i3}, \ldots, v_{id}\}^t \quad (10)$$

Every particle's location in the search space is updated by

$$X_i(t) = X_i(t-1) + V_i(t) \quad (11)$$

where the new velocity

$$V_i(t) = \omega \cdot V_i(t-1) + c_1 \cdot R_1\{P_i(t) - X_i(t-1)\} + c_2 \cdot R_2\{P_g(t) - X_i(t-1)\} \quad (12)$$

where i = 1, 2,…, N (N denotes population size); t = 1, 2,…,T (T denotes a total number of iterations); ω = factor of inertia; R_1 and R_2 are the random values (which between 0 and 1); c_1 and c_2 are the learning or acceleration coefficients. $X_i(t)$ (location of every particle) is calculated by its earlier location $X_i(t-1)$ and its current velocity. $V_i(t)$ (particle's velocity) changes the location of the particles towards a better solution, at every iteration. $V_i(t-1)$ is the velocity from the earlier iteration, P_i is the best location of every particle and P_g is the best position or location ever found by any particle.

The inertia weights of each time interval (or iteration) $\omega(t)$ and acceleration coefficient (c_1 and c_2) are updated with these equations:

$$\omega(t) = \omega_{\max} - \frac{\omega_{\max} - \omega_{\min}}{T} \times t \tag{13}$$

$$c_1 = c_{1,\max} - \frac{c_{1,\max} - c_{1,\min}}{T} \times t \tag{14}$$

$$c_2 = c_{2,\max} - \frac{c_{2,\max} - c_{2,\min}}{T} \times t \tag{15}$$

where T = total number of iterations; $\omega_{\min}$ and $\omega_{\max}$ are the minimum and maximum inertia weights, and their values have been taken as 0 and 0.8, respectively, in the present problem; $c_{1,\max}$ and $c_{2\max}$ = maximum accelerations factors, their values have been taken as 2; $c_{1,\min}$ and $c_{2,\min}$ = minimum accelerations factors, their values have been taken as 0.5.

Particle velocities on every dimension are limited to minimum and maximum velocities.

$$V_{\min} \leq V_i \leq V_{\max} \tag{16}$$

The particle velocities are an important factor. $V_{\max}$ and $V_{\min}$ must be limited. Otherwise, the solution space may not be discovered precisely. $V_{\max}$ is generally considered about 10–20 % of the range of the variable on every dimension [19].

According to the above-mentioned Eqs. (11) and (12), a possible structure of the PSO algorithm is shown below.

1. Initialise a population of particles by randomly assigning initial velocity and location of every particle.
2. Calculate the optimal fitness function for every particle.
3. For every particle, compare the fitness value with the best particle (P_i) fitness value. If the current value is better than P_i, then update the position with the current position.
4. Calculate the best particle of the swarm with the best fitness value, if the best particle value is better than global best (P_g), then update the P_g and its fitness value with the location.
5. Determine new velocities for all the particles using Eq. (12).
6. Update new position of each particle using Eq. (11).
7. Repeat steps 2–6 until the stopping criterion is met.
8. Show the result given by the best particle.

Above-mentioned PSO algorithm deal with both discrete and continuous variables. PSO algorithm with discrete variables is requisite for the design of sewer networks.

4 Optimization of Sewer System

The live example (Sudarshanpura, Jaipur, India sewer network) is considered to check the above-proposed approach. The Sudarshanpura sewer network (Fig. 1) consists of 105 manholes and 104 pipes.

The following steps were used to optimize the component sizing of sewer system using PSO algorithm:

1. Start with first link ($i = 1$) of the first iteration.
2. Calculate constant value K,
   - If $K > 0.305$, then increase diameter.
3. If $K < 0.305$, then calculate sewer hydraulics.
4. Calculate invert levels of upstream and downstream nodes of a particular link.
5. Calculate cost of pipe, cost of manhole and cost of earthwork.
6. Calculate total cost of sewer network (TCOST).
7. Add the respective penalty cost (PC) in TCOST where constraints are violated.
8. Calculate feasible solution using PSO.
9. Check solutions obtained are feasible or not.
10. If feasible solution is not obtained increase iteration by 1 and go to step 1.
11. If feasible solution is obtained, then take output.
12. End.

The cost of pipe (RCC NP4 class), manhole and earth work were taken from Integrated Schedule of Rates, RUIDP [23].

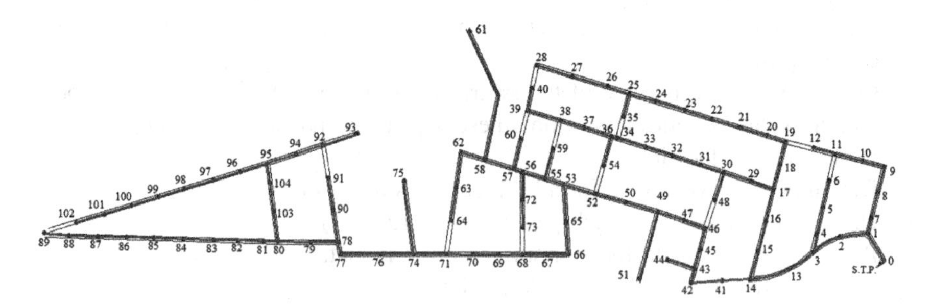

Fig. 1 Sudarshanpura sewer network

5 Results

The performance of the proposed PSO procedure for optimization of the sewer system is now tested against Sudarshanpura sewer network. The optimal results are obtained using 60 iterations and population size of 1000, respectively. The total cost of the sewer system using PSO approach was found to be Rs. 8.505×10^6 and 9.232×10^6 for the traditional design approach. Thus, there is 7.87 % reduction in total cost by applying PSO approach to the present problem. Table 1 shows the solution obtained by PSO approach.

Table 1 Results of the Sudarshanpura sewer network obtained by PSO

Pipe no.	Manhole no.		Pipe length (m)	Diameter (mm)	Slope (1 in.)	Design flow (m^3/s)	V_p (m/s)	d/D	Cover depths (m)	
	Up	Down							Up	Down
6	6	5	30	200	250	0.0004	0.254	0.093	1.120	1.145
13	12	11	20	200	60	0.0003	0.360	0.053	1.577	1.120
30	28	27	30	200	60	0.0004	0.413	0.065	1.350	1.120
38	35	25	12	200	250	0.0004	0.254	0.093	1.657	1.120
43	40	39	14	200	250	0.0004	0.254	0.093	1.120	1.406
48	44	43	30	200	60	0.0004	0.413	0.065	1.120	1.130
52	48	30	24	200	250	0.0005	0.279	0.108	1.874	1.120
55	51	49	72	200	250	0.0009	0.335	0.145	1.237	1.120
59	54	36	24	200	60	0.0005	0.454	0.076	1.120	1.125
62	59	55	30	200	250	0.0006	0.289	0.114	1.120	1.360
66	60	57	32	200	250	0.0006	0.295	0.118	1.120	1.568
69	61	58	143	300	250	0.0518	0.969	0.696	1.220	2.897
72	64	63	33	200	125	0.0007	0.386	0.104	1.120	1.124
79	73	72	30	200	250	0.0006	0.289	0.114	1.125	1.120
85	75	74	76	200	250	0.0010	0.340	0.149	1.120	1.229
99	89	88	30	300	250	0.0504	0.965	0.682	1.220	1.275
101	91	90	33	200	250	0.0008	0.326	0.139	1.120	1.142
103	93	92	36	200	250	0.0005	0.270	0.103	1.120	1.239
112	102	101	30	200	250	0.0008	0.316	0.132	1.120	1.135
5	5	4	30	200	250	0.0008	0.316	0.132	1.145	1.205
12	11	10	30	200	200	0.0008	0.350	0.130	1.130	1.120
29	27	26	30	200	250	0.0008	0.316	0.132	1.120	1.565
42	39	38	30	200	250	0.0006	0.287	0.113	1.406	1.221
71	63	62	33	200	250	0.0011	0.354	0.158	1.124	1.151
78	72	56	21	200	60	0.0011	0.586	0.113	1.120	1.335
98	88	87	30	300	250	0.0508	0.966	0.686	1.275	1.380
100	90	78	33	200	125	0.0013	0.470	0.143	1.161	1.120
104	92	94	30	200	250	0.0008	0.326	0.139	1.239	1.309
105	94	95	26	200	250	0.0012	0.362	0.164	1.309	1.408
111	101	100	30	200	250	0.0011	0.359	0.162	1.135	1.190
4	4	3	10	200	250	0.0009	0.332	0.143	1.205	1.180

(continued)

Table 1 (continued)

Pipe no.	Manhole no.		Pipe length (m)	Diameter (mm)	Slope (1 in.)	Design flow (m^3/s)	V_p (m/s)	d/D	Cover depths (m)	
	Up	Down							Up	Down
11	10	9	20	200	250	0.0013	0.370	0.170	1.120	1.150
28	26	25	27	200	250	0.0011	0.355	0.159	1.565	2.178
41	38	37	30	200	250	0.0011	0.358	0.161	1.380	1.120
70	62	58	24	200	60	0.0014	0.624	0.125	1.151	1.281
97	87	86	30	300	250	0.0511	0.967	0.690	1.380	1.320
110	100	99	30	200	70	0.0015	0.608	0.136	1.190	1.559
10	9	8	30	200	70	0.0015	0.607	0.135	1.391	1.120
27	25	24	30	200	80	0.0019	0.619	0.156	2.178	2.413
40	37	36	16	200	250	0.0013	0.376	0.175	1.151	1.120
68	58	57	33	300	250	0.0536	0.974	0.713	2.897	2.944
96	86	85	30	300	250	0.0515	0.968	0.693	1.465	1.220
109	99	98	30	200	80	0.0019	0.623	0.157	1.559	1.869
9	8	7	30	200	80	0.0019	0.622	0.157	1.830	1.120
26	24	23	30	200	100	0.0022	0.606	0.181	2.413	2.523
39	36	34	7	200	80	0.0019	0.626	0.159	1.125	1.158
65	57	56	8	300	250	0.0545	0.977	0.723	2.944	2.881
95	85	84	30	300	250	0.0519	0.969	0.697	1.220	1.270
108	98	97	30	200	100	0.0023	0.609	0.182	1.869	2.029
8	7	1	9	200	80	0.0020	0.633	0.161	1.273	1.120
25	23	22	30	200	100	0.0026	0.635	0.196	2.523	2.748
36	34	33	18	200	100	0.0023	0.615	0.185	1.158	1.188
64	56	55	25	300	250	0.0560	0.980	0.738	2.881	2.746
94	84	83	30	300	250	0.0523	0.971	0.700	1.275	1.220
107	97	96	30	200	100	0.0027	0.638	0.197	2.029	2.214
24	22	21	30	200	125	0.0030	0.612	0.222	2.748	2.863
35	33	32	30	200	100	0.0027	0.643	0.200	1.188	1.253
61	55	53	20	300	250	0.0570	0.982	0.749	2.746	2.896
93	83	82	30	300	250	0.0527	0.972	0.704	1.220	1.290
106	96	95	30	200	125	0.0030	0.614	0.223	2.214	2.419
23	21	20	30	200	125	0.0034	0.634	0.236	2.863	2.988
34	32	31	30	200	125	0.0031	0.619	0.226	1.253	1.128
92	82	81	30	300	250	0.0530	0.973	0.708	1.290	1.295
116	95	104	27	200	150	0.0046	0.647	0.287	2.419	2.529
115	104	103	27	200	150	0.0049	0.661	0.298	2.529	2.644
22	20	19	18	200	150	0.0036	0.606	0.255	2.988	3.043
33	31	30	30	200	125	0.0035	0.640	0.240	1.230	1.120
91	81	80	10	300	250	0.0532	0.973	0.709	1.295	1.270
114	103	80	27	200	200	0.0052	0.607	0.332	2.644	2.724
21	19	18	12	200	150	0.0040	0.624	0.269	3.043	2.928
32	30	29	22	200	150	0.0043	0.636	0.278	1.223	1.120
90	80	79	31	300	250	0.0588	0.985	0.771	2.724	2.758
20	18	17	30	200	150	0.0043	0.638	0.280	2.928	3.068

(continued)

Table 1 (continued)

Pipe no.	Manhole no.		Pipe length (m)	Diameter (mm)	Slope (1 in.)	Design flow (m^3/s)	V_p (m/s)	d/D	Cover depths (m)	
	Up	Down							Up	Down
31	29	17	30	200	150	0.0047	0.652	0.291	1.120	1.160
89	79	78	31	300	250	0.0592	0.986	0.776	2.758	2.587
19	17	16	30	200	250	0.0096	0.653	0.485	3.068	3.053
88	78	77	13	300	200	0.0606	1.091	0.719	2.587	2.717
18	16	15	30	200	250	0.0099	0.659	0.495	3.053	3.138
87	77	76	38	300	200	0.0611	1.092	0.724	2.717	2.842
17	15	14	30	200	250	0.0103	0.665	0.505	3.138	3.233
86	76	74	38	350	250	0.0616	1.028	0.597	2.842	2.949
84	74	71	34	350	250	0.0630	1.033	0.605	2.949	2.920
83	71	70	26	350	250	0.0635	1.035	0.608	2.920	2.939
82	70	69	26	350	250	0.0638	1.036	0.610	2.939	2.838
81	69	68	26	350	250	0.0642	1.037	0.612	2.838	2.722
77	68	67	22	350	250	0.0644	1.038	0.614	2.722	2.665
76	67	66	22	350	250	0.0647	1.039	0.616	2.665	2.713
75	66	65	30	350	250	0.0651	1.040	0.618	2.713	2.683
74	65	53	30	350	250	0.0655	1.041	0.620	2.683	2.663
60	53	52	30	400	250	0.1229	1.190	0.750	2.896	2.731
57	52	50	30	450	450	0.1234	0.957	0.740	2.731	2.573
56	50	49	30	450	450	0.1238	0.958	0.742	2.573	2.564
54	49	47	26	450	450	0.1251	0.959	0.748	2.564	2.207
53	47	46	26	450	450	0.1254	0.959	0.750	2.207	1.895
50	46	45	20	450	450	0.1258	0.960	0.752	1.895	2.214
49	45	43	20	450	450	0.1261	0.960	0.753	2.214	2.504
47	43	42	11	450	450	0.1266	0.960	0.756	2.504	2.388
46	42	41	30	450	450	0.1270	0.960	0.758	2.388	1.370
45	41	14	30	450	60	0.1274	2.130	0.416	1.625	1.370
16	14	13	30	450	400	0.1381	1.021	0.777	3.233	2.938
15	13	3	30	450	350	0.1384	1.084	0.733	2.938	2.719
3	3	2	23	500	400	0.1396	1.052	0.636	2.719	2.111
2	2	1	23	500	60	0.1399	2.177	0.377	2.407	1.420
1	1	0	30	500	80	0.1423	1.966	0.410	1.420	1.455

6 Conclusion

A particle swarm optimization with new modification was applied in this paper to the optimal solution of sewer system design problems. Using the PSO approach, the total cost of the sewer system was reduced by 7.87 % compared to the traditional design approach. The results indicated that the proposed approach is very promising and reliable, that must be taken as the key alternative to solve the problem of optimal design of sewer system.

References

1. Dajani, J.S., Gemmell, R.S., Morlok, E.K.: Optimal design of urban wastewater collection networks. J. Sanitary Eng. Div. **98**, 853–867 (1972)
2. Elimam, A.A., Charalambous, C., Ghobrial, F.H.: Optimum design of large sewer networks. J. Environ. Eng. **115**(6), 1171–1190 (1989)
3. Price, R.K.: Design of storm water sewers for minimum construction cost. In: 1st International Conference on Urban Strom Drainage. Southampton, UK, pp. 636–647 (1978)
4. Swamee, P.K.: Design of sewer line. J. Environ. Eng. **127**, 776–781 (2001)
5. Walsh, S., Brown, L.C.: Least cost method for sewer design. J. Environ. Eng. Div. **99**(3), 333–345 (1973)
6. Walters, G.A., Templeman, A.B.: Non-optimal dynamic programming algorithms in the design of minimum cost drainage systems. Eng. Optim. **4**, 139–148 (1979)
7. Li, G., Matthew, R.G.S.: New approach for optimization of urban drainage systems. J. Environ. Eng. **116**, 927–944 (1990)
8. Walters, G.A., Lohbeck, T.: Optimal layout of tree networks using genetic algorithms. Eng. Optim. **22**, 27–48 (1993)
9. Afshar, M.H.: Application of a genetic algorithm to storm sewer network optimization. Scientia Iranica **13**, 234–244 (2006)
10. Afshar, M.H.: Partially constrained ant colony optimization algorithm for the solution of constrained optimization problems: application to storm water network design. Adv. Water Resour. **30**, 954–965 (2007)
11. Afshar, M.H.: A parameter free continuous ant colony optimization algorithm for the optimal design of storm sewer networks: constrained and unconstrained approach. Adv. Eng. Softw. **41**, 188–195 (2010)
12. Guo, Y., Walters, G.A., Khu, S.T., Keedwell, E.: A novel cellular automata based approach to storm sewer design. Eng. Optim. **39**, 345–364 (2007)
13. Izquierdo, J., Montalvo, I., Pérez, R., Fuertes, V.S.: Design optimization of wastewater collection networks by PSO. Comput. Math Appl. **56**, 777–784 (2008)
14. Ostadrahimi, L., Mariño, M.A., Afshar, A.: Multi-reservoir operation rules: multi-swarm pso-based optimization approach. Water Resour. Manage **26**, 407–427 (2012)
15. Haghighi, A., Bakhshipour, A.E.: Optimization of sewer networks using an adaptive genetic algorithm. Water Resour. Manage **26**, 3441–3456 (2012)
16. Moeini, R., Afshar, M.H.: Sewer network design optimization problem using ant colony optimization algorithm and tree growing algorithm. In: EVOLVE-A bridge between probability, set oriented numerics, and evolutionary computation IV, pp. 91–105. Springer International Publishing (2013)
17. Saatci, A.: Velocity and depth of flow calculations in partially filled pipes. J. Environ. Eng. **116**, 1202–1208 (1990)
18. Kennedy, J., Eberhart, R.: Particle swarm optimization. In: IEEE International Conference on Neural Networks, IEEE Service Center, Piscataway, NJ, IV, Perth, Australia, pp. 1942–1948 (1995)
19. Shi, Y., Russell, E.: A modified particle swarm optimizer. In: IEEE International Conference on Evolutionary Computation, Anchorage, AK pp. 69–73 (1998)
20. Mu, A., Cao, D., Wang, X.: A modified particle swarm optimization algorithm. Nat. Sci. **1**, 151–155 (2009)
21. Al-kazemi, B., Mohan, C.K.: Multi-phase discrete particle swarm optimization. In: Proceeding of the Fourth International Workshop on Frontiers in Evolutionary Algorithms (2002)
22. Voss, M.S.: Social programming using functional swarm optimization. In: IEEE Swarm Intelligence Symposium, Indiana, USA, pp. 103–109 (2003)
23. Integrated Schedule of Rates: Rajasthan urban infrastructure development project (RUIDP), Government of Rajasthan (2013)

Optimal Tuning of PID Controller for Centrifugal Temperature Control System in Sugar Industry Using Genetic Algorithm

Sanjay Kumar Singh, D. Boolchandani, S.G. Modani and Nitish Katal

Abstract This paper presents optimal tuning of the PID controllers for regulating the temperature in a heat exchanger of centrifugal machines in sugar industry using genetic algorithm. For filtering out sugar from the molasses centrifugal machines are operated at certain temperature and any alterations from the set point will drastically affect the process safety and product quality. The PID controller maintains the temperature of the outgoing fluid at a desired level and that too in a short duration of time, and must be able to adapt to the external disturbances and accept the new set points dynamically. Initially, an oscillatory behavior has been obtained for the PID controller tuned using RTR, followed by optimization using genetic algorithms. The GA optimized PID controller shows better response irrespective of the conventional RTR tuned PID in terms of the performance indices.

Keywords Controller tuning · Industrial systems · Robust time response · Genetic algorithm

1 Introduction

In process industry at control layer, PID controllers are widely used and approximately 90 % of all the controllers used in industry today are PID [1, 2]. Proper PID tuning is an imperial factor as it affects the process safety; so optimal tuning of the PID controllers can be formulated as an optimization problem [3]. In this paper, the considered model maintains the temperature of the outlet fluid at a desired

S.K. Singh (✉)
Department of Electronics & Communication Engineering, Amity University, Jaipur, Rajasthan, India
e-mail: sksingh.mnit@gmail.com

D. Boolchandani · S.G. Modani
Department ECE, Malaviya National Institute of Technology, Jaipur, Rajasthan, India

N. Katal
Department of Electrical Engineering, PEC University of Technology, Chandigarh, India

M. Pant et al. (eds.), *Proceedings of Fifth International Conference on Soft Computing for Problem Solving*, Advances in Intelligent Systems and Computing 437, DOI 10.1007/978-981-10-0451-3_18

temperature and influences the adaptability of the plant to the external noise and changing set points. Heat exchangers find a vast application in process control like petro and chemical processes, sugar industry, paper industry, thermal and nuclear power generation plants, etc.

For industrial system designing of the temperature control loop, PID controller has been used. Initially the tuning of the controller has been carried out using the conventional robust time response (ZN) tuning rules as Ziegler–Nichols method produces a unstable response, and it is observed that a nonoptimal oscillatory response is obtained. So to ensure process safety and product quality, the optimal PID controllers must be obtained [4, 5]. Genetic algorithm has been used for tuning the PID controller with error performance indexes of integral square error (ISE) and integral time square error (ITSE). From the results, GA-based optimized PID controller shows better performance in terms of improved time-domain performance indices.

First section provides the introduction on the role of control in process industry, particularly sugar industry. In Sect. 2, the mathematical model of the heat exchanger used in sugar industry has been obtained using system identification followed by the optimal tuning of PID controllers using RTR and GA in Sect. 3. Section 4 discusses the results obtained by optimal tuning and Sect. 5 concludes the paper.

2 Mathematical Modeling of Centrifugal Heat Exchanger in Sugar Industries

In industrial systems, a heat exchanger tries to maintain the temperature equilibrium by constantly transferring the heat from one body to other. In this paper, for industrial systems a centrifugal machine used in sugar industries has been considered, and the temperature is adjusted at a constant level by varying the amount of steam supplied. Liquid is taken as input from the above inlet as shown in Fig. 1. The flow of steam is regulated using a control valve. The second-order-plus-dead-time model for heat exchanger can be estimated form the open-loop response of the system.

Two values of temperature have been estimated, when the response attains 28.3 and 63.2 % of its final value [6]; the t_1 was found to be 19.90 °C and t_2 to be 34.80 °C. Value of time constant (τ) has been obtained for the calculation of transfer function. The value of dead time (θ) is obtained as $\theta = t_2 - \tau$ and the transfer function has been estimated as [7]

$$G(s) = \frac{e^{-\theta \cdot s}}{\tau \cdot s + 1} = \frac{e^{-15.7 \cdot s}}{22.35 \cdot s + 1} \quad (1)$$

But it does not provide the best fit as can be seen in Fig. 2, so we have used MATLAB-based fitting to estimate the second-order-plus-dead-time transfer

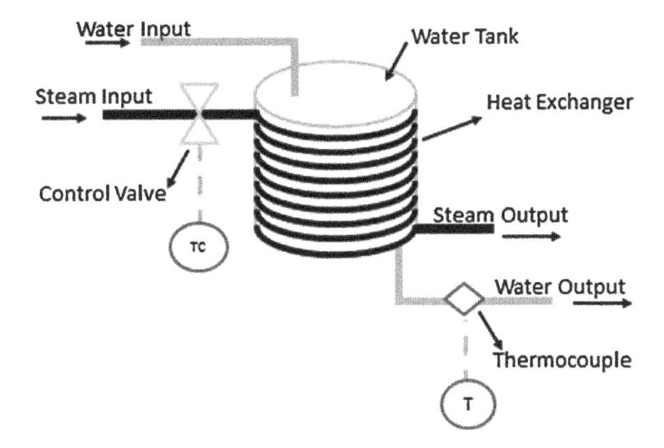

Fig. 1 Schematic representation of heat exchanger system

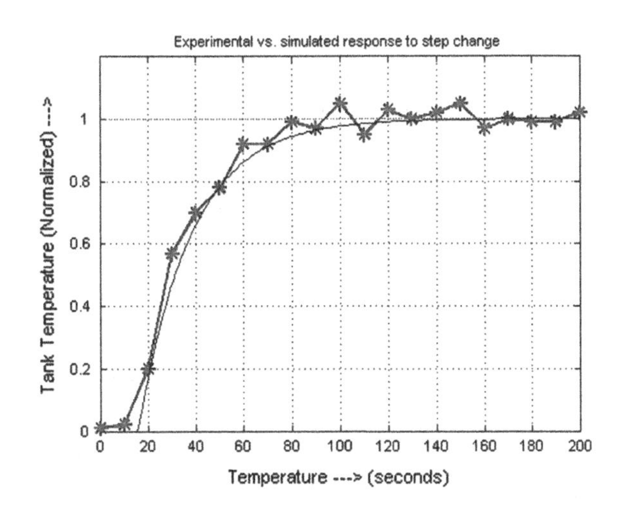

Fig. 2 Curve fitting graphs for experimental and identified system

function which provides much more desired fitting. The estimated transfer function is shown below and Fig. 3 provides the fitted transfer function.

$$G_{\text{new}}(s) = \frac{0.9852}{175.7 \cdot s^2 + 26.51 \cdot s + 1} \cdot e^{-9.7 \cdot s} \tag{2}$$

3 Optimization-Based Design of PID Controllers

PID—Proportional integral and derivative—controllers are the prime controllers used in industry since their inception. Their simplicity and robustness complements their application. The PID controller is shown in Fig. 4. The general equation for a PID controller is [8]

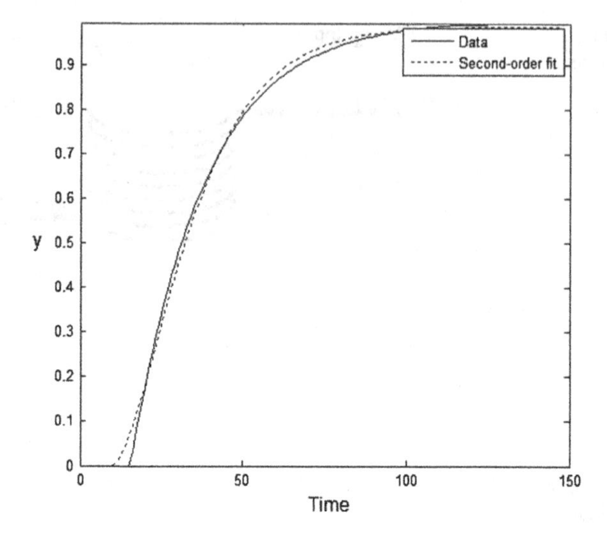

Fig. 3 Response of the second-order-plus-dead-time estimated transfer function versus experimental data

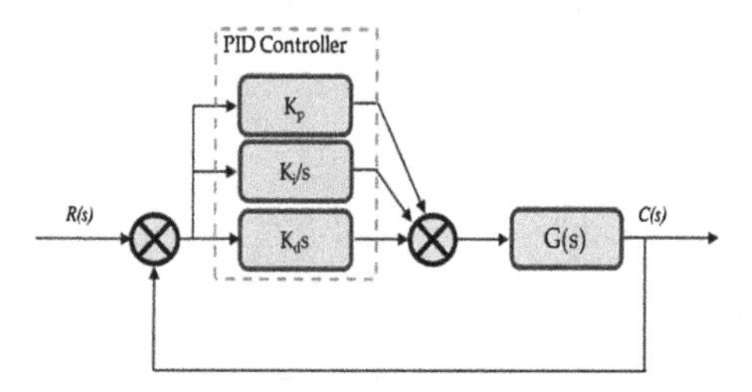

Fig. 4 Block diagram of PID controller with unity feedback

$$C(t) = K_p \cdot e(t) + K_i \int e(t) \cdot \mathrm{d}t + K_d \frac{\mathrm{d}e(t)}{\mathrm{d}t} \tag{3}$$

3.1 *Tuning of PID Controller Using Robust Time Response*

Ziegler–Nichols tuned PID gives a transient response. So the robust time response has been considered here for the initial estimation of the PID gains. In robust time response algorithm, variance of the regulatory signal is minimized under the presence of model uncertainty and the PID tuning is done using least means square-based rules. PID gains obtained are listed in Table 1, and Fig. 5 shows the closed-loop response of the control system.

Table 1 PID parameters obtained by robust time response

PID parameter	Value
K_p	1.9604
K_i	0.10166
K_d	3.471

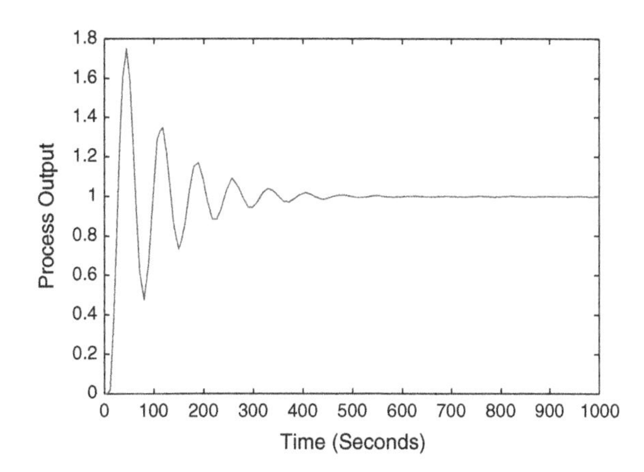

Fig. 5 Closed-loop response of the closed-loop system with RTR–PID controllers

3.2 PID Tuning Using GA

An oscillatory response has been obtained in case of tuning using robust time response, so the parameters are not well suited for direct implementation in the plant. So their organized optimization becomes important taking into consideration the safety of the process. Genetic algorithms are one of the robust optimization algorithms present and their wider adaptability gives a vanguard advantage [9]. Optimization of the PID gains lays emphasis on obtaining the best possible parameters for $[K_p, K_i, K_d]$ by minimizing the objective function stated as integral square error and integral time square error and given as

$$\text{ISE} = \int_0^T e^2(t)\mathrm{d}t \quad \text{and} \quad \text{ITSE} = \int_0^T t \cdot e^2(t)\mathrm{d}t \tag{4}$$

The optimization by genetic algorithms involves the following steps:

- Random generation of initial population for three parameters K_p, K_i, and K_d.
- Evaluation of fitness integral to minimize the performance index of ISE and ITSE; and the selection of the fittest individuals that minimize the objective.
- Reproduction among the members of the population.
- Crossover followed by mutation and the selection of the best individuals, i.e., Survival of the fittest
- Looping (ii) till the predefined stopping criteria are met.

Table 2 PID Parameters obtained by GA

PID parameters	GA (ISE)	GA (ITSE)
K_p	1.7	1.765
K_i	0.046	0.044
K_d	10	10

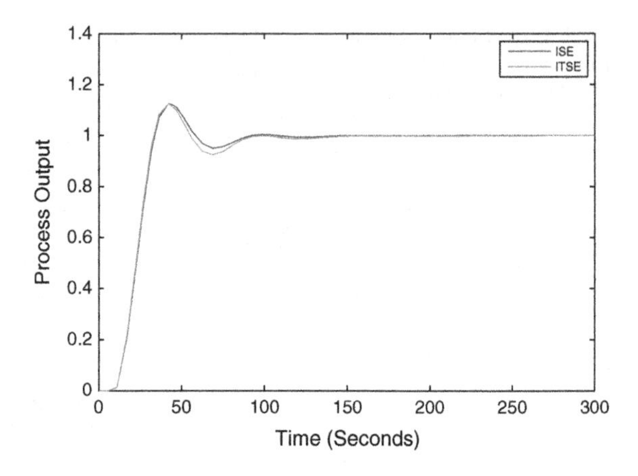

Fig. 6 Closed-loop response of GA–PID Controllers

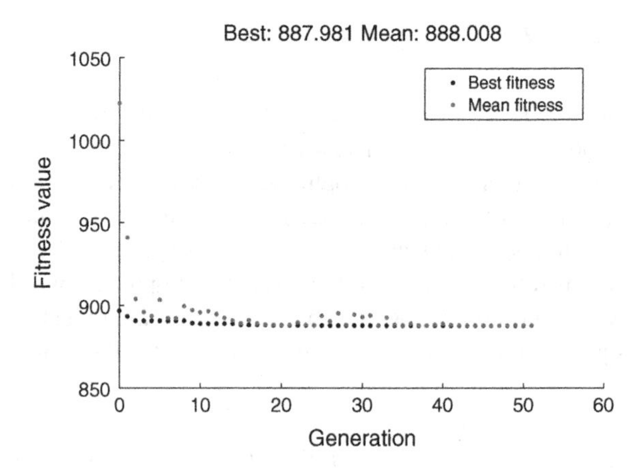

Fig. 7 Best and mean fitness values obtained by optimization using GA (ISE)

The optimization has been carried out with population size of 100 and scattered crossover. GA optimized PID gains have been shown in Table 2, Fig. 6 shows the GA–PID step response of the system, and Figs. 7 and 8 show the plot for best and the mean fitness values obtained.

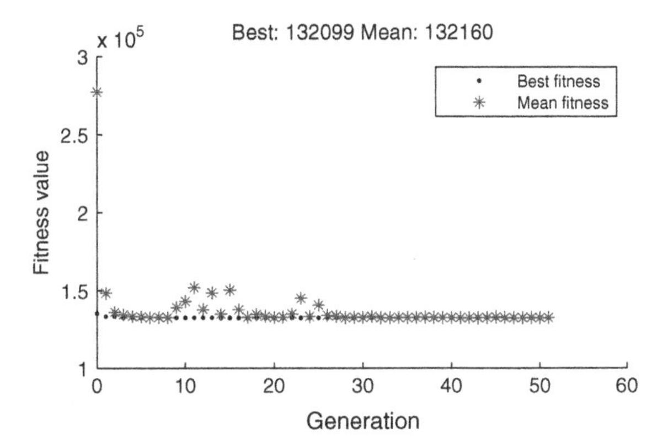

Fig. 8 Best and mean fitness values obtained by optimization using GA (ITSE)

4 Results and Discussions

In this paper, a dynamic model of an industrial system for temperature control loop in centrifugal machines used in sugar industries has been considered and implemented in MATLAB and optimization has been carried out using genetic algorithm. Initial estimates about the bounds of parameters on the search space have been obtained from the RTR–PID gains. The error-based performance indexes of ISE and ITSE have been used for tuning the PID and hence facilitate the determination of the optimum parameters for the industrial systems. It is clearly evident in Fig. 9 that the GA (ISE) solutions present less oscillatory response and reduced rise and settling times in contrast to the robust time response. The results have been presented in Table 3.

Concluding, GA offers superior results in terms of system performance and controller output for the tuning of PID controllers, when the values are compared in Table 3 and Figs. 9 and 10.

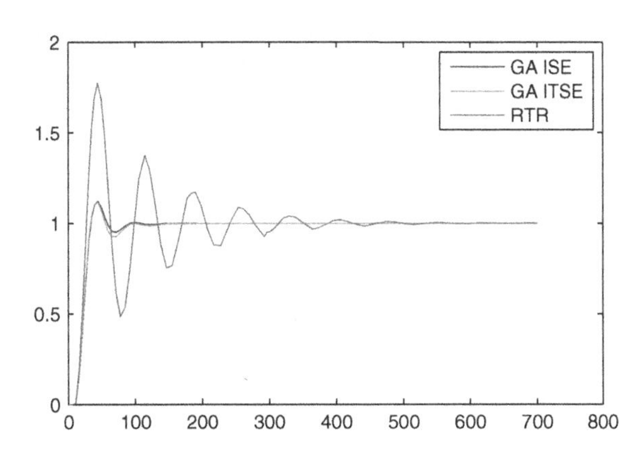

Fig. 9 Compared closed-loop response of the system with RTR, GA (ISE and ITSE), PID controllers

Table 3 Performance Comparison

Method of design	Overshoot %	Rise time	Settling time
Robust time response	71.52	12.36	435.67
GA (ISE)	4.94	15.08	95.98
GA (ITSE)	5.94	14.61	125.61

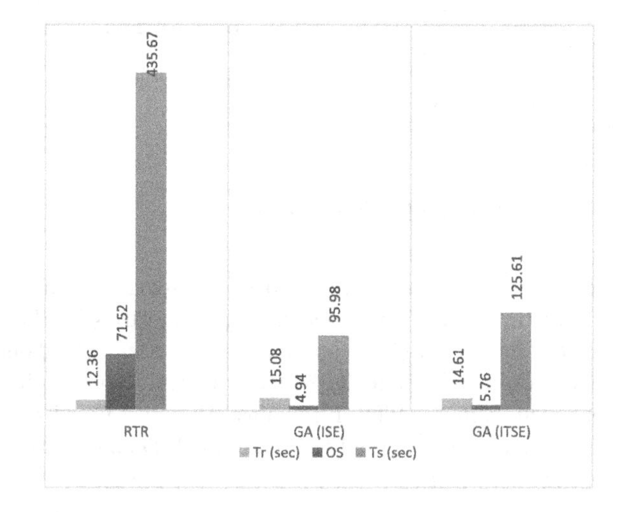

Fig. 10 Compared performance indices of RTR, GA (ISE), and GA (ITSE) tuned PID controllers

5 Conclusion

This paper presents the use of genetic algorithm for optimizing the PID controller, and the obtained controllers have been used for maintaining the temperature control used in centrifugal machines in sugar industry. The optimal tuning of the controllers is crucial as they govern the quality of products and process safety. The result is that, when compared with the other tuning methodologies as presented in this paper, the GA has proved superior in achieving the steady-state response and performance indices.

References

1. Åström, K.J., Hägglund, T.: The future of PID control. Control Eng. Practice, 1163–1175 (2001)
2. Zhao, S.Z., Iruthayarajan, M.W., Baskar, S., Suganthan, P.N.: Multi-objective robust PID controller tuning using two lbests multiobjective particle swarm optimization. Inf. Sci. **181**(16), 3323–3335 (2011)

3. Krohling, R.A.., Rey, J.P.: Design of optimal disturbance rejection PID controllers using genetic algorithms. Evolutionary Computation, IEEE Transactions on **5**(1), 78–82 (2001)
4. Stephanopoulos, G., Reklaitis, G.V.: Process systems engineering: From Solvay to modern bio-and nanotechnology.: A history of development, successes and prospects for the future. Chem. Eng. Sci. **66**(19), 4272–4306 (2011)
5. Nikacevic, N.M., Huesman, A.E., Van den Hof, P.M., Stankiewicz, A.I.: Opportunities and challenges for process control in process intensification. In: Chemical Engineering and Processing: Process Intensification (2011)
6. Optimization Toolbox-Matlab
7. Singh, S.K., Boolchandani, D., Modani, S.G., Katal, N.: Multi-objective optimization of PID controller for temperature control in centrifugal machines using genetic algorithm. Res. J. Appl. Sci., Eng. Technol. **7**(9), 1794–1802 (2014)
8. Nise, N.S.: Control System Engineering, 4th edn. (2003)
9. Larbes, C., Aït Cheikh, S.M., Obeidi, T., Zerguerras, A.: Genetic algorithms optimized fuzzy logic control for the maximum power point tracking in photovoltaic system. Renewable Energy, Elsevier Ltd. **34**(10), 2093–2100 (2009)

Fault Diagnosis of ZSI-Fed 3-Phase Induction Motor Drive System Using MATLAB

Vivek Sharma and Pulkit Garg

Abstract This paper presents the simulation analysis of Z-source inverter-fed induction motor drive system for three-phase operation. Incorporating Z-source inverter enables both buck as well as boost operation. A study of the proposed system is performed under various fault conditions viz. fault at one of the gate terminal of MOSFET, MOSFET blown off fault, and line-to-ground fault which primarily occurs in any drive system. The output voltage and current waveforms with relevant FFT spectrum are presented.

Keywords Boost · Buck · FFT spectrum · Total harmonic distortion (THD) · FFT

1 Introduction

In every application of electrical drives, voltage-fed and current-fed inverters are implemented. But conventional inverters show inherent drawbacks of only one feature of either step up or step-down. This property is due to the gating problem of the respective phases of inverters. This feature has questioned the reliability of conventional inverters and hence an alternative finds its application in the emerging field of drive applications. The disadvantage of EMI or radio interference is also to be eliminated for a pure sine wave output of inverters. Hence, Z-source inverters find their applications for this purpose-source inverters employs LC network in a cross fashion which in turn provides buck or boost output at a desired frequency. The schematic diagram of the proposed system is shown in Fig. 1. It consists of an input DC source which is fed to an LC network and ripple-free DC is then given to

Vivek Sharma (✉) · Pulkit Garg
Department of Electrical & Electronics Engineering,
Graphic Era University, Dehradun, India
e-mail: mail.vivek21@gmail.com

Pulkit Garg
e-mail: pulkitg22@gmail.com

M. Pant et al. (eds.), *Proceedings of Fifth International Conference on Soft Computing for Problem Solving*, Advances in Intelligent Systems and Computing 437, DOI 10.1007/978-981-10-0451-3_19

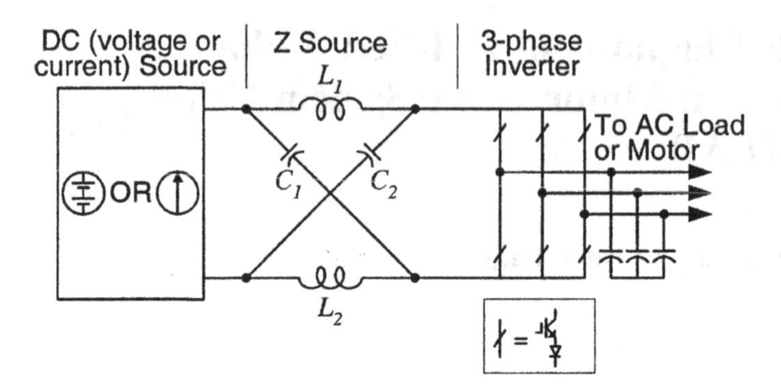

Fig. 1 Schematic diagram of proposed system

three-phase inverter. The inverter provides sinusoidal waveform which is given to the three-phase induction motor load [1].

The proposed scheme inhibits MOSFETs as switching gate circuits for inverter module. The study of fault will be incorporated on the given proposed system as conventional inverters provide AC voltage lower than input voltage. For higher voltage with limited DC voltage applications, implementation of DC-DC boost converter is essential [2]. This in turn increases circuit cost and lowers efficiency. To obtain ripple-free output, an additional LC filter is required in the voltage-source inverter. This increases the cost and complexity of the circuit. In current source inverters, distortions are present since for safe current operation overlap time is needed. Due to interference issues, the reliability of inverters is questionable.

The Z-source inverter arrangement consists an LC network with split inductor and capacitors connected in *X* shape to provide impedance source coupling. The Z-source concept can be applied to all DC-to-AC, AC-to-DC, AC-to-AC, and DC-to-DC power conversions [3]. The unique feature of Z-source inverter is that output voltage can be obtained in any value between zero and infinity regardless of input dc voltage-source inverters are proposed for single-phase DC-to-AC inversion with buck-boost capability. The ratio of peak inverter input voltage to the input dc voltage for ZSI is given by:

$$\frac{V_{\text{pn}}}{V_g} = \frac{1}{1 - 2\alpha} \tag{1}$$

where α denotes shoot through duty ratio.

Due to the above-mentioned novel features shown by ZSI, implementation of ZSI to various applications was done worldwide. This paper deals with the most common faults occurring in the proposed system, i.e., fault at one of the MOSFET gate terminals, MOSFET blown off fault, and line-to ground fault. These faults are most prevailing in every single system in drive applications. Hence, study of these faults becomes prominent [4].

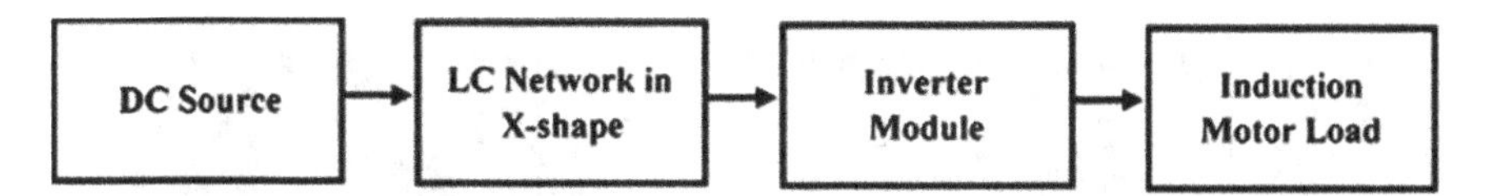

Fig. 2 Block diagram of the proposed system

2 System Design

The block diagram of the proposed system is shown in Fig. 2. The proposed system has DC supply from a source such as battery which is fed to the ZSI-fed induction motor drive system.

The induction motor is connected as load for the 3-phase applications of the proposed system. The fault current waveforms and the FFT spectrum are studied at the load side.

3 Simulation Circuit of ZSI-Fed 3-Phase Induction Motor Drive

In this case single-phase ZSI is simulated with a uniform 3-phase induction motor load. Figure 3 shows the simulation circuit of ZSI operation fed with 3-phase induction motor load. The circuit is driven with an input voltage of 120 V. The output voltage waveform shape is in turn controlled by MOSFETs used in the circuit.

Figure 4 shows the output voltage waveforms of single-phase ZSI fed with 3-phase induction motor for boost operation. Input DC voltage is 120 V.

The THD measured without filter configuration at 50 Hz is 21.76 (Fig. 5).

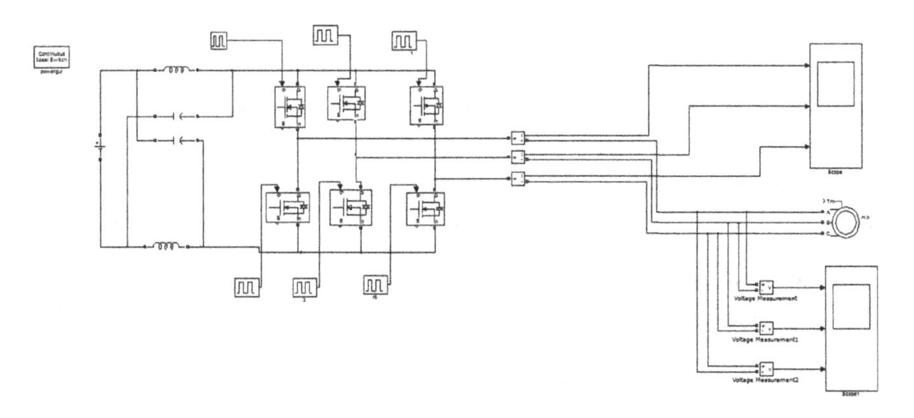

Fig. 3 Simulation circuit of ZSI-fed 3-phase induction motor drive

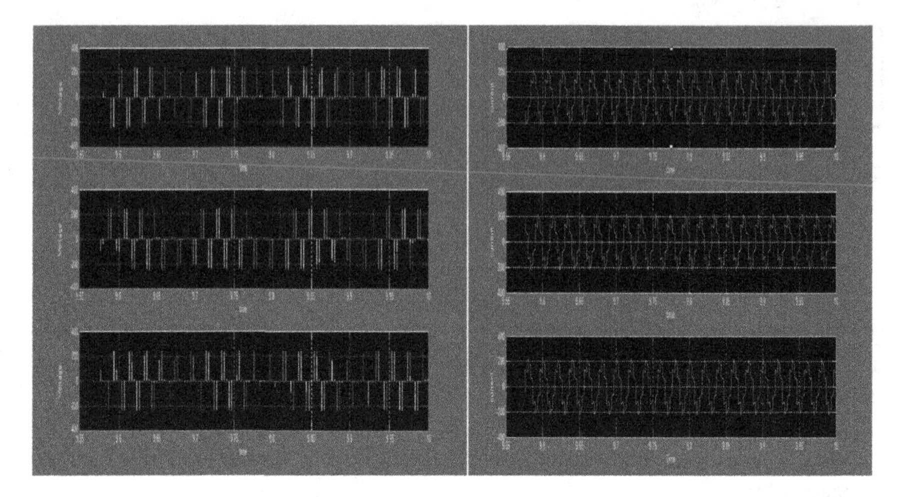

Fig. 4 Output waveform of voltage and current of ZSI-fed 3-phase induction motor drive

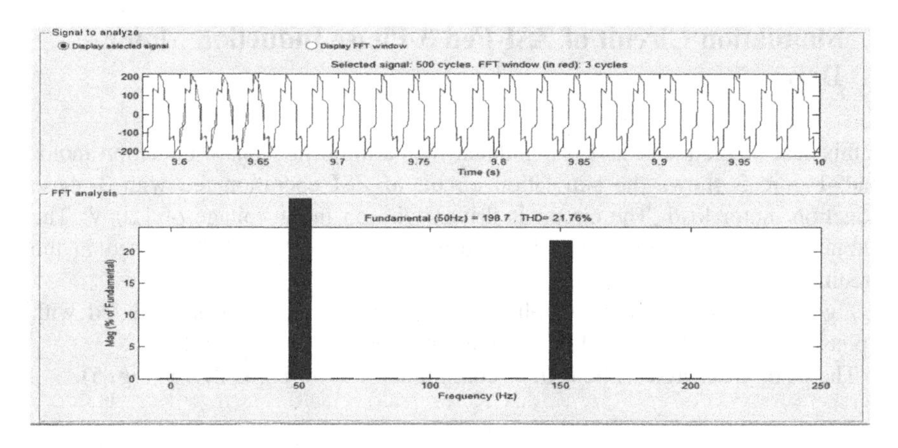

Fig. 5 FFT spectrum of ZSI-fed 3-phase induction motor drive

3.1 Fault at One of the MOSFET Gate Terminals

In this case, the simulation equivalent is achieved by making one of the MOSFET gate terminals grounded. The upper MOSFET of phase A is grounded for the purpose of simulation in the specified condition. Figure 6 shows the simulation circuit for the specified faulty condition. Under the condition of one MOSFET grounded, the output voltage waveform obtained is distorted with only the negative half of the

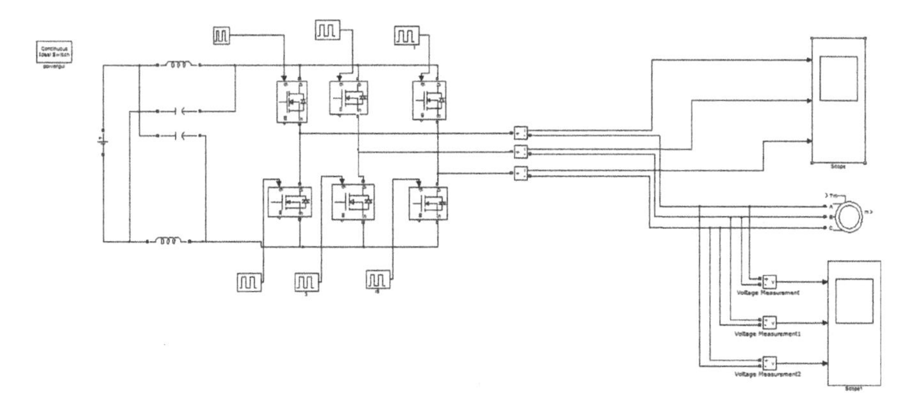

Fig. 6 Simulation circuit of fault at one of the MOSFET gate terminals

Fig. 7 Output voltage and current waveform for fault at one of the gate terminals of MOSFET

waveform obtained. Figure 7 shows the output voltage and the current waveform of ZSI operation under the specified faulty condition.

The THD measured under this faulty configuration at 50 Hz is 72.65 % (Fig. 8).

3.2 *Blowing off One MOSFET in the Simulation Circuit*

This condition is simulated by replacing upper MOSFET by a high resistance. Replacement of MOSFET with high value resistance is equivalent to the open

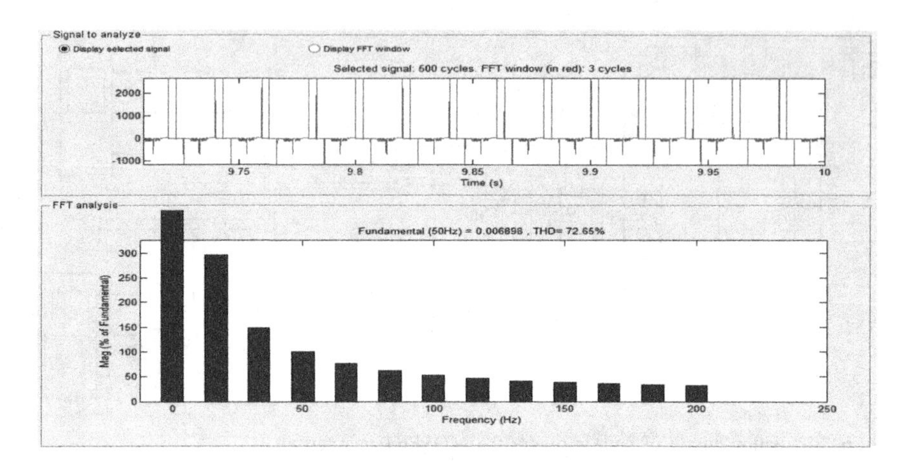

Fig. 8 FFT spectrum of fault at one of the gate terminals of MOSFET

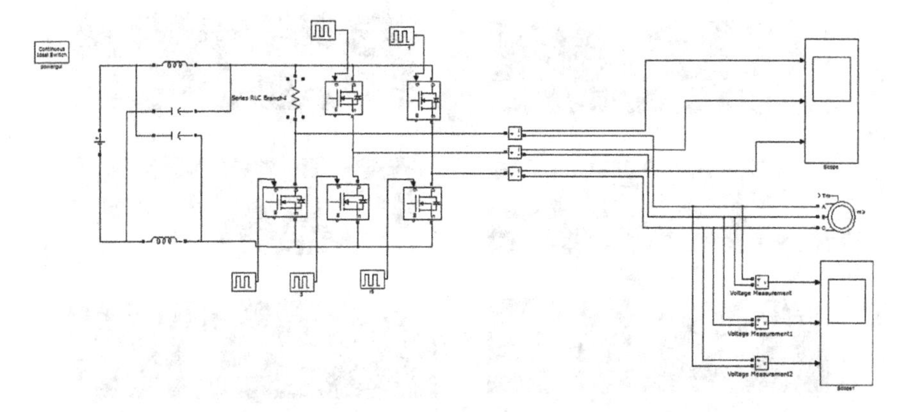

Fig. 9 Simulation circuit of fault due to blowing off one MOSFET

circuiting of one of the MOSFETs. Figure 9 shows the simulation circuit of ZSI boost operation with blowing off one MOSFET fault.

Blowing off one MOSFET inhibits open circuit condition for the whole circuit arrangement. Under open circuit condition, the output voltage waveform obtained is distorted with only the negative half of the waveform obtained. Figure 10 shows output voltage waveform of ZSI boost operation under specified faulty condition.

The THD measured under this faulty configuration at fundamental frequency is 67.74 % (Fig. 11).

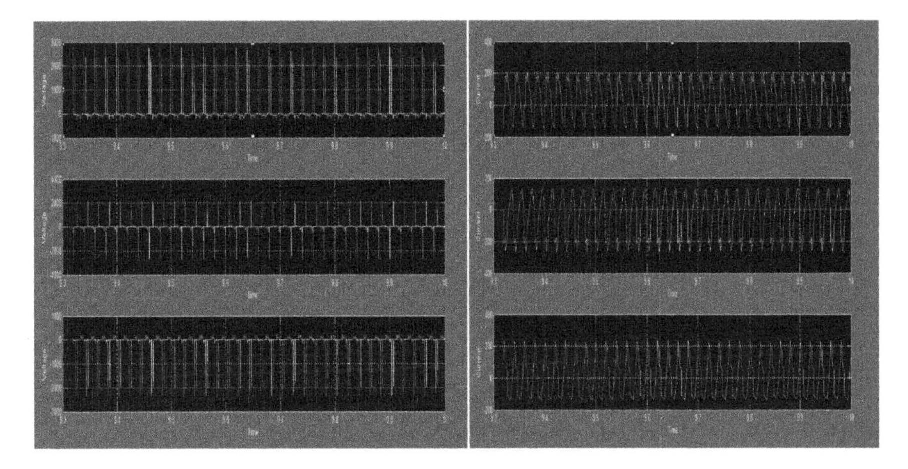

Fig. 10 Output voltage and current waveform with blowing off one MOSFET fault

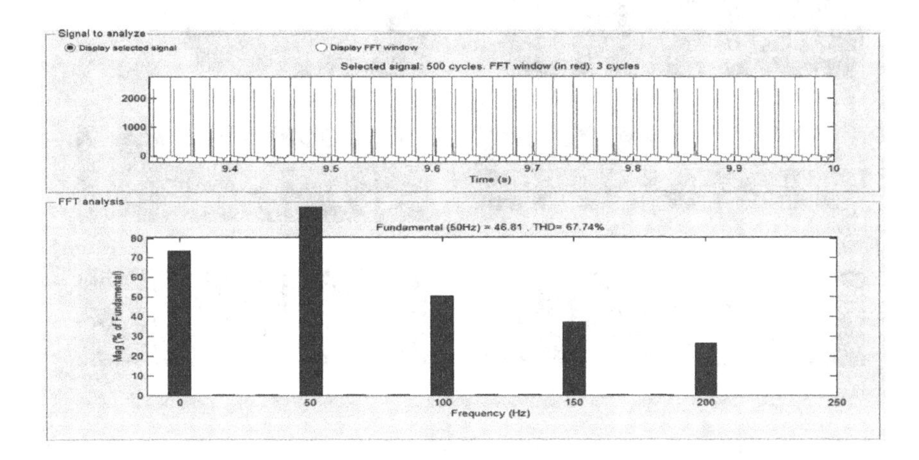

Fig. 11 FFT spectrum of fault with blowing of one MOSFET

3.3 L-G Fault

This condition is simulated by grounding one of the phases of the 3-phase induction motor. Grounding one of the phases of the 3-phase induction motor is equivalent to the L-G Fault Analysis. Figure 12 shows the simulation circuit of ZSI operation-fed 3-phase induction motor with one of the load terminals to be ground.

Grounding of one of the phases of 3-phase induction motor inhibits L-G fault condition for the whole circuit arrangement. Figure 13 shows the output voltage and the current waveform of ZSI operation under specified faulty condition.

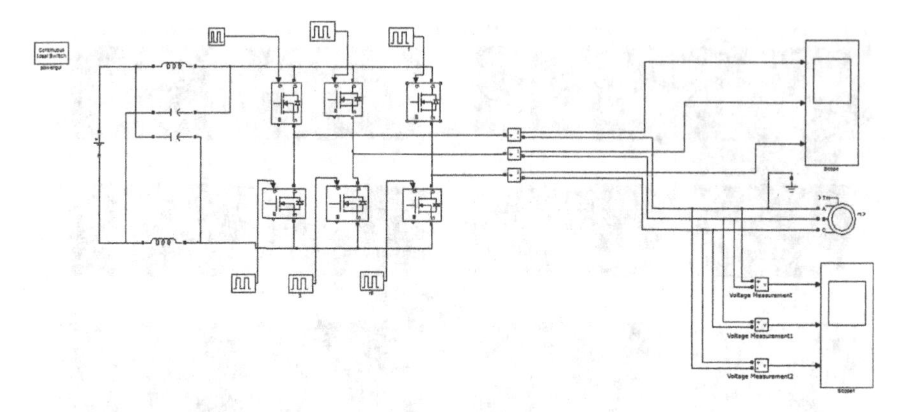

Fig. 12 Simulation circuit of L-G fault analysis

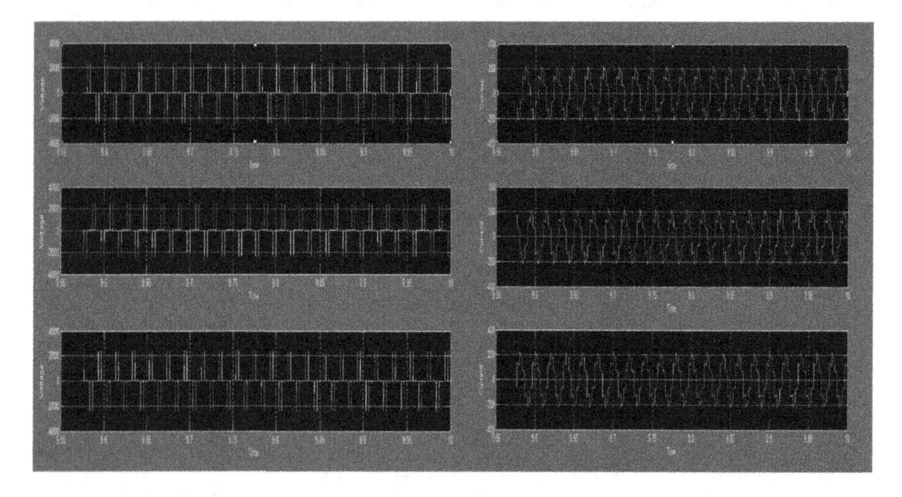

Fig. 13 Output voltage and current waveform with L-G fault

The THD measured under this faulty configuration at fundamental frequency is 32.49 % (Fig. 14).

The summary of FFT analysis of ZSI-fed 3-phase induction motor is shown in Table 1. It can be observed that THD due to MOSFET open circuit fault is increased as compared to the fundamental frequency THD spectrum and MOSFET blown off fault; the current is negative since the value of output current is higher than the value of input current. When faults are introduced, the magnitude of current increases for both the faults.

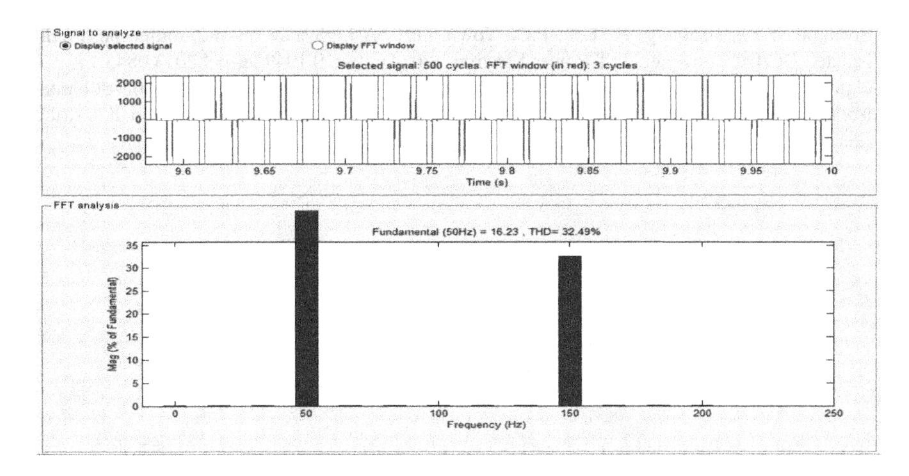

Fig. 14 FFT Spectrum at L-G fault condition

Table 1 Summary of FFT Analysis

Types of simulation	THD (%)	Current (A)
ZSI-fed 3-phase induction motor	21.65	200
Fault at MOSFET gate terminal	72.65	240
MOSFET blown off fault	67.74	−80
L-G fault	32.49	200

4 Conclusion

In this paper, the THDs of single-phase ZSI are evaluated under various fault conditions. Output waveforms are obtained and THD values are tabulated. The result shows that the current harmonics gets introduced upon introduction of faults in rectifier and inverter modules. The THD values increase on introducing faults. Due to faults the current direction gets reversed. The frequency responses with FFT spectrum under three different fault conditions are distinctly different which are precisely presented in this paper.

References

1. Liu, Y., Ge, B., Rub, H.A., Peng, F.Z.: Overview of space vector modulations for three-phase Z-source/quasi Z-source inverters. IEEE Trans. Power Electro. **29**(4), 2098–2108 (2014)
2. Siwakoti, Y.P., Town, G.: Improved modulation technique for voltage fed quasi-Z-source DC/DC converters. In: Proceedings of Applied Power Electronics Conference and Exposition, pp. 1973–1978 (2014)

3. Siwakoti, Y.P., Blaabjerg, F., Loh, P.C., Town, G.E.: A high gain quasi-Z-source push-pull isolated DC/DC converter. IET Power Electron, (2014). doi:10.1049/iet-pel.2013.0845
4. Nguyen, M.K., Phan, Q.D., Nguyen, V.N., Lim, Y.C., Park, J.K.: Trans-Z-source-based isolated DC-DC converters. In: Proceedings of IEEE International Symposium on Industrial Electronics, pp. 1–6 (2013)

OpenMP Genetic Algorithm for Continuous Nonlinear Large-Scale Optimization Problems

A.J. Umbarkar

Abstract Genetic algorithms (GAs) are one of the evolutionary algorithms for solving continuous nonlinear large-scale optimization problems. In an optimization problem, when dimension size increases, the size of search space increases exponentially. It is quite difficult to explore and exploit such huge search space. GA is highly parallelizable optimization algorithm; still there is a challenge to use all the cores of multicore (viz. Dual core, Quad core, and Octa cores) systems. The paper analyzes the parallel implementation of SGA (Simple GA) called as OpenMP GA. OpenMP (Open Multi-Processing) GA attempts to explore and exploit the search space on the multiple cores' system. The performance of OpenMP GA is compared with SGA with respect to time required and cores utilized for obtaining optimal solution. The results show that the performance of the OpenMP GA is remarkably superior to that of the SGA in terms of execution time and CPU utilization. In case of OpenMP GA, CPU utilization is almost double for continuous nonlinear large-scale test problems for the given system configuration.

Keywords Function optimization · Genetic algorithm (GA) · Open multi-processing (OpenMP) · Nonlinear optimization problems · Optimization benchmarks functions

1 Introduction

GAs are inspired from the Darwinian's theory of evolution. It is one of the evolutionary algorithms (EAs). GA is metaheuristic, stochastic search technique, mostly used to solve optimization problems, i.e., to search an optimize solution from a given search space (Set of all feasible solutions) for a given problem with

A.J. Umbarkar (✉)
Department of Information Technology, Walchand College of Engineering, Sangli, MS, India
e-mail: anantumbarkar@rediffmail.com

M. Pant et al. (eds.), *Proceedings of Fifth International Conference on Soft Computing for Problem Solving*, Advances in Intelligent Systems and Computing 437, DOI 10.1007/978-981-10-0451-3_20

certain given constraints. Many continuous, nonlinear, large-scale optimization problems and test problems are solved by EAs.

Today, multicore CPUs (Central processing unit) are easily available as they are cheaper and common, and hence one cannot ignore their importance. Many optimization problems are solved using GA on multicore system but not care about their CPU/core utilizations [1–4]. Uses of system resources like CPU and memory utilization are very critical issues while solving time crucial optimization problems.

The Moore's law says that "functionality of CPU doubles everyone and half year", having said this CPUs are also growing rapidly in their functionality [5]. Currently, all personal systems are multicore processor system like i3, i5, and i7 processors (Intel). These processors boost the performance of application if it is implemented effectively on it.

Advances, computing trends, application, and perspective of parallel genetic algorithm (PGA) given in [1, 6, 7] give a literature survey. Nowadays, the important issues are architecture, OS, topologies, and programming (Libraries). The programming languages are facilitated with set of special system calls or libraries like Linda, OpenMP,[1,2] HPF (High Performance Fortran), Parallel C and OCCAM (both for transputer networks), Java using communication libraries MPI (Message-Passing Interface),[3] Express MPI, PVM (Parallel Virtual Machine), POSIX (Portable Operating System Interface) threads, and Java threads on SMP (Symmetric multiprocessing) machines. There are many parallel programming options available; some are mentioned in footnote.[4]

These languages or libraries are for both, loosely and tightly coupled systems. Some optimization problems are solved using GAs by implementing it over clusters, MPPs (Massively parallel processing) [2, 8], grids [9–16], GPGPU (General-purpose computing on graphics processing units) [3, 4, 17, 18], cloud computing, and multicore/HPC (High-performance computing) [19]. The GAs are experimented on clusters, MPPs, grid, and cloud computing. These GAs can be also called PGAs, as the parallel hardware is used [20]. Even though PGA has the capacity to run in parallel on many computing paradigms, finding suitable hardware and software which give an optimal solution with optimal resource utilization is a critical task. The choice of OpenMP library to implement the SGA on multicore system is due to the wide availability of multicore systems like i3, i5, and i7 processors or HPC.

This paper aims on programming parallel implementation of serial GA with respect to processing speed and CPU utilization. This experimented work is the parallel implementation (OpenMP GA) of SGA on multicore system for unconstrained continuous nonlinear optimization problems (Fig. 1).

The paper is organized as follows. Section 2 explains the OpenMP library and the algorithm with fitness function in detail. Section 3 gives details of test data and experimental results. Section 4 provides conclusions.

---

[1]https://computing.llnl.gov/tutorials/openMP/.

[2]http://en.wikipedia.org/wiki/Openmp.

[3]http://wotug.ukc.ac.uk/parallel/.

[4]http://web.eecs.umich.edu/~sugih/pointers/gprof_quick.html.

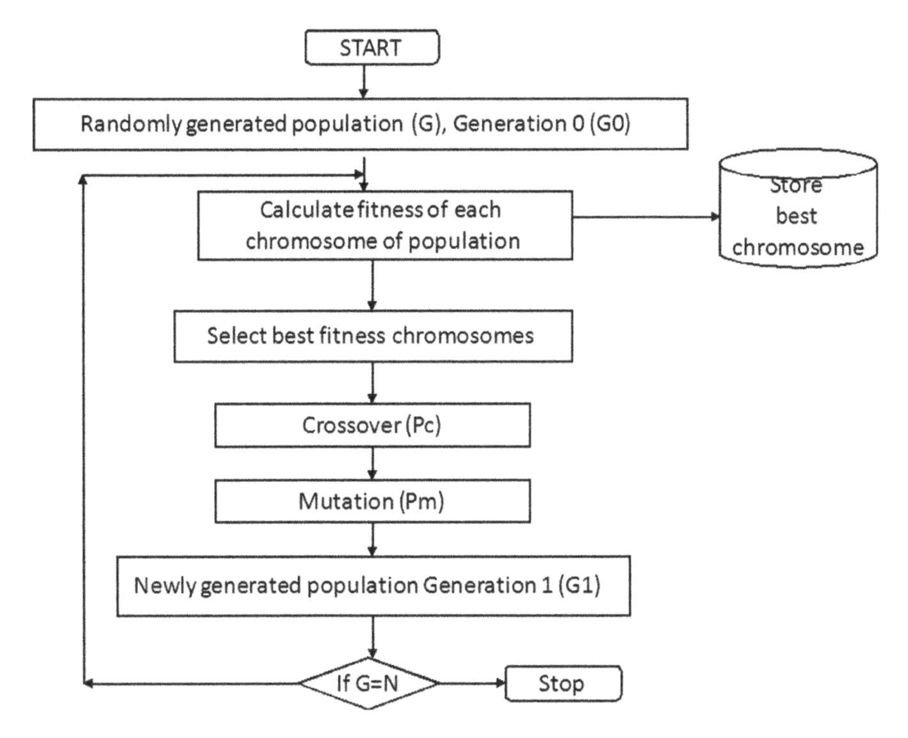

Fig. 1 Schematic diagram of SGA

2 OpenMP GA

OpenMP (Open Multi-Processing) is a C, C++ supported library for parallel programming (see footnote 4). Open MP library works on the concept of thread. The library has many pragmas that parallelize the loops of the algorithm along with known numbers of cores available on a multicore processor. In the implementation, the SGA is loop programmed using OpenMP library. In openMP, the master thread forks off a number of slave threads and divides the task among them. All the threads run concurrently on processors or cores. The section of code that is meant to run in parallel is marked accordingly; preprocessor directive will cause the threads to form before the section is executed. After the execution of parallelized code, all the threads join back into the master thread, which continues onward to the end of the program. Figure 2 shows the concepts of multithreading.

By parallelizing the various operators of SGA such as random number generator, initialization, and fitness calculation, one can reduce the amount of time required to get the optimal solution and can increase the CPU utilization of a multicore system. The flowchart of GA is shown in Fig. 1. This work is experimented on Intel Core i5 CPU and algorithm is a parallel implementation of SGA using OpenMP library.

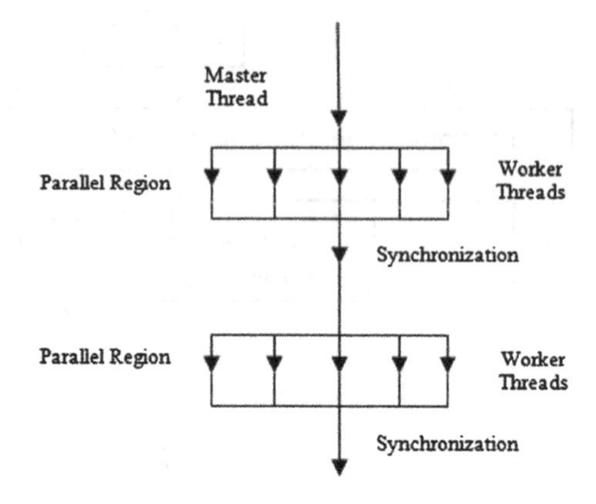

Fig. 2 Concept of multithreading with the master thread forks off a number of threads for executing code in parallel

Table 1 Parameter setting for algorithms

Initial parameters	SGA	OpenMP GA
Main population size	100	100
Crossover method	Single point	Single point
Crossover point K	1	1
Mutation method	Flip Bit	Flip Bit
Crossover probability (Pc)	0.7	0.7
Mutation probability (Pm)	0.03	0.03
Dimension (D)	3	3
No of generation	2000	2000

Fitness functions used in the experimentation are the unimodal and multimodal test functions [7, 21, 22]. The sample Rosenbrock test fitness function F(x) is shown in Eq. (1):

$$f(x) = \sum_{i=1}^{n-1} \left[100(x_{i+1} - x_i^2)^2 + (1 - x_i)^2\right] \tag{1}$$

Test area $-2.048 \leq x_i \leq 2.048$, where $i = 1, \ldots, n$; where n: Dimension. Its global minimum equals F(x) = 0 which is obtainable for x_i, where $i = 1, \ldots, n$ (Table 1).

3 Experimental Results

3.1 Test Data

The algorithm was evaluated on a continuous nonlinear large-scale problem (benchmark problems) given in [7, 21, 22]. The experimentation is carried out with 17 test functions. The (F6, F7, F8, F9, F12, F13, and F17) are unimodal functions and (F1, F2, F3, F4, F5, F10, F11, F14, F15, and F16) are multimodal functions. The test problem set is shown in Table 2.

3.2 Results and Discussion

In the experimentation, SGA and OpenMP GA are implemented using C language over Linux on a personal computer (Intel Core i5 CPU 650 @ 3.20 GHz processor with 2 cores and 4 MB cache, 1 GB of RAM). Table 1 shows the parameter setting for the SGA and OpenMP GA algorithms. Performance of SGA is compared with the performance of OpenMP GA on the basis of execution time taken and speedup obtained. Equation (2) gives the formula used for speedup calculation with respect to SGA:

$$\text{SpeedUp} = \frac{\text{Time taken by SGA} - \text{Time taken by OpenMP GA}}{\text{Time taken by SGA}} \tag{2}$$

CPU utilization is observed using GProf software.[5] It is a performance analyzing and profiling tool, which collects and arranges statistics of programs like execution time, CPU utilization, function calls, function-wise memory, and CPU utilization taken by program using which one can improve the program's efficiency [23].

Table 3 shows the comparison based on best solution obtained, execution time in sec, and CPU utilization in percentage by SGA and OpenMP GA. Table 4 shows the comparison based on best solution obtained and speedup achieved in percentage by OpenMP GA over SGA for multimodal test functions. Figure 3 shows a plot of speedup in percentage versus various multimodal test functions considered. It shows that OpenMP GA outperforms SGA for multimodal test functions. Figure 4 shows a plot of best solution obtained (BSO) versus various multimodal test functions considered.

Table 5 shows the comparison based on best solution obtained and speedup achieved in percentage by OpenMP GA over SGA for unimodal test functions. Figure 5 shows a plot of speedup in percentage versus various unimodal test functions considered. It shows that OpenMP GA performs better in comparison with SGA for unimodal test functions.

Figure 6 shows a plot of execution time in second versus various test functions considered. It shows that the OpenMP GA gives better results in comparison with SGA in terms of execution time taken to obtained best solution. Figure 7 shows a plot of CPU utilization in percentage for BSO versus various test functions considered. It shows that the OpenMP GA doubles the CPU utilization than SGA.

[5]http://www.cs.utah.edu/dept/old/texinfo/as/gprof.html.

Table 2 Problem Set—continuous nonlinear large-scale benchmark problems

Function name	Equation	Range	Optimum value
Rosenbrock (F1)	$f(x) = \sum_{i=1}^{n-1}\left[100(x_{i+1} - x_i^2)^2 + (1 - x_i)^2\right]$	[−2.048, 2.048]	0
Ackley (F2)	$f(x) = -a \cdot \exp\left(-b\sqrt{\frac{1}{n}\sum_{i=1}^{n} x_i^2}\right) - \left(\exp\frac{1}{n}\sum_{i=1}^{n}\cos(cx_i)\right) + a + \exp(1)$	[−32.768, 32.768]	0
Griewangk (F3)	$f(x) = \frac{1}{4000}\sum_{i=1}^{n} x_i^2 - \prod_{i=1}^{n}\cos\left(\frac{x_i}{\sqrt{i}}\right) + 1$	[−600, 600]	0
Schwefel (F4)	$f(x) = \sum_{i=1}^{n}\left[-x\sin\left(\sqrt{\lvert x_i\rvert}\right)\right]$	[−500, 500]	−418.9829
Rastrigin (F5)	$f(x) = 10n + \sum_{i=1}^{n}\left[x_i^2 - 10\cos(2\pi x_i)\right]$	[−5.12, 5.12]	0
De-jong (F6)	$f(x) = \sum_{i=1}^{n} x_i^2$	[−5.12, 5.12]	0
Axis parallel hyper-ellipsoid (F7)	$f(x) = \sum_{i=1}^{n}(i \cdot x_i^2)$	[−5.12, 5.12]	0
Rotated hyper-ellipsoid (F8)	$f(x) = \sum_{i=1}^{n}\sum_{j=1}^{i} x_j^2$	[−65.536, 65.536]	0
Easom (F9)	$f(x_1, f_2) = -\cos(x_1) \cdot \cos(x_2) \exp - \left((x_{1-\pi})^2 - (x_2 - \pi)^2\right)$	[−100, 100]	−1
Michalewiczs (F10)	$f(x) = -\sum_{i=1}^{n}\sin(x_i)\left[\sin\left(\frac{i \cdot x_i^2}{\pi}\right)\right]^{2m}$	[−3.1428, 3.1428]	−4.687

(continued)

Table 2 (continued)

Function name	Equation	Range	Optimum value
Six-hump camel back (F11)	$f(x_1, x_2) = \left(4 - 2.1x_1^2 + \frac{x_1^2}{3}\right)x_1^2 + x_1x_2 + (-4 + 4x_2^2)x_2^2$	[−3, 3]	−1.0316
Goldstein-Price (F12)	$f(x_1, x_2) = \left[1 + (x_1 + x_2 + 1)^2 (19 - 14x_1 + 3x_1^2 - 14x_2 + 6x_1x_2 + 3x_2^2)\right] \left[30 + (2x_1 - 3x_2)^2 (18 - 32x_1 + 12x_1^2 + 48x_2 - 32x_1x_2 + 27x_2^2)\right]$	[−2, 2]	3
Sum of different powers (F13)	$f(x) = \sum_{i=1}^{n} \|x_i\|^{i+1}$	[−1, 1]	0
Shubert (F14)	$f(x_1, x_2) = -\sum_{i=1}^{5} i\cos((i+1)x_1 + 1) \sum_{i=1}^{5} i\cos((i+1)x_2 + 1)$	[−10, 10]	−186.7306
Drop wave (F15)	$f(x_1, x_2) = -\frac{1 + \cos\left(12\sqrt{x_1^2 + x_2^2}\right)}{\frac{1}{2}\left(x_1^2 + x_2^2\right) + 2}$	[−5.12, 5.12]	−2.19334
Fifth De Jong (F16)	$f(x_1, x_2) = \left\{0.002 + \sum_{i=-2}^{2} \sum_{j=-2}^{2} \left[5(i+2) + j + 3^{-1} + (x_1 - 16_j)^6 + (x_2 - 16_i)^6\right]^{-1}\right\}$	[−65.536, 65.536]	0
Branins (F17)	$f(x_1, x_2) = a\left(x_2 - bx_1^2 + cx_1 - d\right)^2 + e(1 - f)\cos(x_1) + e$	[−5, 10] x [0, 15]	0.397887

Table 3 Best solution found, execution time, and CPU utilization for unimodal and multimodal test functions

Test function	Best solution obtained for SGA	Best solution obtained for OpenMP GA	Execution time (s) for SGA	Execution time (s) for OpenMP GA	CPU utilization (%) for SGA	CPU utilization (%) for OpenMP GA
F1	3.36E−01	1.02E+00	8.14	1.55	96.45	189.21
F2	0.00E+00	0.00E+00	12.24	7.55	99.40	189.20
F3	7.82E−01	1.95E−01	8.11	3.49	97.36	186.32
F4	−1.58E +03	−1.25E+03	7.10	3.75	95.21	189.65
F5	6.60E−03	8.20E−03	3.05	3.05	92.34	192.21
F6	2.00E−04	1.00E−04	1.10	0.32	94.28	193.24
F7	3.00E−04	1.00E−04	1.09	0.37	89.66	189.74
F8	2.00E−06	6.49E−04	1.66	0.69	99.87	184.54
F9	9.42E-01	0.00E+00	0.26	0.49	97.45	174.25
F10	−1.93E +00	−1.70E+00	1.58	1.63	89.87	187.26
F11	−9.76E −01	–	0.16	–	98.78	191.24
F12	7.19E+00	–	0.97	–	96.33	189.36
F13	0.00E+00	0.00E+00	0.40	0.29	98.47	187.54
F14	−2.10E +02	−2.10E+02	1.59	1.14	94.21	188.24
F15	9.96E−01	1.00E+00	0.65	0.46	95.74	193.24
F16	5.00E−01	5.00E−01	10.10	12.07	94.28	187.45
F17	7.69E−01	–	0.43	–	97.65	183.21

Table 4 Best solution obtained and speedup gained for multimodal functions

Test function	Best solution obtained for SGA	Best solution obtained for OpenMP GA	Speedup of OpenMP GA compared with SGA (%)
F1	3.36E−01	4.52E−01	50.93
F2	0.00E+00	0.00E+00	27.80
F3	7.82E−01	1.95E−01	42.08
F4	−1.58E+03	−1.25E+03	60.34
F5	6.60E−03	8.20E−03	57.59
F10	−1.93E+00	−1.70E+00	19.91
F11	−9.76E−01	–	20.48
F14	−2.10E+02	−2.10E+02	61.82
F15	9.96E−01	1.00E+00	21.52
F16	5.00E−01	5.00E−01	07.02

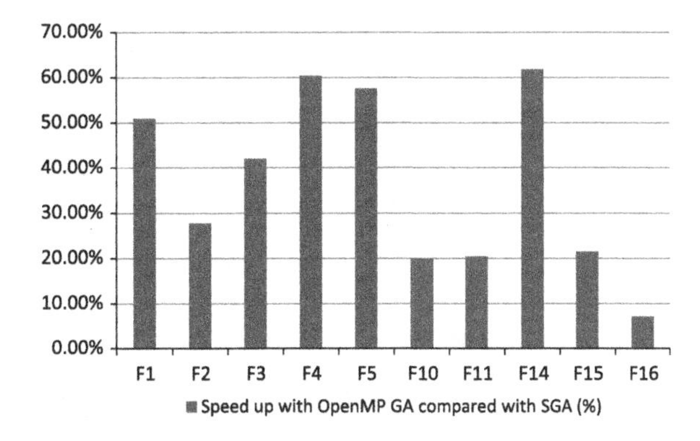

Fig. 3 Speedup gained in OpenMP GA on multimodal test functions

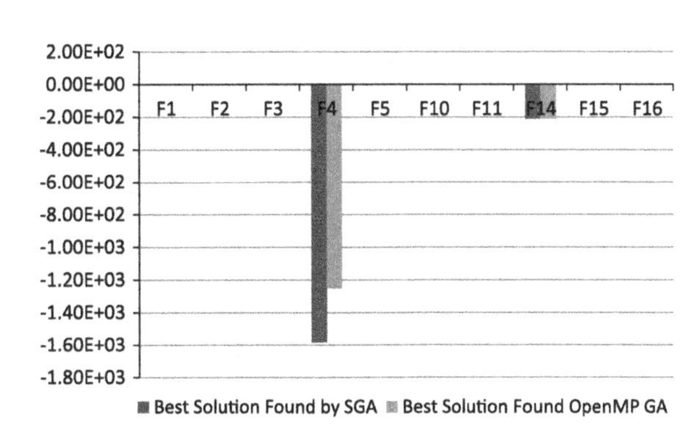

Fig. 4 Best solutions found by OpenMP GA and SGA on multimodal test functions

Table 5 Best solution obtained and speedup gained for unimodal functions

Test function	Best solution obtained for SGA	Best solution obtained for OpenMP GA	Speedup of OpenMP GA compared with SGA (%)
F6	2.00E−04	1.00E−04	51.06
F7	3.00E−04	1.00E−04	27.46
F8	2.00E−06	6.49E−04	49.89
F9	9.42E−01	0.00E+00	17.64
F12	7.19E+00	–	47.60
F13	0.00E+00	0.00E+00	−1.57
F17	7.69E−01	–	02.91

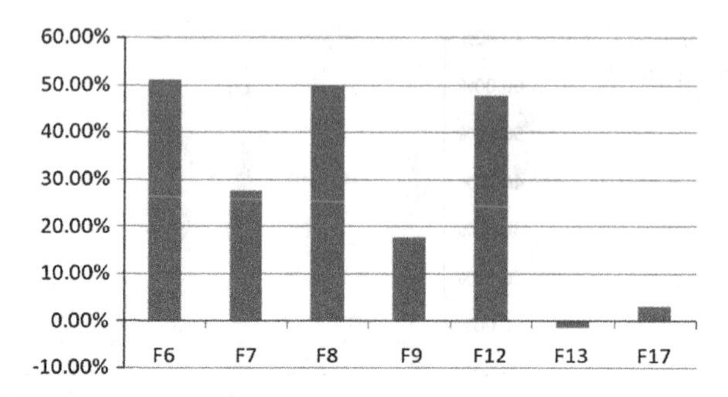

Fig. 5 Speedup gained in OpenMP GA on unimodal test functions

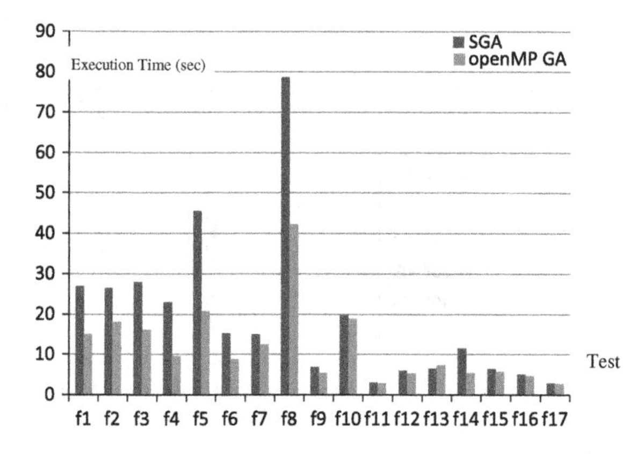

Fig. 6 Execution time in seconds to find BSO versus unimodal and multimodal test functions

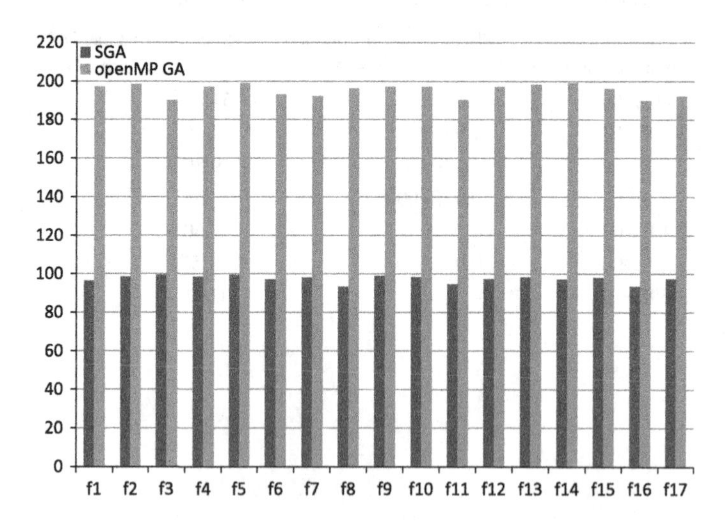

Fig. 7 CPU utilization in percentage to find best solution versus unimodal and multimodal test functions

4 Conclusions

GAs are one of the popular metaheuristics need to improve performance on latest computing trends by implementing or programming them parallel. Results show that the OpenMP GA outperforms over the SGA. OpenMP GA provides the optimal solution in less amount of time (compared to SGA). OpenMP GA doubles the CPU utilization than SGA. OpenMP GA exploits all the available cores efficiently, thereby giving better CPU utilization.

In future scope, researcher can experiment OpenMP GA or similar implementation of OpenMp evolutionary algorithms on multicore system or HPC platform. The analysis of the above-experimented algorithms in terms of CPU utilization, time taken, etc., for benchmark test functions will help to improve the program based on performance improvement of the algorithms for function optimization.

References

1. Konfrst, Z.: Parallel genetic algorithm: advances, computing trends, application and perspective. In: Proceeding of 18th International Parallel and Distributed Processing Symposium [IPDPS'04], IEEE Computer Society (2004)
2. Umbarkar, A.J., Joshi, M.S.: Dual Population Genetic Algorithm (GA) versus OpenMP GA for Multimodal Function Optimization. Int. J. Comput. Appl. **64**(19), 29–36 (2013)
3. Arora, R., Tulshyan, R., Deb, K.: Parallelization of binary and real-coded genetic algorithm on GPU using CUDA. In: Congress on Evolutionary Computation, pp. 1–8 (2010)
4. Vidal, P., Alba, E.: A multi-GPU implementation of a cellular genetic algorithm. In: 2010 IEEE Congress on Evolutionary Computation (2010)
5. August, A.D., Chiou, K.P.D., Sendag, R., Yi, J.J.: Programming multicores: do application programmers need to write explicitly parallel programs?. In: Computer Architecture Debates in IEEE MICRO, pp. 19–32 (2010)
6. Cantú-Paz, E.: Efficient and Accurate Parallel Genetic Algorithms. Kluwer Academic Publishers (2000)
7. Andrei, N.: An unconstrained optimization test functions collection. Adv. Model. Optim. **10** (1), 147–161 (2008)
8. Gagné, C., Parizeau, M., Dubreuil, M.: The master-slave architecture for evolutionary computations revisited. In: Proceedings of the Genetic and Evolutionary Computation Conference, Chicago, IL, 2, pp. 1578–1579 (2003)
9. Hauser, R., Männer, R.: Implementation of standard genetic algorithms on MIMD machines. In: Parallel Problem Solving from Nature [PPSN3], pp. 504–513 (1994)
10. Tanese, R.: Parallel genetic algorithm for a hypercube. In: Proceedings of the Second International Conference on Genetic Algorithms [ICGA2], pp. 177–183 (1987)
11. Tanese, R.: (1989) Distributed genetic algorithms. In: Proceedings of the Third International Conference on Genetic Algorithms [ICGA3], pp. 434–439 (1987)
12. Voigt, H.M., Born, J., Santibanez-Koref, I.: Modeling and simulation of distributed evolutionary search processes for function optimization. In: Parallel Problem Solving from Nature [PPSN1], pp. 373–380 (1991)
13. Voigt, H.M., Santibanez-Koref, I., Born, J., Hierarchically structured distributed genetic algorithm. In: Parallel Problem Solving from Nature [PPSN2], pp. 145–154 (1992)

14. Imade, H., Morishita, R., Ono, I., Ono, N., Okamoto, M.: A grid-oriented genetic algorithm for estimating genetic networks by s-systems. In: SICE 2003 Annual Conference, 3(4–6), pp. 2750–2755 (2003)
15. Herrera, J., Huedo, E., Montero, R., Llorente, I.: A grid oriented genetic algorithm. Adv. Grid Comput. EGC **2005**, 315–322 (2005)
16. Imade, H., Morishita, R., Ono, I., Ono, N., Okamoto, M.: A grid-oriented genetic algorithm framework for bioinformatics. New Gen. Comput. **22**(2), 177–186 (2004)
17. Wong, M., Wong, T.: Parallel hybrid genetic algorithms on consumer-level graphics hardware. In: Congress on Evolutionary Computation, Canada, pp. 2972–2980 (2006)
18. Oiso, M., Matumura, Y.: Accelerating steady-state genetic algorithms based on CUDA architecture. In: 2011 IEEE Congress on Evolutionary Computation, pp. 687–692 (2011)
19. Zheng, L., Lu, Y., Ding, M., Shen, Y., Guo, M., Guo, S.: Architecture-based performance evaluation of genetic algorithms on multi/many-core systems. In: proceeding of 14th IEEE International Conference on Computational Science and Engineering, CSE 2011, Dalian, China, (2011)
20. Cantú-Paz, E.: A Report: A Survey of Parallel Genetic Algorithms. Department of Computer Science and Illinois Genetic Algorithms Laboratory University of Illinois at Urbana-Champaign (2002)
21. Molga, M., Smutnicki, C.: Test functions for optimization needs—2005. Unpublished (2005)
22. Mohan, C., Deep, K.: Optimization Techniques, first edition, New Age International Publication (2009)
23. Susan, L., Graham, P.B., Kessler, M.K., McKusick (2000) gprof: a Call Graph Execution Profiler1, Electrical Engineering and Computer Science Department University of California, Berkeley, California

Combined SCP and Geometric Reconstruction of Spine from Biplanar Radiographs

Sampath Kumar, K. Prabhakar Nayak and K.S. Hareesha

Abstract Three-dimensional reconstruction of the spine is necessary in proper diagnosis of various spinal deformities. This is normally achieved using stereo-radiographic techniques involving biplanar (frontal and lateral) radiographs. Either stereo-corresponding point (SCP) algorithm or non-stereo-corresponding point (NSCP) algorithm is used for this purpose. The NSCP technique suffers from observer variability. Moreover, it is time consuming. Hence, it has restricted usage in clinical environment. Here, a hybrid approach is proposed in which a 3D spine model is reconstructed by applying geometric features to the SCP reconstructed model. The vertebral orientation features are automatically extracted from the calibrated radiographs. The 3D model thus produced is successfully validated. The proposed method has lesser observer variability due to the limited number of anatomical landmarks. Also, the reconstruction errors are within the acceptable limits available in the literature. Thus, the proposed technique can be used in clinical practices for the diagnosis of spinal deformities.

Keywords Scoliosis · Stereo-radiography · 3D reconstruction

Sampath Kumar (✉) · K.P. Nayak
Department of Electronics and Communication, Manipal Institute of Technology, Manipal, India
e-mail: kumar.sampath@manipal.edu

K.P. Nayak
e-mail: kp.nayak@manipal.edu

K.S. Hareesha
Department of Computer Applications, Manipal Institute of Technology, Manipal, India
e-mail: hareesh.ks@manipal.edu

M. Pant et al. (eds.), *Proceedings of Fifth International Conference on Soft Computing for Problem Solving*, Advances in Intelligent Systems and Computing 437, DOI 10.1007/978-981-10-0451-3_21

1 Introduction

Spine is a complex anatomical structure in the human body. Because of this complexity, its pathological deformities like scoliosis are hard to diagnose. Traditionally, they have been assessed in 2D using frontal and lateral view radiographs [1]. The most common 3D imaging modalities such as computed tomography (CT) and magnetic resonance imaging (MRI) are not preferred in the evaluation of spinal deformities [2]. They use supine position for 3D image reconstruction, which can change the actual deformity of the spine. Even though the CT scan reconstruction provides more details of the assessment of spinal deformity, the level of radiation exposure is higher [3]. The MRI is not preferred in spinal deformity assessment as the corrective tools and surgical implants may be present in the patients [4]. The MRI scanning is costly like CT scan reconstruction. Therefore, stereo-radiographic reconstruction is an ideal method for 3D reconstruction of the spine as it uses only biplanar radiographs (frontal and lateral) acquired in standing position.

The earlier approaches of stereo-radiographic reconstruction were based on the triangulation techniques [5, 6]. It uses points that are available in both the radiographs, known as stereo-corresponding points (SCP) for 3D reconstruction. To improve the accuracy, along with the SCPs non-stereo-corresponding points (NSCP) were also used [7, 8]. NSCPs were the points that are available in only one of the radiographs. These approaches involved a lot of human intervention to identify several landmarks on the biplanar radiographs. Hence, they were more appropriate for research purposes and find restricted usage in clinical practices. This is due to the difficulty and delay involved in landmark identification procedure. Also, it involves a complex calibration procedure. To simplify and automate these procedures numerous methods have been developed. Promero et al. [9] used the statistical and geometric approximations derived from the vertebral elements to develop a novel automated process. Humbert et al. [10] added longitudinal inferences to this process to improve the reconstruction accuracy. A semi-automated method developed by Dumas et al. [11] concentrated on interpolation and optimization of the vertebral contours. However, the significant features in the radiographs like vertebral orientations are not fully exploited in these methods.

In recent approaches, the spine midline and vertebral body contour matching techniques were used by Zhang et al. [12] for 3D reconstruction of the spine. Since the epipolar geometry approximations are used, the model is limited to projective instead of Euclidian reconstruction. Also, it is incapable of addressing vertebral deformations. Zhang et al. improved their previous method using pedicle projections for matching the vertebral orientations to reduce the user interaction time [13]. Here, the vertebral centers were manually identified which can cause observer variability and also it lacks in vivo experimentation. Recently, a new uncalibrated procedure was developed by Moura et al. [14] to decrease the occlusions created by calibration objects. It uses average geometrical parameters of the system obtained from a laser range finder for calibration. It assumed that the patient position and radiological settings were fixed during X-ray acquisition. This is not true in a normal clinical environment. Also, the developed model was a six-point model rather than a morpho-realistic model.

To overcome these limitations a calibrated approach is followed for 3D reconstruction, which causes minimal occlusions in the radiograph and produces morpho-realistic model. The occluded landmark positions are estimated using the epipolar geometry. The geometric features like vertebral orientations to the SCP reconstructed model improve the reconstruction accuracy. We call this procedure as combined SCP and geometric (CSCPG) reconstruction method. The accuracy achieved by CSCPG method is nearer to the NSCP method which is considered as the gold standard method for stereo-radiographic reconstruction. The geometric features are automatically extracted from the biplanar radiographs using image processing techniques. Hence, the proposed method reduces observer variability in vertebral landmark identification. Also, a limited number of landmarks need to be identified on the radiographs. Thus a faster, reliable, and accurate 3D spine model can be achieved from the biplanar radiographs using the proposed method.

2 Methodology

Stereo-radiographic method of 3D spine reconstruction presented here is called combined SCP and geometric (CSCPG) reconstruction method. It is based on the deformation of the SCP model according to the automatically extracted geometric features from the radiographs. Figure 1 shows the flowchart of the entire process.

2.1 SCP Reconstruction

Five subjects suffering from idiopathic scoliosis of the age group 11–14 years have volunteered for this study. Frontal and lateral radiographs are obtained from these subjects on a calibration apparatus shown in Fig. 2. It consists of a positioning apparatus to attain an appropriate position and radio-opaque markers for calibrating the radiological environment. The local ethical committee approval has been taken before the experimentation.

Stereo-radiographic reconstruction using stereo-corresponding point (SCP) technique needs calibration of the radiological environment. The technique proposed by Dansereau et al. [15] is used for calibration. It uses direct linear transformation (DLT) [16] to calculate the unknown geometrical parameters. The radio-opaque markers with known 3D coordinates are used for this purpose. DLT implicitly calculates these parameters with the help of these markers. Twenty stereo-corresponding markers are identified on the radiographs to calculate the calibration matrix. DLT algorithm is linear and has less computational complexity, but the error generated by DLT is an algebraic error. This error vector is

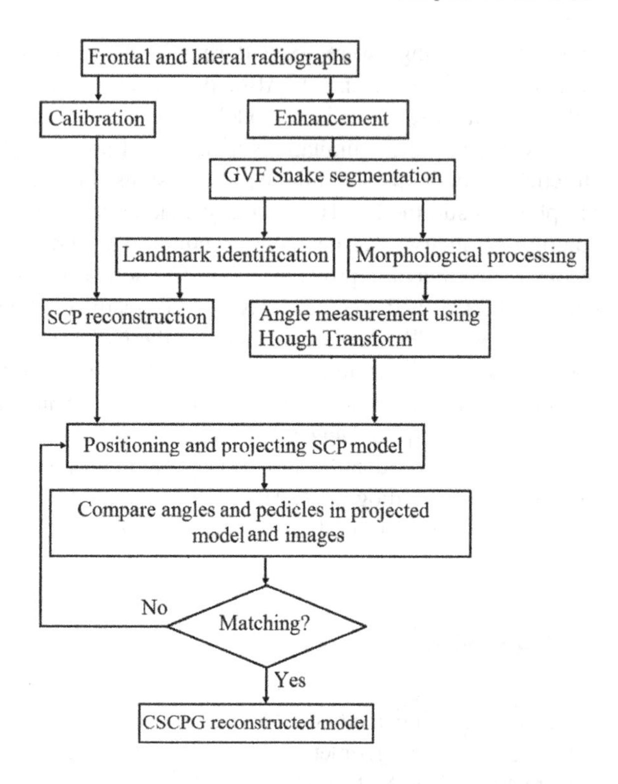

Fig. 1 CSCPG reconstruction procedure

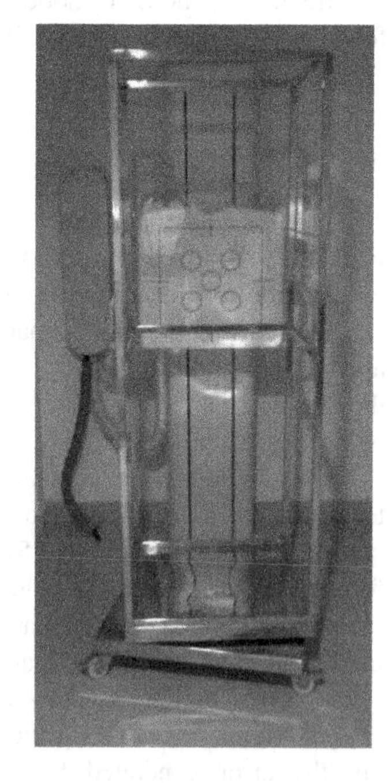

Fig. 2 Calibration apparatus

insignificant and has no clear geometric meaning. Hence, Levenberg–Marquardt algorithm [17] is used to minimize the re-projection error which in turn increases the reconstruction accuracy. This results in an optimized calibration matrix.

For SCP reconstruction six corresponding anatomical landmarks need to be identified on each vertebra. Most visible landmarks are vertebral endplate centers and top and bottom of the pedicles. They are identified using a semi-automatic method, which is explained in detail in our previous work [18]. The radio-opaque markers are very small in size and the calibration apparatus causes minimal occlusions in the radiographs. The occluded landmark positions are estimated using the epipolar geometry constraints related to the acquired biplanar radiographs. With the help of optimized calibration matrix, corresponding landmarks are triangulated to compute their 3D positions. Using this technique, 3D positions of landmarks of all the vertebrae (T1–L5) are computed. They are integrated to obtain the SCP reconstructed model in the form of triangulated mesh. Figure 3 shows an example of SCP reconstruction from the subject-1 radiographs.

SCP reconstruction thus obtained has poor visibility and diagnosing the disease is very difficult. In order to obtain the morpho-realistic model, a generic spine model is developed using a cadaveric X-ray phantom. Deformation of this generic model in accordance with the SCPs can produce a more realistic spine model. The generic model is developed using the method proposed by De Guise et al. [19]. The marching cube algorithm is applied on 1-mm-thick CT scan slices of T1–L5 cadaveric vertebrae. Integration of these vertebrae along the facet joints gives the generic model as shown in Fig. 4a. These models show an accuracy of 1.1 ± 0.8 mm [5].

Dual kriging [20], a free form deformation technique, is used to deform this generic model in accordance with the 3D reconstructed SCPs to obtain a personalized model as shown in Fig. 4b.

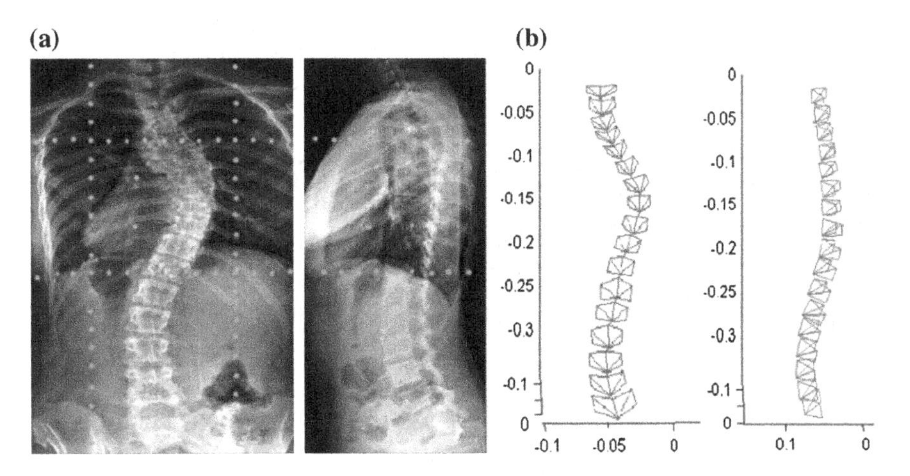

Fig. 3 SCP reconstruction of subject-1. **a** Biplanar radiographs. **b** 3D model

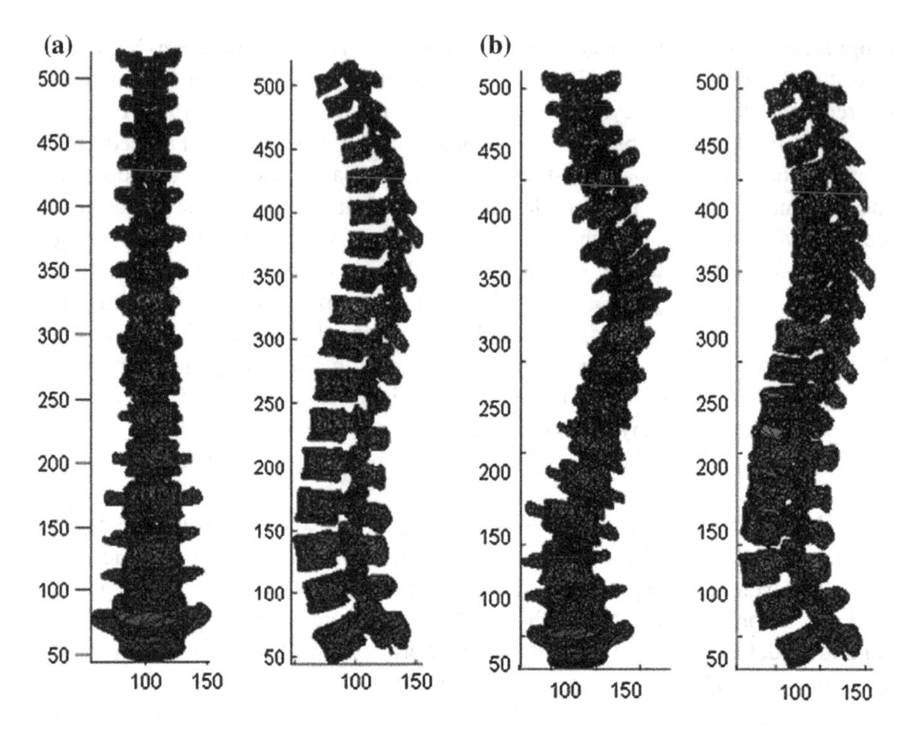

Fig. 4 a Generic spine model. b Personalized SCP model of subject-1 after deformation

2.2 CSCPG Reconstruction

The deformed SCP model shown in Fig. 4b is subject to further deformation in accordance with the automatically extracted geometric features from the radiographs. Initially, the radiographs are enhanced using a local enhancement technique known as multiscale mathematical morphology [21] as shown in Fig. 5a. Then the vertebral contours are automatically extracted from these images using gradient

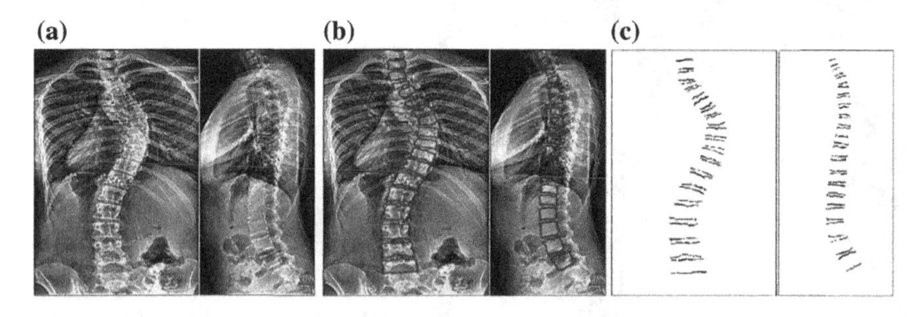

Fig. 5 a Enhancement result. b Segmentation result. c Extraction of horizontal edges

vector flow (GVF) snake algorithm [22] as shown in Fig. 5b. Then morphological operators are applied to retain the vertebral boundaries. The horizontal edges are extracted using Sobel operator as shown in Fig. 5c.

Using the Hough transform, the vertebral orientations are automatically computed from these edges. Similarly, the orientation of vertebrae in the SCP model is also computed by retro-projecting it on frontal and lateral planes. Now, the geometric transformations are applied to the vertebrae in the SCP model till it matches with the automatically extracted geometric features. The pedicle projections are also used for further alignment. When all the vertebrae are aligned along these geometric features, it results in the CSCPG reconstruction. This procedure is repeated to obtain CSCPG models of all the five subjects.

3 Result

Figure 6a shows the CSCPG reconstructed model of subject-1. Both qualitative and quantitative validations are performed to access the reconstruction accuracy. Qualitative validation is performed by retro-projecting the model on the biplanar radiographs as shown in Fig. 6b. A close similarity in shapes is observed in all the five subjects. For quantitative validation, the CT scan reconstruction of the same subject cannot be used, as the supine position of CT can modify the spinal curves. Hence, the NSCP model of the same subject is reconstructed using the method proposed by Mitton et al. [7]. Nineteen more landmarks are identified on the radiographs and 3D reconstructed using this method. This model is considered as

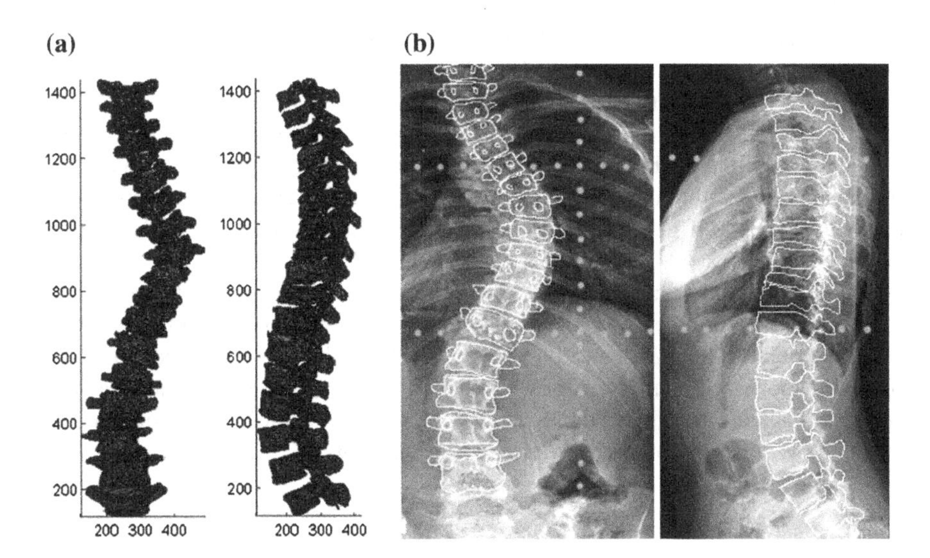

Fig. 6 **a** CSCPG reconstructed model of subject-1. **b** Qualitative validation

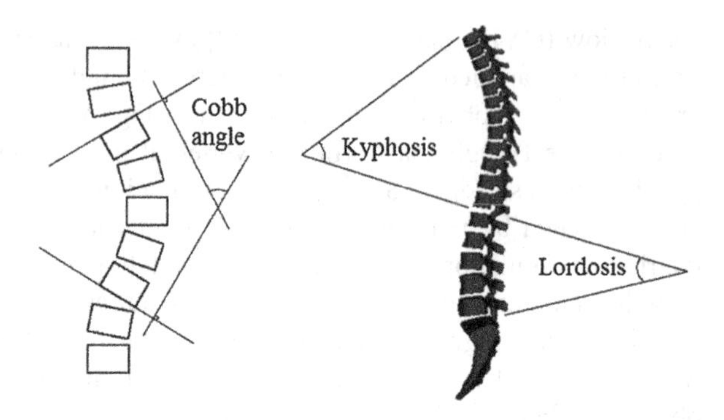

Fig. 7 Clinical indices used for quantitative validation

Table 1 The mean differences between the clinical indices of reference method [7] and the proposed method and the Wilcoxon signed-rank test results (*p* value)

Clinical index	Mean differences	*p*
Cobb angle (°)	0.51	0.81
Kyphosis (°)	0.41	0.78
Lordosis (°)	0.38	0.69
3D spine length (mm)	0.92	0.34

Table 2 Comparison of accuracy of the proposed method with related works

Method	[12]	[13]	[14]	Proposed
Location (mm)	2.4	3.5	3.5	1.93
Orientation (°)	2.8	3.9	3.6	2.51

the reference model for validation. Similarly, the reference 3D reconstructions of all the five subjects are obtained using NSCP algorithm. The clinical indices used for quantitative validation are the Cobb angle, kyphosis, lordosis, and 3D spinal length as shown in Fig. 7. The 3D spinal length is computed by fitting a B-spline curve to the midpoint of the pedicles of each vertebra. The angles are obtained at the plane of maximum curvature in both the methods for all the five subjects. The Wilcoxon test is used to find if the differences between these indices are statistically significant ($p < 0.05$). Table 1 summarizes these results.

The differences in clinical indices are not significant compared to the reference method. To measure the accuracy, the 3D reconstruction error (RMS) is computed using the point-to-point distance between the SCPs in the reference model and the proposed model. Similarly, error in orientation is also computed for all the five subjects. Table 2 summarizes these results. The reconstruction accuracy of CSCPG technique is superior compared to the existing methods.

4 Discussion

Present stereo-radiographic reconstruction of the spine involves identification of six SCPs and twenty or more NSCPs on the biplanar radiographs. The proposed CSCPG method uses only six SCPs and automatically derives geometric features for 3D reconstruction. Thus the proposed method is computationally inexpensive and free from observer variability. The calibration apparatus used here is user-friendly and economic. The minimum occlusions are created by this apparatus on the radiographs. The landmark positions due to the occlusion can be easily estimated using epipolar constraints. The 3D reconstructed model is validated by retro-projecting the model on biplanar radiographs. Qualitative evaluation shows a close similarity between the radiographs and the projections. For quantitative evaluation, a statistical significance test is conducted for the differences in clinical indices between the proposed method and the reference NSCP method. No significant difference is observed in any of these indices. The 3D reconstruction accuracies with respect to the location and orientation are also computed. They are found to be much better than the existing methods.

These results show the validity and benefits of the CSCPG method over current approaches. The uncalibrated approaches require statistical modeling which is difficult to implement in the normal clinical environment. Also, they fail to exhibit the morpho-realistic model of the spine. Hence, NSCP method is considered as the gold standard, but it requires 20 or more landmarks per vertebra which are manually identified. This results in a complex procedure and causes observer variability. It also requires a 3D scanner for validation. However, the proposed method achieves similar results at low cost and lesser observer variability.

5 Conclusion

A novel 3D spine reconstruction technique is proposed by deforming the SCP model using the geometric features. Since these features are automatically extracted, the observer variability is reduced. More geometric features can be incorporated to increase the reconstruction accuracy. The number of points need to be identified on the radiographs is limited. Hence, the entire process is computationally inexpensive. The exclusion of 3D scanner for validation and the low cost calibration apparatus results in an economical procedure. The future work would be spinal deformity quantification using the proposed model. Thus, an economic, robust, and morpho-realistic 3D spine reconstruction technique is proposed. This model can be a used for the diagnosis of the patients with spinal deformities in any clinical environment.

Acknowledgments The authors would like to acknowledge the Department of Science and Technology (DST), Government of India. This project is funded under a SERB-DST, Fast Track Scheme for Young Scientists. The authors also recognize the help extended by the faculty, Kasturba Medical College, Manipal, in data acquisition and expert opinion.

References

1. Weinstein, S., Dolan, L., Cheng, J., Danielsson, A., Morcuende, J.: Adolescent idiopathic scoliosis. The Lancet. **371**(9623), 1527–1537 (2008)
2. Yazici, M., Acaroglu, E., Alanay, A., Deviren, V., Cila, A., Surat, A.: Measurement of vertebral rotation in standing versus supine position in adolescent idiopathic scoliosis. J. Pediatr. Orthop. **21**(2), 252–256 (2001)
3. Moura, D.C., Barbosa, J.G., Reis, A.M., Tavares, J.M.R.S.: A flexible approach for the calibration of biplanar radiography of the spine on conventional radiological systems. Comput. Modell. Eng. Sci. **60**(2), 115–138 (2010)
4. Labelle, H., Belleeur, C., Joncas, J., Aubin, C., Cheriet, F.: Preliminary evaluation of a computer-assisted tool for the design and adjustment of braces in idiopathic scoliosis: a prospective and randomized study. Spine **32**(8), 835–843 (2007)
5. Aubin, C.E., Dansereau, J., Parent, F., Labelle, H., de Guise, J.A.: Morphometric evaluations of personalized reconstructions and geometric models of the human spine. Med. Biol. Eng. Comput. **35**, 611–618 (1997)
6. Andre, B., Dansereau, J., Labelle, H.: Optimized vertical stereo base radiographic setup for the clinical three-dimensional reconstruction of the human spine. J. Biomech. **27**, 1023–1035 (1994)
7. Mitton, D., Landry, C., Verson, S., Skalli, W., Lavaste, F., De Guise, J.A.: 3D reconstruction method from biplanar radiography using non-stereo corresponding points and elastic deformable meshes. Med. Biol. Eng. Comput. **38**, 133–139 (2000)
8. Mitulescu, A., Skalli, W., Mitton, D., De Guise, J.A.: Three-dimensional surface rendering reconstruction of scoliotic vertebrae using a non-stereo-corresponding point's technique. Eur. Spine J. **11**(4), 44–52 (2002)
9. Pomero, V., Mitton, D., Laporte, S., De Guise, J.A., Skalli, W.: Fast accurate stereo-radiographic 3D-reconstruction of the spine using a combined geometric and statistic model. Clin. Biomech. **19**, 240–247 (2004)
10. Humbert, L., De Guise, J.A., Aubert, B., Godbout, B., Skalli, W.: Reconstruction of the spine from biplanar X-rays using parametric models based on transversal and longitudinal inferences. Med. Eng. Phys. **31**, 681–687 (2009)
11. Dumas, R., Blanchard, B., Carlier, R., De Loubresse, C.G., Le Huec, J.C., Marty, C., Moinard, M., Vital, J.M.: A semi-automated method using interpolation and optimization for the reconstruction of the spine from bi-planar radiography. Med. Biol. Eng. Comput. **46**, 85–92 (2008)
12. Zhang, J., Lv, L., et al.: 3D reconstruction of the spine from biplanar radiographs based on contour matching using the Hough transform. IEEE Trans. Biomed. Eng. **60**, 7 (2013)
13. Zhang, J., Shi, X., et al.: A simple approach for 3D reconstruction of the spine from biplanar radiography. In: Proceedings of Sixth International Conference on Digital Image Processing (ICDIP 2014), SPIE Proceedings 9159 (2014)
14. Moura, D.C., Barbosa, J.G.: Real-scale 3D models of the scoliotic spine from biplanar radiography without calibration objects. Comput. Med. Imag. Graph. **38**(7), 580–585 (2014)
15. Dansereau, J., Beauchamp, A., De Guise, J., Labelle, H.: Three-dimensional reconstruction of the spine and rib cage from stereoradiographic and imaging techniques. In: 16th Conference of the Canadian Society of Mech Eng, Toronto. II, pp. 61–64 (1990)

16. Abdel Aziz, Y.I., Karara, H.M.: Direct linear transformation from comparator coordinates into object space coordinates in close-range photogrammetry. In: Proceedings of the Symposium on Close-Range Photogrammetry, Falls Church. VA, American Society of Photogrammetry, pp. 1–18 (1971)
17. Bates, D.M., Watts, D.G.: Nonlinear regression and its applications. Wiley, New York (1988)
18. Kumar, S., Nayak, K.P., Hareesha, K.S.: Semiautomatic method for segmenting pedicles in vertebral radiographs. In: Proceedings of ICCCS'12. Procedia Technology. Elsevier, vol. 6, pp. 39–48 (2012). doi:10.1016/j.protcy.2012.10.006
19. De Guise, J.A., Martel, Y.: 3D-biomedical modeling: merging image processing and computer aided design. In: Proceedings Annual International Conference on IEEE Engineering Medicine Biology Society, New Orleans, pp. 426–427 (1988)
20. Trochu, F.: A contouring program based on dual kriging interpolation. Eng. Comput. **9**, 160–177 (1993)
21. Wei, Z., Hua, Y.: X-ray image enhancement based on multiscale morphology. Sci.Tech. Res. Dev. Progr. Hebei Province, IEEE (2007)
22. Kass, M., Witkin, A., Terzopoulos, D.: Snakes - active contour models. Int. J. Comput. Vision **1**, 321–331 (1998)

Analysis of Network IO Performance in Hadoop Cluster Environments Based on Docker Containers

P. China Venkanna Varma, K.V. Kalyan Chakravarthy, V. Valli Kumari and S. Viswanadha Raju

Abstract Information technology (IT) is creating huge data (big data) everyday. Future business intelligence (BI) can be estimated from the past data. Storing, organizing, and processing big data is the current trend. NoSQL (Moniruzzaman and Hossain, Int J Database Theory Appl 6(4), 2013) [1] and MapReduce (Dean and Ghemawat, MapReduce: simplified data processing on large clusters) [2] technologies find an efficient way to store, organize, and process the big data with commodity hardware using new technologies such as virtualization and Linux containers (LXC) (Sudha et al, Int J Adv Res Comput Sci Softw Eng 4(1), 2014) [3]. Nowadays, all data center services are based on the virtualization and LXC technologies for better resource utilization. Docker (Anderson, Docker software engineering, 2015) [4]-based containers are lightweight virtual machines (VM) being adapted rapidly in hosting big data applications. Docker containers (or simply containers) run inside an operating system (OS) based on Linux Kernel version 2.6.29 and above. Running containers in a virtual machine is a multi-tenant model for scaling in data center services. This leads to higher resource utilization in the data centers and better operational margins. As the number of live containers increases the central processing unit (CPU)'s context switch latency for each live container significantly increases. This will reduce the input and output (IO) throughput of the containers. We observed that the network IO throughput is inversely proportional to the number of live containers sharing the same CPU. The scope of this paper is limited to the network IO

P. China Venkanna Varma (✉)
VistaraIT Inc., Hyderabad, India
e-mail: pc.varma@gmail.com

K.V. Kalyan Chakravarthy
Andhra University, Visakhapatnam, India
e-mail: tokalyankv@gmail.com

V. Valli Kumari
Department of CSSE, Andhra University, Visakhapatnam, India
e-mail: vallikumari@gmail.com

S. Viswanadha Raju
Department of CSE, JNTUH, Hyderabad, India
e-mail: svraju.jntu@gmail.com

M. Pant et al. (eds.), *Proceedings of Fifth International Conference on Soft Computing for Problem Solving*, Advances in Intelligent Systems and Computing 437, DOI 10.1007/978-981-10-0451-3_22

throughput which creates a bottleneck in big data environments. As part of this paper, we studied the working of Docker networks, various factors of CPU context switch latency and how network IO throughput will be impacted with the number of live Docker containers. A Hadoop cluster environment built and executed benchmarks such as TestDFSIO-write and TestDFSIO-read against varying number of the live containers. We observed that Hadoop throughput is not linear with increasing number of live container nodes sharing the same system CPU. The future work of this paper can be extended to analyze the practical implications of network performance and come up with a solution to enhance the performance of the Hadoop cluster environments.

1 Introduction

Raw data are unorganized facts that need to be processed. When data are processed, organized, structured, or presented in a given context so as to make it useful, they are called information. Every business is powered with IT. Future business intelligence can be derived from the past data. Business data grows everyday and there is a potential demand to process the data for rapid and accurate business decisions. Data scientists break big data into five dimensions volume, value, velocity, variety, and veracity (simply big data 5V's). Volume refers to vast amounts of data generated every second. This leads to address how to determine the relevance within the data sets and create value for the business. Velocity means data should be processed at the streaming/gathering speed. This is a challenge for big data correlation and analysis. Variety means that the data comes in different types. It includes structured and unstructured data. Managing, merging, and processing the variety data is again a challenge. Big data veracity refers to the biases, noise, and abnormality in data.

Hadoop [5] is an open-source software framework for storing and processing big data in a distributed fashion. Hadoop does two things, massive data storage and superfast processing. Hadoop is the de facto standard for storing, processing, and analyzing petabytes of data.

At the same time, virtualization technology has become the de facto standard for all public and private cloud requirements. Virtualization has consolidated all the hardware components and created a software redundancy layer on which business applications runs and moves among the hardware hosts for elastic work loads without downtime. Docker technology introduced lightweight virtual machines on top of Linux containers called Docker containers.

Nowadays, data centers are planning to run Hadoop cluster nodes (HCN) on the Docker containers for elastic work loads and multi-tenant service models. Running Docker in production is a beast because it should scale up to the petabytes of data. Hadoop cluster nodes require good amount of CPU cycles, random access memory (RAM) and resource IO for Map-Reduce and data read/write operations. Virtualization solved orchestration of the dynamic resources for data centers, but virtualization overhead is the biggest penalty. "Every single CPU cycle that goes to your Hypervisor is wasted, and likewise every byte of RAM".

As the number of Docker containers increases on a VM or physical machine, there is a significant overhead on the big data operations executed among the Hadoop cluster nodes. This overhead is due to the network IO RTT increase when the CPU is busy. If the CPU is busy, the network IO operation is in pending state and waiting for its scheduled slot to execute; this will increase the RTT of the network IO packet and leads to low network throughput in Hadoop environments. Ideally, Hadoop offers performance linearly by adding new nodes to the cluster. Based on our use cases and test results, we observed that the Hadoop throughput is not linear with the increasing number of nodes sharing the same CPU.

The outline of this paper is as follows: (1) Explore how Docker networking works (Sect. 2). (2) Identify the important factors that effect the system CPU context switch (Sect. 3). (3) Experiment setup, big data environment and test cases (Sect. 4). (4) Analysis of the network IO throughput (Sect. 5). (5) Conclusion (Sect. 6).

2 Docker Networking

Docker has a software ethernet bridge called "docker0" built on top of Linux "bridge-utils" [6]. When Docker starts, it will create a default bridge "docker0" inside the Linux kernel and a virtual subnet to exchange the packets between containers and host. Docker randomly chooses the IP address range not used by host and defined by RFC 1918. Every time a container is created, a pair of virtual interfaces are created that are similar to both the ends of a tube, a packet delivered from one end arrived to another end. One virtual interface is given to the container and another is hooked into docker0 to communicate with host and other containers. Figure 1 describes the Docker's docker0 and virtual interfaces.

Communication between containers depends on the network topology and system firewall (iptables). By default, docker0 bridge allows inter-container communication. Docker never does changes to the iptables to allow a connection to docker0 bridge, it uses a flag—iptables=false, when Docker starts. Otherwise, Docker will add a default rule to the FORWARD chain with a blanket ACCEPT policy if you retain the default—icc=true, else will set the policy to DROP if—icc=false.

We observed that Docker adds a network virtualization layer in the form of a software bridge which adds an extra hop in the network path; it will add some delay in the network packet propagation. Figure 2 describes the network virtualization layers.

Each Hadoop (cluster) node running on a container should communicate with others. Master (NameNode, JobTracker) node should communicate with slave (DataNode, JobTracker) nodes for all data and map/reduce operations. The better network IO throughput leads to better Hadoop cluster throughput.

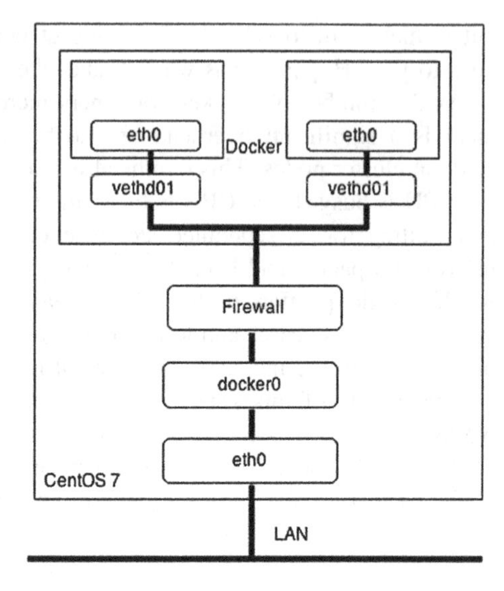

Fig. 1 Describes the Docker's docker0 bridge and virtual interfaces

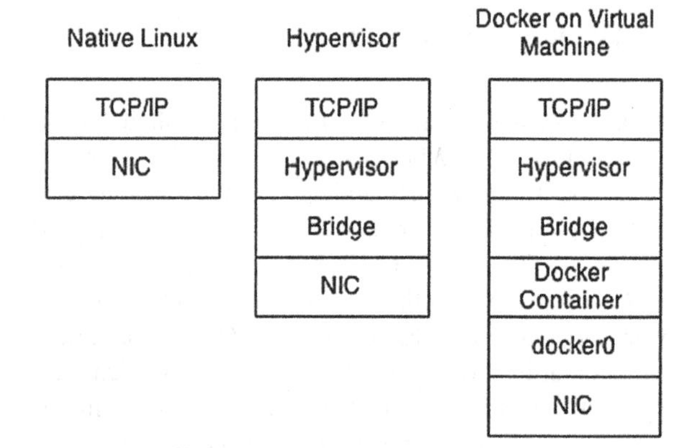

Fig. 2 Describes the network layers for native, hypervisor, and Docker containers

3 Identify the Important Factors that Effects the System CPU Context Switch

"Context switch is the process of storing and restoring the state (context) of a process so that execution can be resumed from the same point at a later time." Context switch is applicable to a set of running processes in the same system CPU. There is a significant time required by the OS kernel to save execution state of the current running processes to the memory, and then resume the state of another process.

3.1 Process Scheduling Algorithms

Process scheduling algorithms play a key role in the context switching. Each algorithm developed is based on a use case. There is no ideal algorithm developed for all the known use cases. There is no choice given to end users to tune the scheduling algorithms. Improving the performance by tuning process scheduling algorithms is not in the scope of this paper.

3.2 Multiple-Processor Scheduling

Multiple-processor scheduling is a complicated concept. Its main aim is to keep all the processors busy, to get the best throughput. But if more and more context switch operations happen, it leads to less throughput. Improving the performance by tuning multiple-processor scheduling algorithms is also not in the scope of this paper.

3.3 Virtualization Extensions

To execute the process that belongs to a virtual machine, hypervisor will add an extra layer on top of the existing process control block (PCB) to identity the virtual machine and process. Coding and decoding of this extra information will add a significant overhead in the processes execution. Recent reports published that round trip time (RTT) latency of a virtual machine process request is around 40 nm.

4 Experiment Setup

VMware ESXi 5.5 x64 OS is installed on a baremetal Dell PowerEdge server with 3.0 GHz Intel Quad Core 2 CPU, 128 GB RAM, 3 TB 7200 RPM hard disk drive and a gigabit ethernet network interface (NIC). Figure 3 illustrates the experimental setup and components.

Provisioned 2 VMs, each with 60 GB RAM, 128 GB Virtual HDD, 2 vNICs and CentOS version 7.x x64 is installed with built-in Docker. A Hadoop cluster using the Ferry Opencore [7] framework was installed. Each Hadoop node will be created and configured with Hadoop application programming interface (API). Yet another markup language (YAML) configuration files are used to increase and decrease the Hadoop nodes to test the network IO throughput of the each node in association with overall system CPU load.

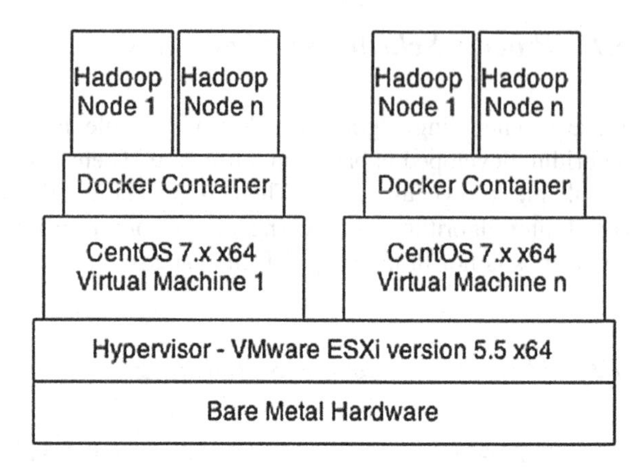

Fig. 3 Experiment setup

Each Hadoop node has one single flat table with enough data for the Hadoop benchmarks. Based on the data volume and variety, the number of Hadoop nodes will be increased/decreased dynamically. Following are the use cases considered to analyze the network IO performance.

1. Varying number of VM sharing the same CPU
2. Varying number of Docker containers sharing the same CPU

4.1 Big Data Environment for Analysis

Installed the Hadoop cluster version 2.7.0 with Hive (version 0.12) using the Opencore Ferry framework. Following is the configuration of the Hadoop cluster. Figure 4 describes the Hadoop environment for analysis.

backend: {storage: {personality: "hadoop", instances: 2 (will be increased), layers: hive}, connectors: {personality: "hadoop-client"}}

Following is the configuration of the each Hadoop node.

Distribution: Ferry Hadoop 2.7.0
NameNode, JobTracker: node0, Secondary NameNode: node1
Workers: All machines/Dockers
dfs.replication = 2, dfs.block.size = 134217728
io.file.buffer.size = 131072
mapred.child.java.opts = -Xmx2098 m Xmn512 m (native)
mapred.child.java.opts = -Xmx2048 m Xmn512 m (virtual)
Cluster topology:
Native, 1 VM per host: all nodes in the default rack 2, 4 VMs per host: each host is a unique rack

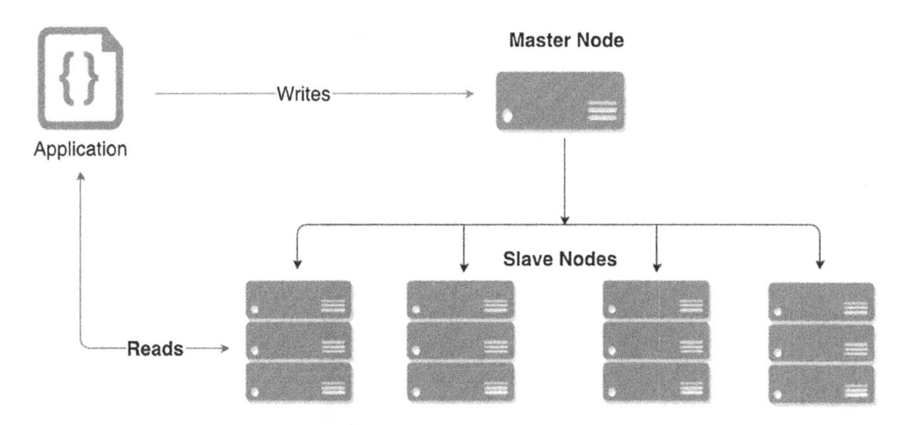

Fig. 4 Big data environment for analysis

5 Analysis of the Network IO Throughput

5.1 CPU Virtualization Layer Overhead

We would like to measure the CPU virtualization layer overhead. There is a tool "nuttcp" to measure the network IO throughput of a unidirectional data transfer over a single TCP connection with standard 1500-byte MTU. We used the nuttcp tool in three use cases.

1. Client-to-server acts as the transmitter.
2. Server-to-client acts as the receiver.
3. Both client-to-server and server-to-client directions since TCP has different code paths for sending and receiving.

All three use cases reach 9.3 Gbps in both the transmit and receive directions, very close to the theoretical limit of 9.41 Gbps due to packet headers. The bottleneck in this test is the network interface card (NIC) leaving other resources mostly idle. In such an I/O-bound scenario, we estimated overhead by measuring the amount of CPU cycles required to transmit and receive data.

Figure 5 shows system-wide CPU utilization for this test, measured using a Linux command "perf stat -a" [5]. The network virtualization in Docker, i.e., Docker's bridging and NAT noticeably increases the transmit path length; Docker containers that do not use NAT have identical performance to the native Linux. In real network-intensive workloads, we expect such CPU overhead will reduce the performance.

Results show that there is a significant overhead added by the virtualization layer. Native application always performs well, compared to the application on the virtualized environments. Both process scheduling and virtual CPU architecture are the important factors for the network IO throughput. If the number of

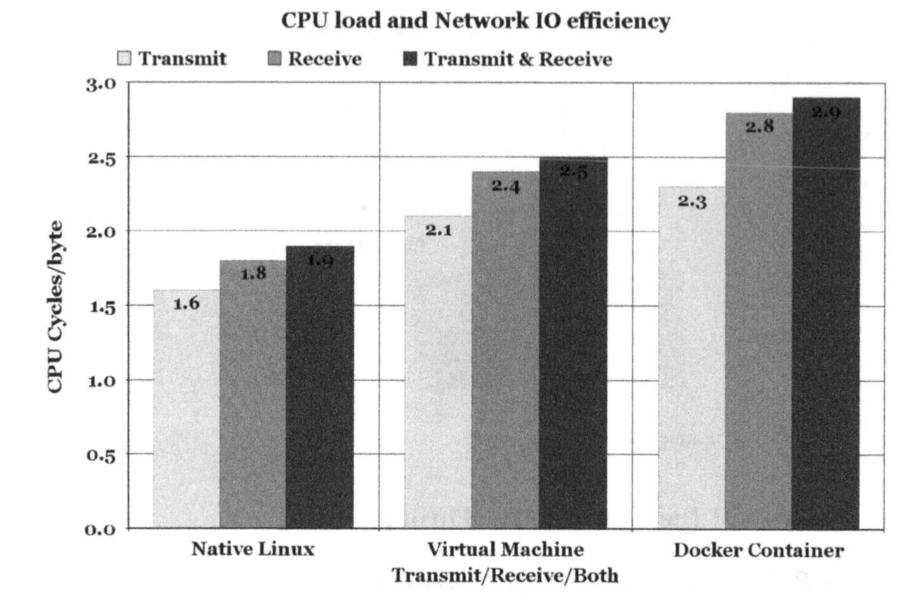

Fig. 5 CPU load and network IO efficiency

process/virtual machines/containers increases on the same CPU, the context switch latency will be increased.

5.2 Network Virtualization Overhead

We again used nuttcp tool to measure network bandwidth between the system under test and an identical machine connected using a direct gigabyte ethernet (GBE). Docker attaches all containers on the host to a bridge and connects the bridge to the network via NAT.

We used the netperf request-response benchmark to measure round-trip network latency using similar configurations. In this case, the system under test was running the netperf server (netserver) and the other machine ran the netperf client. The client sends a 100-byte request, the server sends a 200-byte response, and the client waits for the response before sending another request. Thus, only one transaction is in flight at a time.

Figure 6 describes that the Docker's virtualization layer, increases the latency around 5 s compared to the native and hypervisor environments. Hypervisor adds 30 s of overhead to each transaction compared to the native network stack, an increase of 80 %. TCP and UDP have very similar latency, because in both the cases a transaction consists of a single packet in each direction.

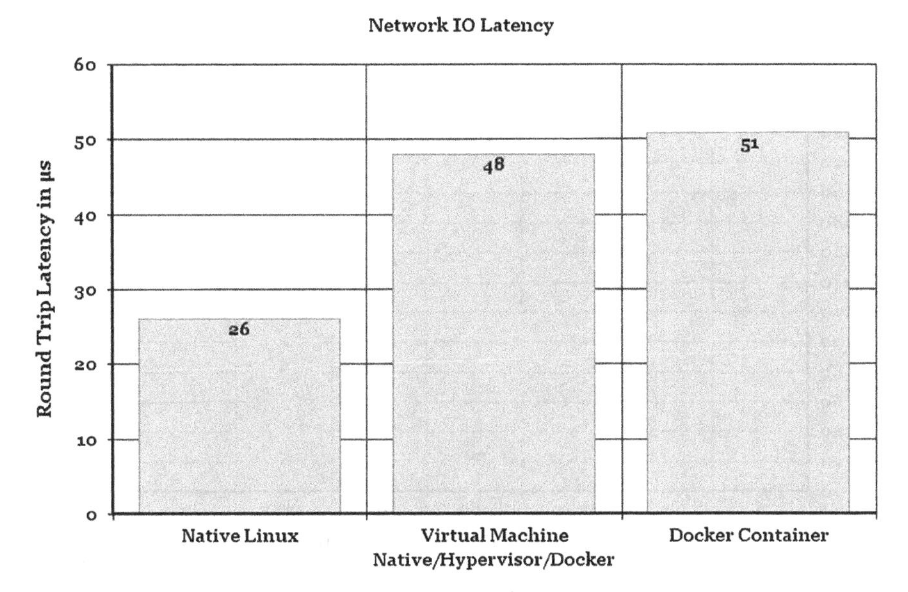

Fig. 6 Describes the transaction latency for native, hypervisor, and Docker container environments

5.3 Number of Docker Containers in the Same Virtual Machine

The same experiment was executed by varying the number of Docker containers within the same VM. Each Docker node is given 4 GB of RAM and performed 1000 Map-Reduce operations per minute from a neighbor container. Figure 7 describes that, as the number of Docker containers increases in the same VM, the RTT of the network IO packets also increases proportionally.

It describes that, adding more nodes to Hadoop will increase the performance of the system. But, adding more nodes (Docker containers) beyond a limit will not increase the performance of the system linearly. We executed the Hadoop benchmarks [5] Pi, TestDFSIO-write, and TestDFSIO-read by varying the number of the containers on the same VM. For all benchmarks we observed the elapsed time in seconds (lower is better). Figure 8 describes that READ operations with TestDFSIO-read benchmark are getting better with number containers. But, WRITE operations with TestDFSIO-write are blocked somewhere and not getting the expected performance. As the number of containers increases, elapsed time also increased. Hadoop nodes on the physical hardware (native) always resulted in good performance.

Finally, we found that Hadoop environment suffers from network IO throughput if the system CPU is busy. But for better data center operational margins, CPU is

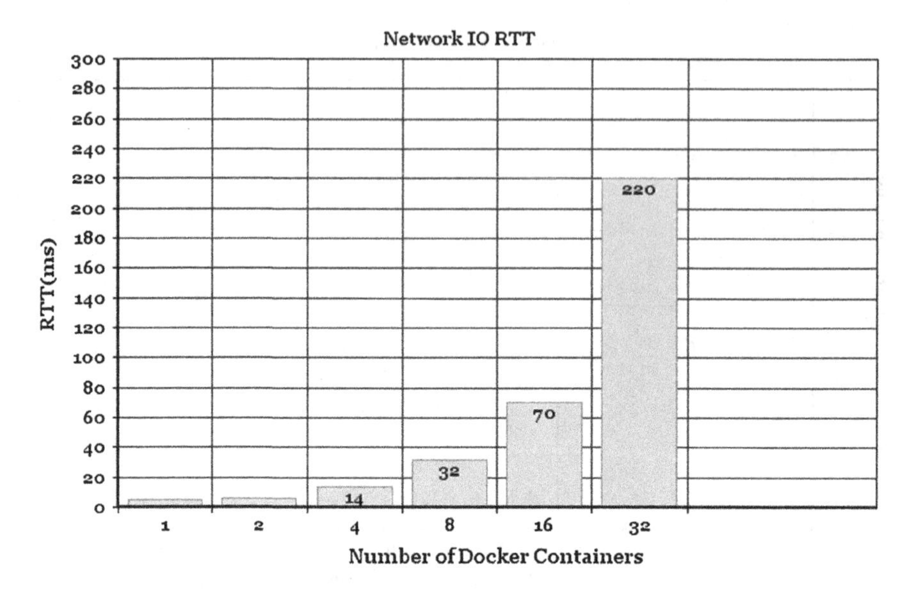

Fig. 7 Describes the increase of RTT with increase of the Docker containers in same virtual machine

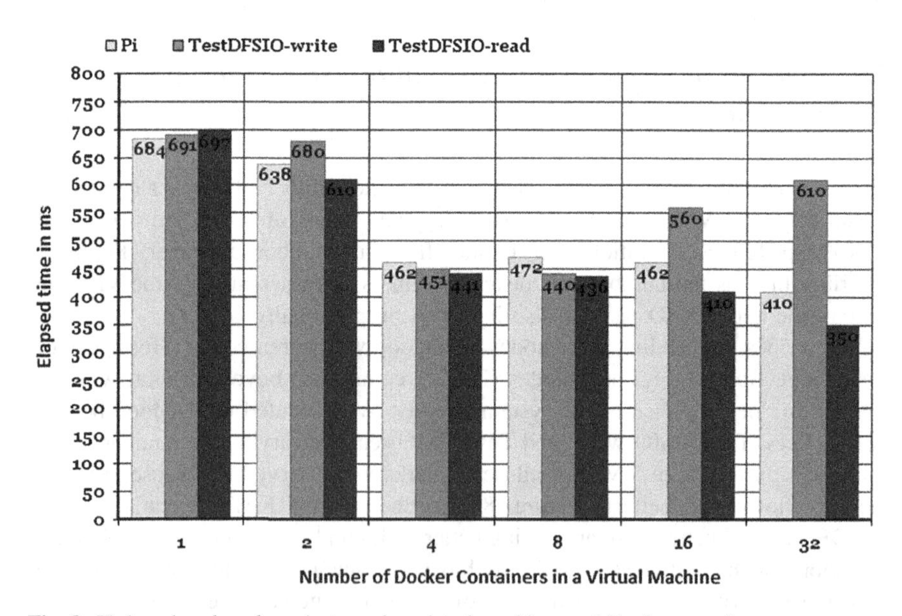

Fig. 8 Hadoop benchmark against number virtual machines and Docker containers

always kept busy. Busy CPU will increase the context switching latency, which leads to increase in the network IO RTT and lower throughput for the big data operations.

6 Conclusion

We have analyzed the factors that effects performance of the Hadoop nodes. We found that the network IO throughput is inversely proportional to the number of live Hadoop nodes on a VM sharing same CPU (Virtual CPU core). If the system CPU is busy in the context switching, it will add a significant latency to the RTT of the network IO packets. Increase in the RTT will reduce the throughput of the entire system. Ideally, adding nodes to the Hadoop cluster will improve the throughput of the system linearly. But, beyond a number, increasing Hadoop nodes will decrease the performance of the data WRITE operations due to increase in the RTT of the network IO packets of the containers that share same virtual CPU.

Future work of this paper would be studying the Docker bridge and to come up with a solution to improve the network IO throughput in big data environments. The Docker network bridge is a software bridge and no hardware components involved. We will study the whole TCP work flow and identify the overhead caused by the hardware components. Then we will try to build a solution to reduce hardware components overhead which will decrease the network IO packet's RTT. The solution would be a counter logic in the Docker network bridge and transparent to the containers as an open-source project.

References

1. Moniruzzaman, A.B.M., Hossain, S.A.: NoSQL database: new era of databases for big data analytics classification, characteristics and comparison. Int. J. Database Theory Appl. **6**(4). http://www.sersc.org/journals/IJDTA/vol6_no4/1.pdf. Accessed Aug 2013
2. Dean, J., Ghemawat, S.: MapReduce: simplified data processing on large clusters, Google, Inc. http://static.googleusercontent.com/media/research.google.com/en/us/archive/mapreduceosdi-04.pdf
3. Sudha, M., Harish, G.M., Usha, J.: Performance analysis of linux containers—an alternative approach to virtual machines. Int. J. Adv. Res. Comput. Sci. Softw. Eng. **4**(1), Jan 2014. http://www.ijarcsse.com/docs/papers/Volume_4/1_January2014/V4I10330.pdf. Accessed Jan 2014
4. Anderson, C.: Docker software engineering. The IEEE Computer Society, 2015. https://www.computer.org/csdl/mags/so/2015/03/mso2015030102.pdf
5. Shvachko, K., Kuang, H., Radia, S., Chansler, R.: The hadoop distributed file system. http://zoo.cs.yale.edu/classes/cs422/2014fa/readings/papers/shvachko10hdfs.pdf
6. Buell, J.: A benchmarking case study of virtualized hadoop performance on VMware vSphere 5. https://www.vmware.com/files/pdf/techpaper/VMWHadoopPerformancevSphere5.pdf
7. Opencore http://ferry.opencore.io/en/latest/

Hough Transform-Based Feature Extraction for the Classification of Mammogram Using Back Propagation Neural Network

Narain Ponraj, Poongodi and Merlin Mercy

Abstract Breast cancer is one of the significant health problems in the world. Early detection can improve the patient recover rate. Mammography is one of the most effective methods for detecting and screening breast cancer. This paper presents Hough transform-based feature extraction method for the classification of mammograms. Hough transform is generally used to separate certain features according to a defined shape within an image. Here, Hough transform is used to detect features of mammogram image and it is classified using back propagation neural network. This method is tested on 208 mammogram images provided by MIAS.

Keywords Hough transform · Breast cancer · SVM classifier · Back propagation neural network

1 Introduction

Cancer is a class of disease characterized by out of control cell growth in a particular location of the body. This cell growth is also referred as tumors. Breast cancer is formed when cancer develops from breast tissue. Breast cancer is one of the major health problems in the world. According to the statistics of world

Narain Ponraj (✉)
Karunya University, Coimbatore, India
e-mail: narainpons@gmail.com

Poongodi
Karpagam College of Engineering, Coimbatore, India
e-mail: poongodiravikumar@gmail.com

Merlin Mercy
Sri Krishna College of Technology, Coimbatore, India
e-mail: emercy86@gmail.com

M. Pant et al. (eds.), *Proceedings of Fifth International Conference on Soft Computing for Problem Solving*, Advances in Intelligent Systems and Computing 437, DOI 10.1007/978-981-10-0451-3_23

healthcare organization (WHO) in the 1960s and 1970s, a rapid increase in breast cancer was registered in the incident rates, adjusted by age, in the cancer records of several countries [1]. Early detection of breast cancer can increase the recovery rate to a great extent. Mammography is one of the standard diagnosis methods, which is used for early detection and treatment of breast lesions [2–4]. Also, it is the most popular diagnostic method to find breast cancer [5]. Breast imaging in mammography is mainly done with the help of low-dose X-rays, high resolution, high contrast, and a system for taking X-ray. Mammography can be used for both screening and diagnosing breast cancer. Full-field digital mammography (FFDM) is mainly used. This method can avoid unnecessary biopsies.

As a result of this, in the past decade many research have been carried out in this particular field to build a computational system to help the radiologist. They are mainly computer-aided detection (CAD) and diagnosis (CADx) systems [6]. This system can provide additional source of information and can increase the correct detection rates for diseases like breast cancer. The performance of the mammogram diagnosis methodologies would be improved if the possible locations of abnormalities are detected efficiently. CAD systems can be used for such situations and they are important and necessary for breast cancer control.

Mammography presents a methodology to help radiologist for detecting the masses in mammogram images and to classify them as normal or abnormal. Nowadays classifiers play a major role in medical diagnosis. Classifiers minimize the possibility of errors and can provide the examination results within a short period of time [6]. But to say that the system is efficient or not mainly depends on the techniques used for segmentation of the mammogram image and feature extraction. Standard enhancement techniques are used to sharpen the interested region of the mammogram image boundaries. Contrast enhancement is done between the region of interest and nearby normal tissue. ROIs are enhanced sharp edges or boundaries of mammogram image [7]. Then segmentation is done on ROI using common statistical, region-based, and morphological approaches. After segmentation Hough transform is done on ROI to extract features. Hough transform is an effective method to recognize complex patterns [8]. Hough transform is an image transform which can be used to isolate features of a particular shape within an image. Gradient decomposed Hough Transform improves accuracy and robustness for extracting eye center [9]. Here, we use generalized Hough transform. It provides us with a systematic way of introducing a priori knowledge about the expected transform [10]. Hough transform transforms the original image into a 2-D function. Feature extraction plays an important role in classification. The feature extracted from Hough transform is given to the classifiers to classify them as normal or abnormal. In this paper, intensity-based modulation is used for segmentation and the resultant output is classified with back propagation neural network and the obtained results were compared with SVM classifiers and GLCM classifier.

2 Methodology

Mammograms are preprocessed in the first stage. This is to increase the difference between needed objects and unwanted background noise. Preprocessing is used to increase the contrast of the mammographic images so that the difference in intensity level can be computed easily for the detection of masses. The block diagram for the entire process is shown in Fig. 1.

The flow of proposed methodology is explained below:

1. Gradient-based thresholding is used for background elimination. The Gradient threshold is found using Otsu's adaptive method. After gradient-based thresholding, the binary image is dilated using a structuring element. For an image of size $M * N$,

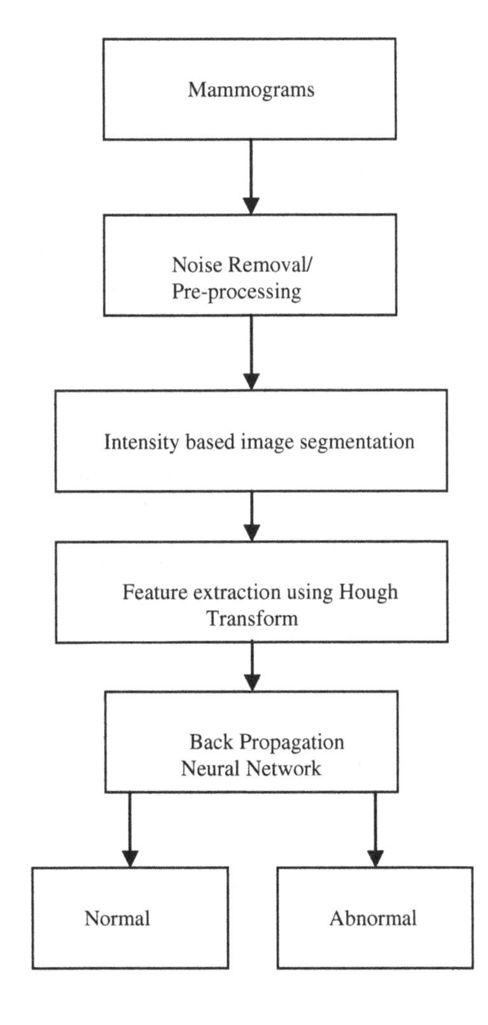

Fig. 1 Block diagram for mammogram classification

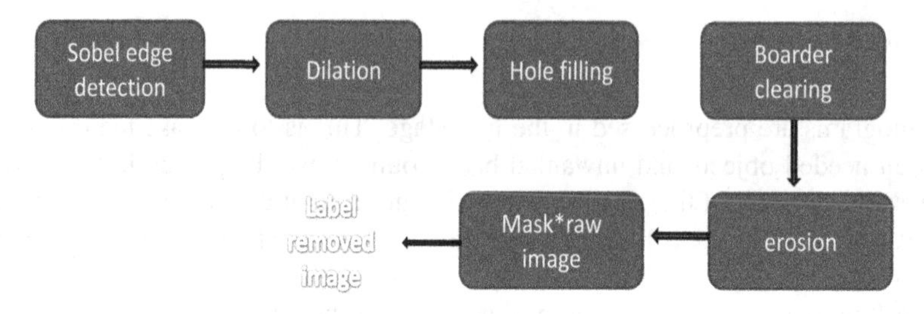

Fig. 2 Block diagram for gradient based thresholding

$$G_{\mathrm{T}} = \frac{1}{MN}\sum_{x=1}^{M}\sum_{y=1}^{N} G_{\mathrm{O}}(x, y) \tag{1}$$

2. The obtained mask is multiplied with the original image. The labels used in digital X-rays are removed when the binary image is multiplied with the original image. The detailed process is shown in Fig. 2.
3. Adaptive histogram equalization technique is then applied to enhance the image.
4. Once the maximum breast region is made available for further processing the pectoral muscle is removed using the proposed intensity-based classification method. This will classify data values based on the user defined intensity classes. For example, if we want to classify the whole image into two intensity classes we will define the input intensity class as two. The equation for intensity-based classifier can be written as

$$I_j = I_{\mathrm{L}} + j(I_{\mathrm{H}} + I_{\mathrm{L}})/K \tag{2}$$

   Here I_j is the jth intensity pixel and j is the intensity class that varies from (0, 1, 2, … j). I_{H} is the high intensity pixel. I_{L} is the low intensity pixel. K is the needed intensity class.
5. For removing the pectoral muscle without affecting the breast region the intensity class will take a value of 4.
6. Local variance and local mean can be done to generalize the intensity-based classification which will improve the accuracy rate of the classification process. This intensity-based classification process is mainly done to separate the pectoral muscle and the breast region into different classes so that segmentation of breast region can easily be obtained.
7. Local binary patterns can necessarily be used to obtain the texture pattern of the segmented image if needed before applying the Hough transform.
8. Hough transform is an image transform which is a derivative of radon transforms [11]. Hough transform calculates image function along the lines. Hough transform is applied on the intensity-based classifier output for feature extraction.

9. Features are calculated from the ROI. The combination of features should be selected carefully so that the classification accuracy will be higher. In this paper mean, variance, standard deviation, entropy are used to extract the features from the image after applying Hough transform.
10. Sample mammograms are taken from MIAS database for all classes like benign, malignant, and normal. After obtaining the features of mammogram image the values are given to a classifier. Back propagation neural network is used to classify the images. Though back propagation neural network requires more number of sample image data compared to other classifiers, it has the ability to approximate complex decision boundaries in a given feature space. The obtained results are compared with SVM and GLCM classifiers.

3 GLCM Feature Set

Apart from the general features extracted from the Hough Transform six other features were also extracted for GLCM classifier. The equation for six features are shown below:

$$\text{contrast} = \sum_{i=0}^{N_m-1} \sum_{j=0}^{N_m-1} (i-j)^2 \cdot \log(m(i,j)) \tag{3}$$

$$\text{Correlation} = \sum_{i=0}^{N_m-1} \sum_{j=0}^{N_m-1} (i-\mu) \cdot (j-\mu) \cdot \frac{m(i,j)}{\sigma^2} \tag{4}$$

$$\text{Indifference Moment} = \sum_{i=0}^{N_m-1} \sum_{j=0}^{N_m-1} \left[\frac{1}{1+(i-j)^2}\right] m(i,j) \tag{5}$$

$$\text{Energy} = \sum_{i=0}^{N_m-1} \sum^{N_m-1} m^2(i,j) \tag{6}$$

$$\text{Entropy} = -\sum_{i=0}^{N_m-1} \sum_{j=0}^{N_m-1} m(i,j) \cdot \log(m(i,j)) \tag{7}$$

$$\text{Variance} = \sum_{i=0}^{N_m-1} \sum_{j=0}^{N_m-1} (i-\mu)^2 \cdot (m(i,j)) \tag{8}$$

4 Result

There are mainly three stages in this work (1) Region of Interest (2) Feature extraction, and (3) Classification using back propagation neural network. The above mentioned stages are applied to the entire 205 mammogram image (101 normal mammogram images and 104 abnormal images). The images were in 8-bit gray resolution format and of size 1024 * 1024 pixels. Figures 3 and 4 show the Hough transformed image of a mammogram.

After applying Hough transform to the mammogram image, features are taken from it. The parameters obtained are listed in the Table 1.

Figure 5 depicts the output of the entire flow process. The raw image is taken and using gradient thresholding the label is removed. Pectoral muscle is removed after applying the new intensity classifier method. Then Hough transform is applied to the mammogram image and features are taken from it. The parameters obtained are listed in the Table 2.

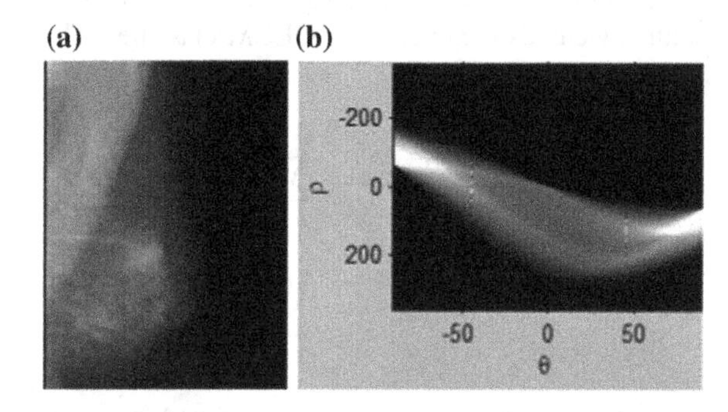

Fig. 3 **a** Mammogram image (Benign). **b** Hough transform image

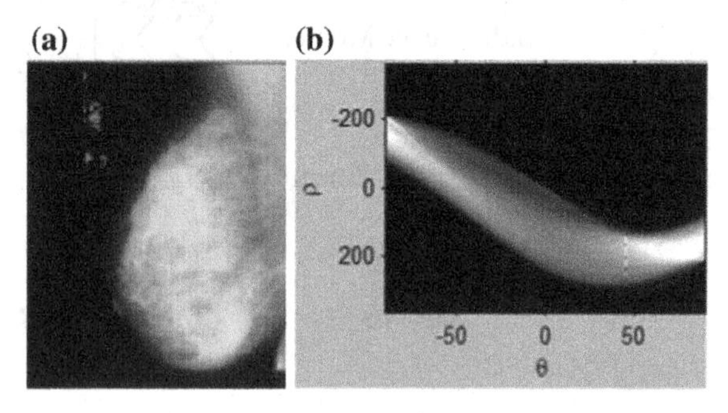

Fig. 4 **a** Mammogram image (Malignant). **b** Hough transform image

Table 1 Parameter table for mammogram image

Parameters	Image 1	Image 2
Mean	3.1675e+003	5.4902e+003
Variance	4.3774e+007	1.0600e+008
Entropy	0.9239	0.9502
Standard deviation	6.6162e+003	1.0296e+004

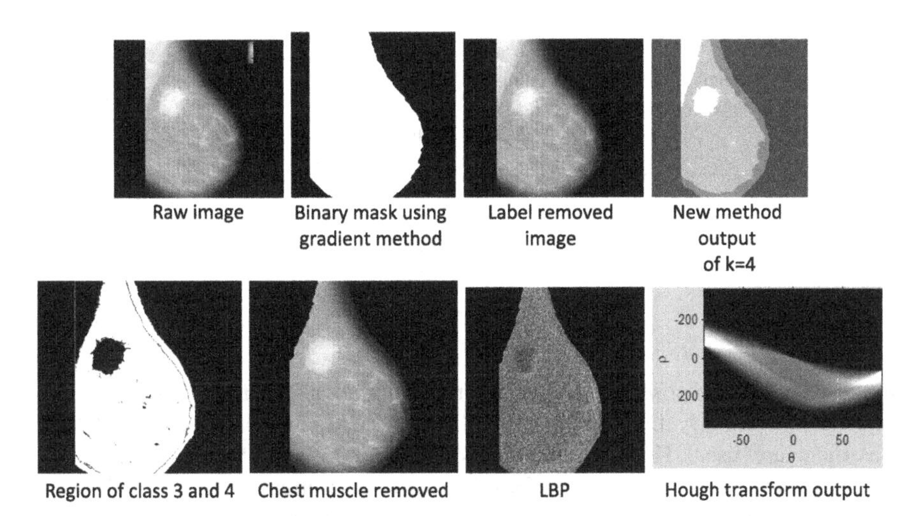

Fig. 5 Outputs obtained using the proposed methodology flow

Table 2 Parameter table for mammogram image with Hough functionals

Parameters	Image 1	Image 2
Mean	3.1675e+003	5.4902e+003
Variance	4.3774e+007	1.0600e+008
Entropy	0.9239	0.9502
Standard deviation	6.6162e+003	1.0296e+004

Table 3 Parameter table for mammogram image using Hough and LBP

Parameters	Image 1	Image 2
Mean	4.4605	5.9196
Variance	73.8886	116.0682
Entropy	0.8774	0.8864
Standard deviation	8.5958	10.7735

For the experimental purpose, LBP-based texture output is obtained and then Hough Transform is applied and is tabulated in Table 3. Using the values in Table 4 recognition rate accuracy false acceptance rate (FAR), and false rejection rate (FRR) are calculated. Recognition rate of 74 %, accuracy of 70.83 %, FAR of 55 %, and of FRR 11 % were obtained using Hough functional. Recognition rate of 93.41 % accuracy of 85.83 % FAR 15 % FRR 11 % is obtained using local variance. Recognition rate of 85.41 % accuracy of 79.83 % FAR of 25 % and FRR of 21 % is obtained by neural network classifier. Recognition rate of 79.11 %, accuracy of 72.83 %, FAR of 35 %, FRR of 21 % are obtained using GLCM.

Table 4 Result obtained

	Trained images	Tested images	Using Hough transform	Using new preprocessing and local variance	Using neural network classifier	Using GLCM
Normal	25	35	27	33	30	31
Abnormal	25	35	24	31	28	32

5 Conclusion

In this paper, we proposed a classification methodology using Hough transform and LBP and to classify the images both SVM, GLCM, and also neural network classifiers are used. Hough transform was used rarely in the classification of mammogram. Earlier, Hough transform is used to discriminate between malignant tumors and nonmalignant masses, and obtained an accuracy of 74 % in detecting malignant tumors using a dataset of 34 computed radiography images including 14 malignant, 9 benign, and 11 normal cases. But we have used Hough transform to extract features and it was tested with digital mammogram images. Based on the results we have obtained, we found out that the proposed methodology is more efficient in classifying the mammogram images when SVM classifier is used. Also GLCM features provide better results than normal features. But as we increase the number of training images, we found that the classifier using back propagation neural network provides better results compared to other classifier types. Under the assumption of having abnormal images more than the normal images in sparse condition, this methodology can be used as one class classifier. The accuracy rate can be increased by using more features that suit Hough transform and using feature ranking criteria. We conclude that more number of training sets should be used to obtain better classification using back propagation neural networks.

References

1. N.C.I. (NCI): Cancer Stat Fact Sheets: Cancer of the Breast. http://seer.cancer.gov/statfacts/html/breast.htmlS (2010)
2. Wang, Y., Gao, X., Li, J.: A feature analysis approach to mass detection in mammography based on RF-SVM. ICIP07, pp. 9–12 (2007)
3. Hassanien, A.: Fuzzy rough sets hybrid scheme for breast cancer detection. Image Vision Comput. **25**(2), 172–183 (2007). ISSN:0262-8856, http://dx.doi.org/10.1016/j.imavis.2006.01.026
4. Suckling, J., Parker, J., Dance, D., Astley, S., Hutt, I., Boggis, C., et al.: The mammographic images analysis society digital mammogram database. Exerpta Med. **1994**, 375–378 (1069)
5. Fenton, J.J., Taplin, S.H., Carney, P.A., Abraham, L., Sickles, E.A., D'Orsi, C., Berns, E.A., Cutter, G., Hendrick, R.E., Barlow, W.E., Elmore, J.G.: Influence of computer-aided detection on performance of screening mammography. N. Engl. J. Med. **356**(14), 1399–1409 (2007). doi:10.1056/NEJMoa066099

6. Huang, C.L., Liao, H.C., Chen, M.C.: Prediction model building and feature selection with support vector machines in breast cancer diagnosis. Expert Syst. Appl. **34**(1), 578–587 (2008)
7. Beghdadi, A., Negrate, A.L.: Contrast enhancement technique based on local detection of edges. Comput. Vision Graphics Image Process. **46**, 162–174 (1989)
8. Hough, P.: Method and means for recognizing complex patterns. U.S. Patent 3069654 (1962)
9. Benn, D.E., Nixon, M.S., Carter, J.N.: Robust eye centre extraction using the hough transform. Audio Video-Based Biometric Person Authent. Lecture Notes Comput. Sci. **1206**, 1–9 (1997)
10. Aggarwal, N., Karl, W.C.: Line detection in images through regularized hough transform. IEEE Trans. Image Process. **15**(3), 582–591 (2006)
11. Kadyrov, A., Petrou, M.: The trace transform and its applications. Ph.D. Dissertation, School Electron. Eng., Inf. Technol. Math., Univ. Surrey, U.K. (2001)

Ant Colony System-Based E-Supermarket Website Link Structure Optimization

Harpreet Singh and Parminder Kaur

Abstract The webgraph link structure of an E-supermarket website is different from static websites. E-commerce companies want to make the browsing experience of their customers hassle free. This can be done by frequently rearranging the link structure of the website according to changing browsing patterns of the users. Only few methods have been developed to optimally design the link structure of an E-supermarket website. In this paper, an ant colony system (ACS) is used to optimize the link structure of E-supermarket website. It has been observed that the ant colony-based method improves the link structure by decreasing the overall weighted distance between webpages. The performance issues of ACS-based method for E-supermarket link structure optimization are also presented.

Keywords Ant colony system · E-supermarket · Optimization · Website link structure

1 Introduction

A lot of companies are moving towards e-commerce and selling their products through websites. The complex link structure of websites and the maze of hyperlinks and webpages encourage prospective customers to leave the website [1]. The website link structure optimization problem is of combinatorial nature [2]. E-supermarket websites fall in the category of dynamic websites where the contents of the webpages change frequently due to addition of new products and removal of old products information from the webpages. The website managers want their

Harpreet Singh (✉)
Department of Computer Science and Engineering, DAV University,
Jalandhar, India
e-mail: harpreet111.singh@gmail.com

Parminder Kaur
Department of Computer Science, Guru Nanak Dev University, Amritsar, India
e-mail: parminderkaur@yahoo.com

M. Pant et al. (eds.), *Proceedings of Fifth International Conference on Soft Computing for Problem Solving*, Advances in Intelligent Systems and Computing 437, DOI 10.1007/978-981-10-0451-3_24

customers to visit the pages that contain information about the products that yield more profit for the company [2]. On the other hand, the website users want to quickly get to the page that contains the information of interest to them. The browsing behaviors of users change frequently because users visit the website with a different target every time [3]. Hence to adapt according to the changing browsing patterns, the website link structure is needed to be modified regularly. While browsing, users face problems such as information overload [4] and more time taken to reach the target page. Information overload occurs when a user at a webpage finds too many hyperlinks to follow and get confused. Users take more time to get to the required page when the hyperlinks that should be present are not present actually.

2 Related Work

Gorafalakis et al. proposed an approach [5] in which pages that receive more hits are moved up in the hierarchy of webgraph by analyzing the user browsing patterns stored in the weblogs. But these methods are not applicable on e-commerce website link graphs because of the constraints that are to be satisfied. Wang et al. [2] developed a mathematical model to represent the link structure improvement problem and also proposed two methods to solve this problem. One method is based on Hopfield neural networks to solve the small-scale problems and another method is based on tabu search to solve the large-scale problems. Chen et al. [6] used data mining algorithms on webserver logs to obtain the paths that are traversed mostly by website users. The information gathered can be used by website managers to add or remove links in the link structure. Lin and Tseng [7] have developed an ant colony system(ACS)-based algorithm that tries to optimize the link structure by using the food foraging behavior [8] of ants. Zhuo et al. [9] proposed a method that finds the association between the webpages to add additional links. Yin and Guo [10] has recently proposed a new mathematical model and enhanced tabu search method for the optimization of webgraph structure. Chen and Ryu [11] have developed a mathematical programming model to facilitate the navigation for users. This model uses a penalty for those pages that break the constraints so that the alterations to the original link structure are minimized. Lin [4] proposed a 0–1 programming models to optimize the website structures. This approach was not applicable on webgraphs of more than hundred nodes. In this paper, ant colony system [7] has been used to optimize the link structure of an E-supermarket website.

3 Formulation of Link Structure Optimization Model

Figure 1 shows a directed graph that represents the link structure of E-supermarket website. It contains two types of webpages: category pages and product pages.

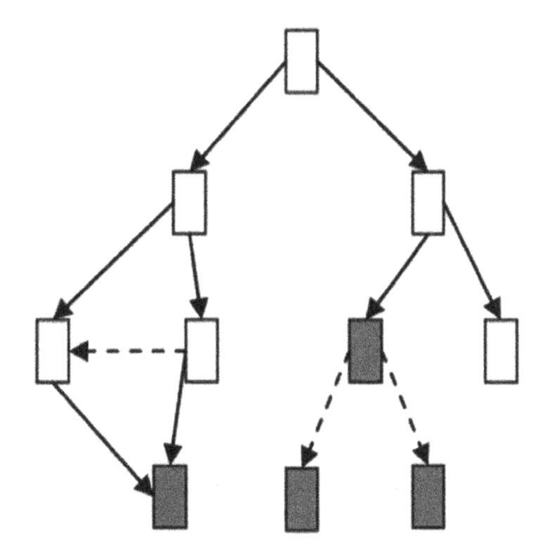

Fig. 1 Website link structure

Category pages are intermediate index pages and product pages contain information about products. Here, dark rectangles represent product pages and rectangles that are not filled represent category pages. The arrows represent the hyperlinks between webpages. Hyperlinks fall into two categories: basic links and add on links. Basic links are in the upper hierarchy level of the link structure. The source nodes of basic links are category pages. Add on links are originated from either category or product pages. Basic links are seldom added or removed but add on links are added or removed frequently according to changing user behavior.

Wang et al. [2] proposed the formulation of the optimization problem for E-supermarket websites. The same formulation has been used in this paper. Some profit factor is attached with every product that represents the profit earned by selling one unit of that product. The symbol p_j represents the profit factor associated with product *j*. For category *j*, symbol p_j represents the average profit factor of all the products with in that category. The relationship between a product/category *i* with another product/category *j* is represented by symbol $u_{i,j}$ where $u_{i,j} \in \{0, 1\}$. If $u_{i,j} = 1$, it means the *i*th product (or category) is associated with *j*th product (or category). The collective sales of the products would help to calculate the correlation coefficient [2] between various products or categories. The correlation coefficient between product/category i and product/category j is represented by symbol $c_{i,j}$ where $0 \leq c_{i,j} \leq 1$. The formulation is described below [2]:

$$\begin{aligned} \mathrm{Min(X)} = & \sum_{i \in V} \sum_{j \in V} c_{i,j} p_j d_{i,j}(X)\left(1 - \theta_{i,j}(X)\right) + \alpha \left[\sum_{i \in V} \sum_{j \in V} \theta_{i,j}(X)\right]^2 \\ & + \omega_1 \left[\sum_{i \in V} \sum_{j \in V} (x_{i,j} - e_{i,j})\right]^+ - \omega_2 \left[\sum_{i \in V} \sum_{j \in V} (e_{i,j} - x_{i,j})\right]^+ \end{aligned} \tag{1}$$

$$\text{st.} \quad x_{i,j} - b_{i,j} \geq 0 \quad i,j \in V \tag{2}$$

$$u_{i,j} - x_{i,j} \geq 0 \quad i,j \in V \tag{3}$$

$$\sum_{j \in V} \left| x_{i,j} - e_{i,j} \right| \leq O_i \quad i \in V \tag{4}$$

$$\sum_{i \in V} \left| x_{i,j} - e_{i,j} \right| \leq I_i j \in V \tag{5}$$

$$\theta_{i,j}(X) = \begin{cases} 1, & d_{i,j}(X) = \infty \\ 0, & \text{otherwise,} \end{cases}$$

If link from node *i* to node *j* exist in the optimized structure then the variable $x_{i,j}$ is set to 1; otherwise $x_{i,j} = 0$. The element in the *i*th row and the *j*th column of adjacency matrix [2] *E* of original structure is denoted by $e_{i,j}$. The link exist between node *i* and *j* in the initial structure if $e_{i,j} = 1$, otherwise not. Parameter b_{ij} represents a basic link, $b_{ij} = 1$ means the basic link (*i*, *j*) exists; otherwise not. The profit factor for product j is represented by p_j. The terms c_{ij} and d_{ij} denote the correlation coefficient and distance between nodes *i* and *j*, respectively. The distance between any two nodes is the length of the shortest path between them. The final objective is to obtain a link structure where the overall weighted distance between the pages is minimized. Equation (1) represents the objective function. The first term in the objective function denotes that pages with larger correlation × profit value should have less distance between them. The second term denotes the punishment for infinite distance. The third term in the objective function denotes the penalty for adding an edge. The last term represents the inducement of removing an edge. Parameters ω_1 and ω_2 represent penalty and inducement factors, respectively. Constraint (2) prevents removal of basic links. Constraint (3) restricts the addition of add-on links that joins the webpages without any relationship. The parameter u_{ij} takes binary values. If $u_{ij} = 1$, it means *i*th product or category is related with *j*th product or category; otherwise not. Constraints (4) and (5) correspond to the out-degree and in-degree variance respectively. The managers want the webpages of products of larger profits to be visited more and users want to get to the webpage of interest quickly. The interest of a customer in various products is reflected by correlation coefficient $c_{i,j}$ and p_j represents the profit generated by different products. The interests of both website managers and customers can be collectively represented by the product of $c_{i,j}$ and p_j known as CP value of webpages *i* and *j*. The CP value for link (*i*, *j*) can be defined as follows:

$$C_{ij}P_j = c_{i,j} \times p_j \tag{6}$$

The final objective is to generate a link structure that is profitable for managers and easily navigable for online customers. The term $C_{ij}P_j$ (CP) represents the degree of benefit congruence [2]. Hence the objective is to add the links with larger CP values under certain constraints.

4 Ant Colony System for E-Supermarket Link Structure Improvement

Ants release pheromone on the paths which they take while moving from nest to food source and on the way back [12]. The paths that get more concentration of pheromones are traveled by ants with more probability. Ant colony system [8] has been used to solve many optimization problems such as traveling salesman problem [8], job scheduling problem [13], quadratic assignment problem [12] etc. The proposed method works in two stages: First a spanning tree is generated to maintain the connectivity constraint [10]. Connectivity constraint specifies that any other webpage node should be reachable from home page. In the second stage, edges are successively added in the solution in decreasing order of their CP value. The stages are described below.

In the first stage, ant colony system [10] method is used to build the spanning tree. Unlike ant colony system for traveling salesman problem [8], every ant starts from the same node that is known as the home page node. In this stage, m ants are used to generate their individual solutions. Two sets are maintained during solution construction: solution set S and candidate set S^1. The solution set is a subgraph that consists of nodes and edges that have been included in the solution being constructed and candidate set consists of edges with source node in the solution set and an end node that is not a part of solution set. First, the solution set is empty. To start the solution generation, home page node is at first inserted in the solution set.

Suppose an ant is present at some node, all the outgoing edges with ending node not in the solution set are the candidate edges that can be included in the solution set. A random number q is generated, if $q \leq q_0$, where q_0 is a predefined parameter between [0, 1] then the ant selects edge (x, y) according to the following rule.

$$C_{xy}P_y = \max_{(i,j)\in c}\{C_{ij}P_j\}, \tag{7}$$

here $C_{ij}P_j$ is the CP value of edge from node i to j.

If $q > q_0$, then an edge is randomly selected from candidate set based on the probability distribution given by following equation.

$$p_{xy} = \begin{cases} \frac{C_{xy}\cdot P_y\cdot \tau(x,y)}{\sum_{(i,j)\varepsilon c} C_{ij}\cdot P_j\cdot \tau(i,j)}, & \text{if } (x,y) \in C, \\ 0, & \text{otherwise} \end{cases} \tag{8}$$

The term $\tau(i,j)$ represents the pheromone level on edge (i, j). The set of candidate edges is represented by $C = \{(i, j) \mid v_i \in S, v_j \notin S, v_i \notin S^1, v_j \in S^1\}$. After the addition of edge (x, y) in the solution, the outdegree of node v_x is checked. If the outdegree of node v_x is less than the specified limit then the outgoing edges of node v_x that does not have ending nodes in the solution set are added to the candidate set. If the outdegree of node v_x becomes equal to the specified limit then the all outgoing edges of node v_x that are present in the candidate set are removed from candidate set and also ruled out for further steps of current solution generation. The generation of spanning tree is stopped when the solution set contains all the nodes and the candidate set becomes empty.

In the second stage, another list of candidate edges in decreasing order according to their CP values is maintained and the edge with the largest CP value is picked from this list and added to the solution set if outdegree and other constraints are satisfied. The local pheromone updating rule is applied on the edges that are included in the current solution by the ant. The updating rule is given below.

$$\tau_{ij} \leftarrow (1-\rho)\tau_{ij} + \rho\Delta\tau \tag{9}$$

The symbol τ_{ij} represents the pheromone level on edge (i, j); $0 < \rho < 1$ is a parameter representing the local pheromone evaporation rate, and $\Delta\tau$ represents the variation in pheromone, which is set to be the initial pheromone level τ_0. The pheromone evaporation rule of Eq. (10) is applied on the edges that are not in the solution set obtained by ant.

$$\tau_{ij} \leftarrow (1-\rho)\,\tau_{ij} \tag{10}$$

Hence every ant generates its own solution using the above mentioned procedure. Once m ants make their own solution link structures, the best out of these m solutions is kept and global updating rule of Eq. (11) is applied on the edges of the globally best link structure obtained so far.

$$\tau_{ij} \leftarrow (1-\beta\tau_{ij}) + \beta\tau_{gb} \tag{11}$$

here $0 < \beta < 1$ represents the global pheromone evaporation rate, and

$$\tau_{\text{gb}} = \begin{cases} 0.1, \text{if edge}(i,j) \in \text{the globally best solution} \\ 0, \qquad\qquad\qquad\qquad \text{otherwise,} \end{cases}$$

4.1 Pseudocode of the Algorithm Is Given Below

```
Begin
Generate a random initial solution
Glb←Initial value of objective function;
lcl←Initial value of objective function;
While( Stopping condition is not met)
 For ant k= 1,..., m            // there are m ants
        $S_k \leftarrow \emptyset$; $S_k^1 \leftarrow \emptyset$; $C \leftarrow \emptyset$;
        $S_k \leftarrow S_k + v_1$;
        $C \leftarrow C + (1,z)$, $\forall(1,z) \in G$; $S_k^1 \leftarrow C + v_1$;
  While($C \neq \emptyset$)  //   first stage started
   If ($q \leq q_0$) //q is a random number generated every time
    The link (x, y) is selected according to equation (7)
   Else
    The link (x, y) is selected according to equation (8)
      $S_k \leftarrow S_k + (x,y) + v_y$;
      $C \leftarrow C - (t,y)$, $\forall(t,y) \in C$;
      $C \leftarrow C + (y,z)$, $\forall(y,z) \in G$;
            $S_k^1 \leftarrow C$; //the candidate solution set is up-
dated with candidate edge set
    If (degree of $v_x$ becomes equal to the maximum allowed
value)
             $C \leftarrow C - (x,g)$, $\forall(x,g) \in C$;
   End  // first stage ends here
    $C \leftarrow$the sorted edges according to decreasing CP value;
   While($C \neq \emptyset$)  // second stage started
    Find the edge (x, y) such that $C_{xy}P_y = \max_{(i,j)\in C}\{C_{ij}P_j\}$,
    If (degree of $v_x$ is less than the maximum allowed
value)
        $S_k \leftarrow S_k + (x,y)$;
        $C \leftarrow C - (x,y)$;
    Else
       $C \leftarrow C - (x,y)$;
   End     // second stage ends here
 If (f($S_k$)≤f(lcl))
   lcl = $S_k$;
Local pheromone update rule of equation (9) applied on
edges of $S_k$;
Local pheromone evaporation rule of equation (10) applied
on links not in $S_k$;
 End // end of for loop
   If(f(Glb)≥f(lcl))
     Glb=lcl;
   Global pheromone update rule using equation (11) ap-
plied on edges of the overall best solution found till
this point;
 End   // end of first while loop
End   // end of begin
```

5 Experiments and Results

All the experiments have been performed on artificially generated website link structures and web user data. The website link structures have been generated the same way as used by Wang et al. [2]. As described by Wang et al. [2], first the total number of nodes and links in the structure are to be decided. Then total number of category pages that are to be generated is decided. Then the number of category pages under every category page is randomly selected from interval of [5, 9]. For category pages that act as leaf nodes this number is 0. The number of product pages under every category page is then decided from the interval [5, 9]. The nodes generated till this time are connected by basic links. The outgoing links out of product pages are always add on links. Hence the number outgoing links out of product pages from the interval of [5, 7] are decided. Out of every category page one or two basic links are converted into add on links. Every category or product page should have one basic link pointing towards it. Now the value of correlation coefficient between every pair of nodes is to be decided. First, a text file containing a dataset of transactions is to be generated. Every transaction represents the products bought by a customer. Overall 1200 transactions have been generated. The total number of products and index numbers of products in each transaction are generated randomly. The concept of association rule mining is used to generate the correlation among webpages. Association rule mining [14] is used to discover all the association rules in the dataset that satisfy the user-specified minimum support (minsup) and minimum confidence constraints. For a rule A→B in a transactional dataset D, support [14] is percentage of transactions in D that contain A and B. It is represented as P(A∪B). The confidence [14] for the same rule is percentage of transactions in D containing A that also contain B. It is represented as the conditional probability P (B/A).

$$\text{Support}(A \to B) = P(A \cup B) = \text{Support}(A \cup B)$$

$$\text{Confidence}(A \to B) = P(B/A) = \text{Support}(A \cup B)/\ \text{Support}(A)$$

The formula for correlation coefficient [2] between nodes A and B is defined below:

$$C_{AB} = \text{Support}(A \to B) \times \text{Confidence}(A \to B)$$

For every product page, the profit factor is generated randomly from the interval of [5, 9]. For a category page, the profit factor is calculated as the average of profit factors of all product pages in that category.

There are two stopping conditions: one is to stop after a certain number of iterations and other is to stop after the least improvement condition [2] is met. Let the lowest objective function value K number of iterations before is Evalue and the lowest value during last K iterations is Last_value. According to the least improvement criteria, the algorithm would terminate if following condition is true.

$$\frac{\text{Evalue} - \text{last\_value}}{\text{Evalue}} \leq \text{percent\_imp}$$

Here the parameter percent_imp represents improvement percentage. In this study, the maximum number of iterations are taken as 1000 and value of K is 100 and of percent_imp is 0.001. The outdegree and indegree variance constraints have not been used in these experiments.

The optimal values of parameters were determined to tune the algorithm. This is done by performing experiments on a graph of 150 nodes and 600 edges. The value of ρ, β, q_0 and m are 0.5, 0.5, 0.6, and 6, respectively. The value of percent_imp is taken as 0.001. The values for parameters α, ω_1 and ω_2 are 0.07, 1.6, and 1.3 respectively. In the experiments, 75 % of pages are used as product pages and 25 % pages are used as category pages.

Figure 2 shows the number of page pairs links with different distances before and after the application of ACS for optimization process in a link structure graph of 100 nodes and 500 links. Figure 3 shows the number of page pairs links with different distances before and after the application of ACS for optimization process in a link structure graph of 200 nodes and 1000 links. In both figures, it can be observed that the curves showing the number of page pairs with different distances have moved to the left as a whole after the application of ACS. The number of page

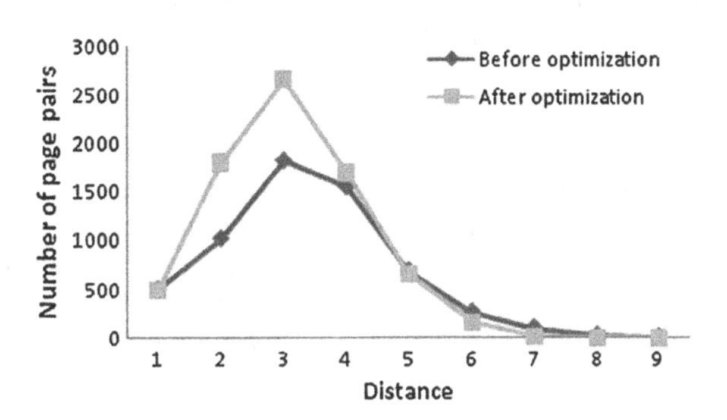

Fig. 2 Number of page pairs with distances on a webgraph of 100 nodes

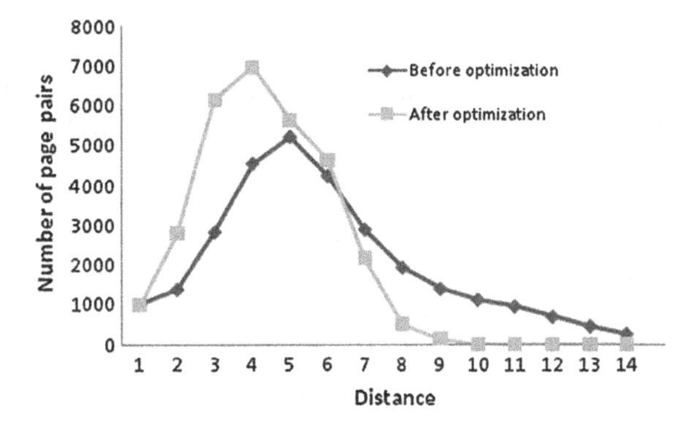

Fig. 3 Number of page pairs with distances on a webgraph of 200 nodes

Table 1 Optimization results of various problems solved by ACS

Number of nodes in the webgraph	Number of links in the webgraph	Outdegree limit	Initial objective value	Objective value of the optimized structure	Time taken to converge (s)
100	500	5	277992.59	127259.36	92.677
150	800	7	615397.37	555102.70	153.893
200	1000	5	2110159.23	1748049.35	265.97
250	1500	6	4563845.34	3506849.28	471.89
300	1800	6	9890575.48	5423545.9	525.33
350	2000	6	20046518.99	10443593.162	705.60
400	2200	6	35383889.94	24451646.37	1138.79
450	2500	7	48435525.34	34177927.55	2019.44
500	3000	6	78389723.16	36654342.92	3298.5

pairs with shorter distances has increased. In Fig. 3, the affect of optimization is more clearly visible. The distance between page pairs have decreased which is considered as a measure of efficient navigation. Microsoft Visual C# has been used to implement the algorithm. The experiments have been performed on a personal computer with core i5-4200U CPU (1.6–2.3 GHZ), 4 GB RAM, and Microsoft Windows 8.1 operating system. Table 1 shows the outcomes of the experiments performed on different webgraphs. The experiments have been performed on the webgraphs of 100, 150, 200, 250, 300, 350, 400, 450, and 500 nodes and the objective function value before and after optimization are presented in Table 1 along with the time taken to converge. It can be observed that the structure is always improved by the ACS. ACS can easily improve the webgraphs of large sizes such as webgraph of 500 nodes. As the number of nodes increases, the time taken by the algorithm also increases. Most of the time is consumed by the second stage in which the node with highest CP value is picked every time. One major problem with ACS is that as the pheromone level on the edges of global best solution increases, the exploration [15] of other possible solutions decreases.

6 Conclusion and Future Work

The link structure of an E-supermarket website needs to be rearranged according to the changing browsing behavior of customers. Website link structure optimization is of combinatorial nature. In this work, ant colony system has been used to optimize the link structure of an E-supermarket website. The distance between webpages is used as a measure of ease of navigation. The experiments on the artificially generated webgraphs show that the ant colony system-based method greatly improves the structure and reduces the average weighted distance between webpages. Future work includes testing this method on complex and larger link graphs with more constraints to discover the limitations and other problems not observed in this paper. The outcomes of experiments on real-world websites also need to be investigated. Future works also include the use of other metaheuristic techniques and hybrid techniques along with ACS for link structure improvement.

References

1. Jahng, J., Jain, H., Ramamurthy, K.: Effective design of electronic commerce environments: A proposed theory of congruence and illustration. IEEE Trans. Syst. Man Cybern.: Part A: Syst. Humans. **30**(4), 456–471 (2000)
2. Wang, Y., Wang, D., Ip, W.: Optimal design of link structure for E-supermarket website. IEEE Trans.: Syst. Man Cybern.: Part A **36**, 338–355 (2006)
3. Singh, H., Kaur, P.: A survey of transformation based website structure optimization models. J. Inf. Optim. Sci. **35**(5 & 6), 529–560 (2014)
4. Lin, C.C.: Optimal web site reorganization considering information overload and search depth. Eur. J. Oper. Res. **173**, 839–848 (2006)
5. Gorafalakis, J., Kappos, P., Mourloukos, D.: Website optimization using page popularity. IEEE Internet Comput. **3**(4), 22–29 (1999)
6. Chen, M.S., Park, J.S., Yu, P.S.: Data Mining for path traversal patterns in a Web Environment. In: 16th IEEE International Conference on Distributed Computing Systems, pp. 385–392 (1996)
7. Lin, C.C., Tseng, L.C.: Website reorganization using an ant colony system. Expert Syst. Appl. **37**, 7598–7605 (2010)
8. Dorigo, M., Gambardella, L.M.: Ant Colony System: a cooperative learning approach to travelling salesman problem. IEEE Trans. Evol. Comput. **1**, 53–66 (1997)
9. Zhou, B., Jinlin, C., Jin, S., Hongjiang, Z., Qiufeng W.: Website link structure evaluation and improvement based on user visiting patterns. In: The 12th ACM Conference on Hypertext and Hypermedia, pp. 241–242 (2001)
10. Yin, P., Guo, Y.: Optimization of multi-criteria website structure based on enhanced tabu search and web usage mining. J. Appl. Math. Comput. **219**, 11082–11095 (2013)
11. Chen, M., Ryu, Y.U.: Facilitating effective user navigation through web site structure improvement. IEEE Trans. Knowl. Data Eng. **99**, 1–18 (2013)
12. Dorigo, M., Blum, C.: Ant colony optimization theory: a survey. Theoret. Comput. Sci. **344**, 243–278 (2005)
13. Gagne, C., Price, W.L., Gravel, M.: Comparing an ACO algorithm with other heuristics for the single machine scheduling problemwith sequence dependent setup times. J. Oper. Res. Soc. **53**, 895–906 (2002)
14. Han, J., Kamber, M.: Data Mining: Concepts and Techniques. Morgan Kaufmann, San Francisco (2001)
15. Glover, F.: Tabu search—Part I. ORSA J. Comput. **1**, 190–206 (1989)

An Aggregation Based Approach with Pareto Ranking in Multiobjective Genetic Algorithm

Muneendra Ojha, Krishna Pratap Singh, Pavan Chakraborty and Sekhar Verma

Abstract Genetic algorithms (GA) have been widely used in solving multiobjective optimization problems. The foremost problem limiting the strength of GA is the large number of nondominated solutions and complexity in selecting a preferential candidate among the set of nondominated solutions. In this paper we propose a new aggregation operator which removes the need of calculating crowding distance when two or more candidate solutions belong to the same set of nondominated front. This operator is computationally less expensive with overall complexity of $O(m)$. To prove the effectiveness and consistency, we applied this operator on 11 different, two-objective benchmarks functions with two different recombination and mutation operator pairs. The simulation was carried out over several independent runs and results obtained have been discussed.

Keywords Multiobjective optimization · Genetic algorithm · Evolutionary algorithms · Aggregation operator · Nondominated sorting

1 Introduction

Evolutionary multiobjective optimization (EMO), especially genetic algorithms (GA), has in recent times become a predominant way for solving multiobjective optimization problem (MOP). The foremost reason for using GA to solve MOPs is the

M. Ojha (✉) · K.P. Singh · P. Chakraborty · S. Verma
Department of Information Technology, Indian Institute of Information Technology, Allahabad, Allahabad, India
e-mail: muneendra@iiita.ac.in

K.P. Singh
e-mail: kpsingh@iiita.ac.in

P. Chakraborty
e-mail: pavan@iiita.ac.in

S. Verma
e-mail: sverma@iiita.ac.in

M. Pant et al. (eds.), *Proceedings of Fifth International Conference on Soft Computing for Problem Solving*, Advances in Intelligent Systems and Computing 437, DOI 10.1007/978-981-10-0451-3_25

approach of a population where each individual is a possible solution candidate. When a new generation is created out of old generation, each individual belonging to this population explores a new possible solution in the space thus giving rise to several parallel explorations. Evolutionary multiobjective optimization algorithm is expected to perform well on two different criteria. First, it is expected to generate a set of solutions as close to the true Pareto as possible, and second, to produce a fairly distributed diverse solutions along the generated Pareto front [1]. For solving first problem, most EMO algorithms focus on generating nondominated solutions. In this approach as a number of generations grow, a large number of nondominated solutions are produced. Thus it becomes difficult to differentiate or prioritize one solution against another. For solving second problem, researchers tend to discard solutions which are very similar and choose the one which has relatively lesser similarity index. Various methods have been suggested in the literature to measure the similarity.

Selection of individuals for next generation plays an important role in genetic algorithms. In case of multiobjective optimization, Pareto dominance defines only a partial order in the objective space. A lot of work has been done in recent times to investigate an ordering mechanism and design of selection operators. For creating ordering among nondominated solutions, the ideas of assigning a ranking similar to the domination rank [2] and domination strength [3] is predominantly common in the literature. Some other approaches include niching and fitness sharing [4], crowding distance [5], k-nearest neighbor method [6], fast sorting [7], gridding, [8] and ϵ-domination method [9]. An extension of Pareto domination has been proposed by Koduru et al. [10], whereas a fuzzy domination approach has been proposed by Kundu et al. [11]. All these approaches have their respective degrees of effectiveness. Another class of methods use scalar techniques over Pareto dominance to assign a fitness value to each candidate solution. A number of approaches can be mentioned under this category, including predefined weighting coefficient such as weighted sum [12–14], weighted min–max [12], and fuzzy weighted product [15–17]. While applying ordering over individuals, a two-stage strategy is implemented. A population is checked for Pareto dominance and based on domination, and ranks are assigned to all individuals such that one with a smaller rank is always the preferred candidate. When several individuals are nondominated and have same rank a further ordering parameter is required which declares a preferred candidate. Such a parameter can be called density [18], a real-valued fitness index, where a lower value is again preferred. Thus ordering $X_1 < X_2$ can be defined as

$$\left(X_1^{\mathrm{rank}} < X_2^{\mathrm{rank}}\right), \tag{1}$$

$$\text{or, } \left(X_1^{\mathrm{rank}} = X_2^{\mathrm{rank}}\right) \text{ and } \left(X_1^{\mathrm{density}} < X_2^{\mathrm{density}}\right) \tag{2}$$

In this paper, we adopt a similar approach and define a new aggregation operator which is applied over the idea of Pareto dominance to find out a candidate for selection. We test the proposed operator on 11 different benchmark functions widely used in the literature: SCH function [5], five functions from the DTLZ test suit [19], and five functions from the ZDT test suit [5]. We have used two pairs of crossover and mutation operators, i.e., blend crossover with non-uniform mutation (BLX-NU) and simulated binary crossover with polynomial mutation (SBX-PN) to check the effectiveness of the proposed operator.

Archiving has been widely used in particle swarm optimization (PSO) algorithms for handling multiobjective problems [8]. Few studies have developed techniques to maintain diversified and Pareto optimal solutions using archiving [20, 21]. In the current work we have used an uninhibited elitist external archive which does not impose any restriction on the way a top ranking solution should be selected for inclusion and stores all rank 1 solutions from each generation in the archive.

2 Proposed Methodology

2.1 Aggregation Based Operator (d)

Suppose,

$$u = (x_1^u, x_2^u, \ldots, x_n^u) \tag{3}$$

and,

$$v = (x_1^v, x_2^v, \ldots, x_n^v) \tag{4}$$

are two solution vectors. We apply the partial order $\prec_n$ defined as

$$u \prec_n v \, \text{if} \, (u_{\text{rank}} < v_{\text{rank}}) \tag{5}$$

$$\text{or,} \; (u_{\text{rank}} = v_{\text{rank}}) \, \text{and} \, d_u < d_v \tag{6}$$

Mathematically, an aggregation operator (d) for any solution vector, is defined as

$$d = \frac{(w_1\delta' + w_2\delta'')}{m+1}, \quad \text{s.t.} \, w_1, w_2 = 1 \tag{7}$$

$$\delta' = \prod_{i=1}^{m} \left\{ \left(y_i^t - y_i^{(T,L)} \right) / \left(y_i^{(T,U)} - y_i^{(T,L)} \right) \right\} \tag{8}$$

and,

$$\delta'' = \sum_{i=1}^{m} \left[\frac{y_i^t - y_i^{(T,L)}}{y_i^{(T,U)} - y_i^{(T,L)}} \right] \tag{9}$$

where,

- m number of objective functions
- t current generation
- T number of generation already passed
- y_i^t value of ith objective in tth generation
- $y_i^{(T,L)}$ minimum value of ith objective till the point of application among generations passed
- $y_i^{(T,U)}$ maximum value of ith objective till the point of application among generations passed

2.2 Implementation

For every objective functions, we maintain two values, a global minimum $y_i^{(T,L)}$ and a global maximum $y_i^{(T,U)}$. These are the minimum/maximum values attained by any function till tth generation.

Algorithm 1: Aggregation Function
Input: $X = (x_1, x_2, \ldots, x_n)$;
Output: Real valued density indicator d;
for all $y_i(X) \in Y(X)$
  if $y_i(X) > y_i^{(T,U)}$
    $y_i^{(T,U)} = y_i(X)$
  if $y_i(X) < y_i^{(T,L)}$
    $y_i^{(T,L)} = y_i(X)$
  $y_i^{'} = \frac{y_i(X) - y_i^{(T,L)}}{y_i^{(T,U)} - y_i^{(T,L)}}$
end for;
  $p = 1.0$
for all $y_i(X) \in Y(X)$
  $p = p * y_i^{'}$
end for
  $s = 0.0$
for all $y_i(X) \in Y(X)$
  $s = s + y_i^{'}$
end for
  $d = \frac{(p+s)}{m+1}$
return d

As per Algorithm 1, as soon as aggregation function is called, the current value of y_i is compared with $y_i^{(T,L)}$ and $y_i^{(T,U)}$ values. If update is required, it is done and $y_i^{'}$ is normalized. The overall complexity of this approach is $O(m)$ pertaining to one loop which iterates over m objective functions. Thus it is computationally cheaper than crowding distance based operator which has complexity to the order of $O(m(2n)\log(2n))$, where m is number of objectives and n is population size [5].

2.3 Selection Operator

Based on dominance values assigned to every individual of population we implement a selection operator as described in Algorithm 2 to generate intermediate population for crossover operation.

```
Algorithm 2: Selection Function
Input: population P
Output: elitist population P′
  n = size(P)
for i = 1 to n
  selected = i  /* generate a random index j such that  j ≤ n
                  and j ≠ i */
  j = random(n)
  if rank(p_j) < rank(p_i)
    selected = j
  else if rank(p_i) == rank(p_j)
    if d(p_j) < d(p_i)
      selected = i
    end if
  end if
    add p_selected to P′
end for
return P′
```

As explained in Algorithm 2, any deadlock of ranking is broken with the aggregation operator. By selecting ith individual as first player of the tournament selection, we provide every member of current population an equal chance to survive, thus maintaining elitism, and by randomly selecting the other player we are able to maintain diversity in the population.

The overall algorithm involves following steps:

1. initialize random population
2. assign rank to each individual
3. perform selection based on proposed aggregation operator
4. perform crossover to produce intermediate generation of population
5. perform mutation to produce final population
6. re-rank individuals
7. select rank 1 individuals to be added to archive which are not yet added
8. if stopping criteria met stop, else return to step 2.

After stopping criteria is met, the elitist external archive is checked for non-dominated solutions. The set of solutions thus obtained after deletion of dominated solutions forms the resulting approximated Pareto set.

3 Experiment Setup, Results, and Discussion

To examine the impact of our proposed operator (*d*) we perform two different sets of experiments where we apply the operator with two pairs of crossover and mutation operators, i.e., BLX-NU and SBX-PN. We then run the algorithm on a set of 11 standard two-objective benchmark problems, i.e., SCH, DTLZ1, DTLZ2, DTLZ3, DTLZ4, DTLZ7, ZDT1, ZDT2, ZDT3, ZDT4, and ZDT6. The parameters required for each problem is listed in Table 1.

All tests were performed on the PC with configuration: System: Windows 7 Professional, 64 bit, CPU: Intel Core i3-3220 @ 3.30 GHz, RAM: 6 GB, Language: Java and MATLAB.

To measure the simulation results, we used the inverted generational distance (IGD) metric [22], which measures the closeness as well as diversity of the approximated Pareto front to the true Pareto front. To measure the consistency of proposed methodology, we conducted 50 independent runs for both algorithms on all test problems. The mean, standard deviation, best case, and worst case IGD values thus obtained for both algorithms are listed in Table 2.

We observe that the implementation of proposed aggregation operator produces good results in terms of convergence to the true Pareto front as well as maintains a fairly good diversity in solutions along the generated Pareto front.

In most of the cases we observe that BLX-NU tends to perform better than SBX-PN algorithm except for SCH, DTLZ3, and DTLZ4 problems. Still in these

Table 1 Listing of parameters taken for every function. These values remain same for different algorithms

Function	No. of generations	Crossover probability	Mutation probability	Min value	Max value
SCH	100	1.0	0.1	−1000	1000
DTLZ1	300	1.0	0.1	0	1
DTLZ2	300	1.0	0.01	0	1
DTLZ3	1000	0.8	0.05	0	1
DTLZ4	100	0.8	0.01	0	1
DTLZ7	500	0.8	0.01	0	2.15
ZDT1	300	1.0	0.01	0	1
ZDT2	500	0.8	0.01	0	1
ZDT3	700	0.8	0.01	0	1
ZDT4	700	1.0	0.01	0	1
ZDT6	1000	0.8	0.01	0	1

Table 2 Mean, standard deviation, best and worst IGD values obtained for BLX-NU and SBX-PN on two-objective DTLZ and ZDT test suits

Problem	Algorithm	Mean	Standard deviation	Best	Worst
SCH	BLX-NU	8.087E−03	1.547E−03	6.037E−03	1.136E−02
	SBX-PN	**4.838E−03**	**7.959E−04**	**3.439E−03**	**5.947E−03**
DTLZ1	BLX-NU	**2.769E−04**	**6.430E−05**	**1.70E−04**	**3.87E−04**
	SBX-PN	5.50E−04	8.658E−05	3.12E−04	6.414E−04
DTLZ2	BLX-NU	**7.634E−04**	**7.472E−05**	**6.006E−04**	**9.011E−04**
	SBX-PN	2.685E−03	4.808E−04	1.296E−03	3.114E−03
DTLZ3	BLX-NU	4.010E−04	1.323E−04	**1.904E−04**	6.847E−04
	SBX-PN	**3.990E−04**	**5.967E−05**	2.730E−04	**4.956E−04**
DTLZ4	BLX-NU	2.933E−03	**3.146E−04**	2.402E−03	3.475E−03
	SBX-PN	**2.747E−03**	3.161E−04	**2.128E−03**	**3.161E−03**
DTLZ7	BLX-NU	**6.967E−03**	**4.687E−04**	**5.972E−03**	**7.659E−03**
	SBX-PN	1.172E−02	1.567E−03	7.620E−03	1.382E−02
ZDT1	BLX-NU	**5.587E−03**	**5.154E−04**	**4.393E−03**	**6.408E−03**
	SBX-PN	2.992E−02	2.601E−03	2.440E−02	3.427E−02
ZDT2	BLX-NU	**4.372E−03**	**5.387E−04**	**3.253E−03**	**5.213E−03**
	SBX-PN	7.441E−03	1.341E−03	4.531E−03	9.276E−03
ZDT3	BLX-NU	**3.104E−03**	**2.632E−04**	**2.487E−03**	**3.445E−03**
	SBX-PN	5.928E−03	6.341E−04	4.768E−03	6.809E−03
ZDT4	BLX-NU	**7.979E−04**	**1.120E−04**	**6.305E−04**	**1.022E−03**
	SBX-PN	2.158E−03	5.417E−04	1.048E−03	2.857E−03
ZDT6	BLX-NU	**1.072E−03**	**7.680E−05**	**8.986E−04**	**1.189E−03**
	SBX-PN	1.225E−03	8.896E−05	1.027E−03	1.369E−03

Best values are shown in *bold*

problems only for SCH, SBX-PN perform well in all categories otherwise, for DTLZ3, the best case IGD value of BLX-NU is better than that of SBX-PN and for DTLZ4 the standard deviation for BLX-NU is marginally better than that of SBX-PN algorithm.

Figures 1 and 2 represent the distribution of IGD values obtained over 50 independent runs for each test problem for BLX-NU and SBX-PN, respectively. With BLX-NU algorithm, the proposed operator produces a near perfect distribution for DTLZ1, DTLZ3, DTLZ7, ZDT1, ZDT2, ZDT3, and ZDT6 where majority of values fall below the median. For problems SCH, DTLZ2, DTLZ4, and ZDT4, it produces better results with SBX-PN algorithm along with problems DTLZ1, DTLZ3, DTLZ4, DTLZ7, ZDT1, ZDT2, ZDT3, and ZDT6, as shown in Fig. 2. This parallel study on two different genetic algorithms suggests that a slight aberration in the efficiency of results can be attributed to the property of algorithm rather than design of aggregation operator. The final Pareto fronts generated in the best case run using the proposed operator (d) with BLX-NU algorithm are shown in Fig. 3.

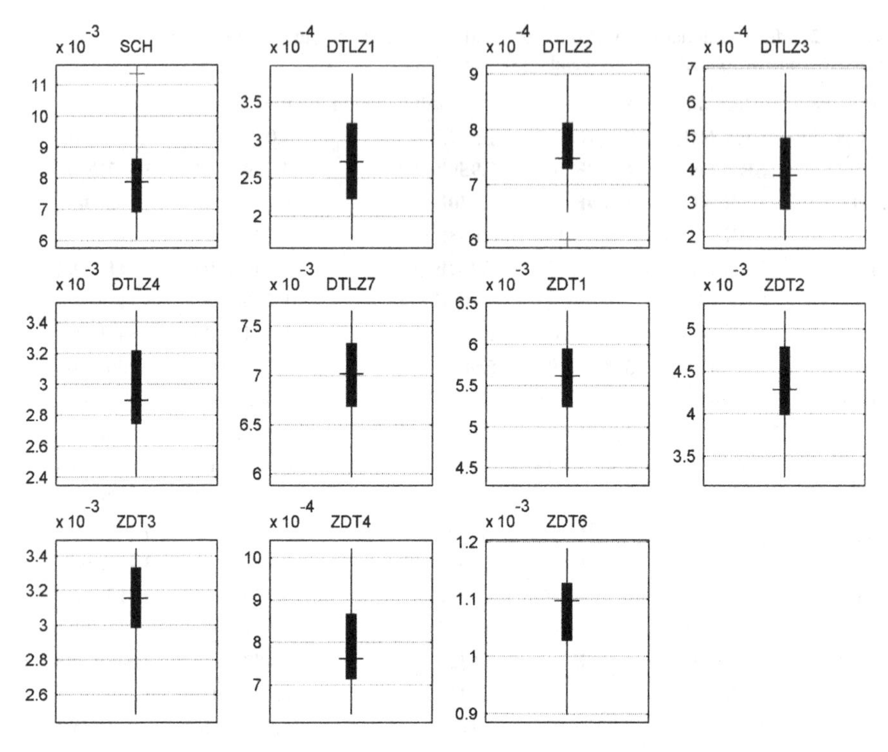

Fig. 1 IGD distribution plot for BLX-NU algorithm for different benchmark functions

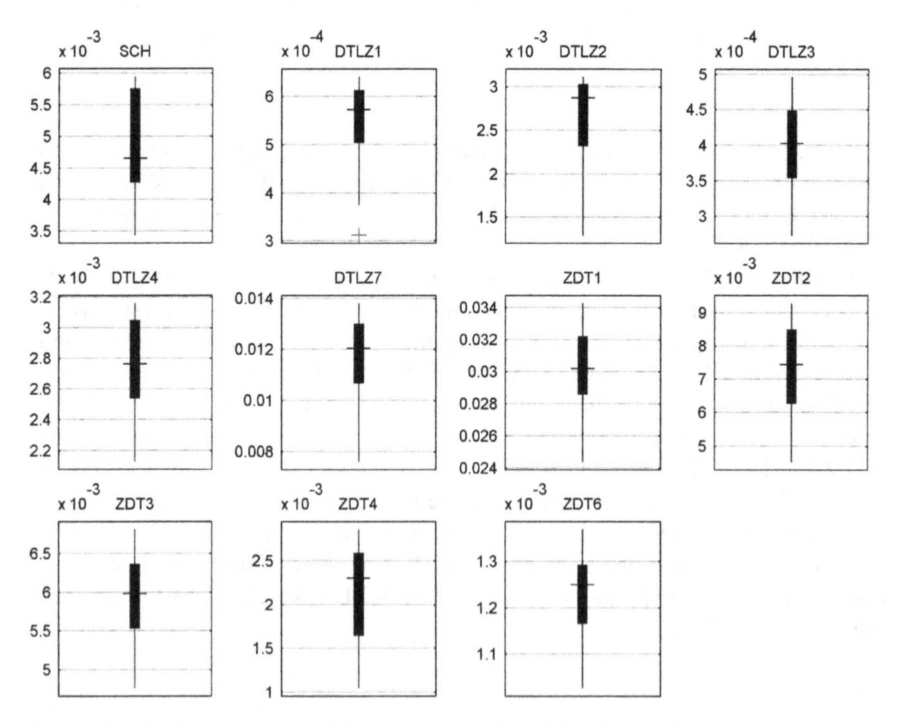

Fig. 2 IGD distribution plot for SBX-PN algorithm for different benchmark functions

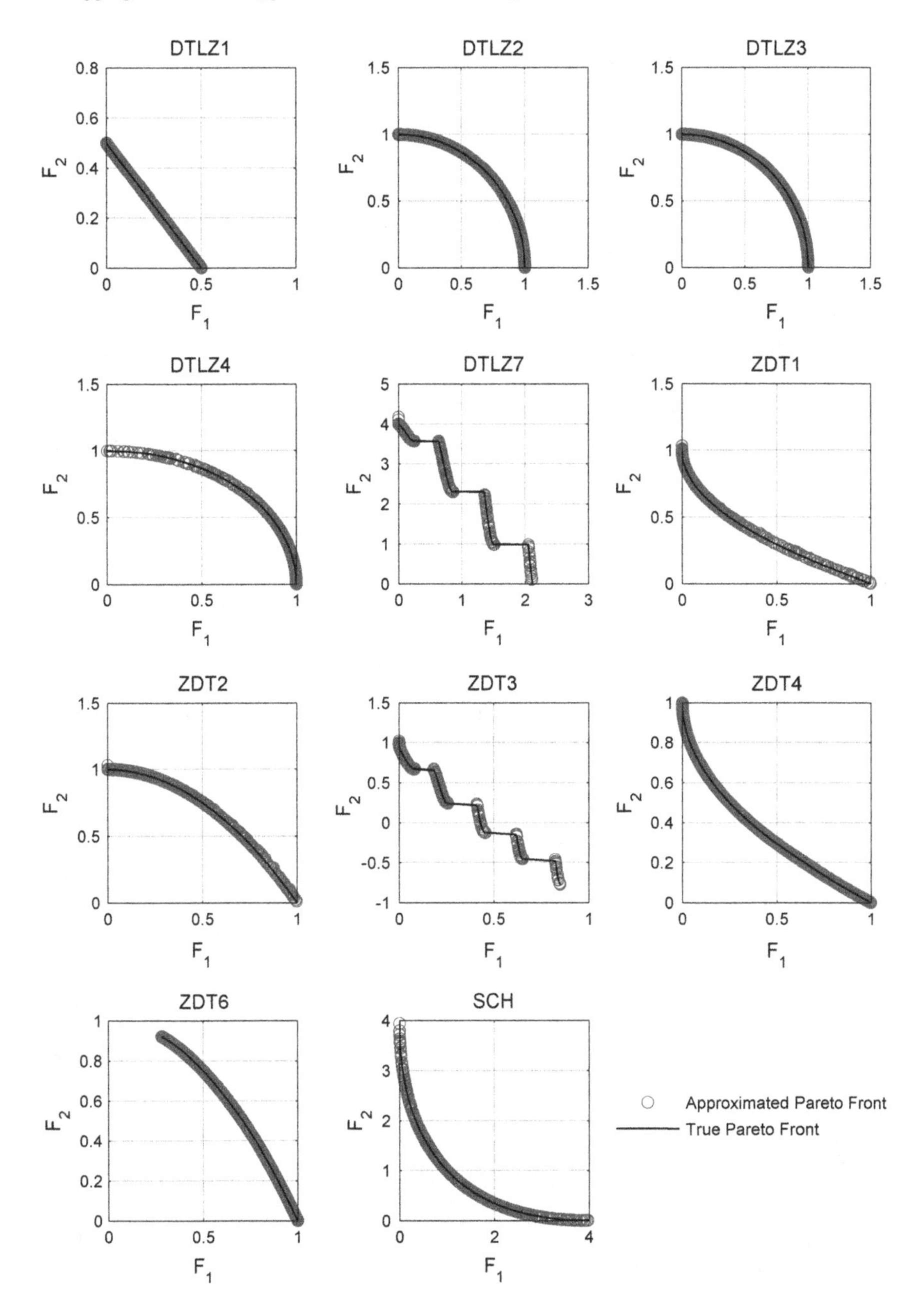

Fig. 3 Plots representing the approximated Pareto front generated for benchmark two-objective multiobjective test problems with BLX-NU algorithm

4 Conclusion and Future Research

The proposed aggregation operator (*d*) performs fairly well on eleven different two-objective standard benchmark problems when implemented with two sets of crossover and mutation operators in genetic algorithm. We have been able to achieve high degree of convergence as well as good diversity as represented by small values of IGD obtained in every test problem. When compared with diversity preserving crowding distance calculation, our operator is much efficient with a complexity of $O(m)$. We also implemented the uninhibited elitist external archive strategy to preserve top ranking solutions in every generation. Since archive is only checked for nondominated solutions at the end of complete algorithm, we are again able to save a lot of time. It remains to be seen as how this operator would fare on test problems with more than two objectives. Thus in future works we intend to apply this approach on many objective as well as real-life test problems.

References

1. Deb, K.: Multi-objective genetic algorithms: problem difficulties and construction of test problems. Evol. Comput. **7**, 205–230 (1999). doi:10.1162/evco.1999.7.3.205
2. Srinivas, N., Deb, K.: Multiobjective optimization using nondominated sorting in genetic algorithms 1 introduction. J. Evol. Comput. **2**, 221–248 (1994)
3. Zitzler, E., Thiele, L.: Multiobjective evolutionary algorithms: a comparative case study and the strength Pareto approach. Evol. Comput. IEEE Trans. **3**, 257–271 (1999). doi:10.1109/4235.797969
4. Fonseca, C.M., Fleming, P.J.: Genetic algorithms for multiobjective optimization: formulation, disscussion and generalization. In: Proceedings of the 5th International Conference on Genetic Algorithms, pp. 416–423 (1993)
5. Deb, K., Pratap, A., Agarwal, S., Meyarivan, T.: A fast and elitist multiobjective genetic algorithm: NSGA-II. IEEE Trans. Evol. Comput. **6**, 182–197 (2002). doi:10.1109/4235.996017
6. Zitzler, E., Laumanns, M., Thiele, L.: SPEA2: improving the strength Pareto evolutionary algorithm. Evol. Methods Des. Optim. Control Appl. Ind. Probl. (2001). doi:10.1.1.28.7571
7. Qu, B.Y., Suganthan, P.N.: Multi-objective evolutionary algorithms based on the summation of normalized objectives and diversified selection. Inf. Sci. (Ny) **180**, 3170–3181 (2010)
8. Coello, C.A.C., Pulido, G.T., Lechuga, M.S.: Handling multiple objectives with particle swarm optimization. Evol. Comput. IEEE Trans. **8**, 256–279 (2004). doi:10.1109/TEVC.2004.826067
9. Zhao, S.-Z., Suganthan, P.N.: Multi-Objective Evolutionary Algorithm With Ensemble of External Archives. Int. J. Innov. Comput. Inf. Control **6**, 1713–1726 (2010)
10. Koduru, P., Dong, Z.D.Z., Das, S., et al.: A multiobjective evolutionary-simplex hybrid approach for the optimization of differential equation models of gene networks. IEEE Trans. Evol. Comput. **12**, 572–590 (2008). doi:10.1109/TEVC.2008.917202
11. Kundu, D., Suresh, K., Ghosh, S., et al.: Multi-objective optimization with artificial weed colonies. Inf. Sci. (Ny) **181**, 2441–2454 (2011)
12. Garza-Fabre, M., Pulido, G.T., Coello, C.A.C.: Ranking methods for many-objective optimization. MICAI 2009 Adv. Artif. Intell. **5845**, 633–645 (2009)

13. Ishibuchi, H., Tsukamoto, N., Nojima, Y.: Empirical analysis of using weighted sum fitness functions in NSGA-II for many-objective 0/1 knapsack problems. In: 2009 11th International Conference on Computer Modeling and Simulation, Uksim, pp. 71–76 (2009)
14. Ishibuchi, H., Doi, T., Nojima, Y.: Incorporation of scalarizing fitness functions into evolutionary multiobjective optimization algorithms. In: Lecture Notes in Computer Science, vol. 4193, Parallel Probl. Solving from Nat.—PPSN IX. pp. 493–502 (2006)
15. He, Z., Yen, G.G., Zhang, J.: Fuzzy-based pareto optimality for many-objective evolutionary algorithms. IEEE Trans. Evol. Comput. **18**, 269–285 (2014). doi:10.1109/TEVC.2013.2258025
16. Deep, K., Singh, K.P., Kansal, M.L., Mohan, C.: An interactive method using genetic algorithm for multi-objective optimization problems modeled in fuzzy environment. Expert Syst. Appl. **38**, 1659–1667 (2011). doi:10.1016/j.eswa.2010.07.089
17. Cheng, F.Y., Li, D.: Multiobjective optimization of structures with and without control. J. Guid. Control Dyn. **19**, 392–397 (1996). doi:10.2514/3.21631
18. Zhou, A., Qu, B.-Y., Li, H., et al.: Multiobjective evolutionary algorithms: A survey of the state of the art. Swarm Evol. Comput. **1**, 32–49 (2011). doi:10.1016/j.swevo.2011.03.001
19. Deb, K., Thiele, L., Laumanns, M., Zitzler, E.: Scalable Test Problems for Evolutionary Multi-Objective Optimization: TIK-Technical Report No. 112 (2001)
20. Laumanns, M., Thiele, L., Deb, K., Zitzler, E.: Combining convergence and diversity in evolutionary multiobjective optimization. Evol. Comput. **10**, 263–282 (2002). doi:10.1162/106365602760234108
21. Knowles, J., Corne, D.: Bounded Pareto archiving: theory and practice. In: Gandibleux, X., Sevaux, M., Sörensen, K., T'kindt, V. (eds.) Metaheuristics multiobjective Optimisation, pp 39–64 (2004)
22. Zhang, Q., Zhou, A., Zhao, S., et al.: Multiobjective optimization Test Instances for the CEC 2009 Special Session and Competition. Report, pp. 1–30 (2009)

Weighted CoHOG (W-CoHOG) Feature Extraction for Human Detection

Nagaraju Andavarapu and Valli Kumari Vatsavayi

Abstract Human recognition techniques are used in many areas such as video surveillance, human action recognition, automobile industry for pedestrian detection, etc. The research on human recognition is widely going on and is open due to typical challenges in human detection. Histogram-based human detection methods are popular because of its better detection rate than other approaches. Histograms of oriented gradients (HOG) and co-occurrence of histogram-oriented gradients (CoHOG) are used widely for human recognition. A CoHOG is an extension of HOG and it takes a pair of orientations instead of one. Co-occurrence matrix is computed and histograms are calculated. In CoHOG, gradient directions alone are considered and magnitude is ignored. In this paper magnitude details are considered to improve detection rate. Magnitude is included to influence the feature vector to achieve better performance than the existing method. In this paper, weighted co-occurrence histograms of oriented gradients (W-CoHOG) is introduced by calculating weighted co-occurrence matrix to include magnitude factor for feature vector. Experiments are conducted on two benchmark datasets, INRIA and Chrysler pedestrian datasets. The experiment results support our approach and shows that our approach has better detection rate.

Keywords Histograms of oriented gradients (HOG) · Co-occurrence histogram of oriented gradients (CoHOG) · Weighted co-occurrence histogram of oriented gradients (W-CoHOG) · Human detection · Pedestrian detection

N. Andavarapu (✉) · V.K. Vatsavayi
Department of Computer Science and Systems Engineering,
Andhra University, Visakhapatnam, India
e-mail: nagraz.a@gmail.com

V.K. Vatsavayi
e-mail: vallikumari@gmail.com

M. Pant et al. (eds.), *Proceedings of Fifth International Conference on Soft Computing for Problem Solving*, Advances in Intelligent Systems and Computing 437, DOI 10.1007/978-981-10-0451-3_26

1 Introduction

Computer vision is a wide and emerging area over the past few years. The analysis of images involving humans comes under computer vision problem. Human detection techniques are used in many areas such as people abnormal behavior monitoring, robots, automobile safety systems, and gait recognition. The main goal of a human detector is to check whether humans are present in the image or not. If human is identified in the particular image then it can be used for further analysis. Human detection is still an open problem. Human detection is one of the active and challenging problems in computer vision, due to different articulations and poses, different types of appearances of clothes and accessories acting as occlusions. In this paper humans are identified in a static image. Identifying humans in a static image is more difficult than in a video sequence because no motion and background information is available to provide clues to approximate human position. In our approach, input of the human detector is an image and output is a decision value finding whether there is a human in a given image or not. In this paper static images are considered to detect humans.

2 Related Works

Many human detection techniques have been proposed so far in different approaches. The implicit shape model (ISM) [1], a part-based object detection algorithm proposed by Leibe et al., uses local features derived from a visual vocabulary or codebook as object parts. Codebooks are generated using SIFT [2] or shape context local feature descriptor. Lu et al. proposed image depth based algorithm [3] to detect humans by taking depth information of given image. Jiaolong et al. proposed a part-based classifier technique [4] to detect humans in a given image window. In this method mixture of parts technique was used for part sharing among different aspects. Andriluka et al. proposed a generic approach for nonrigid object detection and articulated pose estimation based on the pictorial structures framework [5]. Gavrila et al. introduced a template matching approach for pedestrian detection [6]. Template hierarchy of pedestrian silhouettes is built to capture the variety of pedestrian shapes. For identifying shapes, canny edge detector [7] is used.

Gradient orientation based feature descriptors such as SIFT [2], HOG [8], CoHOG [9], etc., are recent trends in human detection. SIFT [2] (scale-invariant feature transform) features proposed by Lowe et al. are used in human body parts detection in [10]. Histogram-based features are popularly used in human recognition and object detection because of their robustness. Histograms of oriented gradients (HOG) [8] is a famous and effective method for human detection. It uses histograms of oriented gradients as a feature descriptor. HOG features are robust towards illumination variances and deformations in objects. Co-occurrence histograms of oriented gradients (CoHOG) [9] is an extensive work of HOG which has more detection rate and lesser miss rate. In recent days, CoHOG used in many computer vision applications such as object recognition [11], image classification [12], and character recognition [13] . In CoHOG, co-occurrence matrices calculated for oriented gradients for making feature descriptor strong. In CoHOG only gradient direction details are considered and gradient magnitude details are ignored. In the proposed method, gradient magnitude components are also considered to bring more accuracy to the existing CoHOG.

The rest of the paper is organized as follows: Sect. 2 gives a brief overview of HOG and CoHOG. Proposed method W-CoHOG is discussed in Sect. 3 in detail. Section 4 contains experimental results and comparison with existing methods. Finally the work concluded in Sect. 5.

3 Background: HOG and CoHOG

3.1 HOG

In HOG, initially gradients are computed on each pixel in a given image and are divided into nine orientations. Next the image is divided into small nonoverlapping regions. Typical regions are of size 8 × 8 or 16 × 16. Then HOGs are calculated for each and every small region. Finally histograms of each region are concatenated using vectorization.

3.2 CoHOG

Co-occurrence histograms of oriented gradients (CoHOG) is an extension to HOG and more robust than HOG. In CoHOG pair of oriented gradients is used instead of single gradient orientation. Co-occurrence matrix is calculated for pair of gradient orientation with different offsets.

$$C_{\Delta x,\Delta y}(p,q)=\sum_{i=1}^{n}\sum_{j=1}^{m}\begin{cases}1 & \text{if } I(i,j)=p \text{ and } I(i+x,j+y)=q\\ \text{None} & \text{Otherwise}\end{cases} \tag{1}$$

Equation (1) shows the calculation of co-occurrence matrix. Figure 1a shows typical co-occurrence matrix histograms of oriented gradients and Fig. 1b shows possible offsets for CoHOG.

In CoHOG, orientation values of gradient are alone considered and magnitude is ignored. In the proposed method magnitude is also considered, as magnitude also contains discriminative information for human detection. Let us consider the following example: Fig. 2a is quite different from Fig. 2b because of different magnitude values even though it has same gradient orientation. Hence magnitude also

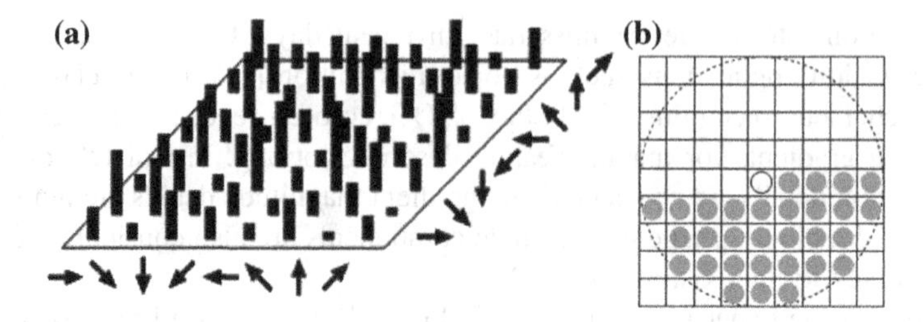

Fig. 1 **a** Typical co-occurrence matrix histogram. **b** Possible offsets

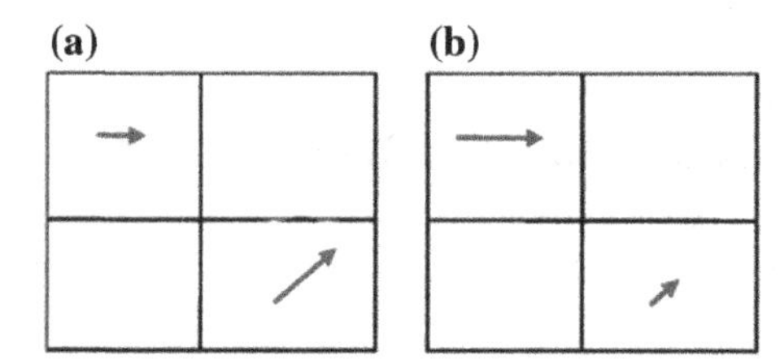

Fig. 2 Magnitudes of two gradients having same orientation. **a** and **b** are in same orientations but they are not the same

describes about what image contains. Existing feature descriptors does not consider magnitude details.

4 Proposed Method (W-CoHOG)

4.1 Overview

In the CoHOG method gradient directions are alone considered and magnitude is ignored. In the proposed method magnitude is also considered to extract more robust feature. Magnitude weighted co-occurrence histograms of oriented gradients (W-CoHOG) is proposed for better feature descriptor. Figure 3 briefly explains the classification process for human detection using W-CoHOG extraction method.

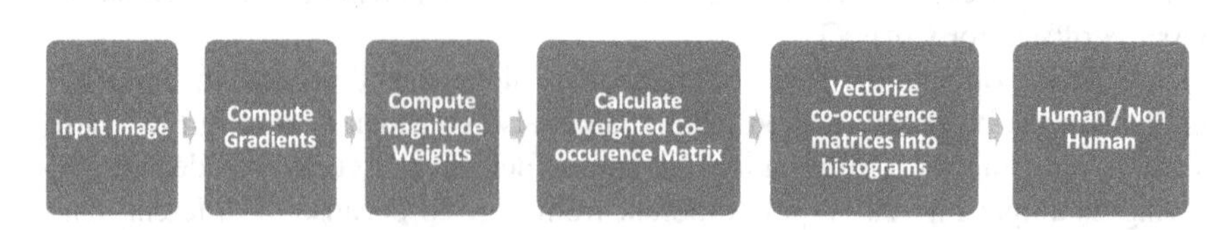

Fig. 3 Our proposed classification process

Initially, gradients of image are computed in magnitude and direction form and converted into oriented gradients. Next, image is divided into 3 × 6 or 6 × 12 sized non-overlapping cells. Then, weighted co-occurrence matrices are computed for each region. After that, all co-occurrence matrices of all regions are combined.

4.2 Feature Extraction

For a given input image, gradients are computed for each pixel. In this method, Sobel and Robert's filters are used to compute gradients of a given input image. Equations (2) and (3) show gradient calculation using Sobel and Robert's filters, respectively, for a given input image I, as shown in below.

Sobel gradient operator

$$\text{(a)}\, G_x = \begin{bmatrix} -1 & 0 & +1 \\ -2 & 0 & +2 \\ -1 & 0 & +1 \end{bmatrix} * I \qquad \text{(b)}\, G_y = \begin{bmatrix} +1 & +2 & +1 \\ 0 & 0 & 0 \\ -1 & -2 & -1 \end{bmatrix} * I \tag{2}$$

Robert's gradient operator

$$\text{(a)}\, G_x = \begin{bmatrix} +1 & 0 \\ 0 & -1 \end{bmatrix} * I \qquad \text{(b)}\, G_y = \begin{bmatrix} 0 & +1 \\ -1 & 0 \end{bmatrix} * I \tag{3}$$

Then, gradients are converted into magnitude and direction using Eq. (4). The gradients directions are converted into eight equal bins with 45^0 intervals.

$$\text{(a)}\, \theta = \tan^{-1}\frac{g_x}{g_y} \qquad \text{(b)}\, m = \sqrt{g_x^2 + g_y^2} \tag{4}$$

After that, magnitude matrix is convoluted with mean mask to eliminate noise which may cause aliasing effect. Equation (5) shows the 7 × 7 mean mask used in the proposed method. Figure 4 shows the overview of W-CoHOG feature calculation process.

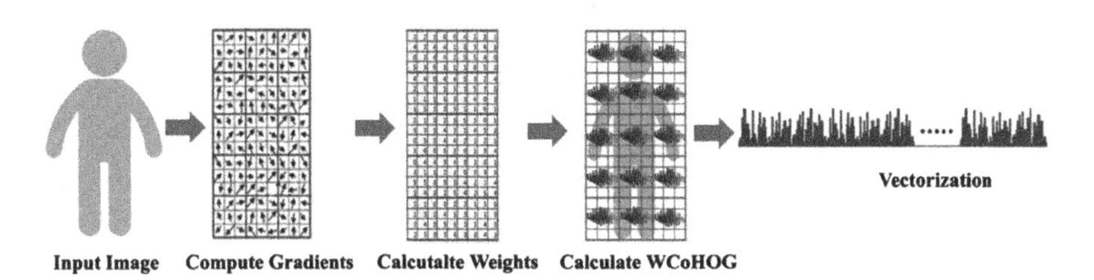

Fig. 4 Overview of W-CoHOG calculation

$$\text{Conv}_{7\times7} = \frac{1}{49}\begin{bmatrix} 1 & 1 & 1 & 1 & 1 & 1 & 1 \\ 1 & 1 & 1 & 1 & 1 & 1 & 1 \\ 1 & 1 & 1 & 1 & 1 & 1 & 1 \\ 1 & 1 & 1 & 1 & 1 & 1 & 1 \\ 1 & 1 & 1 & 1 & 1 & 1 & 1 \\ 1 & 1 & 1 & 1 & 1 & 1 & 1 \\ 1 & 1 & 1 & 1 & 1 & 1 & 1 \end{bmatrix} \tag{5}$$

Weight Function

In this proposed method magnitude component of a gradient is used as weight function to calculate weighted co-occurrence matrix. In order to calculate magnitude weighted co-occurrence matrix, the magnitude weights of each pixel are calculated. Weight function is applied to co-occurrence matrix to influence the co-occurrence matrix using gradient magnitude of each pixel. The weight functions used in this method are described in following paragraph.

Let I be a given input image. i, j are the any two orientations in the given eight orientations and Δx, Δy are offset for co-occurrence. $C_{\Delta x,\Delta y}(i,j)$ is weighted co-occurrence matrix for a given offset Δx, Δy and orientation i, j. The Eqs. (6 and 7) describes the calculation of the weighted co-occurrence matrix.

$$C_{\Delta x,\Delta y}(i,j) = \sum_{p=1}^{n}\sum_{q=1}^{m}\{W_{(p,q),(p+\Delta x,p+\Delta y)} * \alpha \tag{6}$$

where

$$\alpha = \begin{cases} 1 & \text{if } O(p,q) = i \text{ and } O(p+\Delta x, q+\Delta y) = j \\ 0 & \text{Otherwise} \end{cases} \tag{7}$$

Let $m_{p,q}$ be a gradient at a given pixel p, q for a given input image I. $\bar{M}$ and $M_{\max}$ are mean and maximum gradient values in I. The weight calculation was performed with simple operations like mean and division operations. Equations (8 and 9) show two possible weight functions to calculate weight for a given pixel (p, q) and ($p + \Delta x$, $q + \Delta y$). Any of the two functions is preferable to calculate weights for calculating weighted co-occurrence matrix. In this proposed method, Eq. (8) is used to calculate weights for experimental results.

$$W_{(p,q),(p+\Delta x,p+\Delta y)} = \left(\frac{m_{p,q}}{\bar{M}} * \frac{m_{p+\Delta x,p+\Delta y}}{\bar{M}}\right) + \mu \tag{8}$$

$$W_{(p,q),(p+\Delta x,p+\Delta y)} = \left(\frac{m_{p,q}}{M_{\max}} * \frac{m_{p+\Delta x,p+\Delta y}}{M_{\max}}\right) + \mu \tag{9}$$

where, μ is constant and $\mu = 1$.

After computing magnitude weighted co-occurrence matrices for all regions, the matrices are vectorized by simple concatenation of all matrix rows into a single row. There are 31 offsets possible for calculating co-occurrence matrix shown in Fig. 1b. Co-occurrence matrices need not be calculated for all offsets. In calculation of W-CoHOG, two offsets are good enough for pedestrian detection problem.

The size of feature vector is very large in histogram-based feature descriptors. For these types of features linear SVM [04] classifier is suitable. In this proposed method LIBLINEAR classifier [14] is used. LIBLINEAR classifier is an SVM-based classifier which works faster than SVM classifier [15] even for million instances of data. HOG and CoHOG also used SVM classifier for classification.

Algorithm 1: Weighted-CoHOG Calculation

```
1.  Given Image: I
2.  Given offset: x,y
3.    Begin
4.    Calculate O,M of I
5.    Divide image into cells
6.    For each cell in the image
7.      Initialize co-occurrence matrix C←0
8.      For each pair of orientation (i,j)
9.        For each pixel p,q in the cell
10.         Calculate W_(p,q),(p+x,q+y)
11.           If O_p,q =i and O_p+x,q+y =j then
12.             C_p,q = C_p,q + 1*W_(p,q),(p+x,q+y)
13.           end
14.           else
15.             C_p,q = C_p,q + 0*W_(p,q),(p+x,q+y)
16.           end
17.       end
18.     end
19.   end
20.   H ← vectorize all co-occurrence matrices
21.   return H
22. end
```

5 Experimental Results

Experiments are conducted on two datasets Daimler Chrysler [11] and INRIA dataset [8]. These are the familiar benchmark datasets for human detection. In Daimler Chrysler dataset, 4,800 human images and 5,000 nonhuman images are taken for training; and another 4,800 human images for training and 5,000 images

Fig. 5 INRIA sample images in dataset

Fig. 6 Chrysler sample images in dataset

are taken for testing. Each image size in Daimler Chrysler dataset is 48 × 96 pixels. In INRIA dataset, 1208 positive images and 12,180 patches are randomly sampled from person-free image for training and testing.

Figures 5 and 6 show that sample positive and negative examples of INRIA dataset and Chrysler dataset, respectively. Negative images are generated by taking 64 × 128 patches from no person images in INRIA dataset. Simple Sobel filter and Roberts filter were used to calculate the gradients of input image.

ROC curves are used for performance evaluation of binary classification like object detection problems. Sliding window technique is used to detect the humans in the image. A typical scanning window size for INRIA dataset is equal to the same as positive image size 64 × 128. In this paper, true positive rate versus false positive per window (FPPW) was plotted to evaluate the performance of proposed method and to compare with state-of-the-art methods. An ROC curves towards the top-left of the graph means better performance for classification problem. Figures 7 and 8 clearly show that curves obtained by proposed method achieved better

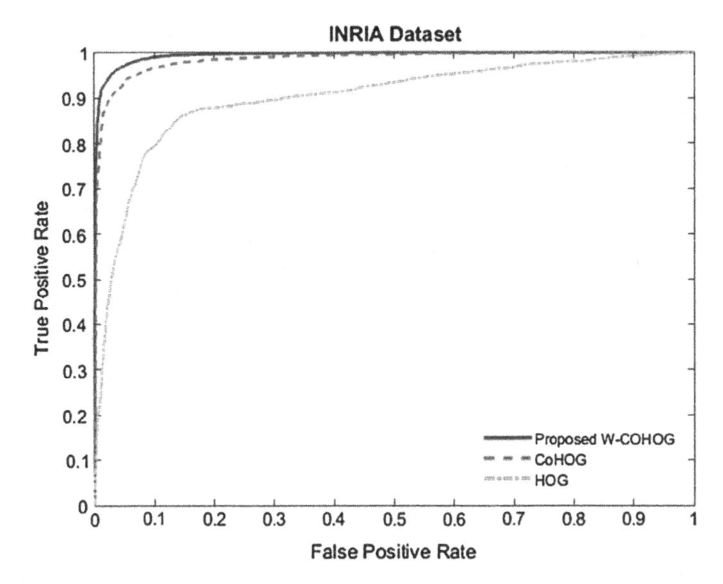

Fig. 7 Comparison of proposed method with other state-of-the-art methods (INRIA Dataset)

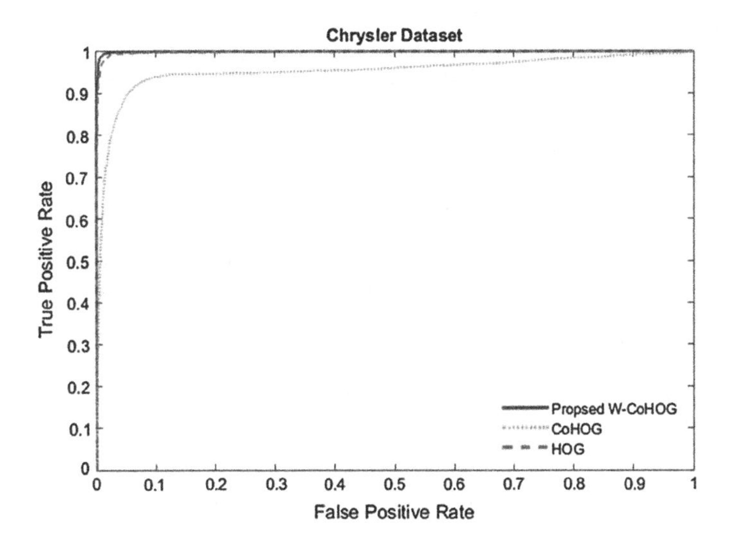

Fig. 8 Comparison of proposed method with other state-of-the-art methods (Chrysler Dataset)

detection rate for all false positive rates than other existing methods or at least comparable. The results clearly show that our method reduced miss rate around 20 % compared with CoHOG. The accuracy of the classifier is also better than other state-of-the-art methods shown in the figure. In the proposed method only two offsets are used instead of all 31 possible offsets, even though good results are acquired by adding gradient magnitude component.

6 Conclusion

In this paper a new method called weighted CoHOG is proposed which is an extension work to CoHOG. Magnitude component is also added to feature vector to improve the classification. The proposed method achieved improvement in accuracy on two benchmark datasets. Experimental results prove that performance of the proposed method is better than the other state-of-the-art methods. Even though calculation of weights adds additional computational complexity, the overall feature vector generation time decreased by reducing the number of offsets to two. Future work involves proposed feature descriptor to be used in other applications such as person tracking.

References

1. Leibe, B., Leonardis, A., Schiele, B.: Combined object categorization and segmentation with an implicit shape model. In: Workshop on statistical learning in computer vision, ECCV, vol. 2. no. 5 (2004)
2. Lowe, David G.: Distinctive image features from scale-invariant keypoints. Int. J. Comput. Vision **60**(2), 91–110 (2004)
3. Lu, X., Chen, C.-C., Aggarwal, J.K.: Human detection using depth information by Kinect. In: 2011 IEEE Computer Society Conference on Computer Vision and Pattern Recognition Workshops (CVPRW), IEEE (2011)
4. Xu, J., et al.: Learning a part-based pedestrian detector in a virtual world. IEEE Trans. Intell. Transp. Syst. **15**(5), 2121–2131 (2014)
5. Andriluka, M., Roth, S., Schiele, B.: Pictorial structures revisited: people detection and articulated pose estimation. In: IEEE Conference on Computer Vision and Pattern Recognition, 2009. CVPR 2009. IEEE (2009)
6. Gavrila, D., Philomin, V.: Real-time object detection for "smart" vehicles. In: The Seventh IEEE International Conference on Computer Vision, vol. 1, pp. 87–93. IEEE Computer Society Press, Los Alamitos (1999)
7. Canny, J.: A computational approach to edge detection. IEEE Trans. Pattern Anal. Mach. Intell. **6**, 679–698 (1986)
8. Dalal, N., Triggs, B.: Histograms of oriented gradients for human detection. In: CVPR 2005 IEEE Computer Society Conference on Computer Vision and Pattern Recognition. vol. 1. IEEE (2005)
9. Watanabe, T., Ito, S., Yokoi, K.: Co-occurrence histograms of oriented gradients for pedestrian detection. In: Advances in Image and Video Technology, pp. 37–47. Springer, Berlin (2009)
10. Shashua, A., Gdalyahu, Y., Hayun, G.: Pedestrian detection for driving assistance systems: single-frame classification and system level performance. In: Intelligent Vehicles Symposium, IEEE (2004)
11. Iwata, S., Enokida, S.: Object detection based on multiresolution CoHOG. Advances in visual computing. In: Springer International Publishing, pp. 427–437 (2014)
12. Kawahara, T., et al.: Automatic ship recognition robust against aspect angle changes and occlusions. In: Radar Conference (RADAR), IEEE (2012)
13. Su, B., et al.: Character Recognition in natural scenes using convolutional co-occurrence HOG. In: 22nd International Conference on Pattern Recognition (ICPR), IEEE (2014)

14. Fan, R.-E., et al.: Liblinear: A library for large linear classification. J. Mach. Learn. Res. **9**, 1871–1874
15. Hearst, M.A., et al.: Support vector machines. In: Intelligent Systems and their Applications, IEEE **13**(4), 18–28 (1998)

Municipal Solid Waste Generation Forecasting for Faridabad City Located in Haryana State, India

Dipti Singh and Ajay Satija

Abstract Faridabad is the largest metropolitan city in Haryana State (India). Solid waste management is one of the biggest environmental issues for the municipal corporation of Faridabad. The Municipal Corporation of Faridabad seems unable to manage the solid waste due to highly increased urbanization and lack of planning, funds, and advanced technology. Hence, various private sector companies and nongovernment organizations are required to work in this sector to resolve such issues. For the success of such useful work, proper planning is required. Successful planning depends on the exact prediction of the amount of solid waste generation. In this paper an artificial neural network model is applied to predict the quantity of solid waste generation in Faridabad city.

Keywords Solid waste management (SWM) · Artificial neural networks (ANN) · Forecasting · Municipal solid waste (MSW)

1 Introduction

Solid waste is the refuse material of daily used items discarded by society. The waste generated from residential sectors, institutional sectors, commercial sectors, and various industries is an environmental issue. People sufferfrom health problems and various atmospheric issues such as foul smell, house flies, cockroaches, or other insects. Hence solid waste management is extremely essential in society. The growing population, migration to urban areas for employment, gross domestic product (GDP) per capita, i.e., living standard parameters, different housing conditions, global longitude and latitude, seasonal conditions, and regional environmental laws

Dipti Singh (✉)
Gautam Buddha University, Greater Noida, India
e-mail: diptipma@rediffmail.com

Ajay Satija
Inderpratha Engineering College, Ghaziabad, India
e-mail: aajaysatija@rediffmail.com

M. Pant et al. (eds.), *Proceedings of Fifth International Conference on Soft Computing for Problem Solving*, Advances in Intelligent Systems and Computing 437, DOI 10.1007/978-981-10-0451-3_27

are the crucial factors affecting solid waste generation. Optimization techniques, soft computing techniques as ANN models, ANFIS (adaptive neurofuzzy inference system) models, expert systems, evolutionary algorithms, etc., have been applied for SWM. Accuracy of waste management forecasting plays an important role in waste management strategy. Various ANN models have been applied to forecast solid waste generation in different cities all over the world. But in India not much work is done in this direction. In this paper ANN model is applied for forecasting of solid waste in Faridabad city. Before introducing the proposed model, a few research papers have been revealed in this direction.

Zade and Noori presented an appropriate ANN model with threshold statistics technique to forecast the solid waste generation in Mashhad city (Iran) [1]. Abdoli et al. suggested a new approach of removing data trend by taking the logarithm of data and creating the stationary condition to predict solid waste generation for the city of Mashhad for the period 2011–2032 [2]. Batinic et al. proposed an ANN model to forecast the waste characteristics (organic waste, paper, glass, metal, plastic, and other waste (output parameters)) in Serbia for the period 2010–2026. The result showed that organic waste will not increase within the period 2010–2016 but within period 2016–2026 there will be expected change. In 2026, 810,000 tons of solid waste will probably need to be disposed in a landfill area in Serbia [3].

Shahabi et al. suggested a feedforward multilayer perception ANN model to forecast waste generation in Saqqez City in Kurdistan Province in northwest Iran. Weekly time series of generated solid waste have been arranged for the period 2004–2007. The authors suggested the Stop Training Approach to solve the problem of increasing error while training and testing phase of ANN [4]. Patel and Meka have proposed the feedforward ANN model to forecast Municipal Solid Waste for the 98 towns of Gujarat for the next 25 years. First, the normalized solid waste data is applied to ANN model and it is found that the predicted data validation follows nearly perfect correlation [5]. Roy et al. proposed a feedforward ANN model to forecast waste generation in Khulna city (Bangladesh). The authors have arranged the city's generated solid waste data weekly with the help of time series. The best neural network model has been selected on the basis of mean absolute errors and regression R of training, testing, validation, etc. Further results of observed and predicted values of solid waste generation have been compared [6]. Falahnezhad and Abdoli proposed a neural network model for solid waste generation in which the effect of preprocessing data is seen. The authors suggest that the logarithm of input and output data to time series provides more accurate results [7]. Shamshiry et al. suggested an ANN model to forecast the solid waste generation in Tourist and Tropical Area Langkawi Island, Malaysia, during 2004–2009. The authors have applied backpropagation neural network for better results. The results are compared with multiple regression analysis but ANN predicted results seem much better [8]. Singh and Satija suggest the concept of incineration process for electricity generation through solid waste in Mumbai and different cities of Gujrat State [9]. Pamnani and Meka propose an ANN model for forecasting solid waste generation for small-scale towns and their neighboring villages of Gujarat State,

India. The authors have validated their results with the help of low value of percentage prediction error [10].

2 Materials and Methods

2.1 Study Area Identification and MSW Management Issues

Faridabad is the main industrial center (ranked 9th in Asia) and a highly populated district in Haryana State established by the Sufi saint, Baba Farid, in 1607 AD. It became the 12th district of Haryana state on 15 August 1979. Its total area is 742.9 km^2 with a dense population of 1,798,954 people (2011 census). There are 91 sectors proposed under the development plan in Faridabad. The city generates per year, per capita 135.72 kg of waste. In Faridabad only 20.49 % of the population lives in rural villages (2011 census). There are about 15,000 small, medium, and large-scale industries, which includes multinational companies and other ISO certified companies. But the waste is not properly treated at these sites. The Municipal Corporation of Faridabad (MCF) was established in 1992. MCF was founded by earlier municipalities of Faridabad Town/New Industrial Township Old Faridabad, Ballabhgarh and its 38 revenue villages. The related solid waste management issues follows: improper short- and long-term strategy to manage solid waste related to MSW Rules 2000, financial issues, high land cost, lack of technical skill in municipal sanitation workers, lack of sufficient numbers of waste gathering carts or advanced vehicles, public unawareness about 4-R principles of environment sciences, sanitation work is not followed on Sundays or on other holidays, etc.

2.2 Case Study and Data

The accurate data plays an important role in waste management strategies. But the collection of reliable and complete solid waste data is a challenging task. The generated solid waste data is arranged from the MCF head office (NIT Faridabad) from January 2010 to December 2014 month-wise. According to MCF officials waste generation (WG) rate is approximately 300 g per day per capita. But there is month-wise fluctuation in waste generation in the city due to the rapid growth of the urban population. In 2014 about 133,871,295 kg waste was collected and disposed by hook loaders in the landfill of village Bandhwari (Gurgaon).

The following are the types of solid waste generated in the city: (a) domestic waste, (b) commercial waste, (d) institutional waste, (e) industrial waste, (f) agriculture waste, (g) energy renewal plant wastes, (h) inert waste, and (i) public place waste (streets, roads, parks etc.)

In the Indian calendar one year is divided into six seasons. There were seasonal variations in solid waste generation per day in Faridabad city in 2014. It is observed that in the monsoon the waste collected is the lowest while in spring the waste collected is the highest.

2.3 *Artificial Neural Network Model (ANN Model)*

Neural network models are similar to models of biological systems. In 1943, Mclloch and Pitts developed the first computational model in which neurophysiology and mathematical logic approaches got merged.

The following model forms the basis of a neural network. Here $\{x_1, x_2, x_3, \ldots, x_n\}$ are the set of inputs to the artificial neurons. The set $\{w_1, w_2, w_3, \ldots, w_n\}$ is the randomly generated weights' set to the input links (synapses). In the biological neuron system the neuron receives its input signals, sums up them, and produces an output. If this sum is greater than a threshold value the input signal passes through the synapse which may accelerate or retard signal. In the ANN model this acceleration or retardation is the reason for creation of the concepts of weights. The weights are multiplied with inputs and transmitted via links (synapses). The larger weight synapse generates a strong signal while a weak weight synapse generates a weak signal. The combined input I (say) is transferred to the soma (nucleus–cell body of neuron in biological system). $I = \sum w_i x_i$, $i = 1, 2, \ldots, n$. The final output is transmitted to a nonlinear filter Φ (Transfer function say), i.e., $y = \Phi(I)$.

A general form of activation function is the threshold function. This sum is equated with the threshold value θ, i.e., $y = \Phi[\sum w_i x_i - \theta]$ (Fig. 1).

This threshold function may be in the form of heaviside function, Signum function, hyperbolic tangent function, or sigmoidal function, i.e., $\Phi(I) = \frac{1}{1+e^{\alpha I}}$. The parameter α is a slope parameter. The three basic classes of ANN models follows:

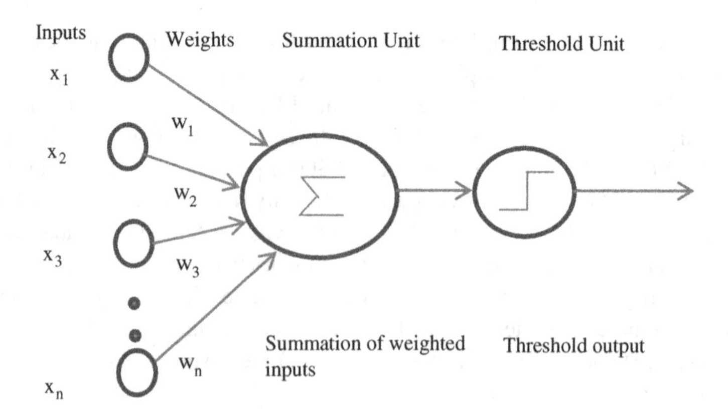

Fig. 1 Simple model of artificial neural network

(a) single layer feedforward network (no hidden layer between input and output), (b) multilayer feed forward network (one or more hidden layers between input and output), (c) recurrent networks [11]. The tansig transfer function can be used in input and hidden layer neurons for summing and nonlinear mapping. The purelin transfer function can be used in linear input–output relationship between the hidden layers and output layer. Both algorithms may be summarized as (i) IO = tansig (IN) (ii) FO = purelin (IO). IN, IO, FO represents input, intermediate outputs, and final output matrices (in Matlab) [5].

There are monthly variations (fluctuation) in estimated solid waste generation. Hence a dynamic time series neural network model is proposed for forecasting waste generation. In Matlab 7.12.0 (R2011a) neural network time series tool (ntstool) is chosen for prediction. Here, nonlinear autoregressive technique is used in which the series $y(t)$ is predicted with given d past values of $y(t)$. The predicted series can be formulated as $y(t) = f(y(t-1), y(t-2), \ldots, y(t-d))$. The input solid waste data is arranged within January 2010 to December 2014 month-wise, i.e, now 60 months solid waste collected data is available for prediction. The solid waste targets (representing dynamic data) have been taken as 1×60 cell array of 1×1 matrices. Now in training data set 70 %, 15 %, 15 % time steps are used for training, validating, and testing respectively. The network is adjusted according to the training time steps. Validation time steps represent the generalization of network. The training stops when generalization stops improving. Testing time steps shows the performance of the network. Autoregressive neural network may be adjusted by the selecting number of hidden neurons in hidden layers and number of time delays. If the network does not perform well, both can be changed. Levenberg–Marquardt backpropagation algorithm (trainlm) is used to fit inputs and targets. Multiple times training will generate different results due to different initial conditions. Mean squared error (MSE) is defined as the average squared difference between outputs and targets. Lower values are considered better. A zero value shows no error. Regression R values show the correlation between outputs and target values. The value $R = 1$ implies a close relationship between output and targets. Finally, testing should be done on large data to decide the network performance.

3 Results and Discussions

The statistical analysis of different ANN models is done. Mean square errors (MSE), root mean square error (RMSE), mean absolute error (MAE), and regression R are measured as performance metrics of ANN models. Different structures of time series neural network models have been investigated with varying values of hidden layer neurons. Table 1 illustrates the ANN model structures, performance metrics MSE, RMSE, and regression values R of training, testing, and validation phases. From table it can be observed that 1-9-1 is the best ANN model for solid waste prediction due to minimum mean square error and maximum correlation R compared to other ANN models.

Table 1 Results of training, testing, and validation of ANN models

ANN model structure	MSE	RMSE	Regression			
			Training	Testing	Validation	All
1-4-1	3681.44	60.67	0.821987	0.493287	0.754089	0.73493
1-5-1	4378.51	66.17	0.81917	0.50451	0.871937	0.79502
1-6-1	2289.66	47.85	0.89636	0.25555	0.580650	0.72636
1-7-1	3737.45	61.13	0.85026	0.83483	0.357105	0.72468
1-8-1	3120.48	55.86	0.85857	0.600234	0.276443	0.67733
1-9-1	**2377.88**	**48.76**	**0.890502**	**0.705548**	**0.634741**	**0.82464**
1-10-1	3271.61	57.19	0.86355	0.56686	0.394007	0.71777
1-11-1	3257.51	57.07	0.85672	0.252753	0.754561	0.64432
1-12-1	6166.94	78.52	0.725112	0.49032	0.597521	0.63368
1-13-1	2786.23	52.78	0.89632	0.802407	0.811832	0.78396
1-14-1	3095.32	55.63	0.82764	0.38631	0.60608	0.62935

Figure 2 illustrates the observed and predicted solid waste generation from training phase of ANN model with structure (1-9-1).

The other performance metric mean absolute error is defined as

$$\mathrm{MAE} = \frac{1}{n}\sum_{1}^{n} \left| w_0 - w_\mathrm{p} \right|$$

Here n is the number of months, w_o is observed solid waste weight, and w_p is predicted solid waste present case. The mean absolute error during training in ANN structure 1-9-1 is 40.91. Figure 3 illustrates the different values of R showing the relationship between observed and predicted values of waste during training, validation, and testing phases of the 1-9-1 ANN model (Fig. 4).

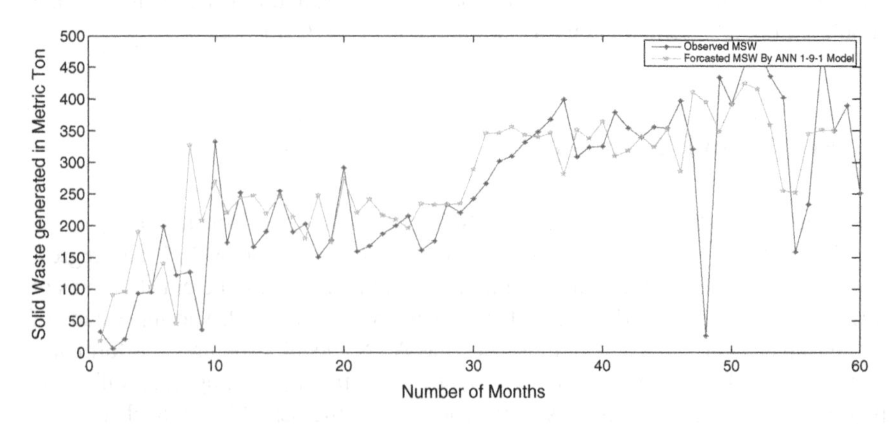

Fig. 2 The observed and predicted solid waste generation from training phase of ANN model with structure (1-9-1)

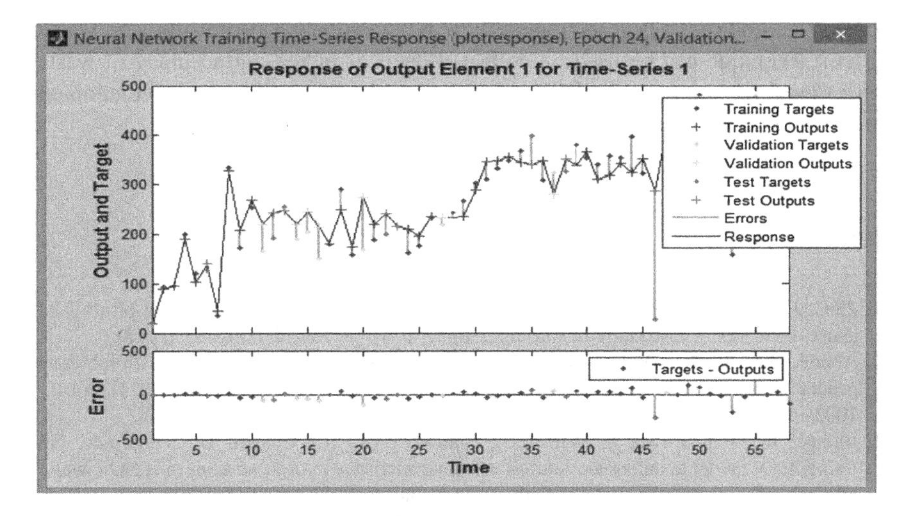

Fig. 3 Overall ANN times series response during training testing and validation

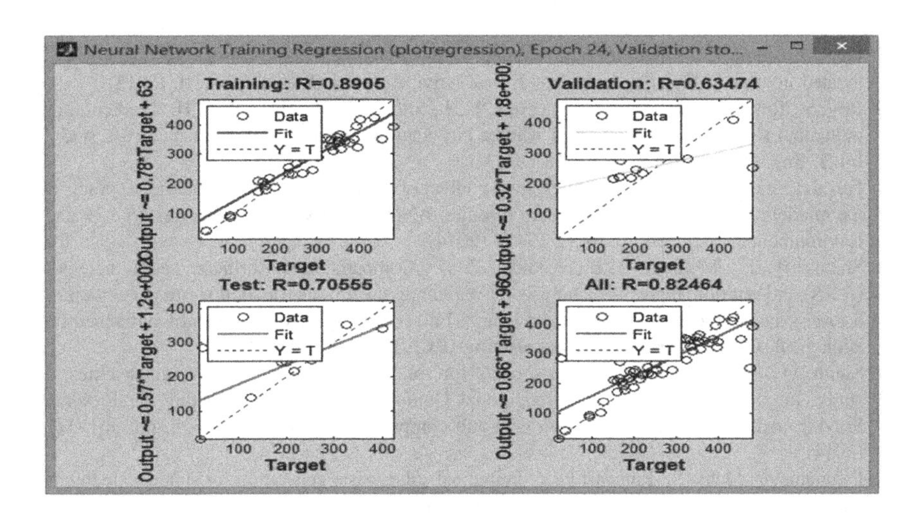

Fig. 4 Regression values of training testing, validating, and ALL for 1-9-1 model

4 Conclusion

The aim of this study is to propose an appropriate model for predicting solid waste for Faridabad city, Haryana (India). Time series neural network tool (Matlab 7.12.0 (R2011a)) has been used for monthly-based solid waste prediction. Different ANN structures have been trained and tested by changing the number of neurons in hidden layers. The ANN structure with 9 hidden layer neurons is selected for its best performance metrics, i.e., low values of MSE, RMSE and high value of R.

The future scope of this work is that the ANFIS (adaptive neuro-fuzzy inference system) technique can be used on collected data for increasing accuracy of waste generation forecasting. Such model may be generalized as solid waste prediction on weekly basis for its fine prediction.

References

1. Zade, J.G.M., Noori, R.: Prediction of municipal solid waste generation by use of artificial neural networks: a case study of Mashhad. Int. J. Environ. Res. **2**(1), 13–22 (2008)
2. Abdoli, M.A., Nazhad, M.F., Sede, R.S., Behboudian, S.: Longterm forecasting of solid waste generation by artificial neural networks. Environ. Progress Sustain. Energy **00** (2011). doi:10.1002/ep.10591
3. Batinic, B., Vukmirovic, S., Vujic, G., Stanisavljevic, N., Ubavin, D., Vumirovic, G.: Using ANN model to determine future waste characteristics in order to achieve specific waste management targets-case study of Serbia. J. Sci. Ind. Res. **70**, 513–518 (2011)
4. Shahabi, H., Khezri, S., Ahmad, B.B., Zabihi, H.: Application of artificial neural network in prediction of municipal solid waste generation (Case study: Saqqez City in Kurdistan Province). World Appl. Sci. J. **20**(2), 336–343 (2012). doi:10.5829/idosi.wasj.2012.20.02.3769
5. Patel, V., Meka, S.: Forecasting of municipal solid waste generation for medium scale towns located in state of Gujarat, India. Int. J. Innovative Res. Sci. **2**(9), 4707–4716 (2013)
6. Roy, S., Rafizul, I.M., Didarul, M., Asma, U.H., Shohel, M.R., Hasibul, M.H.: Prediction of municipal solid waste generation of Khulna city using artificial neural network: a case study. Int. J. Eng. Res. Online **1**(1), 13–18 (2013)
7. Falaahnezhad, M., Abdoli, M.: Investigating effect of preprocessing of data on the accuracy of the modeling solid waste generation through ANNs. Amirkabir J. Sci. Res. (Civil and Environmental Engineering) **46**(2), 13–14 (2014)
8. Shamshiry, E., Mokhtar, M.B., Abdulai, A.M.: Comparision of artificial neural network (ANN) and multiple regression analysis for predicting the amount of solid waste generation of a tourist and tropical area—Langkawi Island. In: proceeding of International Conference on Biological, Civil, Environmental Engineering (BCEE), pp. 161–166 (2014)
9. Singh, D., Satija, A.: Optimization models for solid waste management in indian cities: a study. In: proceeding of fourth International Conference on Soft Computing for Problem Solving, Advances in Intelligent Systems and computing (SocProS), vol. 335, pp. 361–371 (2014)
10. Pamnani, A., Meka, S.: Forecasting of municipal solid waste generation for small-scale towns and surrounding villages located in state of Gujrat, India. International Journal of Current Engineering and Technology **5**(1), 283–287 (2015)
11. Rajashekaran, S., Pai, G.A.V.: Neural networks, fuzzy logic and genetic algorithms. Rajashekaran, S., Pai, G.A.V., Ghosh A.K. (eds,) Fundamentals of Neural Networks, pp. 11–30. Prentice Hall of India (2007)

Secondary Network Throughput Analysis Applying SFR in OFDMA-CR Networks

Joydev Ghosh, Varsha Panjwani and Sanjay Dhar Roy

Abstract In OFDMA femtocell networks, the licensed spectrum of the macro users (MUs) are available to the femto users (FUs), on the condition that they do not spark off notable interference to the MUs. We contemplate wireless data for femto user (FU)/secondary user (SU) in cognitive radio (CR) networks where the frame structure split up into sensing and data transmission slots. Moreover, we consider soft frequency reuse (SFR) technique to improve secondary network throughput by increasing the macrocell edge user power control factor. SFR applies a frequency reuse factor (FRF) of 1 to the terminal located at the cell centre for that all base stations (BSs) share the total spectrum. But for the transmission on each subcarrier the BSs are confined to a certain power level. However, more than 1 FRF uses for the terminals near to the macrocell edge area. In this context, we conceptualize the cognitive femtocell in the uplink in which the femtocell access point (FAP) initially perceive by sensing to find out the availability of MU after that FAP revamps its action correspondingly. Appropriately, when the MU is sensed to be nonexistent, the FU transmits at maximum power. In other respect, the FAP make the best use of the transmit power of the FU to optimize the secondary network throughput concern to outage limitation of the MU. Finally, effectiveness of the scheme is verified by the extensive MATLAB simulation.

Keywords Spectrum sensing · Signal to interference plus noise ratio (SINR) · Average power consumption · Secondary network throughput

Joydev Ghosh (✉)
The New Horizons Institute of Technology, Durgapur-8, West Bengal, India
e-mail: joydev.ghosh.ece@gmail.com

Varsha Panjwani · S.D. Roy
National Institute of Technology, Durgapur-9, West Bengal, India
e-mail: varsha.panjwani2012@gmail.com

S.D. Roy
e-mail: s_dharroy@yahoo.com

M. Pant et al. (eds.), *Proceedings of Fifth International Conference on Soft Computing for Problem Solving*, Advances in Intelligent Systems and Computing 437, DOI 10.1007/978-981-10-0451-3_28

1 Introduction

Radio frequency spectrums are treated as one of the substantial limited resources in wireless communications which should be used effectively. With such stimulation to decrease power consumption and the reuse of radio resources, there is an inescapable shift towards the deployment of femto cell networks by disintegrating traditional single-cell, single-layer network into multi-layer networks with large secondary network throughput [1]. Femtocells are low-power invention that comes up with better coverage to portable users through femtocell access point (FAP) at the indoor scenario. The access node named as FAP, perform as the base station (BS) for the femtocell and take help of internet as backhaul to get connected with MBS. But, it is crucial to emerge effective interference management scheme in order to make sure the coexistence of the two-tier networks [2, 3]. In [4], SFR technique is introduced to keep away from the inter-tier interference between primary and secondary networks [5]. This is obtained by partitioning the total accessible band into a number of subbands and put a limit on the femtocell network's access.

In a CR networks, the SUs are permitted to use the spectrum of PUs when these frequency bands are not been under utilization. To execute this frequency reuse mechanism, the SUs are required to sense the radio frequency scenario, and once the PUs are found to be active, the SUs are needed to evacuate the channel within a certain period of time. Thus, spectrum sensing is of notable importance in CR networks. The variables connected with spectrum sensing are probability of detection (P_d) and probability of false alarm (P_f) [6]. When high the P_d, the PUs are more protected from harmful interference. Moreover, from the SUs point of view, lower the P_f, more chances that the channel can be reutilized when it is accessible, thus larger the secondary network throughput. In this paper, we review the issue of sensing period to optimize the secondary network throughput under the limitation that PUs are sufficiently protected [7]. Consequently, a typical frame structure is contemplated for the SU which composed of the sensing and the data transmission periods [8]. The sensing period and the data transmission period are needed to be incorporated in a unit frame such that (i) the quantity of transmitted data packets become greater and (ii) the number of clashes with the PUs become less.

In Particular, the major contributions of this paper are highlighted below

- In order to construct the sensing-throughput trade-off issue, the objective turns out to be reducing, P_f, under the limitation of P_d. We therefore construct the sensing-throughput trade-off issue from this viewpoint.
- As macrocell edge users' compromise the spectrum with adjacent macrocells, their uplink transmission adapts higher power level to combat cross tier interference. For the adaptation of above power control mechanism, we consider $P_{centre} = P_k$ and $P_{edge} = \alpha P_k$, keeping $\alpha \geq 1$.
- The performance metric of interest, we refer as secondary network throughput, is a complete unit of data that is successfully transmitted in a particular amount of time.

The remainder of this paper is organized as follows. In Sect. 2, we describe system model to define propose network. In Sect. 3, the assumptions considered for execution of exact scenario in the simulation model. In Sect. 4, we present and analyse the numerical results. Ultimately, we finish this work in Sect. 5 with conclusion.

2 System Model

We consider a scenario where femtocells are deployed over the existing macrocell network and share the same frequency spectrum with macrocell. Here, we focus only on downlink scenario. The downlink communications in a network with one macrocell and N_F = 4 number of femtocells is as shown in Fig. 1 in which four femtocells are located at (2, 2), (−2, 2), (−2, −2), (2, −2) within the macrocell coverage. In a macrocell, total N_{MUE} user equipments (UEs) are randomly distributed within its coverage area.

In theoretical analysis, we assume that the OFDMA-based dual tier network consists of N_M number of hexagonal gride macrocell and N_F number of femto cells in each macrocell. Total bandwidth associated with the macrocell edge regions is split up into 3 subbands by applying SFR technique [4, 9]. A subband containing N_{sc} number of subchannels that are available to provide service to the users located at the cell-centre area and the corresponding cell edge area [10, 11]. Besides, we also consider that the channel is slowly time varying and follows the Rayleigh multipath fading distribution. Three kinds of possible links in dual-layer networks are as follows: MBS to outdoor user's link, FBS to indoor user's link, MBS to indoor user's link. Hence, link gain in dual-layer networks can be described as [2].

$$G^{x}_{n,k,i} = d^{-\alpha_p} 10^{\xi_s/10} |h|^2 \quad (1)$$

where $d_{s2s,\,ji}$ is the distance between *j*th FBS and *i*th FBSU. ξ_s (in dB) is a Gaussian random parameter with 0 mean and σ^2, variance, due to shadowing in the channel.

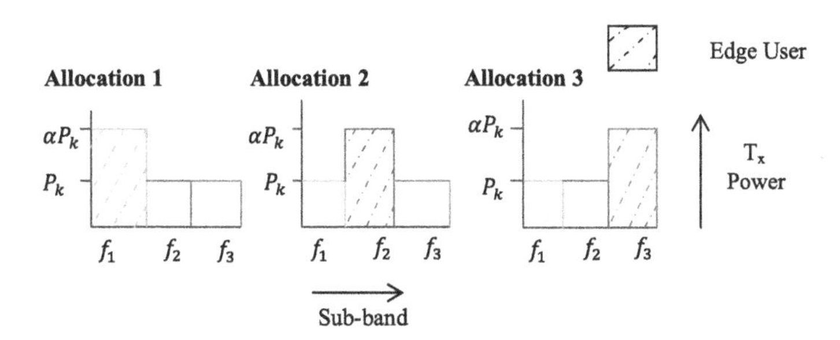

Fig. 1 SFR technique in intra-tier: subbands and transmit power allocations at cell edge area for FRF = 3

Here, $|h|^2$ denote the channel gain between kth MBS and its associated ith FBSU. Moreover, Rayleigh fading gives tractable results which assists understanding of the system response to a particular situation. We use the notation x to denote the serving network entity for a generic user. That is, $x = a$ if the user is associated to a FAP and $x = b$ if the user is associated to a MBS. Without any loss of generic laws, the analysis is conducted on a typical user located at the origin. Therefore, SINR, $\gamma^x_{n,k,i}$ at the typical user located at the origin (which also holds for any generic user) served by an MBS or FAP (MBS/FAP) is given by [3]

$$\gamma^x_{n,k,i} = \frac{P^x_{n,k,i} G^x_{n,k,i}}{I^{x*}_{n,k,i} + I'^{x*}_{n,k,i} + \sigma^2_{n,k,i}} \tag{2}$$

where $G^x_{n,k,i}$ is the wireless link gain between the user to the serving network entity (i.e., an MBS or a FAP) over the nth subchannel. Here, $P^x_{n,k,i}$ is designated as the proportion of total transmit power by an associated serving network entity over the particular subchannel. Likewise, the channel gains from a generic location $x \in \mathbb{D}^2$ to the MBS, b_i and the FAP, a_i are denoted by $h_{b_i} \sim \sqrt{X^2_{b_i} + Y^2_{b_i}}$ and $h_{a_i} \sim \sqrt{X^2_{a_i} + Y^2_{a_i}}$, respectively, where X_x, Y_x are indicated as independent gaussian random variables with zero mean and desired variance, $\sigma^2_{n,k,i}$ is the noise power of zero-mean complex valued additive white Gaussian noise (AWGN).

The energy measuring device is composed of a square law device succeeded with an integrator for finite time. The outcome of the integrator at any instant of t is the energy of the input signal to the square law device on a particular interval (0, T).

The detection is a measure of the following two hypotheses

$\mathcal{H}_o$: The input $c(t)$ is noise alone

(a) $c(t) = n(t)$; $n(t)$ denote zero-mean AWGN with unit variance:

$$n(t) \cong \sigma(0, 1)$$

(b) $E[n(t)] = 0$; where t denotes sample index

(c) noise spectral density = N_{02} (two sided)

(d) noise bandwidth equals to w cycles per seconds

$\mathcal{H}_1$: The input $c(t)$ is signal + noise

$$\begin{aligned} &\text{(a)} \quad c(t) = s(t) + n(t) \\ &\text{(b)} \quad E[s(t)h(t) + n(t)] = s(t) \end{aligned}$$

The tenancy of nth subband is possible to detect with the help of a simple hypothesis test written as [9]

$$V' = \begin{cases} N_n; & \mathcal{H}_{o,n} \\ S_n + N_n; & \mathcal{H}_{1,n} \end{cases} \tag{3}$$

where $n = \{1, 2, \ldots N_{sc}\}$, S_n and N_n indicate the discrete frequency response of $s(t)$ and $n(t)$, respectively.

After proper filtering, sampling, squaring, and integration, the test statistic of an energy detector is

$$T_{\mathrm{v}} = \sum_{i=1}^{2r} |V'|^2 \tag{4}$$

where r is the number of complex signal samples. As described in [4], the probability density function (PDF) of T_{v} follows a central chi-square distribution with $2r$ degrees of freedom (DoF) under $\mathcal{H}_{o,n}$, or a noncentral chi-square distribution with $2r$ DoF and a noncentrality parameter 2γ under $\mathcal{H}_{1,n}$. The test statistic, T_{v}, is compared with a predefined threshold value λ. Hence, the probabilities of detection and false alarm can be written as [6, 12]

$$P_{\mathrm{d}}\left(\gamma^x_{n,k,i}, \lambda\right) = Q_{\mathrm{v}}\left(\sqrt{2\gamma^x_{n,k,i}}, \sqrt{\lambda}\right) \tag{5}$$

$$P_{\mathrm{f}}(\lambda) = \frac{\Gamma\left(r, \frac{\lambda}{2}\right)}{\Gamma(r)} \tag{6}$$

where $Q_{\mathrm{v}}(x) = \frac{1}{\sqrt{2\pi}} \int_x^{\infty} \exp(-t^2/2)\mathrm{d}t$, $\Gamma(a, x) = \int_x^{\infty} t^{a-1} \exp(-t)\mathrm{d}t$, and $\Gamma(a, 0) = \Gamma(a)$. Further, missed detection probability can be calculated as $P_{\mathrm{m}}\left(\gamma^x_{n,k,i}, \lambda\right) = 1 - P_{\mathrm{d}}\left(\gamma^x_{n,k,i}, \lambda\right)$ [13].

In general, a frame made up of a sensing period and a transmission period. The sensing period is represented by τ, while total time span of a frame is T. In the proposed network model, the sensing time is considered as zero from SUs point of view, as SU not an integral part of spectrum sensing. Additionally, typical value of τ differs from 0 to T as the introduced network senses the spectrum maximum to the frame duration. The achievable instantaneous data rate of kth femto user can be calculated by considering two cases, when FAP finds that MU is absent which informs to FU by FAP to transmit at its maximum allowable power ($P^{\max}_{n,k,i}$), and when FAP finds that the MU is present for that FAP optimizes ($P^{*}_{n,k,i}$) and then, inform femto user to transmit to the FAP. The data rates of FU under the above condition can be written as

$$r_{00} = \log_2\left(1 + \frac{P^{\max}_{n,k,i} G^x_{n,k,i}}{\sigma^2_{n,k,i}}\right) \tag{7}$$

when FAP accurately senses that the MU not exist.

$$r_{01} = \log_2\left(1 + \frac{P^{\max}_{n,k,i} G^{x}_{n,k,i}}{I^{x*}_{n,k,i} + I'^{x*}_{n,k,i} + \sigma^2_{n,k,i}}\right) \tag{8}$$

when the FAP inaccurately senses that the MU not exist

$$r_{11} = \log_2\left(1 + \frac{P^{*}_{n,k,i} G^{x}_{n,k,i}}{I^{x*}_{n,k,i} + I'^{x*}_{n,k,i} + \sigma^2_{n,k,i}}\right) \tag{9}$$

when the FAP accurately senses that the MU exists

$$r_{10} = \log_2\left(1 + \frac{P^{*}_{n,k,i} G^{x}_{n,k,i}}{\sigma^2_{n,k,i}}\right) \tag{10}$$

when the FAP inaccurately finds that MU exists while MU not exist. Hence, secondary network throughput can be given by [8]

$$T_{\mathrm{f}} = E\left\{\left(\frac{T - T_{\mathrm{s}}}{T}\right)\left[\begin{array}{c} r_{00}(1 - P_{\mathrm{f}}(\lambda))P(\mathcal{H}_{o,n}) \\ + r_{01}(1 - P_{\mathrm{d}}(\lambda))P(\mathcal{H}_{1,n}) \\ + r_{11}P_{\mathrm{d}}(\lambda)P(\mathcal{H}_{1,n}) + r_{10}P_{\mathrm{f}}(\lambda)P(\mathcal{H}_{0,n}) \end{array}\right]\right\} \tag{11}$$

where $E\{.\}$ is an expectation operator.

$$T_{\mathrm{f}} = \left(\frac{T - T_{\mathrm{s}}}{T}\right)\left[\begin{array}{c} R_{00}(1 - P_{\mathrm{f}}(\lambda))P(\mathcal{H}_{o,n}) \\ + R_{01}(1 - P_{\mathrm{d}}(\lambda))P(\mathcal{H}_{1,n}) \\ + R_{11}P_{\mathrm{d}}(\lambda)P(\mathcal{H}_{1,n}) + R_{10}P_{\mathrm{f}}(\lambda)P(\mathcal{H}_{0,n}) \end{array}\right] \tag{12}$$

where $R_{00} = E\{r_{00}\}$, $R_{01} = E\{r_{01}\}$, $R_{11} = E\{r_{11}\}$, $R_{10} = E\{r_{10}\}$.

3 Simulation Model

For better approximation of numerical results, Rayleigh fading is included with pathloss and shadowing. The simulation testbed model is carried out considering the following steps

1. A fixed number of outdoor users' ($N_{\mathrm{MUE},ku}$) and indoor users' ($N_{\mathrm{FUE},ji}$) is generated and they are randomly distributed within their own coverage area. N_{UE} includes all MBSUs/PUs ($N_{\mathrm{MUE},ku}$) and FBSUs/SUs ($N_{\mathrm{FUE},ji}$) which means $N_{\mathrm{UE}} = N_{\mathrm{MUE},ku} + N_{\mathrm{FUE},ji}$. Here, $j \in \mathcal{N}_{\mathrm{F}} = \{1, 2, \ldots, N_{\mathrm{F}}\}$; $k \in \mathcal{N}_{\mathrm{M}} = \{1, 2, \ldots, N_{\mathrm{M}}\}$; $N_{\mathrm{MUE},ku}\ \forall\ \{1, 2, 3, \ldots, u\mathcal{N}_{\mathrm{M}}\}$, $u \in$ any large integer value; $N_{\mathrm{FUE},ji}\ \forall\ \{1, 2, 3, \ldots, i\mathcal{N}_{\mathrm{F}}\}$, $i \in$ any large integer value.

2. The interference on kth user over the nth subchannel are executed as below [1, 13, 14]

$$I^{f}_{k,i} = \sum_{1=1}^{N_{\mathrm{M}}} P^{m}_{i,l,n} G^{m}_{i,l,n} \quad \forall\ l \in \{1, 2, 3 \ldots N_{\mathrm{M}}\} \tag{13}$$

$$I'^{f}_{k,i} = \sum_{j=1, j \neq i}^{N_{\mathrm{M}} \mathrm{x} N_{\mathrm{F}}} \beta^{n}_{j} P^{f}_{i,j,n} G^{f}_{i,j,n} \quad \forall\ j \in \{1, 2, 3 \ldots\ldots N_{\mathrm{F}}\} \tag{14}$$

$$I^{m}_{k,i} = \sum_{l=1, l \neq i}^{N_{\mathrm{M}}} P^{m}_{i,l,n} G^{m}_{i,l,n} \quad \forall\ l \in \{1, 2, 3 \ldots N_{\mathrm{M}}\} \tag{15}$$

$$I'^{m}_{k,i} = \sum_{j=1}^{N_{\mathrm{M}} \mathrm{x} N_{\mathrm{F}}} \beta^{n}_{j} P^{f}_{i,j,n} G^{f}_{i,j,n} \quad \forall\ j \in \{1, 2, 3 \ldots N_{\mathrm{F}}\} \tag{16}$$

where $P^{m}_{i,l,n}$ and $P^{f}_{i,j,n}$ indicate the transmit signal powers over the nth subchannel of MBS l and FBS j, respectively; $G^{m}_{i,l,n}$ and $G^{f}_{i,j,n}$ indicate the corresponding path gains for MBS l and FBS j, respectively; β^{n}_{j} use as a indicator function for femtocell resource allocation. If $\beta^{n}_{j} = 1$ indicates nth channel is assigned to femtocell j; otherwise $\beta^{n}_{j} = 0$.
3. The received signal strength (RSS) is evaluated from PU/MBSU or SU/FBSU at the reference MBS or FBS.
4. Next, the SINR for a PU/Macro user and/or a SU/Femto user are computed.

4 Results and Discussions

In order to vindicate the excellence predicted by the analytical framework discussed in Sect. 2, we present several applications at different conditions (Fig. 2).

In Fig. 3, average power consumption are shown as a function of user average distance for FRF = 1, 3, and 7. Point to be noted that integer FRF is confined to a set of integer numbers: {1, 3, 4, 7, ….} followed by the equation $i^2 + i \cdot j + j^2 \mid i, j \in N$. By applying FRF = 1 only the users near to the base station (BS) experience better channel quality whereas the users located far apart suffer from poor radio conditions due to sever inter-cell interference (ICI). To accomplish this problem higher FRF employs at macrocell edge area. In general, higher FRF reduces the spectral efficiency of the network, but at the same time it (optimum choice, FRF = 3) reduces power consumptions. For SFR technique, the total spectrum is split up into three subbands: f_1, f_2, and f_3 as shown in Fig. 1. Macrocell edge user are constrained to take service from one of the subbands with large transmit power ($P_{\mathrm{edge}} = \alpha P_{\mathrm{k}}$),

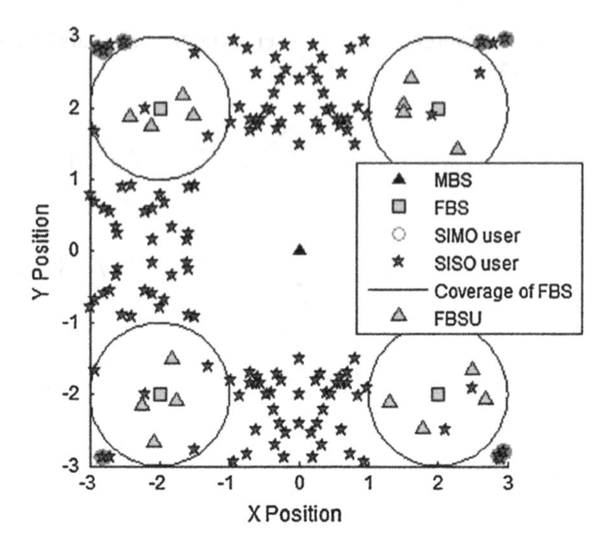

Fig. 2 Co-existance scenario of femtocell and macrocell in our proposed network

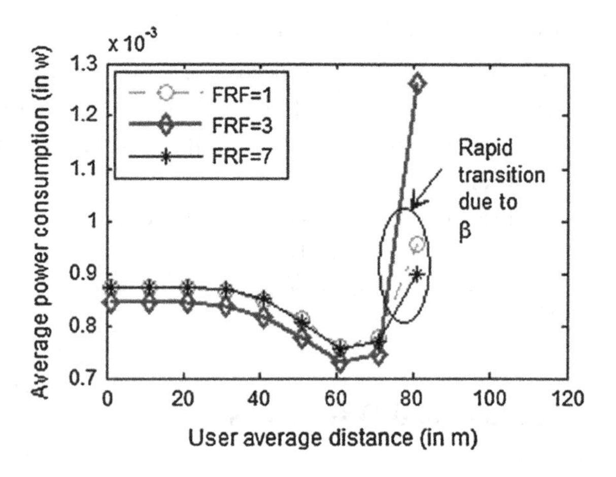

Fig. 3 Average power consumption versus user average distance for different FRF where number of subcarriers (M) = 1200 and subcarrier spacing = 15 kHz

while the centre located users can reuse the whole spectrum with a reduced power ($P_{\mathrm{centre}} = P_{\mathrm{k}}$). When frequency reuse factor is fixed at FRF = 3, the influence of subcarrier spacing and number of subcarrier parameters on the average power consumption is been investigated in Figs. 4 and 5, respectively. It can be seen that, between subcarrier spacing and number of subcarrier parameters, subcarrier spacing is more effective factor by means of less power consumption and stability, particularly in the cell edge area.

In Fig. 6, average network throughput when MUs absent (R_0) and average network throughput when FUs present (R_1) are shown as an increasing and decreasing function of sensing time, respectively, and the resultant curve (R) is becoming secondary network throughput.

Figure 7 plots the secondary network throughput as a function of sensing time for different probability of detection (P_{d}) in imperfect sensing scenarios. However,

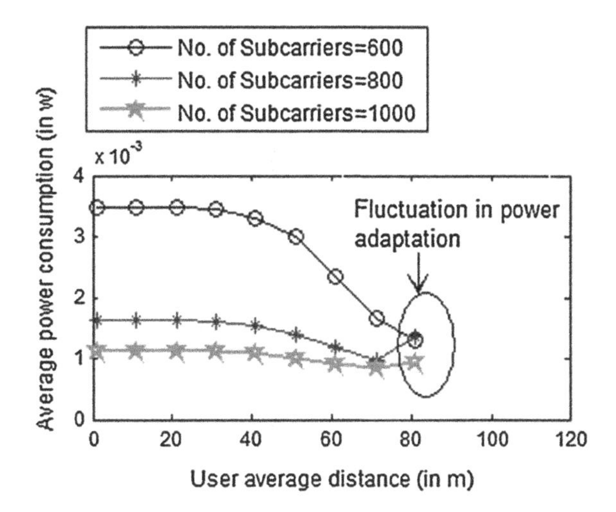

Fig. 4 Average power consumption versus user average distance for different value of M where FRF = 3 and subcarrier spacing = 15 kHz

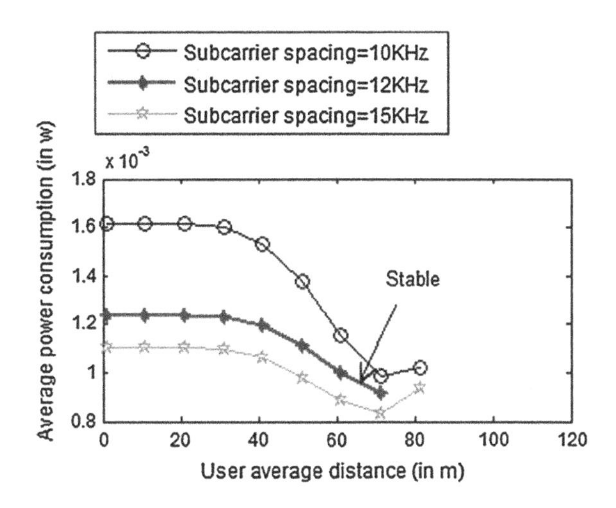

Fig. 5 Average power consumption versus user average distance for different subcarrier spacing where FRF = 3 and M = 1200

there is $\left(\frac{T-T_s}{T}\right)$ in the secondary network throughput expression in Eq. (12), which can rapidly reduce secondary network throughput when the sensing time is too long. P_d is to be set at higher value to make PUs more secure from interference at the cost of secondary network throughput.

Figure 8 shows the secondary network throughput when frame duration are 30, 50, 100 ms, respectively, and P_d is set to 0.9. The issue been raised due to mandatory condition of higher P_d and rapidly reduction in secondary network throughput, particularly in the higher sensing time region, can be resolved up to some extend by increasing the frame duration.

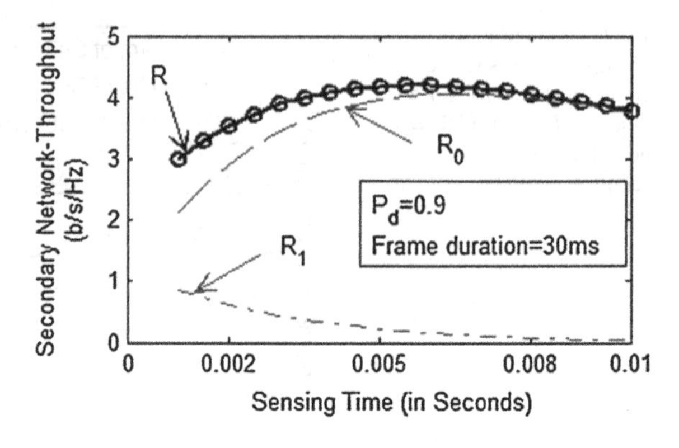

Fig. 6 Secondary network throughput versus sensing time where $P_d = 0.9$ and $T = 30$ ms

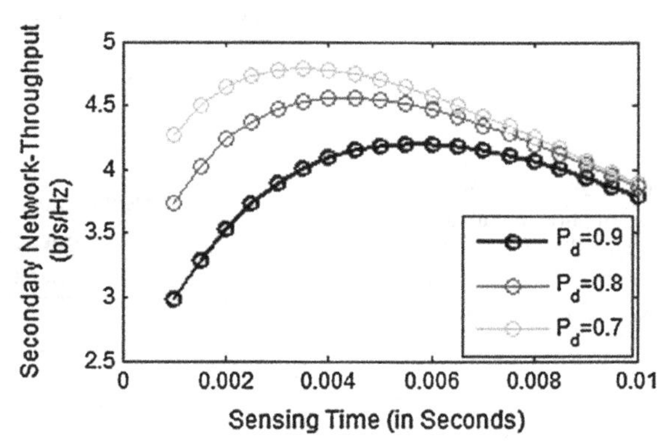

Fig. 7 Secondary network throughput versus sensing time for different P_d where $T = 30$ ms

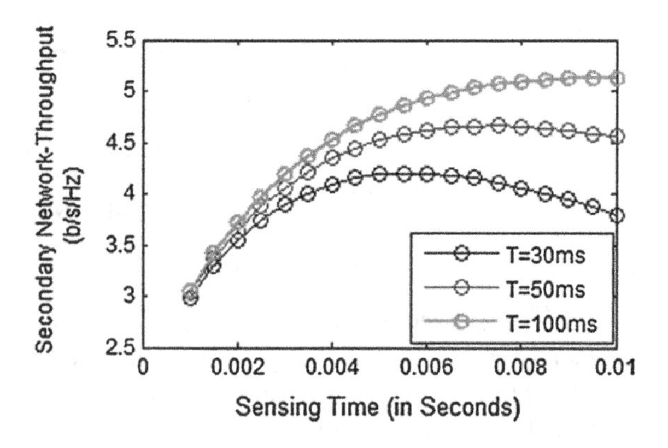

Fig. 8 Secondary network throughput versus sensing time for different T where $P_d = 0.9$

5 Conclusion

In this paper, we develop a novel simulation testbed model to demonstrate various aspects in terms of substantial parameters in connection with imperfect spectrum sensing and secondary network throughput for dual-tier cognitive femtocell networks. It is been found from the results that we have to compromise with stability to deal with the impact of interference at cell edge area by increasing the power level by the factor of α. Tractable analysis of the networks also assist in determining the optimum choice of the parameters which gives insight into the performance tradeoffs of SFR strategies.

References

1. Roy, S.D., Kundu, S., Ferrari, G., Raheli, R.: On spectrum sensing in cognitive radio CDMA networks with beamforming. Phys. Commun. **9**, 73–87 (2013)
2. Lopez-Perez, D., Guvenc, I., de la Roche, G., Kountouris, M., Quek, T.Q.S., Zhang, J.: Enhanced inter-cell interference coordination challenges in heterogeneous networks. IEEE Wireless Commun. Mag. **18**(3), 22–30 (2011)
3. Cheung, W.C., Quek, T.Q.S., Kountouris, M.: Throughput optimization in two-tier femtocell networks. IEEE J. Sel. Areas Commun. (Revised, 2011)
4. Novlan, T.D., Ganti, R.K., Ghosh, A., Andrews, J.G.: Analytical evaluation of fractional frequency reuse for heterogeneous cellular networks. IEEE Trans. Commun. **60**(7) (2012)
5. Li, L., Xu, C., Tao, M.: Resource allocation in open access OFDMA femtocell networks. IEEE Wireless Commun. Lett. **1**(6) (2012)
6. Atapattu, S., Tellambura, C., Jiang, H.: Analysis of area under the ROC curve of energy detection. IEEE Trans. Wire-less Commun. **9**(3) (2010)
7. Xu, Z., Li, G.Y., Yang, C., Zhu, X.: Throughput and optimal threshold for FFR schemes in OFDMA cellular networks. IEEE Trans. Wireless Commun. **11**(8), (2012)
8. Ahmad, A., Yang, H., Lee, C.: Maximizing throughput with wireless spectrum sensing network assisted cognitive radios. Int. J. Distrib. Sensor Netw, Hindawi Publishing Corporation
9. Oh, D.C., Lee, Y.H.: Cognitive radio based resource allocation in femto-cell. J. Commun. Netw. **14**(3) (2012)
10. Misra, S., Krishna, P.V., Saritha, V.: An efficient approach for distributed channel allocation with learning automata-based reservation in cellular networks. SIMULATION: Trans. Model. Simul. Int. **88**(10), 1166–1179 (2012)
11. Krishna, P.V., Misra, S., Obaidat, M.S., Saritha, V.: An efficient approach for distributed dynamic channel allocation with queues for real-time and non-real-time traffic in cellular networks. J. Syst. Softw. (Elsevier) **82**(6), 1112–1124 (2009)
12. Olabiyi, O., Annamalai, A.: Closed-form evaluation of area under the ROC of cooperative relay based energy detection in cognitive radio networks. In: ICCNC 2012, pp. 1103–1107
13. Roy, S.D., Kundu, S.: Performance analysis of cellular CDMA in presence of beamforming and soft handoff. Progr. Electromagn. Res. PIER, **88** 73–89 (2008)
14. Ning, G., Yang, Q., Kwak, K.S., Hanzo, L.: Macro and femtocell interference mitigation in OFDMA wireless systems. In: Globecom 2012-Wireless Communications Symposium, pp. 5290–5295

Unsupervised Change Detection in Remote Sensing Images Using CDI and GIFP-FCM Clustering

Krishna Kant Singh, Akansha Singh and Monika Phulia

Abstract In this paper, an unsupervised change detection method for remote sensing image is proposed. The method takes as input two bi temporal images and outputs a change detection map highlighting changed and unchanged areas. Initially, the method creates a feature vector space by performing the principal component analysis (PCA) of both the images. The first component of PCA of both the images is used to compute the combined difference image. A change map is then created from combined difference image by clustering it into two clusters changed and unchanged using GIFP-FCM clustering technique. The method is applied on Landsat 5 TM images and the results obtained are compared with other existing state of the arts method.

Keywords Change detection · GIFP-FCM · Clustering · Remote sensing

1 Introduction

With the advancement in remote sensing technology, change detection has become a useful technique for identifying various changes between two co-registered remote sensing images of a particular geographical area taken at different time-instants. Satellite imagery has been widely used as a basis for various applications, i.e., urban expansion, cultivation, deforestation monitoring, land use/cover mapping, impact of natural disaster, viz., tsunami, earthquakes, flooding, etc. [1]. Generally, two main schemes are used for such applications: post classification

K.K. Singh (✉) · M. Phulia
EEE Department, Dronacharya College of Engineering, Gurgaon, India
e-mail: krishnaiitr2011@gmail.com

A. Singh
The Northcap University, Gurgaon, India
e-mail: akanshasing@gmail.com

M. Pant et al. (eds.), *Proceedings of Fifth International Conference on Soft Computing for Problem Solving*, Advances in Intelligent Systems and Computing 437, DOI 10.1007/978-981-10-0451-3_29

comparison and direct change detection. Change detection methods are basically of two types: supervised and unsupervised [2]. In supervised classification-based methods, ground-truth data is required for identification of changes in remotely sensed images. Supervised methods have a serious limitation regarding its dependence on ground-truth while unsupervised change detection do not require ground-truth, rather make use of spectral properties of the image for direct comparison and do not require any prior information about land-cover classes [3, 4].

Unsupervised change detection techniques detect the changes of bitemporal images associated with same geographical area. Its procedure can be divided into the following steps: (1) Image preprocessing. In this step, geometric correction and registration are implemented. It is done to align the two bitemporal images in the same coordinate frame. (2) Generation of difference image. Digital ratio image is expressed in a logarithmic or mean scale due to its robustness. Change vector analysis (CVA) is widely used technique for this purpose. (3) Analysis of difference image. This is the most important step which is basically a segmentation problem [3]. A most popular solution for image segmentation is thresholding [5, 6] and clustering [7]. In some cases, feature extraction is done on the difference image prior to image segmentation such as principal component analysis (PCA), stationary wavelet transforms, etc. Bruzzone and Preito proposed an automatic thresholding technique based on Bayes' theory. Optimal threshold was obtained under two assumptions. One was that pixels of the difference image are independent of their spatial location and permits thresholding for reducing error in change detection; the other was that conditional density functions of two classes (changed and unchanged) were modelled by Gaussian functions. In addition, Markov random field (MRF) approach was added to enhance the accuracy of the method [8]. Change detection in the geometry of a city was introduced using panoramic images. The method significantly optimizes the process of updating the 3D model of a time-changing urban environment. Algorithm used in the method specifically detects only structural changes in the environment [9]. Local Gradual Descent Matrix (LGDM) is an efficient change detection technique which analyse the difference image for identifying change detection in remote sensing imagery [10]. Klaric et al. developed a fully automated system namely GeoCDX for change detection of high resolution satellite imagery. GeoCDX performs fully automated imagery co-registration, extracts high-level features from the satellite imagery and performs change detection processing to pinpoint locations of change [11]. A recent technique named multidimensional evidential reasoning (MDER) presented in analyses heterogeneous remote sensing images. The proposed technique is based on a multidimensional (M-D) frame of discernment composed by the Cartesian product of the separate frames of discernment used for the classification of each image wherein results show the potential interest of the MDER approach for detecting changes due to a flood in the Gloucester area in the U.K. from two real ERS and SPOT images [12]. Unsupervised algorithm-level fusion scheme (UAFS-HCD) has also been applied for improving accuracy of PBCD using spatial context information through the preliminary change mask with PBCD to estimate some parameters for OBCD, analysing the unchanged area mask and getting final change map using OBCD [13]. Celik proposed an unsupervised change detection method for multitemporal satellite images using PCA and k-means [14].

Many change detection techniques have been proposed for change detection in remote sensing imagery. A single method does not provide satisfactory result for all types of data. Every method has its own applicability for particular area and particular dataset. In this work, a novel algorithm based on generation of combined difference image (CDI) for unsupervised change detection in heterogeneous remote sensing imagery is proposed. This method performs fusion of two co-registered DI obtained by implementing Principal Component Analysis (PCA), which is used for obtaining feature vector space component from difference images. Finally, change detection is achieved by using clustering using generalized improved fuzzy partitions-FCM (GIFP-FCM).

2 Proposed Method

Let us consider two co-registered bitemporal Landsat satellite images IT1 and IT2 of size $H \times W$ of same geographical area taken at two different times T1 and T2, respectively. The proposed method consists of following steps as shown in Fig. 1:

- Creation of feature vector space using PCA
- Combined Difference Image (CDI)
- Clustering using GIFP-FCM
- Formation of change map

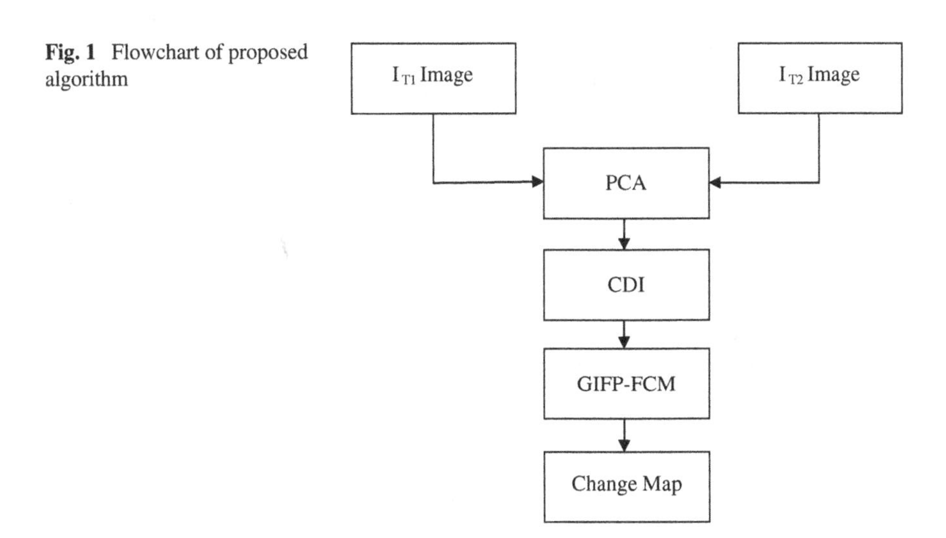

Fig. 1 Flowchart of proposed algorithm

2.1 Principal Component Analysis (PCA)

PCA technique is used for feature extraction in bitemporal digital imagery. The bitemporal satellite images IT1 and IT2 can be shown in a 3D column vector taking band 1, band 2, and band 3 as

$$I(i,j) = \begin{bmatrix} I_{\text{band1}} \\ I_{\text{band2}} \\ I_{\text{band3}} \end{bmatrix} \tag{1}$$

For the images having size $H \times W$, there will be a total of HW vectors consisting of all the pixels in the images. For simpler mathematical notations, vector $I(i, j)$ is denoted by I_h, where h defines an index with $1 \leq h \leq N$ and $N = H \times W$. To extract feature vector using PCA, consider the following equation for mean vector, given by

$$\phi = \frac{1}{N} \sum_{h=1}^{N} I_h \tag{2}$$

Covariance matrix of vector population, Λ has Eigen vector e_k and corresponding Eigen values λ_k. The covariance matrix Λ can be defined as

$$\Lambda = \frac{1}{N-1} \sum_{h=1}^{N} I_h I_h^T - \phi\phi^T \tag{3}$$

Here, we have used $N - 1$, instead of N, due to unbiased estimation of Eigen vectors. Assume that Eigen vector and Eigen values are arranged in decreasing order, i.e. $\lambda_k \geq \lambda_{k+1}$. The mean vector $I(i, j)$ is projected onto Eigen vector to obtain feature extraction at spatial location (x, y) using PCA. Feature vector space is given by

$$I_p(x, y) = \begin{bmatrix} I_{\text{pc1}}(x,y) \\ I_{\text{pc2}}(x,y) \\ I_{\text{pc3}}(x,y) \end{bmatrix} = A[I(i,j) - \phi] \tag{4}$$

Here, let A be a matrix whose rows are formed from the Eigen vectors of Λ such that first row of A is the Eigen vector corresponding to largest Eigen values, and the last row is the Eigen vector corresponding to the smallest Eigen value.

2.2 CDI

First principal component $I_{\text{pc1,T1}}$ and $I_{\text{pc1,T2}}$ are obtained from Principal Component Analysis (PCA) of bitemporal images IT1 and IT2 respectively. In CDI technique [15], absolute differencing and log ratio operation are used for generating difference images namely Δ_{ad} and Δ_{lr} given by

$$\Delta_{\text{ad}} = |\chi_1 - \chi_2| \tag{5}$$

$$\Delta_{\text{lr}} = \left|\log\frac{\chi_2 + 1}{\chi_1 + 1}\right| = |\log\chi_2 + 1 - \log\chi_1 + 1| \tag{6}$$

Log ratio images enhance the low intensity pixels. $\Delta_{\text{ad}}^{'}$ and $\Delta_{\text{lr}}^{'}$ are obtained by applying mean filter and median filter of window size 3 × 3 over Δ_{ad} and Δ_{lr}, respectively. However, combined difference image, Δ by combining image smoothness and edge preservation advantages associated with mean and median filters is given as

$$\Delta = \gamma\,\Delta_{\text{ad}}^{'} + (1 - \gamma)\Delta_{\text{lr}}^{'} \tag{7}$$

where γ is weighting parameter which controls the smoothness of image. Further, $\Delta_{\text{ad}}^{'}$ and $\Delta_{\text{lr}}^{'}$ are the results obtained after applying respective filtering.

2.3 GIFP-FCM

Clustering using GIFP-FCM [16] has been used to partition the difference image Δ from Eq. (7) into two clusters. For better visual effect of clustering, gray level are set to one for changed areas and zero to unchanged areas. Clustering is basically done to show changed and unchanged pixels. It is defined to group up a set of patterns, i.e. $\Delta = \{\Delta_1, \Delta_2, \ldots, \Delta_N\} \subset R^d$ into a set of prototypes $V = \{v_i, 1 \le i \le c\}$, where $u_{ij}(1 \le i \le c, 1 \le j \le N)$ denotes membership degree with reference to the prototype.

Algorithm for GIFP-FCM technique is described as

Step 1: Initialize a membership function constraint $f(u_{ij}) = \sum_{i=1}^{c} u_{ij}(1 - u_{ij}^{m-1})$;
Set $m > 1$ and $\sum_{i=1}^{c} u_{ij} = 1,\ u_{ij} \in [0, 1]$;
Step 2: Minimize the membership function to create an objective function which is given by

$$J_{\text{GIFP-FCM}} = \sum_{i=l}^{c} \cdot \sum_{j=l}^{N} u_{ij}^{m} d^2(x_j, v_i) + \sum_{j=l}^{N} a_j \sum_{i=l}^{c} u_{ij}(1 - u_{ij}^{m-1}) \tag{8}$$

where u_{ij} denotes membership degree for fuzzy partitions;

Step 3: Compute the center and membership using Eqs. 9 and 10 respectively.
Centre updation:

$$\beta_i = \frac{\sum_{j=l}^{N} u_{ij}^m x_j}{\sum_{j=l}^{N} u_{ij}^m} \tag{9}$$

Membership updation:

$$u_{ij} = \frac{1}{\sum_{k=l}^{c} \left(\frac{d^2(x_j, v_i) - a_j}{d^2(x_j, v_k) - a_j} \right)^{\frac{1}{m-1}}}; \tag{10}$$

where

$$a_j = \alpha \cdot \min\{d^2(x_j, v_s) | s \in \{1, \ldots, c\}\}, \ (0 \leq \alpha < 1);$$

Step 4: Compute the membership function u_{ij} and find output of clustering method, otherwise return to step 2.
GIFP-FCM clustering method used with parameter α outrages the other clustering methods, viz., FCM, k-means, etc. From the obtained results, it is found that proposed method has fast convergence speed and low computational complexity O (NcT).

2.4 Formation of Change Map

Each change map is assigned to either changed or unchanged class using Eq. (11).

$$cm(x, y) = \begin{cases} 1 & \text{for } \|I_{pc1}(x, y) - \beta_1\| \leq \|I_{pc1}(x, y) - \beta_2\| \\ 0 & \text{otherwise} \end{cases} \tag{11}$$

where $\|\|$ is Euclidean distance and β_1 and β_2 are clustering centres for changed and unchanged classes, respectively.

3 Result Analysis and Discussion

To investigate the performance of proposed method (Fig. 2), we have taken data from Landsat 5 TM sensor dataset [17] as shown in Table 1. This dataset consists of optical images of a part of Alaska with dimension 1350 × 1520 pixels acquired on 22nd July 1985 and 13th July 2005, respectively. Proposed method is implemented on MatlabR2013a.

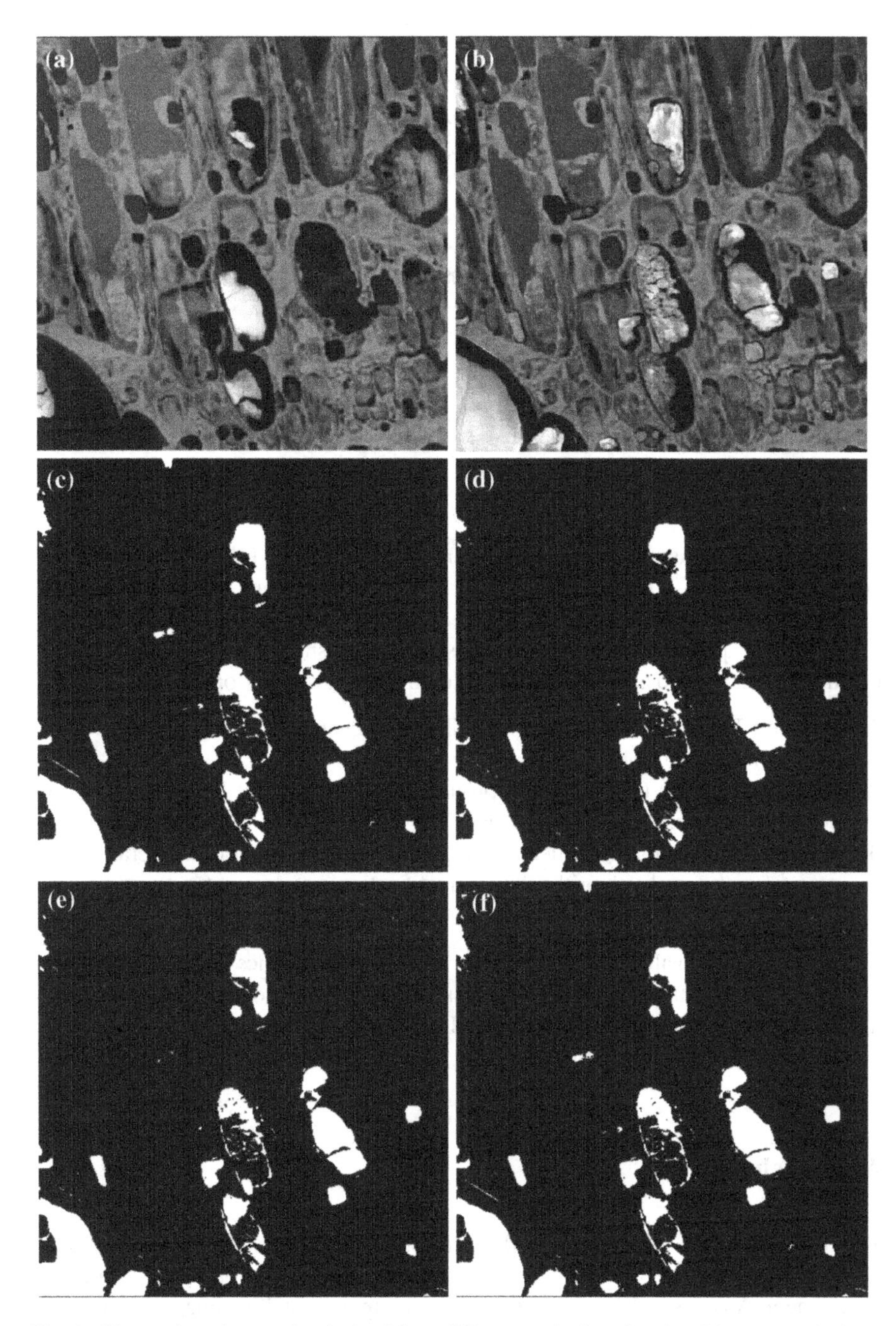

Fig. 2 Change detection results obtained from different methods. **a** Landsat 5 image acquired on July 22, 1985. **b** Landsat 5 image acquired on July 13, 2005. **c** Ground truth of change detection. **d** PCA-based approach. **e** HG-FCM method. **f** Proposed method (PM)

Table 1 Specifications of landsat 5 TM sensor

Band	Wavelength interval (μm)	Resolution (m)
Band 1	0.45–0.52	30
Band 2	0.52–0.60	30
Band 3	0.63–0.69	30
Band 4	0.76–0.90	30
Band 5	1.55–1.75	30
Band 6	10.40–12.50	12
Band 7	2.08–2.35	30

Table 2 Comparison of various parameters calculated for quantitative results

Method	OE (%)	CE (%)	PCC (%)
PCA based	7.10	0.13	99.30
HG-FCM	5.20	0.12	99.40
PM	0.87	0.07	99.78

The specifications of Landsat 5 TM Sensor are shown in Table 1. Results have been analyzed both qualitatively as well as quantitatively and are compared with other detection techniques, i.e., PCA based approach [14] and HG-FCM [2]. It is observed that qualitative results of proposed method show better results as compared to other methods. In this paper, following metrics are adopted:

1. Omission Error (OE) showing probability that changed pixel is wrongly identified as unchanged pixel; OE = FN/(FN + TP)
2. Commission Error (CE) showing probability that unchanged pixel is wrongly identified as changed pixel; CE = FP/(TN + FP)
3. Percentage correction classification (PCC) showing indication of overall accuracy of proposed algorithm in identification of correct change map; PCC = (TP + TN)/(TP +TN + FP + FN)

Table 2 shows values of the various performance metrics used for different methods for quantitative analysis. The results show that proposed algorithm has high percentage correction classification (PCC). Also, OE and CE calculations are quite satisfactory.

4 Conclusion

This paper presents, an unsupervised change detection method based on CDI and GIFP-FCM clustering. Feature vector space is created by applying Principal Component Analysis (PCA) technique on bitemporal satellite images. Absolute differencing and log ratio operation are employed for generating difference image. Finally, GIFP-FCM clustering is applied on the difference image which divides the image into two clusters changed and unchanged. The experiments are performed on

Matlab2013a. The results obtained from the proposed method shows that the method provides more accurate results as omission error is less than that of other methods while PCC is higher. The proposed method has better performance in the preservation of changed areas.

References

1. Khandelwal, P., Singh, K.K., Singh, B.K., Mehrotra, A.: Unsupervised change detection of multispectral images using Wavelet Fusion and Kohonen clustering network. Int. J. Eng. Technol. (IJET) **5**(2), 1401–1406 (2013)
2. Singh, K.K., Mehrotra, A., Nigam, M.J., Pal, K.: Unsupervised change detection from remote sensing images using hybrid genetic FCM. In: IEEE Conference on engineering and systems, pp. 1–5, April 2013
3. Li, Y., Gong, M., Jiao, L., Li, L., Stolkin, R.: Change-detection map learning using matching pursuit. IEEE Trans. Geosci. Remote Sens. 1–12 (2015)
4. Neagoe, V.E., Stoica, R.M., Ciurea, A.I., Bruzzone, L., Bovolo, F.: Concurrent self-organizing maps for supervised/unsupervised change detection in remote sensing images. IEEE J. Selected Topics Appl. Earth Observ. Remote Sens. **7**(8), 1–12 (2014)
5. Bazi, Y., Bruzzone, L., Melgani, F.: An unsupervised approach based on the generalized Gaussian model to automatic change detection in multitemporal SAR images. IEEE Trans. Geosci. Remote Sens. **43**(4), 874–887 (2005)
6. Ronggui, H.M.L.M.W.: Application of an improved Otsu algorithm in image segmentation. J. Electron. Meas. Instrum. **5**, 011 (2010)
7. Coleman, G.B., Andrews, H.C.: Image segmentation by clustering. Proc. IEEE **67**(5), 773–785 (1979)
8. Bruzzone, L., Prieto, D.: Automatic analysis of the difference image for unsupervised change detection. IEEE Trans. Geosci. Remote Sens. **38**(3), 1171–1182 (2000)
9. Taneja, A., Ballan, L., Polleyfeys, M.: Geometric change detection in urban environments using images. J. Latex Class Files **6**(1), 1–14 (2007)
10. Yetgin, Z.: Unsupervised change detection of satellite images using local gradual descent. IEEE Trans. Geosci. Remote Sens. **50**(5), 1919–1929 (2012)
11. Klaric, M.N., Claywell, B.C., Scott, G.J., Hudson, N.J., Li, Y., Barratt, S.T., Keller, J.M., Davis, C.H.: GeoCDX: an automated change detection and exploitation system for high resolution satellite imagery. IEEE Trans. Geosci. Remote Sens. **51**(4), 2067–2086 (2013)
12. Liu, Z., Mercier, G., Dezert, J., Pan, Q.: Change detection in heterogeneous remote sensing images based on multidimensional evidential reasoning. IEEE Geosci. Remote Sens. Lett. **11** (1), 168–172 (2014)
13. Lu, J., Li, J., Chen, G., Xiong, B., Kuang, G.: Improving pixel-based change detection accuracy using an object-based approach in multitemporal SAR flood images. IEEE J. Selected Topics Appl. Earth Observ. Remote Sens. 1–11 (2015)
14. Celik, T.: Unsupervised change detection in satellite images using principal component analysis and k-means clustering. IEEE Geosci. Remote Sens. Lett. 6(4), 772–776 (2009)
15. Zheng, Y., Zhang, X., Hou, B., Liu, G.: Using combined difference image and *k* means clustering for SAR image change detection. IEEE Geosci. Remote Sens. Lett. **11**(3), 691–695 (2014)
16. Zhu, L., Chung, F.L., Wang, S.: Generalized fuzzy c-means clustering algorithm with improved fuzzy partitions. IEEE Trans. Syst. Man Cybern. Part B: Cybern. **39**(3), 578–591 (2009)
17. NASA GSFC, Alaska Circa 2005: Landsat TM Data, Feb. 2011. http://change.gsfc.nasa.gov/alaska.html

[illegible] The results [illegible] method [illegible] proposed method [illegible] preservation of changed areas.

References

[illegible]

Performance Comparison Between InP-ON and SON MOSFET at Different Nanotechnology Nodes Using 2-D Numerical Device Simulation

Shashank Sundriyal, V. Ramola, P. Lakhera, P. Mittal and Brijesh Kumar

Abstract This research paper analyzes and compares the design and electrical behavior of Indium Phosphide-On-Nothing (InP-ON) MOSFET with Silicon-On-Nothing (SON) MOSFET by using Atlas Silvaco two dimensional (2-D) numerical device simulator. The output and transfer characteristics of InP-ON and SON MOSFET are determined at 60, 45 and 35 nm technology nodes while taking the same electrical properties and biasing conditions. Subsequently, the comparison is made between InP-ON and SON MOSFETs in terms of transconductance, leakage current, threshold voltage and drain induced barrier lowering (DIBL) at various technology nodes, i.e., 35, 45 and 60 nm. The advantage of using proposed InP-ON MOSFET is that it gives extremely lower threshold voltage and DIBL effect in comparison with SON MOSFET at same technology nodes. However, both InP-ON and SON MOSFETs works properly for less than 100 nm technology node very efficiently and suppresses short channel effects (SCE). It is difficult to achieve extremely low threshold voltage below 45 nm channel length in SON MOSFET. So, to achieve extremely lower threshold voltage, InP-ON structure is used instead of SON MOSFET. InP-ON MOSFET structure contains different layers that includes source, drain and gate in combination with Indium Phosphide channel layer that is placed above an air tunnel, hence becomes InP-ON. With the advancement in nanotechnology era, it becomes very necessary to have very compact size of semiconductor devices to operate and produce best results at different lower technology nodes. In future, both InP-ON and SON will provide smooth comfort zone to work at lower technology nodes.

Shashank Sundriyal (✉) · V. Ramola
Department of Electronics and Communication Engineering,
Uttarakhand Technical University, Dehradun, India
e-mail: sundriyal.s.1991@ieee.org

P. Lakhera · P. Mittal · Brijesh Kumar
Department of Electronics and Communication Engineering,
Graphic Era University, Dehradun, India
e-mail: brijesh_kumar@ieee.org

M. Pant et al. (eds.), *Proceedings of Fifth International Conference on Soft Computing for Problem Solving*, Advances in Intelligent Systems and Computing 437, DOI 10.1007/978-981-10-0451-3_30

Keywords Indium Phosphide-On-Nothing (InP-ON) · Silicon-On-Nothing (SON) · DIBL effect · MOSFET

1 Introduction

With recent development in microelectronics, mobile computing and communication field, it becomes very necessary to move from sub-micron region to nanometer region for the need of improved performance. To enhance the performance of a device, it is suitable to undergo device scaling [1]. However, continuous scaling of bulk MOSFET meet with many problems such as increased short channel effect (SCE). Besides this, continuous scaling also increases leakage current, DIBL effect and parasitic capacitance between source-substrate and drain-substrate. To overcome these problems, there is a need of new device structure that supports further scaling to few nanometers technology nodes [2]. To obtain such suitable device thin Indium Phosphide layer is placed above an air tunnel and hence becomes Indium Phosphide-On-Nothing device structure as shown in Fig. 1a. The merits of thin air dielectric below the ultrathin InP layer is reduction in parasitic capacitance between source-substrate and drain-substrate due to which short channel effect, transistor scalability, speed and performance of circuit are greatly improved [3, 4].

The most important advantage of InP-ON MOSFET is that electrostatic coupling of the channel w.r.t. source/drain and the substrate through an air layer is greatly

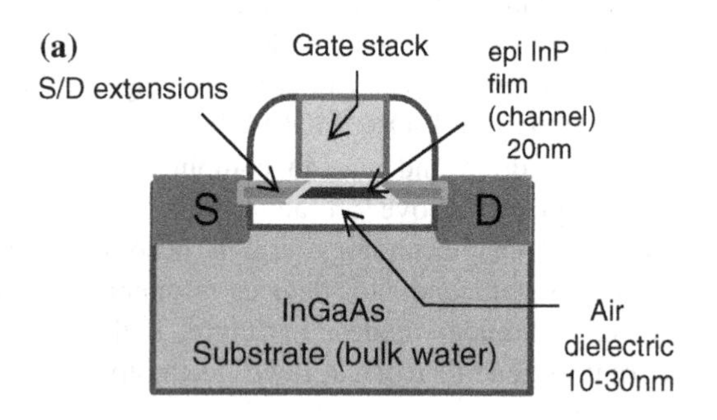

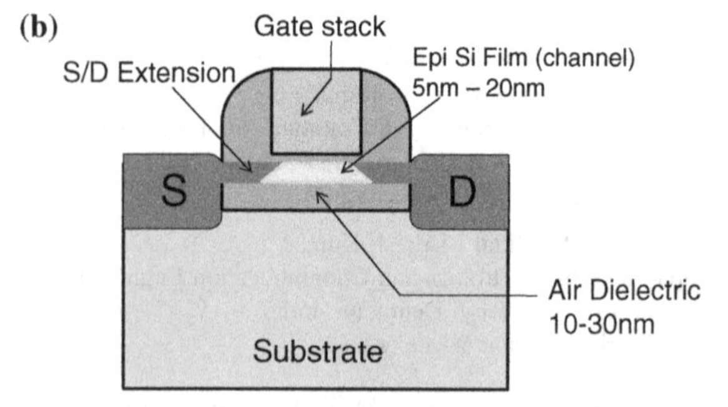

Fig. 1 Simplified design of **a** InP-ON MOSFET, **b** SON MOSFET

reduced. InP-ON has an important feature that it combines the feature of fully depleted devices such as lower sub-threshold slope, higher mobility while controlling Indium phosphide film thickness, no floating body effects [5] and bulk InGaAs substrate that provides better heat dissipation and lower series resistance.

This paper is presented in five sections, including the current introductory Sect. 1. Thereafter, Sect. 2 comprises of structural comparison between Indium Phosphide-On-Nothing MOSFET and Silicon-On-Nothing MOSFET with their advantages and disadvantages. Afterwards, Sect. 3 describes the simulation setup and comparison between InP-ON and SON MOSFET at 35, 45 and 60 nm technology nodes. Atlas Silvaco two dimensional numerical device simulator has been used for the analysis. Section 4 comprises of important results and discussions. Finally, Sect. 5 summarizes the conclusion of the research paper.

2 Structural Comparisons Between InP-ON and SON MOSFETs

The InP-ON MOSFET helps in fabrication of ultra-smaller channel length devices, which in turn helps to reduce supply voltage. To develop InP-ON devices, the thin channel layer used is of Indium Phosphide material while the bulk substrate used in this process is of p-type InGaAs and buried dielectric used is of air whose dielectric constant is unity. The source and drain contacts are made of n-type InP layer while gate used in this process is of polysilicon material. In the structure of InP-ON MOSFET, the thin InP layer is placed above a buried air dielectric which further acts as a barrier between thin InP layer and bulk InGaAs substrate. The advantage of using air as a barrier is that parasitic capacitance is greatly reduced and hence short channel effects reduces to a great extent. In InP-ON structure, the buried oxide is not continuous and is placed only between gate and spacers. Due to which series resistance is greatly reduced. Also extremely shallow and highly doped extensions were made in InP-ON structure.

The design process flow of SON MOSFET is similar to InP-ON MOSFET. The SON MOSFET also has the advantage that it works excellent at nanotechnology nodes as compared to InP-ON MOSFET. In-fact, the current nanotech industry started using SON MOSFET [6] for the fabrication of nanoscale devices due to its enormous advantages. However, in this research paper, for the first time we propose a new device, i.e., Indium Phosphide-On-Nothing MOSFET using InGaAs as a substrate which gives best results for threshold voltage and DIBL as compared to SON MOSFET. In the structure of SON MOSFET, the thin channel layer and bulk substrate is used of p-type silicon material while buried dielectric used is of air. The source and drain contacts are made of n-type silicon material while gate used in this process is of polysilicon material. In the structure of SON MOSFET, the thin silicon layer is placed above a buried air dielectric, which further acts as a barrier between thin silicon layer and bulk silicon substrate.

In this research paper we discuss the performance comparison between SON MOSFET and InP-ON MOSFET at 35, 45 and 60 nm technology nodes [7]. The device comprises of n+ source and drain, a gate formed by polysilicon and an air dielectric below the channel layer as shown in Fig. 2a, b at 35 nm technology node for both InP-ON and SON MOSFET, respectively. The current flowing from source to drain is controlled by gate—depletion region. Current in both the devices is same as that in case of conventional MOSFET, and in saturation mode it is given by:

$$I_{\mathrm{ds}} = \mu_n C_{\mathrm{ox}}(W/L)(V_{\mathrm{gs}} - V_{\mathrm{th}})^2$$

where I_{ds} is the drain to source current in saturation mode, C_{ox} is the gate-oxide capacitance, μ_n is the electron mobility, V_{gs} is the gate-source voltage and V_{th} is the threshold-voltage.

Both depletion region and the gate voltage influence the characteristic behavior of the device; however gate voltage is the more dominating factor to affect the drive current. The width and length of the channel and gate of the device are separately varied to obtain three devices—1, 2 and 3 at 35, 45 and 60 nm channel lengths [8] to study the effect of line width and spacing of the mesh grid.

3 Device Dimensions and Simulation Setup

This section describes the device dimensions and simulation setup used in analyzing the performance comparison between Indium Phosphide-On-Nothing and Silicon-On-Nothing MOSFET at 35, 45 and 60 nm technology nodes in terms of transconductance, threshold voltage, leakage current, drain induced barrier lowering (DIBL) etc. The devices are compared for different channel lengths, gate lengths, line spacing and line widths [9]. But in all cases, the other parameters like gate voltage, drain voltage, source/drain lengths and widths, air dielectric length and width are taken identical for the better comparison. The schematic structures are shown in Fig. 2a, b at 35 nm channel length. The device materials and dimensions of both devices are summarized in Table 1.

For the simulation of both the devices at 35, 45, and 60 nm channel lengths, Silvaco two dimensional numerical device simulator is used. It uses and solves two dimensional Poisson's equations in the 2-D channel region of the ultrathin Indium Phosphide channel for calculating threshold voltage and DIBL. It uses fldmob, srh, conmob, bgn, auger models and uses newton trap [10] method for the excitation of the trapped charge carriers.

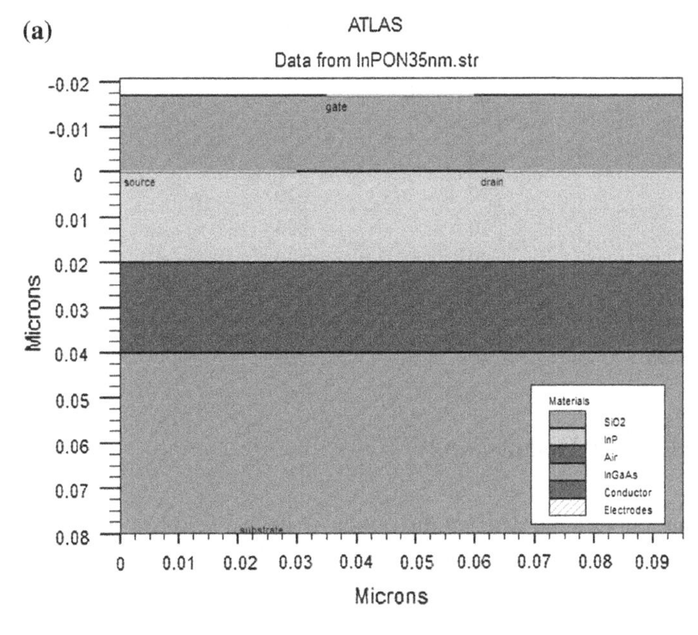

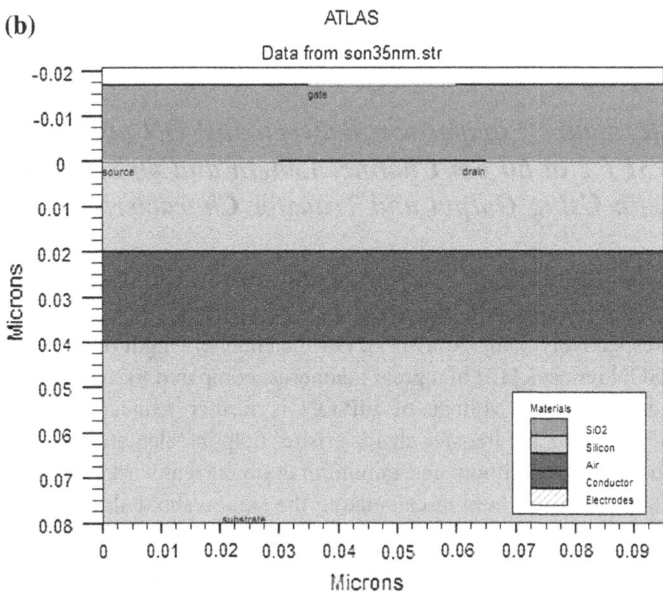

Fig. 2 **a** Schematic of Indium Phosphide-On-Nothing MOSFET device at 35 nm channel length. **b** Schematic of Silicon-On-Nothing MOSFET device at 35 nm channel length

Table 1 Device dimensions and materials for InP-ON MOSFET

Name of parameters (nm)	Device-1 (nm)	Device-2 (nm)	Device-3 (nm)
Channel length	60	45	35
Channel width	20	15	20
Gate electrode	40	25	25
Dielectric	20	20	20
Source (S)	20	20	30
Drain (D)	20	20	30
Device width (W)	80	70	95
Device length (L)	100	85	80
Substrate width	40	35	40

4 Results and Discussions

This section includes the detailed discussion of device behavior in terms of output and transfer characteristics for various gate lengths and channel lengths. Additionally, the transconductance, leakage current, DIBL analysis, threshold voltage are described in depth for device-a and device-b at 35, 45 and 60 nm technology nodes respectively.

4.1 Performance Comparison Between InP-ON and SON MOSFET at 60 nm Channel Length and 40 nm Gate Length Using Output and Transfer Characteristics

The transfer characteristics ($I_{ds} - V_{gs}$) for $V_{ds} = 0.1$ V and output characteristics for calculating sub-threshold slope ($\log I_{ds} - V_{ds}$) for $V_{gs} = -2$ V are shown in Fig. 3a, b, respectively using Atlas [11]. For the channel length of 60 nm, threshold voltage of SON reduces [12] in a great manner as compared to that of conventional MOSFET and threshold voltage of InP-ON is further reduced as compared to SON MOSFET. Also, the transfer characteristics help in calculating the exact value of threshold voltage, maximum and minimum drain current w.r.t gate voltage while the output characteristics help in calculating the subthreshold slope in mV/decade. The transfer characteristics compare the threshold voltage of InP-ON and SON MOSFET. For InP-ON V_{th} is 0.178197 V and for SON V_{th} is 0.3816 V. These results clearly show that InP-ON provides best results in terms of threshold voltage as compared to SON.

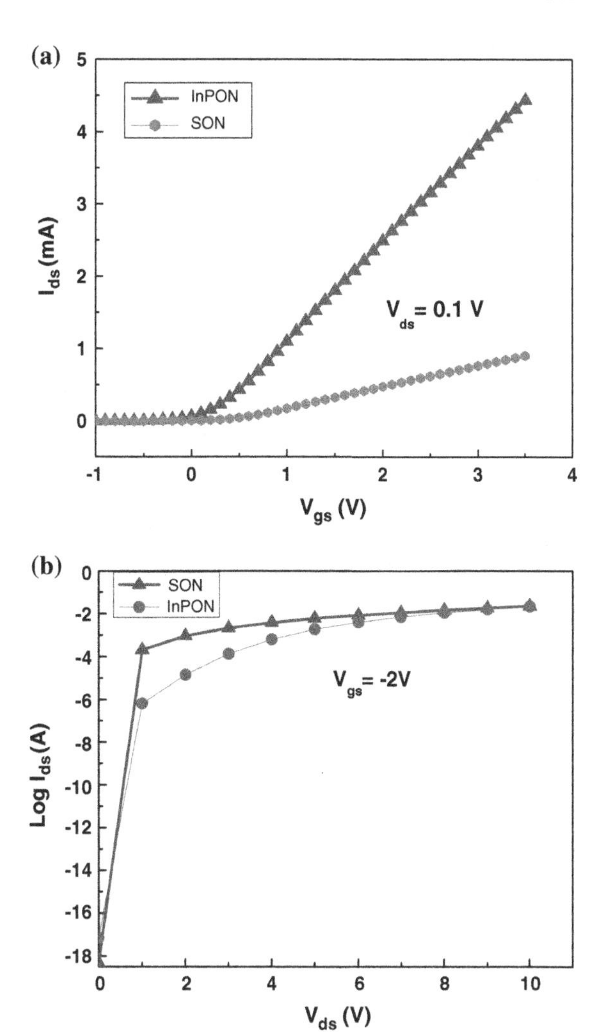

Fig. 3 Characteristics of InP-ON and SON MOSFET at 60 nm technology node. **a** Transfer characteristics ($I_{ds} - V_{gs}$) for $V_{ds} = 0.1$ V. **b** Output characteristics ($I_{ds} - V_{ds}$) for $V_{gs} = -2$ V

4.2 Performance Comparison Between InP-ON and SON MOSFET at 45 nm Channel Length and 25 nm Gate Length Using Output and Transfer Characteristics

The transfer characteristics ($I_{ds} - V_{gs}$) for $V_{ds} = 0.1$ V and output characteristics for calculating sub-threshold slope (log $I_{ds} - V_{ds}$) for $V_{gs} = -2$ V are shown in Fig. 4a, b, respectively. As the channel length further reduces from 60 to 45 nm, threshold voltage also reduces in a great manner. Also the transfer characteristics help in calculating the exact value of threshold voltage, maximum and minimum drain current w.r.t gate voltage while the output characteristics help in calculating the subthreshold slope in mV/decade. The transfer characteristics compare the

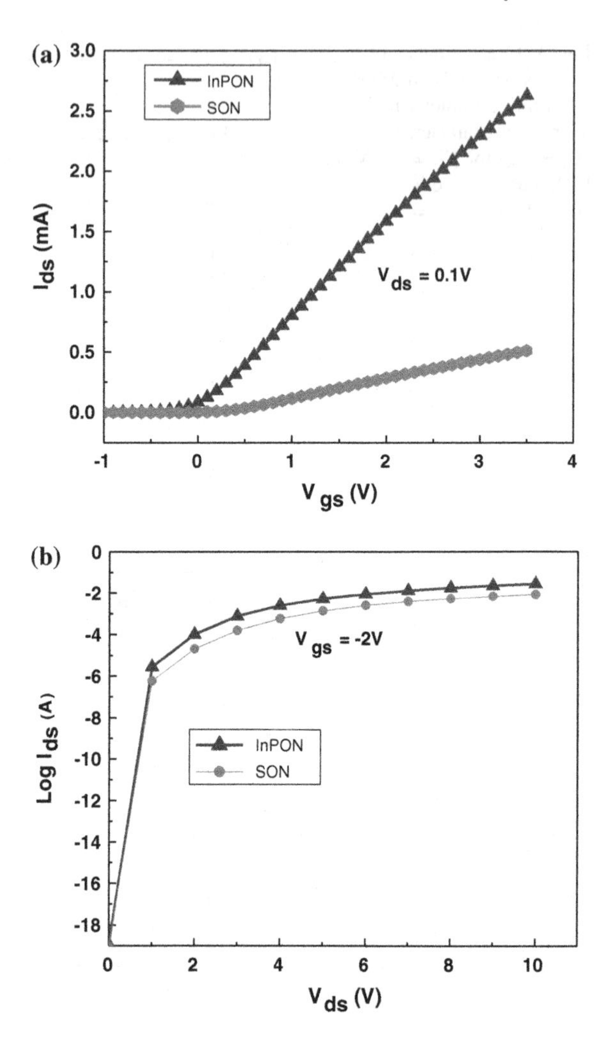

Fig. 4 Characteristics of InP-ON and SON MOSFET at 45 nm technology node. **a** Transfer characteristics ($I_{ds} - V_{gs}$) for $V_{ds} = 0.1$ V. **b** Output characteristics ($I_{ds} - V_{ds}$) for $V_{gs} = -2$ V

threshold voltage of InP-ON and SON MOSFET, for InP-ON $V_{th} = 0.0328366$ V and for SON $V_{th} = 0.341004712$ V. These results clearly show that InP-ON provides better results in terms of threshold voltage as compared to SON.

4.3 Performance Comparison Between InP-ON and SON MOSFET at 35 nm Channel Length and 25 nm Gate Length Using Output and Transfer Characteristics

The transfer characteristics ($I_{ds} - V_{gs}$) for $V_{ds} = 0.1$ V with varying the gate voltage from $V_{gs} = -1$ V to $V_{gs} = 3.5$ V and output characteristics for calculating

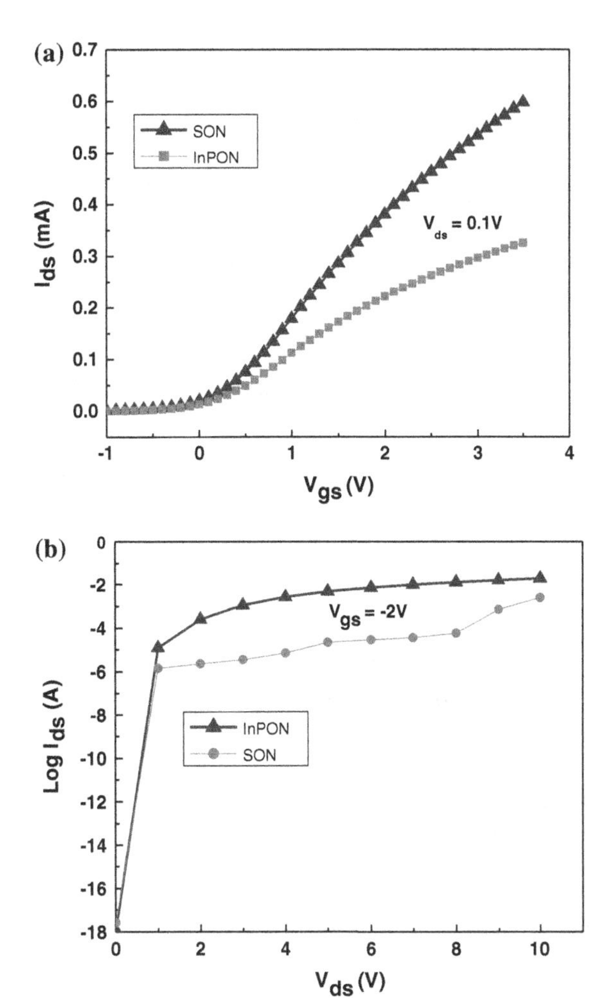

Fig. 5 Characteristics of InP-ON and SON MOSFET at 35 nm technology node. **a** Transfer characteristics ($I_{ds} - V_{gs}$) for $V_{ds} = 0.1$ V. **b** Output characteristics ($I_{ds} - V_{ds}$) for $V_{gs} = -2$ V

subthreshold slope ($\log I_{ds} - V_{ds}$) for $V_{gs} = -2$ V with varying the drain voltage from $V_{ds} = 0$ V to $V_{ds} = 10$ V are shown in Fig. 5a, b, respectively. As the line width increases, the current increases and leakage current reduces, owing to better gate control. As the channel length further reduces from 45 to 35 nm, threshold voltage also reduces. This is because depletion region does not have a significant control over current through the opening. The transfer characteristics compare the threshold voltage between InP-ON and SON MOSFET, for InP-ON $V_{th} = 0.0325232$ V and for SON $V_{th} = 0.144712$ V. These results clearly show that InP-ON provides better results in terms of threshold voltage as compared to SON.

4.4 Performance Comparison Between Parameters of InP-ON and SON MOSFET at 35, 45 and 60 nm Channel Lengths

This section includes the analysis and comparison between parameters of InP-ON and SON MOSFET at 35, 45, and 60 nm technology nodes. The comparison is discussed in terms of threshold voltage, transconductance, and DIBL effect in the following subsections.

4.4.1 Threshold Voltage Comparison at 35, 45 and 60 nm Technology Nodes

Figure 6 shows NMOS InP-ON and SON threshold voltage comparison with different channel lengths, i.e., 35, 45 and 60 nm with constant channel doping concentration of 1.7e16 cm^{-3} at silicon film thickness of 20 nm. The gate work function used in this simulation is 4.5 eV. The source and drain region channel length is 20 nm each for 45 and 60 nm while 30 nm for 35 nm channel length and doping concentration for both source and drain region is taken to be same in all cases. Figure 6 clearly shows that with different channel length, threshold voltage is smaller for InP-ON MOSFET as compared to SON MOSFET.

4.4.2 Transconductance Comparison at 35, 45 and 60 nm Technology Nodes

Figure 7 shows NMOS InP-ON and SON transconductance comparison with different channel lengths of 35, 45 and 60 nm at constant channel doping of

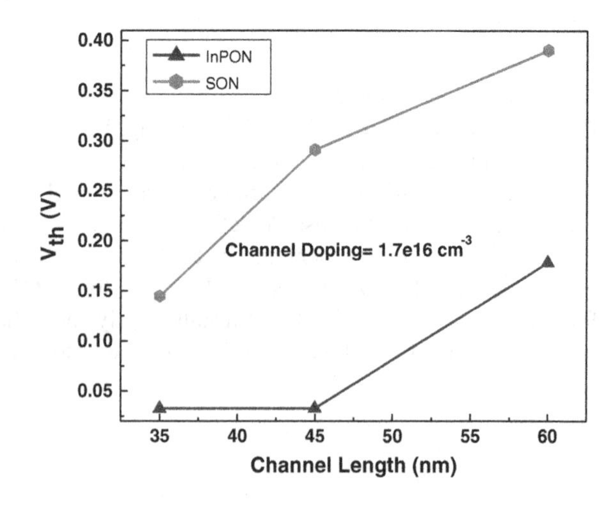

Fig. 6 Characteristic curve for threshold voltage comparison for different channel lengths at constant channel doping concentration of 1.7e16 cm^{-3}

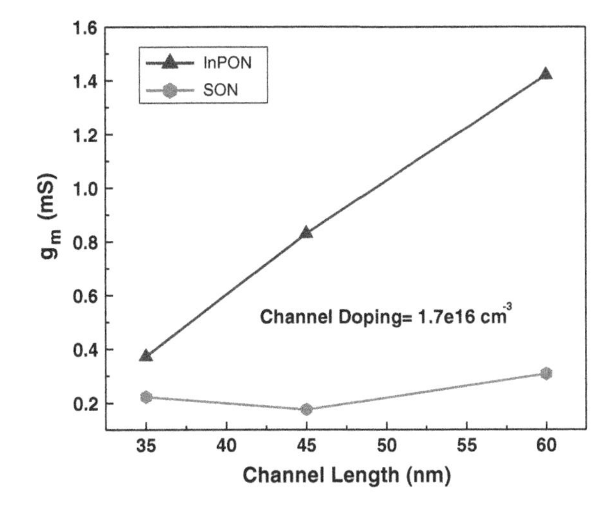

Fig. 7 Characteristic curve for transconductance comparison for different channel lengths at constant channel doping concentration of 1.7e16 cm^{-3}

1.7e16 cm^{-3} and silicon film thickness of 20 nm. The gate work function used in this simulation is 4.5 eV. The source and drain region channel length is 20 nm each for 45 and 60 nm while 30 nm for 35 nm channel length and doping concentration for both source and drain region is taken to be same in all cases. Figure 7 clearly shows that with different channel lengths transconductance is lesser for SON MOSFET as compared to InP-ON MOSFET.

4.4.3 Threshold Voltage Variations for Different Work Functions at 45 nm Channel Length

Figure 8 shows NMOS InP-ON and SON threshold voltage variation for different work function varying from 4.5 to 5.0 eV at constant channel length of 45 nm and doping concentration of channel is 1.7e16 cm^{-3}. The silicon film thickness taken in all cases is 20 nm. The source and drain region channel length is 20 nm each and doping concentration for both source and drain region is taken to be same in all cases. Figure 8 clearly shows that with different work function, threshold voltage varies linearly with work function and is lesser for InP-ON MOSFET as compared to SON MOSFET.

4.4.4 DIBL Variation at 35, 45 and 60 nm Technology Nodes

Figure 9 shows NMOS InP-ON and SON DIBL variation for different channel lengths, i.e., 35, 45 and 60 nm at constant channel doping of 1.7e16 cm^{-3} and silicon film thickness of 20 nm. The gate work function used in this simulation is 4.5 eV [13]. The source and drain region channel length is 20 nm each and doping concentration for both source and drain region is taken to be same in all cases.

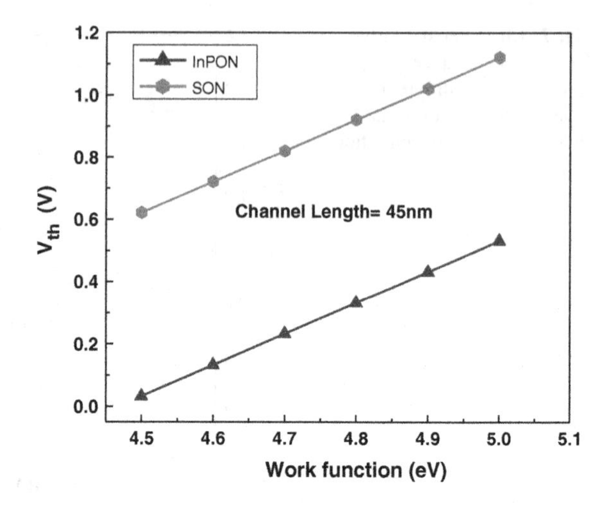

Fig. 8 Characteristic curve for threshold voltage comparison for different work function at 45 nm channel length

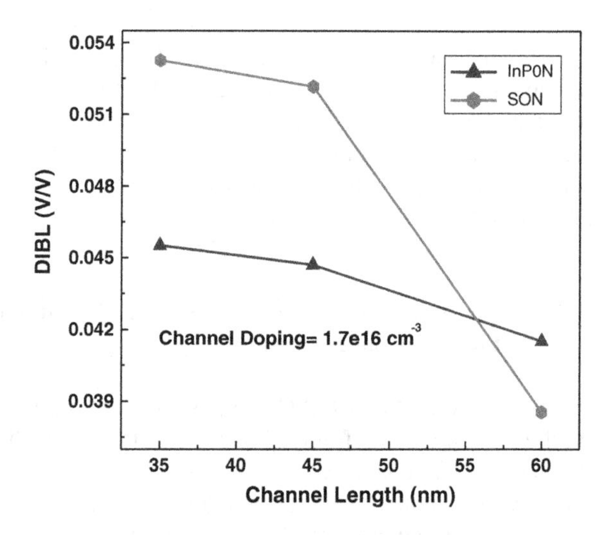

Fig. 9 Characteristic curve for DIBL comparison at different channel lengths

Figure 9 clearly shows that with different channel length, DIBL is less for InP-ON MOSFET at 35 and 45 nm channel length but at 60 nm channel length, DIBL is lesser for SON MOSFET as compared to InP-ON MOSFET.

5 Conclusion

The bulk device n-MOSFET structure gives good results in sub-micron region but when the need of sub-nano region arises it needs to scale down. But when scaling continues in bulk MOSFET the parasitic effects will also appear increasing the threshold voltage, leakage current and DIBL effect. So to avoid these problems a new device SON is developed which suppresses all such parasitic effects and found effective in terms of improved transconductance, threshold voltage, I_{on}/I_{off} ratio, maximum and minimum drain current w.r.t. V_{gs} and V_{ds}. The observed results show that the SON is much better than SOI and bulk MOSFET. However, with the need of day to day applications, it is required that the device should have very less threshold voltage and DIBL effect at sub-nano technology nodes.

This research paper proposes a new device that is InP-ON MOSFET that provides very less threshold voltage and DIBL effect in comparison to SON MOSFET. Additionally, this paper also highlighted the comparison between characteristics and performance parameters of SON MOSFET and InP-ON MOSFET at different technology nodes that includes 60, 45 and 35 nm with variation in gate length, while keeping V_{gs} and V_{ds} same. The performance parameters for individual devices are extracted and analyzed, that further used in circuit applications. The InP-ON MOSFET is preferable in comparison to SON MOSFET at lower than 60 nm technology node because of lower threshold voltage and DIBL effects.

References

1. The international technology roadmap for semiconductor, Emerging Research Devices, 2009
2. Svilicic, B., Jovanovic, V., Suligoj, T.: Vertical silicon-on-nothing FET analytical model of subthreshold slope. Proc. Int. Conf. MIDEM. **1**, 71–74 (2007)
3. Fabrication of silicon-on-nothing (SON) MOSFET fabrication using selective etching of $Si_{1-x}Ge_x$ layer. US 7015147 B2
4. Monfray, S., Skotnicki, T., Morand, Y., Descombes, S., Paoli, M., Ribot, P., Dutartre, D., Leverd, F., Lefriec, Y., Pantel, R., Haond, M., Renaud, D., Nier, M.E., Vizizoz, C., Louis, D., Buffet, N.: First 80 nm SON (Silicon On Nothing) MOSFETs with perfect morphology and high electrical performance. IEDM Tech. Dig. 645–648 (2001)
5. Mizushima, I., Sato, T., Taniguchi, S., Tsunashima, Y.: Empty-space-in silicon technique for fabricating a silicon-on-nothing structure. Appl. Phys. Lett. **77**, 3290–3292 (2000)
6. Pretet, J., Monfray, S., Cristoloveanu, S., Skotnicki, T.: Silicon-on-Nothing MOSFETs: performance, short channel effects and backgate coupling. IEEE Trans. Electron Dev. **51**(2), 240–246 (2004)
7. Monfray, S., Boeuf, F., Coronel, P., et al.: Silicon-on-nothing (SON) applications for low power technologies. In: IEEE International Conference on Integrated Circuit Design and Technology and Tutorial, vol. 1 (2008)
8. Monfray, S., Skotnicki, T., Beranger, C.F., et al.: Emerging silicon-on-nothing (SON) devices technology. Solid-State Electron **48**, 887 (2004)
9. Kilchytska, V., Chang, T.M., Olbrechts, B., et al.: Electrical characterization of true silicon-on-nothing MOSFETs fabricated by Si layer transfer over a pre-etched cavity. Solid-State Electron **51**, 1238 (2007)
10. Brennan, K., Brown, A.: Theory of Modern Electronic Semiconductor Devices. Wiley, New York (2002)
11. Atlas user's manual: device simulation software. Santa Clara: Silvaco International (2010)
12. Deb, S., Singh, N.B., Das, D., De, A.K., Sarkar, S.K.: Analytical model of threshold voltage and sub-threshold slope of SOI and SON MOSFETs: A comparative study. J. Electron Dev. **8**, 300–309 (2010)
13. Sato, T., Nii, H., Hatano, M., et al.: Fabrication of SON (silicon on nothing)-MOSFET and its ULSI applications. IEIC Technical Report, **102**, 178, 99 (2002)

Discrete-Time Eco-epidemiological Model with Disease in Prey and Holling Type III Functional Response

Elizabeth Sebastian, Priyanka Victor and Preethi Victor

Abstract In this paper, we propose a discrete-time eco-epidemiological model with disease in prey incorporating Holling type III functional response and harvesting of prey. Bilinear incidence rate is used to model the contact process. We incorporate harvesting of both the susceptible and infected prey populations. We consider that the predator population will prefer only the infected prey population as the infected ones are more vulnerable. We derive the basic reproduction number of the eco-epidemiological model. We find the equilibrium points of the model and analyse the conditions for their stability. Finally, we carry out simulations to illustrate our analytical findings.

Keywords Difference equations · Eco-epidemiology · Holling type III functional response · Susceptible and infected prey

1 Introduction

Mathematical modelling is used to study and analyse various kinds of systems. Interactions between different species and their habitat can be analysed using ecological models. Outbreak and spread of different diseases can be studied by epidemic models. The dynamics of disease spread in the prey–predator populations is an important area of research. In case of an epidemic, either the prey can be infected or the predator can be infected or both. These problems are challenging to solve but are of great importance. Lotka–Volterra [1, 2] formulated a model to

Elizabeth Sebastian (✉) · Priyanka Victor · Preethi Victor
Auxilium College (Autonomous), Gandhi Nagar, Vellore 632006
Tamil Nadu, India
e-mail: elizafma@gmail.com

Priyanka Victor
e-mail: priyankavictor2@gmail.com

Preethi Victor
e-mail: preethivictor2@gmail.com

M. Pant et al. (eds.), *Proceedings of Fifth International Conference on Soft Computing for Problem Solving*, Advances in Intelligent Systems and Computing 437, DOI 10.1007/978-981-10-0451-3_31

study the predator–prey interactions. Kermack–Mckendrick [3] made some significant contributions to epidemiological models.

Disease in ecological system is a matter of concern because humans coexist with wildlife and any changes in the ecosystem would have an impact on the human population. Eco-epidemiology takes into account both the ecological and epidemiological factors by considering the disease dynamics in the predator–prey populations. Disease in prey–predator population was first studied by Anderson and May [4]. In the last decade, many researchers have analysed the predator–prey models with disease outbreak. Chattopadhyay [5] analysed the predator–prey model with disease in prey and Holling type II functional response. Infection in the prey population is given much importance because it can affect both the prey and predator populations.

Functional response is a function that defines the consumer's food consumption rate. In 1965, Holling [6] formulated three types of functional response functions. Recently, many authors have explored the dynamics of predator–prey systems with Holling-type functional responses. Zhimin He and Bo Li analysed the complex dynamic behaviour of a discrete-time predator–prey system of Holling type III [7]. Sahabuddin Sarwardi, Mainul Haque and Ezio Venturino studied a Leslie–Gower Holling type II eco-epidemic model in continuous time [8]. Holling type III functional response is notable because it considers that the predators develop their efficiency to capture by learning as they hunt. Another important factor which affects the dynamical properties of a prey–predator system is harvesting [9–11]. Harvesting in the predator–prey systems has become an area of interest to economists and ecologists.

Many authors have studied and analysed the eco-epidemic models in continuous time [5, 8, 11–15]. Discrete-time models are more suitable for numerical computing. Because the data that we collect is in discrete times, it would be more appropriate to study our model using difference equations. Until now, there are only few papers on the dynamical behaviours of discrete-time eco-epidemic models. Discrete-time eco-epidemic models were analysed by Hu et al. [16] and Das et al. [17]. Motivated by these works, we have proposed a discrete-time eco-epidemiological model incorporating disease and harvesting in prey, Holling type III functional response, and have analysed its stability conditions.

This paper is organized as follows. In Sect. 2, a mathematical description of our eco-epidemiological model is given and we have constructed our model using a system of difference equations. In Sect. 3, we have listed the possible equilibrium points of our model and have analysed the conditions in which our system is stable or unstable. In Sect. 4, we have analysed the behaviour of the model for different values of h_1 and h_2 using MATLAB. We have discussed about the basic reproduction number based on the harvesting of the prey in Sect. 5. Conclusion of our work is presented in Sect. 6.

2 The Mathematical Model

The following assumptions are made in formulating the discrete-time eco-epidemiological model.

- A predator–prey interaction is considered where the prey species follows the logistic dynamics in the absence of predator and the predator consumes prey following Holling type III response function.The Holling type III functional response is given by

$$g(x) = \frac{ax^2}{x^2 + \alpha}$$

- We assume that the disease spreads among the prey population only and the disease is not genetically inherited. The infected prey population does not recover or become immune. The incidence rate is assumed to be the bilinear mass action incidence βSI.
- Suppose a microparasite infects the prey population and divides it into two disjoint classes, that is, the susceptible (S) and the infected (I) populations, so that the total prey population at any time n is $N_n = S_n + I_n$.
- We assume that the predators consume only the infected preys. This is in accordance with the fact that the infected individuals are less active and can be caught more easily.
- We consider that there is a constant harvesting of both the susceptible and infected prey populations.
- We assume that $h_2 > h_1$, as the infected prey will be more easier to harvest.
- We assume that the parameter values $0<c<1, 0<d<1$, where c is the death rate of the infected prey and d is the death rate of the predator.

From the above assumptions, we formulate the following discrete-time eco-epidemiological model using a system of difference equations:

$$\begin{aligned} S_{n+1} &= S_n + rS_n\left(1 - \frac{S_n + I_n}{K}\right) - \beta S_n I_n - h_1 S_n \\ I_{n+1} &= I_n + \beta S_n I_n - cI_n - \frac{aI_n^2 Y_n}{I_n^2 + \alpha} - h_2 I_n \\ Y_{n+1} &= Y_n - dY_n + \frac{abI_n^2 Y_n}{I_n^2 + \alpha} \end{aligned} \tag{1}$$

where

S_n Susceptible prey population at time n.
I_n Infected prey population at time n.
Y_n Predator population at time n.
r Intrinsic birth rate of the prey population.
K Carrying capacity of the environment for the prey population.

β Transmission coefficient.
c Death rate of the infected prey.
d Death rate of the predator.
a Predation coefficient.
b Conversion efficiency of the predator.
α Half capturing saturation.
h_1 Harvesting rate of the susceptible prey population.
h_2 Harvesting rate of the infected prey population.

System (1) has the following initial conditions:

$$S(0) \geq 0,\ I(0) \geq 0,\ Y(0) \geq 0 \tag{2}$$

In theoretical eco-epidemiology, boundedness of a system implies that a system is biologically well-behaved. Let $S_n + I_n + Y_n = W_n$.

$$S_{n+1} \leq S_n + rS_n\left(1 - \frac{S_n}{K}\right)$$

Then we have

$$\lim_{n \to \infty} S_n \leq K$$

Adding all the equations of system (1), we get

$$\begin{aligned}
W_{n+1} &= W_n + rS_n\left(1 - \frac{S_n + I_n}{K}\right) - h(S_n + I_n) - (cI_n + dY_n) \\
W_{n+1} &\leq W_n + rS_n - (cI_n + dY_n) \\
W_{n+1} &\leq W_n + (r+1)S_n - (S_n + cI_n + dY_n) \\
W_{n+1} &\leq W_n + (r+1)K - (S_n + cI_n + dY_n) \\
W_{n+1} &\leq W_n + (r+1)K - AW_n
\end{aligned}$$

The equilibrium point is

$$W^* = \frac{(r+1)K}{A} \tag{3}$$

where $h = \min(h_1, h_2)$, $A = \min(1, c, d)$. Thus, all the curves of the system (1) will enter the following region:

$$\Omega = \left\{(S, I, Y) \in R_+^3 \backslash S + I + Y \leq \frac{(r+1)K}{A}\right\} \tag{4}$$

3 Equilibria

First, we list all the possible equilibrium points of our model.

System (1) has the following equilibria, namely,

- Trivial equilibrium point: $E_0 = (0,0,0)$
- Existence of susceptible prey: $E_1 = (S^*, 0, 0)$

  where $S^* = K\left(\frac{r-h_1}{r}\right)$ always exists provided

$$r > h_1$$

  E_1 is the disease-free equilibrium.

- Existence of susceptible and infected prey: $E_2 = (S^*, I^*, 0)$

  where $S^* = \frac{c+h_2}{\beta}$, $I^* = \left[\frac{\beta K(r-h_1) - r(c+h_2)}{\beta(K\beta + r)}\right]$

  The existence of this equilibrium point is possible, provided

$$\beta K(r - h_1) - r(c + h_2) > 0$$

E_2 is the equilibrium point in which the predator goes extinct and the disease persists in the prey species.

- Endemic Equilibrium: $E_3 = (S^*, I^*, Y^*)$

  where $S^* = \frac{K(r-h_1) - I^*(K\beta + r)}{r}$, $I^* = \sqrt{\frac{d\alpha}{ab-d}}$, $Y^* = \frac{\alpha\beta b S^* - (c+h_2)b\alpha}{\sqrt{(ab-d)d\alpha}}$

The basic reproduction number R_0 is a fundamental and widely used quantity in the study of epidemiological models. R_0 is the average number of secondary infections produced by one infected individual during the entire course of infection in a completely susceptible prey population. The basic reproduction number of system (1) is given by

$$R_0 = \frac{\beta K(r - h_1)}{r(c + h_2)} \tag{5}$$

Theorem 1 *The Equilibrium point E_0 is locally asymptotically stable if $h_1 < r + 2$, $h_2 < 2 - c$ and $d < 2$. Otherwise it is unstable.*

Proof The Jacobian matrix of system (1) at E_0 is given by

$$J_0 = \begin{pmatrix} 1 + r - h_1 & 0 & 0 \\ 0 & 1 - (c + h_2) & 0 \\ 0 & 0 & 1 - d \end{pmatrix} \tag{6}$$

The eigen values of the above matrix are $\lambda_1 = 1 + r - h_1$, $\lambda_2 = 1 - (c + h_2)$ and $\lambda_3 = 1 - d$. We see that $|\lambda_i| < 1$ for $i = 1, 2, 3$, when the conditions $h_1 < r + 2$,

$h_2 < 2 - c$ and $d < 2$ are satisfied. Therefore, the equilibrium point E_0 of model (1) is locally asymptotically stable. Otherwise unstable. □

Theorem 2 *The Equilibrium point E_1 is locally asymptotically stable if $R_0 < 1$. Otherwise it is unstable.*

Proof The Jacobian matrix of system (1) at E_1 is given by

$$J_1 = \begin{pmatrix} 1 + r - 2(r - h_1) & -\left(\frac{r}{K} + \beta\right)K\left(\frac{(r-h_1)}{r}\right) & 0 \\ 0 & 1 + \beta\left(K\left(\frac{(r-h_1)}{r}\right)\right) - (c + h_2) & 0 \\ 0 & 0 & 1 - d \end{pmatrix} \tag{7}$$

The eigen values of the above matrix are $\lambda_1 = 1 + r - 2(r - h_1)$, $\lambda_2 = 1 + \beta\left(K\frac{(r-h_1)}{r}\right) - (c + h_2)$ and $\lambda_3 = 1 - d$. We see that $|\lambda_i| < 1$, for $i = 1, 2, 3$ if $R_0 < 1$. Therefore, the equilibrium point E_1 of system (1) is locally asymptotically stable. Otherwise unstable. □

Theorem 3 *The Equilibrium point E_2 is locally asymptotically stable if*

$$1 + r < \frac{2rS^* + rI^*}{K} + \beta I^* + h_1 \tag{8}$$

$$1 + \beta S^* < (c + h_2) \tag{9}$$

$$1 + \frac{abI^{*2}}{(I^{*2} + \alpha)} < d \tag{10}$$

Otherwise unstable.

Proof The Jacobian matrix of system (1) at E_2 is given by

$$J_2 = \begin{pmatrix} 1 + r - \frac{2rS^* + rI^*}{K} - \beta I^* - h_1 & -\frac{rS^*}{K} - \beta S^* & 0 \\ \beta I^* & 1 + \beta S^* - (c + h_2) & \frac{-aI^{*2}}{(I^{*2} + \alpha)} \\ 0 & 0 & 1 - d + \frac{abI^{*2}}{I^{*2} + \alpha} \end{pmatrix} \tag{11}$$

The characteristic equation is given by

$$\begin{aligned} \varphi(\lambda) = & \left[1 - d + \frac{abI^{*2}}{I^{*2} + \alpha} - \lambda\right]\left[\lambda^2 - \lambda\left[2 + r - \frac{2rS^* + rI^*}{K} - \beta I^* - h_1\right.\right. \\ & \left. + \beta S^* - (c + h_2)\right] + \left(1 + r - \frac{2rS^* + rI^*}{K} - \beta I^* - h_1\right) \\ & \left. \times [1 + \beta S^* - (c + h_2)] + \left[\frac{rS^*}{K} + \beta S^*\right]\beta I^*\right] = 0 \end{aligned} \tag{12}$$

$$= \left[1 - d + \frac{abI^{*2}}{I^{*2} + \alpha} - \lambda\right]\{\lambda^2 + G\lambda + H\} = 0 \tag{13}$$

where

$$G = \frac{2rS^* + rI^*}{K} + \beta I^* + h_1 + (c + h_2) - 2 - r - \beta S^* \tag{14}$$

$$H = \left[1 + r - \frac{2rS^* + rI^*}{K} - \beta I^* - h_1\right][1 + \beta S^* - (c + h_2)] + \left[\frac{rS^*}{K} + \beta S^*\right]\beta I^* \tag{15}$$

We see that $G > 0, H > 0$ under the given conditions (8) and (9). Consequently, the characteristic Eq. (13) has the following eigenvalues:

$$\lambda_{1,2} = \frac{-G \pm \sqrt{G^2 - 4H}}{2} \tag{16}$$

$$\lambda_3 = 1 - d + \frac{abI^{*2}}{I^{*2} + \alpha} \tag{17}$$

Clearly, $\lambda_{1,2}$ are negative. However, the third eigenvalue λ_3 is negative depending on the condition (10). Therefore, E_2 is asymptotically stable equilibrium point provided that the conditions are satisfied. □

Theorem 4 *The Equilibrium point E_3 is locally asymptotically stable provided that the following conditions are satisfied:*

$$1 + r < \frac{2rS^* + rI^*}{K} + \beta I^* + h_1 \tag{18}$$

$$1 + \beta S^* < (c + h_2) + \frac{2a\beta I^* Y^*}{(I^{*2} + \alpha)^2} \tag{19}$$

$$1 + \frac{abI^{*2}}{I^{*2} + \alpha} < d \tag{20}$$

Otherwise unstable.

Proof The Jacobian matrix of the system (1) at E_3 is given by

$$J_3 = \begin{pmatrix} 1 + r - \frac{2rS^* + rI^*}{K} - \beta I^* - h_1 & -\frac{rS^*}{K} - \beta S^* & 0 \\ \beta I^* & 1 + \beta S^* - (c + h_2) - \frac{2a\alpha I^* Y^*}{(I^{*2} + \alpha)^2} & \frac{-aI^{*2}}{(I^{*2} + \alpha)} \\ 0 & \frac{2ab\alpha I^* Y^*}{(I^{*2} + \alpha)^2} & 1 - d + \frac{abI^{*2}}{I^{*2} + \alpha} \end{pmatrix}$$

The characteristic equation of the above Jacobian matrix is given by

$$\phi(\lambda) = \lambda^3 + \Omega_1\lambda^2 + \Omega_2\lambda + \Omega_3 = 0 \tag{21}$$

where

$$\begin{aligned}\Omega_1 &= -[a_{11} + a_{22} + a_{33}] \\ &= -\left[1 + r - \frac{2rS^* + rI^*}{K} - \beta I^* - h_1\right] - \left[1 - d + \frac{abI^{*2}}{(I^{*2} + \alpha)}\right] \\ &- \left[1 + \beta S^* - (c + h_2) - \frac{2a\alpha I^* Y^*}{(I^{*2} + \alpha)^2}\right]\end{aligned} \tag{22}$$

$$\begin{aligned}\Omega_2 &= a_{11}a_{22} - a_{12}a_{21} + a_{11}a_{33} - a_{13}a_{31} + a_{22}a_{33} - a_{23}a_{32} \\ &= \left[1 + r - \frac{2rS^* + rI^*}{K} - \beta I^* - h_1\right] \\ &\left[\left[1 + \beta S^* - (c + h_2) - \frac{2a\beta I^* Y^*}{(I^{*2} + \alpha)^2}\right] + \left[1 - d + \frac{abI^{*2}}{I^{*2} + \alpha}\right]\right] \\ &+ \left[\frac{rS^*}{K} + \beta S^*\right]\beta I^* + \left[1 + \beta S^* - (c + h_2) - \frac{2a\alpha I^* Y^*}{(I^{*2} + \alpha)^2}\right] \\ &\times \left[1 - d + \frac{abI^{*2}}{I^{*2} + \alpha}\right] + \left[\frac{2a^2 b\alpha (I^*)^3 Y^*}{\left((I^*)^2 + \alpha\right)^2}\right]\end{aligned} \tag{23}$$

$$\Omega_3 = -a_{33}[a_{11}a_{22} - a_{12}a_{21}] - a_{12}a_{23}a_{31} - a_{13}a_{21}a_{32} + a_{13}a_{22}a_{31} + a_{11}a_{32}a_{23} \tag{24}$$

$$\begin{aligned}&= -\left[1 - d + \frac{ab(I^*)^2}{(I^*)^2 + \alpha}\right]\left[\left[1 + \beta S^* - (c + h_2) - \frac{2a\alpha I^* Y^*}{\left((I^*)^2 + \alpha\right)^2}\right]\right. \\ &\times \left[1 + r - \frac{2rS^* + rI^*}{K} - \beta I^* - h_1\right] + \left.\left[\frac{rS^*}{K} + \beta S^*\right]\beta I^*\right] \\ &- \left[1 + r - \frac{2rS^* + rI^*}{K} - \beta I^* - h_1\right]\frac{2a^2 b\alpha (I^*)^3 Y^*}{((I^*)^2 + \alpha)^2}\end{aligned} \tag{25}$$

It follows from the Jury's conditions that the modulus of all the roots of the above characteristic equation is less than 1, if and only if the conditions $\phi(1) > 0, \phi(-1) < 0$ and $|\det J_3| < 1$ hold.

$$\phi(1) = 1 + \Omega_1 + \Omega_2 + \Omega_3 \tag{26}$$

$$\begin{aligned}
&= 1 - \left[1 + r - \frac{2rS^* + rI^*}{K} - \beta I^* - h_1\right] \\
&- \left[1 + \beta S^* - (c + h_2) - \frac{2a\alpha I^* Y^*}{\left((I^*)^2 + \alpha\right)^2}\right] - \left[1 - d + \frac{ab(I^*)^2}{(I^*)^2 + \alpha}\right] \\
&+ \left[1 + \beta S^* - (c + h_2) - \frac{2a\alpha I^* Y^*}{\left((I^*)^2 + \alpha\right)^2}\right]\left[1 + r - \frac{2rS^* + rI^*}{K} - \beta I^* - h_1\right] \\
&+ \left[1 + r - \frac{2rS^* + rI^*}{K} - \beta I^* - h_1\right]\left[1 - d + \frac{ab(I^*)^2}{(I^*)^2 + \alpha}\right] \\
&- \left[1 + r - \frac{2rS^* + rI^*}{K} - \beta I^* - h_1\right]\left[1 + \beta S^* - (c + h_2) - \frac{2a\alpha I^* Y^*}{\left((I^*)^2 + \alpha\right)^2}\right] \\
&\times \left[1 - d + \frac{ab(I^*)^2}{(I^*)^2 + \alpha}\right] + \left[1 + \beta S^* - (c + h_2) - \frac{2a\alpha I^* Y^*}{\left((I^*)^2 + \alpha\right)^2}\right] \\
&\times \left[1 - d + \frac{ab(I^*)^2}{(I^*)^2 + \alpha}\right] + \frac{2a^2 b\alpha (I^*)^3 Y^*}{\left((I^*)^2 + \alpha\right)^2} + \left[\frac{rS^*}{K} + \beta S^*\right]\beta I^* \\
&- \left[\frac{rS^*}{K} + \beta S^*\right]\beta I^*\left[1 - d + \frac{ab(I^*)^2}{(I^*)^2 + \alpha}\right] \\
&- \left[1 + r - \frac{2rS^* + rI^*}{K} - \beta I^* - h_1\right]\frac{2a^2 b\alpha (I^*)^3 Y^*}{\left((I^*)^2 + \alpha\right)^2}
\end{aligned} \tag{27}$$

$$\phi(-1) = -1 + \Omega_1 - \Omega_2 + \Omega_3 \tag{28}$$

$$\begin{aligned}
&= -1 - \left[1 + r - \frac{2rS^* + rI^*}{K} - \beta I^* - h_1\right] \\
&\quad - \left[1 + \beta S^* - (c + h_2) - \frac{2a\alpha I^* Y^*}{\left((I^*)^2 + \alpha\right)^2}\right] - \left[1 - d + \frac{ab(I^*)^2}{(I^*)^2 + \alpha}\right] \\
&\quad - \left[1 + \beta S^* - (c + h_2) - \frac{2a\alpha I^* Y^*}{\left((I^*)^2 + \alpha\right)^2}\right]\left[1 + r - \frac{2rS^* + rI^*}{K} - \beta I^* - h_1\right] \\
&\quad - \left[1 + r - \frac{2rS^* + rI^*}{K} - \beta I^* - h_1\right]\left[1 - d + \frac{ab(I^*)^2}{(I^*)^2 + \alpha}\right] \\
&\quad - \left[1 + r - \frac{2rS^* + rI^*}{K} - \beta I^* - h_1\right]\left[1 + \beta S^* - (c + h_2) - \frac{2a\alpha I^* Y^*}{\left((I^*)^2 + \alpha\right)^2}\right] \\
&\quad \times \left[1 - d + \frac{ab(I^*)^2}{(I^*)^2 + \alpha}\right] - \left[1 + \beta S^* - (c + h_2) - \frac{2a\alpha I^* Y^*}{\left((I^*)^2 + \alpha\right)^2}\right] \\
&\quad \times \left[1 - d + \frac{ab(I^*)^2}{(I^*)^2 + \alpha}\right] - \frac{2a^2 b\alpha (I^*)^3 Y^*}{\left((I^*)^2 + \alpha\right)^2} - \left[\frac{rS^*}{K} + \beta S^*\right]\beta I^* \\
&\quad - \left[\frac{rS^*}{K} + \beta S^*\right]\beta I^* \left[1 - d + \frac{ab(I^*)^2}{(I^*)^2 + \alpha}\right] \\
&\quad - \left[1 + r - \frac{2rS^* + rI^*}{K} - \beta I^* - h_1\right]\frac{2a^2 b\alpha (I^*)^3 Y^*}{\left((I^*)^2 + \alpha\right)^2}
\end{aligned} \tag{29}$$

$$\begin{aligned}
\mathrm{Det}(J_3) &= \left[1 + r - \frac{2rS^* + rI^*}{K} - \beta I^* - h_1\right] \\
&\quad \times \left[1 + \beta S^* - (c + h_2) - \frac{2a\alpha I^* Y^*}{\left((I^*)^2 + \alpha\right)^2}\right]\left[1 - d + \frac{ab(I^*)^2}{(I^*)^2 + \alpha}\right] \\
&\quad + \frac{2a^2 b\alpha (I^*)^3 Y^*}{\left((I^*)^2 + \alpha\right)^2}\left[1 + r - \frac{2rS^* + rI^*}{K} - \beta I^* - h_1\right] \\
&\quad + \left(\frac{rS^*}{K} + \beta S^*\right)\left[1 - d + \frac{ab(I^*)^2}{(I^*)^2 + \alpha}\right]\beta I^*
\end{aligned} \tag{30}$$

Clearly, $\phi(1) > 0$, $\phi(-1) < 0$ and $|\mathrm{Det}\, J_3| < 1$ if the conditions (18), (19) and (20) hold. Hence the proof. □

4 Numerical Simulations

Numerical simulations have been carried out to investigate the dynamics of the proposed model. The simulations have been performed using MATLAB to explain our theoretical results. Taking the following set of parametric values,

$$r = 0.03,\ K = 50,\ \beta = 0.02,\ c = 0.03,$$

$$a = 0.3,\ d = 0.6,\ \alpha = 0.46,\ b = 0.01$$

we assume different values of h_1 and h_2 to investigate the effects of harvesting (Fig. 1). The value of harvesting in figures is as follows: (a) $h_1 = 0.02, h_2 = 0.4$, (b) $h_1 = 0.02$, $h_2 = 0.5$, (c) $h_1 = 0.01, h_2 = 0.8$, (d) $h_1 = 0.01, h_2 = 0.2$, (e) $h_1 = 0.01, h_2 = 0.5$, (f) $h_1 = 0$, $h_2 = 0.08$. For figures (a), (b) and (c), we get $R_0 < 1$ and for the figures (e), (f) and (g), we get $R_0 > 1$.

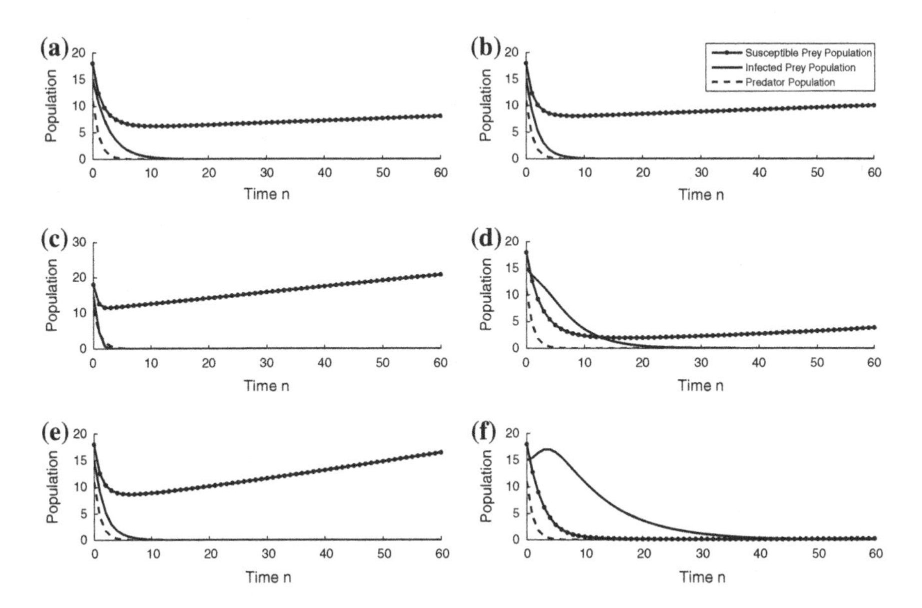

Fig. 1 The dynamical behaviours of the system (1) with $r = 0.03, K = 50, \beta = 0.02$, $c = 0.03, a = 0.3, d = 0.6, \alpha = 0.46, b = 0.01$ and $(S_0, I_0, Y_0) = (18, 15, 11)$

5 Discussion

We have proposed a discrete-time eco-epidemic model with disease and harvesting in prey and consider the case in which the predator predates only on the infected prey. We have considered that both the susceptible and infected prey populations are being harvested which is a natural phenomenon and have applied Holling type III functional response, which considers only high levels of prey density. These assumptions are needed to study an ecological system accurately, which are not done in the previous discrete-time models. We consider the growth rate $r = 0.03$ and find that $h_1 < r$. We analyse the behaviour of the system under different cases of harvesting and see under what conditions the disease spreads or dies out. In the figure, we have shown three cases under which the disease dies out and three cases under which the disease persists. We see that, when $R_0 < 1$, the susceptible prey population is maintained, whereas the infected prey and predator populations decreases. For $R_0 > 1$, the behaviour of the system varies, that is, the susceptible prey population decreases and there is a considerable increase in the infected prey population. We consider a case in which there is no harvesting of susceptible prey population, that is $h_1 = 0, h_2 = 0.08$. In this case, we have an endemic equilibrium. Thus, we can see that harvesting plays an important role in the spread of the disease.

6 Conclusion

In this paper, we have considered a discrete-time prey–predator system with disease and harvesting in the prey population incorporating Holling type III functional response. The prey population is divided into susceptible and infected prey populations. The dynamical behaviour of the system around each equilibrium point has been studied and the basic reproduction number R_0 is computed. Conditions for the existence and stability of prey–predator system around each equilibrium point are obtained. Finally, we have proved the theoretical results using numerical simulations through MATLAB.

References

1. Lotka, A.J.: Elements of Physical Biology. Williams and Wilkins Co., Inc., Baltimore (1924)
2. Volterra, V.: Variazioni e fluttauazionidel numero d individui in specie animals conviventi. Mem. Acad. Lincei **2**, 31–33 (1926)
3. Kermack, W.O., McKendrick, A.G.: Contributions to the mathematical theory of epidemics, part 1. In: Proceedings of the Royal Society of London, Vol. A 115, pp. 700–721 (1927)
4. Anderson, R.M., May, R.M.: The invasion persistence and spread of infectious diseases within animal and plant communities. Philos. Trans. R. Soc. Lond. **314**, 533–570 (1986)

5. Chattopadhyay, J., Arino, O.: A predator-prey model with disease in the prey. Nonlinear Anal. **36**, 747–766 (1999)
6. Holling, C.S.: The functional response of predator to prey density and its role in mimicry and population regulation. Mem. Entomol. Soc. Can. **45**, 1–60 (1965)
7. He, Z., Li, B.: Complex dynamic behavior of a discrete-time predator-prey system of Holling-III type. Adv. Differ. Equ. (2014)
8. Sarwardi, S., Haque, M., Venturino, E.: A Leslie-Gower Holling-type II ecoepidemic model. J. Appl. Math. Comput. **35**, 263–280 (2011)
9. Bairagi, N., Chaudhuri, S., Chattopadhyay J.: Harvesting as a disease control measure in an eco-epidemiological system—a theoretical study. Math. Biosci. 134–144 (2009)
10. Agnihotri, K.B.: The dynamics of disease transmission in a Prey Predator System with harvesting of prey. Int. J. Adv. Res. Comput. Eng. Technol. **1** (2012)
11. Wuaib, S.A., Hasan, Y.A.: Predator-prey interactions with harvesting of predator with prey in refuge. Commun. Math. Biol. Neurosci. (2013)
12. Kooi, B.W., van Voorn, G.A.K., pada Das, K.: Stabilization and complex dynamics in a predator-prey model with predator suffering from an infectious disease. Ecol. Complex. 113–122 (2011)
13. Bahlool, D.K.: Stability of a prey-predator model with sis epidemic disease in prey. Iraqi J. Sci. 52, 484–493(2011)
14. Pal, A.K., Samanta, G.P.: A Ratio-dependent eco-epidemiological model incorporating a prey refuge. Univ. J. Appl. Math. **1**, 86–100 (2013)
15. Wang, S., Ma, Z.: Analysis of an ecoepidemiological model with prey refuges. J. Appl. Math. (2012)
16. Hu, Z., Teng, Z., Jia, C., Zhang, L., Chen, X.: Complex dynamical behaviors in a discrete eco-epidemiological model with disease in prey. Adv. Differ. Equ. (2014)
17. Das, P., Mukherjee, D., Das, K.: Chaos in a Prey-Predator Model with Infection in Predator—A Parameter Domain Analysis. Comput. Math. Biol. (2014)

A Novel Crossover Operator Designed to Exploit Synergies of Two Crossover Operators for Real-Coded Genetic Algorithms

Shashi and Kusum Deep

Abstract In this paper a new crossover operator called the double distribution crossover (DDX) is proposed. The performance of DDX is compared with existing real-coded crossover operator namely Laplace crossover (LX). DDX is used in conjunction with a well-known mutation operator; Power mutation (PM) to obtain a new generational real-coded genetic algorithm called DDX-PM. DDX-PM is compared with the existing LX-PM. The performance of both the genetic algorithms is compared on the basis of success rate, average function evaluation, average error and computational time, and the preeminence of the proposed crossover operator is established.

Keywords Real-coded genetic algorithm · Mutation operator · Crossover operator

1 Introduction

The demand for designing efficient, reliable and robust numerical optimization techniques is very high because of their ability to tackle a number of real-life complex nonlinear optimization problems arising in the field of Engineering, Science, Industry and Finance, which may have constraints associated with them. There are a number of stochastic optimization techniques which are being used to solve complex optimization problems. Among these genetic algorithms (GA) are found to be very promising global optimizers. GAs are population-based heuristics which mimic the Darwin's principal of "survival of fittest". The concept of GA was given by Holland [1]. A detailed implementation of GA could be found in Goldberg

Shashi (✉)
Control and Decision Systems Laboratory, Department of Aerospace Engineering, Indian Institute of Science, Bangalore, India
e-mail: shashibarak@gmail.com

K. Deep
Department of Mathematics, Indian Institute of Technology Roorkee, Roorkee, India
e-mail: kusumfma@iitr.ernet.in

M. Pant et al. (eds.), *Proceedings of Fifth International Conference on Soft Computing for Problem Solving*, Advances in Intelligent Systems and Computing 437, DOI 10.1007/978-981-10-0451-3_32

[2]. The main attribute of GAs is their easy implementation nature, as they do not require any extra information such as continuity and differentiability about objective function and constraints. Although initial versions of GA use binary numbers for encoding of chromosomes, their major drawbacks such as lack of precision and existence of "Hamming-cliff" problem impel to look for another type of encoding scheme. To overcome these difficulties related to binary encoding of continuous parameter optimization problems, real encoding of chromosomes is used. GAs which make use of real encoding of chromosomes are termed as real-coded GA (RCGA).

In GAs crossover is considered to be main search operator. The crossover operator is used to thoroughly explore the search process. After the selection process, the population is enriched with better individuals. Selection makes clones of good strings but does not create new ones. Crossover operator is applied to the mating pool with the hope that it creates better offsprings. In crossover operator the genetic information between two or more individuals is blended to produce new individuals.

Though, along with crossover mutation operator is also essential for thorough search of the function landscape. The most important aspect of mutation operator is that it prevents premature convergence of an algorithm. Mutation operator provides random diversity in the population as described in [3].

A lot of effort has been put into the development of sophisticated real-coded crossover operators [4–8] to improve the performances of RCGAs for function optimization. Deep and Thakur [9] presented Laplace crossover operator which generates a pair of offspring solution from a pair of parent solutions using Laplace distribution. Tutkun [10] proposed a crossover operator based on Gaussian distribution. Kaelo and Ali [11] suggested integration of different crossover rules in the genetic algorithm and recommended some modifications in applying the crossover rules and localization of searches in their study. Recent development in RCGA includes [12].

In an endeavour to define new operators for real-coded genetic algorithms, in this paper, a new crossover operator called double distribution crossover (DDX) is presented. Combining DDX with power mutation (PM) [13], a new generational RCGA called DDX-PM is designed. It is compared with existing RCGA, LX-PM. In order to establish the strength of double distribution crossover computational analysis is performed and discussed.

The paper is organized as follows: the proposed double distribution crossover is defined in Sect. 2. In Sect. 3, new RCGAs based on double distribution crossover are discussed. In Sect. 4, the experimental setup of RCGAs is explained. The computational results and discussion is given in Sect. 5. Finally, in Sect. 6, the conclusions are drawn.

2 The Proposed Crossover Operator

In this section a new crossover operator based on the novel idea of merging of two existing crossover operators is proposed in such a way that it retains the strength of the operators from which it is obtained. The proposed crossover operator is called as double distribution crossover (DDX) as it makes use of two well-established crossover operators: Laplace Crossover (LX) [9] based on Laplace distribution and Weibull Crossover (WX) [14] based on Weibull distribution to produce two off-springs and hence named as double distribution crossover (DDX) operator.

Although DDX could be obtained by merging any two crossover operators, here WX and LX are used because they are well-established potentially effective crossover operators. In DDX two offsprings O_1 and O_2 are generated from a pair of parents P_1 and P_2 obtained after selection, in the following manner:

Generate a uniformly distributed random number $u \in [0, 1]$

if $u \leq \frac{1}{2}$; then

$$\begin{aligned} O_1 &= P_1 + \mu d \\ O_2 &= P_2 + \mu d \end{aligned} \tag{1}$$

else

$$\begin{aligned} O_1 &= P_1 - \mu d \\ O_2 &= P_2 - \mu d \end{aligned} \tag{2}$$

where $d = |P_1 - P_2|$ is the distance between the parents and μ is calculated as $\mu = \frac{r_1 + r_2}{2}$, where r_1 is a random number following Laplace distribution and r_2 is a random number following Weibull distribution. The density function of Weibull distribution is given as follows:

$$f(x) = \begin{cases} ab^{-a}x^{a-1}e^{-(x/b)^a} & \text{if } x > 0 \\ 0 & \text{otherwise} \end{cases} \tag{3}$$

and the distribution function is given by

$$F(x) = \begin{cases} 1 - e^{-(x/b)^a} & \text{if } x > 0 \\ 0 & \text{otherwise} \end{cases} \tag{4}$$

where $a > 0$ is called shape parameter and $b > 0$ is called the scale parameter. And the density function of Laplace distribution is given as follows:

$$f(x) = \frac{1}{2d}\exp\left(-\frac{|x - c|}{d}\right), \quad -\infty < x < \infty \tag{5}$$

the distribution function is given by

$$F(x) = \begin{cases} \frac{1}{2}\exp\left(\frac{|x-c|}{d}\right) & \text{if } x \le c \\ 1 - \frac{1}{2}\exp\left(-\frac{|x-c|}{d}\right) & \text{if } x > c \end{cases} \quad (6)$$

where $c \in R$ is location parameter and $d > 0$ is termed as scale parameter.

DDX preserves the property of both the parent operators, i.e. for fixed values of the crossover index of DDX (which is a two-dimensional vector which includes the crossover indices of both WX and LX), the spread of offsprings is proportional to the spread of parents, i.e. if the parents are near to each other the offsprings are expected to be near to each other and if the parents are far from each other then offsprings are likely to be far from each other.

Also, from Eqs. 1 and 2 it is clear that both the offsprings are placed symmetrical with respect to the position of the parents and if DDX produces an offspring which is not within the bounds of the decision variable then that offspring is assigned a random value within its bounds.

3 The Proposed New RCGA

In this section a new RCGA namely DDX-PM is proposed, which combines the use of Laplace crossover with power mutation [13]. DDX-PM is compared with the existing LX-PM. Both the algorithms use tournament selection with tournament size three. In addition, both the RCGAs are elite preserving with elitism size one. The elitism operator helps to maintain fitness stability to increase the search performance of the proposed algorithm.

Both the RCGAs are terminated if either the pre-specified maximum number of generations (3000) is reached or the best solution obtained by the algorithm lies within specified accuracy (0.01) of known optimum, whichever occurs earlier. The algorithm is tested on standard benchmark problems taken from literature which includes five nonscalable test problems (Table 1) and five scalable test problems of

Table 1 List of nonscalable benchmark functions

S. no.	Function	Mathematical formula	Range	Min
1	Easom 2D	$\text{Min} f(x) = -\cos(x_1)\cos(x_2)\exp\left(-(x_1-\pi)^2-(x_2-\pi)^2\right)$	[−10, 10]	−1
2	Becker and lago	$\text{Min} f(x) = (\lvert x_1\rvert - 5)^2 + (\lvert x_2\rvert - 5)^2$	[−10, 10]	0
3	Bohachevsky 1	$\text{Min} f(x) = x_1^2 + 2x_2^2 - 0.3\cos(3\pi x_1) - 0.4\cos(4\pi x_2) + 0.7$	[−50, 50]	0
4	Eggcrate	$\text{Min } f(x) = x_1^2 + x_2^2 + 25(\sin^2 x_1 + \sin^2 x_2)$	[−2 π, 2 π]	0
5	Periodic	$\text{Min} f(x) = 1 + \sin^2 x_1 + \sin^2 x_2 - 0.1\exp(-x_1^2 - x_2^2)$	[−10, 10]	0.9

size 30 (Table 2) of different levels of complexity and multimodality. The pseudocode of DDX-PM is given below:

Algorithm

Step 1 (Initialization):
- *Initialize population;*
- *Set Generation=0;*

Step 2(Evaluation): Evaluate the fitness for each individual

Step 3(Termination check): Check the termination criteria
If satisfied stop;
else goto 4.

Step 4 (GA Operations)
- *Select individuals according to selection algorithm to build a mating pool*
- *Crossover the population in mating pool with given crossover probability*
- *Mutate the current population with given mutation probability*

Step 5 (Replacement): Replace the old population with new population while retaining the best individual for next generation i.e. apply elitism with size 1

Step 6
- *Evaluate the best fitness and find optimal individual*
- *Increment generation; go to step 3.*

4 Experimental Setup

The experimental setup used for both the RCGAs viz. DDX-PM and LX-PM is given in this section.

Parameter Setting: Population size is taken as 10 times the number of variables. The crossover index for DDX is fixed at (12, 0.20). The final parameter setting is presented in Table 3 where p_c and p_m represent the probabilities of crossover and mutation, respectively. Each GA runs at 100 times with same initial populations while each run is initiated using a different set of initial population.

All the algorithms are implemented in C++ and the experiments are done on a Core Duo Processor with 1.66 GHz speed and 1 GB RAM under WINXP platform.

Performance Evaluation Criteria: A run in which the algorithm finds a solution satisfying $f_{\min} - f_{\text{opt}} \leq 0.01$, where $f_{\min}$ is the best solution found when the algorithm terminates and f_{opt} is the known global minimum of the problem, is considered to be successful.

For each method and problem, the followings are recorded:

- Success Rate $(\text{SR}) = \frac{\text{Number of successful runs}}{\text{Total number of runs}} \times 100$
- Average computational time (ACT) (in seconds).
- Average number of function evaluations (AFE).
- Average Error $(\text{AE}) = \frac{\sum_n (f_{\min} - f_{\text{opt}})}{n}$ where, n is the total number of runs.

Table 2 List of scalable benchmark functions

S. no.	Function	Mathematical formula	Range	Min
1	Paviani	$\min_x f(x) = \sum_{i=1}^{n} \left[(\ln(x_i - 2))^2 + (\ln(10 - x_i))^2\right] - \left(\prod_{i=1}^{n} x_i\right)^{0.2}$	$[2, 10]^{30}$	−997807.705158
2	Rastrigin	$\min_x f(x) = 10n + \sum_{i=1}^{n} \left[x_i^2 - 10\cos(2\pi x_i)\right]$	$[-5.12, 5.12]^{30}$	0
3	Schwefel	$\min_x f(x) = -\sum_{i=1}^{n} x_i \sin\left(\sqrt{\lvert x_i \rvert}\right)$	$[-500, 500]^{30}$	−12569.48
4	Sinusoidal	$\min_x f(x) = -\left[2.5 \prod_{i=1}^{n} \sin\left(x_i - \frac{\pi}{6}\right) + \prod_{i=1}^{n} \sin\left(5\left(x_i - \frac{\pi}{6}\right)\right)\right]$	$[0, \pi]^{30}$	−3.5
5	Zakharov's	$\min_x f(x) = \sum_{i=1}^{n} x_i^2 + \left(\sum_{i=1}^{n} \frac{i}{2} x_i\right)^2 + \left(\sum_{i=1}^{n} \frac{i}{2} x_i\right)^4$	$[-10, 10]^{30}$	0

Table 3 Parameter setting for DDX-PM and LX-PM

Algorithm	Nonscalable		Scalable	
	p_c	p_m	p_c	p_m
DDX-PM	0.70	0.040	0.60	0.010
LX-PM	0.70	0.007	0.60	0.005

Table 4 Computational results for nonscalable and scalable problems

Nonscalable problems					Scalable problems				
Fun	Method	AE	SR	AFE	Fun	AE	SR	AFE	ACT
1	LX-PM	0.00625	100	378	1	0.00865	100	139,364	3.43
	DDX-PM	0.00478	100	332		0.00214	100	57,300	0.98
2	LX-PM	0.00612	100	4059	2	0.01154	77	883,517	31.35
	DDX-PM	0.00583	100	339		0.00634	100	59,476	1.78
3	LX-PM	0.00884	94	649	3	0.00756	100	128,034	2.82
	DDX-PM	0.0048	100	597		0.00632	88	125,463	3.25
4	LX-PM	0.01102	66	12,718	4	0.00805	100	46,486	1.06
	DDX-PM	0.00478	100	387		0.00953	100	17,116	0.42
5	LX-PM	0.02115	44	1394	5	0.00874	100	336,236	6.82
	DDX-PM	0.00598	82	926		0.00966	100	86,225	2.06

5 Results and Discussion

The computational results of both the RCGAs are presented in this section.

Table 4 summarizes the computational results for both the algorithms. As the execution time for nonscalable problems is very less and hence insignificant in most of the cases, it is not included in Table 4. From Table 4 it can be easily observed that the proposed crossover operator DDX outperforms LX in terms of various performance criteria and thus indicates the efficiency of proposed crossover operator for providing suitable exploration for obtaining the global optimal solution.

6 Conclusion

In this paper a new crossover operator called double distribution crossover (DDX) is introduced. By combining DDX with already existing operator, namely PM a new generational RCGA called DDX-PM is proposed. For an analogical comparison it is compared with the existing RCGA based on Laplace crossover and power mutation called LX-PM. Based on the numerical results it is clear that with respect to *reliability*, *efficiency* and *accuracy* measured in terms of success rate, function evaluation and average error, respectively, DDX-PM performed better than LX-PM, though they both use same mutation operator, which clearly signifies the role of crossover operators in the search process for locating global optima.

Thus DDX proved to be an efficient crossover operator when used with power mutation, but whether its performance will remain equally well, when used with other mutation operators for RCGAs, is the question whose answer is to be looked in future studies.

Acknowledgments Shashi thankfully acknowledge the financial assistance from National Board of Higher Mathematics, Department of Atomic Energy, Government of India.

References

1. Holland, J.H.: Adaptation in Natural and Artificial Systems. University of Michigan press, Ann Arbor (1975)
2. Goldberg, D.E.: Genetic Algorithms in Search, Optimization and Machine Learning. Addison-Wesley, New York (1989)
3. Spears, W.M.: Crossover or mutation? In: Whitley, L.D. (ed.) Foundations of Genetic Algorithms 2, pp. 221–238. Morgan Kaufmann, San Mateo, CA (1993)
4. Deb, K., Agrawal, R.B.: Simulated binary crossover for continuous search space. Complex Syst. **9**, 115–148 (1995)
5. Hong, I., Kahng, A.B., Moon, B.R.: Exploiting synergies of multiple crossovers: initial studies. In: Proceedings of Second IEEE International Conference on Evolutionary Computation, IEEE Press, Piscataway, NJ, pp. 245–250 (1995)
6. Yoon, H.S., Moon, B.R.: An empirical study on the synergy of multiple crossover operators. IEEE Trans. Evol. Comput. **6**(2), 212–223 (2002)
7. Herrera, F., Lozano, M., Sanchez, A.M.: Hybrid crossover operators for real coded genetic algorithms: an experimental study. Soft. Comput. **9**(4), 280–298 (2005)
8. Deb, K., Anand, A., Joshi, D.: A computationally efficient evolutionary algorithm for real-parameter evolution. Evol. Comput. J. **10**(4), 371–395 (2002)
9. Deep, K., Thakur, M.: A new crossover operator for real coded genetic algorithms. Appl. Math. Comput. **188**, 895–911 (2007)
10. Tutkun, N.: Optimization of multimodal continuous functions using a new crossover for the real-coded genetic algorithms. Expert Syst. Appl. **36**, 8172–8177 (2009)
11. Kaelo, P., Ali, M.M.: Integrated crossover rules in real coded genetic algorithms. Eur. J. Oper. Res. **176**, 60–76 (2007)
12. Thakur, M.: A new genetic algorithm for global optimization of multimodal continuous functions. J. Comput. Sci. **5**, 298–311 (2014)
13. Deep, K., Thakur, M.: A new mutation operator for real coded genetic algorithms. Appl. Math. Comput. **193**, 229–247 (2007)
14. Deep, K., Shashi., Katiyar, V.K.: Global optimization of Lennard-Jones potential using newly developed real coded genetic algorithms. In: Proceedings of 2011—International Conference on Communication Systems and Network Technologies, (IEEE Computer Society Proceedings), pp. 614–618 (2011)

Classification Techniques for Texture Images Using Neuroph Studio

Priyanka Mahani, Akanksha Kulshreshtha and Anil Kumar Goswami

Abstract With the widespread scope of image processing in fields such as ground classification, segmentation of satellite images, biomedical surface inspection and content-based image retrieval (CBIR), texture analysis is an important domain with a wide scope. The process of texture analysis comprises the following important steps: texture classification, segmentation and synthesis. Texture is a significant property of images which is difficult to define even though the human eye may recognize it with ease. Texture classification has remained an intangible pattern recognition task despite numerous studies. The prime issue in any texture classification application is how to select the features and which features to consider for representing texture. Another major issue remains the type of metric to be used for comparing the feature vectors. The traditional way of texture classification is to convert the texture image into a vector representing the features using a set of filters. This is followed by classification with a few smoothening steps involved in between. This paper outlines a picture of the basic features of an image as texture, colour and shape which are to be extracted to form the feature vector. The concept of applying Machine learning for the purpose of classification is explored and implemented. The use of Soft Computing Technique—Artificial Neural Networks is illustrated for the purpose of classification. The texture features are extracted using the GLCM method and used as input and fed to the neural network. The algorithm used in training the neural networks is the traditional backpropagation algorithm. The results show which configuration of the multi-layer feedforward architecture is best suited according to our experimental set-up.

Priyanka Mahani (✉) · Akanksha Kulshreshtha
Electronics and Computer Engineering Department, Dronacharya College of Engineering, Gurgaon, India
e-mail: priyankamahani@gmail.com

Akanksha Kulshreshtha
e-mail: akanksha.kulshreshtha@gmail.com

A.K. Goswami
DTRL Lab, DRDO, Metcalfe House, New Delhi, India
e-mail: anilkgoswami@gmail.com

M. Pant et al. (eds.), *Proceedings of Fifth International Conference on Soft Computing for Problem Solving*, Advances in Intelligent Systems and Computing 437, DOI 10.1007/978-981-10-0451-3_33

Keywords Content-based image retrieval · Classification · Artificial neural networks · Backpropagation algorithm

1 Introduction

Classification means categorizing any physical entity or object into a predefined set of classes. Image classification is on the whole a difficult task for conventional machine learning algorithms due to the number of images involved and the variety of features that describe an image. As a result, traditional machines are not very stable for the purpose of classifying images from databases. Classification is a two-tier process which involves two tasks, namely learning and testing. The learning stage of classification involves building a prototype for each of the predefined classes based on their characteristics which define the class uniquely. The training data usually comprises images with known class labels. The texture content of these training images may be captured with the help of the various texture analysis methods available. It produces a set of texture features for each image. These features can be represented as empirical distributions, scalar quantities or histograms. These features so obtained help to define the texture of the images such as orientation, roughness, 3-D arrangement, contrast, etc. In the second stage of classification, i.e. the testing or recognition stage, the texture content of an unknown sample is mined with the help of the same analysis technique for texture as was used during the first stage. Then the feature vector so obtained of the unknown sample is contrasted with the feature vector of the training dataset by means of an algorithm used for the purpose of classification. The unknown sample is finally categorized into the class which is closest. Alternatively, if this best case match obtained using the classification algorithm is not sufficiently good on the basis of some predefined criteria, the sample may be discarded altogether. A neural network is defined as a network of nodes which imitates the behaviour of the brain to specific tasks. An artificial neural network (ANN) is a virtual set-up of processing units influenced by the way the human brain functions to process information and get knowledge. An artificial neural network is formed by connection of parallel processing units called neurons which work together coherently to solve specific problems. ANN's just like humans, learn by example. An ANN can be configured to suit specific applications, such as classification or pattern recognition, with the help of an underlying learning process. Learning in biological systems "involves adjustments to the synaptic connections that exist between the neurons" [1].

2 Content-Based Image Retrieval

CBIR or content-based image retrieval is the science of retrieving images based on visual features. These visual features may vary from colour to texture to shape. CBIR is widely used because traditional image indexing methods have proved to be time-consuming, insufficient and laborious in many large image databases. Ancient image indexing methods varied from image database formation and associating it with a keyword or number, to associating it with a categorized description, have become obsolete [2]. Content-based image retrieval meant each image would be stored in the database with its own features extracted using a conventional feature extraction method. These features would then be compared to the features of the query image. Figure 1 shows the working of a CBIR system.

2.1 Texture

Texture can be defined as a measure that defines the spatial arrangement of various features like colour and intensity across an image. Texture is that distinctive feature which defines the visual patterns in an image. Texture helps visually differentiate between surfaces like: fabric, cloud, cemented floor, marble, etc. It also describes the association of the surface to its neighbouring environment. To summarize we can say, texture is that feature which defines the characteristic physical composition of a surface. Properties which define texture include coarseness, regularity, contrast, directionality and roughness (Fig. 2).

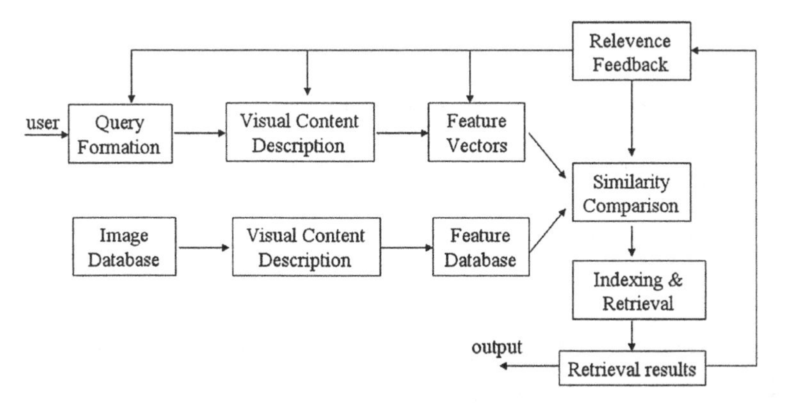

Fig. 1 Steps followed in CBIR

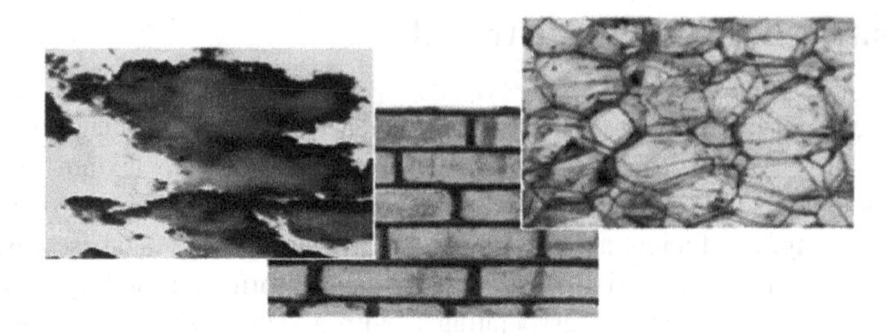

Fig. 2 Examples of texture images taken from Brodatz album

2.2 Texture Analysis Methodologies for Colour Images

2.2.1 Conversion to Grey Scale Equivalent

Since the texture depends on the variations in the intensity values, equivalent grey scale conversion of a colour image is vital. For RGB colour space, the conversion method which is preferred for good results is defined by the formula:

$$\mathrm{Int}_{(i,j)} = \frac{11 * C_{i,j}(\mathrm{R}) + 16 * C_{i,j}(\mathrm{G}) + 5 * C_{i,j}(\mathrm{B})}{32}$$

where the intensity assigned to a pixel (i, j) of the image is given by $\mathrm{Int}_{(i,\ j)}$

$C_{i,j}$ (R) denotes the red colour values of pixel (i, j) of the image,
$C_{i,j}$ (G) denotes the green colour values of pixel (i, j) of the image and
$C_{i,j}$ (B) denotes the blue colour values of pixel (i, j) of the image (Fig. 3).

2.2.2 Formation of Texture Feature Vector by Grey Level Co-occurrence Matrix (GLCM) on Image

Co-occurrence matrix is a widely accepted notation of depicting texture features. It specifies the spatial grey level dependence of pixels in an image. A co-occurrence matrix is technically defined as follows:

Fig. 3 Grey scale conversion of colour images using different methods (colour online)

Consider a position operator POS(i, j). Let MAT represent an m × m matrix. The (i, j)th element of the matrix MAT represents the number of points with grey level (i) w.r.t. (j) in the position specified by the operator POS.

Let CM represent an $m \times m$ matrix. CM is obtained by dividing MAT with the total number of such pairs that satisfy the condition specified by POS. CM [i] [j] thus, represents the joint probability that a pair of point satisfying POS will have values $g[i]$ and $g[j]$.

CM represents the co-occurrence matrix defined by POS [3].

The first stage is to construct the co-occurrence matrix based on the orientation and distance between image pixels as shown in Figs. 4 and 5. The next stage is to extract meaningful statistics from the matrix for the purpose of texture classification. The texture features proposed by Haralick include: Sum Entropy, Contrast, Entropy, Correlation, Angular Second Moment, Variance, Inverse Second Differential Moment, Sum Average, Difference Variance, Sum Variance, Difference Entropy, Measure of Correlation 1, Measure of Correlation 2 and Local Mean [4].

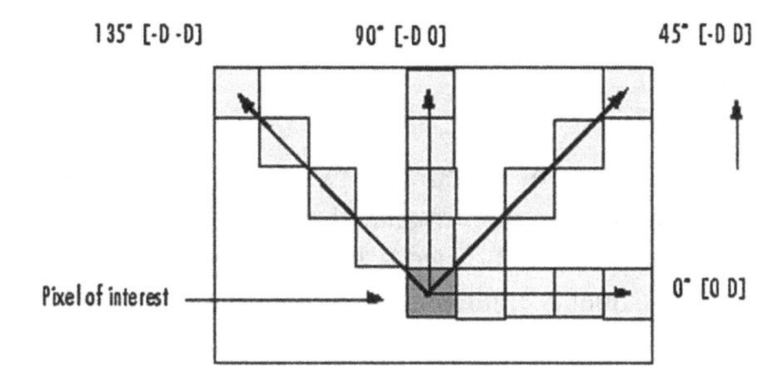

Fig. 4 Directional analysis for GLCM

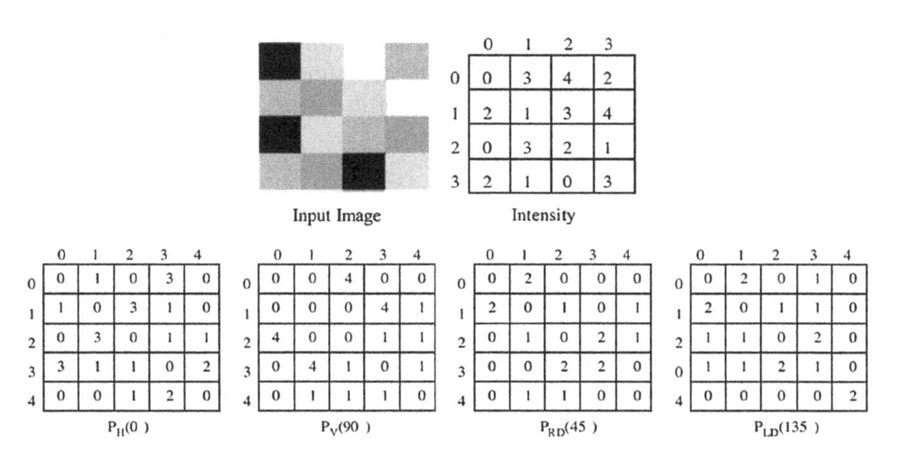

Fig. 5 Formation of classical co-occurrence matrix for the four different directions and a single gap between the pixels

3 Artificial Neural Networks

3.1 *Overview*

Although neural networks were established way before the advent of computers, they may appear to be a modern development. ANN's follow the black box approach since one cannot look inside it to gain knowledge on how it works even though the network may perform well at its job. As stated above, an artificial neural network (ANN) is a virtual set-up of processing units influenced by the way the human brain functions to process information and get knowledge. These processing units which showcase massive parallelism are called neurons. They help to resemble the brain in two respects:

- Knowledge is acquired by the network through a learning process.
- Inter neuron connection strengths known as synaptic weights are used to store the knowledge [1].

3.2 *Learning*

The most important property of a neural network is its ability to learn from its environment and to improve the performance through learning. The procedure used to perform the learning process is called a learning algorithm [5]. Learning is an iterative process of adjustments applied to the synaptic weights of the networks and thresholds.

In the framework of neural networks learning is defined as a process wherein the free parameters of a network are altered and adjusted through a continuing process of stimulation in the host environment of the network. The type of learning is determined by the manner in which the parameter changes take place [6]. A wide variety of learning algorithms are available as shown in Fig. 6. These learning algorithms differ from each other in the way the adjustment to the synaptic weight is formulated. Each algorithm offers advantages of its own.

3.3 *Neural Network Configurations*

Neural networks (NNs) are denoted using directed graphs. Assume an NN represented by graph G'. The ordered pair of 2-tuples (V', E') represents a set of vertices, V' and a set of edges, E' where $V' = \{1, 2, \ldots, n\}$ and arcs $E' = \{<a, b>| \, a \geq 1, b \leq n\}$. The graph $G'(V', E')$ has the following restrictions: V' is partitioned into a set of input nodes, V_I, hidden nodes, V_H, and output nodes, V_O. The vertices are further partitioned into layers. The arc $<a, b>$ represents an edge from node a in layer '$h - 1$' to node b in layer 'h'. Arc $<a, b>$ is labelled with a numeric value w_{ab}, which represents

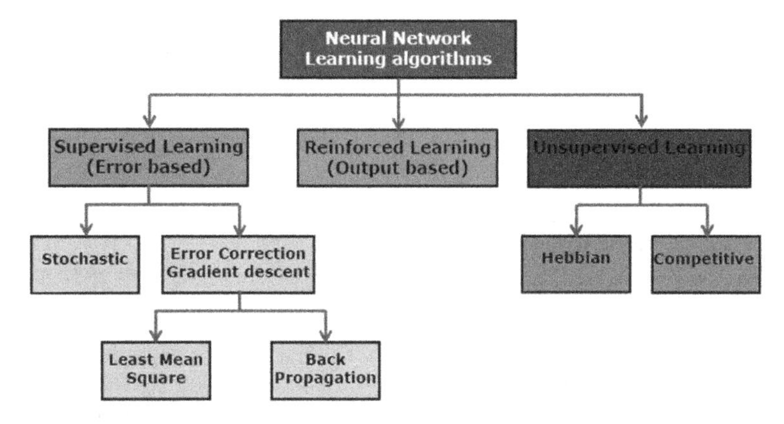

Fig. 6 Classification of learning algorithms

the weight of the connection. Node '*a*' is labelled with a function f_a. When each edge is assigned an orientation, the graph is directed and is called a directed graph or a diagraph. A feedforward network has a directed acyclic graph. The different classes of the networks are single layer feedforward network, recurrent networks, and multilayer feedforward network.

3.3.1 Multilayer Feedforward Network

A multilayer feedforward network as the name suggests comprises multiple layers. It consists of an input and an output layer along with a number of intermediate layers called hidden layers as shown in Fig. 7. The functional units of the input

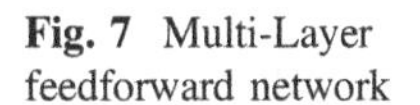

Fig. 7 Multi-Layer feedforward network

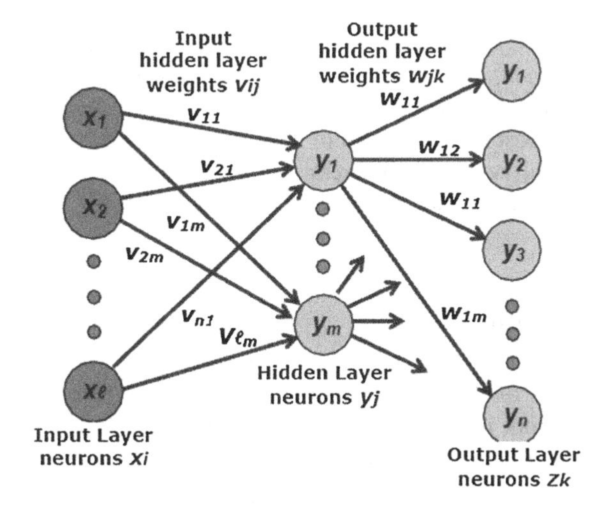

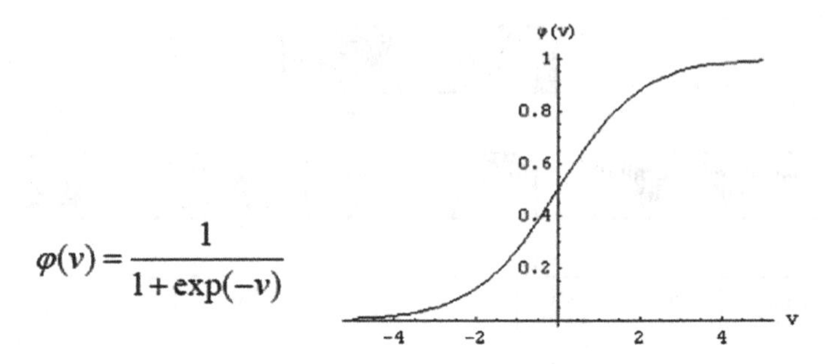

Fig. 8 Sigmoid (logistic) activation function

layer are called input neurons; those of the output layer are called input neurons while those of the hidden layer are known as the hidden neurons [1]. The hidden layer performs the important task of assisting in performing useful transitional computations before directing the input to the output layer. The input layer neurons are linked to the hidden layer neurons. The corresponding weights on these links are referred to as input-hidden layer weights. Similarly, the hidden layer neurons are linked to the output layer neurons and the weights in this case are referred to as hidden-output layer weights.

The nodes in the input layer are fed with the input vector. Necessary computation is performed on the inputs in the input layer and the results are fed to the next layer which is the first hidden layer. The signals are processed further and fed to the next layer. This process continues till the last hidden layer feeds the neurons in the last layer of the network which is the output Layer. This is represented in Fig. 7.

3.4 Activation Function

The various activation functions used in neural networks are threshold activation function, piecewise-linear activation function, hyperbolic tangent function, and sigmoid (logistic) activation function. The sigmoid function as shown in Fig. 8 is the most common form of activation function used in the construction of artificial neural networks. Whereas a threshold function assumes the value of 0 or 1, a sigmoid function assumes a continuous range of values from 0 and 1.

4 Simulation

The experiments are simulated on Neuroph Studio which is a Java-based neural network framework. Texture features extracted from GLCM are energy, contrast, homogeneity and correlation.

$$\text{Energy} = \sum_{i,j=0}^{N-1} (P_{i,j})^2 \qquad \text{Contrast} = \sum_{i,j=0}^{N-1} P_{i,j}(i-j)^2$$

$$\text{Homogeneity} = \sum_{i,j=0}^{N-1} \frac{P_{i,j}}{1+(i-j)^2} \qquad \text{Correlation} = \sum_{i,j=0}^{N-1} P_{i,j} \frac{(i-\mu_r)(j-\mu_c)}{\sigma_r \sigma_c}$$

Here, N is the number of rows/columns of image matrix Q, P_{ij} is the probability that a pair of points in Q will have values (N_i, N_j), μ_r and μ_c are the mean of rows and columns respectively, σ_r and σ_c are the standard deviation of rows and columns respectively. These features are calculated at 0°, 45°, 90° and 135°, each resulting in an input vector of size 16. The neural network architecture we have used for the simulations is multilayer feedforward network. The learning algorithm use is backpropagation network [7]. Some of the sample images which are used in training the neural networks in this work are shown in Fig. 9 [8, 9].

The first stage of simulation was to train the neural network using texture features extracted from the images using GLCM as illustrated in Fig. 10.

The second stage was testing and classification which involved testing unknown images and classifying them into the correct class type (Figs. 11 and 12).

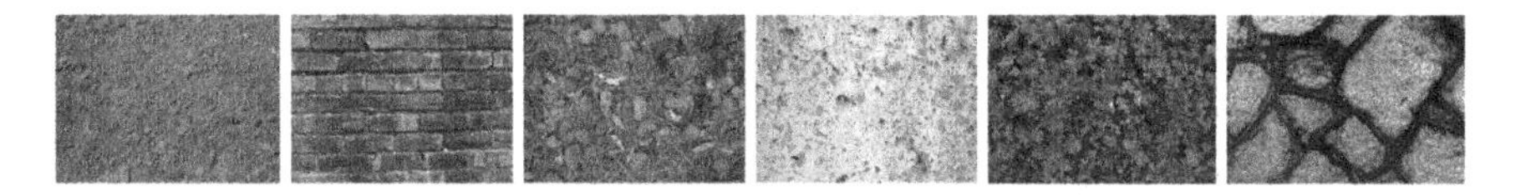

Fig. 9 Sample images used in training the ANN

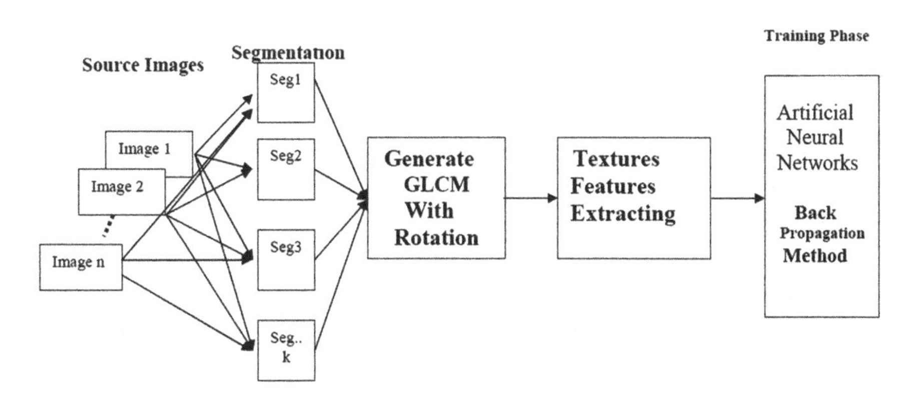

Fig. 10 Experimental set-up—training phase

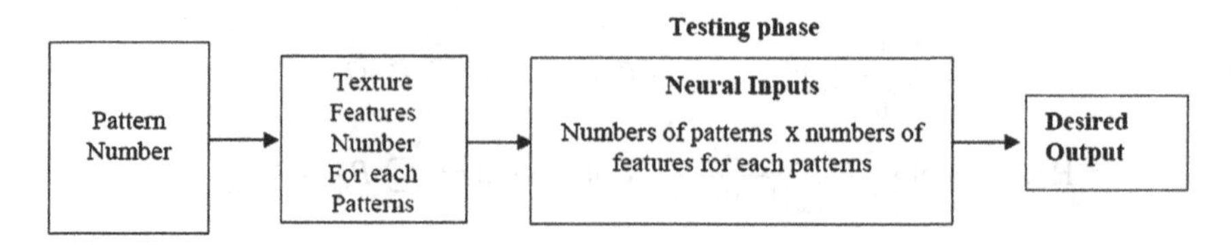

Fig. 11 Experimental set-up—testing phase

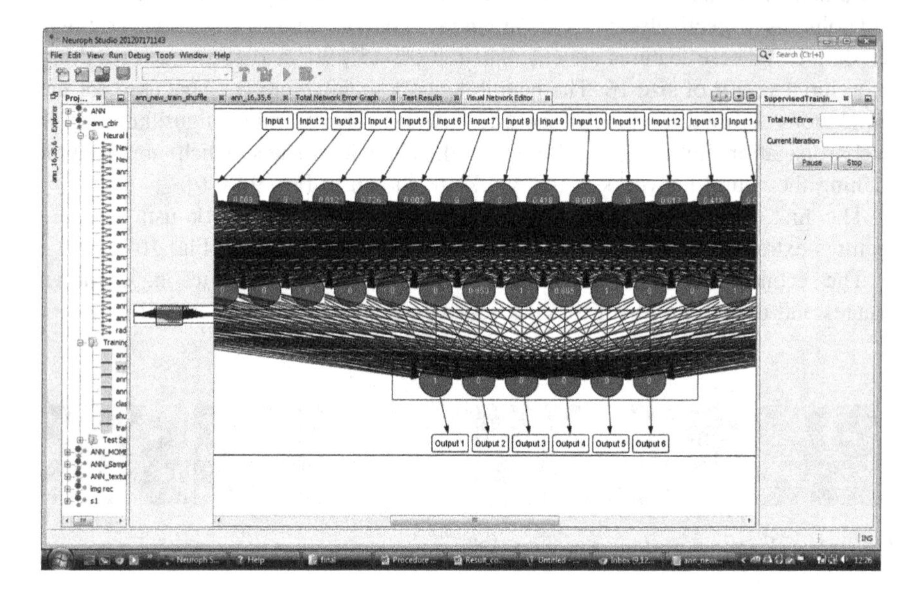

Fig. 12 Snapshot of results generated using Neuroph Studio

5 Results and Discussions

Of the 37 possible configurations tried for Database 1 with 16 inputs and 6 output classes, Tables 1 and 2 show the best results, where the output was obtained within least error limits.

Table 1 Results of database 1 with 6 output classes

S. no.	Architecture	Permissible error (stopping criteria)	η	α	Iterations
1	16, 35, 6	0.05	0.2	0.7	10111
2	16, 35, 6	0.05	0.1	0.7	14199
3	16, 40, 6	0.05	0.2	0.7	11498
4	16, 40, 6	0.05	0.1	0.7	8407
5	16, 45, 6	0.05	0.1	0.7	8834

Table 2 Results of architecture 16, 35, 6

Learning rate	0.2			0.1		
Class	Correctly classified	Incorrectly classified	Class accuracy (%)	Correctly classified	Incorrectly classified	Class accuracy (%)
1	35/60	25/60	58.33	31/60	29/60	51.66
2	50/60	10/60	83.33	46/60	14/60	76.66
3	26/59	33/59	44.06	30/59	29/59	50.84
4	25/60	35/60	41.66	21/60	39/60	35.00
5	40/60	20/60	66.66	46/60	14/60	76.66
6	25/60	35/60	41.66	38/60	22/60	63.33
Total	201/359	158/359		212/359	147/359	
Total accuracy	55.988 %			59.05 %		

From the above results, we can see that the best suited architecture for our experiment was 16, 35, 6, i.e., ANN having 1 hidden layer of 35 hidden neurons and trained using the following parameters: Learning rate, $\eta = 0.1$; Momentum, $\alpha = 0.7$; stopping criteria, error < 0.05 giving an accuracy of 59.05 %.

Since the results obtained for classification for Database 1 were not very efficient, 62 images were removed for the purpose of optimization giving the following results (Table 3).

We clearly observe an improvement in the results by comparing the above results to the previous results obtained without optimization.

Table 3 Optimized results of architecture 16, 35, 6

Learning rate	0.2			0.1		
Class	Correctly classified	Incorrectly classified	Class accuracy (%)	Correctly classified	Incorrectly classified	Class accuracy (%)
1	35/48	13/48	72.91	31/48	17/48	64.58
2	50/60	10/60	83.33	46/60	14/60	76.66
3	26/48	22/48	54.16	30/48	18/48	62.50
4	25/37	12/37	67.56	21/37	16/37	56.75
5	40/55	15/55	72.72	46/55	9/55	83.63
6	25/49	24/49	51.02	38/49	11/49	77.55
Total	201/297	96/297		212/297	85/297	
Total accuracy	67.676 %			71.380 %		

6 Conclusion

This paper illustrates classification of texture images using artificial neural networks. From the above results we can again see that the best suited architecture for our experiment is 16, 35, 6, i.e., ANN having 1 hidden layer of 35 hidden neurons and trained using the following parameters: Learning rate, $\eta = 0.1$; momentum, $\alpha = 0.7$; stopping criteria, error < 0.05 this time giving an accuracy of 71.38 %.

The class that obtains the best accuracy is class 2, i.e., Brick trained at $\eta = 0.2$ with the architecture 16, 40, 6.

It is obvious that the final results of an experiment depend on various factors. The conclusions of these experiments on texture classification are a result in terms of the possible built-in parameters chosen for the texture description algorithm and the various choices made in the experimental set-up. Results of texture classification experiments are suspected to be dependent on individual choices of image acquisition, pre-processing of the acquired data, cleaning of the data, sampling, etc., since no performance characterization has been established in the texture analysis literature. Therefore, all experimental results should be considered to be applicable only to the reported set-up.

References

1. Haykin, S., Networks, N.: A Comprehensive Foundation, 2nd edn., Prentice Hall (1998)
2. Bergman, L.D., Castelli, V., Li, C-S.: Progressive Content-Based Retrieval from Satellite Image Archives, D-Lib Magazine. ISSN: 1082-9873, Oct. 1997
3. Yazdi, M., Gheysari, K.: A new approach for fingerprint classification based on gray-level co-occurrence matrix. World Acad. Sci., Eng. Technol. **47** (2008)
4. Manaa, M.E., Obies, N.T., Al-Assadi, TA.: Department of Computer Science, Babylon University, Object Classification using a neural networks with Graylevel Co-occurrence Matrices (GLCM)
5. Raghu, P.P., Poongodi, R., Yegnanarayana, B.: A combined neural network approach for texture classification. Neural Networks **8**(6), 975–987
6. http://www.scribd.com/doc/267573378/Neural-Network
7. Rady, H.: Reyni's entropy and mean square error for improving the convergence of multilayer backpropagation neural networks: a comparative study
8. http://www-cvr.ai.uiuc.edu/ponce_grp/data/texture_database/samples/
9. http://www.texturewarehouse.com/gallery/

Optimization of Feature Selection in Face Recognition System Using Differential Evolution and Genetic Algorithm

Radhika Maheshwari, Manoj Kumar and Sushil Kumar

Abstract This paper compares the results of the optimization techniques for feature selection of face recognition system in which face as a biometric template gives a large domain of features for optimizing feature selection. We attempt to minimize the number of features necessary for recognition while increasing the recognition accuracy. It presents the application of differential evolution and genetic algorithm for feature subset selection. We are using local directional pattern (LDP), an extended approach of local binary patterns (LBP), to extract features. Then, the results of DE and GA are compared with the help of an extension of support vector machine (SVM) which works for multiple classes. It is used for classification. The work is performed on 10 images of ORL database resulting in better performance of differential evolution.

Keywords Face recognition system · Differential evolution · Local directional pattern (LDP) · Multi-support vector machine (m-SVM) · Genetic algorithm · Feature selection

1 Introduction

Biometric Recognition is the science of studying a person's unique characteristics based on their anatomical and behavioral traits to establish his/her identity [1]. Their application areas range from international border crossings to securing

Radhika Maheshwari (✉) · Manoj Kumar · Sushil Kumar
Department of Computer Science and Engineering, Amity School of Engineering and Technology, Amity University, Noida 201313, Uttar Pradesh, India
e-mail: radhika240990@gmail.com

Manoj Kumar
e-mail: mkumar7@amity.edu

Sushil Kumar
e-mail: skumar21@amity.edu

M. Pant et al. (eds.), *Proceedings of Fifth International Conference on Soft Computing for Problem Solving*, Advances in Intelligent Systems and Computing 437, DOI 10.1007/978-981-10-0451-3_34

information in databases. Face as a biometric though pose a problem of distinctiveness, permanence, or circumvention but its collectability and acceptability makes it the most appropriate for applications that do not require direct collaboration from the observed people. This nonintrusive property makes it reliable to be used in security application [2]. Since face as a biometric is nontransferable and easily allocable, it serves as a convenient solution to both the client and the user.

The study of facial features started with a semi-automated system locating facial features on still images, in 1960s. To automate the recognition in 1970s, Goldstein and Harmon automated recognition using 21 specific subjective markers. Kirby and Sirovich made a face recognition system with the approach of standard linear algebra technique in 1988 [2]. And with this the increasing growth of study in this field is factual from the fact that relating this, we have 2,266 journals and magazines and 18,656 conference publications in IEEE and 131 journals in Elsevier till 2013.

With increasing need of face recognition technology and seeing its comfort with the user, it became imperative to enhance the technique. Face as a biometric trait is a good optimization problem when it comes to selecting the significant facial features for face recognition or detection.

2 Related Work

Face recognition problem was tried to be countered with Genetic Algorithm and compared against particle swarm optimization by Ramadan and Adbel–Kader on ORL database, where both the techniques gave results with same accuracy but PSO using lesser number of features [1, 2].

Again memetic algorithm was compared with genetic algorithm on ORL and YaleB database for face recognition, where we found that memetic algorithm is superior to genetic algorithm. This work was presented by Kumar in [3].

Genetic and Evolutionary Computation (GEC) research community under its subarea Genetic and Evolutionary Biometrics (GEB) evolved Genetic and Evolutionary Methods for the reduction of facial features. There evolved technologies like Genetic Evolutionary Fusion (GEF) and GEFeWS (Genetic Evolutionary Feature Weighting and selection) outperformed on FRGC dataset of 309 subjects with 100 and 99.6 % accuracy [4, 5]. But this work is centric to FRGC dataset only.

Our work is on ORL database and applied differential evolution and genetic algorithm approach, for optimizing feature selection. We used LDP, i.e., local direction pattern which is an extension of LBP, Local Binary Patterns for feature extraction. The classification is being done by multi-support vector machine (m-SVM). The recent work on feature optimization of face recognition is done in the following papers [6–8].

3 Background

3.1 Face Recognition System

Any face recognition system works with two modules [2], Enrolment module: In this module, the database of every individual is fed to make a template out of each which could be further used for verification or validation. And Verification/Validation module: The test image goes through all the steps, and a template is formed which is further matched with the system database in this module (Fig. 1).

3.2 Local Directional Pattern (LDP)

Local dimensionality pattern (LDP) [9] is an extension of local binary patterns (LBP) technique, which is considered best for facial feature extraction application [10, 11]. For any image analysis task, it proves to be highly discriminative and invariant to monotonic and non-monotonic gray-level changes. There is computational efficiency which is an added advantage. It is insensitive to illumination invariance and noise. This LDP contains more information in a stable manner in terms of curves, junctions, and corners as well as other local primitives [9]. It is a robust method which characterize the spatial structure of a local texture in an image.

Any facial image has several regions which are important. Face is treated as a composition of mini patterns which are described with the help of LDP operator. To develop a face descriptor, various features extracted and concatenated by LDP are put into a form of a vector.

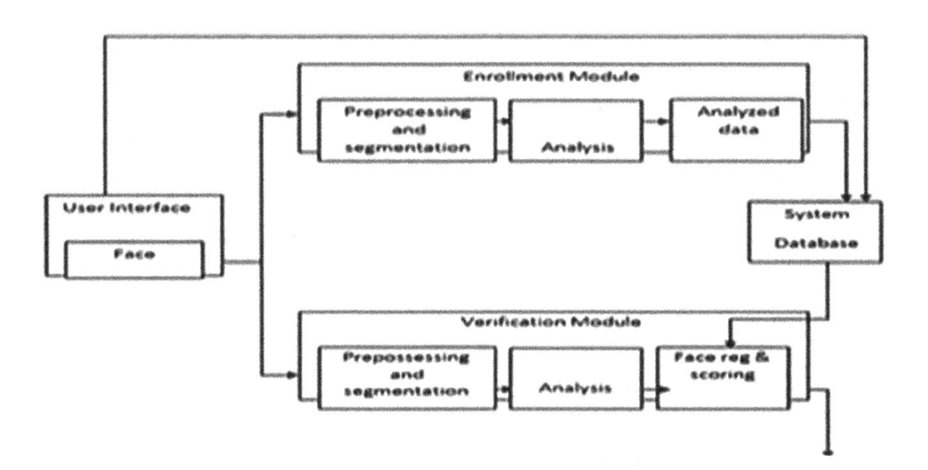

Fig. 1 Face recognition system enrollment and verification module

$$\begin{bmatrix}-3 & -3 & 5\\-3 & 0 & 5\\-3 & -3 & 5\end{bmatrix}\quad\begin{bmatrix}-3 & 5 & 5\\-3 & 0 & 5\\-3 & -3 & -3\end{bmatrix}\quad\begin{bmatrix}5 & 5 & 5\\-3 & 0 & -3\\-3 & -3 & -3\end{bmatrix}\quad\begin{bmatrix}5 & 5 & -3\\5 & 0 & -3\\-3 & -3 & -3\end{bmatrix}$$

East M_0 North East M_1 North M_2 North West M_3

$$\begin{bmatrix}5 & -3 & -3\\5 & 0 & -3\\5 & -3 & -3\end{bmatrix}\quad\begin{bmatrix}-3 & -3 & -3\\5 & 0 & -3\\5 & 5 & -3\end{bmatrix}\quad\begin{bmatrix}-3 & -3 & -3\\-3 & 0 & -3\\5 & 5 & 5\end{bmatrix}\quad\begin{bmatrix}-3 & -3 & -3\\-3 & 0 & 5\\-3 & 5 & 5\end{bmatrix}$$

West M_4 South West M_5 South M_6 South East M_7

Fig. 2 Krissh masks used for LDP

$$LDP_k = \sum_{i=0}^{7} b_i(m_i - m_k) \times 2^i$$

$$b_i(a) = \begin{cases}1 & a \geq 0\\0 & a < 0\end{cases}$$

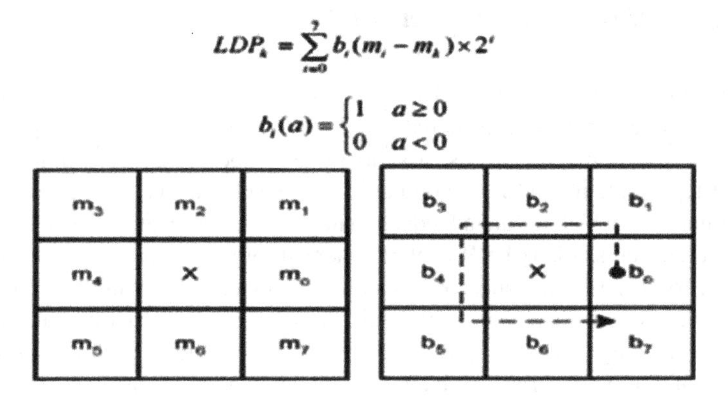

Fig. 3 Evaluation of LDP

Local directional pattern (LDP) computes the edge response values at each position of the pixel from all the eight directions using the Krissh mask shown in Fig. 2.

These masks work with every pixel of the image. And for every X pixel m_1, m_2, m_3, m_4, m_5, m_6, m_7, and m_8 are computed. These values are compared using the equation given below (Fig. 3).

A gray scale texture pattern is generated which is termed as LDP for that pixel and hence a code is generated from the relative strength magnitude which is a binary number—$b_0\ b_1\ b_2\ b_3\ b_4\ b_5\ b_6\ b_7$.

3.3 *Differential Evolution*

When Rainer Storn approached Ken Price with Chebychev polynomial fitting problem, it was the first time differential evolution (DE) approach [12] came into existence. Differential evolution is a simple yet powerful tool for population-based

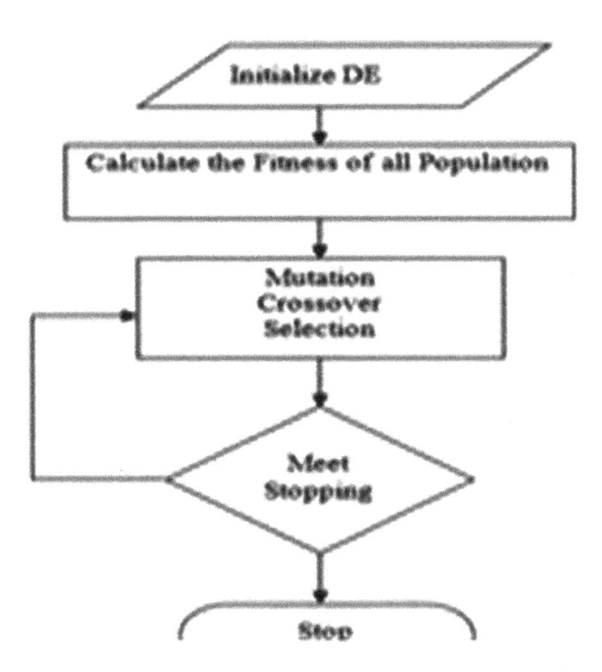

Fig. 4 Flowchart for differential evolution

problems. It is stochastic in nature and is a function minimizer. It is best for genetic type of algorithm. The main concept behind DE is "vector differences for perturbing the vector population." The main idea behind DE is trial parameter vector generalization. This parallel direct search method works for problems having "objective function" which satisfy the following constraints of—non-differentiability, non-continuity, nonlinear, noisy, flat, and multidimensional, and problem whose solution may have many local minima or constraints (Fig. 4).

3.4 Genetic Algorithm

Genetic algorithms (GAs) [13] are randomized searching methods. It imitates the biological model of evolution and natural human genetic system. It is one of the robust and most efficient approach for solving any optimization problem.

The algorithm initially considers a set of random solutions as population. Each member of the population is called as the chromosome and is a particular solution of the problem. These chromosomes are assigned a fitness value on the basis of the goodness of the solution to the problem. After this, there is natural selection on the basis of "survival of the fittest chromosome" theory. And the next generation is prepared with this optimized breed of chromosome. The evolution process includes crossover and mutation, which works iteratively to obtain an optimal or near optimal solution conditioned with the given number of generations (Fig. 5).

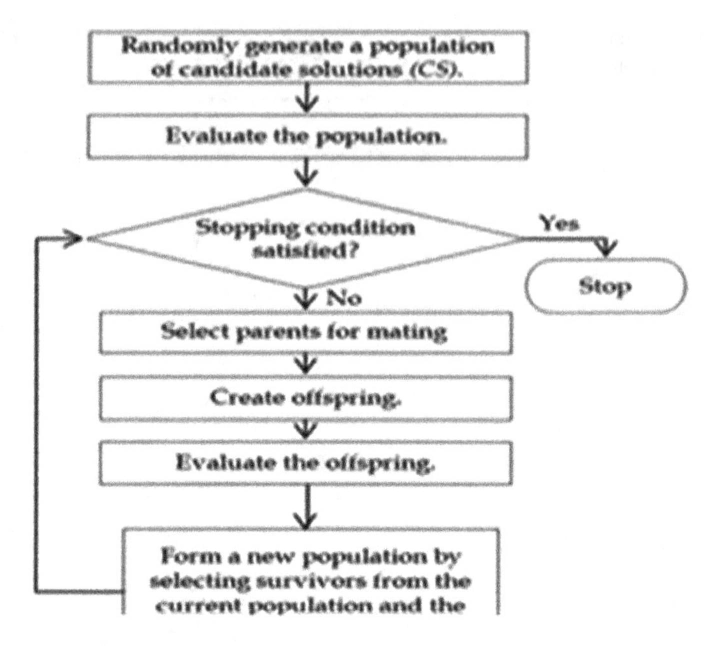

Fig. 5 Flowchart for genetic algorithm

3.5 Support Vector Machine (SVM)

Support vector machine [14] classification works for two classes whereas the face recognition is a K class problem, where k is the number of individuals. SVM has two versions namely, binary classification SVM and multi-class classification SVM.

SVM classifies the objects in classes on the basis of same patterns. It forms a decision surface where we allocate all our training samples in a way that they cover points with maximum distance too. Suppose, the training set have points x_i where i ranges from 1 to N and each x_i belongs either of the class $y_i = \{-1, +1\}$. Suppose it is a linearly separable data, we need to separate this data on the plot in a manner that the distance of support vector is maximized by the hyperplane classification. This optimal separating hyperplane is of the form:

$$f(x) = \sum_{i=1}^{\ell} \alpha_i y_i x_i \cdot x + b. \tag{1}$$

The classification of the third point will then use the following equation for classification:

$$d(x) = \frac{\sum_{i=1}^{\ell} \alpha_i y_i x_i \cdot x + b}{\left\| \sum_{i=1}^{\ell} \alpha_i y_i x_i \right\|} \quad (2)$$

And so multi class classification is done.

For k class problem, one versus all approaches are applied and k SVM are trained to take the result.

4 Proposed Work

The database that we are using is ORL database [15]. There are 40 subjects with 10 images of each. The proposed model works in the following manner:

Enrolment Module

The enrolment module trains the face recognition system in the manner shown in the figure below. For each subject, its each image is first preprocessed and then using LDP we extract its features and then optimizes the result using genetic algorithm and differential evolution. These results are stored and this process runs for all the 10 images of each subject. Then using the support vector machine and these optimized results, the system is trained for every subject (Fig. 6).

Verification Module

The verification module tests the face recognition system in the manner shown in Fig. 7. For an image to be tested, it first preprocesses it and then using LDP extract

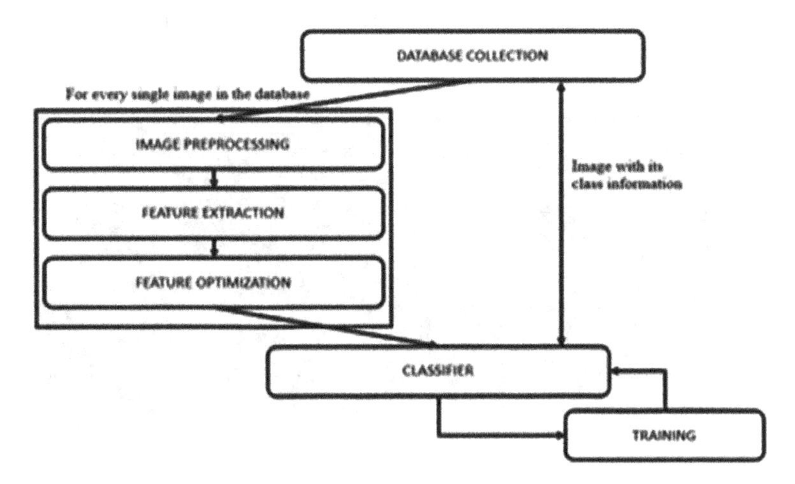

Fig. 6 Enrollment module

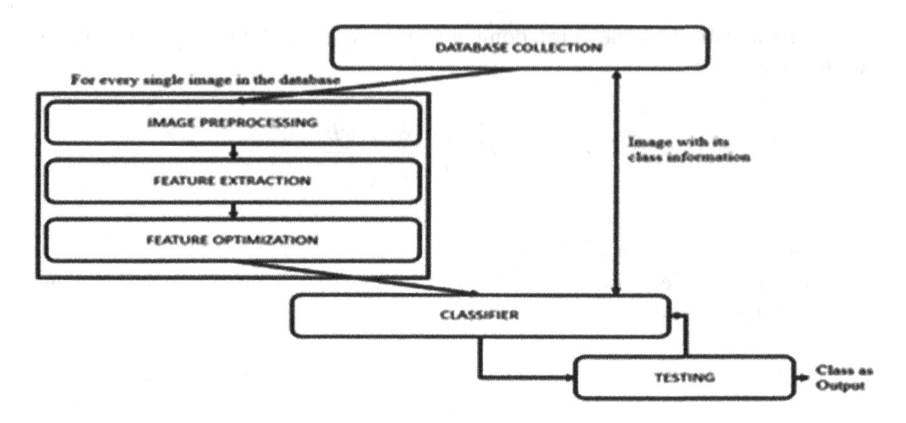

Fig. 7 Verification module

its features and then optimizes the result using genetic algorithm and differential evolution. These results are then fed into the support vector machine which classifies them and gives us the output that any subject similar to that image exists or not in the database.

5 Experiments and Results

The database that we are using is ORL database [15] where we are preprocessing the image by removing noise using filter and then applying histogram equalization technique on it. The experiment is conducted on initial 10 images of the ORL database and the result is obtained in the form of RoC curve, confusion matrix, and recognition rate. The results are obtained first using LDP only. Then, we obtained our result using genetic algorithm and differential evolution technique (Figs. 8, 9, 10, 11, 12, 13 and Table 1).

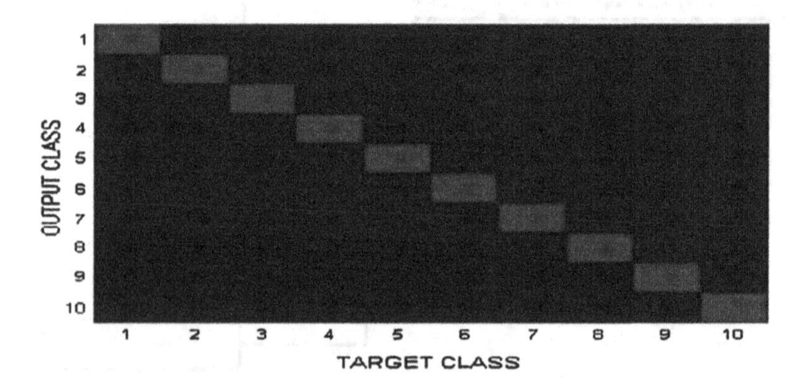

Fig. 8 Confusion matrix for face recognition system using differential evolution

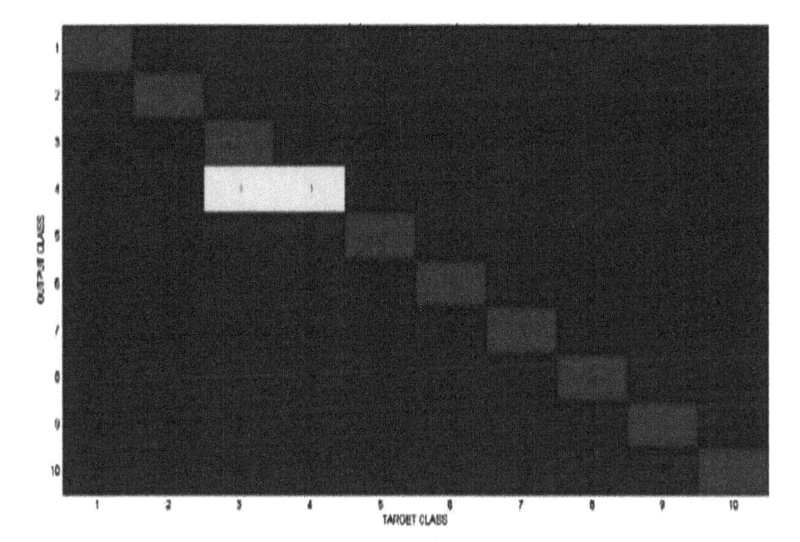

Fig. 9 Confusion matrix for face recognition system using genetic algorithm

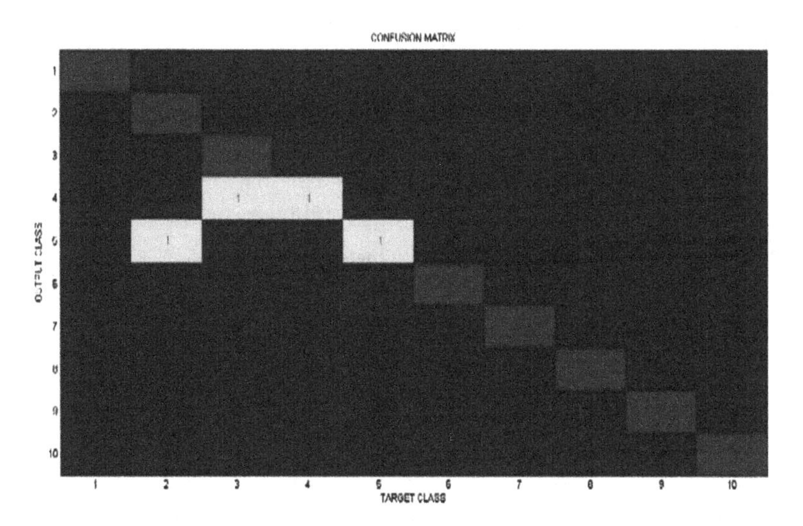

Fig. 10 Confusion matrix for face recognition system using LDP only

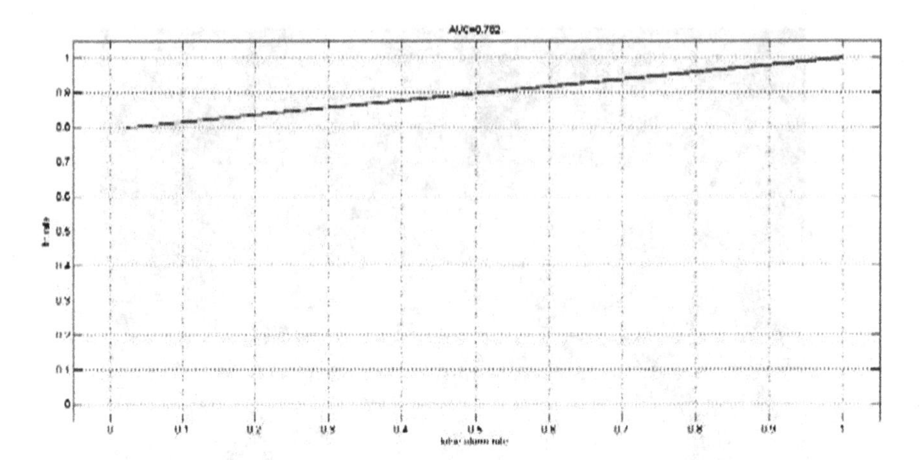

Fig. 11 RoC curve for face recognition system using LDP only

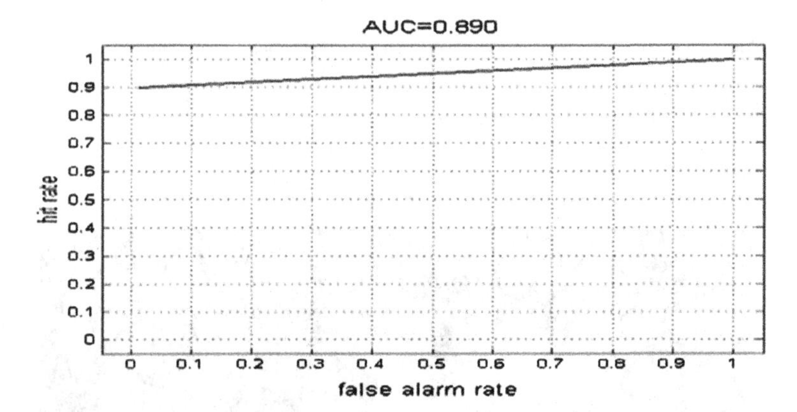

Fig. 12 RoC curve for face recognition system using genetic algorithm

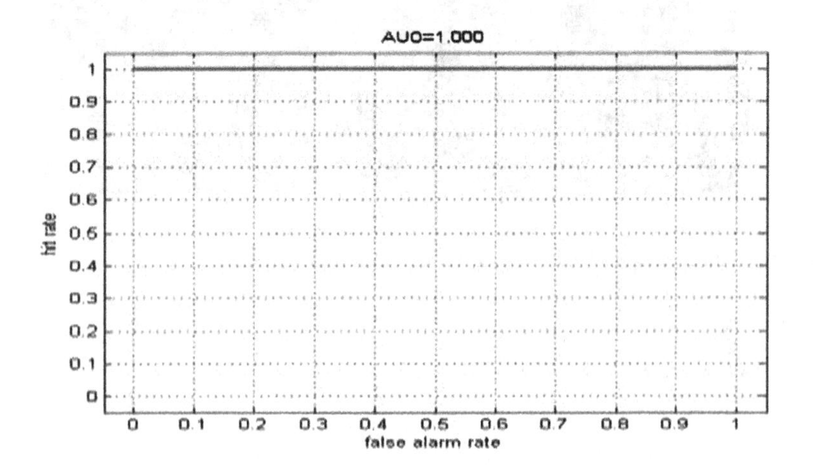

Fig. 13 RoC curve for face recognition system using differential evolution

Table 1 Recognition rate

Technique	No. of features	Recognition rate
LDP	2048	90
Genetic algorithm with LDP	970	95
Differential evolution with LDP	950	100

6 Conclusion

This paper presents the work of differential evolution on face recognition system which outperforms the genetic algorithm approach. Over the ten subjects of the database, it works great but it further requires work on large database with increased number of subjects.

References

1. Jain, A.K., Russ, A., Probhakar, S.: An introduction to biometric recognition. IEEE Trans. Circ. Syst. Video Technol. **14**, 4–20
2. Jain, A.K., Bolle, R., Pankanti, S.: Biometrics: Personal Identification in Networked Society, Kluwer Academic Publishers (1999)
3. Zhao, W., et al.: Face recognition: a literature survey. ACM Comput. Surv. **35**, 399–458 (2003)
4. Ramadan, R.M., Abdel-Kader, R.F.: Face recognition using particle swarm optimization-based selected features. Int. J. Signal Process., Image Process. Pattern Recogn. **2**(2) (2009)
5. Kumar, D., Kumar, S., Rai, C.S.: Feature selection for face recognition: a memetic algorithmic approach. J. Zhejanga Univ. Sci. A **10**(8), 1140–1152 (2009)
6. Jin, Y., Lu, J., Ruan, Q.: Coupled discriminative feature learning for heterogeneous face recognition. IEEE Trans. Inf. Forensics Secur. **10,** 640–652 (2015)
7. Gaynor, P., Coore, D.: Distributed face recognition using collaborative judgement aggregation in a swarm of tiny wireless sensor nodes. In: IEEE conference publication, 1–6 (2015)
8. Radhika, M., Apoorva, G., Nidhi, C.: Secure authentication using biometric templates in kerberos. In: 2nd International Conference on Sustainable Global Development (INDIAcom), IEEE, (2015)
9. Alford, A., Hansen, C., Dozier, G., Bryant, K., Kelly, J., Adams, J., Abegaz, T., Ricanek, K., Woodard, D.L.: GEC-based multi-biometric fusion. In: Proceedings of IEEE Congress on Evolutionary Computation (CEC), New Orleans, LA, (2011)
10. Alford, A., Shelton, J., Dozier, G., Bryant, K., Kelly, J., Adams, J., Abegaz, T., Ricanek, K., Woodard, D.L.: A comparision of GEC-based feature selection and weighting for multimodal biometric recognition. In: Proceedings of IEEE, (2011)
11. Jabid, T., Kabir, M.H., Chae, O.S.: Local directional pattern (LDP) for face recognition. In: IEEE International Conference on Consumer Electronics, Jan 2010
12. Samanta, S.: Genetic algorithm: an approach for optimization (using MATLAB). Int. J. Latest Trends Eng. Technol. (IJLTET) **3** (2014). ISSN: 2278-621X

13. Storn, R., Price, K.: Differential evolution-a simple and efficient heuristic approach for global optimization over continuous spaces. J. Global Optim. **11**, 341–359 (1997)
14. Jabid, T., Kabir, M.H., Chae, O.S.: Facial expression recognition using local directional pattern (LDP). In: IEEE International Conference on Image Processing, Sept 2010
15. Samaria, F., Harter, A.: Parameterization of a stochastic model for human face identification. In: Proceedings of 2nd IEEE Workshop on Applications of Computer Vision, Sarasota FL, Dec 1994

Human Opinion Dynamics for Software Cost Estimation

Ruchi Puri and Iqbaldeep Kaur

Abstract Human opinion dynamics is a novel approach to solve complex optimization problems. In this paper we propose and implement human opinion dynamics for tuning the parameters of the COCOMO model for software cost estimation. The input is coding size or lines of code and the output is effort in person-months. Mean absolute relative error and prediction are the two objectives considered for fine-tuning of parameters. The dataset considered is COCOMO. The current paper demonstrates that use of human opinion dynamics illustrates promising results. It has been observed that when compared with standard COCOMO it gives better results.

Keywords Human opinion dynamics · COCOMO · MARE · Social influence · Update rule

1 Introduction

When the information available is not complete or imperfect with less computation capacity and there is a need to find out the best solution from a large set of sample solutions, then metaheuristic or partial search algorithm is used to provide a solution to an optimization problem in computer science or mathematical optimization.

Ruchi Puri (✉) · Iqbaldeep Kaur
Department of Computer Science Engineering, Chandigarh University, Chandigarh, India
e-mail: Ruchi2192jk@gmail.com

Iqbaldeep Kaur
e-mail: iqbaldeepkaur.cu@gmail.com

M. Pant et al. (eds.), *Proceedings of Fifth International Conference on Soft Computing for Problem Solving*, Advances in Intelligent Systems and Computing 437, DOI 10.1007/978-981-10-0451-3_35

There are various metaheuristics available like particle swarm optimization [1], genetic algorithm, ant colony optimization [2], bacteria foraging optimization, BAT algorithm, memetic algorithm, firefly algorithm, etc. Extensive research has been done on these over a large number of applications. There is no exhaustive research using human opinion dynamics. Human opinion dynamics finds its applications in the areas of social physics. However, recently a researcher has thrown light on the use of human opinion dynamics for solving complex mathematical optimization problems applied on some benchmark mathematical functions and compared the results with PSO [3]. As humans are the highest forms of creation, so the algorithm inspired by human creative problem solving process can be useful and will provide better results. It is used in dynamic social impact theoryfor optimization of an impedance-tongue and the results are compared with genetic algorithm and PSO [3–7].

2 Human Opinion Dynamics

It is a metaheuristic technique to solve complex optimization problems based on human creative problem solving process. Understanding the concept of collective decision making, the study of opinion dynamics and opinion formation is important. It has been one of the most significant areas in social physics. Human interactions give rise to different kinds of opinions in a society [3]. In social networks, the formation of different kinds of opinions is an evolutionary process. There are several models describing human interaction networks such as cultural dynamics, opinion dynamics, crowd behavior, human dynamics, etc., utilized for search strategies and complex mathematical optimization problems. The process of collective intelligence from the tendencies of social influence has effects of individualization escapade for developing search strategies [3]. The algorithm is formed based on the opinion formation structure of individuals. The algorithm is governed by four basic essential elements: social structure, opinion space, social influence, updating rule.

2.1 Social Structure

The interaction between individuals, group of individuals, frequency of interaction, and the way they interact comes under social structure. There are a number of models like cellular automata, small world, random graphs, etc., that have been proposed and simulated in social physics [1].

2.2 Social Influence

It is the influence of individuals on each other and they act according to others' actions or suggestions.

Equation (1) describes social influence $u_{ij}(t)$ of individual j on individual i is given as

$$u_{ij}(t) = \frac{\mathrm{SR}_j(t)}{d_{ij}(t)} \tag{1}$$

where $d_{ij}(t)$ is the Euclidean distance between individuals. Social ranking (SR) is based on the fitness value of individuals.

2.3 Updating Rule

This rule is used to update the position of individuals in the search space. As it is dynamic in nature so change in position according to the best fitness value needs to be updated. In regard to optimization problems it determines the new updated position of individuals. Equation (2) demonstrates the formula for updating rule:

$$\Delta O_i = \frac{\sum_{j=1}^{N} \left(O_j(t) - O_i(t)u_{ij}(t)\right)}{\sum_{j=1}^{N} u_{ij}(t)} + \in_i (t) \tag{2}$$

whereas $O_j(t)$ is the opinion of a number of individuals, $u_{ij}(t)$ represents social influence, $\in_i(t)$ is a normally distributed random noise with mean zero and N is the number of neighbors.

2.4 Opinion Space

There are two types of opinions of individuals, continuous or discrete. Continuous is that which takes real values. Discrete takes values in the given range [0, 1] or [1, −1].

3 Human Opinion Dynamics in Software Cost Estimation

Cost estimation is an important activity and can be done throughout the entire life cycle of the software product to be developed. It is the process of calculation of effort used for the development of project. Time and budget are the two important factors in software project management. The main focus is on time and budget in software project development [4]. There are various models used for effort calculation in software cost estimation. One of the most widely used algorithmic model is COCOMO. The parameters of COCOMO are tuned with the help of metaheuristic techniques. We use human opinion dynamics to optimize the parameters of COCOMO.

3.1 COCOMO

COCOMO model was developed by Boehm and has been widely used for calculation of effort. Effort calculated by COCOMO model is measured in terms of size and constant value parameters *a*, *b*, *c*. We use the Intermediate COCOMO II model in which effort can be calculated using (3) which gives the formula for effort.

$$\text{Effort} = a * (\text{size})^{b} * \text{EAF} + c \tag{3}$$

where size is the size of project measured in LOC (lines of code) or KLOC. EAF are effort multipliers. The value of parameters $a = 3$ $b = 1.2$. These values are fixed for COCOMO model but the parameters vary from organization to organization depending on various factors like environmental factors. So, there is a need to tune the value of parameters to obtain better results in terms of accuracy and less error.

3.2 Fitness Function

Fitness function is the function used to evaluate which opinion is performing the best and gives the best results. Each objective has some weight which is used to combine the two objectives into a single objective. The weights assigned must be equal to one.

$$W1 + W2 = 1 \tag{4}$$

Hence, we use the fitness function where we need to minimize the error and maximize the prediction. Most real-world problems involve optimization of two or more objectives [1]. A multiobjective optimization function involves minimization of one and maximization of the other. The fitness function used in (4) is based on two objectives MARE, i.e., mean absolute relative error and prediction. Equation (5) defines the fitness function we used in our approach.

$$\text{function} = w1(\text{MARE}) + w2(1 - \text{pred.}) \tag{5}$$

where MARE is mean absolute relative error and prediction (n) is the project with n % error. Equation (6) gives the formula to calculate mean absolute error.

$$\%\text{MARE} = \sum \left[\frac{\text{abs.}(\text{mes.effort} - \text{est.effort})}{(\text{mes.effort})} \right] / n \tag{6}$$

4 Proposed Methodology

Step 1 START
Step 2 Initialize the opinions from 1 to 30
Step 3 Assign any random value to a, b and c
Step 4 Apply constraints
Step 5 For every value of opinion i
Step 6 Apply COCOMO formula and calculate effort with the randomly assigned values of parameters a, b, c
Step 7 For these values of a, b, c calculate cost, i.e., equal to fitness function
Step 8 Find social rank for the present value or opinion
Step 9 If social rank is greater than 1 then calculate distance, i.e., Euclidean distance of two individuals
Step 10 Find social influence of the opinion which is based upon social rank and Euclidean distance
Step 11 If best fit number is found, i.e., the opinion which fits aptly in the fitness function then go to Step 12 otherwise go to Step 7
Step 12 Update the value of opinion

And if minimum error < E and maximum iteration = k then stop, otherwise go to Step 5 and perform the optimization process again.

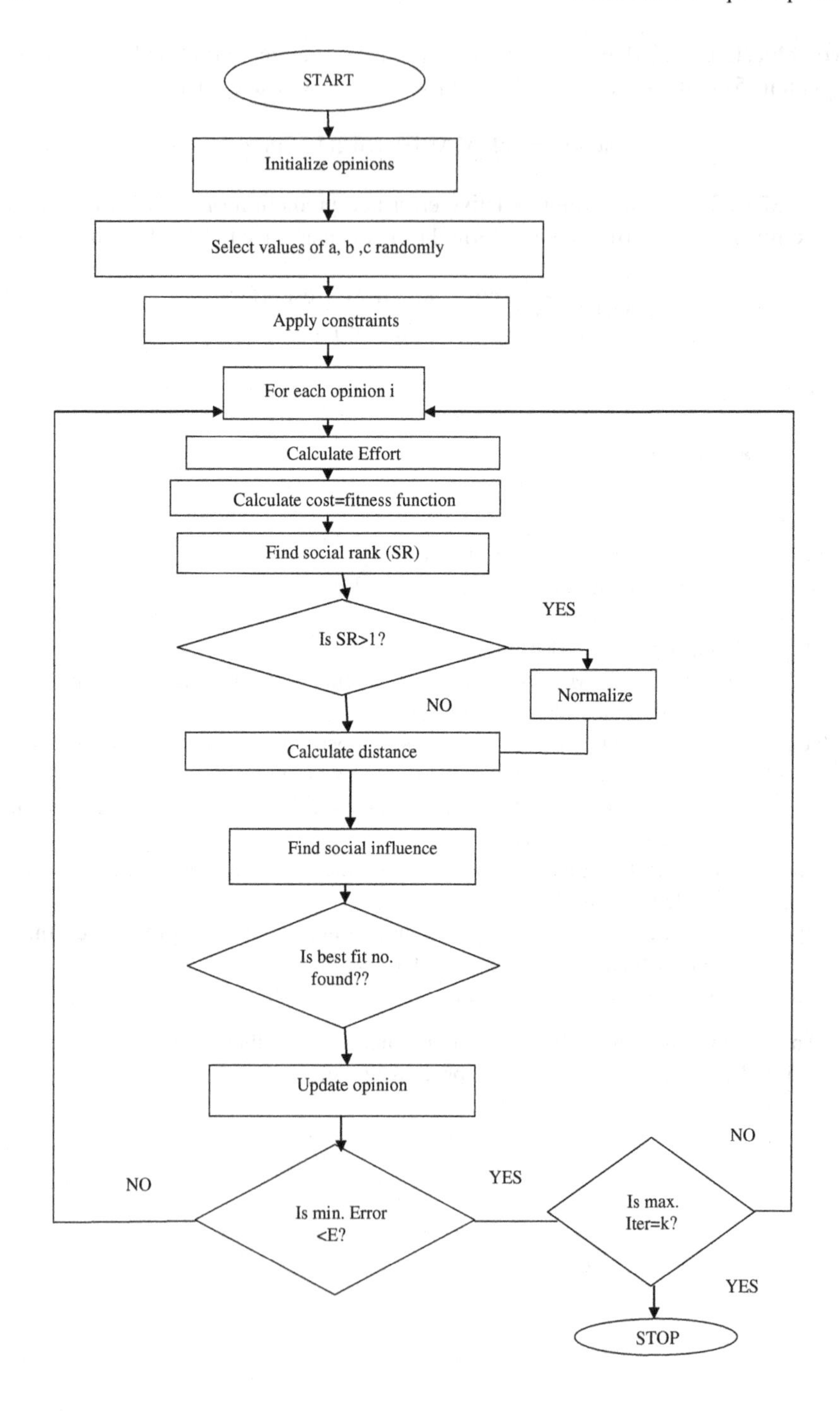
START
Initialize opinions
Select values of a, b ,c randomly
Apply constraints
For each opinion i
Calculate Effort
Calculate cost=fitness function
Find social rank (SR)
Is SR>1?
YES
Normalize
NO
Calculate distance
Find social influence
Is best fit no. found??
Update opinion
NO
Is min. Error <E?
YES
Is max. Iter=k?
NO
YES
STOP

5 Performance Results and Implementation

This section describes the experimentation part. For testing the effectiveness of proposed models, we tested it on COCOMO dataset. Two datasets of 20 projects and 21 projects are considered. Tuned value of parameters are obtained by implementing the above methodology.

Experiment 1

Total 20 projects are considered from the COCOMO dataset. Total number of iterations performed = 100, number of opinions = 30. The optimized values of a, b, c obtained are $a = 4.2$, $b = 1$, $c = 1.3$.

Table 1 shows the values that are already available in dataset from COCOMO dataset (Figs. 1, 2, 3 and 4).

Experiment 2

A total of 21 projects are considered from COCOMO dataset for testing the model. The tuned value of parameters obtained $a = 3.9$, $b = 1.1$, $c = 5.8$.

Table 2 represents the results of comparison of effort calculated by COCOMO, measured and human opinion dynamics (Figs. 5, 6 and 7).

6 Conclusion and Future Scope

The software cost estimation problem was dealt with in this paper, which is an important problem in the SDLC cycle as it influences the decision-making process. A novel metaheuristic approach known as human opinion dynamics-based

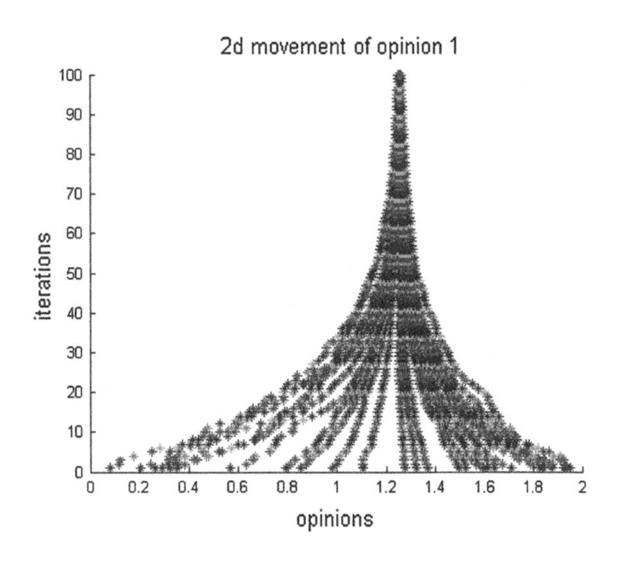

Fig. 1 Convergence of opinions for opinion 1

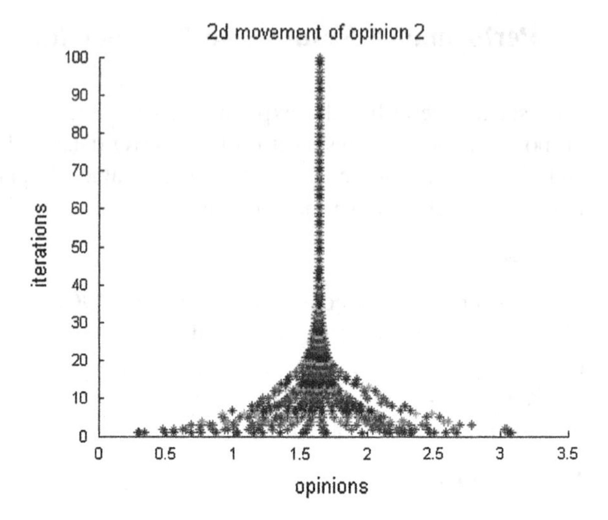

Fig. 2 Convergence of opinion 2

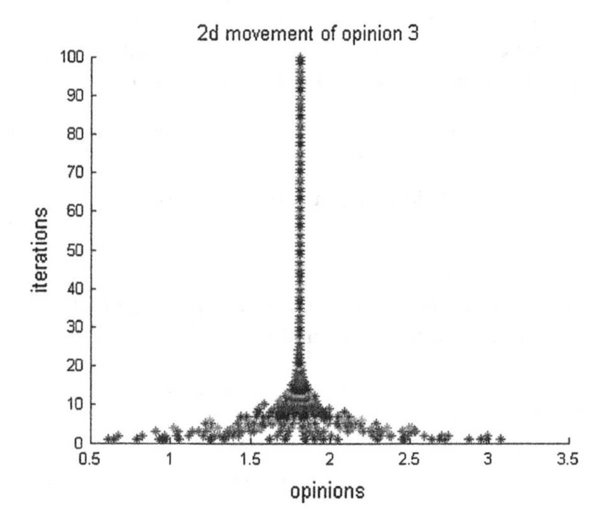

Fig. 3 Convergence of opinion 3

optimization to estimate the software cost using Intermediate COCOMO model. The software costs are predicted and the results are found to be better than that of the normal COCOMO model. The results when compared among HOD and COCOMO infer that HOD's performance is better than COCOMO's results, both in terms of convergence and accuracy.

In the future, other metaheuristic algorithms can be applied and the results can be compared with our proposed methodology. Advanced COCOMO models can also

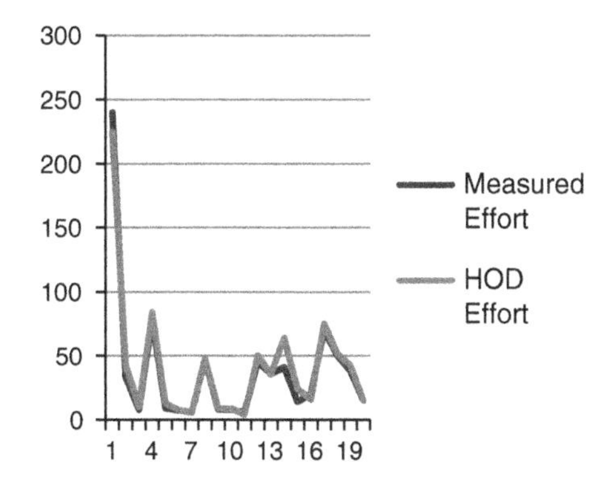

Fig. 4 Comparison of actual and HOD effort

Table 1 Experimental results for the comparison of effort for dataset 1

P. no.	KLOC	EAF	Measured effort	COCOMO effort	HOD effort	COCOMO error	HOD error
1	46	1.17	240	347.2294197	224.744	0.446789249	0.063566667
2	16	0.66	33	55.15808369	43.052	0.671457082	0.304606061
4	6.9	0.4	8	12.18428797	10.292	0.523035997	0.2865
10	24	0.85	79	115.5563114	84.38	0.462738119	0.068101266
14	1.9	1.78	9	11.53574316	12.9044	0.28174924	0.433822222
21	2.14	1	7.3	7.475113353	7.688	0.023988131	0.053150685
22	1.98	0.91	5.9	6.196705766	6.26756	0.050289113	0.062298305
28	34	0.34	47	70.20610386	47.252	0.493746891	0.005361702
30	6.2	0.39	8	10.44854672	8.8556	0.306068341	0.10695
31	2.5	0.96	8	8.648095925	8.78	0.081011991	0.0975
32	5.3	0.25	6	5.548713573	4.265	0.075214405	0.289166667
33	19.5	0.63	45	66.75800317	50.297	0.483511182	0.117711111
38	23	0.38	36	49.08832967	35.408	0.363564713	0.016444444
42	8.2	1.9	41	71.19532527	64.136	0.736471348	0.564292683
43	5.3	1.15	14	25.52408244	24.299	0.823148745	0.735642857
44	4.4	0.93	20	16.51001299	15.8864	0.17449935	0.20568
49	21	0.87	70	100.7635967	75.434	0.439479953	0.077628571
51	28	0.45	50	73.60772685	51.62	0.472154537	0.0324
52	9.1	1.15	38	48.82798088	42.653	0.284946865	0.122447368
53	10	0.39	15	18.54325035	15.08	0.23621669	0.005333333

be utilized and the approach can be utilized to develop an online system which would automatically predict the software cost serving as an automated feedback system for the business analyst.

Table 2 Experimental results of comparison of effort for dataset 2

P. no.	KLOC	EAF	Measured effort	COCOMO effort	HOD effort	COCOMO error	HOD error
1	46	1.17	240	347.2294197	302.010963	0.446789249	0.258379013
2	16	0.66	33	55.15808369	48.5426138	0.671457082	0.470988297
3	4	2.22	43	35.15169074	33.98172143	0.18251882	0.209727409
4	6.9	0.4	8	12.18428797	7.257456815	0.523035997	0.092817898
5	22	7.62	107	933.2187223	884.803766	7.721670302	7.269194075
6	30	2.39	423	424.6827895	387.1125386	0.003978226	0.084840334
7	18	2.38	321	229.1000679	217.2700783	0.286292623	0.323146174
8	20	2.38	218	259.9765682	244.6808678	0.192553065	0.122389302
9	37	1.12	201	255.9654486	226.1020171	0.273459944	0.124885657
10	24	0.85	79	115.5563114	103.5240969	0.462738119	0.310431607
11	3	5.86	73	65.69984976	70.72363706	0.100002058	0.031183054
12	3.9	3.63	61	55.75777155	57.46190997	0.085938171	0.058001476
13	3.7	2.81	40	40.5200193	40.4160499	0.013000482	0.010401247
14	1.9	1.78	9	11.53574316	8.264152613	0.28174924	0.081760821
15	75	0.89	539	474.8809821	395.0870279	0.118959217	0.266999948
16	90	0.7	453	464.8472928	379.52697	0.026152964	0.162192119
17	38	1.95	523	460.1465625	409.9776829	0.120178657	0.216103857
18	48	1.16	387	362.3003293	314.0069941	0.063823438	0.188612418
19	9.4	2.04	88	90.05438122	87.769738	0.023345241	0.002616614
20	13	2.81	98	183.0457279	178.3232425	0.86781355	0.819624924
21	2.14	1	7.3	7.475113353	3.205747324	0.023988131	0.560856531

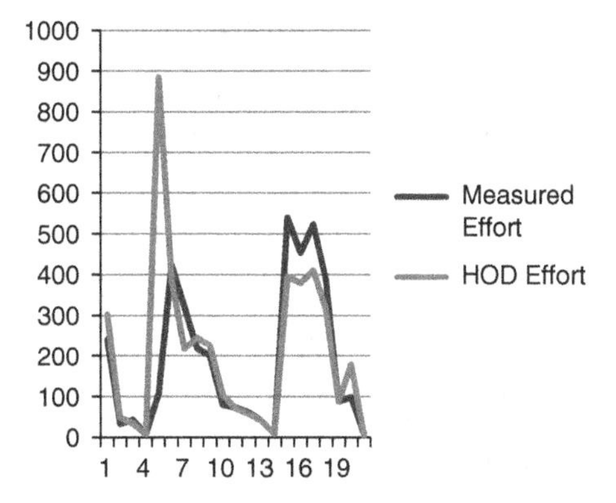

Fig. 5 Comparison of effort dataset 2

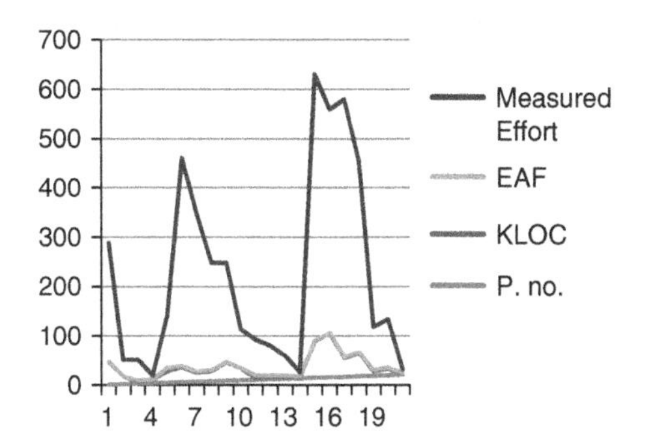

Fig. 6 Values in dataset 2

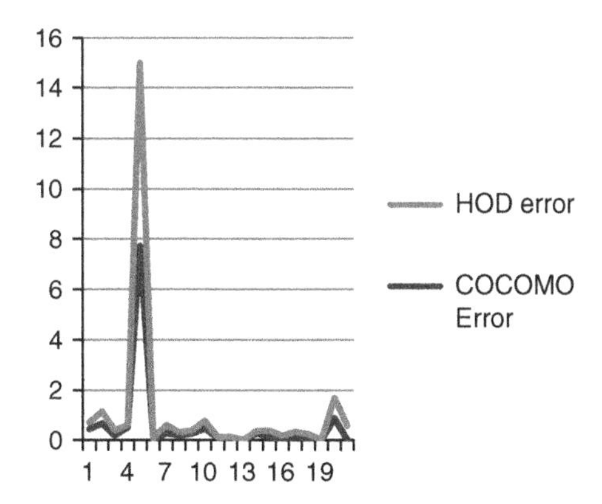

Fig. 7 Comparison of error

References

1. Rao, G.S, Krishna, C.V.P., Rao, K.R.: Multi objective particle swarm optimization for software cost estimation. In: ICT and Critical Infrastructure: Proceedings of the 48th Annual Convention of Computer Society of India, vol. I, pp. 125–132. Springer International Publishing (2014)
2. Maleki, I., Ghaffari, A., Masdari, M.: A new approach for software cost estimation with hybrid genetic algorithm and ant colony optimization. Int. J. Innov. Appl. Stud. **5**(1), 72–81 (2014)

3. Kaur, R., Kumar, R., Bhondekar, A.P., Kapur, P.: Human opinion dynamics: an inspiration to solve complex optimization problems. Scientific reports 3, (2013)
4. Bardsiri, V.K., Jawawi, D.N.A., Hashim, S.Z.M., Khatibi, E.: A PSO-based model to increase the accuracy of software development effort estimation. Software Qual. J. **21**(3), 501–526 (2013)
5. Benala, T.R., Chinnababu, K., Mall, R., Dehuri, S.: A particle swarm optimized functional link artificial neural network (PSO-FLANN) in software cost estimation. In: Proceedings of the International Conference on Frontiers of Intelligent Computing: Theory and Applications (FICTA), pp. 59–66. Springer Berlin Heidelberg (2013)
6. Sheta, A.F., Aljahdali, S.: Software effort estimation inspired by COCOMO and FP models: a fuzzy logic approach. Int. J. Adv. Comput. Sci. Appl. (IJACSA) **4**, 11 (2013)
7. Bhondekar, A.P., Kaur, R., Kumar, R., Vig, R., Kapur, P.: A novel approach using dynamic social impact theory for optimization of impedance tongue (iTongue). Chemometr. Intell. Lab. Syst. **109**(1), 65–76 (2011)

A Comparative Study of Various Meta-Heuristic Algorithms for Ab Initio Protein Structure Prediction on 2D Hydrophobic-Polar Model

Sandhya P.N. Dubey, S. Balaji, N. Gopalakrishna Kini and M. Sathish Kumar

Abstract Ab initio protein structure prediction (PSP) models tertiary structures of proteins from its sequence. This is one of the most important and challenging problems in bioinformatics. In the last five decades, many algorithmic approaches have been made to solve the PSP problem. However, it remains unsolvable even for proteins of short sequence. In this review, the reported performances of various meta-heuristic algorithms were compared. Two of the algorithmic settings—protein representation and initialization functions were found to have definite positive influence on the running time and quality of structure. The hybrid of local search and genetic algorithm is recognized to be the best based on the performance. This work provides a chronicle brief on evolution of alternate attempts to solve the PSP problem, and subsequently discusses the merits and demerits of various meta-heuristic approaches to solve the PSP problem.

Keywords Protein structure prediction · HP model · Meta-heuristic algorithm · NP-complete · Evolutionary algorithm

1 Introduction

In 2003, when the human genome was completely sequenced, researchers were trying to decipher the functional information of it. This endeavor is called structural genomics [1]. Protein structure is crucial to understand its function. The prediction

S.P.N. Dubey (✉) · N. Gopalakrishna Kini
Department of Computer Science & Engineering, MIT, Manipal University, Manipal, India
e-mail: sandhyadubey24@yahoo.co.in

S. Balaji
Department of Biotechnology, MIT, Manipal University, Manipal, India

M. Sathish Kumar
Department of ECE, MIT, Manipal University, Manipal, India

M. Pant et al. (eds.), *Proceedings of Fifth International Conference on Soft Computing for Problem Solving*, Advances in Intelligent Systems and Computing 437, DOI 10.1007/978-981-10-0451-3_36

of native structure of a protein from its sequence is computationally challenging. This has been an open problem for more than 60 years, till today and it is widely considered the 'holy grail' of computational biology [2]. Computationally, there are three different approaches to address the protein structure prediction (PSP) problem: homology (comparative) techniques, threading (fold recognition), and ab initio (de novo) techniques. Homology modeling and threading approaches utilize homologous (similar) sequences and structures. The former uses sequences of known structures from the protein data bank (PDB) to align with the target protein sequence to predict the 3D structure, whereas the latter search is for a homologous structure related to the target sequence [3]. However, both the techniques have two major drawbacks. First, the template structures are obtained through experimental methods that are expensive and time consuming; moreover, only one percent of the structure is known in PDB [3]. The second drawback is that homology and threading approaches do not provide an explanation to the following question: Why the given protein folds into a particular conformation?

Contrariwise, ab initio technique approaches the PSP from scratch. Ab initio PSP involves developing a computer program to predict the 3D structure, based on amino acid's physiochemical properties. It can be used for any protein sequence that does not have homologous structure. Moreover, it overcomes the limitation of comparative and threading approach. Also, it has the potential to accelerate research in the fields of drug design, protein engineering, proteomics, etc. Moreover, ab initio *methods* provide deep insights into the protein folding mechanism. The ab initio PSP is achieved in three stages: (i) Modeling with desirable accuracy, (ii) Defining an energy scoring function, and (iii) Set up a search algorithm to effectively select a energetically favorable conformation.

Successful implementation depends on the availability of a powerful conformation search method that can efficiently find the global minima for a given energy function with complicated energy landscape. Design and development of an efficient search technique can be one of the most prominent progresses to solve the PSP. Hence, in last two decades there are various attempts to find the most efficient search algorithm. Finally, predicted structure is evaluated in terms of its closeness to real structure with root-mean-square deviation (RMSD). RMSD calculates the deviation of predicted structure from original structure by superimposing two structures. In this paper, we consider the PSP from the perspective of developing an efficient search technique, in order to proficiently explore the search space and find the optimum conformation. A variety of algorithms are accumulated over the years with a reasonable speed and accuracy. Some uses very general approaches, extremely rapid but not accurate; others are near accurate and simulate the protein folding process but extremely slow. However, most algorithms try to balance speed and accuracy [4]. Apart from this there are other complexities: the stability of the conformation and the free energy contribution is least understood [5] and there is a difficulty to search in a large conformational space for all possible conformations.

The present work revisits existing wide range of meta-heuristics algorithms and compares them to facilitate the design of new and better search algorithms. The meta-heuristics algorithms used were genetic algorithms (GA), ant colony optimization (ACO), immune algorithm (IA), local search as simulated annealing (SA), hill-climbing (HC), tabu search (TS), and hybrid of local search and evolutionary memetic algorithms.

2 2D Hydrophobic-Polar Model

There are multiple ways to represent a protein structure. Since the conformation space is very huge and complex with many crusts and trough to deal with, this problem requires a simplified model. The best-known representation is the hydrophobic–hydrophilic (HP) model [6], which is based on the observation that hydrophobic contacts are the driving force for the protein to fold. In this model, amino acids are classified as either *hydrophobic* (H) or *polar* (P), but many others do exist. These models are considered to be coarse representation of proteins, as they entail very general principles of protein folding.

2.1 Representation

The amino acid positions in a 2D square lattice are encoded with alphabets of the set $\{F, R, U, D\}$ that indicates the direction viz., forward (F), Reverse (R), Up (U), and Down (D). For example, an optimal 2D structure for sequence "HPHPPH HPHPPHPHHPPHPH" is shown in Fig. 1. The optimal structure has 9 H–H contact with lowest energy value −9.

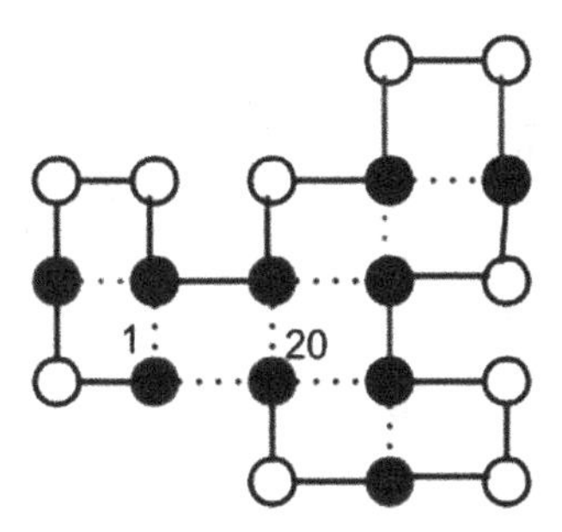

Fig. 1 The optimal 2D structure. *Black* and *white balls* indicate the hydrophobic and polar amino acids, respectively. The hydrophobic contacts (H–H) are represented by *dotted lines*

2.2 Dihedral Angle Space

The protein conformation depends on the nature of amino acids. The atoms in amino acids can take a wide variety of conformations due to main-chain as well as side-chain dihedral angles and bond lengths. However, analyzing the protein entries in databases shows the preponderance of amino acids for certain dihedral angles. Hence, a reduced conformation space is attained by restricting bond lengths and angles preferably by incorporating rotamer or by representing on lattices [5].

2.3 Protein Energy

Modeled conformations are evaluated in terms of Gibbs free energy called as *enthalpy*. It is based on the principle that the chemical compounds attain a state of minimum potential, referred as 'native state.' With HP model, Dill et al. [7] assigned this in terms of H–H interaction. The protein conformation is scored in terms of its free energy by providing the weightage to each type of interactions (Fig. 2). Whenever two H residues reside at one unit distance while folding, it contributes to H–H interaction and scored −1, whereas for H–P, P–H, and P–P it is 0. Further, HP model is extended, in terms of energy matrix as shown in Fig. 2b.

This extended model is named as shifted HP model (or functional model) [8]. *Functional model* is based on experimentally determined structures, where amino acids form the cavity instead of maximally compact structure. This model has considered both attractive and repulsive forces between residues and penalizes the folding by assigning positive weightage to H–P, P–H, and P–P interaction with 1. Even though HP model has reduced the search space, still it is NP-complete problem [9]. Hence, the PSP is attempted with meta-heuristic algorithms to model the protein and find the optimal conformation. This study is constrained to 2D HP model.

(a)

HP Model		
	H	P
H	-1	0
P	0	0

(b)

Functional Model		
	H	P
H	-2	1
P	1	1

Fig. 2 **a** Energy matrix for HP model, **b** energy matrix for functional model

3 Meta-Heuristic Algorithms

Meta-heuristic algorithms are group of algorithms inspired from the natural process such as evolution of species, animal and swarm behavior, and physical annealing process. Meta-heuristic algorithms are widely used [10] to tackle the NP problems such as traveling salesman problem, bin packing problem, scheduling and time-tabling problems, n-queen problem, vehicle routing problems, and graph theoretical problems. These problems have a large solution with many local optima. With Brute force approaches, NP problems result in poor quality solution for practical application. It is presumed that the evolutionary search has the potential for an extensive exploration of the solution space [11]. In this section, we address the use of various meta-heuristic techniques for ab initio PSP problem.

3.1 Evolutionary Algorithm

Evolutionary algorithm (EA) is a superset of procedures, inspired from Darwinian natural selection process. This paradigm includes evolutionary programming (EP), evolution strategies (ES), genetic algorithms (GA), and genetic programming (GP). Each of these procedures involves a population-based approach. EA start with initial population generation and subsequently their fitness calculation. Each individual in the population is called as a *chromosome*. In general, chromosomes are encoded in binary strings of 0s and 1s. For example, the string 1000110111 is binary chromosome of length 10. Collections of chromosomes form a population. EA operates on a pool of possible solutions by means of *selection, crossover, and mutation* operations to attain the global optimal solution. Figure 3 depicts the crossover and mutation operations. At the fundamental level, each of the procedures under EA follows the same flow (Table 1). Details on these procedures can be found in various literatures [11–14].

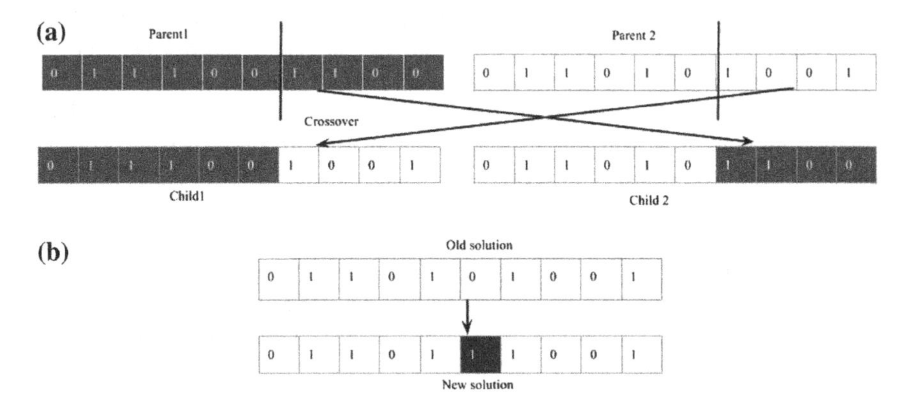

Fig. 3 **a** Crossover and **b** mutation operations

Table 1 Generic framework of evolutionary algorithm

1	Generate initial population//Each chromosome must satisfied all the constrain of particular domain
2	Calculate the fitness of chromosome
3	While Stopping condition not true do
4	Apply selection ()//choose chromosome for crossover
5	Apply crossover ()
6	Calculate fitness of offspring
7	Apply mutation ()
8	Calculate fitness
9	Replace the least fit chromosome with offspring having better fitness
10	Return the best chromosome

Use of evolutionary algorithm for PSP started in early 90s. Unger and Moult first developed the genetic algorithm (GA) combined with Monte Carlo to predict the protein folding. It was represented both in 2d and 3d HP models [15, 16]. In 1995, Patton et al. [17] developed standard GA that outperforms the previous implementation in terms of lowering free energy with less generation.

Later, Konig and Dandekar [18] further improvised the performance of GA with systematic crossover. In last 5 years there is exponential growth in application of GA with local search and other domain-based knowledge for PSP. Further enhancement to GA for PSP was done by Hoque et al. [19] by implementing twin removal to maintain the diversity of the sample space.

3.2 Ant Colony Optimization (ACO)

ACO mimics the behavior of ants; how they find the shortest path to reach the food source from their nest proposed by Dorigo et al. [20]. The first implementation is developed for one of the NP-hard problems namely traveling salesman problem. Ants communicate with each other by means of pheromone trails. Once an ant locates food source, they carries the food to their nest and lay pheromones along the path. Other ants decide the path based on the pheromone concentrations. This indirect communication is referred as *stigmergy*. While following the path, each ant lays the pheromone, resulting in higher pheromone concentration for most followed path. Success of ACO is based on the precise definition of selection probability and pheromone update factor. It follows graph-based probabilistic search where for every node each ant executes a decision policy to obtain the next path to follow. For simple ant colony optimization, transition probability ($P_{ij}^{k}(t)$) of ant k with current location at node i is node j, based on the following condition [11]:

Table 2 General ACO algorithm parameters

Parameter	Meaning
n_k	Number of ants
n_t	Maximum number of iterations
τ_0	Initial pheromone amount
ρ	Pheromone persistence
α	Pheromone intensification
β	Heuristic intensification

$$P_{ij}^{k}(t) = \begin{cases} \frac{T_{ij}^{\alpha}(t)}{\sum_{j \epsilon N_i^k} T_{ij}^{\alpha}(t) n} & \text{if } j \in N_i^k \\ 0 & \text{if } j \notin N_i^k \end{cases} \quad (1)$$

where N_i^k is a set of possible nodes connected to node i, w.r.t. ant k. α is a positive constant used to amplify the influence of pheromone concentrations. Higher value of α may cause rapid convergence to local solution. Solutions of ACO are measured in terms of implicit explicit evaluation. Pheromone deposition ΔT_{ij}^k on link is directly proportional to most followed path, and hence it is used to measure the quality of particular solution. This is termed as explicit evaluation:

$$\Delta T_{ij}^{k}(t) \propto \frac{1}{L^{k}(t)} \quad (2)$$

If all ant deposits equal to amounts of pheromone, then they follow a differential path length, whose solution is called as implicit evaluation (Table 2).

Alena Shmygelska et al. [21, 22] first proposed the use of ACO in protein folding on 2D HP square lattice model. They studied the effect and impact of the above-listed parameters to improvise PSP. It is observed that ACO performance degrades with increase in protein sequence length, but it results in diverse conformation. This suggests making a hybrid of ACO with other meta-heuristics with a local search. Later, Chu et al. [23] implemented ACO in parallel form using distributed programming.

3.3 Immune Algorithm

Immune algorithm (IA) is inspired by mammalian immunology, where immune system differentiates between normal and foreign particle in the body [11]. On encountering of foreign particles, it generates specialized activated cells called as lymphocytes, to destroy them. Lymphocytes and lymphoid organs (for growth, development, and deployment of the lymphocytes) are two important parameters which determine the success of IA. Also, there are various versions of IA based on the behavior of lymphocytes w.r.t. foreign particles. The most widely studied are clonal selection, danger theory, and network theory. For PSP, clonal selection-based

IA is first implemented [24]. In this implementation minimum population size is 10 and maximum is 10^5 with twice duplication rate. Results of this implementation are comparable with the state-of-the-art result.

3.4 Local Search Algorithm

Local search (LS) algorithm is an iterative improvement process, which begins with a single solution and keeps improvising the same by means of random walk, neighborhood search, and random sampling techniques. Hill-climbing, tabu search, and simulated annealing are most used LS approaches to solve the PSP problem. Details on these algorithms can be found in [25]. For PSP, LS starts with a random conformation and keeps improvising it by visiting the vacant neighbor lattice points to attain a better value for a given objective function. TS was first used by Jacek Blazewicz [26], with original HP model for 2D square lattice conformation. In most cases, LS algorithms are implemented adopting evolutionary algorithmic techniques such as hill-climbing GA, which are later classified into a new class of meta-heuristic algorithms called as memetic algorithm.

3.5 Memetic Algorithm

Krasnogor and Smith [27] proposed memetic algorithms (MAs) as an alternate approach for gradient search methods, such as *back-propa*gation. Fundamentally, MAs integrate EA-based stochastic search with problem-specific solvers such as local search heuristics techniques, approximation algorithms or, sometimes, even (partial) exact methods. This hybridization results in better solution, within comparatively less time MAs overcome the premature convergence of EAs and also maintain diversity in population. Intelligent integration of domain knowledge with EAs results in better MAs. To address any optimization problem with MAs, one needs to find out all the possible parts of EAs where it can hybridize the local search. However, this integration increases the complexity of the algorithm. Generic template of MA is presented in Fig. 4, adapted from [27].

In 2002, Krasnogor et al. [28] offered the multi-meme algorithm for PSP. In last few years, there is an increase in the use of MA to solve the PSP problem as compared with other meta-heuristic algorithms. It has been observed that MA handles the diversity and premature convergences problem better. The following year Jiang et al. [29] came up with a hybrid of GA with TS. Further, Islam and Chetty [30] proposed new MA by incorporating the domain knowledge for hydrophobic core construction. Islam and Chetty [30] proposed a non-isomorphic encoding technique to model the conformation, and it results in significant reduction of population size, and reduced the search time and degeneracy of prediction. Camelia et al. [31] proposed EA with HC; in this work they implemented the HC in

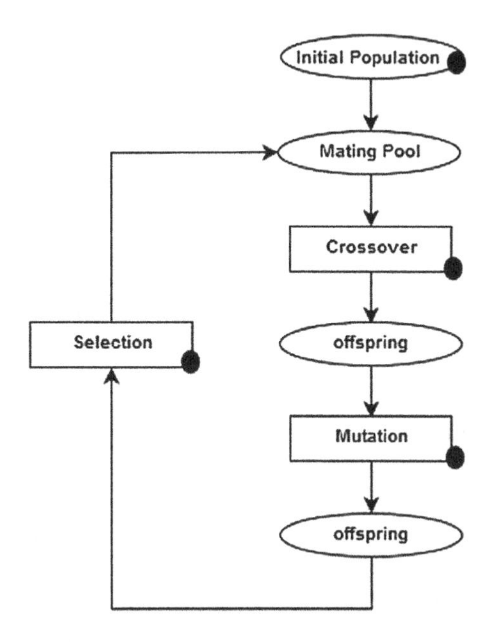

Fig. 4 Generic template of memetic algorithm-*black circle* indicates integration of local search/domain-specific knowledge/intelligent population initialization, etc.

crossover and mutation operator of EA. Su et al. [32] proposed a hybrid of HC and GA on 2D triangular lattice model. In this study, we have considered only a few implementations as the scope is constrained to 2D structure prediction only.

4 Result

The performance of the algorithms is gauged by two parameters, namely (a) free energy and (b) sequence length. Table 3 shows the benchmark sequences used in this study. These sequences were described in [10] and are found at http://www.cs.sandia.gov/tech_reports/compbio/tortilla_hp_benchmarks.html.

Performance Comparison

Free Energy-based comparison
Table 4 shows the comparisons of various state-of-the-art algorithms for the 2D HP models considered in this study. The free energy $\mathbf{E}^*$values for the final models of the bench mark sequences using various algorithms are shown in Table 4.

The hybrid genetic algorithm with twin removal [19] outperformed all other approaches in terms of free energy irrespective of sequence length, whereas the other approaches converge to same energy values (Fig. 5). Although the energy values are same for the modeled proteins, they all differ in their conformation (Fig. 6). This leads to degeneracy issues, where it may require a program to cluster all isomorphic conformations into groups and evaluate further with experimental structure using RMSD to select the biologically relevant conformation.

Table 3. The benchmark instances for 2D HP model

Seq. no.	Length	Protein sequence
1	24	$(HP)^2PH(HP)^2(PH)^2HP(PH)^2$
2	24	$H^2P^2(HP^2)^6H^2$
3	25	$P^2HP^2(H^2P^4)^3H^2$
4	36	$P(P^2H^2)^2P^5H^5(H^2P^2)^2P^2H(HP^2)^2$
5	48	$P^2H(P^2H^2)^2P^5H^{10}P^6(H^2P^2)^2HP^2H^5$
6	50	$H^2(PH)^3PH^4PH(P^3H)^2P^4(HP^3)^2HPH^4(PH)^3PH^2$
7	60	$P(PH^3)^2H^5P^3H^{10}PHP^3H^{12}P^4H^6PH^2PHP$
8	64	$H^{12}(PH)^2((P^2H^2)^2P^2H)^3(PH)^2H^{11}$

Table 4 Comparison on the performances of various algorithms

Algorithms	Bench mark sequences							
	1	2	3	4	5	6	7	8
	Free Energy E* (No. of H–H contact)							
GA [15]	−9	−9	−8	−14	−22	−21	−34	−37
STD GA [17]	−9	−10	−8	−15	N/A	N/A	−34	N/A
Improved GA [18]	−9	−9	−8	−14	−23	−21	−34	−37
TR-GA [19]	N/A	N/A	N/A	N/A	N/A	−21	−34	−37
HGA + TR-80 [19]	N/A	N/A	**−25**	**−51**	**−69**	**−59**	**−117**	**−103**
ACO [21]	−9	−9	−8	−14	−23	−21	−34	−32
Improved ACO [22]	−9	−9	−8	−14	−23	−21	**−36**	**−42**
IA [24]	−9	−9	−8	−14	−23	−21	−35	−39
IA with LRM [24]	−9	−9	−8	−14	−23	−21	−35	−42
TS [26]	−9	−9	−8	−14	−23	−21	−35	−42
MMA [28]	−9	N/A	−8	−14	−22	−21	**−36**	−38
GTS [29]	−9	−9	−8	−14	−23	−21	−35	−39
CMA [30]	−9	−9	−8	−14	−23	−21	**−36**	**−42**
EAHCD [31]	−9	−9	−8	−14	−23	−21	−35	−39
#CMA [32]	**−15**	**−17**	**−12**	**−24**	**−40**	**−40**	**−70**	**−70**
#HHGA [19]	−15	−17	−12	−24	−40	N/A	−70	−50

Represents conformation modeling on 2D triangular lattice; N/A sequences that are not considered for model building; The best E* values are presented in bold face

As the possible conformation space is huge and complex, many approaches get trapped in local optima. Hoque et al. [19] proposed the chromosome correlation factor (CCF), by which similarities among the generated conformations are verified and removed (twin removal) if the conformation is 80 % similar among others. This is used to maintain the diversity in the conformation space and prevent from getting trapped in the local optima. However, this method with GA performed similar to other approaches; in contrast, when GA was combined with a local search and with twin removal, it outperformed others.

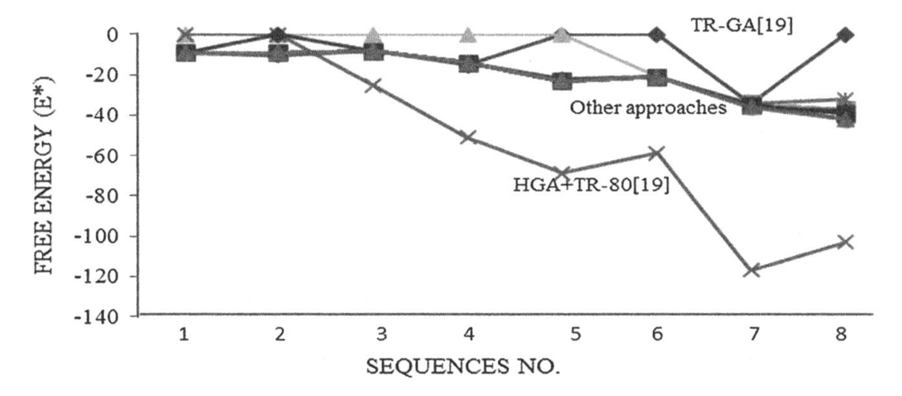

Fig. 5 Free energy obtained for benchmark sequences listed in Table 3 with different approaches

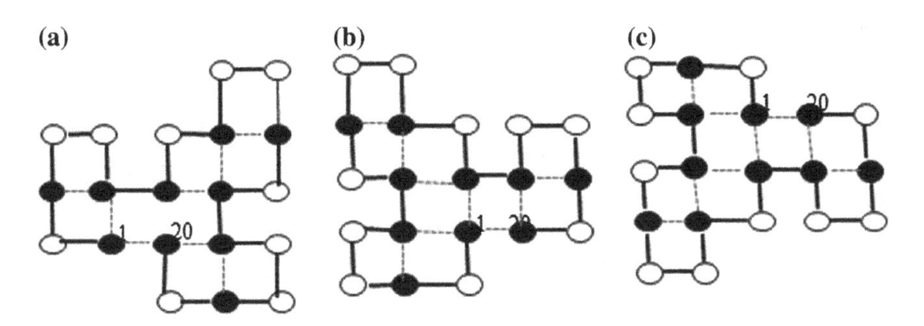

Fig. 6 Isomorphic conformations for the protein string: HPHPPHHPHPPHPHHPPHPH. Each conformation has E* = −9

Sequence length-based comparison

For sequences of length less than 60 residues, the algorithms used for the comparison performed equally better, whereas hybrid approach as [22, 28, 30] outperforms for sequences of length ≥60. It is obvious that the energy values depend on the H–H contacts that in turn depend on the number of hydrophobic residues. This trend is very well reflected in the energy values. The numbers of hydrophobic residues for sequence No. 1–3 were between 9 and 10, and their corresponding energy values are −8 to −9. In case the hydrophobic residues that are between 16 and 24 (Seq. No. 4–6) the energy values is between −14 and −23, calculated by all the algorithms except [19]. At last sequence No. 7 and 8, the numbers of hydrophobic residues are 43 and 42, respectively. This confirms that the increase in the numbers of hydrophobic residues increased the stability of the protein conformation. Also, it is observed that the modeling of conformation on triangular lattice is closer to the experimentally solved protein structure [19, 32]. Moreover, non-isomorphic encoding technique can be employed to maintain the diversity of search space and to reduce the computation time.

5 Conclusion

The performance of various algorithms was compared based on free energy and sequence length. The hybrid of local search and genetic algorithm is recognized to be the best. Although the algorithms undertaken in this study calculated same energy values for the chosen dataset, the final models differ in their conformations. This leads to degeneracy issues and may require an additional program to handle isomorphic conformations. Hence, a new approach with enhanced features is required to predict a stable protein structure with biologically relevant conformation.

Acknowledgments The authors are grateful to the anonymous referees for their constructive comments and suggestions that greatly improved the paper.

References

1. Denise, C.: Structural GENOMICS Exploring the 3D Protein Landscape. Simbios (2010)
2. Lajtha, A.: Handbook of Neurochemistry and Molecular Neurobiology. Springer (2007)
3. Mansour, N.: Search Algorithms and Applications. InTech (2011)
4. Levinthal, C.: Are there pathways for protein folding? J. Chem. Phys. **65**, 44–45 (1968)
5. Helles, G.: A comparative study of the reported performance of ab initio protein structure prediction algorithms. J. R. Soc. Interface **5**(21), 387–396 (2008)
6. Dill, K.A.: Dominant forces in protein folding. J. Biochem. **29**, 5133–7155 (1990)
7. Dill, K.A.: Theory for the folding and stability of globular proteins. Biochemistry **24**(6), 1501–1509 (1985)
8. Blackburne, B.P., Hirst, J.D.: Evolution of functional model proteins. J. Chem. Phys. **115**(4), 1935–1942 (2001)
9. Berger, B., Leighton, T.: Protein folding in the hydrophobic-hydrophilic (HP) is NP-complete. In: Proceedings of the Second Annual International Conference on Computational Molecular Biology, pp. 30–39 (1998)
10. Shmygelska, A., Hoos, H. H.: An improved ant colony optimization algorithm for the 2D HP protein folding problem. In: Proceedings of 16th Canadian Conference Artificial Intelligence, Halifax, Canada, pp. 400–417 (2003)
11. Engelbrecht, A.P.: Computational Intelligence: An Introduction. Wiley (2007)
12. Ashlock, D.: Evolutionary Computation for Modeling and Optimization. Springer (2006)
13. Yu, T., Davis, L., Baydar, C., Roy, R. (eds.) et al.: Evolutionary Computation in Practice. Springer, Berlin Heidelberg (2008)
14. Deb, K.: Multi-Objective Optimization Using Evolutionary Algorithms. Wiley (2001)
15. Unger, R., Moult, J.: Genetic algorithms for protein folding simulations. J. Mol. Biol. **231**(1), 75–81 (1993)
16. Unger, R.: The genetic algorithm approach to protein structure prediction. Struct. Bond. **110**, 153–175 (2004)
17. Patton, A.L., Punch, W.F, Goodman, E.D.: A standard GA approach to native protein conformation prediction. In: Proceedings of the 6th International Conference on Genetic Algorithms, Pittsburgh, PA, USA, July 15–19 (1995)
18. Koing, R., Dandekar, T.: Improving genetic algorithms for protein folding simulations by systematic crossover. BioSystems **50**, 17–25 (1999)

19. Hoque, T., Chetty, M., Lewis, A., Sattar, A.: Twin removal in genetic algorithms for protein structure prediction using low-resolution model. IEEE/ACM Trans. Comput. Biol. Bioinf. **8** (1), 234–245 (2011)
20. Dorigo, M., Maniezzo, V., Colorni, A.: Positive feedback as a search strategy. Technical Report 91–016, Dip. Elettronica, Politecnico di Milano, Italy, (1991)
21. Shmygelska, A., Hoos, H. H.: An improved ant colony optimization algorithm for the 2d hp protein folding problem. In: Xiang, Y., Chaib-draa, B. (eds.) AI 2003, LNAI 2671, pp. 400–417. Springer, Berlin Heidelberg (2003)
22. Shmygelska, A., Hoos, H.H.: An ant colony optimisation algorithm for the 2D and 3D hydrophobic polar protein folding problem. BMC Bioinform. **6**(30), 1471–2105 (2005)
23. Chu, D., Zomaya, A.: Parallel Ant Colony Optimization for 3D Protein Structure Prediction using the HP Lattice Model, Studies in Computational Intelligence (SCI), vol. 22, pp. 177–198. Springer, Berlin Heidelberg (2006)
24. Cutello, V., Nicosia, G., Pavone, M., Timmis, J.: An immune algorithm for protein structure prediction on lattice models. IEEE Trans. Evol. Comput. **11**(1), 101–117 (2007)
25. Lenstra, J. K.: Local Search in Combinatorial Optimization. Princeton University Press (1997)
26. Milostan, M., Lukasiak, P., Dill, K. A., and Blazewicz, J.: A tabu search strategy for finding low energy structures of proteins in HP-model. In: Proceedings of Annual International Conference on Research in Computational Molecular Biology, Berlin, Germany, Poster No. 5-108 (2003)
27. Krasnogor, N., Smith, J.: A tutorial for competent memetic algorithms: model, taxonomy, and design issues. IEEE Trans. Evol. Comput. **9**(5), 474–488 (2005)
28. Krasnogor, N., Blackburne, B.P., Burke, E.K., Hirst, J.D.: Multimeme algorithms for protein structure prediction. In: Proceedings of International Conference Parallel Problem Solving from Nature (PPSN VII), Granada, Spain, pp. 769–778 (2002)
29. Jiang, T., Cui, Q., Shi, G., Ma, S.: Protein folding simulation of the hydrophobic-hydrophilic model by combining tabu search with genetic algorithm. J. Chem. Phys. **119**(8), 4592–4596 (2003)
30. Islam, M.K., Chetty, M.: Clustered memetic algorithm with local heuristics for Ab initio protein structure prediction. IEEE Trans. Evol. Comput. **17**(4), 558–576 (2013)
31. Chira, C., Horvath, D., Dumitrescu, D.: Hill-Climbing search and diversification within an evolutionary approach to protein structure prediction. BioData Min. **4**(23) (2011)
32. Su, S. C., Lin, C.J., Ting, C.K.: An effective hybrid of hill climbing and genetic algorithm for 2D triangular protein structure Prediction. Proteome Sci. **9**(1) (2011)

Population Segmentation-Based Variant of Differential Evolution Algorithm

Pooja, Praveena Chaturvedi and Pravesh Kumar

Abstract Recently, several strategies are proposed for offspring generation and for control parameter adaption to enhance the reliability and robustness of the differential evolution (DE) algorithm, whereas usually the population size is fixed throughout the evolution which results in unsatisfactory performance. In the current study, based on the status of solution-searching, a new variant differential evolution with population segment tuning (DEPT) is proposed to enhance the performance of DE algorithm. The performance of the proposed is validated on a set of eight standard benchmark problems and six shifted problems taken from literature in terms of number of function evaluations, mean error, and standard deviation. Also, the results are compared to DE as well as some state of the art.

Keywords Differential evolution · Mutation · Population size · MRLDE

1 Introduction

Differential evolution (DE) is a variant of evolutionary algorithm (EA) which was proposed by Storn and Price [1] in 1995. DE is a simple and efficient search engine and is used for solving global optimization problems over continuous spaces. It can handle nonlinear, non-differentiable, and multimodal objective functions and a wide range of real-life problems such as engineering design, chemical engineering,

---

Pooja (✉) · Praveena Chaturvedi
Department of Computer Science, Gurukula Kangri Vishwavidyalaya,
Haridwar, India
e-mail: mcapooja.singh2007@gmail.com

Praveena Chaturvedi
e-mail: praveena_c1@rediffmail.com

Pravesh Kumar
Department of Mathematics, Amity University, Haryana, India
e-mail: praveshtomariitr@gmail.com

M. Pant et al. (eds.), *Proceedings of Fifth International Conference on Soft Computing for Problem Solving*, Advances in Intelligent Systems and Computing 437, DOI 10.1007/978-981-10-0451-3_37

pattern recognition, etc. Recently, DE has been applied to multi-level image thresholding [2] and solving integer programming problems [3], and a review of major applications areas of DE can be found in [4].

For enhancing the performance of DE, several variants are proposed in the literature. Some of which are DE with unconventional initialization methods [5], modified DE (MDE) [6], DE with global and local neighborhood (DEGL) [7], Cauchy mutation DE (CDE) [8], self-adaptive DE (SaDE) [9], learning enhance DE (LeDE) [10], DE with self-adaptive control parameter (jDE) [11], adaptive DE with optional external archive (JADE) [12], modified random localization-based DE (MRLDE) [13], DE with adaptive population tuning [14], guiding force strategy-based DE [15], and so on.

Taking the present study from scratch, as suggested by Kaelo and Ali [16], selection of the base vector may help to increase the convergence speed of basic DE and if the base vector is fitter than the difference vectors, the convergence speed of the DE will be better. Keeping this concept in mind, Kumar and Pant [13] proposed an enhanced variant of the DE algorithm, MRLDE, in which the whole search space is divided into three search spaces resulting in an approach which is not totally randomized.

The results produced by MRLDE algorithm are far better than DE algorithm, but the drawback of MRLDE is the selection of the size of the first population segment and the division of the search space while the population is getting dense which is much time consuming. To overcome these problems of DE and MRLDE, a new strategy named DEPT is proposed here for maintaining a balanced approach.

The proposed DEPT algorithm will make use of MRLDE with population segment tuning and when the population will go dense then it will switch to basic DE algorithm as there is no need of partitioning the search space further.

The rest of the paper is organized as follows: Sect. 2 provides a compact overview of DE. Section 3 presents the proposed approach with pseudocode. Benchmark problems and experimental settings are given in Sect. 4. Results and comparisons are reported in Sect. 5, and finally the conclusion derived from the present study is drawn in Sect. 6.

2 Differential Evolution Algorithm

Differential evolution (DE) proposed by Storn and Price [1] is simple, fast, and robust evolutionary algorithm. A brief introduction of the basic DE is given as follows:

DE starts with a population of NP candidate solutions: $\vec{x}_i^g$, $(i = 1, \ldots, \mathrm{NP})$, where it denotes the ith candidate solution of generation g. The three main operators of DE are mutation, crossover, and selection.

2.1 Mutation

The mutation process at each generation begins by randomly selecting three solution vectors from the population set of (say) NP elements. The ith perturbed individual, $\vec{v}_i^g = (v_{1,i}^g, v_{2,i}^g, \ldots, v_{d,i}^g)$, is generated as

$$\text{DE/rand/1/bin}: \qquad \vec{v}_i^g = \vec{x}_{a1}^g + F * (\vec{x}_{a2}^g - \vec{x}_{a3}^g) \tag{1}$$

where $i = 1, \ldots \text{NP}, \text{a1}, \text{a2}, \text{a3} \in \{1, \ldots, \text{NP}\}$ are randomly selected such that $\text{a1} \neq \text{a2} \neq \text{a3} \neq i$, and F is the control parameter such that $F \in [0, 1]$.

2.2 Crossover

The perturbed individual $\vec{v}_i^g = (v_{1,i}^g, v_{2,i}^g, \ldots, v_{d,i}^g)$ is subjected to a crossover operation with target individual $\vec{x}_i^g = (x_{1,i}^g, x_{2,i}^g, \ldots, x_{d,i}^g)$, which finally generates the trial solution, $\vec{u}_i^g = (u_{1,i}^g, u_{2,i}^g, \ldots, u_{d,i}^g)$, as follows:

$$u_{j,i}^g = \begin{cases} v_{j,i}^g & \text{if } \text{rand}_j \leq \text{C}_\text{r} \vee j = j\_\text{rand} \\ x_{j,i}^g & \text{otherwise} \end{cases} \tag{2}$$

where $j = 1, \ldots, d, 1 \leq j\_\text{rand} \leq d$ is a random parameter's index, chosen once for each i. The crossover rate $\text{C}_\text{r} \in [0, 1]$ is set by the user. $0 < \text{rand}_j < 1$ is uniformly distributed random number.

2.3 Selection

The selection scheme of DE also differs from that of other EAs. The population for the next generation is selected from the solution in current population and its corresponding trial solution according to the following rule:

$$\vec{x}_i^{g+1} = \begin{cases} \vec{u}_i^g & \text{if } \; f(\vec{u}_i^g) \leq f(\vec{x}_i^g) \\ \vec{x}_i^g & \text{otherwise} \end{cases} \tag{3}$$

3 Proposed Algorithm

In this section, the working of the proposed algorithm is described. As in MRLDE [13] algorithm, the whole search space is divided into three segments (regions) for maximum exploration and exploitation of the search space, and from each region one individual is selected. The division of the population is done after sorting the population on the basis of the function fitness value. If *R1* represents the region having the fittest individuals, *R2* represents the set of next best individuals and *R3* represents the remaining individuals and the size of these regions are NP*α%, NP*β%, and NP*γ%, respectively.

The base vector is selected from the region *R1*, referred as the base region here. But the selection of the size of base region is the main issue with MRLDE algorithm, as small size will take it toward a greedy approach while large size will make it working like basic DE algorithm which will result in unsatisfactory performance. To overcome this problem, in the proposed approach the size of base region is tuned.

In the proposed approach, the population size of the base region will be adapted by starting from a greedy point and approaching toward the basic DE algorithm. This is done by taking an additional factor, i.e., balancing factor (b_f), which will make a balance between DE and MRLDE.

Initially, the value of α is very small. Here it is taken as 20. The size of the base region *R1* is incremented after every 1000 iterations by incrementing α with b_f, where $b\_f = b\_f +$ NP*5%.

The size of the other two regions is set as defined in MRLDE [13] and decreases accordingly when α increases, making the whole search space more exploratory. This is how the population size of the base region is getting tuned.

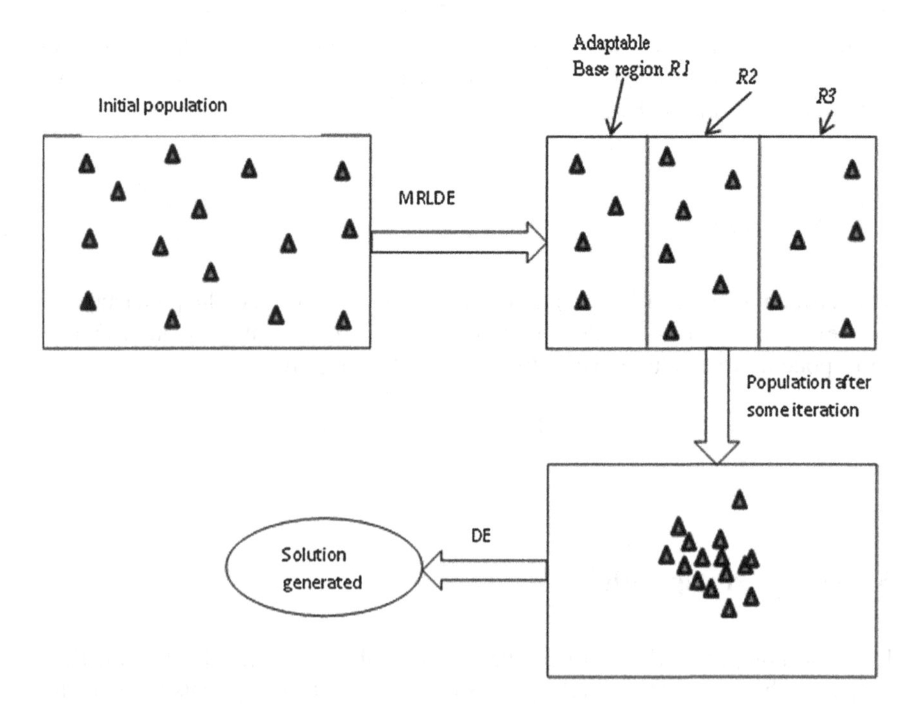

Fig. 1 Graphical representation of working of DEPT algorithm

Simultaneously, this approach is keeping track of the whole population as well. Looking for the stage at which the population converges and the result of the MRLDE simulates with the results of DE algorithm, then instead of partitioning the whole population, the proposed approach uses the basic DE algorithm to generate the new candidate solution as shown in Fig. 1, because there is no need of partitioning the population when all the individuals are moving toward the converging point.

Pseudo Code of the Proposed Approach.

Step 1: Initialize the values of F, Cr and randomly generates initial NP vectors. The jth variable of ith vector is created as:

$x_{j,i} = low_j + rand(0,1) * (up_j - low_j)$, where $rand(0,1)$ is a uniformly generate random number between 0 and 1.

Step 2: Evaluate $f(\vec{x}_i)$.

Step 3: Sort the population.

Step 4: balancing factor *b_f=0.*

Step 5: if (population converged), then goto step 8 else goto step 6.

Step 6: divide population into three regions *R1*, *R2* and *R3* where there sizes are $NP*\alpha\%$, $NP*\beta\%$ and $NP*\gamma\%$ respectively.

Step 7: do{ $\vec{x}_{a1}$ = (int) $rand(0,1) * (\alpha + b\_f)$ }while($\vec{x}_{a1}$ ==i)

do{ $\vec{x}_{a2}$ = (int) ($(\alpha + b\_f) + rand(0,1) * \beta$) }while($\vec{x}_{a2}$ ==i)

do{ $\vec{x}_{a3}$ = (int) ($\beta + rand(0,1) * \gamma$) }while($\vec{x}_{a3}$ ==i)

Step 8: Perform mutation operation using equation-1.

Step 9: Perform crossover operation using equation-2.

Step 10: Evaluate $f(\vec{u}_i)$.

Step 11: Apply selection operation using equation-3.

Step 12: *bf=bf+NP**5%.

Step 13: whether the termination criterion is reached, If yes then stop, else goto step 4.

4 Experimental Settings and Benchmark Problems

4.1 Parameter Settings

Pop size (NP)	100;
Dimension (*d*)	30 (for all Problems);
F, C_r	0.5, 0.9, respectively [5];

α	starts with 20;
β, γ	(NP-α)/2 [13];
Threshold value (θ)	10^{-08} except for f_5 where it is taken as 10^{-02};
#FE	d*10000 for shifted problems [13].

Using DEV-C++ integrated development environment (IDE), all the algorithms are executed 30 times for each problem on Intel Core i3 processor with 4 GB RAM. Each run's termination criteria are defined as when the objective function value is less than the defined threshold value of the respective problem or the maximum number of function evaluations (#FEs) was met.

4.2 Performance Criteria

To evaluate the performance of the algorithms, the performance criteria are selected from the literature [6, 13]. These criteria are as follows:

- Number of function evaluations (#FEs);
- Error;
- Convergence graph;
- Acceleration rate.

4.3 Benchmark Problems

The performance of the proposed algorithm is evaluated on eight traditional benchmark problems listed in Table 1 taken from [10] with functions' names and their corresponding dimension 30. However, six benchmark scalable problems, listed in Table 2, are also taken from [17] to validate the efficiency of the proposed algorithm.

Table 1 Standard benchmark problems

Traditional benchmark functions		Dimension
f_1	Sphere function	30
f_2	Ackley function	30
f_3	Rastrigin function	30
f_4	Rosenbrock function	30
f_5	Noise function	30
f_6	Schwefel 1.2 function	30
f_7	Schwefel 2.22 function	30
f_8	Griewank function	30

Table 2 Shifted benchmark problems

Non-traditional benchmark function		Search space; f_{bias}
f_Sh_1	Shifted sphere function	[−100, 100]; −450
f_Sh_2	Shifted Schwefel 2.21 function	[−100, 100]; −450
f_Sh_3	Shifted Rosenbrock function	[−100, 100]; 390
f_Sh_4	Shifted Rastrigin function	[−5, 5]; −330
f_Sh_5	Shifted Griewank function	[−600, 600]; −180
f_Sh_6	Shifted Ackley function	[−32, 32]; −140

5 Numerical Results and Comparisons

5.1 In Terms of Traditional Benchmark Function

In this section, numerical results are shown for standard benchmark functions and all parametric settings are same for all algorithms as defined above.

Comparison with respect to DE and MRLDE (α = 20, 30, 40, and 50). Here the proposed DEPT algorithm is compared with DE and MRLDE for $\alpha = 20, 30, 40$, and 50 in terms of #FE for 30 runs. The respective results are highlighted in Table 3, which illustrates that the proposed variant performs better than DE and MRLDE (α = 30, 40, and 50) while comparable to MRLDE (α = 20). However, MRLDE (α = 20) shows a greedy approach producing premature solutions. For function f_3, DE and MRLDE algorithms are not working, whereas the proposed variant is working very well. The convergence graph for function f_1 is shown in Fig. 2 and it is clear from the graph too that DEPT produces better results than MRLDE and DE without fixing the value of the base region. Thus, it is true to say that the proposed variant is a balance of DE and MRLDE.

Comparison with respect to some State of the Art. The comparative results of SaDE, jDE, JADE, and LeDE are taken from literature [12]. Keeping all the settings as taken for these functions in literature, the results with the proposed variant are

Table 3 Results and comparison of DEPT with DE and MRLDE in terms of #FE

Function	DE	MRLDE				DEPT
		$\alpha = 20$	$\alpha = 30$	$\alpha = 40$	$\alpha = 50$	
f_1	103,530	40,150	55,350	73,660	91,840	47,150
f_2	163,990	62,050	86,510	113,200	141,600	81,300
f_3	–	–	–	–	–	358,075
f_4	442,780	146,400	183,470	229,000	280,130	283,875
f_5	110,640	58,280	86,020	102,770	117,000	60,075
f_6	410,720	151,390	216,520	283,540	365,460	300,150
f_7	173,740	68,120	94,620	122,910	156,070	91,350
f_8	109,000	42,020	57,960	75,080	94,740	48,325

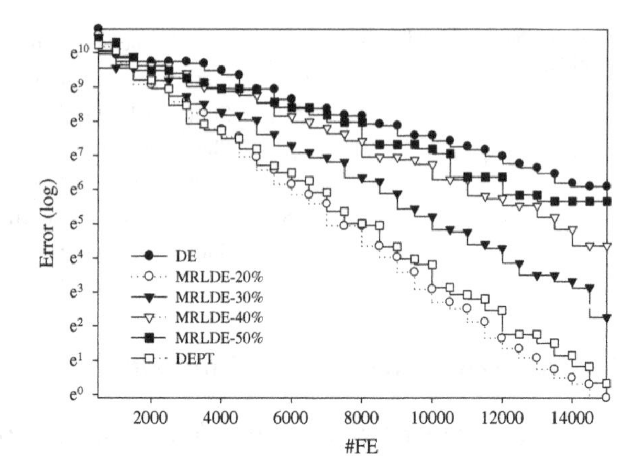

Fig. 2 Convergence graph for f_1 between average error and #FE

Table 4 Results and comparison of DEPT with state of the art in terms of average error and standard deviation

Function	SaDE	jDE	JADE	LeDE	DEPT
f_1	1.03e−37 (1.86e−37)	1.16e−28 (1.24e−28)	**1.30e−54** **(9.20e−54)**	2.19e−34 (1.18e−34)	1.14e−26 (8.15e−27)
f_2	**3.55e−15** **(0.0e+00)**	8.33e−15 (1.86e−15)	4.40e−15 (0.0e+00)	5.89e−15 (0.0e+00)	5.01e−15 (1.99e−15)
f_3	**0.0e+00** **(0.0e+00)**	**0.0e+00** **(0.0e+00)**	**0.0e+00** **(0.0e+00)**	**0.0e+00** **(0.0e+00)**	**0.0e+00** **(0.0e+00)**
f_4	2.66e−01 (1.03e+00)	1.63e−01 (7.89e−01)	3.20e−01 (1.10e+00)	0.0e+00 (0.0e+00)	**2.17e−21** **(3.41e−21)**
f_5	1.21e−03 (3.55e−03)	3.26e−03 (7.74e−04)	6.8e−04 (2.5e−04)	1.1e−03 (5.58e−04)	**1.07e−03** **(2.51e−04)**
f_6	2.42e−35 (6.07e−35)	1.64e−13 (3.65e−13)	**6.0e−87** **(1.9e−86)**	1.16e−38 (2.28e−38)	1.19e−14 (2.96e−13)
f_7	**1.54e−26** **(1.40e−26)**	9.98e−24 (7.59e−24)	3.90e−22 (2.70e−21)	1.09e−24 (4.09e−25)	3.51e−16 (3.72e−15)
f_8	**0.0e+00** **(0.0e+00)**	**0.0e+00** **(0.0e+00)**	2.00e−04 (1.4e−03)	**0.0e+00** **(0.0e+00)**	**0.0e+00** **(0.0e+00)**

produced and listed in Table 4. It can be seen that the proposed variant shows its competence with these state of the art in terms of average error and standard deviation. JADE delivers best results for f_1 while DEPT produces best results for functions f_3, f_4, f_5, and f_8.

Table 5 Comparison of DEPT with DE in terms of mean fitness and standard deviation

Function	DE	DEPT
f_Sh_1	2.2e−14; (2.7e−14)	**0.0e+00; (0.0e+00)**
f_Sh_2	**6.7e−04; (1.7e−04)**	3.21e+00; (1.53e+00)
f_Sh_3	1.8e+01; (1.2e+00)	**8.99e−01; (1.08e+00)**
f_Sh_4	9.4e+01; (8.2e+00)	**5.90e+01; (6.64e+00)**
f_Sh_5	5.6e−15; (1.1e−14)	**0.0e+00; (0.0e+00)**
f_Sh_6	3.9e−14; (1.3e−14)	**2.8e−14; (0.0e+00)**

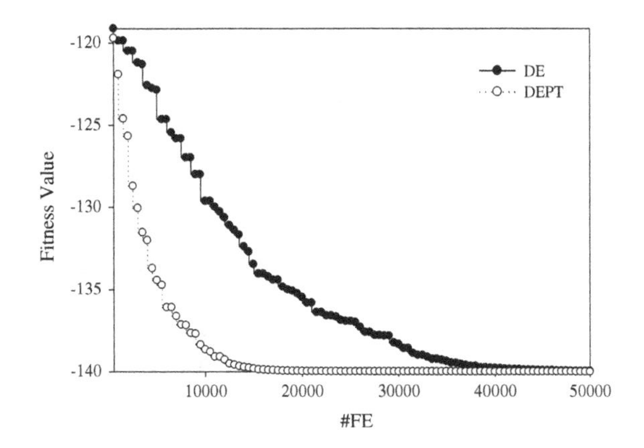

Fig. 3 Convergence graph for f_Sh_6 between mean fitness and #FE

5.2 *In Terms of Non-traditional Benchmark Function*

In this section, numerical results for scalable problems are drawn and comparison is shown with respect to DE algorithm. From Table 5, it is clear that the proposed approach produces much better results than DE algorithm for all functions except f_Sh_2 shown in boldface. Also, the convergence graph between the mean fitness and #FE for function f_Sh_6 is plotted in Fig. 3.

6 Conclusion and Discussion

The proposed algorithms provide a quality solution to global optimization by making use of regional size adaptation in MRLDE algorithm and running both MRLDE and DE algorithms simultaneously. However, the proposed DEPT algorithm depends on the concentration of the population, and when the population gets dense then the MRLDE, which was running with an additional strategy: base region size adaptation, switches to basic DE algorithm. The efficiency of the proposed variant has been verified with traditional as well as non-traditional benchmark functions in terms of some predefined performance criteria. The numerical results indicate that the proposed algorithms perform better than the basic DE and perform either better or at par with the other modified variants proposed in literature.

References

1. Storn, R., Price, K.: Differential evolution—a simple and efficient heuristic for global optimization over continuous spaces. J. Global Optim. **11**, 341–359 (1997)
2. Ali, M., Ahn, C.W., Pant, M.: Multi-level image thresholding by synergetic differential evolution. Appl. Soft Comput. **17**, 1–11 (2014). doi:10.1016/j.asoc.2013.11.018
3. Zaheer, H., Pant, M.: A differential evolution approach for solving integer programming problems. Adv. Intell. Syst. Comput. **336**, 413–424 (2014). doi:10.1007/978-81-322-2220-0_33
4. Plagianakos, V., Tasoulis, D., Vrahatis, M.: A review of major application areas of differential evolution. Adv. Differ. Evol. **143**, 19–238 (2008)
5. Ali, M., Pant, M., Abraham, A.: Unconventional initialization methods for differential evolution. Appl. Math. Comput. **219**, 4474–4494 (2013)
6. Babu, B.V., Angira, R.: Modified differential evolution (MDE) for optimization of non-linear chemical processes. Comput. Chem. Eng. **30**, 989–1002 (2006)
7. Das, S., Abraham, A., Chakraborty, U., Konar, A.: Differential evolution using a neighborhood based mutation operator. IEEE Trans. Evol. Comput. **13**(3), 526–553 (2009)
8. Ali, M., Pant, M.: Improving the performance of differential evolution algorithm using cauchy mutation. Soft. Comput. **15**(5), 991–1007 (2010). doi:10.1007/s00500-010-0655-2
9. Qin, A.K., Huang, V.L., Suganthan, P.N.: Differential evolution algorithm with strategy adaptation for global numerical optimization. IEEE Trans. Evol. Comput. **13**(2), 398–417 (2009)
10. Cai, Y., Wang, J., Yin, J.: Learning enhanced differential evolution for numerical optimization. Soft Comput. **16**(2), 303–330 (2011). doi:10.1007/s00500-011-0744-x
11. Brest, J., Greiner, S., Boskovic, B., Mernik, M., Zumer, V.: Self adapting control parameters in differential evolution: a comparative study on numerical benchmark problems. IEEE Trans. Evol. Comput. **10**(6), 646–657 (2006)
12. Zhang, J., Sanderson, A.: JADE: Adaptive differential evolution with optional external archive. IEEE Trans. Evol. Comput. **13**(5), 945–958 (2009)
13. Kumar, P., Pant, M.: Enhanced mutation strategy for differential evolution. In: IEEE Congress on Evolutionary Computation, pp. 1–6 (2012). doi:10.1109/cec.2012.6252914
14. Zhu, W., Tang, Y., Fang, J.A., Zhang, W.: Adaptive population tuning scheme for differential evolution. Inf. Sci. **223**, 164–191 (2013)
15. Zaheer, H., Pant, M., Kumar, S., Monakhov, O., Monakhova, E., Deep, K.: A new guiding force strategy for differential evolution. Int. J. Syst. Assur. Eng. Manage. 1–14 (2015). doi:10.1007/s13198-014-0322-6
16. Kaelo, P., Ali, M.M.: A numerical study of some modified differential evolution algorithms. Eur. J. Oper. Res. **169**, 1176–1184 (2006)
17. Tang, K., Yao, X., Suganthan, P.N., MacNish, C., Chen, Y.P., Chen, C.M., Yang, Z.: Benchmark Functions for the CEC'2008 Special Session and Competition on Large Scale Global Optimization. Technical Report CEC-08, 1–18 (2008)

PSO Optimized and Secured Watermarking Scheme Based on DWT and SVD

Irshad Ahmad Ansari, Millie Pant and Chang Wook Ahn

Abstract The present study proposed a robust and secure watermarking scheme to authenticate the digital images for ownership claim. The proposed watermarking scheme is making use of 2-level of DWT (Discrete wavelet transform) to provide high capacity of watermark embedding. The SVD (singular value decomposition) is performed on the host and watermark images. Then, principal components are calculated for watermark image. The use of principal components for watermark embedding makes the scheme free from false positive error. PSO (particle swarm optimization) optimized multiple scaling factors along with principal components are utilized for watermark embedding in the singular values of host image. The PSO-optimized scaling factor provides a very good tradeoff between imperceptibility and robustness of watermarking scheme. The scheme is also extended to use for color images. The proposed scheme provides a secured and high data embedding with good robustness toward different signal processing attacks.

Keywords Discrete wavelet transform · Principal components · Secured watermarking · Particle swarm optimization · Robust watermarking

1 Introduction

With the developments in information technology and computer-based communication, the sharing of digital data becomes very easy on the world wide web. Such unprotected sharing of digital data increased the cases of false ownership claims and

I.A. Ansari (✉) · M. Pant
Department of ASE, Indian Institute of Technology Roorkee, Roorkee, India
e-mail: 01.irshad@gmail.com

M. Pant
e-mail: millifpt@iitr.ac.in

C.W. Ahn
Department of CSE, Sungkyunkwan University, Suwon, Republic of Korea
e-mail: cwan@skku.edu

M. Pant et al. (eds.), *Proceedings of Fifth International Conference on Soft Computing for Problem Solving*, Advances in Intelligent Systems and Computing 437, DOI 10.1007/978-981-10-0451-3_38

copyright violations in the recent past. These digital data include images, audios, and videos. Copyrighted images can easily be manipulated with the help of powerful image processing tools in such a way that it leads to financial loss [1]. So there is a need of such tool which can detect/protect these manipulations. Digital image watermarking is one of such tools, which can protect the ownership of digital image by adding some ownership information (watermark) to the host image. Robust watermark is inserted into the host image in such a way that it remains robust to intentional/unintentional attacks and provides the ownership proof, whenever needed [2].

Broadly, image watermarking techniques are divided into visible and invisible image watermarking [1]. Visible image watermarking is a technique in which a logo (watermark) is placed in the original image which helps in identifying the owner of the image. On the other hand, invisible watermarking can again be divided into time domain and transfer domain techniques [1]. Time domain techniques are those, where the watermark is directly embedded by changing the pixel values of the image. On the other hand, original image has to undergo some transformations before the watermark embedding in transfer domain techniques. Transformed domain image watermarking techniques perform better as compared to simple time domain techniques [3]. Transformed domain techniques can further be divided into DCT (discrete cosine transform) [4], FFT (fast fourier transform), DWT (discrete wavelet transform) [5], and SVD (singular value decomposition) [6, 7] based image watermarking. A good watermarking scheme must have a high value of these four features: robustness, imperceptibility, capacity, and security [1, 3]. In this paper, we proposed an optimization of SVD (false positive free) based image watermarking technique in DWT domain using particle swarm optimization so that high value of all the four features can be attained.

Rest of the paper is organized as follows: Sect. 2 gives a brief review of related works and gaps in image watermarking domain. In Sect. 3, basics of DWT, PSO, and false positive solution are discussed, Sect. 4 talks about the proposed watermarking scheme, result and discussions are shown in Sect. 5 and finally Sect. 6 provides a brief conclusion.

2 Related Work

The SVD-based scheme reportedly performs better in the recently developed watermarking algorithms [1, 5, 6]. The main reason for good performance is that the dominant singular values get very less affected by the intentional/unintentional attacks and show very small changes [7]. Generally, SVD-based watermarking is used with other transforms; because used alone increases its computational complexity to quite a high level. Ganic et al. [8] proposed the use of SVD along with DWT to provide robust watermarking scheme. Rastergar et al. [9] suggested shuffling of the host image pixels to increase the security for a 3-level DWT and SVD-based watermarking scheme. Lagzian et al. [10] proposed the use of RDWT at the place of DWT to increase the capacity and then utilized the scheme proposed by Ganic et al. [8] for complete watermarking implementation. This scheme [10] performed better than other schemes in terms of capacity and robustness but suffers with a basic security flaw, which makes the scheme unusable [11]. Lai et al. [12] proposed a secure scheme with DWT. They decompose the watermark into two halves and embedded it into the LH and HL bands of DWT transformed host image. The main drawback of this scheme was low capacity. Makbool et al. [13] proposed a scheme with improved capacity by making use of RDWT. They claimed their scheme to be false positive free but later, it is proved by Ling et al. [14] that it also suffers with a security flaw (false positive problem), which makes their scheme vulnerable to attacks.

This study proposes a new watermarking scheme based on DWT, which is free from false positive error. In order to enhance the robustness further, proposed scheme also makes use of particle swarm optimization for optimal selection of scaling factors during embedding of watermark.

3 Preliminaries

3.1 Particle Swarm Optimization

Soft computing techniques (ANN, GA, PSO, etc.) are used in many fields for efficient problem solving [15, 16]. PSO is an optimization technique proposed by Kennedy et al. [17] in 1995. It is inspired by the swarm behavior of animals like bird flocking, animal herding, fish schooling, etc. Each member in the group learns from its own best performance as well as from the best performance of other members. This kind of property helps it to reach global optimum value in an efficient manner. PSO is a powerful tool for complex and multidimensional search [18, 19]. The very first step is the bounded initialization of swarm particles in the search space. Let each particle P_i have an initial position and velocity of $x_i(t)$ and $v_i(t)$ respectively at the time t. These positions assume to be the local best for the first iteration. PSO Algorithm has these steps:

(1) Initialization of position $x_i(t)$ and velocity $v_i(t)$
(2) Compute the value of fitness function for each position
(3) Compute the local best position, i.e., each particles best position in all the iterations
(4) Compute the global best position, i.e., best particle's position in current iteration
(5) Update the velocity $v_i(t)$ and position $x_i(t)$ of each particle using Eqs. 1 and 2

$$v_i(t+1) = w * v_i(t) + c1 * \text{rand}(l_{\text{best}} - x_i(t)) + c2 * \text{rand}(g_{\text{best}} - x_i(t)) \quad (1)$$

$$\text{and}\, x_i(t+1) = x_i(t) + v_i(t) \quad (2)$$

Here w is inertia weight used to determine the step size for each iteration. $c1$ and $c2$ are the learning factors, which determine the effectiveness of local and global learning and rand function is used to generate a number between (0, 1).

(6) Repeat the steps (2) to (5) till stopping criterion reached.

The stopping criterion can be the maximum number of iterations or change of fitness function below a certain threshold level.

3.2 Discrete Wavelet Transform

DWT provides excellent spatio-frequency localization and this makes it very useful for watermark hiding; as it provides high value of robustness and imperceptibility. Many watermarking schemes have been proposed using this feature of DWT. Then again, DWT is not able to provide a high capacity (large amount of data hiding) due to the shift variance. Shift variance occurs because of down sampling for each level filtering. Even a small variance makes noticeable changes in the wavelet coefficients. This led to inaccurate reconstruction of cover and watermark images.

RDWT is known and implemented by many names like aa trous algorithm, discrete wavelet frames, overcomplete DWT, undecimated DWT, and shift invariant DWT [20]. First time, the implementation of RDWT was done by the aa trous algorithm. It basically removes the down sampling operator from the normal implementation of DWT [20]. To keep the spatial location intact in each sub-band, each band must be of same size as of original image. On the contrary, the size of each level band decreases as the number of level increases in DWT. So, the RDWT kept the same size on each level band and provides more accurate local texture representation [21]. It increases the capacity of RDWT. Then again, any change in RDWT (due to watermark insertion) leads to loss of complete host image because of the band's high correlation (dependence) on each other [14]. In order to get a trade off in this problem (capacity vs. data recovery), DWT is used in proposed work with 2-level of transform.

3.3 Ensuring the False Positive Free Nature of Scheme

The problem of false positive error comes into picture because of complete dependence of extracted watermark on the singular vectors matrices (*U* and *V*), which are supplied by the owner. A supply of false/changed matrices (*U* and *V*) during watermark extraction process led to false/wrong watermark extraction. This dependence need to be removed from the watermarking scheme in order to provide rightful ownership check. Run et al. [22] suggest that one should embed the principal components (PC) of watermark into the host image instead of singular values. The PC values can be obtained by multiplying the singular values to the left singular matrix. The proposed scheme is also embedding the principal components of watermark into the singular values of host image to get rid of false positive error.

4 Watermarking Scheme

The watermarking scheme is divided into three stages: Embedding process, Extraction process, and optimization of scaling factors.

4.1 Embedding Process

The block diagram of proposed scheme is shown in Fig. 1 and it contains following steps:

(1) Perform DWT (2-level) on the host image (H) and obtain (2-level) four bands I_i (i = LL, HL, LH and HH) of transformed host image.
(2) Perform singular value decomposition of these bands using Eq. 3. Here i = LL, HL, LH, HH.

$$I_i = U_i S_i V_i^T \tag{3}$$

(3) Perform DWT (1-level) on the watermark image and obtain four bands W_i (i = LL, HL, LH and HH) of watermark (1-level).
(4) Perform singular value decomposition of the watermark image using Eq. 4 and compute principal components (PC's) using Eq. 5. Here i = LL, HL, LH, HH.

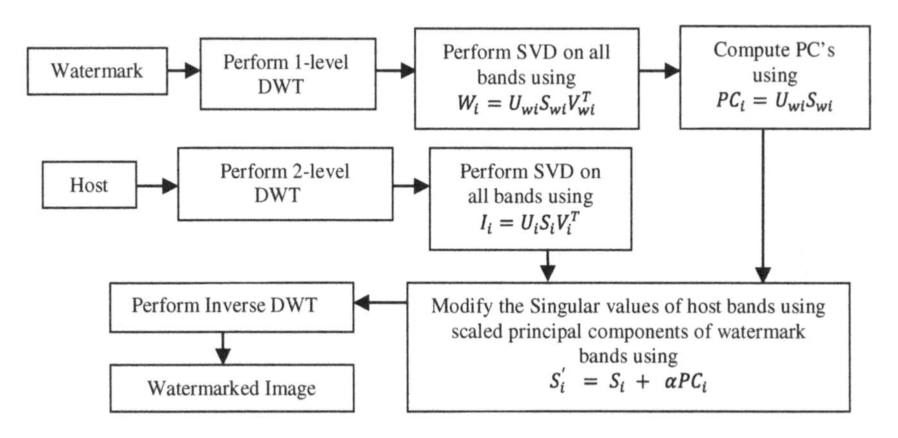

Fig. 1 Block diagram of embedding process

$$W_i = U_{wi}S_{wi}V_{wi}^T \tag{4}$$

$$PC_i = U_{wi}S_{wi} \tag{5}$$

(5) Modify the singular values of host image (each band) with the scaled principal components of watermark (each band) using Eq. 6.

$$S_i' = S_i + \alpha_i PC_i \tag{6}$$

where α is the scaling factor and PC is the principal components.

(6) Use matrix S_i' along with U_i and V_i to create the watermarked sub bands I_{wi} for each band as shown in the Eq. 7.

$$I_{wi} = U_i S_i' V_i^T \tag{7}$$

(7) Perform inverse DWT on I_{wi} to generate watermarked Image H_w.

4.2 Extraction Process

The attacked watermarked image is represented by H_w^*. For proper extraction of watermark, we need original host image. Firstly, DWT is performed on the watermarked image to obtain I_{wi}^*. The extraction process is performed as shown in Fig. 2 and described in following steps:

(1) Perform DWT (1-level) on the original host image and watermarked image to obtain transformed I_i and I_{wi}^* respectively.
(2) Perform subtraction of watermarked bands from the original host bands to obtain P_i using Eq. 8. Here i = LL, HL, LH, HH.

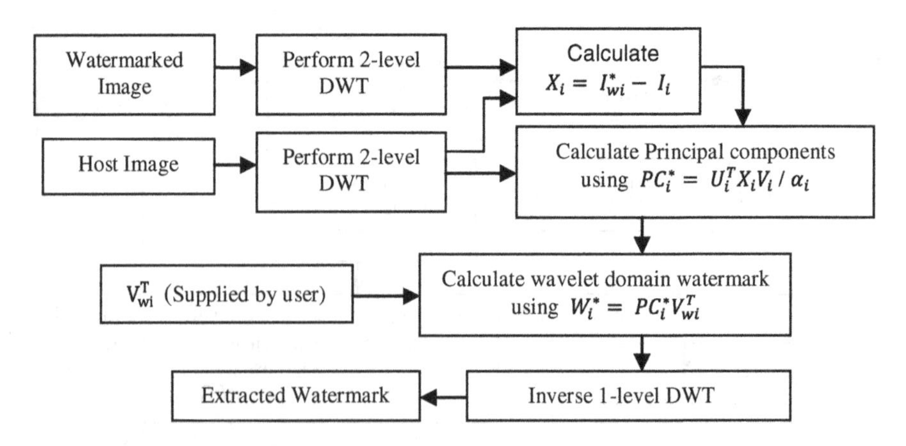

Fig. 2 Block diagram of extraction process

$$X_i = I^*_{wi} - I_i \tag{8}$$

(3) Calculate probably corrupted principal components PC^*_i using Eq. 9.

$$PC^*_i = U^T_i X_i V_i / \alpha_i \tag{9}$$

(4) Use PC^*_i along with user's supplied matrix V^T_{wi} to regenerate watermark image (wavelet domain) with the help of Eq. 10. Here i = LL, HL, LH, HH.

$$W^*_i = PC^*_i V^T_{wi} \tag{10}$$

(5) Perform Inverse DWT to regenerate time domain watermark image W.

4.3 Optimization of Scaling Factors Using PSO

It is quite visible from the embedding and extraction process that the quality of formed watermarked image and extracted watermark greatly depends on the scaling factor α. A low value of scaling factor degrades the robustness of watermark where a high value minimizes the imperceptibility so there is a need to choose an optimal value of scaling factor, which provides a balance between imperceptibility and robustness. These two factors are defined below:

$$\text{Imperceptibility} = \text{correlation}(I, I_w) \tag{11}$$

$$\text{Robustness} = \text{correlation}(W, W^*) \tag{12}$$

Here I is host image, I_w is watermarked image, W is original watermark, and W^* is extracted watermark. Suppose that N type of attacks has been considered then the robustness will become

$$\text{Inverse Robustness} = \frac{N}{\sum_{i=1}^{N} \text{correlation}(W, W^*_i)} \tag{13}$$

As the objective is to maximize both imperceptibility and robustness, the following objective function has been created for minimization:

$$\text{Error} = \frac{N}{\sum_{i=1}^{N} \text{correlation}(W,\ W^*_i)} - \text{correlation}\ (A, A_w) \tag{14}$$

The error function will become a multidimensional search, which cannot be visualized graphically and needs special tool like PSO for optimal value search of scaling factors [α].

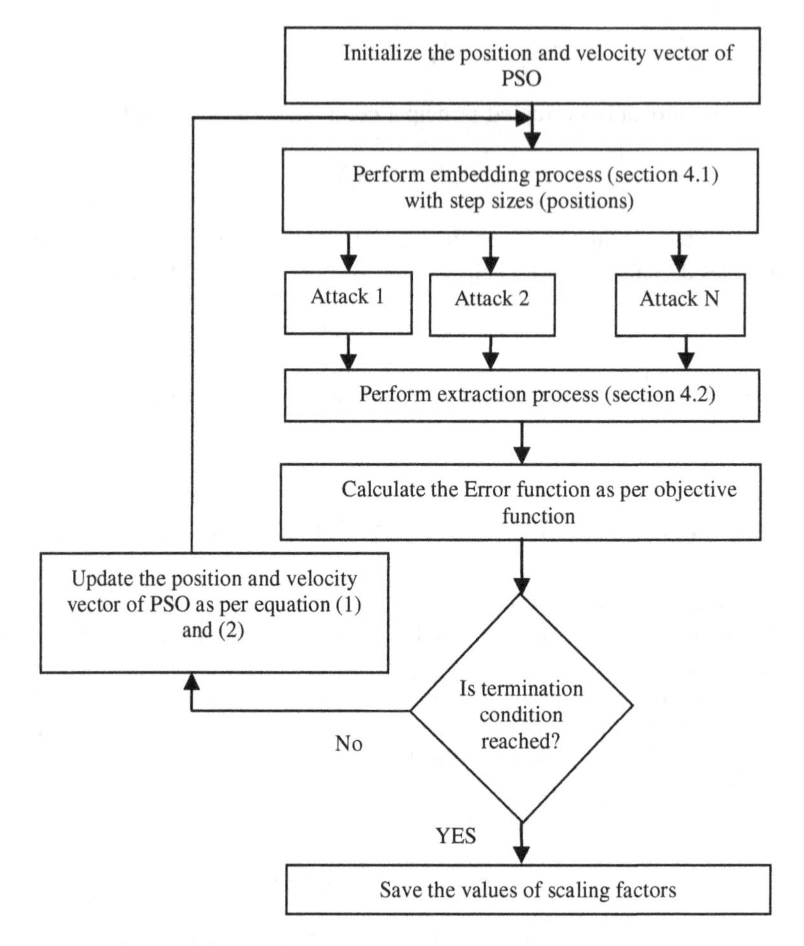

Fig. 3 Block diagram of scaling factor optimization

A bounded initialization of population size of 50 has been done in the range of 0.001–0.8. 200 generations have been used in the PSO optimization. The value of PSO's step size with other parameters has been kept low (w = 0.1, c1 = 0.8, c2 = 0.12) through out the search in order to get better local search as the single optimal value of α was known beforehand. The block diagram of optimization process is shown in Fig. 3.

5 Results and Discussions

Three host images (Lena, Man and Baboon) are used in this study as shown in Fig. 4. All the hosts are of size 512 × 512. The DWT provides the half size of all the bands as that of the size of time domain host image. This makes the capacity of

Fig. 4 Host images and watermark: Lena, Man, Baboon, and Watermark logo (*left* to *right*)

DWT-based watermarking scheme half as compared to the size of host image. So the watermark is also taken as half size of host image, i.e., 256 × 256 and the same is shown in Fig. 4. PSNR (Peak signal to noise ratio) is used to show the similarity between host images and watermarked images. Normal cross-correlation is used to compare the similarity of extracted watermarks with original watermark. Suppose that the size of two images X and X^* is $n \times n$ and they can attain a maximum pixel value as $X_{\max}$. Then PSNR and normalized cross-correlation can be defined as:

$$\mathrm{PSNR} = 10\log_{10}\left(\frac{n \times n \times (X_{\max})^2}{\sum_{i=1}^{n}\sum_{i=1}^{n}(X(i,j) - X^*(i,j))^2}\right) \tag{15}$$

$$\mathrm{correlation}(X, X^*) = \frac{\sum_{i=1}^{n}\sum_{j=1}^{n}\overline{X_{(i,j)}\mathrm{XOR}\,X^*_{(i,j)}}}{n \times n} \tag{16}$$

5.1 *Imperceptibility and Robustness of Scheme*

The quality of watermarked image (Imperceptibility) and extracted watermark (Robustness) with different scaling factors is shown in this section. Table 1 is showing the PSNR of watermarked images with different scaling factors. Table 2 is showing the normalized cross-correlation of watermarked host (Host (W)) and extracted watermark with respect to original host and original watermark, respectively.

Table 1 Variation of PSNR of watermarked images with scaling factors

Host image	PSNR (dB)		
	$\alpha = 0.01$	$\alpha = 0.05$	$\alpha = 0.25$
Lena	38.5334	32.6533	24.4235
Man	37.6545	31.7243	23.7534
Baboon	37.5343	31.2476	23.5464

Table 2 Variation of NCC of watermarked image and extracted watermark with scaling factors

Host image	NCC (normalized cross co-relation)					
	$\alpha = 0.01$		$\alpha = 0.05$		$\alpha = 0.25$	
	Host (W)	Watermark	Host (W)	Watermark	Host (W)	Watermark
Lena	0.9754	0.8743	0.9264	0.9634	0.7943	0.9934
Man	0.9643	0.8856	0.9153	0.9554	0.7865	0.9832
Baboon	0.9676	0.8875	0.9132	0.9564	0.7912	0.9859

5.2 *Optimization of Scaling Factors*

As we can see from the Tables 1 and 2 that the imperceptibility and robustness are dependent on scaling factor and are inversely proportional to each other. So, PSO-optimized multiple scaling factors are used to find an optimal trade off between them; as explained in Sect. 4.3. Figure 5 is showing the watermarked images and Table 3 is showing the imperceptibility and robustness (without any attack) of proposed scheme after the scaling factor optimization.

Table 4 is showing the proposed watermarking scheme's performance under different attacks for three host images.

Fig. 5 Watermarked images: Lena, Man, and Baboon (*left* to *right*)

Table 3 Imperceptibility and robustness (without any attack) of proposed scheme after the scaling factor optimization

Host image	PSNR of watermarked image	NCC of extracted watermark
Lena	34.3563	0.9942
Man	33.5346	0.9874
Baboon	33.6455	0.9894

Table 4 NCC of extracted watermarks under different attacks with respect to original watermark

Attack	NCC of extracted watermarks		
	Lena	Man	Baboon
Average filtering (3 × 3)	0.9454	0.9353	0.9376
Rescaling (512 → 256 → 512)	0.9778	0.9664	0.9687
Gamma correction (0.8)	0.9675	0.9535	0.9543
Median filtering (3 × 3)	0.9784	0.9687	0.9676
Histogram equalization	0.9687	0.9554	0.9576
Gaussian noise (mean = 0, variance = 0.01)	0.8875	0.8754	0.8754
JPEG compression (QF = 50)	0.9885	0.9776	0.9787
Gaussian low-pass filter	0.9902	0.9812	0.9814
Sharpening	0.9054	0.8954	0.8976
Contrast adjustment 20 %	0.9243	0.9165	0.9198

5.3 Applicability of Scheme to Color Images

The proposed scheme is not directly applicable to the color images as the color images contain three channels contrary to single channel of gray host. Figure 6 is showing the color host images used in this study.

The color host is first divided into the RGB (Red-green-blue) channels and then one/multiple channels are selected for watermark embedding. In case of multiple channels, same watermark is embedded in all the channels. PSO-optimized multiple scaling factors are utilized in this embedding. The imperceptibility is shown in Table 5 and robustness of scheme toward different signal processing attacks is shown in Table 6 for host "RGB-Lena." The scheme performs quite similar with the host "RGB-Baboon" also.

Fig. 6 Color host images: RGB-Lena and RGB-Baboon (*left* to *right*)

Table 5 Imperceptibility of color hosts during single and multiple channel embedding

Image name	PSNR (dB)			
	Red channel	Green channel	Blue channel	All channel
RGB-Lena	48.4256	47.9739	48.5734	34.4795
RGB-Baboon	48.5457	47.9374	48.6385	34.5645

Table 6 Robustness of scheme toward different signal processing attacks for host "RGB-Lena"

Attack	NCC of extracted watermarks			
	Red channel	Green channel	Blue channel	All channel
Average filtering (3 × 3)	0.9636	0.9665	0.9644	0.9464
Rescaling (512 → 256 → 512)	0.9869	0.9843	0.9875	0.9763
Gamma correction (0.8)	0.9765	0.9753	0.9743	0.9664
Median filtering (3 × 3)	0.9868	0.9832	0.9886	0.9786
Histogram equalization	0.9822	0.9853	0.9876	0.9665
Gaussian noise (mean = 0, variance = 0.01)	0.9476	0.9432	0.9476	0.8876
JPEG compression (QF = 50)	0.9953	0.9965	0.9987	0.9875
Gaussian low-pass filter	0.9982	0.9978	0.9975	0.9963
Sharpening	0.9554	0.9576	0.9575	0.9065
Contrast adjustment 20 %	0.9443	0.9473	0.9487	0.9274

6 Conclusion

This work utilized embedding to principal components of watermark in the singular values of host to make the watermarking free from false positive error. The watermarking was done in the domain of DWT, so proposed scheme provided decent data embedding. The robustness and imperceptibility of scheme was improved by making use of PSO-optimized multiple scaling factors instead of single scaling factor. After a bit modification in the scheme, it worked almost similar with the color hosts also. In future, more variants of PSO and other soft computing techniques [23–27] will be applied to proposed scheme in order to improve the scheme's performance.

Acknowledgments This work was supported by Ministry of Human Resource Development, India.

References

1. Haouzia, A., Noumeir, R.: Methods for image authentication: a survey. Multimedia Tools Appl. **39**(1), 1–46 (2008)
2. Ansari, I.A., Pant, M.: SVD Watermarking: particle swarm optimization of scaling factors to increase the quality of watermark. In: Proceedings of Fourth International Conference on Soft Computing for Problem Solving, pp. 205–214. Springer, India (2015)
3. Potdar, V.M., Han, S., Chang, E.: A survey of digital image watermarking techniques. In: 3rd IEEE International Conference on Industrial Informatics INDIN'05, pp. 709–716 (2005)
4. Lin, S., Chen, C.F.: A robust DCT based watermarking for copyright protection. IEEE Trans. Consum. Electron. **46**(3), 415–421 (2000)
5. Ansari, I.A., Pant, A., Ahn, C.W.: Robust and false positive free watermarking in IWT domain using SVD and ABC. Eng. Appl. Artif. Intell. **49**, 114-125 (2016)
6. Ansari, I.A., Pant, A., Ahn, C.W.: SVD based fragile watermarking scheme for tamper localization and self-recovery. Int. J. Mach. Learn. & Cyber. 1–15 (2015), doi:10.1007/s13042-015-0455-1
7. Ansari, I.A, Pant, M., Neri, F.: Analysis of gray scale watermark in RGB host using SVD and PSO. In: IEEE Computational Intelligence for Multimedia, Signal and Vision Processing (CIMSIVP), pp. 1–7 (2014)
8. Ganic, E., Eskicioglu, A.M.: Robust embedding of visual watermarks using discrete wavelet transform and singular value decomposition. J. Electron. Imaging **14**(4), 043004–043009 (2005)
9. Rastegar, S., Namazi, F., Yaghmaie, K., Aliabadian, A.: Hybrid watermarking algorithm based on singular value decomposition and radon transform. Int. J. Electron. Commun. (AEU) **65**(7), 658–663 (2011)
10. Lagzian, S., Soryani, M., Fathy, M.: A new robust watermarking scheme based on RDWT–SVD. IJIIP Int. J. Intell. Inform. Process. **2**(1), 22–29 (2011)
11. Guo, J.M., Prasetyo, H.: Security analyses of the watermarking scheme based on redundant discrete wavelet transform and singular value decomposition. AEU-Int. J. Electron. Commun. **68**(9), 816–834 (2014)
12. Lai, C.C., Tsai, C.C.: Digital image watermarking using discrete wavelet transform and singular value decomposition. IEEE Trans. Instrum. Meas. **59**(11), 3060–3063 (2010)
13. Makbol, N.M., Khoo, B.E.: Robust blind image watermarking scheme based on redundant discrete wavelet transform and singular value decomposition. AEU-Int. J. Electron. Commun. **67**(2), 102–112 (2013)
14. Ling, H.-C., Phan, R.C.-W., Heng, S.H.: Comment on robust blind image watermarking scheme based on redundant discrete wavelet transform and singular value decomposition. AEU-Int. J. Electron. Commun. **67**(10), 894–897 (2013)
15. Ansari, I.A., Singla, R., Singh, M.: SSVEP and ANN based optimal speller design for brain computer interface. Comput. Sci. Tech. **2**(2), 338–349 (2015)
16. Kant, S., Ansari, I. A.: An improved K means clustering with Atkinson index to classify liver patient dataset. Int. J. Syst. Assur. Eng. Manage. 1–7 (2015). doi:10.1007/s13198-015-0365-3
17. Eberhart, R.C., Kennedy, J.: A new optimizer using particle swarm theory. In: Proceedings of the sixth international symposium on micro machine and human science, vol. 1, pp. 39–43 (1995)
18. Kiranyaz, S., Ince, T., Yildirim, A., Gabbouj, M.: Fractional particle swarm optimization in multidimensional search space. IEEE Trans. Syst. Man Cybern. B Cybern. **40**(2), 298–319 (2010)
19. Messerschmidt, L., Engelbrecht, A.P.: Learning to play games using a PSO-based competitive learning approach. IEEE Trans. Evol. Comput. **8**(3), 280–288 (2004)
20. Fowler, J.: The redundant discrete wavelet transform and additive noise. IEEE Signal Process. Lett. **12**(9), 629–632 (2005)

21. Cui, S., Wang, Y.: Redundant wavelet transform in video signal processing. In: International conference on image processing, computer vision, pattern recognition, pp. 191–196 (2006)
22. Run, R.-S., Horng, S.-J., Lai, J.-L., Kao, T.-W., Chen, R.-J.: An improved SVD- based watermarking technique for copyright protection. Expert Syst. Appl. **39**(1), 673–689 (2012)
23. Jalhar, S.K., Pant, M., Nagar, M.C.: Differential evolution for sustainable supplier selection in pulp and paper industry: a DEA based approach. Comput. Methods Mater. Sci. **15** (2015)
24. Zaheer, H., Pant, M.: A differential evolution approach for solving integer programming problems. In: Proceedings of Fourth International Conference on Soft Computing for Problem Solving, pp. 413–424. Springer, India (2015)
25. Jauhar, S.K., Pant, M., Abraham, A.: A novel approach for sustainable supplier selection using differential evolution: a case on pulp and paper industry. In: Intelligent Data analysis and its Applications, vol. II, pp. 105–117. Springer International Publishing (2014)
26. Zaheer, H., Pant, M., Kumar, S., Monakhov, O., Monakhova, E., Deep, K.: A new guiding force strategy for differential evolution. Int. J. Syst. Assur. Eng. Manag. 1–14 (2015), doi:10.1007/s13198-014-0322-6
27. Jauhar, S.K., Pant, M.: Genetic algorithms, a nature-inspired tool: review of applications in supply chain management. In: Proceedings of Fourth International Conference on Soft Computing for Problem Solving, pp. 71–86. Springer, India (2015)

Spectral Indices Based Change Detection in an Urban Area Using Landsat Data

Abhishek Bhatt, S.K. Ghosh and Anil Kumar

Abstract This paper proposes a technique to detect the change in some dominantly available classes in an urban area such as vegetation, built-up, and water bodies. Landsat Thematic Mapper (TM) and Landsat 8 imageries have been selected for a particular area of NCR (National Capital Region), New Delhi, India. In this study, three spectral indices have been used to characterize three foremost urban land-use classes, i.e., normalized difference built-up index (NDBI) to characterize built-up area, modified normalized difference water index (MNDWI) to signify open water and modified soil-adjusted vegetation index ($MSAVI_2$) to symbolize green vegetation. Subsequently, for reducing the dimensionality of Landsat data, a new FCC has been generated using above mentioned indices, which consist of three thematic-oriented bands in place of the seven Landsat bands. Hence, a substantial reduction is accomplished in correlation and redundancy among raw satellite data, and consequently reduces the spectral misperception of the three land-use classes. Thus, uniqueness has been gained in the spectral signature values of the three dominant land-use classes existing in an urban area. Further, the benefits of using $MSAVI_2$ as compared with NDVI and MNDWI as compared to NDWI for the highly urbanized area have been emphasized in this research work. Through a supervised classification, the three classes have been identified on the imageries and the change between the image pairs has been found. The overall accuracy (OA) of change detection is 92.6 %. Therefore, the study shows that this technique is effective and reliable for detection of change.

Keywords Urban land use · MNDWI · $MSAVI_2$ · NDBI · Change detection

Abhishek Bhatt (✉) · S.K. Ghosh
Indian Institute of Technology Roorkee, Roorkee, Uttarakhand, India
e-mail: abhishekbhatt.iitr@gmail.com

S.K. Ghosh
e-mail: scangfce@iitr.ac.in

Anil Kumar
Indian Institute of Remote Sensing, Dehradun, Uttarakhand, India
e-mail: anil@iirs.gov.in

M. Pant et al. (eds.), *Proceedings of Fifth International Conference on Soft Computing for Problem Solving*, Advances in Intelligent Systems and Computing 437, DOI 10.1007/978-981-10-0451-3_39

1 Introduction

Urbanization for human society, is a gift, if it is well-ordered, harmonized, and planned. However, unplanned urbanization is a profanity. In 2008, it was found that more than 50 % of the total world's population were urban inhabitants and that this figure is likely to reach 81 % by 2030 [1]. In developing countries, like India, where urbanization rates are high because of mass relocation of individuals from rural places to urban fringes and from smaller to larger urban agglomerations as well. When we go to way back 1970s, the Industrial Revolution seemed as the factor which give thrust to the urbanization process in India. However, during 1990s urbanization further gained momentum due to the globalization. Many mega cities such as Delhi are facing various problems due to rapid urbanization and one of them is extremely high levels of pollution due to a speedy infrastructure development a speedy infrastructure development [2]. Due to global urbanization, there has been an increasing interest in understanding its consequences with respect to ecological factors that includes harm of agricultural land, habitat devastation, and decline in traditional vegetation.

Satellite data provides both spectral and temporal information to such dynamic phenomenon. Using temporal data, change detection at the same geographical location between two images may be carried. If the deviation in spectral response is more than the statistically determined threshold value, then one thematic class is assumed to have been converted into another thematic class [3]. In the context of urban change detection, it is found that dry soil tends to create misidentification of urban areas due to confusing spectral signature of urban feature when using raw data. To resolve such issues with raw data, certain mathematical transformation is available commonly known as spectral indices [4]. These indices tend to normalize and highlight a specific information such as, normalized difference vegetation index (NDVI), which highlights vegetation areas or normalized difference water index (NDWI), which highlights water or normalized difference built-up index (NDBI), which highlights urban areas, etc.

The objective of this research work is to cultivate and investigate the suitability of a new change detection technique using spectral indices based thematic bands for an unevenly urbanized area like New Delhi, India using the spectral indices such as modified normalized difference water index (MNDWI), modified soil-adjusted vegetation index ($MSAVI_2$), and NDBI. This method would appreciate in such a sense that it allows urban planning authorities and policy makers to appropriately update as well as review urban growth pattern with associated land-use changes along with cognizant in terms of justifiable treatment of the precious nature land. The transformed indices havd been proven a powerful tool for analyzing and detection of spatial-frequency characteristics of non-stationary images in a more comprehensive manner [5].

Thus, in this work, some spectral indices for urban change detection have been examined using Landsat 5 TM and Landsat 8 OLI_TIRS dataset have been developed. Further, the comparison of performance of the proposed change detection technique using indiced image and the change detection using raw data have been compared and urban change over a span of 14 years from 2000 to 2014 has been computed.

2 Literature Review

Many studies significantly use satellite imagery to differentiate the built-up class from non-built-up [6]. Ever since multispectral satellite-based land-use and land cover classification had been carried out, it was observed that specifically urban area as a land-use class had been identified. However, due to the non-homogeneity of urban area, since it many include many land cover classes such as vegetation, built-up, and water bodies, it was found that the accuracy of identification of urban area is normally less than 80 %. Results of various research show that there has not been a single technique to classify the built-up area but one can combine the qualities of two techniques in order to create a hybrid approach to enhance the extraction accuracy of urban land use.

Masek et al. [7] applied the NDVI-differencing approach with the support of an unsupervised classification to identify built-up areas of the metropolitan area in Washington D.C., using time series Landsat imagery while achieving an overall accuracy (OA) of 85 % [7]. Xu [8] utilizes a hybrid approach of combining supervised classification. The concept used has been centered around investigation of spectral variation between built-up and non-built land for the Fuqing City in southeastern China [8]. Built-up information was extracted, and finally, incorporated with a land cover classification output to create a final output with a great improvement in results.

Zhang et al. [9] uses spectral data further combined with road density layer and detect the change in Beijing, China using post-classification approach. Hence, it significantly reduced spectral misperception as well as improved the accuracy [9]. Zha et al. [10] proposed NDBI using TM4 and TM5 for extracting urban areas of Nanjing City, China from a Landsat TM image [10]. The thematic map derived through indices contains the vegetation noise and further filtered with the help of NDVI to eliminate the vegetation noise.

Xian and Crane [11] utilizes the regression tree algorithm and unsupervised classification to measure urban land-use development in Florida, Tampa Bay Watershed by mapping urban impervious surfaces and to divulge associated urban land use. The accuracy achieved with such an approach has been greater than 85 % [11].

Li and Liu [12] compared the associations of land surface temperature (LST) with NDBI and NDVI images using moderate resolution imaging spectroradiometer (MODIS) data acquired from four different periods of Changsha–Zhuzhou–Xiangtan metropolitan area, China [13]. However, the binary NDBI and NDVI images are created under the assumption that all positive values of continuous NDBI and NDVI images belong to built-up regions and vegetation regions, respectively. This assumption, however, makes that this approach is unable to accurately identify the built-up regions as well as built-up change regions.

El Gammal et al. [14] have used several Landsat images of different time periods (1972, 1982, 1987, 2000, 2003, and 2008) and processed these images to analyze the changes in the shores of the lake and in its water volume [14].

El-Asmar et al. [15] have applied remote sensing indices, in the Burullus Lagoon, North of the Nile Delta, Egypt for quantifying the change in the water body area of the lagoon during 1973–2011. NDWI and the MNDWI have been used for the study [15].

Bouhennache and Bouden [5] extracted the expansion in urban areas using various indices, i.e., difference soil-adjusted vegetation index (DSAVI), difference normalized difference built-up index (DNDBI). Further post-classification of multispectral and multi-temporal L5 and L7 Landsat satellite using a MLC has been carried out [5].

3 Spectral Indices Used in Study

3.1 $MSAVI_2$-Derived Image for Extraction of Vegetation

NDVI is a widely used index for getting information about the vegetation cover, but it is influenced by the soil background, so SAVI has been suggested by Huete [16]. This index include a constant L, in the denominator of the NDVI calculation and called it factor, in order to minimalize the effects of soil background on the vegetation. Normally SAVI can be represented by the following equation [16]:

$$\text{SAVI} = \frac{(\text{NIR} - \text{RED})(1 + l)}{\text{NIR} + \text{RED} + 1} \ldots \tag{1}$$

where l, signify a soil correction factor, whose value ranges from 0 to 1, where 0 stands for maximum densities to 1 for lowest densities.

Subsequently, Baret et al. [17] suggested TSAVI, which was further amended by Qi et al. [18], and based on the optimal value of the soil adjustment factor of the SAVI. It has been envisioned to provide better correction to the brightness of soil background under diverse vegetation cover situation. This study employed $MSAVI_2$ to represent vegetation due to its benefit over NDVI when applied specially for urban areas where plant cover has been less. $MSAVI_2$ can perform well in the urban area even with a minimum plant cover of 15 %, whereas for NDVI the minimum plant cover required should be above 30 % [18]. The $MSAVI_2$ can be computed using the following equation:

$$\text{MSAVI}_2 = \frac{2 * \text{NIR} + 1 - \sqrt{(2 * \text{NIR} + 1)^2 - 8}(\text{NIR} - \text{RED})}{2} \tag{2}$$

SAVI needs an earlier knowledge about vegetation densities in order to find an optimal value of L, while $MSAVI_2$ automatically adjusts the value of L to optimal one iteratively. It is the quality of $MSAVI_2$ that it raised the vegetation feature impressively and simultaneously reduces the soil-induced variation. Hence, $MSAVI_2$ can be said a more thoughtful indicator for finding degree of vegetation over SAVI as well as other indices.

3.2 *MNDWI-Derived Image for Extraction of Open Water Bodies*

NDWI as proposed by Gao [19] is a measure of liquid water molecules in vegetation canopies that interacts with the incoming solar radiation using two near IR channels.

$$\mathrm{NDWI_{GAO}} = \frac{\mathrm{NIR} - \mathrm{MIR}}{\mathrm{NIR} + \mathrm{MIR}} \tag{3}$$

Subsequently, Mc Feeters, [20] used NDWI to delineate open water features, by replacing IR band with Green band and MIR band with NIR expressed as follows [20]:

$$\mathrm{NDWI} = \frac{\mathrm{GREEN} - \mathrm{MIR}}{\mathrm{GREEN} + \mathrm{MIR}} \tag{4}$$

where GREEN band like TM2, and NIR band like TM4, for Landsat 5 dataset. As a result, positive values have been recoded for water class and zeroed or negative values have been recorded owing to vegetation. Hence, vegetation and soli have suppressed and water class has enhanced in NDWI images.

The applications of the NDWI in highly urbanized areas like New Delhi city, were not as successful as expected because here the open water body exists with built-up land as dominated background. When water bodies are extracted they often gets mixed up with built-up land due to the reason that many built-up areas also contains positive values in the NDWI-derived indiced image. To provide a suitable remedy to this problem, modified NDWI as proposed by Xu [21] has been used. MIR band, such as, TM5 of Landsat 5 has been used in place of the NIR band during calculation. The MNDWI can be articulated as follows [21]:

$$\mathrm{MNDWI} = \frac{\mathrm{GREEN} - \mathrm{MIR}}{\mathrm{GREEN} + \mathrm{MIR}} \tag{5}$$

3.3 *NDBI-Derived Image for Extraction of Built-up Land*

The NDBI has been used to highlight information related to built-up. The evolution of this index based on the exclusive spectral values of built-up lands, which have lower value of reflectance in NIR wavelength range as compared to reflectance in MIR wavelength range. So, NDBI can be represented by the following equation [10]:

$$\mathrm{NDBI} = \frac{\mathrm{MIR} - \mathrm{NIR}}{\mathrm{MIR} + \mathrm{NIR}} \tag{6}$$

It has been observed at some places that, built-up may have lower reflectance in MIR wavelength range, resulting in negative values in NDBI imagery. Whereas, in some conditions, water may reflect MIR waves stronger as compared to NIR spectra, if it contains high suspended matter concentration (SMC). Hence, there is a shift in the peak reflectance near the higher wavelength spectrum and this phenomenon becomes more dominant with intensification in suspended matter [22].

Moreover, the contrast of the NDBI image has not been good when compared to $\mathrm{MSAVI_2}$ and MNDWI images, thus many studies have suggested to make use of combination of indices rather than extraction of the built-up land merely based on an NDBI image independently. Hence, this study proposes to use a combination of NDBI, $\mathrm{MSAVI_2}$ and MNDWI to extract built-up and other dominant classes in an urban fringe. This may also lead to the improvement in the accuracy of classification.

4 Study Area and Dataset Used

The study area selected is located between Latitude 28°15′38″N to 28°54′39″N and longitude 77°26′54″E to 76°57′12″E (Fig. 1) and belongs to New Delhi and its surrounding region commonly known as the National Capital Region (NCR). The altitude of the study area lies between approximately 213–305 m above the *sea level* with 293 m (961 ft) an average altitude.

New Delhi during the past six decades or so has witnessed a phenomenal growth in population. It has increased from 1.74 million in 1951 to 16.7 million in 2011

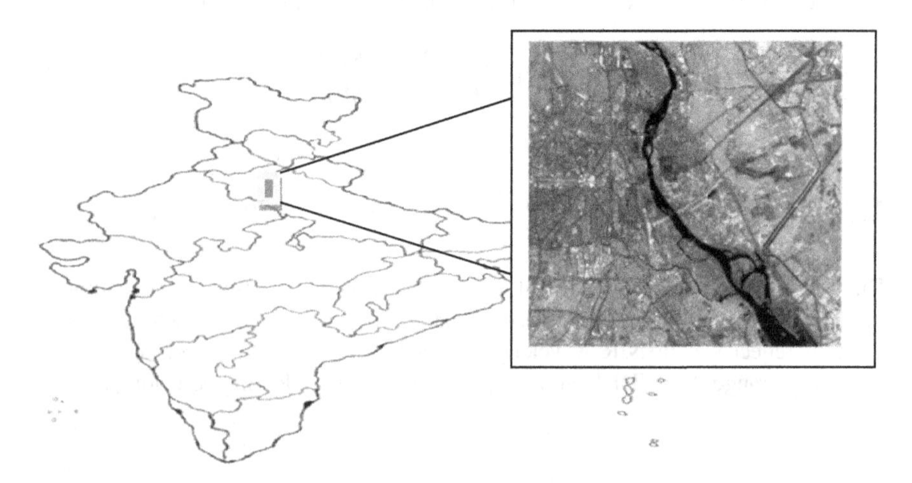

Fig. 1 Study area, i.e., New Delhi and its nearby area, India

Table 1 Summary of dataset used

Satellite	Sensor	Date of image acquisition	Path/row	Datum
Landsat 5	TM	9 May 2000	146/40	WGS 84
Landsat 8	OLI_TIRS	9May 2014	146/40	WGS 84

[Census report 2011]. To detect and analyze the change in urban built-up areas, Landsat data have been used. Satellite data selected for this purpose belong to May 2000 and May 2014 [23]. One of the main reasons for selecting this time of the year is that this is summer time and the foliage of green vegetation is nominal and the water bodies also have the smallest spread. Thus, built-up area is greatly highlighted in comparison to other urban land-use classes. Details of temporal dataset used in this study are summarized in Table 1.

5 Methodology Adopted for Proposed Change Detection

A new method has been proposed in this research work for the detection of change in urban agglomeration (Fig. 2). The change detected has been largely focused on a new FCC image generated using three thematic indices, $MSAVI_2$, MNDWI, and NDBI. The proposed approach has been validated through the detection of change in Delhi city, India using Landsat time series data of past 14 years period (i.e., from 2000 to 2014).

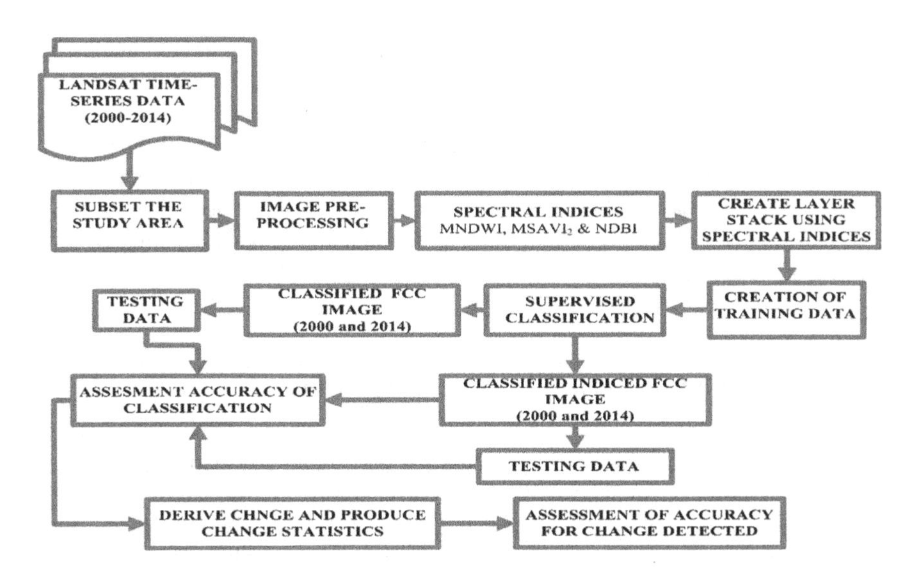

Fig. 2 Flow chart for methodology of change detection

5.1 Preprocessing of Landsat Imagery

Landsat images of have been used in this study being cloud free and of good quality, then radiometric correction was performed [24, 25]. After applying pre-processing, study area has been subset from the whole image.

5.2 Generation of Index-Derived Images

A highly urbanized area can be represented as a complex environment composed of heterogeneous ingredients. However, there have been still some generalizing constituents among these ingredients. Ridd [26] uses three components to describe an urban environment, i.e., vegetation, impervious surface, and bare earth while ignoring open water class [26]. Nevertheless, the open water also has been taken into concern in this research work as it characterizes an important element of the urban surface.

Consequently, the urban area has been categorized into the three generalized classes, i.e., built-up land, green vegetation, and open water body. Three spectral indices have used to represent each of the three dominant class, i.e., NDBI used to signify built-up, $MSAVI_2$ to represent vegetation, and MNDWI is used to characterize open water. Further FCC is generated using the above said three spectral indices.

5.3 Classification Using MLC Classifier

MLC classifier has been used for classification. The maximum likelihood (ML) classifier has turned out to be very popular and widespread in remote sensing society due to its robustness [27, 28]. ML classifier assumes the normal distribution, i.e., each class in each band can be assumed to be normally distributed. Each pixel of image has been allocated to the class with the highest probability among all classes available. For each land-use class training samples have been collected and from these known pixel labels, i.e., training data remaining pixels of images have been assigned labels.

5.4 Assessment of Accuracy for Classification

Assessment of accuracy has been an important part. Land-use map obtained after classification has been compared with the reference information collected in order to evaluate the accuracy of classified image. A total of 256 sample points have been used as the reference data by considering stratified random sampling. Handheld GPS has been used for field verification during the visits. The reference points so obtained were used to authenticate the accuracy of classification.

The three main parameters utilized for assessments of accuracy include OA, user's and producer's accuracy (UA and PA), and kappa coefficient (KC) [29–31].

5.5 *Change Detection*

5.5.1 Theory of Change Detection

A variety of change detection techniques are available for monitoring land-use/land cover change [32]. These techniques can be grouped into two main categories: post-classification comparison techniques and enhancement change detection techniques [33]. The post-classification technique involves the independent production and subsequent comparison of spectral classifications for the same area at two different time periods [34].

The proposed method using spectral indices for change detection used in this study utilize classified indiced image obtained for different years and then compares and analyzes these classified images using a change-detecting matrix and analyzed for 2000–2014 duration.

5.5.2 Assessment of Accuracy for Change Detection

The simplest method to estimate the expected accuracy of change maps is to multiply the distinct classification map accuracies. A more rigorous approach has been used here that consists of selecting the randomly sample areas divided into two categories, i.e., change and no-change and further determine the accuracy of change detected [35].

An error matrix (Table 2) has been built to compute several accuracy indicators based on Eqs. (7) to (10) for identifying change detection [29, 31].

OA is computed for whole of the image dataset and calculated using error matrix. It is the ratio of samples which are detected correctly to the total number of samples taken for assessment of accuracy, given as:

$$\mathrm{OA} = \frac{C_{\mathrm{C}} + U_{\mathrm{U}}}{T} \times 100\,\% \tag{7}$$

When described as a percentage it is nothing but the percentage of correctly changed and unchanged identified pixels to the quantity of total test samples used as reference points.

Kappa coefficient ($\hat{K}$) The kappa indicator integrates the off diagonal components of the error matrices and represents agreement gained after eliminating the amount of agreement that could be estimated to occur by chance. Kappa reveals the

Table 2 Error matrix for change detection accuracy assesment

Detection results	Changed (C)	Unchanged (U)	Total (T)
Detected changed (C)	C_c	C_u	T_c
Detected unchanged (U)	U_c	U_u	T_u
Total (T)	T_c	T_u	T

internal consistency of change detection results. It describes detection accuracy more objectively than OA.

$$\hat{K} = \frac{T(C_c + U_u) - (T_c \times T_c + T_u \times T_u)}{T^2 - (T_c \times T_c + T_u \times T_u)} \times 100\,\% \tag{8}$$

False elarm rate (*commission error*): False alarm rate can be defined as the ratio of false detected changes to total detected changes. It is the percentage of false change so called commission error. It means that the number of pixels that are not actually gone under change but have been detected as changed pixel in the final output.

$$P_F = \frac{C_u}{T_c} \times 100\,\% \tag{9}$$

Miss detection rate (*omission error*): Miss detection rate can be defined as the ratio of undetected changes to total true changes. It means that the number of pixels that are actually gone under change but have been detected as unchanged pixel in the final output. It is also called omission error.

$$P_O = \frac{U_c}{T_c} \times 100\,\% \tag{10}$$

6 Results and Discussion

6.1 Creation of Indiced Image and Classification

Landsat images acquired on two different time period over New Delhi city and its nearby area, India were selected to evaluate the proposed change detection scheme. The size of the image is 417 × 386 pixels. The images were registered using quadratic polynomial with the error less than 0.4 pixels. For each dataset corresponding spectral indices ($MSAVI_2$, MNDWI, and NDBI) are created and using these thematic information an indiced image is generated (Fig. 3).

To create the thematic map for the dominant land cover classes available the MLC classification is carried out for FCC images of years, i.e., 2000 and 2014 (Fig. 4a, b). The same classification is carried out for indiced images created using spectral indices (Fig. 4c, d). Three major classes were identified, i.e., vegetation,

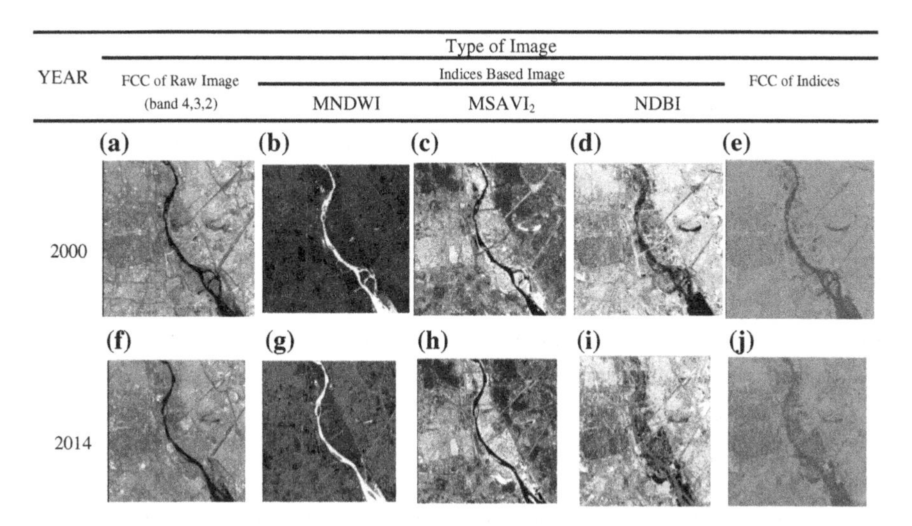

Fig. 3 Landsat FCC image of Delhi area created using band numbers 4, 3, and 2 for different years, i.e., 2000 and 2014 (**a**, **f**), respectively, with their derived MNDWI (**b**, **g**); $MSAVI_2$ (**c**, **h**); and NDBI (**d**, **i**); and the indiced FCC created using all three derived images for each year (**e**, **j**)

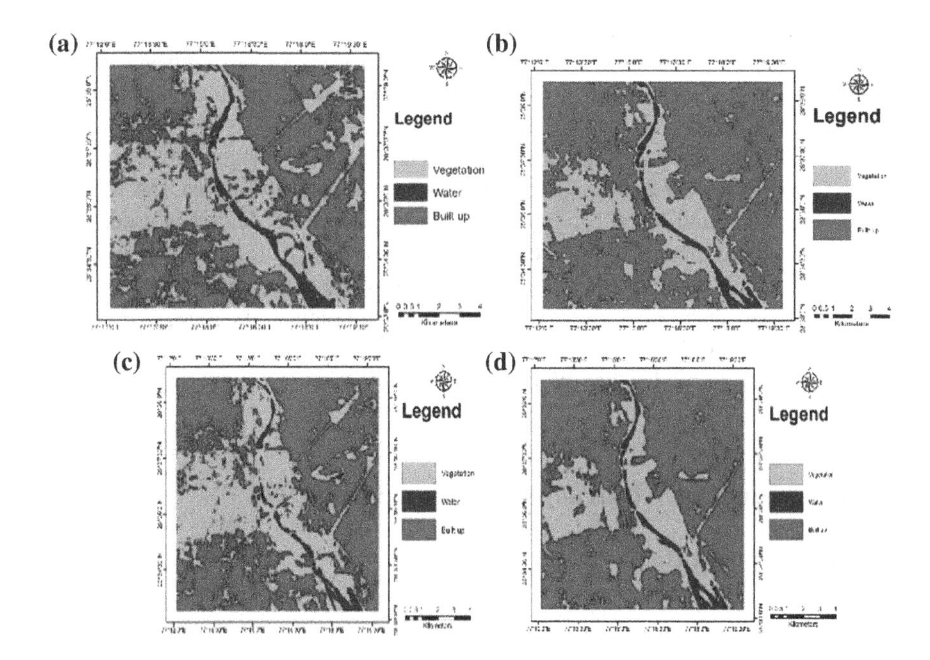

Fig. 4 LULC classified image for different years 2000 (**a**, **c**) and for 2014 (**b**, **d**)

Table 3 Assessment of accuracy for classification

Land cover	Year wise classification accuracy							
	FCC image of RAW data				Indices FCC image			
	2000		2014		2000		2014	
	PA	UA	PA	UA	PA	UA	PA	UA
Vegetation	63	95	52	80	100	83	100	83
Water	100	75	100	100	89	92	100	100
built-up	94	62	91	75	100	88	94	79
OA	76		77		86		89	
KC	0.564		0.507		0.712		0.816	

UA User accuracy (%); *PA* producer accuracy (%); *OA* overall accuracy (%); *KC* kappa coefficient

water, and built-up. The LULC classified images provide the information about the land-use pattern of the study area. The green colour represents the vegetation area; blue colour represents the water bodies whereas built-up area has been represented by red colour.

6.2 Accuracy of Classification

Accuracy is judged in terms of error matrix. A total of 256 sample pixels are chosen as testing data for assessment of accuracy of classification and their summary is given in Table 3.

The OA for the study area has been calculated using FCC image is 76 % for year 2000 and 77 % for year 2014 datasets, whereas for the same study area, OA is 86 % for the year 2000 and 89 % for year 2014 when the indiced dataset is used. This clearly shows that there is a significant rise in the accuracy of classification. Even the kappa coefficient shows a significant rise, thereby indicating that the error due to chance has also decreased. Further, as per USGS classification scheme, the OA exceeds the threshold accuracy of 85 %. Naturally, the base maps with respect to which change is to be detected will be better, leading better change detection.

7 Assessment of Accuracy for Change Detected

7.1 Change Analysis

The analysis for change enumerated in this research work has been centred around the statistics pulled out from the land-use and land cover maps of the given study area for different time period, i.e., 2000–2014. The change map derived from

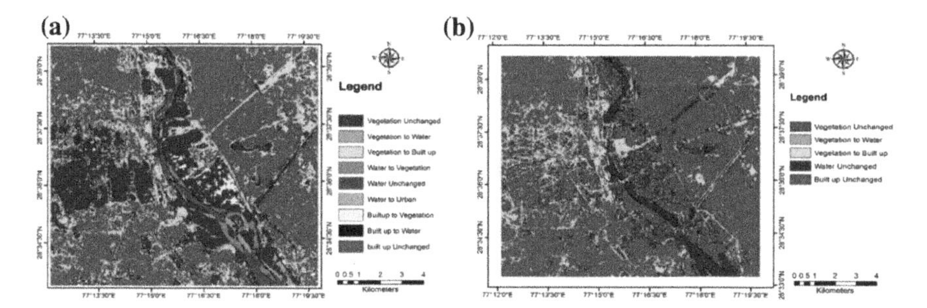

Fig. 5 Change map for the duration 2000–2014. **a** Using FCC. **b** Using indiced image

post-classification change detection are shown in Fig. 5. The change map produced for both the methods, i.e., the normal MLC classification and as well as for the proposed approach.

Using the change map of Fig. 5, the following post-classification change statistics have been created and summarized in the table for study area in the time duration 2000–2014 (Tables 4 and 5).

Table 4 represents the change obtained using the FCC image (Band 4, 3, 2), however, Table 5 represents the change that was obtained by converting the bands of Landsat data into spectral indices and execute the classification process on the indiced image generated using the spectral indices ($MSAVI_2$, MNDWI, and NDBI). Table 4 shows that that water has also been converted to vegetation and some part of the water bodies are also changed to urban whereas in case of using indiced image none of the water pixel has been changed to vegetation the same case has occurred with urban area also. The change results obtained and summarized in Table 5 are more justifiable theoretically as none of the pixel of urban change to vegetation or water as well as there is no change from water to vegetation as well as to urban. However, in both the cases (Tables 4 and 5) the urban class is increasing by compromising in vegetation and water land cover.

Table 4 Post-classification change matrix (using FCC image)

2000	2014			
	Vegetation	Water	built-up	Total
Vegetation	38,220	1555	33,934	74,309
Water	2252	3361	927	6540
Built-up	6119	3	73,991	80,113
Total	47,191	4919	108,852	160,962

Table 5 Post-classification change matrix (using indiced image)

2000	2014			
	Vegetation	Water	built-up	Total
Vegetation	32,099	2615	32,400	67,114
Water	0	4274	0	4274
built-up	0	0	89,574	89,574
Total	32,099	6889	121,974	160,962

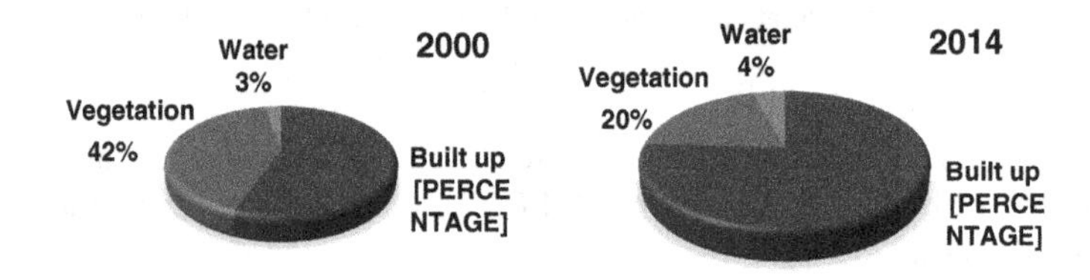

Fig. 6 Land cover for New Delhi and its nearby area (using proposed approach)

7.2 Change Statistics

The figure below shows (Fig. 6) percentage of change that has been occurred in individual land cover class. The built-up area has been changed drastically from 2000 to 2014. Built-up area has been increased by 21 %, vegetation area has also been decreased by 22 %, whereas there is a small change in the area of water bodies and it increases by merely 1 %. The increase in built-up area has many reasons. New Delhi, the capital of India is famous for industrial, educational institutions, IT sector and corresponding infrastructure developments lead to the increase of urban/built-up area.

7.3 Evaluation of the Proposed Change Detection Method

In order to evaluate the accuracy of change detected, a conventional pixel-based method was experimented. The confusion matrix (Table 6) has been assembled with the help of the existing ground truth data, i.e., a group of changed reference data (105 pixels), and a group of unchanged reference data (45 pixels) being used as test samples to evaluate accuracies of change detection methods.

After computation of *OA, kappa coefficient, false alarm rate,* and *miss detection rate* all have been calculated from the values obtained from Table 6. Using the proposed approach of change detection, i.e., FCC of spectral indices the accuracy that obtained has been 92.60 %, whereas for the same duration the accuracy achieved with raw FCC image has been 80.6 %. Table 7 summarizes the various

Table 6 Error matrix for accuracy for change detected

Class	Using FCC of raw image			Using proposed approach		
	Ground truth		Total	Ground truth		Total
	Change	Unchanged		Change	Unchanged	
Change	84	8	92	98	4	102
Unchanged	21	37	68	7	41	48
Total	105	45	150	105	45	150

Table 7 Assessment of accuracy for change detected

Change detection approach	OA	Kappa	Omission rate (%)	Commission rate (%)
Using FCC of raw image	80.6	0.57	22.8	8.7
Proposed approach	92.6	0.82	7.3	4.2

accuracy measures of change detected for the duration 2000–2014 calculated using the Eqs. (7–10) for the study area as shown in Fig. 1.

Table 7 shows the estimation of the various change accuracy measures as previously discussed in Eqs. (7–10). It may be noted that the value of OA has increased from 80.6 to 92.6 % along with there is an improvement in kappa values together with lower commission and omission rates.

It is found that the proposed approach using spectral indices to detect the change yields the best results in terms of OA, kappa, rate of omission and rate of commission in comparison to traditional approaches. Thus, it can be inferred that the proposed approach tends to preserve the change information compared with previously available techniques.

By visual analysis to the change method used, it is observed that for a highly urbanized area like New Delhi the changed pixels are fairly consistent throughout the physical urban expansion tendency. Primarily, the main changes occurred in the peri urban areas, as a result of urban planning and prompt construction

8 Conclusion

Change detection using remote sensing data proved to be a powerful approach for enumerating and investigating the amount and magnitude of landscape changes as a result of urbanization process. Subsequently, it becomes easier to present the outcomes of such evaluation to urban planners and policy-makers as well as the general public in a visually compelling and effective manner. The results of this proposed spectral indices based change detection approach for the given study area reveals that by using the given indices, i.e., MNDWI, NDBI, and $MSAVI_2$, they not only give dimensionality reduction of the data as well as more accurately changed result.

The method proposed has been beneficial and can be used in variable spectral phenomenon as this approach mostly relies on the fact that it takes the advantage of indices that differentiate the classes based on the variation in band reflectance values for each class that can be understood by examining the spectral profile of the classes.

The given approach is a more generalized approach in terms of OA calculation. The OA obtained for the land cover classification has been very much reflected in the accuracy of change detection in turn this proposed approach demonstrates its capabilities to be used for change detection.

Acknowledgments The author would like to thank 'Housing and urban development corporation' (HUDCO, New Delhi) to support this research through 'Rajiv Gandhi HUDCO fellowship,' given to the institutes of National repute in India.

References

1. Unfpa (2009) United Nations Population Fund. Public Health **64**, 25–7
2. Mohan, M., Dagar, L., Gurjar, B.R.: Preparation and validation of gridded emission inventory of criteria air pollutants and identification of emission hotspots for megacity Delhi. Environ. Monit. Assess. **130**, 323–339 (2007). doi:10.1007/s10661-006-9400-9
3. Singh, A.: Digital change detection techniques using remotely sensed data: review article. Int. J. Remote Sens. **10**, 989–1003 (1989)
4. Lyon, J.G., Yuan, D., Lunetta, R.S., Elvidge, C.D.: A change detection experiment using vegetation indices. Photogramm Eng. Remote Sens. **64**, 143–150 (1998). doi: citeulike-article-id:7262520
5. Bouhennache, R., Bouden, T.: Change detection in urban land cover using landsat images satellites, a case study in algiers town. IEEE Xplore **14**, 4799–562 (2014)
6. Guild, L.S., Cohen, W.B., Kauffman, J.B.: Detection of deforestation and land conversion in Rondônia, Brazil using change detection techniques. Int. J. Remote Sens. **25**, 731–750 (2004)
7. Masek, J.G., Lindsay, F.E., Goward, S.N.: Dynamics of urban growth in the Washington DC Metropolitan Area, 1973–1996, from Landsat observations. Int. J. Remote Sens. **21**, 3473–3486 (2000)
8. Xu, H.: Spatial expansion of urban/town in Fuqing and its driving force analysis. Remote Sens. Technol. Appl. **17**, 86–92 (2002)
9. Zhang, Q., Wang, J., Peng, X., Gong, P., Shi, P.: Urban built-up land change detection with road density and spectral information from multi-temporal Landsat TM data. Int. J. Remote Sens. **23**, 3057–3078 (2002)
10. Zha, Y., Gao, J., Ni, S.: Use of normalized difference built-up index in automatically mapping urban areas from TM imagery. Int. J. Remote Sens. **24**, 583–594 (2003)
11. Xian, G., Crane, M.: Assessments of urban growth in the Tampa Bay watershed using remote sensing data. Remote Sens. Environ. **97**, 203–215 (2005). doi:10.1016/j.rse.2005.04.017
12. Li, H., Liu, Q.: Comparison of NDBI and NDVI as indicators of surface urban heat island effect in MODIS imagery. In: International Conference on Earth Observation Data Processing and Analysis (ICEODPA), 10 p, SPIE (2008)
13. Gao, M., Qin, Z., Zhang, H., Lu, L., Zhou, X., Yang, X.: Remote sensing of agro-droughts in Guangdong Province of China using MODIS satellite data. Sensors **8**, 4687–4708 (2008)
14. El Gammal, E.A., Salem, S.M., El Gammal, A.E.A.: Change detection studies on the world's biggest artificial lake (Lake Nasser, Egypt). Egypt J. Remote Sens. Sp. Sci. **13**, 89–99 (2010). doi:10.1016/j.ejrs.2010.08.001

15. El-Asmar, H.M., Hereher, M.E., El Kafrawy, S.B.: Surface area change detection of the Burullus Lagoon, North of the Nile Delta, Egypt, using water indices: A remote sensing approach. Egypt. J. Remote Sens. Sp. Sci. **16**, 119–123 (2013). doi:10.1016/j.ejrs.2013.04.004
16. Huete, A.: A soil-adjusted vegetation index (SAVI). Remote Sens. Environ. **25**, 295–309 (1988)
17. Baret, F., Guyot, G., Major, D.: TSAVI: a vegetation index which minimizes soil brightness effects on LAI and APAR estimation. In: 12th Canadian Symposium on Remote Sensing and IGARSS'90, vol. 4, Vancouver, Canada (1989)
18. Qi, J., Chehbouni, A., Huete, A.R., Kerr, Y.H., Sorooshian, S.: A modified soil adjusted vegetation index. Remote Sens. Environ. **48**, 119–126 (1994). doi:10.1016/0034-4257(94) 90134-1
19. Gao, B.C.: NDWI—a normalized difference water index for remote sensing of vegetation liquid water from space. Remote Sens. Environ. **58**, 257–266 (1996). doi:10.1016/S0034-4257 (96)00067-3
20. Mcfeeters, S.K.: The use of the normalized difference water index (NDWI) in the delineation of open water features. Int. J. Remote Sens. **17**, 1425–1432 (1996). doi:10.1080/ 01431169608948714
21. Xu, H.: Modification of normalised difference water index (NDWI) to enhance open water features in remotely sensed imagery. Int. J. Remote Sens. **27**, 3025–3033 (2006)
22. Xu, H.: Extraction of urban built-up land features from Landsat imagery using a thematic-oriented index combination technique. Photogramm Eng. Remote Sens. **73**, 1381–1391 (2007)
23. Jensen JR (2007) Remote sensing of the environment: an earth resource perspective, 2nd edn. Pearson/Prentice Hall, Upper Saddle River, NJ
24. Dai, X.: The effects of image misregistration on the accuracy of remotely sensed change detection. IEEE Trans. Geosci. Remote Sens. **36**, 1566–1577 (1998). doi:10.1109/36.718860
25. Townshend, J.R.G., Justice, C.O., Gurney, C., McManus, J.: The impact of misregistration on change detection. IEEE Trans. Geosci. Remote Sens. **30**, 1054–1060 (1992). doi:10.1109/36. 175340
26. Ridd, M.K.: Exploring a V-I-S (vegetation-impervious surface-soil) model for urban ecosystem analysis through remote sensing: comparative anatomy for cities†. Int. J. Remote Sens. **16**, 2165–2185 (1995)
27. Mingguo, Z., Qianguo, C., Mingzhou, Q.: The effect of prior probabilities in the maximum likelihood classification on individual classes: a theoretical reasoning and empirical testing. In: 2005 IEEE International Geoscience and Remote Sensing Symposium, pp. 1109–1117 (2009)
28. Strahler, A.H.: The use of prior probabilities in maximum likelihood classification of remotely sensed data. Remote Sens. Environ. **10**, 135–163 (1980)
29. Congalton, R.G.: A review of assessing the accuracy of classification of remotely sensed data. Remote Sens. Environ. **37**, 35–46 (1991). doi:10.1016/0034-4257(91)90048-B
30. Foody, G.M.: Status of land cover classification accuracy assessment. Remote Sens. Environ. **80**, 185–201 (2002). doi:10.1016/S0034-4257(01)00295-4
31. Foody, G.M.: Assessing the accuracy of land cover change with imperfect ground reference data. Remote Sens. Environ. **114**, 2271–2285 (2010). doi:10.1016/j.rse.2010.05.003
32. Lu, D., Mausel, P., Brondízio, E., Moran, E.: Change detection techniques. Int. J. Remote Sens. **25**, 2365–2401 (2004). doi:10.1080/0143116031000139863
33. Nielsen, A.A., Conradsen, K., Simpson, J.J.: Multivariate alteration detection (MAD) and MAF postprocessing in multispectral, bitemporal image data: new approaches to change detection studies. Remote Sens. Environ. **64**, 1–19 (1998). doi:10.1016/S0034-4257(97)00162-4
34. Hansen, M.C., Loveland, T.R.: A review of large area monitoring of land cover change using Landsat data. Remote Sens. Environ. **122**, 66–74 (2012)
35. Fuller, R.M., Smith, G.M., Devereux, B.J.: The characterisation and measurement of land cover change through remote sensing: problems in operational applications? Int. J. Appl. Earth Obs. Geoinf. **4**, 243–253 (2003). doi:10.1016/S0303-2434(03)00004-7

An Approach to Protect Electronic Health Records

B.K. Sujatha, Mamtha Mohan and S.B. Preksha Raj

Abstract A reliable medical image management must ensure proper safeguarding of the electronic health records. Safeguarding the medical information of the patients is a major concern in all hospitals. Digital watermarking is a technique popularly used to protect the confidentiality of medical information and maintaining them which enhances patient health awareness. In this paper, we propose a blend of discrete wavelet transform (DWT) and singular value decomposition (SVD) for watermarking the EHR. The EHR is further encrypted using key-based encryption method for access control. The SVD is applied to the approximate and vertical coefficients of the wavelet transform. The technique improves EHR protection and facilitates in accounting for performance parameters like peak signal-to-noise ratio (PSNR) and mean square error (MSE). Additionally, the properties of gray-level co-occurrence matrix like normalized cross-correlation (NC), wavelet energies, and homogeneity between the pixel pairs of host and the encrypted watermarked EHR are exploited.

Keywords Discrete wavelet transform · Singular value decomposition · Watermarking · Peak signal-to-noise ratio · Mean square error · Normalized cross correlation · Electronic health record (EHR) · Gray-level co-occurrence matrix (GLCM)

1 Introduction

This era has led to major technological innovation, Internet being the forehand in it. The advent of internet being deployed for many applications has made tremendous progress in wide spectrum of fields like medicine, health records, diagnosis, etc. Fast and secure access to patient's records helps to save lives with timely treatment in emergency situations. Therefore, anywhere anytime accessible online health care or medical systems play a vital role in daily life. It taps down this advantage of the need

B.K. Sujatha · Mamtha Mohan (✉) · S.B. Preksha Raj
M S Ramaiah Institute of Technology, Jain University, Bangalore, India
e-mail: mamtha.m@msrit.edu

M. Pant et al. (eds.), *Proceedings of Fifth International Conference on Soft Computing for Problem Solving*, Advances in Intelligent Systems and Computing 437, DOI 10.1007/978-981-10-0451-3_40

for wide scale protected accessibility of health records to provide a safe and efficient way of sharing and accessing the patient's records. The EHRs (electronic health records) is a methodical way of maintaining, securing, and protecting health information of patients. It facilitates high quality patient treatment and awareness. EHR contains case study, diagnosis, blood reports, medication, reports about various other conditions, and other necessary undergoing treatment. Moving to electronic health records is important to the modernization and revamping of our healthcare system, but it poses great challenges in the areas of security, safety, and privacy of patients records. Computerized medical records are prone to potential abuses and threats. In the last few years, thousands of human fraternity have faced compromises of their health information due to the advent of security lapses at hospitals, insurance companies, and government agencies. Various commercial companies make a living, buying, and selling doctor's prescribing habits to the pharmaceutical companies.

Additionally, sensitive electronic data, especially when stored by a third party, is vulnerable to blind subpoena or change in user agreements. The main question here is how to provide a secure way of accessing and sharing of the medical records. Since the internet is prone to snooping by intruders we need to ensure the security of the medical records. Many solutions have been provided for electronic health records. The patient's privacy is assured by access control, which verifies the person's access permission in order to ensure security. Encryption is also proposed along with restricting access. But if the server only holds the decryption key that would be catastrophic. So we propose a design which we refer to as an optimal watermarking technique as a solution to secure and private storage of patient's medical records. The optimal watermarking technique is based on SVD and DWT domain for grayscale images. This has formed from the performance parameters namely peak signal-to-noise ratio (PSNR), mean square error (MSE). PSNR is the ratio between the maximum possible power of a signal and the power of corrupting noise that affects the fidelity of its representation. The high spreading of broadband networks and new developments in digital technology has made ownership protection and authentication of digital data since it makes possible to identify the author of an image by embedding secret information to the host image. The properties of gray-level co-occurrence matrix like normalized cross-correlation (NC), wavelet energies, and homogeneity between the pixel pairs of host and the encrypted watermarked EHR are exploited for enhancement of the EHR security.

The algorithm embeds the watermark by modifying the singular values of the host records. It is followed by singular value decomposition and packing of values and encrypting. Encryption schemes with strong security properties will guarantee that the patient's privacy is protected (assuming that the patient stores the key safely).

The public key encryption is employed to encrypt the records and decrypted using cipher image. This is mainly used to enhance the security. It is an asymmetric cryptographic protocol which is mainly based on the public key. The watermarked EHR is encrypted using the public key and all the users are shared a private key. The user trying to access the EHR decrypts it using his private key. The strength lies in it being impossible for a properly generated private key to be determined from its corresponding public key. Thus, the public key may be published without compromising security, whereas the private key must not be revealed to anyone not authorized to read messages or perform digital signatures. Embedded encrypted patient data is stored in central server of a healthcare provider and shared over the internet which allows anywhere and anytime access of health records at ease. The electronic health records are secured through authenticated access, verifying the identity of the user, and validating their access permissions included for their access. The proposed approach produces watermarked encrypted patient data which protects their copyrights and avoids modifications. This further strengthens the security and protects the patient's records from any sort of compromise and threats.

This Paper is organized as follows: Sect. 2 discusses about the related work, Sect. 3 deals with the proposed methodology involving the algorithms for watermarking and extraction as well as the performance parameters considered. A discussion of the experimental results is done in Sect. 4. Section 5 discusses the conclusion and the future work.

2 Related Work

There are a number of potential applications under the umbrella of privacy-preserving data sharing and processing.

There has been considerable research at de-identification of medical record information [1] and de-identification of clinical records [2]. Various other attempts on de-identification of visit records have been done [3], But these do not include the entire medical records.

Giakoumaki et al. proposed a wavelet transform-based watermarking, the drawback is that it is related only to medical images and not the entire records and medical images are overwritten which may be unacceptable in diagnosis. We plan to work on digital watermarking, which would help ensuring the privacy and security of digital media and safeguard the copyright, and hiding the ownership identification [4]. Watermarking is a process that embeds data into a multimedia object to protect the ownership to the objects [5]. Encryption is a solution to secure and private storage of patient's medical records [6]. The hierarchical encryption system partitions health records into a hierarchical structure, each portion of which is encrypted with a corresponding key. The patient is required to store a root secret key, from which a tree of subkeys is derived [7] (Fig. 1).

3 Methodology

3.1 Discrete Wavelet Transform

Wavelet transform has the capacity of multi-resolution analysis. Embedding of a watermark is made by modifications of the transform coefficients using haar wavelet. The inverse transform is applied to obtain the original record. The host

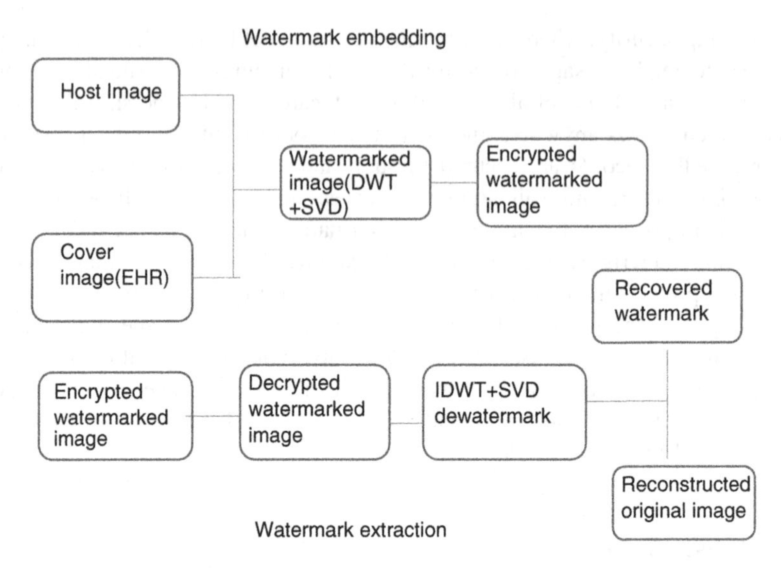

Fig. 1 Block diagram of the proposed methodology

image is decomposed into four subbands namely LL, LH, HL, and HH. A hybrid DWT-SVD-based watermarking scheme which is further encrypted using key-based encryption method is developed that requires less computation effort yielding better performance. Haar wavelet is a sequence of rescaled "square-shaped" functions which together form a family or basis. The Haar wavelet's mother wavelet function $\psi(t)$ can be described as

$$\psi(t) = \begin{cases} 1 & 0 < t < 1/2, \\ -1 & 1/2 \leq t < 1, \\ 0 & \text{otherwise.} \end{cases}$$

Its scaling function $\phi(t)$ can be described as

$$\phi(t) = \begin{cases} 1 & 0 \leq t < 1, \\ 0 & \text{otherwise.} \end{cases}$$

3.2 Singular Value Decomposition (SVD)

Singular value decomposition is a matrix factorization technique. The SVD of a host image is computed to obtain two orthogonal matrices U, V and a diagonal matrix the watermark W is added to the matrix S. Then, a new SVD process is performed on the new matrix $S + kW$ to get Uw, Sw, and Vw, where k is the scale factor that controls the strength of the watermark embedded to the original image.

Then, the watermarked record Fw is obtained by multiplying the matrices U, Sw, and V^T. The steps of watermark embedding are summarized as follows:

i. The SVD is performed on the original image (F matrix).

$$F = USV^T \quad (1)$$

ii. The watermark (W matrix) is added to the SVs of the original image (S matrix).

$$D = S + kW \quad (2)$$

iii. The SVD is performed on the D matrix.

$$D = UwSwV^T \quad (3)$$

iv. The watermarked image (Fw matrix) is obtained using the modified SVs (Sw matrix) (EHR)

$$Fw = USwV^T \quad (4)$$

v. Apply the key-based encryption to the watermarked EHR.

The medical images considered are MRI images. Initially, the watermark is embedded in the image by setting an initial value of scaling factor α. Using gray-level Co-occurrence matrix the properties like wavelet energies, homogeneity, and cross-correlation of the medical document (pixel pairs) are measured.

The original record is reconstructed from the encrypted EHR by extraction of the watermark. Optimum value of scaling factor is found by iteration of the above and tabulating the results to obtain desirable values of PSNR, MSE, NC, wavelet energies, and homogeneity.

3.2.1 Algorithm to Embed Watermark into Cover Image

Steps:

i. Read the cover image and watermark EHR.
ii. Apply DWT to cover image to obtain approximation, horizontal, vertical, diagonal DWT coefficients (LL, HL, LH, HH) Calculate, the approximate DWT coefficient by adding the watermark record using

$$\mathrm{Cal}(i,j) = \mathrm{cal}(i,j) + (\alpha * \mathrm{watermark}) \quad (5)$$

iii. Where Cal and cal are the modified & original approximation coefficients and α is a scaling value as set to 10.
iv. Apply SVD to the decomposed subbands (LL) and (HL) and Encrypt the decomposed record.

v. Find the Inverse DWT and decrypt the watermarked record.
vi. Increment α, apply SVD and inverse DWT.

3.2.2 Algorithm for Watermark Extraction

i. Extract the watermark from vertical & approximation DWT coefficient as per the equation WN = (SN − S)/α; where $\alpha = 10$
ii. Apply two—dimensional DWT, to obtain the first level decomposition of the watermarked image. LL1, HL1, LH1, HH1
iii. Decrypt the watermarked record and calculate the performance parameters.

3.2.3 Performance Parameters

The peak signal-to-noise ratio (PSNR) is used to measure the quality of reconstructed records. PSNR is the ratio between the maximum possible power of a signal and the power of corrupting noise that affects the fidelity of its representation. Because many signals have a very wide dynamic range, PSNR is usually expressed in terms of the logarithmic decible scale. PSNR is most easily defined via the mean squared error (MSE). Given a noise-free $m \times n$ monochrome image I and its noisy approximation K.

The PSNR is defined as

$$\mathrm{PSNR} = 20\,\log_{10}\left[\frac{\mathrm{MAX}_I^2}{\sqrt{\mathrm{MSE}}}\right] \tag{6}$$

Here, MAX_I is the maximum possible pixel value of the image = 255.

MSE is defined as

$$\mathrm{MSE} = \frac{1}{mn}\sum_{i=0}^{m-1}\sum_{j=0}^{n-1}[I(i.j) - k(i,j))] \tag{7}$$

Cross-correlation is a measure of similarity of two waveform as a function of a time lag applied to one of them.

For continuous functions 'f' and 'g', the cross-correlation is defined as

$$(f * g)(\tau) = \int f * (\tau) g(t + \tau)\,\mathrm{d}t \tag{8}$$

where f^* denotes the complex conjugate of 'f' and 'T' is the time lag.

3.3 Grayscale Co-occurrence Matrix (GLCM)

GLCM is an $m \times n \times p$ array of valid gray-level co-occurrence matrices. graycoprops normalizes the gray-level co-occurrence matrix (GLCM) so that the sum of its elements is equal to 1.

The energy of each subbands of the EHR is calculated as

$$\text{Energy} = \frac{1}{M * n} \sum_{i=1}^{M} \sum_{j=1}^{N} |X_{ij}| \tag{9}$$

Homogeneity measures the closeness of the distribution of elements in the GLCM to the GLCM diagonal.

$$\text{homogeneity} = \sum_{i,j} \frac{p(i,j)}{1 + |i - j|} \tag{10}$$

3.3.1 Key-Based Encryption

The public key encryption is employed to encrypt the records and decrypt using cipher image. This is mainly used to enhance the security. It is an asymmetric cryptographic protocol which is mainly based on the public key. The watermarked EHR is encrypted using the public key and all the users are shared a private key. The user trying to access the EHR decrypts it using his private key. It is asymmetric cryptography in terms that where a key used by one party to perform either encryption or decryption is not the same as the key used by another in the counterpart operation.

4 Results

The medical images considered are MRI obtained from the database "MRI Images." The images have been watermarked using EHR of a patient and encrypted.

On experimentation the scaling factors below 8 and above 14 were the scaling factor range is fixed in the range of 8–14. From Table 1 it can be observed that PSNR values are in the range of found to give unrealistic PSNR values of 0 or

Table 1 Performance parameters

Scaling factor	PSNR in dB	MSE	NC
8	80.38	0.00059	0.0213
9	74.984	0.0021	0.0233
10	71.08	0.0051	0.0253
11	67.06	0.0104	0.0264
12	65.07	0.0175	0.0273
14	62.825	0.0339	0.0260

Table 2 Subbands parameters

Subbands	Energy	Homogeneity
LL	0.4949	0.9870
HL	0.5421	0.8628
LH	0.5137	0.8243
HH	0.5461	0.8147

infinity. Hence 62–80 dB while the MSE is in the range of 0.0015–0.048. With increasing values of scaling factor. Wavelet energies of the subbands ranges from 0.4949 to 0.5461 and homogeneity has a decreasing gradient from 0.9870 to 0.8147 portrayed in Table 2. The watermarked EHR is encrypted using the public key and all the users are shared a private key as shown in Fig. 2a. The user trying to access the EHR decrypts it using his private key. This is portrayed in Fig. 2b.

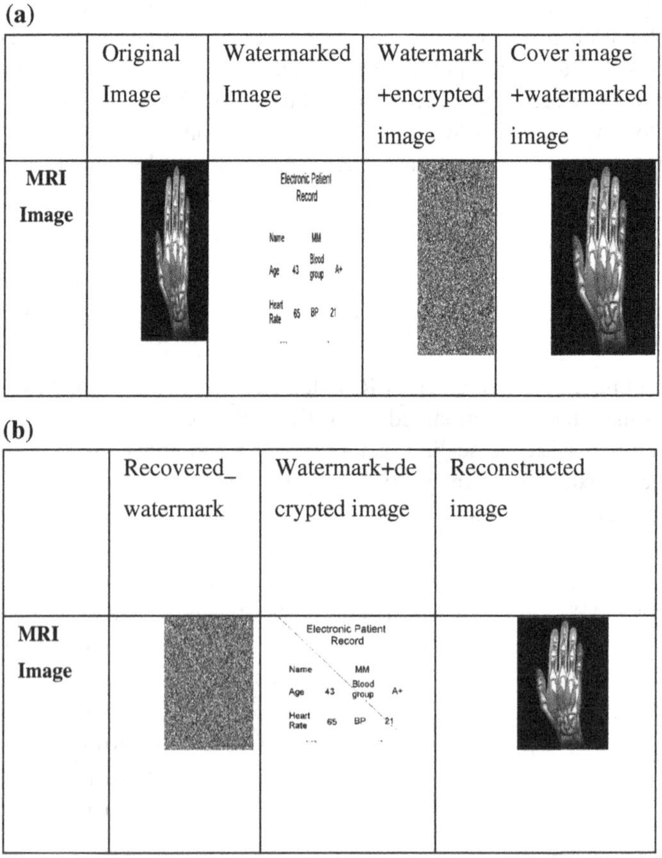

(a)

	Original Image	Watermarked Image	Watermark +encrypted image	Cover image +watermarked image
MRI Image				

(b)

	Recovered_ watermark	Watermark+de crypted image	Reconstructed image
MRI Image			

Fig. 2 **a** Watermark embedding, **b** Watermark extraction

5 Conclusion and Future Work

The watermarking can be used to provide proof of the authenticity of EHR that is to say that the medical information of one patient has been issued from the right source and to the right person. The watermark has been inserted without interfering with the documents usefulness. In this paper we have used DWT + SVD techniques to calculate PSNR, MSE, NC, wavelet energies, and the homogeneity which are the properties of gray level co-occurrence matrix of different medical records. SVD is very aptly used with DWT. It has been tested for different images and PSNR, MSE, NC are calculated for each image. Matlab R2013a has been used. This paper contributes in utilizing SVD generous properties along with hiding, protecting, and safeguarding EHR which is an asset to all the medical fraternity and the most important the patient's awareness regarding their medical health. This paper also introduces new trends and challenges in using SVD and key-based encryption in image processing applications. This paper opens many tracks for future work in using SVD as an imperative tool in signal processing and enhances confidentiality, security, and authenticity of medical health records by encrypting the medical records using public key-based encryption. This assures confidentiality, authenticity, and non-reputability of records.

References

1. Burr, T., Klamann, R., Michalak, S., Picard, R.: Generation of Synthetic BioSense Data. Los Alamos National Laboratory Report LA-UR-05-7841 (2005)
2. Neamatullah, I., Douglass, M., Lehman, L., Reisner, A., Villaroel, M., Long, W., Szolavits, P., Moody, G., Mark, R., Clifford, G.: Automated de-identification of free-text medical records. BMC Biomed. Inform. Decis. Making **8**, 32 (2008)
3. El Emam, K., Dankar, F.K., Issa, R., Jonker, E., Amyot, D., Cogo, E., Corriveau, J.P., Walker, M., Chowdhury, S., Vaillancourt, R., Roffey, T., Bottomley, J.: A globally optimal k-anonymity method for the de-identification of health data. J. Am. Med. Inform. Assoc. **16**(5), 670–682 (2009)
4. Sinha, M.K., Rai, R.K., Kumar, G.: Study of different digital watermarking schemes and its applications. Int. J. Sci. Prog. Res. (IJSPR) **03**(2) 2014
5. Singh, O.P., Kumar, S., Mishra, G.R., Pandey, C., Singh, V.: Study of watermarking. Int. J. Sci. Res. Publ. **2**(10) (2012) 1 ISSN 2250-3153
6. Lavanya, A., Natarajan, V.: Watermarking patient data in encrypted medical images. In: Sadhana Indian Academy of Science Conference on Artificial Intelligence in Computer Science, Malaysia, Vol. 37, pp. 723–729, Part 6, December 2012, pp. 147–156 (2013)
7. Benaloh, J., Chase, M., Horvitz, E., Lauter, K.: Patient controlled encryption: ensuring privacy of electronic medical records, CCSW'09, November 13, Chicago, Illinois, USA (2009)

5 Conclusion and Future Work

[illegible]

Analysis and Design of Secure Web Services

Ambreen Saleem and Ambuj Kumar Agarwal

Abstract This paper presents for the composition of software value a distributed data flow model as it is widely distributed over the internet. These services are ruled by user and they are connected to make a data processing system which is known as the megaservice. This distributed data flow model provides us the exchange of data directly among these services. This is completely different from central data flow model where the central hub for data traffic is the megaservice. This shows that this model that is distributed data flow model is better as compared to central data flow model in the sense of their performance and scalability. In distributed data flow model it fully use the network holding capacity for the free services and it also stops the blockage at the megaservices. The main aim of this paper is to develop web services and then apply the security over them.

Keywords Service appealer · Service dispenser · Prosumer · Web archive

1 Introduction

Web service is simply a means through which we can transmit between electronic devices on any of the network. It is a programming application business logic implemented on the web server at its network address with the service always available for its utility.

Web service is basically a software which is made to provide a support for the machine to machine communication.

Web services model, needs particular implementation of web service as defined by the Web Services Architecture Working Group. Web service interact with all the

A. Saleem (✉) · A.K. Agarwal
College of Computing Sciences and Information Technology,
Teerthanker Mahaveer University, Moradabad, India
e-mail: ambreensaleem4@gmail.com

A.K. Agarwal
e-mail: ambuj4u@gmail.com

M. Pant et al. (eds.), *Proceedings of Fifth International Conference on Soft Computing for Problem Solving*, Advances in Intelligent Systems and Computing 437, DOI 10.1007/978-981-10-0451-3_41

other systems in a way given in its prescription by using SOAP messages, mainly transmitted with an XML serialization in addition with other standards related to web using HTTP. Majority of the web services do not support the described complex architecture.

Web services can be identified into two major categories

- REST complaint web services-REST complaint web services are the web services in which the main aim is to employ portrayal of all web resources by using a similar set of the stateless operations.
- And other are unscientific web services in which the arbitrary set of operations might be disclosed.

Different software system might need to transact data over the internet and this is what we call the web service. It is a way through which it allows the two software systems to transact the data on the internet. There are two one is service appealer and other is service dispenser. Service appealer is the software system that appeals data, in contrast the dispenser is the software system that is use to process the appeal and then dispense the data.

Nowadays using different different programming techniques, different softwares are being designed and implemented so particularly we need a mode through which exchange of data does not depend on any of the particular programming technique. Almost all types of software can understand XML tags. So XML files can be used for data exchange for web services.

The communication between the different system is based on these things:

1. How the neighbour can give data to other system?
2. What specification are needed in the appealing of data?
3. How would be the structure of data that has been extracted?

Basically, the structure of XML file is checked against.xsd file and the exchange of data takes place in XML files. To make troubleshooting easier it is checked that what error messages are being displayed when the given communication rule is not monitored.

2 Types of Web Services

Web services are programmable application logic that can be accessed using Standard Internet Protocols. SOAP (Simple Object Access Protocol) is one of the protocols. SOAP is used to encode and transfer application data. It is a standard technology. It uses extensible markup language, i.e., XML for description of data and for transport it uses HTTP.

The web service users should only understand that how the SOAP messages are send and received, they do not need to understand object model or the language that has been used to implement service.

2.1 List of Web Services in ASP.NET

(A) Business and Commerce.
(B) Standards and Lookup Data.
(C) Graphics and Multimedia.
(D) Utilities.
(E) Value Manipulation/Unit Converter.

2.2 List of Web Services in JAVA

2.2.1 Big Web Services

Big web services, that are following SOAP architecture, text architecture described by XML language and format of messages, functionality of big web services is dispensed by JAX-WS in JAVA EE 6.

The format of SOAP message and Web Services Description Language (WSDL) have been adopted worldwide. The complexity for developing the applications of web service can be reduced by using development tools such as NetBeans IDE.

2.2.2 Restful Web Services

REST (Representational State Transfer) is the one on which our World Wide Web (WWW) is based on. Restful web service is nothing only a way that is based on how our web is working. It was introduced by Roy Fielding in 2000. REST is nor a protocol neither a standard. Web services following client–server architectural style are known as Restful web services.

REST is an easier alternative to SOAP and web services that are based on WSDL, i.e., Web Services Description Language. All set of architectural principles are defined by REST through which we can design web service and it also focuses on system resources. REST in the past few years has emerged as a strongest web service design model. It is having so much growth that it has replaced many SOAP and WSDL-based interface design because it is easy or we can say simple to use.

3 Related Work

Web-based applications and services appealer are rapidly growing and implemented in the tasks where security is a high priority. Most of the security methods depend on two assumptions. First, assumption is made that security sensitive information is free from all kinds of vulnerabilities handled by end point software. Second, point to point communication is presumed by these protocols between the service dispenser and the client.

Complex vulnerable end point software like Internet browser or where there are multiple value adding service dispensers between the original service dispenser and the client. These assumptions are not hold to be true.

To handle this problem it is proposed to develop an architecture which splits the large and complex end point software into three parts: a highly trusted part that fully handles the information which is security sensitive, a medium trusted part that handles the information between two stages and third untrusted part that is used to handle all non sensitive information and it cannot access the secure sensitive information. The proposed architecture decreases the complexity and size of trusted part and helps in making much feasible the formal analysis.

4 Study of Recent Trends in Web Service Security

Marchetto discussed about various standards such as policy of web services, Extensible access control markup language (XAML), XML encryption, XML signature, WS-Trust, WS security policy, and XML key management specification (XKMS) [1].

To control the problem of complex and large web services Satoh and Tokuda have presented the secure information flow architecture that is ISO-WSP [2].

Eggen discussed a execution mechanism for decentralized workflow in order to provide security requirements for the distributed workflows. He ensured that for proper execution of the called operation, each and every web service will access the information only which is needed [3].

Sachs provided us the comparative study about different vulnerability detection tools for web and discussed that programmer of web service must be very careful when they are choosing any vulnerability detection approach [4].

Gruschka and Jensen provided the proper definition of trustworthy web service. He said that trustworthy web service is a web service that is reputed, secure, reliable having less response time as well as having integrated availability [5].

Saledian explained to address the issues of security requirements and evaluation criteria. The various security attacks in web service have been categorized into three main groups that are XML injection, Denail of Service (DO's) and third counterfeit XML fragment [6, 7].

Web service allows different applications to communicate between each other very efficiently and effectively because web service is interoperable that provides absolute connectivity. So using web service in business organization is an advantage only the security risks of it must be perfectly analyzed.

The framework that is use to provide protection to SOAP messages is WS-Security. WS-Security ensures the authenticity, confidentiality and integrity of message [8, 9]. The extension to WS-Security is WS-Trust that is Web Service Trust Language. WS-Trust provides a way to establish trust relationship between different trust domains. Some other features like establishment of secret context and many others are the aim of some other frameworks like WS-Secure Conversation and WS-Security Policy because these framework wants to be better than WS-Trust [10–12].

5 Design of Web Service Platforms

The framework for web service platform or WSP's is specified by web services architecture specification. The interaction between the three entities is transferred by WSP. A support is needed for the three types of functionality: For the exchange of messages, for describing the web services, for publishing and for finding the description of web service (Fig. 1).

This paper mainly focuses on the security required for information exchange between the client and the service dispenser so main focus must be on first functionality, i.e., support for the exchange of message. All the information of web service is stored in web archive.

Apache Axis2 WSP is not widely used, it is the implementation of web services framework. In order to understand its working it is available under an open source license. It dispenses support for implorement, development and management of web services.

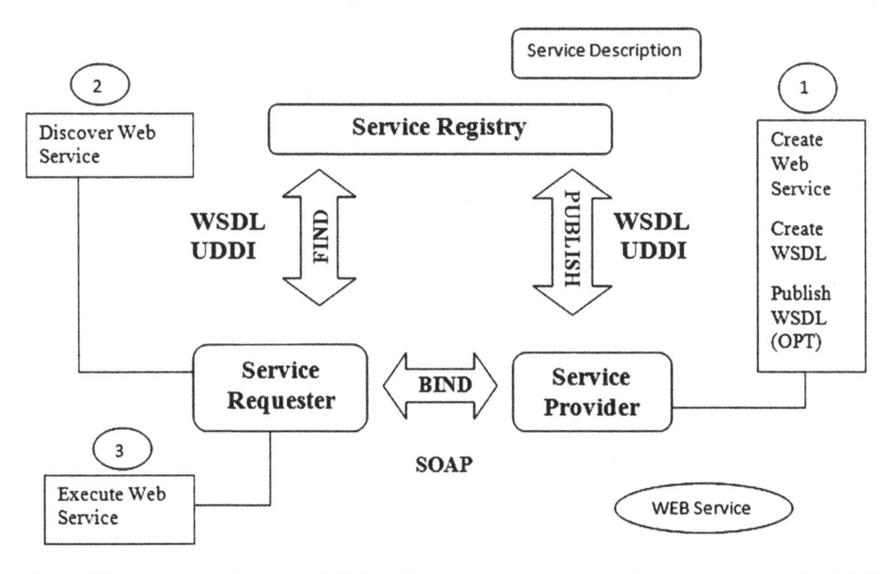

Fig. 1 Web service architecture [13] http://computerscienceimaginarium.blogspot.com/2013/01/web-services-architecture.html

6 Architecture of ISO-WSP

ISO-WSP is classified into two types T-WSP and U-WSP. A difference between the two sequences exist according to their protection domain. In order to prevent the binaries of T-WSP from modifying or configuration files a separation is made between T-WSP and U-WSP by executing the U-WSP with lower privileges in different protection domains. This also helps in preventing the U-WSP from gaining access to secret keys that are used for encryption and decryption.

WSP can be changed to ISO-WSP by making small changes. A fine procedure exist in place of using local calls, i.e., once T-WSP is constructed it is necessary to alter WSP for implorement of T-WSP through remote invocation mechanisms. This remote invocation mechanism involves the identification of messages and results which are exchanged between the two sequence of ISO-WSP. It also adds all the important codes which are required for the process. Users and service dispensers are required to trust the complex and large components for the implementation and designing of web service platform (WSP's). The prototype of ISO-WSP can be implemented on the basis of Apache Axis2 platform. There are some functionalities that poses challenges for the testing and analysis of the WSP's. Configurability and extensibility are some of them. This may cause various security vulnerabilities in WSP's.

The architecture of ISO-WSP has been developed on the basis of App core approach. This approach has been applied in WSP architecture.

There is one weakness in ISO-WSP architecture and that is violation of integrity flow. In this the data dispensed by untrusted part is used by trusted part to perform various operation. So, the problem is this that data is flowing from untrusted part or lower integrity level to trusted part or higher integrity level. Before performing operation the assumption is made the trusted part checks all the inputs that are coming from untrusted part.

Now, attempts are being made to design a new architecture that divides the WSP's into three parts each running in its separate domain.

7 Work Done

Created four web services in ASP.NET

1. HTML Editor.
2. Arithmetic Calculators.
3. GPS Path Finder.
4. SMS Sender.

In order to access the services, we have to go through the SMS verification code security. The code is encrypted and decrypted through the given algorithm.

8 Proposed Algorithm

The Proposed Algorithm will experience through

1. The Key creation.
2. The IV Creation.
3. The format is then encrypted and decrypted.
4. Changing document augmentations.
5. Joining all steps together.

8.1 Step 1

- A string parameter passed to function.
- String parameter passed to function is then changed in array.
- The array is then changed into byte.
- SHA to change over hash the byte.
- Applying MD5 on above step.
- For the key, utilizing initial 256 bits came back from the hashed byte into another byte array (the key).
- The key is returned.

8.2 Step 2

The following 128 bits of the hashed byte to make IV. Thus, the key will be unique in relation to the past key.

8.3 Step 3

- Define the catalog for Encryption and decryption of format.
- Using encrypted/unencrypted document.
- With File Stream object, open and read the document.
- With the Crypto Stream object doing encryption/decryption of format.
- With different file Stream object, write the encrypted/decrypted document.

8.4 Step 4

Changing document Augmentations.

8.5 Step 5

Joining all steps together.

8.6 Step 6

We have also implemented security for link verification and vulnerabilities test as

- Order strategy files from connection specification.
- Checking whether an arrangement of strategy files keep running by any number of senders and recipients effectively executes the objectives of a connection specification, disregarding dynamic aggressors. Strategy driven web services usage are inclined to the typical unobtrusive vulnerabilities connected with cryptographic conventions.

This will help anticipate such vulnerabilities, as we can check strategies when first assembled from connection specifications, furthermore re-confirm arrangements against their unique objectives.

9 Comparison

Different web services were read and analysed. Comparison is being shown by judging their different features.

Features	Services	Other Web Services/WCF
Protocols	Security	Security reliable messaging transactions
Hosting	IIS	IIS, windows, Self-hosting, Windows service
Transports	HTTP, TCP	HTTP, TCP, Namedpipes, MSMQ, P2P
XML	System Xml serialization	System runtime serialization
Encoding	XML 1.0, MTOM, DIME, Custom	XML 1.0, MTOM, Binary, custom
Programming	Attribute has to be added to the class	Attribute has to be added to the class
Operation	One way	One way, appeal-response, duplex
Model	attribute represents the method exposed to client	attribute represents the method exposed to client

10 Conclusion

We have implemented the four web services, we likewise implemented a security feature with algorithm on a SMS response with 256 bit encryption. Got an outline for a guide web services that is sufficiently straightforward for a customer to use without an immediate speculation, and will be exceptionally valuable to be the driver for any number of helpful projects. It is so near to the Web that you can utilize it with a web program. It is a continually meeting expectations and asset situated. It generally be situated in location, with stateless, and joined with neighbors' well.

It is additionally read-just. It expect that customers have nothing to offer except for their voracious for information. There are numerous actualized web services work along these, and there are few of the read-just web services and executed and accessible for the client.

11 Analysis Tool

Website Optimizer is a mosaic testing services provided at google.com that permits you to rotate curious portions of substance on your site to see which areas and position amateur into the most clicks, and by the day's end, the most deals. You can pick what parts of your page you need to test, from the feature to pictures to content, and run tests to see what clients react best to. Furthermore, obviously, with GWS being free (you do not even need Google Analytics to utilize it); it could be the main A/B (a specialized term for various renditions of the site running on the double) and Multivariate (MVT) or complex testing arrangement.

12 Result

An accomplish Web services utilizing a total of components found in IIS and ASP. NET. ASP.NET Web services utilize authorization to react to bids for particular sorts of authentication. Basic/Digest authentication and SSL all have the same obstruction: they want a couple messages to be changed between the SOAP message sender and receiver before the message can be sent safely.

This deal can constrain the speed in which SOAP messages are exchanged. Assist things up is only one of the inspirations driving the WS-Security Specification. WS-Security nonchalance transport convention strategies for a message-driven security model. Until WS-Security is broadly comprehended and sent, HTTP-based security instruments are the most ideal approach to keep your Web services secure.

13 Future Scope

Underneath the term web services we see any imperative of a service that is not inherent in the capacity of the services. For instance, the main aim of a conveyance services would be to pass on bundles between any two subjective areas. This portrayal straightforwardly infers that this services would not mix espresso, so this issue is not a piece of the extension. On the other hand, spasm on the size or weight of the bundles or on the conveyance areas can be communicated as services extension. Formally, web service extension is a situated condition over scope determination which are any of the accompanying three classes of information

1. The data particular of services summons (like shipment location for an electronic shop).
2. Worldwide determination like current time (for instance for working hours).
3. Outer detail allowable for the services summons (the bit rate of the feature to be facilitated, despite the fact that the facilitating services does not advance the bit rate esteem unequivocally as one of its data particular) When extension is expressly characterized in an services portrayal, a potential customer can choose whether it is helpful to conjure this specific services, i.e., whether the services can be relied upon to give positive result; or a customer can see which of the accessible services are in fact presently material.

Web services are anticipated to be the most recent mechanical change that will reform business. Innovation organizations and visionaries have begun their talk. A completely new period gives off an impression of being developing where anybody can distribute their services utilizing standard Internet conventions and customers around the globe can without much of a stretch join these services in any manner to give higher request services. These services system chains will tackle each business issue on the planet and in the process producing incomes to everybody included in the chain.

References

1. Marchetto, A.: Special section on testing and security of Web systems, 14 Oct 2008
2. Satoh, F., Tokuda, T.: Security Policy Composition for Composite Web Services, Oct–Dec 2011
3. Eggen, E.: Norwegian Defense Research Establishment (FFI), Robust Web Services in Heterogeneous Military Networks, IEEE (2010)
4. Sachs, M. (arcus.sachs@verizon.com), Ahmad, D. (drma@mac.com): Splitting the HTTPS Stream to Attack Secure Web Connections, IEEE (2010)
5. Gruschka, N., Jensen, M., Member, IEEE, Luigi Lo Iacono, and Norbert Luttenberger, Server-Side Streaming Processing of WS-Security, Oct–Dec 2011
6. Saiedian, H.: Security Vulnerabilities in the Same-Origin Policy: Implications and Alternatives, University of Kansas, Sept 2011
7. Gaff, B.M., Smedinghoff T.J., Sor, S.: Privacy and Data Security, IEEE, 2012
8. Agarwal, A.: Implementation of Cylomatrix complexity matrix. J. Nat. Inspired Comput. **1**(2), (2013)
9. Ather, D., Agarwal, D.K., Sharma, D., Saxena, D.K.: An Analysis and study of various web development platform in viewpoint of web developer. Acad. J. Online **2**(11), (2013)
10. Fonseca, J., Vieira, M., Madeira, H.: Evaluation of Web Security Mechanisms Using Vulnerability and Attack Injection, Sept/Oct 2014
11. Antunes, N., Vieira, M.: Penetration Testing for Web Services, IEEE, 2014
12. Liu, W.M., Wang, L., Cheng, P., Ren, K.: Senior Member, IEEE, Shunzhi Zhu, and Mourad Debbabi, PPTP: Privacy-Preserving Traffic Padding in Web-Based Applications, Nov/Dec 2014
13. https://www.google.co.in/search?q=web+service+architecture&safe=active&biw=1366&bih=667&tbm=isch&tbo=u&source=univ&sa=X&ved=0CBwQsARqFQoTCIjokLqx0scCFVMFjgodsE0AmQ#imgrc=OIDCGNszqtaKfM%3A

A Fuzzy Clustering Method for Performance Evaluation for Distributed Real Time Systems

Ruchi Gupta and Pradeep Kumar Yadav

Abstract All practical real-time scheduling algorithm in distributed processing environment present a trade-off between their computational intricacy and performance. In real-time system, tasks have to perform correctly and timely. Finding minimal schedule in distributed processing system with constrains is shown to be NP- Hard. Systematic allocation of task in distributed real-time system is one of the major important parameter to evaluate the performance if this step is not execute properly the throughput of the system may be decrease. In this paper, a fuzzy clustering-based algorithm has been discussed.

Keywords Fuzzy logic · Load balancing · Task clustering · Distributed system

1 Introduction

Load balancing is a technique to broaden work among two or more computers, network links, CPUs, hard drives, or other resources, in order to get best possible resource utilization, throughput, or response time. Load balancing attempts through maximizing the system throughput by keeping all available processors active. Load balancing is done by re-allocating the tasks from the overloaded nodes to other unloaded nodes to get better system performance. Load balancing models are in general based on a load index, which provides a measure of the workload at a node relative to some global standard, and four policies, which administrate the actions taken once a load imbalance is detected [1].

We concentrate on the four policies that oversee the action of a load balancing algorithm when a load inequality is noticed compact with information, transfer,

Ruchi Gupta (✉)
Uttarakhand Technical University, Dehradun, India
e-mail: ruchigupta1984@gmail.com

P.K. Yadav
Central Building Research Institute, Roorkee, UK, India
e-mail: pkyadav@cbri.res.in

M. Pant et al. (eds.), *Proceedings of Fifth International Conference on Soft Computing for Problem Solving*, Advances in Intelligent Systems and Computing 437, DOI 10.1007/978-981-10-0451-3_42

location, and selection. The information policy is accountable for keeping the latest load information about each node in the system. A global *Information policy* offers access to the load index of every node, at the cost of added communication for maintaining accurate information [2]. The *Transfer policy* deals with the active facets of a structure. It uses the nodes load information to decide when a node becomes eligible to act as a sender or as a receiver. Transfer policies are typically threshold based. Thus, if the load at a node grows beyond a threshold T, the node becomes an eligible sender.

The *Location policy* selects a companion node for a job transfer operation. If the node is a suitable sender, the location policy pursues out a receiver node to receive the job selected by the selection policy. If the node is a qualified receiver, the location policy looks for an eligible sender node [3]. Once a node becomes a suitable sender, a S*election policy* is used to choice which of the queued jobs is to be transferred to the receiver node. The selection policy procedures some criteria to value the queued jobs. Its goal is to select a job that reduces the local load, incurs as little cost as possible in the transfer, and has good affinity to the node to which it is transferred.

Fuzzy logic is a mathematical tool in representing semantic information and is very useful to solve problems that do not have a precise solution. It includes four components: fuzzifier, inference engine, fuzzy knowledge rule base, and defuzzified [4–7]. Fuzzification converts the crisp input value to a linguistic variable using the membership functions stored in the fuzzy knowledge base. Fuzzy knowledge rule base contains the knowledge on the application domain and the goals of control [8–10]. The dynamic task scheduling approach seeing the load balancing problem is an NP-Hard problem been discussed by Yadav et al. [11, 12]. The inference engine put on the inference tool to the set of rules in the fuzzy rule base to produce a fuzzy set output. Defuzzification transmutes the fuzzy output of the inference engine to inflexible value using membership functions related to the ones used by the Fuzzifier. Recently Yadav et al. [13, 14] discussed the task allocation model for optimization of reliability and cost in distributed system.

2 Objective Function for Task Allocation

Load balancing algorithm aims to distribute the load on resources appropriately to maximize their proficiency while minimizing overall execution time of the tasks. We assume that the number of task is much larger than the number of processor, so the number of processors keeps on idyllic. Sharing sources result in better efficacy to cost ratio, less response time, higher throughput, more reliability, extensibility, and double development and better flexibility in thought.

2.1 Processors Clustering

It be made of a set of loosely connected or tightly connected processor that works together so that in many respects they can be observed as a single system. They identify as group of processor which can perform task instantaneously [15]. We cluster processor on the basic of work expertise and speed. Here we implement one-to-one mapping technique for equal distribution between number of tasks and number of processor [16, 17]. One researcher discusses the desirability force ($B_{ij}/\mu_i + \mu_j$) for the assembling of the processors, here we using B_{ij} is the bandwidth of the channel which connecting two processor P_i and P_j of μ_i and μ_j speed, respectively. The fuzzy function will try to keep processor s of same processing speed in the same masses Membership function is defined as follows: [18].

$$\mu(P_i) = 1/1 + \mathbf{diff}(S_i - S)\ \mathbf{Where\,diff}(S_i - S) = |S_i - S| \tag{1}$$

S_i = sum of communication time of all tasks on the processor P_i

S = maximum value in S_i $\mathbf{1} \leq i \leq n$

By the above Membership function, we formulate the membership value for each processor on the same DRTS, which lies between 0 and 1. Here we categorize the processor cluster in to four categories on the basis or their membership value.

Very High-Speed Processor (VHSP) [0.0, 0.4]
High-Speed Processor (HSP) [0.4, 0.7]
Slow-Speed Processor (SSP) [0.7, 0.9]
Very Slow-Speed Processor (VSSP) [0.9, 1.0]

2.2 Task Clustering

We create cluster of task on the basis of communication time between the tasks. Those clustering task together which is heavily communicated, lightly communicated. When we cluster heavily communicated task together to reduce communication time and applied fuzzy membership function to categories the task cluster. Each task will allocate a membership value defined as [19].

$$\mu(T_i) = 1/1 + \mathbf{diff}(C_i - C)\ \mathbf{Where\,diff}(C_i - C) = |C_i - C| \tag{2}$$

C_i = sum of total communication time if ith task say t_i

C = maximum value of C_i $1 \leq i \leq m$

Here we categorize the task cluster in to four categories on the basis or their membership value.

Very Hard Task (VHT)	[0.0, 0.4]
Hard Task (HT)	[0.4, 0.7]
Easy Task (ET)	[0.7, 0.9]
Very Easy Task (ET)	[0.9, 1.0]

3 Proposed Algorithm

In the proposed model, first of all we determine the L_{avg} (P_i) to be assigned on processor P_j with tolerance factor using the Eq. (1) and T_{load} to be allocated to the system using Eq. (2). The input stage consists of two input variables, i.e., task cluster and processor cluster, as shown in Fig. 1. Task clustering assessed the fuzzy membership value for the all task in reverence to inter-task communication time with first task and categorize them. Processor clustering assessed the fuzzy membership value for all processor starting with first task and categories them on the basis of range.

Here membership functions define the grade to which each input factor represents its connotation (Figs. 2 and 3).

3.1 Membership Functions

$$\mu_{\mathrm{VHT}}(i_1) = \left\{ \begin{array}{ll} 1; & i_1 < 0 \\ \dfrac{0.4 - i_1}{0.4}; & 0 \le i_1 \le 0.4 \\ 0; & i_1 > 0.4 \end{array} \right\}$$

$$\mu_{\mathrm{VHT}}(i_1) = \{1/0 + 0/0.4\}$$

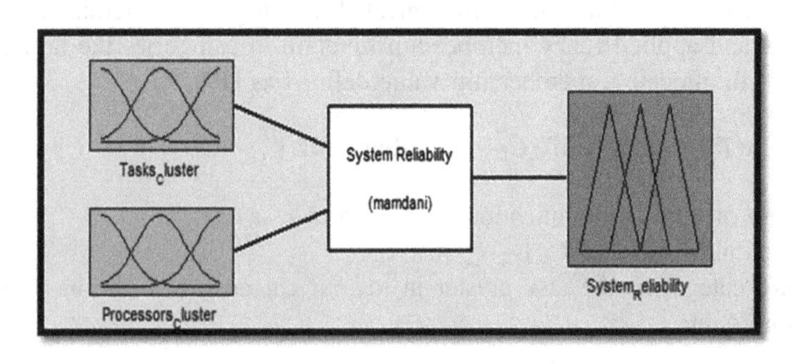

Fig. 1 Proposed model

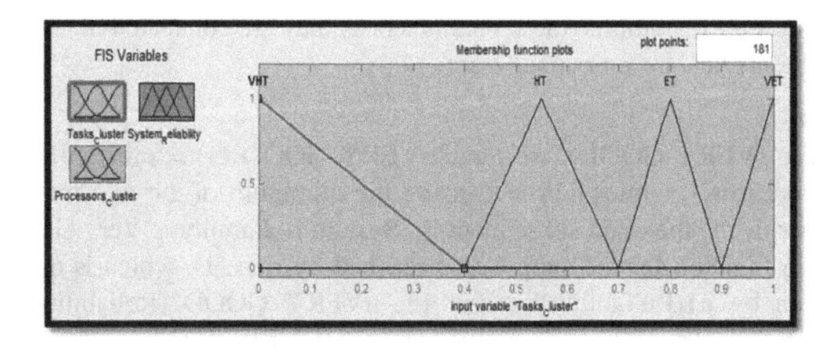

Fig. 2 Fuzzy membership function w.r.t. task cluster

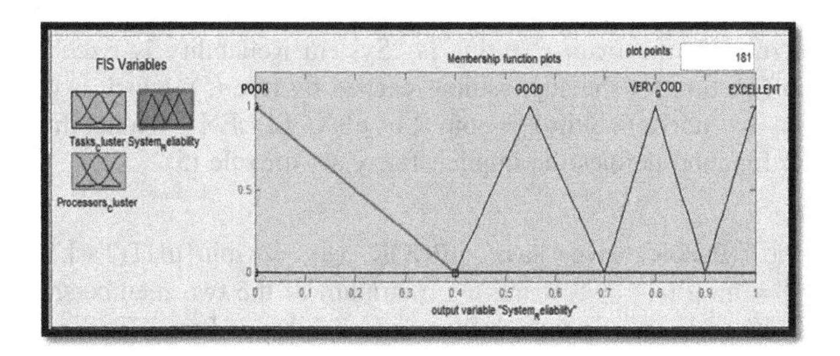

Fig. 3 Fuzzy membership function w.r.t. processor cluster

$$\mu_{\mathrm{HT}}(i_1) = \begin{Bmatrix} \dfrac{i_1 - 0.4}{0.15}; & 0.4 \le i_1 \le 0.55 \\ \dfrac{0.7 - i_1}{0.15}; & 0.55 \le i_1 \le 0.7 \\ 0; & i_1 > 0.7 \end{Bmatrix}$$

$$\mu_{\mathrm{HT}}(i_1) = \{0/0.4 + 1/0.55 + 0/0.7\}$$

$$\mu_{\mathrm{ET}}(i_1) = \begin{Bmatrix} \dfrac{i_1 - 0.7}{0.1}; & 0.7 \le i_1 \le 0.8 \\ \dfrac{0.9 - i_1}{0.1}; & 0.8 \le i_1 \le 0.9 \\ 0; & i_1 > 0.9 \end{Bmatrix}$$

$$\mu_{\mathrm{ET}}(i_1) = \{0/0.7 + 1/0.8 + 0/0.9\}$$

$$\mu_{\mathrm{VET}}(i_1) = \begin{Bmatrix} 0; & i_1 \le 0.9 \\ \dfrac{1 - i_1}{0.1}; & 0.9 \le i_1 \le 1 \\ 1; & i_1 > 1 \end{Bmatrix}$$

$$\mu_{\mathrm{VET}}(i_1) = \{0/0.9 + 1/1\}$$

Fuzzy rules try to combine these parameters as they are connected in real worlds. Some of these rules are mentioned here

For rule 1

The premise (VERY GOOD), we have μVERY GOOD (1) = min{μVHT(Task), μVHSP(Processor)} = min{1,1} = 1, using the minimum of the two membership values. For rule (1) the main subsequent is "System Reliability is Very Good." The membership function for the supposition reached by rule (1), which is denoted as μ1, is given by μ1(Reliability) = min {1, μVERY GOOD (Reliability)}. This membership function defines the implicit fuzzy set for rule (1).

For rule 5

The premise (EXCELLENT), we have μEXCELLENT (5) = min{μHT(Task), μVHSP(Processor)} = min{1,1} = 1, using the minimum of the two membership values. For rule (5) the main resulting is "System Reliability is Excellent." The membership function for the supposition reached by rule (5), which is denoted as μ5, is given by μ5(Reliability) = min {1, μEXCELLENT (Reliability)}. This membership function defines the implicit fuzzy set for rule (5).

For rule 8

The premise (POOR), we have μPOOR (8) = min{μHT(Task), μVSSP (Processor)} = min{1,1} = 1, using the minimum of the two membership values. For rule (8) the main subsequent is "System Reliability is Poor." The membership function for the assumption reached by rule (8), which is denoted as μ8, is given by μ8(Reliability) = min {1, μPOOR (Reliability)}. This membership function defines the implicit fuzzy set for rule (8).

For rule 12

The premise (GOOD), we have μGOOD (12) = min{μET (Task), μVSSP (Processor)} = min{1,1} = 1, using the minimum of the two membership values. For rule (12) the main consequent is "System Reliability is Good." The membership function for the assumption reached by rule (12), which is denoted as μ12, is given by μ12(Reliability) = min {1, μGOOD (Reliability)}. This membership function defines the implicit fuzzy set for rule (12).

In our estimated algorithm as offered under, a recently inwards task will be enter to the input queue. This queue consists of the outstanding tasks from last succession that has not up till now been assigned.

- Loop

  For every processor in the distributed system, prepare the following:

  - For each ready task, provender its processor rate and task size into the inference engine. Contemplate the output of inference module as reliability of the task *T*.
  - Store the values of precedence in an array *P* ( ).
  - Accomplish the task with maximum precedence until a scheduling occurrence.
  - Update the system states.

- End Loop.

Once the effective completion of the algorithm, we shall acquire the task with max value of "reliability" and the parallel will be completed first followed by the next uppermost value of "Reliability." That process repeats itself till all scheduled tasks get executed by the available processors.

When all the tasks get performed by the accessible processors, then we can produce the decision surface to check the value of the output variable under the incorporations of different input parameters. The decision surface accordingly obtained is shown in the Fig. 4.

We can have the 2-D image of grouping of Input variables and their consistent effect on the output "System Reliability." Optimal demonstration of output is portrayed in Fig. 5.

Decide on number of rules and membership functions directly marks system correctness, while performance of the system increases with regulation size decrease. There are specific techniques for changing membership functions. However, simulation results represent that by increasing the speed of system's processor, generation of very easy task increases exceptional system reliability to an acceptable range. It represents that very slow-speed processor and very easy tasks have higher reliability of accomplishment while their deadline would not decrease to critical region. This comportment results in more execution for low-priority tasks in average load cycles. Imitation results show that the model can practicably schedule tasks. When structure load increases and keep system processors loads close to one even at packed times. By examination of task scheduler, periodic tasks age increases robotically by scheduler with respect of their task size and processor speed.

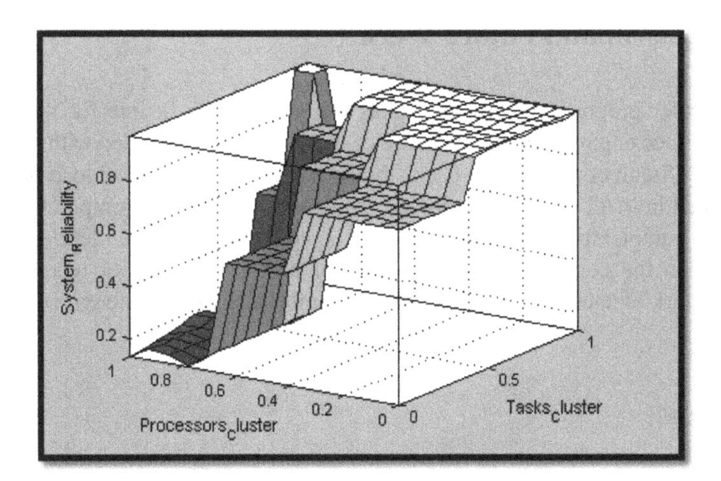

Fig. 4 Decision surface corresponding to inference rules

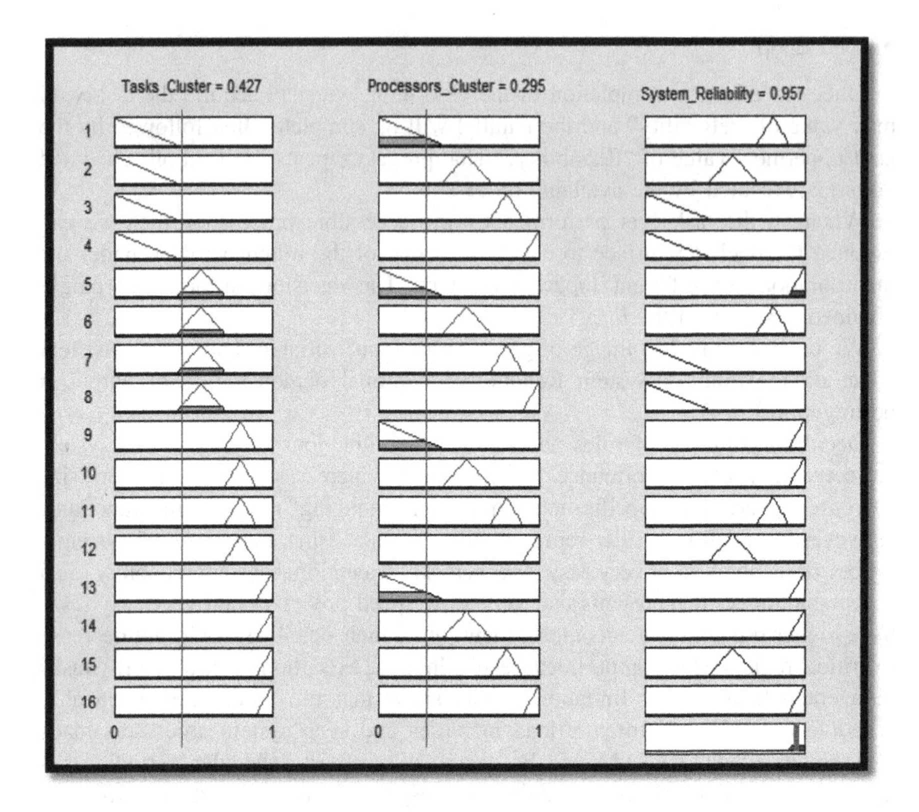

Fig. 5 Task cluster versus processors cluster versus system reliability

4 Conclusion and Future Work

The scheduler proposed in this paper has low complexity in line for the ease of fuzzy inference engine. The performance of the procedure is equated with algorithm reported by Singh et al. [20]. To check the performance of proposed load balancing is calculated in MATLAB simulation environment. Its show the computation of the algorithm complexity and response time is constant and also define that, by increasing in the computation, the number of processors will not increase. This model is work efficient when system has dissimilar tasks with different constraints.

References

1. Shivaratri, N.G., Krueger, P., Singhal, M.: Load distributing for locally distributed systems. Computer **25**, 33–44 (1992)
2. Eager, D.L., Lazowska, E.D., Zahorjan, J.: Adaptive load sharing in homogeneous distributed systems. IEEE Trans. Softw. Eng. **12**, 662–675 (1986)

3. Leinberger, W., Karypis, G., Kumar, V.: Load balancing across near homogeneous multi-resource servers. In: Presented at Proceedings of the 9th Heterogeneous Computing Workshop (HCW 2000) Cancun, Mexico (2000)
4. Karimi, A., Zarafshan, F., Jantan, A.b., Ramli, A.R., Saripan, M.I.: A new fuzzy approach for dynamic load balancing algorithm. Int. J. Comput. Sci. Inf. Sec. **6**(1) (2009)
5. Ally E. El-Bad: Load Balancing in Distributed Computing Systems Using Fuzzy Expert Systems. TCSET'2002. Lviv-Slavsko, Ukraine (2002)
6. Ross, T.J.: Fuzzy Logic with Engineering Applications. McGraw Hill (1995)
7. Huang, M.C., Hosseini, S.H., Vairavan. K.: A Receiver Initiated Load Balancing Method In Computer Networks Using Fuzzy Logic Control. GLOBECOM, 0-7803-7974-8/03. (2003)
8. Chu, E.W., Lee, D., Iffla, B.: A distributed processing system for naval data communication networks. In: Proc. AFIPS Nat. Comput. Conf. **147**, 783–793 (1978)
9. Deng, Z., Liu, J.W., Sun, S.: Dynamic scheduling of hard real-time applications in open system environment. Technical Repeport, University of Illinois at Urbana-Champaign (1993)
10. Buttazzo, G., Stankovic, J.A.: "RED robust earliest deadline scheduling" In: Proceedings of the 3rd International Workshop Responsive Computing Systems, Lincoln, pp. 100–111 (1993)
11. Yadav, P., Kumar, K., Singh, M.P., Harendra, K.: Tasks scheduling algorithm for multiple processors with dynamic reassignment. J. Comput. Syst. Netw. Commun. 1–9 (2008)
12. Bhatia, K., Yadav, P.K., Sagar, G.: A reliability model for task scheduling in distributed systems based on fuzzy theory. CiiT Int. J. Netw. Commun. Eng. (IF 1.953) **4** (11), 684–688 (2012)
13. Yadav, P.K., Singh, M.P., Kuldeep, S.: Task allocation for reliability and cost optimization in distributed computing system. Int. J. Model. Simul. Sci. Comput. (IJMSSC) **2**, 1–19 (2011)
14. Pradhan, P., Yadav, P.K., Singh, P.P.: Task scheduling in distributed processing environment: a fuzzy approach. CiiT Int. J. Programmable Device Circ. Syst. (IF 0.980) **5**(9) (2013)
15. Petters, S.M.: Bounding the execution time of real-time tasks on modern processors. In: Proceedings of the 7th International Conference Real-Time Computing Systems and Applications, Cheju Island, pp. 498–502 (2000)
16. Zhu, J., Lewis, T.G., Jackson, W., Wilson, R.L.: Scheduling in hard real-time applications. IEEE Softw. **12**, 54–63 (1995)
17. Avanish, K., Yadav, P.K., Abhilasha, S.: Analysis of load distribution in distributed processing systems through systematic allocation task. IJMCSIT "Int. J. Math. Comput. Sci. Inf. Technol. **3**(1), 101–114 (2010)
18. Taewoong, K., Heonshik, S., Naehyuck, C.: Scheduling algorithm for hard real-time communication in demand priority network. In: Proceedings of the 10th Euromicro Workshop Real-Time Systems, Berlin, Germany, pp. 45–52 (1998)
19. Yadav, P.K., Prahan, P., Singh, P.P.: A fuzzy clustering method to minimize the inter task communication effect for optimal utilization of processor's capacity in distributed real time system. In Deep, K. et al. (Eds.) Proceeding of the International Conference on SocPros 2011. AISC 130. pp. 151–160. Springer india (2012)
20. Harendra, K., Singh, M.P., Yadav, P.K.: A tasks allocation model with fuzzy execution and fuzzy inter-tasks communication times in a distributed computing system. Int. J. Comput. Appl. (0975–8887) **72**(12), 24–31 (2013)

Image Segmentation Using Genetic Algorithm and OTSU

Jyotika Pruthi and Gaurav Gupta

Abstract Image Segmentation exists as a challenge that aims to extract the information from the image, making it simpler to analyze. There are some major issues associated with the conventional segmentation approaches. To come up with an improvised solution, image segmentation can be modeled as a nonlinear optimization problem which is also very difficult to be solved as global optimization. So to deal with this problem, we present metaheuristic algorithm namely Genetic Algorithm and its combination with OTSU giving the better results. These results have been analyzed with the help of parameters namely Threshold values, CPU Time and Region Non Uniformity.

Keywords Metaheuristic · Genetic algorithm · OTSU · Non linear optimization

1 Introduction

Image Segmentation means to partition or divide the digital image into several smaller parts or segments in order to study the given image in a detailed manner. There have been several challenges in the traditional segmentation techniques like handling of numerous control parameters that interact in a complex, non-linear fashion. The differences between images cause changes in the results of segmentation. To define the objective function is itself a task of big efforts because there is no standard measure to accept or reject the function. In medical imaging segmentation presents few more issues like poor contrast of images. It has been observed in the literature that the medical image segmentation problem is quoted as an optimization problem with suitably represented objective function. The objective

Jyotika Pruthi (✉)
Department of Computer Science, The Northcap University, Gurgaon, India
e-mail: jyotikapruthi@ncuindia.edu

Gaurav Gupta
Department of Applied Sciences, The Northcap University, Gurgaon, India
e-mail: gauravgupta@ncuindia.edu

M. Pant et al. (eds.), *Proceedings of Fifth International Conference on Soft Computing for Problem Solving*, Advances in Intelligent Systems and Computing 437, DOI 10.1007/978-981-10-0451-3_43

function is most of the times multimodal and discontinuous making it complex, also even the search space is usually very noisy with a many local optima's. To improve results, Image Segmentation and Enhancement are modeled as Non Linear Optimization problems but optimization of a nonlinear function is itself a challenge as there may be multiple local minima or local maxima.

Metaheuristics are modern nature inspired algorithms that perform better than heuristics and tend to achieve the global optimization and are becoming popular in dealing with real time problems which are very difficult to be solved by conventional methods. They are inspired from evolutionary approaches and stochastic events. Few of the most popular metaheuritic approaches include Genetic Algorithms (GA) (Holland, J.H. (1975)), Artificial Bee Colony (ABC) (Karaboga, D., (2005)), Differential Evolution (DE) (Storn, R., Price, K.V., (1995)) etc. In this paper we have presented application of genetic algorithm and OTSU Thresholding technique for image segmentation (edge detection) and a variant has also been implemented wherein GA is combined with OTSU.

2 Techniques for Image Segmentation

The segmentation techniques reflected in the literature are broadly classified as: (a) Region Splitting Methods (b) Region Growing Methods (c) Clustering and (d) Edge and Line Oriented Segmentation Methods. In spite of the availability of the multiple methods for segmentation, researchers are still trying hard to develop efficient and robust algorithms that can bring better segmentation results. Due to its simple nature, thresholding has become the most popular way to carry out segmentation. At present many thresholding segmentation algorithms e.g. Otsu, Histogram Thresholding, Maximal entropic thresholding etc. are present in the literature. One of the oldest and most popular method of segmentation for grey images is OTSU (1979) [1]. In this paper we have tried to improve the efficiency of OTSU method by combining it with metaheuristics like Genetic Algorithm.

2.1 OTSU Method

OTSU has been the most simple thresholding algorithm that is being used by researchers on a wide scale. It performs the clustering based image thresholding [2] or in a way reduces image that is gray level into a binary image. This method assumption that the image has two classes of pixels: foreground and background pixels, then the optimized threshold disjoinitng the two classes is calculated, so that their inter-class variance is maximal [3].

OTSU method is described as follows

Assume an image is denoted by M gray levels [0, 1, … $M-1$].

The number of pixels at level i is given by ni and the total number of pixels is given by $n_1 + n_2 + \cdots n_M$.

The probability of i is denoted by:

$$p_i = \frac{n_i}{N},\ p_i \geq 0,\ \sum_{0}^{L-1} p_i = 1 \tag{1}$$

In the two level thresholding method, the pixels of an image are divided into two classes C_1 with gray levels [0, 1, … t] and C_2 with gray levels [0, 1, … $t-1$] by the threshold t.

The gray level probability distributions for the two classes denoted as w_1 and w_2 are given as:

$$w_1 = P_r(C_i) = \sum_{i=1}^{t} p_i \tag{2}$$

$$w_2 = P_r(C_2) = \sum_{i=t+1}^{L-1} p_i \tag{3}$$

The means of C_1 and C_2 are:

$$u_1 = \frac{\sum_{i=0}^{t} iP_i}{w_1} \tag{4}$$

$$u_2 = \frac{\sum_{i=t+1}^{L-1} ip_i}{w_2} \tag{5}$$

The total mean is denoted by u_T

$$u_T = w_1 u_1 + w_2 u_2 \tag{6}$$

The class variances are

$$\sigma_1^2 = \frac{\sum_{i=0}^{t} (i-u_1)^2 p_i}{w_1} \tag{7}$$

$$\sigma_2^2 = \frac{\sum_{i=t+1}^{L-1} (i-u_2)^2 p_i}{w_2} \tag{8}$$

The within-class variance is

$$\sigma_W^2 = \sum_{k=1}^{N} w_k \sigma_k^2 \tag{9}$$

The between-class variance is

$$\sigma_B^2 = w_1(u_1 - u_T)^2 + w_2(u_2 - u_T)^2 \tag{10}$$

The variance of grey levels is

$$\sigma_T^2 = \sigma_W^2 + \sigma_B^2 \tag{11}$$

The optimised threshold t has been chosen by OTSU by maximizing the inter-class variance which holds an equivalence to minimizing the intra-class variance since the total variance is constant for different partitions.

2.2 Genetic Algorithm

GA that is based on theory of survival of the fittest has been applied to many areas of image processing like for image enhancement, image segmentation, image watermarking etc. The basic idea behind GA is natural selection in which fittest entity will survive and remaining will be discarded. GA starts with the chromosome representation of a parameter set $\{x, x, \ldots x_n\}$; generally, the chromosomes are designed as strings of 0's and 1's and uses three operators: Selection, Cross over and Mutation.

Generally, an image can be described by a function $f(x, y)$, where (x, y) denotes the spatial coordinates, and the intensity value at (x, y) is $f(x, y) \in [0{,}255]$.

We present a fitness function that is present in the literature [2] that works for covered and uncovered data. The fitness function includes two terms: covers and uncovers penalty. The point x called covered point if $x \in S_j$, S_j is a region that contains connected points around the centre c_j, and x called uncovered if $x \notin \bigcup_j S_j$

$$\text{fitness} = \propto \sum_{i=1}^{n} \sum_{j=1}^{K} \left\| C_j - R_j(x_i y_i) \right\|^2 + \text{NCR} \tag{12}$$

where, NCR is a penalty for uncover points.

The Euclidian distance term represents the shortest distances between the centroid c_j, $j = 1, 2, \ldots, k$ and all pixels p_i, $i = 1, 2, \ldots, N$ of a region R_j. NCR is used to represent these pixels. NCR can be written as:

$$\text{NCR} = \propto \sum_{i=1}^{m} \|\text{Median} - R(x_i y_i)\|^2 \tag{13}$$

where m is the number of uncovered points.

3 Proposed Methodology for Image Segmentation

The proposed methodology has been stated in a simple manner through the following three steps:

1. **Input the image that is to be segmented and set the initial generation as 0.**
   - Population size (NP) has been taken as the number of pixels for all the metaheuristics used.
   - Grey level images have been taken due to which the initial population has been taken in the range of [0–255].
2. **The fitness function is determined with the help of OTSU method**

$$F(\text{fitness}, i) = S_{\text{between objects}} / S_{\text{within objects}} \tag{14}$$

where the variance between objects, $S_{\text{Between objects}}$, is given as:

$$S_{\text{Between objects}} = \sum_i P_i (m_i - m_g)^2 \tag{15}$$

and the variance within objects is given as:

$$S_{\text{within objects}} = \sum_i \frac{S_i}{S_g} \tag{16}$$

where, M_i is the mean of pixels in the segment whose threshold value is i, m_i is given as:

$$m = \sum_{\text{thsld}\, i}^{\text{thsld}\, i+1} x * \text{hist}(x) \tag{17}$$

The probability of segment i, p_i is given as:

$$p_i = \sum_{\text{thsld}\, i}^{\text{thsld}\, i+1} \text{hist}(x) \tag{18}$$

and the global mean of image is given as:

$$\text{mg} = m_i * p_i \tag{19}$$

3. **The fitness function evaluated has been optimized using metaheuristics (GA + OTSU)**

4 Results

The algorithms have been implemented in Matlab 7.10.0(R2010a) version and experiments are performed on a computer with 2.20 GHz Intel (R) core i5. Test images are from Gonzalez and Woods database of images. The crossover rate for Genetic algorithm is fixed at 0.5. In Fig. 1, image (a) initial image (b) shows image with OTSU being applied (c) segmented image with GA and image (d) shows segmented image with OTSU and GA.

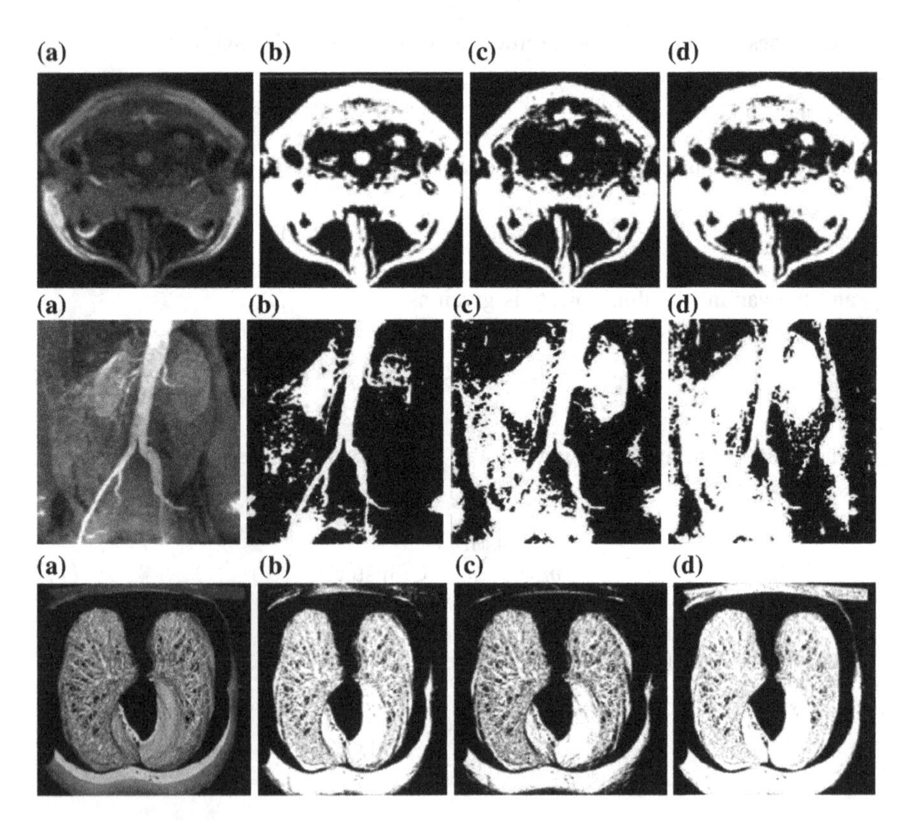

Fig. 1 **a** Original Image. **b** Segmented image with OTSU. **c** Segmented image with GA. **d** Segmented image with OTSU + GA

4.1 Result Analysis on the Basis of Threshold Values and CPU Time

The threshold values obtained by above mentioned algorithms are shown in Table 1. CPU time that each algorithm takes is also given. It can be seen that all hybrid algorithms take more CPU time as compared to OTSU method due to the fact that besides having Otsu method, a metaheuristic technique is also present. However in terms of threshold value, OTSU + GA gives better performance.

4.2 Result Analysis Using Region Non-uniformity Method

The following evaluation criteria are used to analyze the metaheuristics. The segmentation is evaluated using Region Non-uniformity method which was given by

$$\mathrm{NU} = \frac{|F_T|}{|F_T + B_T|} * \frac{\sigma_f^2}{\sigma^2}$$

where
σ^2 represents the variance of whole image
σ_f^2 represents foreground variance
F_T represents foreground threshold value
B_T represents background threshold value for image

The value of NU gives us a measure of the segmentation carried out. If NU value is around ‘0’ then the segmentation is very good. A value close to ‘1’ represents a poor segmentation (Table 2).

Table 1 Table showing threshold value (CPU Time) obtained from OTSU, GA and OTSU + GA

Images	Parameter	OTSU	GA	OTSU + GA
Image 1	Threshold value	0.0123	0.0136	0.0138
	CPU time	0.2745	0.2661	0.3421
Image 2	Threshold time	0.0241	0.0243	0.0246
	CPU time	0.3291	0.4667	0.6846
Image 3	Threshold value	0.0221	0.0288	0.0339
	CPU time	0.4410	0.4314	0.4421

Table 2 Table showing the performance evaluation of OTSU, GA and OTSU + GA

Images	OTSU	GA	OTSU + GA
Image 1	0.877	0.864	0.11
Image 2	0.932	0.822	0.13
Image 3	0.841	0.873	0.21

Table 3 Table showing the statistical analysis

Algorithms	Mean rank
OTSU	3.14
GA	2.45
OTSU + GA	1.35

4.3 Result Analysis Using Mean Rank (Friedman Test)

From the results shown in the above Tables, though it may be seen that GA embedded with OTSU gives a better result, it is difficult to make a final conclusion. In order to provide a more concrete analysis, the algorithms are compared statistically also. These results are given in Table 3. For performance analysis Friedman test is used.

4.3.1 Friedman Test

See Table 3.

5 Conclusion

The segmentation has been performed on images of different domains with OTSU, Genetic algorithm and promising results have been obtained when segmentation is carried through combination of genetic algorithm and OTSU. Thresholding values associated with CPU Time for each image have been calculated and are found to be always higher for OTSU + GA in comparison to only OTSU or only GA which means that though in terms of CPU time, OTSU is fastest but in terms of threshold values the OTSU embedded variants give a better performance with values of 0.0138, 0.0246 and 0.0339. When the hybrid approach is tested with region non uniformity method its value is coming out to be very close to 0 that shows better segmentation as compared to other two algorithms. Mean rank value also proves the hybrid approach to be the better one.

References

1. Otsu, N.: A threshold selection method from gray-level histogram. IEEE Trans. Syst. Man Cybern. **9**, 62–66 (1979)
2. Zanaty, EA, Ghiduk, A.S.: A novel approach based on genetic algorithms and region growing for magnetic resonance image (MRI) segmentation. J. Comput. Sci. Inform. Syst. (ComSIS), **10** (3). doi:10.2298/CSIS120604050Z(2013)
3. Maulik, U.: Medical image segmentation using genetic algorithms. IEEE Trans. Informa. Technol. Biomed. **13**(2) (2009)

On Accepting Hexagonal Array Splicing System with Respect to Directions

D.K. Sheena Christy, D.G. Thomas, Atulya K. Nagar and Robinson Thamburaj

Abstract A new approach to hexagonal array splicing system with respect to directions in uniform and nonuniform ways as an accepting device was proposed in Sheena Christy and Thomas (Proceedings of national conference on mathematics and computer applications 2015). In this paper, we prove the following results (i) $\mathcal{L}(\text{AHexASSD}) - \text{HKAL} \neq \phi$ (ii) $\mathcal{L}(\text{AHexASSD}) - \text{CT0LHAL} \neq \phi$ (iii) $\mathcal{L}_n(\text{AHexASSD}) \subset \mathcal{L}(\text{AHexASSD})$

Keywords Accepting splicing system · Hexagonal array splicing system · Two dimensional languages

1 Introduction

Formal languages have application in the area of DNA computing [1]. In [2], Tom Head defined splicing systems motivated by the behavior of DNA sequences. A specific model of DNA recombination is the splicing operation which consists of "cutting" DNA sequences and then "pasting" the fragments again [3–5].

D.K. Sheena Christy (✉)
Department of Mathematics, SRM University, Kattankulathur 603203, India
e-mail: sheena.lesley@gmail.com

D.G. Thomas
Department of Mathematics, Madras Christian College, Chennai 600059, India
e-mail: dgthomasmcc@yahoo.com

A.K. Nagar
Department of Mathematics and Computer Science, Liverpool Hope University, Liverpool, UK
e-mail: nagara@hope.ac.uk

Robinson Thamburaj
Department of Mathematics, Madras Christian College, Tambaram, Chennai 600059, India
e-mail: robin.mcc@gmail.com

M. Pant et al. (eds.), *Proceedings of Fifth International Conference on Soft Computing for Problem Solving*, Advances in Intelligent Systems and Computing 437, DOI 10.1007/978-981-10-0451-3_44

Mitrana et al. [6] have introduced accepting splicing systems in uniform and nonuniform ways. All the accepting models have been compared with each other with respect to their computational power. Arroyo et al. [7] introduced the generalization of this system and studied its properties.

Another interesting area of research is the study of hexagonal array grammars and their languages [8–10] and application of hexagons in image processing [11]. Thomas et al. [12] have examined the notion of splicing on hexagonal array languages. We continue this study and prove certain results.

2 Accepting Hexagonal Array Splicing System with Respect to Directions

Definition 1 ([13]) An accepting hexagonal array splicing system with respect to directions (AHexASSD) is a pair $\gamma = (S, P)$, where $S = (\Gamma, \mathcal{A})$ is a hexagonal array splicing system with respect to directions and P is a finite subset of Σ^{**H}. Let $w \in \Sigma^{**H}$. The uniform method of the iterated hexagonal array splicing with respect to directions of γ is defined as follows:

$$\Gamma^0(\mathcal{A}, w) = \{w\}; \Gamma^{i+1}(\mathcal{A}, w) = \Gamma^i(\mathcal{A}, w) \cup \left(\bigcup_{D \in \{X,Y,Z\}} \Gamma_D(\Gamma^i(\mathcal{A}, w) \cup \mathcal{A}) \right);$$

$$\Gamma^*(\mathcal{A}, w) = \bigcup_{i \geq 0} \Gamma^i(\mathcal{A}, w).$$

The language accepted by an accepting hexagonal array splicing system with respect to directions γ is $L(\gamma) = \{w \in \Sigma^{**H} / \Gamma^*(\mathcal{A}, w) \cap P \neq \phi\}$.

The class of languages accepted by accepting hexagonal array splicing system with respect to directions is denoted by $\mathcal{L}(\text{AHexASSD})$.

Definition 2 ([13]) Let $\gamma = (S, P)$ be an accepting hexagonal array splicing system with respect to directions and $w \in \Sigma^{**H}$. The nonuniform variant of iterated splicing of γ is defined as follows:

$$\tau^0(\mathcal{A}, w) = \{w\}; \tau^{i+1}(\mathcal{A}, w) = \tau^i(\mathcal{A}, w) \cup \left(\bigcup_{D \in \{X,Y,Z\}} \Gamma_D(\tau^i(\mathcal{A}, w), \mathcal{A}) \right);$$

$$\tau^*(\mathcal{A}, w) = \bigcup_{i \geq 0} \tau^i(\mathcal{A}, w).$$

The language accepted by γ in nonuniform way is $L_n(\gamma) = \{w \in \Sigma^{**H} / \tau^* (\mathcal{A}, w) \cap P \neq \phi\}$.

The class of all languages accepted by AHexASSD in nonuniform way is denoted by $\mathcal{L}_n(\text{AHexASSD})$.

We illustrate a uniform and nonuniform way of accepting hexagonal array splicing system with respect to directions with the following examples:

Example 1 Let $\gamma = (S, P)$ be an AHexASSD, where $S = (\Gamma_Z, \mathcal{A})$, $\Gamma_Z = (\Sigma, R_x^Z, R_y^Z)$ with $\Sigma = \{a\}$, $\mathcal{A} = \{\Lambda\}$, $P = \left\{ \begin{array}{ccccc} & a & & a & \\ a & & a & & a \\ & a & & a & \end{array} \right\}$

$$R_x^Z = \left\{ \begin{array}{|c|} a \\ \hline a \end{array} \# \begin{array}{|c|} \lambda \\ \hline \lambda \end{array} \$ \begin{array}{|c|} \lambda \\ \hline \lambda \end{array} \# \begin{array}{|c|} a \\ \hline a \end{array} \right\}$$

$R_y^Z = \phi$. Let $w = \begin{array}{ccccccc} & a & & a & & a & \\ a & & a & & a & & a \\ & a & & a & & a & \end{array} \in \Sigma^{**}$

$$\begin{aligned} \Gamma^1(\mathcal{A}, w) &= \Gamma^0(\mathcal{A}, w) \cup \Gamma_Z(\Gamma^0(\mathcal{A}, w) \cup \mathcal{A}) \\ &= \left\{ \begin{array}{ccccccc} & a & & a & & a & \\ a & & a & & a & & a \\ & a & & a & & a & \end{array} \right\} \cup \Gamma_Z\left(\left\{ \begin{array}{ccccccc} & a & & a & & a & \\ a & & a & & a & & a \\ & a & & a & & a & \end{array} \right\} \cup \{\Lambda\} \right) \\ &= \left\{ \begin{array}{ccccc} & a & & a & \\ a & & a & & a \\ & a & & a & \end{array}, \begin{array}{ccccccc} & a & & a & & a & \\ a & & a & & a & & a \\ & a & & a & & a & \end{array}, \begin{array}{ccccccccc} & a & & a & & a & & a & \\ a & & a & & a & & a & & a \\ & a & & a & & a & & a & \end{array} \right\}. \end{aligned}$$

Clearly, $\Gamma^1(\mathcal{A}, w) \cap P \neq \phi$. Hence AHexASSD γ accepts the language

$$L = \left\{ \begin{array}{ccccc} & a & & a & \\ a & & a & & a \\ & a & & a & \end{array}, \begin{array}{ccccccc} & a & & a & & a & \\ a & & a & & a & & a \\ & a & & a & & a & \end{array}, \begin{array}{ccccccccc} & a & & a & & a & & a & \\ a & & a & & a & & a & & a \\ & a & & a & & a & & a & \end{array}, \ldots \right\}$$

in uniform way.

Example 2 Let $\gamma = (S, P)$ be a nonuniform variant of AHexASSD where $S = (\Gamma_Z, \mathcal{A})$, $\Gamma_Z = (\Sigma, R_x^Z, R_y^Z)$, $\Sigma = \{1, 2, 3\}$, $\mathcal{A} = \left\{ \begin{array}{ccccc} & 1 & & 1 & \\ 2 & & 2 & & 2 \\ & 3 & & 3 & \end{array} \right\}$ and

$$P = \left\{ \begin{array}{ccccc} & 1 & & 1 & \\ 2 & & 2 & & 2 \\ & 3 & & 3 & \end{array} \right\}, \quad R_x^Z = \left\{ P_1 : \begin{array}{|c|} \lambda \\ \hline \lambda \end{array} \# \begin{array}{|c|} 1 \\ \hline 2 \end{array} \$ \begin{array}{|c|} \lambda \\ \hline \lambda \end{array} \# \begin{array}{|c|} 1 \\ \hline 2 \end{array}, \right.$$

$$\left. P_2 : \begin{array}{|c|} \lambda \\ \hline \lambda \end{array} \# \begin{array}{|c|} 2 \\ \hline 3 \end{array} \$ \begin{array}{|c|} \lambda \\ \hline \lambda \end{array} \# \begin{array}{|c|} 2 \\ \hline 3 \end{array} \right\} \text{ and } R_y^Z = \phi. \, \gamma$$

accepts $L = \{p \in \Sigma^{**H} / p$ is of size (l, m, n) with $l = m = 2$, $n \geq 2$ and the suffix array of p is of the form $\left. \begin{array}{cccc} & & 1 & \\ & 2 & & 2 \\ 3 & & 3 & \end{array} \right\}$

For example, let $w = \begin{array}{ccccccc} & 2 & & 1 & & 1 & \\ 1 & & 1 & & 2 & & 2 \\ & 3 & & 3 & & 3 & \end{array} \in \Sigma^{**H}$. Now $\tau^0(\mathcal{A}, w) = \left\{ \begin{array}{ccccccc} & 2 & & 1 & & 1 & \\ 1 & & 1 & & 2 & & 2 \\ & 3 & & 3 & & 3 & \end{array} \right\}$,

$$\tau^1(\mathcal{A}, w) = \left\{ \begin{array}{ccccccc} & 2 & & 1 & & 1 & \\ 1 & & 1 & & 2 & & 2 \\ & 3 & & 3 & & 3 & \end{array} \right\} \cup \Gamma_Z \left(\left\{ \begin{array}{ccccccc} & 2 & & 1 & & 1 & \\ 1 & & 1 & & 2 & & 2 \\ & 3 & & 3 & & 3 & \end{array} \right\}, \left\{ \begin{array}{ccccc} & 1 & & 1 & \\ 2 & & 2 & & 2 \\ & 3 & & 3 & \end{array} \right\} \right) = \left\{ \begin{array}{ccccc} & 1 & & 1 & \\ 2 & & 2 & & 2 \\ & 3 & & 3 & \end{array}, \begin{array}{ccccccc} & 2 & & 1 & & 1 & \\ 1 & & 1 & & 2 & & 2 \\ & 3 & & 3 & & 3 & \end{array} \right\}.$$

Therefore $\tau^1(\mathcal{A}, w) \cap P \neq \phi$.

3 Main Results

We now compare the generative power of AHexASSD with that of HKAL and CT0LHAL [9]. We show that the class of hexagonal array languages accepted by uniform variant of AHexASSD includes the class of hexagonal array languages accepted by nonuniform variant of AHexASSD.

Theorem 1 *The classes* $\mathcal{L}(\text{AHexASSD})$ *and HKAL are incomparable.*

Proof To show that $\text{HKAL} - \mathcal{L}(\text{AHexASSD}) \neq \phi$, let us consider a HKAG $G_1 = (V, I, P, L, S)$, where $V = V_1 \cup V_2$, $V_1 = \{S\}$, $V_2 = \{a, b\}$, $I = \{X, 0, G\}$, $P = P_1 \cup P_2$, $P_1 = \{S \to (S \textcircled{$\uparrow$} b) \textcircled{$\downarrow$} a\}$, $P_2 = \left\{S \to \begin{array}{ccccc} & x & & x & \\ x & & x & & x \\ & x & & x & \end{array}\right\}$ with $L = \{L_a, L_b\}$, $L_a = \{G^n < G > G^n\}$, $L_b = \{0^n < 0 > 0^n\}$, $n = 1$, S is the start symbol.
The grammar G_1, generates a HKAL

$$L_2 = \left\{ \begin{array}{ccccc} & x & & x & \\ x & & x & & x \\ & x & & x & \end{array}, \begin{array}{ccccccccc} & G & & x & & x & & 0 & \\ G & & x & & x & & x & & 0 \\ & G & & x & & x & & 0 & \end{array}, \begin{array}{ccccccccccccc} & G & & G & & x & & x & & 0 & & 0 & \\ G & & G & & x & & x & & x & & 0 & & 0 \\ & G & & G & & x & & x & & 0 & & 0 & \end{array}, \cdots \right\}$$

The language L_2 cannot be accepted by any uniform variant of AHexASSD, since there does not exist any finite P such that $\Gamma^*(\mathcal{A}, w) \cap P \neq \phi$.

To prove that $\mathcal{L}(\text{AHexASSD}) - \text{HKAL} \neq \phi$, let $\gamma = (S, P)$ be an AHexASSD, where $S = (\Gamma_z, \mathcal{A})$, $\Gamma_z = (\Sigma, R_x^z, R_y^z)$ with $\Sigma = \{1, 2, 3\}$, $\mathcal{A} = \{\Lambda\}$,

$$P = \left\{ \begin{array}{ccccc} & 1 & & 1 & \\ 1 & & 2 & & 1 \\ & 3 & & 3 & \end{array} \right\}; \quad R_x^Z = \left\{ P_1 : \frac{1}{2} \# \frac{\lambda}{\lambda} \$ \frac{\lambda}{\lambda} \# \frac{1}{2}, \quad P_2 : \frac{2}{3} \# \frac{\lambda}{\lambda} \$ \frac{\lambda}{\lambda} \# \frac{2}{3} \right\}$$

$R_y^z = \phi$. Hence the language accepted by γ is

$$L(\gamma) = L_3 = \left\{ \begin{array}{ccccc} & 1 & & 1 & \\ 1 & & 2 & & 1 \\ & 3 & & 3 & \end{array}, \begin{array}{ccccccc} & 1 & & 1 & & 1 & \\ 1 & & 2 & & 2 & & 1 \\ & 3 & & 3 & & 3 & \end{array}, \begin{array}{ccccccccc} & 1 & & 1 & & 1 & & 1 & \\ 1 & & 2 & & 2 & & 2 & & 1 \\ & 3 & & 3 & & 3 & & 3 & \end{array}, \cdots \right\}.$$

But $L_3 \notin$ HKAL, since if we consider the production rule P_2 as $S \rightarrow \begin{array}{ccccc} & 1 & & 1 & \\ 1 & & 2 & & 1 \\ & 3 & & 3 & \end{array}$, it will generate hexagonal arrays which do not belong to the language L_3.

Hence the classes $\mathcal{L}$(AHexASSD) and HKAL are incomparable.

Theorem 2 *The classes $\mathcal{L}$(AHexASSD) and CT0LHAL are incomparable and not disjoint.*

Proof Let $G = (V, H_0, \mathcal{P}, C)$ be a CT0LHAG, where $V = \{M\}$, $H_0 = \left(\begin{array}{ccccc} & M & & M & \\ M & & M & & M \\ & M & & M & \end{array} \right)$; $\mathcal{P} = \{T_1, T_2, T_3\}$, $T_1 = \{M \nearrow MM\}$; $T_2 = \{M \nwarrow MM\}$; $T_3 = \{M \downarrow MM\}$ and $C = \{(T_1T_2T_3)^{2^n - 1} / n \geq 1\}$.

The language produced by the grammar G is the regular hexagonal arrays of sides 2^n.

$$L_4 = \left\{ \begin{array}{ccccc} & M & & M & \\ M & & M & & M \\ & M & & M & \end{array}, \begin{array}{ccccccccccccc} & & & M & & M & & M & & M & & & \\ & & M & & M & & M & & M & & M & & \\ & M & & M & & M & & M & & M & & M & \\ M & & M & & M & & M & & M & & M & & M \\ & M & & M & & M & & M & & M & & M & \\ & & M & & M & & M & & M & & M & & \\ & & & M & & M & & M & & M & & & \end{array}, \cdots \right\}$$

But L_4 cannot be accepted by any AHexASSD since there does not exist any finite P such that $\Gamma^*(\mathcal{A}, w) \cap P \neq \phi$. Hence CT0LHAL $- \mathcal{L}$(AHexASSD) $\neq \phi$. To show that $\mathcal{L}$(AHexASSD) $-$ CT0LHAL $\neq \phi$, consider the language

$$L_5 = \left\{ \begin{array}{ccccc} & 1 & & 1 & \\ 1 & & 2 & & 1 \\ & 3 & & 3 & \end{array}, \begin{array}{ccccccc} & 1 & & 1 & & 1 & \\ 1 & & 2 & & 2 & & 1 \\ & 3 & & 3 & & 3 & \end{array}, \cdots \right\}.$$

We observe that the language L_5 is accepted by a uniform variant of AHexASSD considered in Theorem 1. But it cannot be generated by any regular controlled Table 0L hexagonal array grammar, as either the rule $11 \leftarrow 1$ or $12 \leftarrow 1$ will generate some hexagonal arrays which are not in L_5.

To prove $\mathcal{L}(\text{AHexASSD}) \cap \text{CT0LHAL} \neq \phi$, consider the language

$$L_6 = \left\{ \begin{array}{ccccc} & 1 & & 1 & \\ 2 & & 2 & & 2 \\ & 3 & & 3 & \end{array}, \begin{array}{ccccccc} & 1 & & 1 & & 1 & \\ 2 & & 2 & & 2 & & 2 \\ & 3 & & 3 & & 3 & \end{array}, \cdots \right\}.$$

L_6 is generated by a CT0LHAG in the following way:

Let $G = (V, H_0, \mathcal{P}, C)$ be the CT0LHAG with $V = \{1, 2, 3\}$,

$$H_0 = \begin{array}{ccccc} & 1 & & 1 & \\ 2 & & 2 & & 2 \\ & 3 & & 3 & \end{array}, \ \mathcal{P} = \{T_1, T_2, T_3\},\ T_1 = \{2 \rightarrow 21\},\ T_2 = \{2 \searrow 23\},$$

$T_3 = \{22 \leftarrow 2\}$, $C = \{(T_1 T_2 T_3)^n / n \geq 0\}$. The acceptance of L_6 by a uniform variant of AHexASSD is shown below:

Let $\gamma = (S, P)$, where $S = (\Gamma_z, \mathcal{A})$ with $\Gamma_z = (\Sigma, R_x^z, R_y^z)$; $\Sigma = \{1, 2, 3\}$, $\mathcal{A} = \{A\}$, $P = \left\{ \begin{array}{ccccccc} & 2 & & 1 & & 1 & \\ 1 & & 1 & & 2 & & 2 \\ & 3 & & 3 & & 3 & \end{array} \right\}$; $R_y^z = \phi$ and

$$R_x^Z = \left\{ P_1 : \begin{array}{|c|}\hline 1 \\ \hline 2 \\ \hline\end{array} \# \begin{array}{|c|}\hline \lambda \\ \hline \lambda \\ \hline\end{array} \$ \begin{array}{|c|}\hline \lambda \\ \hline \lambda \\ \hline\end{array} \# \begin{array}{|c|}\hline 1 \\ \hline 2 \\ \hline\end{array}, \right.$$

$$\left. P_2 : \begin{array}{|c|}\hline 2 \\ \hline 3 \\ \hline\end{array} \# \begin{array}{|c|}\hline \lambda \\ \hline \lambda \\ \hline\end{array} \$ \begin{array}{|c|}\hline \lambda \\ \hline \lambda \\ \hline\end{array} \# \begin{array}{|c|}\hline 2 \\ \hline 3 \\ \hline\end{array} \right\}.$$

Notation: If $q = \begin{array}{ccccc} & a & & a & \\ a & & a & & a \\ & a & & a & \end{array}$ and $u = \{(\not{d})^n < \not{d} > (\not{d})^n / n \geq 1\}$ then

$u \circledless q = \begin{array}{ccccccc} & \not{d} & & a & & a & \\ \not{d} & & a & & a & & a \\ & \not{d} & & a & & a & \end{array}$ and $q \circledgtr u = \begin{array}{ccccccc} & a & & a & & \not{d} & \\ a & & a & & a & & \not{d} \\ & a & & a & & \not{d} & \end{array}$. If p is a hexagonal array then we denote the number of rows of p as $|p|_{\ominus}$ and the number of colomns of p as $|p|_{\obar}$. For instance, $|q|_{\ominus} = 3$; $|q|_{\obar} = 5$.

Theorem 3 $\mathcal{L}_n(AHexASSD) \subset \mathcal{L}(AHexASSD)$ *with* $\Gamma_X = \Gamma_Y = \phi$.

Proof Let $\gamma = (S, P)$ be a nonuniform variant of AHexASSD where $S = (\Gamma, \mathcal{A})$ with $\Gamma = (\Gamma_X, \Gamma_Y, \Gamma_Z)$ and $\Gamma_X = \Gamma_Y = \phi$, $\Gamma_Z = (\Sigma, R_x^z, R_y^z)$, where ϕ is the empty set. Let $m = \max\{|k|_{\varobar} \ /k \in \mathcal{A}\}$. We define the finite set $E = \{k' \in L_n(\gamma)/|k'|_{\varobar} \leq m\}$.

Let us consider a uniform variant of AHexASSD $\gamma' = (\Sigma', R_{x'}^z, R_{y'}^z, \mathcal{A}', P')$, where $\Sigma' = \Sigma \cup \{\not d\}$;

$$\mathcal{A}' = \left\{ \begin{array}{ccccccccccccc} & \not d & & a & & b & & c & & \not d & & \\ \not d & & d & & e & & f & & g & & \not d \\ & \not d & & h & & i & & j & & \not d & & \end{array} \Bigg/ \begin{array}{ccccccc} & a & & b & & c & \\ d & & e & & f & & g \\ & h & & i & & j & \end{array} \in \mathcal{A} \right\}.$$

$R_{x'}^z = \{\alpha_2 \oslash \alpha_1 \# \beta_1 \oslash \beta_2 \$ u \oslash r \oslash x' \# y' \oslash s \oslash u / \alpha_1 \# \beta_1 \$ x' \# y' \in R_x^z$ and $p \in \mathcal{A}$, p can be converted into a parallelogram by inserting λ in suitable positions so that p can be written in the form $p_1 \ominus p_2 \ominus \ldots \ominus p_{i_1}$ for $i_1 \geq 1$ with each p_t is a parallelogram of size $2 \times n$, $n \geq 1$, $1 \leq t \leq i_1$ and p_t can be expressed as $r \oslash x' \oslash y' \oslash s$, where $\alpha_1, \beta_1, x', y'$ are 2×1 dominoes of the form $\begin{array}{c} a \\ b \end{array}$, $a, b \in \Sigma$ or $a = b = \lambda$ and β_2, s are parallelograms of sizes $2 \times n$ of the form, $\begin{array}{cccc} a_1 & a_2 & \cdots & a_n \\ b_1 & b_2 & \cdots & b_n \end{array}$ and α_2, r are also parallelograms of sizes $2 \times n$ of the form $\begin{array}{cccc} a_n & \cdots & a_2 & a_1 \\ b_n & \cdots & b_2 & b_1 \end{array}$, where a_j, b_j can be either in Σ or equal to λ, $j \leq n-1$ and $a_n, b_n \in \Sigma \cup \{\lambda\}$ with $|\alpha_2 \oslash \alpha_1 \oslash \beta_1 \oslash \beta_2|_{\varobar} = m+1$ and $u = \{(\not d)^n < \not d > (\not d)^n / n \geq 1\}\} \bigcup \{\alpha_1 \# \beta_1 \$ u \oslash r \oslash x' \# y' \oslash s \oslash u / \alpha_1 \# \beta_1 \$ x' \# y' \in R_y^z$ with $|\alpha_1 \oslash \beta_1|_{\varobar} > m+1$, $p \in \mathcal{A}$ as considered in the earlier case and $u = \{(\not d)^n < \not d > (\not d)^n / n \geq 1\}\}$.

$R_{y'}^z = \{\alpha_2 \obslash \alpha_1 \# \beta_1 \obslash \beta_2 \$ u \obslash r \obslash x' \# y' \obslash s \obslash u / \alpha_1 \# \beta_1 \$ x' \# y' \in R_y^z$, $p' \in \mathcal{A}$, p' can be converted into a parallelogram by inserting λ in suitable positions so that p' can be written in the form $p_1 \ominus p_2 \ominus \ldots \ominus p_{i_1}$ for $i_1 \geq 1$ with each p_t is a parallelogram of size $2 \times n$, $n \geq 1$, $1 \leq t \leq i_1$ and p_t can be expressed as $a, b \in \Sigma$ or $a = b = \{\lambda\}$ and β_2, s are parallelograms of sizes $2 \times n$ of the form $r \obslash x' \obslash y' \obslash s$, where $\alpha_1, \beta_1, x', y'$, are 2×1 dominoes of the form $\begin{array}{c} a \\ b \end{array}$, $a, b \in \Sigma$ or $a = b = \lambda$ and β_2, s are parallelograms of sizes $2 \times n$ of the form $\begin{array}{cccc} a_1 & a_2 & \cdots & a_n \\ b_1 & b_2 & \cdots & b_n \end{array}$, α_2, r are also parallelograms of sizes $2 \times n$ of the form $\begin{array}{cccc} a_n & \cdots & a_2 & a_1 \\ b_n & \cdots & b_2 & b_1 \end{array}$, where a_j, b_j can be either in Σ or equal to λ, $1 \leq j \leq n-1$

and $a_n, b_n \in \Sigma \cup \{\lambda\}$ with $|\alpha_2 \oslash \alpha_1 \oslash \beta_1 \oslash \beta_2|_{\oplus} = m+1$ and $u = \{(\not{q})^n < \not{q} > (\not{q})^n / n \geq 1\}\} \bigcup \{\alpha_1 \# \beta_1 \$ u \oslash r \oslash x' \# y' \oslash s \oslash u / \alpha_1 \# \beta_1 \$ x' \# y' \in R_x^z$ with $|\alpha_1 \oslash \beta_1|_{\oplus} > m+1, p \in \mathcal{A}$ as considered in the earlier case and $u = \{(\not{q})^n < \not{q} > (\not{q})^n / n \geq 1\}\}$.

$$P' = E \cup (I \otimes D) \cup (I \otimes D \otimes I) \cup P \cup (D \otimes I),$$

where $D = E \cup P$; $I = \{(\not{q})^n < \not{q} > (\not{q})^n / n \geq 1\}$.

Claim 1 *For any $w \in \Sigma^{**H}$ with $|w|_{\oplus} > m$;*

(i) $\tau^i_{R^z_x}(A, w) \cap P \neq \phi$ *implies that* $w \in L(\gamma') \cap \Sigma^{**H}$.
(ii) $\tau^j_{R^z_y}(A, w) \cap P \neq \phi$ *implies that* $w \in L(\gamma') \cap \Sigma^{**H}$.

Proof of Claim 1 The statement (i) is trivially true for $i = 0$. If= 1, then let $q \in \Gamma_{R^z_x}(w, Q) \cap P$ for some axiom Q. It follows that $u \otimes q \in \Gamma_{R^z_{x'}}(w, \alpha)$ or $q \otimes u \in \Gamma_{R^z_{x'}}(w, \alpha)$, where $\alpha = (\not{q})^n \, ø \, Q \, ø \, (\not{q})^n \in \mathcal{A}'$ with $|u|_{\ominus} = |q|_{\ominus}$, where $u = \{(\not{q})^n < \not{q} > (\not{q})^n / n \geq 1\}$ and $\mathcal{A}'$ is already defined.

Hence $\Gamma_{R^z_{x'}}(w, \mathcal{A}') \cap P' \neq \phi$. Let us assume that the statement is true for $i \leq j$, i.e., $p \in \Gamma_{R^z_{x'}}(w, Q)$ for some axiom Q and $\tau^j_{R^z_x}(\mathcal{A}, p) \cap P \neq \phi$. From the above consideration, we can have $u \otimes q \in \Gamma_{R^z_{x'}}(w, \alpha)$ or $q \otimes u \in \Gamma_{R^z_{x'}}(w, \alpha)$, where $\alpha = (\not{q})^n \, ø \, Q \, ø \, (\not{q})^n \in \mathcal{A}'$. Now we study about the case $u \otimes q \in \Gamma_{R^z_{x'}}(w, \alpha)$. The other case can be analyzed in a similar way. By the induction hypothesis we have $p \in L(\gamma') \cap \Sigma^{**H}$ which implies that $\Gamma^*_{R^z_{x'}}(\mathcal{A}', p) \cap P' \neq \phi$. It follows that $\Gamma_{R^z_{x'}}(\mathcal{A}', u \otimes p) \cap P' \neq \phi$ (or) $\Gamma_{R^z_{x'}}(\mathcal{A}', p \otimes u) \cap P' \neq \phi$. Hence $w \in L(\gamma') \cap \Sigma^{**H}$. Similarly, we can prove the statement (ii).

Hence $L_n(\gamma) \subseteq L(\gamma') \cap \Sigma^{**H}$.

Claim 2 *For any $w \in \Sigma^{**H}$, $g \geq 0$ and $t \geq 0$,*

(iii) $\Gamma^g_{R^z_{x'}}(\mathcal{A}', w) \cap P' \neq$ *implies that* $w \in L_n(\gamma)$
(iv). $\Gamma^t_{R^z_{y'}}(\mathcal{A}', w) \cap P' \neq$ *implies that* $w \in L_n(\gamma)$.

Proof of Claim 2 If $g = 0$, then as $w \in E \cup P$ we have $w \in L_n(\gamma)$. Assume that the statement (iii) is true for $g \leq j_1$, i.e., let $p \in \Gamma_{R^z_{x'}}(\mathcal{A}' \cup \{w\})$ and $\Gamma^{j_1}_{R^z_{x'}}(\mathcal{A}', p) \cap P' \neq \phi$. Here, p can be either of the form $(u \otimes y_1)$ or $(y_1 \otimes u)$ with $|u|_{\ominus} = |y_1|_{\ominus}$ for some $y_1 \in \Sigma^{**H}$.

We prove the statement for $p = u \otimes y_1$. The other case can be proved similarly. Since $\Gamma^{j_1}_{R^z_{x'}}(\mathcal{A}', p) \cap P' \neq \phi$, we have $\Gamma^{j_1}_{R^z_{x'}}(\mathcal{A}', y_1) \cap P' \neq \phi$. Hence $y_1 \in$

$\Gamma_{R_x^z}(w, \mathcal{A})$ and so $y_1 \in L_n(\gamma)$. The other statement can be proved similarly. Therefore, $L(\gamma') \cap \Sigma^{**H} \subseteq L_n(\gamma)$. Thus we have $L_n(\gamma) = L(\gamma') \cap \Sigma^{**H}$.

To prove the strict inclusion, i.e., $\mathcal{L}(\text{AHexASSD}) - \mathcal{L}_n(\text{AHexASSD}) \neq \phi$, let us consider the language $L_1 = \left\{ \begin{array}{ccccc} & 1 & & 1 & \\ 1 & & 2 & & 1 \\ & 3 & & 3 & \end{array}, \begin{array}{ccccccc} & 1 & & 1 & & 1 & \\ 1 & & 2 & & 2 & & 1 \\ & 3 & & 3 & & 3 & \end{array}, \begin{array}{ccccccccc} & 1 & & 1 & & 1 & & 1 & \\ 1 & & 2 & & 2 & & 2 & & 1 \\ & 3 & & 3 & & 3 & & 3 & \end{array}, \ldots \right\}$. L_1 can be accepted by a uniform variant of AHexASSD $\gamma = (S, P)$, where $S = (\Gamma_Z, \mathcal{A})$, $\Gamma_Z = (\Sigma, R_x^Z, R_y^Z)$, with $\Sigma = \{1, 2, 3\}$, $\mathcal{A} = \{\Lambda\}$, $P = \left\{ \begin{array}{ccccc} & 1 & & 1 & \\ 1 & & 2 & & 1 \\ & 3 & & 3 & \end{array} \right\}$, $R_y^Z = \phi$ and

$$R_x^Z = \left\{ P_1 : \begin{array}{c} 1 \\ 2 \end{array} \# \begin{array}{c} \lambda \\ \lambda \end{array} \$ \begin{array}{c} \lambda \\ \lambda \end{array} \# \begin{array}{c} 1 \\ 2 \end{array} \right.$$

$$\left. P_2 : \begin{array}{c} 2 \\ 3 \end{array} \# \begin{array}{c} \lambda \\ \lambda \end{array} \$ \begin{array}{c} \lambda \\ \lambda \end{array} \# \begin{array}{c} 2 \\ 3 \end{array} \right\}.$$

But L_1 cannot be accepted by any nonuniform variant of AHexASSD, $\gamma' = (S', P')$, where $S' = (\Gamma'_Z, \mathcal{A}')$, $\Gamma'_Z = (\Sigma, R_x^Z, R_y^Z)$ with $\mathcal{A}'$ being a finite subset of Σ^{**H} and since we cannot find any finite P' such that $\tau^*(\mathcal{A}', w) \cap P' \neq \phi$, for some $w \in L_1$. Hence $\mathcal{L}_n(\text{AHexASSD}) \subset \mathcal{L}(\text{AHexASSD})$.

Similarly, we can have the following theorems.

Theorem 4 $L_n(\text{AHexASSD}) \subset L(\mathit{AHexASSD})$ *with* $\Gamma_Z = \Gamma_Y = \phi$.

Theorem 5 $L_n(\text{AHexASSD}) \subset L(\mathit{AHexASSD})$ *with* $\Gamma_X = \Gamma_Z = \phi$.

References

1. Paun, Gh., Rozenberg, G., Salomaa, A.: DNA computing new computing paradigms. Springer, Berlin (1998)
2. Head, T.: Formal language theory and DNA: an analysis of the generative capacity of specific recombinant behaviours. Bull. Math. Biol. **49**, 735–759 (1987)
3. Head, T., Paun, Gh., Pixton, D.: Language theory and molecular genetics: generative mechanisms suggested by DNA recombination. In: Rozenberg, G., Salomaa, A. (eds.) Handbook of Formal Languages, vol. 2, Chap. 7, pp. 296–358. Springer-Verlag (1997)
4. Berstel, J., Boasson, L., Fagnot, I.: Splicing systems and the Chomsky hierarchy. Theoret. Comput. Sci. **436**, 2–22 (2012)
5. Paun, Gh., Rozenberg, G., Salomaa, A.: Computing by splicing. Theoret. Comput. Sci. 168, 321–336 (1996)

6. Mitrana, V., Petre, I., Rogojin, V.: Accepting splicing systems. Theoret. Comput. Sci. **411**, 2414–2422 (2010)
7. Arroyo, F., Castellanos, J., Dasow, J., Mitrana, V., Sanchez-Couso, J.R.: Accepting splicing systems with permitting and forbidding words. Acta Informatica **50**, 1–14 (2013)
8. Dersanambika, K.S., Krithivasan, K., Martin-Vide, C., Subramanian, K.G.: Local recognizable hexagonal picture languages. Int. J. Pattern Recognit Artif Intell. **19**(7), 853–871 (2005)
9. Siromoney, G., Siromoney, R.: Hexagonal arrays and rectangular blocks. Comput. Graph. Image Process. **5**, 353–381 (1976)
10. Subramanian, K.G.: Hexagonal array grammars. Comput. Graph. Image Process. **10**, 388–394 (1979)
11. Middleton, L., Sivaswamy, J.: Hexagonal image processing: a practical approach. Springer-Verlag (2005)
12. Thomas, D.G., Begam, M.H., David, N.G.: Hexagonal array splicing systems. Formal Languages Aspects Natural Comput. 197–207 (2006)
13. Sheena Christy, D.K., Thomas, D.G.: Accepting hexagonal array splicing system with respect to directions. In: Proceedings of National Conference on Mathematics and Computer Applications, pp. 11–22 (2015)

Tracking Movements of Humans in a Real-Time Surveillance Scene

Dushyant Kumar Singh and Dharmender Singh Kushwaha

Abstract Increased security concern has brought up an acute need for being thoughtful in the area of surveillance. The normal trend of surveillance followed is a grid of CCTV cameras with control centralized at a room, which is manually looked upon by a caretaker. Many a times there is no regular watch carried by caretaker, instead logs of video footage are maintained, which are used in the case of any mishaps occurring. This is the practice followed even at major sensitive places. This is a retroactive kind of situation handling. A solution to this could be a system that continuously has a watch using a camera and indentifies a human object and then tracks its movement to identify any uncommon behavior. The sudden responsive action (reaction) made by the caretaker is the expected design objective of the system. In this paper we have proposed a system that analyzes the real-time video stream from camera, identifying a human object anfd then tracking its movement if it tries to go out of the field of view (FoV) of the camera. That is, the camera changes its FoV with the movement of the object.

Keywords Object detection · Haar-like features · Cascade classifier · Surveillance · Camera · Microcontroller · Angular shift · Servo motor

1 Introduction

Object detection and tracking in the real-world scenario has achieved enough focus with amenable challenges in the past few years. Object tracking tends to create a correspondence between consecutive frames and objects/part-of-objects lying within, in order to gather some important information as position, speed, direction, and trajectory. This temporal information could then be used for some secondary calculations. Object detection and tracking in running adjacent frames in a video is significantly a typical job. The first step in video analysis is the moving object

D.K. Singh (✉) · D.S. Kushwaha
Department of CSE, MNNIT Allahabad, Allahabad 211004, UP, India
e-mail: dushyant@mnnit.ac.in

M. Pant et al. (eds.), *Proceedings of Fifth International Conference on Soft Computing for Problem Solving*, Advances in Intelligent Systems and Computing 437, DOI 10.1007/978-981-10-0451-3_45

detection followed by tracking the movements of object and then its behavior, as needed in certain applications. It can be used in many areas such as video surveillance, traffic monitoring, and people tracking. In smart surveillance systems, temporal information about object of focus is extracted and then behavior analysis is done [1].

Processing objects for analyzing and understanding the behavior is a complex task which generally uses visual software tools to produce information. The ability of the human vision is duplicated by some electronic modules with algorithms, working approximately the same as the human eyes and the brain. In robotics and other application areas, Open-Source Computer Vision Library (OpenCV) is now becoming the most used libraries for object detection and processing these objects captured [2].

Not much work has been done in designing such smart/automated systems for surveillance, while there is a good literature on the image processing part of object detection and tracking. Accurate detection with high true positive rate (TPR) and low false positive rate (FPR) is still a challenge [3]. The two directions in which efforts by researchers have been seen are features and classifiers.

Different features for object detection have been discussed so far in history, e.g., Haar features for faces [4], pedestrians and edgelets for pedestrians [5, 6], and HOG features [7]. Haar-like features are used in this and are discussed in subsequent sections. New scope exploited recently by researchers is combining information from more than one source, e.g., color, local texture, edge, motion, etc. [7–9]. This usually increases accuracy in detection at the cost of time of detection.

In terms of classifications, SVM is the prominent classifier used [7]. The object detection systems also rely on other sophisticated machine learning algorithms. The cascade classifier gives good results in detecting faces and pedestrians [4, 5, 9]. Optimization can be done on the size of feature set and the classifiers in the cascade. This is been discussed later in the paper.

2 Proposed Design

The system is developed to work in real time. The camera with a rotating base keeps an eye on the human object in a defined scene periphery. The FoV for camera is always limited compared to the scene periphery, so it is rotated in the left and right directions to capture the complete scene track object. This rotation is manual, or if not manual then random in maximum cases. In this work, camera rotation is not random but depends on the motion of the object in the scene and is automatic, i.e., no human intervention is required. This can be seen in Fig. 1.

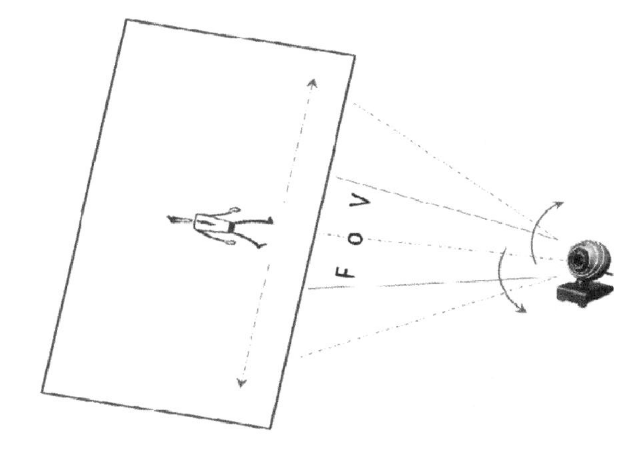

Fig. 1 Surveillance scene

2.1 *System Architecture*

As in Fig. 2, video (stream of images) captured from camera is the input to 'Image Processing' block of the control system. The frames of this real-time video are processed for human object detection and then the position of object to track its motion. The position coordinates calculated are then fed to 'Motor Control' block. Based on the linear motion of the object in any direction, an angular twist calculation of the motor is done. This is then given to the microcontroller to generate the corresponding pulse-width modulated (PWM) signal. PWM signal derives the motor which in turn rotates the camera attached to motor with requisite angle.

The whole system now can be seen as composed of two major components, namely:

- '*Control System*' which is the software control part. It takes video stream from camera as input, and,
- '*Microcontroller*' which is the hardware control part. Its task is deriving motor, with required accuracy.

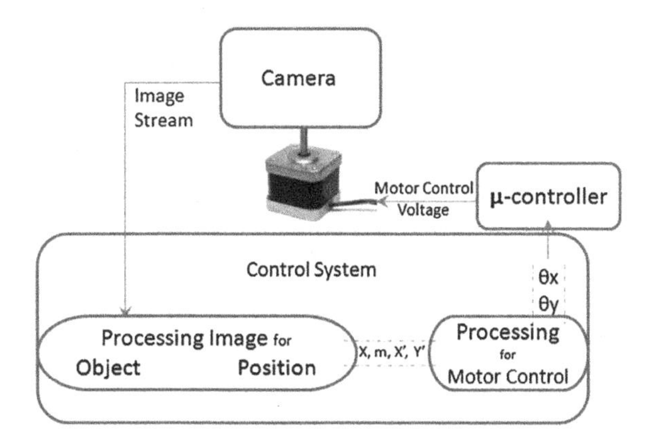

Fig. 2 System architecture

3 Control System

As discussed in the previous subsection, the control system as a whole can be seen as integration of two blocks. First '*Image Processing*' block and the second '*Motor Control*' block.

3.1 Image Processing

This block involves taking video stream as input from camera and then extracting frames. The camera used here in this experimental setup is a webcam with a frame rate of 30 frames per second (fps) for 640 × 480 pixel frame. To make the system faster, not all the frames captured are processed while a key frame selection mechanism is adopted. This can be exploited due to a high degree of correlation between adjacent frames of a high frame rate video. Here in our case, we sampled 6 out of 30 frames at equal interval from one second video, i.e., our frame capture rate was 6 fps. This software part was coded in Python using OpenCV libraries.

The frames sampled are preprocessed by the following mechanism:

Converting RGB image to HSV, as we are using 'Haar Features' for classification of object which uses contrast variation or luminance characteristics.

Erosion and Dilation, it improves the object boundaries in the HSV image.

Noise Removal, removes unwanted noise that hurdles the segmentation and object detection. Here median filter is used for smoothing the image.

Next, the preprocessed image is applied to Haar Cascade Classifier for object detection. Haar features are used in cascade classifier for face detection, given by Viola and Jones [9]. The classifier for face detection is trained for human face. Figure 3 shows the data/control flow for object detection in Image Processing block. The detected face centroid is used to modulate the centroid for whole body. Using a defined aspect ratio of human height and width, and the centroid available, a rectangle is drawn that identifies the human body. Difference in the centroid value for two positions on any kind of movement is calculated as linear left or right shift. This is then given to Motor Control block for calculating the corresponding angular shift.

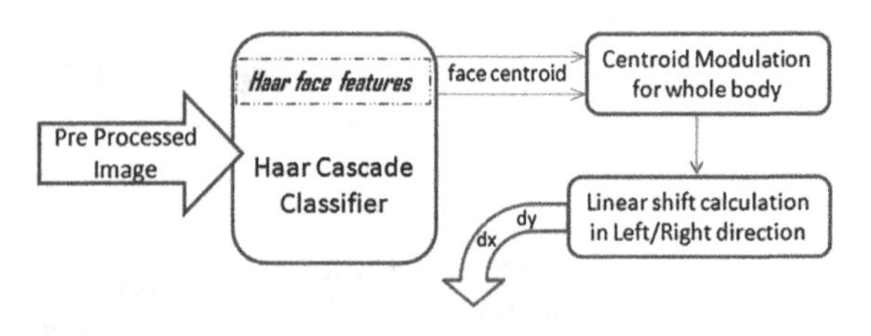

Fig. 3 Data/control flow for object detection

3.2 Human Object Detection

Pixel intensities used directly could work fine for recognizing a specific face and orientation, but would be difficult for detecting a general face independent of any characteristics (as geographical and else). Features are the encoded ad hoc domain knowledge which makes learning easy with a finite quantity of training data. Beside this, systems using features are much faster than a pixel-based system [9].

Haar-like features here are the core basics for Haar classifier-based object detection. The contrast variance between pixels or pixel groups determines the relative light and dark areas in the face portion. Two or three such adjacent groups with relative contrast variance form a Haar-like feature [10].

Calculation of these features do not add extra complexity as features are calculated using integral image, an intermediate representation of image [9], which takes constant time. The integral image can be defined as an array with every element corresponding to one pixel, which is the summation of the pixels' intensity values on the left and above the pixel at (x, y) position. So for any original image $A[x, y]$ the integral image $AI[x, y]$ is given by Eq. 1.

$$AI[x,y] = \sum_{x' \leq x, y' \leq y} A(x', y') \quad (1)$$

Calculating a single feature is easy and efficient but calculating around 180,000 features within a 24×24 subimage is completely impractical [9]. But Menezes et al. in [11] shows that only a limited number of features are needed to determine if a subimage potentially contains the desired object. So, the goal is to eliminate a substantial amount of subimages, around 50 % that do not contain the object.

3.3 Cascade of Classifiers

A cascade of n step with classifiers at each step is a designated decision tree, which rejects all other patterns except that with the objects. Each stage was trained using the Adaboost algorithm. There lies the tradeoff between detection rate and time of computation. With more number of features, cascaded classifiers achieve higher detection rate or low false positive rates, while the time of computation grows high. Possible optimization points in cascade converge to (1) the number of classifier stages, (2) the number of features in each stage, and (3) the threshold of each stage [9].

According to Viola and Jones' face detection classifier model, its cascade has 38 stages with over 6000 features. On the dataset containing 507 faces and 75 million subwindows, faces are detected using an average of 10 feature evaluations per subwindow [9]. Intel has developed an open-source library for computer vision-related programs that includes an implementation of Haar classifier detection and training [2]. The same Haar classifier library of OpenCV is used here for detecting face in the images. Figure 4 shows the detected face in the frame sampled from webcam video. The images in Figs. 4, 5 and 6 were taken in my lab with prior consent of my students.

The actual frame coordinates for rectangle of face are $(x1, y1)$ and $(x2, y2)$. These are the diagonal point coordinates with $x2 > x1$. The coordinates generated for the rectangle marking the whole body are:

Fig. 4 Detected face from sampled frame

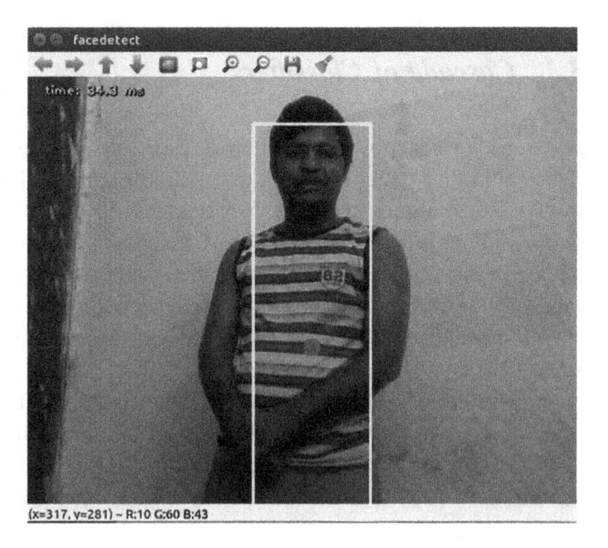

Fig. 5 Left movement

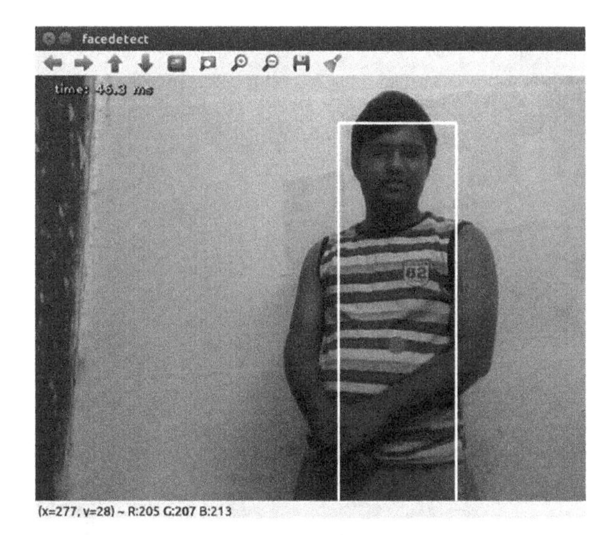

Fig. 6 Right movement

First diagonal coordinate –> ($x1 - 25$, $y1$) and
Second diagonal Coordinate –> ($x2 + 25$, $4 * y2$)

This assumption is based on the aspect ratio of height and width of a normal human body. The results are shown in Figs. 5 and 6.

The red dot is the centroid (cx, cy) of the rectangle or the human object. This is the average of the end coordinates in X and Y directions. Therefore, $cx = (x1 + x2)/2$ and $cy = (y1 + 4*y2)/2$.

3.4 Motor Control

In a frame window of 640 × 480, i.e., 640 on x axis, the deciding criteria's made for checking left/right movement are as …

- If ($280 \leq cx \leq 360$), then centroid and the object is assumed to be in the center, i.e., no motion
- If ($cx < 280$), then it is leftward movement, correspondingly send "L" character serially to the microcontroller to guide the left twist of the motor.
- If ($cx > 360$), then it is rightward movement, correspondingly send "R" character serially to the microcontroller to guide the right twist of the motor.

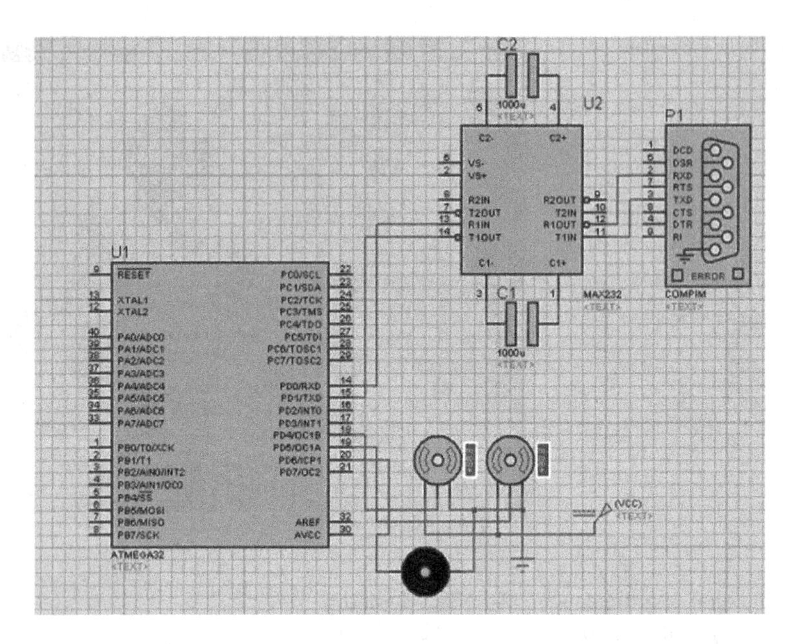

Fig. 7 Microcontroller interfacing circuit

4 Microcontroller

This part of the system is dedicated to rotation of camera fitted on motor base, i.e., controlling the motor rotation either left or right, based on the signal ("L"/"R") received from the software control part. The microcontroller used here is "Atmega32" with "MAX232" as a voltage converter for interfacing microcontroller with serial port of computer. Character "L" or "R" sent by computer is received at microcontroller's receive pin serially through USART, which is Atmega's inbuilt component. The circuit diagram for the microcontroller interfacing well explains the flow in Fig. 7.

Based on the left or right signal received, the angular twist (Θx) is calculated. As in Fig. 8, Θx is twist in X direction, is only considered since human object movement is assumed to be in X direction only. Θx is calculated as in Eq. 2.

$$\Theta x = \tan^{-1}(\Delta x / d) \tag{2}$$

Δx is the difference between the object centroid X coordinate (cx) and frame center X coordinate (320 in our case).

d is the distance of object from camera.

For the calculation for angular twist, a few assumptions are made based on the experimental setup. The value of 'd' is not fixed as for experiment the human object can move anywhere in the arena, that is why 'd' is not used. Δx and hence Θx is not then calculated rather the value of Θx is made fix.

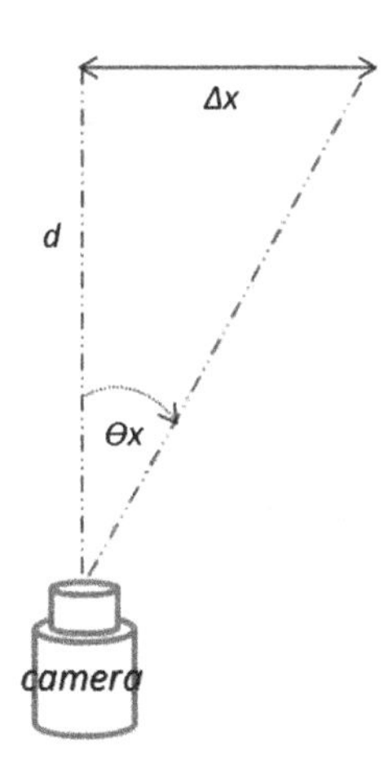

Fig. 8 Angular twist of camera

The degree of twist (Θx) of the motor depends on the PWM signal generated by microcontroller. The PWM signal depends on the timer configuration in Atmega32, i.e., "OCR1A" timer register value. In our setup, the calibrated value for OCR1A to get 90 ° angular twist is 316. So for…

Right twist, OCR1A = OCR1A + 5
and Left twist, OCR1A = OCR1A − 5

This is done at frame sampling rate. This is proposed and experimented by exploiting the speed of human movement, as well as the rpm of the motor accommodating the camera to rotate.

5 Results and Conclusion

Figures 4, 5 and 6 show the real-time tracking of the movements of the human taken for surveillance. The system designed to process a video stream at 30 fps and 640 × 480 pixels resolution works fine in real time. It is been tuned for no delay/glitch in detection, i.e., fluent motor rotation with human object movement in the arena. Yet there are things to be accompanied to make the system more robust and fast with a high resolution and ambient environment. These are discussed as the future work in the next section.

6 Future Work

The classification can be updated to classifier with Haar features for full-body training. Next, the Θx calculation can be made exactly inline with Δx value generated by linear shift of object from image processing. The value of '*d*' can be made variable for flexible movement of object. This is to make the system more real-time and some proximity sensors can be used to find this distance '*d*' of object from camera. In addition, for open environment surveillance, with horizontal linear shift Δx, a vertical linear shift Δy can also be considered for vertical rotation of camera. This will involve an additional motor in vertical direction and will cover whole 3D plane with the use of proximity in Z direction.

References

1. Haering, N., Venetianer, P.L., Lipton, A.: The evolution of video surveillance: an overview. Mach. Vis. Appl. **19**(5-6), 279–290 (2008)
2. Open Computer Vision Library Reference Manual. Intel Corporation, USA (2001)
3. Dollar, P., Wojek, C., Schiele, B., Perona, P.: Pedestrian detection: a benchmark. In: Proceedings of IEEE International Conference on Computer Vision and Pattern Recognition (2009)
4. Viola, P., Jones, M.: Robust real-time face detection. Int. J. Comput. Vision **57**(2), 137–154 (2004)
5. Viola, P., Jones, M., Snow, D.: Detecting pedestrians using patterns of motion and appearance. In: Proceedings of IEEE International Conference on Computer Vision, pp. 734–741 (2003)
6. Wu, B., Nevatia, R.: Detection and tracking of multiple, partially occluded humans by bayesian combination of edgelet based part detectors. Int. J. Comput. Vision **75**(2), 247–266 (2007)
7. Wu, J., Liu, N., Rehg, J.M.: A real-time object detection framework. IEEE Trans. Image Process. (2013)
8. Qiang, Z. et al.: Fast human detection using a cascade of histograms of oriented gradients. In: 2006 IEEE Computer Society Conference on Computer Vision and Pattern Recognition, vol. 2. IEEE (2006)
9. Viola, P., Jones, M.: Rapid object detection using a boosted cascade of simple features. Proceedings of the 2001 IEEE Computer Society Conference on Computer Vision and Pattern Recognition, 2001. *CVPR 2001*. vol. 1. IEEE (2001)
10. Wilson, P.I., Fernandez, J.: Facial feature detection using Haar classifiers. J. Comput. Sci. Coll. **21**(4), 127–133 (2006)
11. Menezes, P., Barreto, J.C., Dias, J.: Face tracking based on haar-like features and eigenfaces. In: IFAC/EURON Symposium on Intelligent Autonomous Vehicles (2004)

An Inventory Model for Deteriorating Items Having Seasonal and Stock-Dependent Demand with Allowable Shortages

S.R. Singh, Mohit Rastogi and Shilpy Tayal

Abstract This paper deals the problem of single deteriorating item, with stock-dependent and seasonal pattern demand rate. It is considered that a constant part of on-hand inventory deteriorates per unit of time. The model is solved for finite time horizon. The aim of this present paper is to give a new height to the inventory literature on stock-dependent and seasonal demand pattern. This paper can be applied in many realistic situations. Shortages are allowed and three different conditions of backlogging are discussed in this model. The purpose of this study is to optimize the overall cost of the system and to find out the optimal ordering quantity. To explain the model and its significant features a numerical illustration and sensitivity analysis with respect to different related parameters is also cited.

Keywords Inventory · Stock and seasonal pattern demand · Shortages · Backlogging

1 Introduction

Many business practices demonstrate that the existence of a larger amount of goods displayed may catch the attention of more customers than that with a smaller amount of goods. This fact implies that the demand may have a positive correlative with stock level. Under such a situation, a firm should seriously consider its pricing and ordering policy.

S.R. Singh · Shilpy Tayal
Department of Mathematics, CCS University, Meerut, India
e-mail: shivrajpundir@gmail.com

Shilpy Tayal
e-mail: agarwal_shilpy83@yahoo.com

Mohit Rastogi (✉)
Centre for Mathematical Sciences, Banasthali University,
Banasthali, Rajasthan, India
e-mail: mohitrastogi84@gmail.com

M. Pant et al. (eds.), *Proceedings of Fifth International Conference on Soft Computing for Problem Solving*, Advances in Intelligent Systems and Computing 437, DOI 10.1007/978-981-10-0451-3_46

As recommended by Lau and Lau [1], in many cases a small change in the demand pattern may result in a large change in optimal inventory decisions. A manager of a company has to examine the factors that affect demand pattern, because customers purchasing behaviour may be affected by the factors such as inventory level, seasonality and so on. Singh and Diksha [2] introduced a supply chain model in a multi-echelon system with inflation-induced demand. Bhunia and Maiti [3] proposed a deteriorating inventory model with linear stock and time-dependent demand. Giri and Chaudhuri [4] introduced an inventory model with power form stock-dependent demand and nonlinear holding cost. Singh and Singh [5] presented an inventory model with stock-dependent demand under inflation in a supply chain for deteriorating items. Hsu et al. [6] presented a deteriorating inventory model with season pattern demand rate. Patel [7] developed an inventory model for deteriorating items with stock-dependent demand under and partial backlogging. Mandal and Maiti [8] focused on a production model with power form stock-dependent demand. Tayal et al. [9] introduced a two echelon supply chain model with seasonal pattern and price-dependent demand for deteriorating items with effective investment in preservation technology. An algorithm for an inventory system with a power form stock-dependent demand presented by Chung [10]. Tayal et al. [11] presented an inventory model for deteriorating items with seasonal products and an option of an alternative market. Teng and Chang [12] investigated a production model with linear stock-dependent demand.

In the existing study of inventory control, the usual inventory models have been developed under the statement that during the storage period, lifetime of an item is infinite. This way that an item once in stock remains unaffected and completely usable for satisfying the future demand in an ideal condition. In actuality, this statement is not always true for some physical goods like wheat, rice or any other kind of food grain, vegetables, fruits, etc., due to their deterioration effect. Deterioration of physical goods is one of the major factors in any inventory and production system. Many researchers had studied deteriorating inventory in the last many years. Singh and Singh [13] developed a production inventory model for deteriorating products with variable demand rate.

Singh and Rastogi [14] investigated an integrated inventory model with amelioration and deterioration under shortages and inflation. Tayal et al. [15] introduced a deteriorating production inventory problem with space restriction in which the extra ordered quantity is returned to the supplier with a penalty cost. Wu et al. [16] illustrated a problem to establish the optimal replenishment policy for non-instantaneous deteriorating items with stock-dependent demand.

It has long been supposed that during the shortage period all happening demand will either totally backlogged or completely lost but in reality the happening demand during stock out is partially backlogged or partially lost. Since it is observed that some customers are willing to wait for the stock up to the next replenishment. Furthermore, the opportunity cost due to lost sales should be considered since some customers would not like to wait for backlogging during the shortage periods. Chang and Dye [17] developed an economic order quantity model for deteriorating items with time-varying demand and partial backlogging. Singh and Saxena [18] presented an optimal returned policy for a reverse logistics inventory model with backorders. Kumar and Rajput [19] suggested a partially backlogging inventory model for deteriorating items with ramp-type demand rate. Tayal et al. [20] developed a multi-item inventory model for deteriorating items with expiration date and allowable shortages. Singh and Singh [21] introduced a perishable inventory model with quadratic demand, partial backlogging and permissible delay in payments.

Here we have developed an inventory model for deteriorating items with stock-dependent and seasonal pattern demand. In this model shortages are permitted and partially backlogged. The different conditions of backlogging are discussed in this model. A numerical illustration and sensitivity analysis with respect to different related parameters are given to demonstrate the model. At last, a conclusion and further extension of this model is mentioned.

2 Assumptions and Notations

2.1 Assumptions

1. The products discussed in this model are deteriorating in nature.
2. The demand for the products is assumed to be a function of time of the season and available stock level.
3. The shortages are allowed and partially backlogged.
4. The shortage period will be less than the length of the season.
5. The warehouse has boundless capacity.
6. The deteriorated items cannot be replaced or repaired.

2.2 Notations

T length of the season
v the time at which inventory level becomes zero
α, β demand coefficients
θ deterioration coefficient
$I_1(0)$ initial stock level
Q_2 backordered quantity during shortages
c purchasing cost per unit

η rate of backlogging
h holding cost per unit
s shortage cost per unit
l lost sale cost per unit
O ordering cost per order

3 Mathematical Model

The inventory time graph for the retailer is depicted in below mentioned Fig. 1. It is shown that I_1 (0) is the initial inventory level for the retailer. During the time period [0, v] the inventory level decreases due to the demand and deterioration. At $t = v$, the inventory level becomes zero and after that during [v, T] shortages occur. Time T is the length of the season. It is assumed that initially the demand for the product is maximum and it decreases as the time for the season increases. The occurring demand during shortages is partially backlogged. The differential equations showing the behaviour of inventory with time are given as follows:

$$\frac{dI_1(t)}{dt} + (\theta+\beta)I_1(t) = -\alpha(T-t) \quad 0 \le t \le v \tag{1}$$

$$\frac{dI_2(t)}{dt} = -\alpha(T-t) \quad v \le t \le T \tag{2}$$

with boundary condition

$$I_1(v) = 0 \text{ and } I_2(v) = 0 \tag{3}$$

Solving these equations:

$$I_1(t) = \alpha\left\{\frac{(T-v)}{(\theta+\beta)} + \frac{1}{(\theta+\beta)^2}\right\}e^{(\theta+\beta)(v-t)} - \alpha\left\{\frac{(T-t)}{(\theta+\beta)} + \frac{1}{(\theta+\beta)^2}\right\} \quad 0 \le t \le v \tag{4}$$

$$I_2(t) = \alpha\left\{T(v-t) - \frac{1}{2}\left(v^2 - t^2\right)\right\} \quad v \le t \le T \tag{5}$$

Total average cost of the system in this case is:

$$\text{T.A.C.} = \frac{1}{T}[\text{Purchasing cost} + \text{Holding cost} + \text{Shortage cost} + \text{Lost sale cost} + \text{Ordering cost}]$$

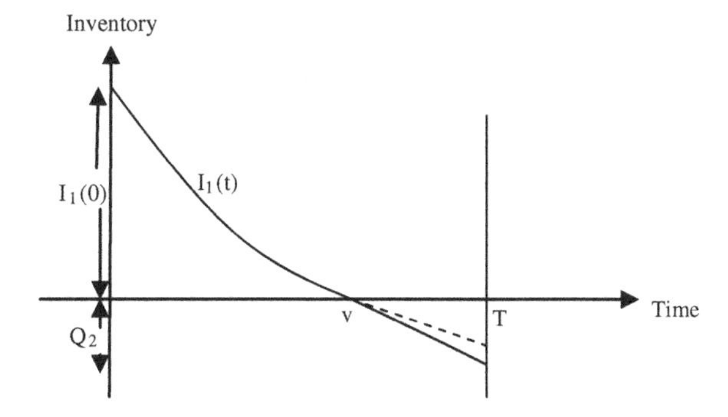

Fig. 1 Inventory time graph for the retailer

$$\text{T.A.C.} = \frac{1}{T}[\text{P.C.} + \text{H.C.} + \text{S.C.} + \text{L.S.C.} + \text{O.C.}] \tag{6}$$

Purchasing Cost

(i) When occurring shortages are partially backlogged:

$$\text{P.C.} = (I_1(0) + Q_2)c$$

where

$$I_1(0) = \alpha\left\{\frac{(T-v)}{(\theta+\beta)} + \frac{1}{(\theta+\beta)^2}\right\}e^{(\theta+\beta)v} - \alpha\left\{\frac{T}{(\theta+\beta)} + \frac{1}{(\theta+\beta)^2}\right\}$$

and

$$Q_2 = \int_v^T \eta\alpha(T-t)\mathrm{d}t$$

$$Q_2 = \frac{\eta\alpha}{2}(T-v)^2$$

Then

$$\text{P.C.} = \left[\alpha\left\{\frac{(T-v)}{(\theta+\beta)} + \frac{1}{(\theta+\beta)^2}\right\}e^{(\theta+\beta)v} - \alpha\left\{\frac{T}{(\theta+\beta)} + \frac{1}{(\theta+\beta)^2}\right\} + \frac{\eta\alpha}{2}(T-v)^2\right]c \tag{7}$$

(ii) When occurring shortages are completely lost:

$$\text{P.C.} = (I_1(0) + Q_2)c$$

where $I_1(0)$ will be the same as in previous case:

$$I_1(0) = \alpha\left\{\frac{(T-v)}{(\theta+\beta)} + \frac{1}{(\theta+\beta)^2}\right\}e^{(\theta+\beta)v} - \alpha\left\{\frac{T}{(\theta+\beta)} + \frac{1}{(\theta+\beta)^2}\right\}$$

In this case Q_2 will be zero, since all the demand during shortages are completely lost.

So in this case purchasing cost will be:

$$\text{P.C.} = \left[\alpha\left\{\frac{(T-v)}{(\theta+\beta)} + \frac{1}{(\theta+\beta)^2}\right\}e^{(\theta+\beta)v} - \alpha\left\{\frac{T}{(\theta+\beta)} + \frac{1}{(\theta+\beta)^2}\right\}\right]c \tag{8}$$

(iii) When occurring shortages are completely backlogged:

$$\text{P.C.} = (I_1(0) + Q_2)c$$

where I_1 (0) will be the same as in previous case:

$$I_1(0) = \alpha\left\{\frac{(T-v)}{(\theta+\beta)} + \frac{1}{(\theta+\beta)^2}\right\}e^{(\theta+\beta)v} - \alpha\left\{\frac{T}{(\theta+\beta)} + \frac{1}{(\theta+\beta)^2}\right\}$$

Since in this case all the occurring demand during shortages is completely backlogged, so in this case Q_2 will be:

$$Q_2 = \int_v^T \alpha(T-t)\mathrm{d}t$$

$$Q_2 = \frac{\alpha}{2}(T-v)^2$$

Thus the purchasing cost in this case will be:

$$\text{P.C.} = \left[\alpha\left\{\frac{(T-v)}{(\theta+\beta)} + \frac{1}{(\theta+\beta)^2}\right\}e^{(\theta+\beta)v} - \alpha\left\{\frac{T}{(\theta+\beta)} + \frac{1}{(\theta+\beta)^2}\right\} + \frac{\eta\alpha}{2}(T-v)^2\right]c \tag{9}$$

Holding Cost

$$\text{H.C.} = h\int_0^v I_1(t)\,\mathrm{d}t$$

$$\text{H.C.} = h\left\{\frac{\alpha}{(\theta+\beta)}\left(\frac{(T-v)}{(\theta+\beta)}+\frac{1}{(\theta+\beta)^2}\right)\left(e^{(\theta+\beta)v}-1\right)-\alpha\left(\frac{(2Tv-v^2)}{2(\theta+\beta)}+\frac{v}{(\theta+\beta)^2}\right)\right\} \tag{10}$$

Shortage Cost

$$\text{S.C.} = s\int_v^T \alpha(T-t)\,\mathrm{d}t$$

$$\text{S.C.} = s\frac{\alpha}{2}(T-v)^2 \tag{11}$$

Lost Sale Cost:

(i) When occurring shortages are partially backlogged:

$$\text{L.S.C.} = l\int_v^T (1-\eta)\alpha(T-t)\,\mathrm{d}t$$
$$\text{L.S.C.} = \frac{l}{2}(1-\eta)\alpha(T-v)^2 \tag{12}$$

(ii) When occurring shortages are completely lost:

$$\text{L.S.C.} = l\int_v^T (\alpha(T-t)\,\mathrm{d}t$$

$$\text{L.S.C.} = \frac{l}{2}\alpha(T-v)^2 \tag{13}$$

(iii) When occurring shortages are completely backlogged:

Since all the demand during stock out is assumed to be completely backlogged so the lost sale cost in this case will be zero.

$$\text{L.S.C.} = 0 \tag{14}$$

Ordering Cost

$$\text{O.C.} = O \tag{15}$$

4 Numerical Example

The numerical is discussed only for the first case of partial backlogging. In order to illustrate the above mentioned model in this case the numerical example is given with respect to below mentioned parameters:

$$T = 10\,\text{days},\ s = 15\,\text{Rs./unit},\ c = 50\,\text{Rs./unit},\ \alpha = 25\,\text{units},\ \eta = 0.9,$$
$$\beta = 0.02,\ \theta = 0.01,\ l = 20\,\text{Rs./unit},\ h = 0.2\,\text{Rs./unit},\ O = 500\,\text{Rs./order}$$

Corresponding to these values the optimal value of 'v' and T.A.C. come out to be 6.86856 days and Rs. 7012.11 (Fig. 2).

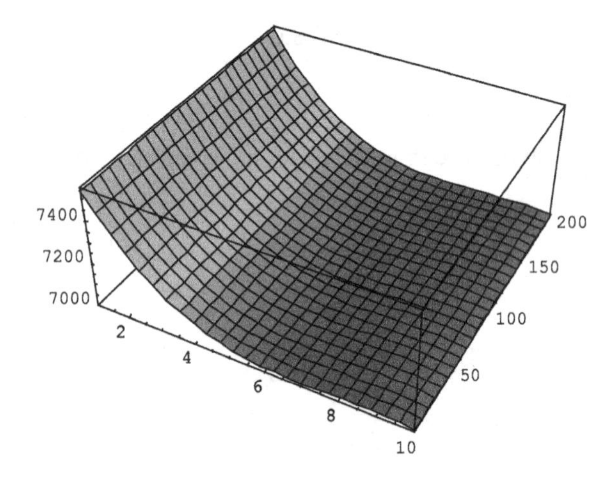

Fig. 2 Convexity of the T.A.C. function

5 Sensitivity Analysis

With respect to different associated parameters a sensitivity analysis is carried out, taking the variation in single parameter and other variables unchanged (Figs. 3, 4, 5, 6 and 7).

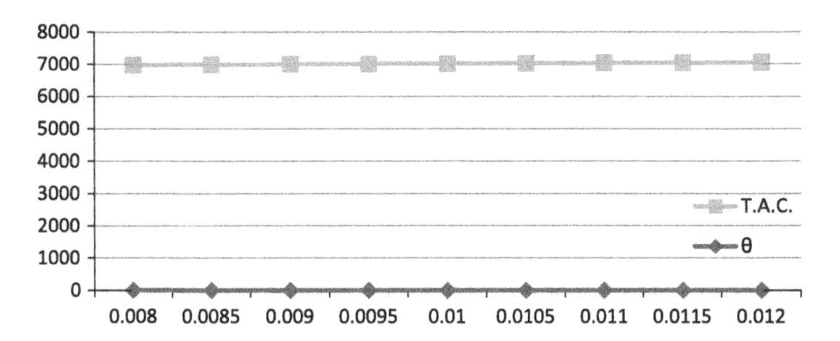

Fig. 3 Variation in T.A.C. function with respect to θ

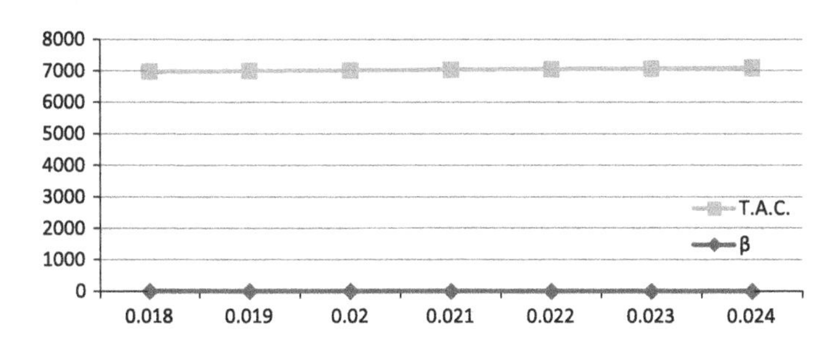

Fig. 4 Variation in T.A.C. function with respect to β

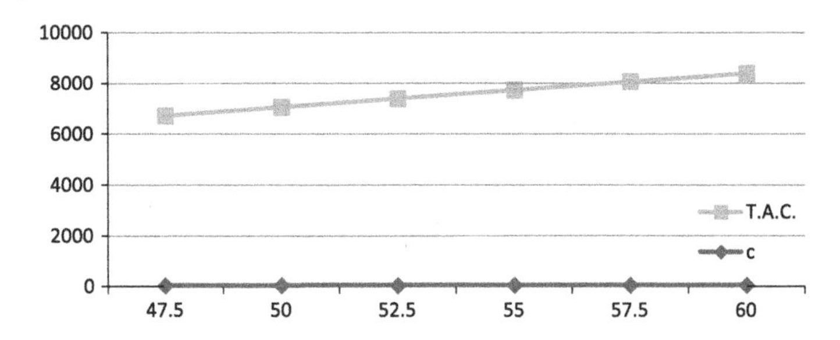

Fig. 5 Variation in T.A.C. function with respect to c

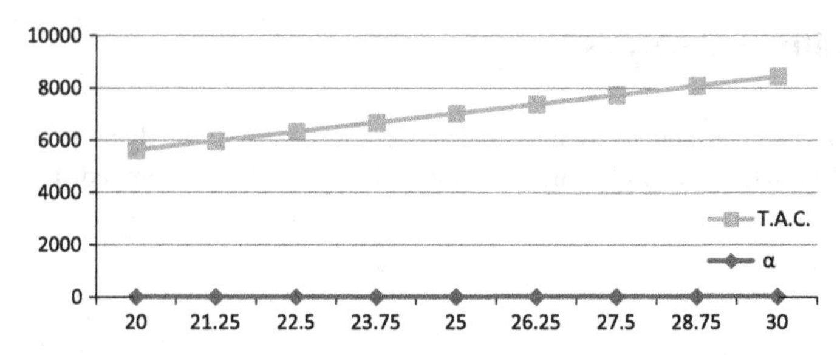

Fig. 6 Variation in T.A.C. function with respect to α

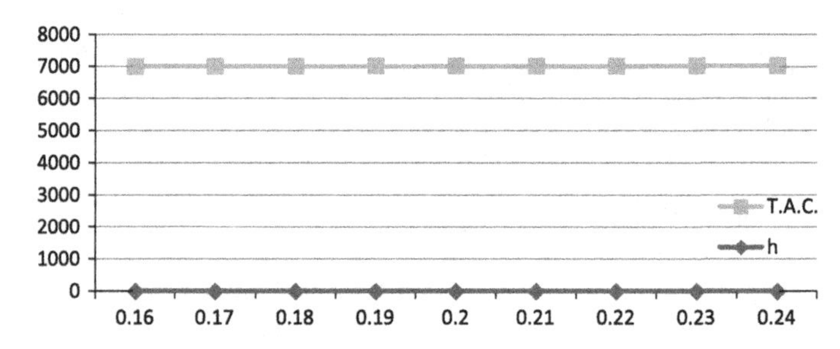

Fig. 7 Variation in T.A.C. function with respect to h

6 Observations

A sensitivity analysis has been carried out with respect to the associated parameters θ, β, c, α and h.

1. From Table 1 we observe that an increment in deterioration parameter θ decreases the time length of positive inventory and increases the T.A.C. of the system.

Table 1 Sensitivity analysis with respect to deterioration parameter θ

% variation in θ	θ	v	T.A.C.
−20	0.008	7.5	6975.94
−15	0.0085	7.31435	6985.28
−10	0.009	7.15185	6994.41
−5	0.0095	7.00459	7003.35
0	0.01	6.86856	7012.11
5	0.0105	6.7414	7020.69
10	0.011	6.62153	7029.1
15	0.0115	6.50788	7037.35
20	0.012	6.39964	7045.43

Table 2 Sensitivity analysis with respect to demand parameter β

% variation in β	β	v	T.A.C.
−10	0.018	7.5	6975.94
−5	0.019	7.15185	6994.41
0	0.02	6.86856	7012.11
5	0.021	6.62153	7029.1
10	0.022	6.39964	7045.43
15	0.023	6.19702	7061.14
20	0.024	6.01005	7076.26

Table 3 Sensitivity analysis with respect to purchasing cost (c)

% variation in c	c	v	T.A.C.
−5	47.5	7.57292	6675.9
0	50	6.86856	7012.11
5	52.5	6.36686	7344.83
10	55	5.94515	7674.45
15	57.5	5.57376	8001.22
20	60	5.2393	8325.31

Table 4 Sensitivity analysis with respect to demand parameter α

% variation in α	α	v	T.A.C.
−20	20	6.86856	5619.69
−15	21.25	6.86856	5967.79
−10	22.5	6.86856	6315.9
−5	23.75	6.86856	6664
0	25	6.86856	7012.11
5	26.25	6.86856	7360.22
10	27.5	6.86856	7708.32
15	28.75	6.86856	8056.43
20	30	6.86856	8404.53

2. Table 2 shows the variation in demand parameter β and other variable unchanged, in this case we observe that with the increment in demand parameter β the value of v decreases and T.A.C. of the system increases.
3. Table 3 lists the variation in purchasing cost c and shows that with the increment in purchasing cost the value of v decreases and T.A.C. of the system increases.
4. From Table 4 it is observed that with the increment in demand parameter α, the value of v remains unchanged and the T.A.C. of the system increases.
5. Table 5 shows that the time v decreases and total average cost of the system increases with the increment in holding cost 'h'.

Table 5 Sensitivity analysis with respect to holding cost h

% variation in h	h	v	T.A.C.
−20	0.16	7.08633	6998.58
−15	0.17	7.02897	7002.01
−10	0.18	6.97371	7005.41
−5	0.19	6.9203	7008.77
0	0.2	6.86856	7012.11
5	0.21	6.81834	7015.42
10	0.22	6.76949	7018.7
15	0.23	6.72192	7021.96
20	0.24	6.67552	7025.18

7 Conclusions

This paper is developed with stock-dependent and seasonal pattern demand. It is shown that the demand for the products decreases as the product is nearer to the end of the season and an increase in stock level increases the demand. The shortages are allowed and three different conditions of backlogging are discussed in this model. Numerical example shows that the model is effective and reflects the real-life situations. A sensitivity analysis is also carried out to check the stability of the model. This model further can be extended for trade credit policy and effect of learning.

References

1. Lau, A.H., Lau, H.S.: Effects of a demand-curve's shape on the optimal solutions of a multi-echelon inventory pricing model. European J. Oper. Res. **147**, 530–548 (2003)
2. Singh, S.R., Diksha, C.: Supply chain model in a multi-echelon system with inflation induced demand. Int. Trans. Appl. Sci. **1**, 73–86 (2009)
3. Bhunia, A.K., Maiti, M.: Deterministic inventory model for deteriorating items with finite rate of replenishment dependent on inventory level. Comput. Oper. Res. **25**, 907–1006 (1998)
4. Giri, B.C., Chaudhuri, K.S.: Deterministic models of perishable inventory with stock-dependent demand rate and nonlinear holding cost. European J. Oper. Res. **105**, 467–474 (1998)
5. Singh, S.R., Singh, C.: Optimal ordering policy for decaying items with stock-dependent demand under inflation in a supply chain. Int. Rev. Pure and Appl. Math. **3**, 189–197 (2008)
6. Hsu, P.H., Wee, H.M., Teng, H.M.: Optimal ordering decision for deteriorating items with expiration date and uncertain lead time. Comput. Ind. Eng. **52**, 448–458 (2007)
7. Patel, S.S.: Inventory model for deteriorating items with stock-dependent demand under and partial backlogging and variable selling price. International. J. Sci. Res. **4**, 375–379 (2015)
8. Mandal, M., Maiti, M.: Inventory of damagable items with variable replenishment rate, stock-dependent demand and some units in hand. Appl. Math. Modelling. **23**, 799–807 (1999)
9. Tayal, S., Singh, S.R., Sharma, R., Chauhan, A.: Two echelon supply chain model for deteriorating items with effective investment in preservation technology. Int. J. Oper. Res. **6**, 78–99 (2014)

10. Chung, K.J.: An algorithm for an inventory model with inventory-level-dependent demand rate. Comput. Oper. Res. **30**, 1311–1317 (2003)
11. Tayal, S., Singh, S.R., Sharma, R.: An inventory model for deteriorating items with seasonal products and an option of an alternative market. Uncertain Supply Chain Manage. **3**, 69–86 (2014)
12. Teng, J.T., Chang, C.T.: Economic production quantity models for deteriorating items with price- and stock-dependent demand. Comput. Oper. Res. **32**, 297–308 (2005)
13. Singh, S.R., Singh, N.: A production inventory model with variable demand rate for deteriorating items under permissible delay in payments. Int. Trans. Math. Sci. Comput. **2**, 73–82 (2009)
14. Singh, S.R., Rastogi, M.: Integrated inventory model with Amelioration and Deterioration under shortages and inflation. Int. J. Trends Comput. Sci. **2**, 831–858 (2013)
15. Tayal, S., Singh, S.R., Chauhan, A., Sharma, R.: A deteriorating production inventory problem with space restriction. J. Inf. Optim. Sci. **35**(3), 203–229 (2014)
16. Wu, K.S., Ouyang, L.Y., Yang, C.T.: An optimal replenishment policy for non-instantaneous deteriorating items with stock-dependent demand and partial backlogging. Int. J. Prod. Econ. **101**, 369–384 (2006)
17. Chang, H.J., Dye, C.Y.: An EOQ model for deteriorating items with time varying demand and partial backlogging. J. Oper. Res. Soc. **50**, 1176–1182 (1999)
18. Singh, S.R., Saxena, N.: An optimal returned policy for a reverse logistics inventory model with backorders. Advances in Decision Sciences. Article ID 386598, p. 21 (2012)
19. Kumar, S., Rajput, U.S.: A Partially Backlogging Inventory Model for Deteriorating Items with Ramp Type Demand Rate. Am. J. Oper. Res. **5**, 39–46 (2015). doi:10.5923/j.ajor.20150502.03
20. Tayal, S., Singh, S.R., Sharma, R.: A multi item inventory model for deteriorating items with expiration date and allowable shortages. Indian J. Sci. Technol. **7**(4), 463–471 (2014)
21. Singh, S.R., Singh, T.J.: Perishable inventory model with quadratic demand, partial backlogging and permissible delay in payments. Int. Rev. Pure Appl. Math. **1**, 53–66 (2008)

Using Differential Evolution to Develop a Carbon-Integrated Model for Performance Evaluation and Selection of Sustainable Suppliers in Indian Automobile Supply Chain

Sunil Kumar Jauhar and Millie Pant

Abstract Automobile industries worldwide are unified in opinion, that successful management of sustainable supply chains is the most important driver to improve both their economic and ecological performances. The significance of sustainable supply chain management (SSCM) is a critical corporate matter in the automobile industries that offers incredible potential for achieving better environmental performance, consumer fulfillment, pull down operating expenditures, reducing inventory investments in addition to achieving better fixed asset usage. The environment concerns, climatic changes, and additional ecological concerns in automobile industries are not only articulated by campaigners or researchers, but also by the common man as well, which has motivated the industries to focus on sustainability. The present research focuses on a DEA-based mathematical model and employs differential evolution to select the competent suppliers providing the utmost fulfillment for the sustainable criteria determined. This study aims to examine the sustainable supplier evaluation and selection practices likely to be adopted by the Indian automobile industry for their products.

Keywords Sustainable supplier selection · Differential evolution · Data envelopment analysis · Multi-criteria decision-making · Sustainable supply chain management · Automobile industry

S.K. Jauhar (✉) · M. Pant
Indian Institute of Technology Roorkee, Roorkee, India
e-mail: suniljauhar.iitr@gmail.com

M. Pant
e-mail: millidma@gmail.com

M. Pant et al. (eds.), *Proceedings of Fifth International Conference on Soft Computing for Problem Solving*, Advances in Intelligent Systems and Computing 437, DOI 10.1007/978-981-10-0451-3_47

1 Introduction

Nowadays much focus is laid on environmental issues by government as well as by common man which has motivated the industries to think in the direction of sustainable supply chain paradigms. As a result, supplier selection problem which is an integral part of supply chain sustainability criterion has been considered [1–6].

Many researchers have discussed the importance of the nature of these decisions of sustainable supplier selection (SSS), which is usually difficult and unstructured [7]. Optimization practices might serve as useful tools for such decision-making problems. During last few years, differential evolution (DE) has risen as a dominant tool used for solving a variety of problems arising in various fields [8–13].

In this paper, we proposed a solution methodology for the SSS based on DE algorithm and data envelopment analysis (DEA), where DE is applied on SSS problem having DEA-based mathematical model. The approach is validated on a hypothetical case taken from Indian automobile industry while considering the environmental issues. The proposed approach is an extension of method introduced by the authors [14], which focuses on developing suitable supplier clusters. In [14] the authors have presented an approach for traditional supplier selection problem. After successfully applied this approach into traditional supplier selection, we further apply this novel approach for SSS problem in automobile industry, which can be said as an extension of the previous work with subtle changes.

The main difference between [14] and present paper is that in the [14] the authors have considered the traditional criteria for supplier selection and have considered a model accordingly. While in this paper the authors have discussed the problem considering the aspect of global sustainability as well. Here the authors have also considered the impact of CO_2 emission, service quality, etc. The input–output criteria are changed accordingly and modified model is considered in addition to the model considered in the previous paper. Further, in this paper the study, including literature review, etc., is done in more detail.

2 SSCM in Indian Automobile Industry

The automobile industry has a significant and complicated role in the worldwide carbon cycle. Automobile industries are massive consumers of energy. The Worldwatch Institute states that the worldwide automotive vehicle has become greater in size as of 50 million units in 1950 in comparison to 550 million automobiles in 2004, with approximations of more than 5 billion automobiles by the year 2050 [15].

The Indian automotive product manufacturing companies are the world's sixth leading manufacturer of vehicles in terms of volume and value and have grown to 14.4 % in the past decade; overall domestic sales of automobiles are led by two-wheelers 77.4 % of total sales in 2012–13, followed by 15.1 % of passenger

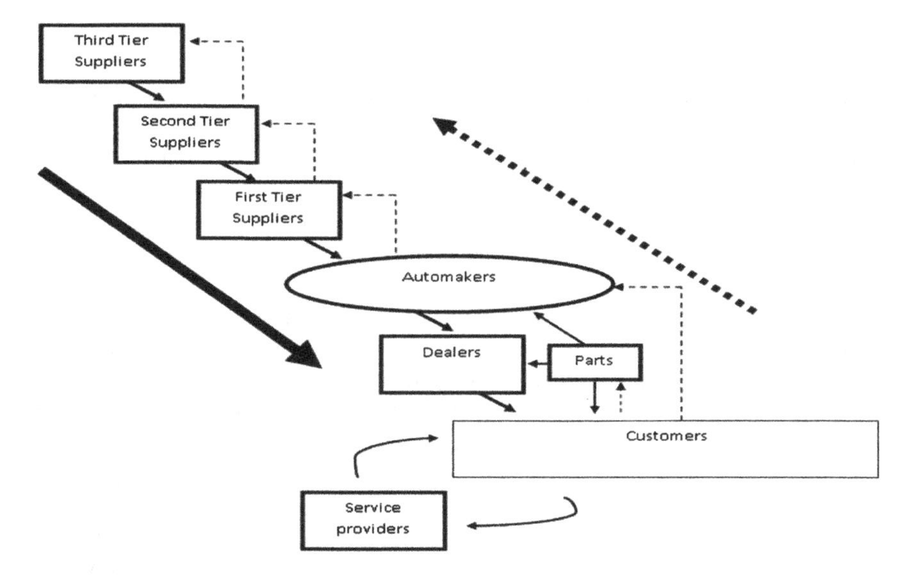

Fig. 1 Supply chain of automobile in India (source [16])

vehicles and 4.45 % of commercial vehicles [16, 17]. In the past 5 years the total growth in Indian automotive product manufacturing units is from 10.85 million vehicles in 2007–08 to 20.63 million vehicles in 2012–13 [18]. Industries in the automotive area have now started to recognize the huge scope and potential future prospects that exist for sustainability in automotive SCM.

Figure 1 presents an illustration of the automobile industry supply chain organization in India with its three tier supplier's involvement. The automotive product manufacturing companies are categorized by an enormous and exceptionally incorporated supply chain. This automotive industry supply chain starts with the three tiers of supplier (procurement network), carries on through automakers (production network), dealers (distribution network), and finishes toward customers (sales network). In this era, concern is by and large for the sector's effect on upstream network (suppliers) of supply chain organization and its emissions of CO_2. The environmental consequences of the present as well as forthcoming development of automobile vehicles have been seen as a major research area [19].

3 Sustainable Supplier Selection in Automobile Industry

Sustainable supplier selection (SSS) is one of the most important factors that assist to build a strong sustainable SCM. Traditionally, only the management aspects like lead time, quality, price, late deliveries, rate of rejected parts, and service quality of the supply chain were considered for selecting a potential supplier, the various multi-criteria decision-making models (MCDM) used extensively for SS were studied by different researches [20–30].

However, with the growing environmental issues researchers are also paying attention to factors like greenhouse effect, reusability, carbon dioxide (CO_2) emission, etc., jointly known as carbon foot printing. The resulting problem is called "sustainable supplier selection," where a balance is maintained between the management and environment concerns. Various sustainable SS methods have been suggested by the researchers to find an optimal solution for the sustainable SS problem [31–42].

Automotive products need high degree of outsourcing to suppliers for their assembly because those are very complicated [43]. At present the automotive sector is one of the most supplier-dependent industries, because its share is between 60 and 80 % of the total manufacturing cost [44–46]. It becomes increasingly evident that the image and public perception of an automotive SCM with regard to sustainability is not only dependent on its own environmental performance, but also on the environmental performance of its supply chain members, and in particular its "suppliers" (primary member of the automotive supply chain) [47], since the SSS is primarily held responsible for the sustainability of the entire automotive SCM.

Authors in [48] state that the development of environmentally friendly automotive paint with the participation of suppliers is vital. Further, [49] examined the role of suppliers in plant-level ecological enrichments in a Canadian printing industry in addition to describing the significance of cooperation for supplier investments in environmentally friendly technologies.

Toward accomplishing a sustainable automobile supply chain, entire associates in the chain from raw material suppliers to topmost administrators must have natural liking in relation to sustainability [50]. Even now, comprehensive SCM study is yet to be accomplished on how corporations can contain suppliers in sustainable management practices and involve them in sustainable activities. Therefore, this study explores the use of DE algorithm to manage this problem and cope with the imprecision that is found in the SSS problem.

4 Methodology

In this research, in order to evaluate the efficiency of sustainable suppliers a methodology was developed, a DEA model, and used DE algorithm to get optimal results. To assess and analyze the efficiency of suppliers for automobile industry, we have followed a four-step methodology:

(1) Designing of input and output criteria.
(2) Selection of an automobile industry for SSS problem.
(3) Choice of DEA-based mathematical model for SSS problem.
(4) Applying DE on mathematical model for performance evaluation and selection of sustainable suppliers.

The methodology presents DE as a tool for evaluating and selecting the optimal suppliers, taking care of sustainability.

5 Choice of DEA-Based Mathematical Model

A variety of DEA models have been discussed in the DEA literature [50–57]. In our case, there are three inputs (inflexible for a year): quality, lead time, and price. Simultaneously, from the output criteria: "Service quality" and "CO_2 emission" cannot be determined in advance. So output-oriented model is suitable for this study. In the present study, we have used the following mathematical model (Eq. 1) developed in our previous research [37].

$$\begin{aligned}
&\text{Max } E_m \sum_{k=1}^{O} w_k \text{Output}_{k,m} \\
&\text{s.t.} \\
&\sum_{l=1}^{I} z_l \text{Input}_{l,m} = 1. \\
&\sum_{k=1}^{O} w_k \text{Output}_{k,n} - \sum_{l=1}^{I} z_l \text{Input}_{l,n} \leq 0, \quad \forall n \\
&w_k, z_l \geq 0; \quad \forall\, k, l
\end{aligned} \tag{1}$$

where a DMU is considered efficient if the efficiency score is 1 otherwise it is considered as inefficient.

6 DE Algorithm

Differential evolution (DE) algorithm is a type of evolutionary algorithm, most efficiently used in optimization problems. Very few previous articles in the literature have mentioned the successful use of the DE algorithm in the automotive industry as well as in mechanical engineering design [58, 59]. The advantage with respect to other techniques and DE algorithm is relevant for the practical application in automotive industry, because DE has emerged as one of the most simple and efficient techniques for solving global optimization problems. DE has some unique characteristics that make it different from many others in the family of evolutionary algorithms. The generation of offspring and selection mechanism of DE is completely different from its counterparts. DE uses a one-to-one spawning and selection relationship between each individual and its offspring. These features though help in strengthening the working of DE algorithm in automobile industries problems.

Table 1 DE (parameter settings)

Pop size (NP)	100
Scale factor (F)	0.5
Crossover rate (Cr)	0.9
Max iteration	3000
DE scheme used	DE/rand/1/bin
Constraint handling	Pareto ranking method [60]

6.1 Experimental Settings

Table 1 shows the parameter settings for DE.

Pop size refers to the population size taken. Scale factor (F) and crossover rate (Cr) are the control parameters of DE defined by the user. Max iteration refers to the maximum number of iterations which are performed. Pareto ranking method is used for solving constrained optimization problems. This method makes use of three criteria to select a candidate: (a) If two solutions are feasible then select the one giving the better fitness function value, (b) If one solution is feasible and one is infeasible then select the one which is feasible, and (c) If two solutions are infeasible, select the one having smaller constraint violation.

7 Case on Indian Automobile Industry

Case: Sustainable supplier selection problem

The problem explored in our paper is the hypothetical data [37] used by an automobile industry in India (*X* Company). The raw material is assumed to be gears, used for transmitting rotational motion in automobiles.

7.1 Designing a Criteria

Here we split the criteria in two manners: the input and output criteria. The input criteria are the traditional supplier selection criteria, such as lead time, price and quality of the delivered goods. The output criteria are the service quality [61] and CO_2 emission [62] of the product and services. Designing of criteria containing the input and output criteria is as follows:

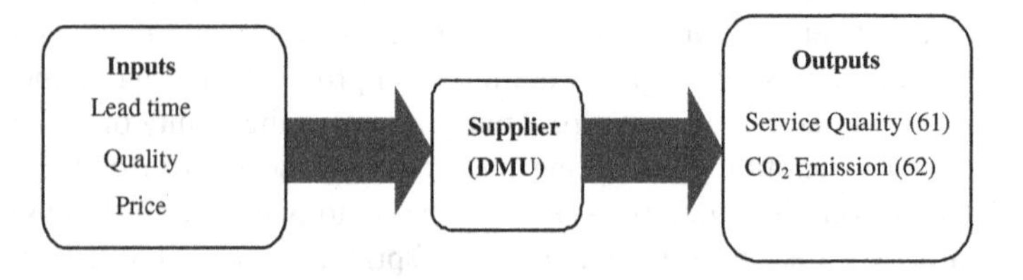

Table 2 Data of an automobile industry

Criteria	Management criteria (inputs)			Environmental criteria (outputs)	
Suppliers	Lead time (L) (Day)	Quality (Q) (%)	Price (P) (Rs.)	Service quality (SQ) (%)	CO_2 emissions (CE) (g)
1	3	75	187	84	40
2	2	77	195	76	30
3	4	85	272	27	25
4	5	86	236	110	22
5	3	74	287	94	38
6	6	62	242	102	10
7	3	73	168	82	24
8	4	92	396	63	38
9	2	77	144	55	26
10	3	69	137	61	18
11	6	54	142	122	24
12	5	57	196	75	30
13	1	77	247	80	55
14	3	61	148	121	39
15	4	69	294	125	8
16	6	94	249	76	6
17	2	88	121	114	55
18	1	78	269	65	48

7.2 Selection of a Problem

The following items provided in the shipment of automobile industry with the supplier's database covering management (input) as well as environmental (output) criteria data are shown in Table 2.

7.3 Formulation of Mathematical Model

Based on the basis of the above data the DEA model of *K*th DMU with the help of Eq. (1) will be as follows:

$$\begin{aligned}
&\text{Max } SQ_m + CE_n \\
&\text{s.t.} \\
&z_1 L_m + z_2 Q_m + z_3 P_m = 1 \\
&w_1 SQ_n + w_2 CE_n - (z_1 L_m + z_2 Q_m + z_3 P_m) \leq 0 \\
&\forall n = 1, \ldots m, \ldots 18
\end{aligned} \tag{2}$$

7.4 Applying DEA on Mathematical Model

In our study, we use CRS assumption for overall performance assessment. DEA solver software is used for all computation associated with the DEA method [63]. Table 3 results show the efficiency score of all DMUs (suppliers).

The results are shown in Table 3. The investigation shows that out of 12 suppliers, only 4 suppliers, namely, 11, 13, 14, and 17 are technically efficient.

7.5 Applying DE on Mathematical Model

After applying DE on sustainable supplier selection problem, in Table 4 average efficiency and weights results of all DMUs are given and from Table 5 we can see that for suppliers 11, 14, and 17, the efficiency score is 1 so these suppliers are assumed to be 100 % efficient while efficiency score for all other suppliers are less than 1. So these suppliers are not as efficient and among these, supplier no. 18 is probably the most inefficient in comparison to all other suppliers.

Table 3 Efficiency scores based on DEA

Supplier	Technical efficiency	Peer supplier	Peer count
1	0.799	14, 13, 17	0
2	0.715	14, 17	0
3	0.412	17, 13	0
4	0.632	11, 14	0
5	0.771	14, 13	0
6	0.749	11, 14	0
7	0.633	14, 17	0
8	0.587	14, 13	0
9	0.525	14, 17, 13	0
10	0.517	11, 17, 14	0
11	1	11	5
12	0.792	14, 13	0
13	1	13	7
14	1	14	12
15	0.896	11, 14	0
16	0.395	11, 14	0
17	1	17	6
18	0.873	13	0
Average	0.739		

Table 4 Average efficiency and weights in 30 runs

Suppliers	Value of input and output weights					Efficiency
	W_1	W_2	Z_1	Z_2	Z_3	
1	0.0093171	0.00422648	0.5864615	0.0021643	0.0077285	0.885125
2	0.0096163	0.08283e−017	1	0	0	0.942402
3	0.0070293	0.33556e−01	0.0668079	0.0088852	0	0.0843525
4	0.0083216	0.00248747	0.0790911	0.010519	0	0.832165
5	0.0112962	0.0219116	0.107364	0.0142795	0	0.734253
6	0.0073778	0.00230969	0.0657469	0.0085730	0.0003544	0.811567
7	0.0089439	0.58488e−01	0.32575	0.0020776	0.0074189	0.822839
8	0.0071902	0.0327347	0.0683372	0.0090886	0.89E−20	0.524889
9	0.0096163	0.0498655	1	0	0	0.721226
10	0.0078839	0.0305688	0.0702566	0.0091611	0.0003788	0.638597
11	0.0089303	0.0031707	0.0321097	0.0136241	0.0008513	1
12	0.0102223	0.29418e−018	0.0984692	0.0092357	0.0015945	0.8689
13	0.0090900	0.0363442	0.0863947	0.0114904	0	0.636303
14	0.0090108	0.0315172	0.0208733	0.0068106	0.0043416	1
15	0.0078651	0.14697e−017	0.0747521	0.0099419	0	0.983144
16	0.0070992	0.0280174	0.0632637	0.0082492	0.0003411	0.468551
17	0.0096163	0.0625062	1	0	0	1
18	0.0081534	0.01723e−014	0.0774918	0.0103063	0.66829	0.448437

Table 5 Suppliers efficiency based on DE

Supplier	Efficiency	Supplier	Efficiency
1	0.885125	10	0.638597
2	0.942402	11	1
3	0.084352	12	0.8689
4	0.832165	13	0.636303
5	0.734253	14	1
6	0.811567	15	0.983144
7	0.822839	16	0.468551
8	0.524889	17	1
9	0.721226	18	0.448437

7.6 Comparison of Efficiency Scores with Two Techniques

Table 6 shows efficiency scores comparison of two techniques and Fig. 2 shows the histogram for comparison of efficiency scores of two techniques, DE-based approach efficiency scores for supplier 11, 14, and 17 is one, except for all other 1, 2, 3, 4, 5, 6, 7, 8, 9, 10, 12, 13, 15, 16, and 18. So these 15 suppliers are inefficient. It is exciting to make a note of that some suppliers, which get comparatively low

Table 6 Efficiency scores comparison of two techniques

Suppliers	DEA technique		DE technique	
	Efficiency	Ranking	Efficiency	Ranking
1	0.799	7	0.885125	6
2	0.715	11	0.942402	5
3	0.412	17	0.084352	18
4	0.632	13	0.832165	8
5	0.771	9	0.734253	11
6	0.749	10	0.811567	10
7	0.633	12	0.822839	9
8	0.587	14	0.524889	15
9	0.525	15	0.721226	12
10	0.517	16	0.638597	13
11	1	1	1	1
12	0.792	8	0.8689	7
13	1	1	0.636303	14
14	1	1	1	1
15	0.896	5	0.983144	4
16	0.395	18	0.468551	16
17	1	1	1	1
18	0.873	6	0.448437	17
Average	0.739		0.7446	

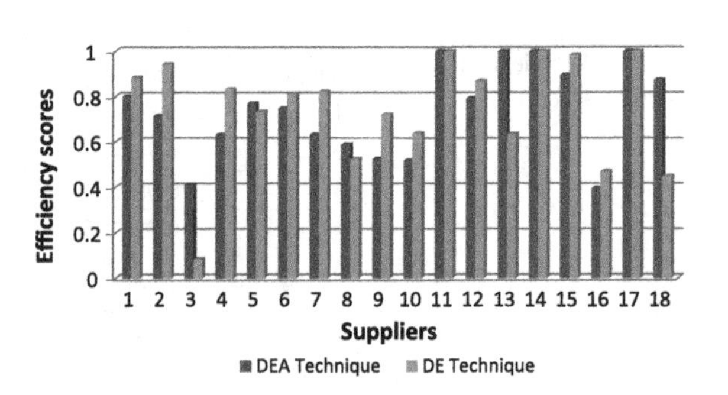

Fig. 2 Histogram for comparison of efficiency scores of two techniques

DEA-based scores, have high DE-based efficiency scores. While 13th supplier has relatively low DE-based efficiency (0.636303), but obtains unit DEA-based score.

This clearly shows that this DE-based approach provides an optimum cluster of sustainable suppliers, which is reduced from four to three suppliers. The results of the case indicate that the DE algorithm can solve the problem effectively. The approach is validated in terms of sustainability, which is an added advantage of the approach.

8 Conclusion

This paper presents how an automotive industry has implemented the DEA model and used DE algorithm for accessing supplier's efficiency in multi-criteria relative to the other supplier's contending in the similar type of market. The main motivation of this research was to access the precision of DE for providing optimal solutions for sustainable supplier selection (SSS).

With the help of this research the purchasing managers of an automobile industry can easily assess the effectiveness of each supplier with respect to their sustainability performance in comparison to "best suppliers" in the marketplace. The findings from this model in terms of results can be used in order to calculate or decide standard values to compare with efficient and inefficient sustainable suppliers. In terms of further researches the use of DE and other soft computing techniques [64–68] for sustainable supplier's evaluation and selection will stimulate by the results of the present paper.

References

1. Jayal, A.D., Badurdeen, F., Dillon, Jr., O.W., Jawahir, I.S.: Sustainable manufacturing: Modeling and optimization challenges at the product, process and system levels. CIRP J. Manufact. Sci. Technol. **2**(3), 144–152 (2010)
2. Floridi, M., Pagni, S., Falorni, S., Luzzati, T.: An exercise in composite indicators construction: Assessing the sustainability of Italian regions. Ecol. Econ. **70**(8), 1440–1447 (2011)
3. Luthe, T., Schuckert, M.: Socially responsible investing–implications for leveraging sustainable development. In: Trends and Issues in Global Tourism 2011 (pp. 315–321). Springer Berlin Heidelberg (2011)
4. Paoletti, M.G., Gomiero, T., Pimentel, D.: Introduction to the special issue: towards a more sustainable agriculture. Crit. Rev. Plant Sci. **30**(1–2), 2–5 (2011)
5. Büyüközkan, G., Çifçi, G.: A novel fuzzy multi-criteria decision framework for sustainable supplier selection with incomplete information. Comput. Ind. **62**(2), 164–174 (2011)
6. Genovese, A., Koh, S.L., Bruno, G., Bruno, P.: Green supplier selection: a literature review and a critical perspective. In: 8th International Conference on Supply Chain Management and Information Systems (SCMIS), 2010 pp. 1–6. IEEE, Oct 2010
7. Purdy, L., Safayeni, F.: Strategies for supplier evaluation: a framework for potential advantages and limitations. IEEE Trans. Eng. Manage. **47**(4), 435–443 (2000)
8. Storn, R., Price, K.: Differential evolution-a simple and efficient adaptive scheme for global optimization over continuous spaces. Berkeley: ICSI, CA, Technical Report TR-95-012 (1995)
9. Plagianakos, V.P., Tasoulis, D.K., Vrahatis, M.N.: A review of major application areas of differential evolution. In: Advances in Differential Evolution, pp. 197–238. Springer, Berlin Heidelberg (2008)
10. Wang, F., Jang, H.J.: Parameter estimation of a bioreaction model by hybrid differential evolution. In: Proceedings of the 2000 Congress on Evolutionary Computation, 2000, vol. 1, pp. 410–417. IEEE (2000)
11. Joshi, R., Sanderson, A.C.: Minimal representation multisensor fusion using differential evolution. IEEE Trans. Syst. Man Cybern. Part A Syst. Hum. **29**(1), 63–76 (1999)

12. Ilonen, J., Kamarainen, J.K., Lampinen, J.: Differential evolution training algorithm for feed-forward neural networks. Neural Process. Lett. **17**(1), 93–105 (2003)
13. Ali, M., Siarry, P., Pant, M.: An efficient differential evolution based algorithm for solving multi-objective optimization problems. Eur. J. Oper. Res. **217**(2), 404–416 (2012)
14. Jauhar, S., Pant, M., Deep, A.: Differential evolution for supplier selection problem: a DEA based approach. In: Proceedings of the Third International Conference on Soft Computing for Problem Solving, pp. 343–353. Springer India, Jan 2014
15. Parker, P.: Environmental initiatives among japanese automakers: new technology, EMS, recycling and lifecycle approaches. Environ.: J. interdiscip. Stud. 29(3) (2001)
16. Kearney, A.T.: Building world class supply chain in India, Conference on Auto Supply Chain Management, 2013. www.atkearney.com. Accessed 23 Nov 2014
17. SIAM India—Society of Indian Automobile Manufacturers (SIAM). www.siamindia.com. Accessed. Accessed 18 Jan 2014
18. Bhattacharya, S., Mukhopadhyay, D., Giri, S.: Supply chain management in indian automotive industry: complexities, challenges and way ahead. Int. J. Managing Value Supply Chains **5**(2) (2014)
19. Nieuwenhuis, P., Wells, P.: The Automotive Industry and The Environment: A Technical, Business and Social Future, Woodhead, Cambridge. 2003, Elsevier (2003)
20. Agarwal, P., Sahai, M., Mishra, V., Bag, M., Singh, V.: A review of multi-criteria decision making techniques for supplier evaluation and selection. Int. J. Ind. Eng. Comput. **2**(4), 801–810 (2011)
21. Weber, C.A., Current, J.R., Benton, W.C.: Vendor selection criteria and methods. Eur. J. Oper. Res. **50**(1), 2–18 (1991)
22. Degraeve, Z., Labro, E., Roodhooft, F.: An evaluation of vendor selection models from a total cost of ownership perspective. Eur. J. Oper. Res. **125**(1), 34–58 (2000)
23. Jauha, S.K., Pant, M.: Recent trends in supply chain management: a soft computing approach. In: Proceedings of Seventh International Conference on Bio-Inspired Computing: Theories and Applications (BIC-TA 2012), pp. 465–478. Springer India, Jan 2013
24. De Boer, L., Labro, E., Morlacchi, P.: A review of methods supporting supplier selection. Eur. J. Purchasing Supply Manage. **7**(2), 75–89 (2001)
25. Holt, G.D.: Which contractor selection methodology? Int. J. Project Manage. **16**(3), 153–164 (1998)
26. Aamer, A.M., Sawhney, R.: Review of suppliers selection from a production perspective. In: Proceedings of the IIE Annual Conference and Exhibition, pp. 2135–2140 (2004)
27. Ho, W., Xu, X., Dey, P.K.: Multi-criteria decision making approaches for supplier evaluation and selection: A literature review. Eur. J. Oper. Res. **202**(1), 16–24 (2010)
28. Tahriri, F., Osman, M.R., Ali, A., Yusuff, R.M.: A review of supplier selection methods in manufacturing industries. Suranaree J. Sci. Technol. **15**(3), 201–208 (2008)
29. Jauhar, S.K., Pant, M.: Genetic algorithms, a nature-inspired tool: review of applications in supply chain management. In: Proceedings of Fourth International Conference on Soft Computing for Problem Solving, pp. 71–86. Springer India, Jan 2015
30. Cheraghi, S.H., Dadashzadeh, M., Subramanian, M.: Critical success factors for supplier selection: an update. J. Appl. Bus. Res. (JABR) **20**(2) (2011)
31. Noci, G.: Designing 'green'vendor rating systems for the assessment of a supplier's environmental performance. Eur. J. Purchasing Supply Manage. **3**(2), 103–114 (1997)
32. Zhu, Q., Geng, Y.: Integrating environmental issues into supplier selection and management. Greener Manage. Int. **2001**(35), 26–40 (2001)
33. Jauhar, S.K., Pant, M., Deep, A.: An approach to solve multi-criteria supplier selection while considering environmental aspects using differential evolution. In: Swarm, Evolutionary, and Memetic Computing, pp. 199–208. Springer International Publishing (2013)
34. Lai, Y.F.: Green supplier evaluation in green supply chain management—examples of printed circuit board suppliers. Unpublished Master's Thesis, Department of Resource Engineering, National Cheng-Kung University, Taiwan (in Chinese) (2004)

35. Seuring, S., Müller, M.: Core issues in sustainable supply chain management—a Delphi study. Bus. Strategy Environ. **17**(8), 455–466 (2008)
36. Awasthi, A., Chauhan, S.S., Goyal, S.K.: A fuzzy multicriteria approach for evaluating environmental performance of suppliers. Int. J. Prod. Econ. **126**(2), 370–378 (2010)
37. Jauhar, S.K., Pant, M., Abraham, A.: A novel approach for sustainable supplier selection using differential evolution: a case on pulp and paper industry. In: Intelligent Data analysis and its Applications, vol. II, pp. 105–117. Springer International Publishing (2014)
38. Buyukozkan, G., Çifci, G.: A novel hybrid MCDM approach based on fuzzy DEMATEL, fuzzy ANP and fuzzy TOPSIS to evaluate green suppliers. Expert Syst. Appl. **39**, 3000–3011 (2012)
39. Kannan, D., Khodaverdi, R., Olfat, L., Jafarian, A., Diabat, A.: Integrated fuzzy multi criteria decision making method and multi-objective programming approach for supplier selection and order allocation in a green supply chain. J. Clean. Prod. **47**, 355–367 (2013)
40. Falatoonitoosi, E., Ahmed, S., Sorooshian, S.: A multicriteria framework to evaluate supplier's greenness. In: Abstract and Applied Analysis, vol. 2014. Hindawi Publishing Corporation, March 2014
41. Yazdani, M.: An integrated MCDM approach to green supplier selection. Int. J. Ind. Eng. Comput. **5**(3), 443–458 (2014)
42. Jalhar, S.K., Pant, M., Nagar, M.C.: Differential evolution for sustainable supplier selection in pulp and paper industry: a DEA based approach. Comput. Methods Mater. Sci. **15** (2015)
43. Simpson, D.F., Power, D.J.: Use the supply relationship to develop lean and green suppliers. Int. J. Supply chain manage. **10**(1), 60–68 (2005)
44. Lee, S.Y., Klassen, R.D.: Drivers and Enablers That Foster Environmental Management Capabilities in Small-and Medium-Sized Suppliers in Supply Chains. Prod. Oper. Manage. **17**(6), 573–586 (2008)
45. Weele, A.J.: Purchasing and Supply Chain Management. Cengage Learning EMEA, Andover, UK (2010)
46. Scannell, T.V., Vickery, S.K., Droge, C.L.: (2000). Upstream supply chain management and competitive performance in the automotive supply industry. J. Bus. Logistics **21**(1)
47. Awaysheh, A., Klassen, R.D.: The impact of supply chain structure on the use of supplier socially responsible practices. Int. J. Oper. Prod. Manage. **30**(12), 1246–1268 (2010)
48. Geffen, C.A., Rothenberg, S.: Suppliers and environmental innovation: the automotive paint process. Int. J. Oper. Prod. Manage. **20**(2), 166–186 (2000)
49. Klassen, R.D., Vachon, S.: Collaboration and evaluation in the supply chain: the impact on plant-level environmental investment. Prod. Oper. Manage. **12**(3), 336–352 (2003)
50. Despotis, D.K., Stamati, L.V., Smirlis, Y.G.: Data envelopment analysis with nonlinear virtual inputs and outputs. Eur. J. Oper. Res. **202**(2), 604–613 (2010)
51. Ramanathan, R. (ed.): A Tool for Performance Measurement. Sage Publication Ltd., New Delhi (2003)
52. Wen, U.P., Chi, J.M.: Developing green supplier selection procedure: a DEA approach. In: 17th International Conference on Industrial Engineering and Engineering Management (IE&EM), 2010 IEEE, pp. 70, 74, 29–31 Oct 2010. doi:10.1109/ICIEEM.2010.5646615
53. Dobos, I., Vörösmarty, G.: Supplier selection and evaluation decision considering environmental aspects. sz. Mőhelytanulmány, HU ISSN 1786–3031, Oct 2012
54. Kumar, P., Mogha, S.K., Pant, M.: Differential evolution for data envelopment analysis. In: Proceedings of the International Conference on Soft Computing for Problem Solving (SocProS 2011) Dec 20–22, 2011, pp. 311–319, Springer India, Jan 2012
55. Srinivas, T.: Data envelopment analysis: models and extensions. In: Production/Operation Management Decision Line, pp. 8–11 (2000)
56. Charnes, A., Cooper, W.W., Rhodes, E.: Measuring the efficiency of decision making units. Eur. J. Oper. Res. **2**(6), 429–444 (1978)
57. Banker, R.D., Charnes, A., Cooper, W.W.: Some models for estimating technical and scale inefficiencies in data envelopment analysis. Manage. Sci. **30**(9), 1078–1092 (1984)

58. Rogalsky, T., Derksen, R.W., Kocabiyik, S.: Differential evolution in aerodynamic optimization. In: Proceedings of 46th Annual Conference, Canadian Aeronautics Space Institute, pp. 29–36 (1999)
59. Joshi, R., Sanderson, A.C.: Minimal representation multi-sensor fusion using differential evolution. IEEE Trans. Syst. Man Cybernetics. Part A **29**, 63–76 (1999)
60. Ray, T., Kang, T., Chye, S.K.: An evolutionary algorithm for constrained optimization. In: Whitley, D., Goldberg, D., Cantu-Paz, E., Spector, L., Parmee, I., Beyer, H.G. (eds.) Proceeding of the Genetic and Evolutionary Computation Conference (GECCO 2000), pp. 771–777 (2000)
61. Shirouyehzad, H., Lotfi, F.H., Dabestani, R.: A data envelopment analysis approach based on the service qualtiy concept for vendor selection. In: International Conference on Computers & Industrial Engineering, 2009. CIE 2009, pp. 426–430. IEEE, July 2009
62. http://www.london2012.com/documents/locog-publications/locog-guidelines-on-carbon-emissions-of-products-and-services.pdf. Accessed 12 Oct 2013
63. Cooper, W.W., Seiford, L.M., Tone, K.: Data envelopment analysis: a comprehensive text with models, applications, references and DEA-Solver Software. Second editions. Springer (2007). ISBN: 387452818, 490
64. Ansari, I.A, Pant, M., Neri, F.: Analysis of gray scale watermark in RGB host using SVD and PSO. In: IEEE Computational Intelligence for Multimedia, Signal and Vision Processing (CIMSIVP), pp. 1–7 (2014)
65. Zaheer, H., Pant, M., Kumar, S., Monakhov, O., Monakhova, E., Deep, K.: A new guiding force strategy for differential evolution. Int. J. Syst. Assur. Eng. Manag. 1–14 (2015), doi:10.1007/s13198-014-0322-6
66. Ansari, I.A., Pant, A., Ahn, C.W.: SVD based fragile watermarking scheme for tamper localization and self-recovery. Int. J. Mach. Learn. Cyber. 1–15 (2015), doi:10.1007/s13042-015-0455-1
67. Zaheer, H., Pant, M.: A differential evolution approach for solving integer programming problems. In: Proceedings of Fourth International Conference on Soft Computing for Problem Solving, pp. 413–424. Springer India (2015)
68. Ansari, I.A., Pant, A., Ahn, C.W.: Robust and false positive free watermarking in IWT domain using SVD and ABC. Eng. Appl. Artif. Intell. **49**, 114–125 (2016)

A New Approach for Rice Quality Analysis and Classification Using Thresholding-Based Classifier

Priyanka Gupta and Mahesh Bundele

Abstract Quality detection and analysis of grains like wheat, rice, etc., is an important process normally followed using manual perceptions. The decision depends and varies from person to person and even involves human error with the same person analyzing at different times. This paper presents a novel approach to provide, simple and efficient classification of rice grains using thresholding method. This approach would be further used to automate the process and maintain the uniformity in decisions. In this research, India Gate basmati rice varieties such as—brown basmati rice, classic, super, tibar, dubar, rozana, and mogra were used for the analysis. The process involved identification of region of interest (ROI) and extraction of six morphological features—area, major axis length, minor axis length, aspect ratio, perimeter, and eccentricity. Further three color features—max hue, max saturation, and max value features from each rice grain of sample image were extracted. Through exhaustive experimentation and analysis the rice qualities were defined into seven classes such as; best, good, fine, 3/4 broken, 5/8 broken, 1/2 broken, and 1/4 broken based on threshold values. Through varying logical conditions and relationships among these quality features, aforesaid rice brands were classified. Three of such varying techniques were used out of which one method is being presented here. The results of the presented approach were 100 % accurate for some classes while 95 % for others.

Keywords Connected component labeling · Thresholding method · Morphological features · SVM · Rice grading · Region of interest

Priyanka Gupta (✉)
Poornima College of Engineering, Jaipur, India
e-mail: Write2priya.abhi@gmail.com

Mahesh Bundele
Computer Engineering, Poornima Foundation, Jaipur, India
e-mail: maheshbundele@poornima.org

M. Pant et al. (eds.), *Proceedings of Fifth International Conference on Soft Computing for Problem Solving*, Advances in Intelligent Systems and Computing 437, DOI 10.1007/978-981-10-0451-3_48

1 Introduction

Enormous increase in population and food grain requirements have led to the need for autonomous systems for grain quality detection and grading. Rice is one of the most consumed cereal grain crops. It is considered an essential food for life in India and is grown on a majority of rural farms. The focus of our research was on rice grain quality detection and grading. There are lot of varieties of rice available in the market with minor changes in shape, size, color, and price. Therefore, it is a tedious job to identify the best quality within price limit. Usually, consumers analyze the quality and variety of rice by just visual inspection. However, the decision-making capability of human inspectors is affected by many external influences such as fatigue, time, vengeance, bias, lack of knowledge, etc. Sometimes due to improper and insufficient knowledge, the consumer relies on branded products only, which is good but not cost-efficient. Hence, it becomes essential to develop an automated system in order to get fast and accurate decisions.

Researchers have been working on food grains quality analysis and classification. Patil et al., worked on different varieties of rice and could identify each of the variety and carried out quality assessment of rice grains based on its size. Based on the length of rice they divided them into three grades—Grade 1, Grade 2, and Grade 3. The average accuracy of classification was 93 % [1]. Ajay et al., worked on milled rice to determine the quality of rice, for this they calculated the quantity of broken rice in milled rice image. If the length of rice was less than 70 % of head rice, it was treated as broken rice [2]. Tanck et al. designed a model for digital inspection of Agmark Standards like broken grain, foreign particle, and admixture for rice grain. For this they measured the area, major axis length, and the perimeter of each rice grain using morphological analysis in MATLAB [3]. Neelam et al., designed an algorithm for identification and classification of four different varieties of rice, using the color and morphological features with the help of neural networks. The average classification and identification accuracy was found 89.7 % [4]. Desai et al., proposed a model to find the percentage of purity in rice grain. They measured area of each rice of sample image and based on area decided the different rice varieties such as basmati, kolam, masoori, etc. After identifying the rice variety sample rice were graded into three classes—good, medium, and poor [5]. Sidnal et al. suggested grading and quality testing of food grains using neural Network-based morphological and color features [6]. Maheshwari et al. designed quality analysis of rice grains by counting normal, long, and small rice seeds and could achieve high degree of accuracy and quality as compared to manual method. [7] Gujjar et al. presented, method to test the purity of rice grain sample. They used the pattern classification technique based on color, morphological, and textural features for rejecting the broken basmati rice grains [8]. Neelamegam et al., performed analysis of basmati rice grains using four morphological features and classified them into three classes—normal, long, and small rice grains [9].

Kaur et al. suggested automated an algorithm for testing the purity of rice grains. They have analyzed head rice, broken rice, and brewers and classified the rice grains as premium, grade A, grade B, and grade C [10]. Aulakh et al., designed an algorithm to find percentage purity of rice samples by image processing. Purity test of grain sample was done based on rice length and broken rice was considered impurity [11]. Gujjar et al., proposed a digital image processing algorithm based on color, morphological, and textural features to identify the six varieties of basmati rice. Purity of Rice grain was checked based on rice grain length, broken rice was considered as impurity [12]. Veena presented an automatic evaluation method to determine the quality of milled rice based on shape descriptors or geometric features [13].

From the review of research works and their comparison about the approaches used for quality analysis and classifications and strengths and weaknesses, it could be found that most of the published research works mainly focuses on broad classification of food grains such as classification of various types food grains, identification of different grains into classes, identification of different varieties of rice, etc. Classified grains into grades like—good/medium/poor or long/medium/broken based on size of grains. From the limited review it could be seen that, little or no work has been done, on the concept of broken rice classification into further classes such as 3/4 broken, 5/8 broken, 1/2 broken, and 1/4 broken on a single variety of grain and compare their result to available varieties in the market. Research papers showed that thresholding-based classifier was not able to classify the rice grains with good classification accuracy.

The basic objective of this research was to design an algorithm for quality analysis and classification of rice grain using thresholding-based method. Various morphological and color features has been used. The morphological features used were—area, major axis length, minor axis length, aspect ratio, perimeter, and eccentricity and the color features—max hue, max saturation, and max value extracted from each rice of sample image. Quality of rice grain was analyzed and graded them into seven categories—best, good, fine, 3/4 broken, 5/8 broken, 1/2 broken, and 1/4 broken based on threshold values. We compared our results by testing different varieties of 'India gate Basmati Rice'. The proposed approach is called as new approach because no researchers in the review could be found considering the number of morphological and color parameters that too with a decision on precise range of consideration and finalization to obtain the desired results.

2 Proposed System

Figure 1 shows the process flow of proposed methodology for rice grain quality analysis and classification using thresholding concept.

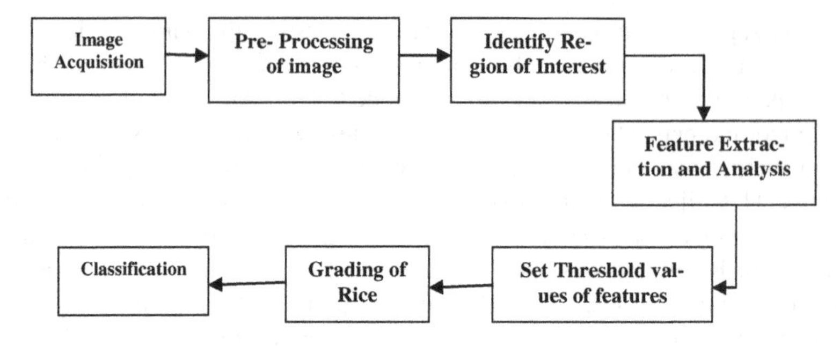

Fig. 1 Process flow of proposed system

2.1 *Image Acquisition*

For collecting the samples of rice grains, we used Cannon EOS 550D digital SLR camera with focal length at 35 mm. Rice grain images were captured on blue background. While capturing images, camera was kept perpendicular to background and distance between camera and background was set to 20 cm in indoor sunlight. Number of rice grains in each sample image was kept around 55–65 grains. Totally, 420 images, 60–60 images of each of seven classes were collected. Size of captured image was 5184 × 3456 pixels in JPEG format. Figure 2 shows seven categories of rice brands selected and their sample images captured.

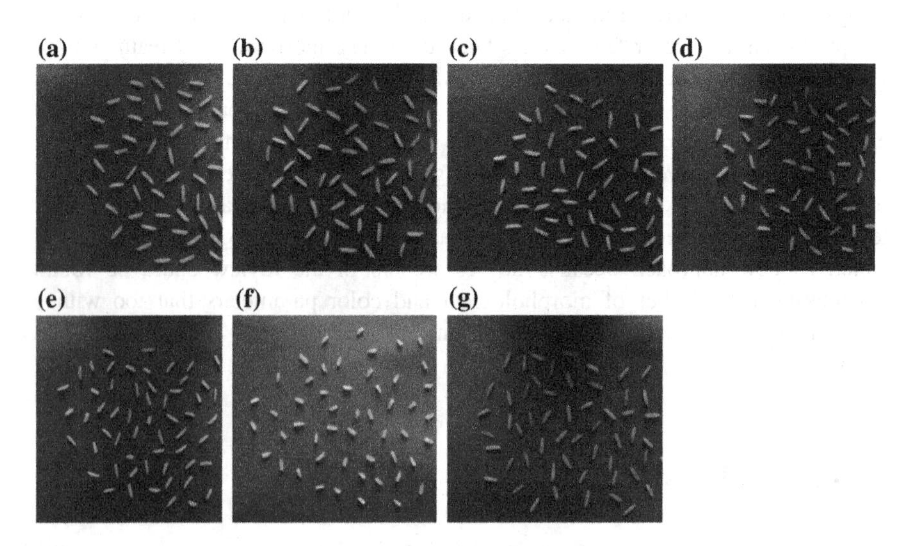

Fig. 2 Sample images of different classes of India Gate basmati rice. **a** Classic India Gate Basmati rice. **b** Super India Gate Basmati rice. **c** Tibar India Gate Basmati rice. **d** India Gate Basmati rice. **e** Rozzana India Gate Basmati rice. **f** Mogra India Gate Basmati rice. **g** Brown India Gate Basmati rice

2.2 Pre-processing

Here all the captured images were resized into 512 × 512 pixels and converted the RGB image into grayscale image. Gaussian filter was applied to remove noise. For solving the nonuniform illumination problem, background image was separated from sample image using morphological opening and closing operations.

2.3 Image Segmentation and Identification of Region of Interest

Image was segmented by threshold values in which background was converted into black (0) and rice grains were converted into white (1) regions. Later, we used connected components labeling technique for identifying the objects (rice) from sample image. This technique scans an image pixel-by-pixel, from top to bottom, left to right to identify connected pixel regions, i.e., regions of adjacent pixels, which share the same set of intensity values, and groups these pixels into components. Once all groups have been determined, each pixel is labeled with a gray level or a color (color labeling) according to its assignment of component. The component with size more than 50 pixels was considered as region of interest (ROI).

2.4 Feature Extraction

Two type of features were extracted from each rice grains of sample image—morphological Features were extracted for geometric analysis of rice grains such as—size, length, shape, width, and the color features were extracted for texture analysis (white or brown) of rice grains. We have extracted six morphological features and three color features for quality analysis and classification process. The extracted features were:

- Morphological Features
  - *Area*: Actual number of pixels present in the region or object is called the area of that region or object.
  - *Major axis length*: The length (in pixels) of the largest diameter of the ellipse that pass through the center and both foci, called major axis length.
  - *Minor axis length*: The length (in pixels) of the shortest diameter of ellipse that pass through the centre called minor axis length.
  - *Aspect ratio*: Ratio of major axis length to minor axis length called aspect ratio.

  - *Perimeter*: The distance around the boundary of the region is called perimeter.
  - *Eccentricity*: The ratio of the distance between the foci of the ellipse and its major axis length is called eccentricity; its value is always between zero and one.

- Color Features

  - *Max hue intensity*: It specifies the 'Hue' value of the pixel with the greatest intensity in the region.
  - *Max saturation intensity*: It specifies the 'Saturation' value of the pixel with the greatest intensity in the region.
  - *Max value intensity*: It specifies the 'Value' value of the pixel with the greatest intensity in the region.

2.5 Setting Threshold

After carrying out repeated experimentation analyses, the ranges of variation for above color features for brown and white rice quality could be fixed as shown in Table 1.

- *Categorization of Morphological Features*: Again with experimental analysis of images captured at initial stage of work, the white rice's morphological features were categorized into seven classes and the range of variation of each morphological feature for these classes were defined as shown in Table 2. These ranges were estimated after doing experimentations through variation in iterative manner.

2.6 Quality Definition for Selected Brand and Grades

Table 3 shows the definition of seven categories of rice brand in terms of quality parameters defined in Table 2.

Table 1 Color feature ranges for brown and white rice

Brown rice	Mean_hue Intensity	0.81–0.63
	Mean_saturation intensity	0.53–0.40
	Mean_value intensity	0.68–0.60
White rice	Mean_hue intensity	0.99–0.70
	Mean_saturation intensity	0.25–0.15
	Mean_value intensity	0.90–0.64

Table 2 Categorization of morphological feature values

Parameters	Best	Good	Fine	3/4 broken rice
Area	360–175	360–150	360–115	340–85
Major axis length	55–45	45–40	40–35	35 – 29
Minor axis Length	10–5	12–5	13–4	10–4
Aspect ratio	10-4.5	10–3	9–2.5	8–1.5
Perimeter	110–85	100–75	95–65	95–55
Eccentrisity	0.995–0.97	0.995–0.95	0.994–0.93	0.992–0.86

Parameters	5/8 broken rice	1/2 broken rice	1/4 broken rice
Area	280–80	250–70	190–50
Major axis length	29–25	25–20	20–10
Minor axis length	14–4	14–4	13–4
Aspect ratio	7–1.5	5–1	4–1
Perimeter	80–50	65–40	55–25
Eccentrisity	0.99–0.85	0.98–0.77	0.97–0.64

Table 3 Rice quality definition based on parameters

Parameters	Classic	Super	Tibar	Dubar
Best	>0 %	≤15 %	≤5 %	≤5 %
Good	≥10 %	>0 %, ≤45 %	>0 and <25 %	>0 % and ≤15 %
Fine	≤65 %	≥15 %and ≤45 %	≥10 % and ≤35 %	≥10 % and ≤30 %
3/4 broken	≤60 %	≥20 % and ≤45 %	≥20 % and ≤40 %	≥20 % and ≤55 %
5/8 broken	Equal to 0 %	>0 % and <= 30 %	>15 % and <= 50 %	≥10 % and ≤50 %
1/2 broken	Equal to 0 %	Equal to 0 %	<10 %	≥5 % and ≤25 %
1/4 broken	Equal to 0 %	Equal to 0 %	Equal to 0	Equal to 0 %

Parameters	Rozzanna	Mogra
Best	Equal to 0 %	Equal to 0 %
Good	≤5 %	Equal to 0 %
Fine	≥5 % and ≤30 %	≤5 %
3/4 broken	≥20 % and ≤40 %	≥0 % and ≤10 %
5/8 broken	≥20 % and ≤50 %	>=6 % and <=35 %
1/2 broken	≥5 % and ≤20 %	≥35 % and ≤60 %
1/4 broken	>0 % and ≤10 %	>10 % and ≤55 %

2.7 Proposed Algorithm

Input Sample rice image in RGB format

Output Classify into India Gate basmati rice variety—brown, classic, super, tibar, dubar, rozana, and mogra

Step 1 Assignment of variable—counter1 = 0 to counter7 = 0

Step 2 Resize the rgbImg image into 512 * 512 size

Step 3 Convert the rgbImg image into 4 images-grayscaleImg, hueImg, SaturationImg, valueImg

Step 4 Apply Gaussian filter and binarization on grayscaleImg to remove noise and nonuniform illumination by subtracting background from binaryImg

Step 5 Apply Connected Component labeling on binaryImg to identify the Region of Interest (ROI) or objects and get total number of rice—NoOfRice in sample image

Step 6 Extract the required features for each rice—area, major_axis_length, minor_axis_length, aspect_ratio, perimeter, eccentricity from grayscaleImg max_hue from hueImg, max_sat from saturationImg, and max_ val

Step 7 Compare these features of a rice by threshold values, identify the grade of rice and increase the counter 1 to counter 7 on the bases of conditions

If condition of best grade satisfy:counter1 = counter1 + 1
elseif condition of good rice satisfy:counter2 = counter2 + 1
elseif condition of fine rice satisfy:counter3 = counter3 + 1
elseif condition of 3/4 broken rice satisfy:counter4 = counter4 + 1
elseif condition of 5/8 broken rice satisfy:counter5 = counter5 + 1
elseif condition of 1/2 broken rice satisfy:counter6 = counter6 + 1
elseif condition of 1/4 broken rice satisfy:counter7 = counter7 + 1
End of if

Step 8 Repeat step 5 and step 6 for all rice of the sample image

Step 9 Calculate the percentage of each counter by dividing the counter to NoOfRice

Step 10 Calculate the mean_Hue, mean_Sat and mean_Val

Mean_hue = mean(Max_hue)
Mean_sat = mean(Max_sat)
Mean_val = mean(Max_val)

Step 11 Classify the sample image on the basis of mean values of hue, saturation, and values into two categories—brown India Gate basmati rice and White India Gate basmati rice

If Condition for brown rice satisfy: Display '**Brown India Gate Basmati Rice**'
If Condition for white rice satisfy: Display '**White India Gate Basmati Rice**'

Step 12 White India Gate basmati rice are divided into six varities on the basis of conditions:

If Condition for Classic rice satisfy: Display '**Classic India Gate Basmati Rice**'
Elseif Condition for super rice satisfy:Display '**Super India Gate Basmati Rice**'

Elseif Condition for tibar rice satisfy:Display ‘**Tibar India Gate Basmati Rice**’
Elseif Condition for dubar rice satisfy:Display ‘**Dubar India Gate Basmati Rice**’
Elseif Condition for rozana rice satisfy:Display ‘**Rozana India Gate Basmati Rice**’
Elseif Condition for mogra rice satisfy:Display ‘**Mogra India Gate Basmati Rice**’
End of If

3 Results and Discussion

The developed model has been tested for seven varieties of India Gate basmati rice—brown rice, classic, super, tibar, dubar, rozana, and mogra using six morphological features and three color features. For testing 60 images, (totally 420 images) of each classes were used.

Confusion matrix: Table 4 shows the results of thresholding-based classifier. Results from thresholding method obtained are 100 % correct classification of brown, classic, super, tibar, rozana, and mogra, while for dubar classification accuracy is 95 %. Therefore, it can be stated that the proposed method-based classification is a good approach to classify the rice in easy and simple manner.

Classification Accuracy: Table 5 shows the percentage of classification accuracy for each class by thresholding approach. For other than Dubar, all the sixty images tested were classified accurately for all six classes of basmati rice. It can be considered, an optimal classification method for rice grains classification.

Table 4 Confusion matrix for thresholding-based classifier

Output/desired	Brown (60)	Mogra (60)	Rozana (60)	Dubar (60)	Tibar (60)	Super (60)	Classic (60)
Brown	60	0	0	0	0	0	0
Mogra	0	60	0	0	0	0	0
Rozana	0	0	60	0	0	0	0
Dubar	0	0	0	57	0	0	0
Tibar	0	0	0	3	60	0	0
Super	0	0	0	0	0	60	0
Classic	0	0	0	0	0	0	60

Table 5 Classification accuracy obtained using thresholding method

Class	Brown	Mogra	Rozana	Dubar	Tibar	Super	Classic
Classification accuracy in %	100	100	100	95	100	100	100

4 Conclusion

A method of rice grain quality analysis and classification was developed using thresholding-based classifier. In this technique, sixmorphological features—area, major axis length, minor axis length, aspect ratio, perimeter, eccentricity and three color features—max hue, max saturation, and max value were used to divide the rice grains into seven grades for quality analysis. Then based on these grades, sample images were classified into seven classes of India Gate basmati rice. Designed classifier was able to classify the India Gate basmati rice quality with 95–100 % accuracy. The accuracy for classic, super, tibar, rozana, mogra, and brown was 100 % but for Dubar the accuracy was 95 %.

5 Future Work

The logic presented can be converted into fuzzy logic method and an automated system can be designed to detect the quality of rice instantly. Further, more features can be added for quality analysis and classification of rice grains such as chalkiness, damage in grain, etc. The method needs to be applied to other rice grains.

References

1. Patil, V., Malemath, V.S.: Quality analysis and grading of rice grain images. IJIRCCE, pp. 5672–5678, June-2015
2. Ajay, G., Suneel, M., Kiran Kumar, K., Siva Prasad, P.: Quality evaluation of rice grains using morphological methods. IJSCE, **2**(6), 35–37 (2013)
3. Tanck, P., Kaushal, B.: A new technique of quality analysis for ricegrading for agmark standards. IJITEE **3**(12), 83–85 (2014)
4. Neelam, Gupta, J.: Identification and classification of rice varieties using neural network by computer vision. IJARCSSE **5**(4), 992–997 (2015)
5. Desai, D., Gamit, N., Mistree, K.: Grading of rice grains quality using image processing. Int. J. Modern Trends Eng. Res. 395–400 (2015)
6. Sidnal, N., Patil, U.V., Patil, P.: Grading and quality testing of food grains using neural network. Int. J. Res. Eng. Technol. **2**(11), 545–548 (2013)
7. Maheshwari, C.V., Jain, K.R., Modi, C.K.: Non-destructive quality analysis of Indian Basmati Oryza Sativa SSP Indica (Rice) using image processing. In: International Conference on Communication Systems and Network Technologies. IEEE, pp. 189–193 (2012)
8. Gujjar, H.S., Siddappa, M.: A method for identification of Basmati rice grain of india and its quality using pattern classification. Int. J. Eng. Res. Appl. (IJERA) ISSN: 2248–9622, vol. 3, Issue 1, pp. 268–273, January–February (2013)
9. Neelamegam, P., Abirami, S., Vishnu Priya, K., Rubalya Valantina, S.: Analysis of rice granules using image processing and neural network. In: IEEE Conference on Information and Communication Technologies, pp. 280–284 (2013)
10. Harpreet, K., Singh, B.: Classification and grading of rice using multi-class SVM. Int. J. Scient. Res. Public. **3**(4), 1–5 (2013)

11. Aulakh, J.S., Banga, V.K.: Percentage Purity of rice samples by image processing. IN: International Conference on Trends in Electrical, Electronics and Power Engineering (ICTEEP'2012). Singapore, pp. 102–104, July 15–16, 2012
12. Gujjar, H.S., Siddappa, M.: Recognition and classification of different types of food grains and detection of foreign bodies using neural networks. In: International Conference on Information and Communication Technologies, pp. 1–5 (2013)
13. Veena, H., Latharani, T.R.: An efficient method for classification of rice grains using Morphological process. Int. J. Innov. Res. Adv. Eng. **1**(1), 118–121 (2014)

Author Biography

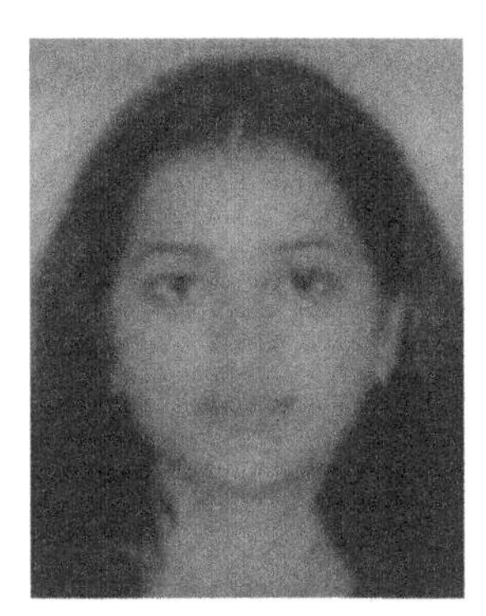

Priyanka Gupta is pursuing her M.Tech. in Computer Science. Her areas of research interest are image processing, pattern recognition, and signal processing. She obtained her Bachelor degree in Computer Science and Engineering in 2006.

Dr. Mahesh Bundele completed his Bachelor's degree in Electronics and Power in 1986, Master's in Electrical Power System, and PhD in Computer Engineering. He has a total of 29 years of teaching experience including 7 years of research. At present, he is working as Dean (R&D) at Poornima University. His areas of research interest are wearable computing, pervasive computing, wireless sensor networks, smart grid issues, etc.

Differential Power Attack on Trivium Implemented on FPGA

Akash Gupta, Surya Prakesh Mishra and Brij Mohan Suri

Abstract Side channel attacks on block ciphers and public key algorithms have been discussed extensively in the literature and a number of techniques have already been developed for recovering the secret information. There is only sparse literature on side channel attacks on stream ciphers. In this paper we introduce an improved DPA technique to apply side channel attacks on Trivium along with the experimental results. This technique is based on the difference of Pearson's Correlation Coefficients, which are calculated for all assumed secret key bit values. Attack is mounted at the resynchronization or re-initialization phase of algorithm. It allows the recovery of secret key information using differential power analysis. To prove the concept, we applied the attack on FPGA implementation of Trivium and are able to extract complete key bits.

Keywords Side channel attack · Pearson's correlation coefficient · Differential power analysis · Stream cipher · FPGA

1 Introduction

Side channel attacks (SCA) [1] are built on the fact that cryptographic algorithms are implemented on a physical device such as FPGA, microcontroller, or ASIC. It can use all types of physical manifestations of the device such as current consumption, electromagnetic radiation, temperature variation, or time variations

Akash Gupta (✉) · S.P. Mishra · B.M. Suri
SAG, DRDO, Metcalfe House, Delhi, India
e-mail: akash.23march@gmail.com

S.P. Mishra
e-mail: spm201@gmail.com

B.M. Suri
e-mail: bmsuri2008@gmail.com

M. Pant et al. (eds.), *Proceedings of Fifth International Conference on Soft Computing for Problem Solving*, Advances in Intelligent Systems and Computing 437, DOI 10.1007/978-981-10-0451-3_49

during the execution of different instructions. These side channels may leak information about secret data. Differential power attack (DPA) is a well-known and thoroughly studied threat for implementations of block ciphers, i.e., DES & AES, and public key algorithms, like RSA. It was introduced by Kocher [2]. In DPA an attacker generates a set of hypotheses (about some secret value or a partial key) and tries to identify the (unique) true hypothesis by finding the highest correlation between power consumption and changes in its internal states during the execution of the algorithm. The true hypothesis provides actual secret information which can be exploited by the attacker. Various other attacks like template attack [3], mutual information attack and fault attack [4] are also discussed in the literature, but DPA is a more generic and device-independent attack. DPA on stream ciphers [5] is a relatively less explored domain due to its complexity and structure. Stream cipher consists of two phases—(i) *key scheduling algorithm (KSA)*: It takes initialization vector (IV) & key (K) to initialize the internal state of algorithm and then states are repeatedly changed for a predefined number of times. (ii) *key sequence generation*: The state is repeatedly updated with the clock and is used to generate the required number of key stream bits. In the case of Trivium the theoretical models of DPA discussed in the existing literature either require more than 10,000 of power traces or are not sufficient(valid for ≤90 nm technology-based FPGA) to retrieve the secret key. Therefore, we introduce a new technique which is based on the difference of correlation traces to retrieve the key in less number of power traces. This is the general approach and is valid for all stream ciphers. It consists of three steps and each gives partial information about secret key. In case of Trivium, our attack methodology targets the initialization phase. We using Pearson's correlation coefficient for correlating the power consumption with hypothetical hamming distance during the execution of algorithm, to recover the actual key of stream cipher.

2 H/W Implementation and Collection of Power Traces

The Trivium algorithm is a hardware-efficient, synchronous stream cipher designed by De Canniere and Preneel [6]. The cipher makes use of 80-bit key and 80-bit IV; its secret state has 288 bits, consisting of three interconnected nonlinear feedback shift registers (NFSR) of length 93, 84, and 111 bits, respectively. The cipher operation consists of two phases: KSA and the key stream generation phase. During KSA shift registers are initialized with the Key & IV and then the algorithm is run for 4×288 steps of the clocking to increase nonlinearity. After that the key stream is generated sequentially. Three bits are computed using nonlinear functions (T_1, T_2 and T_3) in each clock cycle and fed to the NFSRs. The cipher output is generated by xoring the plain text and key stream. NFSRs are a combination of sequential and combinatorial logic. The architectural view of Trivium is shown in Fig. 1.

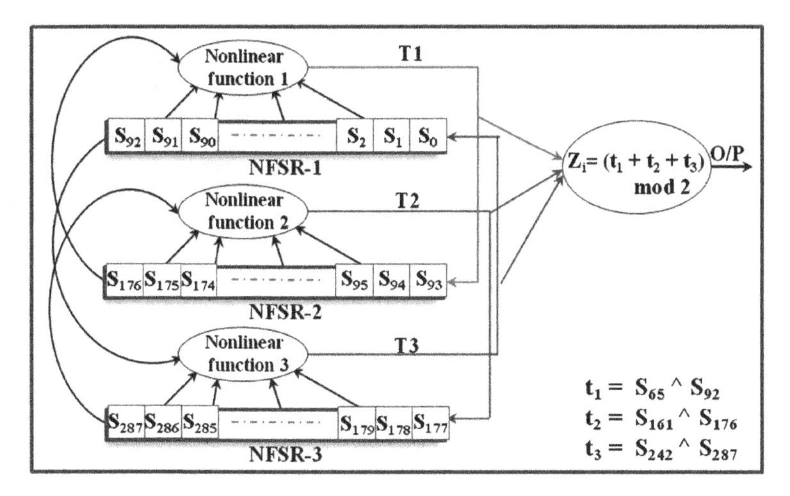

Fig. 1 Trivium architecture

Initialization of Trivium

$$
\begin{array}{lll}
\text{NFSR1} : (S_{92}, \ldots, S_0) & \leftarrow (0, \ldots, 0, K_{79}, \ldots, K_0) & \ldots\text{length} \; - 93 \\
\text{NFSR2} : (S_{176}, \ldots, S_{93}) & \leftarrow (0, 0, 0, 0, \text{IV}_{79}, \ldots, \text{IV}_0) & \ldots\text{length} \; - 84 \\
\text{NFSR3} : (S_{287}, \ldots, S_{177}) & \leftarrow (1, 1, 1, 0, \ldots, 0) & \ldots\text{length} \; - 111
\end{array}
\tag{1}
$$

Nonlinear Function Calculation

$$
\begin{array}{l}
\text{NF-1} : T_1 \leftarrow S_{65} \wedge S_{92} \wedge (S_{90} \,\&\, S_{91}) \wedge S_{170} \\
\text{NF-2} : T_2 161 \wedge S_{176} \wedge (S_{174} \,\&\, S_{175}) \wedge S_{263} \\
\text{NF-3} : T_3 \leftarrow S_{242} \wedge S_{287} \wedge (S_{285} \,\&\, S_{286}) \wedge S_{68}
\end{array}
\tag{2}
$$

After 4 × 288 clock cycle the key stream generation will start:

$$
t_1 = S_{65} \wedge S_{92}; \quad t_2 = S_{161} \wedge S_{176}; \quad t_3 = S_{242} \wedge S_{287}; \quad Z_i = t_1 \wedge t_2 \wedge t_3 \tag{3}
$$

Algorithm was implemented on Xilinx FPGA-Spartan 3E XC3S1600E-4FG320C platform using Verilog. Due to the simplistic nature of the algorithm there is less possibility of different professional implementations for FPGA-based platform (except parallel implementation and countermeasure incorporation). Therefore the attack methodology will remain the same.

Figure 2a, b represents the setup for the acquisition of power traces. Figure 2a shows the detail of signal flow as well as the point of collection of power traces on board while Fig. 2b represents the pictorial view of lab setup. Basically, power traces consist of all instantaneous power consumed by the device during the execution of crypto algorithm and it can be measured with the help of current probe or

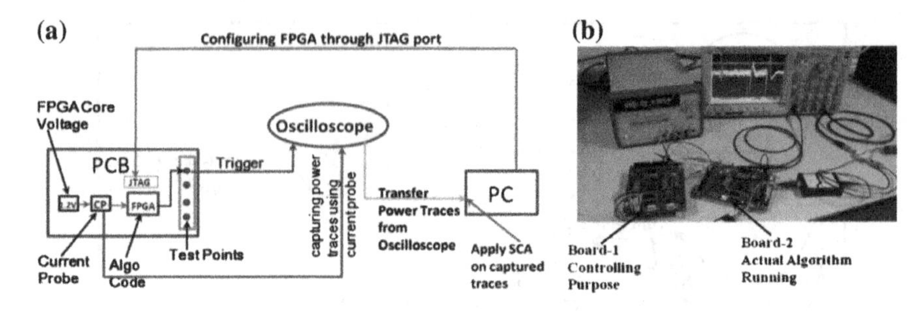

Fig. 2 **a** Signal flow diagram. **b** Lab setup

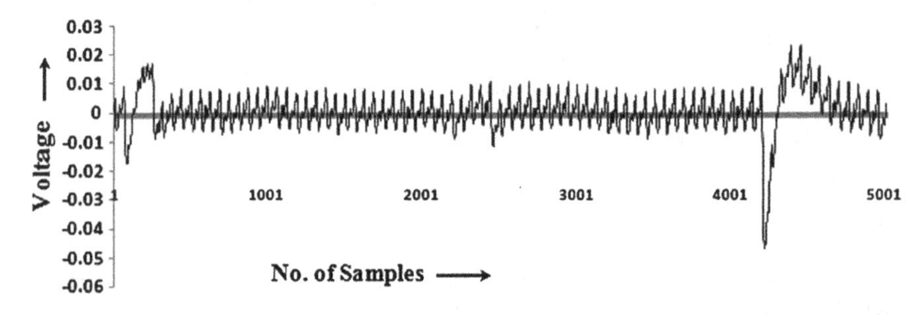

Fig. 3 Power trace of Trivium

differential voltage probe. Two boards are used for generation of control signals and execution of actual algorithm. The power traces are captured during the execution of algorithm inside FPGA at 4 MHz. Figure 3 shows the instantaneous power consumption during execution of the algorithm for random key and IV. Collection of power trace approach is mentioned below:

i. Generation of random IVs in FPGA using RNG: In Trivium, the initialization vector (IV_0–IV_{79}) changes at every resynchronization and key bits K_0–K_{79} are kept constant. So keeping this fact into mind, only the contents of NFSR-2 will change at every resynchronization. Change of IV should be random to avoid other mathematical attacks. To generate random IVs we have implemented 20 stages LFSR with GF2 primitive polynomial $x^{20} + x^{17} + x^9 + x^7 + 1$ as a feedback function.
ii. Reinitialized the algorithm with the same key and random IVs at every 1.5 s and captured power traces by generating trigger signal. Duration of 1.5 s has been taken to ensure that generated power traces are stored properly.
iii. Oscilloscope settings for capturing of power traces are sampling rate—250 MS/s; total number of power traces—5000; number of sample pt/trace—5100; used 20 MHz inbuilt oscilloscope filter;

3 Attack Model and Analysis

FPGA is a CMOS-based device. Whenever the device changes its states, it consumes some amount of power. Power consumption consists of the following two parts:

(a) *Static power consumption*: Static power is required to maintain the CMOS state intact and it is very less.
(b) *Dynamic power consumption*: During the switching, i.e., CMOS state changes from 1→0 or 0→1, FPGA consumes power, and estimation of power consumption is given as

$$\text{Power} = {V_{DD}}^2 * C * f * P_{0\leftrightarrow 1}$$

C Capacitance of device
F Frequency of operation
$P_{0\leftrightarrow 1}$ Number of CMOS state changing from 1→0 or 0→1
V_{DD} Supply voltage

Dynamic power consumption plays an important role in the building of Attack Model and it is large compared to static power consumption.

In case of Trivium the dynamic power consumption occurs at the edge of clock (state of all NFSR's changes with the clock), so total power consumption is

$$P_{total} = P_{NFSR1} + P_{NFSR2} + P_{NFSR3} + \Omega$$

P_{NFSR1}, P_{NFSR2} and P_{NFSR3} are the power consumption of NFRS1, NFSR2, and NFSR3 respectively. Ω is noise content and P_{total} is the total power consumption. Proposed DPA attack framework for Trivium consists of two important steps. First, we model power consumption in such a way that it involves secret parameters to be determined. Once this has been done a suitable efficient methodology is worked out for extraction of secret parameters. Detailed description is as follows.

3.1 Attack Model

Model is based on correlation of power traces $t_{d,j}$ and corresponding hamming distances $h_{d,i}$ of Trivium NFSR's previous and present states. NFSR's stages are changed with respect to the rising edge of clock and feedback is calculated on the basis of nonlinear functions (T_1, T_2 and T_3). *Previous state*: State of NFSRs at 't' clock cycle. *Present states*: States of NFSRs at 't + 1' clock cycle. Analysis is carried out using correlation between hamming distance and power consumed by device. Pearson's coefficient $r_{i,j}$ (where h_i is vector of hamming distances corresponding to different IVs when 'i' is assumed key bit's value and x_j is vector of jth points on all power traces) can be defined as

Given $x_d = \left(x_{d,1}, x_{d,2}, x_{d,3,\ldots}, x_{d,5100}\right)$ where $d = 1$ to N (dth power trace)

$$r_{i,j} = \frac{\sum_{d=1}^{N}\left(h_{d,i} - \bar{H}_i\right)\cdot\left(x_{d,j} - \bar{x}_j\right)}{\sqrt{\sum_{d=1}^{N}\left(h_{d,i} - \bar{H}_i\right)^2 \cdot \sum_{d=1}^{N}\left(x_{d,j} - \bar{X}_j\right)^2}} \tag{4}$$

N Number of power traces (5000 used in our experimentation)

i Assumed key bit/Eq. value

j Position of sample pt within a power trace

$h_{d,i}$ Hamming distance between previous and present states of NFSRs for dth trace as per model for giver 'i'

$\bar{H}_i$ $\sum_{d=1}^{N} h_{d,i}/N$ (Mean of Hamming distances for given value of 'i')

$X_{d,j}$ Value of dth Power trace at jth position

$\bar{X}_j$ $\sum_{d=1}^{N} x_{d,j}/N$ (Mean value of all power trace at jth position)

$r_i = \left(r_{i,1}, r_{i,2}, r_{i,3,\ldots}, r_{i,5100}\right)$ where i = 1 or 2 bit value (*correlation trace*)

Case 1 If 'i' is 1 bit value ('0' or '1'), two possible combinations of correlation traces are required to be computed, i.e., r_0 and r_1.
Case 2 If 'i' is 2 bit value ('00', '01', '10' or '11'), four possible combinations of correlation traces are required to be computed r_0, r_1, r_2 and r_3.

During the execution of Trivium the power consumption mainly depends on the changes in the NFSRs stages. In the whole experimentation we intend to focus on 1 or 2 specific stages among 288 stages of NFSRs involving targeted key bits. The calculation of correlation between total power consumption and the changes in fewer stages (assumed 1 or 2 stages and known stages) is revealing less information, because total power consumption is due to the change in all sequential elements of NFSRs. This increases the number of power traces required for extraction of correct key. It can be verified with Figs. 4a and 5a that correlation traces based on all assumed values are similar to each other. By using the difference of correlation trace technique we can remove the biasness of existing traces and compute the correct values among the assumed values.

3.2 Attack Methodology

In Trivium algorithm, NFSRs to be considered or not for DPA is decided by the presence of IV contents in NFSR's stages. During the execution of Trivium algorithm, the stages of NFSRs may contain information about constant bits, IV bits,

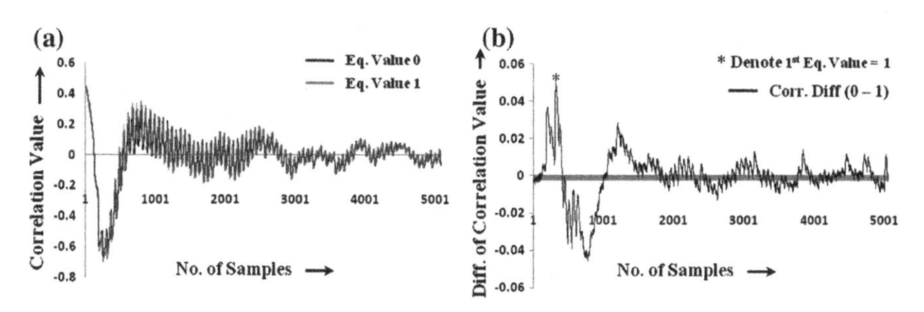

Fig. 4 **a** Correlation traces for Eq. 1 (assumed Eq. value '0' and '1'). **b** Difference of correlation traces for Eq. 1. The value for Eq. 1 is '1'

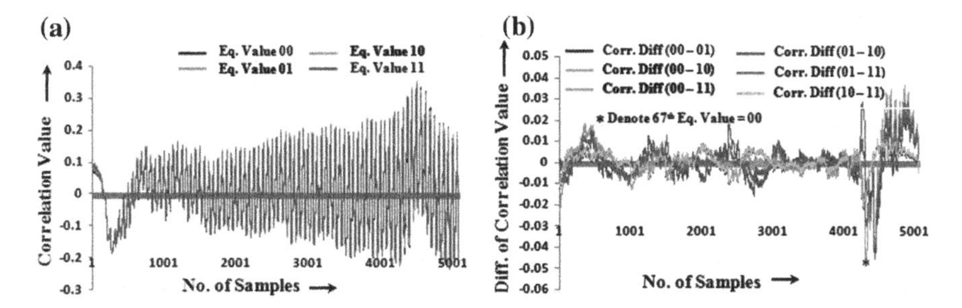

Fig. 5 Correlation traces for Eq. 67 (assumed Eq. values '00', '01', '10' and '11'). **b** Difference of correlation traces for Eq. 67 and the value for Eq. 67 is '00'

key bits, or combination of IV & key bits. But the stages containing only constant bits or key bits cannot be used to mount the attack because interaction of key bits with constant data bits will remain the same over different initializations. If the content of NFSRs stage is a function of key bit/bits and IV bit/bits, then it can leak the side channel information which can be used for mounting attack. This is the basic philosophy behind attack strategy. In our proposed attack determination of key bits is divided into three steps:

Step 1—Extraction of single key bit

Power consumption during initial 12 clock cycles (after initialization of algorithm) leaks information suitable for determining K_{66}–K_{55} bits sequentially. After the initialization of three NFSRs with key and IV the nonlinear functions are:

$$T_1 = K_{66} \wedge S_{92} \wedge (S_{91} \,\&\, S_{90}) \wedge \mathrm{IV}_{78} \quad = K_{66} \wedge \mathrm{IV}_{78} \quad \text{(combination of key, const and IV bits)}$$
$$T_2 = \mathrm{IV}_{69} \wedge S_{176} \wedge (S_{174} \,\&\, S_{175}) \wedge S_{263} \quad = IV_{69} \quad \text{(combination of IV and constant bits)}$$
$$T_3 = S_{242} \wedge S_{287} \wedge (S_{285} \,\&\, S_{286}) \wedge K_{69} \quad = K_{69} \quad \text{(combination of constant bits and key bit)}$$

At first clock states of the NFSRs are changed in the following manner:

NFSR1 :	$(S_{92}, \ldots, S_0)$	$\leftarrow$	$(S_{91}, \ldots, S_0, T_3)$	(all the stages are const and unknown)
NFSR2 :	$(S_{176}, \ldots, S_{93})$	$\leftarrow$	$(S_{175}, \ldots, S_{93}, T_1)$	(all the stage's contents are known)
NFSR3 :	$(S_{287}, \ldots, S_{177})$	$\leftarrow$	$(S_{286}, \ldots, S_{177}, T_2)$	(all the stage's contents are known)

Hamming distances for three shift registers will be given by

$$\{S_{92}(t) \wedge S_{92}(t-1), S_{91}(t) \wedge S_{91}(t-1), \ldots, S_0(t-1) \wedge T_3\} \quad \ldots \text{for NFRS1}$$
$$\{S_{176}(t) \wedge S_{176}(t-1), S_{175}(t) \wedge S_{175}(t-1), \ldots, S_{93}(t-1) \wedge T_1\} \quad \ldots \text{for NFRS2}$$
$$\{S_{287}(t) \wedge S_{287}(t-1), S_{286}(t) \wedge S_{286}(t-1), \ldots, S_{177}(t-1) \wedge T_2\} \quad \ldots \text{for NFRS3}$$

$S_k(t-1)$ represents the value of kth stage at $t-1$ time instance and $S_k(t)$ will denote the value of the same stage of NFSR after one clock cycle. As we can see from the expression of T_3 it is the function of key bit (K_{69}) only. Hence, it is not affected by the change of IV at every reinitialization of Trivium. Its value remains same at every reinitialization for the same key and is fed to NFSR1. It shows that all the states of NFSR1 are constant and are not used to mount the attack. But T_1 and T_2 values keep changing along with the change in IV. So, only NFSR2 and NFSR3 are used to mount DPA. There is only single key bit (K_{66}) involved in computation of T_1. Therefore, corresponding to two different assumed values of K_{66}, i.e., 0 and 1, the value of T_1 will be given by '0^IV_{78}' & '1^IV_{78}'. The values of Pearson's correlation coefficient corresponding to both assumed values are used to determine actual key bit value. The process of computation of value of key bit is mentioned in Algorithm 1.

Algorithm1 : Retrieval of key bits information (Initial 12 bits)	
1. **For** :N= 1 to12	*(N is no. of clock cycles information are required to be solved)*
2. **For** i := 0 to 1	*(based on 2 bit assumed value correspond to T_1 and T_3)*
3. **For** k:=0 to j-1	*(j = total sample points (5100) per trace)*
4. Compute $r_{i,k}$	
5. **End**	
6. **End**	
7. **For** k:=0 to j-1 **Compute**	
8. Absolute Peak or max value **P** in all $r_{0,k} - r_{1,k}$ **and** Sign of Peak val **PS** (+ve or –ve)	
9. **End**	
10. **If : PS** == -ve **then** Key_Bit_D = 0	
11. **Else or PS** == -ve **then** Key_Bit_D = 1	
12. **Use the calculated key bit value for next bit computation**	
13. **End**	

Correlation curves for assumed equation values ('0' and '1') are shown in Fig. 4a. These curves have been drawn using 5000 power traces based on Pearson's correlation coefficient. As we can see in Fig. 4a both correlation traces are similar to each other. During the experimentation it was observed that the peaks in correlation curves were always in negative direction and its position remained the same even when attack was mounted for initial 12 key bits, which indicates that the conclusion about actual key bit is not correct. In our technique, we cancel out the common portion and trying to figure out the instance where difference is maximum in the correlation traces. The instances of maximum difference should be shifted toward the right side on the X-axis because the key bits are determined sequentially. Figure 4b shows the difference of correlation curves ($r_0 - r_1$). The direction of the peak value defines the value of key bit. If the peak is above X-axis, it implies that key bit is '1' otherwise key bit is '0'. This strategy is used to determine 12 key bits, i.e., from K_{64} to K_{53} at different clock cycle. This is done sequentially using previously computed key bits.

Step 2—Extraction of equation value (consists of multi key bits)
After completion of 12th clock cycle nonlinear functions are:

$$
\begin{aligned}
T_1 &= K_{54} \wedge S_{80} \wedge (K_{79} \,\&\, K_{80}) \wedge \mathrm{IV}_{66} && = K_{54} \wedge (K_{79} \,\&\, K_{80}) \wedge \mathrm{IV}_{66} \\
T_2 &= \mathrm{IV}_{57} \wedge \mathrm{IV}_{72} \wedge (\mathrm{IV}_{70} \,\&\, \mathrm{IV}_{71}) \wedge S_{251} && = \mathrm{IV}_{57} \wedge \mathrm{IV}_{72} \wedge (E_{70} \,\&\, \mathrm{IV}_{71}) \\
T_3 &= S_{230} \wedge S_{275} \wedge (S_{273} \,\&\, S_{274}) \wedge K_{57} && = K_{57}
\end{aligned}
$$

Power consumption during 13th to 66th clock cycles (after initialization of algorithm) leaks information about the combination of key bits. Therefore, our methodology for extraction of key bits needs to be changed. It is evident from the expression of T_3 (the feedback for NFSR1) is independent of IV during these clock cycles as shown in Step1. Whereas both T_1 & T_2 are IV dependent, only NFSR2 and NFSR3 are used to mount DPA. As T_1 is a combination of multiple key bits, constant bit and IV bit, therefore, our attack proceeds by assuming the values corresponding to the combination of key bits.

$$
\begin{aligned}
&\text{Case-1 } K_{54} \wedge (K_{79} \,\&\, K_{80}) = \text{“0”} \\
&\text{Case-2 } K_{54} \wedge (K_{79} \,\&\, K_{80}) = \text{“1”}
\end{aligned}
\tag{5}
$$

Therefore, T_1 will be computed as '0 ^ IV_{66}' and '1^ IV_{66}' and corresponding correlation trace for both assumed values r_0 and r_1. The process of computation of equation values in the form of 1's and 0's is mentioned in Algorithm 2.

Algorithm2 : Retrieval of 54 equations information (from 13th to 66th)	
1. **For** :N= 13 to 66	*(N is no. of clock cycles information are required to be solved)*
2. **For** i := 0 to 1	*(based on 2 bit assumed value correspond to T_1 and T_3)*
3. **For** k:=0 to j-1	*(j = total sample points (5100) per trace)*
4. Compute $r_{i,k}$	
5. **End**	
6. **End**	
7. **For** k:=0 to j-1 **Compute**	
8. Absolute Peak or max value **P** in all $r_{0,k} - r_{1,k}$ **and** Sign of Peak val **PS** (+ve or –ve)	
9. **End**	
10. **If : PS** == -ve **then** Eq_Value = 0	
11. **Else or PS** == -ve **then** Eq_Value = 1	
12. **Use the calculated Eq. value for next Eq. computation**	
13. **End**	

The information about combination of key bits for clock cycle 13th to 66th can be easily retrieved in a similar manner, as shown in Fig. 4a, b.

Step 3—*Extraction of* 2 *bits*
After completion of 66th clock cycle nonlinear functions are:

$$\begin{aligned} T_1 &= K_{69} \wedge K_{27} \wedge (K_{25} \,\&\, K_{26}) \wedge \mathrm{IV}_{12} && = K_{69} \wedge K_{27} \wedge (K_{25} \,\&\, K_{26}) \wedge \mathrm{IV}_{12} \\ T_2 &= \mathrm{IV}_3 \wedge \mathrm{IV}_{18} \wedge (\mathrm{IV}_{16} \,\&\, \mathrm{IV}_{17}) \wedge S_{197} && = \mathrm{IV}_3 \wedge \mathrm{IV}_{18} \wedge (\mathrm{IV}_{16} \,\&\, \mathrm{IV}_{17}) \\ T_3 &= \mathrm{IV}_{69} \wedge S_{221} \wedge (S_{219} \,\&\, S_{220}) \wedge K_3 && = \mathrm{IV}_{69} \wedge K_3 \end{aligned}$$

The structure of all three nonlinear functions evidence that power consumption during 67th–78th clock cycles (after initialization of algorithm) leaks two types of information, combination of key bit value in an equation form and single key bit value. We can observe that all nonlinear functions are dependent on IVs. Therefore, all three NFSRs (NFSR2 and NFSR3 completely and NFSR1 partially) are used in the calculation of hamming distance. T_3 contains single bit of key, so as to give 1 bit of key information. T_1 is the combination of key bits so it gives the multiple key bits information in equation form. Four combinations are possible:

$$\begin{array}{llcll} \text{Case-1} & \text{“}K_{69} \wedge K_{27} \wedge (K_{25} \,\&\, K_{26})\text{”} & \text{and} & \text{“}K_3\text{”} = & \text{“00”} \\ \text{Case-2} & \text{“}K_{69} \wedge K_{27} \wedge (K_{25} \,\&\, K_{26})\text{”} & \text{and} & \text{“}K_3\text{”} = & \text{“01”} \\ \text{Case-3} & \text{“}K_{69} \wedge K_{27} \wedge (K_{25} \,\&\, K_{26})\text{”} & \text{and} & \text{“}K_3\text{”} = & \text{“10”} \\ \text{Case-4} & \text{“}K_{69} \wedge K_{27} \wedge (K_{25} \,\&\, K_{26})\text{”} & \text{and} & \text{“}K_3\text{”} = & \text{“11”} \end{array} \quad (6)$$

T_3 is used in the calculation of hamming distance and the previous $T3$ value is also required for this purpose. The previous T_3 (after 65th clock cycle) is

$$T_3 = \underbrace{S_{117} \wedge S_{222} \wedge (S_{220} \,\&\, S_{221})}_{\text{‘0’}} \wedge \underbrace{K_4}_{\text{assume ‘0’ or ‘1’}} \quad (7)$$

After 66th clock cycle, K_4 is the first bit of the NFSR1 and is either assumed to be ‘0’ or ‘1’. In the next clock NFSR1 value becomes $\mathrm{IV}_{69} \wedge K_3$. At that instance, the hamming distance between the previous state and the present state of NFSR2, NFSR3, and first bit of NFSR1 is computed by assuming all four combinations of “$K_{69} \wedge K_{27} \wedge (K_{25} \,\&\, K_{26})$” and “$K_3$”. The correlation traces values are computed for all four combinations (r_0, r_1, r_2 and r_3) using power traces. Multiple key bits information in equation form and 1 bit key can be computed using Algorithm 3.

Algorithm3 : Retrieval of equations information (from 67 to 78) and validation of newly retrieve key bits with the retrieve key bit in Algorithm1

```
1.  Assume K4 = 0
2.  For N:= 67 to 78      (N is no. of clock cycles information are required to be solved)
3.    For i := 0 to 3     (based on 2 bit assumed value correspond to T1 and T3)
4.      For k:=0 to j-1   (j = total sample points (5100) per trace)
5.        Compute r_i,k
6.      End
7.    End
8.    For k:=0 to j-1 Compute
9.      Peak or max value (P1) in all r_0,k – r_1,k and Sign of Peak value (PS1) +ve or –ve
10.     Peak or max value (P2) in all r_0,k – r_2,k and Sign of Peak value (PS2) +ve or –ve
11.     Peak or max value (P3) in all r_0,k – r_3,k and Sign of Peak value (PS3) +ve or –ve
12.     Peak or max value (P4) in all r_1,k – r_2,k and Sign of Peak value (PS4) +ve or –ve
13.     Peak or max value (P5) in all r_1,k – r_3,k and Sign of Peak value (PS5) +ve or –ve
14.     Peak or max value (P6) in all r_2,k – r_3,k and Sign of Peak value (PS6) +ve or –ve
15.   End
16.   Compute absolute value of max(Pn) among P1 to P6 and corresponding PSn
                                                (where n may any value 1 to 6)
17.   If : n==1 then
18.     If : PSn == -ve then   Key_Bit = 0 and Eq_Value = 0
19.     Else   Key_Bit = 1 and Eq_Value = 0
20.   End
21.   If : n==2 then
22.     If : PSn == -ve then   Key_Bit = 0 and Eq_Value = 0
23.     Else   Key_Bit = 0 and Eq_Value = 1
24.   End
25.   If : n==3 then
26.     If : PSn == -ve then   Key_Bit = 0 and Eq_Value = 0
27.     Else   Key_Bit = 1 and Eq_Value = 1
28.   End
29.   If : n==4 then
30.     If : PSn == -ve then   Key_Bit = 1 and Eq_Value = 0
31.     Else   Key_Bit = 0 and Eq_Value = 1
32.   End
33.   If : n==5 then
34.     If : PSn == -ve then   Key_Bit = 1 and Eq_Value = 0
35.     Else   Key_Bit = 1 and Eq_Value = 1
36.   End
37.   If : n==6 then
38.     If : PSn == -ve then   Key_Bit = 0 and Eq_Value = 1
39.     Else   Key_Bit = 1 and Eq_Value = 1
40.   End
```

All four correlation curves of assumed 2 bits equation values ('00', '01', '10' and '11') are shown in Fig. 5a. The differences of correlation curves for all combinations of Eq. 67 are shown in Fig. 5b. and the direction of the peak value also defines the 2 bit equation values. The curve with the absolute peak value among the entire curves represents the 2-bit equation values. For example, the absolute peak value occurred at curve no. 3 ("Corr. Diff (00–11)"), which is the difference of correlation traces when assumed equation values are '00' and '11'. In this case the peak is below 0 level, which means the correct key value is '00'. In the same way, the equation values for Eqs. 68–78 can be retrieved by using the previously computed equation values to compute the next equation values.

4 Experimental Results

Our experimental results are carried out using all three steps. Each step computes the partial key bits information in equation form. All the 80 bits of key can be computed by solving these equations as shown below:

```
for(i=1;i<13;i++)      KEY_Bits[67-i] = Key_Bit_D[i];                                    //K66 to K55 12 bits
for(i=0;i<10;i++)
  KEY_Bits[39-i] = Eq_Value[28+i]^KEY_Bits[66-i]^(KEY_Bits[65-i]&KEY_Bits[64-i]); //K39 to K30 10 bits
for(i=0;i<8;i++)
  KEY_Bits[12-i] = Eq_Value[55+i]^KEY_Bits[39-i]^(KEY_Bits[38-i]&KEY_Bits[37-i]); //K12 to K5 8 bits
for(i=1;i<4;i++)       KEY_Bits[i]   =  Key_Bit [4-i];                                   //K1 to K3 3 bits
KEY_Bits[69]   =   Key_Bit [4];                                                  //K69
KEY_Bits[68]   =   Key_Bit [5]^1;                                                //K68
KEY_Bits[67]   =   Key_Bit [6]^1;                                                //K67
KEY_Bits[42]   =   Eq_Value[25]^KEY_Bits[69]^(KEY_Bits[68] & KEY_Bits[67]);      //K42
KEY_Bits[41]   =   Eq_Value[26]^KEY_Bits[68]^(KEY_Bits[67] & KEY_Bits[66]);      //K41
KEY_Bits[40]   =   Eq_Value[27]^KEY_Bits[67]^(KEY_Bits[66] & KEY_Bits[65]);      //K40
KEY_Bits[15]   =   Eq_Value[52]^KEY_Bits[42]^(KEY_Bits[41] & KEY_Bits[40]);      //K15
KEY_Bits[14]   =   Eq_Value[53]^KEY_Bits[41]^(KEY_Bits[40] & KEY_Bits[39]);      //K14
KEY_Bits[13]   =   Eq_Value[54]^KEY_Bits[40]^(KEY_Bits[39] & KEY_Bits[38]);      //K13
KEY_Bits[16]   =   Eq_Value[78]^KEY_Bits[58]^(KEY_Bits[14] & KEY_Bits[15]);      //K16
KEY_Bits[17]   =   Eq_Value[77]^KEY_Bits[59]^(KEY_Bits[15] & KEY_Bits[16]);      //K17
KEY_Bits[18]   =   Eq_Value[76]^KEY_Bits[60]^(KEY_Bits[16] & KEY_Bits[17]);      //K18
KEY_Bits[19]   =   Eq_Value[75]^KEY_Bits[61]^(KEY_Bits[17] & KEY_Bits[18]);      //K19
KEY_Bits[20]   =   Eq_Value[74]^KEY_Bits[62]^(KEY_Bits[18] & KEY_Bits[19]);      //K20
KEY_Bits[21]   =   Eq_Value[73]^KEY_Bits[63]^(KEY_Bits[19] & KEY_Bits[20]);      //K21
KEY_Bits[22]   =   Eq_Value[72]^KEY_Bits[64]^(KEY_Bits[20] & KEY_Bits[21]);      //K22
KEY_Bits[23]   =   Eq_Value[71]^KEY_Bits[65]^(KEY_Bits[21] & KEY_Bits[22]);      //K23
KEY_Bits[24]   =   Eq_Value[70]^KEY_Bits[66]^(KEY_Bits[22] & KEY_Bits[23]);      //K24
KEY_Bits[25]   =   Eq_Value[69]^KEY_Bits[67]^1^(KEY_Bits[23] & KEY_Bits[24]);    //K25
KEY_Bits[26]   =   Eq_Value[68]^KEY_Bits[68]^1^(KEY_Bits[24] & KEY_Bits[25]);    //K26
KEY_Bits[27]   =   Eq_Value[67]^KEY_Bits[69]^(KEY_Bits[25] & KEY_Bits[26]);      //K27
KEY_Bits[28]   =   Eq_Value[66]^KEY_Bits[1]^(KEY_Bits[26] & KEY_Bits[27]);       //K28
KEY_Bits[29]   =   Eq_Value[65]^KEY_Bits[2]^(KEY_Bits[27] & KEY_Bits[28]);       //K29
KEY_Bits[43]   =   Eq_Value[51]^KEY_Bits[16]^(KEY_Bits[41] & KEY_Bits[42]);      //K43
KEY_Bits[44]   =   Eq_Value[50]^KEY_Bits[17]^(KEY_Bits[42] & KEY_Bits[43]);      //K44
KEY_Bits[45]   =   Eq_Value[49]^KEY_Bits[18]^(KEY_Bits[43] & KEY_Bits[44]);      //K45
KEY_Bits[46]   =   Eq_Value[48]^KEY_Bits[19]^(KEY_Bits[44] & KEY_Bits[45]);      //K46
KEY_Bits[47]   =   Eq_Value[47]^KEY_Bits[20]^(KEY_Bits[45] & KEY_Bits[46]);      //K47
KEY_Bits[48]   =   Eq_Value[46]^KEY_Bits[21]^(KEY_Bits[46] & KEY_Bits[47]);      //K48
KEY_Bits[49]   =   Eq_Value[45]^KEY_Bits[22]^(KEY_Bits[47] & KEY_Bits[48]);      //K49
KEY_Bits[50]   =   Eq_Value[44]^KEY_Bits[23]^(KEY_Bits[48] & KEY_Bits[49]);      //K50
KEY_Bits[51]   =   Eq_Value[43]^KEY_Bits[24]^(KEY_Bits[49] & KEY_Bits[50]);      //K51
KEY_Bits[52]   =   Eq_Value[42]^KEY_Bits[25]^(KEY_Bits[50] & KEY_Bits[51]);      //K52
KEY_Bits[53]   =   Eq_Value[41]^KEY_Bits[26]^(KEY_Bits[51] & KEY_Bits[52]);      //K53
KEY_Bits[54]   =   Eq_Value[40]^KEY_Bits[27]^(KEY_Bits[52] & KEY_Bits[53]);      //K54
KEY_Bits[70]   =   Eq_Value[24]^KEY_Bits[43]^(KEY_Bits[68] & KEY_Bits[69]);      //K70
KEY_Bits[71]   =   Eq_Value[23]^KEY_Bits[44]^(KEY_Bits[69] & KEY_Bits[70]);      //K71
KEY_Bits[72]   =   Eq_Value[22]^KEY_Bits[45]^(KEY_Bits[70] & KEY_Bits[71]);      //K72
KEY_Bits[73]   =   Eq_Value[21]^KEY_Bits[46]^(KEY_Bits[71] & KEY_Bits[72]);      //K73
KEY_Bits[74]   =   Eq_Value[20]^KEY_Bits[47]^(KEY_Bits[72] & KEY_Bits[73]);      //K74
KEY_Bits[75]   =   Eq_Value[19]^KEY_Bits[48]^(KEY_Bits[73] & KEY_Bits[74]);      //K75
KEY_Bits[76]   =   Eq_Value[18]^KEY_Bits[49]^(KEY_Bits[74] & KEY_Bits[75]);      //K76
KEY_Bits[77]   =   Eq_Value[17]^KEY_Bits[50]^(KEY_Bits[75] & KEY_Bits[76]);      //K77
KEY_Bits[78]   =   Eq_Value[16]^KEY_Bits[51]^(KEY_Bits[76] & KEY_Bits[77]);      //K78
KEY_Bits[79]   =   Eq_Value[15]^KEY_Bits[52]^(KEY_Bits[77] & KEY_Bits[78]);      //K79
KEY_Bits[80]   =   Eq_Value[14]^KEY_Bits[53]^(KEY_Bits[78] & KEY_Bits[79]);      //K80
KEY_Bits[4]    =   Eq_Value[63]^KEY_Bits[31]^(KEY_Bits[30] & KEY_Bits[29]);      //K4
```

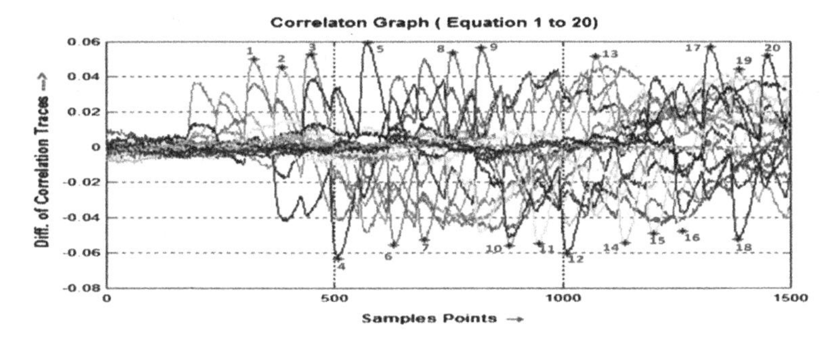

Fig. 6 Difference of correlation traces for initial 20 equations

Figure 6 represents the difference of correlation traces (20 in number using different color) corresponding to the correctly retrieved equation values. The peak for initial 20 clock cycles is represented by the '*' sign and show the maximum correlation with the retrieved equations values. Numbering associated with '*' sign indicates the sequencing of computed equation. In Fig. 6, the peaks are shifted toward the right side. It indicates that the equations values are computed one by one with respect to the different clock cycles. Similarly, the plots for all 78 'difference of correlation traces' can be computed. This technique gives more appropriate results in lesser number of power traces if each power trace is divided in such a manner that every divided trace or local trace contains only single clock cycle information. Then apply DPA on local traces to extract the secret information. But this process requires more information about implementation like frequency of operation of algorithm and this is not a generic approach in the real-world scenario.

The experimentations were conducted for several key and IV pairs. All 80 bits of key were retrieved successfully every time in 78 clock cycles by using 5000 power traces, whereas normal DPA technique requires more than 10,000 traces.

5 Conclusion and Future Work

We have presented differential power analysis on Trivium, which is one of the ciphers in final eSTREAM portfolio. A novel technique, the difference of all Pearson's correlation coefficient traces based on the assumed equations values is introduced to find the secret key bits information. It eliminates the noise of the power traces and provides pure side-channel information. Basically, it computes the maximum gap in the difference of correlation traces corresponding to the correct and wrong assumed value. Shifting in peak values corresponds to different correlation traces with time axis as shown in Fig. 6, which indicates that the correct guess values retrieve with the respective clock cycles, which satisfies our attack methodology. Maximum power traces requirement can be varied with the mechanism of measurement of power traces, implementation of algorithm, and the FPGA family (technology of FPGA, i.e., 90, 65 and 28 nm). The number of power trace requirements can be reduced significantly if each power trace is divided in such a manner that every divided trace or local trace contains only single clock cycle information and applies DPA on local traces to extract the secret information. Difference of correlation traces technique can be used in any nonlinear feedback shift resistor (NFSR)-based stream cipher, especially in lightweight cryptosystems. In lightweight cryptosystems the computational phase is less compared to other stream ciphers. Therefore, SCAs are a serious threat for these kinds of cryptosystems and counter measures are mandatory.

In the future work we will focus on the development of countermeasures for this attack. We will also extend the attack to parallel implementation of Trivium, where 32 or 64 bits of sequences are computed in parallel. We will also try to reduce the number of power traces requirement and the number of sample points per trace using statistical analysis techniques like principle component analysis (PCA) and signal to noise ratio (SNR) analysis; soft computing techniques like neural networks (NN), genetic algorithms, and fuzzy logic.

Acknowledgments We thank Mrs. Amita Malik for providing help in the generation of all the plots used in this paper and also Dr. G. Athithan and Mr. Rajesh Pillai for their guidance, motivation, and valuable suggestions during the duration of paper writing.

References

1. Recberger, Ch., Oswald, E.: Stream Ciphers and Side-Channel Analysis. In: SASC 2004—The State of the Art of Stream Ciphers (Brugge, Belgium, October 14–15, 2004), Workshop Record, pp. 320–326 (2004). http://www.ecrypt.eu.org/stvl/sasc/record.html
2. Kocher, P.C., Jaffe, J., Jun, B.: Differential Power Analysis. In: Wiener M.J. (ed.) Advances in Cryptology—CRYPTO'99, Lecture Notes in CS, vol. 1666, pp. 388–397, Springer-Verlag (1999)
3. Bartkewitz, T., Lemke-Rust, K.: Efficient template attack on probabilistic multi-class support vector machines. Springer (2013)
4. Mohamed, M.S.E., Bulygin, B., Buchmann.: In: COSADE 2011 on the Improved Differential Fault Analysis of Trivium, Darmstadt, Germany, 24–25 Feb 2011
5. Fischer, W., Gammel, M., Kniffler, O., Velten J.: Differential Power Analysis of Stream Ciphers. LNCS 4377, pp. 257–270. Springer-Verlag (2007)
6. De Canniere, C., Preneel, B.: Trivium. In: New Stream Cipher Designs: The eSTREAM Finalists, pp. 244–266 (2008)

Investigations of Power and EM Attacks on AES Implemented in FPGA

Arvind Kumar Singh, S.P. Mishra, B.M. Suri and Anu Khosla

Abstract Side-channel attack is a new area of research which exploits the leakages such as power consumption, execution time, EM radiation, etc., of crypto algorithms running on electronic circuitry to extract the secret key. This paper describes the VHDL implementations of Advanced Encryption Standard (AES) algorithm on Field Programmable Gate Array board (Spartan 3E) employing Xilinx tool and discusses briefly about Correlation Power/EM Analysis attacks. These attacks have been mounted on part of power and EM traces corresponding to tenth round of AES algorithm. Power and EM traces are being acquired using current probe and EM probe station respectively with the help of oscilloscope and PC. Effects of different ways of implementations on these attacks have been explored. Studies have been carried out to find the effect of operating frequencies and number of samples per clock on the computational complexities in terms of number of traces required to extract the key.

Keywords Side Channel Attack · Advanced Encryption Standard · Correlation Power Attack · Correlation EM Attack · Field Programmable Gate Array

1 Introduction

Toward secure communication, a lot of efforts have been made to develop various efficient block and stream cipher systems. This has resulted in the evolvement of very robust cipher systems such as Advanced Encryption Standard (AES) algorithm

A.K. Singh (✉) · S.P. Mishra · B.M. Suri · Anu Khosla
SAG, DRDO, Ministry of Defence, Metcalfe House, Civil Lines,
Delhi 110054, India
e-mail: arvindkums@yahoo.com

S.P. Mishra
e-mail: spm201@gmail.com

B.M. Suri
e-mail: bmsuri2008@gmail.com

M. Pant et al. (eds.), *Proceedings of Fifth International Conference on Soft Computing for Problem Solving*, Advances in Intelligent Systems and Computing 437, DOI 10.1007/978-981-10-0451-3_50

[1, 2], which have been proved to be undefeatable by classical cryptanalysis. However, these devices leak information through execution time [3], power consumption [4], electromagnetic (EM) emission, [5] etc. and it is possible to deduce the secret key by analyzing these leakages using statistical techniques. The extraction of secret information by analyzing these leakages is known as Side Channel Attack (SCA).

Timing attack exploits the variation of execution time with secret data. Power and EM analysis attacks are based on data dependent power consumptions and EM radiations of cipher systems. The use of Complementary Metal Oxide Semiconductor (CMOS) [6] technology in most of the modern electronic circuits has become common practice, where a change in data bit value produces short current pulse. The more current pulses are produced when more states are flipping in the circuit, which results in data dependent power consumption and EM radiation and enables adversary to perform analysis on these leakages to deduce secret information. Both Simple Power Analysis (SPA) [4] and Simple EM Analysis (SEMA) techniques directly interpret power consumption and EM radiation signals, respectively, with detailed knowledge of the cryptographic algorithm of the crypto system to yield information about operations as well as the key. Differential Power Analysis (DPA) [4, 7, 8] attack is a more powerful attack than SPA. It uses statistical techniques and requires less detailed knowledge of the implementation of the cryptographic algorithms. DPA [9] attack divides power traces of multiple encryptions into two groups using some bits of the intermediate results involving part of the key. It computes mean of the two groups of traces and subtracts these mean traces. The resulting differential trace contains peaks for the correct guessed key.

Brier et al. [10] proposed a more efficient version of DPA attack known as Correlation Power Analysis (CPA) attack. This estimates the power consumption of intermediate results, involving part of the guessed key, using hypothetical power model of the cryptographic device. Then the correlations between estimated powers and real power consumptions are computed for all guessed keys. Correctly guessed key shows highest correlation factor. Like power attacks, electromagnetic analysis (EMA) attacks [11–13] also use Differential EM Analysis (DEMA) and Correlation EM analysis (CEMA) attack techniques to extract the secret key by exploiting data dependent EM radiations.

SCA is also known as implementation attack. The success of SCA depends on the way of implementing the cryptographic algorithms and the platform on which these algorithms execute. Field Programmable Gate Array (FPGA) technology based reconfigurable computing systems are being widely used as a platform for designing high-performance cryptographic systems due to their high throughput rates and inherent design flexibility. This inspired researchers to find susceptibility of FPGA based crypto systems to power analysis attack [14–16]. Unfortunately, nearly all prior findings on this topic are based on simulated power consumption models; whereas our analysis and results are based on real power consumptions and EM radiations.

The aim of this task is to study the CPA [10, 15, 17] and CEMA against hardware implementations of 128-bit AES algorithm on FPGA platform as FPGA leaks a significant amount of information about its internal computations through the supply lines [14] and EM radiations. Correlation is computed to find the linear relation between estimated and measured power and EM signals of a cryptographic system to determine the key. This paper also brings in picture some observations, while performing correlation analysis using power and EM traces of AES algorithm running on FPGA, on following points:

- Way of different implementations.
- FPGA board operating frequencies and number of samples/clock.
- Requirement of leakages by CPA and CEMA attacks

The remainder of this paper is organized as follows: Section 2 of this paper describes the way of VHDL implementations of AES on FPGA. The concept of CPA/CEMA attacks is given in Sect. 3. Section 4 presents acquisition setup of power and EM signals. Section 5 contains experimental results of CPA and CEMA attacks mounted on the tenth round of AES algorithm. The conclusions and future works are given in Sect. 6.

2 VHDL Implementations of AES

The 128-bit AES algorithm is a round-based symmetric block cipher, whose processing begins with single Add key operation followed by nine identical rounds of computations involving Substitute byte, Shift rows, Mix Column, and Add Round Key operations. Final (tenth) round does not include Mix Column operation.

AES is implemented on FPGA using VHDL with the help of Xilinx tool. First, 10 round keys are generated in 10 clocks. Then encryption starts and each round completes in 1 clock. The VHDL module of AES encryption on FPGA (Spartan 3E) is shown in Fig. 1. This module runs continuously with sufficient delay (1.5 s) between two consecutive encryptions. A trigger signal is being generated at the start of each encryption process and is fed to the oscilloscope input to start acquisition of leakage. Random plaintext is being generated to be used for next encryption during the delay. The encryption process of AES algorithm is implemented in two different ways. The detail descriptions of these two kinds of implementations are given Sects. 2.1 and 2.2 respectively:

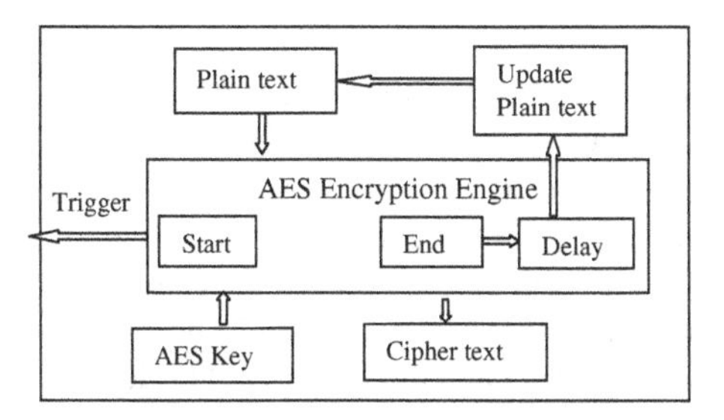

Fig. 1 AES module on FPGA

2.1 Implementation 1: XORing All Plaintext and Key Bits at a Time

The encryption starts with XORing all plaintext and key bits (128 bit) at a time. Then byte-wise substitution is performed. After this, Shift rows operation is performed and bytes are copied in an array considering the Shift row operation. Then the four words (32 bits) are formed from this new array. These words are used in Mix Column operation. Final operation of a round, Add Round Key, is accomplished by XORing 128-bit round key with 128-bit array resulting from concatenation of output bytes of mix column operation. The last (tenth) round of AES algorithm does not include Mix Column operation. Here, the output array bytes of Shift rows operation are concatenated to from 128-bit array, which is being XORed with the 128 bits tenth round key at a time. This implementation is robust to the CPA attack as only few bytes of the tenth round key are being extracted even using very large number of power traces acquired at highest sampling rate of the oscilloscope.

2.2 Implementation 2: Byte-Wise XORing of Plaintext and Key

Here, encryption process starts with byte-wise XORing key with plaintext followed by Substitute byte operation. After the substitute byte operation, the Shift rows operation was performed implicitly (not copying the bytes in another array). The four words (32 bits) are formed from the array of Substitute byte operation, by copying the bytes with index considering the Shift row operation, to be used by Mix Column operation. Add Round Key is accomplished by XORing output bytes of Mix Column operation with the corresponding bytes of round key (byte-wise XORing). The last operation of tenth round is XOR between output bytes of Shift rows operation and corresponding bytes of the round key. This implementation is vulnerable to SCA. All 16 bytes of the tenth round key are being extracted by mounting CPA and CEMA attacks.

3 Correlation Power/EM Analysis Attacks

The SCA exploits the basic concept that is the side channel leakages are statistically correlated to operations and data. It is possible to extract information related to the key by measuring and analyzing leakages such as power consumption and EM radiation, obtained during the execution of cryptographic operations. CPA [10, 16]/ CEMA attacks compare measured power/EM traces with estimated power consumptions/EM radiations. The estimation of leakages requires development of an appropriate model. CMOS devices usually use the Hamming distance

(HD) model that relates switching activity in CMOS devices to the leakages. HD model assumes that the leakages are proportional to the number of 0–1 and 1–0 transitions [17]. HD model also considers that the same amount of leakages are produced by both 0–1 and 1–0 transitions. The estimation of leakage (W) using HD model is given below for two consecutive intermediate values (let X_1 and X_2) of an algorithm.

$$W = \mathrm{HD}(X_1, X_2) = \mathrm{HW}(X_1 \oplus X_2) \tag{1}$$

where HW represents Hamming weight, $\oplus$ denotes exclusive-OR operation. Intermediate values X_1 and X_2 are function of part of key (Ks) and plaintext (PT).

For given N plaintexts, the estimated leakages are derived using (1). Corresponding to each plaintext, real power/EM trace, Lt, are measured at different time t. The correlation coefficient (Pearson's correlation coefficient) between leakage W and Lt is calculated using the formula given as

$$\mathrm{Cs} = \frac{E(\mathrm{W.Lt}) - E(\mathrm{W}).\mathrm{E}(\mathrm{Lt})}{\sqrt{\mathrm{var}(W).\mathrm{var}(\mathrm{Lt})}} \tag{2}$$

where W, Lt are N-dimensional vectors, E denotes the average operation, and var denotes variance. When Ks is not the correct key then the corresponding W and Lt has less correlation and thus obtained correlation factor is small; when Ks is the correct key, the corresponding W and Lt has the highest correlation.

4 Acquisition of Power and EM Signals

The measurement of leakages plays very important role in the success of SCA. These signals are captured using appropriate transducers and they are stored in memory after digitization for analysis purposes. Lab setup to acquire power consumption from FPGA board is shown in Fig. 2. It is comprised of FPGA board, current probe, oscilloscope (DPO 7254, 2.5 GHz, 40 GS/s), and PC/Workstation.

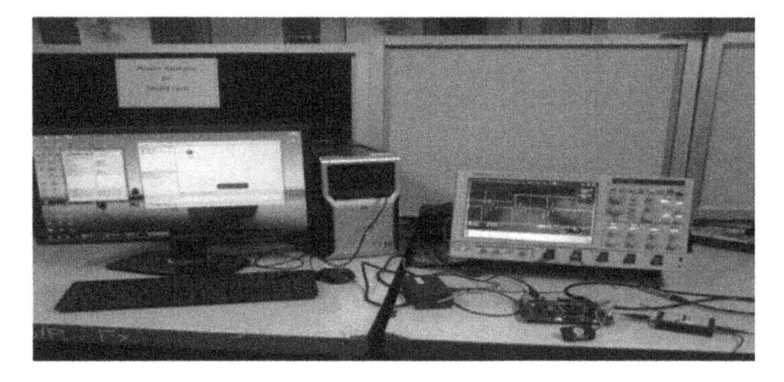

Fig. 2 Lab setup for acquisition of power signal

The FPGA board interacts with PC through JTAG port. Application (bit) file is being ported to the board and executed using Xilinx tool running at PC. The current probe measures the power consumption during the execution of algorithm. This signal is fed to the input of oscilloscope, which acquires it in synchronization with encryption time. Synchronization establishes with the help of a trigger signal generated in FPGA at the start of encryption. Trigger signal is fed to another input of oscilloscope. Oscilloscope acquires data, which are copied to the PC using Tektronix make oscope utility for further analysis. Delay (1.5 s) is inserted between two consecutive encryptions to safely transfer data from oscilloscope to PC. This avoids the overwriting of oscilloscope memory with next encryption data.

However, a trace of multiple encryptions of AES (running continuously) can be acquired and can be splitted into number of traces corresponding to each encryption process. The knowledge of FPGA board operating frequency and sampling rate of oscilloscope makes this task easy. The existence of correlation of plaintext and cipher text with part of power traces corresponding to beginning and end of encryption, respectively, further simplifies this work.

Plot of the power traces of AES executing at different FPGA board operating frequencies are shown in Fig. 3. This figure shows that power trace contains clear patterns of each round of processing at low operating frequency but overlapping between consecutive clock signals increases with the increase of operating frequencies, which increase the requirement of traces to extract the key.

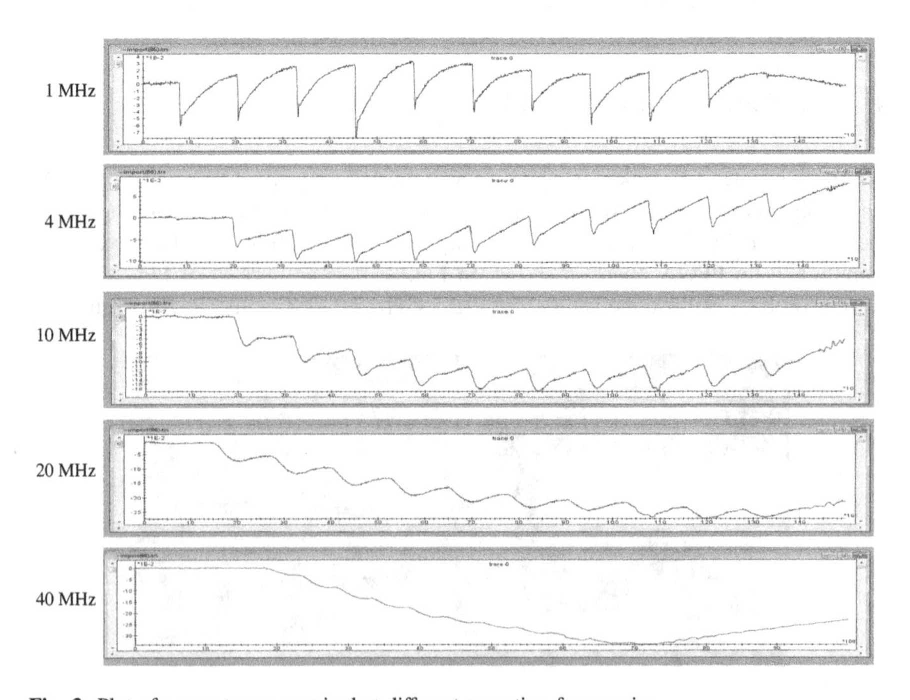

Fig. 3 Plot of power traces acquired at different operating frequencies

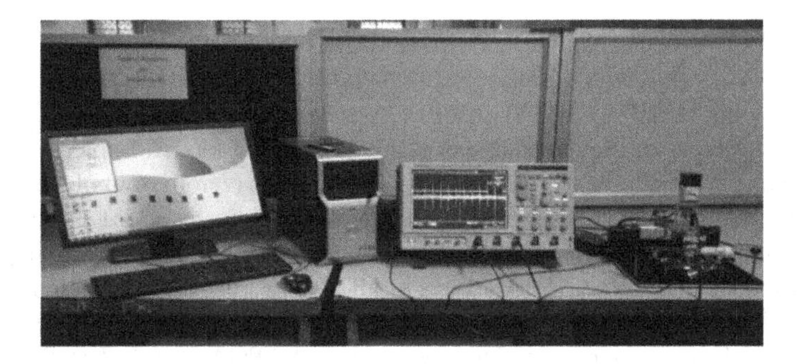

Fig. 4 Lab setup for acquisition of EM signal

The lab setup to acquire EM signal is shown in Fig. 4. It is comprised of all equipments as shown in Fig. 2 except the current probe. In place of current probe, EM probe is mounted on EM probe station to acquire the EM signal. The probe is fixed at the top of the FPGA chip where EM signal is distinct for AES processing. The acquiring process is the same as described above.

Plot of the EM traces of AES executing at different FPGA board operating frequencies are shown in the Fig. 5. This depicts that radiation quality improves with the increase of operating frequency and as a result distinct patterns of each round processing are clearly visible. This is the reason why requirement of traces decreases with the increase of operating frequency to extract the key. Contrary to this, CPA attack requires more traces with increase of operating frequency.

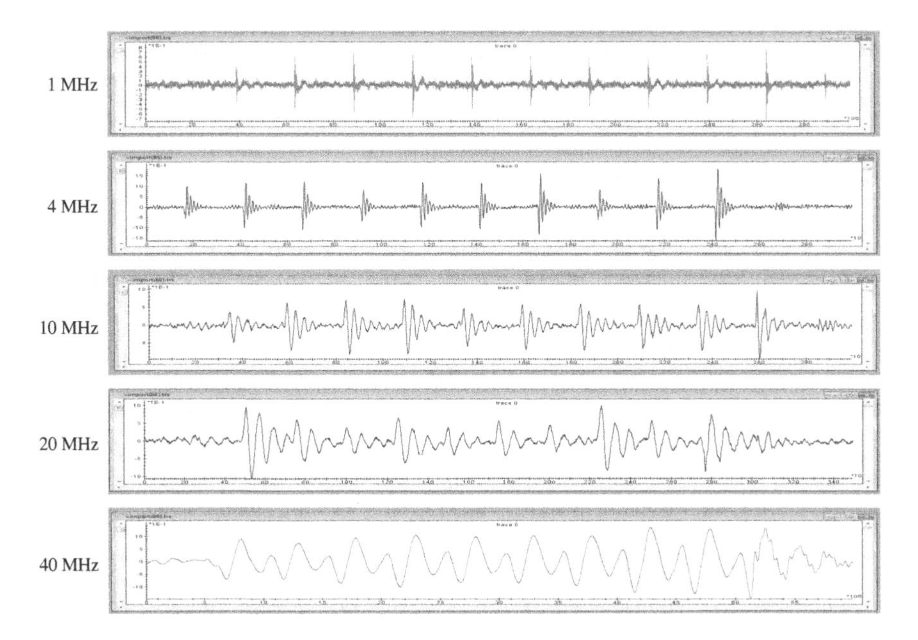

Fig. 5 Plot of EM traces acquired at different operating frequencies

5 Experimental Results

Correlation-based (CPA/CEMA) attacks [15] were mounted on the tenth round of two different types of implementations of AES runing on FPGA. Implementation 1 was found less vulnerable to CPA attack as only 4 bytes (zeroth row) of tenth round key were extracted even using very large number of traces (25 K) acquired at highest sampling rate of oscilloscope. Implementation 2 (byte-wise processing) was found vulnerable to CPA/CEMA attacks as all 16 bytes of AES key were retrieved by mounting these attacks. Experiments were extended to see the effect of sampling rates of oscilloscope and operating frequencies of FPGA board. The Results of power and EM attacks mounted on implementation 2 of AES are given in Sects. 5.1 and 5.2 respectively.

5.1 Results of Power Attack

CPA attack on AES retrieves all 16 bytes of key by considering guessed key byte with highest correlation value as the actual key byte. Filtering using moving average technique reduces the number of traces required. Table 1 contains the

Table 1 Details of CPA mounted on AES

Key bytes	Determination of key bytes using 400 power traces				Determination of key bytes using 600 power traces			
	Rank 1 byte		Rank 2 byte		Rank 1 byte		Rank 2 byte	
	Value	Corr.	Value	Corr.	Value	Corr.	Value	Corr.
0 × A6	0 × A6	0.2861	0 × F3	0.1896	0 × A6	0.2660	0 × 83	0.1541
×0C	×0C	0.2541	0 × 92	0.1770	×0C	0.2276	0 × 19	0.1547
0 × 63	0 × 63	0.2404	0 × 11	0.2067	0 × 63	0.2167	0 × 95	0.1544
0 × B6	0 × 15	0.2036	0 × 90	0.1854	0 × B6	0.1957	0 × 15	0.1556
0 × E1	0 × E1	0.2674	0 × A5	0.1843	0 × E1	0.2144	0 × AC	0.1493
0 × 3F	0 × 3F	0.2838	0 × D6	0.1996	0 × 3F	0.2630	0 × D6	0.1621
0 × 0C	0 × 91	0.2189	0 × 0C	0.2180	0 × 0C	0.2218	×91	0.1719
0 × C8	0 × C8	0.2469	0 × 2B	0.1802	0 × C8	0.2497	0 × 74	0.1473
0 × 89	0 × 89	0.2977	0 × 06	0.1745	0 × 89	0.3039	0 × 0A	0.1492
0 × 25	0 × B5	0.2095	0 × 7D	0.1934	0 × 25	0.1651	0 × B5	0.1582
0 × EE	0 × EE	0.1994	0 × 1D	0.1869	0 × EE	0.2141	0 × BA	0.1616
0 × C9	0 × C9	0.2125	0 × AA	0.1697	0 × C9	0.1797	0 × 5F	0.1429
0 × A8	0 × A8	0.2062	0 × 8E	0.1817	0 × A8	0.2219	0 × FF	0.1577
0 × F9	0 × F9	0.2356	0 × 2F	0.2022	0 × F9	0.2705	0 × FF	0.1670
0 × 14	0 × 14	0.2229	0 × 9C	0.2086	0 × 14	0.2070	0 × 38	0.1553
0 × D0	0 × D7	0.1834	0 × E1	0.1775	0 × D0	0.1567	0 × E1	0.1528
	No. of correct bytes = 12				No. of correct bytes = 16			

Table 2 Effect of sampling rate and operating frequency on success of attack

S. no.	Effect of sampling rate (exp. 1)			Effect of operating frequency (exp. 2)		
	Sampling rate (MS)	Samples per clock	No. of traces	Operating frequency (MHz)	Sampling rate	No. of traces
1	1000	1000	600	1	125 MS	650
2	100	100	600	4	500 MS	1900
3	50	50	600	10	1.25 GS	3450
4	25	25	1500			
5	10	10	1500			
6	5	5	2500			

details of CPA attack on AES when operating frequency was set at 1 MHz and power traces were acquired at 5 GS sampling rate. The tenth round key bytes (Hexadecimal values) along with correlation values are given for different number of power traces. CPA attack extracts correctly 12 and all 16 key bytes with rank 1 candidate using 400 and 600 power traces, respectively. The correlation values associated with key bytes determined as rank 1 candidate are higher than that of rank 2 bytes except few cases. Correlation values of rank 1 and rank 2 bytes are generally competing to each other when key bytes are determined wrongly.

Two experiments were carried out. First experiment was carried out to see the effect of sampling rate (number of samples per clock, i.e., number of samples per round processing) of oscilloscope on the requirement of number of traces to successfully mount the CPA attack and extract all 16 key bytes with rank 1 candidate when board was running at 1 MHz clock frequency. Table 2 contains the experimental details.

From Table 2, it is observed that the optimum number of samples per clock is 50. Acquiring traces with more than optimum number of samples per clock unnecessarily increases the computations. Selecting less number of samples per clock reduces the computations but more number of traces is required to extract the key. This necessitates the availability of cryptosystem to the adversary for longer period of time to run system for more number of encryptions. Therefore, choosing appropriate sampling rate optimizes the complexity of attack.

Experiment 2 was carried out to see the effect of operating frequency of the board on successfully mounting the CPA attack. Here, AES was executed at different operating frequencies and traces were acquired keeping the number of samples (125) per clock constant by varying the sampling rate of the oscilloscope. Table 2 contains the experimental details, which reveals that the requirement of traces and computations increase with the increase of FPGA board operating frequency.

Table 3 Details of CEMA mounted on AES

Key bytes	Determination of key bytes using 6000 EM traces				Determination of key bytes using 8000 EM traces			
	Rank 1 byte		Rank 2 byte		Rank 1 byte		Rank 2 byte	
	Value	Corr.	Value	Corr.	Value	Corr.	Value	Corr.
0 × A6	0 × A6	0.1006	0 × 0F	0.0622	0 × A6	0.0910	0 × 0F	0.0579
×0C	×0C	0.0936	0 × 3F	0.0471	×0C	0.0883	0 × 77	0.0393
0 × 63	0 × 63	0.1827	0 × C6	0.0588	0 × 63	0.1799	0 × C6	0.0507
0 × B6	0 × B6	0.1068	0 × 1F	0.0507	0 × B6	0.1092	0 × 1F	0.0505
0 × E1	0 × C5	0.0457	0 × AF	0.0451	0 × E1	0.0412	0 × 89	0.0381
0 × 3F	0 × 3F	0.0711	0 × 38	0.0442	0 × 3F	0.0790	0 × A9	0.0432
0 × 0C	0 × 0C	0.2403	0 × BC	0.0584	0 × 0C	0.2430	0 × A9	0.0578
0 × C8	0 × 75	0.0423	0 × 8E	0.0420	0 × C8	0.0411	0 × 96	0.0408
0 × 89	0 × 89	0.2226	0 × 2C	0.0720	0 × 89	0.2286	0 × 2C	0.0742
0 × 25	0 × 25	0.0727	0 × 3A	0.0445	0 × 25	0.0732	0 × 8C	0.0406
0 × EE	0 × B9	0.0442	0 × EE	0.0441	0 × EE	0.0535	0 × B9	0.0405
0 × C9	0 × C9	0.0563	0 × AF	0.0455	0 × C9	0.0595	0 × EE	0.0408
0 × A8	0 × 62	0.0475	0 × 6B	0.0431	0 × A8	0.0443	0 × 6B	0.0391
0 × F9	0 × F9	0.1889	0 × 6F	0.0568	0 × F9	0.1940	0 × A8	0.0449
0 × 14	0 × C2	0.0427	0 × 14	0.0421	0 × 14	0.0533	0 × 64	0.0392
0 × D0	0 × D0	0.3363	0 × 39	0.0908	0 × D0	0.3426	0 × 39	0.0903
	No. of correct bytes = 11				No. of correct bytes = 16			

5.2 *Results of EM Attack*

CEMA attack on AES is able to retrieve all 16 bytes of key. Moving average technique reduces the number of EM traces required. EM traces were acquired at 5 GS sampling rate of oscilloscope when FPGA board was running at 1 MHz operating frequency. Table 3 contains the details of CEMA attack.

The tenth round key bytes (Hexadecimal value) along with correlation values are given for different number of EM traces. Here, 11 and all 16 key bytes were determined correctly with rank 1 candidate using 6,000 and 8,000 EM traces respectively. The correlation values associated with key bytes determined as rank 1 candidate are higher than that of rank 2 bytes except few cases. When key bytes are determined wrongly then the correlation values of rank 1 and rank 2 key bytes are found generally competing to each other.

The results of EM attack at different operating frequencies of the board and sampling rates of the oscilloscope are given in Table 4. CEMA attack requires very large number of samples per clock as well as traces to extract the key as compared to the CPA attack. However, these requirements decrease with the increase of FPGA board operating frequency due to improvement in radiation quality with the increase of operation frequency. Success of CEMA attack and requirement of traces also depend on the position and location of the EM probe. EM probe is fixed at the

Table 4 Results of CEMA at different operating frequencies and sampling rates

S. no	Operating frequency (MHz)	Sampling rate (GS)	No. of samples/clock	No. of traces
1	1	5.0	5,000	8,000
2	4	2.5	625	17,500
3	10	2.5	250	7,500

Table 5 Effect of operating frequency on success of attack

S. no.	Operating frequency (MHz)	Sampling rate (GS)	No. of traces
	1	1	11,000
2	4	5	5,400
3	10	10	2,625

top of the FPGA chip, where EM signal contains patterns of AES algorithm processing.

CEMA attack includes two experiments, which were carried out in the case of CPA attack. First experiment was carried out to see the effect of number of samples per clock (Table 2) when operating frequency of the board was set at 1 MHz. Here, CEMA attack could not retrieve all key bytes of AES even using very large number of EM traces (20 K) due to bad radiation quality as illustrated in Fig. 5.

The next experiment was carried out to see the effect of FPGA board operating frequency. In this case, number of samples per clock were kept constant (approximately 1000) by varying the sampling rate of the oscilloscope. The experimental findings are given in Table 5, which shows that the amount of traces decreases with the increase of FPGA operating frequency because of better radiation quality at higher operating frequency. The key bytes were determined by considering first 4 ranks candidates. These experiments prove that CEMA attack requires more number of traces acquired at higher sampling rate than the CPA attack.

6 Conclusions and Future Works

The success of SCA depends on the way of the implementations of cryptographic algorithms on electronic circuitry. AES was implemented in two different ways on FPGA (Spartan 3E) board. Implementation 1 performs XORing all plaintext and key bits at a time while implementation 2 does byte-wise XORing of plaintext and key. CPA attack on implementation 1 of AES was not revealing much secret information even using very large number of traces. The implementation 2 of AES was found vulnerable to CPA and CEMA attacks. All 16 key bytes were extracted by mounting CPA and CEMA attacks on tenth round of AES processing. Our experimental observation reveals that 50 samples per clock is the optimal choice for mounting CPA attack on AES when FPGA board operating frequency is at 1 MHz. In case of EM attack, the correlation values of key bytes with EM traces are considerably less than the correlation values of key bytes with power traces found in power attack. This seems to be the reason why EM attack requires significantly more number of samples per clock and also more number of traces. There is an interesting observation in the case of EM attack. The requirement of traces decreases with the increase of board operating frequency, which is contrary to the power attack. Success of EM attack and requirement of EM traces also depend on positioning of inductive probe and presence of surrounding EM noise signals.

The future works include use of soft computing techniques like Neural Networks [18], Fuggy Logic [19], Genetic Algorithms [20], etc. to improve the efficiency of attacks.

Acknowledgments We are heartily grateful to Dr. G. Athithan, OS and Director, SAG, DRDO, Delhi for his invaluable support, motivation, and informative suggestions. Sincere thanks go to Sh. Devendra Jha, Sc 'F' and SCA team members for their suggestions and help rendered during the execution of this work.

References

1. Advanced Encryption Standard. http://en.wikipedia.org/wiki/Adavanced_Encryption_Standard
2. Stallings, W.: Cryptography and Network Security: Principles and Practice, 3rd edn. Prentice Hall, USA (2003)
3. Kocher, P.C.: Cryptanalysis of Diffie-Hellman, RSA, DSS, and other systems using timing attacks. In: Advances in Cryptology Conference, CRYPTO '95, pp. 171–183 (1995)
4. Kocher, P.C., Jaffe, J., Jun, B.: Differential power analysis. In: Wiener, M., (ed.) CRYPTO 99. LNCS, vol. 1666, pp. 388–397. Springer, Heidelberg (1999)
5. Quisquater, J.-J., Samyde, D.: ElectroMagnetic analysis (EMA): measures and countermeasures for smart cards. In: International Conference on Research in Smart Cards–E-smart 2001. LNCS, vol. 2140, pp. 200–210. Springer, New York (2001)
6. Kang, S.-M., Leblebici, Y.: CMOS Digital Integrated Circuits: Analysis and Design. McGraw Hill, New York (2002)
7. Jaffe, J., Kocher, P.: Introduction to differential power analysis and related attacks. In: Cryptography Research, pp. 1–5 (1998)
8. Mangard, S., Oswald, E., Popp, T.: Power Analysis Attacks. Springer, New York (2007)
9. Han, Y., Zou, X., Liu, Z., Chen Y.: Efficient DPA Attacks on AES Hardware Implementations. Int. J. Commun. Netw. Syst. Sci. **1**(1), 1–103 (2008)
10. Brier, E., Clavier, C., Olivier, F.: Correlation power analysis with a leakage model. In: The Proceedings of CHES 2004. LNCS, vol. 3156, pp. 16–29. Springer, Hiedelberg (2004)
11. Gandolfi, K., Mourtel, C., Oliver F.: Electromagnetic analysis: concrete results. In: The Proceedings of the Workshop on Cryptographic Hardware and Embedded Systems 2001 (CHES 2001), LNCS 2162 Paris, France, pp. 251–261 (2001)
12. Mangard, S.: Exploiting radiated emissions—EM attacks on cryptographic ICs. In: Proceedings of Austrochip (2003)
13. Agrawal, D., Archambeault, B., Rao, J., Rohatgi, P.: The EM side–channel(s): attacks and assessment methodologies. In: Proceedings of Cryptographic Hardware and Embedded Systems—CHES2002. LNCS, vol. 2523, pp. 29–45. Springer, New York (2002)

14. Berna Ors, S., Oswald, E., Preneel, B.: Power-Analysis Attacks on an FPGA–First Experimental Results. CHES 2003, LNCS 2779, pp. 35–50, Springer, Heidelberg (2003)
15. Benhadjyoussef, N., Mestiri, H., Machhout, M., Tourki, R.: Implementation of CPA analysis against AES design on FPGA. In: 2nd International Conference on Communications and Information Technology (ICCIT), pp. 124–128 (2012)
16. Ors, S.B., Gurkaynak, F., Oswald, E., Preneel, B.: Power-analysis attack on an ASIC AES implementation. In: The proceedings of ITCC 2004, Las Vegas, April 5–7 (2004)
17. Mestiri, H., Benhadjyoussef, N., Machhout, M., Tourki, R.: A Comparative study of power consumption models for CPA attack. In: International Journal of Computer Network and Information Security, pp. 25–31 (2013)
18. Tope, K., Rane, A., Rohate, R., Nalawade, S.M.: Encryption and decryption using artificial neural network. Int. Adv. Res. J. Sci. Eng. Techn. **2**(4) (2015)
19. Qaid, G.R., Talbar, S.N.: Encrypting image by using fuzzy logic algorithm. Int. J. Image Process. Vis. Sci. **2**(1) (2013)
20. Ratan, R.: Application of Genetic algorithm in cryptology. Adv. Intell. Syst. Comput. **258**, 821–831 (2014)

Application for Identifying and Monitoring Defaulters in Telecom Network

Aditya Pan, Anwesha Mal and Shruti Gupta

Abstract GGSN is the gateway GPRS support node. Internetworking between external data packets and GPRS networks is the main function of the GGSN. To an outsider GGSN acts like a subnetwork as it conceals GPRS software. The GGSN, when it receives data for a specified user, it first checks if the user is an active user and if the user is active then it forwards data to the SGSN, otherwise it is discarded. The GGSN stores various fields of customer data such as MSNID, the NAT port block, allocation and deallocation time and so on. All telecom companies use the process of natting. Natting is a concept in which each public IP is shared by 1500 private IPs. It is also mandatory for telecom companies in India to follow TRAI regulations for storing of telecom data. TRAI requires data of the past 7 years to be stored. Cyber cells tracking cybercrimes, when investigating a case can only trace back to the public IP of the telecom company and without permission from the telecom company they have no access to the files. Even after gaining access, to go through several million logs to find the correct log is a time-consuming and exhausting process. This paper presents an application which allows fast searching through GGSN logs to produce efficient results in a much shorter period of time. Work which would otherwise take several hours can be accomplished in minutes. The application helps enforce security by making the process of tracking these malicious end users, through a large amount of data, easy and fast.

Keywords Telecom log · Analysis · Security · Map reduce · SPLUNK

Aditya Pan (✉) · Anwesha Mal · Shruti Gupta
Amity School of Engineering and Technology, Amity University, Noida, India
e-mail: pan.aditya93@gmail.com

Anwesha Mal
e-mail: anweshamal@gmail.com

Shruti Gupta
e-mail: sgupta65@amity.edu

M. Pant et al. (eds.), *Proceedings of Fifth International Conference on Soft Computing for Problem Solving*, Advances in Intelligent Systems and Computing 437, DOI 10.1007/978-981-10-0451-3_51

1 Introduction

Tracking and tracing any sort of security breach is an essential component of system security. First it is essential to understand the basic telecom infrastructure. The basic network switching subsystem has the following components as shown in Fig. 1.

BTS stands for base transceiver station. Each telecom tower has a BTS attached to its base. The network could be any type of wireless network like a CDMA (Code Division Multiple Access), GSM (Global System for Mobile Communications), WiMAX (Worldwide Interoperability for Microwave Access) or Wi-Fi [1]. The BTS is part of the cellular network. It is used for the encryption and decryption of communications, spectrum filtering of equipment, transceivers and antennas among others. BTS has several numbers of transceivers which allow it to serve the cell's several different frequencies and sectors. The BSC (base station controller) controls the BTSs via the BCF (base station control function). The BSC manages the operational states of the transceiver and also provides a connection to the NMS (network management system). RNC is the Radio Network Controller. It is a critical component in the UMTS radio access network (UTRAN). It also controls

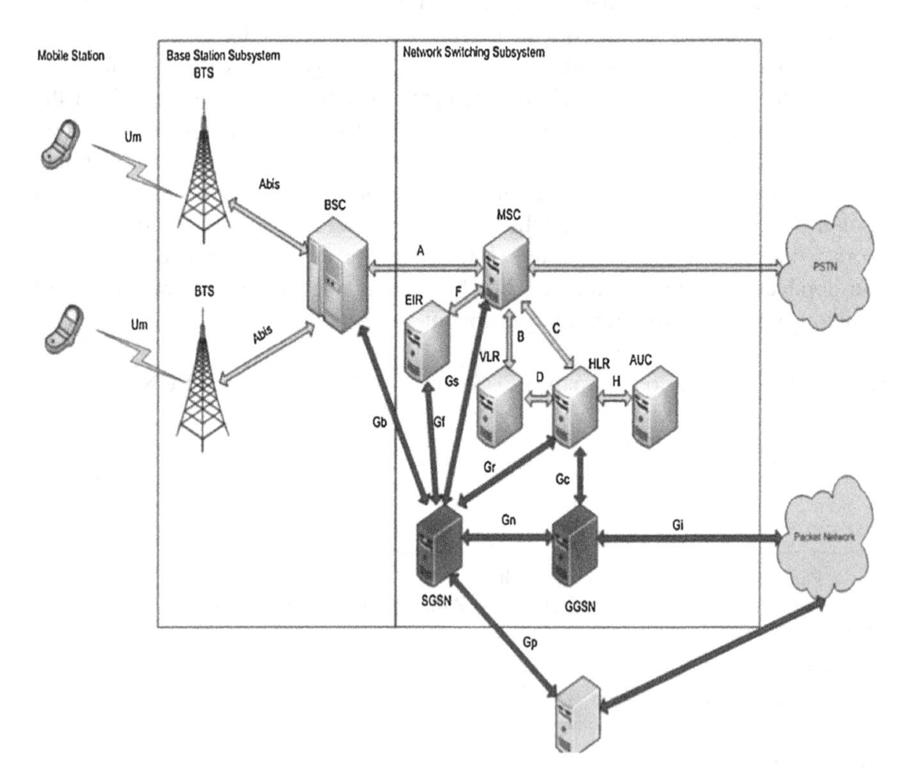

Fig. 1 Ad hoc GSM network

the nodes that are connected to it. Radio Resource Management and a couple of mobility functions are handled by the RNC. It is also the point at which encryption is done before data is sent to and from mobile phones. The RNC is connected via the Media Gateway (MGW) to the circuit switched circuit. It is also connected to the SGSN. MSC is the mobile switching centre and is a central component of the network switching subsystem. It is responsible for switching and routing calls to their destination point. It also has other functions like routing conference calls, SMS messages, service billing and fax. It also interacts with other telephone networks, i.e. it routes call from one service provider to another. The HLR or the home location register is where all customer information is stored like their MSNID, phone number, customer name and other customer related information. The Visitor Location Register contains the exact location of the subscriber. SGSN is called the Serving GPRS support Node and it is the component which is mainly responsible for the GPRS network. The RNC is connected via the MGW to the circuit switched circuit. It is also connected to the SGSN. MSC is the mobile switching centre and is a central component of the network switching subsystem. It is responsible for switching and routing calls to their destination point. It also has other functions like routing conference calls, SMS messages, service billing and fax. It also interacts with other telephone networks, i.e. it routes call from one service provider to another. The HLR or the home location register is where all customer information is stored like their MSNID, phone number, customer name and other customer related information. The Visitor Location Register contains the exact location of the subscriber. SGSN is called the Serving GPRS support Node and it is the component which is mainly responsible for the GPRS network. Transfers data between the GGSN and SGSN. GGSN is the gateway GPRS support node. Internetworking between external data packets and GPRS networks is the main function of the GGSN. To an outsider GGSN acts like a subnetwork as it conceals GPRS software. The GGSN, when it receives data for a specified user, it first checks if the user is an active user and if the user is active then it forwards data to the SGSN, otherwise it is discarded [2].

Figure 1 shows the layout of the GSM network with support for GPRS. It also shows the paths of communication between the different aforementioned components. The public network, which is shown as a cloud, represents the Internet [3]. Several millions of people access the Internet using a telecom provider's Internet services. Tracing back a single individual from a crowd of millions is a very time-consuming task [2]. However, this application returns much faster results and accepts as input the public IP and the session time. It makes use of Splunk's map reduce algorithm implementation to give much faster search results.

This paper first discusses the various security challenges that come across while dealing with large amounts of data, and the way it is changing. Then it moves on to analysis of GGSN logs. Finally the application, its processes and the results are results.

2 Security's New Challenges

With greater security threats over the years, organizations are giving more importance to cyber security. Monitoring data for threats already known is not sufficient any longer. It is necessary for security teams to view activities in order to able to identify and stop attacks. Security teams basically need to leverage four classes of data which are logs, binary data, contextual data and threat and intelligence data. If any one of these data types is missing there is a greater risk that an attack may go unnoticed. These data types help identify what is normal in your system [4, 5].

Taking into consideration the amount of data and the types of data, the most effective security implementation requires that

- It should be able to scale terabytes of data per day without any normalization at collection time and the schema is applied to the data only at search time.
- We should be able to access data from anywhere in the environment such as traditional security sources, databases, time management systems, industrial control systems, and so on.
- It should deliver "fast time-to-answer forensics".
- It should provide a security intelligence platform that is flexible and includes content and apps that are out of the box. In this way we can maximize investments in security infrastructure.

As software, Splunk satisfies all the above requirements. It is scalable and flexible software capable of searching through terabytes of data. Also a forensic investigation once completed can be monitored and saved in real time. SPL is a language that supports those correlations which can generate alerts depending on a combination of conditions and patterns in the system data or when a particular threshold has been reached. Splunk has the ability to extract knowledge from data automatically. We can add additional knowledge by identifying the fields and then by naming and tagging the fields. By correlating information from different sets of data can reduce false positives and provide a greater context and insight. By applying pattern analysis to the activities of users around sensitive data somas to minimize business risks. It is because of these features of Splunk that we have chosen it to develop our platform [6, 7].

The questions that security analysts need to ask regularly are:

- First figuring out who would want to access to my data and what data they would want to access.
- Which are the methods that could be implemented so that I would be able to stealthily spread malware?
- How would I make sure that my activities go undetected?
- How would I ensure that the malware I have installed stays on the target devices?

- What services of the host network would be monitored for any changes.
- How malware data could be distinguished from log-based data depending on location of origin, length of time and the time of the day.

These questions need to be answered so that data can be protected effectively and efficiently. The answers will help in understanding which areas are lacking in security and thereby be able to improve the security in those areas thus creating a more secure and safer environment [8, 9].

3 Experimental Analysis

The application has been deployed on a 32-bit Ubuntu operating system (version 12.04), running on Oracle Virtual Box (version 4.3.20). Virtual box was installed on Ubuntu 14.04 version. Virtualbox was used so as not to corrupt the desktop operating system because of tests. The SPLUNK version used was SPLUNK 4.3.3. Nano text editor was used for editing purposes, and Firefox web browser was used for testing the created application. Plunk has been developed on the cherry-py web framework. It is widely used by various organizations and for various purposes including and not limited to security, monitoring and analysing.

In this paper, set of eight mapper functions—$\{m_1, m_2,\ldots, m_8\}$ Є M, and a set of eight reducer functions—$\{r_1, r_2,\ldots, r_8\}$ Є R, all of them execute in parallel. There is a main control function CF in charge of passing data to mappers, storing the intermediate key-value pairs and passing keys to the reducers [10]. It is responsible for monitoring the status of execution of the mappers and reducers [11]. A barrier is used for the process of synchronization between M and R.

(1) Divide contents of the file F into equally sized blocks $\{f_1, f_2, f_3,\ldots, f_n\}$ such that ${}_1\sum^n{}_f = F$, of 16 MB size.
(2) Each block $\{f_1, f_2, f_3\ldots, f_n\}$ is passed to M, in an ordered manner. Status of M is monitored by CF which is also responsible for delegating blocks to individual mappers in M: $f_x \rightarrow m_y$ where $1 \leq x \leq n$ and $1 \leq y \leq 8$ and yet_to_process (f_x) = True and is executing (m_y) = False

$$\text{Number of blocks } (n) = \text{size of } (F)/(16 * \text{number\_of\_mappers}) \quad (1)$$

where Number of blocks (n) can be equivalent to size of (F)/128
(3) Mappers generate intermediate key, value pairs after every execution is done like <keys, value> count (keys) depends on the f_x passed to it and value = 1.
(4) Synchronization is done through a barrier such that for some m in M if is executing (m) for a TRUE value then for all r in R can execute(r) as a FALSE.
(5) CF is responsible for delegating sets of key values to R, such that if we consider key_set_1 to be any arbitrary intermediate key values, then R: $r_y \rightarrow$ key_set_1 where $1 \leq y \leq 8$ and yet_to_process (key_set_1) = True and is executing (r_y) = False

(6) CF is responsible for collecting outputs generated by all reducers which iterates over the keys provided to it and generates the final <key, value> pairs.

The first step is to set up the Splunk indexer and search head, as shown in Fig. 2. The Splunk indexer is where all the data is stored and the search head is what is used to search through the data. The second step of the process is to add the GGSN logs to Splunk. The GGSN stands for the Gateway GPRS Support Node. The logs generated by the GGSN comprise of 19 fields which include the public and private IP addresses, the port being used, the IMSI, etc. The logs are added by writing an inputs.conf and a props.conf for them. Once the logs have been added to Splunk, it is checked to ensure that the timestamp is extracted correctly. It is necessary to extract the correct time stamp. The GGSN logs have three time entries—one is the allocation time, one is the deallocation time and one is the last activity time. The allocation time is set as the time stamp. This is done using the regex corresponding to the time stamp required to be extracted. If the time stamp is extracted correctly, continue onto the next step, otherwise clear data from the indexer and try to extract the time stamp again. Once the time stamp is extracted correctly, created a saved search and check to make use it runs properly. If the saved search does not execute correctly then delete the data and clear the indexer and restart the process from the third step.

As per Fig. 3, a custom search is created for verification. The application is then designed so that it accepts user input and returns as output allocation time, the radius station calling id, the IMSI, the private IP address of the user, the public IP address and the starting and ending port numbers of the port block being used. Lastly, the application is tested with valid inputs to ensure that the application is working correctly. This is done by entering the public IP address and the time of activity. As an output, the entire log entry for the user is obtained [12].

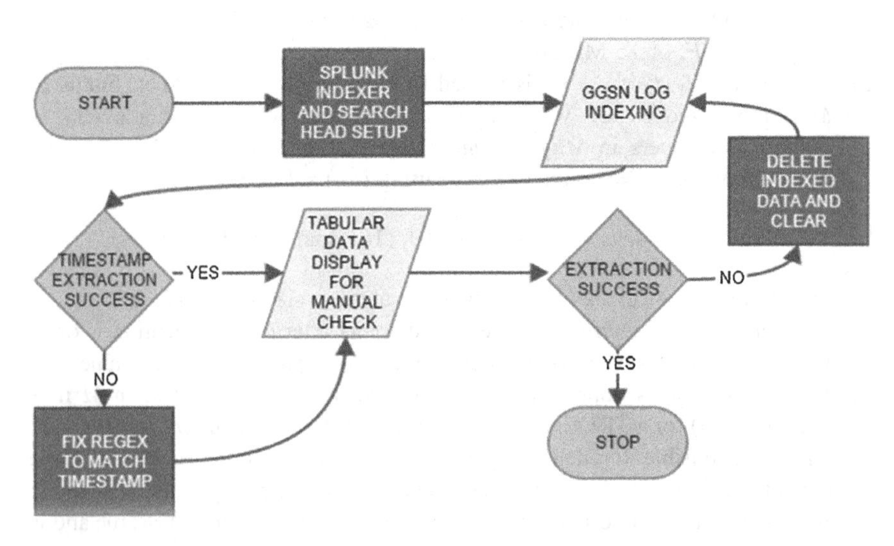

Fig. 2 Input process of log

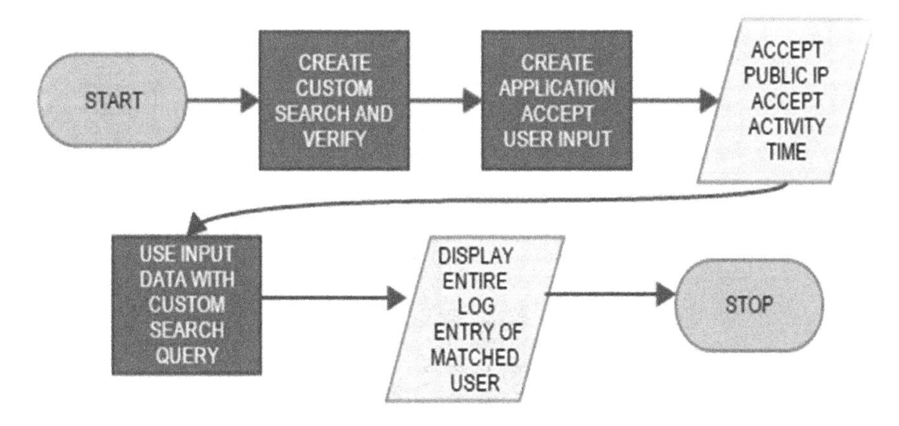

Fig. 3 Analysis process of log

Tables 1, 2 and 3 are one table divided into three parts for the sake of easy representation. It is an excerpt of 10 entries from our collected sample data of 10 entries. The table has 19 columns. The first row explains the content of each column, and each row represents an entry.

The following describes Table 1. The first column represents sn-correlation id. This field contains the identifier for the access network. The second row displays the telecom subscriber internal IP address. As we can see, by the first two octets of the IP address, 10.x.x.x, it is non-routable. The bearer 3gpp charging ID is maintained for GPRS tunnelling purposes (lot of unexplained crap, considering citing something). The bearer-3gpp-sgsn-address is the SGSN it was routed to. The bearer-GGSN-address shows the IP address of the GGSN it was routed to. The bearer-3gpp-imsi number is used for uniquely identify the user of a cellular network. Each cellular network has its own identification connected with it. IMSI stands for International Mobile Subscriber Id. 3gpp stands for third Generation partnership project.

The following describes Table 2. The full form of RADIUS is Remote Authentication Dial In User Service. It is used for authenticating and authorizing any connection requests received by the server. The attribute user name is basically a character string containing the realm name and the user account name which has been supplied by the access client. The radius-calling-station-id is used to store the Access point MAC address or Bridge in the ASCII format. The sn-nat-ip is the public IP which the private IP is finally mapped to. The sn-nat-port-block-start and the sn-nat-port-block-end is the range of network ports allowed per subscriber connection. The sn-nat-binding-timer is what the NAT idle timer maintains. When this timer expires, the public addresses are sent back to the NAT pool. The sn-subscribers-per-ip-address is the number of internal IP address which can be mapped to one public ip address or sn-nat-ip address.

The following describes Table 3. The sn-nat-gmt offset gives the offset of the present time zone from the GMT time which stands for Greenwich meantime. The

Table 1 Sample GGSN logs

#sn-correlation-id	ip-subscriber-ip-address	bearer-3gpp charging-id	bearer-3gpp-sgsn-address	bearer-ggsn-address	bearer-3gpp imsi
i3KqNNDx	10.143.97.15	6.82E+08	203.88.5.96	203.88.3.89	4.06E+14
i3KqMDn6	10.140.44.251	6.82E+08	203.88.2.32	203.88.3.89	4.06E+14
sERSiDsB	10.87.135.205	1.08E+09	203.88.3.33	203.88.3.89	4.04E+14
sESmElLf	10.85.80.151	1.08E+09	203.88.2.176	203.88.3.89	4.06E+14
sES1BneG	10.136.66.57	1.08E+09	203.88.5.96	203.88.3.89	4.06E+14
sESmHb-7	10.68.144.165	1.08E+09	203.88.2.32	203.88.3.89	4.06E+14
sESmEZuC	10.81.5.223	1.08E+09	203.88.2.32	203.88.3.89	4.06E+14
sESmG9aw	10.144.53.117	1.08E+09	203.88.5.96	203.88.3.89	4.06E+14
sESmJJyA	10.69.237.137	1.08E+09	203.88.2.32	203.88.3.89	4.06E+14
sESmIEjK	10.141.116.199	1.08E+09	203.88.3.33	203.88.3.89	4.06E+14

Table 2 Sample GGSN Log (contd.)

Radius-user-name	Radius-calling-station-Id	sn-nat-ip	sn-nat-port-block-start	sn-nat-port-block-end	sn-nat-binding-timer	sn-nat-subscribers-per-ip-address
919,734,360,443@www.40567pre	9.20E+11	112.79.36.65	17,760	17,791	60	1500
919706285859@www.405751pre	9.20E+11	112.79.36.65	8224	8255	60	1500
919830839039@iphone.40430pre	9.20E+11	112.79.37.51	54,944	54,975	60	1500
917797191392@www.40567pre	9.18E+11	112.79.37.51	37,504	37,535	60	1500
919735587877@www.40567pre	9.20E+11	112.79.37.51	56,576	56,607	60	1500
918876018289@www.405751pre	9.19E+11	112.79.37.51	38,208	38,239	60	1500
918794401321@www.405755pre	9.19E+11	112.79.37.51	25,184	25,215	60	1500
919609776757@www.40567pre	9.20E+11	112.79.37.51	33,216	33,247	60	1500
918257801670@www.405755pre	9.18E+11	112.79.37.51	8960	8991	60	1500
919732502654@www.40567post	9.20E+11	112.79.37.51	52,256	52,287	60	1500

Table 3 Sample GGSN Log (contd.)

sn-nat-realm-name	sn-nat-gmt-offset	sn-nat-port-chunk-alloc-dealloc-flag	sn-nat-port-chunk-alloc-time-gmt	sn-nat-port-chunk-dealloc-time-gmt	sn-nat-last-activity-time-gmt
NAT1	530	1	12-05-2013 18:30		
NAT1	530	1	12-05-2013 18:30		
NAT1	530	0	12-05-2013 18:21	12-05-2013 18:30	12-05-2013 18:29
NAT1	530	1	12-05-2013 18:30		
NAT1	530	0	12-05-2013 17:49	12-05-2013 18:30	12-05-2013 18:30
NAT1	530	1	12-05-2013 18:30		
NAT1	530	1	12-05-2013 18:30		
NAT1	530	1	12-05-2013 18:30		

time zone here in India which is five-and-a-half hours ahead of the GMT. The sn-nat-alloc-dealloc flag is used to indicate the sn-nat-port chuck has been allocated or not. If it has been allocated then it is set to 1 else it is set to 0. Sn-nat-port-chunk-alloc-time-gmt gives the time when the port block has been allocated. Sn-nat-port-chunk-dealloc-time-gmt gives the time when the port block has been deallocated. sn-nat-last-activity-time-gmt gives the time stamp of the last performed activity.

4 Methodology for Analysis

4.1 Setting up Splunk Indexer and Search Head

Splunk has to be downloaded and installed before use. After install, it can be started with "./splunk start". On first time setup, the management and web port has to be set. By default these values are 8089 and 8000. However, they can be set to any other port if desired. The web port is used to logging into the web interface of Splunk. After install it can be tested by logging into the Splunk server. In manager of the web interface the server name, session time out times and more can be set [9].

4.2 Adding GGSN Logs to Splunk

The data collected from the GGSN has the following categories: -#sn-correlation-id, ip-subscriber-ip-address, bearer-3gpp charging-id, bearer-3gpp sgsn-address, bearer-3gpp imsi, radius-user-name, radius-calling-station-id, sn-nat-ip, sn-nat-port-block-start, sn-nat-port-block-end, sn-nat-binding-timer, sn-nat-subscribers-per-ip-address, sn-nat-realm-name, sn-nat-gmt-offset, sn-nat-port-chunk-block, alloc-dealloc-flag, sn-nat-port-chunk-alloc-time-gmt, sn-nat-port-chunk-dealloc-time-gmt, sn-nat-last-activity-time-gmt,.

To add the required logs to splunk the following files are created in (installation directory) splunk/etc./system/local-props.conf and inputs.conf files.

To add the required logs to splunk the following files are created in (installation directory)splunk/etc./system/local-props.conf and inputs.conf files. The host is set to ubuntu02, disabled is set to false, followtail to 0. The index is set to main and the source type to GGSN. The inputs.conf file is used for adding GGSN logs to Splunk. This is used to specify the host and the log source type and location and specifying index.

The props.conf file is used in addition with inputs.conf to add the data to Splunk. The file is used for extraction of the correct time stamp from the time format in the log. To verify whether the process, Splunk web summary will show the accessed log data.

The data is stored in CSV format on the local disk. It showed to be noted that the path to the file remains the same at all times. However, the application still works if the contents are modified, since the changes are automatically taken into consideration by Splunk. Since the GGSN logs contain three different time fields it is highly unlikely that splunk will be able to extract the correct time stamp of the logs. For the purpose of this application the sn-nat-port-chunk-alloc-time-gmt time field has to be set as the time stamp. This is being done by specifying the regex of the format in the props.conf file. The TIME_PREFIX field, contains the required regex to extract the correct time stamp. The regex used is ^([^,]*,){16}. The TIME_FORMAT is "%m%d%y %H:%M". It is essential to set the time zone field, especially if the data was being generated by a machine in another time zone. The time zone field helps convert it into local time for easy access. For this case the time zone has been set to Asia/Kolkata [13].

4.3 Verifying Addition of Logs

After saving the files, logging into the web user interface will show a summary of it. The search query, as shown in Table 4, when executed will display the file chronologically, in the defined tablature format as shown in Tables 1, 2 and 3. The GGSN is natted such that a single public IP is shared by 1500 private IPs. This means that 1500 users accessing the Internet from their phones are sharing the same public IP. This thereby makes finding the source of abnormal activity a difficult task. The basic task here is to correlate and filter out the logs so that the attacker can be tracked down efficiently. The logs are saved to the splunk indexer and the map reduce algorithm is then implemented to search through and filter the data. The Map Reduce Algorithm can be thought of as the work being divided between four groups. The mappers, the groupers, the reducer and the master. The mappers and reducers can work independently. The master divides the work and the grouper mainly groups the data based on the key. The job of the mappers and reducers differ depending upon the task being performed [10].

This search is saved for use in the application. The result of the search is shown in Table 5. The MSISDN field has been partially hidden to protect the privacy of the participants [14].

Table 4 Search query

host=KOLGGSN01 | table sn_nat_port_chunk_alloc_time_gmt,radius_calling_station_id,"bearer_3gpp imsi",ip_subscriber_ip_address,sn_nat_port_block_start,sn_nat_port_block_end,sn_nat_ip | rename sn_nat_port_chunk_alloc_time_gmt as "Date/Time", radius_calling_station_id as MSISDN, "bearer_3gpp imsi" as IMSI, ip_subscriber_ip_address as "Private IP", sn_nat_port_block_start as "NAT Start Port", sn_nat_port_block_end as "NAT End Port", sn_nat_ip as "Public IP"

Table 5 Result of search query

Date/time	MSISDN	IMSI	Private IP	NAT start port	NAT end port	Public IP
12-05-2013 18:33	919*********	4.05674E+14	10.71.63.185	26,400	26,431	112.79.36.33
12-05-2013 18:33	918*********	4.05752E+14	10.68.215.72	40,960	40,991	112.79.36.33
12-05-2013 18:33	917*********	4.043E+14	10.143.41.161	10,176	10,207	112.79.36.33
12-05-2013 18:33	918*********	4.05674E+14	10.81.106.205	46,432	46,463	112.79.36.33
12-05-2013 18:33	9175*********	4.05672E+14	10.81.68.115	24,288	24,319	112.79.36.33
12-05-2013 18:33	9196********	4.05752E+14	10.81.106.205	46,432	46,463	112.79.36.33
12-05-2013 18:33	915*********	4.05756E+14	10.80.8.47	46,272	46,303	112.79.36.33
12-05-2013 18:33	917*********	4.05674E+14	10.137.38.118	7232	7263	112.79.36.156
12-05-2013 18:33	9190********	4.05756E+14	10.143.144.252	13,216	13,247	112.79.36.156
12-05-2013 18:33	912*********	4.05674E+14	10.143.144.252	13,216	13,247	112.79.36.156
12-05-2013 18:33	916*********	4.05674E+14	10.81.156.60	49,632	49,663	112.79.36.156
12-05-2013 18:33	919*********	4.05674E+14	10.86.147.42	47,552	47,583	112.79.36.156
12-05-2013 18:33	917*********	4.05674E+14	10.145.3.103	44,256	44,287	112.79.36.156
12-05-2013 18:33	9183********	4.05674E+14	10.68.57.143	52,288	52,319	112.79.36.156
12-05-2013 18:33	9197********	4.05752E+14	10.80.48.141	57,504	57,535	112.79.36.156
12-05-2013 18:33	9184********	4.043E+14	10.71.61.221	53,792	53,823	112.79.36.156
12-05-2013 18:33	9193********	4.05753E+14	10.136.100.41	28,896	28,927	112.79.36.174
12-05-2013 18:33	9190********	4.05752E+14	10.144.116.123	39,936	39,967	112.79.36.174
12-05-2013 18:33	916*********	4.043E+14	10.143.33.83	19,232	19,263	112.79.36.174
12-05-2013 18:33	912*********	4.05753E+14	10.80.207.222	27,104	27,135	112.79.36.174

4.4 The Application Process

The layout of the application, and uses the search query in Table 4, to perform the search operation. The layout has been kept simple to enhance ease of use. The application is created from within the Splunk web interface. The application will accept as input the public IP address and the start and end time of the session. The application will output the details of the person by analysing the logs based on the input information. To build the application an XML script was written.

5 Results

For the purpose of testing the application, data was collected through a survey using surveymonkey.com. The collected log data was analysed to see the most popular service carrier choice based on the survey responses, as shown in Fig. 4. Based on the collected data, a csv file was created as per the format in Tables 1, 2, 3 all combined together.

The survey, graphically shown in Fig. 4, clearly shows Vodafone to be the most popular telecom provider by the survey. Therefore, for the sake of uniformity of data, surveys which were from Vodafone users were used to create a csv file, which acts as input to the application.

The logs were manually scanned through for the purpose of creating a test case. The application required fields for the public IP (or sn_nat_ip) and the start and end time and date of connection. After manual scanning the public IP address of 112.79.37.51, a start time of 12/05/2013 18:21:00 and an end time of 12/05/2013 18:30:00. The concept is since any security monitoring activity will be able to get only these three fields initially from outside the telecom network. Using these fields the application should be able to give the entire user information so that tracking the malicious user becomes easy and fast.

As shown in Table 6, once these fields are provided as input, the application displays the relevant user information, thus enabling the successful tracking of the intended party. Using the fields in the displayed log, the intended party can be uniquely identified.

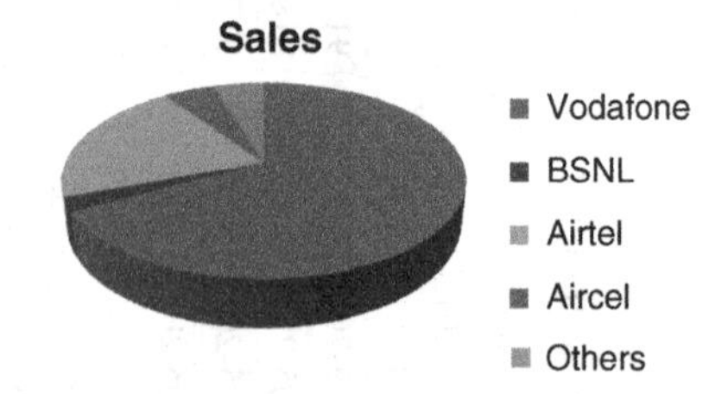

Fig. 4 Pie chart showing Vodafone to be the largest service provider

Table 6 Results after running query

Input			Output						
Public IP	Start time	End time	sn-nat-port-chunk-alloc-time-gmt	Radius-calling-station-id	bearer-3gpp imsi	ip-subscriber-ip-address	sn-nat-port-block-start	sn-nat-port-block-end	sn-nat-ip
112.79.37.51	12/05/2013:18:21:00	12/05/2013:18:30:00	12/05/2013:18:21:00	9.19831E+11	4.04374E+14	10.87.135.205	54,944	54,975	112.79.37.51
112.79.37.51	12/05/2013:18:30:00	12/05/2013:18:31:00	12/05/2013:18:21:00	9.19831E+11	4.043E+14	10.87.135.196	54,934	54,976	112.79.37.51
112.79.37.51	12/05/2013:18:30:00	12/05/2013:18:30:00	12/05/2013:17:49:00	9.19736E+11	4.05672E+14	10.136.66.57	56,576	56,607	112.79.37.51
112.79.36.94	12/05/2013:18:30:00	12/05/2013:18:30:00	12/05/2013:18:25:00	9.18725E+11	4.05752E+14	10.143.82.19	24,960	24,991	112.79.36.94

6 Conclusions with Future Enhancement

The proposed solution provides a fast way to access log data and also the application makes the process of searching much easier. The proposed process requires only a one-time setup. The analyst will only have to interact with the simple GUI, and will only have to deal with the search result of his queries, without worrying about the rest of the data. The use of map reduce algorithm makes the process very fast, which would not have been possible in a traditional database system application based on the setup model, any changes to the currently monitored log file are being automatically updated. However, the location of the log file must not be changed. The application has been setup on a system and the required log files are stored in local disk. However, multiple variations to this can be possible based upon usage scenario. It is possible to forward this data directly from another machine, to the TCP port of the machine where the application has been setup. The application can then be made to work in real time, with the end user not having to worry about where the data is coming from. The process can also be made very secure by providing username and password to authorize only selective users to prevent misuse of the data. A suggested improvement to the application would be to connect it with HLR log data, so that the end user can get the details of the subscriber identified at search time, without worrying about the internal codes and other details. In case of an enterprise use, it is also possible to modify the application so that indexing of the incoming GGSN logs can be performed in a distributed way, with load balancing. The application can then be programmed to access all of these systems simultaneously to provide faster access. It is also possible to set up the application on different machines, so that the functionality may be available on different machines simultaneously, a much required feature in an enterprise setup. Thus it is safe to say that our application is scalable.

References

1. Tang, S., Li, W.: Modelling and evaluation of the 3G mobile networks with hotspot WLANs. Int. J. Wireless Mobile Comput. **2**(4), 303–313 (2007)
2. Mishra, A.: Performance and architecture of SGSN and GGSN of general packet radio service (GPRS). Global Telecommunications Conference, GLOBECOM '01. IEEE, vol. 6, pp. 3494–3498 (2001)
3. Rotharmel, S.: IP based telecom power system monitoring solution in GPRS networks. In: 29th International Telecommunications Energy Conference, INTELEC 2007, pp. 769–774, Sept 30–Oct 4 2007
4. Jacobs, A.: The pathologies of big data. Commun. ACM **52**(8), 36–44 (2009)
5. Laurila, J.K., et al.: The mobile data challenge: Big data for mobile computing research. Pervasive Computing. No. EPFL-CONF-192489 (2012)
6. Oliner, Adam, Ganapathi, Archana, Wei, Xu: Advances and challenges in log analysis. Commun. ACM **55**(2), 55–61 (2012)
7. Ergenekon, E.B., Eriksson, P.: Big Data Archiving with Splunk and Hadoop (2013)

8. Cuzzocrea, A., Song, I.Y., Davis, K.C.: Analytics over large-scale multidimensional data: the big data revolution!. In: Proceedings of the ACM 14th international workshop on Data Warehousing and OLAP. ACM (2011)
9. Chamarthi, Prasad, S., Magesh, S.: Application of splunk towards log files analysis and monitoring of mobile communication nodes. (2014)
10. Dean, J., Ghemawat, S.: MapReduce: simplified data processing on large clusters. Communications of the ACM 51.1. pp. 107–113 (2008)
11. Barrachina, A.D., O'Driscoll, A.: A big data methodology for categorising technical support requests using hadoop and mahout. J. Big Data **1**(1), 1 (2014)
12. Stearley, J., Corwell, S., Lord, K.: Bridging the gaps: joining information sources with splunk. In: Proceedings of the 2010 Workshop on Managing Systems via Log Analysis and Machine Learning Techniques. USENIX Association, 2010
13. Splunk docs. http://docs.splunk.com/Documentation/Splunk/6.2.1/Data/Configureyourinputs#Use_the_CLI
14. Splunk docs. http://docs.splunk.com/Documentation/Splunk/6.2.1/Viz/BuildandeditdashboardswithSimplifiedXML

An Efficient Approach for Mining Sequential Pattern

Nidhi Pant, Surya Kant, Bhaskar Pant and Shashi Kumar Sharma

Abstract There are many different types of data mining tasks such as association rule mining (ARM), clustering, classification, and sequential pattern mining. Sequential pattern mining (SPM) is a data mining topic which is concerned with finding relevant patterns between data where values are delivered in sequence. Many algorithms have been proposed such as GSP and SPADE which work on Apriori property of generating candidates. This paper proposes a new technique which is quite simple, as it does not generate any candidate sets and requires only single database scan.

Keywords Sequential pattern mining · Sequence database

1 Introduction

Data mining refers to discovering of interesting patterns from large databases. Sequential pattern mining (SPM) process finds sets of data items that are present together frequently in some sequences of sequential database. In real world, huge amount of data continuously being collected and stored, many industries are becoming interested in mining sequential patterns from these databases.

SPM is a process of extracting those patterns, whose support exceeds predefined minimum support, i.e., it helps to extract patterns which reflect most frequent behavior in sequence database. To reduce very large number of sequence into most interesting sequential pattern, so as to fulfill the different user requirements, minimum support is introduced which prunes the sequential patterns with no interest.

Nidhi Pant · S.K. Sharma
Graphic Era Hill University, Dehradun, India

Surya Kant (✉)
IIT Roorkee, Roorkee, India
e-mail: suryak111@gmail.com

Bhaskar Pant
Graphic Era University, Dehradun, India

M. Pant et al. (eds.), *Proceedings of Fifth International Conference on Soft Computing for Problem Solving*, Advances in Intelligent Systems and Computing 437, DOI 10.1007/978-981-10-0451-3_52

SPM is used in several domains such as in business organization to study customer behavior and also in area of web mining.

Many SPM algorithms have been proposed and prior algorithms among them are based on property of Apriori algorithms proposed by [1]. The property states that frequent pattern containing sub-pattern are also frequent. Based on this assumption many algorithms were proposed. Additionally, the Apriori property-based horizontal formatting method, generalized sequential pattern (GSP), has been presented in 1996 by same authors [2]. GSP is the exact algorithm to find frequent sequence in original database according to user-defined minimum support. It has the property that all subsequence of a frequent sequence must also be frequent and the property is as anti-monotone property. GSP works according to Apriori downward principle, where the candidate sequence length is generated from shorter length subsequence and thereafter candidates who have infrequent subsequence are pruned.

GSP makes multiple databases scan. In first scan, items of 1-sequence are recognized, eg: a_1, a_2....a_m. From recognized frequent items, candidate 2-sequence sets are generated from already identified first-level frequent sequences excluding rare subsequences from the first level. The algorithm consists of two important steps: 1. *Candidate generation*—Suppose a frequent set of Fk-1 sequence is given, performing a join of Fk-1 with itself will generate candidate sets for next scan and then pruning of database is done that eliminates sequences whose subsequences are not in frequent order. 2. *Support counting*—The search based on heap is used for efficient support counting. Finally, non-minimal sequences are removed.

Apriori property-based vertical formatting method (SPADE) was presented [3]. SPADE utilizes an equivalence class that decomposes initial problem into small subproblem, so that it can independently be solved in memory, using simple joins. In very first step spade generates 1-sequence items, items with single element, and this is done with one database scan. Second step, which consists of counting of 2-sequence, is performed by changing vertical database representation to horizontal database representation in the memory. It uses vertical representation of database; each row contains events uniquely identified by sequence id and event id. All sequences are recognized in three database scans. Many more algorithms like PrefixSpan, FreeSpan, and Sequential pattern mining using Bitmap Representation were also proposed for mining sequential patterns.

2 Literature Survey

The Apriori algorithm [1] has set a basis for those algorithms that largely depend on property of apriori and uses its Apriori-generate method. Apriori property says that all non-empty subsets of frequent item set should also be frequent and this property is called anti-monotonic property. Algorithms having priori property have disadvantage of maintaining frequency of each and every subsequence. To overcome from this problem, algorithms have discovered a way of calculating frequency and pruning sequences without maintaining count in each iteration.

A sequence database, where every single sequence is a set of transactions, is ordered by transaction time. The problem was analyzed by algorithm [2]. The problem of sequential mining is to identify all those patterns having support equal to more than a predefined minimum support and presented GSP algorithm. Algorithm [4] proposed a PSP algorithm. This algorithm uses a prefix tree that holds candidate sequence for each sequence along with their support count. When support threshold is very low algorithm performs very inefficiently.

Algorithm [3] introduced SPADE algorithm discovered for fast generation of sequential patterns. All existing algorithm of sequential mining makes multiple scans of database but SPADE discovers all frequent patterns in just three scans of database.

FreeSpan algorithm [5] uses projected databases for generation of database annotation that helps to quickly find frequent patterns. Shrinking factor of projected database is less than PrefixSpan.

In PrefixSpan algorithm [6], during scan of original database projected database is created. Ordering of lexicographic is maintained and is important as well. A lot of memory is occupied to store projections, especially when use of recursion is there.

Algorithm [7] introduces, generates, and tests feature, which are efficient for mining of sequences. Memory consumption is decreased to a magnitude order in this case.

Another technique [8] uses a rough set theory technique for finding local pattern from sequences, which is efficient for mining sequential patterns. Then [9] proposes the sequential pattern mining method specialized for analysis of learning histories of programming learning. [10] proposed a technique for identifying temporal relationships between medications and accurately predicting the next medication likely to be prescribed for a patient.

3 Proposed Algorithm

Given sequential database, scan the database ones, take the first sequence id, and from that generate all possible combinations with the help of following algorithm:

- Scan the database D.
- Set $i = 1$;
- For $i = 1$ to n;
- Select sequence Si
- Generate all possible combinations.
- Store them in tree.

GSP and SPADE algorithms of sequential pattern mining make multiple database scans and generate candidate sequences, which takes lot of time. But this new technique is far better as our technique makes only one database scan and also it does not generate candidate sequences. This proposed technique work on simple combination method. On a single database scan, a sequence id is taken and all possible combination of sequences is generated. Once the sequences are generated, the sequences are stored in a tree. The nodes of the tree also contain frequency of that sequence. If a same sequence exists in single sequence id, its count will not be increased. The count will only be increased if that sequences will be present in some other sequence id. A sequence can be searched with the help of the following steps:

- Take the sequence which you want to search.
- Take first item of the sequence and search it in the tree.
- If the first item is matched, then search in downward child nodes for the sequence.

After we are done with all sequence ids, we will look for nodes that will satisfy the minimum support. The nodes which will not satisfy the minimum support will be pruned and we will be with nodes which are frequent.

4 Illustrative Example

First, we take sequence id S1 and make all possible sequences. For example, suppose we take first sequence id S1 of sequence database given in Fig. 1.

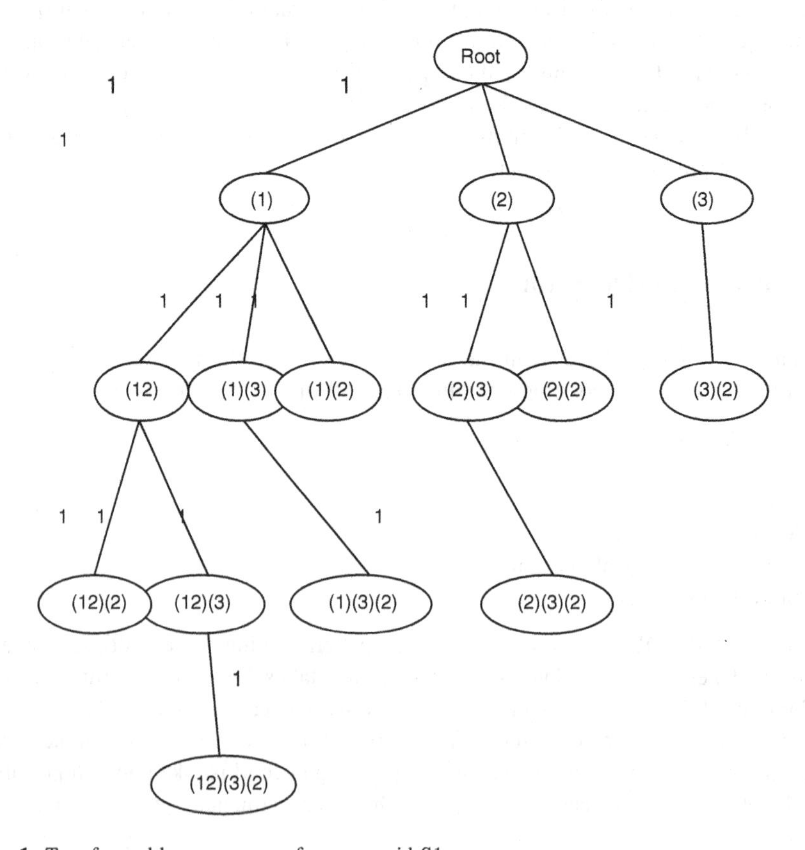

Fig. 1 Tree formed by sequences of sequence id S1

S1 (1 2) (3) (2)

The sequences generated from this sequence are as follows:

(1), (2), (3), (1 2), (1)(3), (1)(2), (2)(3), (2)(2), (3)(2), (1)(3)(2), (2)(3)(2), (1 2)(3), (1 2)(2), (1 2)(3)(2).

The sequences that are generated are placed in the form of tree given in Fig. 1.

Root will be at level 0, 1-sequence sequences will be stored at level-1, and so on. Figure 1 shows the tree formed by the sequences of S1. The count 1 indicates the frequency of all the sequences. Now we will take sequence S2, generate all sequences, and place them in the same tree shown in Fig. 1. If the sequence is already present in the tree, then count of the node will be increased; otherwise the sequence will be placed in the tree.

After S1 is stored in tree, now we take S2. All the sequences formed by S2 are S2 (3)(2 3)

(3), (2), (3)(2), (3)(3), (2 3), (3)(2 3).

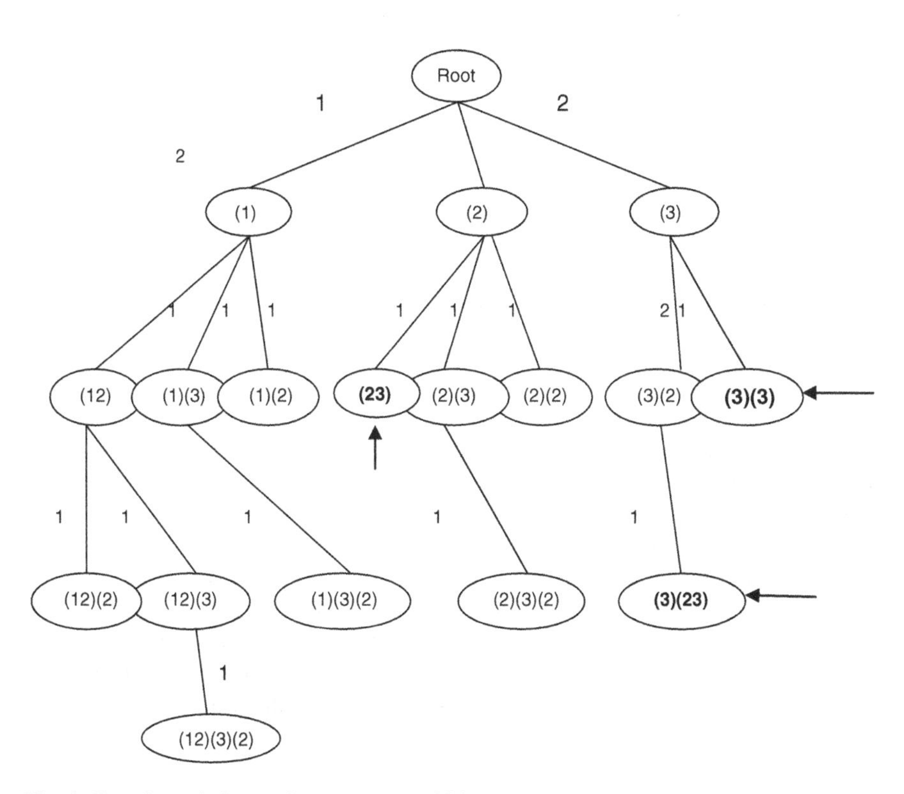

Fig. 2 Tree formed after storing sequences of S2

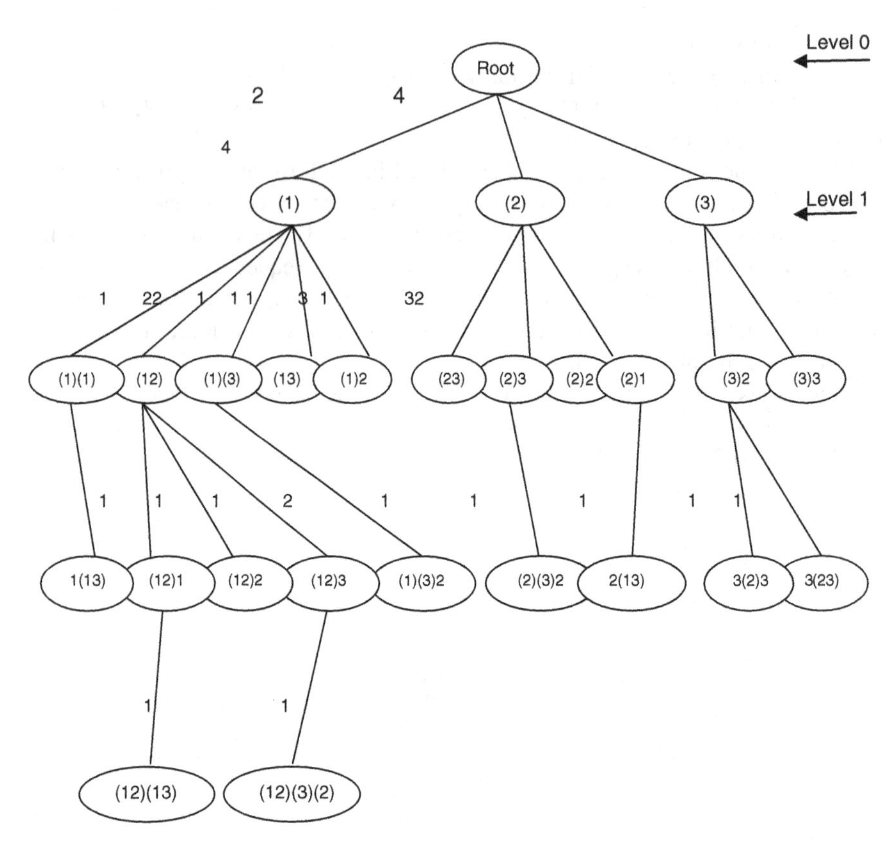

Fig. 3 Final tree formed by sequence database given in Table 1

Sequences such as (2), (3), and (3)(2) already exist in tree in Fig. 1, so their counts are increased to 2, whereas sequences (3)(3), (2 3), and (3)(2 3) are stored as shown in Fig. 2. Same procedure is applied for all sequences (Fig. 3).

Searching any sequence in the tree is also very easy and simple, since all the nodes save the path that makes easy for us to know all the nodes above that particular node. Now suppose we have to search sequence (1 2)(3). The sequence (1 2)(3) is a 3-sequence, so it will be at level-3. According to the searching technique, since the first item of them sequence (1 2)(3) is 1, we will look for the nodes under node (1) in the tree. Afterward, we will look for the second item of the sequence, i.e., (1 2).

So we will look for (1 2) within same bracket, and finally we will look for nodes under node (1 2).

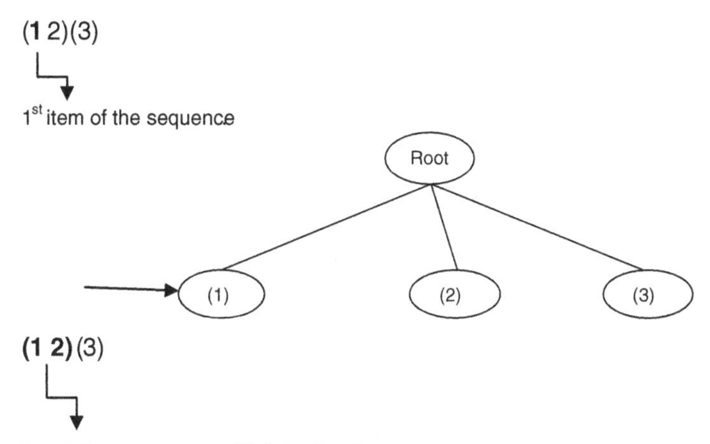

Look for sequence (12) in the tree.

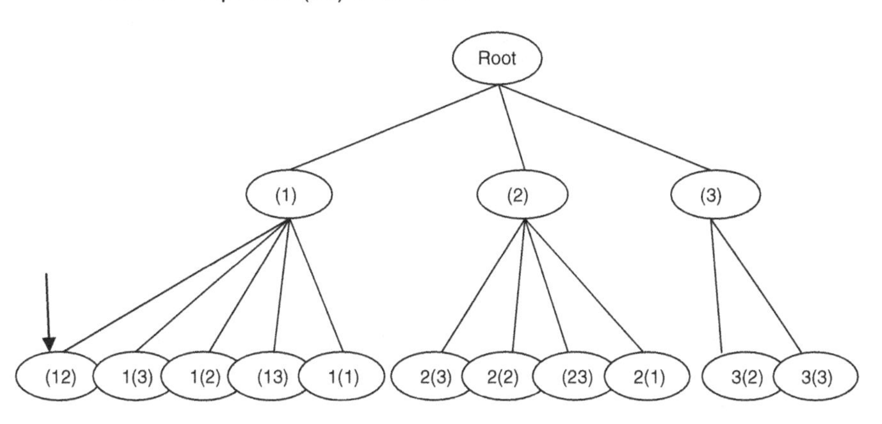

After that look for (12)(3) node below the node (12).

After that look for (12)(3) node below the node (12).

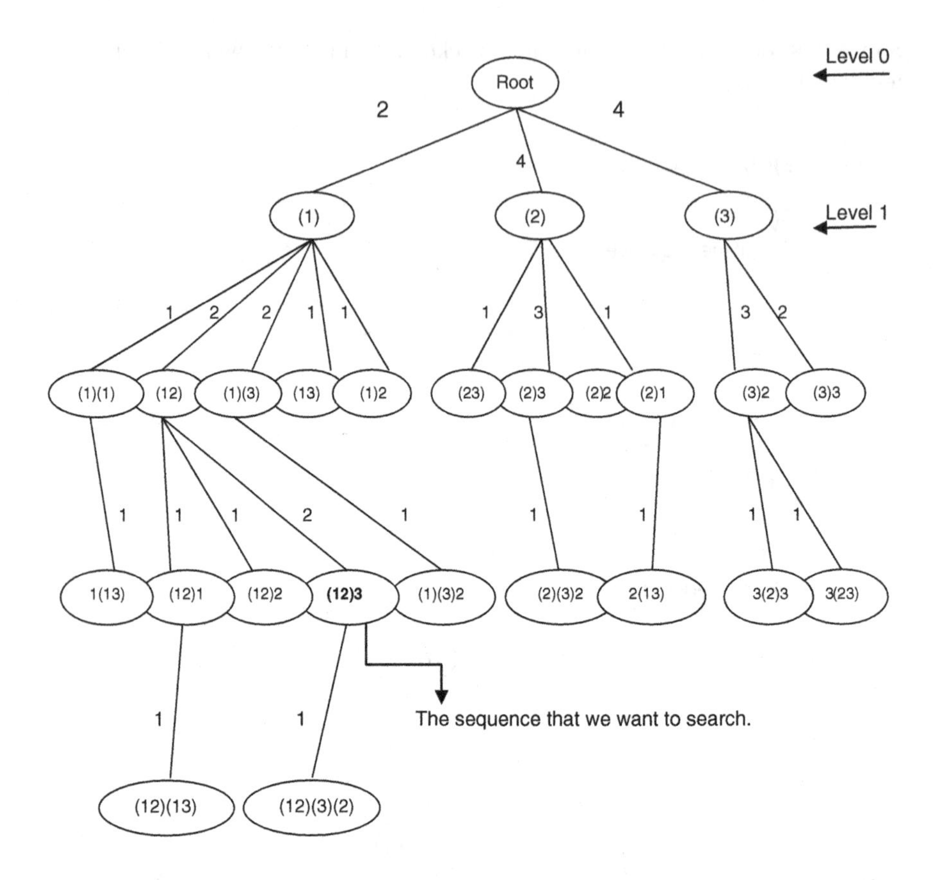

By this way storing complete previous path on each and every node makes the searching of sequence very easy; since we do not have to backtrack to search for a particular, this provides us an idea of where a particular sequence might be stored.

So the frequent sequential patterns of sequence database are shown in Tables 1 and 2.

Table 1 Sequence database

Sequence Id	Sequences
S1	(1 2)(3)(2)
S2	(3)(2 3)
S3	(1 2)(1 3)
S4	(3)(2)(3)

Table 2 All frequent sequential patterns

Sequence	Count
(1)	2
(2)	4
(3)	4
(1 2)	2
(1)(3)	2
(2)(3)	3
(3)(3)	2
(3)(2)	3
(1 2)(3)	2

5 Conclusion

The algorithm proposed in this paper generates the complete set of frequent sequential patterns without generating any candidates who reduce both the times. It also reduces the effort of repeated database scanning, thereby improving the performance of algorithm. The algorithm stores both frequent as well as non-frequent items. Although storing of non-frequent items require little more memory, but in future better performance will be achieved for incremental mining because of stored non-frequent items. We can ignore memory usage issue for performance enhancement benefits as nowadays memory is not so expensive, and will also be shown later.

References

1. Agrawal, R., Srikant, R.: Fast algorithms for mining association rules in large databases. In: Proceedings of 20th International Conference on Very Large Databases (1994)
2. Srikant, R., Agrawal, R.: Mining sequential patterns: generalizations and performance improvements. In: Proceedings of the 5th International conference on Extending Database Technology (EDBT'96), Avignon, France, September, pp. 3–17 (1996)
3. Zaki, M.J.: SPADE: an efficient algorithm for mining frequent sequences. Mach. Learn. **42**(1–2), 31–60 (2001)
4. Masseglia, F., Poncelet, P., Teisseire, M.: Using data mining techniques on web access logs to dynamically improve hypertext structure. ACM Sig. Web Lett. **8**(3), 13–19 (1999)
5. Han, J., et al.: FreeSpan: frequent pattern-projected sequential pattern mining. In: Proceedings of the Sixth ACM SIGKDD International Conference on Knowledge discovery and data mining, ACM (2000)
6. Pei, J., et al.: PrefixSpan: mining sequential patterns efficiently by prefix-projected pattern growth. In: 2013 IEEE 29th International Conference on Data Engineering (ICDE), IEEE Computer Society (2001)
7. Ayres, J., et al.: Sequential pattern mining using a Bitmap Representation. In: Proceedings of Conference on Knowledge Discovery and Data Mining, pp. 429–435 (2002)
8. Kaneiwaa, K., Kudob, Y.: A sequential pattern mining algorithm using rough set theory. Int. J. Approximate Reasoning **52**(6), 881–893 (2011)

9. Nakamura, S., Nozaki, K., Norimoto, Y., Miyadera, Y.: Sequential pattern mining method for analysis of programming learning history based on learning process. In: The International Conference on Educational Technologies and Computers (ICETC), IEEE, ISBN: 978-1-4799-647-1, September 2014, pp. 55–60 (2014)
10. Wright, A.P., Wright, A.T., McCoy, A.B., Sittig, D.F.: The use of sequential pattern mining to predict next prescribed medications. J. Biomed. Inf. **53**, 73–80 (2015)

A Better Approach for Multilevel Association Rule Mining

Priyanka Rawat, Surya Kant, Bhaskar Pant, Ankur Chaudhary and Shashi Kumar Sharma

Abstract Finding frequent item sets is an important problem for developing association rule in data mining. Several techniques have been developed for solving this. One of them is Pincer approach. Pincer approach is a bidirectional search, which makes use of both bottom-up and top-down process. But it consumes a lot of time on finding the set of maximum frequent itemset. This paper proposes an optimized approach which makes use of And-Or graph search technique. In this we make a tree at each level and we prune that node which occurs in the minimum cost path. This approach is simple and effective because it uses Ao* search which is a heuristic search and finds optimal solution.

Keywords Association rule mining · Data mining · Multilevel association rule mining · And-Or graph

1 Introduction

The role of data mining and knowledge discovery is simple and has been described as extracting precious information from huge databases. Data Mining has become an essential technology for information management, query processing, decision making, process control, etc. [1].

Association rule mining aims to extract the interesting associations and co-relations among sets of items in large datasets. Association rule mining is one of the dominating technique of data mining research was first introduced by Agrawal. Traditionally, there are two basic measures used in association rule mining support

Priyanka Rawat · S.K. Sharma
Graphic Era Hill University, Dehradun, India

Surya Kant (✉)
IIT Roorkee, Roorkee, India
e-mail: suryak111@gmail.com

Bhaskar Pant · Ankur Chaudhary
Graphic Era University, Dehradun, India

M. Pant et al. (eds.), *Proceedings of Fifth International Conference on Soft Computing for Problem Solving*, Advances in Intelligent Systems and Computing 437, DOI 10.1007/978-981-10-0451-3_53

and confidence. Support is a measure that defines the percentage/fraction of records/entries in the dataset that contain X to the total number of records/entries, where X and Y are two different sets of items from the given database [1, 2].

$$\text{Support} = \frac{\text{The number of records/entries that contain } X}{\text{The total number of records/entries}}$$

Confidence is a measure that defines the percentage/fraction of records/entries in the dataset that contain X and Y to the total number of records/entries that contain just X.

$$\text{Confidence} = \frac{\text{The number of records/entries that contain } X \text{ and } Y}{\text{The total number of records/entries that contain } X}$$

Here the aim is to find out all rules that satisfy user-specified minimum support and confidence values. Example of association rule is "80 % of customer who buy itemset X also buy itemset Y" [1]. Basically association rule follows apriori principle, apriori is an iterative approach which generates and tests candidate set step-by-step, it takes more number of scans and time for generating association rule, resulting in low performance [3]. Many applications of association rule mining needs; mining is to be done at multilevel of abstraction [4]. One of the major arguments for the use of multilevel mining is that it has the potential for undiscovered knowledge to be discovered. Single-level approach could not found such knowledge and this new knowledge may be highly relevant or interesting to a given user. For example 80 % of people that purchase computer may also purchase antivirus, it is fascinating to let users to drill-down and display that 75 % of customer purchase quick hill antivirus if they purchase laptop [5]. For traversing multilevel association rule mining, two things are necessary: (1) Data should be organized in the form of concept hierarchy and (2) Effective methods for multilevel rule mining. Maximum frequent set (MFS) is the set of all maximal frequent itemsets. It uniquely determines the entire frequent set, the union of its subsets form the frequent set.

2 Literature Survey

An algorithm was proposed in [6] the problem of mining generalized association rules. Transactions dataset is given, where each transaction contains set and hierarchy of items, association rule is relation between items at any level of the hierarchy. The author mentioned that the hierarchy of transaction did not consider in previous work of association rules and restricted the items in the association rules to only single-level items in the hierarchy. It has also mentioned that a clear solution for this problem is to consider all the sets of items in the actual transaction as well as all the antecedents of each item in the actual transaction. Author has proposed two algorithms, Cumulate and EstMerge, whose running time is two to five times better than Basic.

A top-down progressive deepening method used to generate multilevel association rules from datasets having huge amount of transaction by extending some earlier techniques of association rule mining proposed in [7]. Based on ways of sharing intermediate results various algorithms were discovered, and the relative performance of these algorithms tested on different types of data. This method also discussed relaxation of the rule conditions for finding "level-crossing" association rules. Earlier works on multilevel association rule mining ignore the fact that the sets of item in a transaction can be dynamic new sets of item and transactions are continuously adding into the actual dataset.

A new technique for multilevel association rules mining in large datasets was proposed in [8]. In this concept of counting, inference approach is used that allows performing few support counts as possible. In this method, number of database scan is less at each level as compared to other existing methods for multilevel association rule mining from large datasets.

A new approach based on Boolean matrix used to find frequent item sets was developed in [9]. It uses Boolean relational calculus to discover maximum frequent itemsets at lower level. It scans the database at every level to produce the association rules. In this method, property of Apriori is used to prune the itemsets. It used top-down progressive deepening method and also used Boolean logical operation to generate the multilevel association rules.

To find association rules from given transactional database a new technique proposed in [10], authors employed partition and Boolean concepts, multiple level taxonomy, and different minimum supports. This method works well with problems involving uncertainty in data relationships, which are presented by Boolean concepts. Level-by-level frequent itemsets can be generated by this algorithm and then produce association rules from transaction dataset. It also examines the method of partitioning to limit the total memory requirement.

Mining multilevel association rule using fuzzy logic approach was proposed in [11]. In this approach, a fuzzy multilevel association rule mining algorithm extract implicit knowledge from multilevel dataset. This approach uses a top-down progress and also incorporates fuzzy boundaries instead of sharp boundary interval to derive large itemsets.

Another algorithm removal of duplicate rule for association rule mining from multilevel dataset was proposed which removes hierarchical duplicity in multilevel [12], thus reducing the size of the rule set which improve the quality and usefulness.

Multilevel apriori algorithm based on genetic algo (MAAGA) was proposed [13]. This algorithm combines a concept hierarchy and a genetic algo with apriori algorithm to extract association rule from construction defect databases. In which the rule is processed to remove insignificant ones.

3 Proposed Technique

In this paper, we have used search property of AO* algorithm for mining multilevel association rule and finding MFS in large databases. AO* is an artificial intelligence technique. It always finds minimum cost solution; it is a heuristic search and uses heuristic value. But in this algorithm, we prefer maximum cost path rather than minimum cost path.

3.1 Ao Algorithm*

Ao* which is also known as And-or graph, an optimal solution graph, which finds an optimal solution without considering every problem state. It solves a state-space problem by gradually building a solution graph, beginning from start state. It also finds the minimum cost path from the start state to the goal state.

3.2 Proposed Algorithm

Input Multilevel Transactional database D containing all transactions
Output Maximum Frequent Itemsets

Procedure:

1. Set user-defined min-support and heuristic function cost.
2. Scan the database to count the occurrence of items present in the database.
3. Prune the item that does not qualify the min-support.
4. For qualifying items generate AND-OR graph.
5. The items in max cost path are 1-frequent itemsets.
6. For $k = 2$
   (i) Generate pairs from $k - 1$ frequent itemset.
   (ii) Prune those items that do not qualify min-support.
   (iii) Form the AND-OR graph and prune the itemset which occur in the minimum cost path.
   (iv) Check for stopping condition, i.e.,

   If

   Only 1 itemset is generated, and then no more trees can be formed.

   Go to step 7.
   Else
   Increment k.

7. Stop.

4 An Illustrative Example

An example is used to describe the concept of the proposed algorithm. Transactional dataset is given which consists of 10 transactions as shown in Table 1.

Examine the above transactional dataset. According to this technique, first we scan the database and find the frequency of each items present in thedatabase as shown in Table 2 to construct the AND-OR graph. Minimum Support for this example is 2 and heuristic value is 1.

Those items whose frequency is greater than or equal to minimum support are used to generate the AND-OR graph shown in Fig. 1. And prune the item that does not qualify the minimum support. The item having highest frequency will be at the top of the tree. If there are two or more than two items with the highest frequency, then root of the tree will be null.

Table 1 Transaction table

Transaction no	Items
T1	A11, B11, E11
T2	B11, D11
T3	B11, C11
T4	A11, D11, B22, C22
T5	A22, C22
T6	B11, C11, D11
T7	A22, C22, D11
T8	A11, B11, C11, E11
T9	A11, B11, C11, E11
T10	E22, D22

Table 2 Frequency table

Item	Frequency
A11	4
B11	6
C11	4
D11	4
E11	3
A22	2
B22	1
C22	3
E22	1
D22	1

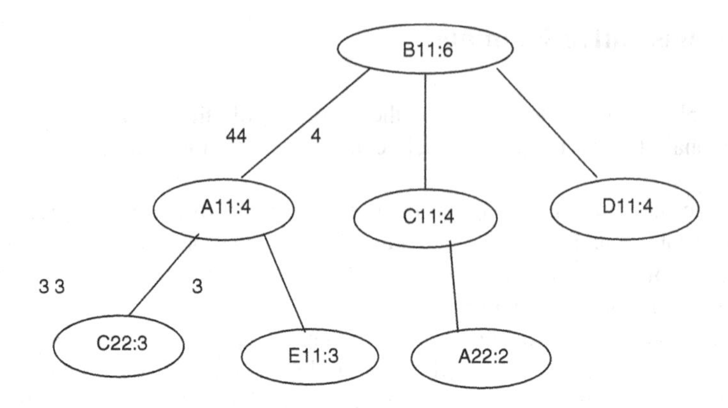

Fig. 1 1-Itemset AND-OR graph

According to this algorithm, we find all the maximum cost path and prune those with minimum cost. Items which occur in the maximum cost path are used to generate the *k*-item sets.

Maximum cost path P1 = (4 + 3 + 1 + 1) = 9, P2 = (4 + 3 + 1 + 1) = 9, P3 = (4 + 2 + 1 + 1) = 8, P4 = (4 + 1) = 5.

P4 is the minimum cost path and we prune the minimum cost path.

According to Fig. 1 B11, A11, C22, E11, C11, A22 items are used to generate the 2-itemsets because they occur in the maximum cost path.

B11A11 = 3, B11C22 = 0, B11E11 = 3, B11C11 = 4, B11A22 = 0, A11C22 = 1, A11E11 = 3, A11C11 = 2, A11A22 = 0, C22E11 = 0, C22C11 = 0, C22A22 = 2, E11C11 = 2, E11A22 = 0, C11A22 = 0

B11A11, B11C11, B11E11, A11E11, A11C11, C22A22, E11C11 are those set which qualify the minimum support are used to generate tree in the next level.

According to Fig. 2 P1 = (3 + 2 + 1 + 1) = 7, P2 = (3 + 2 + 1 + 1) = 7, P3 = (3 + 2 + 1 + 1) = 7, P4 = (3 + 1) = 4.

P4 is minimum cost path we prune it. And 3-item sets which are formed using maximum cost path items are:

(B11C11A11) = 2, (B1C11E11) = 2, (C22A22B11) = 0, (C22A22C11) = 0, (C22A22E11) = 0, (C22A22A11) = 0

B11C11A11, B11C11E11 are that sets which qualify the minimum support and used to generate tree in next level (Fig. 3).

P1 = (2 + 1) = 3, P2 = (2 + 1) = 3, both P1 and P2 are equal in cost we consider both path items. And 4-Item sets which are formed are: B11C11A11E11 = 2.

This set qualifies the minimum support. According to proposed approach, if only one item set qualify the minimum support, no tree is generated this itemset is considered as the final result. Finally B11C11A11E11 is a maximum frequent itemset in this database.

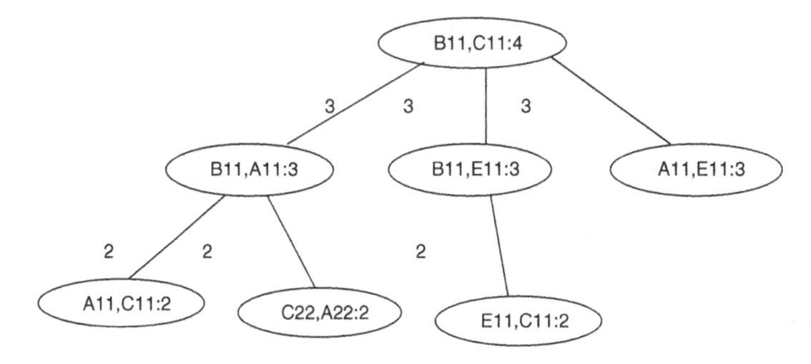

Fig. 2 2-Itemset AND-OR graph

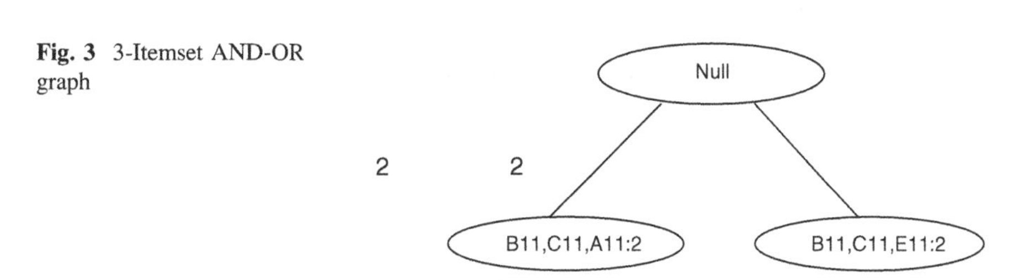

Fig. 3 3-Itemset AND-OR graph

5 Conclusion

Apriori is an important and classical algorithm for association rule mining it works iteratively but it increases number of databases scans, therefore in case of large databases it is not efficient. In this paper, we have developed a multilevel association rule mining algorithm based on AO* search. This makes it possible to find out optimal solution as efficiently as possible without evaluating the entire space. In this example, the result indicates that the proposed approach can find the maximum frequent items simply and effectively.

References

1. Han, J., Kamber, M.: Data mining: concepts and techniques. The Morgan Kaufmann Series (2001)
2. Agrawal, R., Srikant, R.: Fast algorithms for mining association rules. In: Proceedings of the 20th VLDB Conference, pp. 487–499 (1994)
3. Pujari, A.K.: Data Mining Techniques. Universities press, (India), Hyderabad (2001)
4. Rajkumar, N., Karthik, M.R., Sivananda, S.N.: Fast algorithm for mining multilevel association rules. IEEE Trans. Knowl. Data Eng. **2**, 688–692 (2003)
5. Kaya, M., Alhajj, R.: Mining multi-cross-level fuzzy weighted association rules. In: Second IEEE International Conference on Intelligent System, vol. 1, pp. 225–230 (2004)

6. Srikant, R., Agrawal, R.: Mining generalized association rules. In: Proceedings of the 21th International Conference on Very Large Data Bases, pp. 407–419 (1995)
7. Han, J., Fu, Y.: Mining multiple-level association rules in large databases. IEEE TKDE **1**, 798–805 (1999)
8. Thakur, R.S., Jain, R.C., Pardasani, K.R.: Fast algorithm for mining multilevel association rule mining. J. Comput. Sci. **1**, 76–81 (2007)
9. Gautam, Pratima, Pardasani, K.R.: A fast algorithm for mining multilevel association rule based on boolean matrix. (IJCSE) Int. J. Comput. Sci. Eng. **2**, 746–752 (2010)
10. Gautam, P., Pardasani, K.R.: Efficient method for multiple-level association rules in large databases. J. Emerg. Trends Comput. Inf. Sci. **2** (2011)
11. Usha Rani, Vijaya Prakash R., Govardhan, A.: Mining multi level association rules using fuzzy logic. Int. J. Eng. **3**, 8 August. ISSN 2250-2459 (2013)
12. Chandanan, A.K., Shukla, M.K.: Removal of duplicate rule for association rule mining from multilevel dataset. In: International Conference on Advanced Computing Technologies and Applications (ICACTA), vol. 45, pp. 143–149 (2015)
13. Cheng, Yusi, Wen-der, Yu., Li, Qiming: GA-based multilevel association rule mining approach for defect analysis in the construction industry. Autom. Constr. **1**, 78–91 (2015)

Prediction of Market Power Using SVM as Regressor Under Deregulated Electricity Market

K. Shafeeque Ahmed, Fini Fathima, Ramesh Ananthavijayan, Prabhakar Karthikeyan Shanmugam, Sarat Kumar Sahoo and Rani Chinnappa Naidu

Abstract This paper proposes a methodology to utilize support vector machines (SVM) as a regressor tool for predicting market power. Both the companies, i.e., Generation (Gencos) and the Distribution (Discos), can utilize this tool to forecast market power on their perspective. Attributes and criterion are to be chosen properly to classify market power. In this paper, the effectiveness of SVM technique in predicting market power is formulated. Independent system operator (ISO) can also use this tool as regressor and it is discussed elaborately. Both linear and nonlinear kernels are compared. Nodal must run share (NMRS) is used as an index for predicting market power. A sample of three-bus system consisting of two generators and one load/two loads is used to illustrate the study.

Keywords Support vector machine · Market power · Regressor · Kernels · Nodal must run share

1 Introduction

Market power

Must Run Generation (*MRG*): To analyze the load variation and transmission line constraints, market power indices like MRS and nodal must run share (NMRS) are considered [1]. Taking account on generation and transmission constraints, MRG is

K. Shafeeque Ahmed · Fini Fathima · R. Ananthavijayan · P.K. Shanmugam (✉)
S.K. Sahoo · R.C. Naidu
School of Electrical Engineering, VIT University, Vellore 632014, Tamil Nadu, India
e-mail: sprabhakarkarthikeya@vit.ac.in

S.K. Sahoo
e-mail: sksahoo@vit.ac.in

R.C. Naidu
e-mail: crani@vit.ac.in

M. Pant et al. (eds.), *Proceedings of Fifth International Conference on Soft Computing for Problem Solving*, Advances in Intelligent Systems and Computing 437, DOI 10.1007/978-981-10-0451-3_54

the contribution of individual generator to the total load. MRG can be obtained by solving the following linear optimization equations:

$$\text{Minimize}\, P_{\text{gk}} \tag{1}$$

$$\text{Subject to}\quad e^{T} = (\text{Pg} - \text{Pd}) = 0 \tag{2}$$

$$0 \leq \text{Pg} \leq \text{Pg}_{\text{max}} \tag{3}$$

$$-\text{Pl}_{\text{max}} \leq F(\text{Pg} - \text{Pd}) \leq \text{Pl}_{\text{max}} \tag{4}$$

where

e Vector with all one;
Pg Power dispatch vector;
Pd Demand vector;
Pl_{max} Line limit vector; and
F Distribution factors matrix [1]

Equation (2) denotes power balance equation; Eqs. (3) and (4) imply output limits of generator and limits of transmission lines, respectively.

Must Run Share (*MRS*): MRS indicates the variations in the market power with respect to dynamic load. MRS_k of generator k in a power market is given by

$$\text{MRS}_k = \text{Pg}_k^{\text{must}}/\text{Pd} \tag{5}$$

where Pd is the total demand in the power market. The generator is said to have market power only if the MRS is greater than zero.

Nodal Must Run Share (*NMRS*): $\text{NMRS}_{k,i}$ is the minimum contribution of the must run generator k to feed the load at node i and to analyze the geographical difference of market power

$$\text{NMRS}_{k,i} = \text{Pg}_{k,i}^{\text{must}}/\text{Pd}_i\, i = 1, 2 \ldots N \tag{6}$$

The NMRS is determined from the following expression:

$$\text{NMRS}_{k,i} = \frac{\text{Pg}_{k,i}^{\text{must}}}{\text{Pd}_i} = \frac{[M^{-1}]_{ik}\text{Pg}_k^{\text{must}}}{\sum_{j \in N} [M^{-1}]_{ij} P_{gj}} \tag{7}$$

where $\text{Pg}_{k,i}^{\text{must}}$ is the contribution of the must run generator k to Pd_i, N is the number of buses, Pd_i is the load at bus i, j is the bus which is directly connected to bus i through transmission lines, and $[M]$ is the distribution matrix. This matrix is used to show how the power supplied at a node is contributed from all the generators in system and the complete discussion is available in [1]. A complete survey on various indices, used to measure the market power, is available in [2].

2 Support Vector Machines

SVM has been extensively used as a classifier and regressor in various power system domains like fault diagnosis, load forecasting, stability assessment, security assessment, market clearing price estimation/prediction, and so on.

In [3], Correa-Tapasco et al. have combined SVM as a regressor and Chu-Beasley genetic algorithm (CBGA) for locating faults in power distribution systems. Authors have limited their scope with single-phase fault as this fault is predominant in distribution systems. Similarly in [4], the authors have studied power system faults using SVM. SVM multiple classifiers are used to identify all types of power system faults by employing HHT to extract the instantaneous amplitude and Hilbert marginal spectrum of the current signal. The parameters are optimized by the method of cross validation [5]. Another efficient and computationally economical method to classify the power quality events and to detect complex perturbations is based on wavelet transforms [6].

Kalyani et al. [7] proposed multi-class SVM for the application of classifying and assessing the status of a power system during security studies. Different security modes namely secure, critically secure, insecure, and highly insecure are used as parameters to classify the given system status. The results were also compared with the conventional method like least-squares, probabilistic neural network, etc. In this paper, SVM is used as a regressor to predict the value of market power for a set of training points. Here, for a Genco, its generation in active power for a given load level is taken as an attribute and the amount of market power it is exercising and it can exercise (for other loads) is predicted. The prediction is also extended from Disco's and ISO's point of view.

Nowadays, inevitable support is given to the development of large-scale production of green energy. The research hence focuses on predicting the power market patterns incorporated with large-scale wind power [8]. The analysis of SVM algorithm employed for prediction of wind speed shows that prediction time is closer to transaction time boosting accuracy of forecasting speed. Results of work done by Arunkumaran et al. point out that using SVM technique switching losses and harmonics are controlled which results in very low THD level compared to SPWM method [9].

Galactic researches were done for optimal AGC regulators, in which LS SVM interfaced with standard MATLAB is used for AGC of a two-area interconnected power system [10]. Without any efficient training, Kernel function σ and adjustable function γ are found with better performance. Optimum values of SVM are solved using genetic algorithms to improve accuracy and to shrink the training time for online voltage stability monitoring by estimating the voltage stability margin index [11].

3 SVM for Regression

The following discussions about SVM are made available from [1, 12, 13] to facilitate the readers in understanding the concept. Our data of form $\{x_i, y_i\}$ *is* analyzed to predict real valued output y_i where $i = 1, 2, 3....L$,

$$y_i \in R, x \in R^d, \quad y_i = w \cdot x_i + b$$

The regression SVM will use a more sophisticated penalty function than before, not allocating a penalty if the predicted value y_i is less than a distance ξ away from the actual value t_i, i.e., if $|t_i - y_i| < \xi$. Referring to Fig. 1, the region bounded by $y_i \pm \xi \forall i$ is called an ξ-insensitive tube. The other modification to the penalty function is that output variables which are outside the tube are given one of two slack variable penalties depending on whether they lie above (ξ^+) or below (ξ^-) the tube (where $\xi^+ > 0; \xi^- > \forall i$): $t_i \leq y_i + \xi + \xi^+$ and $t_i \geq y_i - \xi - \xi^-$.

The error function for SVM regression can then be written as shown in Eq. (8):

$$\frac{1}{2}||w||^2 + C\sum_{i=1}^{L} \xi_i^+ + \xi_i^- \tag{8}$$

This needs to be minimized subject to the constraints $\xi^+ \geq 0; \xi^- \geq 0 \forall i$. In order to do this we introduce Lagrange multipliers $\alpha_i^+ \geq 0, \alpha_i^- \geq 0, \mu_i^+ \geq 0, \mu_i^- \geq 0, \forall i$.

Now the Lagrangian equation L_p in terms of α is given in Eq. 9:

$$\begin{aligned} &\frac{1}{2}||w||^2 + C\left[\sum_{i=1}^{L} (\xi_i^+ + \xi_i^-)\right] - \sum_{i=1}^{L} \alpha_i^+ (y_i + \xi + \xi^+ - t_i) \\ &\quad - \sum_{i=1}^{L} \alpha_i^- (y_i + \xi + \xi^- - t_i) - \left(\sum_{i=1}^{L} \xi_i + \mu_i^+ + \xi_i - \mu_i^-\right) \end{aligned} \tag{9}$$

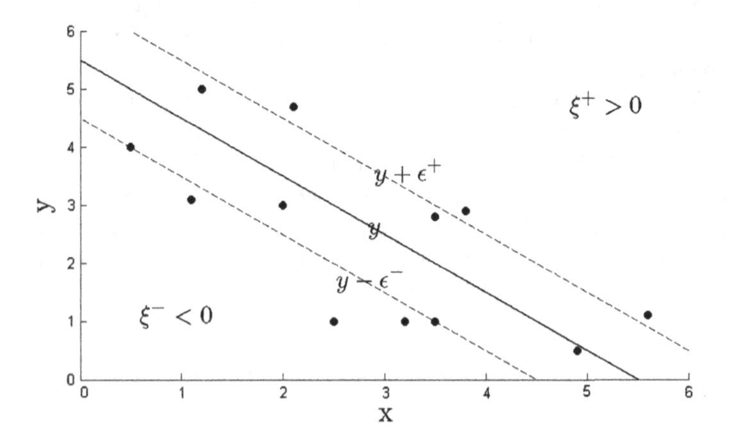

Fig. 1 Regression with ξ-insensitive tube

Substituting for y_i, differentiating with respect to w, b, ξ^+, and ξ^-, and setting the derivatives to 0, we get

$$\frac{dL_p}{dw} = 0 \approx w = \sum\nolimits_{\downarrow} (i=1)^{\uparrow} L \cong (\alpha_i^+ - \alpha_i^-) x_i \tag{10}$$

$$\frac{dL_p}{db} = 0 \approx \sum\nolimits_{\downarrow} (i=1)^{\uparrow} L \cong (\alpha_i^+ - \alpha_i^-) = 0 \tag{11}$$

$C = \alpha_i^+ + \mu_i^+$ and $C = \alpha_i^- + \mu_i^-$.

Substituting in L_p we get the Lagrangian as

$$\begin{aligned} L_p = \sum\nolimits_{\downarrow} (i=1)^{\uparrow} L \cong (\alpha_i^+ - \alpha_i^-) t_i - \xi \sum\nolimits_{\downarrow} (i=1)^{\uparrow} L \\ \cong \left(\alpha_i^+ - \alpha_i^-\right) \\ -\frac{1}{2} \sum\nolimits_{\downarrow} (i,j=1)^{\uparrow} L \cong (\alpha_i^+ - \alpha_i^-)(\alpha_i^+ - \alpha_i^-) x_i x_j \end{aligned} \tag{12}$$

such that

$0 \le \alpha_i^+ \le C$ and $0 \le \alpha_i^- \le C$ and $\sum_{\downarrow} (i=1)^{\uparrow} L \cong (\alpha_i^+ - \alpha_i^-) = 0 \forall i$

And new predictions y' can be found using $y' = \sum_{\downarrow} (i=1)^{\uparrow} L \cong (\alpha_i^+ - \alpha_i^-) x_i x' + b$ where x' is the test point and b is given in Eq. 13:

$$b = \frac{1}{N_s} \left(\sum\nolimits_{\downarrow} (i=1)^{\uparrow} M \equiv t_s - \xi - \sum\nolimits_{\downarrow} (j=1)^{\uparrow} M \cong (\alpha_j^+ - \alpha_j^-) x_j x_i \right) \tag{13}$$

4 Simulation and Results

A three-bus system which is shown in Fig. 2 consists of two generators and one consumer. Each generator can be considered as an individual generation company. The load is considered to be inelastic in nature. All line reactances are of j0.2p.u.

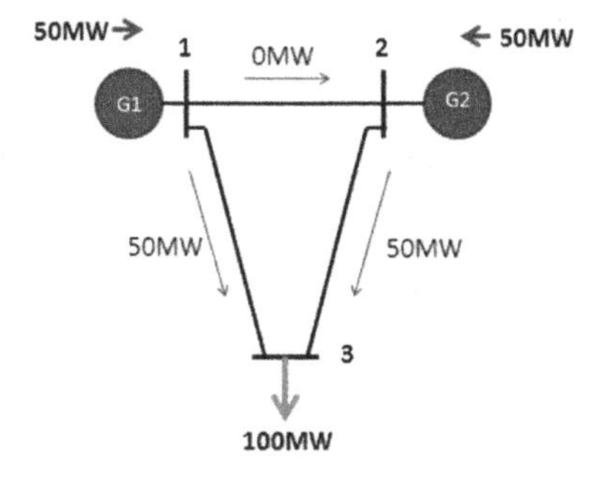

Fig. 2 Sample three-bus system

The cost functions of the generators are assumed to be $F(\mathrm{Pg}_i) = 0.02\mathrm{Pg}_i^2 + 2\mathrm{Pg}_i$ (where $c_i = 0$) as discussed in [14]. The said sample system is sufficient enough to explain the work carried out by the authors.

4.1 Without Any Constraints

The maximum and the minimum generations are taken as 100 and 0 MW. For this system, as both the generators are located at an equal distance from the load, both have the same chance of supplying the load.

It is inferred that as the maximum generation of both the generators is 100 MW, one generator can run without the other, i.e., MRG values of both the generators are zero. Therefore, both MRS and NMRS are also zero. It is clear that no generators enjoy any market power. Hence, there exists perfect competition in this market. Unfortunately, HHI indicates some market power which is not true.

4.2 Case 1: SVM as a Regressor—Discos Perspective for the Sample System

As a regressor, SVM can be used to predict the value of the market power for a given set of training data. For the sample system shown in Fig. 2, the Discos can control or change the information on the load required that is submitted to the ISO. 40 random training points are generated with a maximum value of 100 MW and the corresponding NMRS values pertaining to generator 2 are found. Because of the transmission constraint on the line which is close to generator 1, the NMRS is always zero. Hence, NMRS of generator 1 cannot be taken as a criterion for regression. The random test points generated are shown in Table 1. Since the points lie on a curve that starts from zero on a point not from the origin, it could not be fitted on a linear or a polynomial kernel. Hence, an rbf kernel is used for this data.

The order of the kernel chosen is 10. The result of the rbf kernel regression using the training data is given in Table 1. The type and the order of the kernel have to be chosen manually by the user in order to fit the training data effectively. Ten test points are chosen and the predicted values, their original values, and the percentage error are calculated and shown in Table 2. The regression plot using rbf kernel of order 10 is also shown below in Fig. 3.

From the results shown, it is inferred that the maximum percentage error encountered is around 5.4 %, i.e., the regression process using SVM is quite efficient and from Fig. 3, it is found that the Discos can infer that up to a load of 67 MW, no generator can impose market power on them.

Table 1 Training points for case 1

Pd (*X*)	NMRS 2 (*Y*)	Pd (*X*)	NMRS 2 (*Y*)	Pd (*X*)	NMRS 2 (*Y*)	Pd (*X*)	NMRS 2 (*Y*)	Pd (*X*)	NMRS 2 (*Y*)
43.8744	0	70.9365	0.0969	54.7216	0	11.8998	0	81.4285	0.3421
38.1558	0	75.4687	0.2112	13.8624	0	49.8364	0	24.3525	0
76.5517	0.2365	75.1267	0.203	14.9294	0	95.9744	0.5934	92.9264	0.5472
79.52	0.3023	25.5095	0	25.7508	0	34.0386	0	34.9984	0
18.6873	0	50.5957	0	27.6025	0	58.5268	0	19.6595	0
48.9764	0	69.9077	0.0689	67.9703	0.0138	22.3812	0	25.1084	0
44.5586	0	89.0903	0.4847	65.5098	0	84.0717	0.3942	61.6045	0
64.6313	0	95.9291	0.5927	16.2612	0	25.4282	0	47.3289	0

Table 2 Sample results for case 1

X_{test}	Y_{pred}	Y_{orig}	% error	X_{test}	Y_{pred}	Y_{orig}	% error
32	0	0	0	77	0.2441	0.2468	1.094
47.5	0	0	0	82	0.3559	0.3537	0.6219
68.5	0.0307	0.0292	5.136	83.5	0.3838	0.3832	0.1565
69	0.0419	0.0435	3.678	94.5	0.5645	0.5714	1.2075
71	0.0973	0.0986	1.1318	97	0.5769	0.6082	5.4255

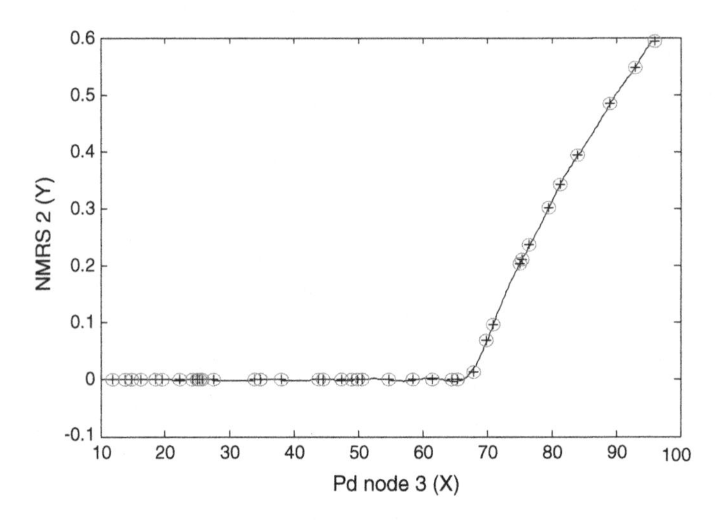

Fig. 3 SVM as a regressor plot using rbf kernel of order 10 for case 1

4.3 *Case 2: SVM as a Regressor—Gencos Perspective on the Sample System*

As a Genco, from their perspective, they can control or change the information on their maximum generation available and the bids that is submitted to the ISO. Due to the presence of the transmission constraint in the line connecting buses 1 and 3, the NMRS of generator 1 is always zero as discussed earlier. Thus, generator 1 can go for minimizing the market power of generator 2 and to obtain profit. Since the bids do not affect the market power, they are not taken as attributes. 40 training points are generated with the maximum value of P_{max} as 100 MW. Here P_{max} is taken as one attribute and $NMRS_{23}$ is taken as Y. And for estimating the profit, generator 1's profit is taken as Y with P_{max} as the attribute. The training points generated are given below in Table 3.

The profits of generator 1 and $NMRS_{23}$ are generated as two different plots using the SVM regressor using rbf kernel of order 10. As mentioned before the nature and the order of the kernel are decided by the user in order to fit the data as efficiently as

Table 3 Training points for case 2

P_{max1} (*x*)	NMRS 2 (*Y*1)	Profit (*Y*2)	P_{max1} (*x*)	NMRS 2 (*Y*1)	Profit (*Y*2)	P_{max1} (*x*)	NMRS 2 (*Y*1)	Profit (*Y*2)	P_{max1} (*x*)	NMRS 2 (*Y*1)	Profit (*Y*2)
10.6653	0.8933	2.275	43.1414	0.65	24.5	85.3031	0.65	24.5	41.7267	0.65	24.5
96.1898	0.65	24.5	91.0648	0.65	24.5	62.2055	0.65	24.5	4.9654	0.9503	0.4931
0.4634	0.9544	0.0043	18.1847	0.8182	6.6137	35.0952	0.65	24.5	90.2716	0.65	24.5
77.491	0.65	24.5	26.3803	0.7362	13.9184	51.325	0.65	24.5	94.4787	0.65	24.5
81.7303	0.65	24.5	14.5539	0.8545	4.2363	40.1808	0.65	24.5	49.0864	0.65	24.5
86.8695	0.65	24.5	13.6069	0.8639	3.7029	7.5967	0.924	1.1542	48.9253	0.65	24.5
8.4436	0.9156	1.4259	86.9292	0.65	24.5	23.9916	0.7601	11.512	33.7719	0.6623	22.8109
39.9783	0.65	24.5	57.9705	0.65	24.5	12.3319	0.8767	3.0415	90.0054	0.65	24.5
25.9787	0.7401	13.5065	54.986	0.65	24.5	18.3908	0.8161	6.7644	36.9247	0.65	24.5
80.0068	0.65	24.5	14.4955	0.855	4.2024	23.9953	0.76	11.5154	11.1203	0.8888	2.4732

Table 4 Sample results for case 2

$P_{\max 1}$ (*X*)	Profit pred	Profit orig	% error	$P_{\max 1}$ (*X*)	Profit pred	Profit orig	% error
9	1.562	1.62	3.5	35	23.8102	24.5	2.815
15	4.592	4.5	2.04	37	24.5795	24.5	0.325
25	12.4841	12.5	0.1272	42	24.5022	24.5	0.008
28	16.354	15.68	4.29	48	24.299	24.5	0.82
32	21.3688	20.48	4.339	52	24.6205	24.5	0.49
$P_{\max 1}$ (*X*)	NMRS pred	NMRS orig	% error	$P_{\max 1}$ (*X*)	NMRS pred	NMRS orig	% error
9	0.9104	0.91	0.04	35	0.655	0.65	0.769
15	0.8493	0.85	0.08	37	0.6492	0.65	0.123
25	0.7501	0.75	0.013	42	0.6501	0.65	0.015
28	0.7152	0.72	0.667	48	0.6515	0.65	0.231
32	0.6739	0.68	0.897	52	0.6491	0.65	0.138

possible. Linear and polynomial kernels cannot fit this set of data as the curve almost looks exponential. The predicted values, original values, and the % error of both the above plots are shown in Table 4.

It is found that the maximum % error in predicting the profit of generator 1 is 4.4 % and in predicting NMRS of the generator 2 on the load at bus 3 is 0.9 %. It is proved that the SVM as a regressor is quite accurate. From Figs. 4 and 5, generator

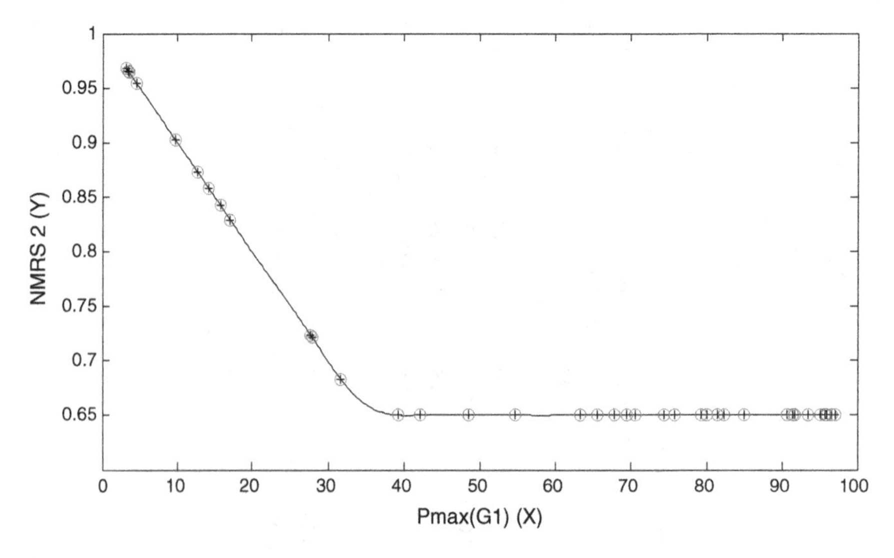

Fig. 4 SVM as a regressor-$P_{\max 1}$ versus NMRS_{23} (using rbf kernel—order 10) for case 2

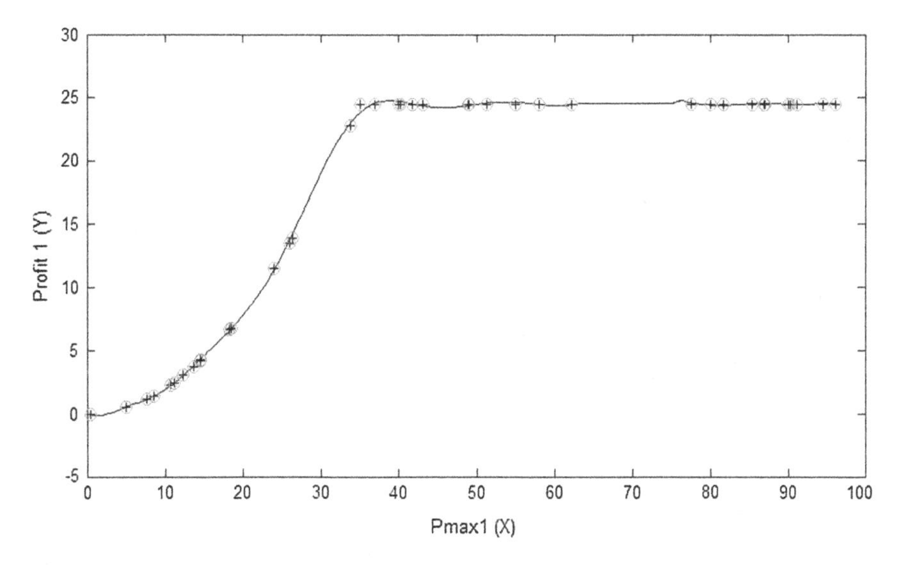

Fig. 5 SVM as a regressor-P_{max1} versus Gen_1's Profit (rbf kernel—order 10) for case 2

1 can infer that both its profit and $NMRS_{23}$ stabilize at maximum and minimum values beyond P_{max} value of 35 MW. Hence, it is enough if generator 1 allocates and submits a P_{max}value of 35 MW to the ISO.

4.4 *Case 3: SVM as a Regressor—ISO's Perspective on the Sample System*

It is observed that in the previous cases generator 2 enjoys market power due to the presence of a transmission constraint. Hence in order to curb its market power, the transmission constraint on that line has to be nullified. More power can be pushed through the transmission line by adding FACTS device in the system. SVM can be used as a regressor by the ISO to know the amount of power that can be allowed through the transmission lines at which the market remains in equilibrium. In other words, this tool helps in considering the market power as a constraint in deciding the location of the FACTS devices under congestion. Random values of transmission constraints are generated with 45 MW as the maximum value and their corresponding values of $NMRS_{23}$ are generated. Hence, the P_{max} of $line_{13}$ is taken as an attribute and $NMRS_{23}$ is taken as *Y*. The generated training points are shown in Table 5 and the regression plot is shown in Fig. 6.

From Fig. 6, it is inferred that the market power imposed by generator 2 on bus 3 becomes null beyond $Pmax_{13}$ of 67 MW. Hence, for example, the ISO can install a FACTS device that can bring up the maximum power allowable in line 1–3 up to 67 MW in order to curb generator 2's market power.

Table 5 Training points for case 3

P_{max23} (*X*)	NMRS 2 (*Y*)	P_{max23} (*X*)	NMRS 2 (*Y*)	P_{max23} (*X*)	NMRS 2 (*Y*)	P_{max23} (*X*)	NMRS 2 (*Y*)
48.3156	0.5505	35.4166	0.9375	48.3156	0.5505	37.7361	0.8679
54.8738	0.3538	46.7993	0.596	54.8738	0.3538	50.4937	0.4852
37.6549	0.8704	40.6764	0.7797	37.6549	0.8704	38.7328	0.838
36.8883	0.8934	62.8	0.116	36.8883	0.8934	68.6664	0
53.5779	0.3927	45.8925	0.6232	53.5779	0.3927	35.1622	0.9451
62.2709	0.1319	53.4987	0.395	62.2709	0.1319	62.1219	0.1363
67.6904	0	40.7977	0.7761	67.6904	0	63.6056	0.0918
39.5467	0.8136	56.0694	0.3179	39.5467	0.8136	65.4043	0.0379
54.9088	0.3527	44.204	0.6739	54.9088	0.3527	37.9553	0.8613
51.4287	0.4571	57.8928	0.2632	51.4287	0.4571	48.9924	0.5302

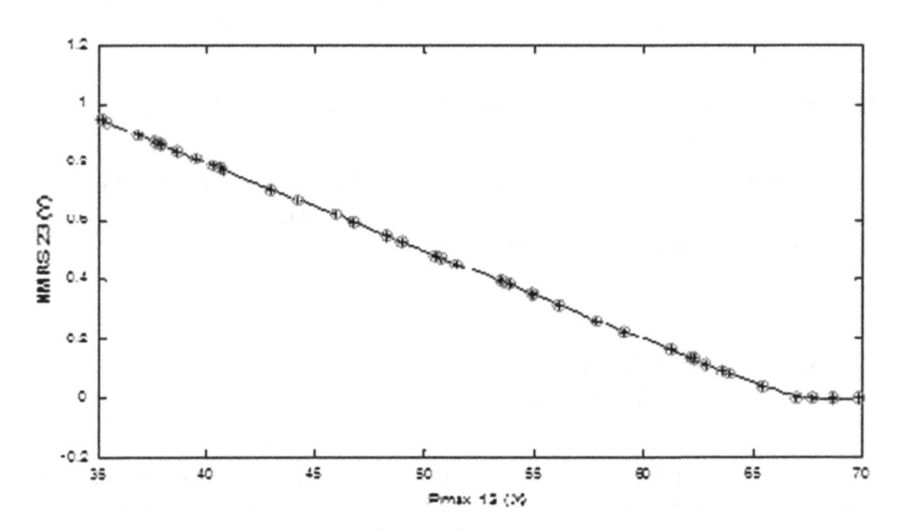

Fig. 6 SVM as a regressor-P_{max13} versus $NMRS_{23}$ (using rbf kernel of order 10) for case 3

5 Conclusions

In this paper, SVM is used as a tool to predict market power of the generators shown in the sample system. Only one attribute is chosen so that the results were plotted in two-dimensional graph. The sample system taken for illustration is made simple to understand how SVM tool can be used as regressor. It is found that prediction is more accurate when nonlinear kernels are used. The only limitation is the requirement of more number of training points. The nature and the order of the kernels have to be chosen manually by the user in order to fit the data points in an effective manner.

References

1. Wang, P., Xiao, Y., Ding, Y.: Nodal market power assessment in electricity markets. IEEE Trans. Power Syst. **19**(3), 1373–1379 (2004)
2. Prabhakar Karthikeyan, S., Jacob Raglend, I., Kothari, D.P.: A review on market power in deregulated electricity market. Int. J. Electr. Power Energy Syst. **48**, 139–147 (2013)
3. Correa-Tapasco, E., Perez-Londono, S., Mora-Florez, J.: Setting strategy of a SVM regressor for locating single phase faults in power distribution systems. In: IEEE/PES Transmission and Distribution Conference and Exposition, Latin America, pp. 798–802 (2010)
4. Wang, Y., Wu, C., Wan, L., Liang, Y.: A study on SVM with feature selection for diagnosis of power systems **2**, 173–176 (2010)
5. Guo, Y., Li, C., Li, Y., Gao, S.: Research on the Power system fault classification based on HHT and SVM using wide-area information. Energy Power Eng. **5**, 138–142 (2013)
6. De Yong, D., Bhowmik, S., Magnago, F.: An effective power quality classifier using wavelet transform and support vector machines. Expert Syst. Appl. **42**, 6075–6081 (2015)
7. Kalyani, S., Shanti Swarup, K.: Classification and assessment of power system security using multiclass SVM. IEEE Trans. Syst. Man Cybernetics-Part C: Appl. Rev. **41**(5), 753–758 (2011)
8. Hua, D., Zhang, Y.T., Wang, Z., Dang, X.: The research of power market patterns incorporation of large-scale wind power. In: IEEE PES Asia-Pacific Power and Energy Engineering Conference (APPEEC), pp. 1–4 (2013)
9. Arunkumaran, B., Raghavendra Rajan, V., Ajin Sekhar, C.S., Tejesvi, N.: Comparison of SVPWM and SVM controlled smart VSI fed PMSG for wind power generation system. In: IEEE National Conference on Emerging Trends in New & Renewable Energy Sources and Energy Management (NCET NRES EM), pp. 221–226 (2014)
10. Sharma, G., Ibraheem, R.: Application of LS-SVM technique based on robust control strategy to AGC of power system. In: Proceedings of IEEE International Conference on Advances in Engineering & Technology Research (ICAETR-2014) (2014)
11. Sajan, K.S., Kumar, V., Tyagi, B.: Genetic algorithm based support vector machine for on-line voltage stability monitoring. Electr. Power Energy Syst. **73**, 200–208 (2015)
12. Shawe-Taylor, J., Cristianini, N.: Support Vector Machines and other kernel-based learning methods. Cambridge University Press (2000)
13. Support Vector Machines Explained—Tristan Fletcher, Tutorial paper-(Ph D 2008). www.cs.ucl.ac.uk/staff/T.Flecher (2009)
14. Prabhakar Karthikeyan, S., Jacob Raglend, I., Sathish Kumar, K., Sahoo, S.K., Priya Esther, B.: Application of SVM as classifier in estimating market power under deregulated electricity market, Lecture Notes in Electrical Engineering. In: 2015 Proceedings of Power Electronics and Renewable Energy Systems: ICPERES 2014

References

Design of Boost Converter for PV Applications

M. Balamurugan, S. Narendiran, Sarat Kumar Sahoo, Prabhakar Karthikeyan Shanmugam and Rani Chinnappa Naidu

Abstract This paper presents a detailed design of portable boost converter. The input is obtained by using photovoltaic (PV) system along with the maximum power point tracking algorithm (MPPT) to pull out maximum power. The boost converter is used to increase the fluctuating PV array voltage to high constant DC voltage. ATmega328 microcontroller is used to design the algorithm and able to track the PV voltage and current with the help of sensors and provides the appropriate pulse width modulation (PWM) signal to control the MOSFET in the boost converter. The main objective of this work is to implement perturb and observe (P&O) method as MPPT to control the boost converter. A detailed hardware design of the converter and MPPT method is carried out with the very good result shows that this design can effectively realize the actual practical Boost converter.

Keywords Boost converter · MPPT · PWM · P&O

M. Balamurugan (✉) · S. Narendiran · S.K. Sahoo · P.K. Shanmugam · R.C. Naidu
School of Electrical Engineering, VIT University, Vellore, India
e-mail: balamurugan.m27@gmail.com

S. Narendiran
e-mail: narengaag@gmail.com

S.K. Sahoo
e-mail: sksahoo@vit.ac.in

P.K. Shanmugam
e-mail: sprabhakarkarthikeya@vit.ac.in

R.C. Naidu
e-mail: crani@vit.ac.in

M. Pant et al. (eds.), *Proceedings of Fifth International Conference on Soft Computing for Problem Solving*, Advances in Intelligent Systems and Computing 437, DOI 10.1007/978-981-10-0451-3_55

1 Introduction

The use of the maximum power point tracking (MPPT) algorithms [1] has led to enhance the competence of function of the PV modules, and thus, it is valuable in the field of Nonconventional energy.

Due to the variation in atmospheric temperature and irradiation the power extracted from the panel is not constant [2]. There are quite a lot of methods to solve this problem. One such is mechanical tracking which proves to be more expensive as it requires powerful motors to rotate the solar panels installed iron rods along with the sun throughout the day [3]. The challenge is to develop an efficient and inexpensive mechanism which would serve the purpose. Over the last decades, Distributed energy resources are focused greatly on the photovoltaic system as it holds massive potential in energy extraction. The factors which lead to such a change in government incentive plan are due to reduction of cost, rising electricity demand, and enhancements in the power electronics industry.

The PV cell output voltage and current are dependent on environmental conditions, i.e., radiation and temperature. Keeping this in mind, to aim for optimized power flow by efficient energy conversion. For high efficient energy conversion, the switch mode power supply (SMPS) switch must have high on and off speed. The switching losses associated with it should also be minimum [4].

The primary intention of this paper is to develop DC–DC step up converter hardware for solar applications and to find a suitable maximum power point tracking algorithm. Modeling of PV cell, boost converter, and interface with the MPPT algorithm to perform the maximum power point operation has been done in Simulink. Then, the hardware implementation of the converter is processed. One common problem with solar technology is the efficiency of solar power extraction. The output power of the PV panel is dependent on temperature variations of the atmosphere. It also varies with loading conditions and hence it is required to continuously monitor it. Perturb and observe (P&O) method [5] is the most commonly used algorithm in PV systems to obtain the maximum power. This scheme has several advantages such as simple implementation, high reliability, low cost, and tracking efficiency. This algorithm could result in oscillations of power output.

2 Modeling of PV System

PV cell is a semiconductor piece of equipment that converts the illumination of solar energy into electrical energy by absorbing sunlight. PV cells are arranged in parallel and series to form the PV array. Energy produced by the PV cell depends on parameters such as temperature and irradiance.

The PV Module is connected to boost converter where the sensor is used to sense voltage and current from the PV module in order to optimize the DC power. The block diagram of the proposed system is shown in Fig. 1.

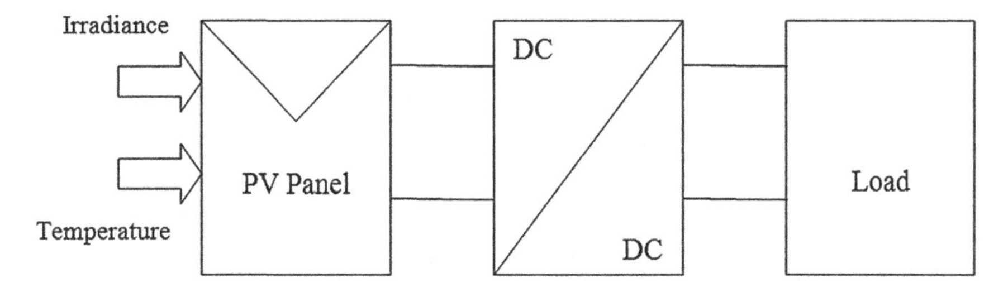

Fig. 1 Block diagram of the proposed system

2.1 Modeling of PV Array

The PV Array is made up of connecting the solar cell in parallel and series.

The panel current I_{av} can be stated by Eq. (1)

$$I_{\mathrm{av}} = I_{\mathrm{rr}} - I_{\mathrm{ss}} * \left[\exp\left(\frac{e(V_{\mathrm{av}} + I_{\mathrm{av}} * R_{\mathrm{se}})}{N_{\mathrm{ss}} A_s k T_s}\right) - 1\right] - \frac{V_{\mathrm{av}} + I_{\mathrm{av}} * R_{\mathrm{se}}}{R_{\mathrm{pl}}} \tag{1}$$

The Irradiance current I_{rr} can be put across by Eq. (2)

$$I_{\mathrm{rr}} = I_{\mathrm{scc}} + (K_e(T_s - T_{\mathrm{refe}}))\frac{G_s}{G_{\mathrm{refe}}} \tag{2}$$

The diode saturation current I_{ss} can be articulated by Eq. (3)

$$I_{\mathrm{ss}} = I_{\mathrm{rc}}\left(\frac{T_s}{T_{\mathrm{refe}}}\right)^3 \left[\exp\left(\frac{eE_a}{A_s k}\right) * \left(\frac{1}{T_{\mathrm{refe}}}\right) - \left(\frac{1}{T_s}\right)\right] \tag{3}$$

The reverse saturation current I_{rc} can be written as Eq. (4)

$$I_{\mathrm{rc}} = \frac{I_{\mathrm{scc}}}{\left(\exp\left(\frac{e * V_{\mathrm{ocv}}}{N_{\mathrm{ss}} A_s k T_s}\right) - 1\right)} \tag{4}$$

where

I_{av} Current of PV array
V_{av} Voltage of PV array
V_{ocv} Open-circuit voltage of PV panel
I_{scc} Short-circuit current of PV panel
I_{rr} Photocurrent or irradiance current of PV panel
I_{ss} Diode saturation current of PV panel
I_{rc} Reverse saturation current of PV panel
R_{se} Series resistance

Table 1 PV panel datasheet

PV cell characteristics	Values
Open circuit voltage (V_{ocv})	21.239 V
Short circuit current (I_{scc})	4.735 A
Irradiation (I_{ra})	1000 W/m^2
Reference temperature (T_{refe})	25 °C
Peak power current (I_{ms})	4.54 A
Maximum power voltage (V_{ms})	21.24 V

R_{pl} Parallel resistance of PV array
T_s Operating temperature of PV panel
T_{refe} Reference temperature (25 °C at STC)
G_s Solar iradiance
G_{refe} Reference solar irradiance (1000 W/m$^{2)}$
A_s Ideality factor of diode (A_s = 1.3 for silicon diode)
K Boltzmann constant (1.380*10^{-23} J/K)
E Charge of electron (1.6*10^{-19} C)
N_{ss} Number of cells in series in one module
N_{pl} Number of cells in parallel in one module
E_a Bandgap energy of semiconductor (E_a = 1.1 eV for silicon)

Although most of the works cited in the literature uses double diode model for better operation. The single diode model is utilized in this work for the reason that it offers a superior balance between accuracy and ease.

Most of the constraints in the equations should be taken from data sheet offered by company. The PV panel requirement of the proposed system is displayed in Table 1.

2.2 MPPT

The efficiency of a PV panel is very low. In order to improve the effectiveness, several methods are to be carrying out to match the source and load impedance. Maximum power point tracking (MPPT) is one such method is to extract maximum power [6]. The parameters of the PV panel are shown in Table 1.

The MPPT is devised to track the working point by sensing the output power from the PV panel. This technique is employed to acquire the utmost power under changeable atmospheric conditions. In PV systems, the I–V characteristic graph is extremely nonlinear, thereby making it intricate to be used to power a firm load [7]. This can be done by using boost converter where duty signal is tuned by using the algorithm.

The PV array is coupled to the boost converter in order to produce the maximum power by tuning the electrical operating point. MPPT charge regulator is mostly a

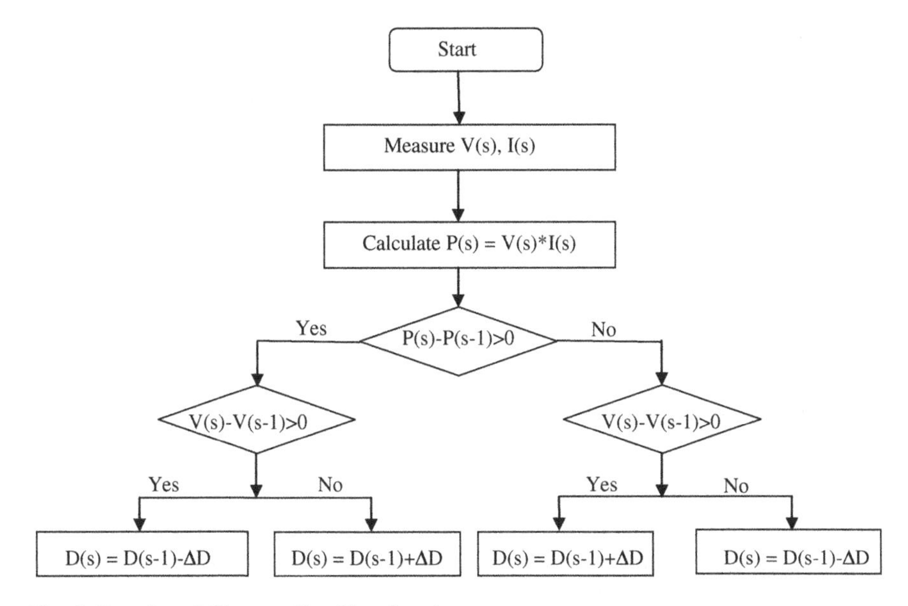

Fig. 2 Perturb and Observe Algorithm flowchart

DC–DC boost converter. The converter alters the operating voltage of the PV panel to facilitate and it is operated at a maximum voltage to produce most power. The output voltage is regulated by varying the duty signal of the converter. The PWM signal is given to the gate of the MOSFET in the boost converter and the control indicator is routinely tuned by using an algorithm.

The function of perturb and observe algorithm is to control the duty signal of the boost converter. The work flow of the processed algorithm is shown in Fig. 2. In perturb and observe algorithm, the voltage, and current are tracked with the help of sensors from the PV array and the utmost power is calculated. By comparing the power drawn from the PV array of each cycle, the new perturbation path can be determined.

If the power is increased then the previous cycle perturbation path will be go after in the new control cycle. Otherwise, the perturbation path will alter. By this method, the maximum peak point of the PV array can get closer to the maximum power and then reaches a steady state.

3 Hardware Design of Boost Converter

DC–DC converter acts as a crossing point between the load and PV array. By changing the duty signal of the converter, the load resistance is adjusted and matched the peak power to transfer the maximum power. The circuit illustration of the boost converter is shown in Fig. 3.

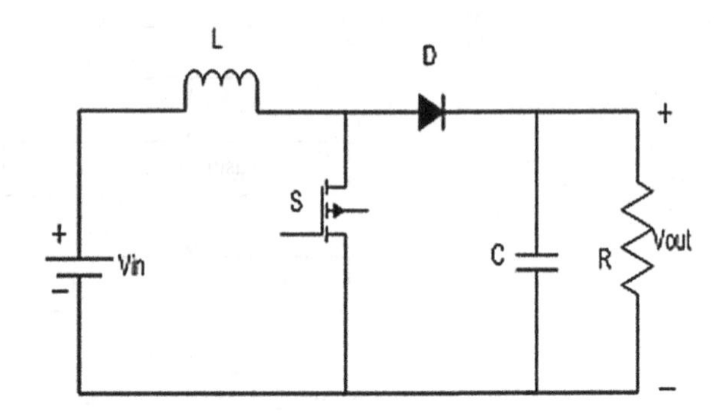

Fig. 3 Circuit diagram of boost converter

The output voltage equation of boost converter is:

$$\frac{V_o}{V_i} = \frac{1}{1 - d_c} \quad (5)$$

The duty cycle, inductor, and the capacitor values of the boost converter is determined by the following equation,

$$d_c = 1 - \frac{V_i}{V_o} \quad (6)$$

$$L_m = (1 - d_c)^2 * d_c * R_l / 2f_s \quad (7)$$

$$C_m = V_o * d_c / R_l * f_s * \Delta V_{re} \quad (8)$$

where,

V_i Input Voltage of converter
L_m Inductance
C_m Capacitance
R_l Load resistor
V_o Output voltage of converter
d_c Duty cycle of converter
f_s Switching frequency of converter
ΔV_{re} Ripple Voltage

The parameter constraint of the boost converter is shown in Table 2.

3.1 Design of Inductor

To calculate the inductance of air core single-layer coil the following formula is used (Fig. 4):

Table 2 Parameter constraint of the boost converter

Parameter	Values
Inductor	100 uH
Capacitor	1000 uF
Resistor	500 Ω

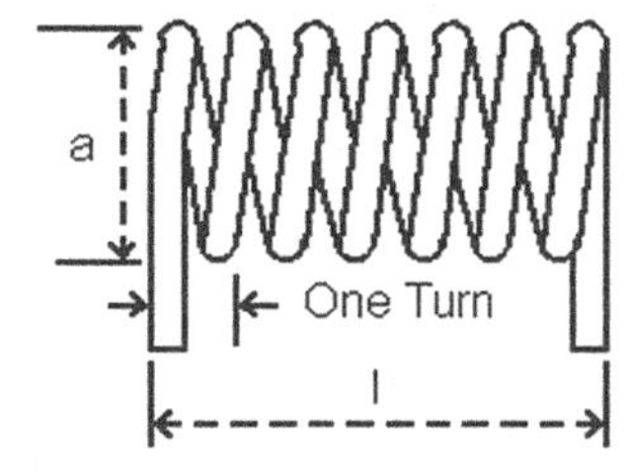

Fig. 4 Core diagram of single-layer air-core coil

$$L = a^2 \times N^2/[(18 * a) + (40 * l)] \quad (9)$$

where,

L Inductance
N Total no of turns
L Length of the coil
A Coil outside diameter

The parameters of designed inductor with its ratings are shown in Table 3.

3.2 Hardware Setup

The hardware realization of the boost converter is shown in Fig. 5. ATmega328 microcontroller [8, 9] is used for designing the algorithm by sensing the PV voltage

Table 3 Design specifications for inductor

Parameter	Ratings
Core material type	Hollow cylinder
Turns	100
Diameter (cm)	3
Length (cm)	7.5
Coil wire (AWG)	18
Temperature rating (°C)	150

Fig. 5 Experimental setup of boost converter

Table 4 Boost converter experimental results

Operating point	Voltage	Current	Power	Duty cycle
MPP	5	2.32	11.58	0.5
Below MPP	3.17	2.32	7.54	0.53
	3.34	2.32	7.75	0.56
Above MPP	4.5	2.32	10.43	0.65
	4.33	2.32	10.02	0.62

and current using sensors and generate the suitable pulse width modulation (PWM) signal to control the MOSFET in the boost converter. It has a very high power handling capability and was tested with a load of 20 W, two 1000uF capacitors connected in series to increase voltage handling capability.

The experimental results of the boost converter for different operating conditions are shown in Table 4. When the operating point is at MPP the duty cycle is constant, there is no change in voltage and power.

When the peak point of the PV panel is less than MPP the power is increasing in successive iterations. When the operating point is above MPP the power is decreasing in successive iterations.

4 Conclusion

In this paper, the hardware implementation of a boost converter model is developed. This model can work well under varying atmospheric condition. The maximum power of the PV panel is traced with a P&O MPPT algorithm by using boost converter. The experimental results have confirmed that the proposed design shows good performance of the boost converter and it can be used for high power applications.

Acknowledgments The authors gratefully acknowledge support from Department of Science and Technology (DST), Government of India, Project No. DST/TSG/NTS/2013/59 and VIT University, Vellore, India.

References

1. Xiao, W., Dunford, W.G.: A modified adaptive hill climbing MPPT method for photovoltaic power systems. In: Proceedings of IEEE PESC, pp. 1957–1963, 2004
2. Zhang, C., Zhao, D., Wang, J., Chen, G.: A modified MPPT method with variable perturbation step for photovoltaic system. In: Power Electronics and Motion Control Conference, pp. 2096–2099, 2009
3. Green, M.A.: Photovoltaics: coming of age. In: Photovoltaic Specialists Conference, 1990
4. Piegari, L., Rizzo, R.: Adaptive perturb and observe algorithm for photovoltaic maximum power point tracking. IET Renew. Power Gener. **4**(4), 317–328, July 2010
5. Femia, N., Petrone, G., Spagnuolo, G., Vitelli, M.: Optimization of perturb and observe maximum power point tracking method. IEEE Trans. Power Electr. **20**(4), 963–973, July 2005
6. Femia, N., Petrone, G., Spagnuolo, G., Vitelli, M.: Optimizing sampling rate of P&O MPPT technique. In: Proceedings of IEEE PESC, pp. 1945–1949, 2004
7. Nguyen, T.L., Low, K.-S.: A global maximum power point tracking scheme employing DIRECT search algorithm for photovoltaic systems. IEEE Trans. Ind. Electr. **57**(10), 3456–3467 (2010)
8. Surek, T.: Progress in U.S. photovoltaics: Looking back 30 years and looking ahead 20. In: World Conference on Photovoltaic Energy Conversion. Osaka, Japan, 2003
9. Koutroulis, E., Kalaitzakis, K., Member, IEEE, Voulgaris, N.C.: Development of a microcontroller-based, photovoltaic maximum power tracking control system. IEEE Trans. Power Electr. **16** (2001)

A Review of Soft Classification Approaches on Satellite Image and Accuracy Assessment

Ranjana Sharma, Achal Kumar Goyal and R.K. Dwivedi

Abstract Classification is a widely used technique for image processing and is used to extract thematic data for preparing maps in remote sensing applications. A number of factors affect the classification process. But classification is only half part of image processing and incomplete without accuracy assessment. Accuracy assessment of classification tells how accurately the classification process has been carried out. This research paper presents a review study of image classification through soft classifiers and also presents accuracy assessment of soft classifiers using entropy. Soft classifiers help in the development of more robust methods for remote sensing applications as compared to the hard classifiers. In this paper, two supervised soft classifiers, FCM, and PCM have been used to demonstrate the improvement in the classification accuracy by membership vector, RMSE, and also it has tried to generate fraction output from FCM, PCM, and noise with entropy.

Keywords Fuzzy C-Means · Fuzzy possibilistic C-Means · K-means · Soft classification · Maximum likelihood classifier (MLC) · Entropy

1 Introduction

Remote sensing [1, 2] has thus become an important data source for providing effective land use, land cover information particularly at regional to global scales. The conventional hard classification methods, which assume that the pixels [3] are

Ranjana Sharma (✉)
Uttarakhand Technical University, Dehradun, Uttarakhand, India
e-mail: sharmaranjana04@gmail.com

A.K. Goyal
Computer Centre, Gurukul Kangri University, Haridwar, India
e-mail: dr_achalk@yahoo.com

R.K. Dwivedi
CCSIT, Teerthanker Mahaveer University, Moradabad, India
e-mail: r_dwivedi2000@yahoo.com

M. Pant et al. (eds.), *Proceedings of Fifth International Conference on Soft Computing for Problem Solving*, Advances in Intelligent Systems and Computing 437, DOI 10.1007/978-981-10-0451-3_56

pure, force the mixed pixels [4] to be allocated to one and only one class. This may result in a loss of pertinent information present in a pixel. Mixed pixels [5] may thus be treated as error, noise, or uncertainty in class allocation for hard classification methods. Over and underestimate the actual area covered by these classes on ground. The land use land cover areal estimates obtained from hard classification, go as input to any other remote sensing or geographical information system (GIS)-based application, may affect the accuracy of the end product Clearly, the conventional hard classification methods may not be appropriate for classification of images fraught with mixed pixels. This problem may be resolved either by ignoring or removing the mixed pixel from the classification process. This may be undesirable due to loss of class information hidden in these pixels. Due to this problem, scientists and practitioners have made great effort in developing advanced classification approaches and technique for improving classification accuracy [6, 7].

The conventional use of hard classification methods, which assume that the pixels are pure force, is that the mixed pixel to be allocated to one and only one class. This may result in a loss of pertinent information present in a pixel. Mixed pixels may thus be treated as error, noise, or uncertainty in class allocation for hard classification methods. There are two types of classification; one is supervised and the other is unsupervised. Used in supervised classification is fuzzy-based classification; [8] also soft classification, data elements can belong to more than one class and associated with each element is a set of membership level and then using them to assign a data element to one or more class. In mixed pixel, they are assigned to the class with the highest proportion of coverage of yielding a hard classification due to which considerable amount of information is lost. To overcome this problem, this loss soft classification was introduced.

A software classification assigns a pixel to different class according to the area, it represent, inside the pixel [9]. This software classification [10] yield a number of fraction images equal to the no of land cover classes. The use of fuzzy set-based classification methods in remote sensing has received growing interest for their particular value in situations where the geographical phenomena are inherently fuzzy. So there are two soft classifier FCM [11] and PCM. The basic difference between FCM and PCM is the objective function in FCM includes an additional term called as regularizing PCM clustering is similar to FCM but it does not have probabilistic constraint of FCM. Therefore, term. Previously, FCM and PCM Clustering worked in unsupervised mode. But in this paper we have discussed soft classifiers FCM and PCM that are essentially unsupervised classifiers, but in this study these classifiers are applied in supervised mode.

"This paper is focused on advanced classification approaches and techniques used for improving classification accuracy".

2 Literature Review

The extraction of land cover from remote sensing Images [12, 13] has traditionally been viewed as a classification problem where each pixel in the image is allocated to one of the possible classes. So, remotely sensed data of the earth may be analyzed sensing has thus become an important data source for providing effective land use land cover information particular at regional to global scales. Digital image classification [14] is usually performed to retrieve this information using a range of statistical pattern recognition or a classification technique (supervised and unsupervised) such as maximum like hood classifier, k-mean classifier, the minimum distance to mean classifier etc. There classifiers allocate each pixel of the remote sensing image to a single land use land cover class. Thus producing hard classification. However, often the images are dominated by mixed pixels, which do not represent a single land use, land cover class, but contain two or more classes in a single pixel area [15]. The conventional use of hard classification methods that allocate one class to a pixel may tend to over- and underestimate the actual aerial extends of the classes on the ground and thus may provide erroneous results [15]. For assessing the accuracy of soft classification output mixed pixels have been incorporated at the testing stage. Recently, the concept of fuzzy error matrix (FERM) has been put forth to assess the accuracy of soft classification [16]. Utilization of soft reference data, created either from actual class proportions are based on a linguistic scale allows for incorporation of mixed pixels in the testing stage of classification. For measuring the uncertainty using entropy of MSS data [17] in order to evaluate the classification performance. Another widely used method to determine the class composition of mixed pixel in the allocation stage is the linear mixture model (LMM), which has been used in a number of studies for various applications [18]. Fuzzy set theory-based classifier, fuzzy, c-means (FCM) clustering, an unsupervised approach has been most widely used in remote sensing studies [19, 20, 16]. The FCM, however, is based on the probabilistic constraint that class memberships of a pixel across the classes sum to one.

3 General Review

In hard classification, class is assigned to pixel is crisp i.e. pixel belongs to one of the class from all classes. The classified pixel is either completely belongs to a class or not. This is called hard classification [21]. Although in real world the pixel has some spatial resolution and can cover a mixture of two or more class features on ground. The pure pixels are rare. Most likely, boundaries of classes have the mix pixel. Therefore the soft classification approach was developed [22]. Soft classification is used to produce class proportions within a pixel ill order to increase the classification accuracy [23] and to produce meaningful and appropriate land cover composition [24, 25]. One of the most popular fuzzy clustering [10] methods are the fuzzy c-means (FCM) [11] which is an unsupervised classifier that in an iterative process assigns class membership values to pixels of an image by minimizing an objective function. Although, a few studies on the use of FCM have been reported, the major limitations of FCM are the probabilistic sum to one constraint. Therefore, besides using this classifier, another fuzzy set clustering method, namely possibilistic c-means (PCM) [26], which relaxes this constraint so as to be robust to the noise (i.e., Pixels with a high degree of class mixtures) present in the dataset, has also been implemented.

3.1 Fuzzy c-Means (FCM) Clustering

FCM is a method of clustering which allows one piece data to belong to two or more clusters that may be employed to partition pixels of remote sensing images into different class membership values [27]. The key to represent that similarly that a pixel shares with each cluster with a function whose value lie between zero and one. The objective function FCM is

$$j_m(U,V) = \sum_{i=1}^{N}\sum_{j=1}^{C} (\mu_{ij})^m ||x_i - v_j||_A^2 \quad [21,28] \tag{1}$$

Subject to constraints

$$\begin{aligned} &\sum_{j=1}^{c} \mu_{ij} = 1 \text{ for all } I \\ &\sum_{j=1}^{c} \mu_{ij} > 0 \text{ for all } j \\ &0 \leq \mu_{ij} \leq 1 \text{ for all } i,j \end{aligned} \tag{2}$$

where x_i is the vector denoting spectral response I (i.e., a vector of spectral response of a pixel), V is the collection of vector of cluster centres, and v_j, μ_{ij} are class membership values of a pixel (members of fuzzy c-partition matrix), c and n are number of cluster and pixels respectively, m is a weighting exponent ($1 < m < \infty$), $||x_i - v_j||_A^2$ is the squared distance (d_{ij}) between x_i and v_j, and is given by,

$$d_{ij}^2 = ||x_i - v_j||_A^2 = (x_i - v_j)^T A(x_i - v_j) \quad [21,15] \tag{3}$$

where A is the weight matrix

3.2 Possibilistics C-Mean (PCM) Clustering

The formulation of PCM is based on a modified FCM objective function, whereby an additional term called is regularizing term is also included. PCM is also an iterative process where the class membership values are obtained by minimizing the generalized least-square error objective function [11], given by

$$j_m(U,V) = \sum_{i=1}^{N}\sum_{j=1}^{C}\left(\mu_{ij}\right)^m \|x_i - v_j\|_A^2 + \sum_{i=1}^{c}\left(1-\mu_{ij}\right)^m \quad [11,21] \tag{4}$$

Subject to constraints

$$\begin{aligned} &\max_j \mu_{ij} > 0 \text{ for all } i \\ &\sum_{i=1}^{N} \mu_{ij} > 0 \text{ for all } j \\ &0 \leq \mu_{ij} \leq 1 \text{ for all i, } j \quad [11,21] \end{aligned} \tag{5}$$

where η_j is the suitable positive number. And η_j depends on the shape and the average size of the cluster j and its value may be computed as

$$\eta = K\frac{\sum_{i=1}^{N}\mu_{ij}^m d_{ij}^2}{\sum_{i=1}^{N}\mu_{ij}^m} \quad [11,21] \tag{6}$$

where K is a constant and is generally kept as 1. The class memberships, μ_{ij} are

$$\mu_{ij} = \frac{1}{1+\left(\frac{d_{ij}^2}{\eta_j}\right)^{\frac{1}{(m-1)}}} \quad [11,21] \tag{7}$$

4 Remote-Sensing Classification Process

Accuracy assessment process of sub-pixel classification is also differing from hard classification as hard classification have single land cover information with each pixel where as in soft classification, each pixel have land cover proportion information for every land cover classes in it. Accuracy assessment of soft classified data has to handle multiple land cover information associated with pixel. The major factors of remote sensing are selection of suitable classification approaches, post-classification process, and accuracy assessment.

4.1 Selection of Data in Remotely Sensed

Understanding the strengths and weaknesses of different types of sensor data is essential for the selection of suitable remotely sensed data for image classification. Some previous literature has reviewed the characteristics of major types of remote-sensing data [29, 18].

4.2 Selection of a Classification Method and Training Sample for Remotely Sensed Data

A suitable classification system and a sufficient number of training samples are prerequisites for a successful classification [18]. Identified three major problems when medium spatial resolution data are used for vegetation classifications defining adequate hierarchical levels for mapping, defining discrete land-cover units discernible from selected remote-sensing data, and selecting representative training sites

5 Uncertainty Visualization

There are different ways in literatures to assess accuracy of soft classified data depending upon availability and type of reference data. Hamid Dehghan in 2006 [30], proposed that the classification process result in classified pixel which can be divided into three groups, given as (a) pixel whose classification is certain and correct, some (b) pixels whose classification are certain, but incorrect, and the (c) and others, whose classification is uncertain. Correctness of the classification results are evaluated by some criteria such as the accuracy and reliability. There are several methods available to access accuracy of soft classified data, such a Membership vector, Mean relative error, RMSE and Entropy

5.1 Membership Vector

Membership vectors method is used to measure the uncertainty for the classification results. The elements of membership vector for a pixel are the actual membership value to one of the "*M*" classes. This vector for pixel "*x*" is defined as

$$\mu(W/x) = \begin{bmatrix} \mu(w_1/x) \\ \cdot \\ \cdot \\ \mu(w_k/x) \\ \cdot \\ \cdot \\ \mu(w_M/x) \end{bmatrix} \tag{8}$$

where μ "(w_k/x)" is the estimated membership function of class "k" for pixel "x".

5.2 Mean Relative Error (MRE)

Mean relative error for a pixel is defined based on all actual classifier outputs and desired outputs. If the desired membership function for class "k" is "md(w_k/x)", the relative error (RE) will be defined as

$$\mathrm{RE} = \frac{|\mu d(w_k/x) - \mu(w_k/x)|}{|\mu d(w_k/x)|} \quad [30] \tag{9}$$

The mean relative error for a pixel is defined

$$\mathrm{MRE} = \frac{1}{M}\left[\sum_{k=1}^{m}\frac{|\mu d(w_k/x) - \mu(w_k/x)|}{|\mu d(w_k/x)|}\right] \quad [30] \tag{10}$$

5.3 Root Mean Square Error (RMSE)

This accuracy measure is based on membership value difference between actual and desired data. It is calculated by square root of average squared difference between these two values. It is calculated by

$$\mathrm{MRE} = \frac{1}{M}\left[\sum_{k=1}^{m}|\mu d(w_k/x) - \mu(w_k/x)|\right]^{1/2} \tag{11}$$

6 Proposed Criteria of Assessment of Accuracy: Entropy

Different classification algorithms will be investigated for extracting information of interest at sub-pixel level. These algorithms will be used in supervised classification mode; it means reference data will be generated from input multi spectral image [31, 21, 30]. The reference data will be used for generating statistical parameters of different algorithms. The output generated from these algorithms in the form of land use and land cover map will be evaluated [32].

The accuracy [33–35] of the output generated in this methodology will be evaluated in different ways. In the first step, accuracy of the output will be checked using higher resolution multi-spectral data of same date as date of classification data resolution from Resource-Sat-1 (P6) satellite [36]. In the second step the accuracy will be checked using ground reference data collected from Mobile GIS/GPS/Geodetic Single Frequency GPS. We are measuring the assessment accuracy using entropy [33, 34].

$$H = -\sum_{i=1}^{C} \mu_{ij} \log_2(\mu_{ij}) \quad [30] \tag{12}$$

where, μ_{ij} is the membership value and it denotes the class proportion of class j in pixel i of the image and c is the number of classes. The entropy of each pixel is summed over all pixels to compute average entropy of whole classification.

7 Study Area and Data Used

Study area for this research work has been identified, Sitarganj Tehsil (Udham Singh Nagar District, Uttarakhand). Surroundings of Aurangabad, Maharashtra.

In this research work soft classification approach will be applied while preparing land use and land cover map using multi-spectral remote sensing data sets (Recourse Sat data). The data sets to be used in this research work will be of two types. One data set will be fine spatial resolution data, to check classifier algorithms toward coarser spatial resolution data set and second data with coarser to medium coarser resolution for classification (Fig. 1).

The AWIFS versus LISS-III [21] data set have been used for the purpose of classification, whereby LISS-III versus LISS-IV data set has been used to generate reference data and AWIFS versus LISS-IV data is used for the purpose of similarity analysis (Table 1).

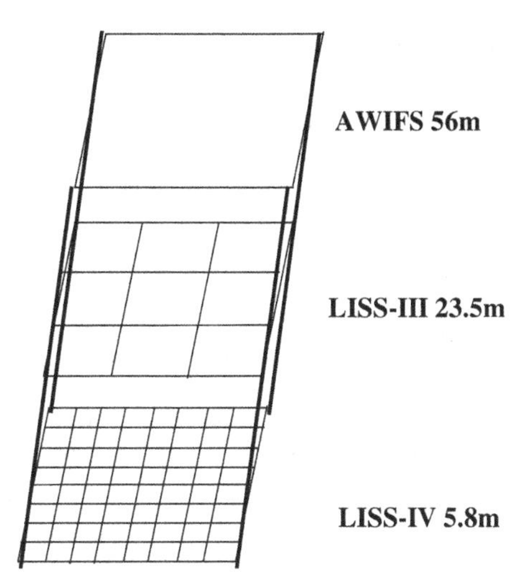

Fig. 1 IFOV of different sensors of resource Sat-1 (IRS-P6) Satellite with respect to single pixel coverage of AWIFS payload

Table 1 Resource Sat-1 payload characteristic

Sensor	Bands	Resolution (m)	Swath (km)	Quantization (bits)
LISS-IV Mono mode	Red	5.8	70.3	7
LISS-IV MX mode	Green red NIR	5.8	70.3	7
LISS-III	Green red NIR SWIR	23	141	10
AWiFS	Green red NIR SWIR	56 (nadir) … 70 (edge)	740	10

8 Discussion and Conclusion

The expected outcomes from this research work would be as follows:

i. By this paper, understand and illustrate the efficiency and capability of FCM and PCM.
ii. In this paper, we addressed the uncertainty problem for classification applications and introduce a new entropy-based criterion.
iii. The effect of learning parameters on the output classified image.

References

1. Eastman, J.R., Laney, R.M.: Bayesian soft classification for sub-pixel analysis: a critical evaluation. In: Photogram. Eng. Remote Sens. **68**, 1149–1154 (2002)
2. Foody, G.M.: Approaches for the production and evaluation of fuzzy land cover classifications from remotely sensed data. Int. J. Remote Sens. **17**(7), 1317–1340 (1996)
3. Shi, W.Z., Ehlrs, M., Temphli, K.: Analysis modeling of positional and thematic uncertainty In: Integration of Remote Sensing And GIS. In: Transaction in GIS, vol. 3, pp. 119–136, (1999)
4. Foody, G.M.: Approaches for the production and evaluation of fuzzy land cover classifications from remotely sensed data. Int. J. Remote Sens. **17**(7), (1996)
5. Foody, G.M., Arora, M.K.: Incorporating mixed pixels in the training, allocation and testing stages of supervised classification. Pattern Recogn. Lett. **17**, 1389–1398 (1996)
6. Adams, J.B., Smith, M.O, Gillespie, A.R.: Imaging spectroscopy: interpretation based on spectral mixture analysis. In: Pieters, C.M., Englert, P.A.J. (eds.) Remote Geochemical Analysis: Elemental and Mineralogical Composition, pp. 145–166 (1993)
7. Pal, M., Mather, P.M.: An assessment of the effectiveness of decision tree methods for land cover classification. Remote Sens. Environ. **86**, 554–565 (2003)
8. Lewis, H.G., Brown, M.: A generalized confusion matrix for assessing area estimates from remotely sensed data. Int. J. Remote Sens. **22**, 3223–3235 (2001)
9. Cracknell, A.P.: Synergy in remote sensing-what's in a Pixel? Int. J. Remote Sens. **19**(11), 2025–2047 (1998)
10. Dunn, J.C.: A fuzzy relative of the ISODATA process and its use in detecting compact well-separated clusters. J. Cybern. **3**, 32–57 (1973)
11. Bezdek, J.C.: Pattern recognition with fuzzy objective function algorithms. In: Plenum, New York, USA (1981)
12. Jensen, J.R.: Introductory Digital Image Processing: A Remote Sensing Perspective, 2nd edn. Prentice Hall PTR, USA (1996)
13. Mather, P.M.: Preprocessing of training data for multispectral image classification proceedings of advances in digital image processing. In: 13 Annual Conference of the. Remote Sensing Society, Remote Sensing Society, Nottingham, pp. 111–120 (1987)
14. Yannis, S.A., Stefanos, D.K.: Fuzzy image classification using multi-resolution neural with applications to remote sensing. In: Electrical Engineering Department, National Technical University of Athens, Zographou 15773, Greece, (1999)
15. Foody, G.M.: The role of soft classification techniques in refinement of estimates of ground control points location. Photogram. Eng. Remote Sens. **68**, 897–903 (2002)
16. Wei, W., Jerry, M.M.: A fuzzy logic method for modulation classification in non ideal environment. IEEE Trans. Fuzzy Syst. **7**, 333–344 (1999)
17. Dehghan, et al.: For Measuring the Uncertainty Using Entropy on MSS Data in Order to Evaluate the Classification Performance (2005)
18. Erbek, F.S., Ozkan, C., Taberner, M.: Comparison of maximum likelihood classification method with supervised artificial neural network algorithms for land use activities. Int. J. Remote Sens. **25**, 1733–1748 (2004)
19. Aplin, P., Atkinson, P.M.: Sub-pixel land cover mapping for per-field classification. Int. J. Remote Sens. **22**(14), 2853–2858 (2001)
20. Lark, R.M., Bolam, H.C.: Uncertainty in prediction and interpretation of spatially variable data on soils. Geoderma **77**, 263–282 (1997)
21. Shalan, M.A., Arora, M.K., Ghosh, S.K.: An evaluation of fuzzy classification from IRS IC LISS III data. Int. J. Remote Sens. **23**, 3179–3186 (2003)

22. Kumar, A., Ghosh, S.K., Dadhwal, V.K.: A comparison of the performance of fuzzy algorithm versus statistical algorithm based sub-pixel classifier for remote sensing data. In: Proceedings of Mid-term Symposium ISPRS, ITC-The Netherlands (2006)
23. Aziz, M.A.: Evaluation of soft classifiers for remote sensing data. In: Unpublished Ph.D thesis, IIT Roorkee, Roorkee, India
24. Lu, D., Mausel, P., Batistella, M., Moran, E.: Comparison of land cover classification methods. In the Brazilian amazon basin. Photogram. Eng. Remote Sens. **70**, 723–732 (2004)
25. Mertens, K.C., Verbeke, L.P.C., Ducheyne, E.I., Wulf, R.R. De.: Using genetic algorithms in sub-pixel mapping. Int. J. Remote Sens. **24**(21), 4241–4247 (2003)
26. Krishnapuram, R., Keller, J.M.: A possibilistic approach to clustering. IEEE Trans. Fuzzy Syst. **1**(2), 98–110 (1993)
27. Foody, G.M., Lucas, R.M., Curran, P.J., Honzak, M.: Non-linear mixture modelling without end-members using an artificial neural network. Int. J. Remote Sens. **18**(4), 937–953 (1997)
28. Congalton, R.G., Plourde, L.: Quality assurance and accuracy assessment of information derived from remotely sensed data. In: Bossler, J. (ed.) Manual of Geospatial Science and Technology, pp. 349–361. Taylor and Francis, London (2002)
29. Binaghi, E., Brivio, P.A., Ghezzi, P., Rampini, A.: A fuzzy set accuracy assessment of soft classification. Pattern Recogn. Lett. **20**, 935–948 (1999)
30. Dehghan, H., Ghassemian, H.: Measurement of uncertainty by the entropy: application to the classification of MSS data. In: Int. J. Remote Sens. 1366–5901, **27**(18), 4005– 4014 (2006)
31. Kumar, G.J.: Automated interpretation of sub-pixel vegetation from IRS LISS-III images. Int. J. Remote Sens. **25**, 1207–1222 (2004)
32. Kuzera, K., Pontius, R.G. Jr.: Categorical coefficients for assessing soft-classified maps at multiple resolutions. In: Proceedings of the Joint Meeting of the 15th Annual Conference of the International Environmental Society and the 6th Annual Symposium on Spatial Accuracy Assessment in Natural Resources and Environmental Sciences Portland ME (2004)
33. Congalton, R.G.: A review of assessing the accuracy of classification of remotely sensed data. Remote Sens. Environ. **37**, 35–47 (1991)
34. Foody, G.M.: Cross-entropy for the evaluation of the accuracy of a fuzzy land cover classification with fuzzy ground data. ISPRS J. Photogram. Remote Sens. **50**, 2–12 (1995)
35. Foody G.M., Arora, M.K.: An evaluation of some factors affecting the accuracy of classification by an artificial neural network. Int. J. Remote Sens. **18**, 799–810 (1997)
36. Evaluation of the Application Potential of IRS-P6 (Recourse Sat-1). In: Action Plan Document, NRSA (2003)

23. [illegible] A., [illegible] ... structural algorithm based on ... set classification ... [illegible]
... Symposium ... [illegible]
... [illegible] classifier for remote sensing [illegible]
[illegible]
... [illegible] ... method ... [illegible]
25. [illegible]
26. [illegible]

A New Collaborative Approach to Particle Swarm Optimization for Global Optimization

Joong Hoon Kim, Thi Thuy Ngo and Ali Sadollah

Abstract Particle swarm optimization (PSO) is population-based metaheuristic algorithm which mimics animal flocking behavior for food searching and widely applied in various fields. In standard PSO, movement behavior of particles is forced by the current bests, global best and personal best. Despite moving toward the current bests enhances convergence, however, there is a high chance for trapping in local optima. To overcome this local trapping, a new updating equation proposed for particles so-called extraordinary particle swarm optimization (EPSO). The particles in EPSO move toward their own targets selected at each iteration. The targets can be the global best, local bests, or even the worst particle. This approach can make particles jump from local optima. The performance of EPSO has been carried out for unconstrained benchmarks and compared to various optimizers in the literature. The optimization results obtained by the EPSO surpass those of standard PSO and its variants for most of benchmark problems.

Keywords Metaheuristics · Particle swarm optimization · Extraordinary particle swarm optimization · Global optimization

1 Introduction

Nature is rich source for inspiring of optimization algorithms. Nowadays, most of new algorithms have been developed by mimicking natural systems. These meta-heuristic algorithms, so-called nature-inspired algorithms can be classified

J.H. Kim (✉) · T.T. Ngo · Ali Sadollah
School of Civil, Environmental, and Architectural Engineering, Korea University, Seoul 136-713, South Korea
e-mail: jaykim@korea.ac.kr

T.T. Ngo
e-mail: ngothuy@korea.ac.kr

Ali Sadollah
e-mail: sadollah@korea.ac.kr

M. Pant et al. (eds.), *Proceedings of Fifth International Conference on Soft Computing for Problem Solving*, Advances in Intelligent Systems and Computing 437, DOI 10.1007/978-981-10-0451-3_57

emphasizing on the sources into three main categories, namely, biology, physics, and chemistry [1]. Those algorithms have been widely applied for many areas in both designing and operation optimization problems. Some of the most well-known bio-inspired algorithms are genetic algorithms (GAs) [2], ant system [3], and particle swarm optimization (PSO) [4]. These algorithms can solve optimization problems with large dimensions; however, no specific algorithm could be suited for all problems [5].

Nowadays, the PSO is one of the most common optimizer applied for problems in different areas. The PSO is a population-based optimization technique originally developed by Kennedy and Eberhart [4] inspiring the movement behavior of bird and fish flock. The PSO mimics the society of animals (i.e., bird or fish) in which each individual so-called particle shares its information with the others.

An individual in the standard PSO tends to move toward the best particle among them (global best, *gbest*) also with considering their best experience (personal best, *pbest*). By this motion approach, the PSO has several advantages: fast convergence, easy implementation, and simple computation. In the other hand, the PSO exhibits some drawbacks in trapping local optima and premature convergence. To overwhelm these disadvantages, some variants of PSO are proposed.

In general, variants of PSO can be classify into four main aspects: improve moving environment such as quantum-behaved particles swarm [6, 7]; modify swarm population such as nonlinear simplex initializing [8], partitioning the population [9], or using Gaussian mutation [10]; alter parameters of inertia weight [11–18]; and improving the intelligence level of particles using learning strategies such as adaptive learning PSO (ALPSO) [19], self-learning PSO (SLPSO) [20], and PSO with improved learning strategy (PSO-ILS) [21].

In this paper, a new approach is proposed to the standard PSO focusing on movement behavior of particles. Instead of the two coefficients presenting cognitive and social components, a combined operator is used. Particles in the new movement strategy can exchange information with any other particles, such as global best, local bests, or even the worst particle. This approach may help the PSO to escape from the local optima. The particles may not fly toward currents bests and have opportunities to break the moving rule used in the PSO and become extraordinary particles. Also, for the mutation part in this improved PSO, the uniform random search is utilized.

The remaining parts of this paper are given as follows: Sect. 2 explain brief descriptions of standard PSO. Section 3 describes detailed approach and governing equations of the proposed improved PSO. Efficiency and performance of proposed PSO is investigated using several benchmarks in Sect. 4 accompanied by reporting optimization results compared with other optimizers. Finally, conclusions and future research are drawn in Sects. 5 and 6, respectively.

2 Particle Swarm Optimization

The PSO originally developed by Eberhart and Kennedy [4] inspired by collective behavior of animal flocking. The particles in PSO move through the search space following direction of global and personal bests. Each particle in the PSO at iteration t is evaluated by fitness function and characterized by its location and velocity [22].

The PSO starts with an initial group of random particles in the D-dimensional search space of the problem. The particle ith at iteration t is represented by position $X_i^t = \left(x_i^1, x_i^2, \ldots, x_i^D\right)$ and velocity $V_i^t = \left(v_i^1, v_i^2, \ldots, v_i^D\right)$. At every iteration, location of each particle is updated by moving toward the two best locations (i.e., global and personal bests) as given in the following equation [22]:

$$\vec{V}_i(t+1) = w(t)\vec{V}_i(t) + r_1 C_1\left(\vec{pbest}_i(t) - \vec{X}_i(t)\right) + r_2 C_2\left(\vec{gbest}(t) - \vec{X}_i(t)\right), \quad (1)$$

$$\vec{X}_i(t+1) = \vec{X}_i(t) + \vec{V}_i(t+1), \quad (2)$$

where r_1 and r_2 are uniform distributed random numbers in [0,1], C_1 and C_2 are cognitive and social coefficients known as acceleration constants, respectively; $pbest_i(t)$ denotes the best personal position of the particle ith and $gbest$ is the best position among all particles at iteration t; $w(t)$ is an inertia weight a user parameter between zero and one. The velocity of each particle is limited by the range of $[v_{\min}^d, v_{\max}^d]$. In entire paper, notations having vector sign corresponds vector values, otherwise the rest of notations and parameters consider as scalar values.

3 Extraordinary Particle Swarm Optimization

In the PSO, balance of exploitation and exploration is obtained by combining local and global searches as well as inertia weight coefficient (see Eq. (1)). The inertia weight represents exploitation–exploration tradeoff [12]; larger inertia weight is preferred for global search. However, this tradeoff cannot always successfully apply [18].

The PSO simulates movement behaviors of animal flocking when searching food. By information sharing mechanism among particles, each individual is forced to move following the current best particles. Although, this process represents some advantages, however, there are several drawbacks in terms of being trapped in local optima and fast immature convergence rate. Directions oriented by current bests of particles may not be always the best way and moving toward the worst particles, in fact, may neither be the worst way to the global optima. This fact therefore motivates a new movement strategy for particles to escape from local optima and speed up the mature convergence.

Particles in EPSO, so-called extraordinary particles, fly to their determined targets which are updated at each iteration. The target of each particle is stochastically defined and could be any particle in population, the global best, the local best, or normal particle with different cost/fitness. These extraordinary particles tend to break the movement rule applied in standard PSO. The location and velocity of particle *i*th is updated by iteration as the following equation:

$$\vec{V}_i(t+1) = C\big(\vec{X}_{Ti}(t) - \vec{X}_i(t)\big), \tag{3}$$

$$\vec{X}_i(t+1) = \vec{X}_i(t) + \vec{V}_i(t+1), \tag{4}$$

where $X_{Ti}(t)$ is the determined target T of particle *i*th at iteration t; C is combined component representing both cognitive and social factors. The target particle could be the best among entire particles or the best that extraordinary particle has experienced or just a normal particle in the swarm population. If the chosen target is global or personal bests, C coefficient, therefore is cognitive or social coefficient, respectively.

The random search is involved in EPSO by limiting the potential range of targets. If the target chosen by particle *i*th belong to the range, the particle will be guided by its target, otherwise random search is taken part. The range of potential target is defined by the upper bound T_{up} which calculated by user-defined parameter α and population size N_{pop}, as follows: T_{up} = round ($\alpha \times N_{pop}$). Sorting particle by cost/fitness is consisted before determining targets. It means if α is 0.9, 90 % of best particles in population can be randomly chosen as target solution, more exploitation considering worst and best solutions and less exploration using the random search. In the other hand, if α is 0.1, 10 % of best particles in population can be selected as target solution and more exploration using random search and more greedy movements to the other solutions.

The detailed procedure of EPSO is described in the following steps:

Step 1: Setting initial parameters: combined coefficient C [0, 2], and α indicates the upper bound of a potential target in entire population [0, 1]

Step 2: Initializing the swarm population: Randomly generate new particles between lower and upper bounds of a given problem

Step 3: Sorting particles based on their cost/fitness

Step 4: Selecting the target: Select the target for each particle at each iteration among all particles using the determined index (T) of particles

$$T = \text{round}\big(\text{rand} \times N_{\text{pop}}\big), \tag{5}$$

where rand is uniform distributed random number [0, 1]. The index of target (T) ranges from zero to N_{pop}

Step 5: Updating particles to their new locations using the following equation:

$$\vec{X}_i(t+1) = \begin{cases} \vec{X}_i(t) + C\left(\vec{X}_{Ti}(t) - \vec{X}_i(t)\right) & \text{if } T \in \left(0, T_{\text{up}}\right) \\ \vec{LB}_i + \text{rand} \times \left(\vec{UB}_i - \vec{LB}_i\right) & \text{Otherwise} \end{cases}, \quad (6)$$

where LB and UB are lower and upper bounds of the search space, respectively

Step 6: Check the stopping criterion. If the stopping condition is met, stop. Otherwise, return to Step 3.

4 Optimization Results and Discussions

To validate performance of EPSO, MATLAB programming software has been used for various unconstrained benchmark problems. To compare the ability of EPSO with other optimizers, 13 unconstrained benchmark functions are investigated in this paper. Statistical indicators have been obtained from 30 independent runs, and collected from the literatures for other optimization methods. For all benchmark functions, we used the recommended parameters with $N_{\text{pop}} = 20$, $C = 0.3$, and $\alpha = 0.8$.

The unconstrained benchmark problems used in our experimental study are extracted from [23]. A comparison of EPSO with other variants of PSO and other population-based metaheuristic algorithms has been given in the following.

The optimization results of EPSO applied for 13 functions (F_1 to F_{13}) are represented in Table 1 and compared to other algorithms including the real-coded genetic algorithm (RGA) [24] and the gravitational search algorithm GSA [23].

Table 1 Comparison of EPSO with the considered optimizers for reported benchmarks

Test function	RGA		GSA		EPSO	
	Average	SD	Average	SD	Average	SD
F_1	2.31E+01	1.21E+01	6.80E−17	2.18E−17	7.77E−18	6.01E−18
F_2	1.07E+00	2.66E−01	6.06E−08	1.19E−08	6.78E−12	3.00E−12
F_3	5.61E+02	1.25E+02	9.42E+02	2.46E+02	2.12E−01	5.46E−01
F_4	1.17E+01	1.57E + 00	4.20E+00	1.12E+00	9.94E−03	9.85E−03
F_5	1.18E+03	5.48E+02	4.79E+01	3.95E+00	1.78E−02	2.13E−02
F_6	2.40E+01	1.01E+01	9.31E−01	2.51E+00	0.00E+00	0.00E+00
F_7	6.75E−02	2.87E−02	7.82E−02	4.10E−02	6.47E−04	4.54E−04
F_8	−1.24E+04	5.32E+01	−3.60E+03	5.64E+02	−1.25E+04	3.85E−12
F_9	5.90E+00	1.17E+00	2.94E+01	4.72E+00	2.27E−14	2.83E−14
F_{10}	2.14E+00	4.01E−01	4.80E−09	5.42E−10	1.28E−09	7.28E−10
F_{11}	1.16E+00	7.95E−02	1.66E+01	4.28E+00	2.53E−08	1.31E−07
F_{12}	5.10E−02	3.52E−02	5.04E−01	4.24E−01	6.05E−20	8.97E−20
F_{13}	8.17E−02	1.07E−01	3.40E+00	3.68E+00	9.37E−16	1.80E−15

Table 2 Comparison of EPSO with several variants of PSO for reported functions

Test function	PSO		CPSO		CLPSO	
	Average	SD	Average	SD	Average	SD
F_1	5.19E−40	1.13E−74	5.14E−13	7.75E−25	4.89E−39	6.78E−78
F_2	2.07E−25	1.44E−49	1.25E−07	1.17E−14	8.86E−24	7.90E−49
F_3	1.45E+00	1.17E+00	1.88E+03	9.91E+06	1.92E+02	3.84E+03
F_5	2.54E+01	5.90E+02	8.26E−01	2.34E+00	1.32E+01	2.14E+02
F_6	**0.00E+00**	**0.00E+00**	**0.00E+00**	**0.00E+00**	**0.00E+00**	**0.00E+00**
F_7	1.23E−02	2.31E−05	1.07E−02	2.77E−05	4.06E−03	9.61E−07
F_8	−1.10E+04	1.37E+05	−1.21E+04	3.38E+04	−1.25E+04	4.25E+03
F_9	3.47E+01	1.06E+02	3.60E−13	1.50E−24	**0.00E+00**	**0.00E+00**
F_{10}	1.49E−14	1.86E−29	1.60E−07	7.86E−14	**9.23E−15**	**6.61E−30**
F_{11}	2.16E−02	4.50E−04	2.12E−02	6.31E−04	**0.00E+00**	**0.00E+00**

Talking about parameter selection, in the GSA, the initial gravitation constant (G_0) and reduction rate (α) were set to 100 and 20, respectively. Regarding the RGA, crossover and mutation probabilities were set to 0.3 and 0.1, accordingly.

Also, a comparative study with other variants of PSO for several benchmarks is given in Table 2. The reported variants included in this study are the standard particle swarm optimization (PSO) [22], the cooperative PSO (CPSO) [25], the comprehensive learning PSO (CLPSO) [26], the fully informed particle swarm (FIPS) [27] and the Frankenstein's PSO (F-PSO) [28]. The user parameters of these PSO's variants were set to the recommended values used in the literatures [22, 25–28].

The results from EPSO for 30D problems are obtained after 30 independent runs with maximum number of function evaluations (NFEs) of 50,000 for Table 1 and 200,000 for Tables 2 and 3. In the entire paper, the values lower than 1.00E−324 (i.e., defined zero in MATLAB) are considered as 0.00E+00. The optimization

Table 3 Comparison of EPSO with several variants of PSO for reported functions (cont.)

Test function	FIPS		F-PSO		EPSO	
	Average	SD	Average	SD	Average	SD
F_1	4.58E−27	1.95E−53	2.40E−16	2.00E−31	**1.66E−74**	2.76E−74
F_2	2.32E−16	1.14E−32	1.58E−11	1.03E−22	**1.90E−47**	2.15E−47
F_3	9.46E+00	2.59E+01	1.73E+02	9.15E+03	**2.01E−03**	1.93E−03
F_5	2.67E+01	2.00E+02	2.81E+01	2.31E+02	**2.82E−05**	3.65E−05
F_6	**0.00E+00**	**0.00E+00**	**0.00E+00**	**0.00E+00**	**0.00E+00**	**0.00E+00**
F_7	3.30E−03	**8.66E−07**	4.16E−03	2.40E−06	**2.58E−04**	1.87E−04
F_8	−1.10E+04	9.44E+05	−1.12E+04	2.77E+05	**−1.25E+04**	**2.48E−12**
F_9	5.85E+01	1.91E+02	7.38E+01	3.70E+02	**0.00E+00**	**0.00E+00**
F_{10}	1.38E−14	2.32E−29	2.17E−09	1.71E−18	1.21E−14	3.10E−15
F_{11}	2.47E−04	1.82E−06	1.47E−03	1.28E−05	**0.00E+00**	**0.00E+00**

results of other algorithms are extracted from [18, 23], except for GSA which had been implemented by authors.

In Table 1, the EPSO shows the far better performance rather than the RGA and GSA in terms of the best solution, the average of solutions and standard deviation (SD) for all considered functions. The consequences of 30 independent runs with 200,000 NFEs for the EPSO and several variants of PSO shows the superiority of the EPSO over the other reported improved versions as shown in Tables 2 and 3. In most of functions (9 out of 10 functions) the EPSO found the lower average solution than the others. In particular, the EPSO found the global optimal solution of multimodal functions F_9 and F_{11}, while the others except CLPSO failed.

Significance of this research is given in the following: an improved version of PSO considering rule-breaking particles has been proposed and applied for optimization purposes. The new improved PSO shows its supervisory over other PSO variants and the other reported optimization methods. Therefore, using the proposed improved PSO may be considered as an alternative optimizer for solving optimization problems.

5 Conclusions

This paper proposed a new cooperative approach for particle swarm optimization (PSO) using extraordinary particles which tend to interrupt the defined rules of standard PSO and move toward their own targets. The algorithm therefore is called extraordinary PSO (EPSO). The proposed improved PSO enhances the advantages of standard PSO with two user-defined parameters (C and α) and overcomes the drawbacks of standard PSO by its approach for finding optimal solution. The performance of EPSO through unconstrained problems has surpassed the standard PSO and the other reported variants of PSO.

6 Future Research

As further studies, multi-objective version of the proposed improved method (EPSO) can be developed and applied to solve multi-criteria optimization problems. Tackling real-life engineering problems arising in industry may be suitable target area for future research. Comparison with other optimizers and application to large-scale benchmarks show better efficiency and reliability of the proposed PSO for solving complex optimization problems.

Acknowledgments This work was supported by the National Research Foundation of Korea (NRF) grant funded by the Korean government (MSIP) (NRF-2013R1A2A1A01013886).

References

1. Fister, I.Jr., Yang, X.S., Fister, I., Brest, J., Fister, D.: A Brief Review of Nature-Inspired Algorithms for Optimization, CoRR, 1–7. arXiv:abs/1307.4186 (2013)
2. Holland, J.H.: Adaptation in natural and artificial systems: an introductory analysis with applications to biology, control, and artificial intelligence. University of Michigan Press, Michigan (1975)
3. Dorigo, M., Maniezzo, V., Colorni, A.: Ant system: optimization by a colony of cooperating agents. IEEE Trans. Syst. Man Cybernatics **26**(1), 29–41 (1996)
4. Kennedy, J., Eberhart, R.C.: Particle swarm optimization. In: IEEE International Conference on Neural Networks, pp. 1942–1948. Piscataway, NJ (1995)
5. Wolpert, D.H., Macready, W.G.: No free lunch theorems for optimization. IEEE Trans. Evol. Comput. **1**, 67–82 (1997)
6. Sun, J., Feng, B.: Particle swarm optimization with particles having quantum behavior. IEE Proc. Con. Evolut. Comput. **1**, 325–331 (2004)
7. Chen, W., Zhou, D.: An improved quantum-behaved particle swarm optimization algorithm based on comprehensive learning strategy. J. Control Decis. 719–723 (2012)
8. Parsopoulos, K.E., Vrahatis, M.N.: Initializing the particle swarm optimizer using the nonlinear simplex method. In: Grmela, A., Mastorakis, N.E. (eds.) Advances in Intelligent Systems, Fuzzy Systems, Evolutionary Computation, pp. 216–221. World Scientific and Engineering Academy and Society Press, Stevens Point, WI, U.S.A. (2002)
9. Jiang, Y., Hu, T., et al.: An improved particle swarm optimization algorithm. Appl. Math. Comput. **193**(1), 231–239 (2007)
10. Higashi, N., Iba, H.: Particle swarm optimization with gaussian mutation. In: Proceedings of the IEEE Swarm Intelligence Symposium, pp. 72–79, Indiana (2003)
11. Shi, Y.H., Eberhart, R.C.: A modified particle swarm optimizer. In: IEEE International Conference on Evolutionary Computation. Anchorage, Alaska, pp. 69–73 (1998)
12. Eberhart, R., Shi, Y.: Comparing inertia weights and constriction factors in particle swarm optimization. In: IEEE Conference on Evolutionary Computation, pp. 84–88 (2000)
13. Chatterjee, A., Siarry, P.: Nonlinear inertia weight variation for dynamic adaption in particle swarm optimization. Comput. Oper. Res. **33**(3), 859–871 (2006)
14. Lei, K., Qiu, Y., He, Y.: A new adaptive well-chosen inertia weight strategy to automatically harmonize global and local search ability in particle swarm optimization. In: ISSCAA (2006)
15. Yang, X., Yuan, J., et al.: A modified particle swarm optimizer with dynamic adaptation. Appl. Math. Comput. **189**(2), 1205–1213 (2007)
16. Arumugam, M.S., Rao, M.V.C.: On the improved performances of the particle swarm optimization algorithms with adaptive parameters, cross-over operators and root mean square (RMS) variants for computing optimal control of a class of hybrid systems. Appl. Soft Comput. **8**(1), 324–336 (2008)
17. Panigrahi, B.K., Pandi, V.R., Das, S.: Adaptive particle swarm optimization approach for static and dynamic economic load dispatch. Energ. Convers. Manage. **49**(6), 1407–1415 (2008)
18. Nickabadi, A., Ebadzadeh, M.M., Safabakhsh, R.: A novel particle swarm optimization algorithm with adaptive inertia weight. Appl. Soft Comput. **11**, 3658–3670 (2011)
19. Li, C., Yang, S.: An adaptive learning particle swarm optimizer for function optimization. In: Proceedings of Congress on Evolutionary Computation, pp. 381–388 (2009)
20. Li, C., Yang, S., Nguyen, T.: A self-learning particle swarm optimizer for global optimization problems. IEEE Trans. Syst. Man Cybernatics—Part B Cybernetics **43**(3), 627–646 (2012)
21. Lim, W., Isa, N.: Particle swarm optimisation with improved learning strategy. J. Eng. Sci. **11**, 27–48 (2015)
22. Eberhart, R.C., Kennedy, J.: A new optimizer using particle swarm theory. In: Proceedings of the International Symposium on Micro Machine and Human Science. Nagoya, Japan, pp. 39–43 (1995)

23. Rashedi, E., Nezamabadi-pour, H., Saryazdi, S.: GSA: a gravitational search algorithm. Inf. Sci. **179**(13), 2232–2248 (2009)
24. Sarafrazi, S., Nezamabadi-pour, H., S. Saryazdi,: Disruption: A new operator in gravitational search algorithm. Scientia Iranica 539–548 (2011)
25. Bergh, F.V.D., Engelbrecht, A.P.: A Cooperative approach to particle swarm optimization. IEEE Trans. Evolut. Comput. **8**(3), 225–239 (2004)
26. Liang, J.J., Qin, A.K.: Comprehensive learning particle swarm optimizer for global optimization of multimodal functions. IEEE Trans. Evolut. Comput. **10**(3), 281–295 (2006)
27. Mendes, R., Kennedy, J., Neves, J.: The fully informed particle swarm: simpler, maybe better. IEEE Trans. Evolut. Comput. **8**(3), 204–210 (2004)
28. Oca, M.A., Stutzle, T.: Frankenstein's PSO: a composite particle swarm optimization algorithm. IEEE Trans. Evolut. Comput. **13**(5), 1120–1132 (2009)

Reliability Optimization for the Complex Bridge System: Fuzzy Multi-criteria Genetic Algorithm

Michael Mutingi and Venkata P. Kommula

Abstract System reliability optimization often involves multiple fuzzy conflicting objectives, for instance, reducing system cost and reliability improvement. This paper presents a system reliability optimization problem for the complex bridge system. First, the problem is formulated as a fuzzy multi-criteria nonlinear program. Second, we propose a fuzzy multi-criteria genetic algorithm approach (FMGA) to solve the problem. Fuzzy evaluation techniques are used to handle fuzzy goals and constraints, resulting in a flexible and adaptable approach that provides high-quality solutions within reasonable computation times. Using fuzzy theory concepts, the preferences of the decision maker on the cost and reliability objectives are judiciously incorporated. Third, computational experiments results are presented based on benchmark problems in the literature. The computational results obtained show that the proposed method is encouraging.

Keywords System reliability optimization · Multi-criteria optimization · Genetic algorithm · Fuzzy optimization

1 Introduction

Reliability is central to productivity and effectiveness of industrial systems [1–3]. To maximize productivity, the systems should always be available. However, it is difficult for an industrial system, comprising several complex components to

Michael Mutingi (✉)
School of Engineering, Namibia University of Science & Technology, Windhoek, Namibia
e-mail: mmutingi@gmail.com

Michael Mutingi
Faculty of Engineering and the Built Environment, University of Johannesburg, Johannesburg, South Africa

V.P. Kommula
Mechanical Engineering Department, University of Botswana, Gaborone, Botswana
e-mail: pramkv@gmail.com

M. Pant et al. (eds.), *Proceedings of Fifth International Conference on Soft Computing for Problem Solving*, Advances in Intelligent Systems and Computing 437, DOI 10.1007/978-981-10-0451-3_58

survive over time since its reliability directly depends on the characteristics of its components. Failure is inevitable, such that system reliability optimization has become a key subject area in industry. Developing effective system reliability optimization is imperative. Two approaches for system reliability improvement are: (i) using redundant elements in subsystems and (ii) increasing the reliability of system components.

Reliability-redundancy allocation maximizes system reliability via redundancy and component reliability choices [1], with restrictions on cost, weight, and volume of the resources. The aim is to find a trade-off between reliability and other resource constraints [2]. Thus, for a highly reliability system, the main problem is to balance reliability enhancement and resource consumption. However, real-life reliability optimization problems are complex: (i) management goals and the constraints often imprecise; (ii) problem parameters as understood by the decision maker may be vague; and (iii) historical data is often imprecise and vague. Uncertainties in component reliability may arise due to variability and changes in the manufacturing processes that produce the system component. Such uncertainties in data cannot be addressed by probabilistic methods which deal with randomness. Therefore, the concept of fuzzy reliability is more promising [4–9]. Contrary to probabilistic, fuzzy theoretic approaches address uncertainties that arise from vagueness of human judgment and imprecision [2, 10–15].

A number of methods and applications have been proposed to solve fuzzy optimization problems by treating parameters (coefficients) as fuzzy numerical data [16–20]. In a fuzzy multi-criteria environment, simultaneous reliability maximization and cost minimization require a trade-off approach. Metaheuristics are a potential application method for such complex problems [21–23]. Population-based metaheuristics are appropriate for finding a set of solutions that satisfy the decision maker's expectations. This calls for interactive fuzzy multi-criteria optimization which incorporates preferences and expectations of the decision maker, allowing for expert judgment. Iteratively, it becomes possible to obtain the most satisfactory solution.

In light of the above issues, the aim of this research is to address the system reliability optimization problem for a complex bridge system in a fuzzy multi-criteria environment. Specific objectives of the research are:

1. to develop a fuzzy multi-criteria decision model for the problem;
2. to use an aggregation method to transform the model to a single-criteria optimization problem; and
3. to develop a multi-criteria optimization method for the problem.

We use the max–min operator to aggregate the membership functions of the objective functions, incorporating expert opinion. To this end, we define notations and assumptions.

Assumptions

1. The availability of the components is unlimited;
2. The weight and product of weight and square of the volume of the components are deterministic;
3. The redundant components of individual subsystems are identical;
4. Failures of individual components are independent;
5. All failed components will not damage the system and are not repaired.

Notation

m The number of subsystems in the system
n_i The number of components in subsystem i, $1 \leq i \leq m$
n $\equiv(n_1, n_2, \ldots, n_m)$, the vector of the redundancy allocation for the system
r_i The reliability of each component in subsystem i, $1 \leq i \leq m$
r $\equiv(r_1, r_2, \ldots, r_m)$, the vector of the component reliabilities for the system
q_i $=1 - r_i$, the failure probability of each component in subsystem i, $1 \leq i \leq m$ $R_i(n_i) = 1 - q_i^{n_i}$, the reliability of subsystem i, $1 \leq i \leq m$
R_s The system reliability
g_i The ith constraint function
w_i The weight of each component in subsystem i, $1 \leq i \leq m$
v_i The volume of each component in subsystem i, $1 \leq i \leq m$
c_i The cost of each component in subsystem i, $1 \leq i \leq m$
V The upper limit on the sum of the subsystems' products of volume and weight
C The upper limit on the cost of the system
W The upper limit on the weight of the system
b The upper limit on the resource

2 Reliability Optimization of the Complex Bridge System

The complex bridge system comprises five subsystems (Fig. 1) [1, 24]. The aim is to maximize system reliability, subject to multiple constraints.

The model can be expressed as a mixed integer nonlinear program as follows:

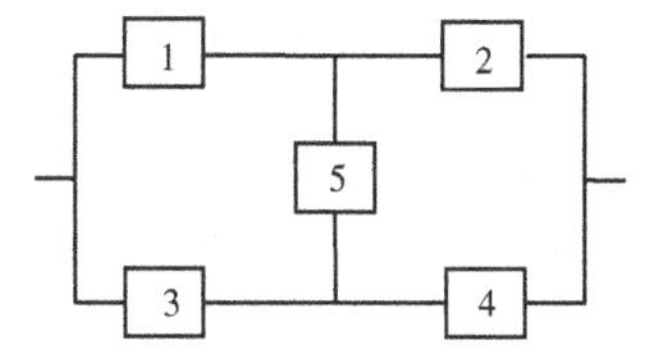

Fig. 1 The schematic diagram of the complex bridge system

$$\left.\begin{aligned}
&\mathrm{Max} f(r,n) = R_1R_2 + R_3R_4 + R_1R_4R_5 + R_2R_3R_5 + 2R_1R_2R_3R_4R_5 \\
&\qquad - R_1R_2R_3R_4 - R_1R_2R_3R_5 - R_1R_2R_4R_5 - R_1R_3R_4R_5 - R_2R_3R_4R_5 \\
&\text{Subject to} \\
&g_1(r,n) = \sum_{i=1}^{m} w_i v_i^2 n_i^2 \le V \\
&g_2(r,n) = \sum_{i=1}^{m} \alpha_i(-1000/\ln r_i)^{\beta_i}(n_i + \exp(n_i/4)) \le C \\
&g_3(r,n) = \sum_{i=1}^{m} w_i n_i \exp(n_i/4) \le W \\
&0 \le r_i \le 1,\ n_i \in Z^+,\ 1 \le i \le m \\
&0 \le r_i \le 1,\ n_i \in Z^+,\ 1 \le i \le m
\end{aligned}\right\} \tag{1}$$

We present the proposed fuzzy multi-criteria optimization approach, based on genetic algorithms in the next section.

3 Fuzzy Multi-criteria Optimization Approach

In a fuzzy environment, the aim is to find a trade-off between reliability, cost, weight, and volume. A common approach is to simultaneously maximize reliability and minimize cost. Constraints are transformed into objective functions, so that reliability and other cost functions can be optimized jointly. This is achieved through the use of membership functions, which are easily applicable and adaptable to the real-life decision process. In general, the fuzzy multi-criteria optimization problem can be represented by the following:

$$\left.\begin{aligned}
&\mathrm{Min}\ \tilde{f}(x) \\
&\text{Subject to:} \\
&g_z(x) \lesssim \text{or} \equiv \text{or} \gtrsim 0 \qquad z = 1, 2\ldots, p \\
&x_q^l \le x_q \le x_q^u \quad q = 1, 2, \ldots, Q
\end{aligned}\right\} \tag{2}$$

where $x = (x_1, x_2, \ldots, x_Q)^T$, is a vector of decision variables that optimize a vector of objective functions, $\tilde{f}(x) = \{\tilde{f}_1(x), \tilde{f}_2(x), \ldots, \tilde{f}_d(x)\}$ over decision space X; $f_1(x), f_2(x), \ldots, f_d(x)$ are d individual objective functions; x_q^l and x_q^u are lower and upper bounds on the decision variable x_q, respectively.

3.1 Membership Functions

Fuzzy set theory permits gradual assessment of membership, in terms of a suitable function that maps to the unit interval [0, 1]. Membership functions such as Generalized Bell, Gaussian, Triangular, and Trapezoidal can represent the fuzzy membership. Linear membership functions can provide good quality solutions with much ease, including the widely recommended triangular and trapezoidal membership functions [4, 8, 17]. Thus, we use linear functions to define fuzzy membership of objective functions.

Let a_t and b_t denote the minimum and maximum feasible values of each objective function $\tilde{f}_t(x)$, $t = 1, 2,\ldots, h$, where h is the number of objective functions. Let μ_{f_t} denote the membership function corresponding to the objective function f_t. Then, the membership function corresponding to minimization and maximization is defined based on satisfaction degree. Figure 2 illustrates the linear membership functions defined for minimization and maximization.

For minimization, the linear membership function is represented by the following expression:

$$\mu_{f_t}(x) = \begin{cases} 1 & f_t(x) \leq b_t \\ \frac{b_t - f_t(x)}{b_t - a_t} & a_t \leq f_t(x) \leq b_t \\ 0 & f_t(x) \geq b_t \end{cases} \tag{3}$$

The function is monotonically decreasing in $f_t(x)$. For maximization, the membership function is as follows:

$$\mu_{f_t}(x) = \begin{cases} 1 & f_t(x) \geq b_t \\ \frac{f_t(x) - a_t}{b_t - a_t} & a_t \leq f_t(x) \leq b_t \\ 0 & f_t(x) \geq a_t \end{cases} \tag{4}$$

The function is a monotonically increasing function in $f_t(x)$. The next step is to formulate the corresponding crisp model. Fuzzy evaluation enables FMGA to cope with infeasibilities which is otherwise impossible with crisp formulation. This gives the algorithm speed and flexibility, which ultimately improves the search power of the algorithm.

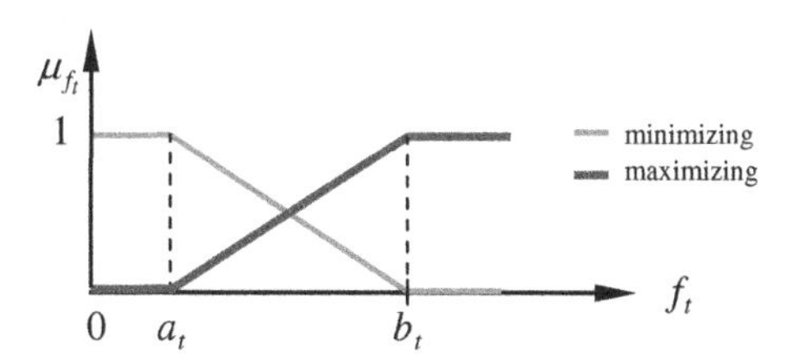

Fig. 2 Fuzzy membership function for $f_t(x)$

3.2 Corresponding Crisp Model

To improve flexibility and to incorporate the decision maker's preferences into the model, we introduce user-defined weights, $w = \{w_1, w_2,\ldots, w_h\}$. The multi-criteria system reliability optimization model is converted into a single-objective optimization model [18]:

$$\left.\begin{array}{l} \text{Max } \left(1 \wedge \frac{\lambda_1(x)}{w_1}\right) \wedge \left(1 \wedge \frac{\lambda_2(x)}{w_2}\right) \wedge \cdots \wedge \left(1 \wedge \frac{\lambda_h(x)}{w_h}\right) \\ \text{Subject to:} \\ \lambda_t(x) = \mu_{f_t}(x) \quad w_t \in [1,0] \quad t = 1,\ldots,h \\ x_q^l \leq x_q \leq x_q^u \quad q = 1,\ldots,Q \end{array}\right\} \tag{5}$$

Here, $\mu_{f_t}(x) = \{\mu_{f_1}(x), \mu_{f_2}(x), \ldots, \mu_{f_h}(x)\}$ signifies a set of fuzzy regions that satisfy the objective functions; λ_t is the degree of satisfaction of the tth objective, x is a vector of decision variables, w_t is the weight of the tth objective function, and symbol "$\wedge$" is the aggregate min operator. Thus, expression $(1 \wedge \lambda_1(x)/w_1)$ gives the minimum between 1 and $\lambda_1(x)/w_1$. Though $\lambda_1(x)$ are in the range [0, 1], the value of $\lambda_1(x)/w_1$ may exceed 1, howbeit, the final value of $(1 \wedge \lambda_1(x)/w_1)$ will always lie in [0, 1]. A FMGA approach is used to solve the model.

4 The Fuzzy Multi-criteria Genetic Algorithm

The proposed FMGA is a developed from the classical genetic algorithm (GA) that was developed by Holland [25, 26]. Based on the concepts of GA, the FMGA is a global optimization method that stochastically evolves a population of candidate solutions based on the philosophy of survival of the fittest. Over the generations, quality solutions, coded into a string of digits (called chromosomes) are given more preference to survive. However, to maintain diversity, a few low-quality solutions are allowed to survive in the population. Probabilistic group genetic operators such as selection, crossover, mutation, and inversion are used to generate new offspring. Using fuzzy evaluation, new and old (parent) candidates are compared in terms of their respective finesses, so that best performing candidates are retained into the next generation. Thus, good traits of candidate solutions are passed from generation to generation. The general structure of the FMGA is presented in Fig. 3.

Algorithm 1: FMGA pseudo code

1: Input parameters, pop, crossover probability, mutation probability
2: Randomly generate initial population
3: **Repeat**
4: Evaluation of fitness, objective: $f(x)$, $x = (x_1, x_2, \ldots, x_h)$
5: Selection strategy
6: Crossover
7: Mutation
8: Replacement
9: Advance population; *oldpop* = *newpop*
10: **Until** (Termination criteria is satisfied)

Fig. 3 The general structure of the overall FMGA

4.1 Chromosome Coding

In this problem, FMGA uses the variable vectors n and r. An integer variable n_i is coded as a real variable and transformed to the nearest integer value upon objective function evaluating.

4.2 Initialization and Evaluation

An initial population of size *pop* is randomly generated. The algorithm then calculates the objective function value for each chromosome. Thus, the chromosome is evaluated according to the overall objective function in model (5).

4.3 Selection and Recombination

Several selection strategies have been suggested by Goldberg [26]. The 'remainder stochastic sampling without replacement' is preferred; each chromosome j is selected and stored in the mating pool according to the expected count e_j;

$$e_j = \frac{f_j}{\sum_{j=1}^{\text{pop}} f_j / \text{pop}} \tag{6}$$

where f_j is the objective function value of the jth chromosome. Each chromosome receives copies equal to the integer part of e_i, while the fractional part is treated as success probability of obtaining additional copies of the same chromosome into the mating pool.

4.4 Crossover

Genes of selected parent chromosomes are partially exchanged to produce new offspring, to form a selection pool. We use an arithmetic crossover operator which defines a linear combination of two chromosomes [23]. Assume a crossover probability of 0.45 in this study. Let s_1 and s_2 be the selected parents, and r represent a random number in [0, 1]. Then, the resulting offspring, c_1 and c_2, are given by the following expression:

$$\begin{aligned} c_1 &= rs_1 + (1 - r)s_2 \\ c_2 &= (1 - r)s_1 + rs_2 \end{aligned} \tag{7}$$

4.5 Mutation

The mutation operator is applied to every new chromosome, at a very low probability, so as to maintain diversity of the population. A uniform mutation with rate 0.035 is used in this study.

4.6 Replacement

In every generation, newly created offspring may be better or worse. As such, nonperforming chromosomes should be replaced. A number of replacement strategies exist in the literature, e.g., probabilistic replacement, crowding strategy, and elitist strategy [23]. Here, a combination of these was used. By successively comparing chromosomes in the current population and the selection pool, better performing chromosomes are advanced into the next generation. Additional chromosomes from the selection pool are probabilistically accepted to ensure a complete population.

4.7 Termination Criteria

Two termination conditions are used to stop the FMGA iterations: when the number of generations exceeds maximum iterations, and when the average change in the fitness of the best solution over specific generations is less than a small number, which is 10^{-5}.

5 Numerical Illustrations

This section presents the comparative results of our FMGA computations based on benchmark problems [1, 24, 27, 28]. We use the parameter values in [3] and define the specific instances of the problems as shown in Table 1.

From the computational experience and fine tuning, the parameters of the FMGA were as follows: The crossover and mutation were set at 0.44 and 0.036, respectively. A two-point crossover was used in this application. The population size was set to 20. The maximum number of generations was set at 150. Termination is controlled by either the maximum number of iterations or the order of the relative error set at 10^{-5}, whichever is earlier. Specifically, whenever the best fitness f^* at iteration t is such that $|f_t - f^*| < \varepsilon$ is satisfied, then three best solutions are selected; where ε is a small number equal to 10^{-5}. The FMGA was implemented in JAVA, and the program was run 25 times, while selecting the best three solutions out of the converged population.

The formulation in (5) is used to solve benchmark problems in [1]. A fuzzy region of satisfaction is constructed for each objective function, that is, objective functions corresponding to system reliability, cost, volume, and weight, denoted by λ_1, λ_2, λ_3, and λ_4, respectively. Using the constructed membership functions together with their corresponding weight vectors, we obtain the equivalent crisp optimization formulation

$$\left.\begin{array}{ll}
\text{Max } \left(1 \wedge \frac{\lambda_1(x)}{\omega_1}\right) \wedge \left(1 \wedge \frac{\lambda_2(x)}{\omega_2}\right) \wedge \left(1 \wedge \frac{\lambda_3(x)}{\omega_3}\right) \wedge \left(1 \wedge \frac{\lambda_4(x)}{\omega_4}\right) & \\
\text{Subject to:} & \\
\lambda_t(x) = \mu_{f_t}(x) & t = 1, \ldots, 4 \\
0.5 \le r_i \le 1 - 10^{-6} & r_i \in [0, 1] \\
1 \le n_i \le 10 & n_i \in Z^+ \\
0.5 \le R_s \le 1 - 10^{-6} & R_s \in [0, 1]
\end{array}\right\} \quad (8)$$

The weight set $\omega = \{\omega_1, \omega_2, \omega_3, \text{and } \omega_4\}$ was selected in the range [0.2, 1], where the values of the weights indicate the bias toward specific objectives as specified by the decision maker. For instance, the weight set $\omega = [1, 1, 1, 1]$ implies that the expert user expects no preference at all, while the weight set $\omega = [1, 0.5,$

Table 1 Basic data used for the bridge complex system

i	$10^5\alpha_i$	β_i	$w_i v_i^2$	w_i	V	C	W
1	2.330	1.5	1	7	110	175	200
2	1.450	1.5	2	8	110	175	200
3	0.541	1.5	3	8	110	175	200
4	8.050	1.5	4	6	110	175	200
5	1.950	1.5	2	9	110	175	200

Table 2 Minimum and maximum values of objective functions

	Complex bridge system			
	f_1	f_2	f_3	f_4
b_i	1	180	190	110
a_i	0.6	60	70	20

0.5, 0.5], indicates preferential bias toward the region that is closer to the objective corresponding to reliability than to the rest of the objectives that are equally ranked with weight value of 0.5. Therefore, the decision-making process takes into account the decision maker's preferences and choices based on expert opinion.

FMGA interactively provides a set of good solutions, rather than prescribing a single solution. The algorithm enables the decision maker to specify the minimum and maximum values of objective functions in terms of reliability, cost, volume, and weight, denoted by f_1, f_2, f_3, and f_4, respectively. Table 2 provides a list of selected minimum and maximum values of the objective functions for the complex bridge system. Results and discussions are presented in the next section.

6 Computational Results and Discussions

This section presents the comparative results of the numerical experiments. The best three FMGA solutions are compared with the results obtained by algorithms in the literature [3, 24, 28].

Table 3 shows comparative results in which the best three solutions of each problem are compared against solutions from the literature. The results indicate that each of the three solutions is better than previous solutions, in terms of system reliability. Cost wise, the three solutions are no better than the previously reported solutions. However, the difference in cost is quite small. Overall, FMGA provides better solutions than the previous approaches.

Table 3 Comparative results for the complex bridge system

	Best 3 MGA solutions			Wu et al. [3]	Chen [28]
	$(r_i{:}n_i)$	$(r_i{:}n_i)$	$(r_i{:}n_i)$	(3,2,2,3,3)	$(r_i{:}n_i)$
1	(0.82868361:3)	(0.82808733:3)	(0.81059326:3)	(0.82868361:3)	(0.812485:3)
2	(0.85802567:3)	(0.85780316:3)	(0.85436730:3)	(0.85802567:3)	(0.867661:3)
3	(0.91364616:2)	(0.91423977:3)	(0.88721528:3)	(0.91364616:2)	(0.861221:3)
4	(0.64803407:4)	(0.64814829:3)	(0.72126594:3)	(0.64803407:4)	(0.713852:3)
5	(0.70227595:1)	(0.70416346:1)	(0.71732358:1)	(0.70227595:1)	(0.756699:1)
R_s	0.999889631	0.999889640	0.999958830	0.99988963	0.99988921
C_s	174.9999960	175.0000000	175.0000000	174.999996	174.998506
W_s	198.4395341	198.4395340	195.7352300	198.439534	175.735230
V_s	105.0000000	105.0000000	92.00000000	105.000000	81.0000000

The FMGA approach offers a number of practical advantages to the decision maker, that is

- FMGA method addresses conflicting multiple objectives, giving a trade-off between the objectives;
- The approach accommodates the decision maker's fuzzy preferences;
- The method gives a population of alternative solutions, rather than prescribing a single solution;
- The method is practical, flexible, and easily adaptable to problem situations.

In view of the above advantages, FMGA is a useful decision support tool for the practicing decision maker in system reliability optimization.

7 Conclusion

Practicing decision makers in system reliability optimization seek to find a judicious trade-off between maximizing reliability and minimizing cost to an acceptable degree. In a fuzzy environment, management goals and constraints are imprecise and conflicting. One most viable and useful option is to use a fuzzy satisfying approach that includes the preferences and expert judgments of the decision maker. This study provided a multi-criteria nonlinear mixed integer program for reliability optimization of a complex bridge system. The fuzzy multi-criteria model is transformed into a single-objective model which uses fuzzy multi-criteria evaluation. FMGA uses fuzzy evaluation to evaluate the fitness of individuals in each population at every generation. Numerical results demonstrate that FMGA approach is able to provide high-quality solutions.

This work is a useful contribution to practicing decision makers in the field of system reliability design. FMGA provides a trade-off between management goals, contrary to single-objective approaches which optimize system reliability only. Oftentimes, at design stage, the information required for system reliability design is imprecise and incomplete. With such ill-structured problems, reliance on expert information is inevitable. Using the FMGA approach, the vagueness and imprecision of the expert knowledge, at the design stage, can be addressed effectively while taking into account the multiple conflicting objectives. Furthermore, FMGA provides a population of good alternative solutions, which offers the decision maker a wide choice of practical solutions and an opportunity to consider other practical factors not included in the formulation. Therefore, the approach gives a robust method for system reliability optimization, specifically for the complex bridge system.

References

1. Kuo, W., Prasad, V.R.: An annotated overview of system-reliability optimization. IEEE Trans. Reliab. **49**(2), 176–187 (2000)
2. Huang, H.Z., Tian, Z.G., Zuo, M.J.: Intelligent interactive multi-criteria optimization method and its application to reliability optimization. IIE Trans. **37**(11), 983–993 (2005)
3. Wu, P., Gao, L., Zou, D., Li, S.: An improved particle swarm optimization algorithm for reliability problems. ISA Trans. **50**, 71–78 (2011)
4. Sakawa, M.: Fuzzy Sets and Interactive Multi-criteria Optimization. Plenum Press, New York (1993)
5. Onisawa, T.: An application of fuzzy concepts to modeling of reliability analysis. Fuzzy Sets Syst. **37**(3), 267–286 (1990)
6. Cai, K.Y., Wen, C.Y., Zhang, M.L.: Fuzzy variables as a basis for a theory of fuzzy reliability in the possibility context. Fuzzy Sets Syst. **42**, 145–172 (1991)
7. Bellman, R., Zadeh, L.: Decision making in a fuzzy environment. Manage. Sci. **17**, 141–164 (1970)
8. Chen, L.: Multi-objective design optimization based on satisfaction metrics. Eng. Optim. **33**, 601–617 (2001)
9. Chen, S.M.: Fuzzy system reliability analysis using fuzzy number arithmetic operations. Fuzzy Sets Syst. **64**(1), 31–38 (1994)
10. Bing, L., Meilin, Z., Kai, X.: A practical engineering method for fuzzy reliability analysis of mechanical structures. Reliab. Eng. Syst. Saf. **67**(3), 311–315 (2000)
11. Mohanta, D.K., Sadhu, P.K., Chakrabarti, R.: Fuzzy reliability evaluation of captive power plant maintenance scheduling incorporating uncertain forced outage rate and load representation. Electr. Power Syst. Res. **72**(1), 73–84 (2004)
12. Duque, O., Morifiigo, D.: A fuzzy Markov model including optimization techniques to reduce uncertainty. IEEE Melecon **3**(1), 841–844 (2004)
13. Bag, S., Chakraborty, D., Roy, A.R.: A production inventory model with fuzzy demand and with flexibility and reliability considerations. J. Comput. Ind. Eng. **56**, 411–416 (2009)
14. Garg, H., Sharma, S.P.: Stochastic behavior analysis of industrial systems utilizing uncertain data. ISA Trans. **51**(6), 752–762 (2012)
15. Garg, H., Sharma, S.P.: Multi-criteria reliability-redundancy allocation problem using particle swarm optimization. Comput. Ind. Eng. **64**(1), 247–255 (2013)
16. Slowinski, R.: Fuzzy sets in decision analysis. Operations Research and Statistics. Kluwer Academic Publishers, Boston (1998)
17. Delgado, M., Herrera, F., Verdegay, J.L., Vila, M.A.: Post optimality analysis on the membership functions of a fuzzy linear problem. Fuzzy Sets Syst. **53**(1), 289–297 (1993)
18. Huang, H.Z.: Fuzzy multi-criteria optimization decision-making of reliability of series system. Microelectron. Reliab. **37**(3), 447–449 (1997)
19. Huang, H.Z., Gu, Y.K., Du, X.: An interactive fuzzy multi-criteria optimization method for engineering design. Eng. Appl. Artif. Intell. **19**(5), 451–460 (2006)
20. Mahapatra, G.S., Roy, T.K.: Fuzzy multi-criteria mathematical programming on reliability optimization model. Appl. Math. Comput. **174**(1), 643–659 (2006)
21. Coit, D.W., Smith, A.E.: Reliability optimization of series-parallel systems using genetic algorithm. IEEE Trans. Reliab. **R-45**(2), 254–260 (1996)
22. Chen, T.C., You, P.S.: Immune algorithm based approach for redundant reliability problems. Comput. Ind. **56**(2), 195–205 (2005)
23. Michalewicz, Z.: Genetic Algorithms + Data Structures = Evolutionary Programs. Springer (1996)
24. Hsieh, Y.C., Chen, T.C., Bricker, D.L.: Genetic algorithm for reliability design problems. Microelectron. Reliab. **38**(10), 1599–1605 (1998)
25. Holland, J.H.: Adaptation in Natural and Artificial System. University of Michigan Press, Ann Arbor, MI (1992)

26. Goldberg, D.E.: Genetic Algorithms: In Search, Optimization & Machine Learning. Addison-Wesley Inc, Boston, MA (1989)
27. Hikita, M., Nakagawa, Y., Harihisa, H.: Reliability optimization of systems by a surrogate constraints algorithm. IEEE Trans. Reliab. **R-41**(3), 473–480 (1992)
28. Chen, T.-C.: IAs based approach for reliability redundancy allocation problems. Appl. Math. Comput. **182**, 1556–1567 (2006)

26. [illegible] Addison-Wesley, Boston, MA [illegible]
27. [illegible] algorithm. IEEE Trans. [illegible]
28. [illegible] based [illegible] Comput. [illegible]

Laser Power Distribution for Detection Performance in Laser System

Sunita and V.N. Singh

Abstract Laser ranging system has issues of low detection sensitivity and poor ranging accuracy in the adverse field environment. In this paper, analysis has been carried out for laser power distribution at the plane of semi-active laser (SAL) terminal guidance seeker. For the calculation of power distribution, the model is based on laser beam's transmission path and its power loss. The Beer–Lambert Law is used to compute power loss due to atmospheric attenuation. Lambert's Cosine Law and designator-seeker-target geometry are used to determine the portion of the power that the seeker receives after it reflects off the target. The model is simulated for different weather conditions and different target reflectivities to measure the seeker receiving power reflected from a rectangular parallelepiped at different ranges. Under the specific known conditions, the relationship between the emitting laser wavelengths and atmospheric attenuation and the relationship between the detection distances of detection system and receiving laser power of detector are shown through simulations.

Keywords Semi-active laser · Power distribution · Beer–Lambert law

1 Introduction

There has been large-scale proliferation of lasers and optoelectronic devices and systems for applications such as range finding, target acquisition, target tracking, and designating. The unique characteristics of lasers offer significant improvements in the detection limits up to several kilometers. Highly divergent beams can give

---

Sunita (✉) · V.N. Singh
IRDE, DRDO, Ministry of Defence, Raipur Road, Dehradun, India
e-mail: sunnidma@gmail.com

M. Pant et al. (eds.), *Proceedings of Fifth International Conference on Soft Computing for Problem Solving*, Advances in Intelligent Systems and Computing 437, DOI 10.1007/978-981-10-0451-3_59

pinpoint accuracy in the detection of target of interest from very long distances. Laser guidance is one of the preferred methods among guidance systems, used against both stationary and moving targets. Laser guidance gives pinpoint accuracy. There are two types of laser guidance; semi-active laser guidance and laser beam riding, which is a type of command guidance [1, 2].

In a laser beam riding guidance, the system consists of a missile control/launch unit equipped with electro-optical means to track and illuminate targets with laser, and a missile equipped with a detector or receiver at the back of the missile to detect the incoming laser energy form the launcher. The laser designator always sends the laser beam on the target or the proper intercept point till impact. Detector is used to sense the difference between the laser energy coming on its quadrants and corrective action is taken to align the missile with the laser beam.

The other and most widely used laser guidance method is the semi-active laser guidance. Semi-active laser (SAL) terminal guidance typically consists of a scout illuminating a target with short, high-energy laser pulses in a near-infrared (IR) wavelength. Some of the energy from each pulse reflects off the target and into the seeker's quadrant detector. By comparing the power measured in each channel of the quadrant detector, the seeker approximates the location of the reflected power within its field of view (FOV). Using this information, the seeker returns control signals to the projectile's maneuver system to steer toward the target.

In this paper simulation is carried out to estimate the power received at semi-active laser (SAL) terminal guidance seeker plane in different scenarios. Detection range of the seeker is also simulated based on received power at seeker's plane for various designator powers.

2 Working of Laser Guided Weapons

In order to employ laser guided weapons, two main components are necessary. A laser designator and a laser guided weapon. The designator is aimed at the target by means of operator optics. The laser beam strikes the target surface and reflected. The reflected light enters the seeker as a collimated large beam due to the small aperture of the seeker and the large volume of reflected energy in space. Optics collects the incoming laser beam and directs it on the surface of the detector where the falling energy causes a voltage or current formation on the detector. Since there is a line of sight angle between the weapon and the target, when refracted by the lenses most of the energy will fall into one region of the detector, giving information about the line of sight angle between the target and weapon. This is a necessary knowledge for the guidance system to operate. Depending on the sensitivity and structure of the seeker it is possible to extract the LOS (line of sight) angle, lead angle, or LOS rate from the seeker. This information may be used at various guidance methods, along with some additional sensors.

Laser designators and seekers use a pulse coding system to ensure that a specific seeker and designator combination works in harmony. By setting the same code in both the designator and the seeker, the seeker will track only the target illuminated by the designator. The pulse coding is based on PRF (pulse repetition frequency). The laser beam and its behavior at the atmosphere, along with the weather conditions are modeled in this paper, target types and their response to incoming laser are also analyzed briefly. A làser model based on minimum detectable power constraint is formed in light of all this information.

3 Analysis of Received Power at Seeker

In laser-aided weapon delivery systems, a laser designator is used to illuminate the target. The designator produces a train of very short duration, high peak power pulses of light which are collimated in a very narrow beam and directed to the target. These laser pulses are reflected off the target and are detected by a laser receiver. In this paper for simulation, ground laser designators (1064 nm wavelength) with pulsed laser designation (20 ns pulse width) and divergence 0.5 mils have been taken.

3.1 *Laser Attenuation in the Atmosphere*

Laser beams travel in the atmosphere according to Beer's law, which states that [3]

$$T = e - \sigma R \tag{1}$$

$$\sigma = \frac{3.91}{V}\left(\frac{550}{\lambda}\right)^{q} \tag{2}$$

where R is range in kilometers, V is visibility (km), λ is wavelength (nm), and q is the size distribution of the scattering particles (1.6 for high visibility for $V > 50$ km, 1.3 for average visibility for 6 km $< V <$ 50 km).

If the visual range is less than 6 km due to haze, the exponent q is related to the visual range by the following empirical formula [3]:

$$q = 0.585V^{1/3} \tag{3}$$

where V is expressed in kilometers.

3.2 Target Type and Reflection Pattern

An important factor in laser reflection is the target reflectivity. Each target has a different reflectivity at a certain wavelength, depending on its material properties. The reflectivity, γ, at the laser wavelength of different targets can vary from less than 1 % to almost 100 %. When the reflectivity is not known and cannot be estimated, a value of 20 % or 0.2 (absolute number) is generally used [1, 4]. For simulation the target taken is Russian T-72 MBT (Main Battle Tank) having dimensions of about 2.3 × 2.3 m with different reflectivity 0.2, 0.3 and 0.5. The tank is equipped with ERA (Explosive Reactive Armour) packages on turret, hull front, and skirt sides. It is also assumed that all the incoming laser energy falls and is reflected from the target, without any spillover. A diffuse reflection characterized by the Lambertian scattering is taken for simulation.

3.3 Measurement of Detection Range

The output power of the laser designator is taken as 5 MW, considering the specifications of ground designators. Since the entire laser beam is assumed to be reflected from the target's projected area (no spillover), the power reflected from the target will be attenuated only by the atmospheric attenuation and target reflectance parameters.

The power at the target P_tar can be found by multiplying the designator output power P_des with the transmission coefficient T [5]:

$$P_{\_tar} = P_{\_des}\,T \tag{4}$$

The transmission coefficient T can be calculated using Beer's law in Eq. (1). The atmospheric attenuation coefficient is found with the help of Eq. (2). Assuming that there is no spillover, the reflected power P_ref can be found by considering the target reflectivity as

$$P\_ref = P\_tar\,\gamma \tag{5}$$

where γ is the target reflectivity.

The power collected by the seeker P_rec, is equal to P_ref multiplied by the atmospheric transmission, and the ratio of the seeker optics to the area of a hemisphere with a radius equal to range as shown below.

$$P_{\_rec} = P_{\_ref} \cdot T \cdot \frac{r^2_{\text{seeker}}}{2 \cdot R^2} \tag{6}$$

Equations from (1) to (6) determined the envelop at which the seeker of the laser guided weapon will acquire target. This is achieved by use of minimum detectable power by the laser seeker and the corresponding range at which power reaches to the seeker at a given designator-target geometry, within specified atmospheric conditions. The minimum power detectable by the weapon seeker determines the range at which the reflected energy can be sensed and projectile guidance command generation starts.

4 Simulation Results

Different scenarios are performed with a ground designator at different ranges from the target in high visibility conditions and aircraft attack bearing is assumed the same as the designator-target line at the time of firing. Graph 3 shows atmospheric attenuation coefficient versus visibility. For reflectivity 0.2, assuming that there is no spillover, the power at target P_tar, reflected power P_ref and power sensed by the seeker P_rec versus visibility for 5 km range to target have shown in Graph 1. Graph 2 shows the effect of target reflectivity coefficient (0.2, 0.3, 0.5) on seeker acquisition range for designator target range of 3 km as function of visibility. The target acquisition range is heavily dependent on the target reflectivity and minimum detectable power value of the weapon seeker. The designator range to target is less effective on detection range. The laser reflection model given above does not include the effects caused by screening aerosols, snow, etc., or countermeasure systems like creation of false targets (decoy) by laser replicators.

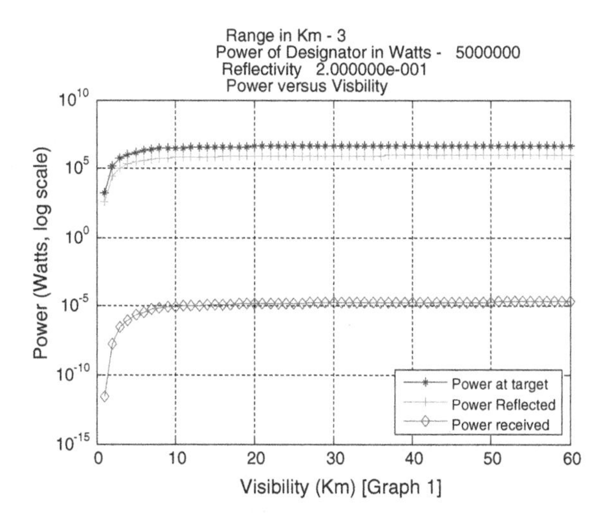

Visibility (Km) [Graph 1]

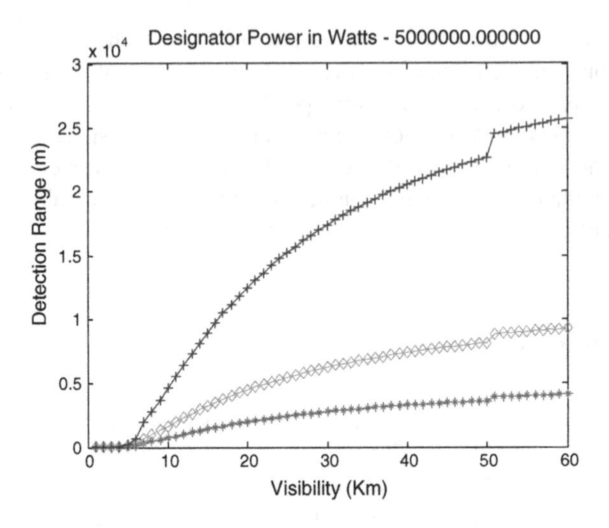

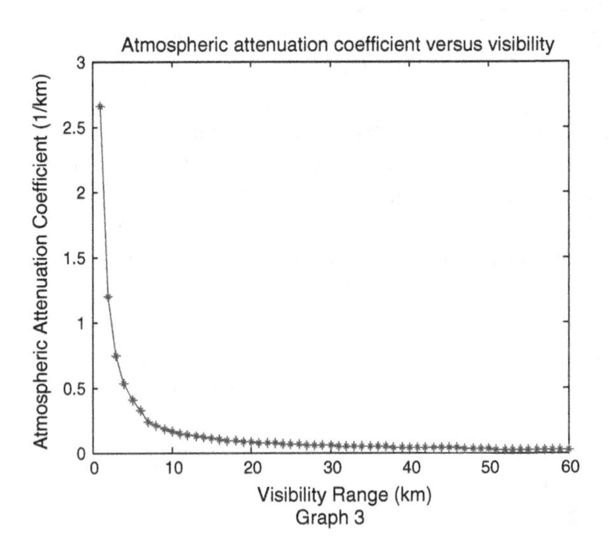

Graph 3

5 Conclusion

In this paper, the laser reflection from a target is modeled. The reflection scheme of the laser is mainly used to determine the envelope at which the seeker of the laser guided weapon will acquire target. This is achieved by the use of minimum detectable power by the laser seeker and the corresponding range at which power reaches the seeker at a given designator-target geometry, within specified atmospheric conditions. The minimum power detectable by the weapon seeker determines the range at which the reflected energy can be sensed and projectile guidance command generation starts.

References

1. Maini, A.K., Verma, A.L.: Pre-flight functionality check to enhance mission efficacy of precision guided munitions. Defence Sci. J. **59**(5), 459–4 (2009)
2. Maini, N., Sabharwal, A., Sareen, K., Singh, A., Kumar, P.: A user programmable electro-optic device for testing of lasers seekers, Amity University Noida. Defence Sci. J. **64**, 1 (2014)
3. Sanyal, S., Mandal, R., Sharma, P.K., Singh, I.: Optics for Laser Seeker. In: International conference on Optics and Photonics (ICOP 2009), CSIO, Chandigarh, India, 30 Oct–1 Nov (2009)
4. The New Seekers Defence Helicopter, vol. 34 no. 1, www.rotorhub.com/Helicopter (2015)
5. Airborne Laser Systems Testing and Analysis, RTO AGARDograph 300, Flight Test Techniques Series, vol. 26 (2010)

Numerical Solution of a Class of Singular Boundary Value Problems Arising in Physiology Based on Neural Networks

Neha Yadav, Joong Hoon Kim and Anupam Yadav

Abstract In this paper, a soft computing approach based on neural networks is presented for the numerical solution of a class of singular boundary value problems (SBVP) arising in physiology. The mathematical model of artificial neural network (ANN) is developed in a way to satisfy the boundary conditions exactly using log-sigmoid activation function in hidden layers. Training of the neural network parameters was performed by gradient descent backpropagation algorithm with sufficient number of independent runs. Two test problems from physical applications have been considered to check the accuracy and efficiency of the method. Proposed results for the solution of SBVP have been compared with the exact analytical solution as well as the solution obtained by the existing numerical methods and shows good agreement with others.

Keywords Neural networks · Singular boundary value problem · Gradient descent algorithm

1 Introduction

The main aim of this paper is to introduce a neural network approach for the solution of the following class of singular boundary value problems arising in physiology [1]:

Neha Yadav · J.H. Kim (✉)
School of Civil, Environmental and Architectural Engineering,
Korea University, Seoul 136-713, South Korea
e-mail: jaykim@korea.ac.kr

Neha Yadav
e-mail: nehayad441@yahoo.co.in

Anupam Yadav
Department of Sciences and Humanities, National Institute
of Technology, Uttarakhand, Srinagar 246174, Uttarakhand, India
e-mail: anupam@nituk.ac.in

M. Pant et al. (eds.), *Proceedings of Fifth International Conference on Soft Computing for Problem Solving*, Advances in Intelligent Systems and Computing 437, DOI 10.1007/978-981-10-0451-3_60

$$u'' + \frac{\alpha}{x} u' = f(x, u(x)), \quad x \in (0, \beta) \tag{1}$$

subject to the following boundary conditions:

$$u'(0) = 0 \quad \text{and} \quad a\,u'(\beta) + b\,u'(\beta) = \eta \tag{2}$$

Let us suppose that $f(x, u)$ is continuous, first order partial derivative $\frac{\partial f}{\partial x}$ exists, and is continuous over the domain $[0, \beta]$. SBVP defined in Eq. (1) arises in various applications, depending on the value of α and on certain functions $f(x, u)$. Particularly for the case when $\alpha = 2$ and $f(x, u) = -pe^{-\text{ply}}, p > 0, l > 0$, represents the formulation of the distribution of heat sources in human head [2, 3].

Singular boundary value problems arising in physiology have been solved by many researchers in recent years. In [4] a finite difference method based on a uniform mesh is presented for solution of SBVPs, which arises in physiology. Different numerical methods like Chebyshev economization [5], B-spline functions [6], cubic extended spline methods [7], Adomian decomposition method [8], etc., have been applied for the numerical solution of SBVP arising in human physiology. A technique based on the Green's function and Adomian decomposition method has been developed by Singh and Kumar in [9]. Existence and uniqueness of the solution of SBVP have also been discussed in [10] and the references therein for different applications where it arises.

Although the SBVP has been solved using different numerical schemes, most of the techniques require discretization of the domain into two parts and then provides a solution for them. The problem becomes difficult when we have more complex boundaries for SBVP, thus the need for meshless techniques arises. Neural networks are one of the meshless techniques used for the solution of differential equations nowadays. A sufficient amount of research is dedicated to the solution of boundary value problem with ODEs as well as PDEs [11–15]. Neural networks have been applied to effectively solve the linear and nonlinear differential equations, e.g., Riccati differential equation [11], first Painleve equation [12], and Beam column equation [13], etc. The survey article [14] presents the history and development in the same field. Keeping in view the advantages and recent work of neural network methods over other existing numerical methods, in this paper we present a neural network method to handle singularity point at the origin.

The rest of the paper is organized as follows. In Sect. 2, a brief overview of the neural network method is presented for solving singular boundary value problems. Problem formulation based on neural networks is discussed in Sect. 3. Section 4 contains numerical simulation of the test problems along with results and discussion. Finally, Sect. 5 presents conclusion and summarizes the numerical results.

2 Neural Networks for SBVP

To illustrate the neural network technique for the solution of singular boundary value problem, let us consider a general SBVP of the following form:

$$x^n u''(x) = F(x, u(x), u'(x)) \tag{3}$$

subject to the boundary conditions imposed, where $n \in Z$, $x \in R$ $D \subset R$ represents the domain and $u(x)$ is the solution to be computed. The strategy to solve SBVP is to first construct an approximate trial solution using neural network that will satisfy the boundary conditions and then train the neural network to obtain the solution over whole domain.

In [15] McFall et al. defined a concept of length factor to construct a trial solution that exactly satisfies the boundary condition and the strategy for defining the boundary conditions will be of the form:

$$u_T(x, \vec{\theta}) = A(x) + L(x)\, N(x, \vec{\theta}) \tag{4}$$

where $L(x)$ represents length factor, which is a measure of distance from the define boundary. The first term $A(x)$ is designed in a way that exactly satisfies the boundary conditions and $N(x, \vec{\theta})$ represents a feedforward neural network with input x and network parameters $\vec{\theta}$. The defined length factor function $L(x)$ should fulfill the following conditions: (i) it should return zero value for all x on boundary, (ii) return nonzero value for all x within the domain, and (iii) there exists at least one nonzero partial derivative $\partial L/\partial x_k$.

After construction of the trial solution Eq. (4) ANN output N must be optimized in a way that it approximately satisfies the differential equation in Eq. (3) with less approximate error. The residual error then can be defined as

$$G = \left[u_T(x_i, \vec{\theta}) - u(x_i)\right] \tag{5}$$

for $i = 1,\ 2, \ldots, n$ represents the collocation points inside the domain.

The mean absolute error MAE then calculated serves as the error function for optimizing the neural network parameters. Updation of neural network parameters then can perform by any optimization algorithm as the local or global search technique. Neural network method for solution of SBVP can be defined as the following algorithm given in Fig. 1.

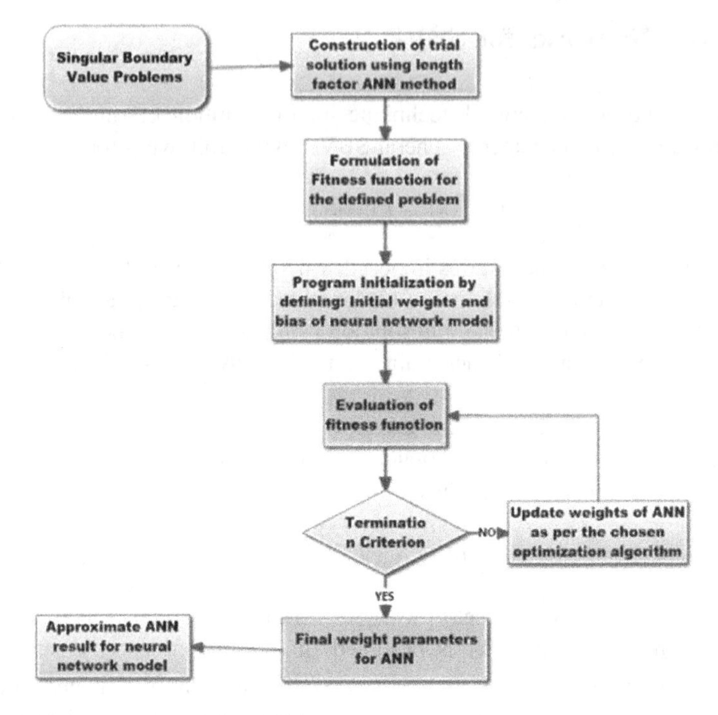

Fig. 1 ANN algorithm for solving SBVP

3 Problem Formulation

In this section, we consider the following two test problems for tackling SBVP arising in the application of human physiology. We also present the neural network formulation for these test problems in the following subsection.

3.1 Test Problem 1

Let us consider the following special case of Eq. (1) defined as follows:

$$u'' + \frac{1}{x}u' = -\exp(u),\ x \in (0, 1) \tag{6}$$

subject to the boundary conditions:

$$u'(0) = 0 \quad \text{and} \quad u(1) = 0. \tag{7}$$

The exact solution for test problem 1 can be written as

$$u(x) = 2\ln\left(\frac{K+1}{Kx^2+1}\right) \tag{8}$$

where $K = 3 - 2\sqrt{2}$. The problem defined in Eq. (6) is also known as Emden–Fowler equation of the second kind.

The trial solution for Eq. (6) can be written as

$$u_T(1,\vec{\theta}) = (2L^{-4} - 4L^{-3} + 2L^{-1})\left(\frac{N_0' - N_L'}{2L} - N_0' + N\right) = 0 \tag{9}$$

which satisfies the desired boundary conditions as

$$\begin{aligned} u_T'(x,\vec{\theta}) &= (8L^{-4}x^3 - 12L^{-3}x^2 + 4L^{-1}x)\left(\frac{N_0' - N_L'}{2L}x^2 - N_0'x + N\right) + \cdots \\ &\quad + (2L^{-4}x^4 - 4L^{-3}x^3 + 2L^{-1}x^2)\left(\frac{N_0' - N_L'}{L}x - N_0' + N'\right) \\ &\text{so,} \quad u_T'(0,\vec{\theta}) = 0. \end{aligned} \tag{10}$$

3.2 *Test Problem 2*

Let us consider another special case of Eq. (1) as

$$u'' + \frac{2}{x}u' = -pe^{-\text{plu}}, p > 0, l > 0 \tag{11}$$

where $0 < x \leq 1$, and subject to the following boundary conditions:

$$u'(0) = 0 \quad \text{and} \quad 0.1u(1) + u'(1) = 0. \tag{12}$$

Since the exact solution of test problem 2 is not known, we investigate the trial solution by satisfying it to the original nonlinear SBVP Eq. (11). The trial solution of Eq. (11) can be written similar to Eq. (9), as it satisfies the boundary condition $u'(0) = 0$ (Eq. 10) and another boundary condition as

$$u'_T(1,\vec{\theta}) = (8L^{-4} - 12L^{-3} + 4L^{-1})\left(\frac{N'_0 - N'_L}{2L} - N'_0 + N\right) + \cdots$$
$$+ (2L^{-4} - 4L^{-3} + 2L^{-1})\left(\frac{N'_0 - N'_L}{L} - N'_0 + N'\right) = 0 \tag{13}$$

and $u_T(1,\vec{\theta}) = (2L^{-4} - 4L^{-3} + 2L^{-1})\left(\frac{N'_0 - N'_L}{2L} - N'_0 + N\right) = 0$

Hence, it satisfies another boundary condition $0.1u(1) + u'(1) = 0$.

4 Numerical Simulation

In this section, numerical simulations are performed for both the test problems and their results are presented for detailed analysis. A three-layer multilayer perceptron neural network has been chosen to solve the above-mentioned SBVP. As the success of ANN method in providing the more accurate solution lies in how well the neural network parameters are chosen, hence, for the solution of SBVP we simulate the neural network for different number of neurons in the hidden layer $h = 5, 10, 20, 25$, number of training points $n = 100$, and 100 different starting weights. The numerical experimentation of each neural network models is carried out on the basis of the chosen number of neurons in the layer and the error function is constructed for both the test problems for inputs $x \in (0, 1)$ respectively as

$$G = \sum_{i=1}^{n} x_i u''_T(x_i,\vec{\theta}) + u'_T(x_i,\vec{\theta}) + x_i \times \exp\left(u_T(x_i,\vec{\theta})\right) \tag{14}$$

$$G = \sum_{i=1}^{n} x_i u''_T(x_i,\vec{\theta}) + 2u'_T(x_i,\vec{\theta}) + p \times \exp\left(-\mathrm{pl}u_T(x_i,\vec{\theta})\right) \tag{15}$$

The accuracy of the neural network models is based on the strength of trial solution and the level of tolerance achieved. The value of the absolute error in the ANN solution has been also computed as $\left|u_{\text{Exact}}(x_i) - u_T(x_i,\vec{\theta})\right|$ for different number of neurons in the hidden layer and are reported in Tables 1 and 2 for both the test problems respectively.

From Table 1 it can be seen that by increasing the number of neurons in the hidden layer from 5 to 25, better accuracy is achieved in neural network solution. Hence, we solved the SBVP defined in test problem 1 and test problem 2 using $h = 25$ and $n = 100$ training points inside the domain (0, 1). Solution obtained by

Table 1 Absolute error in the solution for test problem 1

x	$h = 5$	$h = 10$	$h = 20$	$h = 25$
0	4.36829E−04	2.96485E−04	4.12623E−05	4.06719E−05
0.1	6.05678E−04	3.08246E−04	3.40862E−04	1.97039E−05
0.2	4.18236E−03	3.93142E−03	4.08266E−04	4.39122E−05
0.3	3.124455E−03	3.66574E−04	2.64221E−04	1.53577E−04
0.4	4.99105E−03	2.86428E−04	2.22549E−04	3.19082E−04
0.5	4.32891E−03	3.33574E−03	3.24663E−03	5.62493E−04
0.6	5.34025E−03	3.48624E−03	2.01843E−03	9.25803E−04
0.7	2.82451E−02	4.10562E−03	4.008426E−03	1.47987E−03
0.8	1.95746E−02	3.83521E−03	3.260421E−03	2.334114E−03
0.9	1.54277E−02	4.574425E−03	4.654440E−03	3.646379E−03
1.0	1.12336E−02	1.063445E−02	5. 928112E−03	5.39616E−03

Table 2 Numerical solution for test problem 2

x	Method in [1] for step length (1/20)	Method in [6] for step length (1/60)	ANN solution for $h = 25$
0	1.14704079	1.147039936	1.14670012
0.1	1.14651141	1.146510559	1.146159777
0.2	1.14492228	1.144921418	1.144539109
0.3	1.14227034	1.142269478	1.141837995
0.4	1.13855053	1.138549660	1.138056436
0.5	1.13375570	1.133754812	1.133194431
0.6	1.12787656	1.127875662	1.127251981
0.7	1.12090166	1.120900762	1.120229085
0.8	1.11281731	1.112816415	1.112125743
0.9	1.10360749	1.103606592	1.102941956
1.0	1.09325371	1.093252826	1.093752767

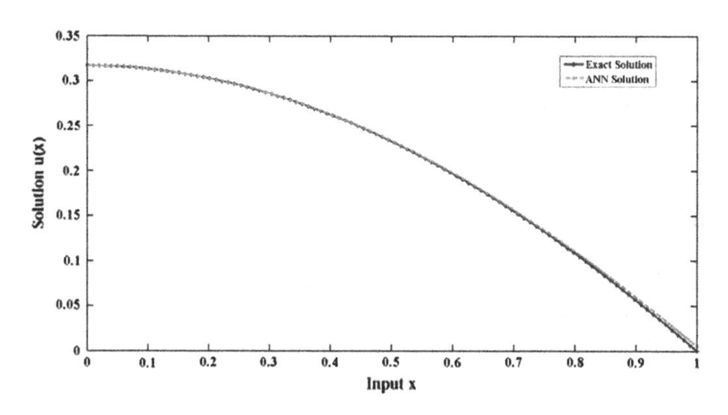

Fig. 2 ANN and exact solution for test problem 1 for $h = 25$ and $n = 100$

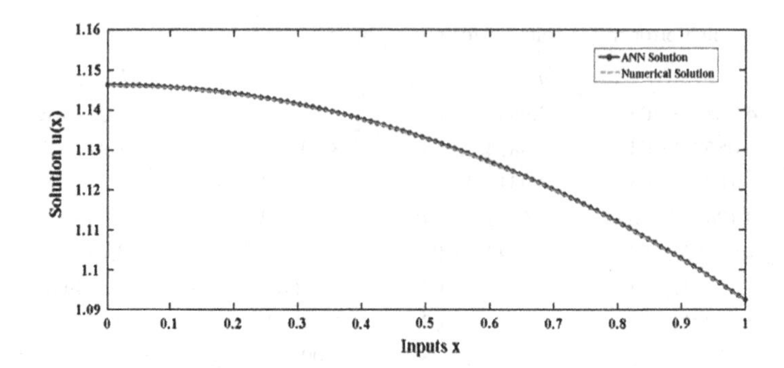

Fig. 3 ANN and numerical solution [1] for test problem 2 for $h = 25$ and $n = 100$

ANN method is also compared to the exact solution for test problem 1 as depicted in Fig. 2, and for test problem 2, ANN solution is compared to the solution obtained by other numerical methods in the existing literature and is presented in Fig. 3.

5 Conclusion

In this paper, an approach based on neural network has been presented for the numerical solution of a class of SBVP arising in physiology. For the construction of trial solution, the concept of length factor is used in ANN to satisfy the boundary conditions. The results reported in the paper show that the neural network method provides better estimation to the solution and is also comparable to the solution of other existing methods. It can also be concluded that the method is effective and applicable for a wide range of SBVPs arising in any engineering and science applications. Moreover, the presented method does not require discretization of domain and the solution is not dependent on the mesh size. Once the network is trained, it provides continuous evaluation of solution inside the domain.

Acknowledgments This work was supported by National Research Foundation of Korea (NRF) grant funded by the Korean government (MSIP) (NRF-2013R1A2A1A01013886) and the Brain Korea 21 (BK-21) fellowship from the Ministry of Education of Korea.

References

1. Khuri, S.A., Sayfy, A.: A novel approach for the solution of a class of singular boundary value problems arising in physiology. Math. Comput. Model. **52**(3–4), 626–636 (2010)
2. McElwain, D.L.S.: A re-examination of oxygen diffusion in a spherical cell with Michaelis-Menten oxygen update kinetics. J. Theor. Biol. **71**(2), 255–263 (1978)
3. Flesch, U.: The distribution of heat sources in the human head: a theoretical consideration. J. Theor. Biol. **54**(2), 285–287 (1975)

4. Pandey, R.K., Singh, A.K.: On the convergence of a finite difference method for a class of singular boundary value problem arising in physiology. J. Comput. Appl. Maths. **166**(2), 553–564 (2004)
5. Ravi Kanth, A.S.V., Reddy, Y.N.: A numerical method for singular two point boundary value problems via Chebyshev economization. Appl. Math. Comput. **146**(2–3), 691–700 (2003)
6. Caglar, H., Caglar, N., Ozer, M.: B-spline solution of non-linear singular boundary value problems arising in physiology. Chaos Soliton. Fract. **39**(3), 1232–1237 (2009)
7. Gupta, Y., Kumar, M.: A computational approach for solution of singular boundary value problem with applications in human physiology. Natl. Acad. Sci. Lett. **353**(3), 189–193 (2012)
8. Singh, R., Kumar, J., Nelakanti, G.: Approximate series solution of nonlinear singular boundary value problems arising in physiology. Sci. World J. **945872**, 10 (2014)
9. Singh, R., Kumar, J.: An efficient numerical technique for the solution of nonlinear singular boundary value problems. Comput. Phys. Commun. **185**, 1282–1289 (2014)
10. Hiltmann, P., Lory, P.: On oxygen diffusion in a spherical cell with Michaelis-Menten oxygen update kinetics. Bull. Math. Biol. **45**(5), 661–664 (1983)
11. Raja, M.A.Z., Khan, J.A., Qureshi, I.M.: A new stochastic approach for solution of Riccati differential equation of fractional order. Ann. Math. Artif. Intell. **60**(3–4), 229–250 (2010)
12. Raja, M.A.Z., Khan, J.A., Siddiqui, A.M., Behloul, D.: Exactly satisfying initial conditions neural network models for numerical treatment of first Painleve equation. Appl. Soft Comput. **26**, 244–256 (2015)
13. Kumar, M., Yadav, N.: Buckling analysis of a beam-column using multilayer perceptron neural network technique. J. Frank. Inst. **350**(10), 3188–3204 (2013)
14. Kumar, M., Yadav, N.: Multilayer perceptrons and radial basis function neural network methods for the solution of differential equations: a survey. Comput. Math. Appl. **62**(10), 3796–3811 (2011)
15. McFall, K.S., Mahan, J.R.: Artificial neural network method for solution of boundary value problems with exact satisfaction of arbitrary boundary conditions. IEEE Trans. Neural Net. **20** (8), 1221–1233 (2009)

Prototype Evidence for Estimation of Release Time for Open-Source Software Using Shannon Entropy Measure

H.D. Arora, Ramita Sahni and Talat Parveen

Abstract Software systems are updated either to provide more comfort to the users or to fix bugs present in the current version. Open-source software undergoes regular changes due to high user-end demand and frequent changes in code by the developer. Software companies try to make the interface more comfortable and user friendly, which requires frequent changes and updation at a certain period of time and addition of new features from time to time as per customer's demand. In this paper, we have considered the bugs recorded in various bugzilla software releases and calculated the Shannon entropy for the changes in various software updates. We have applied multiple linear regression models to predict the next release time of the software. Performance has been measured using goodness-of-fit curve, different residual statistics and R^2.

Keywords Open-source software · Shannon entropy · Next release time · Software repositories

1 Introduction

Open-source software systems are becoming increasingly popular these days due to the advancement in Internet technologies and communication and thus are proven to be highly successful in various fields. With this growing demand for open-source software in today's generation, changes in the software are inescapable. In order to

H.D. Arora (✉) · Ramita Sahni · Talat Parveen
Department of Applied Mathematics, Amity Institute of Applied Sciences, Amity University, Sector-125, Noida, Uttar Pradesh, India
e-mail: hdarora@amity.edu

Ramita Sahni
e-mail: smiles_ramita@yahoo.co.in

Talat Parveen
e-mail: talat.tyagi@gmail.com

M. Pant et al. (eds.), *Proceedings of Fifth International Conference on Soft Computing for Problem Solving*, Advances in Intelligent Systems and Computing 437, DOI 10.1007/978-981-10-0451-3_61

meet customer expectation, changes occur in the software frequently. Bug repair, feature enhancement and addition of new features are the three main types of code changes that occur in the software source code. These changes are being made continuously to remove the bugs, enhance the features and implement new features. Massive changes in the code make the software complex over a time period and results in introduction of bugs and at times cause failure of the software. It is of great significance to record these changes in a proper manner. For this purpose, we have source code repositories which record these changes properly. It also contains various details such as bugs present, record of every change occurred, creation data of a file and its initial content, communication between users and developers, etc. These storage locations, viz. repositories, offer a great opportunity to study the code change process. By code change process we mean to study the patterns of source code modifications. A complex code in the software system is undesirable as this complexity leads to many faults and ultimately faces many delays. Thus, it is a necessary task to manage and measure complexity of source code software.

Project managers strive hard to control the complexity of the open-source software. The paramount goal is to release the software on time and with an appropriate budget. To achieve this, researchers have proposed many metrics to reduce the enhancing entropy with the regular change in code of software due to customer demand, feature enhancement, development process and increasing competition. There is delay in the release of a new version of the software if the software developer does not have a proper understanding of the code change process.

The entropy-based estimation proves to be important in studying the code change process. Entropy is an important branch of information theory, which is a probabilistic approach that focuses on measuring uncertainty relating to information. Entropy is a measure of randomness/uncertainty/complexity of code change. It depicts how much information is contained in an event. Entropy is evaluated taking into consideration the number of changes in a file for a defined time period with respect to the total number of changes in all files. This period can be considered as a day, week, month, year, etc. We have used a sound mathematical concept from information theory, viz., Shannon entropy [1].

The concept of entropy in information theory was developed by Shannon [1]. A formal measure of entropy, called Shannon entropy [1] was defined by him which is as follows:

$$S = -\sum_{i=1}^{n} p_i \log_2 p_i \tag{1}$$

where p_i denotes the probability of occurrence of an event.

The above Eq. (1) is known as 'measure of uncertainty'. This uncertainty measure is used as a measure of equality, disorder or uniformity. The base of logarithm is taken to be 2 for discrete case and otherwise it is taken arbitrary (>1). Shannon entropy [1] possesses interesting properties as mentioned below:

1. It is permutationally symmetric.
2. It is additive in nature.
3. It is a decreasing function and a concave function.
4. It is maximal if all outcomes are equally likely.
5. If one outcome is certain then it is zero.
6. It is continuous in $0 < p_i < 1$.
7. Shannon entropy is never negative as $0 < p_i < 1$.

In this paper, we have developed a method to determine the next release time of the Bugzilla software using Shannon entropy [1]. We have applied multiple linear regression model using SPSS (Statistical Package for Social Sciences) in predicting the time of the next release of the product. We have also measured the performance using goodness-of-fit curve and R^2 and residual statistics.

2 Literature Review

It is of great importance to release a software version on time to fulfil the demands of the customers. Researchers are working hard to determine the suitable time for the release of the software. A few open-source projects release their software versions frequently using milestone wise, requirement wise, important patch wise, using fixed time, etc. Hassan [2] proposed information theory in measuring the amount of uncertainty or entropy of the distribution to quantify the code change process and to predict the bugs based on past defects using entropy measures. Singh and Chaturvedi [3] calculated the complexity of code change and used it for bug prediction. Ambros and Robbes [4] presented a benchmark for defect prediction and provided an extensive comparison of well-known bug prediction approaches. Singh and Chaturvedi [5] proposed to predict the future bugs of software based on current year entropy using vector regression and compared with simple linear regression. Garzarelli [6] attempted to explain the working of open-source software development.

The 'Next Release Problem'(NRP) term was coined by Bagnall et al. [7] They formulated NRP to find the set of requirements within resource constraint. Geer and Ruhe [8] proposed an approach to optimize software releases using genetic algorithm. Baker et al. [9] considered both ranking and selection of components to which search-based approach is automated using greedy and simulated annealing algorithm to address next release problem. Garey and Johnson [10] estimated that no exact algorithm can determine the number of requirements for the next release in polynomial time. Cheng and Atlee [11] concluded that it is difficult to take decisions about the next release of the product to attain maximum profit as user needs go on increasing. Ngo-The and Ruhe [12] proposed a two-phase optimization by combining integer programming to relax the search space and genetic programming to reduce search space. Jiang et al. [13] proposed a hybrid ant colony optimization algorithm (HACO) for next release problem (NRP). Xuan et al. [14] proposed backbone-based multilevel algorithm (BMA) to address the large-scale NRP. BMA reduces the problem scale and constructs a final optimal set of customers. BMA can achieve better solutions to large-scale NRP. Kapur et al. [15] proposed a method for release time problem based on reliability, cost and bugs fixed. Chaturvedi et al. [16] proposed a complexity of code change-based method to address the next release problem using multiple linear regression model and measured the performance using different residual statistics, goodness-of-fit curve and R^2.

3 Methodology

Bugzilla (www.bugzilla.org) [17] is the world's leading free software bug tracking system. It tracks bugs, communicates with teammates and manages quality assurance. Several organizations use bugzilla such as Mozilla, Linux Kerner, GNOME, KDE, Apache Project, LibreOffice, Open Office, Eclipse, Red Hat, Mandriva, Gentoo, Novell, etc. The bugzilla project was started in 1998 and till now it has had 144 releases. In our paper, we have considered the bugs present in several releases of the Bugzilla project from 2001 to May 2015. We have taken into consideration only the major releases of the project over this period of time. Data are organized with some consideration and are pre-processed. If there is more than one release in a particular month, all the bugs fixed are merged in a single release as a latest release in that month. Data are organized on yearly basis and time along the release is taken in months. Entropy is used to quantify the changes as suggested by Hassan [2] over the releases in terms of complexity of code change. This method of predicting next release date using complexity of code change was used by Kapur et al. and Chaturvedi et al. [15, 16]. We have used Shannon entropy [1] to predict the release time of the next version.

The following methodology has been proposed for data preparation of the project:

1. Release dates and information about previous Bugzilla versions are noted.
2. Bugs fixed in each release are recorded and organized along each release.
3. Exclude the first release of the project as this release does not contain any source code changes and bugs fixed and also excludes releases which do not contain any bugs.
4. From the release history, if there are multiple releases in a single month, merge all releases as a single release and also merge corresponding bugs.
5. Arrange the releases on a monthly basis and then merge the releases belonging to a particular year with the latest release of that year.
6. Shannon entropy/complexity of code change is calculated.
7. Multiple linear regression model is fitted with SPSS while taking time as dependent and entropy and number of bugs as independent variables.

We have calculated the entropy change in various software releases of bugzilla and thus predicted the next release time of software. Multiple linear regression is applied using SPSS to predict the next release time.

Multiple linear regression is a statistical tool that uses several variables (independent) to predict dependent variable. When there are two or more independent variables it is called a multiple linear regression. It is applied to predict the release time of the software.

For multiple linear regression we have the following notation:

X_1 : Shannon entropy/Complexity of code changes

X_2 : Number of bugs present

Y : Release time in months

The multiple linear regression [18] has been used for the independent variable X_1 and X_2 and dependent variable Y using Eq. (2)

$$Y = a + bX_1 + cX_2 \quad (2)$$

where a, b and c are regression coefficients

The values of a, b and c are calculated using the SPSS. After estimation of regression coefficients using the historical data of number of bugs fixed and complexity of code changes/entropy, we construct a system with which we can predict the next release time of software. Table 1 depicts Shannon entropy of bugs fixed for each release along with predicted months for each release.

SPSS (Statistical Package for Social Sciences) has been used for analysis. It is a comprehensive and flexible statistical data management and analysis tool.

Table 1 Shannon entropy of bugs fixed along with predicted months of different release

Shannon entropy	Number of bugs	Predicted value of months	Number of months	Software versions
0.000	7	4	3	4.4.8
0.99403	22	8	6	4.4.4
1.36504	35	9	11	4.4.1
2.30502	54	13	13	4.2.4
1.91039	40	12	12	4.0.3
2.26518	45	13	11	3.6.3
2.27649	37	13	12	3.4.4
1.73701	25	11	13	3.2
0.9449	18	8	9	3.0.2
1.52057	34	10	11	2.22.1
1.8496	16	12	13	2.2
0.88129	10	8	6	2.16.7
0.92621	41	7	8	2.16.4
1.22105	17	9	9	2.16
0.65002	12	7	9	2.14

4 Results and Discussions

R defines the correlation between the observed and the predicted value. It measures the strength and the direction of a linear relationship between two variables and it lies between −1 and 1. R^2 is known as coefficient of determination which is the square of R. The coefficient of determination is the proportion of variability of the dependent variable accounted for or explained by the independent variable. The value of R^2 is between 0 and 1. Adjusted R^2 adjusts the R^2 statistic based on the number of independent variable in the model. It gives the most useful measure to model development.

The results of the multiple linear regression are shown in Table 2. This table depicts the value of R, the value of R^2 and adjusted R^2 and standard error of estimate.

The difference between the observed and predicted value is termed as residual. The residual statistics are given in Table 3.

Goodness-of-fit curve has been plotted between observed and predicted value. Goodness-of-fit for multiple linear regression shows the fitting of the next release observed value with the next release predicted value. Figure 1 depicts this goodness-of-fit curve.

Table 2 Parameters of multiple linear regression

R	R^2	Adjusted R^2	Standard error of estimate
0.876	0.768	0.729	1.55439

Table 3 Residual statistics

Statistics	Minimum	Maximum	Mean	Std deviation	N
Predicted value	4.316	13.25783	9.723921	2.614574	15
Residual	−2.17674	2.048765	0.009413	1.439097	15

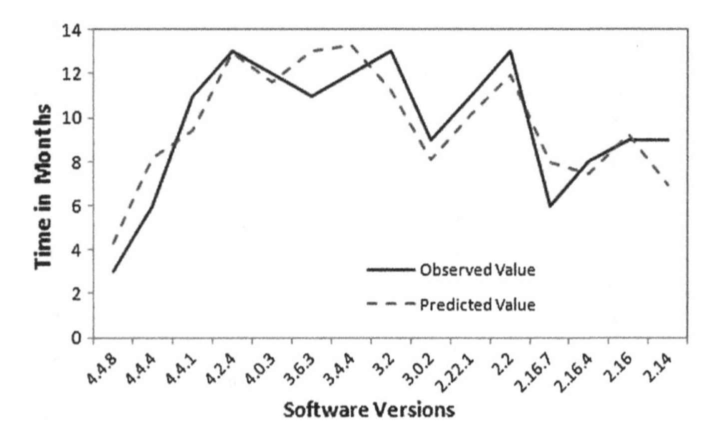

Fig. 1 Goodness-of-fit curve for different software versions

In the releases, the predicted time of next release is similar to the observed value of the next release. The regression coefficients of the multiple linear regression helps in determining the predicted value of the next release in months. These coefficients can be easily evaluated using the number of bugs present in the current year and Shannon entropy [1]/complexity of code changes observed for the same period.

5 Conclusion

In this paper, an approach has been developed to determine the predicted time of next release of the open-source software Bugzilla using multiple linear regression model. In multiple linear regression, we have considered the number of bugs recorded and Shannon entropy calculated both as independent variables and time in months as dependent variable for the statistics record for Bugzilla software (www.bugzilla.org). For this study we have collected our data from the website www.bugzilla.org and arranged it with some consideration. We have applied information theoretic concept given by Shannon [1] to calculate the entropy change for various releases of the Bugzilla software. Comparison between the observed and the predicted value is depicted using goodness-of-fit curve. From the graph it could be depicted that there is no significant difference between the observed and the predicted value of all the releases. We have not considered the latest release 4.4.9 because it does not contain the complete revised data. R^2 for Shannon entropy [1] is 0.768, which means that 76.8 % of the variance in dependent variable is predictable from independent variables In this study, the results are limited to only open-source software systems. We can further extend our study by applying this method in predicting the next release time of other software projects.

References

1. Shannon, C.E.: A mathematical theory of communication. Bell Syst. Tech. J. **27**(379–423), 623–656 (1948)
2. Hassan, A.E.: Predicting Faults based on complexity of code change. In: Proceedings of 31st International Conference on Software Engineering, pp. 78–88 (2009)
3. Singh, V.B., Chaturvedi, K.K.: Improving the quality of software by quantifying the code change metric and predicting the bugs. In: Murgante, B. et al., (eds.), ICCSA 2013, Part II, LNCS 7972, pp. 408–426, 2013. © Springer-Verlag Berlin Heidelberg (2013)
4. D'Ambros, M., Lanza, M., Robbes, R.: An extensive comparison of bug prediction approaches. In MSR'10: Proceedings of the 7th International Working Conference on Mining Software Repositories, pp. 31–41 (2010)
5. Singh, V.B., Chaturvedi, K.K.: Entropy based bug prediction using support vector regression. In: Proceedings of 12th International Conference on Intelligent Systems Design and Applications during 27–29 Nov 2012 at CUSAT, Kochi (India). ISBN: 978-1-4673-5118-8_c 2012. IEEE Explore, pp. 746–751 (2012)

6. Garzarelli, G.: Open source software and the economics of organization. In: Bimer, J., Garrouste, P. (eds.) Markets, information and communication, pp. 47–62. Routledge, New York (2004)
7. Bagnall, A., Rayward-Smith, V., Whittley, I.: The next re-lease problem. Inf. Softw. Technol. **43**(14), 883–890 (2001). doi:10.1016/S0950-5849(01)00194-X
8. Greer, D., Ruhe, G.: Software release planning: an evolutionary and iterative approach. Inf. Softw. Technol. **46**(4), 243–253 (2004). doi:10.1016/j.infsof.2003.07.002
9. Baker, P., Harman, M., Steinhofel, K., Skaliotis, A.: Search based approaches to component selection and prioritization for the next release problem. In: Proceedings 22nd International Conferrence Software Maintenance (ICSM '06), pp. 176–185, Sept. 2006. doi:10.1109/ICSM.2006.56
10. Garey, M.R., Johnson, D.S.: Computers and intractability: a guide to the theory of NP-completeness, pp. 109–117. W.H. Freeman, New York (1979)
11. Cheng, B.H.C., Atlee, J.M.: Research directions in requirements engineering. In: Proceedings International Conference on Software Engineering Workshop Future of Software Engineering (FoSE '07), pp. 285–303, May 2007. doi:10.1109/FOSE.2007.17
12. Ngo-The, A., Ruhe, G.: Optimized resource allocation for software release planning. IEEE Trans. Softw. Eng. **35**(1), 109–123 (2009). doi:10.1109/TSE.2008.80
13. Jiang, H., Zhang, J., Xuan, J., Ren, Z., Hu, Y.: A hybrid ACO algorithm for the next release problem. In: Proceeding International Conference Software Engineering and Data Mining (SEDM '10), pp. 166–171, June 2010
14. Xuan, J., Jiang, H., Ren, Z., Luo, Z.: Solving the large scale next release problem with a backbone based multilevel algorithm. IEEE Trans. Softw. Eng. **38**(5), 1195–1212 (2012)
15. Kapur, P.K., Singh, V.B., Singh, O.P., Singh, J.N.P.: Software release time based on multi attribute utility functions. Int. J. Reliab. Qual. Safety Eng. **20**(4), 15 (2013)
16. Chaturvedi, K.K., Bedi, P., Mishra, S., Singh, V.B.: An empirical validation of the complexity of code changes and bugs in predicting the release time of open source software. In: IEEE 16th International Conference on Computational Science and Engineering (2013)
17. The bugZilla project. http://www.bugzilla.org
18. Weisberg, S.: Applied Linear Regression. John Wiley and Sons (1980)

A Broad Literature Survey of Development and Application of Artificial Neural Networks in Rainfall-Runoff Modelling

Pradeep Kumar Mishra, Sanjeev Karmakar and Pulak Guhathakurta

Abstract Rainfall-Runoff (R-R) modelling is one of the most important and challenging work in the real and present world. In all-purpose, rainfall, temperature, soil moisture and infiltration are highly nonlinear and complicated parameters. These parameters have been used in R-R modelling and this modelling requires highly developed techniques and simulation for accurate forecasting. An artificial neural network (ANN) is a successful technique and it has a capability to design R-R model but selection of appropriate architecture (model) of ANN is most important challenge. To determine the significant development and application of artificial neural network in R-R modelling, a broad literature survey last 35 years (from 1979 to 2014) is done and results are presented in this survey paper. It is concluded that the architectures of ANN, such as back propagation neural network (BPN), radial basis function (RBF), and fuzzy neural network (FNN) are better evaluated over the conceptual and numerical method and worldwide recognized to be modelled the R-R.

Keywords ANN · BPN · Feed forward · RBF · Rainfall-Runoff

P.K. Mishra (✉)
Research Center, Bhilai Institute of Technology,
Durg, CG, India
e-mail: pradeepmishra1883@gmail.com

S. Karmakar (✉)
Department of Master of Computer Application, Bhilai Institute of Technology,
Durg, CG, India
e-mail: dr.karmakars@gmail.com

P. Guhathakurta
India Meteorological Department (IMD), Pune, India
e-mail: pguhathakurta@rediffmail.com

M. Pant et al. (eds.), *Proceedings of Fifth International Conference on Soft Computing for Problem Solving*, Advances in Intelligent Systems and Computing 437, DOI 10.1007/978-981-10-0451-3_62

1 Introduction

Accurate forecasting of Rainfall-Runoff (R-R) over a river basin through modelling has been challenging for scientists and engineers of hydrology for ages and centuries. To meet this challenge the mathematical modelling and computation may play an important role. To get exact runoff estimates, various techniques have been used. However, the R-R analysis is quite difficult due to presence of complex nonlinear relationships in the transformation of rainfall into runoff. However, runoff analyses are extremely significant for the prediction of natural calamities like floods and droughts. It also plays a crucial role in the design and operation of various components of water resources projects like barrages, dams, water supply schemes, etc. Runoff analyses are also needed in water resources planning, development and flood mitigations. Every year floods to a loss of several lives, damage of property and crops of millions of dollars only because there is no assessment of runoff forecasting. Various types of modelling tools had been used to estimate runoff. These techniques consist of distributed physically based models, stochastic models, lumped conceptual models and black box (time series) models. Conceptual and physically based models although try to account for all the physical processes involved in the R-R process, their successful use is limited mainly because of the need of catchment specific parameters and simplifications involved in the governing equations. On the other hand the use of time series stochastic models is complicated due to non-stationary behaviour and nonlinearity in the data. These models require the modellers to be expert and experienced.

From 1986, artificial neural networks (ANNs) have emerged as a powerful computing system for highly complex and nonlinear systems (chaos), such as climate, runoff, etc. ANNs belongs to the black box time series models and offers a relatively flexible and quick means of modelling. These models can treat the nonlinearity of system to some extent due to their parallel architecture. However, various architecture of ANNs is used in nonlinear system. It is found that the architecture of ANNs depends on the problem space.

The aim of this study is to categorize the ANNs in R-R modelling and their applicability without any scientific controversy. To achieving the aim, the objectives of this study are to identify all methods including ANNs for R-R modelling up to till date and their performances and evaluate the performance of ANNs. These objectives are considered via comprehensive review of literature from 1979 to 2014. It is found that, several methods are used including ANNs. Although, ANNs are found suitable without any controversy. However, detail of discussion concerning the architecture of ANN for the same is rarely visible in the literature, while various applications of ANN are available.

The paper has been constructed with the sections. Section 2 discussed year wisely comprehensive review of world-wide contribution from 1979 to 2014. ANNs methods to R-R modelling those are indisputably accepted and no scientific disagreements are discussed in Sect. 3. Results of the review are discussed in Sect. 4 and finally conclusions are discussed in the Sect. 5.

2 Literature Review

The significant contributions in the field of R-R modelling from 1979 to 2014 are reviewed and identified very vital methodologies. The major contributions are discussed in this section.

Initially in this research area, Kitanindis and Bras have applied ridge regression technique and this technique provides parameter estimates that seem more intuitive and it handles co-linearity problems in a consistent objective manner and without endangering regression accuracy [1]. After that Kitanindis and Bras have developed 'kalman filter' which is used in real-time river discharge forecasting [2].

Martinec has found deterministic approach require measurements of the snow covered area provided by remote sensing for short-term discharge forecasts [3]. Baumgartner et al. derived the main input variable for a deterministic snowmelt runoff model (SRM), it is assured snow cover information for a specific area and time [4]. Vandewiele and Yu have used statistical methodology for calibrating the models of given catchments is described, it reduces effectively to regression analysis, as well as residual analysis, sensitivity to calibration period and extrapolation test. The results of implementing the new models are sufficient from a statistical point of view [5].

In the major involvement in 1993, Shashikumar et al. have developed snowmelt runoff model (SRM) using variables like temperature, precipitation and snow covered area along with some externally derived parameters like temperature lapse rate, degree-day factor and this snowmelt runoff model is useful for snow fed river basins, especially where meteorological and hydrological data are not sufficient [6]. Kember et al. have applied nearest neighbour method (NNM), the NNM model is found to improve forecasts as compared to auto-regressive integrated moving average (ARIMA) models [7].

Seidel et al. have described that the Runoff is computed by the SRM model with snow covered areas as well as temperature and precipitation forecasts as input variable. These runoff forecasts can be utilized, among other purposes, for optimizing the hydropower production and for gently Decisions on the electricity market [8]. Shi et al. have used polarimetric model and this model shows that scattering mechanisms control the relationship between snow wetness and the co-polarization signals in data from a multi-parameter synthetic aperture radar [9]. Franchinia et al. have observed that the stochastic storm transposition (SST) approach have been converted to a range of possible flood peak values by a R-R model (the ARNO model) and a probabilistic disaggregation method of cumulative storm depths to hourly data. The SST approach has been extended to a probabilistic procedure for assessment of yearly possibilities of flood peaks by coupling with a R-R model [10].

After the development of linear model as discussed above, Franchini et al. have Compared several genetic algorithm (GA) schemes, GA is connected with a conceptual R-R model, is such as to make the GA algorithm relatively inefficient [11].

The forecast system uses hydrological model is developed by Bach et al. [12], applied for the translation of rainfall into runoff and this model has been modified to allow remote sensing inputs in an automated way using geographical information system (GIS) tools. This technique improves flood forecast through a better estimation of spatial input parameters [12].

The important contributions in 1999, Sajikumar et al. have implemented a simple "black box" model to identify a direct functioning between the inputs and outputs without detailed consideration of the internal structure of the physical process and compared with a conceptual R-R model, it was slightly better for a particular river flow-forecasting problem [13]. During the growth of ANN concept, Dibike and Solomatine have analyzed two types of ANN architectures that is multi-layer perceptron network (MLP) and radial basis function network (RBN) and have concluded that ANN-based forecast model is better than conceptual model [14].

In the context of hybrid modelling 2000, Toth et al. have used linear stochastic auto-regressive moving average (ARMA) models, ANN and the non-parametric nearest-neighbours method as well as analyze and compare the relative advantages and restriction of each time series analysis technique, used for predicting rainfall for lead times varying from 1 to 6 h. The results also indicate a significant improvement in the flood forecasting accuracy [15].

In the considerable contributions in 2001, Sivapragasam et al. singular spectrum analysis (SSA) coupled with support vector machine (SVM) techniques have been used and SSA–SVM technique found 58.75 % improvement over nonlinear prediction (NLP) in the runoff prediction for catchment [16]. Chang et al. applied and modify radial basis function (RBF) neural network (NN) and the modified RBF NN is capable of providing arbitrarily good prediction of flood flow up to 3 h ahead [17].

In 2002, Randall and Tagliarini have used feed-forward neural networks (FFNNs) technique and this technique provides better forecasting results rather than auto-regressive moving average (ARMA) techniques [18]. Cannas et al. have Used neural network-based approach with feed-forward, multi-layer perceptrons (MLPs) and this neural models applied to the R-R problem could provide a useful tool for flood prediction water system [19]. Breath et al. have developed time series analysis technique for getting better the real-time flood forecast by deterministic lumped R-R model and they have concluded that aside from ANNs with adaptive training, all the time series analysis techniques considered allow significant enhancements if flood forecasting accuracy compared with the use of empirical rainfall predictors [20].

In the significant contributions in 2003, Mahabir et al. have applied fuzzy logic modelling techniques and this technique provides more accurate quantitative forecast [21]. Slomatine and Dulal have used ANNs and model trees (MTs) technique and concluded that both techniques have almost similar performance for 1 h ahead prediction of runoff, but the result of the ANN is a little superior than the MT for higher lead times [22]. Gaume and Gosset have employed feed-forward artificial neural network (FFNN) and the FFNN model appear to be better forecasting tools than linear models. The FFNN model can be efficient only if those function are suitable for the process to be simulated [23].

In the year of 2004, Hossain et al. have analyzed framework modelling and studied satellite rainfall error model (SREM) to simulate passive microwave (PM) and infrared (IR) satellite rainfall and it is useful for the design, planning and application evaluation of satellite remote sensing in flood and sudden flood forecasting [24]. In the significant contributions in 2005, Corani and Guariso have applied neural networks and fuzzy logic technique and performance improvements have been found by comparing the projected framework approach about traditional ANN for flood forecasting [25]. Khan and Coulibaly have introduced a Bayesian learning approach for ANN modelling (BNN) of daily stream flows implemented with a multi-layer perceptron (MLP) and have concluded that BNN model is better than the non-Bayesian model for forecasting [26]. In 2006, Ghedira et al. have used ANN system and conclude that the performance of neural network reaches 82 %, 10 % higher than the filtering algorithm for the test data sets which is not used in the neural network training process [27]. Liu et al. have studied time series transfer function noise (TFN) of time series, the grey system (GM), along with the adaptive network-based fuzzy inference system (ANFIS) and found that adaptive network-based fuzzy inference system (ANFIS) is better [28]. Tayfur et al. have applied ANN fuzzy logic (FL) and kinematic wave approximation (KWA) models, found ANN and FL models are used to predict runoff at a very small scale of a flume 6.9 m^2 and a larger scale of a plot 92.4 m^2, and also predict a small scale of a watershed 8.44 m^2. It is easier to construct ANN and FL models than KWA model [29].

In the important contributions in 2007, Cheng et al. have proposed bayesian forecasting system (BFS) framework, with back-propagation neural network (BPN) and this technique not only increases forecasting precision greatly but also offers more information for flood control [30]. Jiang et al. have introduced Fletcher-Reeves algorithm in BPN model and found that this model can enhance the convergence rate without increasing its complexity, so as to improve the forecasting precision of the BPN model [31]. Ju et al. have used BPN model and studied comparison of with Xinanjiang model indicates that BPN performs well on the stream flow simulation and forecasting [32]. After using BPN, Broersen, has introduced auto-regressive moving average (ARMA) time sseries models and its performed well for small samples [33]. Lohani et al. have used fuzzy logic technique and the fuzzy modelling approach is slightly better than the ANN [34].

In the noteworthy contributions in 2008, Li and Yuan have used data mining technology with ANN as well as this approach needs less input data requisite, not as much of maintenance and performs more simple forecasting process and Good precision of forecasting [35]. After that in terms of designing of model, Liu et al. have conceptualized adaptive-network-based fuzzy inference system (ANFIS) and concluded that ANFIS was better than ARMA Model [36]. Pei and Zhu have used fuzzy inference technique and found that the results specify that the model can efficiently select the forecast factors and the forecast precision is improved [37]. Xu et al. BPN model, distributed hydrological model had been used and the simulated results of the BPN model indicate a acceptable performance in the daily-scale simulation than distributed hydrologic model [38]. Sang and Wang have applied stochastic model for mid-to-long-term runoff forecast is built by combining wavelet analyze (WA), ANN and hydrologic frequency analysis and compared with traditional methods, this model is of higher accuracy, high eligible rate and improve the performance of forecast results about the runoff series [39]. Guo et al. have proposed wavelet analysis and artificial neural network (WANN) model and the result shows that this model can get a good result in simulating and predicting monthly runoff [40]. Remesan et al. have applied two system models (i) neural network auto-regressive with exogenous input (NNARX) and (ii) adaptive neuro-fuzzy inference system (NIFS). ANFIS and NNARX models can be applied successfully in R-R modelling to achieve highly accurate and reliable forecasting results [41]. Aytek et al. have analyzed genetic programming (GP) technique, ANN techniques: (i) the BPN and (ii) generalized regression neural network (GRNN) methods and found that (GP) formulation performs very well compared to results achieved by ANNs. Gene expression programming (GEP) can be proposed as an alternative to ANN models [42]. Solaimani, has used artificial neural network (ANN) with feed-forward back propagation and found the ANN method is more appropriate and well organized to predict the river runoff than typical regression model [43].

In the significant contributions in 2009, Ping, has combined technique of ANN. Wavelet transformation and solving the non-stationary time series problem. This method is feasible and effective [44]. Yan et al. have used BP neural network model. BP model has high accuracy and can be used to forecast runoff and the sediment transport volume during the flood period and non-flood period [45]. In the duration of development of R-R modelling, Yan et al. have proposed RBF and compared the RBF emulating results with the field data, the forecasting error is analyzed and the methods to improve the forecast precision [46]. Luna et al. have applied fuzzy inference system (FIS) approach and have concluded that FIS approach is better than soil moisture accounting procedure (SMAP) approach, it is promising for daily runoff forecasting [47]. Xu et al. have used BPN and compared between BPN models and four process-based models, namely a lumped model (Xinanjiang), a semi-distributed model (XXT), and ESSI and (SWAT) two distributed models (SWAT). It is found that all the four process-based models showed poor performance as compared to BPN models for I-day lead time to 20-day lead time [48]. Feng and Zhang have used back propagation neural network and found that the ANN technology is a comparatively efficient way of resolving problems in forecast of surface runoff [49]. Hundecha et al. have developed fuzzy logic-based R-R model and have concluded that Fuzzy logic-based R-R model is better than

HBV model. This approach is easier and faster to work with [50]. Hung et al. have utilized ANN technique and found that the superiority in performance of the ANN model over that of the persistent model and the accuracy and efficiency of forecast has been tremendously improved by ANN [51].

In 2010, Xu et al. have utilized support vector machine (SVM)-based R-R models and it is found that both the process-based models TOPMODEL and Xinanjiang model perform poorer compared with SVM model [52]. Deshmukh and Ghatol have applied that ANN-based short-term runoff forecasting system, Jordan neural network model, Elman neural network model and have concluded that Jordan neural Network is performing better as compared to Elman neural network in forecasting runoff for 3 h lead time [53]. Wang and Qiu have applied adaptive network-based fuzzy inference system (ANFIS) model and found ANFIS model has better forecasting performance than artificial neural network (ANN) model [54]. Wang et al. have proposed SVM model with chaotic genetic algorithm (CGA) and results is compared with the ANN performance. It is found that the SVM-CGA model can give good prediction performance and more accurate prediction results [55]. Liu et al. have designed optimal subset regression and back propagation (OSR-BP) neural network model and the result shows that the stability of model is good and accuracy is satisfactory for long-term runoff prediction [56]. Yan et al. have compared between BPN model and RBFN model and found that the BP model has high accuracy and can be used to forecast runoff and the RBF model should be improved by treating the input value reasonably, deciding the data number suitably [57]. Liu et al. have proposed shuffled complex evolution metropolis algorithm (SCEM-UA) and have concluded that the SCEM-UA is much better than that by GA [58]. Huang and Tian have applied multi-layer perception (MLP) with BPN, or simply as MLP-BPN model and ANN-based forecasting model components aimed to enhance modelling efficiency of interactive runoff forecasting [59]. Jizhong et al. have designed Adaptive regulation ant colony system algorithm (ARACS) and RBF neural network combined to form ARACS-RBF hybrid algorithm. This method improves forecast accuracy and improves the RBF neural network generalization capacity, it has a high computational precision and in 98 % of confidence level, the average percentage error is not more than 6 % [60].

In a considerable contributions in 2011, Linke et al. have applied conceptual R-R model based on statistical black-box models and this model is applied to get high quality weather forecasts and flood warning systems [61]. Ilker et al. have applied ANN along with coefficient of determination (R2) and root-mean square error (RMSE) and found that the performance of the best model using ANN [62]. Huang and Wang have Proposed novel hybrid NN–GA (neural network—genetic algorithm) and it shows that it increased the R-R forecasting accuracy more than any other model [63]. In this duration, Zhenmin et al. have combined the neural network rainfall forecast model with genetic algorithm and GIS and found that the model efficiently improved forecast precision and speed [64]. Zhang and Wang have applied wavelet-artificial fuzzy neural network (ANFIS) model and the result shows that, the prediction accuracy is greatly enhanced and it is fit to be used in daily runoff prediction [65]. Aiyun and Jiahai have combined the wavelet analysis and artificial neural network (BP algorithm) and found that this model has a better capability of simulation for the process of monthly runoff and the model designed for predict with a higher accuracy [66]. Wu and Chau have developed ANN, modular artificial neural network (MANN) with singular spectrum analysis (SSA) and the ANN R-R model coupled with SSA is more promising compare to MANN and LR models [67].

In 2012, Shah et al. have applied hybrid approach of possibilistic fuzzy C-mean and Interval type-2 Fuzzy logic methods and this approach can help in planning and reforming the agriculture structure, storm water management, runoff and pollution control system as well as the accurate rainfall prediction [68]. Bell et al. have combined support vector machine with sequential minimum optimization (SMO) algorithm with a RBF kernel function and this technique yield a comparative absolute error 48.65 % against 63.82 % for the human ensemble forecast, this hybrid model optimize forecasts wet and dry years equally [69]. Li et al. have proposed a model based on BPN algorithm and found that this technique deal with nonlinear hydrological series problems and provides a new idea to mid and long-term runoff forecast of reservoirs [70]. Mittal et al. have developed a dual (combined and paralleled) artificial neural network (D-ANN) and concluded that the D-ANN model performs very well and better than the feed-forward ANN (BPN) model [71].

In the noteworthy contributions in 2013, Chen et al. have studied BPN and conventional regression analysis (CRA) and found that BPN perform better than the conventional regression analysis method [72]. Phuphong and Surussavadee have applied ANN-based runoff forecasting model and results show good forecast accuracy and useful forecasts 12 h in advance [73]. Patil et al. have reviewed the different forecasting algorithms. ANN, FL, GA of R-R modelling and have concluded that if fuzzy logic and ANN are combine together it will provide better result which can be used for prediction [74]. Ramana et al. have combined the wavelet technique with ANN and found that the performances of wavelet neural network models are more efficient than the ANN models [75]. In 2014, Karmakar et al. described that the BPN in identification of internal dynamics of chaotic motion is observed to be ideal and suitable in forecasting various natural phenomena [76].

At the end of this review following methods are identified in suitability of R-R modelling as shown in Table 1. However, the different ANNs and fuzzy-based system are found better evaluated over the other conceptual methods. In the next section, the performances of ANN methods those are significantly used in R-R modelling are discussed.

Table 1 Identified methods of R-R modelling in the literature (1979–2014)

No.	Year	Methods	Contributor(s)
1	1979	Ridge regression technique	Kitanidis and Bras
2	1980	Kalman filter technique	Kitanidis and Bras
3	1982	Deterministic approach	Martinec.
4	1987	SRM	Baumgartner et al.
5	1993	NNM	Chember et al.
7	1992	Statistical model	Vandewiele and Yu
8	1993	SRM	Shashikumar et al.
9	1993	NNM	Kember et al.
10	1994	SRM	Seidel et al.
11	1995	Polarimetric model	Shi et al.
12	1996	SST	Franchinia et al.
13	1997	GA	Franchini et al.
14	1999	GIS tools	Bach et al.
		Black box model	Sajikumar et al.
		BPN, RBF	Dibike and Solomatine
15	2001	SSA–SVM	Sivapragasam et al.
		RBFN	Chang et al.
		BPNs	Randall and Tagliarini
		BPN	Cannas et al.
		Time series analysis technique	Breath et al.
16	2003	Fuzzy logic modelling	Mahabir et al.
		MLP with BP algorithm	Slomatine and Dulal
		BPN	Gaume and Gosset
17	2004	SREM	Hossain et al.
18	2005	Neural networks and fuzzy logic technique	Corani and Guariso
		BPN and MLP	Khan and Coulibaly
		ANNs	Ghedira et al.
		TFN and ANFIS	Liu et al.
		BPN, FL, KWA	Tayfur et al.
19	2006	Motes-based sensor network	Chang and Guo
		ANNs	Ghedira et al.
		TFN and ANFIS	Liu et al.
		Fuzzy neural network	Li et al.
		DDM	Khan and See
		Hydro CA routing model	Huan et al.
		DEM	Du et al.
		BPN, FL, KWA	Tayfur et al.
20	2007	BFS,BPN	Cheng et al.
		BPN	Jiang et al.
		BPN	Ju et al.
		ARMA time series model	Broersen
		Fuzzy based model	Lohani et al.

(continued)

Table 1 (continued)

No.	Year	Methods	Contributor(s)
21	2008	BPN	Li and Yuan
		ANFIS	Liu et al.
		Fuzzy inference technique	Pei and Zhu
		BPN	Xu et al.
		WA, ANN	Sang and Wang
		WANN model	Guo et al.
		ANFIS, NNARX	Remesan et al.
		GP, FFBP, GRNN,	Aytek et al.
		ANN	Solaimani
22	2009	ANN with wavelet transformation	Ping
		BPN	Yan et al.
		RBF	Yan et al.
		FIS	Luna et al.
		BPN	Xu et al.
		TOPMODEL, DEM	Xui et al.
		BPN	Feng and Zhang
		Fuzzy logic-based rainfall-runoff model	Hundecha et al.
		ANN	Hung et al.
23	2010	SVM	Xu et al.
		ANN-based model	Deshmukh and Ghatol et al.
		SVM with CGA	Wang et al.
		OSR-BPNN	Liu et al.
		BPN, RBF	Yan et al.
		SCEM-UA	Liu et al.
		MLP with BP	Huang and Tian
		ARACS-RBF	Jizhong et al.
		SVM	Huang and Tian
24	2011	Black-box models	Linke et al.
		ANN with RMSE	Ilker et al.
		NN–GA	Huang and Wang
		ANN,GA,GIS	Zhenmin et al.
		Wavelet-ANFIS	Zhang and Wang
		Wavelet analysis-BPNN	Aiyun and Jiahai
		MANN with SSA	Wu and chau
25	2012	Fuzzy logic	Shah et al.
		SVM with SMO	Bell et al.
		BPN	Li et al.
		D-ANN	Mittal et al.
		XXT	Jingwen et al.

(continued)

Table 1 (continued)

No.	Year	Methods	Contributor(s)
26	2013	BPN	Chen et al.
		ANN	Phuphong and Surussavadee
		ANN, FL, GA	Patil et al.
		Wavelet technique with ANN	Ramana et al.
27	2014	BPN	Karmakar et al.

3 ANNs and Their Performance

Two main architectures of ANN known as BPN, RBF and Fuzzy-based system are found sufficiently suitable to R-R modelling. Very significant contributions based on the performance of the model are discussed as follows.

3.1 Back Propagation Neural Network (BPN)

In 1999, Dibike and Solomatine have found performance of the conceptual model corresponds to model efficiencies of 95.89 % for calibration and 80.23 % for verification periods, respectively. The results show that while model efficiency on training data were 98.4 and 91.1 % on the verification data were obtained using ANN with the appropriate input pattern, found corresponding values of 95.9 and 80.2 % with the properly calibrated conceptual tank model [14].

In 2007; 2008, Ju et al. have found that the prediction error of the BPN model is lower than these of the Xinanjiang model—its coefficient of efficiency (E) is 10 % or lower than these of the Xinanjiang model and the context of the comparative error is higher than these of the Xinanjiang model. On the other hand, the BPN and the Xinanjiang models still have a high error in predicting some flood measures as well as the maximum error arises during the same flood measures. This leads to the conclusion the input data has to be processed more efficiently and some new attributes should be added to improve the prediction accuracy [32].

Year 2008, Li and Yuan, proposed a approach, In this approach, a long-term integrative ANN runoff forecasting model is developed with data mining technology and BPN. This forecasting model represent following results (Table 2).

When compared this BPN model with traditional runoff forecasting method, BPN model needs less input data requirement, less maintenance and performs more simple forecasting process. The test result shows that it is fit for engineering and forecasting area [35].

In 2009, Yan et al. have compared with the field data, during the flood period in 1991, the forecast error of runoff is 1.3 million m^3 (the relative error is 0.03 %) and the forecast error of the sediment transport volume is 3.6 million t (the relative error is 4.5 %). During the non-flood period, the forecast error of runoff is 42 million m^3

(the relative error is 0.05 %) and the forecast error of the sediment transport volume is 8 million t (the relative error is 4.73 %) [45].

In 2009, Hung et al. developed BPN model for real-time rainfall runoff forecasting and flood management in Bangkok, Thailand. The performance ANN model shows in Table 3.

In this table, for 1 h lead forecasts at some stations the value of EI reached up to 98 %, while the lowest EI value of all stations was 74 %. Similarly, the maximum and minimum correlation coefficient values of 0.99 and 0.74 shows highly satisfactory results. For 2 h ahead forecast, the results also quite satisfactory with maximum EI of 87 %, and minimum EI of 63 %, and maximum and minimum R2 of 0.92 and 0.63, respectively. Forecasting results of 3 h ahead are less satisfactory, results for 4–6 h ahead may be considered to be poor with the mean EI varying between 45 and 36 % and mean R2 in the range from 0.78 to 0.71. This ANN model was compared to the convenient approach namely simple persistent method. The results prove that ANN forecasts are much better than over the ones obtained by the persistent model [51].

Table 2 Performance of BPN model [35]

Date	Forecasting value/(m^3/s)	Real value/(m^3/s)	Error/(%)
21	610	901	32.29
22	487	411	18.49
23	469	422	11.13
24	466	523	10.89
25	455	378	20.37
26	401	801	49.93
27	431	766	43.73
28	476	1023	53.47
29	509	683	25.47
30	531	672	20.98
31	512	620	17.42

Table 3 Performance of BPN model with lead time [51]

Lead time (h)	Efficiency index (%)			Correlation coefficient			RMSE (mm/h)		
	Max	Min	Mean	Max	Min	Mean	Max	Min	Mean
1	98	74	86	0.99	0.74	0.88	1.48	0.42	0.87
2	87	63	69	0.92	0.63	0.77	2.16	0.73	1.36
3	68	42	54	0.84	0.55	0.67	2.55	1.06	1.72
4	62	35	45	0.78	0.48	0.64	2.82	1.11	1.85
5	58	30	41	0.73	0.46	0.62	2.72	1.16	1.88
6	48	29	36	0.71	0.36	0.60	2.75	1.24	1.93

In 2010, Deshmukh and Ghatol have found Jordan and Elman model shown excellent performance which is applied for R-R modelling for the upper area of Wardha River in India. An ANN-based short-term runoff forecasting system is developed in this work. A assessment between Jordan and Elman neural network model is made to investigate the performance of the two different approaches. The performance is shown as following Table 4 [53].

In 2010, Liu et al. developed BPN and trained this model with the help of collection of data from 2001 to 2008. Prediction analysis and accuracy analysis on Runoff show in Table 5.

The BPN model test was examined for 6 years (year 2001–2008) and it was found that the accuracy runoff in the month of September was 75 % further. The accuracy reached 85 % in October. On this basis, it can be concluded that BPN model is excellent for training and performance is highly accurate for flood forecast [56].

In 2010, Yan et al. introduced BPN and RBF model for the Yellow River Mouth during the flood and non-flood period in 1990 and 1991. In BPN model during the flood period, found the comparative forecast error of runoff is 0.03 % and the relative forecast error of the sediment transport volume is 4.5 % and during the non-flood period, the relative forecast error of runoff is 0.05 % and the relative forecast error of the sediment transport volume is 4.73 %. Which indicate the BPN model is higher accurate than RBF model and it can be used to accurate runoff forecasting [57].

Table 4 Comparison of performances [53]

S. no.	Model	Performance measure	
		MSE	Co-coefficient
1	Jordan	0.0187	0.86
2	Elman	0.0192	0.81

Table 5 Training and actual performance of BPN model [56]

Objection year	September			October		
	Actual value	Prediction value	Accuracy evaluation	Active value	Prediction value	Accuracy evaluation
2001	25.7	22.5	Pass	25.3	28.4	Pass
2002	11.6	10.2	Pass	8.7	6.6	Pass
2003	170.2	145.7	Pass	86.7	22.7	Fail
2004	60.5	67.0	Pass	40.4	44.2	Pass
2005	51.6	11.5	Fail	117.6	128.4	Pass
2006	27.8	19.8	Pass	26.1	19.1	Pass
2007	35.6	12.7	Fail	15.2	21.1	Pass
2008	40.1	36.3	Pass	40.7	42.8	Pass
Pass rate	75.0 %			87.5 %		

Ilker et al. have studied, various developed ANN models by different input combinations for Kizilirmak River in turkey. The best-fit network structure is determined according to the model performance criteria for testing data set. The best coefficient of correlation (R) values of the best model are 0.98 and 0.97 for training and testing set, correspondingly [62].

Recently in 2012, Li et al. have observed the forecasting errors analysis as shown in table that model 1 and 2 are back propagation neural network-based model as showed in Table 6, by introducing frequencies, the average forecasting error has decreased from 12.28 to 9.00 % in 2005 and from 13.13 to 10.58 % in 2006, respectively; the maximum forecasting errors have gone down −21.9 to 15.43 % in 2005 and from 20.4 to −15.3 % in 2006, respectively, the forecasting qualified rate (relative error is below 20 %) has both improved from 83 to 100 %. Thus, in premise that the influencing runoff factors (such as temperature, sunspot activity, etc.) are complicated to achieve, by introducing frequencies from the data of many years historical runoff can improve the forecasting accuracy to some extent [70].

In 2012, Mittal et al. proposed Dual-ANN (D-ANN) model and applied on Kolar river in India. This model has been compared with that of the feed-forward neural network in the context of statistical criteria coefficient of correlation (*R*), coefficient of efficiency (*E*), root-mean-square error (RMSE). Comparison of D-ANN and BPN model on the basis of statistical criteria is presented in Table 7.

The performance of the D-ANN and BPN model is similar to coefficient of correlation (R) and coefficient of efficiency (E) but BPN has improved the result on root-mean-square (RMSE) and it's a better technique for prediction of high flow in R-R forecast [71].

The research work in 2013, Chen et al. developed BPN R-R model for Linbien River basin located in the low-lying areas of Pingtung County in the southern part of Taiwan and compared with conventional regression analysis (CRA).

Table 6 Forecasting errors analysis of BPN model 1 and 2 [70]

Year	Model	Average forecasting error (%)	Maximum forecasting error (%)	Number of the month that forecasting error is above 20 %	Forecasting qualified rate (%)
2005	1 (4-10-1)	12.28	−21.9	2	83
	2 (8-10-1)	9.00	15.43	0	100
2006	1	13.13	20.4	2	83
	2	10.58	−15.3	0	100

Table 7 Performance comparison between D-ANN and BPN [71]

	D-ANN	BPN
Coefficient of correlation (R)	0.99	0.99
Coefficient of efficiency (E)	0.98	0.98
Root-mean-square error (RMSE)	27.16	23.24

They studied that 1745 datasets collected were divided into 1221 sets for linear regression and 524 sets were for regression model testing. It is found that the determination coefficient (*R*2) performed by the regression model is only 0.284 applied the same data to the BPN to obtain the determination coefficient (*R*2) for is 0.969. It can easily conclude that the ANN approach performed better than the conventional statistical method [72].

In 2013 Phuphong and Surussavadee, developed BPN configuration have three hidden layers with 10, 5, and 1 neurons. This model had been applied for two hydrological stations. The correlation coefficients between forecasts and observations are higher than 0.92 and 0.86 for station 1 and station 2 and root mean square (RMS) errors are within 1.92 and 6.67 %, respectively. The result shows high forecast accuracy for runoff forecasting model-based BPN and this model useful forecast 12 h in advance and can be adapted for other river basins where good quality rainfall data are not available [73].

3.2 Radial Basis Function (RBF)

In 2001, Chang et al. have applied and developed RBF model. They discovered that the RBF provides arbitrarily good prediction of flood flow up to 3 h ahead [17]. In 2009, Yan et al. have proposed RBF; they build RBF able to forecast the runoff and the sediment transport volume of Lijin section during the flood period and non-flood period in 11th year according to the previous 10 years field data. They compared the RBF emulating results with the field data. The forecasting error of the runoff and the sediment transport volume is too obviously. They analyzed forecasting error and found RBF is sufficiently suitable [46].

4 Result and Discussion

It is observed that BPN and RBF model is significant. However, in 2010, Yan et al. have compared between BPN model and RBF model and found that the BPN has high accuracy and can be used to forecast runoff and the RBF model should be improved by treating the input value reasonably, choosing the data number suitably [57]. By BPN model, for the duration of the flood period in 1991, the relative forecast error of runoff is 0.03 % and the comparative forecast error of the sediment transport volume is 4.5 %. During the non-flood period, the relative forecast error of runoff is 0.05 % and the comparative forecast error of the sediment transport volume is 4.73 %, which indicate the BPN model has high accuracy and can be used to forecast runoff and the sediment transport volume during the flood period and non-flood period in the Yellow River Mouth. However, RBF model should be improved by treating the input value reasonably, choosing the data number suitably, etc.

In 2014, Karmakar et al. explained that the BPN in identification of internal dynamics of chaotic motion is observed to be ideal and suitable in forecasting various natural phenomena. However, selection of its parameters likes, number of input vectors (n), number of hidden layers (m), number of neurons in hidden layers (p), number of output neurons (y), weights and biases, learning rate (α), momentum factors (μ) seems crucial during design time. Specially, for chaos forecast, no author has provided optimum value of these parameters. Also during the training process, it is found that, a high learning rate (α) leads to rapid learning but the weights may oscillate, while a lower value of α leads to slower learning process in weight updating. Identification of a correct value of α maintaining a higher learning process needs special attention of the researchers in this area. The main purpose of the momentum factor (μ) is to accelerate the convergence of error during the training. It is the most complicated and experimental task to select appropriate value of 'α' and 'μ' during the training process of BPN as well. They identified these parameters to identify internal dynamics of monsoon rainfall data time series through deterministic forecast. They concluded that, with $\alpha = 0.3$ and $\mu = 0.9$, the BPN is trained properly and also found efficient enough. Particularly for this problem $\alpha = 0.3$ and $\mu = 0.9$ is found optimum and these values have produced exceptional performance (standard deviation = 7.3; mean absolute deviation = 2.841; CC = 0.88) with a high level of mean square error MSE = 4.99180426869658E-04 during the training process. And their optimized parameters are $m_1 = 3$, $m_2 = 1$, $n = 11$, $p = 3$, $e = 15 \times 10^5$, $y_k = 1$, $\alpha = 0.3$, $\mu = 0.9$ and $f(x)$ is sigmoid. It is noted that BPN parameters values may diverge for other problems. Accordingly, optimization of BPN parameters is extremely useful and therefore their optimum values must be chosen carefully through experiments only. Finally, it can be observed that the BPN model can be applied to forecast R-R modelling, however, required superiority to select its parameters like $v_{ij}s$, $w_{ij}s$, v_0, w_0, m_1, m_2, n, p, $f(x)$, e, y_k, α, and μ is vital. Thus, a careful experimentation is suggested before applying the BPN model in R-R modelling [76].

5 Conclusion

R-R modelling over river basin is a major challenge in the context of accuracy and time. Through broad study of different techniques, it has been found that (i) statistical model (ii) lumped conceptual models (iii) distributed physically based models (iv) stochastic models (v) and black box (time series) models are not sufficient in R-R modelling for accurate forecasting because coefficient of correlation (R), coefficient of efficiency (E) and root mean square error (RMSE) rate is high. In the last 10 years ANN emerged as a powerful technique for R-R modelling that provide accurate, adequate forecast timely, but selection and designing of ANN architecture is very difficult. ANN basically classified in BPN and RBF models. Research survey of these ANN models show that accurate forecast result of R-R modelling but when BPN model has been compared with RBF model BPN has high accuracy and can be used to forecast runoff and the RBF model should be improved by treating the input value all most practically. BPN model can be hybridized with fuzzy logic for better R-R modelling in the future expectations.

Acknowledgments We are thankful to India Meteorological Department Pune, India, State Data Center, Raipur, Chhattisgarh, and Bhilai Institute of Technology, Durg, Chhattisgarh, India.

References

1. Kitanidis, P.K., Bras, R.L.: Collinearity and stability in the estimation of rainfall-runoff model parameters. J. Hydrol. **42**, 91–108 (1975)
2. Kitanidis, P.K., Brass, R.L.: Adaptive filtering through detection of isolated transient error in rainfall-runoff models. Water Resour. Res. **16**, 740–748 (1980)
3. Martinec, J.: Runoff modeling from snow covered area. IEEE Trans. **20**(3), 259–262 (1982)
4. Baumgartner, M.F., Seidel, K., Martinec, J.: Toward snowmelt runoff forecast based on multisensor remote-sensing information. IEEE Trans. **25**(6), 746–750 (1987)
5. Vandewiele, G.L., Yu, C.: Methodology and comparative study of monthly water balance models in Belgium, China and Burma. J. Hydrol. **134**(1–4), 315–347 (1992)
6. Shashiumar, V., Paul, P.R., Ramana Rao, Ch.L.V., Haefner, H., Seibel, K.: JAHS Publication, 218, 315–347 (1993)
7. Kember, G., Flower, A.C.: Forecasting river flow using nonlinear dynamics. Stochast. Hydrol. Hydraulics **7**, 205–212 (1993)
8. Seidel, K., Brusch, W., Steinmeier, C.: Experiences from real time runoff forecasts by snow cover remote sensing. IEEE Trans. 2090–2093 (1994)
9. Shi, J., Dozier, J.: Inferring snow wetness using C-band data from SIR-C's polarimetric synthetic aperture radar. IEEE Trans. **33**(4), 905–914 (1995)
10. Franchinia, M., Helmlinger, T.K.R., Foufoula-Georgioub, E., Todini, E.: Stochastic storm transposition coupled with rainfall-runoff modeling for estimation of exceedance probabilities of design floods. J. Hydrol. **175**, 511–532 (1996)
11. Franchini, M., Galeati, G.: Comparing several genetic algorithm schemes for the calibration of conceptual rainfall-runoffmodels. Bydrol. Sci. J. Des Sci. Hydrol. **42**(3), 357–379 (1997)
12. Bach, H., Lampart, G., Strasser, G., Mauser, W.: First results of an integrated flood forecast system based on remote sensing data. IEEE Trans. **6**(99), 864–866 (1999)
13. Sajikumar, N., Thandavewara, B.S.: A non-linear rainfall–runoff model using an artificial neural network. J. Hydrol. **216**(1–2), 32–55 (1999)
14. Dibike, Y.B., Solomatine, D.P.: River flow forecasting using artificial neural networks. Elsevier **26**(1), 1–7 (1999)
15. Toth, E., Brath, A., Montanari, A.: Comparison of short-term rainfall prediction models for real-time flood forecasting. J. Hydrol. **239**, 132–147 (2000)
16. Sivapragasam, C., Liong, S.Y., Pasha, M.F.K.: Rainfall and runoff forecasting with SSA–SVM approach. J. Hydroinformatics 141–152 (2001)
17. Chang, F.J., Liang, J.M., Chen, Y.C.: Flood forecasting using radial basis function neural networks. IEEE Trans. **31**(4), 530–535 (2001)
18. Randall, W.A., Tagliarini, G.A.: Using feed forward neural networks to model the effect of precipitation on the water levels of the northeast Cape fear river. IEEE Trans. 338–342 (2002)
19. Cannas, B., Fanni, A., Pintusb, M., Sechib, G.M.: Neural network models to forecast hydrological risk. IEEE Trans. 623–626 (2002)

20. Brath, A., Montanari, A., Toth, E.: Neural networks and non-parametric methods for improving realtime flood forecasting through conceptual hydrological models. Hydrol. Earth Syst. Sci. **6**(4), 627–640 (2002)
21. Mahabir, C., Hicks, F.E., Fayek, A.R.: Application of fuzzy logic to forecast seasonal runoff. Hydrol. Process **17**, 3749–3762 (2003)
22. Slomatine, D.P., Dulal, K.N.: Model trees as an alternative to neural networks in rainfall–runoff modelling. Hydrol. Sci. J. Des Sci. Hydrol. **48**(3), 399–411 (2003)
23. Gaume, E., Gosset, R.: Over-parameterisation, a major obstacle to the use of artificial neural networks in hydrology. Hydrol. Earth Syst. Sci. **7**, 693–706 (2003)
24. Hossain, F., Anagnostou, E.N., Dinku, T.: Sensitivity analyses of satellite rainfall retrieval and sampling error on flood prediction uncertainty. IEEE Trans. **42**(1), 130–139 (2004)
25. Corani, G., Guariso, G.: Coupling fuzzy modeling and neural networks for river flood prediction. IEEE Trans. **35**(3), 382–390 (2005)
26. Khan, M.S., Coulibaly, P.: Streamflow forecasting with uncertainty estimate using bayesian learning for ANN. IEEE Trans. 2680–2685 (2005)
27. Ghedira, H., Arevalo, J.C., Lakhankar, T., Azar, A., Khanbilvardi, R., Romanov, P.: The effect of vegetation cover on snow cover mapping from passive microwave data. IEEE Trans. 148–153 (2006)
28. Liu, C.H., Chen, C.S., Su, H.C., Chung, Y.D.: Forecasting models for the ten-day streamflow of Kao-Ping river. IEEE Trans. 1527–1534 (2006)
29. Tayfur, G., Vijay, Singh, V.P., Asce, F.: ANN and fuzzy logic models for simulating event-based rainfall-runoff. J. Hydraul. Eng. 1321–1330 (2006)
30. Cheng, C.T., Chau, C.W., Li, X.Y.: Hydrologic uncertainty for bayesian probabilistic forecasting model based on BP ANN. IEEE Trans. (2007)
31. Jiang, G., Shen, B., Li, Y.: On the application of improved back propagation neural network in real-time forecast. IEEE Trans. (2007)
32. Ju, Q., Hao, Z., Zhu, C., Liu, D.: Hydrologic simulations with artificial neural networks. IEEE Trans. (2007)
33. Broersen, P.M.T.: Error correction of rainfall-runoff models with the ARMAsel program. IEEE Trans. **56**(6), 2212–2219 (2007)
34. Lohani, A.K., Goel, N.K., Bhatia, K.K.: Deriving stage–discharge–sediment concentration relationships using fuzzy logic. Hydrol. Sci. J. **52**(4), 793–807 (2007)
35. Li, C., Yuan, X.: Research and application of data mining for runoff forecasting. IEEE Trans. 795–798 (2008)
36. Liu, C.H., Chen, C.S., Huang, C.H.: Revising one time lag of water level forecasting with neural fuzzy system. IEEE Trans. 617–621 (2008)
37. Pei, W., Zhu, Y.Y.: A multi-factor classified runoff forecast model based on rough fuzzy inference method. IEEE Trans. 221–225 (2008)
38. Xu, Q., Ren, L., Yu, Z., Yang, B., Wang, G.: Rainfall-runoff modeling at daily scale with artificial neural networks. IEEE Trans. 504–508 (2008)
39. Sang, Y., Wang, D.: A Stochastic model for mid-to-long-term runoff forecast. IEEE Trans. 44–48 (2008)
40. Guo, H., Dong, G.Z., Chen, X.: WANN model for monthly runoff forecast. IEEE Trans. 1087–1089 (2008)
41. Remesan, R., Shamim, M.A., Han, D., Mathew, J.: ANFIS and NNARX based rainfall-runoff modeling. IEEE Trans. 1454–1459 (2008)
42. Aytek, A., Asce, M., Alp, M.: An application of artificial intelligence for rainfall–runoff modeling. J. Earth Syst. Sci. **117**(2), 145–155 (2008)
43. Solaimani, K.: Rainfall-runoff prediction based on artificial neural network (a case study: Jarahi Watershed). Am. Eurasian J. Agric. Environ. Sci. **5**(6), 856–865 (2009)
44. Ping, H.: Wavelet neural network based on BP algorithm and its application in flood forecasting. IEEE Trans. (2009)
45. Yan, J., Liu, Y., Wang, J., Cao, H., Zhao, H.: BP model applied to forecast the water and sediment fluxes in the yellow river mouth. IEEE Trans. (2009)

46. Yan, J., Liu, Y., Wang, J., Cao, H., Zhao, H.: RBF model applied to forecast the water and sediment fluxes in Lijin section. IEEE Trans. (2009)
47. Luna, I., Soares, S., Lopes, J.E.G., Ballini, R.: Verifying the use of evolving fuzzy systems for multi-step ahead daily inflow forecasting. IEEE Trans. (2009)
48. Xu, J., Zha, J., Zhang, W., Hu, Z., Zheng, Z.: Mid-short-term daily runoff forecasting by ANNs and multiple process-based hydrological models. IEEE Trans. 526–529 (2009)
49. Feng, L.H., Zhang, J.Z.: Application of ANN in forecast of surface runoff. IEEE Trans. (2009)
50. Hundecha, Y., Bardossy, A., Werner, H.: Development of a fuzzy logic-based rainfall-runoff model. Hydrol. Sci. J. **46**(3), 363–376 (2009)
51. Hung, N.Q., Babel, M.S., Weesakul, S., Tripathi, N.K.: An artificial neural network model for rainfall forecasting in Bangkok, Thailand. Hydrol. Earth Syst. Sci. **13**, 1413–1425 (2009)
52. Xu, J., Wei, J., Liu, Y.: Modeling daily runoff in a large-scale basin based on support vector machines. IEEE (Int. Conf. Comput. Commun. Technol. Agric. Eng.) 601–604 (2010)
53. Deshmukh, R.P., Ghatol, A.: Comparative study of Jorden and Elman model of neural network for short term flood forecasting. IEEE Trans. 400–404 (2010)
54. Wang, W., Qiu, L.: Prediction of annual runoff using adaptive network based fuzzy inference system. IEEE Trans. (2010)
55. Wang, W., Xu, D., Qiu, L.: Support vector machine with chaotic genetic algorithms for annual runoff forecasting. IEEE Trans. 671–675 (2010)
56. Liu, Y., Chen, Y., Hu, J., Huang, Q., Wang, Y.: Long-term prediction for autumn flood season in Danjiangkou reservoir basin based on OSR-BP neural network. IEEE Trans. 1717–1720 (2010)
57. Yan, J., Chen, S., Jiang, C.: The application of BP and RBF model in the forecasting of the runoff and the sediment transport volume in Linjin section. IEEE Trans. 1892–1896 (2010)
58. Liu, J., Dong, X., Li, Y.: Automatic calibration of hydrological model by shuffled complex evolution metropolis algorithm. IEEE Trans. 256–259 (2010)
59. Huang, M., Tian, Y.: Design and implementation of a visual modeling tool to support interactive runoff forecasting. IEEE Trans. 270–274 (2010)
60. Jizhong, B., Biao, S., Minquan, F., Jianming, Y., Likun, Z.: Adaptive regulation ant colony system algorithm—radial basis function neural network model and its application. IEEE Trans. (2010)
61. Linke, H., Karimanzira, D., Rauschenbach, T., Pfutzenreuter, T.: Flash flood prediction for small rivers. IEEE Trans. 86–91 (2011)
62. Ilker, A., Kose, M., Ergin, G., Terzi, O.: An artificial neural networks approach to monthly flow estimation. IEEE Trans. 325–328 (2011)
63. Huang, G., Wang, L.: Hybrid neural network models for hydrologic time series forecasting based on genetic algorithm. IEEE Trans. 1347–1350 (2011)
64. Zhen-min, Z., Xuechao, W., Ke, Z.: Rainfall-runoff forcast method base on GIS. IEEE Trans. 2406–2409 (2011)
65. Zhang, R., Wang, Y.: Research on daily runoff forecasting model of lake. IEEE Trans. 1648–1650 (2011)
66. Aiyun, L., Jiahai, L.: Forecasting monthly runoff using wavelet neural network model. IEEE Trans. 2177–2180 (2011)
67. Wu, C.L., Chau, K.W.: Rainfall-runoff modelling using artificial neural network coupled with singular spectrum analysis. J. Hydrol. **399**(3–4), 394–409 (2011)
68. Shah, H., Jaafar, J., Ibrahim, R., Saima, H., Maymunah, H.: A hybrid system using possibilistic fuzzy C-mean and interval type-2 fuzzy logic for forecasting: a review. IEEE Trans. 532–537 (2012)
69. Bell, B., Wallace, B., Zhang, D.: Forecasting river runoff through support vector machines. IEEE Trans. 58–64 (2012)
70. Li, K., Ji, C., Zhang, Y., Xie, W., Zhang, X.: Study of mid and long-term runoff forecast based on back-propagation neural network. IEEE Trans. 188–191 (2012)
71. Mittal, P., Chowdhury, S., Roy, S., Bhatia, N., Srivastav, R.: Dual artificial neural network for rainfall-runoff forecasting. J. Water Resour. Prot. **4**, 1024–1028 (2012)

72. Chen, S.M., Wang, Y.M., Tsou, I.: Using artificial neural network approach for modelling rainfall–runoff due to typhoon. J. Earth Syst. Sci. **122**(2), 399–405 (2013)
73. Phuphong, S., Surussavadee, C.: An artificial neural network based runoff forecasting model in the absence of precipitation data: a case study of Khlong U-Tapao river basin, Songkhla Province, Thailand. IEEE Trans. 73–77 (2013)
74. Patil, S., Walunjkar: Rainfall-runoff forecasting techniques for avoiding global warming. Robertson (2013)
75. Ramana, R.V., Krishna, B., Kumar, S.R., Pandey, N.G.: Monthly rainfall prediction using wavelet neural network analysis. Springer (Water Resour. Manage.) **27**, 3697–3711 (2013)
76. Karmakar, S., Shrivastava, G., Kowar, M.K.: Impact of learning rate and momentum factor in the performance of back-propagation neural network to identify internal dynamics of chaotic motion, Kuwait. J. Sci. **41**(2), 151–174 (2014)

Development of iMACOQR Metrics Framework for Quantification of Software Security

Arpita Banerjee, C. Banerjee, Santosh K. Pandey and Ajeet Singh Poonia

Abstract With the advent of new technologies, software has become relatively more interactive and has extended support for multiple users in a distributed as well as collaborative environment. Though the extensive use of software by the global players have surely improved productivity and efficiency, but at the same time has also provided ample opportunity for the attackers to exploit it. The software development team has been inspired by the idea of strengthening the software against such attacks. Many techniques are available for security implementation during its development and among them OO techniques like use case, misuse case, and abuse case due to their simplicity are mostly favored. Since security is a qualitative feature of software, and mechanism should be in place to provide its quantification so that it can be measured and controlled. This paper extends the previous work done by the researchers using misuse case modeling and integrates it with abuse case modeling and proposes iMACOQR (improvised Misuse and Abuse Case Oriented Quality Requirements) metrics framework. It was found that after applying the proposed iMACOQR metrics framework as per the recommended implementation mechanism, the security team of the software development process

---

A. Banerjee
St. Xaviers College, Jaipur, India
e-mail: arpitaa.banerji@gmail.com

C. Banerjee (✉)
Amity University, Jaipur, India
e-mail: chitreshh@yahoo.com

S.K. Pandey
DoEIT, Ministry of Communications & IT, New Delhi, India
e-mail: Santo.panday@yahoo.co.in

A.S. Poonia
Department of CSE, Government College of Engineering & Technology,
Bikaner, India
e-mail: pooniaji@gmail.com

M. Pant et al. (eds.), *Proceedings of Fifth International Conference on Soft Computing for Problem Solving*, Advances in Intelligent Systems and Computing 437, DOI 10.1007/978-981-10-0451-3_63

may eliminate vulnerability, induce proper mitigation mechanism, and specify improvised security requirements during requirements elicitation phase and thus more secure software could be built.

Keywords Software security · Software security metrics · Misuse cases · Use cases · Abuse cases

1 Introduction

Today's software systems, keeping pace with the current and future technologies like cloud computing, are capable of providing online and real-time computing services which have equipped the organizations to hold command in the global marketplace [1]. Every day, there are number of attacks on the software, of which, some are reported and some are not noticed and hence are not reported and documented [2, 3].

Strengthening the software against these attacks has inspired the software development team to build secure software which could withstand the malicious attacks and continue to operate uninterruptedly. Though the work in the field of software security is being carried out both by the research community as well as the industry people, but still for providing a comprehensive and pragmatic solution there is a long way to go [4].

Now, there are two aspects to the protection of software, of which, one aspect advocates the use of certain tools and techniques like antivirus, malware detection, firewalls, penetration testing, etc., for this purpose and the second aspect suggests that the software itself should be made self-sufficient to protect itself for such attacks. The central objective is to implement security in the software right from the beginning, i.e., the requirements engineering phase of the software development process. The first aspect is reactive in nature and the second aspect is proactive in nature. Hence, second aspect needs more exposure from research point of view [5, 6].

How secure is software, during development and when developed and fully implemented has always been a matter of concern and an issue of debate. This is because security is a quality attribute of software and needs quantification for its measurement. Researchers worldwide have always supported the idea that a process can never be indicative, can never provide estimation, and can never be improved and controlled if it cannot be measured. The same fact applies to the security attribute and hence some methods and techniques need to be devised to measure the security of the software under development and the software implemented for a reasonable period of time [7].

Although many methods, standards, and techniques have been proposed by researchers from time to time but there seems a lack of comprehensive approach toward implementation of software security during its development process. The methods are too generalized, standards are too broad, and the techniques are too irrational in nature. Further, a clear cut synchronization of these methods, standards, and techniques with the various phases of software development process is also a limitation [8].

Since the research community and the industry person have collectively emphasized on the implementation of security during the early phases of software development process, hence, some techniques need to be explored which could be properly applied and synchronized with the requirements engineering phase which is also the first phase of software development process. A number of techniques are available like attack trees, use case, and misuse case modeling, threat modeling, risk analysis, etc., which could be a potential source of security implementation. Out of the available techniques mentioned above, OO technique like use case and misuse case modeling could be a potential contender for the implementation of security in software during the software development process [9, 10].

In extension to the work carried out earlier, in this paper, we intend to propose a framework named iMACOQR using misuse case and abuse case modeling which may be practically applied from the beginning of SDLC to implement security in the software. Rest of the paper is organized as follows: in Sect. 2 we discuss about the concept of misuse case and abuse case, Sect. 3 covers the work done so far in the field, Sect. 4 focus on the proposed iMACOQR framework for quantitative assessment of software security during its development process, Sect. 5 throws light on "Implementation Mechanism," and Sect. 6 focuses on "Validation and Experimental Results" with conclusion and future work is given in Sect. 7.

2 Use Case, Misuse Case, and Abuse Case

Use cases, in today's software development scenario, are considered as a standard practice for gathering the requirements of software [11]. A use case proves to be a potential tool to capture functional as well as nonfunctional requirements. Hence, for specifying the security requirements which are a nonfunctional requirement of software, use case may be very useful [12].

Using use cases to elicit security requirements may have limitations; hence, the concept of use cases can be extended to include misuse cases. Misuse cases may be used to elicit security requirements as they can integrate more closely with the use cases [10].

McDermott and Fox defined abuse case as an interaction between the system, actors, and stakeholders which results in causing harm to the system. The researchers quoted that abuse cases are those harmful attacks which are reported after the software is developed, implemented, and operated for a sufficient period of time [13].

3 Related Work

McDermott and Fox, in the research work carried out, adapted an object oriented technique called use case to capture and analyze the security requirements of software in the process of development. The research work proposes the use of abuse case modeling technique for the specification of security requirements which is easier to understand and implement than any security mathematical model [13].

Abdulrazeg. et al. proposed a security metrics model which is based upon the GQM (Goal Question Metric) approach. The proposed security metrics model focuses on the design of the misuse case model. According to the researchers, the proposed security metrics model may assist the security team in examining the misuse case model which in turn may help in the discovery of defects and vulnerabilities and timely development of mitigation mechanism of known security risk and vulnerabilities. The proposed security metrics is based on the OWASP top 10-2010 and misuse case modeling antipattern [14].

Okubo et al. proposed an extension of misuse case model called MASG (Misuse Case with Assets and Security Goals) by incorporating new model elements, assets, and security goals. According to the researchers, the proposed model may be used to elicit and analyze the security features of a software system [15].

Banerjee et al. proposed MACOQR (Misuse and Abuse Case Oriented Quality Requirements) metrics from defensive perspective. As per the researchers, the purpose of the proposed metrics is to measure the predicated and observed ratio of flaw and flawlessness in misuse case modeling. It was further stated that the more defect free misuse case modeling is made, the better security requirements could be specified. This in turn will ensure development of secured software. The researchers further demonstrated that the measures and ratios obtained by the implementation of proposed metrics may help the security team involved during the requirements engineering phase plan and eliminate the defects of misuse case modeling [16].

Banerjee et al. proposed MACOQR (Misuse and Abuse Case Oriented Quality Requirements) metrics from the attacker's perspective. According to the researcher, the aim of the proposed metrics is to measure the ratio of internal and external attacks targeted toward the software. The proposed metrics uses the OO technique like misuse case modeling and abuse case modeling during requirements engineering phase to obtain indicators and estimators. The measures and ratios obtained may help the security analyst team to ensure proper and timely action for development and implementation of effective countermechanism especially against the internal attacks [17].

The limitation of the abuse case modeling proposed by John McDermott and Chris Fox in defining security requirements is that it seems unfit for higher assurance and can be used as a complementary tool [13]. The limitation of the security metrics model proposed by Abdulrazeg et al. is that further experiments are to be planned to ensure and demonstrate its usefulness in the software development process. Moreover, its applicability in all types of vulnerability classification is a matter of concern which further needs to be examined [14].

According to Okubo et al., one of the limitations of the proposed MASG model is that it is not comprehensive in nature and further extension of the diagram is possible. Another limitation is a lack of tool support for automation purpose. The resolution of the conflict aspect of the various security features is also a limitation which needs to be addressed [15].

The MACOQR metrics from defensive perspective proposed by Banerjee et al. need to be validated using larger samples for standardization. Further, more security dimensions could be explored and the MACOQR metrics could be further decomposed into more granular form to ensure a comprehensive approach toward specification of security requirements [16].

The limitation of the MACOQR metrics from an attacker's perspective proposed by Banerjee et al. is that it lacks insight of more security dimensions and certain factors which may have an influence on the internal attack. Further, factor analysis could be performed for priority setting based on the type of software being developed [17].

4 Proposed *i*MACOQR (*improvised* Misuse and Abuse Case Quality Requirements) Metrics Framework

Banerjee et al. have proposed MCOQR Metrics Framework for quantification of software security during the requirements elicitation phase of software development process. Though the researchers have significantly contributed to the initial measurement process of software security during the early phases of SDLC, still the limitation of the study is that, confinement of the framework boundaries within misuse case modeling.

Moreover, the study lacks the insight of the number of misuse cases which were actually targeted during the implementation and operation of the software for a reasonable period of time. It also lacks the insight of how many misuse cases were not predicted beforehand during the software development process [18]. These indicators and estimators if properly planned and implemented will contribute toward defect free misuse case modeling. The proposed and existing MCOQR metrics framework is shown in Fig. 1.

Based on the study carried out during the research work, it was found that MCOQR Metrics Framework is suitable for obtaining the security indicators and estimators during the requirements elicitation phase of software development process using misuse case modeling. These indicators and estimators are further used by the security team to remove defect from the misuse case modeling and specify better security specifications with sound counter measures.

But in order to have a complete and comprehensive view of the security implementation, it is recommended that the software after development can be made operational for a suitable period of time to note and analyze the various misuse cases which were predicted using MCOQR Metrics Framework (Fig. 3).

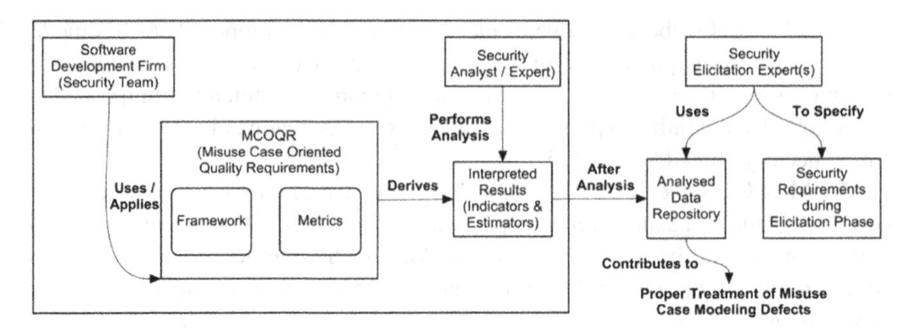

Fig. 1 Existing MCOQR metrics framework

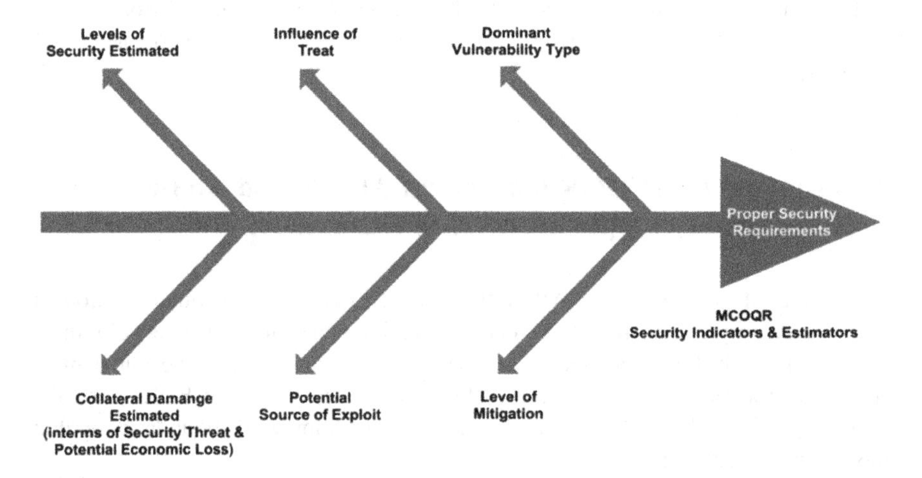

Fig. 2 Identified security indicators and estimators

Now, based on the research work carried out by McDermott and Fox [13] those misuse cases which provided harm to the system and its resource be termed as predicted misuse cases or "Abuse Cases." Further, some attacks may also be reported which were not predicted earlier using the MCOQR Metrics framework, they be termed as unpredicted misuse cases or "Abuse Cases."

Hence this coordinated and synchronized idea of integrating the use cases, misuse cases, and abuse cases could be used to provide a comprehensive approach of estimating and specifying the security requirements of the similar software which an organization intends to develop in future.

Hence, after carefully studying the limitation of MCOQR, we extended it and proposed its improvised version with inclusion of abuse cases and named the proposed framework as iMACOQR (improvised Misuse and Abuse Case Oriented Quality Requirements) metrics framework. The proposed iMACOQR metrics framework is shown in Fig. 2.

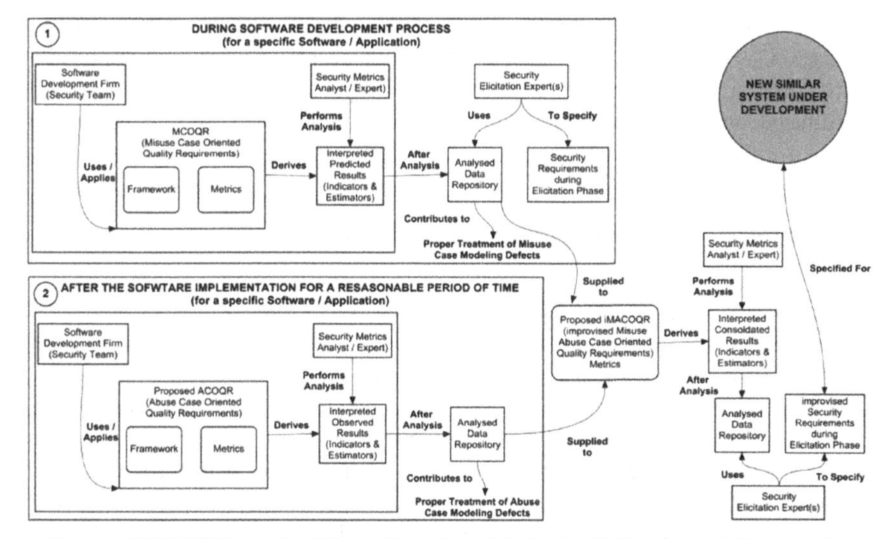

Fig. 3 Proposed iMACOQR metrics framework

5 Implementation Mechanism

The implementation of proposed iMACOQR (improvised Misuse and Abuse Case Oriented Quality Requirements) metrics framework may be done at two levels:

Level 1: During software development process using MCOQR Metrics Framework.

Level 2: After the software implementation for a reasonable period of time using ACOQR Metrics Framework.

The process for both level 1 and level 2 is as follows:

(1) Security team will apply MCOQR/proposed ACOQR metrics framework and metrics to derive and interpret predicted/observed results (indicators and estimators).
(2) Security metrics analyst/expert will perform analysis on the predicted/observed results and after analysis store data in individual respective repositories.
(3) Analyzed data from the respective repositories will be supplied to proposed iMACOQR Metrics which in turn may be used to derive interpreted consolidated results in form of indicators and estimators.
(4) Security metrics analyst/expert will perform analysis on the interpreted consolidated results and after analysis store data in central repository.
(5) The central repository may be used by security elicitation experts to specify improvised security requirements during the elicitation phase for the new similar system under development.

Note: For this, a central repository will be used containing CVSS metrics database, CVSS document, CVE database, misuse case database, and application specific misuse case database. The central repository needs to be well synchronized, coordinated, and updated on regular basis with CVSS and CVE external repository.

6 Validation and Experimental Results

The proposed iMACOQR (improvised Misuse Case and Abuse Case Quality Requirements) metrics framework was applied to a real life project from industry (on the request of the company, identity is concealed). The final results of security assessment were calculated according to the recommended implementation mechanism. Further, the level of security implementation was compared with the security assurance of other project in which the proposed framework was not applied.

The investigation which is carried out further showed that the level of security implementation was very high when applied using iMACOQR metrics framework and thus the risk was minimized up to 43.5 %. Due to the page limit constraint, we are not providing the details of validation results in this paper; we will discuss in our next paper.

7 Conclusion and Future Work

The studies carried out using misuse case and abuse case modeling and the results thus obtained clearly shows that these OO techniques can very well be integrated with the requirements engineering phase of SDLC to specify improvised security requirements which aid in the development of a secured software. The research paper tried to present some concrete work done on software security metrics using misuse case modeling and abuse case modeling and hence iMACOQR metrics framework is proposed. Validation results show the applicability of the proposed iMACOQR metrics framework during the development life cycle to develop secure software.

Future work may include applying the proposed iMACOQR metrics framework on a large sample for ensuring the accuracy of the work. Future work may also involve exploring the various subdimensions of security for making the proposed framework more comprehensive and pragmatic in nature. The research work carried out will certainly help the security team and the software development team in general to quantify security aspects so as to develop a more secure software.

References

1. Wang, B., Zheng, Y., Lou, W., Hou, Y.T.: DDoS attack protection in the era of cloud computing and software-defined networking. Comput. Netw. **81**, 308–319 (2015)
2. Zhang, D., Liu, D., Csallner, C., Kung, D., Lei, Y.: A distributed framework for demand-driven software vulnerability detection. J. Syst. Softw. **87**, 60–73 (2014)
3. McMahon, J.: An analysis of the characteristics of cyber attacks. Discov. Invention Appl. (1) (2014)
4. Banerjee, C., Banerjee, A., Murarka, P.D.: Evaluating the relevance of prevailing software metrics to address issue of security implementation in SDLC. Int. J. Adv. Stud. Comput. Sci. Eng. **3**(3), 18 (2014)
5. Banerjee, C., Pandey, S.K.: Software Security Rules, SDLC Perspective (2009). arXiv:0911. 0494
6. McGraw, G.: Software Security: Building Security in (Vol. 1). Addison-Wesley Professional (2006)
7. Fenton, N., Bieman, J.: Software Metrics: A Rigorous and Practical Approach. CRC Press (2014)
8. Brotby, W.K., Hinson, G.: PRAGMATIC Security Metrics: Applying Metametrics to Information Security. CRC Press (2013)
9. Schumacher, M., Fernandez-Buglioni, E., Hybertson, D., Buschmann, F., Sommerlad, P.: Security Patterns: Integrating Security and Systems Engineering. John Wiley & Sons (2013)
10. Sindre, G., Opdahl, A.L.: Eliciting security requirements with misuse cases. Requirements Eng. **10**(1), 34–44 (2005)
11. Kulak, D., Guiney, E.: Use Cases: Requirements in Context. Addison-Wesley (2012)
12. Wiegers, K., Beatty, J.: Software Requirements. Pearson Education (2013)
13. McDermott, J., Fox, C.: Using abuse case models for security requirements analysis. In: Proceedings of the 15th Annual Computer Security Applications Conference (ACSAC'99), pp. 55–64. IEEE (1999)
14. Abdulrazeg, A., Norwawi, N.M., Basir, N.: Security metrics to improve misuse case model. In: 2012 International Conference on Cyber Security, Cyber Warfare and Digital Forensic (CyberSec), pp. 94–99. IEEE (2012, June)
15. Okubo, T., Taguchi, K., Kaiya, H., Yoshioka, N.: Masg: advanced misuse case analysis model with assets and security goals. J. Inf. Process. **22**(3), 536–546 (2014)
16. Banerjee, C., Banerjee, A., Murarka, P.D.: Measuring software security using MACOQR (misuse and abuse case oriented quality requirement) metrics: defensive perspective. Int. J. Comput. Appl. **93**(18), 47–54 (2014)
17. Banerjee, C., Banerjee, A., Murarka, P.D.: Measuring software security using MACOQR (misuse and abuse case oriented quality requirement) metrics: attackers perspective. Int. J. Emerg. Trends Technol. Comput. Sci. **3**(2), 245–250 (2014)
18. Banerjee, C., et al.: MCOQR (misuse case oriented quality requirements) metrics framework. In: Deepti (ed.) Problem solving and uncertainty modeling through optimization and soft computing applications. IGI Global Publishers (2016)

Miniaturized, Meandered, and Stacked MSA Using Accelerated Design Strategy for Biomedical Applications

Love Jain, Raghvendra Singh, Sanyog Rawat and Kanad Ray

Abstract In this paper, a strategy to design implantable microstrip antenna used in biomedical telemetry has been suggested. A skin-implanted multilayer microstrip antenna containing ground, and meandered lower and upper radiating patch covered with a superstrate layer has been designed and simulated. A material with high permittivity such as Rogers 3010 has been used for substrate and superstrate layer. The volume of proposed antenna is 230 mm^3. The size of proposed MSA is small enough to implant into human head-scalp. The designed antenna is resonating in the MICS (402–405 MHz) band and the 10 dB bandwidth is 23 MHz. The return loss is found as −18.274 dB at 402 MHz.

Keywords Implantable microstrip antenna · Superstrate layer · MICS · Multilayer · Biomedical telemetry

L. Jain (✉)
Department of Electronics & Communication Engineering,
JK Lakshmipat University, Jaipur, Rajasthan, India
e-mail: lovejain.eck@gmail.com

R. Singh
JK Lakshmipat University, Jaipur, Rajasthan, India
e-mail: planetraghvendra@gmail.com

S. Rawat · K. Ray
Department of Electronics & Communication, Amity University,
Jaipur, Rajasthan, India
e-mail: sanyog.rawat@gmail.com

K. Ray
e-mail: kanadray00@gmail.com

M. Pant et al. (eds.), *Proceedings of Fifth International Conference on Soft Computing for Problem Solving*, Advances in Intelligent Systems and Computing 437, DOI 10.1007/978-981-10-0451-3_64

1 Introduction

In the past few years, tremendous growth has been noticed in the area of development of microstrip antennas (MSA) for biomedical telemetry applications. Biomedical telemetry is receiving continuous attention of researchers. Biomed-telemetry deals with the transmission of physiological information of the patient to the nearby receiving station with the help of in-body and on-body active nodes or sensors. The collected information may be further utilized for diagnosis purposes [1–4].

While the technology is still in its primeval phase it is being extensively researched and once taken up, is expected to be a burst through discovery in healthcare, leading to concepts like mobile health monitoring. The development in microelectromechanical systems (MEMS), electrically small antenna and body centric communication systems will play a dominant role in futuristic medical/health care system [4].

The bidirectional communication between in-body and external base station uses licensed Medical Implant Communication Service (MICS) band ranging from 402 to 405 MHz [5]. The MICS band has been standardized by European Radiocommunications Committee (ERC) [6] and the United States Federal Communications Commission (FCC) [7].

In-body nodes or implanted devices (ID) must be biocompatible in order to ensure patients safety and avoid direct contact between ID and body tissue, because there is a possibility of shortening due to conductive nature of body tissue. To avoid this problem, the metallic radiator is usually covered with a dielectric layer of biocompatible material having high dielectric constant [8]. Commonly used biocompatible materials found in the literature are Macor ($\varepsilon_r = 6.1, \tan\delta = 0.005$), Teflon ($\varepsilon_r = 2.1$, $\tan\delta = 0.001$), ceramic alumina ($\varepsilon_r = 9.4$, $\tan\delta = 0.006$) [4], and polydimethylsiloxane (PDMS) [2].

Miniaturization, patient safety, low power consumption, and far-field gain are the main constraints of the implanted MSA to embed in ID. In [1], it is reported that the electrical length is reduced by almost seven times as in free space, when it is implanted inside the human body. Miniaturization of MSA by increasing the length of the current path by meandering using spiral and waffle-type, and hook-slotted-shaped are shown in [9]. Spiral-shaped [4], serpentine-shaped [4], circular-meandered [9, 10], stacked radiating patch [10] microstrip antenna, electrically doubling the size of antenna with the use of shortening between ground and radiating patch [4, 10] recently have been reported for biomedical telemetry operating in MICS band.

The paper is divided into five sections. A design strategy has been proposed for designing implantable antenna in Sect. 2. Design strategy speedups the design cycle time by designing the antenna into tissue model rather than designing for free space first and retuning again for tissue model. Section 3 explains about the proposed MSA design and chosen skin-tissue model. Section 4 discusses the results of FDTD numerical simulation. Conclusion is drawn in Sect. 5.

2 Proposed Strategy

The flow chart of the proposed strategy for designing implantable antenna is shown in Fig. 1. The proposed flow graph is specified for designing implanted MSA for biomedical telemetry applications, keeping in view that implanted MSA is surrounded by biological tissues. It has been reported by Gabriel [11] in 1996 that dielectric property of human tissue varies with frequency. Tissue electrical properties at 402 MHz are considered and approximated as constant inside the 300–500 MHz frequency range for simulation purpose [12].

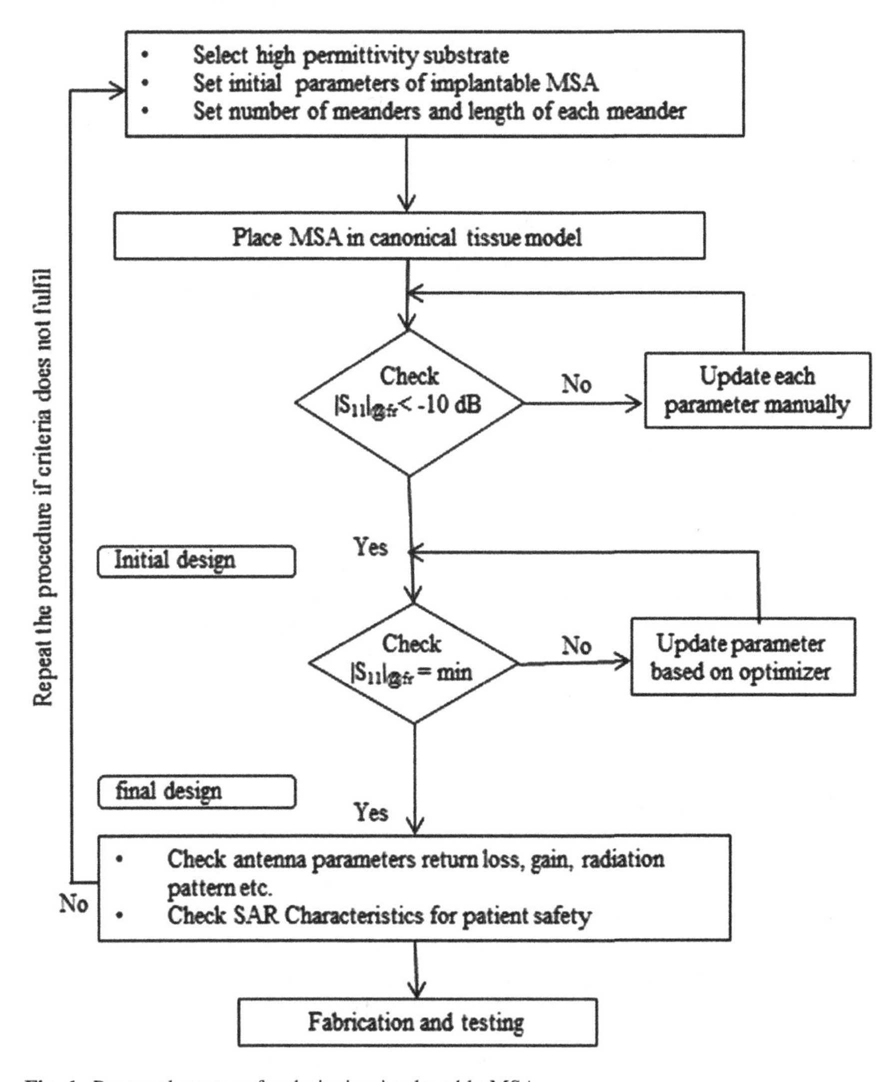

Fig. 1 Proposed strategy for designing implantable MSA

The proposed design strategy shortens the design cycle time by designing the antenna into tissue model rather than designing for free space first and retuning again for tissue model.

In the first step, initial parameters of MSA are set manually. Designer can set number of meanders and their length initially. To miniaturize the design further, radiator patches can be stacked separated by substrate. These patches can have spiral [4] or meandered [9] type rectangular [4]/circular [9, 10] shape. All design parameters of the antenna can be set with respect to the reference antenna quoted in the literature. The high permittivity dielectric material should be chosen to get ultra-small-sized design. The substrate material should be chosen with care.

The implantable device is numerically simulated in artificially modeled human body environment and it is called as body phantoms or simply phantoms. A large number of models are available in the literature to predict the effects of the human body on the characteristics of implanted antennas. For ease and simplicity, canonical models are often being used. These may be single layer (see Fig. 5(a) of [3], Fig. 3(a) of [13]) or multilayer (see Fig. 5(b) of [3], Fig. 6 of [1]). To obtain a more realistic result, anatomical tissue model [3, 13] is used. These models are developed by computer tomography and magnetic resonance imaging techniques.

The values of MSA model are manually varied in an iterative manner to get the return loss less than −10 dB at the desired resonating frequency (f_r).

$$|S_{11@fr}| < -10\,\mathrm{dB} \tag{1}$$

Once the initial design is met, the MSA model parameters are further optimized with the help of optimizer available in simulation tool. The cost function [13] of the optimizer is defined as

$$\mathrm{Cost} = |S_{11@fr}| \tag{2}$$

Once the final design is obtained, other antenna parameters such as radiation pattern, gain, directivity are verified. Then SAR characteristics are also verified to meet the compliance with international safety guidelines [14, 15] for human exposure to electromagnetic radiation.

3 Proposed MSA and Simple Tissue Model

The proposed design presented in this paper is inspired from the MSA design of Soontornpipit et al. [4]. The reference design consists of only radiating patch and has a volume of $19.6 \times 29.4 \times 6 = 3457.44\,\mathrm{mm}^3$.

The proposed antenna structure is shown in Fig. 2. The antenna consists of a rectangular ground plane of size $7.5\,\mathrm{mm} \times 16.14\,\mathrm{mm}$, and two vertically stacked

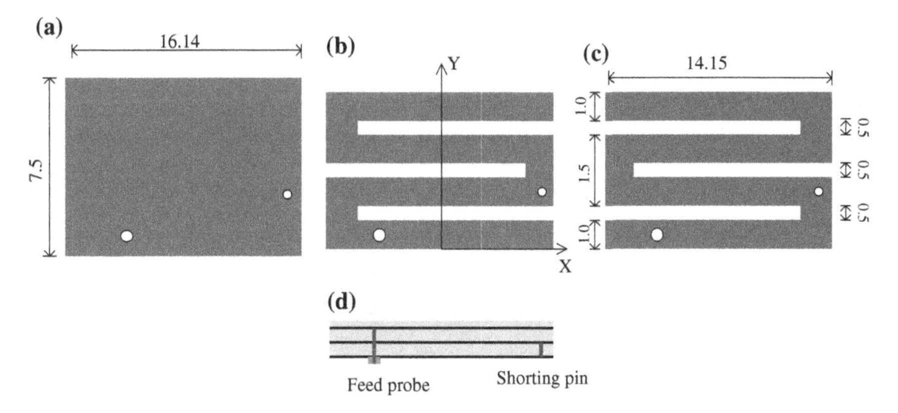

Fig. 2 Geometric view of the proposed microstrip antenna: **a** Ground plane. **b** Lower radiating patch. **c** Upper radiating patch, and **d** Side view (all dimensions are in millimeters)

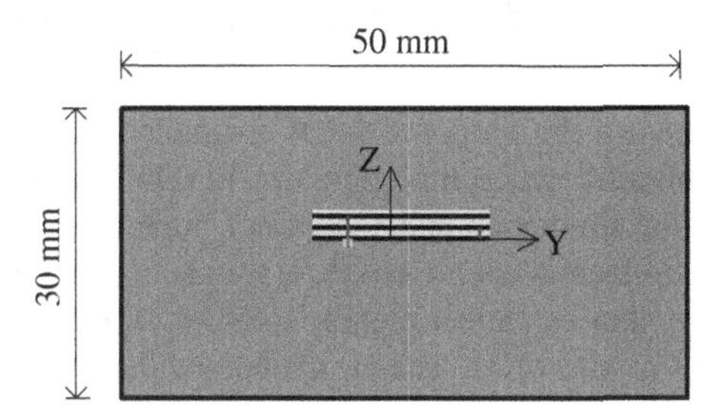

Fig. 3 Simulation setup: MSA inside modeled skin-tissue

identical radiating meandered rectangular patches, relatively rotated by 180°. All dimensions are depicted in Fig. 2. Meandering and stacking of patches are done to increase the length of current flow path, which results in reduction in the size of the MSA [16]. A shorting pin with radius of 0.3 mm connects lower patch to ground, resulting in further reduction in size [4]. Each of the radiating patches is printed on 0.635 mm thick Rogers RO3010 ($\varepsilon_r = 10.2$) substrate. A superstrate layer of same thickness and material is applied on MSA for the compliance of robustness and biocompatibility. Throughout this study, the origin is assumed at the midpoint of the lowest edge of radiating patch as shown in Fig. 2. The location of feed point and shorting pin are (−4, 0.5) and (6.5, 2), respectively.

Since the MSA is intended for implantation in human skin-tissue; the proposed MSA is simulated in simple cuboid-shaped skin canonical model as shown in Fig. 3. The electrical properties of skin-tissue ($\varepsilon_r = 46.74, \sigma = 0.689S/m$) are considered at 402 MHz and approximated as constant inside 300–500 MHz [11].

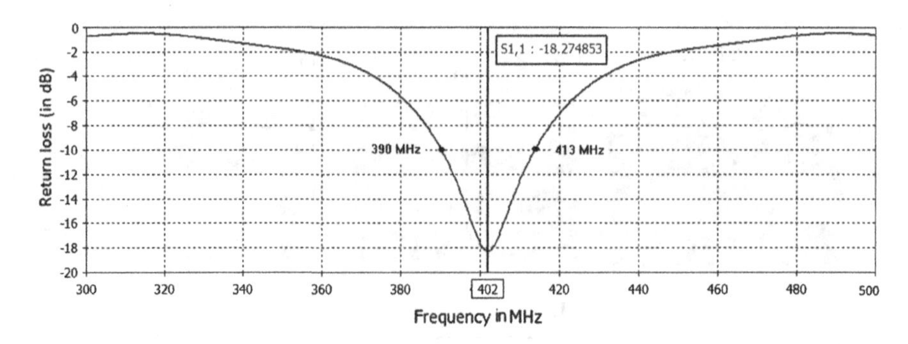

Fig. 4 Reflection coefficient frequency response of the proposed MSA inside a simple cuboid skin canonical model

4 Numerical Analysis and Discussion

4.1 The Reflection Coefficient Frequency Response

The numerical analysis of the proposed MSA surrounded by skin-tissue model is carried out using finite-difference time-domain (FDTD) method, which has been used extensively for the simulation and analysis of implanted MSA in the literature [1, 3, 4, 10, 13]. The reflection coefficient (S_{11}) frequency response of the proposed MSA inside a simple skin canonical model is shown in Fig. 4. The return loss is found as −18.274 dB at 402 MHz. The 10 dB bandwidth of the proposed MSA is found as 23 MHz.

4.2 The Far-Field Radiation Pattern

The simulated far-field radiation pattern of the proposed implanted MSA inside the single layer canonical skin model at 402 MHz is shown in Fig. 5. The dimensions of the antenna are very small, that is why nearly omnidirectional radiation pattern is achieved. The obtained maximum gain is −46.24 dB at $\emptyset = 100°$. Low gain value is attributed to the miniaturized MSA dimensions and high permittivity tissue absorption.

5 Conclusion

An antenna performance evaluation for body-implanted device is more difficult than for antennas in the free space, as there are limitations caused due to the environment and challenge of reducing the size also. A simple strategy has been proposed to design implantable microstrip antenna used in biomedical telemetry. Size/volume of

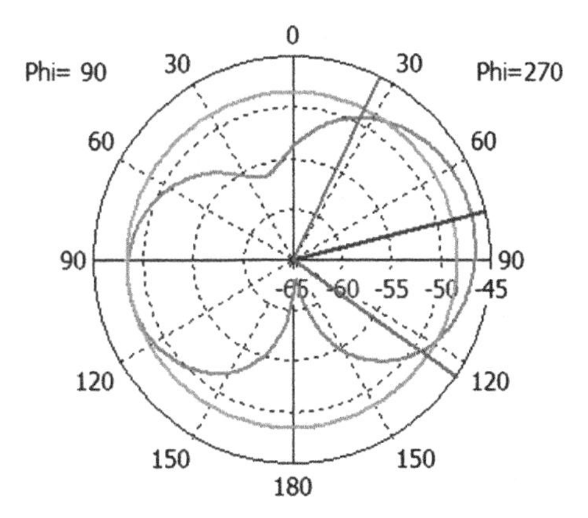

Fig. 5 Far-field pattern for 402 MHz

implanted microstrip antenna can be drastically reduced with the use of meandering and stacking of radiating patches. The size of the proposed MSA is small enough to implant into human head-scalp.

Future work may include refining of MSA dimension in multilayer canonical and anatomical model. SAR characteristics can be accessed to meet the compliance with international safety guidelines for human exposure to electromagnetic radiation.

References

1. Lin, H.Y., Takahashi, M., Saito, K., Ito, K.: Performance of implantable folded dipole antenna for in-body wireless communication. IEEE Trans. Antennas Prop. **61**(3), 1363–1370 (2013)
2. Scarpello, M.L., Kurup, D., Rogier, H., Vande Ginste, D., Axisa, F., Vanfleteren, J., Joseph, W., Martens, L., Vermeeren, G.: Design of an implantable slot dipole conformal flexible antenna for biomedical applications. IEEE Trans. Antennas Prop. **59**, 3556–3564 (2011)
3. Kim, J., Rahmat-Samii, Y.: Implanted antennas inside a human body: simulations, design and characterization. IEEE Trans. Microwave Theory Tech. **52**(8), 1934–1943 (2004)
4. Soontornpipit, P., Furse, C.M., Chung, Y.C.: Design of implantable microstrip antenna for communication with medical implants. IEEE Trans. Microwave Theory Tech. **52**(8), 1944–1951 (2004)
5. Establishment of a Medical Implant Communications Service in the 402–405 MHz Band. Fed. Reg.: Rules Regulations **64**(240), 69926–69934 (1999)
6. European radio-communications commission recommendation 70-03 relating to the use of short range devices. In: European Conference of Postal and Telecommunications Administration, CEPT/ERC 70-03, Annex 12 (1997)
7. FCC guidelines for evaluating the environmental effects of radio frequency radiation. FCC, Washington (1996)
8. Karacolak, T., Hood, A.Z., Topsakal, E.: Design of a dual-band implantable antenna and development of skin mimicking gels for continuous glucose monitoring. IEEE Trans. Microw. Theory Tech. **56**(4), 1001–1008 (2008)
9. Kiourti, A., Nikita, K.S.: A review of implantable patch antennas for biomedical telemetry: challenges and solutions. IEEE Antennas Prop. Mag. **54**(3), 210–228 (2012)

10. Kiourti, A., Christopoulou, M., Kouloudis, S., Nikita, K.S.: Design of a novel miniaturized PIFA for biomedical telemetry. In: International ICST Conference on Wireless Mobile Communication and Healthcare. Ayia Napa, Cyprus (2010)
11. Gabriel, C., Gabriel, S., Corthout, E.: The dielectric properties of biological tissues. Phys. Med. Biol. **41**, 2231–2293 (1996)
12. Kiourti, A., Christopoulou, M., Nikita, K.S.: Performance of a novel miniature antenna implanted in the human head for wireless biotelemetry. In: Proceedings of IEEE International Symposium Antennas Propagation, Spokane, WA, pp. 392–395 (2011)
13. Kiourti, A., Nikita, K.S.: Miniature scalp-implantable antennas for telemetry in the MICS and ISM bands: design, safety considerations and link budget analysis. IEEE Trans. Ant. Prop. **60** (8), 3568–3575 (2012)
14. IEEE, Standard for Safety Levels with Respect to Human Exposure to Radio Frequency Electromagnetic Fields, 3 kHz to 300 GHz. IEEE Standard C95.1–1999 (1999)
15. IEEE, Standard for Safety Levels with Respect to Human Exposure to Radio Frequency Electromagnetic Fields, 3 kHz to 300 GHz. IEEE Standard C95.1–2005 (2005)
16. Wong, K.L.: Compact and Broadband Microstrip Antennas. Wiley, New York (2002)

Artificial Bee Colony-Trained Functional Link Artificial Neural Network Model for Software Cost Estimation

Zahid Hussain Wani and S.M.K. Quadri

Abstract Software cost estimation is forecasting the amount of developmental effort and time needed, while developing any software system. A good volume of software cost prediction models ranging from the very old algorithmic models to expert judgement to non-algorithmic models have been proposed so far. Each of these models has their advantage and disadvantage in estimating the development effort. Recently, the usage of meta-heuristic techniques for software cost estimations is increasingly growing. So in this paper, we are proposing an approach, which consists of functional link ANN and artificial bee colony algorithm as its training algorithm for delivering most accurate software cost estimation. FLANN reduces the computational complexity in multilayer neural network, and does not has any hidden layer, and thus has got fast learning ability. In our model, we are using MRE, MMRE and MdMRE as a measure of performance index to simply weigh the obtained quality of estimation. After an extensive evaluation of results, it showed that training a FLANN with ABC for the problem of software cost prediction yields a highly improved set of results. Besides this, the proposed model involves less computation during its training because of zero hidden layers and thus is structurally simple.

Keywords Artificial neural network (ANN) · Functional link artificial neural network (FLANN) · Artificial bee colony (ABC) · Software cost estimation (SCE)

Z.H. Wani (✉) · S.M.K. Quadri
Department of Computer Sciences, University of Kashmir, Srinagar, India
e-mail: zahid.uok@gmail.com

S.M.K. Quadri
e-mail: quadrismk@kashmiruniversity.ac.in

M. Pant et al. (eds.), *Proceedings of Fifth International Conference on Soft Computing for Problem Solving*, Advances in Intelligent Systems and Computing 437, DOI 10.1007/978-981-10-0451-3_65

1 Introduction

Software cost estimation simply defines the prophecy of sum of effort and calendar time needed to develop any software project. Software development effort covers the number of work hours and the workers count in terms of staffing levels needed to develop a software project. Every single project manager is always in a quest of finding better estimates of software cost so that he can evaluate the project progress, has potential cost control, delivery accuracy, and in addition can gave the organization a better insight of resource utilization and consequently, will land the organization with a better schedule of futuristic projects. The software effort prediction facet of software development project life cycle is always made at an early stage of software development life cycle. However, estimating the software development project at this time is most difficult to obtain because in addition to availability of sparse information about the project and the product at this stage, the so presented information is likely to be vague for the estimation purpose. So, predicting development effort for any software remained an intricate problem, thus accordingly continues to be the centre of focus for many researchers. In this connection, a range of software cost estimation techniques have been projected and accordingly have been classified into various broad categories which will be discussed in later sections. So far variations of a range of ANN models have also been developed for this purpose in addition to previously used algorithmic models.

In this paper, our focus will be on a specially varied ANN called as functional link artificial neural network (FLANN). FLANN is infact a high order artificial neural network model and in-particular its training algorithm which will be an artificial bee colony algorithm based on the purpose of achieving a better software cost prediction model.

The rest of the paper is structured as in Sect. 2, the evolution of algorithmic and non-algorithmic software cost estimation models will be discussed. In addition, characteristics of COCOMO II model will be presented here. FLANN architecture and the implementation of training algorithm will be presented in Sect. 3. In Sect. 4 datasets and evaluation criteria will be discussed. Section 5 reports the experimentation results and their analysis. Finally, Sect. 6 will conclude the study along with future work.

2 Software Cost Estimation Models

Software cost estimation is an uncertain task, as it is based on large set of factors which are normally vague in nature and is thus hard to model them mathematically. A successful planning as well as tracking of any software development project is potentially sustained by precise cost estimation process [1]. The steadfast and precise software cost estimation process of any software development project in software engineering is always an ongoing challenge because of the reason that it allows for any organization's substantial financial and tactical planning. Software cost estimation techniques are classified into various categories and they include:

2.1 Parametric models or algorithmic models are the models which lay down their foundation based on the numerical analysis of past data. These models use arithmetic formula to forecast software cost rooted in the approximations of project dimensions, digit count of software engineers, plus some other factors [2]. Models of this nature are developed by simple cost analysis and other necessary properties of already concluded projects and then discovering the formula fitting closest to real experience. Boehm's COCOMO'81 II [2], Albrecht's function point [3], and Putnam's SLIM [1] represent some of the famous algorithmic models. These models normally demand a precise estimate of specific attributes such as line of code (LOC), number of user screen, complexity and so on, which normally are difficult to obtain at this early stage of software development process. In addition to this, the other limitations of these models include the presence of built-in complex relationships between their related attributes, so comprehending and calculations of these models are much difficult. Because of their less comprehensive nature, these models are less capable of gripping over the categorical data and are thus deficient in analytic abilities. Thus, their limitations lead to the exploration of the non-algorithmic techniques which are soft computing based. Among a good number of algorithms also called parametric models, COCOMO uses mathematical formulae which is the most commonly used SCE model for predicting the development effort of any software project at diverse stages [1] of software development life cycle. Thus, COCOMO because of its simplicity is taken as a base model for SCE and thus necessitates an overview in the following section.
The COCOMO model was developed by Barry Boehm [2] in 1981. It is a regression-based model, used to determine the sum of effort as well as time plan of software projects. It is considered to be the generally cited and highly credible model among all conventional algorithmic models. Although COCOMO 81 is considered to be the most stable software cost prediction model of the time, it fails to compete against the late 1990s development standards and environment. So, COCOMO II got a chance to get printed in 1997 which gave solution to most of the problems lying with COCOMO 81. COCOMO II has three models which include:

(1) Application composition model—This model lays down its foundation on new object points and thus fits to projects that are developed using modern GUI-builder tools.
(2) Early design model—It is an unadjusted function-points-based model and is used to generate jagged approximations of project's cost and time period prior to the determination of projects complete structural design.

This model works by including a novel but little collection of cost drivers, and also explores novel equations for estimation purpose.

(3) Post-architecture model—Though normally used once the overall structural design of a software project is completed, this model is the detailed one among all the three and uses LOC or FP's as a measure of size estimate and engrosses the real development as well as maintenance of the final software product.

COCOMO II considers 17 cost drivers. These cost drivers are rated on a scale from very low to extra high just as they are followed in COCOMO 81. This model is given below

$$\text{Effort} = A \times [\text{SIZE}]^B \times \prod_{i=1}^{17} \text{Effort Multiplier}_i \quad (1)$$

$$\text{where } B = 1.01 + 0.01 \times \sum_{j=1}^{5} \text{Scale Factor}_j$$

In Eq. (1), A means multiplicative constant and SIZE depicts the size of the software calculated as selection of scale factors (SF).

2.2 Expert judgment is a unstructured process where experts normally formulate their decisions for problem solving. They do not need any historical data [4–6] for this purpose rather use their knowledge, wisdom and experience to give more accurate results.

2.3 Thus in the 1990s, nonparametric models began to exist and were planned to work for software cost estimation. These nonparametric models are based on FL, ANN and evolutionary computation (EC). These models assemble together a group of techniques that later symbolize some of the aspects of common human brain [3], and examples include fuzzy systems, rule induction, regression trees, genetic algorithms, Bayesian networks, evolutionary computation and ANNs. However, tools based on ANN are progressively gaining attractiveness because of their inherent capacity and capability of approximating any nonlinear function to a high level of precision. Artificial Neural networks have the capability of generalizations from trained data. A well-defined training dataset along with a precise training algorithm delivers a procedure that fits the data in a rational manner [7]. A lot of research has already been done in developing models which are based on a variety of pointed soft computing techniques from the past more than a couple of decades. Early ones use the MLP architecture with backpropagation as their training scheme for the purpose, while a major portion of new work is totally derived from evolutionary optimization techniques like genetic algorithms.

3 Proposed FLANN Model and Its Training via ABC Algorithm

The FLANN model used for the purpose of estimating software development effort is very simple. It is a single-layered feedforward ANN containing an input layer plus an output layer. This model computes output as development effort by intensifying cost drivers presented as initial inputs and then processes them onto the concluding output layer. Each input neuron symbolizes an element of input vector. There is a single output neuron within an output layer. Once the outputs of the input layer undergo their linear weighted sum, the output neuron figures out the software development effort [8]. The reason of using FLANN contrary to multilayer feed-forward network (MLP) is that ANN model in software cost estimation utilizes a supervised training scheme called backpropagation for the purpose of training the network where as the multi-layered structure in a MLP lets down its training speeds to a minimum [3]. Thus, problems of the nature of overfitting and interference of weights let the training of an MLP network difficult [9]. Thus, Pao [10] has introduced FLANN architecture to overcome such hurdles. His approach takes away the hidden layer from ANN architecture, thus reduces the complexity within the architecture of the neural network and this way provides them with an enhancement in representation of input nodes for the network to be able to perform a better accuracy of software cost estimations. His network can handily be occupied when one is in demand of functional approximations of higher convergence rates than any MLP architecture. Because of its flatness and simplicity, the computations are simple, functional expansion of the input very efficiently amplifies the size of the input domain and accordingly the hyper-planes thus produced by FLANN yield better biassed means in input pattern space [11]. A good number of research proposals on system identification, noise cancellation and channel equalization have recently been reported [12]. In all these cases, FLANN has offered acceptable results, thus can be applied to problems containing exceedingly nonlinear and dynamic data.

3.1 The FLANN architecture: In addition to minimizing the computational complexity, FLANN is known for its key role in functional estimations. Further when coming to its learning aspect, the FLANN architecture is supposed to be faster than other networks. Using FLANN to predict software development effort needs a good fortitude of its architectural parameters as per the features of COCOMO. General structure of FLANN and its detailed architecture are shown in Figs. 1 and 2, respectively.
The details of the FLANN structure shown in Fig. 2 are as follows: It is a nonlinear neural network with just one layer and k number of input pattern duos. All these input pattern duos need to be trained by the FLANN. Let X_k be the input vector pattern and is n dimensional. Let Y_k be the output and a scalar. So the training patterns will accordingly be represented as $\{X_k\ Y_k\}$. To functionally expand the input $X_k = [x_1(k)\ x_2(k)\ \ldots\ x_n\ (k)]^T$, let N,

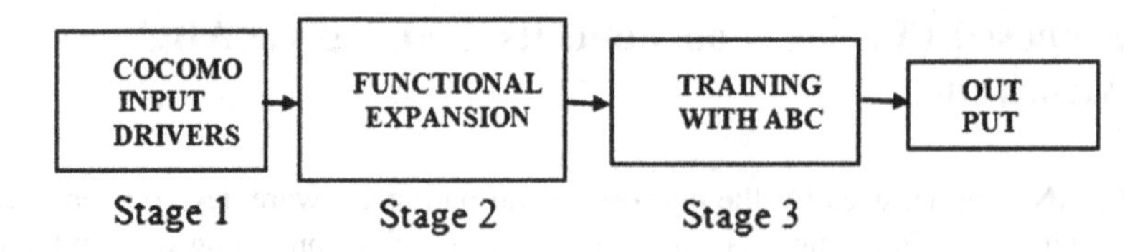

Fig. 1 General sructure of FLANN

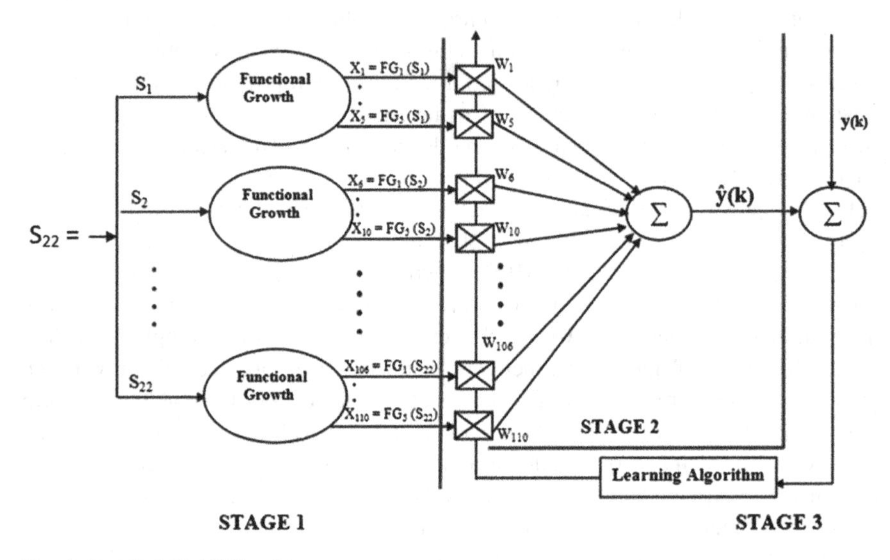

Fig. 2 Detailed FLANN architecture

$Ø(X_k) = [Ø(X_k)\ Ø(X_k)\ \ldots\ Ø(X_k)]^T$, be the number of basis functions used where $N > n$. The purpose of using these linearly independent N functions is to get the $R^n \rightarrow R^N$; $n < N$ mapping.

We can put the linear grouping of the above function values in their matrix form as $S = WØ$, where for any $S_{k,}$ $S_k = [S_1(k)\ S_2(k)\ \ldots\ S_m(k)]^T$ and W is the weight matrix of the order of $m \times N$. Finally, for getting the output $\hat{Y} = [\hat{y}_1\ \hat{y}_2\ \ldots\ \hat{y}m]^T$, $\hat{y}_j = \rho(S_j)$, $j = 1, 2, \ldots, m$, we use a group of nonlinear functions $\rho(\bullet) = \tan h(\bullet)$ onto matrix S_k as input.

The various steps involved here are as follows:

Step-1: An input for the network is first of all chosen from a validated dataset. The input here is simply a set of 22 cost factors. These cost factors are then functionally expanded using the formulas.

In FLANN, there are three functional expansions namely, Legendre, Chebyshev and power series of input pattern. Their respective networks include L-FLAN, C-FLANN and P-FLANN.

Let $c(j)$, $I < j < S$ be an individual element of the input pattern prior to its expansion. Here in our study, S represents the total number of characteristics of the chosen dataset $c_n(j)$, $1 < n < N$ is the functional expansion of every $c(j)$, where N is the total count of expanded points for every input element. Expansion of an input pattern can be done as

$$x_0(c(j)) = 1, x_1(c(j)) = c(j), x_2(c(j)) = 2c(j)^2 - 1,$$
$$x_3(c(j)) = 4c(j)^3 - 3c(j), x_4(c(j)) = 8c(j)^3 - 8c(j)^2 + 1.$$

where $c(j)$, $1 < j < d$, d = characteristics set present in dataset.
Upon multiplication of these nonlinear outputs by a set of weights which are arbitrarily initialized from the range [−0.5, 0.5] and then summing them up we get the approximated output $y(k)$ later summing up all those $y(k)$s, we get the final $\hat{y}(k)$.

Training with Artificial Bee Colony Algorithm
Backpropagation algorithm holds good number of disadvantages in training a FLANN but it is the most widely used. However, on the other hand, the strengths and plasticity offered by population-based optimization algorithms directs us to go for the ABC algorithm as a good training scheme. The sole purpose of this training scheme is to beat the drawbacks influenced by backpropagation while training a FLANN. The training algorithm is presented below. In the first step, the FLANN architecture including bias as well as weight plus the training dataset is converted into an objective function. For the purpose of searching optimal weight parameters, the objective function in the next step is fed to the ABC algorithm. Considering error calculation as a yard stick, the weight changes are tuned by the ABC algorithm. Here in ABC algorithm, every single bee corresponds to a solution with an exacting weight vector collection. The ABC algorithm [13] used is given as below.

0. Cycle 1.
1. Choose a population of scout bees and initialize them with any random solution.

$$x_i, i = 1, 2 \ldots \mathrm{SN}$$

2. Evaluating fitness function of all x_i's. Initialize FLANN including both its bias as well as weight.
3. Cycle 2: while cycle is less than maximum cycle, iterate through step 4 to step 10.
4. Generate fresh population v_i for the employed bees as

$$v_{ij} = x_{ij} + \emptyset_{ij}(x_{ij} - x_{kj})$$

where k represents the solution in i's neighbourhood, $\emptyset$ representing an arbitrary number from [−1, 1] and evaluate them.

5. Evaluate fitness function of fresh population as

$$\mathrm{fit}_i = \begin{cases} \frac{1}{1+f_i}, & \text{if } f_i^3 0 \\ 1+\mathrm{abs}(f_i), & \text{if } f_i < 0 \end{cases}$$

6. Apply the greedy selection process between x_{ij} and v_{ij}.
7. Generate probability values (p_i) for the solutions X_i as

$$p_i = \frac{\mathrm{fit}_i}{\sum_{n=1}^{\mathrm{SN}} \mathrm{fit}_n}$$

8. Check onlookers for their distributed nature and if yes go to step 11.
9. Otherwise, for onlookers repeat iteration through step 4 until step 7.
10. Use greedy selection process for onlookers, $v_{i.}$
11. Conclude the discarded solution for the scout and if exists, substitute it with a fresh arbitrarily generated solution $_{xi}$ using:

$$x_i^j = x_i^j + \mathrm{rand}(0,1)(x_{\max}^j - x_{\min}^j)$$

12. Remember the top solution. Accordingly upgrade FLANN's fresh weights and bias.
13. Increment cycle counter as cycle+ = 1.
14. Stop once cycle reaches to maximum cycle.
15. Finally, save updated FLANN along with its fresh bias and weight set.

4 Datasets and Evaluation Criteria

In our study, we approached PROMISE repository of software engineering [14] for datasets purpose as it is publicly available for research purpose. In our study, we have used NASA 93, COCOMO 81 and COCOMO_SDR datasets. There are 63 projects within the COCOMO 81 dataset, each of them is using COCOMO model as already described in earlier section. Each project is thus described by 17 cost drivers, 5 scale factors, software size in KDSI, tactual effort, total defects and the development time in months. The NASA 93 dataset contains the data of 93 projects from different centres of NASA over many years. It contains 26 attributes which include 17 cost drivers of standard COCOMO II and 5 scale factors ranging from very low to extra high, measure of LOC, total defects, actual effort and the development time. The COCOMO_SDR dataset is from Turkish software industry. It consists of data from 12 projects and 5 different software companies in various domains. It has 24 attributes: 22 attributes from COCOMO II model, one being KLOC and the final one is actual effort (in person months). The entire dataset is classified into two classes namely a training set and a validation set in the ratio of

80:20. The only purpose of dividing the dataset this way is to get more accuracy in estimation. The proposed model is thus accordingly trained with the presented training data and tested with presented the test data.

The evaluation is based on the accuracy comparison of estimated effort against the actual effort. A general measure of evaluation of different cost estimation models is the magnitude of relative error (MRE) and is stated as below

$$\text{MRE} = \frac{|\text{Actual Effort} - \text{Estimated Effort}|}{\text{Actual Effort}}$$

For each project in the validation set, its MRE value is calculated. Similarly, MMRE computes the average MRE of n no. of projects as given below

$$\text{MMRE} = \frac{1}{n}\sum_{x=1}^{n} \text{MRE}$$

Since MRE is the most common evaluation criteria, it is thus adopted as the performance evaluator in the present paper. In addition to the mean magnitude of relative error (MMRE) and the Median of MRE (MdMRE) for entire validation set is also calculated and compared with COCMO model in the present paper.

5 Experimentation Results and Their Analysis

This section presents and confers about the results obtained while applying the proposed neural network model to the COCOMO 81, NASA 93 and COCOMO_SDR datasets. For the implementation purpose of the model, we use MATLAB here. The MRE value is calculated for the projects in the validation set for all the three datasets and is later compared against the COCOMO model.

The results as well as comparison based on COCOMO dataset is depicted in Table 1. It is clear there that the MRE in percentage for project 5 comes to out to be 7.44 while using COCOMO model and effectively and more efficiently 4.01 while using the proposed model. Similarly, the MMRE calculated for COCOMO model on entire validation set is 15.165 and 3.406 for the proposed model. Likewise, the calculated Median of MRE (MdMRE) for COCOMO model equivalent to 12.4 and 3.22 % for the proposed model.

Figure 3 illustrates the graphical demonstration of the two models based on their MRE values when executed on COCOMO 81 dataset. It is clear from there that there exists an upright fall in the relative error value while delivering with proposed model. Hence the obtained results thus recommend that proposed model can be best applied for accurate software cost prediction.

The results as well as comparison based on NASA 93 dataset is depicted in Table 2. Here also exists a marginal decrease in relative error value while applying the proposed model. It is clear from there that the MRE in percentage for project 30

Table 1 Comparison of various SCE models against the proposed FLANN

S. no.	Project ID	MRE (%) using COCOMO model	MRE (%) using proposed ABC based FLANN model
1	5	7.44	3.87
2	12	19.83	3.94
3	30	6.49	1.62
4	38	50.98	7.22
5	40	12.4	4.10
6	45	5.35	3.91
7	47	16.4	2.87
8	59	8.66	3.23
9	61	13.10	3.10
10	62	6.22	2.68
11	63	19.95	3.27

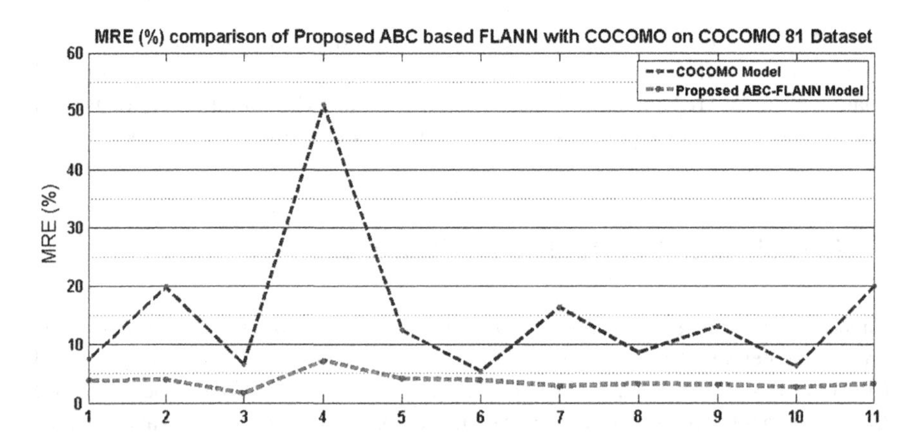

Fig. 3 Comparison of COCOMO model against proposed model on COCOMO dataset

comes to out to be 8.81 while using COCOMO model and effectively and more efficiently 3.34 while applying the proposed model. Same is true for project ID 62 which yields 13.20 incase of COCOMO and 4.70 while using the proposed one. Similarly, the MMRE calculated for COCOMO model on overall validation set is 12.746 and 4.194 for the proposed model. Likewise the calculated Median of MRE (MdMRE) for COCOMO is equivalent to 13.43 and 4.46 % for the proposed model.

Figure 4 illustrates the graphical demonstration of the two models based on their MRE values when validated on NASA 93 dataset. The marginal decrement again in the value of relative error suggests the very original magnitude of the proposed model.

Table 2 Comparison of MRE on NASA 93

S. no.	Project ID	MRE (%) using COCOMO model	MRE (%) using proposed ABC based FLANN model
1	1	9.33	3.80
2	5	8.84	3.29
3	15	16.75	4.45
4	25	14.09	3.77
5	30	8.81	3.22
6	42	13.9	5.13
7	54	13.67	4.49
8	60	11.78	5.21
9	62	13.2	4.70
10	75	17.09	3.88

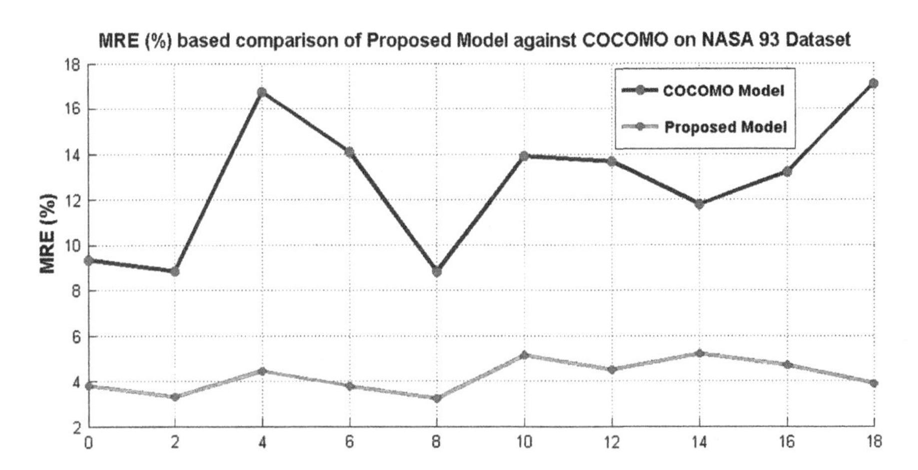

Fig. 4 Comparison of COCOMO against proposed model on NASA93 dataset

Table 3 shows the estimated effort and their MRE values using the proposed model on COCOMO_SDR dataset. MMRE value for the estimated effort is 6.34. The value of MdMRE incase of proposed model when applied on entire validation set is 4.62 %.

Figure 5 shows the bar graph representation of actual effort values and estimated effort values done with the proposed model for COCOMO_SDR. The bar graph shows that the estimated effort is very close to the actual effort. The results obtained thus, suggest that the proposed model outperformed the COCOMO model in terms of all the discussed evaluation criteria, i.e. MRE, MMRE and MdMRE and thus can be applied for accurate prediction of software costs.

Table 3 Comparison of effort on COCOMO_SDR

S. no.	Project Id	Actual effort	Estimated effort	MRE (%)
1	1	1	1.22	0.22
2	2	2	1.67	0.16
3	3	4.5	4.29	0.046
4	4	3	2.94	0.02
5	5	4	3.67	0.082
6	6	22	20.12	0.085
7	7	2	1.93	0.035
8	8	5	3.67	0.266
9	9	18	16.73	0.070
10	10	4	3.84	0.04
11	11	1	1.06	0.06
12	12	2.1	2.12	0.009

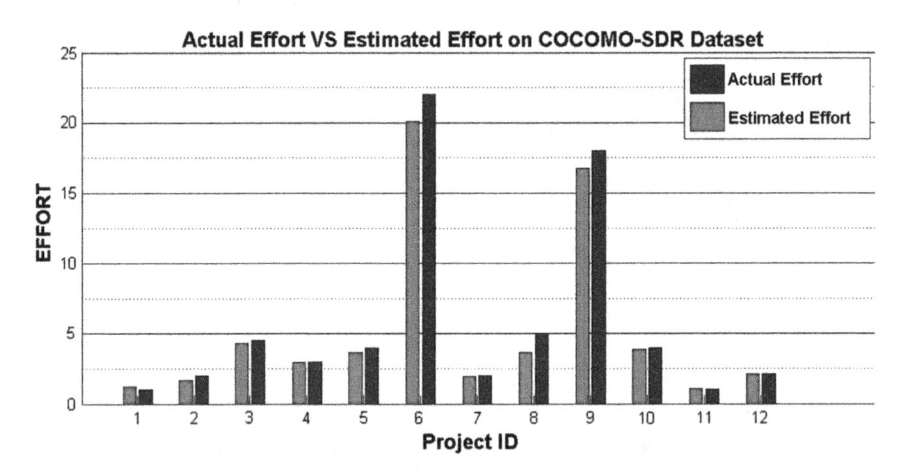

Fig. 5 Comparison of actual versus estimated effort on COCOMO_SDR dataset

6 Conclusion and Future Work

In the proposed work, we have evaluated the ABC-based FLANN model for the task of software cost prediction. Our study has confirmed that FLANN-ABC carries out the prediction task fairly well. The obtained results prove that the proposed ABC learning algorithm effectively trains the FLANN for cracking prediction problems with improved accurateness even under the availability of vague data. An important facet of this research work is to train the FLANN with an improved training scheme as a replacement of standard BP learning algorithm for the purpose of getting an accurate software development cost estimation. The major benefit of the proposed model is that it trims the computational complexity without any forfeit on its performance. The results obtained from the proposed method have been evaluated using a number of criteria such as MRE, MMRE and MdMRE. According to the obtained results, it can be said that FLANN increases the accuracy of the cost estimation and especially when this model is trained with ABC algorithm the accuracy of the estimated cost became even much better.

In future, we can combine meta-heuristic algorithms with ANNs and other data mining techniques to obtain better results in SCE field. In addition, the accuracy of SCE can be enhanced by applying other particle swarm optimization methods.

References

1. Putnam, L.H.: A general empirical solution to the macro software sizing and estimating problem. IEEE Trans. Soft. Eng. **4**(4), 345–361 (1978)
2. Boehm, B.W.: Software Engineering Economics. Prentice Hall, Englewood Cliffs (1981)
3. Patra, J.C., Pal, R.N.: A functional link artificial neural network for adaptive channel equalization. Sig. Process. **43**, 181–195 (1995)
4. Booker, J.M., Meyer, M.M.: Elicitation and Analysis of Expert Judgement. Los Alamos National Laboratory
5. Zull, J.E.: The Art of Changing the Brain: Enriching the Practice of Teaching by Exploring the Biology of Learning. Stylus Publishing (2002)
6. Shepperd, M., Cartwright, M.: Predicting with sparse data. IEEE Trans. Soft. Eng. **27**, 987–998 (2001)
7. Venkatachalam, A.R.: Software cost estimation using artificial neural networks. In: Proceedings of the 1993 International Joint Conference on Neural Networks, pp. 987–990 (1993)
8. Rao, B.T., Dehuri, S., Rajib, M.: Functional link artificial neural networks for software cost estimation. Int. J. Appl. Evol. Comput. 62–68 (2012)
9. Dehuri, S., Cho, S.B.: A comprehensive survey on functional link neural networks & an adaptive PSO–BP learning for CFLNN. Neural Comput. Appl. 187–205 (2010)
10. Pao, Y.H., Takefuji, Y.: Functional-link net computing: theory, system architecture, and functionalities. Computer **25**(5), 76–79 (1992)
11. Pao, Y.H., Phillips, S.M., Sobajic, D.J.: Neural net computing and intelligent control systems. Int. J. Control **56**(2), 263–289 (1992)
12. Patra, J.C., Pal, R.N., Chatterji, B.N., Panda, G.: Identification of non-linear & dynamic system using functional link artificial neural network. IEEE Trans. Syst. Man Cybern. B Cybern. **29**(2), 254–262 (1999)
13. Karaboga, D., et al.: Artificial bee colony (ABC) optimization algorithm for training feed-forward neural networks. In: Torra, V., et al. (eds.) Modelling Decisions for Artificial Intelligence, vol. 4617, pp. 318–329. Springer, Berlin/Heidelberg (2007)
14. www.promisedata.org

Improvised Cyber Crime Investigation Model

Ajeet Singh Poonia, C. Banerjee and Arpita Banerjee

Abstract Crime is a major social and legal problem in the world we live in. Crime changes with time and circumstances. Criminals use new technologies to facilitate and maximize criminal activities. With increased use of Computers and Internet, there has been an explosion of new crime called the Cybercrime which is also known as 'Internet crimes' or 'e-crime'. Cybercrime is the most complicated problem and it has become the major concern for the global community. It is hard to detect. The nature of cybercrime shows that investigations are often technically complex, requiring access to specialist skill. For detection of cybercrime many investigation techniques and methods are there in the system. Also many investigation models have been proposed. Till date the investigation models proposed by various researchers have some or the other shortfalls. An improvised cybercrime investigation model with detail phases has been proposed in this paper.

Keywords Cybercrime · e-crime · Cybercrime investigation model

1 Introduction

With the rise of digital era and merger of computing and communication, unmatched and unparalleled job opportunities have been created which has prompted the ordinary citizens with criminal minds. The continued development of digital technology

A.S. Poonia (✉)
Department of CSE, Government College of Engineering and Technology,
Bikaner, Rajasthan, India
e-mail: pooniaji@gmail.com

C. Banerjee
Amity University, Jaipur, Rajasthan, India
e-mail: chitreshh@yahoo.com

Arpita Banerjee
St. Xavier's College, Jaipur, Rajasthan, India
e-mail: arpitaa.banerji@gmail.com

M. Pant et al. (eds.), *Proceedings of Fifth International Conference on Soft Computing for Problem Solving*, Advances in Intelligent Systems and Computing 437, DOI 10.1007/978-981-10-0451-3_66

has contributed in creation of novel ways of criminal operation. It is solely the responsibility of users and policy making agencies to furnish themselves with the knowledge that will permit an effective response against such crime and criminal.

As we know that, at one end, cyber system provides incomparable opportunities to communicate and learn and at the other end some individuals or community exploit its power for criminal purposes [1]. Cybercrime as we say differs from the conventional crimes in a number of ways. Cybercrime is any illegal activity arising from one or more Internet components done by the cyber criminals. It may be created or committed for a no of reasons like for the sake of recognition, making quick money, global expansion, less chances of being caught, lack of proper controlling rules, lack of reporting and standards, limited media exposure, and due to infrequent official investigation and criminal prosecution.

Complete security is unattainable but it can be made comprehensive so as to minimize the harm caused due to its exploitation. Cyber criminals can be any one of the following like motivated criminals, organized hackers, organized hackers, discontented employees, cyber terrorists, etc. However, it is possible to obtain effective results by involving good management, enforced procedures, and the adequate and updated technical tools, with an appropriate and effective policy framework [2].

It is now the duty of the society to fight against cybercrime by adopting some ethics, guidelines, models, etc. At present to safe guard against the cybercrime and to investigate it there is no generalized model available. Some of the available model like Casey focuses on processing and examining digital evidence and comprises of steps like recognition, preservation, collection, documentation, classification, comparison, individualization, and reconstruction [3].

Lee et al. reflected upon crime scene of scientific nature and its procedural investigation [4]. The first Digital Forensics Research Workshop by Palmer introduced a working model which highlighted the various steps to be taken during digital forensic analysis in a linear fashion, viz., Identification, Preservation, Collection, Examination, Analysis, Presentation, and Decision [5].

Reith et al. in their research paper described a model which to some extent is a resultant of Digital Forensic Research Workshop (DFRWS) model [6]. Ciardhuáin proposed a cybercrime model which is comprehensive in nature and contains 13 steps which highlights the importance of awareness, authorization, planning, notification. The proposed model also emphasized on procedures like search and identifies evidence, evidence collection, transportation, storage and examination, hypothesis, and distribution of information [7].

Carrier and Spafford suggested a process model which is based on investigative procedures. In addition to investigation, the proposed model through a phase also ensures that the investigation is supported by procedures like operation and infrastructure [8]. Ceresini suggested and stressed upon methods of implementation and maintaining forensic viability of log files supported by policies and procedures [9]. Pollitt [10] and Selamat et al. [11] studies concluded that most of the existing models are improvised version of existing models with the aim to make them comprehensive and robust.

2 Phases of Cyber Crime Investigation Model

Many current methods are simply too technology-specific. The proposed model attempts to improve upon existing models through the combination of common techniques while trying to ensure method shortfalls are addressed. The inclusion of information flow, as well as the investigative activities, makes it more complete than other models. The proposed improvised model have various phases as: Realization phase, Authorization phase, Audit planning, Auditing, Managing evidence, Hypothesis, Challenge analysis, Final report presentation, Updating polices, Report abstraction and Dissemination. The details discussion of each phase is mentioned below [1]:

1. Realization phase: In realization phase the particular organization/system/individual realizes that some form of cybercrime has occurred. This realization is a phase prior to the actual investigation process. The end results depend on the initial phase so it should be carried out carefully and vigilantly. This realization invokes the person/organization to carry out cybercrime investigation to find out the amount of loss occurred in any form, i.e., whether in the form of information or monetary or physically or personally or in any other form and to find out the culprit. In this phase the Local investigation plays an important role where the organization/person itself carries out investigation at its own level to identify the crime and its reasons and effects.
   This phase is optional and sometimes abandoned to be carried out in order to conserve the evidences and protecting the crime scene from any form of loss. But in some organization, local investigation may be present as a protocol to be followed in their crime policy. The investigation depends upon the type of the crime or the criminal, i.e., whether the case is related to outside or inside attack. Outsiders are hackers/crackers, etc., and Insiders could be employees of the organization or users of a particular application(s) (Fig. 1).
2. Authorization phase: The authorization phase is concerned with the synchronization of investigation team's policy, procedures and protocols with the policies, procedures and protocols of the suffered party/organization/individual and granting of necessary rights to the investigation team by the organization/person and the law. So in this authorization phase all possible

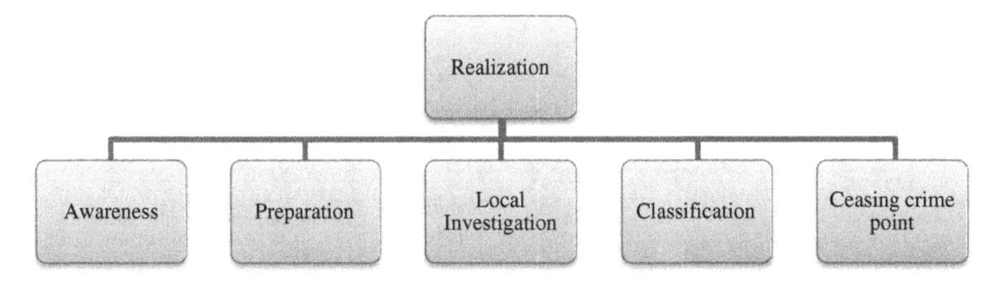

Fig. 1 Shows the sub-phases of realization phase

rights are granted by the organization/person to the investigation team to carry out investigation to any extent possible. This phase is important because in the cybercrime investigation process, the investigating team may require support from the organization/person and need to check and gather confidential data.
Authorization may be complex as it requires interaction with both external and internal entities to obtain the necessary permission. Structure of authorization may vary depending on the type of investigation. At one level it may only require a simple verbal approval from person/organization to carry out a detailed investigation, at the other extreme advanced technology, high tech tools, policies, methodology along with law enforcement agencies. All the communication between the organization/person and the investigating team is documented to deal with controversies and to ensure openness and transparency in the investigation process (Fig. 2).

3. Audit Planning: In audit planning all the necessary preparations are done before reaching the location of crime, by any of the mode or medium. In such preparations, the auditing team gathers all sorts of queries that need to be put to the organization/person in order to get maximum information about the background of the system suffering from crime and about the crime itself. It is the important phase where the questionnaire (e.g., the basic question to be asked at the site can be, who first noticed the incident? At what time(s) did the incident occur? Type of hardware platforms, who was the last person on the system? When the system was last backed up? Is the attacker still online? Are there any suspects?
Do people share passwords or user IDs? What type of work is this organization involved with. Any new applications recently added. New contractors, employees, etc., hired in the past months. Any patches or operating system upgrades recently done.), technical tools (e.g., the basic audit tools are Data Recovery Programs, Honey pots, IP Address Tracking, Chat Room Monitors, log files, etc.), information extraction (for extracting log files, deleted files, password files, user session information, etc.) and documentation tools (Some of the basic tools that can be referenced are DFD, ERD, Decision Tree, Decision Table, Flowcharts, etc.) are prepared as per the case, i.e., as per the gravity of the cybercrime held (Fig. 3).
4. Auditing: Auditing is one of the vital phase which is concerned with procedures related to investigation of cybercrime. The outcome of this phase reveals the real 'cause' and the associated area(s) which has been impacted by the attack.

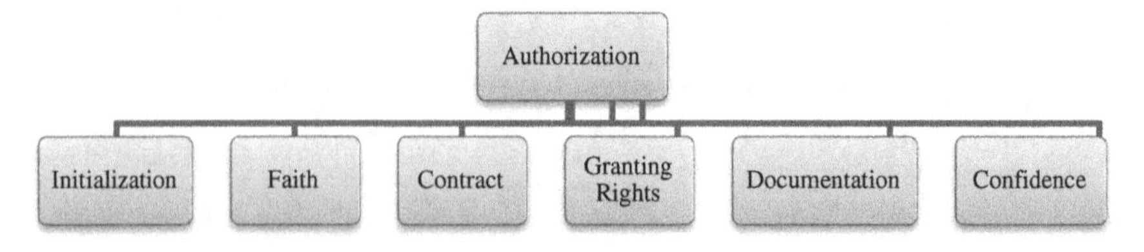

Fig. 2 Shows the sub-phases of authorization phase

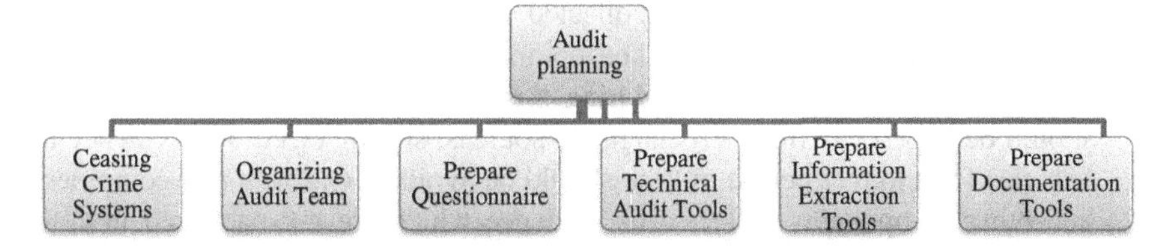

Fig. 3 Shows the sub-phases of the audit planning phase

Supporting material like references, records, etc., containing reports noted down from the previous cases happened inside and outside the organization helps the investigating team. Investigative procedures could be as simple as identifying the computer(s) manipulated by a suspect or it could be as complex as tracing the computers used for malicious purpose through multiple ISPs based on knowledge of an IP address. The depth of investigation is solely depended on the gravity of crime held.
Material for investigation may include record, log, and document checking. e-mail messages and associated email addresses could also be potential material for the investigative team along with the information related to sender, content of the communications, IP addresses, date and time, user information, attachments, passwords, and application logs that may highlight the evidence related to spoofing, etc. In the initial step, the auditor solely focuses on review of answers given by the client associated with questions concerning the crime. The step forms Audit Planning phase. The devices like routers, firewalls, etc., are required and used by the investigative team to collect potential evidences. They are supported by software like ERP packages, Intrusion Detection used for monitoring malicious activities, packet sniffers, keyboard loggers, and content checker, etc. Logs like general logs, such as access logs, printer logs, web traffic, internal network logs, etc., also forms a potential material in the auditing process (Fig. 4).

5. Managing Evidences: After the auditing phase is over, it is followed by managing evidences phase. The materials collected during auditing phase are used to identify the crime and criminal. This phase focused on scrutinizing the potential and relevant evidences for providing the crime's effect and to reach

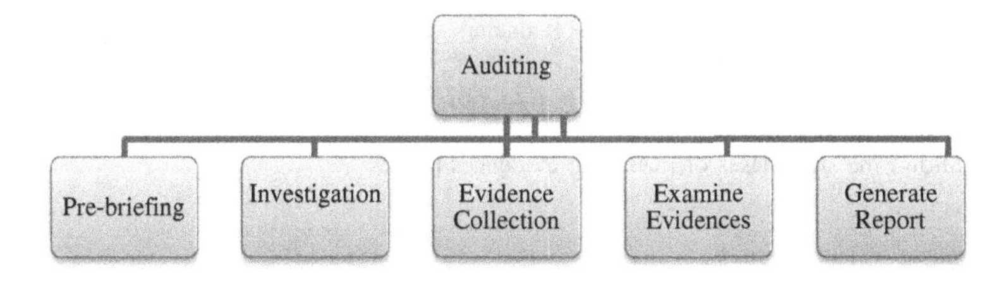

Fig. 4 Shows the sub-phases of the auditing phase

the criminal. All the evidences collected together are stored, packaged, and transported. They are further backed up and analyzed.
Few points to be taken care of during this phase is that; electronic evidence should be kept away from any magnetic sources, storing electronic evidence in vehicles for a longer period of time should be avoided, ensuring that computers and other components collected as evidence which are not packaged in containers are secured in the vehicle to avoid shock and excessive vibrations, maintaining the chain of custody on all evidence transported, and inventory management should be done properly (Fig. 5).

6. Hypothesis: It describes how the data acquired during the collection and reporting phase's. It is important that the report only describe facts that were observed, analyzed, and thus documented. This is the major and the most important phase of the cybercrime investigation. Based on the examination of the evidence, the investigators/auditors construct a hypothesis of what occurred. The degree of formality of this hypothesis depends on the type of investigation which varies from case to case. These hypotheses are helpful in generating the proofs for the investigating team. These hypotheses also help in reaching the actual criminal and to prove its crime (Fig. 6).
7. Challenge analysis: This is the phase in which the investigating team tries to go against their own built hypotheses and try to prove their own hypotheses wrong of irrelevant. If the investigating team becomes successful in proving any of its hypotheses absolutely irrelevant than it has to revisit the evidences available as well as the hypotheses in order to correct the shortcomings. Moreover, it should collect and generate stronger hypotheses that could not be easily rejected by the opposition in the court. This cycle keeps on repeating till all the hypotheses made by the investigating team are strong enough not to be beaten or rejected by the opposition's hypotheses and proofs (Fig. 7).
8. Final report presentation: In this phase, a final report of the whole investigation process is generated and presented well in front of the relevant person or organization and the opposition party. This report documents all the evidences

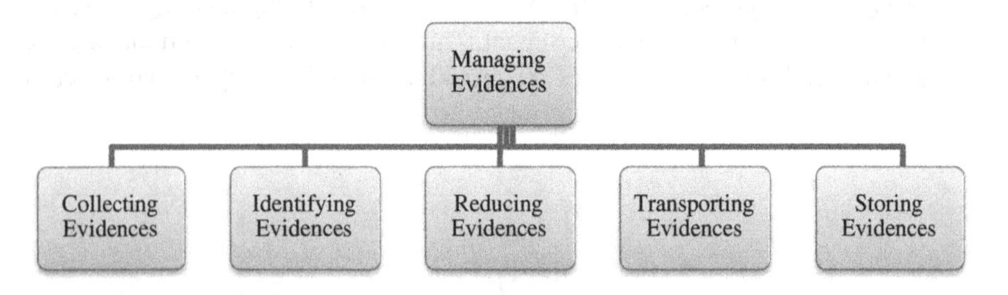

Fig. 5 Shows the sub-phases of managing evidences phase

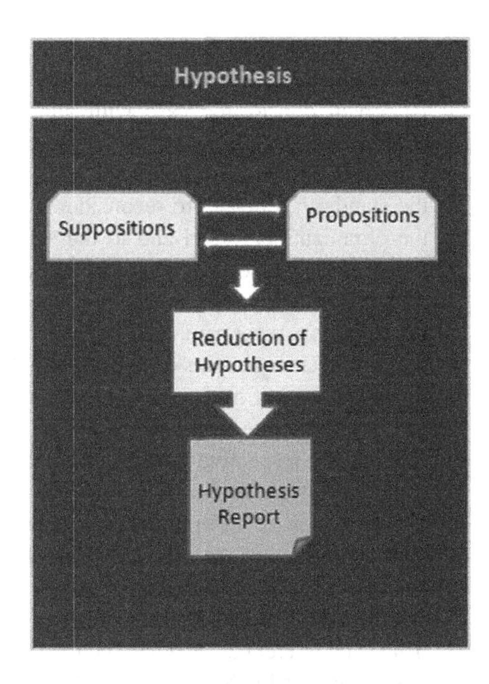

Fig. 6 Shows the sub-phases of hypothesis phase

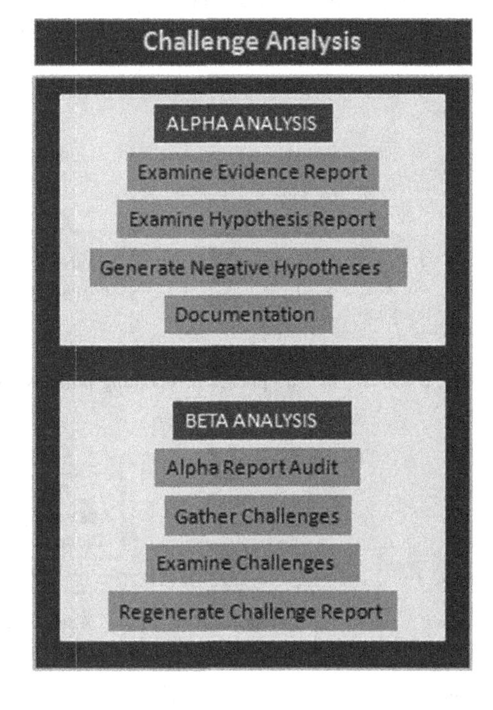

Fig. 7 Shows the sub-phases of challenge analysis phase

and the hypotheses results. This is a detailed report and documents of all the investigating results collected and generated by the investigating team related to the given cybercrime case with all the positive and negative points of the evidences and then facing the challenges if there are any so.

9. Updating policies: In this phase, the security policies of the organization and the standard policies are reconsidered and updated if required with respect to the organizational level and to the standard levels if required. New protocols meant for the security may also be developed if it is found important. Sometimes we have to develop the sophisticated forensic techniques for investigating cybercrime and begin to track the cybercrime, because every day the new techniques are developed for doing the cybercrime, after that, sometimes we are not having the investigating technique required for tracking that newly cybercrime technique, these techniques are developed according to the rules of the respective country, they should be according to the law (Fig. 8).
10. Report Abstraction and Dissemination: The final activity in the model is the dissemination of information from the investigation. In this phase, the final report containing results is abstracted at different levels of abstraction depending upon the sensitivity and requirement of publishing the results to corresponding users. Few reports may be highly abstracted and may be published for reference for other such cases happening in future while few reports may be negligible abstracted and kept by the organization for future assistance and policy updation. This report is kept safe for future use, if same type of case is performed again then we can track the victim easily (Fig. 9).

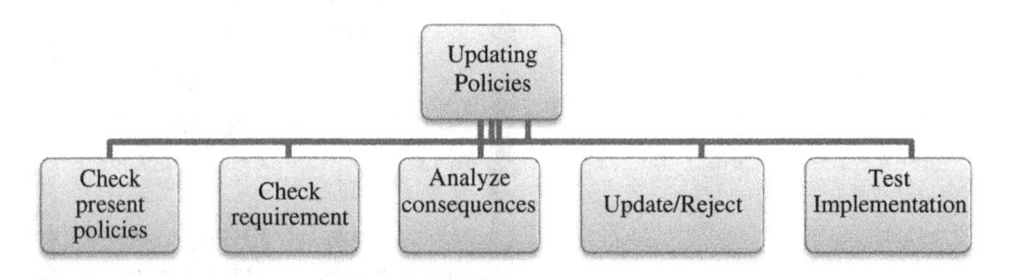

Fig. 8 Shows the sub-phases of updating policies phase

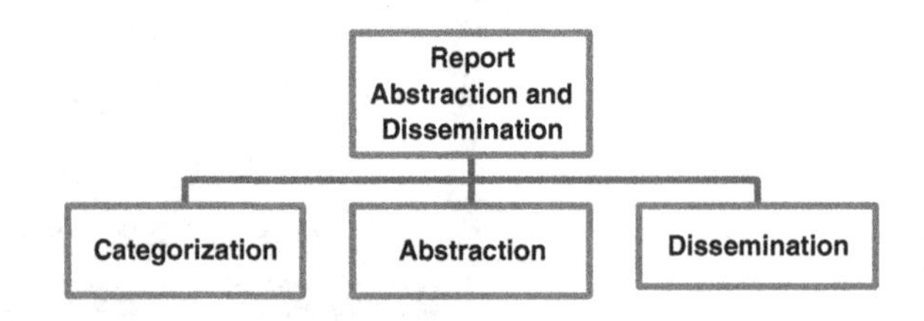

Fig. 9 Shows the sub-phases of report abstraction and dissemination phase

3 Conclusion

The above described improvised Cybercrime Investigation Model has its own importance as it provides a reference, an abstract, a framework, a technique, a procedure, a method, an environment independent of any particular type of digital crime for any system or any organization, for supporting the work of any investigators/auditors. The improvised model can be used in a practical way to identify opportunities for the development and operation of technology to support the work of investigators, and to provide a framework for the capture and analysis of requirements for investigative tools.

This Cybercrime Investigation Model is useful not just for law enforcement but is also beneficial for IT managers, security practitioners, and auditors. The model will not only help in the investigation part but will also support the development of tools, techniques, training and the certification/accreditation of investigators, auditors, system and tools.

References

1. Poonia, A.S.: Managing digital evidences for cyber crime investigation. Int. J. Adv. Stud. Comput. Sci. Eng. (IJASCSE) **3**(11), 22–26 (2014). ISSN: 2278 7917
2. Poonia, A.S.: Investigation and Prevention of Cyber Crime in Indian Perspective, Ph.D Thesis, MNIT, Jaipur (2013)
3. Casey, E.: Error, uncertainty, and loss in digital evidence. Int. J. Digital Evid. **1**(2), 1–45 (2002)
4. Lee, H.C., Palmbach, T., Miller, M.T.: Henry Lee's Crime Scene. Academic Press (2001)
5. Palmer, G. (ed.): A Road Map for Digital Forensic Research: Report from the First Digital Forensic Workshop, 7–8 Aug 2001. DFRWS Technical Report DTR-T001-01, 6 Nov 2001
6. Reith, M., Carr, C., Gunsch, G.: An examination of digital forensic models. Int. J. Digital Evid. **1**(3) (2002)
7. Ciardhuáin, S.O.: An extended model of cybercrime investigation. Int. J. Digital Evid. **3**(1) (2004)
8. Carrier, B., Spafford, E.H.: Getting Physical with the investigative process. Int. J. Digital Evid. **2**(2) (2004)
9. Ceresini, T.: Maintaining the forensic viability of log files. http://www.giac.org/practical/gsec/Tom_Ceresini_GSEC.pdf (2001)
10. Pollitt, M.: An ad-hoc review of digital forensics models. In: Higgins, G. (ed.) Cybercrime: An Introduction to an Emerging Phenomenon. McGraw Hill
11. Selamat, S.: Mapping process of digital forensic investigation framework. Int. J. Comput. Sci. Netw. Secur. **8**(10) (2008)

Performance of Wideband Falcate Implantable Patch Antenna for Biomedical Telemetry

Raghvendra Singh, Love Jain, Sanyog Rawat and Kanad Ray

Abstract The biomedical telemetry provides diagnostic inspection and monitoring of physiological signals at a far distance. Nowadays implanted medical devices are recent advances in this field. Patch antennas are main devices to send these signals at far distance and attain high attention for integration into implantable medical devices. Implantable patch antennas can be designed for transfer of high-speed data rate for transferring high quality video from implanted patch to outside telemetry. Dedicated MICS band (402–405 MHz) is not sufficient enough for transferring the real-time videos. High data rate for transferring real-time videos can be achieved in IEEE 802.11 (2.4 GHz) ISM band (5.725–5.825 and 2.4–2.5 GHz). The objective of this study is to analyze the performance of implanted falcate-shaped antenna for high-speed transfer rate achieving impedance bandwidth 1.2978 GHz. Proposed antenna is useful for transferring 54 mbps (802.11g) and 600 mbps (802.11n), theoretically, for particular usage in biotelemetry.

Keywords Industrial scientific and medical (ISM) band · Wideband · IEEE 802.11: medical telemetry · Falcate-shaped antenna · Implantable antenna

Raghvendra Singh (✉)
JK Lakshmipat University, Jaipur, Rajasthan, India
e-mail: planetraghvendra@gmail.com

Love Jain
Department of Electronics & Communication Engineering,
JK Lakshmipat University, Jaipur, Rajasthan, India
e-mail: lovejain.eck@gmail.com

Sanyog Rawat · Kanad Ray
Department of Electronics & Communication Engineering,
Amity University, Jaipur, Rajasthan, India
e-mail: sanyog.rawat@gmail.com

Kanad Ray
e-mail: kanadray00@gmail.com

M. Pant et al. (eds.), *Proceedings of Fifth International Conference on Soft Computing for Problem Solving*, Advances in Intelligent Systems and Computing 437, DOI 10.1007/978-981-10-0451-3_67

1 Introduction

Wireless communication is receiving much interest nowadays in biomedical field. In biomedical field, wireless transmitters and receivers are very much helpful for experts to find exact data of the body of patient in same way it is easy to use and comfortable for patient. There are a variety of medically implanted devices being used in the human body for sensing, monitoring and the wireless implanted communication is required for adequate working of these devices [1, 2].

There are a variety of applications for medical care and many more are coming in the future. Wireless biotelemetry allows doctors to establish a reliable and high-speed link for health monitoring [3]. High-speed data transfer, images, and video transfer is not possible in MICS band (402–405 MHz), but there is another band as ISM (5.725–5.825 and 2.4–2.5 GHz) band and IEEE 802.11 (2.4 GHz) range. This band is sufficient for high-speed data transfer and video transfer. However, the MICS band with many advantages is not suited well for transferring the image and huge data from implanted device in human body.

It is obvious that, any time a person uses a wireless communication device, the device will have an antenna to provide the communication link between transmitter and receiver. Therefore, the key element for any wireless implanted communication is an antenna, and there are many parameters to be considered while designing biotelemetry system for human body such as antenna size, impedance matching, and the unique radio frequency transmission challenges [4, 5].

The biotelemetry systems use implanted antennas which are covered by the lossy body tissues. So the implanted antennas should be able to operate in lossy dielectrics (electrical analogy of human tissues) following the variable electrical properties appropriately with frequency [6]. The implanted antenna properties may vary inside the different tissue layers of human body.

Human body is considered here as a medium for wave propagation for creating reliable wireless communication links. Hence electrical analogy of human body is developed, in which different tissues and bones are considered as stacked dielectric layers [7]. The electrical properties of different layers should be known for radio wave propagation at a particular frequency of operation [8]. In this paper, an implanted antenna for the human body for wideband radio frequency propagation, 1.8–3.1 GHz is presented. CST® Microwave studio, a high frequency simulator is used for antenna design and optimization.

2 Antenna Design Schemes

There are several strategies for implantable antenna design for human body but mostly are stated by the fact that antennas are to operate inside the human tissue instead of open atmosphere. According to first strategy, the antenna should be designed in free space then refined for covered tissue model. In the second strategy, an antenna can be designed directly in an environment surrounded by human tissues [9].

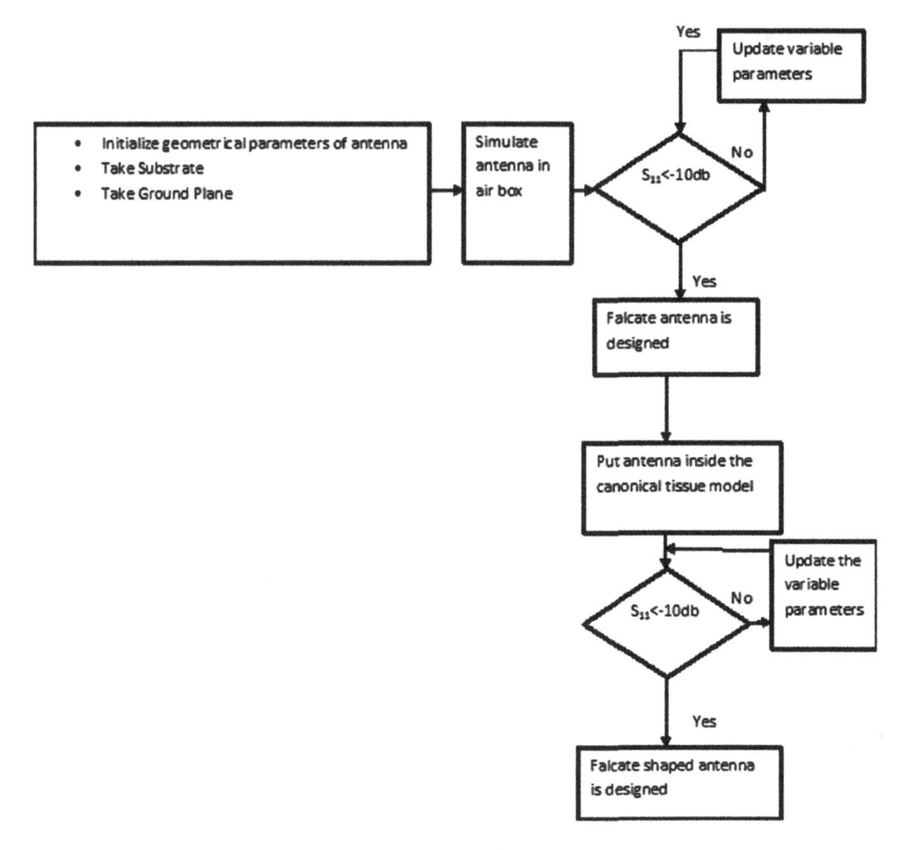

Fig. 1 Flow chart of design of falcate implantable patch antenna

Here wideband falcate implantable patch antenna is simulated first in free space and then it is verified in tissue model also, flowchart is shown in Fig. 1. The tissue model used here consists of two layers such as skin and bone. The wavelength in the tissues is shorter, since the velocity of wave propagation is lowered [2]. Hence, the geometrical size of implantable antenna is usually less than 10 % of wavelength in open environment.

3 Antenna Design

The proposed geometry and constraints related to the design of wideband falcate-shaped implantable patch antenna are as follows:

Here the substrate is FR-4 of dimensions 58 mm × 40 mm and patch is printed on one side of substrate. These dimensions are calculated after applying the formula of dimensions calculations of patch antenna. The radiator dimensions are 18 × 1 mm

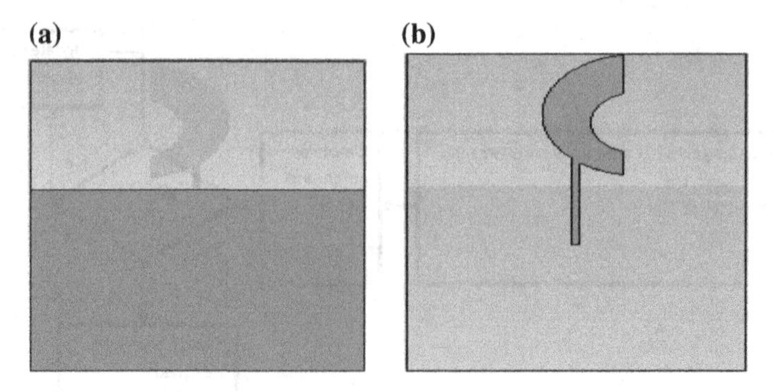

Fig. 2 **a** Back view. **b** Front view of implanted falcate antenna simulated in CST®

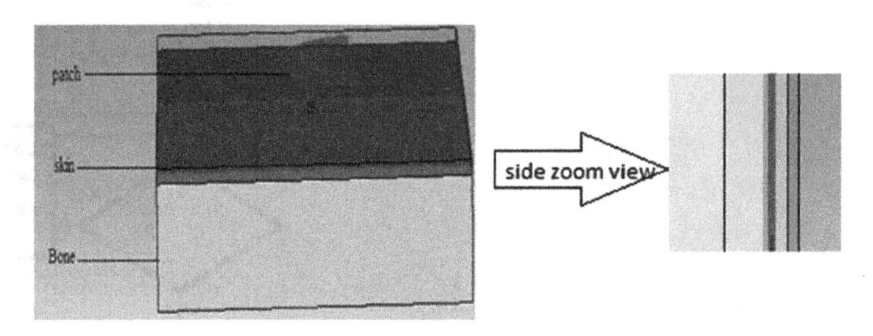

Fig. 3 Side view and side zoom view of implanted falcate antenna simulated in CST®

for achieving 50 Ω impedance. This will achieve accurate impedance matching with the input line and transfer the maximum power into the radiator. The dielectric constant of substrate is 4.3 and the loss tangent is 0.025. The ground plane also plays the vital role for impedance matching as its dimensions are 38 mm × 40 mm.

Falcate shape is generated by two circles of radius 11.3 and 5.5 mm, respectively Front view and back view of the proposed antenna is shown in Fig. 2 (Fig. 3).

4 Tissue Models

Falcate-shaped implantable patch antenna is analyzed inside the inhomogeneous lossy dielectric that is an electrical equivalent of biological tissues. Usually, the biological tissues contain their own permittivity (εr), conductivity (σ), and mass-density values. Canonical model approach of biological tissues is used here to make fast simulations and for easier design of implantable antennas. Here multi-layer canonical model (e.g. Fig. 4) is used which is providing a simplified model of falcate-shaped antenna implanted inside the biological tissues.

Fig. 4 Two layer canonical tissue model [10]

Table 1 Properties of canonical model tissue structure

Tissue	Permittivity	Conductivity
Skin	46.7	0.69
Bone	13.1	0.09

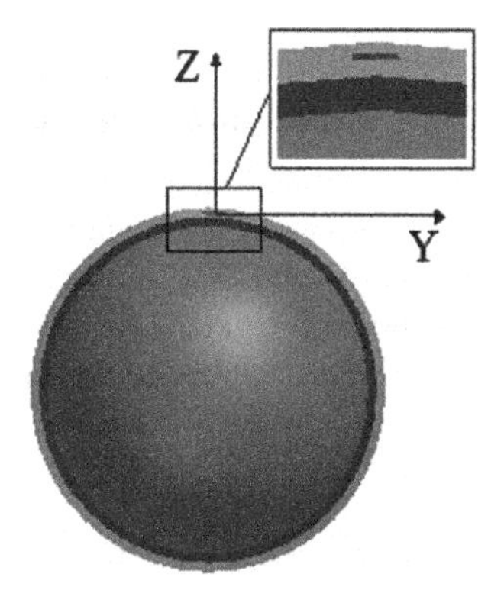

Fig. 5 Skin layer of two layer spherical head

The canonical model is inspired from Gabriel and Gabriel [10, 11]. Properties of canonical model tissue structure are given in Table 1 [10].

The proposed antenna is used for implanting inside the human head model. Here two layers are considered as skin and bone. Falcate antenna is implanted and sandwiched in between skin and bone as shown in Fig. 5 [10].

5 Antenna Performance

The antenna used for the implantable device should be able to radiate outside the body and yet the power associated with the device must comply with the radiation requirements [10]. Most tissues have significant attenuation and, therefore, there will be losses, when signal passes through different tissue layers or organs [12]. Increased power levels might cause heating of the tissue and consequently permanent tissue damage. S-parameter of the proposed falcate antenna is shown in Fig. 6.

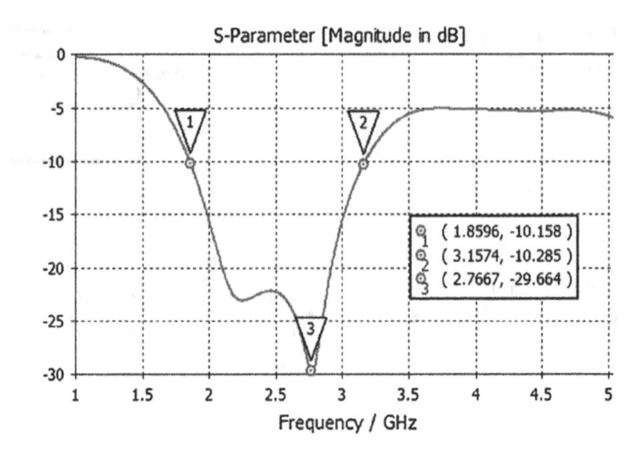

Fig. 6 Antenna return loss in canonical tissue model

5.1 *Reflection Coefficient Versus Frequency Response*

There is a clear wide band 1.8596–3.1574 GHz in Fig. 6. This band is supported by both WiFi and ISM. So this antenna is useful for transferring the 54 mbps (802.11g) and 600 mbps (802.11n) theoretically for particular usage. Impedance bandwidth is 1.2978 GHz which can be used for high data rate transfer.

5.2 *Far-Field Patterns*

Far-field patterns of falcate antenna are observed at 1.9, 2.4, and 5.7 GHz as in Figs. 7, 8, and 9. Maximum directivity is on 0° and 180°. With the frequency increment, the directivity is shifting in the lower half of the plot. It means at higher frequency the antenna will be more directive in a particular direction.

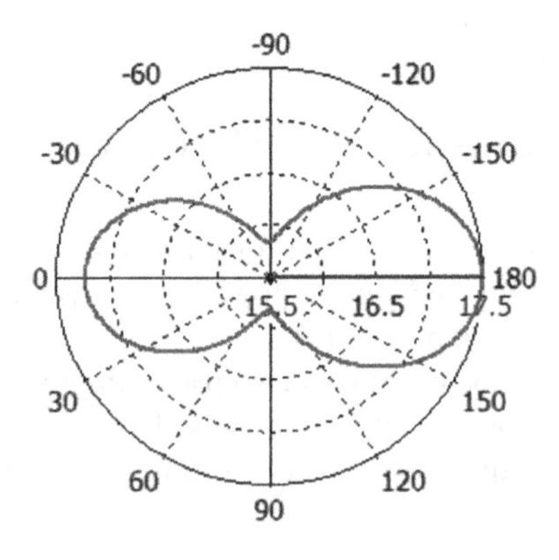

Fig. 7 Far-field pattern at 1.9 GHz in *YZ* plane

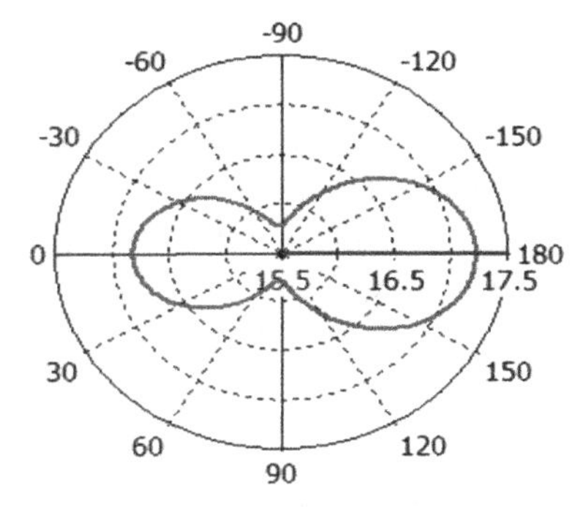

Fig. 8 Far-field pattern at 2.4 GHz in *YZ* plane

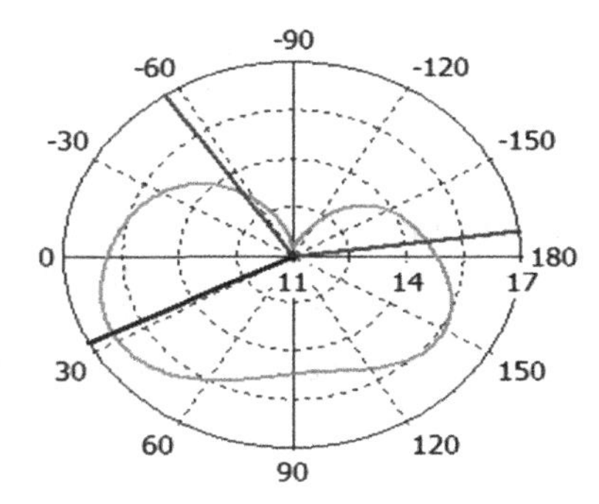

Fig. 9 Far-field pattern at 5.7 GHz in *YZ* plane

5.3 Gain Versus Frequency Plot

Gain versus frequency plot is shown in Fig. 10 which has a maximum gain of 4.6 at 3 GHz frequency. After 3 GHz, the gain is decreasing and the last value is 3. Gain is increasing up to 4.5 from the frequency range 1 to 3 GHz.

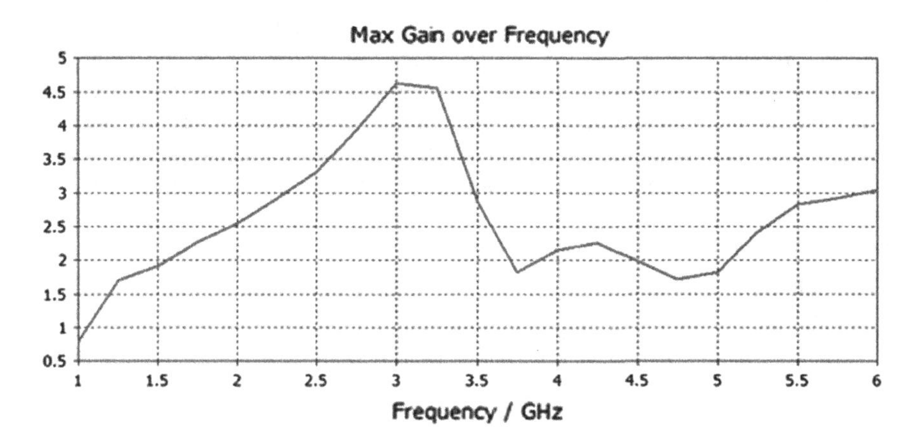

Fig. 10 Gain versus frequency plot

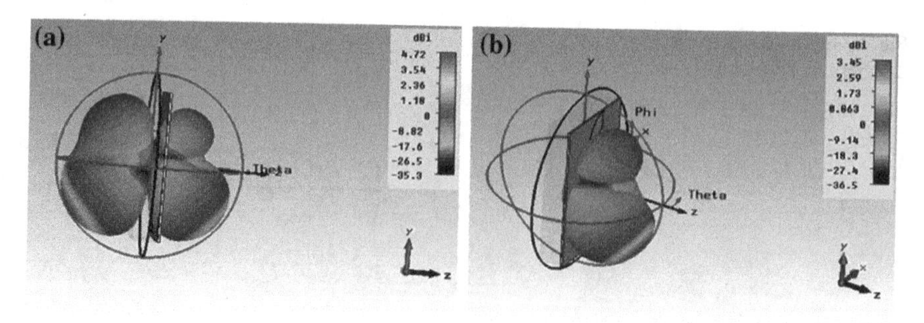

Fig. 11 **a** Radiation pattern in three-dimensional view (θ = 0°–90°) at 3.5 GHz. **b** θ = 90° at 3.5 GHz

5.4 Three-Dimensional Far-Field Patterns

Radiation Pattern in three-dimensional view (θ = 0°–180°) at 3.5 GHz is shown in Fig. 11a and (θ = 90°) in Fig. 11b. Maximum directivity is on 0° and 180°. It is clear from far-field radiation pattern that this antenna can be implanted horizontally in the body tissue and performs well in both the direction as shown in Fig. 11.

6 Conclusion

Conductivity and permittivity of the surrounded tissues in implanted devices generate restrictions along with the geometrical size of antenna due to which the antenna performance analysis is more complex than the antennas in open environment. The proposed falcate-shaped antenna is with higher impedance matching, 1.2978 GHz impedance bandwidth, and higher gain value with frequency which can support both WiFi and ISM band. So this antenna is useful for transferring 54 mbps (802.11g) and 600 mbps (802.11n) theoretically for particular usage.

The proposed falcate-shaped antenna dimensions are small, fulfilling the requirements of medical implant devices. Electrical parameters of human body are considered to meet appropriate performance from implanted antenna. The antenna performance was carried out using CST® Microwave studio with different dielectrics in canonical model so that it would perform best inside the human body.

References

1. Cao, Y.F., Cheung, S.W., et al.: A multiband slot antenna for GPS/WiMAX/WLAN systems. IEEE Trans. Antennas Propag. **63**(3), 952–958 (2015)
2. Kiourti, A., Tsakalakis, M., et al.: Parametric study and design of implantable PIFAs for wireless biotelemetry. In: Proceedings of the 2nd ICST® International Conference on Wireless Mobile Communication and Healthcare (MobiHealth 2012). Kos Island, Greece (2011)

3. Dissanayake, T., et al.: Dielectric loaded impedance matching for wideband implanted antennas. IEEE Trans. Microw. Theory Tech. **57**(10), 2480–2487 (2009)
4. Skrivervik, A.K., et al.: Design strategies for implantable antennas. In: Proceedings of the Antennas and Propagation Conference. Loughborough, UK (2011)
5. Kiourti, A., Christopoulou, M., et al.: Performance of a novel miniature antenna implanted in the human head for wireless biotelemetry. In: IEEE International Symposium on Antennas and Propagation. Spokane, Washington (2011)
6. Karacolak, T., Hood, A.Z., et al.: Design of a dual-band implantable antenna and development of skin mimicking gels for continuous glucose monitoring. IEEE Trans. Microw. Theory Tech. **56**(4), 1001–1008 (2008)
7. Kim, J., Rahmat-Samii, Y.: Implanted antennas inside a human body: simulations, designs, and characterizations. IEEE Trans. Microw. Theory Tech. **52**(8), 1934–1943 (2004)
8. Kiourti, A., et al.: A review of implantable patch antennas for biomedical telemetry, challenges and solutions. In: IEEE Trans. Antennas Propag.
9. Medical implant communications service (MICS) federal register. Rules Reg. **64**(240), 69926–69934 (1999)
10. Gabriel, S., et al.: The dielectric properties of biological tissues: II. Measurements in the frequency range 10 Hz to 20 GHz. Phys. Med. Biol. **41**, 2251–2269 (1996)
11. Gabriel, C., Gabriel, S., et al.: The dielectric properties of biological tissues: I. Lit. Surv. Phys. Med. Biol. **41**, 2231–2249 (1996)
12. Tang, Z., et al.: Data transmission from an implantable biotelemetry by load-shift keying using circuit configuration modulator. IEEE Trans. Biomed. Eng. 524–528 (1995)
13. Kiourti, A., Nikita, K.S.: Meandered versus spiral novel miniature PIFAs implanted in the human head, tuning and performance. In: Proceedings of the 2nd ICST® International Conference on Wireless Mobile Communication and Healthcare (MobiHealth 2012). Kos Island, Greece (2011)
14. Yekeh Yazdandoost, K., Sayrafian-Pour, K.: channel model for body area network. In: IEEE P802.15 Working Group for Wireless Personal Area Networks, IEEE P802.15-08-0780-12-0006 (2010)
15. Warty, R., Tofighi, M.R., et al.: Characterization of implantable antennas for intracranial pressure monitoring, reflection by and transmission through a scalp phantom. IEEE Trans. Microw. Theory Tech. **56**(10), 2366–2376 (2008)
16. Zhou, Y., Chen, C.C., Volakis, J.L.: Dual band proximity-fed stacked patch antenna for tri-band GPS applications. IEEE Trans. Antennas Propag. **AP-55**(1), 220–223 (2007)
17. Yu, H., Irby, G.S., Peterson, D.M., et al.: Printed capsule antenna for medication compliance monitoring. Electron. Lett. **43**(22), 41–44 (2007)
18. Soontornpipit, P., Furse, C.M., et al.: Miniaturized biocompatible microstrip antenna using genetic algorithm. In: IEEE Trans. Antennas Propag. **AP-53**(6), 1939–1945 (2005)
19. Kiziltas, G., Psychoudakis, D., Volakis, J.L., et al.: Topology design optimization of dielectric substrates for bandwidth improvement of a patch antenna. IEEE Trans. Antennas Propag. **AP-51**(10), 2732–2743 (2003)
20. Kiourti, A., Nikita, K.S.: Miniature scalp-implantable antennas for telemetry in the MICS and ISM bands, design, safety considerations and link budget analysis. IEEE Trans. Antennas Propag.
21. Yazdandoost, K.Y.: UWB antenna for body implanted applications. In: Proceedings of the 42nd European Microwave Conference

Mining Frequent Itemset Using Quine–McCluskey Algorithm

Kanishka Bajpayee, Surya Kant, Bhaskar Pant, Ankur Chaudhary and Shashi Kumar Sharma

Abstract This paper presents an approach which uses Quine–McCluskey algorithm in order to discover frequent itemsets to generate association rules. In this approach, the given transaction database is converted into a Boolean matrix form to discover frequent itemsets. After generating the Boolean matrix of given database Quine–McCluskey algorithm is applied. Quine–McCluskey algorithm minimizes the given Boolean matrix to generate the frequent itemset pattern. This method requires less number of scans compared to other existing techniques.

Keywords Association rule · Data mining · Frequent itemset

1 Introduction

Data mining [1] can be described as a process which is used to extract useful information from the given large database. Association rule mining [2] is an important part of data mining. Association rule mining finds relation between the different attributes present in the database.

There are two measures, support and confidence, which are used to verify the validity of an association rule. The support can be defined as number of times an item X occurs in the given database.

Kanishka Bajpayee · S.K. Sharma
Graphic Era Hill University, Dehradun, India

Surya Kant (✉)
IIT Roorkee, Roorkee, India
e-mail: suryak111@gmail.com

Bhaskar Pant · Ankur Chaudhary
Graphic Era University, Dehradun, India

M. Pant et al. (eds.), *Proceedings of Fifth International Conference on Soft Computing for Problem Solving*, Advances in Intelligent Systems and Computing 437, DOI 10.1007/978-981-10-0451-3_68

$$\text{Support} = \frac{(\text{Number of transactions which contains } X)}{(\text{Total number of transaction})}$$

Confidence is used to measure the accuracy of the generated association rule. It is defined as number of transactions that contains itemset X and itemset Y together divided by number of transactions that contain only X.

$$\text{Confidence} = \frac{(\text{Number of transactions which contains } X \text{ and } Y)}{(\text{Total number of transaction that contain only } X)}$$

For example, if a database contains 100 records out of which 20 records contains X and 8 of which also contains Y then the association rule $X \geq Y$ will have a support of 0.8 (support = 8/100) and confidence of 40 % (confidence = (8/20) = 40). Both the measures, support and confidence can be defined by the user to eliminate any rules which are not useful to the user.

2 Literature Survey

The first approach that was proposed for association rule mining was Agarwal, Imielinski and Swami (AIS) algorithm. AIS algorithm was proposed by Agarwal et al. [2]. Its main aim was to improve the quality of database and to find the relationship between different items which are present within the database.

The AIS approach had several drawbacks. These drawbacks were overcome by Apriori algorithm which was proposed in [3, 4]. Apriori algorithm is a popular algorithm used for finding association rules. Apriori algorithm has two major drawbacks; one of which is that it scans the database multiple times to produce frequent itemsets. The second major drawback is that it consumes a lot of time and resources.

FP-tree [5] was proposed to overcome the drawbacks of Apriori algorithm. FP-tree algorithm unlike Apriori generates frequent itemset without generating any candidate itemset. Also, FP-tree requires only two scans of database for generating frequent itemset. FP-tree performs better than Apriori as it requires less number of scans compared to Apriori but still there were some disadvantages of FP-tree.

Later K-Apriori approach was proposed [6], which used K-map properties to generate frequent itemsets pattern. This approach reduced numbers of scans of the database to one, thereby increasing the performance of the algorithm in large databases.

Further enhancement was done in this algorithm by applying partition algorithm to generate frequent itemset pattern [7].

Another algorithm was proposed for association rule mining, which was based on K-map and genetic algorithm [8]. In this algorithm, support is calculated using K-map and rules are generated using genetic algorithm.

Efficient and optimized positive–negative association rule mining (EO-ARM) algorithm [9] was proposed. It used a concept similar to K-map to find the positive and negative association rule by scanning the database only once.

The limitation of K-Apriori is that it is efficient for small number of variables but its performance decreases with the increase in number of variables.

3 Proposed Algorithm

In order to remove the drawback of K-Apriori, an algorithm is proposed that uses Quine–McCluskey algorithm for finding frequent itemsets from the database. Quine–McCluskey algorithm is also used to solve Boolean functions but its main advantage is that it can handle more number of variables than K-map algorithm.

3.1 Quine–McCluskey

Quine–McCluskey algorithm [10, 11] provides solution for Boolean function minimization. A fundamental concept used in Quine–McCluskey method of reduction is the concept of a prime implicant. The Quine–McCluskey method is implemented in three stages

- Record the canonical terms in increasing order of number of 1's that appear when the minterm numbers are expressed in binary
- Reduce the number of terms by combining terms, starting with the smallest of 1's
- Select the essential prime implicants, starting with the reduced terms containing the fewest number of literals.

3.2 Algorithm for Generation of Frequent Itemset

This paper proposes an algorithm using Quine–McCluskey algorithm to generate the frequent itemset from the given database.

ALGORITHM:

Input: Transactional database D containing all transactions.
Output: Set of Frequent Itemset
Procedure:
START

1. Scan the database D
2. Represent each transaction in boolean form
3. If,

   An item exist in the current transaction mark 1 below that item

   Else,

   If item is not present in the current transaction mark 0 below that item
4. Arrange the table obtained from above step according to the number of 1's in them, i.e. Group 0 will contain terms with no 1's, then Group 1 will contain terms with one 1, and so on.
5. Set i=0.
6. Pick up each item with index i and i+1 to see if they differ in exactly one position.
7. If, true
8. Write the single term which results from the combination in 1,0,- notation in new column
9. Else,
10. Continue with other pairs unit all the pairs with i and i+1 have been compared.
11. Set i=i+1. Repeat step 4 unit all terms have been covered.
12. Repeat Steps 3,4, and 5 on the new list to form another list
13. Terminate the process when no new lists are formed.
14. After the paring process terminates a Prime Implicant Chart is formed in order to find the final sum-of-product expression.

END

4 Illustrative Example of Proposed Algorithm

The algorithm proposed is applied on the following database which consists of following transactions. The transaction consists of four items A, B, C and D.

The transaction table is then converted in Boolean form. The item which is present in a particular transaction is marked as 1 in that transaction otherwise it is marked as 0. Following is the Boolean representation of the given database (Table 1).

Group the transactions from 1, 2, … *N*, where *N* represents the number of items in each transaction. For example transaction 3 in Table 2 contains only 1 item so it should be placed in group 1 and so on.

Table 1 Transaction table

Transaction no.	Items
1	B, C, D
2	A, D
3	B
4	A, C, D
5	A, B
6	C
7	A, B, C
8	A, B, C, D
9	A, C

Table 2 Truth table of transaction table

Trans. no.	A	B	C	D
1	0	1	1	1
2	1	0	0	1
3	0	1	0	0
4	1	0	1	1
5	1	1	0	0
6	0	0	1	0
7	1	1	0	1
8	1	1	1	1
9	1	0	1	0

In column 1 of Table 3 start with group 1 and search for rows in the next group which differ in only 1 bit position. Place these pairs of rows in next column and mark the rows with a check sign in front of both terms to show that they have been paired and covered.

The terms in each group in column 1 are compared with each other and the pairs generated are placed in column 2. Again the rows are matched to locate rows which differ in only one bit position. The –'s must also match in the pairs of rows. This

Table 3 Comparison table

	Minterm	Column 1	Column 2	Column 3
Group 1	0010 0100	2✓ 4✓	−100 (4, 12) −010 (2, 10)	
Group 2	1001 1010 1100	9✓ 10✓ 12✓	10−1 (9, 11) 101− (10, 11)✓ 1−10 (10, 14)✓ 11-0 (12, 14)	
Group 3	0111 1011 1110	7✓ 11✓ 14✓	−111 (7, 15) 1−11 (11, 15)✓ 111−(14, 15)✓	1−1−(10, 11, 14, 15)
Group 4	1111	15✓		

Table 4 Prime implicant chart

		2	4	7	9	10	11	12	14	15
(10, 11, 14, 15)	1–1–					x	x		x	x
(2, 10)	–010	x				x				
(4, 12)	–100		x					x		
(9, 11)	10–1				x		x			
(12, 14)	11–0							x	x	
(7, 15)	–111			x						x

process continues till no pairs can be generated. In column 3 no more reduction is possible, therefore the procedure gets terminated. The next step is to reduce these prime implicants to the set of essential prime implicants. The reduction to the set of essential prime implicants is carried out by forming a prime implicant chart Table 4. This is done by tabulating the minterms against the prime implicant, where the minterm included in the prime implicant is indicated by an *X*.

If there is only one *X* in a column then there is only one prime implicant covering that minterm and that prime implicant is an essential prime implicant. When a term is included in the final sum of products expression its row is crossed out and the column covering it is also crossed out.

The prime implicants are BCD + AB′D + BC′D′ + B′CD′ + AC.

The items with a "′" on them are eliminated as they are non-frequent items. A "′" over an item represents the absence of that particular item from that frequent itemset. For example, in AB′D, B′ represents the absence of item B when A, D appear together. Also the frequent itemset whose superset is frequent, its subset will also be frequent. For example BCD is frequent then its subset {B}, {C}, {D}, {BC}, {CD}, {BD} will also be frequent.

The final set of frequent items is: A, B, C, D, AD, AC, BC, CD, BD, BCD.

5 Conclusion

The algorithm discussed in this paper attempts to overcome the shortcoming of K-Apriori algorithm. The algorithm proposed in this paper uses Quine–McCluskey algorithm to find the frequent itemsets. The benefit of mining frequent itemset pattern using Quine–McCluskey over K-Apriori is that it can handle more number of variables than k-map approach. Another advantage of this approach is that the transaction table is that it only requires one scan of the database to generate frequents itemsets which increases the efficiency of this algorithm.

References

1. Han, I., Kamber, M.: Data Mining concepts and Techniques, pp. 335–389. M. K. Publishers (2000)
2. Agrawal, R., Imielinski, T., Swami, T.: Mining association rules between sets of items in large databases. In: Proceedings of the ACM SIGMOD Conference on Management of Data, pp. 207–216. Washington, D.C. (1993)
3. Agrawal, R., Srikant, R.: Fast algorithms for mining association rules. In: Proceedings of the 20th VLDB Conference. Santiago, Chile (1994)
4. Srikant, R., Agrawal, R.: Mining Generalized Association Rules, pp. 407–419 (1999)
5. Han, J., Pei, J., Yin, Y.: Mining frequent patterns without candidate generation. In: ACM SIGMOD International Conference on Management of Data, pp. 1–12. ACM Press (2000)
6. Lin, Y.C., Hung, C.M., Huang, Y.M.: Mining ensemble association rules by Karnaugh Map. In: World Congress on Computer Science and Information Engineering, pp. 320–324 (2009)
7. Sharma, N., Singh, A.: K-partition for mining frequent patterns in larger databases. Int. J. Comput. Sci. Eng. (IJCSE). **4**(09), (2012). ISSN:0975-3397
8. Mandrai, P., Baskar, R.: A novel algorithm for optimization of association rule using k-map and genetic algorithm. Int. Conf. Comput. Commun. Netw. Technol. 1–7 (2013)
9. Ravi, C., Khare, N.: EO-ARM: an efficient and optimized k-map based positive-negative association rule mining technique. Int. Conf. Circuits Power Comput. Technol. 1723–1727 (2014)
10. McCluskey, E.J.: Minimization of boolean functions. Bell Syst. Tech. J. **35**(5), 1417–1444 (1956)
11. Quine, W.V.: The problem of simplifying truth functions. Am. Math. Montly **59**(8), 521–31 (1952)

Robotics and Image Processing for Plucking of Fruits

Om Prakash Dubey, Rani Manisha, Kusum Deep and Pankaj Kumar Singh

Abstract This electronic document is a description giving the various components of robotics, basics, and its varieties. It also describes its different forms and usages. Here image processing and its basics are also discussed. Here the main stress is laid on a particular application of robotics. Thus it tries to redefine a robot in such a way that it suits that particular application for plucking fruits and medicinal plants along with its protection issues.

Keywords Actuators · AVR microcontroller · ATmega16 · ATmega32 · Autonomous

1 Introduction

Robotics is a vast field. It is basically divided into three categories:

1. Manual: It consists of a bot (robot) which comprises manual parts such as wheels, actuators, relays, etc. It is also handled manually.
2. Autonomous: It mainly consists of parts like microcontroller board and parts which can be handled autonomously and is used mainly in places like space.

O.P. Dubey (✉)
Department of Mathematics, Bengal College of Engineering and Technology, Durgapur, West Bengal, India
e-mail: omprakashdubeymaths@gmail.com

Rani Manisha
Wipro Limited, Main Office, Gopanpally, Hyderabad, Telangana 501301, India
e-mail: ranimanisha28@gmail.com

K. Deep
Department of Mathematics, IIT Roorkee, Roorkee, Uttarakhand, India
e-mail: kusumfma@iitr.ac.in

P.K. Singh
Department of Mathematics, GGS Engineering & Technical Campus, Bokaro, Jharkhand, India
e-mail: pksingh9431580612@gmail.com

M. Pant et al. (eds.), *Proceedings of Fifth International Conference on Soft Computing for Problem Solving*, Advances in Intelligent Systems and Computing 437, DOI 10.1007/978-981-10-0451-3_69

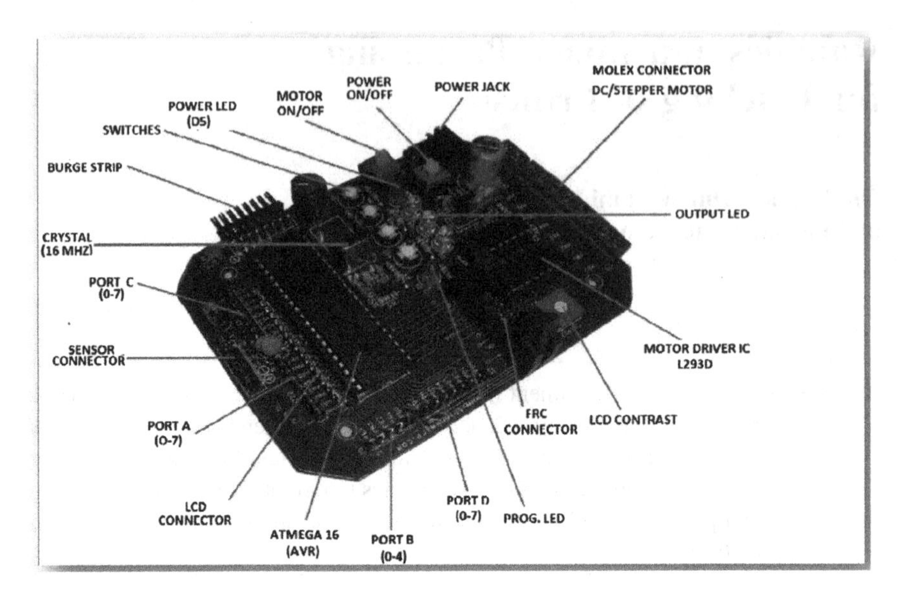

Fig. 1 Microcontroller board

3. Semi-autonomous: It is a combination of manual and autonomous, i.e. it can be handled both manually and autonomously.

Here we will be dealing mainly on the autonomous portion. The simplest autonomous robot consists of the following parts: wheels, actuators, chassis, microcontroller board and sensors as required.

The microcontroller board looks as follows and the following are its parts (Fig. 1).

- Sensors: There are different types of sensors available; some of them are bump sensors, line sensors, light sensors using IR diode and IR detectors, obstacle detectors, temperature sensor, shaft encoder and gyroscope.
- Actuators/Motors: The different types of actuators are DC motor, stepper motor, servo motors and linear actuators.
- Motor Drivers: Motor driver using relay, motor driver using transistor, solid-state drivers, H-bridge and IC drivers.
- AVR Microcontroller: It is mainly an IC named ATmega16 or ATmega32. It is basically a 40 pin IC.

Functions of the various parts are that the sensors sense the environment and give the related feedback, for example, the obstacle sensor senses the obstacle and gives the related output signal. The output signal of the sensor is taken as input to the microcontroller. This input is processed in the microcontroller according to the programme or instructions fed in the microcontroller and then actions are taken accordingly which is given as output through the output port of the microcontroller. For example, if there is an obstacle the bot should move backward. This instruction of moving backward is given by the output port of the microcontroller which is given as input to the motor driver IC (l293d) which then enables the motors to move which are connected to the output port of the motor driver IC. It can provide a maximum of 36 V to the motors [1, 2].

There are different ways in which robots can be designed and controlled; some of them are control robots using mobile phones, control robots using IR, design your own IR communication protocol, interface your TV remote with your robot, design your own IR remote control, control robots using RF, design your own RF communication protocol, control robots using PC, design your own robotic control station, control robots using sound, control robots using internet and control robots using keyboard and mouse.

Thus robotics having a wide range of usages using various methodologies gives a varied application form. One such application of robotics is image sensing and image processing [3, 4].

There are mainly four types of images in the image processing toolbox: (Figs. 2, 3, 4 and 5).

1. Binary images: In binary images the images are stored in binary form, i.e. in 0 and 1, where 0 for white and 1 for black are used.
2. Intensity images: In intensity images the images are black and white. Here, the value is between 0 and 1 representing the shades of black and white.
3. Indexed image: In indexed image, the image contains two arrays, colour map and image matrix. The colour map contains set of values to represent the colour in the image. For each image pixel, the image matrix contains a value that is an index into the colour map
4. RGB images: The only difference here is that these intensity values are stored directly in the image array, not indirectly in a colour map.

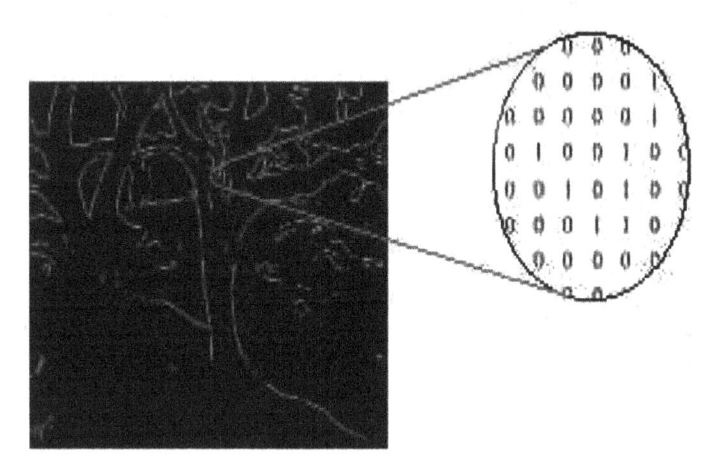

Fig. 2 Binary image

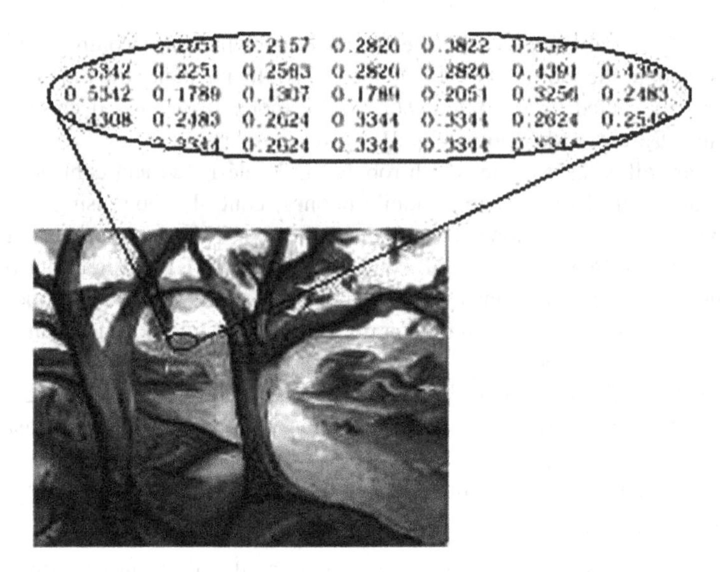

Fig. 3 Intensity image

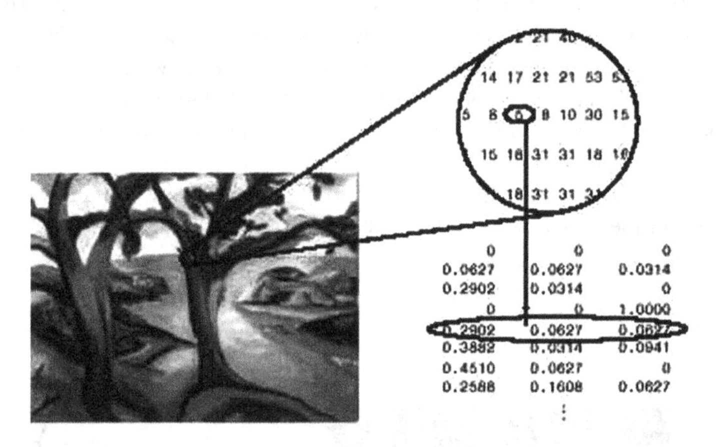

Fig. 4 Indexed image

2 General Commands

The locations in the coordinate system can be represented in two forms—the pixel coordinate system and the spatial coordinate system. The difference between the two is that the pixel coordinate system is discrete, while spatial coordinate system is continuous in nature (Table 1).

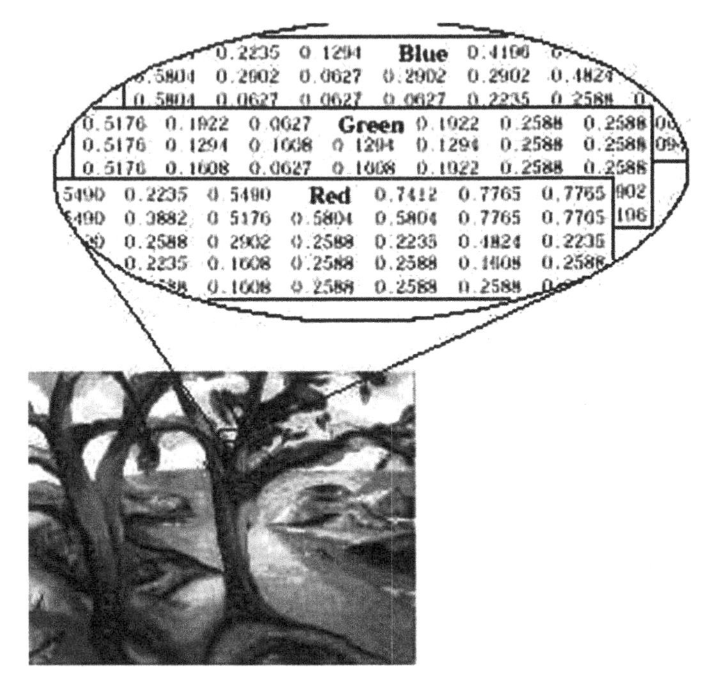

Fig. 5 RGB image

Table 1 Matlab command

Operation	Matlab command
Read an image (Within the parenthesis you type the name of the image file you wish to read. Put the file name within single quotes ' '.)	imread();
Write an image to a file (As the first argument within the parenthesis you type the name of the image you have worked with. As a second argument within the parenthesis you type the name of the file and format that you want to write the image to. Put the file name within single quotes ' '.)	imwrite();
Creates a figure on the screen	Figure;
Displays the matrix g as an image	imshow(g);
Turns on the pixel values in our figure	pixval on;
The command returns the value of the pixel (i, j)	impixel(i, j);
Information about the image	iminfo;
Zoom in (using the left and right mouse button)	zoom on;
Turn off the zoom function	zoom off;

3 Process

Read an image, validate the graphic format (bmp, hdf, jpeg, tiff) and store it in array:

I = *imread('pout.tif'); [X, map] = imread('pout.tif');*
Display an image- *imshow(I)*
Histogram Equalizations - Distribution of intensities
figure,imhist(I)
Equalize Image
I2=histeq(I); figure, imshow(I2)
figure, imhist(I2)

Write the image

Imread(I2,'pout2.png');
imwrite(I2,'pout2.png', 'BitDepth', 4);

Adding Images

I = imread('rice.tif'); J = imread('cameraman.tif'); K = imadd(I,J); imshow(K)
Substracting an image –This gives the background of a scene

rice = imread('rice.tif');
background = imopen(rice, strel('disk', 15));
rice2 = imsubtract(rice, background);

imshow(rice), figure, imshow(rice2);

Multiplying images

I = imread('moon.tif');
J = immultiply(I, 1.2);
imshow(I);

figure, imshow(J)

Dividing Images

I = imread('rice.tif');
background = imopen(I, strel('disk', 15));
Ip = imdivide(I, background);

imshow(Ip)

Resizing images

I = imread('ic.tif');
J = imresize(I, 1.25);
K = imresize(I, [100 150]);
figure, imshow(J)

figure, imshow(K)

Rotating images

I = imread('ic.tif');
J = imrotate(I, 35, 'bilinear');
imshow(I)

figure, imshow(J)

Cropping images

imshow ic.tif

I = imcrop;

4 Causes

With the help of concepts of robotics, image processing and some other features, various devices can be generated to pluck fruits especially ripe fruits of different categories. First, we will discuss the various problems which we face in getting a fresh ripe fruit. The various problems are as follows:

1. The fruits are either broken or plucked before they ripe and then put into artificial ripening agents like calcium carbide. Such agents are really harmful for health, can cause damage to kidney, etc. It is defined as a carcinogen that has potentiality to convert normal human cells into cancer cells.
2. The ripened fruits fall down and get wasted. No one picks it up and uses because it gets smashed up as it falls from great height.
3. The branches of many trees are old and weak; therefore, there is a probability that the branches may break leading to damage the person who climbs the tree.

5 Procedure

Thus we can design a robot which can pluck ripe fruits from the trees, especially the tall ones. This can be done in two ways, i.e. (a) either by controlling the robot through a mobile or a remote or by some other gadget in which manual interference is required or (b) by using such robots which have embedded artificial intelligence in it and need no human interference.

We will be concerning with the second type of robot. For this we require an autonomous robot with image sensing and image processing mechanism. It should also contain pressure sensors to touch and detect whether the fruits have ripened or not. It should contain a container to keep the plucked fruits. The container should be a soft one so that the fruits do not get smashed up when they fall in the container. There should be a mechanism to pluck the fruits; this can be done mainly in two ways: (a) first, by plucking it simply by breaking the fruit stalk. This is mainly done for fruits which are bigger in size like mango, custard apple, etc., and (b) second, it can be done by simply waving the stem or branch containing the fruits. This is mainly done for fruits smaller in size and larger in number, i.e. in bunch like Indian blackberry (jamun), date palm (khajur), etc. [5–9].

One of the most important things for such robots is that it should have the image of the fruits fed into it so that the fruits are plucked properly and easily, especially without human interference or human presence. It has two very important aspects of plucking procedure which are as follows: first, it should be able to identify and locate the fruit easily; and second, the robot should have an access to the fruit in the tree easily as fruits are generally located at high ends in the bushy regions. For this a very efficient image processing system should be developed so that the fruits can be easily located and identified.

6 Design

It is just a theoretical and imaginative form which suggests the different ways of building the robot. Here, we will be discussing the different ways and tactics of making the robot and its different parts according to the necessities of the circumstances.

The external design of the robot can be of two types: (a) One, which has its base on the ground and from there it can move to different parts of the tree and branches; and (b) second, it can have an arial movement like a helicopter or quad copter. The design of the first kind of robot which has its base on the ground can be as follows:

It can have a movable stick which can increase or decrease in size as per the need. It may have 5–6 degrees of freedom and 4–5 hinge points which can move in all directions. Such robots will be useful in places where human beings can reach easily like the plants planted in our locality or anywhere over a plain area, whereas the second one is useful in places where human beings cannot reach or do not have proper accessibility like the mountains, hills, etc.

Following is a prototype which suggests some of the features as to how the arm of the robot should be and how it should work. It should also have the feature of increasing or decreasing in size (Fig. 6).

Till now, we discussed the overall design of the robot but now we shall look into the aspects as to how the plucking mechanism should work. As previously discussed for plucking, we need to either break the fruit stalk or wave the stem containing the bunch of fruits. One of the best mechanisms of plucking can be through hand mechanism which means that plucking procedure is similar to the method in which we pluck fruits through our hands. This can be made possible

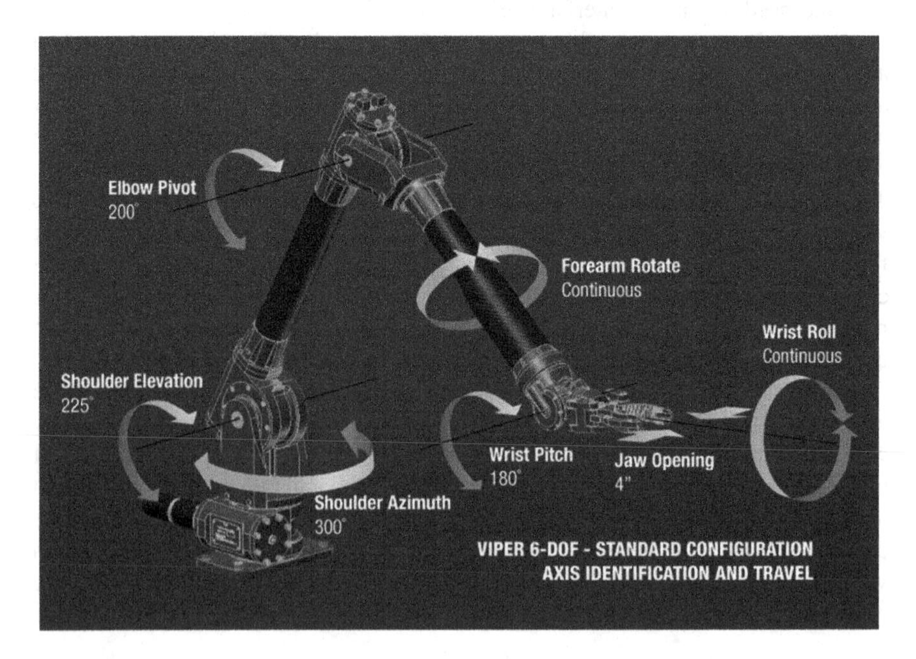

Fig. 6 Robotic arm

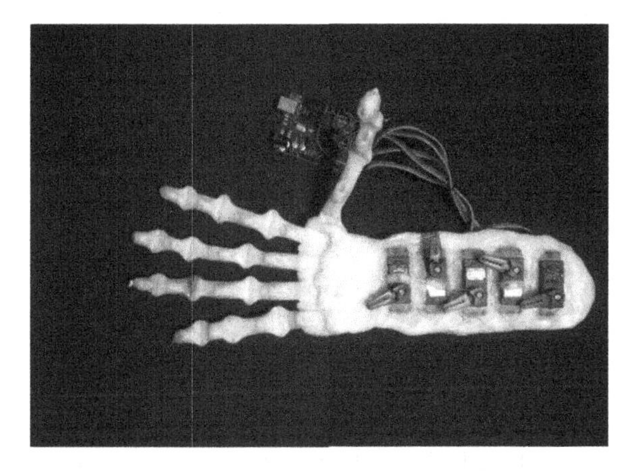

Fig. 7 Robotic hand

through robotic hand or arm controlled through gestures [10]. One can do two things: either each time the robot encounters a fruit, the hand gesture is made such as to pluck the fruit or what should be the hand movement for plucking is fed in the robot beforehand. The second option is the better one because each time the fruit is plucked, one need not change the gestures of his/her hand regularly (Fig. 7).

These robots will not be useful only in plucking fruits but also be useful in getting important medicinal herbs from places like Himalayan ranges where human accessibility is difficult. It can bring a revolution in ayurveda because there are many such plants which are neither available nor can be grown around us in the plains. There can also be a third design of robot which can both move on the ground as well as climb the tree and move along the branches till the end with proper grip. In this kind of robots, the algorithm fed into the microcontroller should be such that once the robot has moved across a branch it should not go to the same branch end repeatedly. It should be light weight so that the branches may not break. It may get heavy with load of the fruits plucked; hence, there should be a mechanism to put the fruits down into a bigger container placed at the ground. For this it would be quite repetitive, slow, and hectic process to bring back the robot frequently at the bottom; therefore, there must be a pulley sort of system to swiftly and carefully drop the fruits at the bottom without damaging it.

7 Other Aspects

The other aspects which ought to be dealt with are the security factors. There are many agents from which the fruits need to be protected. These are especially the fruit thieves. This concept is not that prevalent in foreign countries USA, UK, etc. but is quite prevalent in India. There are many reasons why people do this (Fig. 8):

(1) They have easy accessibility to plants or trees located in small plots or regions.
(2) They may do it just for fun.
(3) Due to sheer need of money.
(4) Due to momentary hunger.

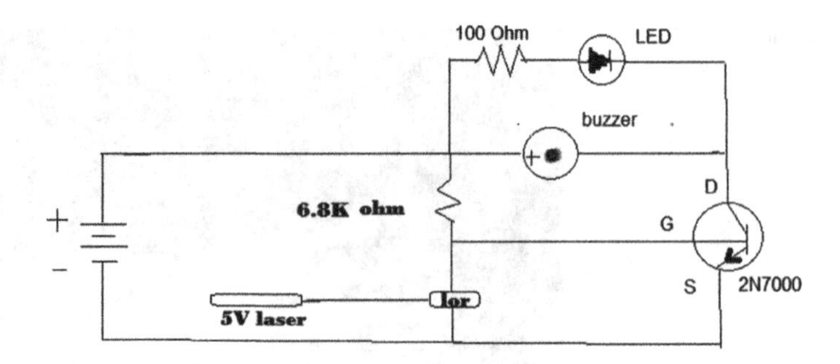

Fig. 8 Laser security alarm

For this an efficient security system must be developed in order to protect the fruits. One can draw a laser security system for this. The materials required for this are laser light, battery, resistor, transistor, LED and buzzer.

This is the miniature form of the laser detector which can be advanced to form a proper system so as to make it possible to detect any unwanted activities. Thus the fruits can be protected from the thieves. A problem which can occur due to this is that the sensor may also detect the flying birds. Therefore, the sensors should be placed at lower levels where generally birds may not come.

8 Conclusion

In this paper, an effort is made to draw the attention towards the basics of robotics and one of the applications of robots using image processing which is not so prevalent, i.e. robotics used for plucking fruits and medicinal plants along with its protection issues, henceforth preventing the damage caused to the fruits because of falling and the destruction caused due to falling down from trees. It also provides proper usage of medicinal plants which till now had been an unexploited resource.

References

1. Gammell, J.D., Srinivasa, S.S., Barfoot, T.D.: Batch informed trees (BIT*): Sampling-based optimal planning via the heuristically guided search of implicit random geometric graphs (2015) [http://arxiv.org/pdf/1405.5848.pdf]
2. Rusk, Natalie, Resnick, Mitchel, Berg, Robbie, Pezalla-Granlund, Margaret: New pathways into robotics: strategies for broadening participation. J. Sci. Educ. Technol. **17**(1), 59–69 (2008)
3. Grover, N., Garg, N.: Image processing using MATLAB. Int. J. Res. **2**(5), 160–161 (2015)
4. Agrawal, A.: Elementary introduction to image processing based robots: IIT, Kanpur (2009). http://students.iitk.ac.in/roboclub/old/data/tutorials/Elementary%20Introduction%20to%20Image%20Processing%20Based%20Robots.pdf
5. Cha, E., Dragan, A.D., Srinivasa, S.S.: Perceived robot capability, (2015). http://www.ri.cmu.edu/pub_files/2015/8/roman2015_137.pdf

6. Robotic hand movement through gestures. http://www.instructables.com/member/acali/
7. http://www.rakeshmondal.info/L293D-Motor-Driver
8. http://www.kanda.com/blog/microcontrollers/avr-microcontrollers/
9. http://robotix.in/rbtx13/tutorials/category/imageprocessing/ip_matlab#222
10. Kaura, H.K., Honrao, V., Patil, S., Shetty, P.: Gesture controlled robot using image processing. In: Int. J. Adv. Res. Artif. Intell. **2**(5), 69–77 (2013)

A Review on Role of Fuzzy Logic in Psychology

Shilpa Srivastava, Millie Pant and Namrata Agarwal

Abstract The process of medical diagnosis, like many other fields, has to pass through various stages of uncertainty, especially in cases where the data is mostly available in linguistic format. Under such circumstances of vague data, application of fuzzy logic concepts can play an important role in extracting approximate information which in turn may help in reaching to a particular diagnosis. This study is devoted to the application of fuzzy logic in the psychological domain. The paper provides a detailed literature review on the use of fuzzy logic rules in analyzing the different aspects of psychological behavior of human beings. Further, it also provides some suggestions to make the system more effective.

Keywords Fuzzy rules · Linguistic · Psychological behavior · Fuzzy logic

1 Introduction

Fuzzy logic has emerged as a powerful tool for solving complex real-life problems where the data is based on approximation rather than being fixed and exact. It can be considered as a powerful tool for decision-making systems and other artificial intelligence applications. In medical field, fuzzy logic concepts play an important role as the diagnosis of an ailment involves several levels of uncertainty and

Shilpa Srivastava (✉)
RKGIT, Ghaziabad, India
e-mail: shri.shilpa03@rediffmail.com

M. Pant
IIT Roorkee, Saharanpur Campus, Roorkee, India
e-mail: millidma@gmail.com

Namrata Agarwal
NIFM, Faridabad, India
e-mail: namrata_agrawal@hotmail.com

M. Pant et al. (eds.), *Proceedings of Fifth International Conference on Soft Computing for Problem Solving*, Advances in Intelligent Systems and Computing 437, DOI 10.1007/978-981-10-0451-3_70

Table 1 Use of Fuzzy in different situations

Heart [11–20]	Brain [21–23]
MRI [24–26]	Diabetes [27–30]
Cancer [31–33]	Anesthesia [34, 35]
Classification of fibromyalgia syndrome [36]	Blood pressure [37]
Pain assessment [38]	Neuro muscular [39]
Color blindness [40]	Obstetrics [41]
Alzhiemer [42]	Muscle fatigue [43]

imprecision [1]. The best and most precise description of disease entities use linguistic terms that are usually imprecise and vague. Generally, an expert physician tends to specify his experience in rather fuzzy terms, which is more natural to him (and to the patient) than trying to cast his knowledge in rigid rules having abrupt premises. For example, words like 'severe pain' are difficult to formalize and to measure.

Consequently, fuzzy logic suffices the need for a tool that can appropriately deal with vague and linguistic data. It introduces partial truth values between true and false. Fuzzy logic plays an important role in medicine [2–8]. Fuzzy set theory has also been used in some medical expert systems [9]. Through fuzzy logic we can develop intelligent systems in diagnosing the diseases. Table 1 cites a few references where fuzzy logic has been used for different ailments.

This study throws light on the application of fuzzy logic in the psychological domain. The paper is structured into four sections. Section 1 illustrates the different concepts of fuzzy logic, Sect. 2 describes the aspects of psychological behavior of human beings, Sect. 3 provides detailed literature review in the chronological order, and Sect. 4 concludes the chapter besides giving some future directions.

In the next section, we describe some basic concepts of psychology.

2 Psychological Behavior

Psychology can be defined as a systematic and scientific study of mental processes, experiences, and behaviors which provide a scientific understanding of human nature, how we think, feel, act, and interact individually and in groups. Study of psychology helps us to develop a basic understanding about human nature and facilitates dealing with a number of personal and social problems. The various goals of studying psychology can be summarized as follows:

- Describing human thought and behavior and its reason.
- Predicting how, why, and when these behaviors will occur again in the future.
- Modifying and improving behaviors to make the lives of individuals better.

The various psychological processes that are involved in human behavior are sensation, attention, perception, learning, memory, and thinking. It relies on scientific methods to investigate questions and arrive at conclusions. It may involve the use of different experiments, case studies, observations, and questionnaire.

3 Literature Review

The concept of vagueness in natural language had been identified in the 1970s but actual interest of psychologists in fuzzy logic has visibly been growing since mid-1980s [10]. Psychologists have known for some time that people use categories with blurred edges and gradations of membership; the field has been slow to take up ideas from fuzzy set theory and fuzzy logic. By the early 80s, several conceptual and theoretical critiques of fuzzy set from the psychological community had appeared. Fuzzy set theory was alleged to suffer from theoretical incoherence or paradoxes, mismatches with human intuition, and a lack of measurement foundations. Despite critiques and controversies in the late 80s, fuzzy set achieved a degree of legitimacy in psychology and fuzzy theories which have come to be accepted as a regular part of the psychological landscape researches.

The following review shows the application of fuzzy logic in the different psychological aspects like emotional adjustment, perception, social relationship, children's lying behavior, motivation, risk behavior, intellectual disability, etc. Table 2 presents the work done in this field in the chronological order. The literature survey shows that fuzzy logic has a remarkable contribution in the psychological domain and the application of fuzzy logic makes the evaluation process robust and consistent. We have tried to cover maximum possible papers for our literature review. However, there is a possibility that some interesting studies might have been skipped.

Table 2 Literature review

S. no.	Objective
1	Presents the vagueness in the natural language by the fuzzy sets. Fuzzy set theory is applied to the phrases like very small, sort of large, etc. [44]
2	Different decision-making and information analysis models are presented in the fuzzy environment [45]
3	Presents tools for applying fuzzy set concepts to the social and behavioral sciences with examples [46]
4	A framework based on fuzzy logic and possibility theory was designed compatible to the psychological explanations having partial and uncertain constraints [47, 48]
5	Discusses the role of fuzzy sets in social sciences [47, 48]
6	Discusses the need of fuzzy measurement in different fields of psychology. The paper also describes a fuzzy graphic rating scale, which can be used in various fields

(continued)

Table 2 (continued)

S. no.	Objective
7	Uses fuzzy set theory to design a matching model and discusses its application for analyzing an asymmetric similarity matrix
8	Investigates the application of fuzzy graphic rating scale for measuring preferences for occupational sex type, prestige, and interests by considering L.S. Gottfredson's circumscription and compromise theory and the concept of social space [49]
9	Discusses a framework designed by combining the concepts of probability and fuzzy set and also throws light on the two major sources of imprecision in human knowledge, linguistic inexactness, and stochastic uncertainty [50]
10	Tests two competing theories of career compromise through a fuzzy graphic rating scale [51]
11	Tests different components of the Theory of Work Adjustment on 170 bank personnel using a fuzzy graphic rating scale to elicit work preferences and job perceptions [52]
12	Uses Fuzzy Cognitive Mapping (FCMs) for providing qualitative information about complex models in social and behavioral science research [53]
13	Presents an analysis of the fuzzy logic model of perception (FLMP) through a measurement-theoretic perspective [54]
14	Introduce a model of truth qualification in fuzzy logic, into the process of the subjective judgment so that the behavior of context can be understood ideally and easily. The study was developed into a practical psychological response model including the concept of randomness [55]
15	Describes a rating system consisting of two parts: one is to estimate the standard of judgment in every second by the use of pairs of data observed in the past; and the other plays a role of reasoning the psychological impression by stimulus using if-then rules based on the method of fuzzy successive categories [56]
16	A model FLAME—Fuzzy Logic Adaptive Model of Emotions—is proposed to produce emotions and to simulate the emotional intelligence process. It uses fuzzy logic in capturing the fuzzy and complex nature of emotions [57]
17	Compares the concept of perception and measurement and describes the relation of perception and fuzzy logic [58]
18	Represents emotions with the logical knowledge. The emotional simulator for an autonomous mobile robot which is able to express various emotions is developed with simplified fuzzy rules [59]
19	Review developments and state of fuzzy set theory and applications in psychology. It surveys the uses of fuzzy sets in the operationalization and measurement of psychological concepts. Also provides a description of psychological theories and frameworks that use fuzzy sets [60]
20	Fuzzy rules have been applied to track the user's interest in web navigation pattern and then analyzing about his change in priorities [61]
21	In this paper, fuzzy dynamic Bayesian network, a modification of the Dynamic Bayesian Network (DBN) is adopted for capturing the intrinsic uncertainty of human common-sense reasoning and decision-making during their communications [62]
22	Uses fuzzy logic to develop a method to simultaneously measure the ratings of the patient's Global Assessment of Functioning (GAF), by both the therapist and patient in psychotherapy [63]

(continued)

Table 2 (continued)

S. no.	Objective
23	Introduces STEM-D, a fuzzy logic model for assessing depression in OSA (obstructive Sleep Apnea) patients by considering the multifactorial nature of depression [64]
24	The authors have presented a novel method for continuously modeling emotion using physiological data [65]
25	Issues related to recognition of music emotions based on subjective human emotions and acoustic music signal features have been discussed. Later, fuzzy inference system is used to present an intelligent music emotion recognition system [66]
26	Assessment of operator functional state (OFS) based on a collection of psychophysiological and performance measures is presented with the help of two models, namely ANFIS (adaptive network-based fuzzy inference system) and GA (genetic algorithm)based on Mamdami fuzzy model [67]
27	Investigates the application of fuzzy logic and gas in psychopathological field [68]
28	A model in the fuzzy environment has been designed for describing the individual nature of mind. Later the model is extended for groups, couples, and social relationship dealing with both perfect and approximate reasoning [69]
29	Presents an emotion model based on fuzzy inference [70]
30	Propose a new analytical method based on fuzzy logic for psychological research. It also compares and discusses fuzzy and traditional approaches [10]
31	Presents analysis of the fuzzy rules based systems that make the transitions between emotional states [71]
32	Psychological theory of consciousness have been studied in this paper and on the basis of the Maslow's Hierarchy of Needs theory fuzzy comprehensive evaluation-based artificial consciousness model is constructed [72]
33	It shows that fuzzy logic can improve the research of consumer ethnocentrism and can illuminate uncovered sides in the area of consumer ethnocentrism [73]
34	Applies fuzzy logic for designing a stress detection system [74]
35	Discusses fuzzy logic approach for examining consumer ethnocentrism level for Jordanians based on their socio-psychological variables [75]
36	Shows the role of fuzzy logic in the linguistic description of emotions, the description of a cognitive rule base and in the mapping between the core affect level and the cognitive level [76]
37	Discusses a new software technique based on fuzzy logic theory for determining emotional tension on humans. The membership functions and the rules were designed for the psychological tests [77]
38	The study applies fuzzy logic theory in exploring the children's lying behavior. The authors suggest that fuzzy logic is an improved method for detecting the lying behavior [78]
39	The authors suggest the use of fuzzy numbers for accessing the behavior of customers who want to purchase the technological goods like mobiles or smart phone in the retail outlets in the provinces of Pescara and Chieti [79]
40	A fuzzy logic approach for bilateral negotiation (winwin solutions for both the parties) has been suggested in this paper. The suggested approach allows agents to improve the negotiation process [80]

(continued)

Table 2 (continued)

S. no.	Objective
41	A fuzzy emotional behavior model is proposed to assess the behaviors of mine workers. According to personalized emotion states, internal motivation needs and behavior selection thresholds the proposed emotion model can generate more believable behaviors for virtual miners [81]
42	The paper introduces a framework for extracting the emotions and the sentiments expressed in the textual data. The emotions are based on the Minsky's conception of emotions, which consists of four affective dimensions, each one with six levels of activations. Sentiments and emotions are modeled as fuzzy sets and the intensity of the emotions has been tuned by fuzzy modifiers, which act on the linguistic patterns recognized in the sentences [82]
43	The authors have presented a fuzzy methodology approach in designing a computational model of the elicitation and the dynamics of affective states in autonomous agents [83]
44	In this paper, the classroom behaviors of college students have been modeled through fuzzy logic approach. The proposed fuzzy logic system helps educators quantitatively analyze and understand students' conformity behavior in classroom [84]
45	Application of fuzzy logic to map the a fuzzy logic model was established to map the psychophysiological signals to a set of key emotions, namely, frustration, satisfaction, engagement, and challenge of engineers performing CAD activities [85]
46	Throws light on use of fuzzy expert systems for the assessment of intellectual disability. A detailed review of the psychological assessment is presented by means of theoretical studies and practical experiments with data collected from patients affected by different levels of intellectual disability [86]
47	In this paper, Z-number-based fuzzy approach is applied for modeling the effect of Pilate exercises on motivation, attention, anxiety, and educational achievement [87]
48	This paper is based on the well-known appraisal theory of emotions and targets five primary emotions. Simultaneous emotion elicitation is also supported by the model. The design of model is using the concept of linguistic variables and if-then rules [88]

4 Conclusion and Suggestions

Fuzzy logic provides a simple and effective methodology for diagnosing a wide range of diseases. In this study, we provide a detailed literature review on the application of fuzzy logic in the psychological domain in the chronological order. We have analyzed that the fuzzy logic has been applied to the different subfields of psychology like perception, emotional, motivational, anxiety, frustration, stress, etc. Although the acceptance of fuzzy logic in the psychological field has been accepted, we still have to go a long way for its wider acceptability in developing nations. Followings are the few suggestions which can make the system more effective:

1. Exposure to computer/Internet technology: Internet technology can provide a diverse array of online resources for elder, disabled, etc. Internet awareness can help them to maintain social connections and online counseling can also be done if required.

2. User-friendly mobile applications: Customized packages focusing on the different psychological problems related to older generation/disabled/pregnant women/adolescents.
3. Audio video interactive programs at school level for solving the adolescence problems: According to WHO, internet and mobile communications have significant potential in providing health services to the school at an affordable cost. They can, for example, provide confidential and anonymous interactions, easy access 24 h a day, and in some cases should also provide personalized interaction http://apps.who.int/adolescent/second-decade/section6/page4/dchool-health-E-health.html

References

1. Madkour, M.A., Roushdy, M.: Methodology for medical diagnosis based on fuzzy logic. Egypt. Comput. Sci. J. **26**(1), 1–9 (2004)
2. Abbod, M.F., von Keyserlingk, D.G., Linkens, D.A., Mahfouf, M.: Survey of utilization of fuzzy technology in medicine and healthcare. Fuzzy Sets Syst. **120**(2), 331–349 (2001)
3. Barro, S., Marín, R. (eds.): Fuzzy Logic in Medicine, vol. 83. Springer Science & Business Media (2001)
4. Boegl, K., Adlassnig, K.P., Hayashi, Y., Rothenfluh, T.E., Leitich, H.: Knowledge acquisition in the fuzzy knowledge representation framework of a medical consultation system. Artif. Intell. Med. **30**(1), 1–26 (2004)
5. Mahfouf, M., Abbod, M.F., Linkens, D.A.: A survey of fuzzy logic monitoring and control utilisation in medicine. Artif. Intell. Med. **21**(1–3), 27–42 (2001)
6. Mordeson, J.N., Malik, D.S., Cheng, S.-C.: Fuzzy Mathematics in Medicine. Physica, Heidelberg, Germany (2000)
7. Steimann, F.: On the use and usefulness of fuzzy sets in medical AI. Artif. Intell. Med. **21**(1–3), 131–137 (2001)
8. Szczepaniak, P.S., Lisoba, P.J.G., Kacprzyk, J.: Fuzzy Systems in Medicine. Physica, Heidelberg, Germany (2000)
9. Phuong, N.H., Kreinovich, V.: Fuzzy logic and its applications in medicine. Int. J. Med. Inf. **62**(2), 165–173 (2001)
10. Kushwaha, G.S., Kumar, S.: Role of the fuzzy system in psychological research. Eur. J. Psychol. **2**, 123–134 (2009)
11. Malhotra, V.K., Kaur, H., Alam, M.A.: A spectrum of fuzzy clustering algorithm and its applications. In: International Conference on Machine Intelligence and Research Advancement (ICMIRA), pp. 599–603. Katra, 21–23 Dec 2013
12. Narasimhan, B., Malathi, A.: A fuzzy logic system with attribute ranking technique for risk-level classification of CAHD in female diabetic patients. In: International Conference on Intelligent Computing Applications (ICICA), pp. 179–183. Coimbatore, 6–7 March 2014
13. Arief, Z., Sato, T., Okada, T., Kuhara, S., Kanao, S., Togashi, K., Minato, K.: Radiologist model for cardiac rest period determination based on fuzzy rule. In: Annual International Conference of the IEEE Engineering in Medicine and Biology Society (EMBC), pp. 4092–4095. Buenos Aires. Aug 31 2010–Sept 4 2010
14. Gouveia, S., Bras, S.: Exploring the use of Fuzzy Logic models to describe the relation between SBP and RR values. In: Annual International Conference of the IEEE Engineering in Medicine and Biology Society (EMBC), pp. 2827–2830. San Diego, CA, Aug 28 2012–Sept 1 2012

15. Abbasi, H., Unsworth, C.P., Gunn, A.J., Bennet, L.: Superiority of high frequency hypoxic ischemic EEG signals of fetal sheep for sharp wave detection using wavelet-type 2 fuzzy classifiers. In: 36th Annual International Conference of the IEEE Engineering in Medicine and Biology Society (EMBC), pp. 1893–1896. Chicago, IL, 26–30 Aug 2014
16. Abbasi, H., Unsworth, C.P., McKenzie, A.C., Gunn, A.J., Bennet, L.: Using type-2 fuzzy logic systems for spike detection in the hypoxic ischemic EEG of the preterm fetal sheep. In: 36th Annual International Conference of the IEEE Engineering in Medicine and Biology Society (EMBC), pp. 938–941. Chicago, IL, 26–30 Aug 2014
17. Orrego, D.A., Becerra, M.A., Delgado-Trejos, E.: Dimensionality reduction based on fuzzy rough sets oriented to ischemia detection. In: Annual International Conference of the IEEE Engineering in Medicine and Biology Society (EMBC), pp. 5282–5285. San Diego, CA, Aug 28 2012–Sept 1 2012
18. Tsipouras, M.G., Karvounis, E.C., Tzallas, A.T., Goletsis, Y., Fotiadis, D.I., Adamopoulos, S., Trivella, M.G.: Automated knowledge-based fuzzy models generation for weaning of patients receiving Ventricular Assist Device (VAD) therapy. In: Annual International Conference of the IEEE Engineering in Medicine and Biology Society (EMBC), pp. 2206–2209. San Diego, CA, Aug 28 2012–Sept 1 2012
19. Shamsi, H., Ozbek, I.Y.: Heart sound localization in chest sound using temporal fuzzy c-means classification. In: Annual International Conference of the IEEE Engineering in Medicine and Biology Society (EMBC), pp. 5286–5289. San Diego, CA, Aug 28 2012–Sept 1 2012
20. Becerra, M.A., Orrego, D.A., Delgado-Trejos, E.: Adaptive neuro-fuzzy inference system for acoustic analysis of 4-channel phonocardiograms using empirical mode decomposition. In: 35th Annual International Conference of the IEEE Engineering in Medicine and Biology Society (EMBC), pp. 969–972. Osaka, 3–7 July 2013
21. Rabbi, A.F., Aarabi, A., Fazel-Rezai, R.: Fuzzy rule-based seizure prediction based on correlation dimension changes in intracranial EEG. In: Annual International Conference of the IEEE Engineering in Medicine and Biology Society (EMBC), pp. 3301–3304. Buenos Aires, Aug 31 2010–Sept 4 2010
22. Liu, R., Xue, K., Wang, Y.X., Yang, L.: A fuzzy-based shared controller for brain-actuated simulated robotic system. In: Annual International Conference of the IEEE Engineering in Medicine and Biology Society, EMBC, pp. 7384–7387. Boston, MA, Aug 30 2011–Sept 3 2011
23. Aymerich, F.X., Sobrevilla, P., Montseny, E., Rovira, A.: Fuzzy approach toward reducing false positives in the detection of small multiple sclerosis lesions in magnetic resonance. In: Annual International Conference of the IEEE Engineering in Medicine and Biology Society, EMBC, pp. 5694–5697. Boston, MA, Aug 30 2011–Sept 3 2011
24. Gambino, O., Daidone, E., Sciortino, M., Pirrone, R., Ardizzone, E.: Automatic skull stripping in MRI based on morphological filters and fuzzy c-means segmentation. In: Annual International Conference of the IEEE Engineering in Medicine and Biology Society, EMBC, pp. 5040–5043. Boston, MA, Aug 30 2011–Sept 3 2011
25. Peng, Ke., Martel, S.: Preliminary design of a SIMO fuzzy controller for steering micro particles inside blood vessels by using a magnetic resonance imaging system. In: Annual International Conference of the IEEE Engineering in Medicine and Biology Society, EMBC, pp. 920–923. Boston, MA, Aug 30 2011–Sept 3 2011
26. Pham, T.D.: Australia Brain lesion detection in MRI with fuzzy and geostatistical models. In: Annual International Conference of the IEEE Engineering in Medicine and Biology Society (EMBC), pp. 3150–3153. Buenos Aires, Aug 31 2010–Sept 4 2010
27. Narasimhan, B., Malathi, A.: Fuzzy logic system for risk-level classification of diabetic nephropathy. In: International Conference on Green Computing Communication and Electrical Engineering (ICGCCEE), pp. 1–4. Coimbatore, 6–8 March 2014
28. San, P.P., Ling, S.H., Nguyen, H.T.: Intelligent detection of hypoglycemic episodes in children with type 1 diabetes using adaptive neural-fuzzy inference system. In: Annual International Conference of the IEEE Engineering in Medicine and Biology Society (EMBC), pp. 6325–6328. San Diego, CA, Aug 28 2012–Sept 1 2012

29. Ranamuka, N.G., Meegama, R.G.N.: Detection of hard exudates from diabetic retinopathy images using fuzzy logic. Image Processing, IET **7**(2), 121–130 (2013)
30. Ling, S.H., Nuryani, N., Nguyen, H.T.: Evolved fuzzy reasoning model for hypoglycaemic detection. In: Annual International Conference of the IEEE Engineering in Medicine and Biology Society (EMBC), pp. 4662–4665. Buenos Aires, Aug 31 2010–Sept 4 2010
31. Zhang, M., Adamu, B., Lin, C., Yang, P.: Gene expression analysis with integrated fuzzy C-means and pathway analysis. In: Annual International Conference of the IEEE Engineering in Medicine and Biology Society, pp. 936–939. Boston, MA, EMBC, Aug 30 2011–Sept 3 2011
32. Karemore, G., Mullick, J.B., Sujatha, R., Nielsen, M., Santhosh, C.: Classification of protein profiles using fuzzy clustering techniques: an application in early diagnosis of oral, cervical and ovarian cancer. In: Annual International Conference of the IEEE Engineering in Medicine and Biology Society (EMBC), pp. 6361–6364. Buenos Aires, Aug 31 2010–Sept 4 2010
33. Schaefer, G., Nakashima, T.: Hybrid cost-sensitive fuzzy classification for breast cancer diagnosis. In: Annual International Conference of the IEEE Engineering in, Medicine and Biology Society (EMBC), pp. 6170–6173. Buenos Aires, Aug 31 2010–Sept 4 2010
34. Pawade, D.Y., Diwase, T.S., Pawade, T.R.: Designing and implementation of fuzzy logic based automatic system to estimate dose of anesthesia. In: Confluence 2013: The Next Generation Information Technology Summit (4th International Conference), pp. 95–102. Noida, 26–27 Sept 2013
35. Mirza, M., GholamHosseini, H., Harrison, M.J.: A fuzzy logic-based system for anesthesia monitoring. In: Annual International Conference of the IEEE Engineering in Medicine and Biology Society (EMBC), pp. 3974–3977. Buenos Aires, Aug 31 2010–Sept 4 2010
36. Arslan, E., Yildiz, S., Köklükaya, E., Albayrak, Y.: Classification of fibromyalgia syndrome by using fuzzy logic method. In: 15th National Biomedical Engineering Meeting (BIYOMUT), pp. 1–5. Antalya, 21–24 April 2010
37. Honka, A.M., van Gils, M.J., Parkka, J.: A personalized approach for predicting the effect of aerobic exercise on blood pressure using a fuzzy inference system. In; Annual International Conference of the IEEE Engineering in Medicine and Biology Society, EMBC, pp. 8299–302. Boston, MA. Aug 30 2011–Sept 3 2011
38. Araujo, E., Miyahira, S.A.: Tridimensional fuzzy pain assessment. In: IEEE International Conference onFuzzy Systems (FUZZ), pp. 1634–1639. Taipei, 27–30 June 2011
39. Roshani, A., Erfanian, A.: Fuzzy logic control of ankle movement using multi-electrode intraspinal microstimulation. In: 35th Annual International Conference of the IEEE Engineering in Medicine and Biology Society (EMBC), pp. 5642–5645. Osaka, 3–7 July 2013
40. Leec, J.: Fuzzy-based simulation of real color blindness. In: Annual International Conference of the IEEE Engineering in Medicine and Biology Society (EMBC), pp.6607–6610. Buenos Aires, Aug 31 2010–Sept 4 2010
41. Stylios, C.S., Georgopoulos, V.C.: Fuzzy cognitive maps for medical decision support—a paradigm from obstetrics. In: Annual International Conference of the IEEE Engineering in Medicine and Biology Society (EMBC), pp. 1174–1177. Buenos Aires, Aug 31 2010–Sept 4 2010
42. Lee, C.S., Lam, C.P., Masek, M.: Rough-fuzzy hybrid approach for identification of bio-markers and classification on Alzheimer's disease data. In: IEEE 11th International Conference on Bioinformatics and Bioengineering (BIBE), pp. 84–91. Taichung, 24–26 Oct 2011
43. Lalitharatne, T.D., Hayashi, Y., Teramoto, K., Kiguchi, K.: Compensation of the effects of muscle fatigue on EMG-based control using fuzzy rules based scheme. In: 35th Annual International Conference of the IEEE Engineering in Medicine and Biology Society (EMBC), pp. 6949–6952. Osaka, 3–7 July 2013
44. Hersh, H.M., Caramazza, A.: A fuzzy set approach to modifiers and vagueness in natural language. J. Exp. Psychol. Gen. **105**(3), 254–276 (1976)

45. Alexeyev, A.V., Borisov, A.N., Krumberg, O.A., Merkuryeva, G.V., Popov, V.A., Slyadz, N. N.: A linguistic approach to decision-making problems. Fuzzy Sets Syst. **22**(1–2), 25–41 (1987)
46. Smithson, Michael: Applications of fuzzy set concepts to behavioral sciences. Math. Soc. Sci. **2**(3), 257–274 (1982)
47. Smithson, M.: Possibility theory, fuzzylogic, and psychological explanation. Adv. Psychol. **56**, 1–50 (1988)
48. Smithson, M.: Fuzzy set theory and the social sciences: the scope for applications. Fuzzy Sets Syst. **26**(1), 1–21 (1988)
49. Hesketh, B., Pryor, R., Gleitzman, M., Hesketh, T.: Practical applications and psychometric evaluation of a computerised fuzzy graphic rating scale. Adv. Psychol. **56**, 425–454
50. Zwick, R., Walisten, T.S.: Combining stochastic uncertainty and linguistic inexactness: theory and experimental evaluation of four fuzzy probability models. Int. J. Man Mach. Stud. **30**, 69–111 (1989)
51. Hesketh, B., Elmslie, S., Kaldor, W.: Career compromise: an alternative account to Gottfredson's theory. J. Couns. Psychol. **37**(1), 49–56 (1990)
52. Hesketh, B., Mclachlan, K., Gardner, D.: Work adjustment theory: an empirical test using a fuzzy rating scale. J. Vocat. Behav. **4**, 318–337 (1992)
53. Craiger, P., Coovert, M.D.: Modelling dynamic social and psychological processes with fuzzy cognitive maps. In: Proceedings of the Third IEEE Conference on Fuzzy Systems, IEEE World Congress on Computational Intelligence, vol. 3, pp. 1873–1877. Orlando, FL, 26–29 Jun 1994
54. Crowther, C.S., Batchelder, W.H., Hu, X.: A measurement-theoretic analysis of the fuzzy logic model of perception. Psychol. Rev. **102**(2), 396–408 (1995)
55. Kato, Y., Yamaguchi, S., Oimatsu, K.: A study of context effects based on fuzzy logic in the case of psychological impression caused by noise stimulus. In: Proceedings of International Joint Conference of the Fourth IEEE International Conference on Fuzzy Systems and the Second International Fuzzy Engineering Symposium on Fuzzy Systems 1995, vol. 2, pp. 569–576, 20–24 Mar 1995
56. Kato, Y., Yamaguchi, S., Oimatsu, K.: A proposal for a dynamic rating system based on the method of fuzzy successive categories in the case of psychological impression caused by noise stimulus. In: Proceedings of the Sixth IEEE International Conference, vol. 3, pp. 1607–1613, 1–5 Jul 1997
57. El-Nasr, M.S., Yen, J.: Agents, emotional intelligence and fuzzy logic. In: Conference of the North American Fuzzy Information Processing Society-NAFIPS, pp. 301–305. Pensacola Beach, FL, 20–21 Aug 1998
58. Lotfi, A., Zadeh, A.: New direction in fuzzy logic-toward a computational theory of perceptions. In: 18th International Conference of the North American Fuzzy Information Processing Society, NAFIPS, pp. 1–4. New York, NY, Jul 1999
59. Maeda, Y.: Fuzzy rule expression for emotional generation model based on subsumption architecture. In: 18th International Conference of the North American Fuzzy Information Processing Society NAFIPS, pp. 781–785. New York, NY, July 1999
60. Smithson, M., Oden, G.C.: Fuzzy set theory and applications in psychology. Practical Applications of Fuzzy Technologies, Part IV, pp. 557–585
61. Agarwal S., Agarwal, P.: A fuzzy logic approach to search results' personalization by tracking user's web navigation pattern and psychology. In: Proceedings of the 17th IEEE International Conference on Tools with Artificial Intelligence (ICTAI'05), pp. 318–325. Hong Kong, 14–16 Nov 2005
62. Hua, Z., Rui, L., Jizhou, S.: An emotional model for non verbal communication based on fuzzy dynamic Bayesian network. In: IEEE Canadian Conference on Electrical and Computer Engineering 2006, pp. 1534–1537. Ottawa, May 2006
63. Sripada, B., Jobe, T.H.: Fuzzy measurements, consensual and empathic validation of the active observations of multiple observers of the same psychotherapeutic event. In: Annual meeting of

the North American Fuzzy Information Processing Society, NAFIPS 2006, pp. 576–585. Montreal, Que, 3–6 June 2006
64. McBurnie, K., Kwiatkowska, M., Matthews, L., D'Angiulli, A.: A multi-factor model for the assessment of depression associated with obstructive sleep apnea: a fuzzy logic approach. In: Fuzzy Information Processing Society, 2007. NAFIPS'07. Annual Meeting of the North American. IEEE, 2007, pp. 301–306
65. Mandryk, R.L., Atkins, M.S.: A fuzzy physiological approach for continuously modeling emotion during interaction with play technologies. Int. J. Hum. Comput. Stud. **65**, 329–347 (2007)
66. Jun, S., Rho, S., Han, B., Hwang, E.: A fuzzy inference-based music emotion recognition system. In: 5th International Conference on Visual Information Engineering, VIE 2008, pp. 673–677. Xian China, July 29 2008–Aug 1 2008
67. Zhang, J.H., Wang, X.Y., Mahfouf, M., Linkens, D.A.: Fuzzy logic based identification of operator functional states using multiple physiological and performance measures. In: International Conference on BioMedical Engineering and Informatics, pp. 570–574. Sanya, 27–30 May 2008
68. Di Nuovo, A.G., Catania, V., Di Nuovo, S., Buono, S.: Psychology with soft computing: an integrated approach and its applications. Appl. Soft Comput. **8**, 829–837 (2008)
69. Araujo, E.: Social relationship explained by fuzzy logic. In: IEEE International Conference on Fuzzy Systems FUZZ-IEEE 2008, 2129–2134. Hong Kong, 1–6 June 2008
70. Liu, L., He, S.H., Xiong, W.: A fuzzy logic based emotion model for virtual human. In: International Conference on Cyberworlds 2008, 284–288. Hangzhou, 22–24 Sept 2008
71. Eisman, E.M., López, V., Castro, J.L.: Controlling the emotional state of an embodied conversational agent with a dynamic probabilistic fuzzy rules based system. Expert Syst. Appl. **36**, 9698–9708 (2009)
72. Zhu, C., Wang, Z.: Fuzzy comprehensive evaluation-based artificial consciousness model. In: Proceedings of the 8th World Congress on Intelligent Control and Automation, pp. 1594–1598. Jinan, China, July 7–9 2010
73. Ganideh, S.F.A., El Refae, G.: Socio-psychological variables as antecedents to consumer ethnocentrism: a fuzzy logic based analysis study. In: Annual Meeting of the North American Fuzzy Information Processing Society (NAFIPS), 2010, pp. 1–6. Toronto, ON, 12–14 July 2010
74. de Santos Sierra, A., Ávila, C.S., Casanova, J.G., Pozo, G.B.D.: A stress-detection system based on physiological signals and fuzzy logic. IEEE Trans. Ind. Electr. **58**(10), 4857–4865 (2011)
75. Ganideh, S.F.A., El Refae, G.A., Aljanaideh, M.: Can fuzzy logic predict consumer ethnocentric tendencies? an empirical analysis in Jordan. In: Annual Meeting of the North American Fuzzy Information Processing Society (NAFIPS), 2011, pp. 1–5. El Paso, TX, 18–20 March 2011
76. van der Heide, A., Sánchez, D., Trivino, G.: Computational models of affect and fuzzy logic. In: Proceedings of the 7th Conference of the European Society for Fuzzy Logic and Technology Published by Atlantis Press 2011, pp. 620–627
77. Al-Kasasbeh, R.T.: Biotechnical measurement and software system controlled features for determining the level of psycho-emotional tension on man–machine systems by fuzzy measures. Adv. Eng. Softw. **45**, 137–143 (2012)
78. Chen, C.C., Chang, D.F., Wu, B.: Analyzing children's lying behaviors with fuzzy logics. Int. J. Innovative Manage. Inf. Prod. ISME **3**(3), (2012). ISSN 2185-5455
79. Maturo, A., Rosiello, M.G.: Psychological and social motivations to the purchase of technological goods: fuzzy mathematical models of interpretation. Procedia-Soc. Behav. Sci. **84**(9), 1845–1849 (2013)
80. Dalel, K.: Fuzzy psychological behavior for computational bilateral negotiation. In: International Conference on Computer Applications Technology (ICCAT), pp. 1–5. Sousse, 20–22 Jan 2013

81. Cai, L., Yang, Z., Yang, S.X., Qu, H.: Modelling and simulating of risk behaviours in virtual environments based on multi-agent and fuzzy logic. Int. J. Adv. Rob. Syst. **10**(387), 1–14 (2013)
82. Loia, V., Senatore, S.: A fuzzy-oriented sentic analysis to capture the human emotion in Web-based content. Knowl. Based Syst. **58**, 75–85 (2014)
83. Schneider, M., Adamy, J.: Towards modelling affect and emotions in autonomous agents with recurrent fuzzy systems. In: IEEE International Conference on Systems, Man, and Cybernetics, pp. 31–38. Oct 5–8 2014, San Diego, CA, USA
84. Wang, G.P., Chen, S.Y., Yang, X., Liu, J.: Modeling and analyzing of conformity behavior: a fuzzy logic approach. Optik **125**, 6594–6598 (2014)
85. Liu, Y., Ritchie, J.M., Lim, T., Kosmadoudi, Z., Sivanathan, A., Sung, R.C.W.: A fuzzy psycho-physiological approach to enable the understanding of an engineer's affect status during CAD activities. Comput. Aided Des. **54**, 19–38 (2014)
86. Di Nuovo, A., Di Nuovo, S., Buono, S., Cutello, V.: Benefits of fuzzy logic in the assessment of intellectual disability. In: IEEE International Conference on Fuzzy Systems (FUZZ-IEEE), pp. 1843–1850. Beijing, 6–11 July 2014
87. Aliev, R., Memmedova, K.: Application of-number based modeling in psychological research. Comput. Intell. Neurosci. Article ID 760403 (in press)
88. Jain, S., Asawa, K.: EmET: emotion elicitation and emotion transition model. In: Proceedings of Second International Conference INDIA 2015, Information Systems Design and Intelligent Applications, vol. 1, pp. 209–217

Modified Single Array Selection Operation for DE Algorithm

Pravesh Kumar and Millie Pant

Abstract In this study, a modified selection operation is proposed for differential evaluation (DE) algorithm. The proposed selection strategy called information utilization (IU) strategy and the proposed DE variant called IUDE reuse redundant trial vectors embedded with single array selection strategy. The proposed selection strategy is implemented on DERL and MRLDE, the enhanced DE variants and the corresponding algorithms are termed IU-DERL and IU-MRLDE. Six traditional functions are taken for experiments. Results confirm that the proposed selection strategy is helpful in amplifying the convergence speed.

Keywords Optimization · Differential evolution · Selection · Single array selection

1 Introduction

Differential evolution (DE) algorithm proposed by Storn and Price [1] is a stochastic, vigorous, and direct search-based algorithm for solving global optimization problems. According to frequently reported experimental studies, DE has imposed its robustness on other evolutionary algorithms (EAs) in terms of convergence speed over numerous benchmark and real-world problems [2]. Some of the primary and salient features of DE include small code, simple to employ and requires few control parameters. It is capable of handling non-differentiable, nonlinear, and multi-modal objective functions and has been successfully demonstrated to a wide range of real-life problems such as engineering design, chemical engineering, mechanical engineering, power engineering, pattern recognition, image processing, and so on [2–8].

Pravesh Kumar (✉)
Amity University, Haryana, India
e-mail: praveshtomariitr@gmail.com

M. Pant
IIT Roorkee, Roorkee, India
e-mail: millidma@gmail.com

M. Pant et al. (eds.), *Proceedings of Fifth International Conference on Soft Computing for Problem Solving*, Advances in Intelligent Systems and Computing 437, DOI 10.1007/978-981-10-0451-3_71

In order to get the better performance of DE, a lot of research has been carried out in the past few years. Some interesting variants of DE include trigonometric mutation (TDE) [9], fuzzy adaptive DE (FADE) [10], DE with random localization (DERL) [11], DE with self-adaptive control parameter (jDE) [12], DE with simplex crossover local search (DEahcSPX) [13], opposition-based DE (ODE) [14], self-adaptive DE (SaDE) [15], adaptive DE with optional external archive (JADE) [16], cauchy mutation DE (CDE) [17], MRLDE [18], dynamic parameter selection-based DE [19], cultivated DE [20], and so on. A comprehensive literature survey of DE variants can be found in [21–23].

In this paper, we have proposed a modified single array selection operation for DE and call it information utilization (IU) selection. The corresponding variant is called IUDE. Furthermore, proposed selection strategy is also implemented on two other DE variants, DERL [11] and MRLDE [18]. Both are improved mutation operation-based DE variants and the corresponding variants are called as IU-DERL and IU-MRLDE respectively. The significance of new selection operation is discussed later in the paper.

The rest of the paper is structured as follows; In Sect. 2, introduction of basic DE is given. Proposed modification in DE selection is described in Sect. 3. Experimental settings and numerical results are discussed in Sect. 4 and the paper is concluded in Sect. 5.

2 Basic Differential Evolution (DE)

Let $P = \{X_i^{(g)}, i = 1, 2, \ldots, N\}$ be the population at any generation t which contains N, D-dimensional vectors, i.e., each vector has the form as $X_i^{(g)} = (x_{1,i}^{(g)}, x_{2,i}^{(g)}, \ldots, x_{D,i}^{(g)})$. For basic DE (DE/rand/1/bin) [1] mutation, crossover and selection operations are defined as below:

i. **Mutation** For each target vector $X_i^{(g)}$ select three different vector say $X_a^{(g)}$, $X_b^{(g)}$ and $X_c^{(g)}$ from the population P such that $a \neq b \neq c \neq i$ then the mutant vector $V_i^{(g)} = (v_{1,i}^{(g)}, v_{2,i}^{(g)}, \ldots, v_{D,i}^{(g)})$ is defined as

$$V_i^{(g)} = X_a^{(g)} + F(X_c^{(g)} - X_c^{(g)}) \tag{1}$$

Here, $X_a^{(g)}$ is a base vector, F is a real and constant factor having value between [0, 2] and controls the amplification of differential variation $(X_b^{(g)} - X_c^{(g)})$.

ii. **Crossover** Perform crossover operation to create a trial vector $Y_i^{(g)} = (y_{1,i}^{(g)}, y_{2,i}^{(g)}, \ldots, y_{D,i}^{(g)})$ as

$$y_{j,i}^{(g)} = \begin{cases} v_{j,i}^{(g)}, \text{ if } \mathrm{Cr} < \mathrm{rand}_j \forall j = j_{\mathrm{rand}} \\ x_{j,i}^{(g)} \quad \text{otherwise} \end{cases} \tag{2}$$

$rand_j$ is the uniform random number between 0 and 1; Cr is the crossover constant that takes values in the range [0, 1] and $j_{\text{rand}} \in 1, 2, \ldots, D;$ is the randomly chosen index.

iii. **Selection** During the selection operation we generate a new population $Q = \{X_i^{(g+1)}, i = 1, 2, \ldots, N\}$ for next generation $g + 1$ by choosing the best vector between trial vector and target vector.

$$X_i^{(g+1)} = \begin{cases} Y_i^{(g)}, \text{ if } f(Y_i^{(g)}) < f(X_i^{(g)}) \\ X_i^{(g)} \quad \text{otherwise} \end{cases} \tag{3}$$

3 Proposed Modification

In the basic DE selection strategy, the fitness of the trial vector is compared to the corresponding target vector and is selected for the next generation only if it is better than the latter. Due to the unique one-to-one selection strategy of DE, some of the trial vectors though having a good fitness value may be discarded resulting in loss of some important information. In this study, we have done a slight modification in the selection strategy of DE. Here, instead of completely discarding the trial vector we compare it with the individual having the worst fitness (say x_w) and retain the trial vector if it is better than x_w.

By doing so, we try to utilize the significant information of the trial vector. This strategy condenses the search domain area in each generation and, consequently, increases the convergence speed.

Further, in order to reduce the redundant computer memory, we have implemented on a single array [24]. In [24], a single array was used to update the population for the next generation. Also, there was only a single way to update this array in case of acceptance of trial vector. However, the proposed IUDE updates the single array in case of rejection of trail vectors also.

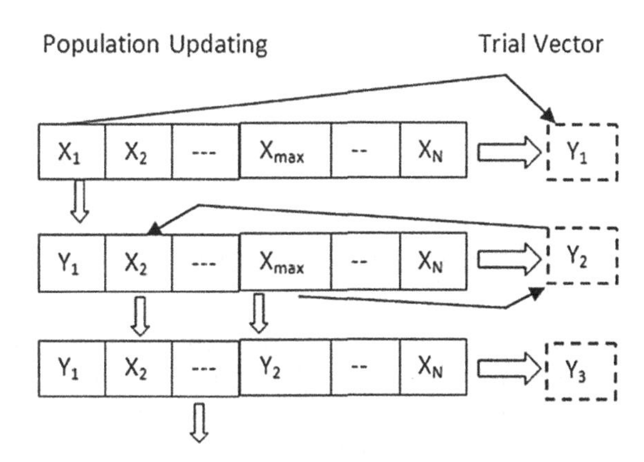

Fig. 1 Population updating by modified single array strategy

Population updating by proposed IUDE is expressed graphically in Fig. 1.

Pseudo Code of IUDE Algorithm

```
BEGIN
Generate uniformly distributed random population
P= {X_i^(g), i=1,2,...N}.
      X_i^(0) = X_lower +(X_upper -X_lower)*rand(0,1),
      where i =1, 2,..,NP
Evaluate f(X_i^(g))
while (Termination criteria is met )
{
 for i=1:N
  {
   Perform mutation operation
   Perform crossover operation and generate trial
   vector Y_i^(g)
   Evaluate f(Y_i^(g))

//** Modified Selection Operation with Single
Array strategy***////

   If (f(Y_i^(g))< f(X_i^(g)))
    {
     X_i^(g+1)= Y_i^(g)
    }
     Else
    {
     If (f(Y_i^(g))< f(X_max^(g)))
      X_max^(g)= Y_i^(g)
    }
 }/* end for loop*/
} /* end while loop*/
END
```

4 Experimental Setting and Simulated Results

To evaluate the performance of proposed algorithms six standard benchmark functions taken from [11, 12, 16] are used in this study. Next in this section, algorithms used for comparison, performance criteria, experimental settings, and numerical results are given.

4.1 Performance Criteria

We have used four different criteria to evaluate the proposed algorithm:

Error The error of a solution X is defined as $f(X) - f(X^*)$, where X^* is the global optimum of the function. The minimum error is recorded when the Max NFEs is reached in 50 runs. Also, the average and standard deviation of the error values are calculated.

NFEs We set the termination criteria as $|f_{\text{opt}} - f_{\text{globqal}}| \leq \text{VTR}$ and record the average NFE of successful run over 50 runs, where VTR is *Value TO Reach* and NFE is the *Number of Function Evaluations*.

Convergence graphs The convergence graphs show the mean fitness performance of the total runs, in the respective experiments.

Acceleration rate (AR) This criterion is used to compare the convergence speeds between algorithms A and B. It is defined as follows:

$$\text{AR} = \frac{\text{NFE}_\text{A} - \text{NFE}_\text{B}}{\text{NFE}_\text{A}} \%$$

4.2 Experimental Setting

All algorithms are implemented in Dev-C++ and the experiments are conducted on a computer with 2.00 GHz Intel (R) core (TM) 2 duo CPU and 2 GB of RAM. All results are calculated as an average result over 50 independent runs for every algorithm.

The parameters used are set as follows [12, 18].

4.3 Simulated Result and Comparisons

(a) Comparison of IUDE with DE and MDE

In this section, numerical results for DE, MDE, and IUDE are given in Table 2 in terms of average NFEs of 50 runs. Here, MDE states modified DE with single array as proposed by Babu and Angira [24]. Parameter settings are taking identical for each variant as discussed in Table 1.

From Table 2, it is clear that IUDE takes less NFEs for all benchmark problems compared to both DE and MDE. Total NFEs obtained by DE and MDE are 1,128,860, and 970,030, while IUDE takes 887,100 which gives average acceleration rate of IUDE against DE and MDE as 21.42 and 8.55, respectively. Hence we can conclude that the proposed modification in selection helps to boost the convergence speed of DE.

Table 1 Parameter setting

Pop size (NP)	100
Dimension (D)	30
Scale factor (F) and crossover rate (Cr)	0.5 and 0.9 respectively
Value to reach (VTR)	10^{-08} except for noisy function where VTR is 10^{-02}
Max NFE	300,000

Table 2 Experimental results and comparison of DE, MDE, and IUDE in terms of average NFEs over 50 runs

Fun	DE	MDE	IUDE	AR	
				IUDE versus DE	IUDE versus MDE
Sphere	106,500	92,400	86,600	18.69	6.28
Schwefel 2.22	177,200	156,500	142,500	19.58	8.95
Rosenbrock	437,300	375,600	351,900	19.53	6.31
Noise	136,400	102,630	82,400	39.59	19.71
Griewank	109,200	97,400	87,400	19.96	10.27
Ackley	162,260	145,500	136,300	16.00	6.32
Total	1,128,860	970,030	887,100	21.42	8.55

(b) Effect of Proposed Modification on MRLDE and DERL

In this section, we have implemented the proposed modification IU on other enhanced DE variants such as DERL [11] and our previously proposed variant MRLDE [5, 8, 18]. We named these variants IU-DERL and IU-MRLDE respectively.

The results are simulated in Table 3 in terms of average NFEs of 50 runs. We can observe the effect of proposed modification in each variant DERL and MRLDE. Both variants perform quickly compared to their old versions. Also, we can see that IU-MRLDE takes less NFEs compared to all other variants and therefore high acceleration rate compared to IU-DERL.

Figure 2 represents the bar chart corresponding to each NFEs taken by DE, IUDE, MDE, MRLDE, and IU-MRLDE for each function. We can easily analyze

Table 3 Experimental results and comparison of DERL, IU-DERL, MRLDE, and IU-MRLDEin term of average NFEs over 50 runs

Fun	DERL	IU-DERL	MRLDE	IU-MRLDE	AR
					IU-MRLDE versus IU-DERL
Sphere	54,400	44,000	38,400	32,200	26.82
Schwefel 2.22	96,700	72,300	68,800	54,500	24.62
Rosenbrock	276,100	179,100	145,300	124,200	30.65
Noise	88,800	52,400	47,900	28,500	45.61
Griewank	58,900	45,900	40,200	35,000	23.75
Ackley	87,200	67,500	63,300	52,200	22.67

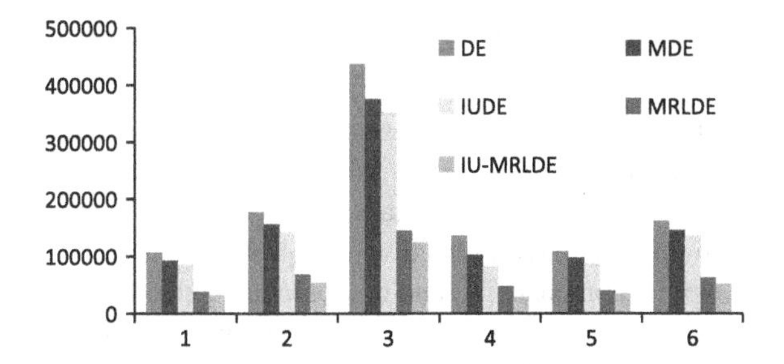

Fig. 2 Results analysis of DE, IUDE, MDE, MRLDE, and IU-MRLDE on all benchmark problems

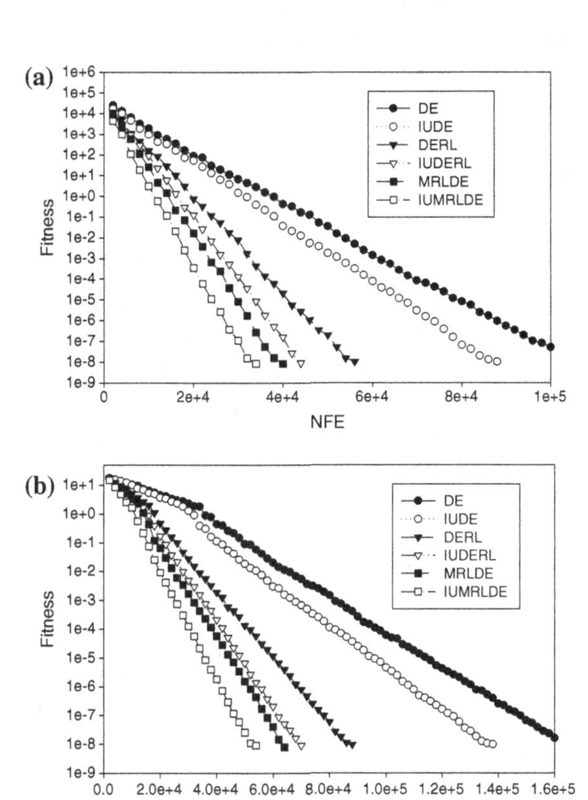

Fig. 3 Convergence graph of **a** sphere function and **b** Ackley function

the complete effect of the proposed modification in the corresponding variant from Fig. 2.

Figure 3 represents the convergence graphs of DE, IUDE, DERL, IU-DERL, MRLDE, and IU-MRLDE for Sphere and Ackley function in terms of NFE and fitness values.

5 Conclusions

In this study, a modified selection strategy of DE algorithm is proposed and is applied on a single array selection strategy. The proposed strategy utilized the significant information regarding rejected trial vector and hence called it information utilization (IU) selection operation. Initially, the proposed IU selection strategy is applied on basic DE and compared its performance with MDE and DE. Later, the proposed IU selection strategy is implemented on two other enhanced DE variants, DERL and MRLDE and named these IU-DERL and IU-MRLDE respectively. Numerical results prove the effectiveness of proposed IU selection and conclude that it can be adapted as an influential selection operation for DE algorithm.

References

1. Storn, R., Price, K.: Differential evolution—a simple and efficient adaptive scheme for global optimization over continuous. Spaces. Berkeley, CA, Tech. Rep. TR-95-012 (1995)
2. Vesterstrom, J., Thomsen, R.: A comparative study of differential evolution, particle swarm optimization and evolutionary algorithms on numerical benchmark problems. In: Congress on Evolutionary Computation, pp. 980–987 (2004)
3. Plagianakos, V., Tasoulis, D., Vrahatis, M.: A review of major application areas of differential evolution. Adv. Diff. Evol. **143**, 197–238 (2008). Springer, Berlin
4. Kumar, P., Kumar, S., Pant, M.: Gray level image enhancement by improved differential evolution algorithm. Proc. BICTA **2**(2012), 443–453 (2012)
5. Kumar, S., Kumar, P., Sharma, T.K., Pant, M.: Bi-level thresholding using PSO, artificial bee colony and MRLDE embedded with Otsu method. Memetic Comput. **5**(4), 323–334 (2013)
6. Kumar, P., Pant, M.: Noisy source recognition in multi noise plants by differential evolution. Proc. SIS **2013**, 271–275 (2013)
7. Ali, M., Ahn, C.W., Pant, M.: Multi-level image thresholding by synergetic differential evolution. Appl. Soft Comput. **17**, 1–11 (2014)
8. Kumar, P., Pant, M., Singh, V.P.: Modified random localization based DE for static economic power dispatch with generator constraints. Int. J. Bio-Inspired Comput. **6**(4), 250–261 (2014)
9. Fan, H., Lampinen, J.: A trigonometric mutation operation to differentia evolution. J. Global Opt. **27**, 105–129 (2003)
10. Liu, J., Lampinen, J.: A fuzzy adaptive differential evolution algorithm. Soft Comput. Fus. Found Methodol. Appl. **9**(6), 448–462 (2005)
11. Kaelo, P., Ali, M.M.: A numerical study of some modified differential evolution algorithms. Eur. J. Oper. Res. **169**, 1176–1184 (2006)
12. Brest, J., Greiner, S., Boskovic, B., Mernik, M., Zumer, V.: Self adapting control parameters in differential evolution: a comparative study on numerical benchmark problems. IEEE Trans. Evol. Comput. **10**(6), 646–657 (2006)
13. Noman, N., Iba, H.: Accelerating differential evolution using an adaptive local Search. IEEE Trans. Evol. Comput. **12**(1), 107–125 (2008)
14. Rahnamayan, S., Tizhoosh, H., Salama, M.: Opposition based differential evolution. IEEE Trans. Evol. Comput. **12**(1), 64–79 (2008)
15. Qin, A.K., Huang, V.L., Suganthan, P.N.: Differential evolution algorithm with strategy adaptation for global numerical optimization. IEEE Trans. Evol. Comput. **13**(2), 398–417 (2009)

16. Zhang, J., Sanderson, A.: JADE: adaptive differential evolution with optional external archive. IEEE Trans. Evol. Comput. **13**(5), 945–958 (2009)
17. Ali, M., Pant, M.: Improving the performance of differential evolution algorithm using cauchy mutation. Soft Comput. (2010). doi:10.1007/s00500-010-0655-2
18. Kumar, P., Pant, M.: Enhanced mutation strategy for differential evolution. In: Proceeding of IEEE Congress on Evolutionary Computation (CEC-12), pp. 1–6 (2012)
19. Sarker, R.A., Elsayed, S.M., Ray, T.: Differential evolution with dynamic parameters selection for optimization problems. IEEE Trans. Evol. Comput. **18**(5), 689–707 (2014)
20. Pooja, C.P., Kumar, P.: A cultivated differential evolution variant for molecular potential energy problem. Procedia Comput. Sci. **57**, 1265–1272 (2015)
21. Neri, F., Tirronen, V.: Recent advances in differential evolution: a survey and experimental analysis. Artif Intell. Rev. **33**(1–2), 61–106 (2010)
22. Das, S., Suganthan, P.N.: Differential evolution: a survey of the state-of-the-art. IEEE Trans. Evol. Comput. **15**(1), 4–13 (2011)
23. Gonuguntla, V., Mallipeddi, R., Veluvolu, K.C.: Differential evolution with population and strategy parameter adaptation. In: Mathematical Problems in Engineering (2015), doi:http://dx.doi.org/10.1155/2015/287607 (2015)
24. Babu, B.V., Angira, R.: Modified differential evolution (MDE) for optimization of non-linear chemical processes. Comput. Chem. Eng. **30**, 989–1002 (2006)

Application of Laplacian Biogeography-Based Optimization: Optimal Extraction of Bioactive Compounds from Ashwgandha

Vanita Garg and Kusum Deep

Abstract In this paper, the problem of extraction of bioactive compounds from the plant Ashwagandha is formulated as a nonlinear unconstrained optimization problem with more than one objective. Further this problem is solved using Laplacian Biogeography-Based Optimization. This paper focuses on the maximum extraction of two bioactive compounds (withaferin A and withanolide-A) from the roots of Ashwgandha (*Withania somnifera*). There are two objective functions each representing the maximizing of two bioactive compounds, i.e., withaferin A and withanolide A. The output of the yields is dependent on two factors namely, concentration of methanol and extraction temperature. This nonlinear optimization problem is converted into single objective problem using weighted sum approach and then solved using Laplacian Biogeography-Based Optimization (LX-BBO). Comparing these results with the previously reported results, it can be concluded that LX-BBO works comparatively well for these kind of optimization problems.

Keywords Biogeography-Based Optimization · Extraction of compounds · Real coded Genetic Algorithm · Laplacian BBO

1 Introduction

Ashwagandha (*Withania somniferia*) is a herb found in abundance in many parts of India and China. The dried roots of Ashwgandha are used for many medicinal cures. It has many therapeutic benefits. It is used in many ayurvedic medicines for relieving stress and even for prevention of cancer. It is well-known for its medicinal values, and hence the pharmaceutical industries use its extracts in the preparation of vital drugs.

Vanita Garg (✉) · K. Deep
Department of Mathematics, Indian Institute of Technology, Roorkee, India
e-mail: vanitagarg16@gmail.com

K. Deep
e-mail: kusumdeep@gmail.com

M. Pant et al. (eds.), *Proceedings of Fifth International Conference on Soft Computing for Problem Solving*, Advances in Intelligent Systems and Computing 437, DOI 10.1007/978-981-10-0451-3_72

Ashwagandha is rich in iron, antioxidants, tannis, glucose, potassium, nitrate, somnina, somniferine, anferina and withanolides, fatty acids, and numerous other substances. It has two influent bioactive compounds namely, withaferin A and withanolide A.

Extraction process of these two ingredients is costly and a timing consuming process. Due to various benefits and lack of availability of resources, optimization of extraction of bioactive compounds becomes very important. Extraction optimization contains two parts: First, the problem of extraction is formulated in an optimization problem and second, this optimization problem is solved using some optimization method.

Many methods are available for solving complex and difficult optimization problems. Traditional methods need additional information of optimization function or constraints. This makes traditional methods less useful than meta-heuristic methods for solving optimization problems. Meta-heuristic approaches are derivative free methods and are used by many scientists and researchers for solving numerous optimization problems. Meta-heuristic techniques like genetic algorithm (GA), ant colony optimization (ACO), particle swarm optimization (PSO), artificial bee colony (ABC), glow swarm optimization (GSO), biogeography-based optimization (BBO), etc., could be used for extraction optimization. In this paper, genetic algorithm and biogeography-based optimization are used for extraction optimization of bioactive compound of ashwgandha. The same problem has been solved already using real coded genetic algorithm in [1]. In this paper, a novel attempt is made to solve the same problem by Laplacian BBO proposed by Garg and Deep [2].

In the optimal extraction of bioactive compounds from ashwgandha problem, yields of namely, withaferin A and withanolide A are considered as the dependent variables. These two yields are calculated upon methanol concentration (MeOH) and extraction temperature. Thus, these are considered as independent variables in the problem. The yield of the two compounds is given as a function of two independent variables. The problem is to extract maximum yields of two bioactive compounds simultaneously. Thus, optimization problem is to maximize both the yield functions and thus formulate a multi-objective optimization problem.

In the second phase of the problem, this multi-objective optimization problem is converted into single-objective problem by weighted sum method and then Laplacian biogeography-based optimization is applied to solve the problem.

The paper is organized as follows: The extraction of bioactive compounds is described in Sect. 2. Section 3 gives the summary of the Laplacian BBO algorithm. Section 4 gives the numerical and computational results of the problem. The conclusion and future scope are given in Sect. 5.

2 Model of Problem of Optimal Extraction of Compounds from Ashawgandha

Problem of extraction of compounds from ashwgadha roots is formulated in [1], where yields of two bioactive compounds namely, withaferin A (Y_1) and withanolide-A (Y_2) are affected by the two independent variables namely, concentration of methanol (X_1) and extraction temperature (X_2).

The data used for formulation of the problem are produced by Shashi [1] using HPLC (high-pressure liquid chromatography). HPLC is based upon the principal of chromatography. Chromatography is the process, which is used to separate the mixtures on the basis of their absorbency of individual components [1]. The data produced are fitted into the model using least square fitting method. Same kind of problem is defined in [3] for gardenia plant.

The output of the model is defined as the two yields withaferin A (Y_1) and withanolide A (Y_2), which are represented as nonlinear function of two independent variables X_1 and X_2. These functions are defined in the form of second-order polynomial equation as follows

$$Y_n = b_0 + \sum_{i=1}^{2} b_i X_i^2 + \sum_{i=1}^{2} b_{ij} X_i X_j$$

where Y_n is the nth yield, b_0 is a constant, b_i and b_{ij} are the quadratic and interactive coefficients of the model, respectively. Using least square method, the resultant equations for yields Y_1 and Y_2 are calculated as follows:

$$\begin{aligned} Y_1 = {} & 0.0320790574 + 0.0001292883 X_1 + 0.0012707700 X_2 - 0.0000016421 X_1^2 \\ & - 0.0000195955 X_2^2 + 0.0000097853 X_1 X_2 \end{aligned}$$

$$\begin{aligned} Y_2 = {} & -0.0457594117 + 0.002021757 X_1 + 0.0017892681 X_2 - 0.0000148519 X_1^2 \\ & - 0.000016451 X_2^2 - 0.0000164518 X_1 X_2 \end{aligned}$$

where $X_i \in [30, 80]$ for $i = 1, 2$.

After choosing the model of the problem, the problem presents a multi-objective optimization problem where both the yields Y_1 and Y_2 need to be maximized simultaneously. Weighted sum approach is used to convert this multi-objective optimization problem into single-objective optimization problem. Same weights $w_1(1/2)$ and $w_2(1/2)$ are given because we need to maximize both the yields simultaneously. LX-BBO is a recently proposed algorithm [4] which is used for solving this problem.

LX-BBO has already been used for such extraction optimization for gardenia in [4]. LX-BBO has outperformed real coded genetic algorithm (RCGA) in extraction optimization problem of gardenia. Thus, LX-BBO is applied in the problem of the extraction of the bioactive compounds from ashwgandha.

Mathematically, for given yields Y_1, and Y_2 the single objective function is to solve the following function:

$$\max g = w_1 Y_1 + w_2 Y_2$$

where w_1 and w_2 are user-defined weights given to different yields [1].

3 Laplacian BBO (LX-BBO)

Biogeography-based optimization (BBO) is a nontraditional optimization method which was proposed by Dan Simon in 2008 [5]. BBO derives its inspiration from biogeography. Species tend to move from one habitat to another on the basis of its suitability which is characterized by weather, vegetation, natural climate, etc.

In BBO algorithm, a solution is considered as a habitat. Each solution is replaced by another solution on the basis of HSI (habitat suitability index). This replacement is decided by emigration and immigration rates. BBO algorithm has two main operators and are described as follows:

Migration: Migration is the key operator of BBO algorithm. The information shared between two solutions or habitats is known as migration. The information shared between two solutions is determined probabilistically with the help of emigration rate and immigration rate. Emigration rate is used to decide which habitat or solution is to be replaced and immigration rate decides which habitat or solution will take the place of the replaced solution. These rates are calculated based on habitat suitability index (HSI) of the solutions. The immigration rate and emigration rate are calculated using the following formula:

$$\lambda_i = I\left(1 - \frac{k(i)}{n}\right)$$

$$\mu_i = E\left(\frac{k(i)}{n}\right)$$

λ_i is the immigtaion rate of ith solution and I is the maximum possible immigration rate.
$K(i)$ is the rank of the solution which is decided by its HSI.
μ_i is the emigtaion rate of ith solution and E is the maximum possible emigration rate.
N is the total number of solutions.
HSI is usually taken as objective function value [5].

Mutation: A sudden change in natural calamity or disease, etc., can change the course of species migration. Hence it can affect the number of species living in the habitat. This phenomenon is termed as mutation. Mutation helps in improving the exploration of the solution.

BBO has become very popular since 2008. BBO was first proposed by Dan Simon on discrete optimization problems. But many improved version of BBO for continuous, constrained optimization problems are available in the literature [6]. The latest addition in these different versions is Laplacian biogeography-based optimization (LX-BBO). LX-BBO is proposed by Garg and Deep [2] for unconstrained optimization problems. LX-BBO is an attempt to improve migration operator of BBO. It introduces Laplacian operator of real coded genetic algorithm in migration operator. Laplace crossover is well tested on different sets of benchmark problems.

In LX-BBO, the migration operator of BBO is replaced by new migration operator where two habitats x_1 and x_2 are chosen on the basis of its migration rates. Immigration and emigration rates are calculated using the Eq. (1) as in BBO. These habitats or solutions are not exchanged like in BBO. x_1 and x_2 further give rise to two new habitats or solutions y_1 and y_2 using Laplace crossover [2]. Since Laplcace crossover is a parent-centric approach, so necessary information is not lost and further it improves the exploration of the solution. Thus including Laplace crossover in BBO makes BBO more exploring and exploiting.

Two new habitats y_1 and y_2 are generated by the following equation:

$$y_1^i = x_1^i + \beta(x_1^i - x_2^i)$$
$$y_2^i = x_2^i + \beta(x_1^i - x_2^i)$$

where random number which follows Laplace distribution is generated given by the equation:

$$\beta = \begin{cases} a - b * \log(u), & u \leq 1/2 \\ a + b * \log(u), & u > 1/2 \end{cases}$$

$a \in R$ is called location parameter and $b > 0$ is called scale parameter.
$u_i \in [0, 1]$ is a uniform random number.

Then, y_1 and y_2 are blended with the help of a blended parameter and a new solution z is obtained which is given in following equation:

$$z = \gamma y_1^i + (1 - \gamma) y_2^i$$

γ is the blended parameter. The values of all these parameters are given in [2].

Random mutation operator is used in LX-BBO and its mutation rate is set equal to 0.005. This random mutation operator is same as used in genetic algorithm.

Migration and mutation operator are used and repeated until some predefined criteria is satisfied.

To understand LX-BBO, flowchart of the LX-BBO algorithm is given in Fig. 1.

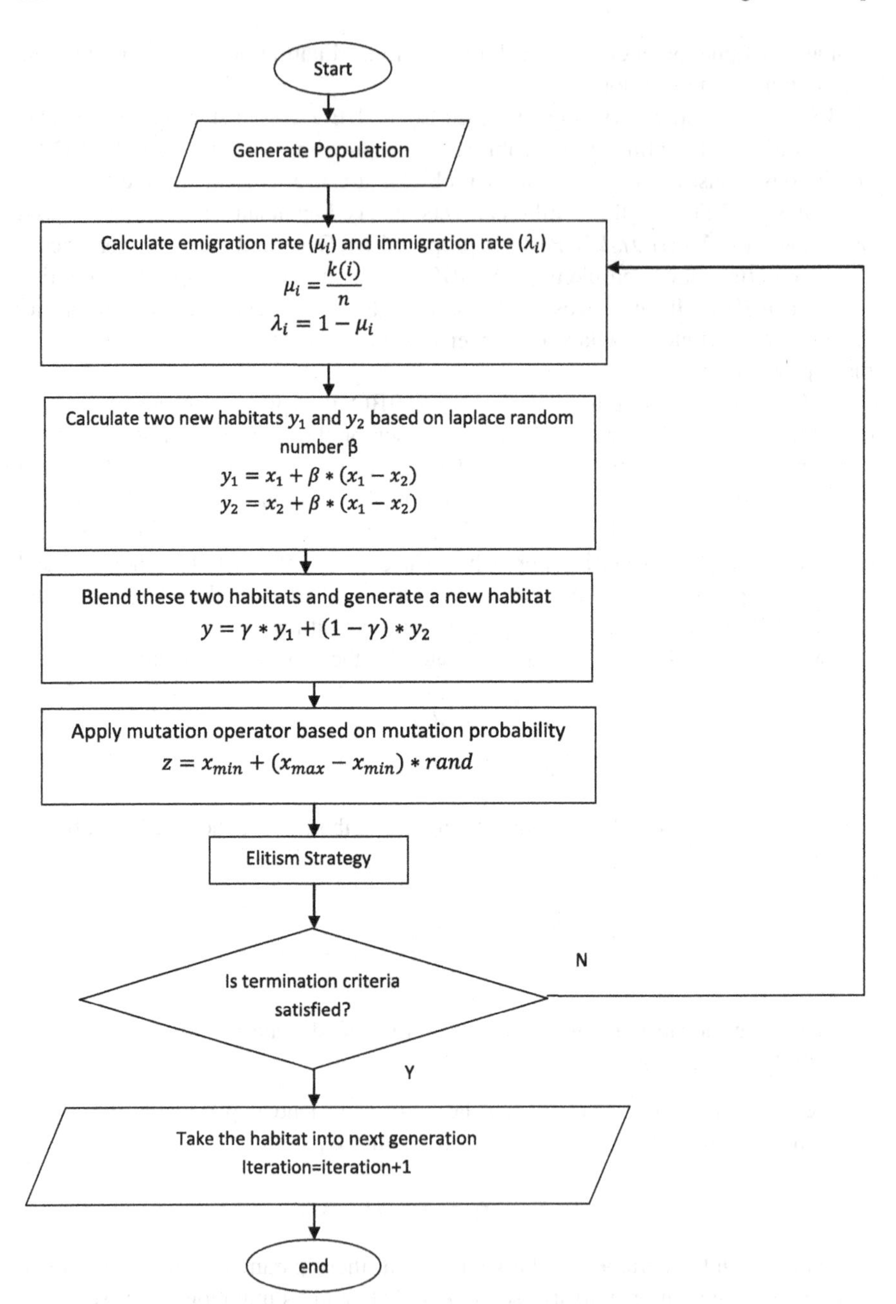

Fig. 1 Flowchart of the LX-BBO algorithm

Table 1 Yields obtained by different algorithms

Algorithm/yield	Y_1 (withaferin A)	Y_2 (withanolide A)
LX-BBO	0.0636	0.0631
DDX-LLM	0.0633	0.0628

4 Computational Results

Numerical results of the solution of the problem obtained by Laplacian Biogeography-Based Optimization are presented in Table 1. In the presented nonlinear unconstrained optimization problem, both the yields (outputs) needto be maximized. Hence, both the yields are equally important. Thus, both yields are given equal weights. Weights w_1 and w_2 are set equal to 0.5 and 0.5 for both objective functions.

The extraction optimization from ashwgandha is a two variable problem. Population size for Laplacian Biogeography-Based Optimization is set equal to 30. The maximum number of generations is set equal to 300. Totally, 30 independent runs are acted. The best result out of 30 independent run is recorded. Random mutation is used in LX-BBO.

The problem is solved by real coded genetic algorithm (RCGA) in [1]. The solution obtained by LX-BBO is compared with the solution by earlier recorded results. LX-BBO and DDX-LLM are compared on the basis of the solution of extraction optimization problem. DDX-LLM is a real coded genetic algorithm proposed by Shashi in [1]. In DDX-LLM log-logistic mutation was used. The results are given in the form of Table 1.

Table 1 shows the objective function value of the problem, i.e., the yields of two bioactive compounds

withaferin A and withanolide A. These are the results produced by both the algorithms DDX-LLM and LX-BBO. Now, it is clearly shown that LX-BBO is superior than DDX-LLM in solving optimal extraction problem of Ashwgandha. LX-BBO is also a winner in obtaining maximum output of the yields for Gardenia also [4]. Thus, LX-BBO can be helpful in solving extraction of compounds problem. The value of the independent variable X_1 (concentration of ethanol) and X_2 (extraction temperature) is 68.3917 and 49.4324, respectively.

The results obtained by LX-BBO are marginally well from results obtained by DDX-LLM. But it can be seen that LX-BBO outperformed DDX-LLM in extraction optimization of gardenia problem also. Thus, LX-BBO should be given preference over RCGA for solving extraction optimization problems.

5 Conclusion

Ashwagandha has been used widely in ancient Indian medicines. Its medicinal properties cannot be ignored in modern world too. The extraction of bioactive compounds from Ashwagandha roots is formulated as a multi-objective optimization problem. This multi-objective optimization problem is converted into single-objective by giving equal weights to all the objectives. This single-objective problem is solved by using Laplacian biogeography-based optimization. The results obtained by LX-BBO are compared with the results obtained by real coded genetic algorithm. It is concluded that Laplacian biogeography-based optimization provides superior results as compared to the earlier published results. A real-life problem of extraction of bioactive compounds from other plants may also contain some constraints. LX-BBO has been proposed for unconstrained optimization problems. But it can also be extended for constrained optimization problems.

References

1. Shashi, B.: New real coded genetic algorithms and their application to biorelated problem. Ph. D. thesis, Indian Institute of Technology Roorkee, India (2011)
2. Garg, V., Deep, K.: Performance of Laplacian biogeography-based optimization algorithm on CEC 2014 continuous optimization benchmarks and camera calibration problem. (Accepted in Swarm and Evolutionary Computation)
3. Deep, K., Katiyar, V.K.: Multi objective extraction optimization of bioactive compounds from gardenia using real coded genetic algorithm, 6th world. Congr. Biomech. **31**, 1436–1466 (2010)
4. Garg, V., Deep, K.: Optimal extraction of bioactive compounds from gardenia using Laplacian biogeography based optimization. In: The proceedings of 2nd ICHSA 2015, Seoul, South Korea
5. Simon, D.: Biogeography-based optimization. IEEE Trans. Evol. Comput. **12**(6), 702–713 (2008)
6. Garg, V., Deep, K.: A state of the art of biogeography based optimization. In: The proceedings of 4th SocPros 2014, NIT Silchar, Assam

Implementation of 'Fmincon' Statistical Tool in Optimizing the SHG Efficiency of Polymeric Films

Renu Tyagi, Millie Pant and Yuvraj Singh Negi

Abstract In this paper, we have studied nonlinear optical properties like second harmonic generation of organic composite polymeric films. Composite films are prepared using organic host polymer polyethersulfone and organic guest material metanitroaniline. Films are poled by contact poling method to attain the desired noncentrosymmetry. Poled films are characterized by Nd:YAG laser to analyze the second harmonic generation efficiency. Second harmonic intensity of films are evaluated applying 'fmincon' statistical tool, as a function of laser input power and metanitroaniline doping concentration.

Keywords Fmincon · Second harmonic generation · Polymeric films · Electrical poling

1 Introduction

Nonlinear optics (NLO) deals with the interactions of applied electromagnetic fields in various materials to generate new electromagnetic field altered in phase, frequency, amplitude, or other physical properties [1, 2]. The fields of NLO and photonics are rapidly emerging as the technology for the twenty-first century [3]. Photonics is the technology in which a photon instead of an electron is used to acquire, store, process, and transmit information. A photonic circuit is equivalent to electronic circuit, in which photons are conducted through channels. Light can be switched from one channel to another at certain junction points. For optical

Renu Tyagi (✉) · Y.S. Negi
Department of Polymer Science and Process Engineering, IIT Roorkee,
Saharanpur Campus, Roorkee, India
e-mail: renutyagi80@gmail.com

M. Pant
Department of Applied Science and Engineering, IIT Roorkee,
Saharanpur Campus, Roorkee, India
e-mail: milliepant.iitr@gmail.com

M. Pant et al. (eds.), *Proceedings of Fifth International Conference on Soft Computing for Problem Solving*, Advances in Intelligent Systems and Computing 437, DOI 10.1007/978-981-10-0451-3_73

switching at these junctions, one needs to use a material that allows the manipulation of an electric field or laser pulse. The materials which manipulate the light at these junction points are known as NLO materials and these are gaining importance in technologies such as optical communication, optical computing, and dynamic image processing [4, 5].

Before the advent of the lasers, optics assumed that optical parameters of the medium are independent of the intensity of the light propagating in these mediums. The reason is that the electric field strength generated by the nonlaser light sources of the order of 10^3 v/cm is very much smaller than the interatomic fields, i.e., 10^7–10^{10} v/cm of the medium, which is unable to affect the atomic fields of the medium and there by the optical properties of the medium. Lasers have drastically changed the situation as they generate electric field strength varying from 10^5 to 10^9 v/cm, which is able to commensurate to that of the atomic electric fields of the medium and there by affect the optical properties of the medium and thus generate new electromagnetic fields altered in phase, frequency, and amplitude. This is the domain of NLO [6].

1.1 Second Harmonic Generation

The process of transformation of light with frequency into light with double frequency is referred as second harmonic generation (SHG). The process is illustrated in Fig. 1. The p_1 and p_2 are the momenta of the absorbed photons, and p is the momentum of the emitted one. The process is spontaneous process and involves three-photon transitions, i.e., two photons with energy of hv per photon are absorbed spontaneously to emit a photon with an energy ($2hv$). The dashed (- - - - -) line corresponds to the virtual level [1, 7].

1.2 Nonlinear Constrained Minimization (Fmincon)

'Fmincon' finds a minimum of a constrained nonlinear multivariable function, and by default is based on the SQP (sequential quadratic programming) algorithm [8, 9]. It solves problems of the form:

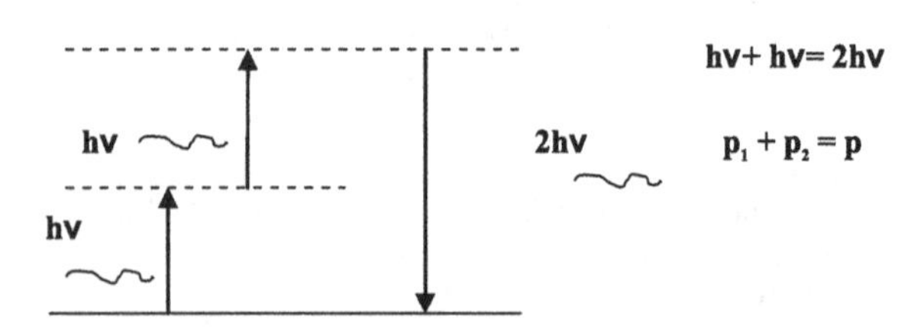

Fig. 1 Second harmonic generation (SHG)

min $f(x)$ subject to

$$\left.\begin{array}{r} A * x \leq B \\ \mathrm{Aeq} * x = \mathrm{Beq} \end{array}\right\} \text{linear constraints}$$

$$\left.\begin{array}{r} C(x) \leq 0 \\ \mathrm{Ceq}(x) = 0 \end{array}\right\} \text{nonlinear constraints}$$

$$\mathrm{LB} \leq x \leq \mathrm{UB} \rightarrow \text{bounding of variables}$$

1.2.1 Common Fmincon Syntax [10]

x = fmincon (fun, x_0, A, b)
Starts at x_0 and finds a minimum x to the function described in fun subject to the linear inequalities $A * x \leq b$. x_0 can be a scalar, vector, or matrix.

x = fmincon (fun, x_0, A, b, Aeq, Beq)
Minimizes fun subject to the linear equalities $\mathrm{Aeq} * x = \mathrm{beq}$ as well as $A * x \leq b$. Set $A = []$ and $B = []$ if no inequalities exist.

x = fmincon (fun, x_0, A, b, Aeq, Beq, lb, ub)
Defines a set of lower and upper bounds on the design variables, x, so that the solution is always in the range $\mathrm{lb} \leq x \leq \mathrm{ub}$. Set $\mathrm{Aeq} = []$ and $\mathrm{Beq} = []$ if no equalities exist.

x = fmincon (fun, x_0, A, b, Aeq, Beq, lb, ub, nonlcon)
Subjects the minimization to the nonlinear inequalities $c(x)$ or equalities $\mathrm{ceq}(x)$ defined in nonlcon. fmincon optimizes that $c(x) \leq 0$ and $\mathrm{ceq}(x) = 0$. Set $\mathrm{lb} = []$ and/or $\mathrm{ub} = []$ if no bounds exist.

x = fmincon (fun, x_0, A, b, Aeq, Beq, lb, ub, nonlcon, options)
Minimizes with the optimization parameters specified in the structures options.

1.3 *Objective of Present Work*

In the present study, we have developed polymeric film having different compositions through guest–host system. Metanitroaniline (MNA) is incorporated into Polyethersulfone (PES) as doping material from 2 to 18 % by weight of PES. Thin and transparent films of developed composite are prepared by solution casting method. These films are poled by contact poling method under external field (6 kV/cm) at 120 °C temperature for half an hour to align the MNA molecules in electric field direction. Poled films are characterized using Q-switched Nd:YAG laser. Frequency doubling efficiency of films is studied as a function of incident laser beam power and MNA concentration. It is observed that how nonlinear optical properties of films are affected by applying external DC electric field and by different doses of substituted aniline (MNA). The behavior of NLO property (SHG) is optimized mathematically using fmincon (a MATLAB tool).

2 Materials Used and Polymeric Film Preparation

Commercial grade polyethersulfone (PES) [11] purchased from Solvay Chemicals is used as host material. Optical grade metanitroaniline (MNA) [12] (Merck) is used as guest material. The total solid weight is kept 1gm to make all film samples and totally nine film samples are prepared. The amount of MNA is added in the range of 2–18 % by weight % of PES, taking the step size as 2. The mixture of PES and MNA in appropriate composition is dissolved in dimethyl sulfoxide (Merck) solvent and stirred until a clear solution is obtained. The viscous solution thus obtained is filtered to remove the impurities and filtered solution is spread on a glass plate by solution casting method. The films are allowed to dry at room temperature to evaporate solvent. The thin films are then peeled off and kept in vacuum oven to remove the residual solvent.

3 Poling Procedure

Doped MNA guest molecules will occupy random position into polymer matrix relative to each other. In order to impart such randomness of guest molecules the electric field poling technique is applied [13]. Studied poling parameters are stated in Table 1.

The developed polymeric composite films are poled by contact electrode poling method. The poling setup has been shown in Fig. 2. The DC electric field (6 kV/cm) is applied to the films for 30 min at 120 °C temperature. After half an hour the temperature is switched off and films are allowed to cool down up to room temperature in the presence of DC electric field. This step freezes the alignment of MNA molecules into composite films.

Table 1 Poling parameters

Poling time (min)	30
Poling temperature (°C)	120
Poling field strength (kV/cm)	6

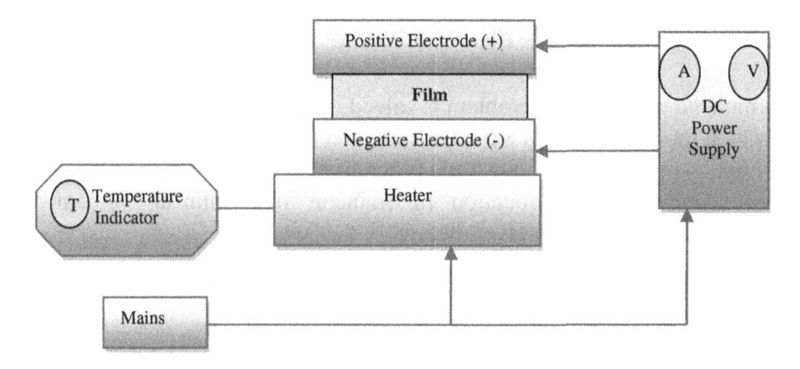

Fig. 2 Poling setup

4 Optical Characterization (SHG) Procedure

For second harmonic (SH) signal, poled doped polymer film samples are illuminated with a compact diode-pumped solid state, Q-switched Nd:YAG laser [1] (wavelength 1064 nm, pulse width 40 ns and 1 pps) beam. SHG setup is shown in Fig. 3. The pump spot size was measured 150 μm. The spot size was experimentally estimated using a beam profiler. The passed out beam from the film samples is filtered using an IR filter to eliminate the fundamental wavelength (1064). The second harmonic generation was confirmed by the emission of green radiation (532 nm) emitted by the film sample.

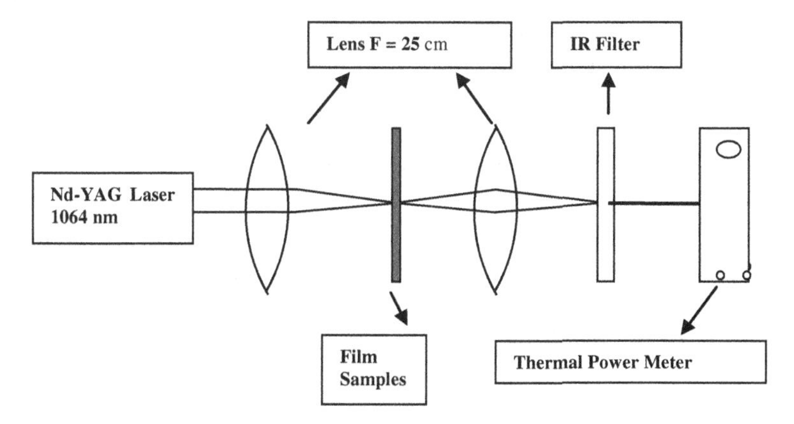

Fig. 3 SHG setup

5 Methods Used for Optimization

The formulated optimization problem is solved using the MATLAB 'fmincon' [14] optimization tool box (version 8.0). This function is based on local gradient method which implements sequential quadratic programming to obtain a local minimum of a constrained multivariable function. In 'fmincon' the nonlinear problem can be represented using the syntax given below

$$x = \text{fmincon}(\text{fun}, x_0, \text{lb}, \text{ub}, \text{options})$$

where objective function 'fun' of a multidimensional design vector x is the scalar vector. x_0 stands for starting point (initial design point). 'lb' and 'ub' are the lower bounds and upper bounds of design valuables.

6 Results and Discussions

6.1 SHG Analysis of Polymeric Films

Prepared PES-MNA films are poled with the method explained in Sect. 3. After poling, the films are characterized by Nd:YAG laser to analyze the second harmonic generation efficiency. Generated SH signal is recorded as a function of laser input energy and doping concentration of chromophores.

Measured SHG intensity is plotted with input laser power. Graph is shown in Fig. 4. It is observed that generated SHG intensity through the developed films increases with increasing laser input power and with doping NLO chromophore concentration. Initially at low laser input power, the enhancement in SHG is observed quickly but at high laser power (6 W onwards), the gain in SHG is minor.

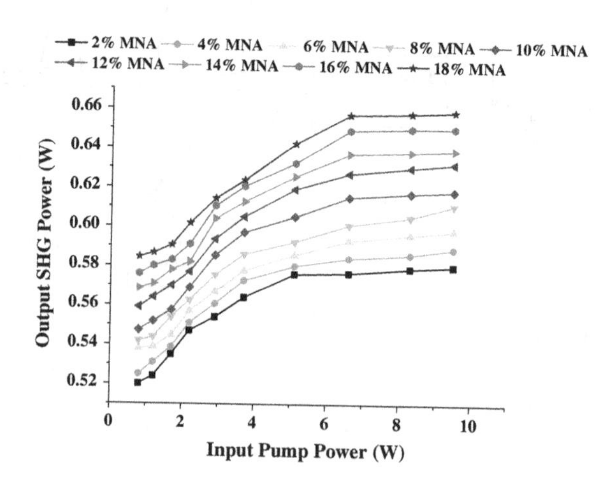

Fig. 4 Variation of output SHG power versus input laser pump power for PES-MNA composite films

From the graph, it can be observed that addition of MNA molecules in PES generates the nonlinear polarizability through the films.

6.2 Optimization of SHG Intensity of Polymeric Films

SHG intensity of developed composite films is mathematically formulated in the form of second-order polynomial equation (Eq. 1) using *least square fitting* method.

$$Y_k = a_0^k + \sum_{i=i}^{2} a_i^k X_i + \sum_{i=1}^{2} a_{ii}^k X_i^2 + \sum_{i \neq j=1}^{2} a_{ij}^k X_i X_j \quad (1)$$

In the above equation, Y_k represents the kth SHG intensity such that $k = 1, 2, 3$; a_0^k is a constant, a_i^k, a_{ii}^k, and a_{ij}^k are the linear, quadratic, and interactive coefficients, respectively. X_i and X_j are the levels of the independent variables, viz., laser input power and MNA concentration. Objective function F is constructed as follows:

$$F = \min \sum_{i=0}^{n} \left(y_i^{\text{fit}} - y_i^{\text{obs}}\right)^2$$

where n indicates the total number of observations, y_i^{fit} and y_i^{obs} are the fitted value and the observed value of the ith observation for SHG output power.

The resulting value of the constants a_0^k, and the coefficients a_i^k, a_{ii}^k, and a_{ij}^k corresponding to observed SHG power (Y) is presented in Table 2. The initial solutions are obtained using fmincon. Predictive capability of the model and statistical validity of the model are determined by calculating coefficient of determination (R^2) and absolute average deviation (AAD), respectively. In Table 3, a comparison in terms of absolute average deviation (AAD) and coefficient of

Table 2 Coefficient of fitted polynomials for SHG power (Y)

Coefficients	a_0	a_1	a_2	a_{11}	a_{12}	a_{22}
Value obtained by fmincon (Y_1)	0.4976	0.01986	0.001865	−0.001256	0.000159	0.0000386

Table 3 Value of R^2 and AAD for SHG response

R^2	AAD
0.90906	1.4933

determination (COD) is given. The result indicates that fmincon optimization technique produces satisfactory performance in terms of COD and AAD. The resultant equation for SHG power (Y) is given below.

$$\begin{aligned} Y &= 0.4976 + 0.01986X_1 + 0.001865X_2 \\ &\quad - 0.001256X_1^2 + 0.000159X_1X_2 + 0.0000386X_2^2 \end{aligned}$$

7 Conclusions

Second harmonic intensity has been studied in detail in this chapter for all developed PES-MNA composite films. SHG power enhances with doping both chromophoric material MNA as well as falling input laser power. We developed a mathematical model to optimize SHG intensity and solved it by applying statistical tool 'fmincon'. The nonlinear constrained optimization method is observed satisfactory because of high value of R^2 (0.91) and low value of AAD (1.49).

References

1. Calvert, P.D., Moyle, B.D.: Second harmonic generation by polymer composites. MRS Online Proc. Libr. **109** (1987)
2. Prasad, P.N., Williams, D.J.: Introduction to Nonlinear Optical Effects in Molecules and Polymers. Wiley, New York (1991)
3. Prasad, P.N.: Polymers as multi-role materials for photonics technology. In: Polymers and Other Advanced Materials, pp. 441–449, Springer, (1995)
4. Kuang, L., Chen, Q., Sargent, E.H., Wang, Z.Y.: [60] Fullerene-containing polyurethane films with large ultrafast nonresonant third-order nonlinearity at telecommunication wavelengths. J. Am. Chem. Soc. **125**(45), 13648–13649 (2003)
5. Agarwal, G.: Applications of Nonlinear Fiber Optics. Academic Press, New York (2001)
6. Boyd, R.W.: Nonlinear Optics. Academic press, New York (2003)
7. Arivuoli, D.: Fundamentals of nonlinear optical materials. PRAMANA **57**, 871–884 (2001)
8. Sharma, R.: Computational Approach to Function Minimization and Optimization with Constraints. Digital Repository @ Iowa State University. http://lib.dr.iastate.edu/undergradresearch_symposium/2014/presentations/30/ (2014)
9. Quancai, L., Xuetao, Q., Cuirong, W., Xingxing, W.: The study on gear transmission multi-objective optimum design based on SQP algorithm. SPIE-International Society for Optics and Photonics (2011)
10. Ku, H.-M.: MATLAB Syntax for fminbnd, fminunc, fminsearch, fmincon. www.chemeng.kmutt.ac.th/cheps/MATLAB_Optimization_Syntax
11. Ballet, W., Picard, I., Verbiest, T., Persoons, A., Samyn, C.: Chromophore-functionalized poly (ether sulfone)s with high poling stabilities of the nonlinear optical effect. Macromol. Chem. Phys. **205**(1), 13–18 (2004)
12. Valluvan, R., Selvaraju, K., Kumararaman, S.: Studies on the growth and characterization of meta-nitroaniline single crystals. Mater. Lett. **59**(10), 1173–1177 (2005)

13. Zhang, X., Lu, X., Wu, L., Chen, R.T.: Contact poling of the nonlinear optical film for polymer-based electro-optic modulator. In: SPIE-International Society for Optics and Photonics (2002)
14. López, C.P.: Optimization techniques via the optimization toolbox. In: MATLAB Optimization Techniques, Springer, pp. 85–108 (2014)

Curve Fitting Using Gravitational Search Algorithm and Its Hybridized Variants

Amarjeet Singh, Kusum Deep and Aakash Deep

Abstract In many experimental studies in scientific applications a set of given data is to be approximated. This can be performed either by minimizing the least absolute deviation or by minimizing the least square error. The objective of this paper is to demonstrate the use of gravitational search algorithm and its recently proposed hybridized variants, called LXGSA, PMGSA and LXPMGSA, to fit polynomials of degree 1, 2, 3, or 4 to a set of N points. It is concluded that one of the hybridized version namely, LXPMGSA outperform all other variants for this problem.

Keywords Gravitational search algorithm · Method of least square · Least absolute deviation

1 Introduction

Curve/Surface reconstruction is the problem of recovering the shape of a curve/surface [1]. It has received much attention in the few years. In the literature, the problem could be studied in two different orientations. In the first one, the problem of obtaining the curve or surface modal from a set of given cross-section or points. This is a problem in most of the research and application areas where an object (acquired from 3D laser scanning, ultrasound imaging, magnetic resonance

Amarjeet Singh (✉) · Kusum Deep
Department of Mathematics, Indian Institute of Technology, Roorkee,
Roorkee 247667, Uttarakhand, India
e-mail: amarjeetiitr@gmail.com

Kusum Deep
e-mail: kusumdeep@gmail.com

Aakash Deep
Department of Applied Science and Engineering, Indian Institute of Technology,
Roorkee, Roorkee 247667, Uttarakhand, India
e-mail: aakash791@gmail.com

M. Pant et al. (eds.), *Proceedings of Fifth International Conference on Soft Computing for Problem Solving*, Advances in Intelligent Systems and Computing 437, DOI 10.1007/978-981-10-0451-3_74

imaging, computer tomography, etc.) is a sequence of 2D cross-sections, such as CAD/CAM, biomedical, and medical science. Other one is, reconstructing curve/surface from a given set of data points. Based on the nature of the data points, there are two different approaches of fitting (i) interpolation (ii) approximation. In interpolation, parametric curve is constrained to pass through all the given set of data points. It is suitable when the data points describing the curve are sufficiently accurate and smooth. In approximation, parametric curve passes close to the given set of data points, but not necessarily through them. It is recommended when data is noisy or not exact. Approximation technique requires less computational effort to obtain a surface in comparison to interpolation on an infinite number of data. Approximation curve/function minimize the total error. Data fitting plays an important role in geometric modeling and image analysis. In most of the cases, geometric designers are asked to create curves or surfaces from measured data points, received from a digitizing device (i.e., coordinate measuring machine or a laser scanning device).

This paper discusses the determination of parameter values in order to perform least absolute deviation and least squares fitting of a polynomial using gravitational search algorithm and its variants. Numerous papers have been published on the selection of such parameters in which heuristic techniques are used to fit polynomial curve, B-spline, NURBS. Ülker and Arslan [1] proposed a method to determine the appropriate location of knots automatically and simultaneously using an artificial immune system for B-spline curve approximation. Siriruk [2] employed the particle swarm optimization to approximate a curve by optimizing the number of segments as well as their knot locations with fixed initial and final points. Pittman and Murthy [3] used genetic algorithms to fit piecewise linear functions into two-dimensional data. Cang and Le [4] used non uniform B-spline (NUB) surface fitting through a Genetic Algorithm technique. Sun et al. [5] used particle swarm optimization for B-spline curve fitting. Gálvez and Iglesias [6] used particle swarm optimization to reconstruct a nonuniform rational B-spline (NURBS) surface of a certain order from a given set of 3D data points. Gálvez et al. [7] applied genetic algorithm iteratively to fit a given set of data points in two steps: the first one determines the parametric values of data points; the later computes surface knot vectors. Gálvez and Iglesias [8] used firefly algorithm to compute the approximating explicit B-spline curve to a given set of noisy data points. Yoshimoto et al. [9] proposed a method for determining knots of data fitting with a spline using a genetic algorithm. Kalaivani et al. [10] proposed a procedure for single curve piecewise fitting stiction detection method and quantifying valve stiction in control loops based on ant colony optimization. Islam et al. [11] proposed an optimization technique for microstrip patch antenna using curve fitting-based particle swarm optimization (CFPSO). Xiao et al. [12] used a modified ant colony optimization algorithm to estimate the weight and knot by minimizing the sum square error between the fitted and target curve and surface. Garcia-Capulin et al. [13] used a hierarchical genetic algorithm to tackle the B-spline surface approximation of smooth explicit data.

The paper is organized as follows: In Sect. 2 gravitational search algorithm and its variants, namely LXGSA, PMGSA, LXPMGSA are described. In Sect. 3, Method of least squares is presented. Experimental results and analysis are given in Sect. 4. The paper concludes with Sect. 5.

2 Gravitational Search Algorithm

Gravitational search algorithm (GSA) [14] is a newly developed Nature Inspired Algorithm based on the metaphor of gravity and mass interactions. In this algorithm, the solution of the problem is represented by the position of the particle at the specified dimension and quality of the solution is represented by the mass of the particle, higher the mass better solution.

The continuous nonlinear optimization problem is defined as:

$$\begin{array}{ll} \text{Minimize} & f(x) \\ \text{subject to} & x_{\text{lower}} \leq x \leq x_{\text{upper}} \end{array} \quad (1)$$

Each iteration of GSA passes through three steps: (a) Initialization (b) Force calculation (c) Motion. Consider a system of N particle in which the position of ith particle is represented by

$$X_i = \left(x_i^1, x_i^2, \ldots, x_i^d, \ldots, x_i^m\right) \quad \text{for } i = 1, 2, \ldots, N \quad (2)$$

where x_i^d is the position of ith particle in dth direction.

In Initialization, a population of N particles is generated randomly in the search space. The velocities of each particle are initialized to zero (it could be nonzero if desired). Fitness value is evaluated using the objective function fit (X_i). The position of the best particle at step t is denoted by X_{best}.

In Force calculation, first mass of each particle is calculated using a function of fitness of particle, i.e., $M_i = g(f(X_i))$. Where $M_i \in (0, 1]$ and $g(.)$ is bounded and monotonically decreasing. The function g is defined in such a way that the best particle has the largest value (normalized) and worst particle has smallest value. Thus, after evaluating the current population fitness, the gravitational mass, and inertia mass of each particle are calculated as follows:

$$M_{ai} = M_{pi} = M_{ii} = M_i \quad (3)$$

$$m_i = \frac{\text{fit}_i(t) - \text{worst}(t)}{\text{best}(t) - \text{worst}(t)}, \quad i = 1, 2, \ldots, N. \quad (4)$$

$$M_i(t) = \frac{m_i(t)}{\sum_{j=1}^{N} m_j(t)} \quad (5)$$

where M_{ai} is the active gravitational mass, M_{pi} is the passive gravitational mass, M_{ii} is the inertia mass of particle *i*, $fit_i(t)$ is the fitness value of the *i*th particle at time t. Also, best(t) and worst(t) are the best and worst particle with regard to their fitness value.

For minimization problem

$$\text{best}(t) = \min_{j \in \{1,2,\ldots,N\}} \text{fit}_j(t) \quad \text{and worst}(t) = \max_{j \in \{1,2,\ldots,N\}} \text{fit}_j(t) \tag{6}$$

For maximization problem

$$\text{best}(t) = \max_{j \in \{1,2,\ldots,N\}} \text{fit}_j(t) \quad \text{and worst}(t) = \min_{j \in \{1,2,\ldots,N\}} \text{fit}_j(t) \tag{7}$$

Then, the force acting on mass 'i' from 'j' is evaluated by

$$F_{ij}^d(t) = G(t) \frac{M_{pi}(t) \times M_{aj}(t)}{R_{ij}(t) + \varepsilon} \left(x_j^d(t) - x_i^d(t) \right) \tag{8}$$

where $M_{aj}(t)$ is the active gravitational mass related to particle j, $M_{pi}(t)$ is the passive gravitational mass related to particle i. ε is a small value.

$G(t)$ is the gravitational constant and it is calculated by

$$G(t) = G_0 \exp(-\alpha t / \text{max\_iter}) \tag{9}$$

$R_{ij}(t)$ is the Euclidean distance between i and j particles and it is defined as follows:

$$R_{ij}(t) = \left\| X_i(t), X_j(t) \right\|_2 \tag{10}$$

The total force acting on *i*th particle in dimension d is calculated by

$$F_i^d(t) = \sum_{j \in \text{Kbest}, j \neq i} \text{rand}_j F_{ij}^d(t) \tag{11}$$

where rand_j is randomly distributed random number in interval $(0, 1]$, Kbest is the set of first k particles with the best fitness value and k is a decreasing function with time. Initially k is set to the number of particles in the system and it decreases linearly in such a way that at the last iteration $k = 1$.

In Motion, first acceleration of each particle is calculated by

$$a_i^d(t) = \frac{F_i^d(t)}{M_i(t)} \tag{12}$$

where $a_i^d(t)$ is the acceleration of particle i in the dimension d at time t. Then, the velocities and next position of particles i in the dth dimension are updated by

$$v_i^d(t+1) = \text{rand}_i \times v_i^d(t) + a_i^d(t) \tag{13}$$

$$x_i^d(t+1) = x_i^d(t) + v_i^d(t+1) \tag{14}$$

where rand_i is randomly distributed random number in the interval $(0, 1]$.

in the initial population, a particle having best fitness value is set to Lbest and in successive iteration the fitness of Lbest is compared with the best particle's fitness in each iteration, if it has better fitness than Lbest is updated otherwise Lbest remains same. Figure 1 show the pseudo code of GSA.

Laplace Crossover [15] and Power Mutation [16] are two well-known real coded genetic algorithms operators, introduced by Deep and Thakur. The pseudo code of

```
Set number of particles = N
Set dimension of the problem = m
Set parameter value: G_0, α
Deploy N particles randomly in the search space
Let x_i(t) = (x_i^1(t),...,x_i^d(t),...,x_i^m(t)) be the position of particle i at iteration t
Set maximum number of iteration = max_iter
t=0
while (t ≤ max_iter ) do:
  { Evaluate fitness f of each particle
    G(t) = G_0 exp(−α t/max_iter)
    best(t) = min_{j∈{1,...,N}} f(x_j(t)), worst(t) = max_{j∈{1,...,N}} f(x_j(t)),  % for minimization
    msum=0;
    for i =1 to N
       { m_i(t) = (f(x_i(t)) − worst(t)) / (best(t) − worst(t));  msum=msum + m_i(t); }
    for i =1 to N
      { M_i(t) = m_i(t)/msum ; }
    for each particle i = 1 to N do:
      {  for d = 1 to m do:
         { F_i^d(t) = Σ_{j∈kbest, j≠i} rand_j G(t) (M_pi(t)×M_aj(t))/(R_ij(t) + ε) (x_j^d(t) − x_i^d(t))
           a_i^d(t) = F_i^d(t)/M_ii(t)                % M_pi(t) = M_ai(t) = M_ii(t) = M_i(t)
           v_i^d(t+1) = rand_i v_i^d(t) + a_i^d(t)
           x_i^d(t+1) = x_i^d(t) + v_i^d(t+1)
         }
      }
    t=t+1
  }
```

Fig. 1 Pseudo code of GSA

```
let x_1 = (x_1^1, x_1^2, ..., x_1^m) and x_2 = (x_2^1, x_2^2, ..., x_2^m) are a pair of parents.
y_1 = (y_1^1, y_1^2, ..., y_1^m) and y_2 = (y_2^1, y_2^2, ..., y_2^m) are generated offspring
for i=1 to m
  {
  Generate two uniformly distributed random numbers r_i, s_i ∈ [0,1]
```

$$\beta_i = \begin{cases} a - b\log_e(r_i), & s_i \le 0.5 \\ a + b\log_e(r_i), & s_i > 0.5 \end{cases}$$

$$y_1^i = x_1^i + \beta_i |x_1^i - x_2^i|,$$

$$y_2^i = x_2^i + \beta_i |x_1^i - x_2^i|,$$

If $y_1^i, y_2^i \notin (x_{lower}^i, x_{upper}^i)$ then y_1^i, y_2^i be the random value in $[x_{lower}^i, x_{upper}^i]$.

}

Fig. 2 Pseudo code of Laplace crossover

let $x = (x^1, x^2, \ldots, x^m)$ is a parent.

$y = (y^1, y^2, \ldots, y^m)$ is a generated offspring

for i=1 to m

{ Generate a uniformly distributed random number $r, v \in [0,1]$

$$w = r^{1/p},$$

$$t = \frac{x^i - x_{lower}}{x_{upper} - x_{lower}}$$

$$y^i = \begin{cases} x^i - w(x^i - x_{lower}), & if\ t < v \\ x^i + w(x_{upper} - x^i), & if\ t \ge v \end{cases}$$

}

Fig. 3 Pseudo code of power mutation

Laplace Crossover is shown in Fig. 2 and the pseudo code of Power Mutation is shown in Fig. 3.

To enhance the exploration and exploitation ability of GSA, Singh and Deep [17] proposed an improved GSA based on Lbest update mechanism. Singh and Deep [18] also hybridized GSA with Laplace Crossover and Power Mutation and proposed LXGSA, PMGSA, and LXPMGSA. In LXGSA, after the completion of each iteration of GSA, the Lbest particle and a random particle from current population are selected as parents and Laplace crossover is applied to produce two offsprings called y_1 and y_2 as shown in Fig. 2. If fitness of y_1 is better than the fitness of worst particle then, worst is replaced by y_1 and worst is updated. In either case, if fitness of y_2 is better than the fitness of worst then, worst is replaced by y_2 and Lbest is updated if offsprings have better fitness.

Similarly, in PMGSA, after the completion of each iteration of GSA, the Lbest particle is selected and the Power Mutation is applied to produce a mutated offspring called y as shown in Fig. 3. If fitness of y is better than the fitness of worst, then worst is replaced by y and Lbest is updated if offspring has better fitness.

In LXPMGSA, After the completion of each iteration of GSA, the Lbest particle and a random particle from the current population are selected as parents and Laplace crossover is applied to produce two offsprings called y_1 and y_2. If fitness of y_1 is better than the fitness of worst then, worst is replaced by y_1 and worst is updated. In either case, if fitness of y_2 is better than the fitness of worst then, worst is replaced by y_2 and Lbest is updated if offsprings have better fitness. Then, Lbest particle is selected and the Power Mutation is applied to produce a mutated offspring called y. If fitness of y is better than the fitness of worst, then worst is replaced by y and Lbest is updated if offspring has better fitness.

3 Curve Fitting

Curve fitting is the process of constructing a curve, or mathematical function that has the best fit to a series of data points, possibly subject to constraints. A wide variety of methods are available for fitting an equation to data such as interpolation, least absolute Deviation, method of Least Squares, extrapolation etc. but Method of Least Squares is widely used in statistical estimation procedure. In this method the sum of the squared deviations is minimized.

3.1 The Method of Least Squares

Least square method (LSM) fits a curve to a set of approximate data points. Through LSM a large data set can be summarized adequately by a simple analytic function that has a few adjustable parameters.

Let $(x_1, y_1), (x_2, y_2), \ldots, (x_N, y_N)$ be the N data points and a polynomial of degree n, $y = f(x) = a_0 + a_1 x + a_2 x^2 + \cdots + a_n x^n$ is used to fit the data points where $N \geq n$.

Then, the sum of absolute error

$$E_1 = \sum_{i=1}^{N} |y_i - f(x_i)| = \sum_{i=1}^{N} |y_i - (a_0 + a_1 x_i + a_2 x_i^2 + \ldots + a_n x_i^n)| \quad (15)$$

The objective is to find the coefficients of the polynomial $f(x)$(i.e., a_i's) such that Eq. (15) is minimized. This method is known as Least Absolute Deviation or L_1 estimators. Since, E_1 is not derivable so it is not possible to find minima using classical methods of calculus. Therefore, Instead of using absolute values, square all the values are used and if necessary a square root is taken at the end.

$$E_2 = \sqrt{\sum_{i=1}^{N} (y_i - f(x_i))^2} = \sqrt{\sum_{i=1}^{N} (y_i - (a_0 + a_1 x_i + a_2 x_i^2 + \cdots + a_n x_i^n))^2} \quad (16)$$

Now, the objective is to find a_i's such that equation is (16) minimized. Since E_2 is differentiable therefore calculus method is applicable on Eq. (16). Here, E_2 is a function of a_i.

For minimum E_2

$$\frac{\partial E_2}{\partial a_i} = 0, \; i = 0, 1, \ldots, n \quad (17)$$

A system of $n + 1$ variables and $n + 1$ equations are obtained from Eq. (17) and by solving it a_i's are obtained. This method is known as the Method of Least Squares. Least Absolute Deviation has the advantage of not being as sensitive to outliers as the Method of Least Squares.

4 Simulation and Results

Consider a data set given in Table 1. The aim is to fit a polynomial of degree $n = 1, 2, 3, 4$ on the given set of data [19].

On the above data set, the Method of Least Squares is applied to fit a polynomial of degree $n = 1, 2, 3, 4$ and error E_2 are shown in Table 2. The determined polynomials are shown in Fig. 4. In the Fig. 4, the data points are shown by 'o'.

In the present study, Gravitational Search Algorithm and its variants proposed by Singh and Deep namely LXGSA, PMGSA and LXPMGSA [18] are applied on a curve fitting problem. Two type of experiments are performed. In the experiment I,

Table 1 Data for curve fitting

x_i	0.05	0.11	0.15	0.31	0.46	0.52	0.70	0.74	0.82	0.98	1.17
y_i	0.956	0.890	0.832	0.717	0.571	0.539	0.378	0.370	0.306	0.242	0.104

Table 2 Coefficients obtained by method of least square and error E_2

n	a_0	a_1	a_2	a_3	a_4	E_2
1	0.95227687	−0.76040691				0.09563441
2	0.99796838	−1.01804246	0.22468213			0.04321473
3	1.00369670	−1.07944638	0.35137487	−0.06893509		0.04258073
4	0.98809985	−0.83690198	−0.52679830	1.046131891	−0.45635100	0.04057287

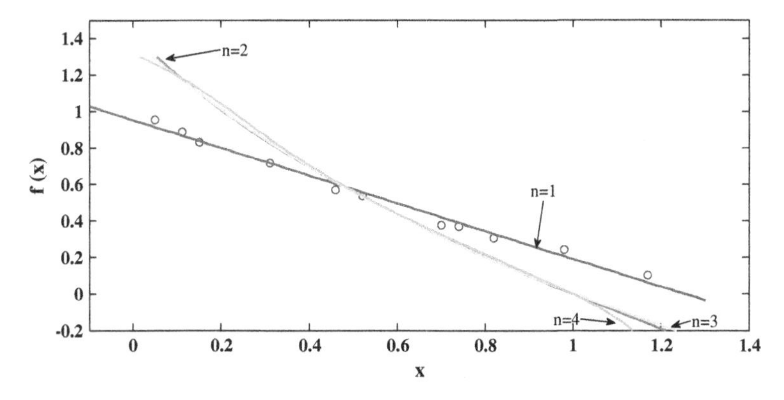

Fig. 4 Plot of data point and approximated function when n = 1, 2, 3, 4

error E_1 formulated by Least Absolute Deviation is considered and coefficients of the polynomial are determined by GSA, LXGSA, PMGSA and LXPMGSA. In the experiment II, error E_2 formulated by the Method of Least Squares is considered and coefficients of the polynomial are determined by GSA, LXGSA, PMGSA and LXPMGSA. The environment for running program of the experiments is processor: Intel(R) Xeon(R) CPU E5645 @ 2.40 GHz 2.39 GHz RAM: 12 GB, operating system window, integrated development environment: MATLAB 2008. The parameters of the algorithm are population size = 50 and the termination criteria is 5000 iterations for both the experiments. The GSA and its variants are run 30 times each. For a fair comparison among the all variants the first randomly generated population is used for the first run of all variants, second randomly generated population is used for the second run of all variants algorithm, and so on.

4.1 Analysis the Results of Experiment I

The best value of error E_1 and corresponding over 30 runs are shown in Table 3. From the Table 3, it is concluded that for n = 1, all variants have same error E_1 and a_i's. For n = 2, PMGSA has best E_1, for n = 3, LMPMGSA has best E_1 and for n = 4, LXGSA has better E_1.

The best, worst, median, average, and standard deviation (STD) of error E_1 are evaluated for experiment I and reported in Table 4. From the Table 4, it is concluded that for n = 1, the results obtained by LXGSA and LXPMGSA are approximately same and better than GSA and PMGSA. For n = 2, PMGSA has better Best but LXPMGSA has better worst, average, and median. For n = 3, LXPMGSA has better Best but LXGSA has better Worst, Average, and Median. For n = 4, LXPMGSA has better Worst but LXGSA has better Best,aAverage and median.

The best coefficients obtained by GSA, LXGSA, PMGSA, and LXPMGSA in experiment I are used to calculate the error E_1 and E_2 and these errors are shown in

Table 3 Coefficients of best solution obtained by GSA, LXGSA, PMGSA, and LXPMGSA and error E_1

Deg.	Algo.	a_0	a_1	a_2	a_3	a_4	Error E_1
$n = 1$	GSA	**0.962000**	**−0.800000**				**0.259000**
	LXGSA	**0.962000**	**−0.800000**				**0.259000**
	PMGSA	**0.962000**	**−0.800000**				**0.259000**
	LXPMGSA	**0.962000**	**−0.800000**				**0.259000**
$n = 2$	GSA	1.001349	−1.037718	0.231413			0.112185
	LXGSA	1.001352	−1.037754	0.231441			0.112184
	PMGSA	**1.001360**	**−1.037829**	**0.231499**			**0.112184**
	LXPMGSA	1.001358	−1.037805	0.231481			0.112184
$n = 3$	GSA	0.980913	−0.816443	−0.161356	0.186815		0.122328
	LXGSA	1.013039	−1.166974	0.522066	−0.161297		0.106535
	PMGSA	1.013189	−1.177025	0.544326	−0.173135		0.108026
	LXPMGSA	**1.011707**	**−1.161052**	**0.513945**	**−0.157851**		**0.106413**
$n = 4$	GSA	1.019162	−1.333687	1.493712	−1.735483	0.736480	0.117386
	LXGSA	**0.997552**	**−0.970871**	**−0.128284**	**0.632886**	**−0.317876**	**0.107383**
	PMGSA	0.978110	−0.717051	−0.926869	1.562165	−0.676857	0.109869
	LXPMGSA	0.998592	−0.991751	−0.003014	0.428173	−0.221937	0.108391

Table 4 Best, worst, average, median and STD of error E_1 over 30 runs

Deg.	Algorithm	Best	Worst	Average	Median	STD
$n = 1$	GSA	**0.25900000**	0.25906479	0.25900873	0.25900000	1.7676E−05
	LXGSA	**0.25900000**	**0.25900000**	**0.25900000**	**0.25900000**	4.4890E−12
	PMGSA	**0.25900000**	0.25903944	0.25900898	0.25900264	1.2089E−05
	LXPMGSA	**0.25900000**	**0.25900000**	**0.25900000**	**0.25900000**	**4.0012E−12**
$n = 2$	GSA	0.11218505	0.11232597	0.11221740	0.11220909	3.0538E−05
	LXGSA	0.11218444	0.11228979	0.11221602	0.11221367	**2.6255E−05**
	PMGSA	**0.11218400**	0.11230230	0.11222422	0.11221357	3.2831E−05
	LXPMGSA	0.11218414	**0.11226197**	**0.11221298**	**0.11220698**	2.6471E−05
$n = 3$	GSA	0.12232824	0.85405123	0.36236411	0.29067794	0.20656451
	LXGSA	0.10653491	**0.14833581**	**0.11281951**	**0.11045577**	**0.00783828**
	PMGSA	0.10802649	0.81467347	0.40312524	0.37133778	0.22265093
	LXPMGSA	**0.10641284**	0.23564771	0.12368039	0.11196100	0.03349783
$n = 4$	GSA	0.11738564	0.99312864	0.37157843	0.27893552	0.27124099
	LXGSA	**0.10738301**	0.27496756	**0.14299596**	**0.12178736**	**0.04490740**
	PMGSA	0.10986909	0.75233285	0.34134886	0.30266555	0.18166492
	LXPMGSA	0.10839068	**0.26557707**	0.15350507	0.12994224	0.05054518

Table 5. From Table 5, it is concluded that for $n = 1$, GSA, LXGSA, PMGSA, and LXPMGSA have same E_1 and same E_2. For $n = 2$, PMGSA has better E_1 but GSA has better E_2. For $n = 3$, LXPMGSA has better E_1 and better E_2. For $n = 4$, LMGSA has better E_1 and better E_2.

Table 5 Error E_1 and E_2 when coefficient (a_i's) are taken of best fitness function E_1

Deg.	Error	GSA	LXGSA	PMGSA	LXPMGSA
$n = 1$	E_1	**0.25900000**	**0.25900000**	**0.25900000**	**0.25900000**
	E_2	**0.11337989**	**0.11337989**	**0.11337989**	**0.11337989**
$n = 2$	E_1	0.11218505	0.11218444	**0.11218400**	0.11218414
	E_2	**0.04790531**	0.04791172	0.04792451	0.04792033
$n = 3$	E_1	0.12232824	0.10653491	0.10802649	**0.10641284**
	E_2	0.05555218	0.04406728	0.04433521	**0.04394772**
$n = 4$	E_1	0.11738564	**0.10738301**	0.10986909	0.10839068
	E_2	0.05918353	**0.04161843**	0.04186075	0.04164449

Table 6 Coefficients of best solution obtained by GSA, LXGSA, PMGSA and LXPMGSA and error E_2

Deg.	Algo	a_0	a_1	a_2	a_3	a_4	Error E_2
$n = 1$	GSA	**0.952277**	**−0.760407**				**0.095634**
	LXGSA	**0.952277**	**−0.760407**				**0.095634**
	PMGSA	**0.952277**	**−0.760407**				**0.095634**
	LXPMGSA	**0.952277**	**−0.760407**				**0.095634**
$n = 2$	GSA	**0.997968**	**−1.018042**	**0.224682**			**0.043214**
	LXGSA	**0.997968**	**−1.018042**	**0.224682**			**0.043215**
	PMGSA	**0.997968**	**−1.018042**	**0.224682**			**0.043215**
	LXPMGSA	**0.997968**	**−1.018042**	**0.224682**			**0.043215**
$n = 3$	GSA	1.015060	−1.193882	0.579246	−0.189884		0.044555
	LXGSA	**1.003697**	**−1.079446**	**0.351375**	**−0.068935**		**0.042581**
	PMGSA	1.005166	−1.094346	0.381107	−0.084736		0.042615
	LXPMGSA	**1.003697**	**−1.079446**	**0.351375**	**−0.068935**		**0.042580**
$n = 4$	GSA	1.014497	−1.310141	1.351131	−1.476369	0.614312	0.051782
	LXGSA	0.999280	−1.003137	0.053093	0.329467	−0.168804	0.041410
	PMGSA	0.961787	−0.441443	−1.915968	2.771061	−1.150753	0.045188
	LXPMGSA	**0.988631**	**−0.844800**	**−0.499254**	**1.012114**	**−0.442712**	**0.040575**

4.2 Analysis the Results of Experiment II

The best value of error E_2 and corresponding a_i's over 30 runs are shown in Table 6. From the Table 6, it is concluded that for $n = 1$ all variants have same error E_2 and a_i's and these results are same as obtained by method of least squares. For $n = 2$, all variants have same error E_2 and a_0, a_2 but PMGSA and LXPMGSA have better a_1 and the results obtained by PMGSA and LXPMGSA are same as obtained by method of least squares. For $n = 3$, LXGSA and LMPMGSA have same error E_2 and equal to error obtained by method of least squares and a_i's are approximately same. For $n = 4$, LXPMGSA has better error E_2 and approximately equal to the error obtained by method of least squares.

Table 7 Best, worst, average, median, and STD of error E_2 over 30 runs

Deg.	Algorithm	Best	Worst	Average	Median	STD
$n = 1$	GSA	**0.09563441**	**0.09563441**	**0.09563441**	**0.09563441**	**4.1554E−17**
	LXGSA	**0.09563441**	**0.09563441**	**0.09563441**	**0.09563441**	**4.1554E−17**
	PMGSA	**0.09563441**	**0.09563441**	**0.09563441**	**0.09563441**	4.1951E−17
	LXPMGSA	**0.09563441**	**0.09563441**	**0.09563441**	**0.09563441**	4.1951E−17
$n = 2$	GSA	**0.04321473**	**0.04321473**	**0.04321473**	**0.04321473**	1.1233E−17
	LXGSA	**0.04321473**	**0.04321473**	**0.04321473**	**0.04321473**	1.7901E−17
	PMGSA	**0.04321473**	**0.04321473**	**0.04321473**	**0.04321473**	1.6299E−17
	LXPMGSA	**0.04321473**	**0.04321473**	**0.04321473**	**0.04321473**	1.7667E−17
$n = 3$	GSA	0.04455544	0.55372083	0.25282426	0.20776573	0.17872075
	LXGSA	**0.04258073**	0.05293810	0.04292836	**0.04258073**	0.00189056
	PMGSA	0.04261510	0.56902528	0.25022636	0.23105848	0.15076386
	LXPMGSA	**0.04258073**	**0.04328426**	**0.04261617**	**0.04258073**	**0.00013188**
$n = 4$	GSA	0.05178171	0.50751414	0.18717237	0.17354562	0.10523103
	LXGSA	0.04141000	**0.14714254**	**0.08027723**	0.07360802	**0.03291153**
	PMGSA	0.04518858	0.55896381	0.18255382	0.16333776	0.11968256
	LXPMGSA	**0.04057478**	0.20421831	0.08771791	**0.06817811**	0.04683150

The best, worst, median, average, and standard deviation (STD) of error E_2 are evaluated for experiment II and reported in Table 7. From the Table 7, it is concluded that for $n = 1$, 2 GSA, LXGSA PMGSA, and LXPMGSA have approximately same error E_2. For $n = 3$, LXPMGSA performs better and get better results in comparison to GSA, LXGSA and PMGSA. For $n = 4$, LXPMGSA has better best and median but LXGSA has better Worst and Average.

The best coefficients obtained by GSA, LXGSA, PMGSA and LXPMGSA in experiment II are used to calculate the error E_1 and E_2 and these errors are shown in Table 8. From Table 8, it is concluded that for $n = 1$, LXGSA has slightly better E_1 and GSA, LXGSA, PMGSA and LXPMGSA have same E_2. For $n = 2$, GSA, LXGSA and LXPMGSA have same E_1 and GSA, LXGSA, PMGSA and LXPMGSA have same E_2. For $n = 3$, GSA has better E_1 and LXGSA and

Table 8 Error E_1 and E_2 when coefficient (a_i's) are taken of best fitness function E_2

Deg.	Error	GSA	LXGSA	PMGSA	LXPMGSA
$n = 1$	E_1	0.27976355	**0.27976354**	0.27976355	0.27976355
	E_2	**0.09563441**	**0.09563441**	**0.09563441**	**0.09563441**
$n = 2$	E_1	**0.12127551**	**0.12127551**	0.12127552	**0.12127551**
	E_2	**0.04321473**	**0.04321473**	**0.04321473**	**0.04321473**
$n = 3$	E_1	**0.11468141**	0.11599869	0.11475191	0.11599869
	E_2	0.04455544	**0.04258073**	0.04261510	**0.04258073**
$n = 4$	E_1	0.13170661	**0.11313097**	0.11366527	0.11334359
	E_2	0.05178171	0.04141000	0.04518858	**0.04057478**

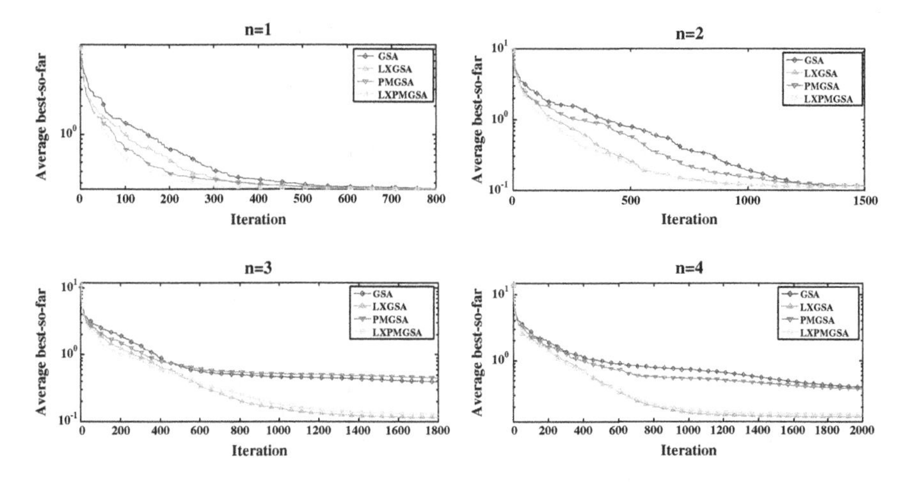

Fig. 5 Iteration wise convergence plot of average best-so-far for error function of experiment I

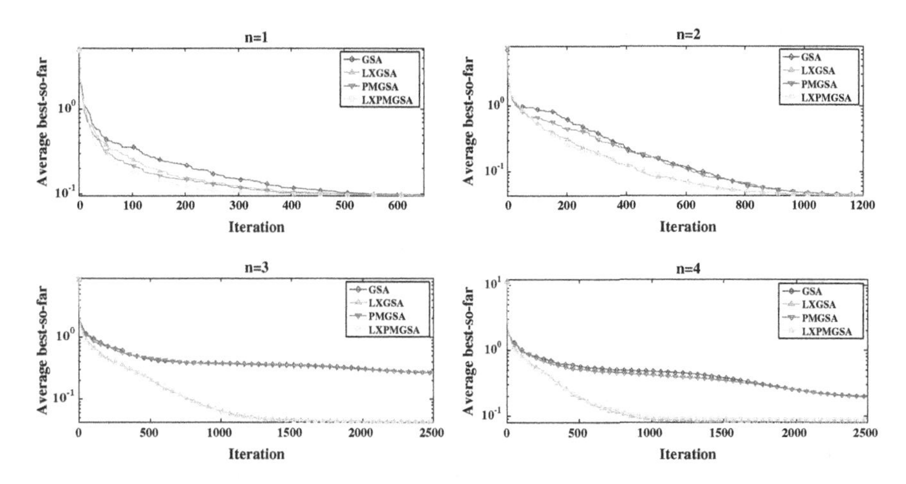

Fig. 6 Iteration wise convergence plot of average best-so-far for error function of experiment II

LXPMGSA have better E_2. For n = 4, LXGSA has better E_1 and LXPMGSA has better E_2.

To study the convergence behavior of the GSA, LXGSA, PMGSA ,and LXPMGSA toward optima, iteration wise convergence plots of average best-so-far are plotted and shown in Fig. 5 for experiment I and in Fig. 6 for experiment II. On the horizontal axis the iterations are shown, whereas on the vertical axis the average best-so-far is shown. Average best-so-far is the average of the best objective function value in each iteration over 30 runs. In the Fig. 5, LXPMGSA converge faster than GSA, LXGSA and PMGSA for n = 1 and 2 but for n = 3 and 4, in the initial iterations LXPMGSA converges faster and in latter iterations LXGSA. In the Fig. 6, LXPMGSA converge faster than GSA, LXGSA, and PMGSA for $n = 1$ but for $n = 2$ and 3, in the initial iterations the convergence of LXPMGSA is better but in latter iterations LXPMGSA and LXGSA have same behavior. For $n = 4$, the convergence of LXPMGSA is better in the initial iterations and in latter iterations LXGSA has better convergence.

5 Conclusion

Gravitational Search Algorithm is a recently proposed heuristic algorithm and successfully applied in many applications. In the present paper, Gravitational Search Algorithm and its recently proposed variants namely LXGSA, PMGSA, LXPMGSA are applied to determine the coefficient of a polynomial of degree 1, 2, 3 and 4. A number of numerical and graphical analysis is performed and it is concluded that the variant, in which Laplace Crossover and Power mutation is incorporated has better performance in comparison to other variants.

Acknowledgments The first author would like to thank Council for Scientific and Industrial Research (CSIR), New Delhi, India, for providing him the financial support vide grant number 09/143(0824)/2012-EMR-I and ICC, Indian Institute of Technology Roorkee, Roorkee for computational facility.

References

1. Ülker, E., Arslan, A.: Automatic knot adjustment using an artificial immune system for B-spline curve approximation. Inf. Sci. **179**(10), 1483–1494 (2009)
2. Siriruk, P.: Fitting piecewise linear functions using particle swarm optimization. Suranaree J. Sci. Technol. **19**(4), 259–264 (2012)
3. Pittman, J., Murthy, C.A.: Fitting optimal piecewise linear functions using genetic algorithms. IEEE Trans. Pattern Anal. Mach. Intell. **22**(7), 701–718 (2000)
4. Cang, V.T., Le, T.H.: Modeling of ship surface with non uniform B-spline. In: Proceedings of the International Multiconference of Engineers and Computer Scientists (vol. 2) (2011) www.iaeng.org/publication/IMECS2011/IMECS2011_pp1129-1132.pdf
5. Sun, Y.H., Tao, Z.L., Wei, J.X., Xia, D.S.: B-spline curve fitting based on adaptive particle swarm optimization algorithm. Appl. Mech. Mater. **20**, 1299–1304 (2010)
6. Gálvez, A., Iglesias, A.: Particle swarm optimization for non-uniform rational B-spline surface reconstruction from clouds of 3D data points. Inf. Sci. **192**, 174–192 (2012)
7. Gálvez, A., Iglesias, A., Puig-Pey, J.: Iterative two-step genetic-algorithm-based method for efficient polynomial B-spline surface reconstruction. Inf. Sci. **182**(1), 56–76 (2012)
8. Gálvez, A., Iglesias, A.: Firefly algorithm for explicit B-spline curve fitting to data points. Mathematical Problems in Engineering, 2013, (2013). Article ID 528215, 12 pp., doi:10.1155/2013/528215
9. Yoshimoto, F., Harada, T., Yoshimoto, Y.: Data fitting with a spline using a real-coded genetic algorithm. Comput. Aided Des. **35**(8), 751–760 (2003)

10. Kalaivani, S., Aravind, T., Yuvaraj, D.: A single curve piecewise fitting method for detecting valve stiction and quantification in oscillating control loops. In: Proceedings of the Second International Conference on Soft Computing for Problem Solving (SocProS 2012), Springer India December 28–30, 2012, pp. 13–24 (2014)
11. Islam, M.T., Moniruzzaman, M., Misran, N., Shakib, M.N.: Curve fitting based particle swarm optimization for UWB patch antenna. J. Electromagn. Waves Appl. **23**(17–18), 2421–2432 (2009)
12. Xiao, R.R., Zhang, J., Liu, H.Q.: NURBS fitting optimization based on ant colony algorithm. Adv. Mater. Res. **549**, 988–992 (2012)
13. Garcia-Capulin, C.H., Cuevas, F.J., Trejo-Caballero, G., Rostro-Gonzalez, H.: Hierarchical genetic algorithm for B-spline surface approximation of smooth explicit data. Math. Probl. Eng. (2014). doi:10.1155/2014/706247
14. Rashedi, E., Nezamabadi-Pour, H., Saryazdi, S.: GSA: a gravitational search algorithm. Inf. Sci. **179**(13), 2232–2248 (2009)
15. Deep, K., Thakur, M.: A new crossover operator for real coded genetic algorithms. Appl. Math. Comput. **188**(1), 895–911 (2007)
16. Deep, K., Thakur, M.: A new mutation operator for real coded genetic algorithms. Appl. Math. Comput. **193**(1), 211–230 (2007)
17. Singh, A., Deep, K., Atulya, N.: A new improved gravitational search algorithm for function optimization using a novel "best-so-far" update mechanism, 2015 international conference on soft computing and machine intelligence (accepted)
18. Singh, A., Deep, K.: Real coded genetic algorithm operators embedded in gravitational search algorithm for continuous optimization. International journal of intelligent systems and applications (in press)
19. Gerald, C.F., Wheatley, P.O.: Applied numerical analysis with MAPLE. Addison-Wesley, New York (2004)

Investigation of Suitable Perturbation Rate Scheme for Spider Monkey Optimization Algorithm

Kavita Gupta and Kusum Deep

Abstract Spider Monkey Optimization (SMO) is a new metaheuristic whose strengths and limitations are yet to be explored by the research community. In this paper, we make a small but hopefully significant effort in this direction by studying the behaviour of SMO under varying perturbation rate schemes. Four versions of SMO are proposed corresponding to constant, random, linearly increasing and linearly decreasing perturbation rate variation strategies. This paper aims at studying the behaviour of SMO technique by incorporating these different perturbation rate variation schemes and to examine which scheme is preferable to others on the benchmark set of problems considered in this paper. A benchmark set of 15 unconstrained scalable problems of different complexities including unimodal, multimodal, discontinuous, etc., serves the purpose of studying this behaviour. Not only numerical results of four proposed versions have been presented, but also the significance in the difference of their results has been verified by a statistical test.

Keywords Metaheuristics · Spider monkey optimization · Control parameters · Perturbation rate

List of symbols

Dim	No. of dimensions
$R(a, b)$	Uniformly generated random number between a and b
NG	Number of groups in current swarm
Pr	Perturbation rate
$G[k]$	Number of members in the kth group
$I[k][0]$	Index of first member of the kth group in the swarm
$I[k][1]$	Index of last member of the kth group in the swarm

Kavita Gupta (✉) · K. Deep
Department of Mathematics, Indian Institute of Technology Roorkee, Roorkee 247667, Uttarakhand, India
e-mail: gupta.kavita3043@gmail.com

K. Deep
e-mail: kusumdeep@gmail.com

M. Pant et al. (eds.), *Proceedings of Fifth International Conference on Soft Computing for Problem Solving*, Advances in Intelligent Systems and Computing 437, DOI 10.1007/978-981-10-0451-3_75

S_i	Position vector of *i*th spider monkey in the swarm
s_{ij}	*j*th coordinate of the position of *i*th spider monkey
S_{new}	A trial vector for creating a new position of a spider monkey
S_{r}	Position vector of randomly selected member of group
ll_{kj}	*j*th co-ordinate of the local leader of the *k*th group
gl_j	*j*th co-ordinate of the global leader of the swarm
$s_{\text{min}j}$	Lower bound on the *j*th decision variable
$s_{\text{max}j}$	Upper bound on the *j*th decision variable
fitness (S_i)	Fitness of the position of *i*th spider monkey
LLC_k	Limit count of the local leader of the *k*th group

1 Introduction

Last few decades witness the successful application of metaheuristics in solving real-world optimization problems [1]. Genetic Algorithms (GA) [2], Ant Colony Optimization (ACO) [3], Particle Swarm Optimization (PSO) [4], Differential Evolution (DE) [5], Artificial Bee Colony (ABC) [6], etc., are just a few names in the emerging list of metaheuristics which have been successfully applied to find solutions of complex real-world optimization problems. Metaheuristics are placed in the category of modern optimization techniques. The main reason of increasing popularity of metaheuristics is the problem independent nature of them. Whereas different traditional optimization techniques require mathematical model of the optimization problems which are to be solved to satisfy different mathematical properties like continuity, differentiability, separability, etc., metaheuristics are not bound to follow any such restrictions. Though metaheuristics do not guarantee to find optimal solution like traditional optimization methods, yet they provide a satisfactory solution to those problems at reasonable computational cost for which no problem specific algorithms exist. Increasing applicability of metaheuristics also demand for the theoretical development of metaheuristics. Every metaheuristic technique has some parameters which play dominant role in its performance. Performance of a metaheuristic largely depends on the setting of these parameters. For theoretical development, we need to find out the parameters that effect the performance of a metaheuristic and methods to control them. Some parameters are static while others are dynamic in nature. Static parameters are those which remain constant from the beginning of the optimization technique till the termination criterion is met and dynamic parameters take different values in different iterations based on the strategy used to alter them.

SMO is a new metaheuristic which falls in the category of swarm intelligent techniques [7]. Though there are numerous established metaheuristics, yet all the claims about the superiority of any metaheuristic over the others is shattered by the no free lunch theorem [8] and it makes room for more invention and discoveries in the field of metaheuristics. This paper deals in discussing the behaviour of Spider Monkey Optimization (SMO) technique by varying one of its control parameters namely perturbation rate. Section 2 gives brief introduction of Spider Monkey Optimization technique. Section 3 describes the role of perturbation rate in SMO and its variation schemes proposed in this paper. Section 4 describes experimental setup. Section 5 deals with the experimental results and discussion followed by conclusion and future scope in Sect. 6.

2 Spider Monkey Optimization Technique

SMO technique finds its source of inspiration in food searching strategy of spider monkeys. Spider Monkeys belong to the class of Fission Fusion Structure-based animals. These type of animals live in a group of 40–50 individuals. Scarcity of food compels these animals to divide themselves into smaller groups and search for their food in different directions. Later, having enough food, they all unite together and share their food. SMO technique has been introduced by Bansal et al. in [7]. Like all other metaheuristics, it starts with a collection of randomly generated solutions called swarm (in case of swarm intelligent techniques) which are intended to be improved through iterations to converge towards optimal solution. It also shares features like ease of implementation, few control parameters, efficiency in solving non-linear optimization problems with other swarm intelligent techniques. Premature convergence and stagnation are two of the main problems of a metaheuristic technique. Even initial versions of well-established algorithms like PSO, ABC, DE, etc., face these problems, but SMO has been designed in such a way that it deals with these two problems very efficiently. Also, the results shown in [7] depicts that SMO is a competitive member of the family of metaheuristics.

SMO has four control parameters namely Local Leader Limit, Global Leader Limit, Perturbation rate and maximum number of groups. Spider monkeys update their position based on linear equations. These update equations are guided by two types of solutions which are global leader and local leaders. SMO starts with the swarm of randomly generated solutions and passes through six iterative steps namely Local Leader Phase, Global Leader Phase, Local Leader Learning Phase, Global Leader Learning Phase, Local Leader Decision Phase, Global Leader Decision Phase. Detailed description of each phase can be found in [7]. Pseudocode of SMO has been given in Fig. 1.

3 Perturbation Rate and Its Variation Schemes

Perturbation rate controls the amount of change in the current position of a spider monkey in SMO technique. It is used in Local Leader Phase and Local Leader Decision Phase. Pseudocode for Local Leader Phase and Local Leader decision

```
begin:
        Initialization of the swarm
        Set local leader limit, global leader limit, perturbation rate, maximum number of groups
        Iteration = 0
        Evaluate fitness value of each spider monkey in the swarm
        Apply greedy selection to find global and local leaders
        while(termination criterion is not satisfied)do
                //Local Leader Phase
                //evaluate selection probability of each spider monkey
                //Global Leader Phase
                //Global Leader Learning Phase
                //Local Leader Learning Phase
                //Local Leader Decision Phase
                //Global Leader Decision Phase
                Iteration = Iteration+1
        end while
end
```

Fig. 1 Pseudocode of SMO

Phase have been given in Figs. 2 and 3 respectively. Local Leader Phase provides chance to every individual of the swarm to update itself and Local Leader Decision Phase is responsible for the re-initialization of a particular group if its local leader gets stagnated means not improving for specified number of iterations. From Figs. 2 and 3, usage of perturbation rate in these two phases can be seen. Also, it can be observed that in both the phases, it has been used in different ways. In Local Leader Phase, it has been used to decide whether a particular dimension of the solution should be updated or not. Here, it can be observed that lower the perturbation rate, more number of dimensions of a solution get the chance to be updated as compared to that of higher perturbation rate. Thus, lower perturbation rate facilitates

```
begin:
        for k = 1: NG do
                for i = I[k][0]:I[k][1] do
                        for j = 1: Dim do
                                if R(0,1)≥ pr then
                                  s_newj = s_ij + R(0,1) × (ll_kj − s_ij) + R(−1,1) × (s_rj − s_ij)
                                else
                                  s_newj = s_ij
                                end if
                        end for
                        if (fitness(S_new) > fitness(S_i))
                                S_i = S_new
                        end if
                end for
        end for
end
```

Fig. 2 Pseudocode for local leader phase

```
begin:
    For k = 1: NG do
        If (LLC_k > local leader limit) then
            LLC_k = 0
            for i = I[k][0]:I[k][1] do
                for j = 1: Dim do
                    if R(0,1)≥ pr then
                    s_ij = s_minj + R(0,1) × (s_maxj − s_minj)
                    else
                    s_ij = s_ij + R(0,1) × (gl_j − s_ij) + R(0,1) × (s_ij − ll_kj)
                    end if
                end for
            end for
        end if
    end for
end
```

Fig. 3 Pseudocode of local leader decision phase

exploration while higher perturbation rate facilitates exploitation in this phase. In local leader decision phase, perturbation rate has been used to decide how a particular dimension of the solution should be updated. It decides whether it should be randomly generated between the bounds or it should be updated using an equation which is a linear combination of its own position, its local leader position and global leader position. Thus perturbation rate plays central role in updating the swarm.

In this paper, we have used four variation schemes for perturbation rate. These are constant, random, linearly increasing and linearly decreasing. Corresponding to four perturbation rate varying schemes, there are four versions of SMO has been proposed in this paper. These are PRC (constant), PRR (random), PRLI (linearly increasing) and PRLD (linearly decreasing). Perturbation rate varying scheme has been mentioned in round parentheses in the previous line. Original SMO [7] adopts linearly increasing perturbation rate scheme. Formula for generating perturbation rate in each iteration in PRR, PRLI and PRLD has been given in Table 1. In PRR, perturbation rate is generated from a function of random numbers. In PRLI and PRLD, perturbation rate is generated from a formula such that the value of perturbation rate is increasing and decreasing respectively with iterations.

Table 1 Formulas for perturbation rate (pr)

Serial no.	Perturbation rate varying strategy	Corresponding version of SMO	Formula for perturbation rate
1	Constant	PRC	$\mathrm{pr} = 0.5$
2	Random	PRR	$\mathrm{pr} = 0.5 + \frac{\mathrm{rand}()}{2}$ Here, rand() is the uniform random number generated in (0, 1)
3	Linearly increasing (form 0.1 to 0.4 as mentioned in [1])	PRLI	$\mathrm{pr}(\mathrm{current}) = \mathrm{pr}(\mathrm{old}) + \frac{(0.4-0.1)}{\text{maximum number of iterations}}$
4	Linearly decreasing (form 0.4 to 0.1)	PRLD	$\mathrm{pr}(\mathrm{current}) = \mathrm{pr}(\mathrm{old}) - \frac{(0.4-0.1)}{\text{maximum number of iterations}}$

4 Experimental Setup

Experiments have been performed with 15 scalable benchmark problems. List of benchmark problems along with their search range and optimal value is given in Table 2. All the problems are of same dimension, unconstrained in nature and of minimization type.

Except perturbation rate, rest of the parameters are same for each version of SMO. Parameter setting has been provided in Table 3.

Since initial swarm heavily effects the performance of a metaheuristic, results from 100 independent runs have been considered to reduce any kind of biasness. Also, to make fair comparison among different versions, same run in every algorithm starts with the same initial seed.

5 Experimental Results and Discussion

Results of four versions of SMO namely PRC, PRR, PRLI and PRLD have been compared on the basis of reliability and convergence speed of the algorithm. For this purpose, two things of all the versions have been recorded. One is number of

Table 2 List of benchmark problems

Test problems	Objective function	Search range	Optimal value
Sphere function	$f_1(x) = \sum_{i=1}^{D} x_i^2$	[−5.12, 5.12]	0
De Jong's F4	$f_2(x) = \sum_{i=1}^{D} ix_i^4$	[−5.12, 5.12]	0
Griewank	$f_3(x) = \sum_{i=1}^{D} \frac{x_i^2}{4000} - \prod_{i=1}^{D} \cos\left(\frac{x_i}{\sqrt{i}}\right) + 1$	[−600, 600]	0
Rosenbrock	$f_4(x) = \sum_{i=1}^{D} \left[100(x_i^2 - x_{i+1})^2 + (x_i - 1)^2\right]$	[−100, 100]	0
Rastrigin	$f_5(x) = \sum_{i=1}^{D} (x_i^2 - 10\cos(2\pi x_i) + 10)$	[−5.12, 5.12]	0
Ackley	$f_6(x) = -20\exp\left(-0.2\sqrt{\frac{1}{D}\sum_{i=1}^{D} x_i^2}\right) - \exp\left(\frac{1}{D}\sum_{i=1}^{D} \cos(2\pi x_i)\right) + 20 + e$	[−30, 30]	0
Alpine	$f_7(x) = \sum_{i=1}^{D} \lvert x_i \sin(x_i) + 0.1x_i \rvert$	[−10, 10]	0
Michalewicz	$f_8(x) = -\sum_{i=1}^{D} \sin(x_i)\left[\frac{\sin(ix_i^2)}{\pi}\right]^{20}$	[0, π]	−9.66015
Cosine mixture	$f_9(x) = \sum_{i=1}^{D} x_i^2 - 0.1\sum_{i=1}^{D} \cos(5\pi x_i)$	[−1, 1]	−D * 0.1
Exponential	$f_{10}(x) = -\exp(-0.5\sum_{i=1}^{D} x_i^2)$	[−1, 1]	−1
Zakharov	$f_{11}(x) = \sum_{i=1}^{D} x_i^2 + (\frac{1}{2}\sum_{i=1}^{D} ix_i)^2 + (\frac{1}{2}\sum_{i=1}^{D} ix_i)^4$	[−5.12, 5.12]	0
Cigar	$f_{12}(x) = x_1^2 + 100000\sum_{i=2}^{D} x_i^2$	[−10, 10]	0
Brown 3	$f_{13}(x) = \sum_{i=1}^{D-1} \left[(x_i^2)^{(x_{i+1}^2+1)} + (x_{i+1}^2)^{(x_i^2+1)}\right]$	[−1, 4]	0
Schewel prob 3	$f_{14}(x) = \sum_{i=1}^{D} \lvert x_i \rvert + \prod_{i=1}^{D} \lvert x_i \rvert$	[−10, 10]	0
Salomon problem	$f_{15}(x) = 1 - \cos\left(2\pi\sqrt{\sum_{i=1}^{D} x_i^2}\right) + 0.1\sqrt{\sum_{i=1}^{D} x_i^2}$	[−100, 100]	0

Table 3 Parameter setting and termination criterion

Parameter	Value
Swarm size	150
Dimension	30
Maximum number of groups	5
Local leader limit	100
Global leader limit	50
Total number of independent runs	100
Maximum number of iterations	4000
Acceptable error (difference between obtained global optima and optimal value)	10^{-5}
Termination criterion	Either maximum number of iterations performed or acceptable error is achieved

successful runs out of 100 independent runs and the other is average number of function evaluations for successful runs. Reliability is measured in terms of number of successful runs and convergence speed is measured in terms of average number of function evaluations for successful runs.

First, all the versions of SMO have been compared on the basis of number of successful runs out of 100 independent runs. The version having highest number of successful runs is the winner. If all the versions have same number of successful runs, then the results are compared on the basis of function evaluations for successful runs and the algorithm with lowest number of function evaluations for successful runs is the winner. Also, to see whether there is a significant difference in the results, a statistical test namely t-test at a significance level 0.05 has been employed. The sign "=" is used where there is no significant difference in results. Signs "+" and "−" are used where the best proposed version takes less or more number of function evaluations, respectively, for successful runs in comparison to other proposed versions. N.A. is used where t-test is not applicable, i.e. the cases where all the four versions of SMO do not have same number of successful runs and comparison is made on the basis of number of successful runs out of total number of independent runs instead of number of function evaluations.

From Table 4, it can be seen that out of 15 problems, there are 12 problems where all the versions have 100 successful runs. Problem no. 4 and 15 remain completely unsolved by all the versions while for problem no. 11, PRLD has highest number of successful runs.

Further out of 12 problems, where all the algorithms have same number of successful runs, it can be seen from Table 5 that there are 9 problems where PRLI has lowest number of function evaluations for successful runs. So, from the results, it can be concluded that PRLI performs best among all the versions of SMO proposed in this paper on the basis of criterion decided for comparison among them.

Table 4 Number of successful runs out of 100 independent runs

Fun	PRC	PRR	PRLI	PRLD
1	100	100	100	100
2	100	100	100	100
3	100	100	100	100
4	0	0	0	0
5	100	100	100	100
6	100	100	100	100
7	100	100	100	100
8	100	100	100	100
9	100	100	100	100
10	100	100	100	100
11	10	0	47	**68**
12	100	100	100	100
13	100	100	100	100
14	100	100	100	100
15	0	0	0	0

Table 5 Number of function evaluations for successful runs

Fun	PRC	PRR	PRLI	PRLD
1	44001	59,579	**33,192**	40,494
2	33,389	41,863	**26,566**	31,305
3	**82,569**	98,972	103,625	83,820
4	N.A.	N.A.	N.A.	N.A.
5	222,874	**205,089**	229,495	227,081
6	84,357	115,227	**63,917**	77,588
7	207,462	238,851	**189,827**	198,791
8	22,035	**20,659**	24,027	21,042
9	45,451	60,724	**37,701**	41,809
10	33,718	45,929	**25,421**	31,078
11	N.A.	N.A.	N.A.	N.A.
12	76,714	103,146	**58,502**	70,549
13	43,047	58,532	**32,220**	39,519
14	80,386	108,356	**62,664**	74,499
15	N.A.	N.A.	N.A.	N.A.

Further to judge the significance in difference of number of function evaluations for successful runs of PRLI with other versions, pairwise t-test is conducted among PRLI and other versions and the results are presented in Table 6. Note that t-test has been conducted only for those problems where all the versions of SMO has same number of successful runs in order to see the significance in the difference of

Table 6 Outcome of t-test

Fun	PRLI versus PRC	PRLI versus PRR	PRLI versus PRLD
1	+	+	+
2	+	+	+
3	−	=	+
4	N.A.	N.A.	N.A.
5	+	−	=
6	+	+	+
7	=	+	=
8	=	−	+
9	+	+	+
10	+	+	+
11	N.A.	N.A.	N.A.
12	+	+	+
13	+	+	+
14	+	+	+
15	N.A.	N.A.	N.A.

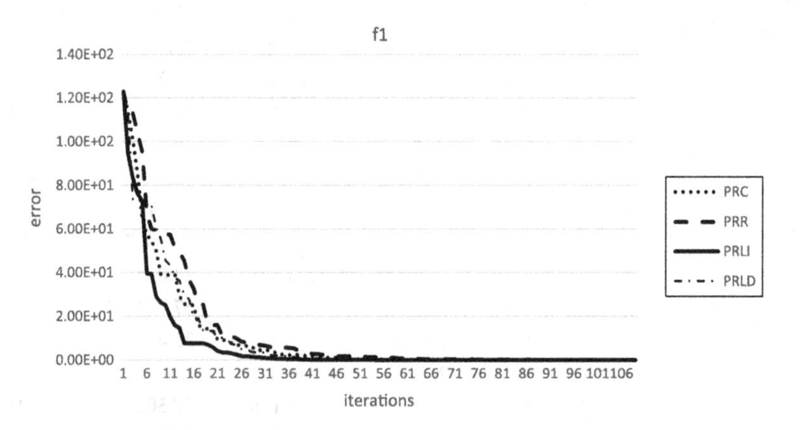

Fig. 4 Convergence graph of problem 1

function evaluations. From Table 6, it can be concluded that PRLI performs significantly better than other versions on most of the problems.

Also, the convergence of problem nos. 1, 6 and 12 have been shown in Figs. 4, 5 and 6 respectively. Boxplot of average number of function evaluations for successful runs has been provided in Fig. 7.

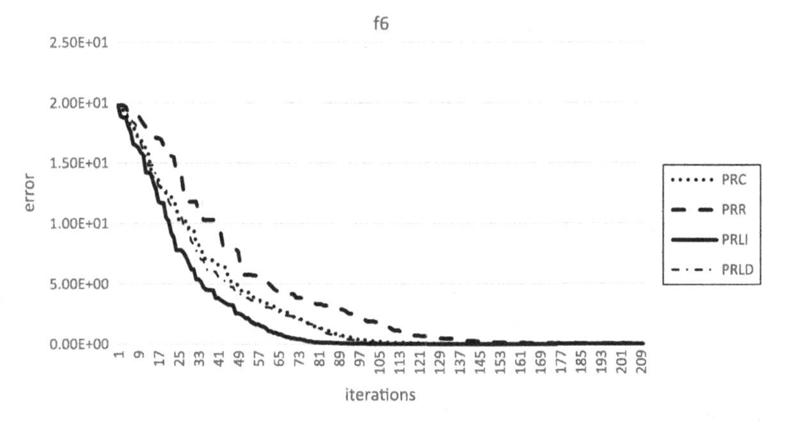

Fig. 5 Convergence graph of problem 6

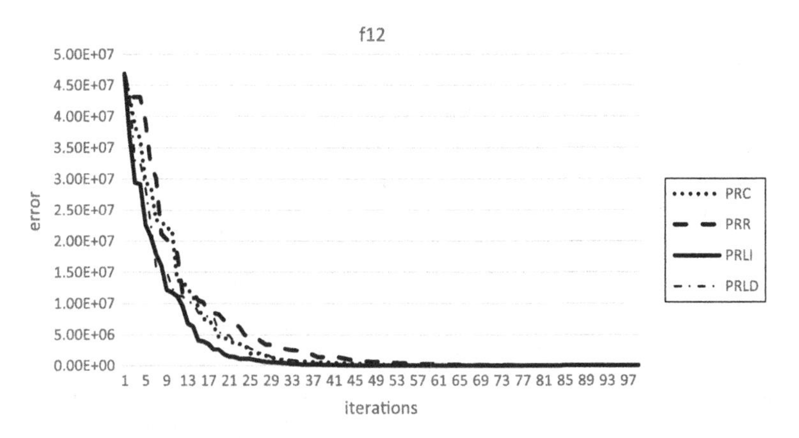

Fig. 6 Convergence graph of problem 12

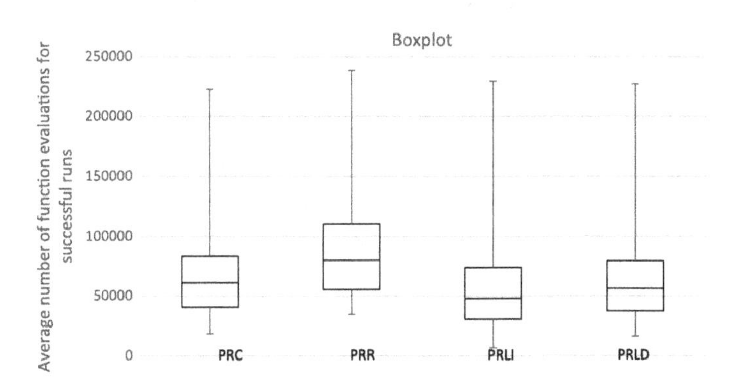

Fig. 7 Boxplot graph of average number of function evaluations of successful runs

6 Conclusion and Future Scope

In this paper, four versions of SMO namely PRC, PRR, PRLI and PRLD are proposed which adopt constant, random, linearly increasing and linearly decreasing perturbation rate strategies respectively. It is concluded from the results that the linearly increasing perturbation rate which has been used in original SMO is preferable to other three perturbation rate strategies, but these conclusions are made on experimental results only. No theoretical proof and discussion has been provided to make any claim regarding the superiority of one strategy over the other. In future, more perturbation rate varying strategies like adaptive, non-linear, chaotic, etc., will be studied and theoretical details of their performance will be attempted to discuss.

Acknowledgments The first author would like to acknowledge the Ministry of Human Resource Development, Government of India for financial support.

References

1. Gonga, A., Tayal, A.: Metaheuristics: review and application. J. Exp. Theor. Artif. Intell. **25**(4), 503–526 (2013)
2. Holland, J.H.: Adaptation in natural and artificial systems. University of Michigan press, AnnArbor (1975)
3. Dorigo, M.: Optimization, Learning and Natural Algorithms. Ph.D. thesis, Politecnico di Milano, Italy (1992)
4. Kennedy, J., Eberhart, R.: Particle swarm optimization. Proc. IEEE Int. Conf. **4**, 1942–1948 (1995)
5. Storn, R., Price, K.: Differential evolution—a simple and efficient heuristic for global optimization over continuous spaces. J. Global Optim. **11**(4), 341–359 (1997)
6. Karaboga, D.: An idea based on honeybee swarm for numerical optimization. In: Technical Report TR06, Erciyes University, Engineering Faculty, Computer Engineering Department (2005)
7. Bansal, J.C., Sharma, H., Jadon, S.S., Clerc, M.: Spider monkey optimization algorithm for numerical optimization. Memetic Comput. **6**(1), 31–47 (2014)
8. Wolpert, D.H., Macready, W.G.: No free lunch theorems for optimization. IEEE Trans. Evol. Comput. **1**, 67–82 (1997)

A Portfolio Analysis of Ten National Banks Through Differential Evolution

Hira Zaheer, Millie Pant, Oleg Monakhov and Emilia Monakhova

Abstract Portfolio optimization guides about the management of assets. Among several investment offers the task is to choose the plan so as to attain the maximum financial benefit. The investor requires a thorough comparative study to decide the best possible option where the return is maximum and the risk is minimum. Portfolio optimization can help in the process of decision-making by bringing out the selected options beneficial for the investor. In this paper we have considered the mean semivariance portfolio optimization model given by Markovitz and the data is taken from National Stock Exchange (NSE), Mumbai. Ten years data of ten different banks is taken from national stock exchange. The model is solved with the help of differential evolution which is a population-based metaheuristics.

Keywords Portfolio optimization · Differential evolution · Mean semivariance model

1 Introduction

Optimization models have a long history in number of financial domains. Many computational financial problems such as resource portion, risk management, from alternative estimating to model adjustment can be tackled effectively using optimization models. In 1950s Markowitz [1] distributed his spearheading work where

Hira Zaheer (✉) · M. Pant
Indian Institute of Technology, Roorkee, Roorkee, India
e-mail: hirazaheeriitr@gmail.com

M. Pant
e-mail: millifpt@iitr.ernet.in

Oleg Monakhov · Emilia Monakhova
Institute of Computational Mathematics and Mathematical Geophysics, Novosibirsk, Russia
e-mail: monakhov@rav.sscc.ru

Emilia Monakhova
e-mail: emilia@rav.sscc.ru

M. Pant et al. (eds.), *Proceedings of Fifth International Conference on Soft Computing for Problem Solving*, Advances in Intelligent Systems and Computing 437, DOI 10.1007/978-981-10-0451-3_76

he proposed a straightforward quadratic project for selecting a differentiated arrangement of securities by considering two aspects: expected return and portfolio risk. His model for portfolio choice can be formed mathematically either as maximization or as a minimization problem. In maximization problems, return is maximized with bounded risk and in minimization problems risk is minimized with some fixed return.

Markowitz's model has become an important tool for portfolio selection. However, its practical application is not much popular because of the complex quadratic nature of model having a large number of variables. Thus, a portfolio will be efficient if it is giving the maximum possible return for a given level of risk and if it provides minimum risk for given return. The group of such portfolios is called efficient frontier.

2 Portfolio Optimization

A portfolio is collection of two or more risky assets held by an institution or an individual and represented in an ordered n-tuple, i.e., $(x_1, x_2, x_3 \ldots x_n)$ where x_i denotes the amount of fund invested in *i*th asset. Portfolio management is basically a method of guiding how to buy, hold, sell, and invest for both risky and riskless asset. Portfolio theory tells about the allocation of wealth and management of liabilities.

Portfolio optimization problem is basically the problem which seeks to maximize the profitable returns through optimally allocating capital into a large number of assets under the minimization constraint of risk. Modern portfolio theory can be traced back to the work proposed by Markowitz in 1952. Since then the portfolio optimization problem has been widely studied by utilizing many optimization methods, for example, the stochastic optimization method, fuzzy optimization method, and robust optimization method [1].

Although many portfolio optimization models have been extensively proposed, most of them only focus on the solutions of the concentrated investment which contradicts the risk diversified property as the requirement in practical applications.

Its solution methods include quadratic programming, nonlinear programming, mixed integer programming, and metaheuristic techniques such as genetic algorithm, genetic programming, differential evolution, simulated annealing, particle swarm optimization, ant colony optimization, Tabu search, etc. Portfolio optimization approaches are mainly based on both the stochastic optimization and fuzzy optimization. In this paper, we present a constraint nonlinear problem based on fuzzy optimization approach. Here, variables can take any fuzzy value between 0 and 1, i.e., a person may invest in one asset or in fractions too.

Bonami and Lejeune [2] proposed a solution approach in which ambiguity of expected return and integer dealing restrictions both are considered simultaneously. Their method is based on branch and bound algorithm having two more rules of branching. The first rule is static in nature which is known as idiosyncratic risk branching and the second one is dynamic, known as portfolio risk branching. Liu [3] discusses a fuzzy portfolio optimization problem in which the return is taken as fuzzy number. In his approach bounds (upper and lower) of portfolio return are calculated with the help of a couple of two-step mathematical program and then that pair of dual-step mathematical program is converted into a pair of simple one-step linear programs by some transformation techniques and duality theorems. A linear portfolio management model is presented by Konno and Yamazaki [4] known as mean absolute deviation (MAD) model. In this model, absolute deviation risk is considered in place of quadratic risk function (variance) of MV model. Speranza [5] presented a semi-absolute deviation model for measuring risk. Mansini and Speranza [6] developed an algorithm by constructing subproblems of mixed integer type with minimum transaction lots and took semi-absolute deviation model for solving the problem. Oritowe et al. [7] introduce a method to initialize population using the extreme point of Hessian approach and used this approach in GA to solve portfolio optimization problems. Yen and Yen [8] developed a coordinatewise descent algorithm in which assets are normed by l_q norms for $1 \leq q \leq 2$, applied to solve a minimum variance portfolio (mvp) optimization problem which is tested on two data sets. Chen et al. [9] applied artificial bee colony algorithm for solving this type of problems and tested the efficiency of their algorithm on four global stock market indexes from OR-Library. Tuba and Bacanin [10] applied firefly algorithm to solve this type of problems and do not find appropriate results, so, modified their algorithm to achieve better exploration/exploitation and checked on benchmark data sets from OR-Library. Skolpadungket et al. [11] applied different variants of multi-objective GA to solve portfolio optimization problem. Aranha et al. [12] and Chang et al. [13] also used GA techniques for solving portfolio optimization problem.

3 Differential Evolution

DE [14] is population-based technique for solving global optimization problem. DE can solve single objective, multi-objective, linear, nonlinear, continuous, noncontinuous, and even more complicated problems. It contains mainly three steps: mutation, crossover, and selection. Pseudo code of DE is given in Fig. 1. A very few work have been done using DE for solving portfolio optimization problem. Differential evolution is not directly applied to portfolio optimization problems. Ardia et al. [15] worked on a DE optim package which was developed in 2005, and solve nonconvex portfolio optimization problem. De Optim follows the steps of Differential evolution. Ma et al. [16] proposed a hybridized DE approach by combing penalty function method with DE. The model which they have chosen was minimum of value-at-risk with cardinality constraint. The problem was taken from Shanghai and Shenzhen Stock Exchange. Korczak and Roger [17] introduced a new method of share portfolio optimization and solved with the help of differential evolution. Zaheer et al. [18] gave a new mutation strategy of differential evolution which can be used further for solving portfolio optimization problem.

Fig. 1 Pseudo code for DE

```
/* Initialize parameters */
Set mutation scale factor Fϵ [0,2],
crossover parameter CRϵ (0,1),
population size NP, counter g =0
/* Initialize a population */
while stopping criteria do
g=g+1
for i=1 to NP do
  /*Differential Mutation */
  Select randomly 3 individuals
  X^g_p1, X^g_p2, X^g_p3 ϵ p^g
V_i = X^g_p1 + FX (X^g_p2 + X^g_p3)
  p1,p2,p3 ϵ { 1,2,....NP}/{i}
 /* probability Binomial Crossover */
  Generate randomly J_rand ϵ {1,2,.....,D}
for j=1 to D do
generate randomly rand(0,1)
if rand(0,1) ≤ CR or j=j_rand
u_1^j,g = x^j_i
end
end for
/*Greedy Selection */
If f(u_i^g) < f(X^j_g)
else
u_1^j,g = x^g_i
end if
end for
end while
```

4 Mean Semivariance Model

The assessment of risk by variance is just average, when returns are typically circulated or financial specialists show quadratic inclinations. Various studies prove that the majority of hedge funds is with asymmetric returns, mean variance model becomes less appropriate [19], that is why there is a need to replace variance with a downside risk measure, i.e., the measure which considers only the negative deviations from the level of reference return. So, Markovitz introduced semivariance as it is one of the most popular downside risk measure and named that as men semivariance model [19–22]. It is preferred over variance because it does not consider value beyond the critical values. If all return distributions are symmetric or have same degree of asymmetry then both models will give same set of efficient portfolios. In previous work, Zaheer et al. [23] worked over mean variance model of portfolio optimization.

Markovitz theory helps in deciding which portfolio is best among different portfolios in the market. In this model, he explained how to maximize the return for a given level of risk and how to minimize the risk for the fixed return. Markowitz mean semivariance optimization model is single objective model with constraints. Semivariance is the expected value of squared negative deviations of possible outcomes from expected return. Mathematical formulation of mean variance optimization model is as follows:

Let us suppose that the return rate of ith variable is denoted by random variable R_i (i varies from 1 to m), x_i is the amount of fund invented in ith asset.

Asset return is the rate of return which can be calculated for a given period of time. Mathematically,

Return = (Closing price of current period − closing price of previous period + dividend collect during the period)/(closing price of previous period)

$$r_{\text{in}} = \frac{(p_{\text{in}}) - (p_{\text{in}-1}) + (d_{\text{in}})}{(p_{\text{in}-1})}$$

where p_{in} is the closing price of ith asset for the duration n,

$p_{\text{in}-1}$ is the closing price of ith asset for period $n-1$,
d_{in} is the dividend collected during the period.

The expected return of portfolio is given by

$$r(x_1, x_2, \ldots, x_m) = E\left[\sum_{i=1}^{m} R_i x_i\right] = \sum_{i=1}^{m} E[R_i] = \sum_{i=1}^{m} r_i x_i$$

where $E[R_i]$ is the expected value of random variable R_i. $E[R_i] = r_i$.

Portfolio risk measured as semivariance is denoted by $s(x_1, x_2, \ldots, x_n)$ and is defined as follows:

$$\begin{aligned} s(x_1, x_2, \ldots, x_n) &= E\left[\left[\sum_{i=1}^{m} R_i x_i - E\left[\sum_{i=1}^{m} R_i x_i\right]\right]^{-}\right]^2 \\ &= \frac{1}{T}\sum_{i=1}^{T}\left\{\left[\sum_{i=1}^{m}(r_{it} - r_i)x_i\right]^{-}\right\}^2 \end{aligned}$$

where

$$\left[\sum_{i=1}^{m} R_i x_i - E\left[\sum_{i=1}^{m} R_i x_i\right]\right]^{-} = \begin{cases} \sum_{i=1}^{m} R_i x_i - E\left[\sum_{i=1}^{m} R_i x_i\right], & \text{if} \sum_{i=1}^{m} R_i x_i - E\left[\sum_{i=1}^{m} R_i x_i\right] < 0 \\ 0, & \text{if} \sum_{i=1}^{m} R_i x_i - E\left[\sum_{i=1}^{m} R_i x_i\right] \geq 0 \end{cases}$$

In mean semivariance model, it is not required to compute variance covariance matrix, only the joint distribution of assets is necessary. Mathematical formulation of Marcowitz mean semivariance model with risk minimization is as follows:

$$\min E\left[\left[\sum_{i=1}^{m} R_i x_i - E\left[\sum_{i=1}^{m} R_i x_i\right]\right]^{-}\right]^2$$

Subject to

$$\sum_{i=1}^{m} r_i x_i = r_0$$

$$\sum_{i=1}^{n} x_i = 1$$

$$x_i \geq 0, i = 1, 2, \ldots, m$$

The above model can be written as follows:

$$\min \frac{1}{T} \sum_{i=1}^{T} p_t^2$$

Subject to

$$p_t \geq -\sum_{i=1}^{m} (r_{it} - r_i) x_i, \quad t = 1, 2, \ldots, T$$

$$\sum_{i=1}^{m} r_i x_i = r_0,$$

$$\sum_{i=1}^{m} x_i = 1,$$

$p_t \geq 0$ for $t = 1, 2, \ldots, T$.
$x_i \geq 0$ for $i = 1, 2, \ldots, m$.

5 Numerical Illustration

The mathematical model of the problem is taken from Gupta et al. [24] where data is taken from National Stock Exchange (NSE), Mumbai which is the ninth largest stock exchange in the world. Ten reputed national banks randomly selected from

NSE are considered as 10 different assets for the financial year 2004–2014. Average monthly return is calculated on the basis of daily time data for calculating variance and covariance of the assets. Mathematical model is formulated on the basis of monthly return and variance covariance matrix.

With the help of above tables and equations following mathematical model is formed in which risk is minimized in the objective function.

$$\min f(X) = \frac{1}{10}(p_1^2 + p_2^2 + p_3^2 + p_4^2 + p_5^2 + p_6^2 + p_7^2 + p_8^2 + p_9^2 + p_{10}^2)$$

subjected to

$$p_1 - 0.032940x_1 + 0.012646x_2 + 0.017183x_3 + 0.028531x_4 + 0.144380x_5 + 0.015385x_6 - 0.025730x_7 + 0.201983x_8 - 0.00192x_9 + 0.001805x_{10} \geq 0$$

$$p_2 + 0.537731x_1 + 0.003292x_2 + 0.004294x_3 + 0.022608x_4 + 0.010180x_5 - 0.034410x_6 - 0.000070x_7 - 0.187960x_8 - 0.020350x_9 + 0.011434x_{10} \geq 0$$

$$p_3 - 0.018020x_1 + 0.003386x_2 + 0.015762x_3 - 0.013720x_4 + 0.023990x_5 + 0.004529x_6 + 0.013755x_7 - 0.035177x_8 + 0.021163x_9 + 0.031460x_{10} \geq 0$$

$$p_4 - 0.091690x_1 - 0.008060x_2 - 0.031530x_3 - 0.067680x_4 - 0.054650x_5 - 0.032360x_6 - 0.058740x_7 - 0.249090x_8 + 0.001350x_9 + 0.008644x_{10} \geq 0$$

$$p_5 - 0.003400x_1 + 0.081151x_2 + 0.046475x_3 + 0.093674x_4 + 0.0767560x_5 + 0.074118x_6 + 0.069783x_7 + 0.269243x_8 + 0.053773x_9 + 0.000399x_{10} \geq 0$$

$$p_6 - 0.044600x_1 + 0.012560x_2 + 0.003601x_3 + 0.004162x_4 - 0.003520x_5 + 0.027065x_6 + 0.042311x_7 + 0.116613x_8 + 0.005672x_9 - 0.001470x_{10} \geq 0$$

$$p_7 - 0.087610x_1 - 0.022270x_2 - 0.070150x_3 - 0.023230x_4 - 0.028800x_5 - 0.029080x_6 - 0.027620x_7 - 0.060090x_8 - 0.034950x_9 - 0.005700x_{10} \geq 0$$

$$p_8 - 0.067540x_1 - 0.018050x_2 + 0.001173x_3 + 0.005572x_4 - 0.007280x_5 - 0.021640x_6 - 0.021480x_7 - 0.066180x_8 - 0.005340x_9 - 0.001070x_{10} \geq 0$$

$$p_9 - 0.070220x_1 + 0.015091x_2 + 0.001612x_3 + 0.010224x_4 - 0.000130x_5 - 0.027520x_6 - 0.014670x_7 - 0.066680x_8 - 0.031020x_9 - 0.001820x_{10} \geq 0$$

$$p_{10} - 0.121720x_1 - 0.079730x_2 + 0.011574x_3 - 0.060140x_4 - 0.030980x_5 + 0.023911x_6 + 0.022457x_7 + 0.006979x_8 + 0.011637x_9 - 0.043670x_{10} \geq 0$$

$$0.072001x_1 + 0.005743x_2 + 0.016592x_3 + 0.013213x_4 + 0.022602x_5 + 0.014836x_6 + 0.01199x_7 + 0.105402x_8 + 0.012896x_9 + 0.040609x_{10} = r_0$$

$$x_1 + x_2 + x_3 + x_4 + x_5 + x_6 + x_7 + x_8 + x_9 + x_{10} = 1,$$

$$p_t \geq 0, \quad i = 1, 2 \ldots 10$$

$$x_i \geq 0, \quad i = 1, 2 \ldots 10$$

Here, $f(X)$ represents the risk and x_i's represents different assets, i.e., ten banks which are shown on the left-hand side of Table 1. The problem consists of twelve

Table 1 Return of assets for the period April 01, 2004 to March 31, 2015

	Yearly returns									
State Bank of India (SBI)	0.039057	0.609732	0.053982	−0.01969	0.068606	0.027397	−0.0156	0.004464	0.001783	−0.04972
Punjab National Bank (PNB)	0.018389	0.009035	0.009129	−0.00232	0.086895	0.018303	−0.01653	−0.01231	0.020834	−0.07399
HDFC Bank (HDFC)	0.033774	0.020885	0.032353	−0.01494	0.063067	0.020193	−0.05355	0.017765	0.018204	0.028165
ICICI Bank (ICICI)	0.041743	0.035821	−0.00051	−0.05447	0.106887	0.017375	−0.01002	0.018785	0.023437	−0.04693
Axis Bank (Axis)	0.037040	0.032781	0.046592	−0.03205	0.099357	0.019083	−0.0062	0.015323	0.022468	−0.00838
Canara Bank (Canara)	0.030221	−0.01958	0.019366	−0.01752	0.088954	0.041901	−0.01425	−0.00680	−0.01268	0.038747
Vijaya Bank (Vijaya)	−0.01374	0.011919	0.025745	−0.04675	0.081773	0.054301	−0.01563	−0.00949	−0.00268	0.034447
Syndicate Bank (SYND)	0.307384	−0.08255	0.140579	−0.14369	0.374645	0.222015	0.045314	0.039218	0.038725	0.112381
Union Bank (UNI)	0.010971	−0.00745	0.034059	0.014245	0.066669	0.018567	−0.02206	0.007553	−0.01813	0.024532
Bank of India (BOI)	0.042415	0.052043	0.072069	0.049253	0.041008	0.039137	0.034906	0.039536	0.038789	−0.00306

constraints in which there are ten inequality constraints and two equality constraints.

For solving the above problem, the value of r_0 should be known. The problem is solved in two phases,

In phase I, the value of r_0 is calculated. The value of r_0 lies between r_{min} and r_{max} where r_{max} is the maximum return and r_{min} is the return corresponding to minimum risk, and can be calculated by omitting the return constraint and then minimizing the objective function subjected to all other constraints. Thus, we get the range in which r_0 lies.

In phase II, the objective function is minimized for different values of r_0 between r_{min} and r_{max}. By doing so, we get a solution set with different values of r_0 which is known as efficient frontier.

We applied standard differential evolution algorithm to solve this problem using 0.5 as mutation factor and crossover probability is set as 0.9 and Dev CPP as developer.

By solving for phase I, we get the objective function value as 0.335263 (risk). r_{min} is the return for minimum risk which is obtained as 0.030609 and from Table 2 it can be seen that r_{max} is 0.1054.

The result obtained from phase I is shown below in Table 3.

Following is the efficient frontier graph obtained from the data given in Table 4. On x-axis there are different values of return r_0, on y-axis there is risk. It can be seen from graph, that by increasing risk, the value of return is also increasing.

Table 2 Average return of assets

Bank	Return
SBI	0.072000
PNB	0.005743
HDFC	0.016591
ICICI	0.013212
Axis	0.022601
Canara	0.014836
Vijaya	0.011989
SYND	0.105401
UNI	0.012895
BOI	0.040609

Table 3 Summary result of portfolio selection using semivariance

Portfolio risk	Allocations				
	SBI	PNB	HDFC	ICICI	Axis
0.335263	0.239561714	0.119301143	0.002615871	0.087300286	0.150970571
	Canara	Vijaya	SYND	UNI	BOI
	0.013419429	0.246005429	0.004630086	0.039708857	0.096486614

Table 4 Portfolio set for different values of r0

Return	×1	×2	×3	×4	×5	×6	×7	×8	×9	×10	Risk
0.031221	0.593325	0.000000	0.016445	0.005748	0.005654	0.002255	0.000007	0.327412	0.046007	0.003147	0.085101
0.035520	0.592066	0.000578	0.015869	0.004849	0.003548	0.002268	0.000087	0.336548	0.040887	0.003300	0.095682
0.036948	0.580054	0.001918	0.010521	0.005798	0.001589	0.002870	0.003669	0.335548	0.054723	0.003310	0.100044
0.040434	0.570012	0.000383	0.004599	0.004889	0.001378	0.002298	0.017598	0.349778	0.045741	0.003324	0.106406
0.047648	0.566230	0.000047	0.004236	0.008844	0.001250	0.002399	0.002547	0.369547	0.036544	0.008356	0.125262
0.051706	0.559875	0.000285	0.003985	0.004549	0.001000	0.002413	0.009348	0.362465	0.023689	0.032391	0.138897
0.056911	0.530006	0.000351	0.003654	0.005464	0.001005	0.002500	0.070041	0.369999	0.013570	0.003410	0.152356
0.059510	0.521436	0.000445	0.003448	0.003248	0.000995	0.002574	0.065877	0.389005	0.009547	0.003425	0.165815
0.062108	0.500310	0.000718	0.003164	0.006950	0.000921	0.002710	0.090478	0.375412	0.015890	0.003447	0.181941
0.066628	0.492210	0.000058	0.002897	0.005698	0.000832	0.002770	0.082300	0.399755	0.009987	0.003493	0.198067
0.071147	0.488009	0.053813	0.002795	0.003155	0.000800	0.015004	0.000785	0.423445	0.008699	0.003495	0.220524
0.073895	0.485326	0.013385	0.002713	0.003186	0.000740	0.015472	0.000547	0.466542	0.008541	0.003548	0.240490
0.076643	0.473256	0.004015	0.002699	0.003201	0.000731	0.016423	0.000587	0.487562	0.007856	0.003670	0.249981
0.080524	0.412356	0.006491	0.002648	0.003266	0.000723	0.017547	0.025478	0.521249	0.006547	0.003695	0.265472
0.085164	0.400578	0.000230	0.002547	0.003276	0.000719	0.018560	0.000048	0.569745	0.000597	0.003700	0.280648
0.091008	0.404547	0.008670	0.005136	0.003999	0.000752	0.020958	0.002417	0.545879	0.003155	0.004488	0.310800
0.096484	0.378974	0.008838	0.004549	0.003219	0.000999	0.021925	0.003645	0.569598	0.003750	0.004504	0.353044
0.099250	0.360110	0.000700	0.001474	0.003240	0.000872	0.021479	0.003950	0.601531	0.002654	0.003990	0.381458
0.102445	0.355752	0.000723	0.001467	0.003460	0.000687	0.027798	0.000894	0.600804	0.003899	0.004516	0.411756
0.104245	0.349998	0.000986	0.001971	0.003570	0.000983	0.032478	0.000795	0.604005	0.002008	0.003206	0.436471

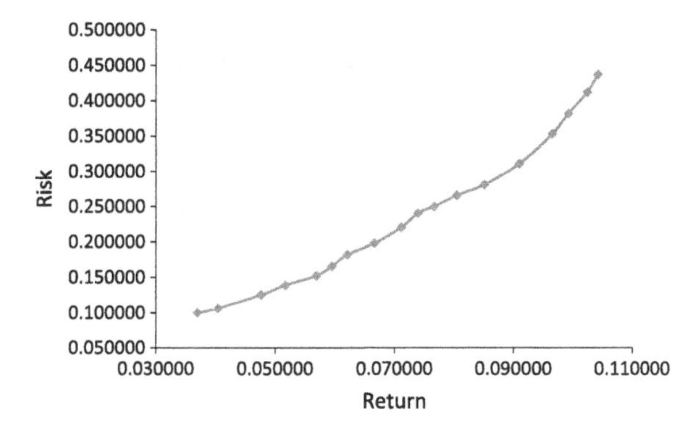

Fig. 2 Efficient Frontier

In solving phase II results are taken for 20 different values between $r_{\min}$ and $r_{\max}$ which are treated as 20 portfolios shown in Table 4. We can see from the table that by increasing the level of return the risk also increases.

6 Conclusion

It has been found that in the literature of financial mathematics, mean Markowitz model had not solved before with the help of DE. We applied DE to solve the minimization form of the mean variance model, and found reasonably good results, which are shown in Table 4 and in Fig. 2 (portfolio curve). This model can be used in two ways, if the investor knows that how much the return should, i.e., if there is any expected desired return then the portfolio objective problem can be solved directly to find out the portfolio. Second, if the expected portfolio is not known, then we may produce a set of efficient portfolios using the method explained above, from which investor can choose any one according to his/her preference.

Acknowledgments The reported study was partially supported by DST, research project No. INT/RFBR/P-164, and RFBR, research project No. 14-01-92694 IND-a.

References

1. Markowitz, H.: Portfolio selection*. J. Finance **7**(1), 77–91 (1952)
2. Bonami, P., Lejeune, M.A.: An exact solution approach for portfolio optimization problems under stochastic and integer constraints. Oper. Res. **57**(3), 650–670 (2009)
3. Liu, S.T.: Solving portfolio optimization problem based on extension principle. In: Trends in Applied Intelligent Systems, pp. 164–174. Springer, Berlin (2010)
4. Konno, H., Yamazaki, H.: Mean-absolute deviation portfolio optimization model and its applications to Tokyo stock market. Manag. Sci. **37**(5), 519–531 (1991)
5. Angelelli, E., Mansini, R., Speranza, M.G.: A comparison of MAD and CVaR models with real features. J. Bank. Finance **32**(7), 1188–1197 (2008)

6. Mansini, R., Speranza, M.G.: An exact approach for portfolio selection with transaction costs and rounds. IIE Trans. **37**(10), 919–929 (2005)
7. Orito, Y., Hanada, Y., Shibata, S., Yamamoto, H.: A new population initialization approach based on bordered hessian for portfolio optimization problems. In: 2013 IEEE International Conference on Systems, Man, and Cybernetics (SMC), pp. 1341–1346. IEEE (2013)
8. Yen, Y.M., Yen, T.J.: Solving norm constrained portfolio optimization via coordinate-wise descent algorithms. Comput. Stat. Data Anal. **76**, 737–759 (2014)
9. Chen, A.H.L., Liang, Y.C., Liu, C.C.: An artificial bee colony algorithm for the cardinality-constrained portfolio optimization problems. In: 2012 IEEE Congress on Evolutionary Computation (CEC), pp. 1–8. IEEE (2012)
10. Tuba, M., Bacanin, N.: Upgraded firefly algorithm for portfolio optimization problem. In: Proceedings of the 2014 UKSim-AMSS 16th International Conference on Computer Modelling and Simulation, pp. 113–118. IEEE Computer Society (2014)
11. Skolpadungket, P., Dahal, K., Harnpornchai, N.: Portfolio optimization using multi-objective genetic algorithms. In: IEEE Congress on Evolutionary Computation, CEC 2007, pp. 516–523, IEEE (2007)
12. Aranha, C., Iba, H.: Modelling cost into a genetic algorithm-based portfolio optimization system by seeding and objective sharing. In: IEEE Congress on Evolutionary Computation, 2007. CEC 2007, pp. 196–203. IEEE (2007)
13. Chang, T.J., Yang, S.C., Chang, K.J.: Portfolio optimization problems in different risk measures using genetic algorithm. Expert Syst. Appl. **36**(7), 10529–10537 (2009)
14. Storn, R., Price, K.: Differential evolution–a simple and efficient heuristic for global optimization over continuous spaces. J. Global Optim. **11**(4), 341–359 (1997)
15. Ardia, D., Boudt, K., Carl, P., Mullen, K., Peterson, B.G.: Differential evolution with DEoptim: an application to non-convex portfolio optimization. R J. **3**(1), 27–34 (2011)
16. Ma, X., Gao, Y., Wang, B.: Portfolio optimization with cardinality constraints based on hybrid differential evolution. AASRI Procedia **1**, 311–317 (2012)
17. Korczak, J., Roger, P.: Portfolio optimization using differential evolution. PraceNaukoweAkademiiEkonomicznej we Wrocławiu **855**, 302–319 (2000)
18. Zaheer, H., Pant, M., Kumar, S., Monakhov, O., Monakhova, E., Deep, K.: A new guiding force strategy for differential evolution. Int. J. Syst. Assur. Eng. Manag. 1–14 (2014)
19. Chunhachinda, P., Dandapani, K., Hamid, S., Prakash, A.J.: Portfolio selection and skewness: evidence from international stock market. J. Bank. Finance **21**, 143–167 (1997)
20. Grootveld, H., Hallerbach, W.: Variance vs downside risk: is there really muchdifference? Eur. J. Oper. Res. **114**, 304–319 (1999)
21. Markowitz, H., Todd, P., Xu, G., Yamane, Y.: Computation of mean-semivarianceefficient sets by the critical line algorithm. Ann. Oper. Res. **45**, 307–317 (1993)
22. Rom, B.M., Ferguson, K.W.: Post-modern portfolio theory comes of age. J. Invest. **2**, 27–33 (1994)
23. Zaheer, H., Pant, M., Monakhov, O., Monakhova, E.: Solving a real life portfolio optimization problem: a differential evaluation approach. In: XI International Conference on Innovation and Business Management (ICIBM-15), 978-93-85000-294, pp. 435–444 (2015)
24. Gupta, P., Mehlawat, M.K., Inuiguchi, M., Chandra, S.: Fuzzy Portfolio Optimization. Springer (2014)

Applications of Harmony Search Algorithm in Data Mining: A Survey

Assif Assad and Kusum Deep

Abstract The harmony search (HS) is a music-inspired algorithm that appeared in the year 2001. Since its introduction HS has undergone a lot of changes and has been applied to diverse disciplines. The aim of this paper is to inform readers about the HS applications in data mining. The review is expected to provide an outlook on the use of HS in data mining, especially for those researchers who are keen to explore the algorithm's capabilities in data mining.

Keywords Harmony search · Data mining · Classification · Clustering · Prediction

1 Harmony Search

Harmony search (HS) is a musicians behaviour inspired evolutionary algorithm developed in 2001 by Geem et al. [1]; although it is a comparatively new meta-heuristic algorithm, its efficiency and advantages have been demonstrated in various applications like structural design [2], traffic routing [3], satellite heat pipe design [4], etc.

In order to explain HS, let us first consider the improvisation process by a musician. During improvisation process a skilled musician has three possible choices:

Assif Assad (✉) · K. Deep
Department of Mathematics, Indian Institute of Technology Roorkee, Roorkee 247667, Uttarakhand, India
e-mail: assifassad@gmail.com

K. Deep
e-mail: kusumdeep@gmail.com

M. Pant et al. (eds.), *Proceedings of Fifth International Conference on Soft Computing for Problem Solving*, Advances in Intelligent Systems and Computing 437, DOI 10.1007/978-981-10-0451-3_77

(1) Play any famous piece of music exactly from his memory.
(2) Play something similar to a known piece (thus adjusting the pitch slightly).
(3) Compose new or random notes.

Geem et al. [1] in 2001 formalized these three choices into quantifiable optimization processes and the three corresponding components became harmony memory (HM) usage, pitch adjusting and randomization. The HM usage is similar to the choice of selecting best fit individuals in genetic algorithms(GAs). This will guarantee that the best harmonies are passed over to the New Harmony Memory. Associated with HM is a parameter called harmony memory considering rate (HMCR $\in$ [0, 1]), so that it can be used efficiently. If this rate is too low (near 0), only few best harmonies are selected and thus convergence of algorithm may be too slow and if HMCR is very high (near 1), it results in exploitation of the harmonies in the HM and hence other harmonies are not explored, thus leading to wrong solutions. Typically, it is suggested to use HMCR $\in$ [0.7, 0.95]. The second component called pitch adjustment is determined by pitch adjusting rate (PAR) and pitch bandwidth (BW). Pitch adjustment refers to generating a slightly different solution. Pitch adjustment can be implemented linearly or nonlinearly, however, linear adjustment is most commonly used. Thus we have

$$X_{\text{new}} = X_{\text{old}} + BW \times r$$

where X_{old} is some existing harmony from the HM and X_{new} is the new harmony produced after pitch adjustment. This produces a new solution in the vicinity of existing solution by varying the solution marginally. Here r is a random number generated in the range [−1, 1]. The PAR controls the degree of adjustment. A narrow bandwidth with low PAR can slow down the convergence of HS because of the limitation in the exploration of only a small subspace of the whole search space. On the other hand, an extremely high PAR with a wide bandwidth may cause the solution to swing around some potential optimal solution. Thus most commonly used value of PAR $\in$ [0.1, 0.5]. The third component is the randomization, its use is to increase the exploration of the search space. Although pitch adjustment has a similar role, it is limited to close neighbourhood of harmony, thus it corresponds to local search. The use of randomization can push the system further to explore various diverse solutions, so as to find the global optima. The pseudo code of HS is shown in Fig. 1. It is evident from the pseudo code that the probability of randomization is 1-HMCR and the probability of pitch adjustment is HMCR × PAR. In the algorithm, H represents a potential solution or harmony.

```
HARMONY SEARCH ALGORITHM

Begin
        Define Objective function f(H)
        Define HMCR, PAR, BW
        Initialize Harmony Memory (HM)
        While ( Stopping Criteria Not Reached )
        Find current WORST and BEST harmony in HM
                If (rand < HMCR)
                        Generate new harmonics by using harmony memory and name it H.
                        If (rand < PAR)
                                Adjust pitch of H to get new harmony (solutions)
                        End if
                Else
                        Generate new harmony H using randomization.
                End if
                If (H is better than worst Harmony in HM)
                        Update HM by replacing WORST harmony by H
                End if
        End While
End
```

Fig. 1 Pseudo code of harmony search algorithm

2 Data Mining

"We are living in the information age" is a popular saying, however, actually we are living in the data age. There is a well-known paradox that more data means less information. The hardware cost is decreasing day by day and at the same time its efficiency is increasing; there has been a revolution in the Internet technology in the past decade and now access to Internet is very easy. The mobile phone industry has witnessed a revolution in the last couple of years with the result that smart phones have become easily available. Due to the above-mentioned reasons the data generating, collecting and storing rate has exploded in the past decade. Computerization of our society has resulted in explosive growth of data. Business worldwide generates enormous volumes of data in the form of sales transaction, product description, stock trading records, sales promotion, company profiles, customer information, customer feedback and performance. Petabytes of data are generated from scientific and engineering experiments, remote sensing and environmental surveillance. The medical and health industry generate volumes of data from medical imaging, medical records and patient monitoring. Millions of interactive websites produce large volumes of data. Communication and social media have become very important data generating sources producing videos, pictures and text. The list of sources producing volumes of data is endless. This explosively growing and extensively available enormous body of data makes our period truly a data age. Thus sophisticated, automated, efficient and effective tools are required that can produce meaningful information from these data tombs. This necessity has led to the birth of what we call data mining. Data mining also known as knowledge discovery in databases (KDD) "*is the non-trivial process of identifying valid, novel, potentially useful, and ultimately understandable patterns in data*". Data mining is also called knowledge extraction, knowledge mining from data, data archaeology, data/pattern analysis and data dredging. Hence data mining is concerned with developing algorithms and computational tools to help people extract patterns from data. Data mining can be applied to any kind of data like data warehouse, database data, transactional data, ordered/sequential data, data streams, text data, spatial data, multimedia data, experimental data, simulation data or data that is flooded into the system dynamically. Data mining has found its applications in business, science and engineering, spatial data analysis, sensor data mining and pattern analysis.

3 Applications of Harmony Search in Data Mining

Around mid-2015, approximately 35 publications were available related to HS and data mining, since data mining covers a vast area and includes many functionalities like classification, clustering, association rule analysis, prediction, feature selection, etc. In this section we summarize the applications of HS in different areas of data mining. Figure 2 shows the current approximate distribution of HS applications, as measured by the number of publications devoted to that topic. Table 1 shows the applications of HS in data mining, discipline wise.

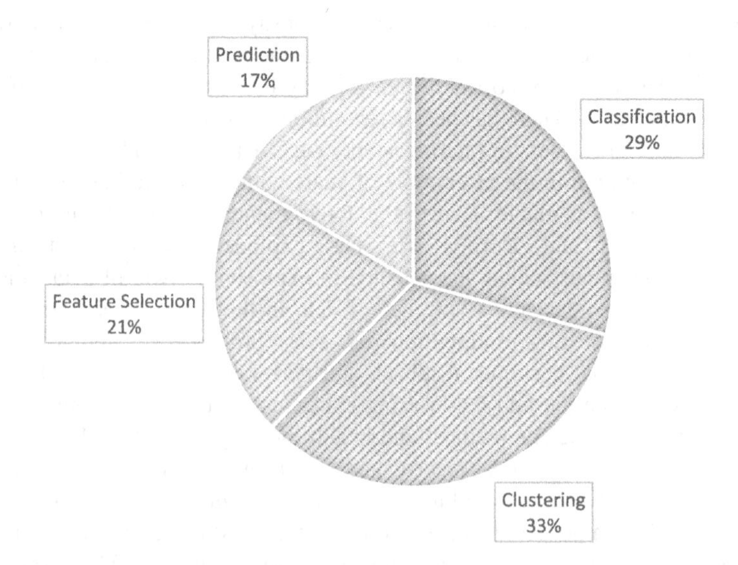

Fig. 2 Approximate breakdown of HS applications in data mining by discipline area as on July 2015

Table 1 Applications of HS in various areas of data mining

Area	Algorithm(s)	Application	Author	Year	Reference
Classification	HS	Classify emails as spam or legitimate	Wang et al.	2015	[5]
	Self-adaptive HS	Music genre classification	Huang et al.	2014	[6]
	Hybridized clonal selection algorithm and HS	Classification of Fisher Iris and wine dataset.	Wang et al.	2009	[7]
	Variants of HS	Train multilayer feedforward neural networks	Kulluk et al.	2012	[14]
	Self-adaptive Global best HS	Training of feedforward neural networks	Kulluk et al.	2011	[15]
	HS	Malware detection	Sheen et al.	2013	[16]
	HS	Sound classification in digital hearing aids	Alexandre et al.	2009	[17]
Clustering	HS	Clustering of four artificial and five real datasets	Amiri et al.	2010	[18]
	Hybridized HS and K-means algorithm	Document clustering	Mahdavi et al.	2009	[19]
	Hybridized global best HS, frequent term sets and Bayesian information criteria	Document clustering	Cobos et al.	2010	[20]
	HS and K-mean	Clustering of datasets	Moh'd et al.	2011	[21]
	Modified harmony search-based clustering (HSC)	Clustering of datasets	Kumar et al.	2014	[22]
	HS	Clustering of images	Ibtissem & Hadria	2013	[23]
	HS	Web page clustering	Forsati et al.	2008	[24]
	HS	Web page clustering	Forsati et al.	2008	[25]
	HS	Minimizing intra-cluster distance to optimize energy consumption in wireless sensor networks (WSN)	Hoang et al.	2010	[26]

(continued)

Table 1 (continued)

Area	Algorithm(s)	Application	Author	Year	Reference
Feature Selection	Hybrid HS and optimal path forest algorithm	Non-technical losses detection	Ramos et al.	2011	[27]
	HS	Classification	Diao and Shen	2012	[28]
	HS	Classification	Diao and Shen	2010	[29]
Prediction	HSA and extreme learning machines	Medium term fashion sale forecast	Wong and Guo	2010	[30]
	Differential HS	Stock market volatility prediction	Dash et al.	2015	[31]
	HS and artificial neural network	Predict optimal cutting pattern leading to minimal surface roughness in face milling of stainless steel	Razfar et al.	2011	[32]

Classification Classification deals with organization of data in classes. It is also known as supervised classification or supervised learning. Classification approaches usually use a training set where all objects are already associated with known class labels. The classification algorithm learns from the training set and builds a model, which is used to classify new unseen objects.

Wang et al. [5] used HS to develop an algorithm, namely Document Frequency and Term frequency combined feature selection (DTFS) to classify emails as ham/legitimate or spam. In order to evaluate the performance of DTFS algorithm the authors have applied it on six corpuses: PU2, CSDMC2010, PU3, Lingspam, Enron-spam and TREC2007 and compared the results with other state-of-the art classification algorithms like fuzzy support vector machine (FVSM) and Naïve Bayesian classifiers and have concluded that DTFS significantly outperforms other classification algorithms.

Huang et al. [6] used self-adaptive harmony search (SAHS) algorithm to develop music genre-classification system that is based on local feature selection strategy. Initially five audio characteristics, namely pitch, rhythm, intensity, timbre and tonality are extracted to create original feature set. Then using SAHS algorithm a feature selection model is employed for each pair of genres to derive a corresponding local feature set. Finally, each one against one support vector machine (SVM) classifier is fed with the corresponding local feature set and, finally, majority voting method is used to classify each musical recording. In order to carry out experiments the authors have used GTZAN dataset and concluded that the proposed algorithm performs well.

Wang et al. [7] proposed a hybrid optimization method based on the fusion of the clonal selection algorithm (CSA) and harmony search technique. The hybrid optimization algorithm is used to optimize Sugeno fuzzy classification systems for the Fisher Iris data and wine datasets. Computer simulation results demonstrate the remarkable effectiveness of this new approach.

Neural networks have proved to be very efficient tools for classification, however, their performance depends on the performance of training algorithm. Backpropagation is one of the most commonly used algorithm for training neural networks, however, it has the disadvantage of converging to local optimal. Stochastic algorithms like GAs [8, 9], particle swarm optimization (PSO) [10], differential evolution (DE) [11], ant colony optimization (ACO) [12, 13], Harmony Search (HS), etc., have been used in training neural networks. Substantial literature is available on the use of HS to train neural networks. In [14] five different variations of HS algorithm have been used to train multilayer feedforward neural networks. In order to test the efficiency of the proposed algorithms six benchmark classification datasets and a real-world dataset have been used. Real-world dataset is associated to the classification of frequently encountered quality defect in a textile company. The classification parameters like training time, sum of squared errors, etc., of self-adaptive global best Harmony Search (SGHS) algorithm is compared with other HS algorithm variations and standard backpropagation algorithm. The experiments showed that the SGHS algorithm is highly competitive with other methods. Kulluk et al. [15] used SGHS algorithm for the supervised training of feedforward neural networks. In order to test the efficiency of the SGHS algorithm the authors have applied it on two real datasets, Iris and WBC.

Malware detection using data mining techniques has been studied comprehensively.

Structural features dependent techniques for malware detection are based on anomaly identification in the structure of executable files. API call sequences, portable executable, byte n-grams and strings are some of the structural attributes of an executable file that can be extracted. In order to detect malware, Sheen et al. [16] extracted various features from executable files and applied it on an ensemble of classifiers. A huge collection of windows-based malware have been analysed and some interesting patterns related to PE header and API calls which will distinguish the two categories of files as benign and malware were identified. Ensemble of classifiers usually gives better performance than single learners, an ensemble approach for malware detection has been used in [16]. With the increase in the number of classifiers the storage requirement also increases, thus ensemble pruning is used to find the most optimum subset of classifiers that can perform at least as good as the original ensemble. In [16] the authors have used Harmony search Algorithm to perform ensemble pruning.

HS algorithm has been used to the problem of selecting appropriate features for sound classification in digital hearing aids [17]. Implementing sound classification algorithms embedded in hearing aids is a very difficult task because hearing aids work at very low clock frequency so that they can minimize power consumption and hence maximize the battery life. The computational load must be reduced and at the same time, a low error rate must be maintained. Feature extraction process is one of the most time-consuming tasks in classification. Alexandre et al. [17] used music-inspired harmony search algorithm for feature extraction so that sound classification in hearing aids can efficiently be performed.

Clustering Clustering also called unsupervised classification is the organization of data in classes. However, unlike classification, in clustering, class labels are unknown and it is up to the clustering algorithm to discover suitable classes. All clustering approaches all built on the principle of maximizing intra class similarity and minimizing inter-class similarity.

In [18], HS has been used for clustering. The authors introduce some deterministic clustering algorithms like hierarchical clustering, partition base clustering, density base clustering, artificial intelligence base clustering and K-means clustering. Then they highlight that though K-means clustering algorithm is very powerful, it has two main shortcomings, one being significantly sensitive to the initial randomly selected cluster centres and second convergence to local optimal. The authors propose stochastic algorithm, namely HS to solve the classification problem and compare the results with other scholastic algorithms like GA, simulating annealing (SA), taboo search (TA) and ACO. For conducting experimentation they have used four artificial and five real datasets (Vowel, Wine, Iris, Crude Oil and Thyroid Disease Data) and have concluded that HS algorithm performs better than other stochastic algorithms both in terms of solution accuracy and computational efficiency.

Fast and high quality document clustering is a crucial task in information retrieval, enhancing search engine results, web crawling and organization of information. Studies have shown that K-means algorithm is more suitable for large datasets, however, it can generate a local optimal solution. In [19], harmony K-means algorithm (HKA) based on HS optimization method and K-means algorithm has been developed that deals with document clustering. By using the concept of Markov chain theory it has been proved that HKA converges to the global optimum. To demonstrate the effectiveness of HKA algorithms it has been applied on some standard datasets. The authors have compared HKA with other metaheuristics like GA and PSO. Experimental results show HKA algorithm converges faster to the best known optimum method than other methods like GA and PSO.

By the hybridization of the global-best harmony search (GHS), frequent term sets and bayesian information criterion (BIC), a novel algorithm has been developed for web document clustering [20]. The new algorithm automatically defines the number of clusters. The Global Best HS globally searches in the solution space, K-means algorithm refines the solution by concentrating on local search space. FP-growth is used to reduce dimensionality in the vocabulary and BIC is used as a fitness function. The resulting algorithm is called IGBHSK and has been tested with datasets based on DMOZ and Reuters-21578.

The C-mean clustering algorithm has two types, namely hard C mean algorithm (HCM) and fuzzy C-mean algorithm (FCM). The quality of the result produced by C-mean clustering algorithm depends on the initially guessed cluster centres. Alia et al. [21] proposed a clustering algorithm consisting of two stages. In the first stage, HS algorithm explorers the search space to determine the near optimal cluster centres. In the second stage, the best cluster centres found in the first stage are used as initial cluster centres for the C-mean clustering algorithm. In order to carry out experimentation the authors have applied the algorithm on nine datasets, two of them are artificial and seven are real. The real datasets used are Iris, BUPA liver disorder, Haberman, Glass, MAGIC gamma telescope, Breast cancer Wisconsin and Diabetes and have compared the results with C-mean clustering algorithm (in which the initial cluster centres are picked randomly). The authors concluded that the proposed algorithm performs significantly better than C-mean particularly when the dataset contains multiple extremes.

Kumar et al. [22] modified the improved HS algorithm and named the algorithm as modified harmony search based clustering (MHSC). The proposed algorithm has been implemented and tested on five real-life datasets: Wine, Haberman, BUPA liver disorders, CMC, Breast Cancer. On comparing the results of proposed algorithm, it has been found that MHSC resulted in better cluster quality metrics than GA-based clustering (GAC), K-Means, HS-based clustering (HSC), Improved Harmony search-based clustering (IHSC). Experimental results reveal that except the wine dataset, MHSC yields better intra-cluster distance than other techniques for all the datasets used. It has been experimentally confirmed that MHSC provides highly cohesive clusters than the other techniques. HS has also been used to perform unsupervised clustering of images [23]. The authors have used both artificial and satellite images to demonstrate the efficiency of HS in image clustering. In [24, 25], HS has been used for web page clustering. The authors have first used HS algorithm to find approximate clusters, then they have hybridized the HS and K-means algorithm to fine-tune the results.

In [26], HSA has been used for minimizing the intra-cluster distance and optimizing the energy consumption of the wireless sensor networks (WSN).

Feature Selection In machine learning, feature selection is also known as variable selection, attribute selection or variable subset selection. It is the process of selecting a subset of relevant features (variables, predictors) for use in model construction.

Non-technical losses in power distribution systems are related to the consumed but not billed energy and it represents a problem to the electric power companies, since these losses are strongly related with frauds and energy theft. Ramos et al. [27] proposed a hybrid algorithm to accomplish the task of feature selection by using HS and Optimum-Path Forest (HS–OPF) for non-technical loss detection. In order to validate the proposed algorithm the authors have used two labelled datasets obtained from a Brazilian electric power company, one constituted by industrial consumers and the other by commercial consumers.

Diao et al. [28] used HS for the task of feature selection (FS), in an effort to identify more compact and better quality subsets of features for classification. Then they compared the algorithm with hill climbing, GAs and PSO. In [29], HS along with fuzzy-rough has been used for feature evaluation of feature selection problem.

Prediction Prediction has attracted significant attention given the potential implications of successful forecasting in business, science and the engineering context.

Wong and Guo [30] used HS algorithms and extreme learning machines to tackle the medium term fashion sales forecasting problem. The authors carried out experiments on real fashion retail data and public benchmark datasets to evaluate the performance of the proposed model. Results demonstrate that the performance of the proposed algorithm is far superior to the traditional autoregressive integrated moving average (ARIMA) algorithm.

Forecasting and modelling volatility with a reasonable accuracy plays a key role in achieving gain in different financial applications. Volatility provides a measure of fluctuation in a financial security price around its expected value. In [31] the authors used HS for stock market volatility prediction.

In [32] artificial neural network (ANN) and HS are coupled to predict optimal cutting parameters leading to minimum surface roughness in face milling of stainless steel.

4 Conclusion

In this paper an introduction is provided about data mining and music-inspired HS algorithm followed by a detailed description about applications of HS algorithm in different areas of data mining, particularly clustering, classification, neural network training, feature selection and prediction. In some papers HS has been proven to outperform other well-established heuristic algorithms like GA and PSO. Based on the review it is observed that even though HS has been recently developed it has been extensively used in data mining due to its efficiency and ease of implementation.

References

1. Geem, Z.W., Kim, J.H., Loganathan, G.V.: A new heuristic optimization algorithm harmony search. Simulation **76**(2), 60–68 (2001)
2. Lee, K.S., Geem, Z.W.: A new structural optimization method based on the harmony search algorithm. Comput. Struct. **82**(9–10), 781–798 (2004)
3. Geem, Z.W., Lee, K.S., Park, Y.: Application of harmony search to vehicle routing. Am. J. Appl. Sci. **2**(12), 1552–1557 (2005)
4. Geem, Z.W., Hwangbo, H.: Application of harmony search to multi-objective optimization for satellite heat pipe design. In: Proceedings of, pp. 1–3 (2006, August)
5. Wang, Y., Liu, Y., Feng, L., Zhu, X.: Novel feature selection method based on harmony search for email classification. Knowl.-Based Syst. **73**, 311–323 (2015)
6. Huang, Y.F., Lin, S.M., Wu, H.Y., Li, Y.S.: Music genre classification based on local feature selection using a self-adaptive harmony search algorithm. Data Knowl. Eng. **92**, 60–76 (2014)
7. Wang, X., Gao, X.Z., Ovaska, S.J.: Fusion of clonal selection algorithm and harmony search method in optimisation of fuzzy classification systems. Int. J. Bio-Inspired Comput. **1**(1–2), 80–88 (2009)
8. Montana, D.J., Davis, L.: Training feedforward neural networks using genetic algorithms. IJCAI **89**, 762–767 (1989)

9. Leung, F.H., Lam, H.K., Ling, S.H., Tam, P.K.: Tuning of the structure and parameters of a neural network using an improved genetic algorithm. IEEE Trans. Neural Netw. **14**(1), 79–88 (2003)
10. Mendes, R., Cortez, P., Rocha, M., Neves, J.: Particle swarms for feedforward neural network training. Learning **6**(1), (2002)
11. Ilonen, J., Kamarainen, J.K., Lampinen, J.: Differential evolution training algorithm for feed-forward neural networks. Neural Process. Lett. **17**(1), 93–105 (2003)
12. Socha, K., Blum, C.: An ant colony optimization algorithm for continuous optimization: application to feed-forward neural network training. Neural Comput. Appl. **16**(3), 235–247 (2007)
13. Blum, C., Socha, K.: Training feed-forward neural networks with ant colony optimization: An application to pattern classification. In: Fifth International Conference on Hybrid Intelligent Systems, 2005. HIS'05. (pp. 6–pp). IEEE (2005, November)
14. Kulluk, S., Ozbakir, L., Baykasoglu, A.: Training neural networks with harmony search algorithms for classification problems. Eng. Appl. Artif. Intell. **25**(1), 11–19 (2012)
15. Kulluk, S., Ozbakir, L., Baykasoglu, A.: Self-adaptive global best harmony search algorithm for training neural networks. Procedia Comput. Sci. **3**, 282–286 (2011)
16. Sheen, S., Anitha, R., Sirisha, P.: Malware detection by pruning of parallel ensembles using harmony search. Pattern Recogn. Lett. **34**(14), 1679–1686 (2013)
17. Alexandre, E., Cuadra, L., Gil-Pita, R.: Sound classification in hearing aids by the harmony search algorithm. In: Music-Inspired Harmony Search Algorithm, pp. 173–188. Springer, Berlin (2009)
18. Amiri, B., Hossain, L., Mosavi, S.E.: Application of harmony search algorithm on clustering. In: Proceedings of the World Congress on Engineering and Computer Science, vol. 1, pp. 20–22, (2010, October)
19. Mahdavi, M., Abolhassani, H.: Harmony K-means algorithm for document clustering. Data Min. Knowl. Disc. **18**(3), 370–391 (2009)
20. Cobos, C., Andrade, J., Constain, W., Mendoza, M., León, E.: Web document clustering based on global-best harmony search, K-means, frequent term sets and Bayesian information criterion. In: 2010 IEEE Congress on Evolutionary Computation (CEC), pp. 1–8. IEEE (2010, July)
21. Moh'd Alia, O., Al-Betar, M.A., Mandava, R., Khader, A.T.: Data clustering using harmony search algorithm. In: Swarm, Evolutionary, and Memetic Computing, pp. 79–88. Springer, Berlin (2011)
22. Kumar, V., Chhabra, J.K., Kumar, D.: Clustering using modified harmony search algorithm. Int. J. Comput. Intell. Stud. **2**, **3**(2–3), 113–133 (2014)
23. Ibtissem, B., Hadria, F.: Unsupervised clustering of images using harmony search algorithm. Nature **1**(5), 91–99 (2013)
24. Forsati, R., Mahdavi, M., Kangavari, M., Safarkhani, B.: Web page clustering using harmony search optimization. In: Canadian Conference on Electrical and Computer Engineering, 2008. CCECE 2008, pp. 001601–001604, IEEE (2008, May)
25. Forsati, R., Meybodi, M., Mahdavi, M., Neiat, A.: Hybridization of K-means and harmony search methods for web page clustering. In: Proceedings of the 2008 IEEE/WIC/ACM International Conference on Web Intelligence and Intelligent Agent Technology, vol. 01, pp. 329–335, IEEE Computer Society (2008, December)
26. Hoang, D.C., Yadav, P., Kumar, R., Panda, S.K.: A robust harmony search algorithm based clustering protocol for wireless sensor networks. In: 2010 IEEE International Conference on Communications Workshops (ICC), pp. 1–5, IEEE (2010, May)
27. Ramos, C.C., Souza, A.N., Chiachia, G., Falcão, A.X., Papa, J.P.: A novel algorithm for feature selection using harmony search and its application for non-technical losses detection. Comput. Electr. Eng. **37**(6), 886–894 (2011)
28. Diao, R., Shen, Q.: Feature selection with harmony search. IEEE Trans. Syst. Man Cybern. B Cybern. **42**(6), 1509–1523 (2012)

29. Diao, R., Shen, Q.: Two new approaches to feature selection with harmony search. In: 2010 IEEE International Conference on Fuzzy Systems (FUZZ), pp. 1–7, IEEE (2010, July)
30. Wong, W.K., Guo, Z.X.: A hybrid intelligent model for medium-term sales forecasting in fashion retail supply chains using extreme learning machine and harmony search algorithm. Int. J. Prod. Econ. **128**(2), 614–624 (2010)
31. Dash, R., Dash, P.K., Bisoi, R.: A differential harmony search based hybrid interval type 2 fuzzy EGARCH model for stock market volatility prediction. Int. J. Approximate Reasoning **59**, 81–104 (2015)
32. Razfar, M.R., Zinati, R.F., Haghshenas, M.: Optimum surface roughness prediction in face milling by using neural network and harmony search algorithm. Int. J. Adv. Manuf. Technol. **52**(5–8), 487–495 (2011)

Enriched Biogeography-Based Optimization Algorithm to Solve Economic Power Dispatch Problem

Vijay Raviprabhakaran and Coimbatore Subramanian Ravichandran

Abstract This article offers an enriched biogeography-based optimization (EBBO) technique to crack the economic power dispatch (EPD) problem of coal-fired generating units. The considered EPD involves the complex limitations including valve point loading effects, transmission line losses, and ramp rate limits. The geographical smattering of biological species is the vital scope of this algorithm. The proposed EBBO describes the arousal, enhanced migration of species from one environment to another. The algorithm has two main steps specifically, migration and mutation. These steps are involved in searching the global optimum solution. The EBBO's efficiency has been verified on 13 and 40 generating test systems. The proposed technique produces superior results when compared with the conventional biogeography-based optimization (BBO) and other prevailing techniques. Also, it gives the quality and promising results for solving the EPD problems. Further, it can be applied for practical power system.

Keywords Enriched biogeography-based optimization · Enhanced migration operator · Thermal generating units · Economic power dispatch problem

1 Introduction

Economic power dispatch (EPD) is a recent problem in the power system operation and control. EPD is the optimal allocation of the thermal generating system to the generating stations, which satisfies the power demand and losses incurred in the

Vijay Raviprabhakaran (✉)
Department of Electrical & Electronics Engineering,
Anna University Regional Centre, Coimbatore, India
e-mail: vijai.mtp@gmail.com

C.S. Ravichandran
Department of Electrical & Electronics Engineering,
Sri Ramakrishna Engineering College, Coimbatore, India
e-mail: eniyanravi@gmail.com

M. Pant et al. (eds.), *Proceedings of Fifth International Conference on Soft Computing for Problem Solving*, Advances in Intelligent Systems and Computing 437, DOI 10.1007/978-981-10-0451-3_78

transmission system. It also finds the minimum generating cost of the allocated generating units [1]. The nonlinear EPD optimization problem considered in this article deliberates the real-world operational limitations comprising valve point loading effects, ramp rate limits, etc.

The nonlinear characteristics need to remain estimated to encounter the necessities of conventional optimization techniques. Owing to this estimation, the final solution is merely suboptimal, henceforth, an enormous expense occurs in that period. The extreme nonlinear characteristics of these generating unit power demands for solution methods that ensure no limitations on the contour of the thermal cost characteristics. The calculus-built technique is unsuccessful in determining the EPD problem [2]. The dynamic programming method does not inflict slight limitation on the thermal cost characteristics [3]. It explains the linear and nonlinear EPD problems. However, this technique struggles from the dimensionality of the problem and software runtime rises promptly through the growth of the generating system size [4].

Recently, several optimization techniques such as genetic algorithm (GA) [5], particle swarm optimization (PSO) [6], artificial immune system (AIS) [7], evolutionary programming (EP) [8], differential evolution (DE) [9], ant colony optimization (ACO) [10], bacterial foraging optimization (BFO), [11], etc., have been applied efficiently to crack the EPD problems.

Although these algorithms have efficaciously offered to explain this problem, this technique has certain disadvantages. The main hindrances are solving the problems that lead to lag in convergence rate, the solutions fail to attain the global optimal point, the solution of the problem progresses extremely quickly on the way to the optimal solution in the primary phase, increase in system constraints result in complexity, etc.

In recent years, a new optimization technique, namely biogeography-based optimization (BBO) is proposed to solve the problems in the engineering domain [12]. In this algorithm, biogeography is defined as nature's approach of allocating species (plant or living organism). In BBO, the island (land mass) of habitat through a high habitat suitability index (HSI) is compared to the good (best optimal) solution and the landmass by means of a low HSI solution to a poor (worst) solution. The high HSI solutions fight to convert better than low HSI solutions. The low HSI solutions motivate to counterfeit worthy characters from high HSI solutions. Collective characters persist in the high HSI solutions, even though they simultaneously act as innovative characters in the low HSI solutions. This occurs when particular representatives (agents) of species are specified toward an environment, whereas other representatives persist in their indigenous habitat. The solutions with poor characteristics approve many innovative features from the worthy solutions. These accumulations of innovative characters on low HSI solutions could promote the superiority of the optimal results [13].

Further, the BBO technique has assured some distinctive qualities resulting in numerous drawbacks of the typical approaches as revealed below. In the GA owing to the crossover process the worthy solution attains initially, occasionally the solution may fail to attain the fitness in later iterations. Similarly, BBO does not have any crossover technique and due to migration process the solution is modified progressively. The most important process in this algorithm is elitism. This process retains the best solution and makes the proposed BBO technique more competent with the other techniques. Considering the PSO algorithm, the solutions are further likely to be grouped together into analogous groups to search the optimal solution, though in BBO algorithm the solutions do not group owing to its mutation operation. Simultaneously, the limitations handling is considerably accessible compared to the BFO technique. Although the conventional BBO has an edge over the other algorithms, it suffers from poor convergence characteristics when considering complex problems. In the BBO algorithm, the deprived solution admits some new features from worthy ones, this progresses the superiority of problem solutions. Comparably, this is another distinctive component of BBO technique when associated with alternative methods.

In this paper, an enhanced migration-based BBO (EBBO) technique is developed. There are many migration models available for the BBO technique [14, 15]. In the projected technique the migration step is modified for the best optimal solution. The proposed EBBO technique is authenticated on 13 and 40 thermal generating system to verify its efficiency. The results attained with the projected EBBO technique are related with the traditional BBO and other optimization techniques.

2 EPD Problem Formulation

The main goal of EPD problem is to estimate the optimum arrangement of thermal power generations in order to minimize the entire generation fuel cost subjected to the constraints. The constraints may be about equality and inequality. The fuel cost function of every online generator is characterized by a quadratic function specified as follows:

The goal function EPD problem is composed as follows:

$$\text{Min}\, F_t = \sum_{i=1}^{n} F_i(P_i) \tag{1}$$

where $F_i\,(P_i)$ is the total thermal fuel cost of the ith thermal online generator (in $/h) and is given below as

$$F_i(P_i) = a_{i_i} P_i^2 + b_i P_i + c_i \tag{2}$$

where thermal generated fuel cost coefficients of the ith generator is a_i, b_i and c_i.

The generic EPD problem deliberates the quadratic cost function additionally with system load demand and operational limitations. In this paper the practical EPD with valve point loading and generator ramp rate operating limits are examined.

2.1 Nonsmooth Cost Function with Valve Point Loading Effect

The steam powered turbo generator with multiple valves has dissimilar input–output characteristics associated with the smooth cost function. Usually, the valve point loading consequences and the steam turbine valve starts to open and close simultaneously, this causes ripple-like effects. The cost function comprises non-linearity of higher order. Because of valve point effect, the objective function F_t of EPD problem is rewritten as follows:

$$\text{Min}\, F_i(P_i) = \sum_{i=1}^{n} a_i P_i^2 + b_i P_i + c_i + \sum_{i=1}^{n} \left| e_i \times \sin\left(f_i \times \left(P_{i,\min} - P_i\right)\right)\right| \quad (3)$$

The practical nonsmooth EPD problem minimizes the valve point objective function subject to the following constraints.

2.2 Power Balance Limitations

$$\sum_{i=1}^{n} P_i - P_{\text{Dd}} - P_{\text{Lo}} = 0 \quad (4)$$

where P_i is the output for unit i, n is generator number participating in the problem, P_{Dd} is the total load demand, P_{Lo} is the transmission loss.

2.3 Generator Capacity Limitations

$$P_i^{\min} \leq P_i \leq P_i^{\max} \quad (5)$$

where $P_i^{\min}$ and $P_i^{\max}$ are minimum and maximum operating output of the online generation unit i respectively.

The transmission loss P_{Lo} may be expressed using B-coefficients equation as

$$P_{\text{Lo}} = \sum_{i=1}^{n} \sum_{j=1}^{n} P_i B_{ij} P_j + \sum_{i=1}^{n} B_{0i} P_i + B_{00} \quad (6)$$

Here B_{ij} represents the i, jth component of transmission loss coefficient matrix, loss coefficient vector of ith element is represented by $B_{0\text{i}}$, and the coefficient of transmission loss is symbolized by B_{00}.

2.4 Generator Ramp Rate Limits

The real governing scope of all the thermal generators is regulated by their resultant ramp rate limits. The ramp-up and ramp-down limitations are presented as follows:

$$P_i - P_i^0 \leq \mathrm{UR}_i \quad P_i^0 - P_i \leq \mathrm{DR} \tag{7}$$

where P_i^0 is the earlier power generated by the ith generating unit. UR_i and DR_i are the up-ramp and down-ramp restrictions of ith generator correspondingly.

To examine the ramp rate and generated power bounds the above equation is redrafted as an inequality constraint as shown below.

$$\max\{P_{i,\min}, P_i^0 - \mathrm{DR}_i\} \leq P_i \leq \min\{P_{i,\max}, P_i^0 + \mathrm{UR}_i\} \tag{8}$$

3 Steps Involved in Enriched Biogeography-Based Optimization (EBBO) Algorithm

Biogeography algorithm describes the biological species migration from one island to another. The proposed EBBO algorithm comprises of two main steps, namely the enriched migration and mutation and is described as follows.

3.1 Enriched Migration Model

The fitness of the projected solution is increased by a monotonic downturn and increases in degrees of immigration and emigration, which determines that as an increase in the species count results in more fitness of the solution, the uniting characteristic probability from additional solutions decreases. Though recently biogeography is observed in certain pioneer plant species, in this a primary rise in species numbers results in a primary increase and decrease of immigration and emigration rates. The new unfavorable circumstantial surroundings of the island have been enhanced by the first explorers, who categorized it as being extra generous to additional species, i.e., progressive outcome of augmented diversity owing to the first immigration restricts the negative influence of enlarged size of species populations. In BBO this effect causes a primary growth in immigration rate, in this the appropriate deprived candidate solution primarily progresses the fitness of solution [16]. Such a step is perceived as a momentary progressive feedback method in BBO. The most deprived aspirant solution admits the characters from further solutions. When fitness of the solution is increased, consequently it increases the probability of accommodating additional features from the other solutions. This motivation comes from biogeography. The immigration rate α and emigration rate β of the proposed EBBO algorithm is projected as follows:

$$\alpha = \frac{I}{2}\left(\cos\left(\frac{(S_h * \pi) + \xi}{S_{max}}\right) + 1\right) \quad (9)$$

$$\beta = \frac{E}{2}\left(-\cos\left(\frac{(S_h * \pi) + \xi}{S_{max}}\right) + 1\right) \quad (10)$$

where S_h is the species count of each habitat and element of S, E and I is the maximum emigration and immigration rate (generally assumed as 1), ξ is the degree of temporary positive immigration rate feedback usually between $[-\pi/2, 0]$, S_{max} represents the maximum number of species.

Using this enhanced model, the fitness is normalized to $[0, 1-\beta/\alpha]$. In the projected methodology, the immigration primarily escalates with the fitness of the solution. It generally contributes refining solutions, i.e., the momentum that they need to remain improving. When the immigration primarily increases, the solution remains to develop appropriately. The immigration rate originates the decrease to contribute lesser fitness solutions comparatively. It recommends immigrating worthy solutions to the problem.

3.2 Mutation of Species

The mutation process inclines to intensification among the species. Each species probability is deliberated by means of the differential equation given as

$$\dot{P}_s = \begin{Bmatrix} -(\alpha_s + \beta_s)P_s + \beta_{s+1}P_{s+1}\ldots S = 0 \\ -(\alpha_s + \beta_s)P_s + \alpha_{s-1}P_{s-1} + \beta_{s+1}P_{s+1}\ldots 1 \leq S \leq S_{max} - 1 \\ -(\alpha_s + \beta_s)P_s + \alpha_{s-1}P_{s-1}\ldots S = S_{max} \end{Bmatrix} \quad (11)$$

After the adjustment, the extremely possible solutions are inclined to succeed more in the considered species. The mutation approach creates the low HSI solutions which are likely to alter; this offers it a process of refining.

Also, the mutation is applied to high HSI solutions, which progresses beyond that. The elitism process is used to retain the worthy solution in the EBBO technique. If the mutation remains, their HSI is conserved and returned later, if required. Mutation operation is needed for both poor and good solutions. In this the average solutions are optimistically improved already and thus the mutation process is evaded.

4 Implementation of EBBO to EPD Problem

In this section, the step-by-step implementation of EBBO technique to solve the EPD problems is described below.

4.1 *Initialization*

Initialize the number of generating units which indicate the number of suitability index variable (SIV) as *m*, the habitats number is specified as *N*, generating unit maximum and minimum capacity, power demand at each dispatch hour, B-loss matrix for calculating loss in the power transmission line. Also, initialize the other EBBO algorithm parameters.

The variables for determining the EPD problems are power generators, which represent the specific habitat. The power generated by whole thermal generating units is characterized as the SIV of the habitat.

The habitat in matrix form is signified as

$$\mathrm{HB} = [\mathrm{HB}^1, \mathrm{HB}^2, \mathrm{HB}^3, \ldots, \mathrm{HB}^i, \ldots, \mathrm{HB}^N] \quad (12)$$

where $i = 1, 2, \ldots, N$ represents the generator number from 1 to *N* and *j* signifies the SIV of habitat from 1 to *m*.

From the above habitat's matrix, HB^i is the *i*th habitat position vector. Each habitat represents the possible solutions of the EPD problem. The habitat size is the same as the amount of EBBO population. The component HB^{ij} signifies *j*th SIV of *i*th habitat. SIV^{ij} denotes the power generated of the *i*th habitat group.

4.2 *Generation of Habitat*

Every SIV of an assumed habitat matrix is fixed as HB. It is set randomly in the limits of actual power generation. The initialization is created on generating capacity of generators and ramp rate limits. Every single habitat has to satisfy these constraints.

4.3 *Compute the HIS of Every Habitat*

The HSI denotes the thermal fuel cost of the online thermal generating units for a specific power demand. The *i*th individual HB^i is defined as follows:

$$\begin{aligned}\mathrm{HB}^{i} = \mathrm{SIV}^{ij} &= \left[\mathrm{SIV}^{i1}, \mathrm{SIV}^{i2}, \ldots \mathrm{SIV}^{im}\right] \\ &= \left[P_{\mathrm{g}}^{i1}, P_{\mathrm{g}}^{i2}, P_{\mathrm{g}}^{i3}, \ldots P_{\mathrm{g}}^{im}\right]\end{aligned} \tag{13}$$

$$i = 1, 2, \ldots, N, \ldots, S \text{ and } j = 1, 2, \ldots, m$$

where $S \times m$ is the whole habitat matrix set. Entire elements of each individual are signified as real values.

4.4 Apply Elitism

This operation is created with the value of HSI (Thermal fuel cost) to recognize the best habitats. The expression of elite is applied to designate these habitat groups of output generating power. This inclusion offers the superior fuel cost. These habitat sets retained the same after separate iteration without formulating a few adjustments. These habitat fitness values, specifically the finite HSI values, are deliberated as effective species S in EPD problem.

4.5 Perform Enhanced Migration Operation

Apparently accomplish enhanced migration process at the SIVs of every nonelite habitat nominated for immigration. The selection of a few SIVs for this process is explained as follows:

(i) Primarily, immigration rate of lower α_{lower} and upper value α_{upper} are selected. Then determine the species count.
(ii) Estimate the rate of α and β for every habitat group using Eqs. (9) and (10). Subsequently, determine the habitat and when migration occurs, then for newly generated habitat the SIV is to be designated. After the migration process, the fresh habitat set is created. Concerning EPD problems, these habitats signify the fresh improved generation values of online generation units (P_{g}).
(iii) The limitations are fulfilled using the subsequent method as follows:

If ith generator's power is greater than the maximum generation of ith generator, i.e., $P_{\mathrm{g}}^{i} \geq P_{\mathrm{gmax}}^{i}$
then set $P_{\mathrm{g}}^{i} = P_{\mathrm{gmax}}^{i}$
end
If the output of ith generator is less than the maximum generation of ith generator, i.e., $P_{\mathrm{g}}^{i} \leq P_{\mathrm{gmin}}^{i}$
then set $P_{\mathrm{g}}^{i} = P_{\mathrm{gmin}}^{i}$
end

If output of *i*th online generating unit P_{gi} is within the maximum and minimum generation restrictions
then fix $P_{\mathrm{g}}^{i} = P_{\mathrm{g}}^{i}$
end

Equality constraint shown in Eq. (4) is fulfilled by the model of slack generator.

4.6 Execute the Mutation Operation

Mutation process is achieved on the nonelite habitat. It is basically substituted by an additional randomly produced novel habitat group that satisfies the constraints of EPD problem. The every new habitat group is computed again based on HSI value, i.e., the thermal fuel cost of every generating unit.

4.7 Termination

This termination loop of the algorithm is ended when a predefined amount of iterations or subsequently an optimal solution has been produced. Otherwise, go to Step 4.3. After each habitat is improved (in Steps 4.5 and 4.6), the problem solution practicability is substantiated, specifically, each SIV would fulfill the constraints as specified above.

5 Simulation Results and Analysis

The subsequent optimum traits of EBBO limitations are included for simulating the EPD problem in Table 1.

The projected EBBO technique has been tested on EPD problems for three distinct experimental cases, i.e., 13 and 40-generator system to verify its practicality. The software for the system is programmed using the numerical computing environment called MATLAB.

Table 1 EBBO parameters considered for EPD problem

Parameters	Optimum values
Size of the habitat	30
Habitat conversion probability	1
Limits of immigration probability	[0, 1]
Step length of the habitat	0.1
Absolute immigration and emigration amount for every landmass	1
Probability of mutation	0.005

Table 2 Analyzing power dispatch results for 13-generator system (1800 MW)

Gen. Power (MW)	HQPSO [16]	HGA [18]	PS [19]	BBO (Props.)	EBBO (Props.)
G1	628.32	628.32	538.56	628.32	628.31
G2	149.11	222.75	224.64	222.75	148.11
G3	223.32	149.60	149.85	149.59	224.32
G4	109.87	109.87	109.87	60.00	109.86
G5	109.86	109.87	109.87	109.87	109.86
G6	109.87	109.87	109.87	109.87	109.86
G7	109.79	109.87	109.87	109.87	109.86
G8	60.00	60.00	109.87	109.87	60.00
G9	109.87	109.87	109.87	109.87	109.87
G10	40.00	40.00	77.47	40.00	40.00
G11	40.00	40.00	40.22	40.00	40.00
G12	55.00	55.00	55.03	55.00	55.00
G13	55.00	55.00	55.03	55.00	55.00
TPD[a]	1800.00	1800.00	1800.00	1800.00	1800.00
TGC[b]	17963.95	17963.83	17969.17	17963.82	17963.80

[a]*TPD* Total power demand (MW)
[b]*TGC* Total generation cost ($/h)

5.1 Test Case 1–13 Thermal Generating Units

In this case 13 thermal generating units [2] using a practical valve point loading is deliberated. Table 2 illustrates the improved dispatch solutions accomplished through the proposed EBBO technique over the load demand of 1800 MW.

The convergence profile for EBBO method is presented in Fig. 1. The total generation cost generated over 50 trail runs using the HQPSO, HGA, PS, and the projected BBO, EBBO are specified in Table 3. The results reveal that the projected technique is better than the other techniques in rate of superior optimal solution.

5.2 Test Case 2–40 Thermal Generating Units

A test case with 40 thermal generating units is considered. In this paper, for each generating system the valve point loading effect is deliberate. The data for the system are taken from [17]. The power demand met by all the 40 thermal generating systems is 10,500 MW. The loss in power transmission has not been included. The result acquired from the projected EBBO method is shown in power dispatch Table 4.

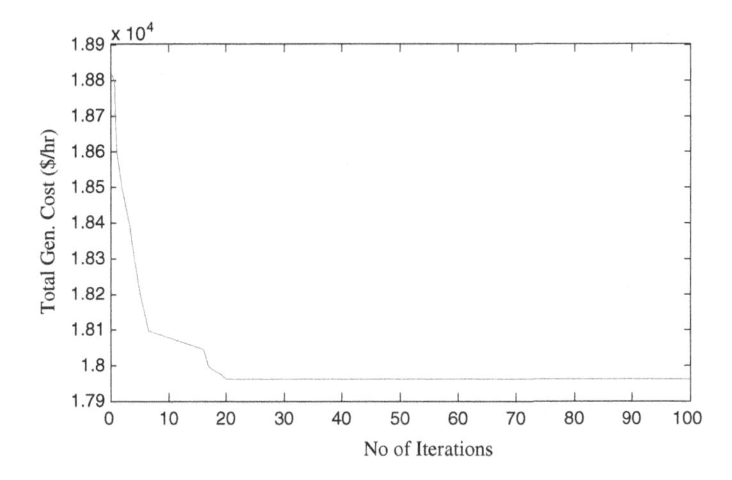

Fig. 1 Convergence curve of proposed EBBO technique for 13-generating system (1800 MW)

Table 3 Convergence results comparison over 50 trials (13-generating system including 1800 MW demand)

Technique	Generation fuel cost ($/hr)		
	Minimum	Average	Maximum
HQPSO [20]	17963.95	18273.86	18633.04
HGA [17]	17963.83	17988.04	NO[a]
PS [21]	17969.17	18088.84	18233.52
BBO (Props.)	17963.80	18087.23	18220.35
EBBO (Props.)	17962.74	18080.93	18198.76

[a]*NO* Not offered in the literature

The generation cost acquired from the PS, EP, PSO, EP-PSO, PSO-SQP, IFEP, SPSO, BBO, and the proposed method for 50 trail runs are compared in Table 5.

It is clear from Table 5 that the projected EBBO technique has produced the minimum fuel generation cost among all the techniques in the test case. Therefore, the proposed technique finds the optimum solution even for large generating systems in the power system. The convergence curve of proposed EBBO technique for 40 thermal generating systems is shown in Fig. 2.

6 Conclusion

In this paper, the enriched biogeography-based optimization (EBBO) technique has effectively shown to solve 13- and 40-thermal generating system of EPD considering the multiple valve loading effects and generator ramp limits. The proposed

Table 4 Result of power dispatch for 40-generator system (10,500 MW)

Generator	Gen. power (MW)	Generator	Gen. power (MW)	Generator	Gen. power (MW)
G1	110.71	G14	394.27	G27	10.00
G2	110.71	G15	394.27	G28	10.00
G3	97.16	G16	394.25	G29	10.00
G4	180.03	G17	489.24	G30	88.43
G5	89.79	G18	489.24	G31	190.00
G6	139.92	G19	511.21	G32	190.00
G7	259.49	G20	511.21	G33	190.00
G8	284.01	G21	523.12	G34	165.61
G9	284.46	G22	523.12	G35	196.46
G10	130.05	G23	523.26	G36	200.00
G11	94.00	G24	523.24	G37	109.80
G12	94.00	G25	523.24	G38	109.80
G13	212.36	G26	523.24	G39	109.80
				G40	510.41
TPD = 10500.00		TGC = 121398.87			

[a] *TPD* Total power demand (MW)
[b] *TGC* Total generation cost ($/h)

Table 5 Evaluating convergence results evaluation over 50 trials (40-thermal generating units including 10,500 MW demand)

Technique	Generation fuel cost ($/h)		
	Minimum	Average	Maximum
PS [21]	121415.14	122332.65	125486.29
EP [22]	122624.35	123382.00	NO
PSO [22]	123930.45	124154.49	NO
EP-SQP [22]	122323.97	122379.63	NO
PSO–SQP [22]	122094.67	122245.25	NO
SPSO [20]	124091.16	122049.66	122327.36
IFEP [2]	125740.63	122624.35	123382.00
BBO [16]	121588.66	121426.95	121508.03
EBBO (Props.)	121398.87	121424.29	121452.58

[a] *NO* Not offered in the literature

EBBO technique acquires the competence to estimate the best superior solution, efficiency, and robustness for the considered test cases. It is evident from the dispatch results that the projected EBBO technique has a good convergence rate. It evades the premature convergence, when related to other techniques presented in the literature. Thereby, the projected technique may apply to practical complex systems for accurate dispatch results.

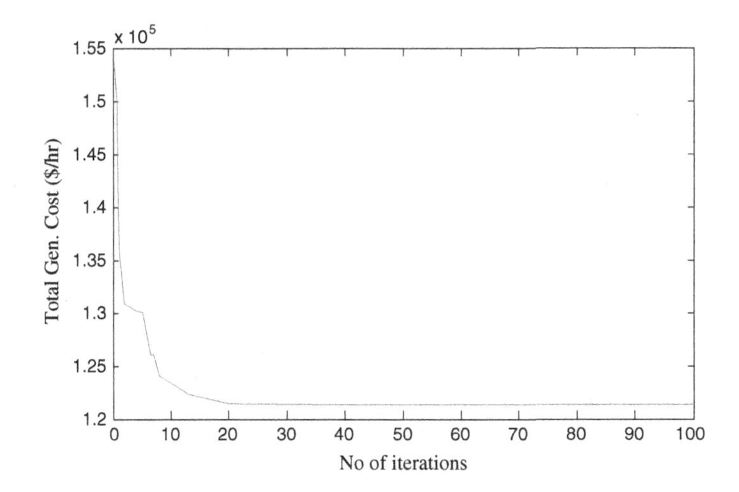

Fig. 2 Convergence curve of proposed EBBO technique for 40-unit system

References

1. El-Keib, A., Ma, H., Hart, J.L.: Environmentally constrained economic dispatch using the Lagrangian relaxation method. IEEE Trans. Power Syst. **9**(4), 1723–1729 (1994)
2. Sinha, N., Chakrabarti, R., Chattopadhyay, P.K.: Evolutionary programming techniques for economic load dispatch. IEEE Trans. Evolut Comput. **7**(1), 83–94 (2003)
3. Liang, Z.X., Glover, J.D.: A zoom feature for a dynamic programming solution to economic dispatch including transmission losses. IEEE Trans. Power Syst. **7**(5), 544–549 (1992)
4. Chowdhury, B.H., Rahman, S.: A review of recent advances in economic dispatch. IEEE Trans. Power Syst. **5**(4), 1248–1259 (1990)
5. Wang, J., Huang, W., Ma, G., Chen, S.: An improved partheno genetic algorithm for multi-objective economic dispatch in cascaded hydropower systems. Int. J Electr. Power. **67**, 591–597 (2015)
6. Basu, M.: Modified particle swarm optimization for nonconvex economic dispatch problems. Int. J. Electr. Power. **69**, 304–312 (2015)
7. Rahman, T.K.A., Suliman, S.I., Musirin, I.: Artificial immune-based optimization technique for solving economic dispatch in power system. Neural Nets. **3931**, 338–345 (2006)
8. Jayabarathi, T., Sadasivam, G., Ramachandran, V.: Evolutionary programming based economic dispatch of generators with prohibited operating zones. Elect. Power Syst. Res. **52**, 261–266 (1999)
9. Yegireddy, N.K., Panda, S., Kumar Rout, U., Bonthu, R.K.: Selection of control parameters of differential evolution algorithm for economic load dispatch problem. Comput. Intell. Data Min. **3**, 251–260 (2015)
10. Secui, C.D.: A method based on the ant colony optimization algorithm for dynamic economic dispatch with valve-point effects. Int Trans. Electr. Energy **25**(2), 262–287 (2015)
11. Vijay, R.: Intelligent bacterial foraging optimization technique to economic load dispatch problem. Int. J. Soft Comput. Eng. **2**, 55–59 (2012)
12. Simon, D.: Biogeography-based optimization. IEEE Trans. Evolut. Comput. **12**(6), 702–713 (2008)
13. Roy, P.K., Ghoshal, S.P., Thakur, S.S.: Biogeography-based optimization for economic load dispatch problems. Electr. Power Compon. Syst. **38**(2), 166–181 (2009)

14. Wang, G.G., Gandomi, A.H., Alavi, A.H.: An effective krill herd algorithm with migration operator in biogeography-based optimization. Appl. Math. Model. **38**(9), 2454–2462 (2014)
15. Ma, H.: An analysis of the equilibrium of migration models for biogeography-based optimization. Inf. Sci. **180**(18), 3444–3464 (2010)
16. Bhattacharya, A., Chattopadhyay, P.K.: Biogeography-based optimization for different economic load dispatch problems. IEEE Trans. Power Syst. **25**(2), 1064–1077 (2010)
17. Dakuo, H., Fuli, W., Zhizhong, M.: A hybrid genetic algorithm approach based on differential evolution for economic dispatch with valve-point effect. Electr. Power Energy Syst. **30**, 31–38 (2008)
18. He, D.K., Wang, F.L., Mao, Z.Z.: Hybrid particle swarm optimization algorithm for non-linear, genetic algorithm for economic dispatch with valve-point effect. Elect. Power Syst. Res. **78**, 626–633 (2008)
19. Chaturvedi, K.T., Pandit, M., Srivastava, L.: Self-organizing hierarchical particle swarm optimization for nonconvex economic dispatch. IEEE Trans. Power Syst. **23**(3), 1079 (2008)
20. Coelho, L.D.S., Mariani, V.C.: Particle swarm approach based on quantum mechanics and harmonic oscillator potential well for economic load dispatch with valve-point effects. Energ. Convers Manag. **49**, 3080–3085 (2008)
21. Alsumait, J.S., Al-Othman, A.K., Sykulski, J.K.: Application of pattern search method to power system valve-point economic load dispatch. Electr. Power Energy Syst. **29**, 720–730 (2007)
22. Aruldoss Albert Victoire, T., Ebenezer Jeyakumar, A.: Hybrid PSO-SQP for economic dispatch with valve-point effect. Electr. Power Syst. Res. **71**(1), 51–59 (2004)

Modified Artificial Bee Colony Algorithm Based on Disruption Operator

Nirmala Sharma, Harish Sharma, Ajay Sharma and Jagdish Chand Bansal

Abstract Artificial bee colony (ABC) algorithm has been proven to be an effective swarm intelligence-based algorithm to solve various numerical optimization problems. To improve the exploration and exploitation capabilities of ABC algorithm a new phase, namely disruption phase is introduced in the basic ABC. In disruption phase, disrupted operator in which the solutions are attracted or disrupted from the best solution based on the their respective distance from the best solution, is applied to all the solutions except the best solution. Further, the proposed strategy has been evaluated on 15 different benchmark functions and compared with basic ABC and two of its variants, namely modified ABC (MABC) and best so for ABC (BSFABC).

Keywords Artificial bee colony · Swarm intelligence · Optimization · Disruption operator

1 Introduction

The swarm intelligence-based algorithms use hit and trial methods with their ability to learn from individual and neighbor's intelligence to solve complex problems those were tedious to solve by other existing numerical methods. Artificial bee colony (ABC) algorithm is a popular swarm intelligence-based algorithm, developed

Nirmala Sharma (✉) · Harish Sharma
Rajasthan Technical University, Kota, Rajasthan, India
e-mail: nirmala_rtu@yahoo.com

Harish Sharma
e-mail: harish.sharma0107@gmail.com

Ajay Sharma
Government Engineering College, Jhalawar, Rajasthan, India
e-mail: ajay_2406@yahoo.com

J.C. Bansal
South Asian University, New Delhi, India
e-mail: jcbansal@gmail.com

M. Pant et al. (eds.), *Proceedings of Fifth International Conference on Soft Computing for Problem Solving*, Advances in Intelligent Systems and Computing 437, DOI 10.1007/978-981-10-0451-3_79

by taking inspiration from the intellectual behavior and the ability of communication of honey bees [1, 2]. But there is always a possibility that ABC also may sometimes stop proceeding toward global optimum although the solutions have not converged to local optimum [3]. Researchers are continuously working to improve the performance of the ABC algorithm [4–10].

In order to enhance the exploration and exploitation capabilities of ABC algorithm, a new operator namely disruption operator which is based on astrophysics is applied with ABC. When a group of gravitationally bound particles approaches very near to a massive object M, the group of particles tends to be torn apart. The same phenomenon is applied with the basic ABC. Here, the best candidate solution is chosen as a star and all the other candidate solutions disrupt under the gravity force of the star solution. The proposed algorithm is named as disrupt artificial bee colony algorithm (DiABC).

The rest of the paper is structured as follows: In Sect. 2, disruption phenomenon is described. The proposed disruption operator and its characteristics are described in Sect. 3. A comparative study is presented in Sect. 4. Finally, Sect. 5 includes a summary and conclude the work.

2 Disruption Phenomenon in Nature

When a group of gravitationally bound particles (having total mass m), approaches very near to a massive object M, the group of particles tends to be torn apart. This is called disruption phenomenon in nature [11]. The same thing occurs in the solid body also. Disruption phenomenon in astrophysics is defined as *The sudden inward fall of a group of particles that are gravitationally bound under the effect of the force of gravity* [11, 12]. It occurs when all the forces fail to supply a sufficient high pressure to counterbalance the gravity and keep the massive body in equilibrium [13].

3 Simulation of Disruption Phenomenon in Artificial Bee Colony Algorithm

Metaheuristic algorithms must have a proper balance between exploitation and exploration properties to achieve efficiency in local and global search. Since each operator has its own ability of performing the exploration and exploitation of the search space, new operators are simulated with various heuristic algorithms to add specific capabilities and enhancing the performance [14].

For simulating the disruption phenomenon, a new phase called disruption phase is introduced within the ABC. In the proposed phase, the disruption operator initially explores the search space and as time passes it switches to the exploiting conditions. The proposed strategy is named as disruption phase and the algorithm is named as disrupt artificial bee colony algorithm (DiABC). In DiABC, it is assumed that the solution having best fitness value in the swarm is nominated as a star and rest of the candidate solutions are scattered in the search space under the gravity force of the star solution.

The disruption phase is described as follows: For all the candidate solutions except the star solution (having best fitness value) following disruption condition is checked.

$$\frac{R_{i,j}}{R_{i,\text{best}}} < C \tag{1}$$

where $R_{i,j}$ and $R_{i,\text{best}}$ are the Euclidean distances between the *i*th and *j*th candidate solution and between *i*th and the best solution, respectively. Here, j is the nearest neighbor of i. C is a threshold and it is defined as

$$C = C_0\left(1 - \frac{t}{T}\right) \tag{2}$$

The solutions that satisfy Eq. 1 are disrupted under the vicinity of the star (best) solution. The threshold C is a variable which is used to make the operator more meaningful. Initially, when the solutions are not converged the value of C is kept to be large that leads to increase the exploration of the search space and as the solutions get closer to each other, C has to be small for exploitation of the search space. The position update equation for the candidate solutions that satisfy Eq. 1 is defined as

$$x_i(t+1) = x_i(t) \times D \tag{3}$$

Here, $x_i(t)$ and $x_i(t+1)$ are the positions of the *i*th candidate solution during the iteration t and $t+1$, respectively.

The value of D is defined as

$$D = \begin{cases} R_{i,j} \times \text{U}(-0.5, 0.5), & \text{if } R_{i,\text{best}} \geq 1 \\ (1 + \rho \times \text{U}(-0.5, 0.5), & \text{otherwise.} \end{cases} \tag{4}$$

In the above Eq. 4, $U(-0.5,0.5)$ is a uniformly distributed random number in the interval $[-0.5,0.5]$. ρ is a small number used for the purpose of exploitation of the search space. Depending upon the value of D, the exploration and exploitation of search space is performed during the phase. When the value of $R_{i,\text{best}} \geq 1$ means that the solutions are not converged then the value of D can be less than or greater than 1 (using Eq. 4) and by multiplying this value with previous value of the *i*th solution, the dimension of the *i*th solution is changed randomly and it can be smaller or larger than the previous solution value. So this leads to explore the search space. On the other part, when $R_{i,\text{best}} < = 1$ means that *i*th solution is close to the best solution, then the value of D is set to very small and it will update the position of the candidate solution near to old position. This helps in exploitation of the search space. This implies that when the solutions are not converged the disruption operator explores and as the solutions are converging and getting close to the star solution, the operator exploits the search space. C and D are the two new control parameters in the disruption phase. Fine tuning of both the parameters is required for proper implementation of the strategy.

The DiABC algorithm is divided into four phases, namely employed bee phase, onlooker bee phase, scout bee phase, and disruption phase. The first three phases are same as in the basic ABC and after these three phases a new phase, namely disruption phase is added to implement the above discussed strategy. Initially, all the solutions are randomly initialized in the given search space through Eq. 5:

$$x_{ij} = x_{\mathrm{minj}} + U[0,1](x_{\mathrm{maxj}} - x_{\mathrm{minj}}) \tag{5}$$

where x_i represents the ith food source in the swarm, x_{minj} and x_{maxj} are bounds of x_i in jth dimension and $U[0,1]$ is a uniformly distributed random number in the range [0, 1]. Based on the above discussion, the pseudocode of the proposed algorithm is shown in the Algorithm 1.

4 Results and Discussions

To analyze the performance of the proposed DiABC algorithm, 15 different global optimization problems (f_1–f_{15}) are chosen as shown in Table 1.

Algorithm 1 Disrupt Artificial Bee Colony Algorithm($DiABC$)

```
Initialize the parameters: MCN (Maximum number of cycles), D (Dimension of the
problem), SN (Swarm Size), C_O, ρ;
Initialize the swarm having solutions, x_i where (i=1,2,.....,SN) by using equation
(1.5) cycle = 1;
while cycle <> MCN do
  Step 1: Employed bee phase for generating new food sources;
  Step 2: Onlooker bees phase for updating the food sources depending on their
  nectar amounts;
  Step 3: Scout bee phase for discovering the new food sources in place of aban-
  doned food sources;
  Step 4: Memorize the best food source found so far (Consider as star Solution);
  Step 5: Disruption Phase: /* Explained as follows:*/
  for each solution do
    Check the condition using equation (1.1) for all the candidate solutions except
    the star solution; here R_{i,j} represents the Euclidean distance between the i^th
    solution and its neighbor j, while R_{i,best} is the Euclidean distance between
    the i^th solution and the star solution; C is a threshold calculated using the
    equation (1.1);
    if (R_{i,j} / R_{i,best} < C) then
      D is calculated using the equation 1.4;
      Change the position of the solutions using equation 1.3;
    end if
  end for
  Memorize the best food source found so far;
  cycle=cycle+1;
end while
Output the best solution found so far.
```

Table 1 Test problems

Objective function	Search range	OV	D	AE
$f_1(x) = -20 + e + \exp\left(-\frac{0.2}{D}\sqrt{\sum_{i=1}^{D} x_i{}^3}\right)$	[−1, 1]	$f(0) = 0$	30	1.0E −05
$f_2(x) = \sum_{i=1}^{D} \lvert x_i \rvert^{i+1}$	[−1, 1]	$f(0) = 0$	30	1.0E −05
$f_3(x) = -\sum_{i=1}^{D-1}\left(\exp\left(\frac{-(x_i^2 + x_{i+1}^2 + 0.5x_i x_{i+1})}{8}\right) \times I\right)$ where, $I = \cos\left(4\sqrt{x_i^2 + x_{i+1}^2 + 0.5x_i x_{i+1}}\right)$	[−5, 5]	$f(\mathbf{0}) = -D + 1$	10	1.0E −05
$f_4(x) = \frac{\pi}{D}(10\sin^2(\pi y_1) + \sum_{i=1}^{D-1}(y_i - 1)^2 \times \left(1 + 10\sin^2(\pi y_{i+1})) + (y_D - 1)^2\right)$, where $y_i = 1 + \frac{1}{4}(x_i + 1)$	[−10, 10]	$f(\mathbf{-1}) = 0$	30	1.0E −05
$f_5(x) = 0.1(\sin^2(3\pi x_1) + \sum_{i=1}^{D-1}(x_i - 1)^2 \times (1 + \sin^2(3\pi x_{i+1})) + (x_D - 1)^2(1 + \sin^2(2\pi x_D))$	[−5, 5]	$f(\mathbf{1}) = 0$	30	1.0E −05
$f_6(x) = a(x_2 - bx_1^2 + cx_1 - d)^2 + e(1 - f)\cos x_1 + e$	$-5 \le x_1 \le 10, 0 \le x_2 \le 15$	$f(-\pi, 12.275) = 0.3979$	2	1.0E −05
$f_7(x) = \sum_{i=1}^{D} z_i^2 + f_{\text{bias}}$, $z = x - o$, $x = [x_1, x_2, \ldots, x_D]$, $o = [o_1, o_2, \ldots, o_D]$	[−100, 100]	$f(o) = f_{\text{bias}} = -450$	10	1.0E −05
$f_8(x) = \sum_{i=1}^{D} \frac{z_i^2}{4000} - \prod_{i=1}^{D} \cos\left(\frac{z_i}{\sqrt{i}}\right) + 1 + f_{\text{bias}}$, $z = (x - o)$, $x = [x_1, x_2, \ldots, x_D]$, $o = [o_1, o_2, \ldots, o_D]$	[−600, 600]	$f(o) = f_{\text{bias}} = -180$	10	1.0E −05
$f_9(x) = -20\exp\left(-0.2\sqrt{\frac{1}{D}\sum_{i=1}^{D} z_i^2}\right) - \exp(\frac{1}{D}\sum_{i=1}^{D}\cos(2\pi z_i)) + 20 + e + f_{\text{bias}}$, $z = (x - o)$, $x = (x_1, x_2, \ldots, x_D)$, $o = (o_1, o_2, \ldots, o_D)$	[−32, 32]	$f(o) = f_{\text{bias}} = -140$	10	1.0E −05
$f_{10}(x) = (1 + (x_1 + x_2 + 1)^2 \times (19 - 14x_1 + 3x_1^2 - 14x_2 + 6x_1x_2 + 3x_2^2)) \times (30 + (2x_1 - 3x_2)^2 \times (18 - 32x_1 + 12x_1^2 + 48x_2 - 36x_1x_2 + 27x_2^2))$	[−2, 2]	$f(0, -1) = 3$	2	1.0E −14

(continued)

Table 1 (continued)

Objective function	Search range	OV	D	AE
$f_{11}(x) = (4 - 2.1x_1^2 + x_1^4/3)x_1^2 + x_1x_2 + (-4 + 4x_2^2)x_2^2$	[−5, 5]	$f(-0.0898, 0.7126) = -1.0316$	2	1.0E −05
$f_{12}(x) = 10^5x_1^2 + x_2^2 - (x_1^2 + x_2^2)^2 + 10^{-5}(x_1^2 + x_2^2)^4$	[−20, 20]	$f(0, 15) = f(0, -15) = -24777$	2	5.0E −01
$f_{13}(x) = \sin(x_1 + x_2) + (x_1 - x_2)^2 - \frac{3}{2}x_1 + \frac{5}{2}x_2 + 1$	$x_1 \in [-1.5, 4]$, $x_2 \in [-3, 3]$	$f(-0.547, -1.547) = -1.9133$	30	1.0E −04
$f_{14}(x) = \sum_{i=1}^{5}\left(\frac{x_1x_3t_i}{1 + x_1t_i + x_2v_i} - y_i\right)^2$	[−10, 10]	$f(3.13, 15.16, 0.78) = 0.4 \times 10^{-4}$	3	1.0E −03
$f_{15}(x) = -\sum_{i=1}^{5} i\cos((i+1)x_1 + 1)\sum_{i=1}^{5} i\cos((i+1)x_2 + 1)$	[−10, 10]	$f(7.083, 4.858) = -186.7309$	2	1.0E −05

OV Optimum value, *D* Dimension, *AE* Acceptable error

To prove the performance of DiABC, a comparative analysis is carried out among DiABC, ABC, and two of its variants, namely best so far ABC (BSFABC) [15] and modified ABC (MABC) [16]. To test DiABC, ABC, BSFABC, and MABC over the considered problems, following experimental setting is adopted:

- $C_0 = 60$
- $\rho = 10^{-10}$
- The number of simulations/run = 100,
- Colony size NP $=$ 50 and number of food sources SN $=$ NP$/2$,
- $\phi_{ij} = \text{rand}[-1, 1]$ and limit = $D \times$ SN [16],
- Parameter setting for the algorithms BSFABC and MABC are similar to their original research paper.

The experimental results of the considered algorithms are shown in Table 2. Table 2 provides a report about the standard deviation (SD), mean error (ME), average number of function evaluations (AFE), and success rate (SR). Results in Table 2 reflect that most of the time DiABC outperforms in terms of reliability, efficiency, and accuracy as compared to the ABC, BSFABC, and MABC.

Further, boxplots analysis of AFE is carried out to compare all the considered algorithms in terms of consolidated performance as it can efficiently represent the empirical distribution of data graphically. The boxplots for DiABC, ABC, BSFABC, and MABC are shown in Fig. 1. The results reveal that interquartile range and medians of DiABC are comparatively low. The considered algorithms are also compared by giving weighted importance to the ME, SR, and AFE. This comparison is measured using the performance indices which is described in [5, 17]. The values of PIs for the DiABC, ABC, BSFABC, and MABC are calculated and corresponding PIs graphs are shown in Fig. 2.

The graphs corresponding to each of the cases, i.e., giving weighted importance to AFE, SR, and ME (as explained in [5, 17]) are shown in Fig. 2a–c, respectively. In these figures, horizontal axis represents the weights while vertical axis represents the PI.

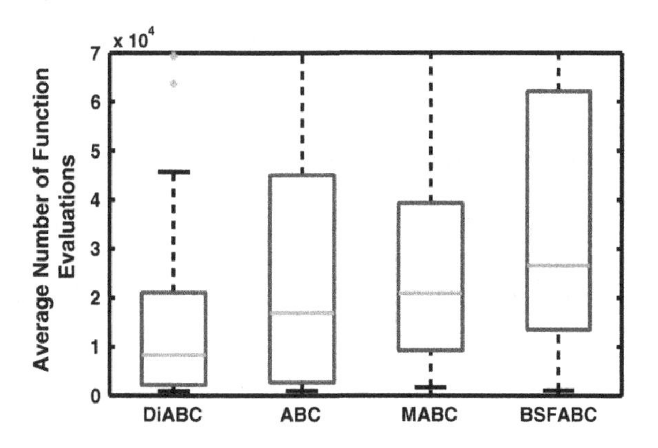

Fig. 1 Boxplots graphs for average number of function evaluation

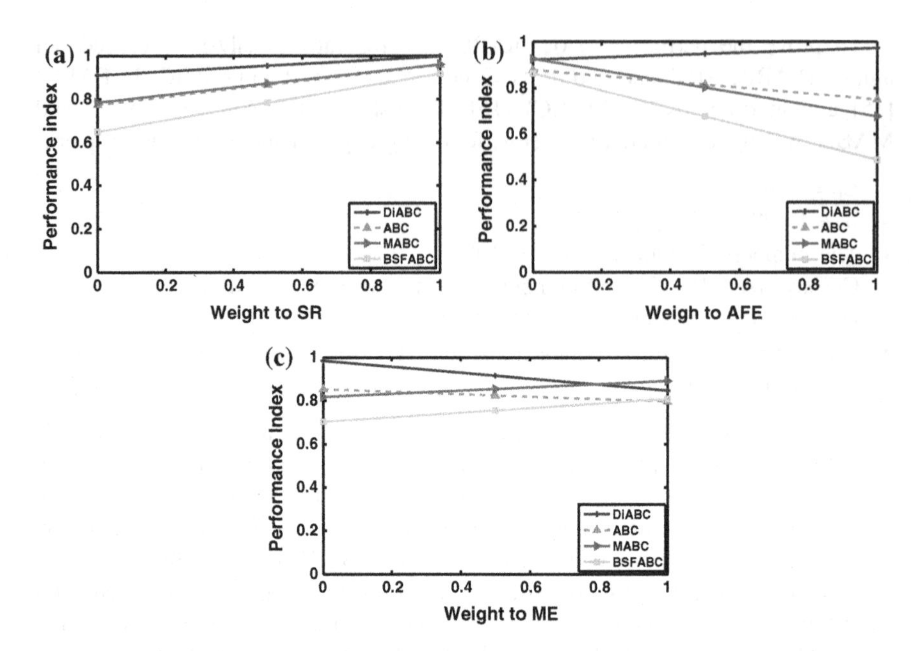

Fig. 2 Performance index for test problems; **a** for weighted importance to SR, **b** for weighted importance to AFE, and **c** for weighted importance to ME

It is clear from Fig. 2 that PI of DiABC is higher than the considered algorithms in each case, i.e., DiABC performs better on the considered problems as compared to the ABC, BSFABC, and MABC. Further, we compare the convergence speed of the considered algorithms by measuring the average function evaluations (AFEs). A smaller AFE means higher convergence speed. In order to minimize the effect of the stochastic nature of the algorithms, the reported function evaluations for each test problem is the average over 100 runs. In order to compare convergence speeds, we use the acceleration rate (AR) which is defined as follows, based on the AFEs for the two algorithms ALGO and DiABC:

$$AR = \frac{AFE_{ALGO}}{AFE_{DiABC}}, \tag{6}$$

where ALGO $\in$ {ABC, MABC, BSFABC} and AR > 1 means DiABC is faster. In order to investigate the AR of the proposed algorithm compared to the basic ABC and its variants, results of Table 2 are analyzed and the value of AR is calculated using Eq. 6. Table 3 shows a clear comparison between DiABC and ABC, DiABC and MABC, and DiABC and BSFABC in terms of AR. It is clear from Table 3 that convergence speed of DiABC is faster among all the considered algorithms.

Table 2 Comparison of the results of test problems, *TP* Test problem

TP	Algorithm	SD	ME	AFE	SR
f_1	DiABC	9.06E−07	8.97E−06	45690.75	100
	ABC	1.59E−06	8.31E−06	49188.5	100
	BSFABC	1.74E−06	8.28E−06	72368.5	100
	MABC	4.23E−07	9.53E−06	43630	100
f_2	DiABC	2.55E−06	6.11E−06	13896.75	100
	ABC	2.77E−06	5.57E−06	16888.5	100
	BSFABC	2.72E−06	5.84E−06	14434	100
	MABC	2.06E−06	7.31E−06	9365	100
f_3	DiABC	5.21E−02	5.25E−03	69407.79	99
	ABC	1.05E−01	2.36E−02	90991.91	93
	BSFABC	1.99E−01	6.09E−02	123141.06	85
	MABC	1.53E−06	8.29E−06	68906.06	100
f_4	DiABC	1.59E−06	8.13E−06	19557	100
	ABC	2.51E−06	6.93E−06	20037	100
	BSFABC	2.52E−06	6.72E−06	26570.5	100
	MABC	7.12E−07	9.10E−06	22702.5	100
f_5	DiABC	2.01E−06	7.74E−06	21552.75	100
	ABC	2.43E−06	7.49E−06	21671.5	100
	BSFABC	2.57E−06	6.98E−06	28887.5	100
	MABC	7.93E−07	9.16E−06	20881	100
f_6	DiABC	6.62E−06	5.71E−06	1715.32	100
	ABC	6.05E−06	5.08E−06	1944.54	100
	BSFABC	7.19E−06	6.09E−06	29551.41	86
	MABC	6.73E−06	5.97E−06	22809.08	90
f_7	DiABC	2.10E−06	7.40E−06	8315.25	100
	ABC	2.42E−06	7.11E−06	9109	100
	BSFABC	2.49E−06	6.94E−06	18117	100
	MABC	1.63E−06	7.97E−06	8688.5	100
f_8	DiABC	1.26E−03	2.27E−04	63675.67	97
	ABC	2.57E−03	7.90E−04	65692.24	91
	BSFABC	6.18E−03	4.58E−03	118467.79	58
	MABC	1.03E−03	1.55E−04	73173.09	97
f_9	DiABC	1.24E−06	8.54E−06	13875	100
	ABC	1.76E−06	8.02E−06	16585.5	100
	BSFABC	1.93E−06	8.13E−06	31326.5	100
	MABC	1.04E−06	8.79E−06	14231.02	100
f_{10}	DiABC	4.41E−15	4.86E−15	5856.04	100
	ABC	1.30E−05	2.33E−06	123320.73	53
	BSFABC	4.74E−15	6.36E−15	13202.48	100
	MABC	4.43E−15	5.40E−15	14370.25	100

(continued)

Table 2 (continued)

TP	Algorithm	SD	ME	AFE	SR
f_{11}	DiABC	1.07E−05	1.38E−05	883.5	100
	ABC	1.03E−05	1.15E−05	973.02	100
	BSFABC	1.51E−05	1.74E−05	104395.46	48
	MABC	1.49E−05	1.58E−05	94866.77	53
f_{12}	DiABC	5.29E−03	4.89E−01	1180.5	100
	ABC	5.52E−03	4.89E−01	1473.56	100
	BSFABC	5.28E−03	4.91E−01	2800.72	100
	MABC	5.46E−03	4.91E−01	2347.19	100
f_{13}	DiABC	6.51E−06	8.81E−05	909	100
	ABC	6.83E−06	8.85E−05	1203.04	100
	BSFABC	6.44E−06	8.71E−05	1013.58	100
	MABC	6.79E−06	8.77E−05	1641.17	100
f_{14}	DiABC	2.70E−06	1.95E−03	7493.71	100
	ABC	2.75E−06	1.95E−03	32554.88	100
	BSFABC	2.64E−06	1.95E−03	17641.71	100
	MABC	2.96E−06	1.95E−03	9296.69	100
f_{15}	DiABC	5.89E−06	5.28E−06	3575.37	100
	ABC	5.85E−06	5.13E−06	4599.74	100
	BSFABC	5.76E−06	5.11E−06	9396.07	100
	MABC	5.37E−06	4.65E−06	26573.93	100

Table 3 Acceleration rate (AR) of DiABC compared to the basic ABC, MABC, and BSFABC

Test Problems	ABC	MABC	BSFABC
f_1	1.076552694	0.954897873	1.583876386
f_2	1.215284149	0.673898573	1.038660118
f_3	1.310975468	0.992771273	1.774167712
f_4	1.024543642	1.160837552	1.358618398
f_5	1.005509738	0.968832284	1.340316201
f_6	1.133631043	13.29727398	17.22792832
f_7	1.095457142	1.044887406	2.178767926
f_8	1.031669396	1.149153044	1.86048753
f_9	1.195351351	1.025659099	2.257765766
f_{10}	21.05872398	2.453919372	2.254506458
f_{11}	1.101324278	107.3760838	118.161245
f_{12}	1.248250741	1.988301567	2.372486235
f_{13}	1.323476348	1.805467547	1.115049505
f_{14}	4.344294081	1.240599116	2.354202391
f_{15}	1.28650741	7.432497895	2.62799934

5 Conclusion

In this paper, a new phase, namely disruption phase is introduced with ABC and the proposed strategy is named as disrupt ABC (DiABC). In disruption phase, exploration and exploitation capabilities of ABC algorithm are balanced by disruption operator. In the proposed phase, the best solution among all the solutions is considered as a star and all the remaining solutions are disrupted under the gravity force of the star solution. This property helps to improve the exploration and exploitation capability of the ABC algorithm. The proposed strategy is evaluated through extensive experimental analysis and found that the DiABC may be a good choice for solving the continuous optimization problems.

References

1. Karaboga, D., Basturk, B.: On the performance of artificial bee colony (abc) algorithm. Appl. Soft Comput. **8**(1), 687–697 (2008)
2. Bansal, J.C., Sharma, H., Jadon, S.S.: Artificial bee colony algorithm: a survey. Int. J. Adv. Intell. Paradigms **5**(1), 123–159 (2013)
3. Karaboga, D., Akay, B.: A comparative study of artificial bee colony algorithm. Appl. Math. Comput. **214**(1), 108–132 (2009)
4. Bansal, J.C., Sharma, H., Arya, K.V., Deep, K., Pant, M.: Self-adaptive artificial bee colony. Optimization **63**(10), 1513–1532 (2014)
5. Bansal, J.C., Sharma, H., Arya, K.V., Nagar, A.: Memetic search in artificial bee colony algorithm. Soft. Comput. **17**(10), 1911–1928 (2013)
6. Bansal, J.C., Sharma, H., Nagar, A., Arya, K.V.: Balanced artificial bee colony algorithm. Int. J. Artif. Intell. Soft Comput. **3**(3), 222–243 (2013)
7. Jadon, S.S., Bansal, J.C., Tiwari, R., Sharma, H.: Expedited artificial bee colony algorithm. In: Proceedings of the Third International Conference on Soft Computing for Problem Solving, pp. 787–800. Springer (2014)
8. Sharma, H., Bansal, J.C., Arya, K.V.: Opposition based lévy flight artificial bee colony. Memetic Comput. **5**(3), 213–227 (2013)
9. Sharma, H., Bansal, J.C., Arya, K.V.: Power law-based local search in artificial bee colony. Int. J. Artif. Intell. Soft Comput. **4**(2/3), 164–194 (2014)
10. Sharma, H., Bansal, J.C., Arya, K.V., Deep, K.: Dynamic swarm artificial bee colony algorithm. Int. J. Appl. Evol. Comput. (IJAEC) **3**(4), 19–33 (2012)
11. Sarafrazi, S., Nezamabadi-Pour, H., Saryazdi, S.: Disruption: a new operator in gravitational search algorithm. Scientia Iranica **18**(3), 539–548 (2011)
12. Ding, G.Y., Liu, H., He, X.Q.: A novel disruption operator in particle swarm optimization. Appl. Mech. Mater. **380**, 1216–1220 (2013)
13. Liu, H., Ding, G., Sun, H.: An improved opposition-based disruption operator in gravitational search algorithm. In: 2012 Fifth International Symposium on Computational Intelligence and Design (ISCID), vol. 2, pp. 123–126. IEEE (2012)
14. Chen, T.-Y., Chi, T.-M.: On the improvements of the particle swarm optimization algorithm. Adv. Eng. Softw. **41**(2), 229–239 (2010)
15. Banharnsakun, A., Achalakul, T., Sirinaovakul, B.: The best-so-far selection in artificial bee colony algorithm. Appl. Soft Comput. **11**(2), 2888–2901 (2011)

16. Akay, B., Karaboga, D.: A modified artificial bee colony algorithm for real-parameter optimization. Inf. Sci. **192**, 120–142 (2012)
17. Bansal, J.C., Sharma, H.: Cognitive learning in differential evolution and its application to model order reduction problem for single-input single-output systems. Memetic Comput. **4**(3), 209–229 (2012)

DG Integration with Power Quality Improvement Feature for Smart Grid

Archana Sharma, Bharat Singh Rajpurohit and Lingfeng Wang

Abstract The scheme proposed in the paper aims to maximize the utilization of a grid interfacing inverter used for distributed generation (DG) integration. To maximize the utilization, inverter is controled to integrate DG unit with active power filtering features. The inverter is controlled in a way to (1) inject power generated from DG to grid, and (2) compensation of grid current harmonics, load current balance, and reactive power compensation. Control strategy used is a learning based Anti-Hebbian (AH) algorithm. AH algorithm extracts fundamental active and reactive component of load side power and continuously tune these components through feedback principle with respective changes in input. Furthermore, closed loop power control is done without using phase-locked loop (PLL). Thus, power is controlled accurately with harmonic compensation function. In addition, ability of this scheme to work in condition of frequency deviation is also discussed. Simulation of the scheme with all possible condition variation is done to show effectiveness of the scheme.

Keywords Distributed generation (DG) · Power quality improvement · Anti-Hebbian

Archana Sharma (✉) · B.S. Rajpurohit
School of Computing and Electrical Engineering, IIT Mandi, Mandi, India
e-mail: archana_sharma@students.iitmandi.ac.in

B.S. Rajpurohit
e-mail: bsr@iitmandi.ac.in

Lingfeng Wang
College of Engineering and Applied Sciences, University of Wisconsin, Milwaukee, Wisconsin, USA
e-mail: wang289@uwn.edu

M. Pant et al. (eds.), *Proceedings of Fifth International Conference on Soft Computing for Problem Solving*, Advances in Intelligent Systems and Computing 437, DOI 10.1007/978-981-10-0451-3_80

1 Introduction

Nowadays, renewable energy sources become significant part of electricity generation. Renewable energy sources with storage facilities like batteries and micro-turbines, forms a DG unit. DG integration with grid is required to increase the reliability of power supply to end users. DG integration to grid requires a power electronic converter, which makes the DG power compatible to grid power [1]. Concentration of DG sources increases the number of converters and hence increases the complexity in power system. Presence of nonlinear loads further degrades distribution system power quality [2]. From past few decades, a number of filtering methods have been developed to compensate harmonics. These methods require an additional filter (power electronics converter based) installation, which is not cost effective in present scenario. Alternatively, DG integration with power quality improvement together provides a solution. DG needs an interfacing device to integrate with grid and active filters (voltage source converters) are used for harmonic filtering/reactive power compensation. By modifying control scheme used in converter for harmonic filtering, DG can be integrated with power quality improvement feature together.

For load current harmonic compensation, an accurate detection of fundamental component of load current is required. Various techniques for harmonic compensation has been reported from many years, such as synchronous reference frame theory based compensation [3], fourier transform based current detection, second order generalized integrator, and many more. Some computationally intelligent control algorithms such as neuro-fuzzy and fuzzy genetic controllers had also reported [4]. Some of the conventional control schemes use phase locked loop (PLL) to synchronize the fundamental reference current with grid current. In PLL, for synchronization it is assumed that grid voltage is ripple free and has constant magnitude. However, point of common coupling (PCC) voltage varies due to power flow fluctuations. The variation in voltage can cause non trivial power control errors [5].

This paper proposes an Anti-Hebbian-based control scheme for integration of DG with grid. This scheme is a neural network based total least square technique which allows tuning of fundamental active and reactive power component of load current in case of input parameters variations. With integration of DG, harmonic elimination as well as reactive power compensation is done by the same converter. A balanced power supply distribution between grid and DG is achieved.

2 System Description with Results and Discussions

2.1 System Description

The scheme proposed consists of DG unit connected to the grid as shown in Fig. 1. For interfacing DG to the grid a voltage source inverter (VSC) is used which requires a control scheme to work. The control scheme takes system parameters

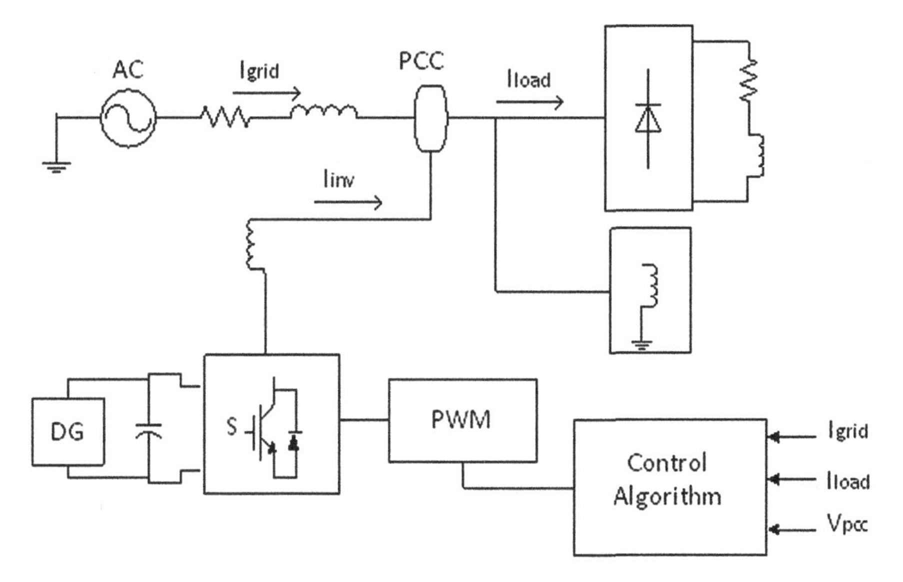

Fig. 1 Schematic of DG integration with grid

(current, voltages, etc.) as inputs and generates a reference current signal to bring about pulses for the VSC. VSC (specifically inverter) and loads (both non-linear and linear) are connected to grid at the point of common coupling (PCC). Power from DG source is not uniform because renewable sources are always nature dependent. The dc-link capacitor is used to interface DG, plays an important role to deals with the variable nature of DG input power. Power transferred from DG to grid is controlled so that inverter always supplies reactive power needed by connected load and extra reactive power to the grid, and active power to the load, if generated power is equal to the load power and to the grid and load, if generated power is more than required by load [6].

2.2 *Control Algorithm Design for VSC*

AH-based learning algorithm is used as a control scheme for grid interfacing inverter. AH algorithm is based on classical Hebbian rule [7], which makes linear neurons of weighted input signals to converge the eigenvectors without much complexity. The simplest neural equation (linear) is given as Eq. (1)

$$\eta(t) = \sum_{i=1}^{n+1} \phi_i(t)\xi_i(t) \tag{1}$$

In Eq. (1), $\eta(t)$ is output of neuron, $\xi_i(t)$ is the input vector and $\phi_i(t)$ is the weight vector. In basic Hebbian algorithm, weight vector is recalculated by adding to it a small increment that is proportional to the product of input and output signal. In AH, instead of adding the incremented weight, it is subtracted from the original weight. The output signal for $\phi_i(t)$ is calculated as Eq. (2)

$$\widetilde{\phi_i}(t+1) = \widetilde{\phi_i}(t) - \mu\eta(t)\xi_i(t) \tag{2}$$

where, μ is scaling factor. It is a constant term with value 1.

By performing the scaling function on the weight vector

$$\widetilde{\phi_i}(\mathrm{t}+1) = -1 \times \frac{\widetilde{\phi_i}(t+1)}{\widetilde{\phi_{n+1}}(t+1)} \tag{3}$$

Combining Eqs. (2) and (3) together, we get Eq. (4)

$$\widetilde{\phi_i}(\mathrm{t}+1) = \frac{\widetilde{\phi_i}(t) - \mu\eta(t)\xi_i(t)}{1 + \mu\eta(t)\xi_{n+1}(t)} \tag{4}$$

AH learning algorithm requires weight normalization to prevent weight vectors from collapsing to zero. Oja's active decay rule [8] is a popular local approximation to explicit weight normalization used here.

$$\Delta \mathrm{W} = \eta(xy - wy^2) \tag{5}$$

The first term in parentheses (xy) represents the standard Hebb rule, while the second (wy^2) is the active decay. The Eq. (5) is used to calculate the power component of load current, will be given later.

The procedure for implementation of this algorithm for power quality enhancement is described below:

First, in phase unit voltage templates are calculated. For this amplitude of three-phase terminal phase voltages (v_{soa}, v_{sob} and v_{soc}) at PCC is computed as Eq. (6)

$$V_m = \mathrm{sqrt}(\{(2/3)(v_{\mathrm{soa}}^2 + v_{\mathrm{sob}}^2 + v_{\mathrm{soc}}^2)\}) \tag{6}$$

The in-phase and quardrature unit voltage templates are calculated as Eqs. (7–10), respectively.

$$u_{\mathrm{pa}} = \frac{v_{\mathrm{soa}}}{V_m},\ u_{\mathrm{pb}} = \frac{v_{\mathrm{sob}}}{V_m},\ u_{\mathrm{pc}} = \frac{v_{\mathrm{soc}}}{V_m} \tag{7}$$

$$u_{\mathrm{qa}} = -\frac{u_{\mathrm{pb}}}{\sqrt{3}} + \frac{u_{\mathrm{pc}}}{\sqrt{3}} \tag{8}$$

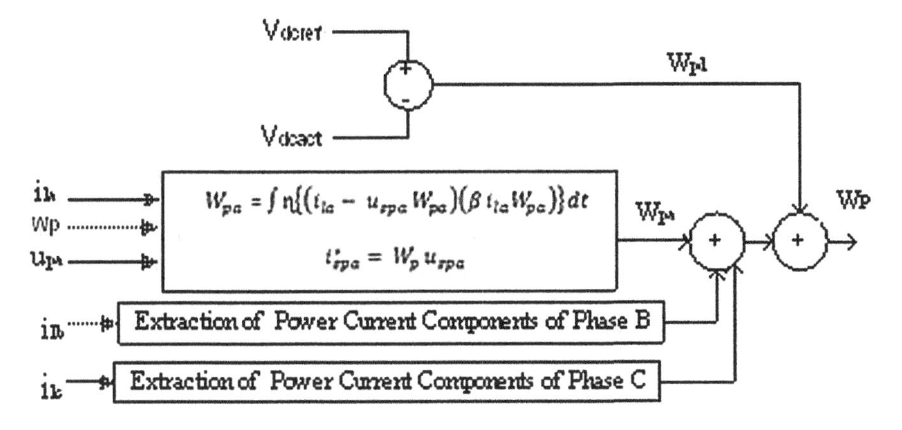

Fig. 2 Weight estimation of active power of load using Anti-Hebbian

$$u_{\mathrm{qb}} = \frac{\sqrt{3}u_{\mathrm{pa}}}{2} + \frac{(u_{\mathrm{pb}} - u_{\mathrm{pc}})}{2\sqrt{3}} \tag{9}$$

$$u_{\mathrm{qc}} = -\frac{\sqrt{3}u_{\mathrm{pa}}}{2} + \frac{(u_{\mathrm{pb}} - u_{\mathrm{pc}})}{2\sqrt{3}} \tag{10}$$

Next is the estimation of weight of active power taken by load; this can be done by using Eq. (5) as shown in Fig. 2. Instantaneous values of load currents (i_{la}, i_{lb} and i_{lc}) are sensed and fed to the controller, which calculates the weights of active power components (W_{pa}, W_{pb} and W_{pc}) of load current for a phase as given in Eq. (11) [9].

$$W_{\mathrm{pa}} = \int \eta\{(i_{\mathrm{da}} - u_{\mathrm{pa}}W_{\mathrm{pa}})(\beta i_{\mathrm{la}}W_{\mathrm{pa}})\}\mathrm{d}t \tag{11}$$

Equation (11) given for weight estimation of active power component of load, it contains two terms 'η' and 'β'. 'η' is an iterative feedback factor used for closed loop control. Estimation accuracy and convergence rate depends on 'η'. High value of 'η' leads to low estimation accuracy and large error in updated weight. This makes choice of low value for 'η' to minimize error of weighted signal. AH is based on TLS so a constant 'β' is used with a value of 1. A factor 'i_{da}' is there in calculation of weight of active power component. This factor depends on instantaneous value of load current and can be calculated from Eq. (12).

$$i_{\mathrm{da}} = (i_{\mathrm{la}} - u_{\mathrm{spa}}W_{\mathrm{pa}}) \tag{12}$$

Similarly, for other phases the weights (W_{pb}, W_{pc}) can be calculated from their respective load current components.

The regulation of dc link voltage gives information about the exchange of active power from grid to load Eq. (13).

$$v_d(n) = v_{\text{dcref}}(n) - v_{\text{dcact}}(n) \tag{13}$$

$$W_{\text{dp}}(n) = W_{\text{dp}}(n-1) + K_{\text{pd}}(v_d(n) - v_d(n-1)) + K_{\text{id}} v_d(n) \tag{14}$$

Thus output of PI controller is a current value which is given to regulate weight of active power. Weight component of output current component of PI controller is given in Eq. (14). Average of weights components of three phases gives the fundamental reference current as Eq. (15)

$$W_{\text{PA}} = (W_{\text{pa}} + W_{\text{pb}} + W_{\text{pc}})/3 \tag{15}$$

Total fundamental active power component of reference current can be calculated by adding weight component of load current and dc link voltage active power component Eq. (16)

$$W_p = W_{\text{pA}} + W_{\text{dp}} \tag{16}$$

The three fundamental component of three phase source current is calculated by multiplying weight component by in-phase unit voltage template Eq. (17).

$$i^*_{\text{sfap}} = W_P * u_{\text{pa}},\; i^*_{\text{sfbp}} = W_P * u_{\text{pb}},\; i^*_{\text{sfcp}} = W_P * u_{\text{pc}} \tag{17}$$

Similar calculation is done to calculate reactive power component of reference current using quadrature unit voltage templates Eqs. (8–10), using Eq. (18) as reactive power weight calculation for load current.

$$W_{\text{qa}} = \int \eta\{(i_{\text{qa}} - u_{\text{qa}} W_{\text{qa}})(\beta\, i_{\text{la}} W_{\text{qa}})\}\text{d}t \tag{18}$$

To maintain terminal voltage constant a PI controller is used which convert the voltage difference between reference and actual terminal voltage to reactive current component Eq. (19).

$$W_{\text{qt}}(n) = W_{\text{qt}}(n-1) + K_{\text{pq}}(v_{\text{te}}(n) - v_{\text{te}}(n-1)) + K_{\text{iq}} v_{\text{te}}(n)) \tag{19}$$

where, $v_{\text{te}}(n) = v_{\text{tref}}(n) - v_t(n)$, v_t is sensed terminal voltage, v_{tref} is reference terminal voltage, K_{pq} is proportion gain factor and K_{iq} is integral gain factor.

Total fundamental reactive power reference current weight component is given by Eq. (20) and current is given by Eq. (21).

$$W_q = W_{\text{qA}} + W_{\text{qt}} \tag{20}$$

$$i^*_{\text{sfaq}} = W_q * u_{\text{qa}},\ i^*_{\text{sfbq}} = W_q * u_{\text{qb}},\ i^*_{\text{sfcq}} = W_q * u_{\text{qc}} \tag{21}$$

Total reference current is addition of active and reactive components of reference current Eq. (22).

$$i^*_{\text{sfa}} = i^*_{\text{sfap}} + i^*_{\text{sfaq}},\ i^*_{\text{sfb}} = i^*_{\text{sfbp}} + i^*_{\text{sfaq}},\ i^*_{\text{sfa}} = i^*_{\text{sfap}} + i^*_{\text{sfaq}} \tag{22}$$

To generate pulses for voltage source inverter the difference of grid current and fundamental component of reference current is calculated and PWM technique is used.

2.3 Results and Discussions

This section contains the response of system with proposed scheme under various unfavorable conditions. A three-phase VSC is used to interface DG with grid and to make grid current unity power factor. The conditions included for scheme verification are change in DG input power, load power, fault condition at load and deviation in frequency from nominal value. DG power output can be allowed to vary for a very small percentage (<3 %), as large variation in DG power makes grid power unbalance. Next condition is sudden change in load power. After that the impact of single phase and three phase fault is considered.

The frequency variation (<4 %) is considered so that system can work under worst condition.

Table 1 shows the system parameters under different conditions. Supply frequency is deviated 2 Hz for considering the worst condition, nowadays fast acting devices are used in power systems which does not allow frequency to deviate much. An inductor is used to interface to grid and to remove some higher order harmonics from inverter.

A complete response of the system under all variations is shown in Fig. 3. The figure shows the waveform of grid voltages (v_{soa}, v_{sob} and v_{soc}), grid currents (i_{sa}, i_{sb} and i_{sc}), load currents (i_{la}, i_{lb} and i_{lc}), inverter current (i_{inva}, i_{invb} and i_{invc}),

Table 1 System parameters

Parameters	Simulation
AC line voltage, frequency	415 V(L-L), 50 Hz, 52 Hz (deviation)
Source impedance	$R_s = 0.06\ \Omega$, $L_s = 3$ mH
Load	$R = 20\ \Omega$, $L = 200$ mH (non-linear) $Q_L = 1000$ VAR (linear)
DC bus voltage	830 V
Interfacing inductor	2.1 mH
Max. switching frequency	20 kHz
DG power	35 kW, 18 kVA

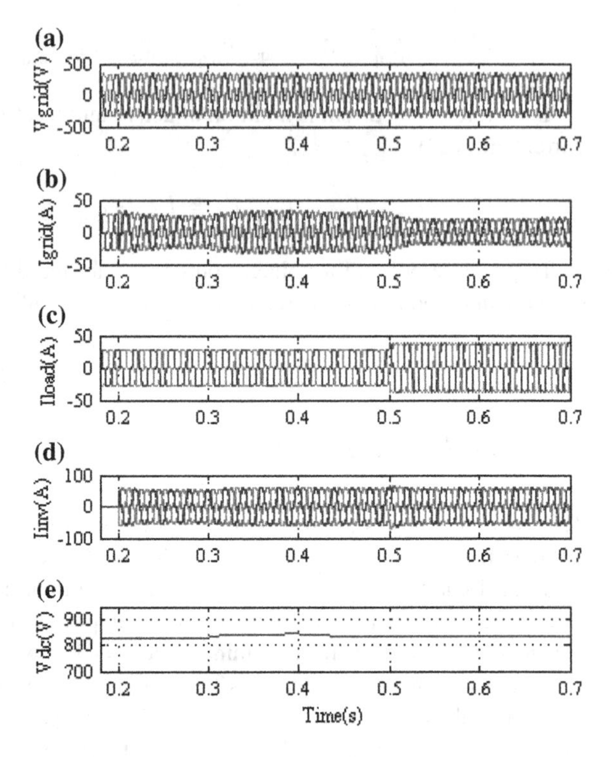

Fig. 3 Complete response: **a** grid voltages, **b** grid currents, **c** load currents, **d** inverter currents, **e** DC link voltage

and dc link voltage (Vdc). A detailed description of Fig. 3 is given later with enhanced waveform figures.

Figure 4 shows exchange of active and reactive power between grid (P_{grid}, Q_{grid}), DG (P_{inv}, Q_{inv}), and load (P_{load}, Q_{load}). This figure shows the variation in power exchange between grid, load, and inverter at different conditions.

Variable Output from DG. DG source output depends on various nature dependent factors. At $t = 0.3$ s, DG output increases, this causes grid current to rise to some higher value due to constant nature of load as shown in Fig. 5.

Variation in Load. System behavior for load change is shown in Fig. 6. At $t = 0.5$ s, load changes to some higher value. This value of load change is less than the total capacity of DG power. DG supplies the power due to load change, but the grid current decreases due to decrease in amount of power supplied by DG as shown in Fig. 4.

In Condition of Fault. To check the performance of the system under faults both single line and three phase faults are considered. At $t = 0.8$ s, a single line fault encounterd the system as shown in Fig. 7, this fault has a tendency to make system unbalance. DG power supply makes the system balanced under the fault. At $t = 0.9$ s, three phase fault occur in the system, this makes load current zero and power generated from DG is supplied to grid. At $t = 1$ s, fault is cleared and system reached its initial condition back. This shows that system can take momentarily fault condition properly.

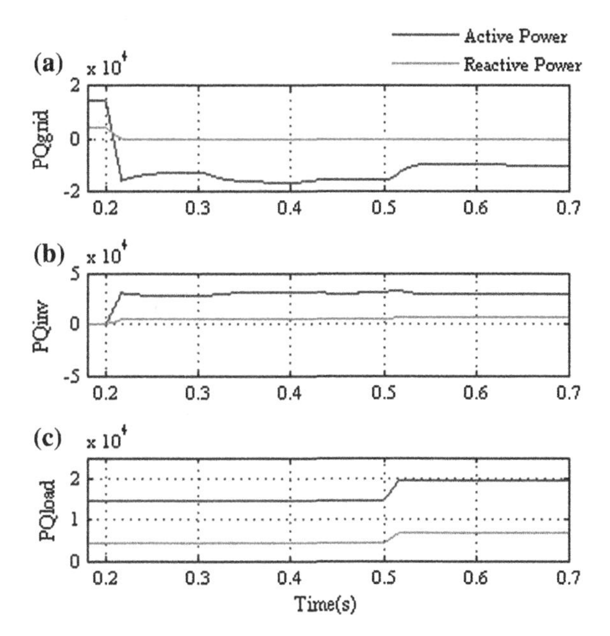

Fig. 4 Complete response: **a** PQ-grid, **b** PQ-inverter, **c** PQ-load

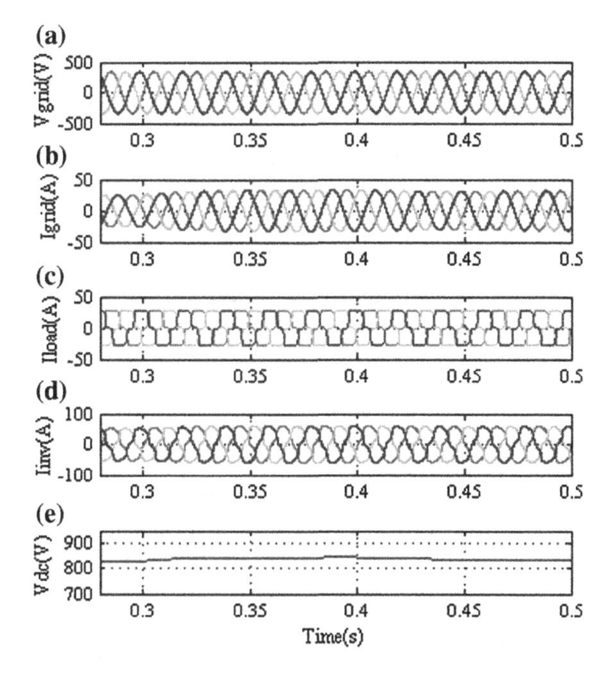

Fig. 5 Response under DG output variation: **a** grid voltages, **b** grid currents, **c** load currents, **d** Inverter currents, **e** DC link voltage

Variable Reactive Power supply. Figures 8 and 9 shows the performance of system if excess reactive power is supplied to grid. The scheme proposed has a feature that we can control the supply of reactive power, if reactive power supply is needed for

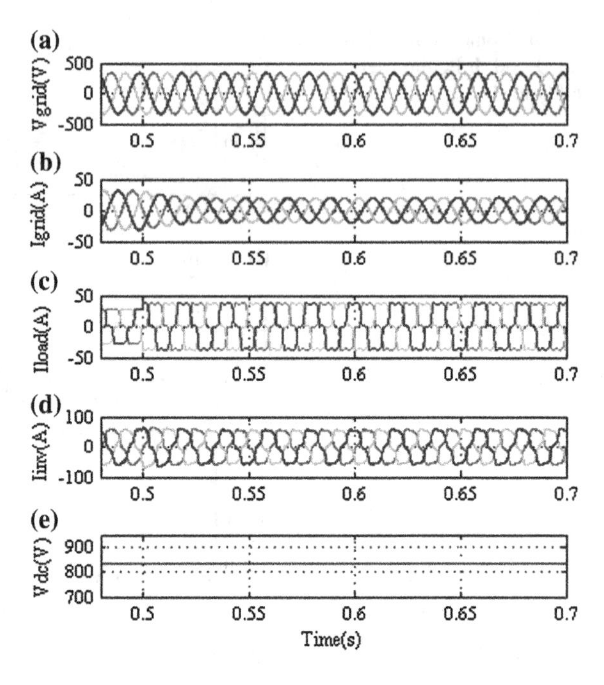

Fig. 6 Response under load variation: **a** Grid voltages, **b** grid currents, **c** load currents, **d** inverter currents, **e** DC link voltage

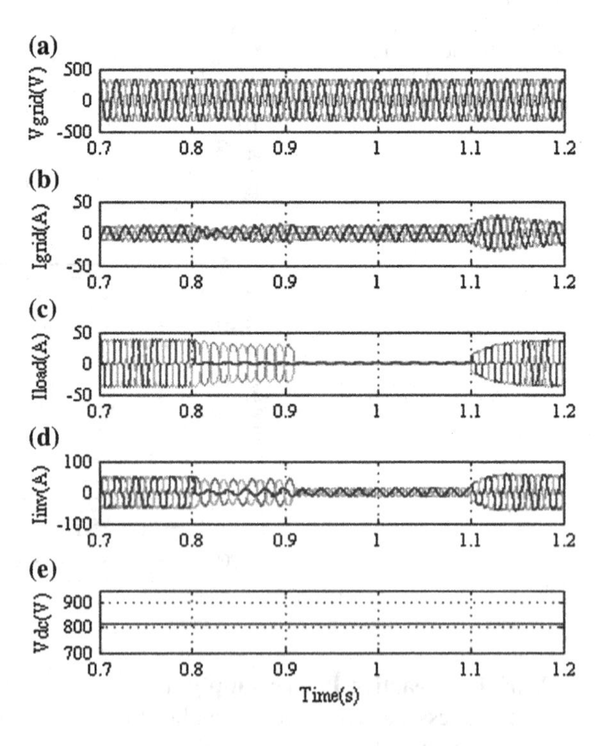

Fig. 7 Response under fault condition: **a** grid voltages, **b** grid currents, **c** load currents, **d** inverter currents, **e** DC link voltage

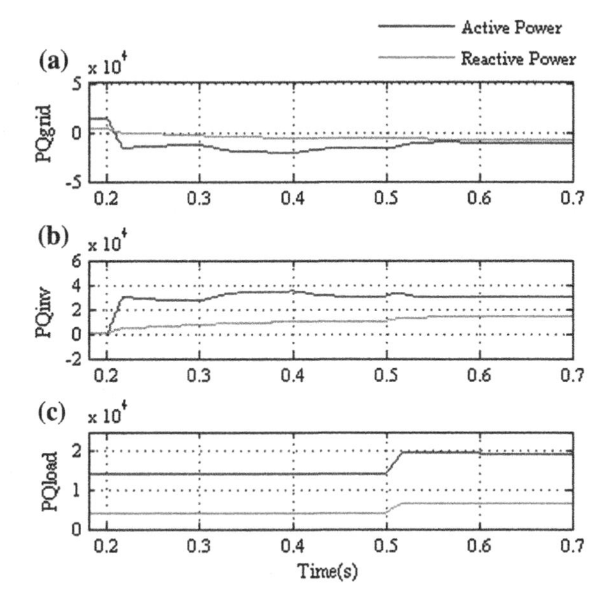

Fig. 8 Complete response under reactive power supply: **a** PQ-grid, **b** PQ-inverter, **c** PQ-load

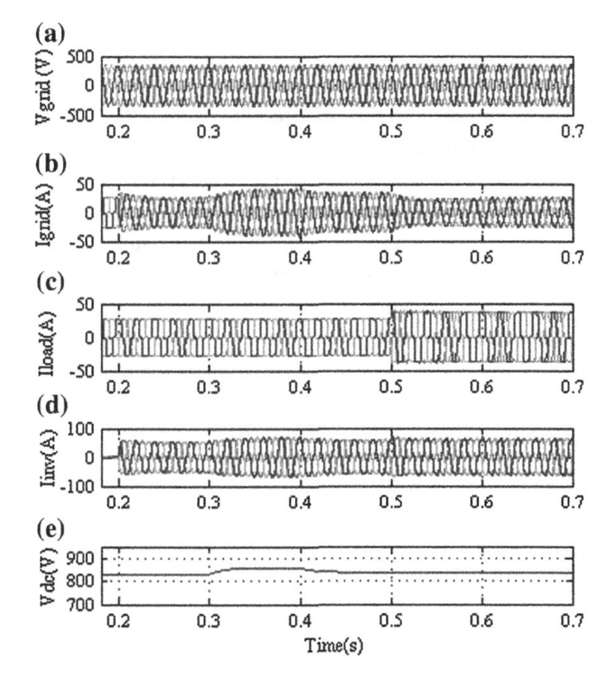

Fig. 9 Response under reactive power supply: **a** grid voltages, **b** grid currents, **c** load currents, **d** inverter currents, **e** DC link voltage

the connected load only, at that time reactive power component calculation loop should not contribute for total fundamental component current calculation. To allow the DG to supply reactive power to grid a fundamental reactive component calculation is incorporated. It hardly affects other parameters of system.

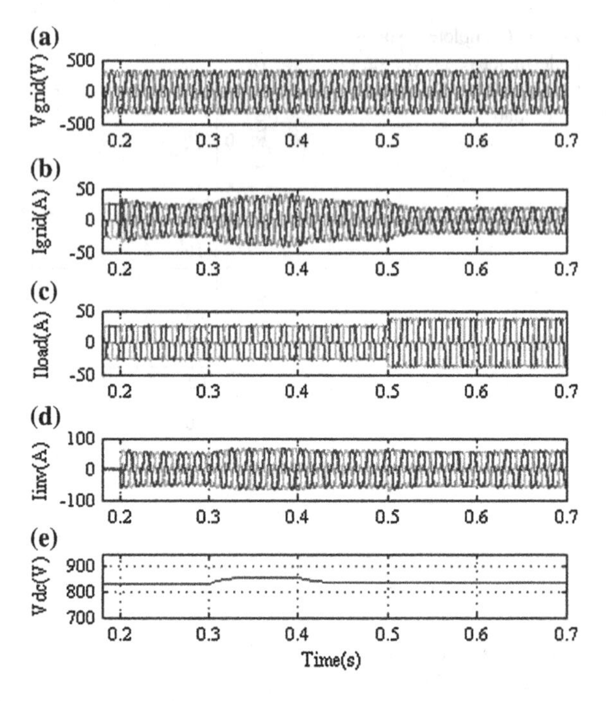

Fig. 10 Response under frequency variation: **a** grid voltages, **b** grid currents, **c** load currents, **d** inverter currents, **e** DC link voltage

Frequency Deviation. Frequency deviation is a rare phenomenon because of many regulating devices used to correct frequency, but severity of frequency deviation is very high. To make system able to momentarily work in condition of frequency deviation, system performance is checked for a deviation of 2 Hz from its nominal value. All the conditions remains same as earlier, change in frequency does not show any major difference in system performance. Figure 10 shows the effectiveness of algorithm to make system work during frequency deviation.

Total Harmonic Distortion (THD). Harmonic distortion of grid current is 1.51 %, which is very less compared to load current 21.02 %. Grid voltage THD is only 0.32 %.

3 Conclusion

The scheme presented in the paper has been verified for possible variations in power system operation while a DG is connected to the system. The scheme is able to supply active and reactive power from DG to grid in conditions of increase in DG power output, change in load, and few severe conditions like single phase, three-phase fault and frequency deviation. THD of load current decreases from 21.02 to 1.51 %, it shows the capability of control scheme in harmonic elimination. Voltage THD is 0.32 %, which can be neglected.

References

1. Wang, F., Duarte, J.L., Hendrix, M.A.M., Ribeiro, P.F.: Modeling and analysis of grid harmonic distortion impact of aggregated DG inverters. IEEE Trans. Power Electron. **26**(3), 786–797 (2011)
2. Divan, D.M., Bhattacharya, S., Banerjee, B.: Synchronous frame harmonic isolator using active series filter. In: European Power Electronics and Conference, pp. 3030–3035 (1991)
3. Asiminoaei, L., Blaabjerg, F., Hansen, S.: Detection is key—harmonic detection methods for active power filter applications. IEEE Ind. Appl. Mag. **13**(4), 22–33 (2007)
4. Kumar, P., Mahajan, A.: Soft computing techniques for the control of an active power filter. IEEE Trans. Power Del. **24**(1), 452–46 (2009)
5. He, J., Li, Y.W., Blaabjerg, F., Wang, X.: Active harmonic filtering using current controlled grid connected dg units with closed-loop power control. IEEE Trans. Power Elect. **29**(2), 642–652 (2014)
6. Singh, M., Khadkikar, V., Chandra, A., Verma, R.K.: Grid interconnection of renewable energy sources at the distribution level with power quality improvement features. IEEE Trans. Power Del. **26**(1), 307–315 (2011)
7. Douglas, S.C.: Analysis of an anti-Hebbian adaptive FIR filtering algorithm. IEEE Trans. Circ. Syst. I1 Analog Digit. Sig. Process. **43**(11), 777–780 (1996)
8. Oja, E.: A simplified neuron model as a principle component analyzer. J. Math. Biol., 267–273 (1982)
9. Arya, S.R., Singh, B., Chandra, A., Al-Haddad, K.: Learning-based Anti-Hebbian algorithm for control of distribution static compensator. IEEE Trans. Ind. Elect. **61**(11) 6004–12 (2014)

Asynchronous Differential Evolution with Convex Mutation

Vaishali and Tarun Kumar Sharma

Abstract Asynchronous differential evolution (ADE) is recently introduced variant of differential evolution (DE). In ADE the mutation, crossover, and selection operations are performed asynchronously whereas in DE these operations are performed synchronously. This asynchronous process helps in good exploration and well suited for parallel optimization. In this study the strength of ADE is enhanced by incorporating convex mutation. Convex mutation efficiently utilizes the information of the parents which assists in faster convergence. The proposal is named ADE–CM. The potential of the proposal is evaluated and compared with state-of-the-art algorithms over a selected noisy benchmark functions consulted from the literature.

Keywords Differential evolution · Asynchronous differential evolution · Convex mutation · Optimization · Convergence

1 Introduction

Differential evolution (DE) was proposed by Storn and Price in 1995 [1, 2]. It is a simple but robust global optimization algorithm [3]. It has less number of control parameters as compared to other evolution strategies. In most cases it efficiently converges to global minimum. It has proven its efficiency in various fields of science and engineering [4–6].

This study is concentrated on asynchronous differential evolution algorithm (ADEA), a simple and efficient metaheuristic algorithm. ADE is derived from DE. ADE concept given by Zhabitskaya and Zhabitsky [7] have managed to seek the attention of many researchers to solve optimization problems.

Vaishali (✉) · T.K. Sharma
Amity University, Jaipur, Rajasthan, India
e-mail: vaishaliyadav26@gmail.com

M. Pant et al. (eds.), *Proceedings of Fifth International Conference on Soft Computing for Problem Solving*, Advances in Intelligent Systems and Computing 437, DOI 10.1007/978-981-10-0451-3_81

The process of ADE is similar to DE except the evolution strategy. In general, synchronous evolution strategy (based on generation) is employed in DE [2] whereas in case of ADE an asynchronous evolution strategy (based on individual target vectors) is employed [7]. In ADE the mutation, crossover, and selection operations are performed asynchronously. The performance of ADE especially in case of parallel optimization is competitive. ADE, a variant of DE also operates in the same manner as DE until the termination criteria is met. But there are no generation increments in ADE [7]. As soon as we find a better vector, it becomes a part of the population without any delay and will participate in evolution. The generation concept which was one of the most important parts of DE has no meaning in ADE. So, in ADE the population is updated without time lag on the basis of fitness value of the target and trial vector in contrast to DE where a vector with better fitness has to wait until all the vectors in a particular generation are evaluated and the better vector is able to participate in the next generation only. ADE also outperforms many variants of DE. The control parameters of ADE are same as in classical DE.

Various enhancements have been proposed to ADE. A synchronization degree (SD) parameter in ADE was introduced by Milani and Santucci [8] which regulates the synchrony of the current population evolution, i.e., how fast a better candidate vector becomes the population member. SD regulates the number of mutant vectors produced by mutation and given as input to the crossover selection phase. So, SD ranges from 1 to NP. The authors discussed SD ≤ NP and SD ≥ NP where SD = NP corresponds to SDE in each case. Zhabitskaya and Zhabitsky [9] formulated asynchronous strategy with good explorative ability and fast convergence rate by modifying the method of choosing target vector. Instead of Cr adaptation, a modified crossover operator is used by Zhabitskaya and Zhabitsky [10] and this crossover is based on adaptive correlation matrix, which is identified by the symbol 'acm' as DE/rand/rand/1/acm. Ntipteni et al. [11] realized the parallelization using an asynchronous approach with the help of master–slave architecture. The algorithm was based on the fact that each individual in the population is compared only against its counterpart not to all individuals in current population. In [12] ADE is shown to perform better than simplex and Migrad method to find the solution of optimization problem by Zhabitskaya et al.

In this study, the mutation process of ADE is modified by incorporating convex mutation. The information of the parents is efficiently used in convex mutation which helps in accelerating convergence. The proposal is named as ADE–CM and validated on a set of five noisy benchmark functions [13] consulted from the literature.

Rest of the paper is structured as follows: Sect. 2 presents a brief introduction of ADE. The proposed algorithm, ADE–CM is detailed in Sect. 3. Section 4 demonstrates experimental settings with performance measures and results analysis. The conclusions drawn from the study are summarized in Sect. 5.

2 ADE: An Overview

In ADE, asynchronous strategy is incorporated during mutation, crossover, and selection operations. The working of ADE is similar to classical DE (described below) except in choosing the target vector and inclusion of vector found with better fitness in the population. The working pseudocode of ADE is discussed in Fig. 1. In ADE the target vectors are not taken into loop where as it is (target vector) chosen from the current population to generate the mutant vector. Then crossover is performed to generate the trial vector. If this resulted trial vector provides better fitness function value then it replaces the target vector chosen from the population. This process is repeated until the termination criterion is reached. To get a clear picture, the flow graphs of DE and ADE are depicted in Figs. 2 and 3, respectively.

2.1 *Basic DE*

Initialization

Population (S) (solutions) of NP candidate solutions is initialized:

$$X_{i,G} = \{x_{1,i,G}, x_{2,i,G}, x_{3,i,G}, \ldots, x_{D,i,G}\} \tag{1}$$

where the index i represents ith individual in the population and i = 1, 2, …, NP; G is generation; and D is dimension of the problem.

```
// Initialize a population:
    X_{i,G} = {x_{1,i,G}, x_{2,i,G}, x_{3,i,G},..., x_{D,i,G}}  (Eq. 1)
do {
i = target_vector_selection();
//Perform Mutation:
    V_{i,G} = X_{r1,G} + F×(X_{r2,G} − X_{r3,G})  (Eq. 3)
// Perform Crossover: (trial vector is generated)
for (j = 0; j < D; j = j + 1)
    u_{j,i,G} = { v_{j,i,G}   if rand_{i,j}[0,1] ≤ Cr ∨ j = j_rand
                { x_{j,i,G}   otherwise                           (Eq. 4)
//Perform Selection:
    X_{i,G+1} = { U_{i,G}   if f(U_{i,G}) ≤ f(x_{i,G})
                { X_{i,G}   Otherwise                             (Eq. 5)
} while termination criterion is not satisfied.
```

Fig. 1 Pseudocode of ADE

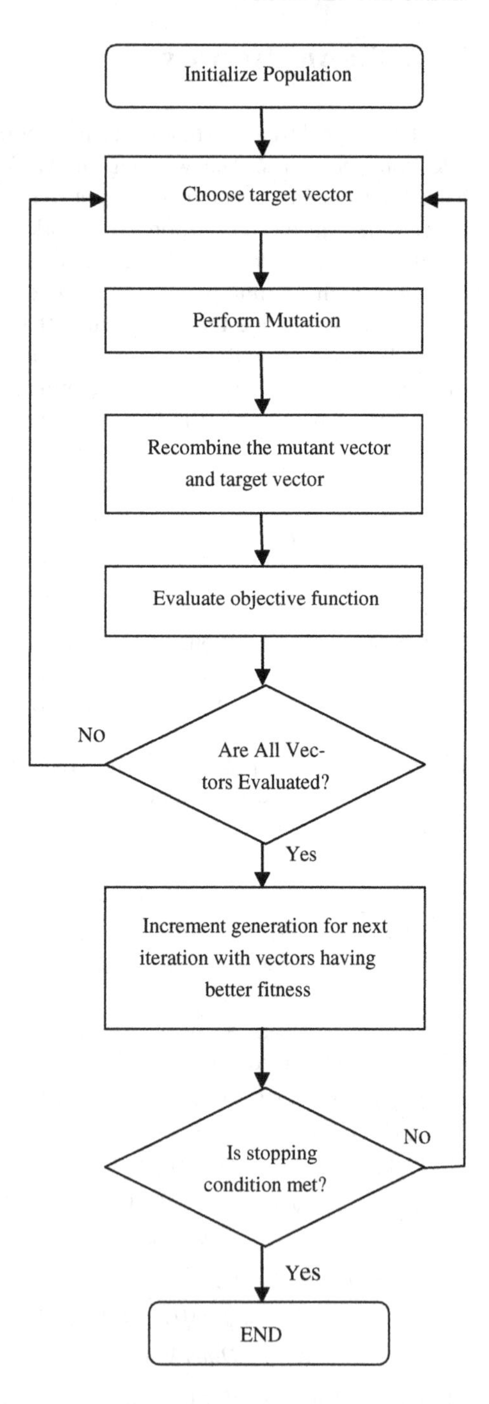

Fig. 2 DE flowchart

Fig. 3 ADE flowchart

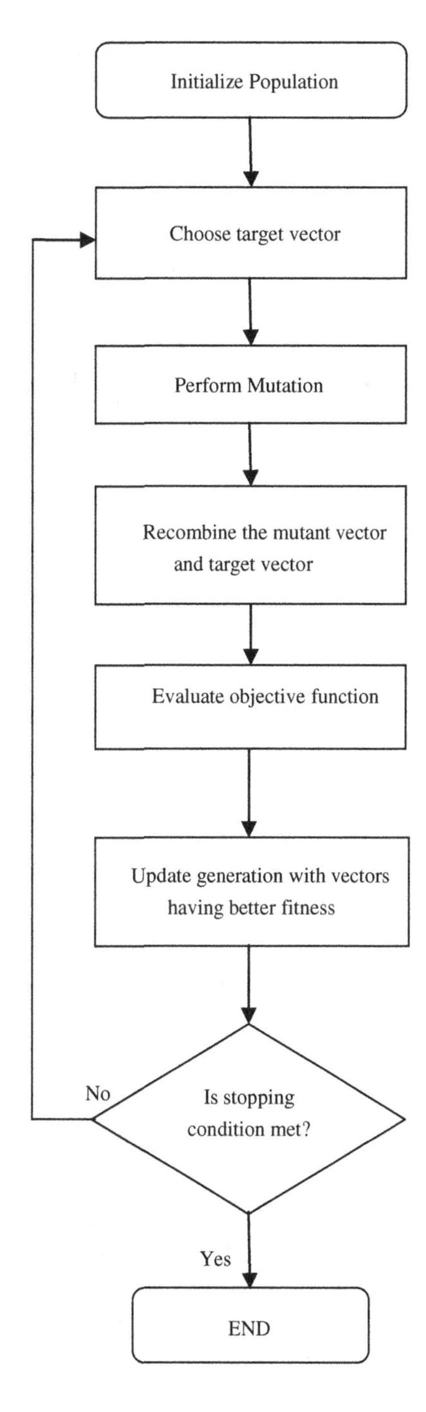

Uniformly randomizing individuals within the search space constrained by the prescribed lower and upper bounds: $l = \{l_1, l_2, \ldots, l_D\}$ and $u = \{u_1, u_2, \ldots, u_D\}$ are initialized.

jth component of the ith vector may be initialized as follows:

$$x_{j,i} = l_j + \text{rand}_{i,j}[0,\ 1] \times (u_j - l_j) \tag{2}$$

where $\text{rand}_{i,j}$ [0, 1] is uniformly distributed random number between 0 and 1.

Mutation

In the current generation (G), corresponding to each target vector ($X_{i,j}$) a donor (perturbed) vector ($V_{i,G}$) is created. There are several schemes defined in DE for this. Here, the scheme called DE/rand/bin is used throughout the study and defined as

$$V_{i,G} = X_{r1,G} + F \times (X_{r2,G} - X_{r3,G}) \tag{3}$$

where $r1$, $r2$, $r3$ are random numbers taken randomly from [1, NP] in such a way that $r1 \neq r2 \neq r3 \neq i$. F denotes scaling (amplification) factor and is a positive controlling parameter which lies between 0 and 1.

Crossover

This phase helps in increasing diversity of the population and gets activated after the mutation phase where the donor or perturbed vector

$$V_{i,G} = \{v_{1,i,G}, v_{2,i,G}, v_{3,i,G}, \ldots, v_{D,i,G}\}$$

exchanges its components with that of target vector

$$X_{i,G} = \{x_{1,i,G}, x_{2,i,G}, x_{3,i,G}, \ldots, x_{D,i,G}\}$$

to introduce trial vectors

$$U_{i,G} = \{u_{1,i,G}, u_{2,i,G}, u_{3,i,G}, \ldots, u_{D,i,G}\}$$

The binomial crossover is performed to produce trail vectors.

$$u_{j,i,G} = \begin{cases} v_{j,i,G} & \text{if } \text{rand}_{i,j}[0,\ 1] \leq \text{Cr} \vee j = j_{\text{rand}} \\ x_{j,i,G} & \text{otherwise} \end{cases} \tag{4}$$

where $j = 1, \ldots, D$ and $j_{\text{rand}} \in \{1, \ldots, D\}$. Cr is second positive control parameter and known as crossover rate and is $\text{Cr} \in [0,\ 1]$. Comparing Cr to $\text{rand}_{i,j}$ [0,1] determines the source for each remaining trial parameter. If $\text{rand}_{i,j}$ [0,1] ≤ Cr, then the parameter comes from the mutant vector; otherwise, the target vector is the source.

Selection

Selection is the final phase of DE, where the population for the next generation is selected from the individual in the current population and its corresponding trail vector based on greedy mechanism and is given as

$$X_{i,G+1} = \begin{cases} U_{i,G} & \text{if} f(U_{i,G}) \leq f(x_{i,G}) \\ X_{i,G} & \text{Otherwise} \end{cases} \quad (5)$$

3 ADE–CM (ADE with Convex Mutation): Proposal

The convex mutation process is performed in ADE instead of original mutation as given in Eq. 3. In convex mutation the information provided by the parents is utilized more efficiently [14]. The mutation strategy in ADE performs non-convex search, where the mutants are aligned in a line.

Because of which the information provided by the parents is not used efficiently by basic mutation of ADE. The population in convex mutation is scattered uniformly in the search space. The convex mutation that performs convex search is performed using Eq. 6 below.

$$V_{i,G} = a_1 X_{r1,G} + a_2 X_{r2,G} + a_3 X_{r3,G} \quad (6)$$

where $a_1 = 1.0$, $a_2 = F$ and $a_3 = -F$ and $a_1 + a_2 + a_3 = 1$ and $r1$, $r2$, and $r3$ are three mutually distinct individuals randomly extracted from the population (Fig. 4).

```
// Initialize a population:
    X_{i,G} = {x_{1,i,G}, x_{2,i,G}, x_{3,i,G}, ..., x_{D,i,G}}  (Eq. 1)
do {
i = target_vector_selection();
//Perform Mutation:
    V_{i,G} = a1 X_{r1,G} + a2 X_{r2,G} + a3 X_{r3,G}      (Convex Mutation) (Eq. 6)
// Perform Crossover: (trial vector is generated)
for (j = 0; j < D; j = j + 1)
    u_{j,i,G} = { v_{j,i,G}   if rand_{i,j}[0,1] <= Cr v j = j_rand
                { x_{j,i,G}   otherwise                              (Eq. 4)
//Perform Selection:
    X_{i,G+1} = { U_{i,G}   if f(U_{i,G}) <= f(x_{i,G})
                { X_{i,G}   Otherwise                                (Eq. 5)
} while termination criterion is not satisfied.
```

Fig. 4 Pseudocode of ADE with convex mutation (ADE–CM)

4 Experimental Setup, Parameter Settings, and Results

The efficiency of the proposed ADE–CM is tested and validated on five benchmark optimization problems consulted from [13] with the following settings:

PC configuration: Microsoft Windows XP Professional (2002); CPU: Celeron (R) Dual Core CPU T3100@1.90GHz; RAM: 1.86 GB; Simulated in MATLAB.

The value to reach (VTR) with lower and upper bounds for search space for each function is described in Table 1. Other parameter for experiment like scaling factor (*F*) was set to 0.5. Experiments were carried out for different combinations of NP (population size), *D* (dimension), and Cr (Crossover rate) and are mentioned clearly in each table of results. Every function is optimized for 25 optimization executions. All functions are tested in MATLAB.

The average NFE, if it reaches the global minimum (VTR), standard deviation (Std. dev.), and success rate (S_r) is reported. The results are also compared with that of ADE and basic DE at same environmental and parameter settings.

Tables 2 and 3 summarize the values of average NFE, Std when the function reaches the specified global minimum. Cr is set to 0.5 and *D* is 10 and 20 for Tables 2 and 3, respectively. All the parameters are reported for NP = 50 and 100.

Table 1 The benchmark functions used in the experiment

Test function	Search space	VTR
f_{01}: Sphere function	−100 to 100	10^{-6}
f_{02}: Hyper ellipsoid	−50 to 50	10^{-6}
f_{09}: Restrigin function	−5 to 5	10^{-3}
f_{10}: Ackley function	−30 to 30	10^{-6}
f_{11}: Griewank function	−30 to 30	10^{-6}

Table 2 Results of DE, ADE, and ADE–CM with *D* = 10 and Cr = 0.5

Cr = 0.5		*D* = 10					
		DE		ADE		ADE with CM	
F	NP	Avg. NFE	Std. dev.	Avg. NFE	Std. dev.	Avg. NFE	Std. dev.
f_{01}	50	1.15E+04	6.92E-08	1.12E+04	6.34E−08	**1.11E+04**	**4.54E−08**
	100	2.36E+04	5.54E-08	2.29E+04	4.46E−08	**2.27E+04**	**3.95E−08**
f_{02}	50	1.19E+04	6.68E−08	**1.14E+04**	7.52E−08	1.15E+04	**3.80E−08**
	100	2.43E+04	4.52E−08	2.34E+04	2.65E−08	**2.32E+04**	**3.24E−08**
f_{09}	50	4.40E+04	5.20E−05	4.32E+04	**4.44E−05**	**4.26E+04**	7.21E−05
	100	9.48E+04	4.83E−05	9.53E+04	3.46E−05	**9.15E+04**	**2.58E−05**
f_{10}	50	1.20E+04	3.70E−09	1.49E+04	6.18E−09	**1.00E+04**	**3.40E−10**
	100	2.30E+04	1.10E−08	2.50E+04	5.70E−09	**2.12E+04**	**5.12E−09**
f_{11}	50	4.54E+04	7.22E−08	4.06E+04	**5.82E−08**	**4.00E+04**	6.01E−08
	100	8.77E+04	3.98E−08	8.47E+04	6.45E−08	**8.36E+04**	**3.52E−08**

Table 3 Results of DE, ADE, and ADE–CM with D = 20 and Cr = 0.5

Cr = 0.5		D = 20					
		DE		ADE		ADE with CM	
F	NP	Avg. NFE	Std.dev.	Avg. NFE	Std.dev.	Avg. NFE	Std.dev.
f_{01}	50	2.37E+04	**1.04E−07**	2.26E+04	1.41E−07	**2.25E+04**	1.25E−07
	100	4.86E+04	**1.05E−07**	4.71E+04	3.91E−08	**4.70E+04**	1.06E−07
f_{02}	50	2.52E+04	1.05E−07	2.45E+04	1.37E−07	**2.44E+04**	**1.05E−07**
	100	5.19E+04	1.19E−07	5.05E+04	5.27E−08	**5.04E+04**	**5.13E−08**
f_{09}	50	**6.43E+05**	1.54E−04	6.54E+05	1.08E−04	7.44E+05	**7.50E−05**
	100	NA	NA	NA	NA	NA	NA
f_{10}	50	2.42E+04	3.70E−08	2.34E+04	3.07E−08	**2.11E+04**	**1.08E−08**
	100	4.53E+04	4.17E−08	4.46E+04	4.86E−08	**4.29E+04**	**4.05E−08**
f_{11}	50	2.28E+04	9.60E−08	2.37E+04	1.16E−07	**2.25E+04**	**9.19E−08**
	100	4.93E+04	8.17E−08	4.97E+04	6.86E−08	**4.85E+04**	**8.98E−08**

Table 4 Results of DE, ADE, and ADE–CM with D = 10 and Cr = 0.9

Cr = 0.9		D = 10					
		DE		ADE		ADE with CM	
F	NP	Avg. NFE	Std	Avg. NFE	Std	Avg. NFE	Std
f_{01}	50	1.22E+04	1.17E−07	1.08E+04	9.61E−08	**1.02E+04**	**7.30E−08**
	100	2.62E+04	**3.71E−08**	2.44E+04	5.08E−08	**2.41E+04**	5.08E−08
f_{02}	50	1.21E+04	5.45E−08	1.12E+04	6.93E−08	**1.02E+04**	**4.82E−08**
	100	2.71E+04	6.22E−08	2.46E+04	**2.81E−08**	**2.42E+04**	5.01E−08
f_{09}	50	6.11E+04	6.96E−03	**6.08E+04**	**1.05E−03**	6.80E+04	8.75E−03
	100	3.33E+05	**7.61E−05**	**2.59E+05**	3.15E−03	2.89E+05	3.15E−04
f_{10}	50	1.29E+04	4.61E−08	1.16E+04	4.60E−08	**1.10E+04**	**3.57E−08**
	100	2.23E+04	7.51E−08	2.14E+04	5.66E−08	**2.00E+04**	**5.24E−08**
f_{11}	50	3.56E+04	1.64 E−02	3.14E+04	9.201 E−03	**3.10E+04**	**1.68 E−03**
	100	1.85E+05	1.01 E−02	1.74E+05	5.438 E−03	**1.71E+05**	**4.20 E−03**

Tables 4 and 5 conclude the results of same parameters for Cr = 0.9. In all the tables the best results are highlighted in boldface.

Tables 6, 7, 8, and 9 summarize the acceleration rate (AR) of ADE–CM with respect to DE and ADE.

Figures 5 and 6 show the convergence of f_{01} and f_{02}, respectively with NFE (x-axis) and fitness value (y-axis).

Table 5 Results of DE, ADE, and ADE–CM with D = 20 and Cr = 0.9

Cr = 0.9		D = 20					
		DE		ADE		ADE with CM	
F	NP	Avg. NFE	Std	Avg. NFE	Std	Avg. NFE	Std
f_{01}	50	2.33E+04	**1.03E−07**	2.09E+04	1.51E−07	**2.07E+04**	1.75E−07
	100	5.96E+04	1.49E−07	5.32E+04	**6.88E−08**	**5.29E+04**	1.13E−07
f_{02}	50	2.50E+04	1.77E−07	2.22E+04	1.84E−07	**2.23E+04**	**1.27E−07**
	100	6.36E+04	8.53E−08	5.80E+04	1.18E−07	**5.71E+04**	**7.79E−08**
f_{09}	50	1.21E+05	**2.79E−03**	1.27E+05	3.22E−02	**1.16E+05**	2.46E−02
	100	**5.65E+05**	9.61E−03	6.21E+05	1.64E−03	6.32E+05	**1.15E−03**
f_{10}	50	2.19E+04	3.05E−07	2.39E+04	1.81E−07	**2.00E+04**	**2.89E−07**
	100	5.76E+04	6.27E−08	5.20E+04	5.14E−07	**5.00E+04**	**4.64E−08**
f_{11}	50	2.08E+04	5.81E−03	1.76E+04	3.12 E−03	**1.59E+04**	**1.20E−03**
	100	5.86E+04	3.78 E−03	5.28E+04	3.67E−03	**4.87E+04**	**2.35E−03**

Table 6 Acceleration rate (AR) of ADE–CM with respect to DE and ADE for D = 10 and Cr = 0.5

Cr = 0.5		D = 10					
		Avg. NFE		Acceleration rate (AR) (%)	Avg. NFE		Acceleration rate (AR) (%)
F	NP	DE	ADE–CM		ADE	ADE–CM	
f_{01}	50	1.16E+04	1.11E+04	4	1.12E+04	1.11E+04	5
	100	2.36E+04	2.27E+04	4	2.29E+04	2.27E+04	1
f_{02}	50	1.19E+04	1.15E+04	3	1.14E+04	1.15E+04	1
	100	2.43E+04	2.33E+04	4	2.34E+04	2.33E+04	–
f_{09}	50	4.41E+04	4.26E+04	3	4.33E+04	4.26E+04	1
	100	9.49E+04	9.16E+04	3	9.54E+04	9.16E+04	1
f_{10}	50	1.21E+04	1.01E+04	17	1.50E+04	1.01E+04	4
	100	2.31E+04	2.13E+04	8	2.51E+04	2.13E+04	32
f_{11}	50	4.55E+04	4.00E+04	12	4.07E+04	4.00E+04	15
	100	8.78E+04	8.36E+04	5	8.47E+04	8.36E+04	1

4.1 Results Analysis

Performance Measures

To evaluate the performance of chosen benchmark functions, four measures are adopted from CEC2005 [13].

- **Average NFE** Average NFE is calculated as the summation of NFEs of total runs divided by successful runs. NFE specifies how many times the fitness is evaluated to reach global minimum.
- **Success Rate (S_r)** A successful run is the run when the specified global minimum is reached before the population hits specified maximum NFE value. So,

Table 7 Acceleration rate (AR) of ADE–CM with respect to DE and ADE for D = 20 and Cr = 0.5

Cr = 0.5		D = 20					
		Avg. NFE		Acceleration rate (AR) (%)	AVG. NFE		Acceleration rate (AR) (%)
F	NP	DE	ADE–CM		ADE	ADE–CM	
f_{01}	50	2.37E+04	2.25E+04	5	2.26E+04	2.25E+04	1
	100	4.86E+04	4.70E+04	3	4.71E+04	4.70E+04	0.00
f_{02}	50	2.52E+04	2.44E+04	3	2.45E+04	2.44E+04	0.00
	100	5.19E+04	5.04E+04	3	5.05E+04	5.04E+04	0.00
f_{09}	50	6.43E+05	7.44E+05	–	6.54E+05	7.44E+05	–
	100	NA	NA	NA	NA	NA	NA
f_{10}	50	2.42E+04	2.11E+04	13	2.34E+04	2.11E+04	10
	100	4.53E+04	4.29E+04	5	4.46E+04	4.29E+04	4
f_{11}	50	2.28E+04	2.25E+04	1	2.37E+04	2.25E+04	5
	100	4.93E+04	4.85E+04	2	4.97E+04	4.85E+04	2

Table 8 Acceleration rate (AR) of ADE–CM with respect to DE and ADE for D = 10 and Cr = 0.9

Cr = 0.9		D = 10					
		Avg. NFE		Acceleration rate (AR) (%)	Avg. NFE		Acceleration rate (AR) (%)
F	NP	DE	ADE–CM		ADE	ADE–CM	
f_{01}	50	1.22E+04	1.02E+04	16	1.08E+04	1.02E+04	6
	100	2.62E+04	2.41E+04	8	2.44E+04	2.41E+04	1
f_{02}	50	1.21E+04	1.02E+04	16	1.12E+04	1.02E+04	9
	100	2.71E+04	2.42E+04	11	2.46E+04	2.42E+04	1
f_{09}	50	6.11E+04	6.80E+04	–	6.08E+04	6.80E+04	–
	100	3.33E+05	2.89E+05	13	2.59E+05	2.89E+05	–
f_{10}	50	1.29E+04	1.10E+04	14	1.16E+04	1.10E+04	5
	100	2.23E+04	2.00E+04	10	2.14E+04	2.00E+04	7
f_{11}	50	3.56E+04	3.10E+04	13	3.14E+04	3.10E+04	1
	100	1.85E+05	1.71E+05	7	1.74E+05	1.71E+05	2

the success rate (S_r) is calculated by dividing the number of successful runs by the total number of runs.

- **Acceleration Rate (AR)** The AR of ADE–CM is calculated with respect to DE and ADE and is represented in percentage. The AR is calculated as

$$\frac{\mathrm{NFE}_{\mathrm{DE\,or\,ADE}} - \mathrm{NFE}_{\mathrm{ADE-CM}}}{\mathrm{NFE}_{\mathrm{ADE-CM}}}$$

Table 9 Acceleration rate (AR) of ADE–CM with respect to DE and ADE for D = 20 and Cr = 0.9

Cr = 0.9		D = 20					
		Avg. NFE		Acceleration rate (AR) (%)	Avg. NFE		Acceleration rate (AR) (%)
F	NP	DE	ADE–CM		ADE	ADE–CM	
f_{01}	50	2.33E+04	2.07E+04	11	2.09E+04	2.07E+04	1
	100	5.96E+04	5.29E+04	11	5.32E+04	5.29E+04	1
f_{02}	50	2.50E+04	2.23E+04	11	2.22E+04	2.23E+04	0.00
	100	6.36E+04	5.71E+04	10	5.80E+04	5.71E+04	1
f_{09}	50	1.21E+05	1.16E+05	4	1.27E+05	1.16E+05	9
	100	5.65E+05	6.32E+05	–	6.21E+05	6.32E+05	–
f_{10}	50	2.19E+04	2.00E+04	9	2.39E+04	2.00E+04	16
	100	5.76E+04	5.00E+04	13	5.20E+04	5.00E+04	4
f_{11}	50	2.08E+04	1.59E+04	24	1.76E+04	1.59E+04	10
	100	5.86E+04	4.87E+04	17	5.28E+04	4.87E+04	8

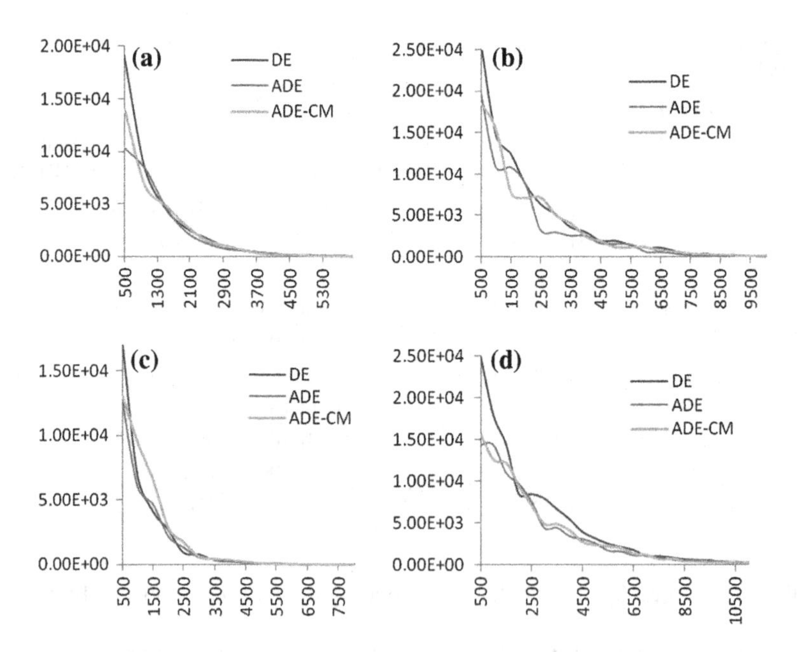

Fig. 5 **a–d** Convergence graphs of DE, ADE, and ADE–CM for f_{01} (*Sphere* function). **a** Cr = 0.5; DE = 20; NP = 50. **b** Cr = 0.5; DE = 20; NP = 100. **c** Cr = 0.9; DE = 20; NP = 50. **d** Cr = 0.9; DE = 20; NP = 100

- **Convergence Graph** The convergence graphs show the performance of DE, ADE, and ADE–CM to reach the global minimum over total runs for the chosen functions.

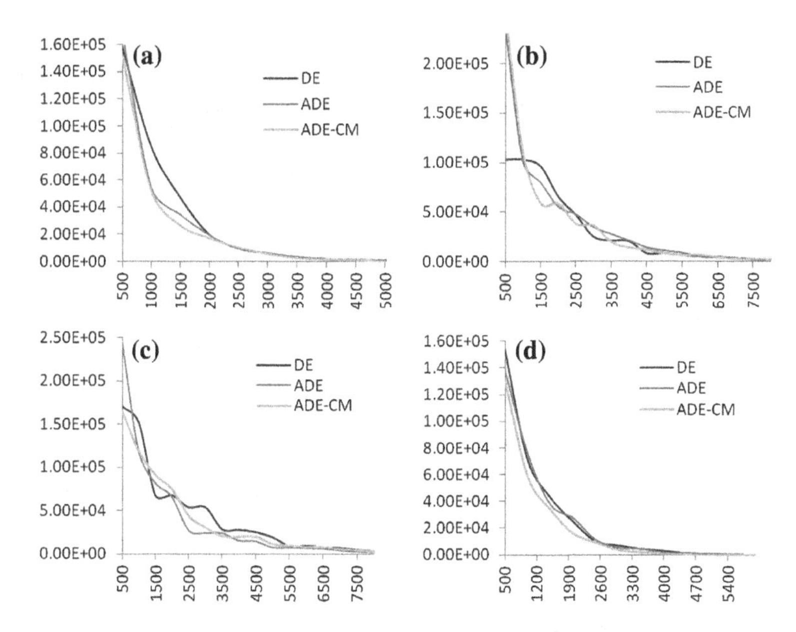

Fig. 6 **a–d** Convergence graphs of DE, ADE, and ADE–CM for f_{02} (*Hyper Ellipsoid*). **a** Cr = 0.5; DE = 20; NP = 50. **b** Cr = 0.5; DE = 20; NP = 100. **c** Cr = 0.9; DE = 20; NP = 50. **d** Cr = 0.9; DE = 20; NP = 100

4.2 Performance of ADE–CM

As shown in Tables 2, 3, 4, and 5 ADE–CM outperforms DE and ADE for most of the functions in terms of NFEs and standard deviation. Only for the function f_{09}, DE or ADE perform better than ADE–CM. The convergence graphs for f_{01} and f_{02} are presented in Figs. 5 and 6 which demonstrate that ADE–CM converges faster for most of the values which can also be seen in Tables 6, 7, 8, and 9 where acceleration rate (AR) is reported. Except for some values of f_{09}, AR of ADE–CM is increased or same for other functions. The values for which AR of ADE–CM is lesser are not presented in the tables. The success rate (S_r) for each function is noticed as 100 % except f_{09} ($D = 20$, Cr = 0.5, NP = 100) where the function does not reach to global minimum within specified NFE.

5 Conclusion

This study presents a modification in mutation phase of ADE. The modification is intended to improve performance of ADE in terms of convergence rate. The proposal uses the concept of convex mutation that enables to use the information provided by the parents efficiently. The convex mutation performs convex search.

The proposal is named as asynchronous differential evolution with convex mutation (ADE–CM). ADE–CM is tested and validated on a small test bed of five benchmark problems. The comparative result analysis proves the efficiency of the proposal in most of the cases in terms of convergence speed.

In future, we will try to test it on more multimodal benchmark and real-world problems with single objective.

References

1. Storn, R., Price, K.: differential evolution—a simple and efficient adaptive scheme for global optimization over continuous spaces. Berkeley, CA, Tech. Rep. TR-95-012, 1995
2. Differential evolution—a simple and efficient heuristic for global optimization over continuous spaces. J. Global Optim. **11**, 341–359 (1997). Norwell, MA: Kluwer
3. Price, K., Storn, R.M., Lampinen, J.A.: Differential evolution: a practical approach to global optimization (Natural Computin Series), 1st edn. Springer, New York (2005). ISBN 3540209506
4. Rogalsky, T., Derksen, R.W., Kocabiyik, S.: Differential evolution in aerodynamic optimization. In: Proceedings of the 46th Annual Conference of Canadian Aeronautics and Space Institute, Montreal, QC, Canada (May 1999)
5. Ilonen, J., Kamarainen, J.-K., Lampinen, J.: Differential evolution training algorithm for feed-forward neural networks. Neural Process. Lett. **7**(1), 93–105 (2003)
6. Storn, R.: On the usage of differential evolution for function optimization. In: Proceedings of Biennial Conference of the North American Fuzzy Information Processing Society, pp. 519–523, Berkeley, CA, (1996)
7. Zhabitskaya, E., Zhabitsky, M.: Asynchronous differential evolution. Lect. Notes in Comp. Sci. **7125**, 328–333 (2012)
8. Milani, A., Santucci, V.: Asynchronous differential evolution. In: Proceedings of the 2010 IEEE Congress on Evolutionary Computation, pp. 1210–1216 (2010)
9. Zhabitskaya, E., Zhabitsky, M.: Asynchronous differential evolution. Lect. Notes in Comp. Sci. **7125**, 328–333 (2012)
10. Zhabitskaya, E., Zhabitsky, M.: Asynchronous differential evolution with adaptive correlation matrix. In: GECCO'13, Amsterdam, The Netherlands (July 2013)
11. Ntipteni, M.S., Valakos, I.M., Nikolos, I.K.: An asynchronous parallel differential evolution algorithm
12. Zhabitskaya, E.I., Zemlyanaya, E.V., Kiselev, M.A.: Numerical analysis of SAXS-data from vesicular systems by asynchronous differential evolution method. Matem. Mod. **27**(7), 58–64 (2015) (Mi mm3623)
13. Suganthan, P.N., et al.: Problem definitions and evaluation criteria for the CEC05 special session on real-parameter optimization. Technical report, Nanyang Technological University, Singapore (2005). http://www.ntu.edu.sg/home/epnsugan/indexfiles/CEC-05/CEC05.htm
14. Yan, J., Guo, C., Gong, W.: Hybrid differential evolution with convex mutation. J. Softw. **6**, 11 (2011)

Equilibrium Cable Configuration Estimation for Aerostat Using Approximation Method

A. Kumar, S.C. Sati and A.K. Ghosh

Abstract This paper presents estimation technique for equilibrium cable parameters applicable to aerostat balloon. The direct integration expressions are first expressed for uniform wind speed and density with height. Two types of polygonal approximations are then derived for numerical estimation which can directly be applied to varying wind speed as well as air density with height. A comparison of polygonal approximations is then carried out with direct integration for a small size aerostat assuming uniform wind speed and air density with height. Then the polygonal approximation is applied to a medium size aerostat considering the air density variation with height and thereby proposing a tether length factor for aerostat design. Finally, the method is applied for a practical case of both wind speed and air density variation with height. The results suggest that with a reasonable number of elements, the polygonal approximations can successfully be applied for estimation of equilibrium cable parameters for the aerostat balloon.

Keywords Tether cable · Polygonal approximation · Tether length factor

List of symbols

B	Buoyancy force
C_D, C_L	Drag and lift coefficients
C_{D_c}	Tether cable drag coefficient
d_c	Tether cable diameter

A. Kumar (✉) · S.C. Sati
Aerial Delivery Research & Development Establishment, Agra 282001, India
e-mail: ajit.kpc@gmail.com

S.C. Sati
e-mail: scsati@rediffmail.com

A.K. Ghosh
Department of Aerospace Engineering, Indian Institute of Technology, Kanpur, Kanpur 208016, India
e-mail: akg@iitk.ac.in

M. Pant et al. (eds.), *Proceedings of Fifth International Conference on Soft Computing for Problem Solving*, Advances in Intelligent Systems and Computing 437, DOI 10.1007/978-981-10-0451-3_82

D_A	Aerodynamic drag force
dl	Small/polygonal cable element
g	Acceleration due to gravity
l	Tether cable length from anchor point
L	Total tether cable length
L_A	Aerodynamic lift force
n_d	Cable drag per unit length for cable normal to the wind
$\overline{p}$	$w_c/2n$
S	Characteristic area of balloon, $V_b^{2/3}$
T_0, T_1	Tensions of tether cable at lower and upper ends, respectively
V_∞	Steady wind velocity
V_b	Volume of balloon hull (i.e., gas bag)
W_S	Structural weight of balloon (including bridle, payload, and test instruments)
w_c	Tether cable weight per unit length
x, z	Coordinates on tether cable w.r.t. anchor point
x_1, z_1	Coordinates of balloon confluence point with respect to tether cable anchor point
α	Trim angle of attack and sideslip angle
δ_n	Angle between nth cable element and horizontal
γ_0, γ_1	Angles between the horizontal and tether cable at lower and upper ends
ρ_h	Atmospheric air density at height
ρ_g	Helium density
τ_0, τ_1	τ at lower and upper ends, respectively, of tether cable

1 Introduction

Aerostats are lighter than air systems, usually filled with helium gas and are tethered to the ground. They provide useful platform for payload mounting due to increased line of sight. There are many advantages of using aerostat platforms, some of which are long endurance, low vibrations and noise, and low cost. The major applications of aerostat include surveillance and communication intelligence. During its operation, an aerostat encounters wind which causes its tether to take a particular shape. It is of vital importance to estimate the tether cable shape and tension variation along its length under the operating wind conditions. Commendable tether cable modeling for uniform wind speed and uniform air density to estimate cable shape and tension has been done in reference [1–3]. For the case where velocity magnitude varies with vertical position, Berteaux [4] suggested an approximate step-wise change of velocity with vertical position. An optimization technique has also been proposed [5] for accounting velocity variation. Dynamic tether has been considered in [6, 7].

In this paper, Neumark's method [2] has been used to express equilibrium cable parameters for aerostat assuming uniform wind speed and density followed by derivation of two types of expressions for polygonal approximations. As is the case with medium and large size aerostats, wind speed and density variations with height cannot be neglected. Since the tether can be divided into finite number of elements from anchor point to confluence point, wind speed and air density variations with height can be easily accounted for in polygonal approximations. Once the tether tension and angle at the confluence point are known, the method proceeds downward with approximation of these parameters for the next lower element. An error analysis has also been carried out for the case of uniform wind speed and density. The polygonal approximations are then applied for medium size aerostat considering the wind speed and density variations.

2 Tether Tension

The aerostat configuration establishes its equilibrium under given wind conditions with a certain value of pitch angle and a blow-by. The aerostat in this condition is acted upon by its weight, buoyancy force, aerodynamic force, and tether tension [8] as shown in Fig. 1. Balancing the forces in horizontal and vertical directions, tether tension (T_1) and angle with horizontal (γ_1) can be estimated as below:

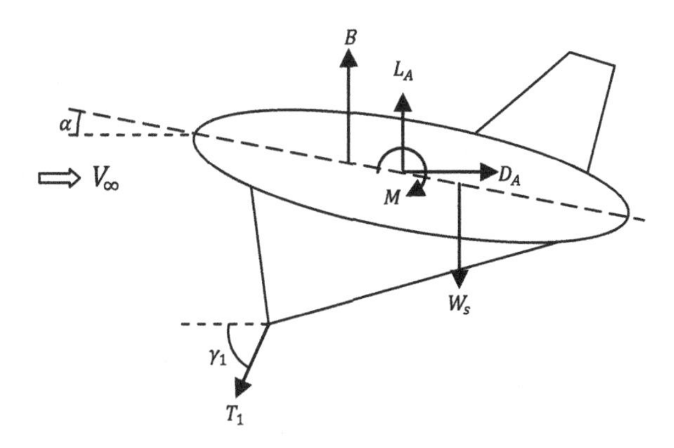

Fig. 1 Forces and moments acting on aerostat in equilibrium condition

$$T_1 = \sqrt{D_A^2 + (B + L_A - W_s)^2} \quad (1)$$

$$\gamma_1 = \tan^{-1}\left(\frac{B + L_A - W_s}{D_A}\right) \quad (2)$$

where $B = V_b(\rho_h - \rho_g)g$; $L_A = \frac{1}{2}\rho_h V_\infty^2 SC_L$ and $D_A = \frac{1}{2}\rho_h V_\infty^2 SC_D$.

Using Eqs. (1) and (2), the tether parameters at the balloon confluence point are evaluated.

3 Cable Configuration for Uniform Wind Speed and Density

Once the tether tension and angle with horizontal are known at the confluence point, it is required to estimate these parameters at the anchor point and also the cable shape from confluence point to anchor point. The relevant expressions are presented here for the case of uniform wind speed and air density as adapted from reference [2]. The coordinate system and the forces acting on the cable are shown in Fig. 2.

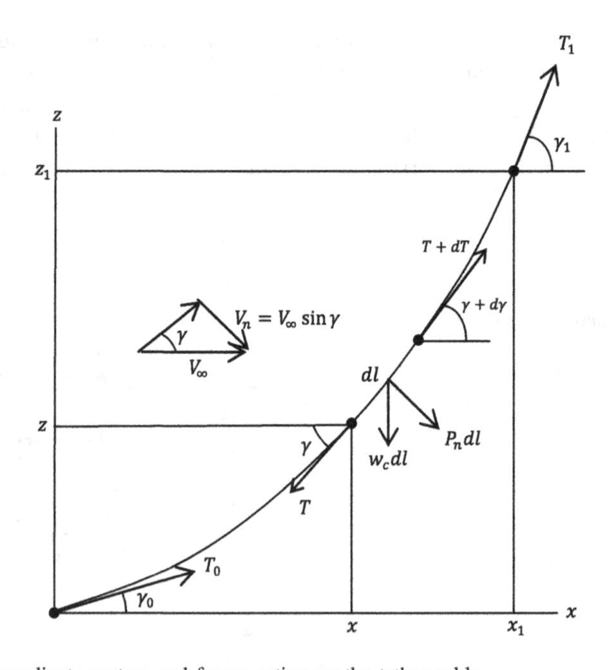

Fig. 2 The coordinate system and forces acting on the tether cable

The forces acting on an element of the cable of length dl (see Fig. 2) are the tension T, cable weight $w_c \mathrm{d}l$, and cable drag $P_n \mathrm{d}l$ normal to the cable. Drag along the cable is neglected in this analysis. The normal drag force per unit length P_n depends on the component of wind velocity normal to the cable V_n, the cable drag coefficient C_{D_c}, and cable diameter d_c as follows:

$$P_n = C_{D_c} d_c \frac{1}{2} \rho_h V_n^2$$

Retaining only first-order terms in the infinitesimals, the drag on the element is

$$P_n \mathrm{d}l = n_d \mathrm{d}l \sin^2 \gamma \text{ , where } n_d = C_{D_c} d_c \frac{1}{2} \rho_h V_\infty^2$$

is the cable drag per unit length for a cable normal to the wind. Summing the x- and z-forces and retaining only first-order terms yields

$$P_n \mathrm{d}l \sin\gamma - T\mathrm{d}\gamma \sin\gamma + \mathrm{d}T \cos\gamma = 0 \tag{3}$$

$$-w_c \mathrm{d}l - P_n \mathrm{d}l \cos\gamma + T\mathrm{d}\gamma \cos\gamma + \mathrm{d}T \sin\gamma = 0 \tag{4}$$

Equations (3) and (4) may be combined to give

$$\mathrm{d}T = w_c \mathrm{d}l \sin\gamma \tag{5}$$

$$T\mathrm{d}\gamma = \left(n \sin^2\gamma + w_c \cos\gamma\right)\mathrm{d}l \tag{6}$$

Dividing Eq. (5) by (6) gives

$$\frac{\mathrm{d}T}{T} = \frac{2\overline{p}\sin\gamma}{\sin^2\gamma + 2\overline{p}\cos\gamma}\mathrm{d}\gamma \tag{7}$$

where $\overline{p} = \frac{w_c}{2n}$. Using the substitution $f = \cos\gamma$ and partial-fraction decomposition in Eq. (7) to obtain

$$\frac{\mathrm{d}T}{T} = -\frac{\overline{p}}{\overline{q}}\left(\frac{\mathrm{d}f}{\overline{q}+\overline{p}-f} + \frac{\mathrm{d}f}{\overline{q}-\overline{p}+f}\right) \tag{8}$$

where $\overline{q} = \sqrt{1+(\overline{p})^2}$, and integrating Eq. (8) from the upper end of the cable yields

$$T = \frac{T_1 \tau}{\tau_1} \tag{9}$$

where

$$\tau(\gamma) = \left(\frac{\overline{q} + \overline{p} - \cos\gamma}{\overline{q} - \overline{p} + \cos\gamma}\right)^{\overline{p}/\overline{q}} \tag{10}$$

For present purposes it is assumed that the cable length l, drag per unit length n, weight per unit length w_c, tension at the upper end T_1, and angle of the upper end γ_1 are known. The procedure required to determine the coordinates of the upper end of the cable x_1 and z_1, the tension at the lower end T_0, and the angle of the lower end γ_0 is outlined as follows: Substituting Eqs. (9) and (10) in Eq. (6) gives

$$\mathrm{d}l = \frac{T_1}{n\tau_1} \frac{\tau}{\sin^2\gamma + 2\overline{p}\cos\gamma} \mathrm{d}\gamma \tag{11}$$

Integration of Eq. (11) from the lower end to upper end yields

$$l = \frac{T_1}{n\tau_1}\left(\overline{\lambda_1} - \overline{\lambda_0}\right) \tag{12}$$

where

$$\overline{\lambda}(\gamma) = \int_0^{\gamma} \frac{\tau(\gamma)}{\sin^2\gamma + 2\overline{p}\cos\gamma} \mathrm{d}\gamma \tag{13}$$

$\overline{\lambda_0} = \overline{\lambda}(\gamma_0)$, and $\overline{\lambda_1} = \overline{\lambda}(\gamma_1)$. Angle γ_0 is unknown, but may be obtained by solving Eq. (12) for
$\overline{\lambda_0} = \overline{\lambda_1} - \frac{n\tau_1 l}{T_1}$ and using this value with Eq. (13) to obtain

$$\overline{\lambda_0} = \int_0^{\gamma_0} \frac{\tau(\gamma)}{\sin^2\gamma + 2\overline{p}\cos\gamma} \mathrm{d}\gamma \tag{14}$$

This equation is solved for the unknown limit of integration γ_0 by Newton iteration [9]. With γ_0 known, Eqs. (9) and (10) are used to find

$$T_0 = T_1 \frac{\tau_0}{\tau_1} \tag{15}$$

where $\tau_0 = \tau(\gamma_0)$. From Fig. 2, $\mathrm{d}x = \mathrm{d}l\cos\gamma$.Using Eq. (11) in this expression yields

$$\mathrm{d}x = \frac{T_1}{n\tau_1}\mathrm{d}\sigma \tag{16}$$

where $\mathrm{d}\sigma = \frac{\tau\cos\gamma}{\sin^2\gamma + 2\overline{p}\cos\gamma}\mathrm{d}\gamma$. Equation (16) may be integrated numerically to give

$$x_1 = \frac{T_1}{n\tau_1} \int_{\gamma_0}^{\gamma_1} \mathrm{d}\sigma \tag{17}$$

Finally, from Fig. 2 and Eq. (5)

$$\mathrm{d}z = \mathrm{d}l \sin \gamma = \frac{\mathrm{d}T}{w_c} \tag{18}$$

which is integrated to give

$$z_1 = \frac{T_1 - T_0}{w_c} \tag{19}$$

4 Approximation for Wind Speed and Density Variation

The expressions presented in the previous section for tether cable tension, angle with horizontal, and profile are valid only for constant wind speed and density with height, and hence are applicable only for the cases where these variations are negligible, e.g., for a small size aerostat due to low height of operation. Hence, it is required to develop relations which are applicable to variable wind speed and density with height. To account for such variations, Berteaux [4] suggested an approximate step-wise change of velocity with vertical position. Following this assumption, expressions for two types of polygonal approximation have been developed in this section. The basic idea is to represent the cable through finite number of elements and assume constant wind speed and density over a particular element. Using the equilibrium of a particular cable element, cable tension and angle with horizontal for the next cable element are estimated in terms of current element parameters. If we start from the confluence point of the cable at the top, the tether tension and angle are known using Eqs. (1) and (2), and hence the cable tension and angle for the second element can be estimated and so on.

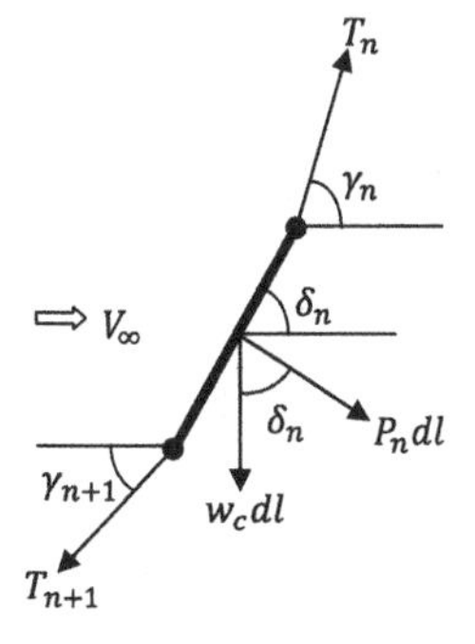

Fig. 3 Polygonal approximation of tether cable element

Figures 2 and 3 show the equilibrium of a cable element and the forces acting on it. Balancing the forces in *X*- and *Z*-direction in Fig. 3 which represents *n*th element, we get

$$\sum F_X = 0 \Rightarrow P_n \mathrm{d}l \sin\delta_n + T_n \cos\gamma_n - T_{n+1}\cos\gamma_{n+1} = 0 \tag{20}$$

$$\sum F_Z = 0 \Rightarrow T_n \sin\gamma_n - T_{n+1}\sin\gamma_{n+1} - P_n \mathrm{d}l \cos\delta_n - w_c \mathrm{d}l = 0 \tag{21}$$

Based on angle δ_n, two types of approximations are proposed as below.

4.1 Polygonal Approximation 1

In this case, it is assumed that δ_n is the mean of angle γ_n and γ_{n+1} i.e., $\delta_n = \frac{\gamma_n + \gamma_{n+1}}{2}$ Then, using Eqs. (20) and (21), we have

$$\tan\gamma_{n+1} = \frac{T_n \sin\gamma_n - w_c \mathrm{d}l - 0.5 C_{\mathrm{D_c}} d_c \rho_h V_\infty^2 \sin^2\left(\frac{\gamma_n + \gamma_{n+1}}{2}\right) \mathrm{d}l \cos\left(\frac{\gamma_n + \gamma_{n+1}}{2}\right)}{T_n \cos\gamma_n + 0.5 C_{\mathrm{D_c}} d_c \rho_h V_\infty^2 \sin^2\left(\frac{\gamma_n + \gamma_{n+1}}{2}\right) \mathrm{d}l \sin\left(\frac{\gamma_n + \gamma_{n+1}}{2}\right)} \tag{22}$$

Since γ_{n+1} is the only unknown parameter here, Eq. (22) can be solved for γ_{n+1} using Newton iteration method [9]. Then using Eq. (20), tension in the lower element can be estimated as below:

$$T_{n+1} = \frac{T_n \cos\gamma_n + 0.5 C_{\mathrm{D_c}} d_c \rho_h V_\infty^2 \sin^2\left(\frac{\gamma_n + \gamma_{n+1}}{2}\right) \mathrm{d}l \sin\left(\frac{\gamma_n + \gamma_{n+1}}{2}\right)}{\cos\gamma_{n+1}} \tag{23}$$

Using Eqs. (22) and (23), tension and angle for the next lower element can be approximated, thereby evaluating these values for all the lower elements. Summation of *x*- and *z*-components of element length provides *x*- and *z*-distance.

4.2 Polygonal Approximation 2

Further simplification can be achieved and tension and angle for the next element can directly be written in terms of previous values if $\delta_n = \gamma_n$ is assumed. Under this assumption, using Eqs. (20) and (21), we have

$$\tan\gamma_{n+1} = \frac{T_n \sin\gamma_n - w_c \mathrm{d}l - \left(0.5 C_{\mathrm{D_c}} d_c \rho_h V_\infty^2 \sin^2\gamma_n\right) \mathrm{d}l \cos\gamma_n}{T_n \cos\gamma_n + \left(0.5 C_{\mathrm{D_c}} d_c \rho_h V_\infty^2 \sin^2\gamma_n\right) \mathrm{d}l \sin\gamma_n} \tag{24}$$

Hence, γ_{n+1} is directly obtained as all the parameters in the right side of Eq. (24) are known for current element. Further, using Eq. (20), we get

$$T_{n+1} = \frac{T_n \cos\gamma_n + 0.5 C_{D_c} d_c \rho_h V_\infty^2 \sin^2\gamma_n dl \sin\gamma_n}{\cos\gamma_{n+1}} \tag{25}$$

Using Eqs. (24) and (25), tension and angle for the next lower element can be directly written in terms of previous values. As for the previous case, summation of x- and z-components of element length provides x- and z-distance.

It is clear that the expressions for polygonal approximation 2 are simpler. As will be presented in the results, for large number of elements both approximations give similar results.

5 Results and Discussions

The results are presented for one small size aerostat (*Aerostat* 1: Volume ~ 250 m^3 and Operating Height ~ 100 m) and one medium size aerostat (*Aerostat* 2: Volume ~ 2000 m^3 and Operating Height ~ 1000 m). For *Aerostat* 1, since operating height is small, a constant air density has been assumed from ground to operational height. But for *Aerostat* 2, air density variation with height has been taken into account while using polygonal approximation as below [10]:

$$\rho_h = \rho_0 \left(\frac{T_0 - 0.0065h}{T_0}\right)^{4.254}, \text{ where } T_0 = 288.15 \text{ K and } \rho_0 = 1.225 \text{ kg/m}^3.$$

For the aerostat parameters, the tether tension and angle with horizontal are evaluated using Eqs. (1) and (2) and are presented in Table 1.

Table 1 Tether tension and angle values at confluence point for two aerostats

Wind speed (m/s)	Aerostat 1		Aerostat 2	
	Tether tension (N)	Angle (degree)	Tether tension (N)	Angle (degree)
5	1118.0	88.43	7461.2	89.14
10	1324.6	84.69	8210.8	86.87
15	1677.4	80.54	9482.7	83.88
20	2180.3	77.01	11295.0	80.85
25	2834.2	74.33	13657.0	78.14

Table 2 Tether parameters for the two aerostats

Parameters	Aerostat 1	Aerostat 2
Tether unit weight (kg/m)	0.1	0.3
Tether diameter (mm)	10	17

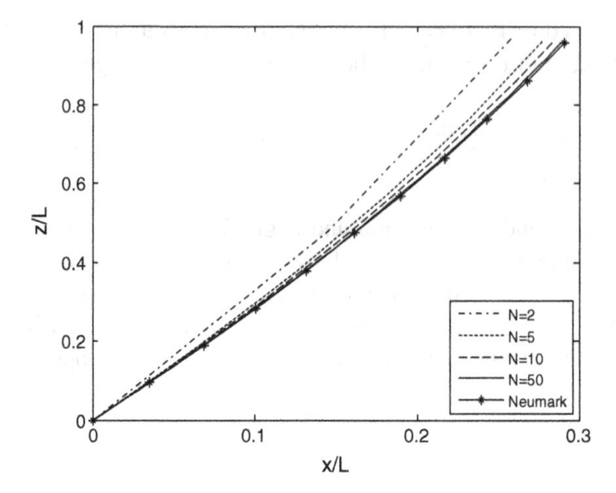

Fig. 4 Non-dimensional cable profile for *Aerostat* 1 using polygonal approximation 1 and V_∞ = 20 m/s

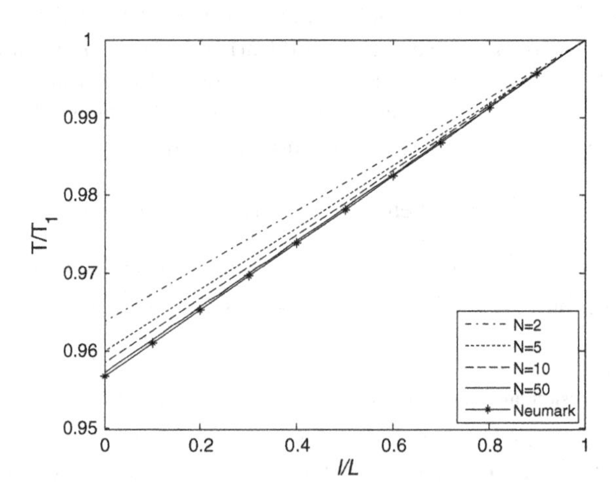

Fig. 5 Non-dimensional cable tension for *Aerostat* 1 using polygonal approximation 1 and V_∞ = 20 m/s

The drag coefficient of tether cable element is assumed as 1.2 [11] for the flow considered. The tether parameters of the two aerostats are assumed as in Table 2.

First, comparison of Neumark's method [2] is done with polygonal approximations for varying number of elements. Since a constant density is assumed in Neumark's method, this comparison is done only for *Aerostat* 1 with 20 m/s uniform wind speed. The results are presented in Figs. 4, 5, 6, 7, 8, 9, and 10. Figures 4, 5, and 6 present the comparison of Neumark's method with polygonal

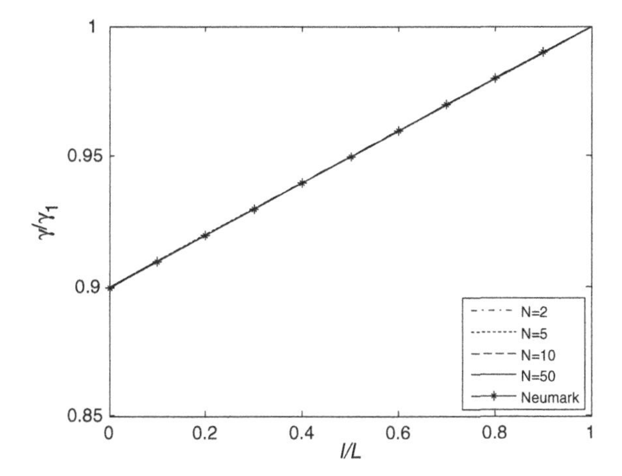

Fig. 6 Non-dimensional cable angle with horizontal for *Aerostat* 1 using polygonal approximation 1 and V_∞ = 20 m/s

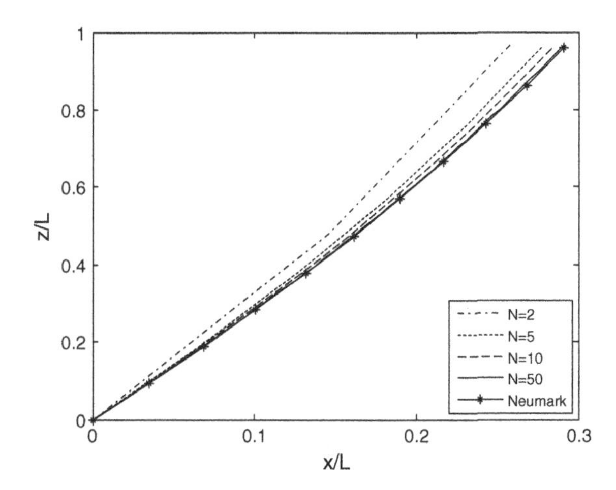

Fig. 7 Non-dimensional cable profile for *Aerostat* 1 using polygonal approximation 2 and V_∞ = 20 m/s

approximation 1 for non-dimensional position, tether tension, and angle with horizontal. Similar results for polygonal approximation 2 have been presented in Figs. 7, 8, and 9. The percentage errors for these two approximations with increasing number of elements have been plotted in Fig. 10. These plots suggest that for large number of elements both approximations give similar results. But for small number of elements, angle with horizontal is better estimated using polygonal approximation 1, whereas tether tension is better estimated using polygonal approximation 2. Also, the percentage error for position estimation becomes less

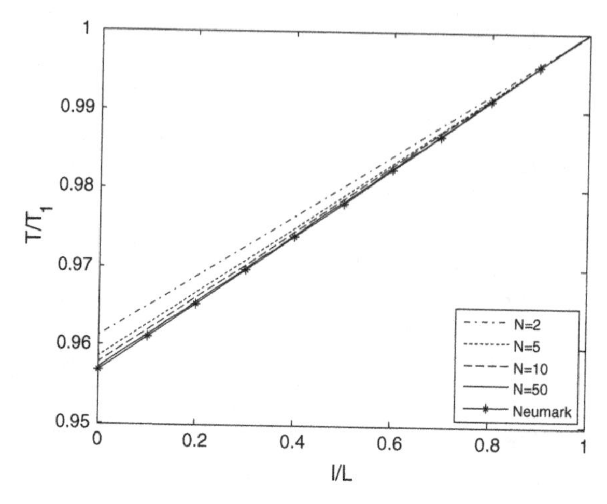

Fig. 8 Non-dimensional cable tension for *Aerostat* 1 using polygonal approximation 2 and V_∞ = 20 m/s

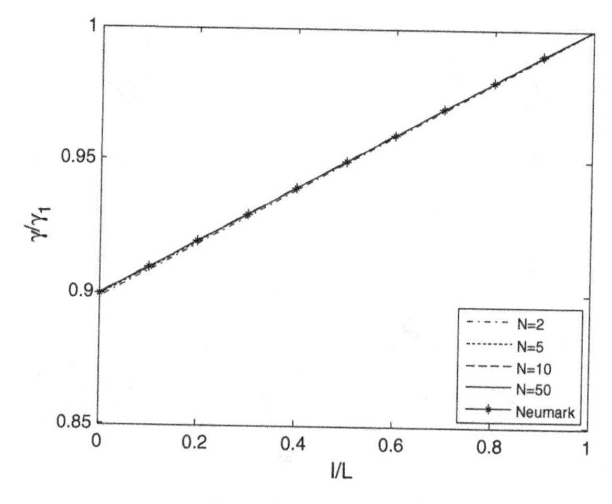

Fig. 9 Non-dimensional cable angle with horizontal for *Aerostat* 1 using polygonal approximation 2 and V_∞ = 20 m/s

than 1 % if 50 or more elements are used and hence for these many numbers of elements, both the approximations can be used to generate practical results.

Figures 11, 12, and 13 show the equilibrium tether cable non-dimensional position, tension, and angle of *Aerostat* 1 for uniform wind speed varying from 5 to 25 m/s. Since the air density variation with height has been neglected in this case, Neumark's method has been used. As can be seen from the figure, the operating height decreases and blow-by increases with the increasing wind speed as expected.

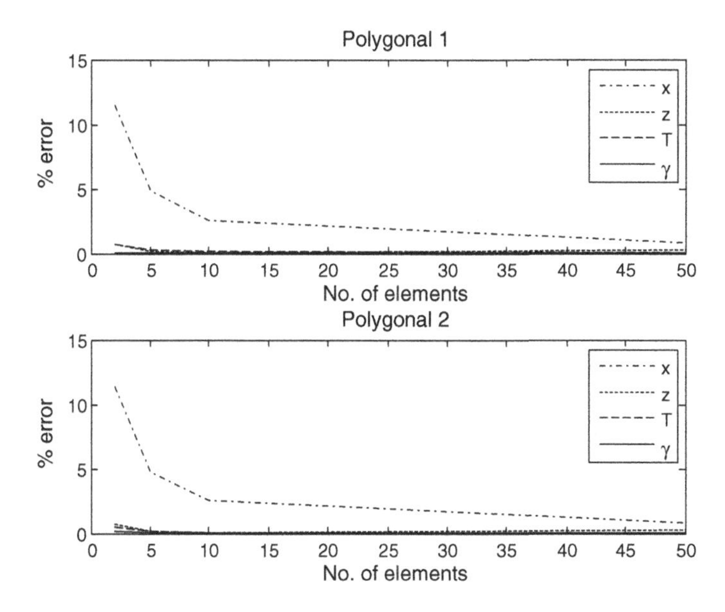

Fig. 10 Percentage error with number of elements for *Aerostat* 1 for V_∞ = 20 m/s

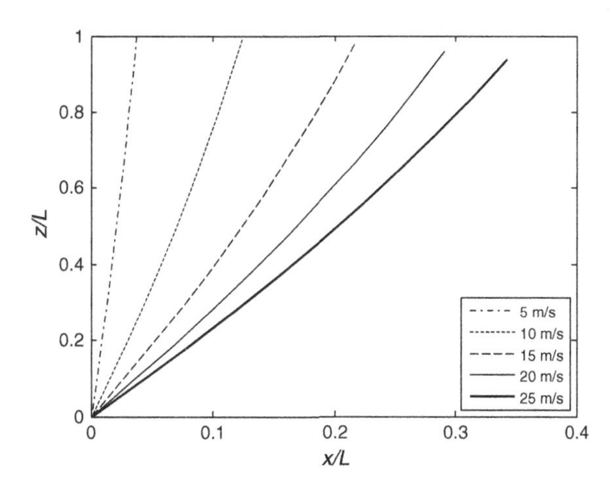

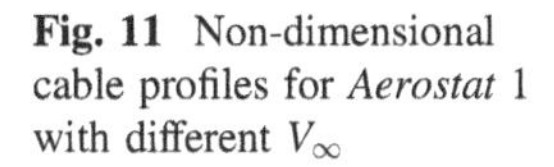
Fig. 11 Non-dimensional cable profiles for *Aerostat* 1 with different V_∞

For a wind speed of 25 m/s, the operating height decreases to 93.8 m and blow-by increases to 34.2 m. Also, with increase in wind speed, tether tension increases and angle with horizontal decreases as expected.

Figures 14, 15, and 16 show the equilibrium tether cable non-dimensional position, tension, and angle of *Aerostat* 2 for uniform wind speed varying from 5 to 25 m/s. Since the air density variation with height has been taken into account, polygonal approximation 2 has been used for calculations with 100 elements. Here,

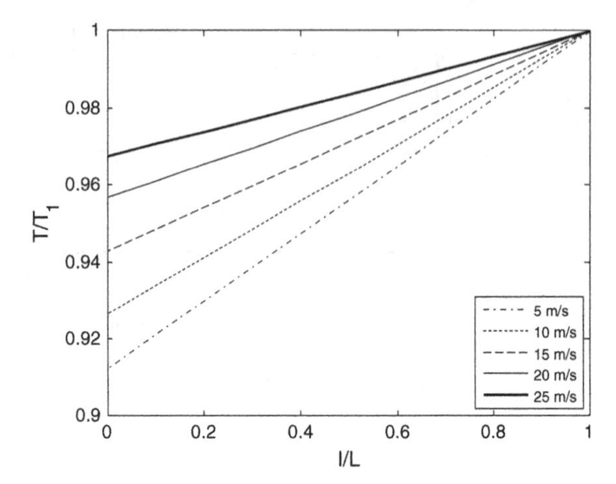

Fig. 12 Non-dimensional cable tensions for *Aerostat* 1 with different V_{∞}

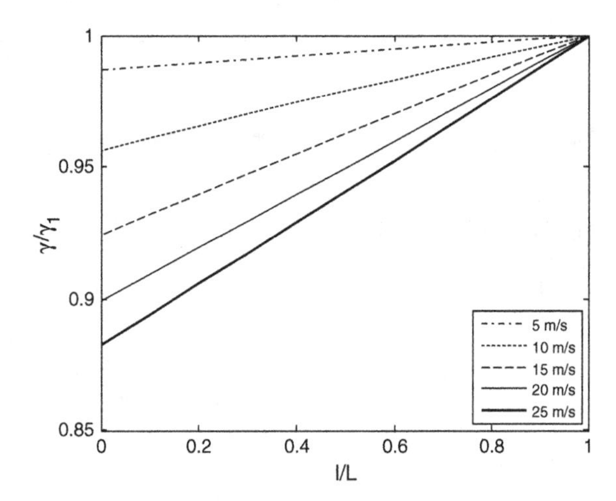

Fig. 13 Non-dimensional cable angles with horizontal for *Aerostat* 1 with different V_{∞}

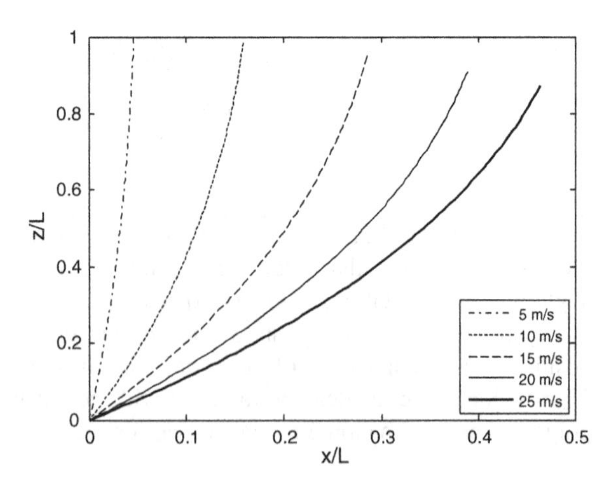

Fig. 14 Non-dimensional cable profiles for *Aerostat* 2 with different V_{∞}

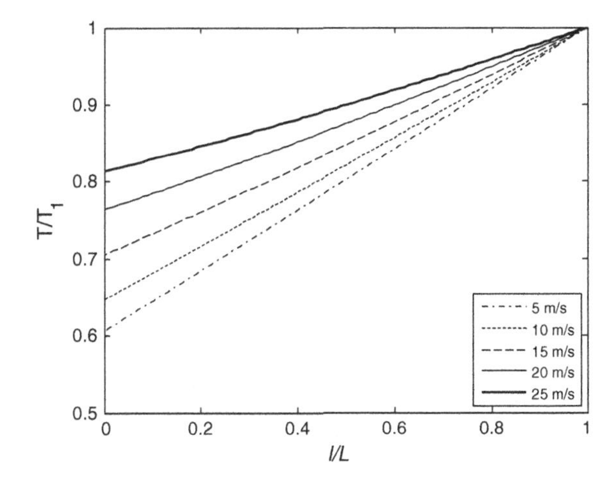

Fig. 15 Non-dimensional cable tension for *Aerostat* 2 with different V_∞

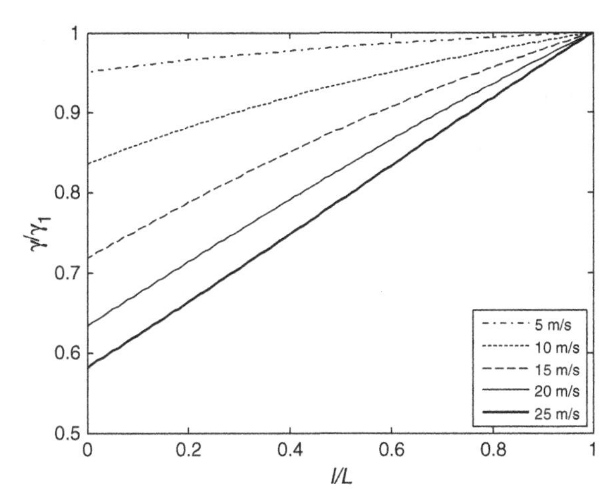

Fig. 16 Non-dimensional cable angles with horizontal for *Aerostat* 2 with different V_∞

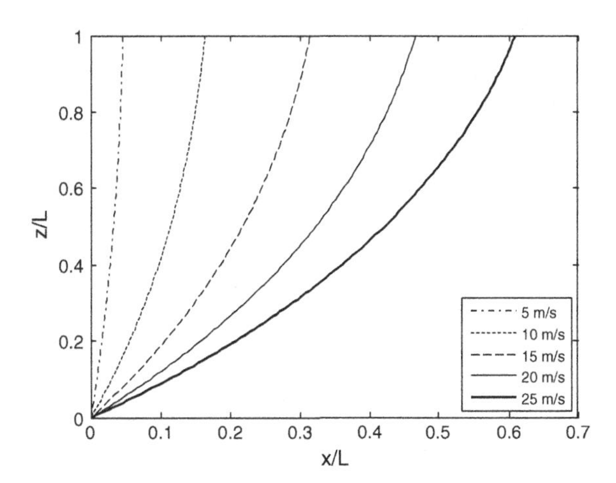

Fig. 17 Non-dimensional cable profiles for *Aerostat* 2 with different V_∞ and fixed operational height

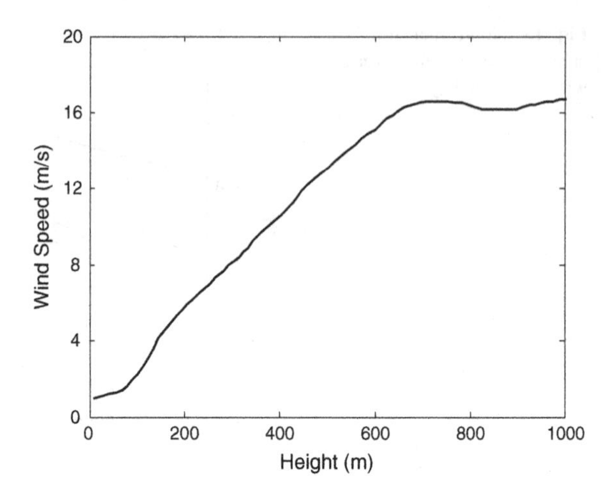

Fig. 18 A typical measured wind speed with height

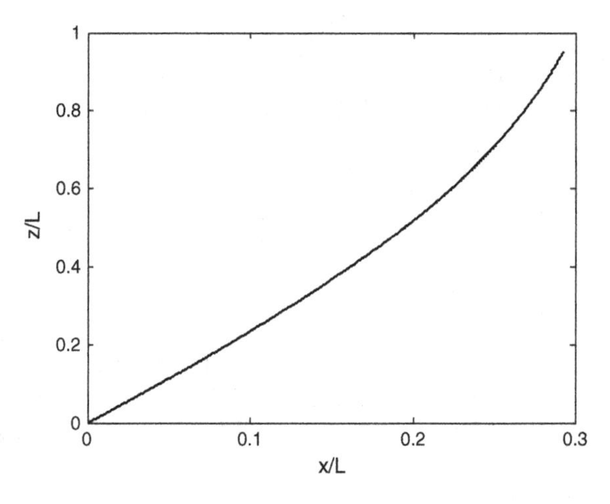

Fig. 19 Non-dimensional cable profile for *Aerostat* 2 for a measured wind speed with height and taking density variation with height

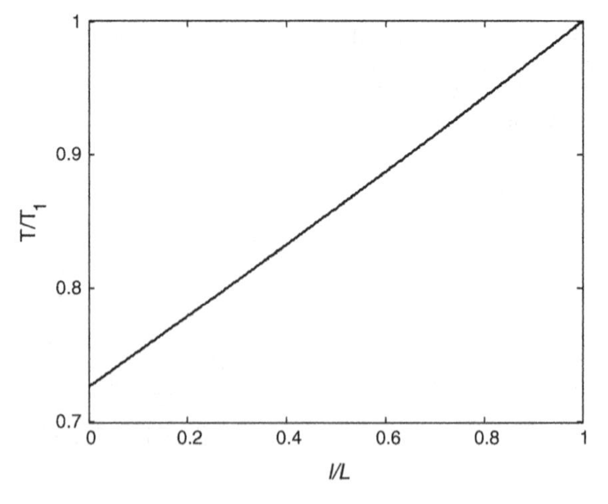

Fig. 20 Non-dimensional cable tension for *Aerostat* 2 for a measured wind speed with height and taking density variation with height

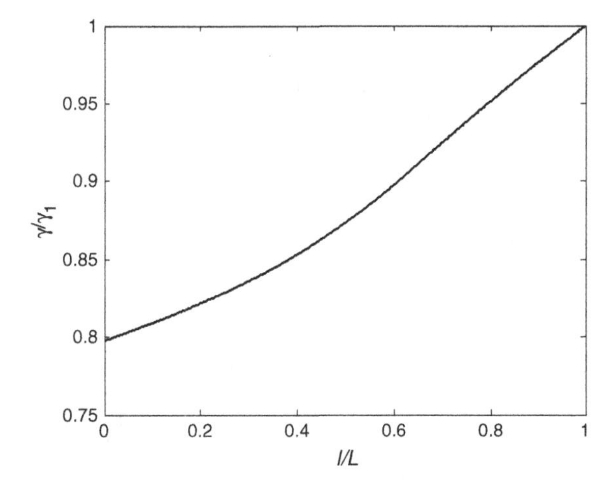

Fig. 21 Non-dimensional cable angle with horizontal for *Aerostat* 2 for a measured wind speed with height and taking density variation with height

the operating height decreases and blow-by increases with the increasing wind speed as expected. For a wind speed of 25 m/s, the operating height decreases to 871 m and blow-by increases to 463 m. If it is required to maintain height of 1000 m, as is generally the case for full coverage, the tether non-dimensional position is shown in Fig. 17. The blow-by has increased here to 608 m for 25 m/s wind speed. In this case, it is required to release extra tether of about 200 m to maintain height, and hence extra tether weight has to be taken into account for payload calculations of aerostat. Since the upper wind speed limit for aerostat operation is generally about 25 m/s, a tether length factor of 1.2 is proposed for payload calculations of medium size aerostat.

Finally, the case of both wind speed and air density variation with height is considered where the numerical method is most suited. Figure 18 shows a typical measured wind speed with height in a coastal region. The wind speed varies from about 1 m/s near surface to about 17 m/s at a height of 1000 m as per pattern shown. Polygonal approximation 2 has been used with 500 elements to estimate the tether cable position, tension, and angle and the result is presented in Figs. 19, 20, and 21. Here, the operating height is reduced to 953 m with an increase in blow-by to 291 m.

6 Conclusion

The estimation techniques for equilibrium cable configurations applicable to aerostat balloon have been presented. The direct integration expressions have been presented and compared with polygonal approximations for a small size aerostat for which wind speed and air density variations with height were neglected. The results are in agreement with a reasonable number of finite elements of cable. The polygonal expressions are then applied for a medium size aerostat including a practical case for which wind speed and density variations with height have been taken into account. A tether length factor for medium size aerostat design has also been proposed. The method presented is only applicable to static equilibrium case for a flexible and inextensible tether cable.

References

1. Pode, L.: Tables for computing the equilibrium configuration of a flexible cable in a uniform current. DTMB report 687, March (1951)
2. Neumark, S.: Equilibrium configurations of flying cables of captive balloons, and cable derivatives for stability calculations. R. and M. No. 3333, Brit. A.R.C. (1963)
3. Redd, L.T., Benett, R.M., Bland, S.R.: Stability analysis and trend study of a balloon tethered in wind, with comparisons. NASA TN D-7272 (1973)
4. Berteaux, H.O.: Buoy Engineering. Wiley, New York, Chap. 4 (1976)
5. Maixner, M.R., McDaniel, D.R.: Preliminary calculations for a flexible cable in steady, non-uniform flow. Aerosp. Sci. Technol. **18**(1), 1–7 (2012)
6. Rajani, A., Pant, R.S., Sudhakar, K.: Dynamic stability analysis of a tethered aerostat. J. Aircraft, Vol. 47, No. 5, Sep-Oct (2010)
7. Hembree B, Slegers, N.: Tethered aerostat modeling using an efficient recursive rigid-body dynamics approach. J. Aircr., **48**, 2, March–April (2011)
8. Krishnamurthy, M., Panda, G.K.: Equilibrium analysis of a tethered aerostat. Flight Experiment Division, NAL, Bangalore, India (1998)
9. Chapra, S.C.: Applied Numerical Methods with MATLAB for Engineers and Scientists, 3rd edn. New York, McGraw-Hill Higher Education (2011)
10. Anderson Jr., J.D.: Introduction to Flight, 5th edn. Tata McGraw-Hill Publishing Company Limited, New Delhi (2008)
11. Hoerner, S.F.: Fluid-dynamic drag-theoretical, experimental and statistical information. Published by the author (1965)

Differential Evolution Approach to Determine the Promotion Duration for Durable Technology Product Under the Effect of Mass and Segment-Driven Strategies

Arshia Kaul, Anshu Gupta, Sugandha Aggarwal and P.C. Jha

Abstract Promotion plays an important role for the success of a product in the market and accounts for a large portion of the firm's expenditure. This calls for judicious planning so that the promotion resources can be used efficiently at the same time creating maximum effectiveness. Firms use both mass and segment-driven promotion strategies to promote their product in a segmented market. Mass promotion spreads wider awareness of a product in the whole market with a spectrum effect while the segment-driven promotion is targeted to the distinct potential customers of the segments. This study proposes a mathematical model to determine the optimal length of a promotion campaign for a durable technology product, marketed in a segmented market under the joint effect of mass and segment-driven promotions. An application of the proposed model is demonstrated using real-life data. Differential evolution algorithm is used to solve the model.

Keywords Promotion duration · Mass promotion · Segment-driven promotion · Differential evolution

Arshia Kaul (✉) · Sugandha Aggarwal · P.C. Jha
Department of Operational Research, University of Delhi, Delhi, India
e-mail: arshia.kaul@gmail.com

Sugandha Aggarwal
e-mail: sugandha_or@yahoo.com

P.C. Jha
e-mail: jhapc@yahoo.com

Anshu Gupta
School of Business, Public Policy and Social Entrepreneurship, Ambedkar University, Delhi, India
e-mail: anshu@aud.ac.in

M. Pant et al. (eds.), *Proceedings of Fifth International Conference on Soft Computing for Problem Solving*, Advances in Intelligent Systems and Computing 437, DOI 10.1007/978-981-10-0451-3_83

1 Introduction

Markets today are very volatile and change rapidly with time. To survive in the market firms need to project their products to the customers such that awareness of product is increased. Through promotion, firms try to spread awareness of their products among customers and subsequently convert the awareness into a purchase decision. It is important for a firm to plan an appropriate promotion strategy to promote a new product in the market or to increase the market for an existing product. The strategy of the firm should be developed such that it is able to achieve maximum profit while utilizing the limited financial resources judiciously.

While developing a promotion strategy, firms need to keep in mind a number of aspects pertaining to the nature of the market. In today's world of tough competition and rapid growth in technology with several choices available to customers, they tend to show different needs and preferences making the potential market for the product heterogeneous. A uniform promotion strategy is not effective for all individuals in such a market. Firms, nowadays, have moved away from pure mass promotion strategy and develop a mixed strategy. A heterogeneous market is divided into homogeneous groups sharing similar characteristics called segments on the basis of geographic, demographic, psychographic, or behavioristic characteristics [1]. Promotion strategy is planned to target the individual segments specifically such that promotion resources can be concentrated in the segments. Firms often also use mass promotion in addition to promotion specific to the segment. While with the segment-driven promotion strategy firms target the potential segments and gain the competitive advantage; mass promotion is carried out to ensure their presence in the mass market and create wider awareness for their products. Mass promotion reaches all segments jointly representing the firms total market potential and the effect is sieved through to each segment (spectrum effect) [2–4].

Promotion strategy of a firm is not only limited by the budget but also by the constraint on the duration of promotion. All promotion campaigns launched to promote a product are time bound. According to the theory of diffusion of innovations in marketing [5, 6], adopters of a new product are categorized as innovators and imitators, who are influenced by two types of communications, viz., external and interpersonal (word-of-mouth) channels to adopt the product. External influences that spread due to promotion activities influence the innovators while the word-of-mouth communications influence the imitators. During the initial phases of product life cycle (PLC) only innovators adopt a product on being persuaded by the promotion activities and other factors. In the later phases as the word-of-mouth communication spreads, imitators start adopting the product. As time progresses the influences due to interpersonal communication channels increases and dominates the promotion activities with a decreasing marginal effect of promotion on adoption. The decrease in marginal effect on adoption due to promotion affects the profitability of the firm [7] as the expenditure on promotion may exceed the profit earned due to additional product adoption with respect to promotion activities. Thus, while planning for a promotion campaign the firm must decide the time period for the promotion campaign. The research aim of this paper is to develop a mathematical model to determine the optimal duration of promotion campaign for a new durable technology product marketed in a segmented market using both segment-driven and mass promotion strategies.

Several scholars have carried out studies related to the duration of a promotion campaign [8–11]. Most of the studies in the literature in this area assumed a homogeneous market targeted with a uniform promotion strategy. Market segmentation is considered as an important tool in marketing to gain competitive advantage. If a product is promoted in a segmented market using segment-driven promotion strategies, duration of the campaign should also be determined considering the segmentation aspect. Segments differ from one another in their adoption behavior under the effect of promotion efforts, the total segment potential and segment characteristics. This difference appears as variation in shape, size of adopter categories ([5] categorized adopters into early adopters, early majority, late majority, and laggards over the PLC), height, rate of growth, and the life span of the PLC curve. Owing to the rapid changing technology, short life cycle of durable technology products and changing customer demands, firms comes up with new product or new version of the existing product (also called technology substitution products) to remain competitive in the market. Most firms plan to terminate an existing product and introduce a new product in the market at a same time in all the segments. Continuing product for different time periods in different segments affects the market of the previous as well as future products and hence the profitability. In order to synchronize the PLC in all the segments promotion campaigns should terminate at the same time in all the segments. A recent study due to [11] has considered the segmentation aspect to determine the campaign duration for a segmented market assuming that pure segment-driven promotion strategy. A firm may choose combined mass and segment-driven promotion strategy considering the differential benefits of each type of strategy. If adoption is governed by both types of promotion strategies, promotion duration should also be computed based on the joint effect. Our study addresses this gap in the literature and proposes a profit maximization mathematical model that finds application to determine the optimal duration of the promotion campaign for a durable technology product marketed in a segmented market promoted under the joint effect of segment-driven and mass promotion strategies. An application of the proposed model is also presented through a real-life case solved using differential evolution.

The study is presented as follows: Section 2 discusses the literature review. Model development is presented in Sect. 3. A real life case study is discussed to validate and illustrate the application of the proposed model in Sect. 4. Results of data analysis, discussions, and conclusions are also drawn in Sect. 4.

2 Literature Review

The theory of communication proposed in [5] defined diffusion of an innovation as a process in which an innovation is communicated by means of some channels of communication over time to the potential market. Focus of the theory was on the two types of communication channels that support the diffusion, viz., promotion campaign and interpersonal communication channels. The theory has been extensively studied and applied by several researchers from various fields in the diverse situations. In the marketing literature researches have studied the implications of hypothesis of this theory for targeting and developing marketing strategies for new products [12, 13] and proposed analytical models to describe and forecast the diffusion process of technological innovations (for detail review see [14]). The analytical model proposed by [6] based on diffusion theory is recognized as the pioneering scholarly study. Large-scale number of applications of the model has resulted in considerable research on the model and development of the literature in the area.

Scholars have also studied the impact of two types of communication channels on the diffusion process. These studies discuss the fact that the marginal returns due to promotion increases in the beginning of the diffusion process and decreases with time over the product life cycle [7, 15]. Some scholarly studies have also focused on the development of analytical models to study the diffusion process of successive technological innovations of a firm [16, 17]. They have also analyzed the market to determine the time to introduce newer generations of the product. Findings from these two types of studies suggest that while a firm plans for its promotion strategy it should also decide the optimal length of its promotion campaign such that promotion is continued as long as it is profitable.

The topic of optimal duration for promotion has also attracted the interest of scholars in academic marketing literature. The model proposed by [18] was to determine the optimum duration between insertions in a long run advertisement campaign such that minimum effectiveness level of advertising is maximum at any point in time considering the economic axioms of non-satiation and diminishing marginal returns. Studying the advertising data of five Korean mobile companies, [8] proposed a regression model to identify the advertising lifetime (time between start and withdrawal of TV advertising) distribution. Authors suggest that lognormal distribution provides the best fit to the data. In the study by [19], the impact of time duration for promotion on retail sales based on primary data was examined. The findings suggests that time-limited promotions have more impact than the promotions conducted for longer time durations. [2] discusses an optimal control model to determine the advertising plan for launching a new product that maximizes the product goodwill at the launch time, minimizing the time duration of campaign and the advertising expenditure. The study due to [9] considered two opposing forces, viz., awareness and urgency that determine the response to an email advertising campaign. The authors proposed a model based on decision calculus that maximizes profit to determine the optimal time limit of campaign assuming exponential form of awareness and urgency functions. The findings suggest longer time limits allow greater awareness but may reduce the urgency of the purchase and vice versa in case of shorter time limits. [20] proposed a model to evaluate the impact of time restriction on purchase intensions driven by deal evaluation, anticipated regret, and urgency during a promotion campaign. [10] proposed a profit maximization model to determine the optimal duration of advertising using a pure external influence diffusion model. The authors ignored the role of interpersonal communications in the diffusion process.

In [21], the effects of short- and long-term promotion durations for the restaurant industry were discussed. The authors defined autoregressive distributed lag model-based error correction model with net sales as the dependent variable and advertisement expenditure and franchising sales ratios as independent variables. The study concluded that the promotion had short-term positive effect on sales growth and does not have an impact in the long run. The long-term effects were found significant only when advertising is interplayed jointly with franchising. [22] extended the work of [9] assuming hyperbolic S-shaped functional form of awareness and urgency functions. In [23] authors have discussed analysis of covariance to examine the relationship between sales performance of tourism and hospitality group buying deals and promotion time frame. Extending the work of [10], [24] proposed a profit maximization model for determining promotion duration of technological innovations based on mixed influence model [6] of innovation diffusion to describe the diffusion process. In a recent study by [11], it is stated that the studies in area of promotion duration largely assume a uniform market for product targeting. Given the heterogeneous nature of potential markets in the present times and tough competition, market segmentation and segment-driven promotion strategies are recognized as an important tool for better product positioning. They proposed profit maximization models to determine promotion duration for a durable product considering that the product is marketed in a segmented market. Authors suggested alternative decision policies to determine promotion time frame jointly for all segments.

Innovation diffusion in marketing as a field has evolved over the years, but nonetheless with the new market conditions new research in this field is inevitable. Researchers have also discussed the combined use of mass media and segment-driven promotion strategies to promote products in segmented markets [2–4]. Mass media (uniform market promotion) spreads awareness among the wider market with a spectrum effect on the segments while promotion strategies tailored with respect to segments target the segment potential and gain competitive advantage. [4] proposed an analytical diffusion model to describe the path of diffusion of a durable product over its life cycle under the joint influence of mass and segment specific promotion. None of the existing studies in the literature consider the combined influence of mass and segment-specific external communication channels while analyzing the promotion time frame. In this study, a mathematical model is developed with an objective to maximize profit to determine the time duration of promotion campaign for a durable product marketed in a segmented market. It is assumed that the potential market is targeted using both mass and segment specific promotion strategy. The diffusion model given in [4] is used to describe the diffusion process.

3 Model Development

3.1 Notations

S	Total number of market segments $(i=1, 2, \ldots, S)$
$\bar{N}_i$	Expected number of potential adopters of the product in ith segment
p_i	Coefficient of innovation in ith segment
q_i	Coefficient of imitation in ith segment
$x_i(t)$	Instantaneous rate of promotion in the ith segment
$X_i(t)$	Cumulative segment specific promotional effort in the ith segment by time t
$X(t)$	Cumulative mass promotion effort by time t
$N_i\begin{pmatrix} Xi(t), \\ X(t) \end{pmatrix}$	Number of adopters expected to adopt product under the joint influence of segment-specific and mass promotion in the ith segment by time t
$R_i(t)$	Revenue earned from sales by time t in the ith segment
$C_i(t)$	Variable cost incurred by time t in the ith segment
FC_i	Fixed cost incurred by time t in the ith segment
a_i	Fixed cash flow on promotion per unit time in the ith segment
A	Fixed cash flow on mass promotion per unit time
ϖ_i'	Unit sale price of the product in the ith segment
ϖ_i''	Unit cost price of the product in the ith segment
ϖ_i	Unit profit from product sale in the ith segment; $\varpi_i' - \varpi_i''$
α_i	Factor of spectrum effect of mass promotion in the ith segment; $0 \leq \alpha_i \leq 1$
r	Discount factor (present value factor)
Z	Upper bound on promotion expenditure
$\phi(t)$	Expected profit attained by time t

The mathematical model to determine the promotion campaign length is formulated in this study with a profit maximization objective. The profit earned by the firm depends on the adoption level of the product, per unit profit margin from sales, expenditure on promotion and other fixed costs incurred. To measure the adoption level, we use an innovation diffusion model proposed by [4].

3.2 The Diffusion Model [4]

Considering that the potential market of a durable technology product is segregated into S independent market segments, such that each segment has finite market potential $\bar{N}_i$; the model of first time purchase behavior is developed under the following assumptions

1. The consumer decision process is binary.
2. Adopters are categorized into innovators and imitators.
3. Parameter of external and interpersonal communications are time independent.
4. Mass and segment-driven promotion jointly act as external influence communication channel.
5. The rate of purchase with respect to promotional effort intensity is proportional to the remaining number of non-purchasers of the product.

Mathematically, the expected number of adopters by time t in the ith segment is defined as

$$N_i(X_i(t), X(t)) = \bar{N}_i\left(\frac{1 - e^{-(p_i + q_i)(X_i(t) + \alpha_i X(t))}}{1 + ((q_i/p_i)e^{-(p_i + q_i)(X_i(t) + \alpha_i X(t))})}\right)\forall(i = 1, 2, \ldots, S) \quad (1)$$

In the literature [4] exponential $(\mu_i(1 - \exp(-\gamma_i t)))$; Rayleigh $(\mu_i(1 - \exp(-\gamma_i t^2/2)))$; Weibull $(\mu_i(1 - \exp(-\gamma_i t^{m_i})))$ and logistic $(\mu_i/(1 + m_i \exp(-\gamma_i t)))$ forms are used to define the promotion effort functions (PEF) $X_i(t)$. Diffusion model (1) is used in the profit model of this study to describe the diffusion process using the best fit form of PEF.

3.3 Optimisation Model for Promotion Duration

Assuming that both mass and segment specific promotion in segments starts at the same time (t = 0), there is no restriction on the supply side, and at each time unit there is a fixed cash flow on promotion in segments (say a_i; (i= 1,2,…, S)) and mass market (say A), the present value of total profit is defined as

$$\begin{aligned}\max\phi(t) &= \sum_{i=1}^{S} e^{-rt}(R_i(t) - C_i(t) - a_i X_i(t)) - \mathrm{FC}_i - AX(t)e^{-rt} \\ &= \sum_{i=1}^{S} e^{-rt}(\varpi_i N_i(X_i(t), X(t)) - a_i X_i(t)) - \mathrm{FC}_i - AX(t)e^{-rt}\end{aligned} \quad (2)$$

where $R_i(t) = \omega_i'(t)N_i(X_i(t), X(t))$ $C_i(t) = \omega_i'' N_i(X_i(t), X(t))$ are the present values of revenue and variable cost of the product, respectively and $\omega_i = \omega_i' - \omega_i''$, $a_i X_i(t)e^{-rt} \forall (i= 1,2,\ldots, S)$ is the present value of segment specific promotion and $AX(t)e^{-rt}$ is the present value of expenditure on mass promotion and where $N_i(X_i(t), X(t))$ is defined by (1).

Concavity of the profit function (2) is not defined thus mathematical programming methods could not be applied here to obtain an optimal length of promotion time. Here, we use nature inspired evolutionary algorithm Differential Evolution (DE) [25, 26] to solve the model to obtain a near optimal solution.

Substituting the values of the model parameters in the profit model (2), setting $t \geq 0$ and solving using DE we obtain the length of time of the promotion campaign (say $(0, t^*)$). Using t^* in (1) and best fit PEF we can calculate the expected market size attainable by the time promotion winds up and the expenditure on each type of promotion. Firms usually set a fixed budget for promotion. If solution obtained using (2) suggests expenditure on promotion above the limit, the firm is willing to spend on promotion model (2) can be solved under the budget constraint

$$\sum_{i=1}^{S} a_i X_i(t) + AX(t) \leq Z \tag{3}$$

Given the budgetary limitation, the firm may also want to achieve a certain proportion of market share in each segment and/or minimum amount of total market share combined from all the segments. The profit objective in this case will be restricted by the following constraints depending on the firm's requirements

$$N_i(X_i(t), X(t)) \geq N_i^* \forall i = 1, 2, \ldots, S \tag{4}$$

$$\sum_{i=1}^{S} N_i(X_i(t), \mathrm{X}(t)) \geq N^* \tag{5}$$

where N_i^*; $(i = 1, 2, \ldots, S)$ and N^* are market share aspirations in segment and total market, respectively. Combining all constraints the profit maximization mathematical model to determine promotion duration is defined as follows:

(P1) Maximize

$$\phi(t) = \sum_{i=1}^{S} \mathrm{e}^{-rt}(\varpi_i N_i(X_i(t), X(t)) - a_i X_i(t)) - FC_i - AX(t)e^{-rt}$$

Subject to

$$\sum_{i=1}^{S} a_i X_i(t) + \mathrm{AX}(t) \leq Z$$

$$N_i(X_i(t), X(t)) \geq N_i^* \forall\, i = 1, 2, \ldots, S$$

$$\sum_{i=1}^{S} N_i(X_i(t), X(t)) \geq N^*$$

$$t \geq 0$$

To determine the length of the promotion campaign using model (P1), where values of the sales price and the variable cost parameters are provided by the firm, data of fixed cash flows per unit time on promotion can be collected from media agencies/firm and parameters of diffusion model and PEF can be estimated using the adoption data for some initial periods or similar past products sales data.

Budget and minimum market shares restrictions are contradictory in nature and may result in infeasibility if imposed simultaneously on the model. Solving the model using DE yields a compromised solution in case of infeasibility. Alternative solutions can be obtained by carrying sensitivity analysis on budget limitations and market share aspirations.

4 Data Analysis and Model Validation

Data Description Validity of the model is established on real life adoption data with respect to promotion for 24 months taken from an ABC automobile company after appropriate rescaling. The target market of the firm is segmented into four geographical segments. The firm's promotion strategy is to promote the product at the mass level in the total potential market as well as at the segment level to target the segment customers distinctly. Due to commercial confidentiality company's name, target market information and data is not shared here.

The same adoption data has previously been used in the study [4], the authors have reported the result of parameter estimation of this data set while establishing the validity of the diffusion model used in our study. The estimation was carried using nonlinear regression using statistical software support SPSS. Mean square error (MSE) is reported to validate the goodness of fit. Here, we adopt the estimates of diffusion model directly from their study (Table 1). However [4] has not established the mathematical nature of the promotion effort functions. Using the forms of the promotion effort functions discussed in the literature [4] we estimate the best fit form of the promotion effort function. The estimation is again carried using nonlinear regression in SPSS. Nonlinear least square estimates are calculated for different functions and that which gives the minimum MSE is chosen. The promotion function is estimated for mass promotion as well for segment-level promotion for each of the four segments. The parameter values of the best fit promotion effort functions and corresponding MSE are tabulated in Table 1.

Data related to cost parameters (sales price and variable cost of the car in each segment, per time unit fixed cash flow on mass and segment targeted promotion, fixed costs) (see Table 1), upper bound on budget and market share aspirations are also given by the firm. Promotion cost per time unit on mass promotion is given as Rs. 4,000,000.00 and fixed cost for each segment is Rs. 40,000,000.00. It is assumed that present value factor $r = 0.02$. Rs. 160,000,000.00 (160 million) is specified as the upper bound on promotion expenditure.

Table 1 Cost parameters and diffusion model parameter estimates

Segment		S1	S2	S3	S4	Mass
Sales price	In Rs.	457,000	468,000	446,000	453,700	
Variable Cost		388,000	403,000	378,200	393,500	
Unit promotion cost		1,300,000	1,020,000	1,820,000	1,720,000	
Diffusion model parameters	$\bar{N}_i$	287,962	156,601	106,977	223,291	774,831
	p_i	0.000671	0.001128	0.001344	0.000621	
	q_i	0.132113	0.470658	0.566036	0.331665	
	α_i	0.373	0.198	0.166	0.264	
	MSE	196105.48	13832.28	11994.30	173423.74	
Best fit PEF		Weibull	Exponential	Exponential	Exponential	Exponential
PEF parameters	μ_i	41.56	14.66	12.27	33.41	77.89
	γ_i	0.0034	0.0265	0.0274	0.0155	0.01485
	m_i	1.72	1	1	1	1
	MSE	0.07229	0.351279	0.711139	0.741132	

Data Analysis The data in Table 1 is used to solve the profit model (2) using DE algorithm. The DE algorithm is not discussed in the study due to page limitation of the manuscript. Reader may refer to [26] for algorithm. Values of the parameters of DE are chosen as suggested in the literature [25, 27, 28], i.e., F (the scaling factor) ϵ [0.5, 1]; CR (the crossover probability) ϵ [0.8, 1]; NP (the population size) = 10 * D; where D is the dimension of the model and is equal to 1 here, corresponding to the decision variable time (promotion length). The values of DE parameters chosen in this paper are $F = 0.5$, $C_r = 0.9$ and base vector selection method selected is roulette wheel. A high value of crossover probability speeds up the convergence rate. Population size, NP is taken = 15 > 10*D; (D = 1). Large population size allows better exploration of the search space. The algorithm is programmed and executed on DEV C++, with five runs. Every run is executed with a desired accuracy of 0.0001 between maximum and minimum values of fitness function. The final solution reported here is the mean of the solution obtained in 5 runs. The execution time varies between 3 and 5 min to solve the models on DEV C++ platform. The goodness of fit curves are given for segment 1 only (Fig. 1) due to page limitation.

4.1 Results and Discussion

First, the unconstrained model (2) is solved. The solution suggests time to stop for promotion to be t^* = 37.76 months. If the product is promoted for 37.76 months according to the PEF estimated from data obtained from the company (function

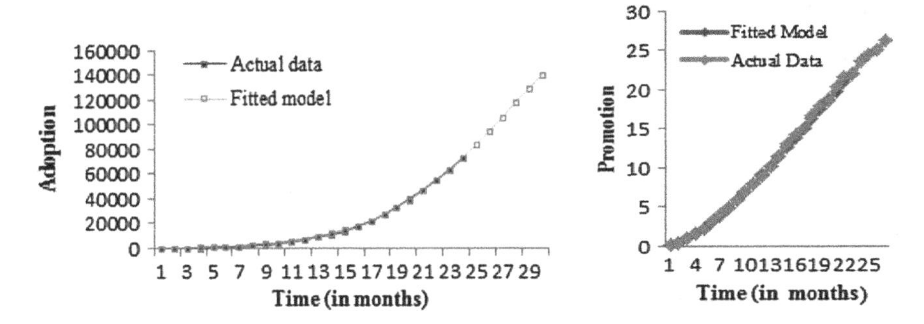

Fig. 1 Goodness of fit curves

parameters and form are tabulated in Table 1), it is expected that 82.30 % total market potential can be captured with a budget of Rs. 227,722,672.00 (detailed results shown in Table 2). The expenditure on promotion exceeds the given limit of Rs. 160 million with unconstrained maximization of profit function. The optimization model is then run with budget restriction (3). With a budget restriction of Rs. 160 million the optimal time to stop promotion is obtained to be 24.19 months. The expected total market share to be captured by 24.19 months reduces to 43.71 %. Corresponding to this solution, the expected market potential that can be attained by the time promotion winds up in segments (S1, S2, S3, S4) is 24.33%, 84.18%, 66.02% and 29.625 %, respectively. From this we infer that the propensity of adoption is slow in segments S1 and S4 compared to segments S2 and S3. Firms want to achieve certain minimum proportion of their potential markets before they wind up promotion activities. Soon after termination of promotion of a product, companies indulge into launching activities for subsequent generation products. The current product continues to diffuse in the market due to interpersonal communication channels and price promotions. When a new product is introduced, cannibalization is observed for the previous period affecting their market potential and hence the profitability and goodwill. In the view of these facts firms wants to reach at a minimum acceptable adoption level before terminating the promotion activities.

Since the expected adoption levels to be reached by the stopping time 24.19 months is quite low in some segments, we impose restrictions of 60 % minimum adoption level in each segment. Solving the optimization model with constraints (3), (4) and setting $t \geq 0$, same solution is obtained using DE algorithm, i.e., $t^* = 24.19$ as the constraints (3) and (4) act as opposing forces on the model and a compromised solution is obtained. The solution remains same even if market potential aspirations on segments S1 and S4 is set to 50 % due to their slow propensity of adoption and 60 % on S2, S3 and total potential. Since in all these cases the DE algorithm compromises the market potential keeping the budget at the constant level of Rs. 160 million, we further conducted sensitivity on the model for different levels budget. Solutions are summarized in Table 2. On the basis of the management preferences, one the alternative solutions can be chosen for implementation.

Table 2 Results for model validation

Optimal time to stop promotion, t^*	Segment	Expected market potential to be reached (%) (aspirations)	Expected profit (in Rs.)	Promotion budget (in Rs.)
Results of unconstrained model				
37.7621	S1	69.57	19,317,800,000	227,722,672
	S2	99.05		
	S3	96.04		
	S4	80.39		
	Total	82.30		
Results of sensitivity analysis				
24.19	S1	24.33 (50)	13,411,166,872	160,000,000
	S2	84.18 (60)		
	S3	66.03 (60)		
	S4	29.63 (50)		
	Total	43.71 (60)		
27.79	S1	37.43 (50)	16,200,510,332	180,000,000
	S2	92.68 (60)		
	S3	80.29 (60)		
	S4	44.69 (50)		
	Total	56.61 (60)		
29.70	S1	44.68 (50)	17,335,899,504	190,000,000
	S2	95.14 (60)		
	S3	85.51 (60)		
	S4	52.93 (50)		
	Total	62.89 (60)		
31.69	S1	51.94 (50)	18,241,140,771	200,000,000
	S2	96.81 (60)		
	S3	89.54 (60)		
	S4	61.12 (50)		
	Total	68.85 (60)		

4.2 Conclusion, Limitations, and Scope of Future Research

In this study, a mathematical model is proposed that finds application in planning the time frame of promotion activities for durable technology products. The model is developed on the background of diffusion theory of marketing communications and uses a diffusion model to describe the adoption process under the joint influence of mass and segment-driven promotion strategies. A case study is presented to establish the validity of the model and is solved using differential evolution algorithm. A marketing practitioner can use the proposed model for planning promotional activities of technological products. The model can be used to examine how the profit and adoption level grows with respect to promotion spending and decide the level of promotion spending and length of time of promotion activities.

The models find limitations in application if the customer segments are not independent and interact with each other, effect of changes in price, competitor's promotion and other marketing variable is not included in the model. The model is developed assuming a continuous promotion strategy and finds limitation if pulsating strategy is adopted by the firm. The study can be extended in future considering the limitations of the model and expanding the scope of its application.

References

1. Weinstein, A.: Handbook of Market Segmentation: Strategic Targeting for Business and Technology Firms, 3rd edn. Routledge, New York (2013)
2. Buratto, A., Viscolani, B.: New product introduction: goodwill, time and advertising cost. Math. Methods Oper. Res. (ZOR) **55**, 55–68 (2002)
3. Saak, A.: Dynamic informative advertising of new experience goods. J. Ind. Econ. **60**, 104–135 (2012)
4. Jha, P., Aggarwal, S., Gupta, A., Kumar, U., Govindan, K.: Innovation diffusion model for a product incorporating segment-specific strategy and the spectrum effect of promotion. J. Stat. Manag. Syst. **17**, 165–182 (2014)
5. Rogers, E.: Diffusion of Innovations. Free Press, New York (1962)
6. Bass, F.: A New Product growth for model consumer durables. Manag. Sci. **15**, 215–227 (1969)
7. Dubé, J., Hitsch, G., Manchanda, P.: An empirical model of advertising dynamics. Quant. Market Econ. **3**, 107–144 (2005)
8. Sohn, S., Choi, H.: Analysis of advertising lifetime for mobile phone. Omega **29**, 473–478 (2001)
9. Hanna, R., Berger, P., Abendroth, L.: Optimizing time limits in retail promotions: an email application. J. Oper. Res. Soc. **56**, 15–24 (2004)
10. Çetin, E.: Determining the optimal duration of an advertising campaign using diffusion of information. Appl. Math. Comput. **173**, 430–442 (2006)
11. Jha, P., Aggarwal, S., Gupta, A.: Optimal duration of promotion for durable technology product in a segmented market. J. Promot. Manag. (in press)
12. Talke, K., Hultink, E.: Managing diffusion barriers when launching new products. J. Prod. Innov. Manag. **27**, 537–553 (2010)
13. Iyengar, R., Van den Bulte, C., Valente, T.: Opinion leadership and social contagion in new product diffusion. Mark. Sci. **30**, 195–212 (2011)
14. Peres, R., Muller, E., Mahajan, V.: Innovation diffusion and new product growth models: a critical review and research directions. Int. J. Res. Mark. **27**, 91–106 (2010)
15. Lee, S., Trimi, S., Kim, C.: Innovation and imitation effect s dynamics in technology adoption. Ind. Manag. Data Syst. **113**, 772–799 (2013)
16. Mahajan, V., Muller, E.: Timing, diffusion, and substitution of successive generations of technological innovations: the IBM mainframe case. Technol. Forecast. Soc. Change. **51**, 109–132 (1996)
17. Jun, D., Park, Y.: A choice-based diffusion model for multiple generations of products. Technol. Forecast Soc. Chang. **61**, 45–58 (1999)
18. Balakrishnan, P.S., Hall, N.: A maximin procedure for the optimal insertion timing of ad executions. Eur. J. Oper. Res. **85**, 368–382 (1995)

19. Aggarwal, P., Vaidyanathan, R.: Use it or lose it: purchase acceleration effects of time-limited promotions. J. Consum. Behav. **2**, 393–403 (2003)
20. Swain, S., Hanna, R., Abendroth, L.: How time restrictions work: The roles of urgency, anticipated regret, and deal evaluations. Adv. Consum. Res. **33**, 523 (2006)
21. Park, K., Jang, S.: Duration of advertising effect: considering franchising in the restaurant industry. Int. J. Hosp. Manag. **31**, 257–265 (2012)
22. Chiang, C., Lin, C., Chin, S.: Optimizing time limits for maximum sales response in internet shopping promotion. Expert. Syst. Appl. **38**, 520–526 (2011)
23. Lo, A., Wu, J., Law, R., Au, N.: Which promotion time frame works best for restaurant group-buying deals? Tourism Recreat. Res. **39**, 203–219 (2014)
24. Aggarwal, S., Gupta, A., Singh, Y., Jha, P.: Optimal duration and control of promotional campaign for durable technology product. In: Proceedings of IEEE IEEM 2012, pp. 2287–2289, Dec 2012
25. Storn, R., Price, K.: Differential evolution—a simple and efficient heuristic for global optimization over continuous spaces. J. Glob. Optim. **11**, 341–359 (1997)
26. Price, K., Storn, R., Lampinen, J.: Differential Evolution. Springer, Berlin (2006)
27. Ali, M., Törn, A.: Population set-based global optimization algorithms: some modifications and numerical studies. Comput. Opera. Res. **31**, 1703–1725 (2004)
28. Vesterstrom, J., Thomsen, R.: A comparative study of differential evolution, particle swarm optimization, and evolutionary algorithms on numerical benchmark problems. In: IEEE Congress on Evolutionary Computation, CEC2004, vol. 2, pp. 1980–1987 (2004)

A Fuzzy Optimization Approach for Designing a Sustainable RL Network in a Strategic Alliance Between OEM and 3PRL

Vernika Agarwal, Jyoti Dhingra Darbari and P.C. Jha

Abstract Increased competition, higher profits, and drive toward sustainability are motivating third party revere logistics providers (3PRLs) to seek strategic alliances with original equipment manufacturers (OEMs) for recovering maximum value from product returns. With emphasis on combining technological capabilities, enhancing performance levels, developing a fast redistribution system, and addressing common sustainable concerns, a successful partnership can be achieved between the 3PRLs and OEMs. An alliance between the OEM and 3PRL is such that either the OEM outsources all its reverse logistics (RL) activities to the provider or both parties collaborate for jointly planning and executing the RL operations through mutually shared infrastructure and facilities while sharing profits. However, from 3PRL provider's perspective, a long-term view is needed to anticipate the economic viability of a collaborative system before entering into a logistics relationship with the manufacturers. As our contribution to the existing literature, we propose a fuzzy integer nonlinear programming model which can be used as a decision tool by 3PRLs for making this key decision of enhancing sustainability of the product recovery design. The proposed model is designed with the main objective of maximizing the profit generated by the multi-echelon RL network while analyzing the nature of the alliance to be formed between the OEMs and 3PRLs for each product to be recovered. The model also determines which clients (manufacturers) and how many clients to be acquired for maximum profitability. The core focus of the research is to exploit the maximum recoverable potential of the product returns and achieve a socially and environmentally sound

Vernika Agarwal (✉) · J.D. Darbari · P.C. Jha
Department of Operational Research, University of Delhi, New Delhi, India
e-mail: vernika.agarwal@gmail.com

J.D. Darbari
e-mail: jydbr@hotmail.com

P.C. Jha
e-mail: jhapc@yahoo.com

M. Pant et al. (eds.), *Proceedings of Fifth International Conference on Soft Computing for Problem Solving*, Advances in Intelligent Systems and Computing 437, DOI 10.1007/978-981-10-0451-3_84

and yet profitable RL system. The proposed model is validated by considering case study of a 3PRL offering product recovery services such as collection, repair, refurbishing, dismantling, and redistribution to manufacturers of electronic equipments.

Keywords Third party reverse logistics (3PRL) · Collaborative alliance · Product recovery services · Sustainability

1 Introduction

In recent past, strict enforcement of take back laws and environmental regulation along with emphasis on social development has stimulated manufacturers to be responsible toward the entire life cycle of their products and examine their reverse logistical operations. In accordance with the National Environmental Policy (NEP) and to address sustainable development concerns, there is a need to facilitate the recovery and/or reuse of useful materials from waste generated from a process and/or from the use of any material thereby, reducing the wastes destined for final disposal and to ensure the environmentally sound management of all materials [1]. Reverse logistics (RL) activities refer to the activities involved in product returns, source reduction, conservation, recycling, substitution, reuse, disposal, refurbishment, repair, and remanufacturing [2, 3]. Increased online purchases, stricter environmental standards, higher quality standards, and more lenient return policies have dramatically increased the volume of returned products and managing such a huge volume is a mounting task for original equipment manufacturers (OEMs) [4]. Since implementation of an effective RL system is a costly endeavor which requires additional capital investment, labor, training, new facilities, access to new resources, etc., therefore most of the manufacturing firms are reluctant to take control of all their RL activities. Handling product returns is not considered their core "value creating" business and returned products are passively handled which diminishes the value recovery process [5]. To manage the product returns in a more cost-efficient manner, OEMs prefer utilizing the services of third party reverse logistics providers for implementation of a fast and cost-effective product recovery network, thus creating a great demand for third party reverse logistic (3PRL) providers. While there is tremendous growth opportunity for these RL providers and encouragement from the government (in terms of subsidies), the major strategic challenge for them is what to offer and how much to offer in order to successfully position themselves in this continuously expanding market and be economically and socially responsible toward society. In addition to simply offering transportation services, a survey conducted by [6] suggests that a vast majority (91 %) of the 3PLs offer value-added services, such as repair, remanufacturing, repackaging, and relabeling for managing product returns. Offering these product recovery services allows for profits in terms of reusable products, components and material and gives a competitive advantage to 3PRLS.

Electronic product returns in particular have high residual value in terms of precious metals, reusable parts, and components. Recovery activities, such as collection, dismantling, sorting-segregation, are mostly done manually in India and predominantly handled by an informal sector based on a network existing among the collectors (kabadiwalas), traders, and recyclers/dismantlers [7–9]. In this regard, 3PRLs can also play a crucial role for providing a legal and well-organized channel to all the stakeholders involved in the recovery system, while creating jobs and adding value at every point of the chain. Seeking high potential for financial gains and maintaining a sustainable image, OEMs of electronic equipments are also being driven to collaborate with 3PRL providers in order to recover maximum benefit from the returns. Implementation of reverse logistics for product returns allows for savings in inventory carrying cost, transportation cost, and waste disposal cost due to returned products, and improves customer loyalty and future sales [10]. 3PRL provider on the other hand, would prefer to enter a strategic alliance with OEMs if it yields him substantial cost benefits and enhances his competitive position. Other convincing factors such as level of commitment of OEMs, percentage of profit sharing, nature of the services utilized, flexibility in coordination, and similar sustainability goals also play a crucial role in 3PRL's involvement in a collaborative network. Therefore, from the 3PRL provider's perspective, a long-term view is needed to anticipate the economic viability of a collaborative system before entering into a logistics relationship with the manufacturers. As a new contribution to the existing literature, the intention of this paper is to propose an analytical model to understand the nature of the alliance to be formed between the OEMs and 3PRLs for each product to be recovered. We propose a decision-making mathematical model for a third party reverse logistics provider which determines that for each type of product to be recovered for the OEMs, whether the company should collaborate with OEMs (collaborative clients) to jointly perform the operations of RL and share the profits or pay a fixed buyback cost to manufacturers (regular clients) for the returned products collected and take full responsibility of the all the RL activities while keeping the profits generated in the process. The multi-product, multi-echelon network is developed as an integer non linear programming (INLP) problem under uncertainty with the objective of maximizing the profit generated by the network while also determining which clients and how many clients to be acquired for maximum utilization of the facilities. The model can serve as a decision tool for 3PRL provider for understanding the nature and viability of a strategic alliance with electronic manufacturers for attaining economic sustainability and social benefits.

The rest of the paper is structured as follows: Sect. 2 discusses the analytical model; Sect. 3 presents the numerical model and discussion based on results; conclusion and suggestions for future research are presented in Sect. 4.

2 Analytical Model

The extended producer responsibility (EPR) extends the OEMs responsibility to the post consumer stage of the product life cycle. Owing to this, many OEMs outsource this responsibility to a 3PRL company which optimally extracts maximum value recovery from the discarded products. 3PRLs are specialized companies which offer comprehensive RL solutions to cater to OEMs sustainable requirements. 3PRL provide technical capabilities, expertise, and techniques to improve efficiency and process velocity which provides the highest value for the return products. The proposed model considers the design of a multi-echelon reverse logistics network from the perspective of a 3PRL provider company. The network consists of collection point, treatment sites (processing center, disassembly site, component fabrication) and final sites (scrap facility, secondary market for products and components). The reverse network starts with the collection of the returned products from collaborative or regular clients. These products are consolidated at the collection center from which they are sent to treatment sites (processing center, disassembly site, component fabrication). At the processing center, the returned products are classified into repairable or non repairable products. The repairable products are repaired to working condition and are sold to secondary market based on their demand. The non repairable products are sent to disassembly site. At the disassembly site, the reusable components are sent for component fabrication. Final sites encompass scrap facility, secondary market for products and components. The waste components from disassembly site are discarded as scrap. At the component fabrication center, the fabricated components of the returned products are sold in the secondary market or can be reused at the service center of the clients.

Monitory factors play a crucial role while designing a network for a 3PRL provider company. In this proposed framework, we have considered various costs as well as the revenue from the final sites. In case of collaboration with OEMs the company is under revenue sharing contract, thus, the revenue obtained from selling the repaired products and scrap in the secondary market is shared between the two parties. The refurbished components are retained by the collaborative clients for repairing and remanufacturing purposes. In such a partnership there is no cost of buying back the returned product while the other cost, such as repair cost, dismantling cost, transportation cost are borne by both companies involved in the partnership. In case of regular clients, the buyback cost (transportation cost is included in the buyback cost), operational cost of the facility, and various processing cost for repair and refurbishment activities is borne by the 3PRL and the provider benefits from revenue generated by selling the repaired products and refurbished components, and from the material recovered at scrap yard. In case of collaborative clients, an integrated facility or provision could be made for allocating space for a dismantling unit while in case of regular clients the cost for setting up a dismantling center is borne by the 3PRL. The configuration of the network shown in Fig. 1 is mainly dependent on the strategic decision of whether to collaborate with clients or to choose independent/regular clients for each product. The 3PRL

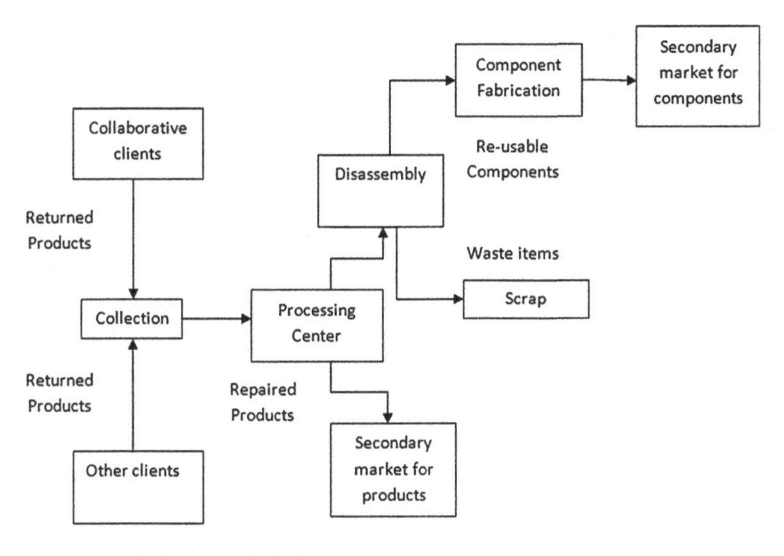

Fig. 1 Reverse logistics network design

company wants to determine whether to collaborate with clients or to have regular clients based on economic incentives for each type of returned product, the number of clients with whom the company wants to work in a given planning horizon while calculating the amount of flow of products and components across the network to maximize the profit of the network by enhancing the value recovery from the returned products. In addition, it is required not to exceed the permissible capacity of the facilities. To configure the above network, we propose a fuzzy optimization model.

Assumptions

In the creation of the model, the following assumptions were considered:

1. The time frame considered is one planning horizon.
2. The capacities of the processing center and the fabrication center are limited.
3. All the products refurbished by the 3PRL are sold in the secondary market no inventory is considered.
4. Under revenue sharing considerations, the revenue generated is shared between the 3PRL and the collaborative clients.

Sets

Set of collaborative clients with cardinality I indexed by i, Set of regular clients with cardinality J indexed by j, Product Set with cardinality K is indexed by k, Components with cardinality R is indexed by r.

Parameters

ϕ_{ik} denotes the fraction of the total profit allocated to the 3PRL provider by the ith collaborative clients for kth product while $\hat{\phi}_{\text{ik}}$ is the fraction of the total cost beared to the 3PRL provider by the *i*th collaborative clients for *k*th product. Fraction of *k*th returned products by clients to be repaired is given by α_k. β_r is the fraction of *r*th components of the clients to be fabricated. u_{rk} is a binary variable equal to '1' if *r*th component is present in *k*th product, '0' otherwise. S_k^{rep} is the per unit selling cost of kth product at the secondary market, S_k^{scrap} and S_k^{comp} are the per unit selling cost of *r*th component at the scrap facility and spare market respectively. C_{ik}^d, C_{ik}^p, $C_{ik}^{buyback}$ and T_{ik}^p per unit dismantling cost, repair cost, buyback cost, and transportation cost of *k*th product. C_r^f is the per unit fabrication cost of rth product. U_i^p and U_i^d are the maximum capacity (in terms of number of units of products) of processing and dismantling center for collaborative clients. U^P and U^d are the maximum capacity (in terms of number of units of products) of processing and dismantling center for other clients. D_k is the demand of kth product (in terms of number of units of products). C_k^{eff} is the per unit operational cost calculated as an average of the total setup cost over a planning horizon of T months. M and $\hat{M}$ are the maximum number of collaborative and regular clients that can be selected. RX_{ik}^{min} and RX_{ik}^{max} are the minimum and maximum number of returns by *i*th clients of *k*th product.

Decision Variables

Y_k is the binary variable equal to '1' if the 3PRL prefers collaborating for *k*th product, '0' otherwise. Z_{ik} is the binary variable equal to '1' if the *i*th client is selected for *k*th product, '0' otherwise. X_{ik} are the number of *k*th products returned by *i*th client. X_{ik}^p is the number of *k*th products repaired at processing center for *i*th client, X_{ik}^d is number of *k*th products dismantled for *i*th client. X_{ir}^f and X_{ir}^{rec} are the number of *r*th components fabricated and send to scrap, respectively for *i*th client.

Objective

Maximize Profit

$$\begin{aligned} F \cong \sum_i \sum_k \Bigg(& S_k^{rep} X_{ik}^p Z_{ik}(\phi_{ik} Y_k + (1 - Y_k)) + \sum_i \sum_k \sum_r Z_{ik} \Big(S_r^{comp} X_{ir}^f \\ & + S_r^{scrap} X_{ir}^{rec} \Big) (\phi_{ik} Y_k + (1 - Y_k)) - \sum_i \sum_k \Big[\Big(C_{ik}^{buyback} + T_{ik}^p \Big) X_{ik} Z_{ik} (1 - Y_k) \\ & - C_{ik}^p X_{ik}^p Z_{ik} \Big(\hat{\phi}_{ik} Y_k + (1 - Y_k) \Big) - C_{ik}^d X_{ik}^d Z_{ik} \Big(\hat{\phi}_{ik} Y_k + (1 - Y_k) \Big) - Z_{ik} C_k^{eff} (1 \\ & - Y_k) \Big] - \sum_i \sum_k \sum_r C_r^f X_{ir}^f Z_{ik} \Big(\hat{\phi}_{ik} Y_k + (1 - Y_k) \Big) \Bigg) \end{aligned} \tag{1}$$

In this model, we are taking into account fuzziness in the objective as well as in the constraints by considering decision makers aspiration levels. The symbol $\cong$ reflects the vagueness in the objective and can be interpreted as 'essentially equal to.' The objective of the model is to maximize the total profit generated by the model.

For each kth type of return, the model selects whether the 3PRL company should collaborate with its clients or work independently. As in case of collaboration, the company has to undergo a revenue sharing contract but the cost incurred is also less as part of it is borne by the collaborative clients. On the other hand, the entire profit as well as the cost will be on the companies account in case of regular clients. Based on the selection for each type of return, the cost associated with either the collaborative clients or the regular clients will be picked up by the model. The first term of the equation gives the revenue generated by selling of repaired products in the secondary market. The second term is associated with the revenue generated by selling of components at the secondary market for components and scrap facility. The third term gives the buyback cost and transportation of the returned products from regular clients. The fourth term signifies the cost incurred at the processing center for repairing of products. The fifth term signifies the dismantling cost. The sixth term signifies the operational cost calculated as an average of the total setup cost over a planning horizon of T months. The seventh term gives the cost of fabrication at the component fabrication center.

Subject to

$$R_{\mathrm{ik}}^{\min} \leq X_{ik} \leq R_{\mathrm{ik}}^{\max} \ \forall i, k \tag{2}$$

$$X_{\mathrm{ik}}^{p} = X_{\mathrm{ik}} \alpha_k Z_{\mathrm{ik}} \ \forall i, k \tag{3}$$

$$X_{\mathrm{ik}}^{d} = X_{\mathrm{ik}} (1 - \alpha_k) Z_{\mathrm{ik}} \ \forall i, k \tag{4}$$

$$X_{\mathrm{ir}}^{f} = \sum_k X_{\mathrm{ik}}^{d} u_{\mathrm{rk}} \beta_r \ \forall i, r \tag{5}$$

$$X_{\mathrm{ir}}^{\mathrm{rec}} = \sum_k X_{\mathrm{ik}}^{d} u_{rk} (1 - \beta_r) \ \forall i, r \tag{6}$$

$$\sum_i Z_{\mathrm{ik}} + \sum_i Z_{\mathrm{ik}} \leq M Y_k + \widehat{M}(1 - Y_k) \qquad \forall k \tag{7}$$

$$\sum_i X_{\mathrm{ik}}^{p} \tilde{\leq} D_k \ \forall k \tag{8}$$

$$\sum_k X_{\mathrm{ik}}^{p} Y_k \tilde{\leq} U_i^{p} \ \forall i \tag{9}$$

$$\sum_k X_{\mathrm{ik}}^{d} (1 - Y_k) \tilde{\leq} U_i^{d} \ \forall i \tag{10}$$

$$\sum_i \sum_k X_{\mathrm{ik}}^{p} (1 - Y_k) \tilde{\leq} U^{p} \tag{11}$$

$$\sum_i \sum_k X^d_{ik}(1 - Y_k) \tilde{\leq} U^d \tag{12}$$

$$Z_i, Y_k \in \{0, 1\} \tag{13}$$

$$X_{ik}, X^p_{ik}, X^d_{ik}, X^f_{ir}, X^{rec}_{ir} \geq 0 \text{ and integers } \forall i, k \tag{14}$$

Constraint (2) limits the number of returns by the collaborative or regular clients within a range. Constraint (3) determines the total number of repairable returns at the processing center. The number of units sent to dismantling center is shown by constraint (4). Constraints (5) and (6) determine the total number of components fabricated at the component fabrication center and the components sent to scrap. Constraint (7) insures that the number of collaborative or regular clients selected should be less than a specified number. Constraint (8) limits the repaired products to be less than or equal to the demand at the secondary market. Constraints (9) and (10) limit the total number of products returned and dismantled to be less than the allotted capacity of the facility for each collaborative client. Constraints (11) and (12) insure that the total number products repaired and dismantled for all the regular clients should be less than the allotted capacity of that center. Constraint (13) assures the binary integrality of decision variables Z_{ik}, Y_k. Constraint (14) assures non negativity of the decision variable $X_{ik}, X^p_{ik}, X^d_{ik}, X^f_{ir}, X^{rec}_{ir}$.

The above fuzzy INLP has been converted into crisp programming problem as follows [11]:

Maximize θ

Subject to:

Constraints (2)–(7), (13) and (14)

$$\sum_i \sum_k \Big(S^{rep}_k X^p_{ik} Z_{ik} (\phi_{ik} Y_k + (1 - Y_k)) + \sum_i \sum_k \sum_r Z_{ik} \Big(S^{comp}_r X^f_{ir} + S^{scrap}_r X^{rec}_{ir} \Big) (\phi_{ik} Y_k + (1 - Y_k)) - \sum_i \sum_k \Big[\Big(C^{buyback}_{ik} + T^p_{ik} \Big) X_{ik} Z_{ik} (1 - Y_k) - C^p_{ik} X^p_{ik} Z_{ik} \Big(\hat{\phi}_{ik} Y_k + (1 - Y_k) \Big) - C^d_{ik} X^d_{ik} Z_{ik} \Big(\hat{\phi}_{ik} Y_k + (1 - Y_k) \Big) - Z_{ik} C^{eff}_k (1 - Y_k) \Big] - \sum_i \sum_k \sum_r C^f_r X^f_{ir} Z_{ik} \Big(\hat{\phi}_{ik} Y_k + (1 - Y_k) \Big) \Big) \geq Z_0 - (1 - \theta) p^0$$

$$\sum_i X^p_{ik} \leq D_k + (1 - \theta) p^1_k \;\; \forall k$$

$$\sum_k X^p_{ik} Y_k \leq \big(U^p_i + (1 - \theta) p^2_i \big) \;\; \forall i$$

$$\sum_k X^d_{ik} (1 - Y_k) \leq \big(U^d_i + (1 - \theta) p^3_i \big) \;\; \forall i$$

$$\sum_i \sum_k X^p_{ik} (1 - Y_k) \leq U^p + (1 - \theta) p^4$$

$$\sum_i \sum_k X^d_{ik} (1 - Y_k) \leq U^d + (1 - \theta) p^5$$

$$\theta \varepsilon \{0, 1\}$$

where Z_o is the aspiration level of the decision maker, p_0 is the permissible tolerance from the objective function and p_k^1, p_i^2, p_i^3, p^4 and p^5 are the permissible tolerance for the constraints. $\theta \epsilon \{0, 1\}$ represents the degree of satisfaction of the decision maker.

3 Numerical Illustration

E-waste Management and Handling Rules, 2011 came into effect from May 1, 2012. This law mandates producers of electronic equipment to ensure that e-waste is collected, transported to specific collection, dismantling and processing units, and safely disposed under the principle of extended producer responsibility or EPR. Due to this law, a number of OEMs are turning toward 3PRL providers to achieve sustainability in their RL network by maximizing the recovery. The 3PRL companies help in retrieving maximum value while ensuring environmentally sound recovery option as per the existing government regulations. The 3PRL company considered in this case study wants to expand its existing RL network for product recovery by opening new facilities and offering their services to more clients. The major focus of the company is to increase its economic sustainability by enhancing the value recovery and strengthen the social impact of its business by creating more job opportunities. The 3PRL company considered for the validation of the model deals with three major types of product returns namely personal computers, laptops, and printers of varying quality. The 3PRL provider has the option of collecting the returned products from collaborative clients or regular clients. In case of collaborative clients, the 3PRL has to undergo a revenue sharing contract in which the profit as well as the cost are divided among both the 3PRL provider as well as the clients. In case of regular clients, the entire profit as well as the cost will be on the company's account. If the collaborative clients are selected then the company gets only certain percent of revenue from selling of repaired products, components and scrap while all the other costs are borne by both the company as well as its collaborative partner. In case of regular clients the company can incur higher profits as the entire revenue from selling of products, components as well as the scrap is its own. But the cost associated with processing as well as the setup cost is higher. Another important decision is the ideal number of collaborative or regular clients that the company can tie up with, so as not to exceed the limited resources of the organization. So a strategic decision has to be made for the planning horizon of few years so that the trade-off between cost and revenue is achieved. The proposed model helps the company in finding the optimal alternative with the main objective of maximizing profit.

There are six potential clients for collaboration and six regular potential clients. $\alpha_k = 0.5, 0.6, 0.37$ denote the fraction of personal computers (k1), laptops (k2), and printers (k3) returned by clients, to be repaired, respectively. The demand of these products in secondary market is 300, 350, and 200 units; the selling price is

Rs. 4500, Rs. 5000, and Rs. 3200. Operational cost calculated as an average of the total setup cost over a planning horizon of 10 years or 120 months is Rs. 200,000, Rs. 200,000 and Rs. 100,000, respectively.

There are 22 major components that can be retrieved by disassembly of personal computers, laptops, and printers viz. LCD (r1), RAM (r2), rear input slot and support (r3), screws (r4), keyboard (r5), outer body (r6), CPU circuit board (r7), chaises and cables (r8), metal frame (r9), motherboard (r10), fan (r11), main circuit board (r12), hard disk drive (r13), adaptors (r14), cd drive (r15), mouse (r16), battery (r17), front tray (r18), printer upper and bottom housing subassembly (r19), ribbon wire (r20), printer cartridges (r21), and printer circuit board (r22) [12–14]. The utilization rate of these components in products is given in Table 1.

Table 2 gives the minimum and maximum number of returned products and the revenue and cost sharing parameters of the clients and Table 3 gives the value of various costs associated with product returns. U_i^p= 100, 100, 100, 100, 100, 100, U_i^d = 80, 80, 80, 70, 70, 80 are maximum capacity of processing and dismantling centers for collaborative clients while 600 and 700 are the maximum capacity (in terms of number of units of products) of processing and dismantling center for regular clients. A minimum of 3 collaborative or regular clients can be selected. The fractions of *r*th component to be fabricated are 0.5, 0.56, 0.3, 0, 0.7, 0, 0.4,0, 0, 0.3, 0.4, 0.35, 0.5, 0.2, 0.5, 0.8, 0.6, 0, 0.23, 0, 0.4, 0.45, respectively.

The per unit selling costs of *r*th component at the scrap and spare market are Rs. 80, Rs. 40, Rs. 50, Rs. 10, Rs. 60, Rs. 70, Rs. 150, Rs. 50, Rs. 40, Rs. 100, Rs. 30, Rs. 120, Rs. 200, Rs. 30, Rs. 70, Rs. 50, Rs. 100, Rs. 50, Rs. 55, Rs. 40, Rs. 100, Rs. 80 and Rs. 1000, Rs. 1500, Rs. 200, Rs. 0, Rs. 380, Rs. 0, Rs. 300, Rs. 0, Rs. 0, Rs. 2200, Rs. 200, Rs. 700, Rs. 1500, Rs. 300, Rs. 500, Rs. 300, Rs. 1200, Rs. 0, Rs. 500, Rs. 0, Rs. 600, Rs. 700, respectively. Rs. 600, Rs. 400, Rs. 60, Rs. 0, Rs. 50, Rs. 0, Rs. 100, Rs. 0, Rs. 0, Rs. 500, Rs. 50, Rs. 300, Rs. 550, Rs. 100, Rs. 200, Rs. 150, Rs. 400, Rs. 0, Rs. 300, Rs. 0, Rs. 170, Rs. 220 are the per unit fabrication costs of *r*th product.

3.1 Result

The problem is solved using LINGO 11.1 with the help of the above parameters, and the model is solved achieving 88.29 % aspiration of the decision makers. The fuzzy objective and fuzzy constraints are converted into crisp forms. The aspiration level Z_o of the decision maker is Rs. 4,200,000, the permissible tolerance from the objective function p^0 is given as 600,000 and the permissible tolerance p_k^1, p_i^2, p_i^3, p^4 and p^5, for the constraints are 50, 25, 50, 120 and 120, respectively. The model selects collaborative partnership for desktop (k1) and laptops (k2) while in case of printers, the company prefers to perform the RL operations independently. A total profit of Rs. 4,129,779 is obtained. Table 4 shows the clients selected for *k*th type of product. Table 5 gives the optimal quantity of products from selected clients.

Table 1 The utilization rate (u_{rk}) of *r*th component in *k*th product

	r1	r2	r3	r4	r5	r6	r7	r8	r9	r10	r11	r12	r13	r14	r15	r16	r17	r18	r19	r20	r21	r22
k1	1	1	1	1	1	1	1	1	1	1	1	1	1	1	1	1	0	0	0	0	0	0
k2	1	1	1	1	1	1	1	1	1	1	1	1	1	1	1	0	1	0	0	0	0	0
k3	1	0	0	0	0	1	0	0	0	0	0	0	0	1	0	0	0	1	1	1	1	1

Table 2 Number of minimum and maximum returned products and the revenue and cost sharing parameters of clients

	Minimum expected return			Maximum expected return			ϕ_{ik}			$\hat{\phi}_{ik}$		
	K1	K2	K3	K1	K2	K3	K1	K2	K3	K1	K2	K3
i1	90	40	20	150	100	80	0.6	0.6	0.6	0.45	0.5	0.6
i2	40	110	30	100	170	90	0.5	0.5	0.5	0.45	0.4	0.5
i3	30	80	40	90	140	100	0.7	0.7	0.7	0.62	0.65	0.58
i4	110	130	40	170	190	100	0.5	0.6	0.58	0.5	0.5	0.5
i5	50	20	40	110	80	100	0.5	0.56	0.4	0.5	0.44	0.4
i6	30	150	50	90	210	110	0.6	0.56	0.56	0.6	0.5	0.56

Table 3 Various costs associated with product returns

	Cost at processing center			Cost at dismantling center			Transportation cost			Buyback cost		
	K1	K2	K3	K1	K2	K3	K1	K2	K3	K1	K2	K3
i1	1000	800	400	200	300	150	800	500	500	2000	3000	1500
i2	1100	750	350	220	320	140	900	550	450	2200	3200	1400
i3	900	600	300	200	280	170	600	500	200	2000	2800	1700
i4	1000	800	400	190	250	180	900	600	350	1900	2500	1800
i5	1100	750	350	190	240	150	800	550	300	1900	2400	1500
i6	900	750	350	210	300	150	750	600	250	2100	3000	1500

Table 4 Selection of *i*th client for *k*th product (Z_{ik})

	Products		
Clients	K1 (collaboration)	K2 (collaboration)	K3 (regular)
I1	Yes	No	Yes
I2	Yes	Yes	No
I3	Yes	Yes	Yes
I4	No	Yes	No
I5	Yes	No	Yes
I6	No	Yes	Yes

The optimal numbers of components fabricated are 63, 42, 22, 0, 53, 0, 30, 0, 0, 23, 30, 26, 38, 25, 37, 60, 0, 0, 12, 0, 20 and 22; 40, 45, 24, 0, 57, 0, 32, 0, 0, 24, 32, 28, 40, 16, 40, 29, 26, 0, 0, 0, 0 and 0; 73, 46, 25, 0, 58, 0, 33, 0, 0, 25, 33, 29, 41, 29, 41, 36, 23, 0, 14, 0, 25 and 28; 34, 38, 20, 0, 48, 0, 27, 0, 0, 20, 27, 24, 34, 13, 34, 0, 41, 0, 0, 0, 0 and 0; 59, 30, 16, 0, 38, 0, 22, 0, 0, 16, 22, 19, 28, 23, 28, 44, 0, 0, 14, 0, 25 and 28; 68, 38, 20, 0, 48, 0, 27, 0, 0, 20, 27, 24, 34, 27, 34, 0, 41, 0, 16, 0, 27 and 31 for clients 1, 2, 3, 4, 5 and 6, respectively. The optimal quantity of components sent for scrap are 63, 33, 53,75, 23, 125, 45, 75, 75, 53, 45, 48, 37,

Table 5 Optimal quantity of products from selected clients

		Clients					
	Products	I1	I2	I3	I4	I5	I6
X_{ik}	K1	150	73	90	0	110	0
	K2	0	110	96	171	0	172
	K3	80	0	100	0	100	110
X_{ik}^{p}	K1	75	37	45	0	55	0
	K2	0	66	58	103	0	102
	K3	30	0	37	0	37	41
X_{ik}^{d}	K1	75	36	45	0	55	0
	K2	0	44	38	68	0	70
	K3	50	0	63	0	63	69

100, 38, 15, 0, 50, 39, 50, 30 and 28; 40, 36, 57, 81, 24, 81, 49, 81, 81, 57, 48, 53, 40, 65, 40, 7, 17, 0, 0, 0, 0 and 0; 73, 37, 59, 84, 25, 147, 50, 84, 84, 59, 50, 54, 42, 117, 42, 9, 15, 63, 49, 63, 38 and 34; 34, 30, 48, 69, 21, 69, 41, 69, 68, 48, 41, 44, 34, 54, 34, 0, 27, 0, 0, 0, 0 and 0; 59, 24, 39, 55, 16, 118, 33, 55, 55, 38, 33, 35, 28, 94, 27, 11, 0, 63, 48, 63, 37 and 34; 68, 30, 48, 68, 20, 138, 41, 68, 68, 48, 41, 45, 34, 110, 34, 0, 27, 69, 53, 69, 42 and 38 for clients 1, 2, 3, 4, 5, and 6 respectively.

4 Conclusion

With the aim of increasing profit, strengthening market presence, and achieving sustainability, third party reverse logistics providers are constantly trying to improve upon their existing supply chain network by offering innovative services. Providers are also venturing into the possibility of collaborating with OEMs for establishing an efficient and valuable product recovery network. However, a long-term view is needed before exploring the possibilities so that strategic alliance between the providers and OEMs is mutually beneficial and has a positive impact on the environment. An alliance can be formed between the OEM and 3PRL such that either the OEM outsources all its RL activities to the provider or both parties collaborate for jointly planning and executing the reverse logistics operations through mutually shared infrastructure and facilities while sharing profits. As our contribution to the existing literature, we propose a mathematical model which can be used as a decision tool by 3PRLs for understanding the nature and viability of a strategic alliance with electronic manufacturers for attaining economic sustainability and social benefits and making key decisions to enhance sustainability of their product recovery design. The multiproduct, multi-echelon network is developed as an integer nonlinear programming (INLP) problem under uncertainty with the objective of maximizing the profit generated by the network while also determining which clients (manufacturers) and how many clients to be acquired for maximum utilization of the facilities and to attain maximum value recovery.

References

1. http://www.cpcb.nic.in/
2. Stock, J.R.: Reverse logistics: white paper. Counc. Logist. Manag. (1992)
3. Mutha, A., Shaligram, P.: Strategic network design for reverse logistics and remanufacturing using new and old product modules. Comput. Ind. Eng. **56**(1), 334–346 (2009)
4. Min, H., Ko, H.-J.: The dynamic design of a reverse logistics network from the perspective of third-party logistics service providers. Int. J. Prod. Econ. **113**(1), 176–192 (2008)
5. Guide, V.D.R., Harrison, T.P., Van Wassenhove, L.N.: The challenge of closed-loop supply chains. Interfaces **33**(6), 3–6 (2003)
6. Lieb, R.C.: Third-party Logistics: A Manager's Guide. JKL Publications, Houston (2000)
7. Wath, S.B., Vaidya, A.N., Dutt, P.S., Chakrabarti, T.: A roadmap for development of sustainable E-waste management system in India. Sci. Total Env. **409**(1), 19–32 (2010)
8. Manomaivibool, P.: Extended producer responsibility in a non-OECD context: The management of waste electrical and electronic equipment in India. Resour. Conserv. Recycl. **53**(3), 136–144 (2009)
9. Dwivedy, M., Mittal, R.K.: An investigation into e-waste flows in India. J. Clean. Prod. **37**, 229–242 (2012)
10. Ko, H.J., Evans, G.W.: A genetic algorithm-based heuristic for the dynamic integrated forward/reverse logistics network for 3PLs. Comp. Oper. Res. **34**(2), 346–366 (2007)
11. Bector, C.R., Chandra, S.: Fuzzy mathematical programming and fuzzy matrix games (Vol. 490). Springer, Berlin (2005)
12. Shrivastava, P., Zhang, H.C., Li, J., Whitely, A.: Evaluating obsolete electronic products for disassembly, material recovery and environmental impact through a decision support system. In: Proceedings of the 2005 IEEE International Symposium on Electronics and the Environment, pp. 221–225, May 2005
13. Mishima, K., Mishima, N.: A study on determination of upgradability of laptop PC components. In: Functional Thinking for Value Creation, pp. 297–302. Springer Berlin, Heidelberg (2011)
14. Das, S.K., Yedlarajiah, P., Narendra, R.: An approach for estimating the end-of-life product disassembly effort and cost. Int. J. Prod. Res. **38**(3), 657–673 (2000)

Fuzzy Bicriteria Optimization Model for Fault Tolerant Distributed Embedded Systems Under Reuse-or-Build-or-Buy Strategy Incorporating Recovery Block Scheme

Ramandeep Kaur, Stuti Arora and P.C. Jha

Abstract A practical approach for component selection of hardware and software components in embedded systems has been considered in this paper. Redundancy has been incorporated using RB/1/1 architecture to make the system fault tolerant. Software industry has been relying on CBSD approach for adopting components with reuse attribute. Software engineers are constantly facing a challenge to select a right mix of components from ready-to-use COTS, in-house built, and fabricated components. Such a decision is reuse-build-buy decision, based upon several factors. Through this paper, an optimization model has been proposed for selection of right mix of components using reuse-build-buy decision, maximizing the overall reliability of the embedded system simultaneously minimizing cost under RB/1/1 fault tolerant scheme. Imprecision can be induced in the parameters due to ambiguity human judgment for which fuzzy optimization approach has been considered.

Keywords RB/1/1 · Reuse · COTS · Fabricated components · CBSD

1 Introduction

Embedded systems have encircled mankind with enumerable applications. Their use can be attributed to global domains of life-critical system such as avionics, aeronautics, medical appliances, etc. These systems tend to fit in the requirements

Ramandeep Kaur (✉)
Department of Computer Science, Banasthali Vidyapith, Vanasthali, Rajasthan, India
e-mail: rrdk_07@yahoo.com

Stuti Arora · P.C. Jha
Department of Operational Research, University of Delhi, Delhi, India
e-mail: stu.aro@gmail.com

P.C. Jha
e-mail: jhapc@yahoo.com

M. Pant et al. (eds.), *Proceedings of Fifth International Conference on Soft Computing for Problem Solving*, Advances in Intelligent Systems and Computing 437, DOI 10.1007/978-981-10-0451-3_85

of aforesaid industries seamlessly. But, dependability on the embedded systems is a major concern which generally related to reliability of systems and ability to tolerate faults. The factors for achieving dependability consist of fault prevention and fault removal techniques, but usually fault tolerance (FT) techniques cover the transient, hardware permanent, and residual software faults [1]. Fault tolerance is a means of achieving a continuous system service in the presence of active faults [2]. The advent of more sophisticated embedded systems supporting powerful functions, while relying on profound submicron process technologies for their fabrication, has raised concerns pertaining to reliability [3]. The overall reliability of the system can be enhanced by incorporating redundancy at hardware and/or software level to make them fault tolerant [4]. Single software version-based techniques such as checkpoint and restart [5] and triple modular redundancy [6] were used, with their effect on only hardware and transient software faults. Popular fault tolerant architectures for embedded systems are N-Version Programming (NVP) and Recovery Block (RB) schemes [7–10], with widely implemented variants as NVP/0/1 Architecture, NVP/1/1 architecture and RB/1/1 Architecture [11, 12]. Modularized embedded system can be realized as series/parallel distributed system with no of subsystems [13] (Fig. 2).

Studies were conducted based on genetic algorithm (GA) for redundancy allocation problem for the series-parallel system [14]. Figure 1 depicts RB/1/1 architecture as an extension of RB scheme, which can be realized in a general form of *X/i/j*, with *X* as a fault tolerant technique, *i* be the number of hardware faults tolerated and *j* be the number of software faults tolerated [11]. Models for optimizing overall reliability of distributed systems within a specified budget so as to incorporate fault tolerance were proposed [15–17]. However, improving software reliability, using redundancy requires additional resources [18]. The scheme allows subsystem to tolerate one hardware or/and one software fault. It consists of an adjudication module (or acceptance test (AT)), with at least two software components or alternates. The system initiates by rendering the output of the primary alternate for AT. In the event of failure, the process rolls back to the initial state, and tests the output

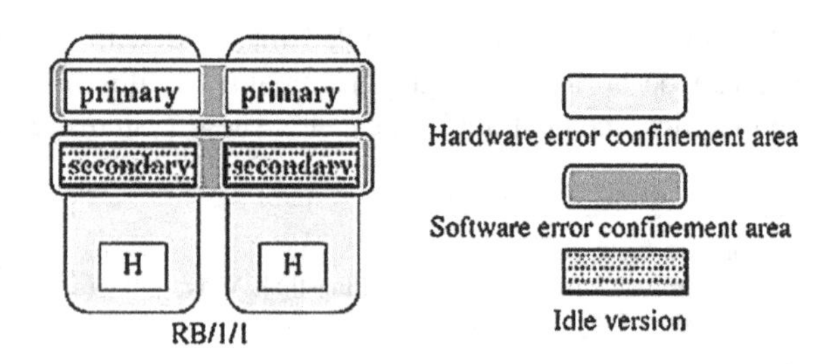

Fig. 1 RB/1/1 fault tolerant architecture [10]

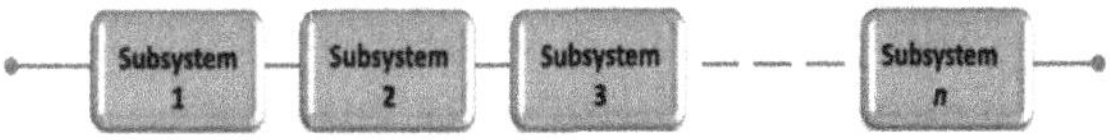

Fig. 2 Framework of subsystem-based modularized embedded system

of secondary alternative for acceptance and continues until an output from an alternate is accepted or outputs of all alternates are rejected. Under the occurrence of any of the conditions, i.e., existence of related faults among both software versions; failure of AT; faults in specifications; failure of both hardware components; independent failure of both software versions; the distributed system is entitled to fail (Fig. 2).

Hardware components manufactured by different vendors, are assembled by a third-party to form a complete system. The componentization of a software into manageable chunks has eased off the engineers from developing entire software in-house. Components are segregated and developed by vendors on the basis of their functionality and specialization, and integrated by the software engineers as per user needs. Such an approach is known as component-based software development (CBSD) [19], focusses on developing components that can reused endlessly rather than coding from the scratch. These ready-to-use components can be procured from the vendors as commercial off-the-shelf (COTS) components [19, 20], are economical as well as more reliable. CBSD has roots in both software and hardware and involves the purchase and integration of pre-made components [21, 22]. More than one functionally equivalent COTS component may exist in the market with variance in cost, reliability, delivery time, and the number of source lines of code (SLOC). Certain components are in-house build, which depends upon several factors. In addition, pre-existing in-house developed components can also be reused after fabrication [23, 24], which entails fabrication cost and time. It becomes challenging for the developer to decide optimally whether to fabricate the existing component, or develop in-house from scratch, or procure it as COTS. This decision is referred to as reuse-or-build-or-buy decision [25] and depends upon parameters such as expertise availability, requirement specificity, execution time, cost, etc. This paper formulates an optimization model, each with the objective of maximizing the reliability of fault tolerant-embedded system, simultaneously minimizing the cost under the constraints of execution time, delivery time, and SLOC. Very Often, human judgments about abovementioned parameters are ambiguous and inaccurate. Therefore, a fuzzy approach has been used based on reuse-or-build-or-buy scheme incorporating RB/1/1 FT architecture. Aim of the paper is to select the hardware and software components optimally, at the same time optimize redundancy levels for each subsystem.

2 Framework for Component Selection Using Reuse-or-Build-or-Buy Strategy

An optimization model has been proposed in this paper, an extension of the model discussed in [4]. The model is based on RB/1/1 architecture as seen in Fig. 1 incorporating reuse-build-buy decision. The formulation is done with an assumption that each subsystem can have at most one reusable component.

2.1 Notation

N	Number of subsystems within the embedded system
m_i	Number of hardware component choices for subsystem i; $i = 1,\ldots,n$
p_i	Number of COTS software alternatives for subsystem i; $i = 1,\ldots,n$
C_{ij}^{hw}	Cost of using hardware component j for subsystem i; $i = 1,\ldots,n$; $j = 1,\ldots,m_i$
C_{ik}^{COTS}	Cost of kth COTS software component for ith subsystem; $i = 1,\ldots n$; $k = 1,\ldots p_i$
C_i^{ihd}	Net cost incurred if in-house developed software component is deployed in subsystem i; $i = 1,\ldots,n$
C_i^{fab}	Net cost of fabricated reusable component available for subsystem i; $\forall\, i$
c_i	Unitary development cost of in-house developed software component for subsystem i; $i = 1,\ldots,n$
t_i^{ihd}	Estimated development time of in-house developed software instance for subsystem i; $i = 1,\ldots,n$
τ_i^{ihd}	Average time required to perform a test case on in-house developed component for ith subsystem; $\forall\, i$
N_i^{tot}	Total number of tests performed on in-house developed software component for ith subsystem; $\forall\, i$
N_i^{suc}	Number of successful tests performed on in-house developed component for ith subsystem; $\forall\, i$
π_i^{ihd}	Testability probability that single execution of in-house developed software instance fails on a test case from a certain input distribution; $\forall\, i$
R_i^{ihd}	Probability that the in-house developed software component for ith subsystem is failure free during a single run given that N_i^{suc} test cases have been successfully performed; $\forall\, i$
t_i^{fab}	Estimated fabrication time for reusable software component of subsystem i; $\forall\, i$
τ_i^{fab}	Average time required to perform a test case on fabricated reusable software component for subsystem i; $\forall\, i$
Nf_i^{tot}	Total number of tests performed on fabricated reusable software component for subsystem i; $\forall\, i$
Nf_i^{suc}	Number of successful tests performed on fabricated reusable software component for subsystem i; $\forall\, i$

π_i^{fab}	Probability that single execution of the fabricated reusable software instance fails on a test case chosen from a certain input distribution; $\forall\ i$
R_i^{fab}	Probability that the fabricated software component for ith subsystem is failure free during a single run given that Nf_i^{suc} test cases have been successfully performed; $\forall\ i$
R_{ik}^{COTS}	Reliability of *k*th COTS software component for *i*th subsystem; $\forall\ i, k$
R_{ij}^{hw}	Reliability of *j*th hardware component for *i*th subsystem; $\forall\ i, j$
R_i	Reliability of subsystem *i*; $\forall\ i$
l_{ik}^{COTS}	Number of source lines of code for *k*th COTS software component available for *i*th subsystem; $\forall\ i, k$
l_i^{ihd}	Number of source lines of code for in-house developed software component for subsystem *i*; $\forall\ i$
l_i^{fab}	Number of source lines of code for fabricated software component for subsystem *i*; $\forall\ i$
L	Limit set on total source lines of code of the system
d_{ij}^{hw}	Delivery time for *j*th hardware instance of *i*th subsystem $\forall\ i, j$
d_{ik}^{COTS}	Delivery time for *k*th COTS software instance of *i*th subsystem; $\forall\ i, k$
D	Bound set on overall delivery time of the system
T_{ik}^{COTS}	Execution time for *k*th COTS software instance of *i*th subsystem; $\forall\ i, k$
T_i^{ihd}	Execution time for in-house developed component of *i*th subsystem; $\forall\ i$
T_i^{fab}	Execution time for fabricated software instance of *i*th subsystem; $\forall\ i$
E_i	Bound set on execution time of the *i*th subsystem
P_i^{rv}	Probability of failure from the related faults between software versions in subsystem *i*, which has RB/1/1 redundancy; $i = 1,\ldots,n$; $Q_i^{\mathrm{rv}} = 1 - P_i^{\mathrm{rv}}$
$P_{i,t,t'}^{\mathrm{rv}}$	Probability of failure of subsystem *i* from related faults between COTS software versions *t* and *t'*; $i=1,\ldots, n$; $t, t' = 1,\ldots, p_i$
$P_{i,\mathrm{ihd},t}^{\mathrm{rv}}$	Probability of failure of subsystem *i* from related faults between in-house developed and COTS software component *t*; $i = 1,\ldots,n$; $t = 1,\ldots p_i$
$P_{i,\mathrm{fab},t}^{\mathrm{rv}}$	Probability of failure of subsystem *i* from related faults between fabricated reusable software and COTS software component *t*; $i = 1,\ldots,n$; $t = 1,\ldots p_i$
$P_{i,\mathrm{ihd,fab}}^{\mathrm{rv}}$	Probability of failure of *i*th subsystem from related faults between fabricated software component and in-house developed software instance; $\forall\ i$
P^d	Probability of failure of Acceptance Test (AT); $Q^d = 1 - P^d$
P_i^{fs}	Probability of failure from faults in specification, that leads to failure of all software versions in subsystem *i*, which has RB/1/1 redundancy; $i = 1,\ldots, n$; $Q_i^{\mathrm{fs}} = 1 - P_i^{\mathrm{fs}}$
P_i^{hw}	Probability of failure of a hardware in subsystem *i*, which has RB/1/1 redundancy; $i = 1,\ldots,n$
P_i^{fbv}	Probability that both software versions of subsystem *i*, which has RB/1/1 redundancy, fail independently; $\forall\ i$

x_{ij} Decision variable; $i = 1,\ldots,n$; $j = 1,\ldots,m_i$

$$x_{ij} = \begin{cases} 1; & \text{if } j\text{th hardware component selected for subsystem } i\text{, without redundancy} \\ 2; & \text{if } j\text{th hardware component selected for subsystem } i\text{, with redundancy} \\ 0; & \text{if } j\text{th hardware component not selected for subsystem } i \end{cases}$$

y_{ik} Binary decision variable; $i = 1,\ldots,n$; $k = 1,\ldots,p_i$

$$y_{ik} = \begin{cases} 1; & \text{if } k\text{th COTS component of } i\text{th subsystem is selected} \\ 0; & \text{otherwise} \end{cases}$$

v_i Binary decision variable; $i = 1,\ldots,n$

$$v_i = \begin{cases} 1; & \text{if in-house developed software component is selected for } i\text{th subsystem} \\ 0; & \text{otherwise} \end{cases}$$

w_i Binary decision variable

$$w_i = \begin{cases} 1; & \text{if fabricated reusable component is selected for } i\text{th subsystem} \\ 0; & \text{otherwise} \end{cases}$$

z_i Binary decision variable; $i = 1,\ldots,n$

$$z_i = \begin{cases} 1; & \text{if } i\text{th subsystem uses RB/1/1 redundancy technique} \\ 0; & i\text{th subsystem does not use any redundancy technique} \end{cases}$$

2.2 Assumptions

1. Several but finite number of subsystems connected in series are considered in modularized system.
2. Reliability and cost is known for hardware and software components.
3. Component are non-repairable with two states, functional or failed.
4. Nonidentical nature of software and hardware failures. System failure occurs when all software versions or hardware components fail.
5. COTS, in-house developed and reusable software components are available for a SS.
6. Vendor provides information pertaining to Reliability, cost, SLOC, execution time, and delivery time of COTS components.
7. Cost and reliability of an in-house built and fabricated component can be specified based on the parameters of intricate development process.
8. At most one component can be developed in-house or fabricated for each SS, while abundant COTS software versions for a SS.
9. Each subsystem either has RB/1/1 redundancy or has no redundancy.
10. Subsystem without any redundancy has exactly one hardware and software component. SS with RB/1/1 redundancy has two identical hardware components and two distinct software components.
11. Failure of individual component are s-independent [15].
12. Number of hardware units purchased has no effect on delivery time for hardware component(s) of a subsystem.
13. Time taken for integration of components is negligible.

2.3 Model Formulation

The optimization model for component selection can be written as follows:
Problem (P1)

$$\text{Max}\, R \cong \prod_{i=1}^{n} R_i \tag{1}$$

$$\text{Min}\, C \cong \sum_{i=1}^{n}\left[\sum_{j=1}^{m_i} C_{ij}^{\text{hw}} x_{ij} + \sum_{k=1}^{p_i} C_{ik}^{\text{cots}} y_{ik} + C_i^{\text{ihd}} v_i + C_i^{\text{fab}} w_i\right] \tag{2}$$

subject to

$X \in S = \{x_{ij}, y_{ik}, v_i, w_i, z_i$ are decision variables

$$C_i^{\text{ihd}} = c_i\left(t_i^{\text{ihd}} + \tau_i^{\text{ihd}} N_i^{\text{tot}}\right);\ \ i = 1, \ldots, n \tag{3}$$

$$C_i^{\text{fab}} = c_i\left(t_i^{\text{fab}} + \tau_i^{\text{fab}} N f_i^{\text{tot}}\right);\ \ i = 1, \ldots, n \tag{4}$$

$$\left(\sum_{k=1}^{p_i} \tilde{\text{T}}_{ik}^{\text{cot s}}\, y_{ik}\right) + \tilde{\text{T}}_i^{\text{ihd}}\, v_i + \tilde{\text{T}}_i^{\text{fab}}\, w_i \le E_i;\ \ i = 1, \ldots, n \tag{5}$$

$$\sum_{i=1}^{n}\left[\sum_{k=1}^{p_i} l_{ik}^{\text{cot s}} y_{ik} + l_i^{\text{ihd}} v_i + l_i^{\text{fab}} w_i\right] \le L \tag{6}$$

$$\sum_{j=1}^{m_{,i}} x_{ij} - z_i = 1;\ \ i = 1, \ldots, n \tag{7}$$

$$\sum_{k=1}^{p_i} y_{ik} + v_i + w_i - z_i = 1;\ \ i = 1, \ldots, n \tag{8}$$

$$x_{ij} \times x_{il} = 0;\ \ i = 1, \ldots, n;\ \ j, l = 1, \ldots, m_i;\ \ j \ne l \tag{9}$$

$$\underset{\substack{i = 1, \ldots, n \\ j = 1, \ldots, m_i}}{\text{Max}} \frac{\tilde{d}_{ij}^{\text{ hw}} \cdot x_{ij}}{(z_i + 1)} \le D \tag{10}$$

$$\underset{\substack{i = 1, \ldots, n \\ k = 1, \ldots, p_i}}{\text{Max}} \tilde{d}_{ik}^{\text{ cot s}} \cdot y_{ik} \le D \tag{11}$$

$$\operatorname*{Max}_{i=1,\ldots,n}\left(t_i^{\text{ihd}} + \tau_i^{\text{ihd}} N_i^{\text{tot}}\right) v_i \le D \tag{12}$$

$$\operatorname*{Max}_{i=1,\ldots,n}\left(t_i^{\text{fab}} + \tau_i^{\text{fab}} N f_i^{\text{tot}}\right) w_i \le D \tag{13}$$

$$R_i^{\text{ihd}} = \frac{1 - \pi_i^{\text{ihd}}}{(1 - \pi_i^{\text{ihd}}) + \pi_i^{\text{ihd}}(1 - \pi_i^{\text{ihd}})^{N_i^{\text{suc}}}};\ \ i = 1, \ldots, n \tag{14}$$

$$R_i^{\text{fab}} = \frac{1 - \pi_i^{\text{fab}}}{(1 - \pi_i^{\text{fab}}) + \pi_i^{\text{fab}}(1 - \pi_i^{\text{fab}})^{Nf_i^{\text{suc}}}};\ \ i = 1, \ldots, n \tag{15}$$

$$N_i^{\text{suc}} = (1 - \pi_i^{\text{ihd}}) N_i^{\text{tot}};\ \ i = 1, \ldots, n \tag{16}$$

$$Nf_i^{\text{suc}} = (1 - \pi_i^{\text{fab}}) Nf_i^{\text{tot}};\ \ i = 1, \ldots, n \tag{17}$$

$$R_i = (1 - z_i)\left[\left(\sum_{j=1}^{m_i} R_{ij}^{\text{hw}} x_{ij}\right) \cdot \left(\sum_{k=1}^{p_i} R_{ik}^{\text{cot s}} y_{ik} + R_i^{\text{ihd}} v_i + R_i^{\text{fab}} w_i\right)\right] + z_i(1 - P_i); \forall i \tag{18}$$

$$P_i = P_i^{\text{rv}} + Q_i^{\text{rv}} P_d + Q_i^{\text{rv}} Q_d P_i^{\text{fs}} + Q_i^{\text{rv}} Q_d Q_i^{\text{fs}} (P_i^{\text{hw}})^2 + Q_i^{\text{rv}} Q_d Q_i^{\text{fs}} \left(1 - (P_i^{\text{hw}})^2\right) P_i^{\text{fbv}};\quad i = 1, \ldots, n \tag{19}$$

$$P_i^{\text{rv}} = \sum_{t=1}^{p_i - 1} \sum_{t'=t+1}^{p_i} P_{i,t,t'}^{\text{rv}} y_{it} y_{it'} + \sum_{t=1}^{p_i} P_{i,\text{ihd},t}^{\text{rv}} y_{it} v_i + \sum_{t=1}^{p_i} P_{i,\text{fab},t}^{\text{rv}} y_{it} w_i + P_{i,\text{ihd},\text{fab}}^{\text{rv}} v_i w_i;\quad i = 1, \ldots, n \tag{20}$$

$$P_i^{\text{hw}} = \sum_{j=1}^{m_i} \frac{\left(1 - \widetilde{R}_{<ij}^{\text{hw}}\right) x_{ij}}{(z_i + 1)};\ \ i = 1, \ldots, n, \tag{21}$$

$$\begin{aligned} P_i^{\text{fbv}} = & \sum_{t=1}^{p_i - 1} \sum_{t'=t+1}^{p_i} \left[(1 - \widetilde{R}_{it}^{\text{cot s}}) y_{it}\right] \cdot \left[(1 - \widetilde{R}_{it'}^{\text{cot s}}) y_{it'}\right] + \sum_{t=1}^{p_i} \left[(1 - \widetilde{R}_{it}^{\text{cot s}}) y_{it}\right] \cdot \left[(1 - R_{<i}^{\text{ihd}}) v_i\right] \\ & + \sum_{t=1}^{p_i} \left[(1 - \widetilde{R}_{it}^{\text{cot s}}) y_{it}\right] \cdot \left[(1 - R_{<i}^{\text{fab}}) w_i\right] + \left[(1 - R_{<i}^{\text{ihd}}) v_i\right] \cdot \left[(1 - R_{<i}^{\text{fab}}) w_i\right];\ \ i = 1, .., n \end{aligned} \tag{22}$$

$$x_{ij} \in \{0,1,2\};\ \ i = 1,\ldots,n;\ j = 1,\ldots,m_i \tag{23}$$

$$y_{ik} \in \{0,1\};\ \ i = 1,\ldots,n;\ k = 1,\ldots,p_i \tag{24}$$

$$v_i \in \{0,1\};\ i = 1,\ldots,n \tag{25}$$

$$w_i \in \{0,1\};\ i = 1,\ldots,n \tag{26}$$

$$z_i \in \{0,1\};\ i = 1,\ldots,n \tag{27}$$

($\sim$) represents the fuzziness in data. Here, the objective function (1) maximizes system reliability and objective function (2) minimizes overall cost. Equations (3) and (4) give estimated in-house developed and fabrication cost for ith subsystem. Equations (5) and (6) compute the execution time for each subsystem and total SLOC of COTS, in-house developed and fabricated components, respectively, with a total bound on it. Constraint (7) represents the selection of exactly one hardware component in case of no redundancy, and exactly two components if the subsystem has redundant architecture. Similarly selection of exactly one software component in case of no redundancy, and selection of two components if redundancy exists, is shown in (8). Equation (9) guarantees selection of only one type of hardware component for a subsystem. Thus in case of redundancy in ith subsystem, two hardware components selected will be identical. Since there is a bound on delivery time of the system, constraints (10)–(13) ensure that all selected hardware and software components will be delivered within D time units. Equations (14) and (15) evaluate reliability for in-house developed component and fabricated component, respectively, with the help of testability and number of successful test cases. Number of successful test cases can be evaluated as shown in (16) and (17). Reliability of ith subsystem is computed in (18), as product of reliabilities of selected hardware component and software component if subsystem has no redundancy. Otherwise, it is computed on the basis of the reliability for RB/1/1 architecture. If ith subsystem has RB/1/1 redundant architecture, then its unreliability is given by (19) which can be computed using Eqs. (20), (21) and (22). Equations (23)–(27) represent decision variables.

3 Fuzzy Approach for Finding Solution

For fuzzy parameters, crisp equivalent is computed using defuzzification function $f = (a_1 + 2a_2 + a_3)/4$, where a_1, a_2, and a_3 are triangular fuzzy numbers (TFNs). (−) on top of notations represent notations in defuzzified form. Resulting problem is: Problem (P2)

$$\operatorname{Max} R \cong \prod_{i=1}^{n} R_i \tag{28}$$

$$\operatorname{Min} C \cong \sum_{i=1}^{n} \left[\sum_{j=1}^{m_i} C_{ij}^{\text{hw}} x_{ij} + \sum_{k=1}^{p_i} C_{ik}^{\text{cots}} y_{ik} + C_i^{\text{ihd}} v_i + C_i^{\text{fab}} w_i \right] \tag{29}$$

subject to

$$X \in \overline{S} = \{x_{ij}, y_{ik}, v_i, w_i, z_i \text{ are decision variables}\}$$

Constraint (3), Constraint (4), Constraints (6)–(9),
Constraints (12)–(17), Constraints (23)–(27)

$$\left(\sum_{k=1}^{p_i} \overline{T}_{ik}^{\text{cots}} y_{ik} \right) + \overline{T}_i^{\text{ihd}} v_i + \overline{T}_i^{\text{fab}} w_i \leq E_i;\ i = 1, \ldots, n \tag{30}$$

$$\underset{\substack{i=1,\ldots,n \\ j=1,\ldots,m_i}}{\operatorname{Max}} \ \frac{\overline{d}_{ij}^{\text{hw}} \cdot x_{ij}}{(z_i + 1)} \leq D \tag{31}$$

$$\underset{\substack{i=1,\ldots,n \\ k=1,\ldots,p_i}}{\operatorname{Max}} \ \overline{d}_{ik}^{\text{cots}} \cdot y_{ik} \leq D \tag{32}$$

$$R_i = (1 - z_i) \left[\left(\sum_{j=1}^{m_i} \overline{R}_{ij}^{\text{hw}} x_{ij} \right) \cdot \left(\sum_{k=1}^{p_i} \overline{R}_{ik}^{\text{cots}} y_{ik} + R_i^{\text{ihd}} v_i + R_i^{\text{fab}} w_i \right) \right] + z_i (1 - P_i); \forall\, i \tag{33}$$

$$\begin{aligned} P_i &= P_i^{\text{rv}} + Q_i^{\text{rv}} P_d + Q_i^{\text{rv}} Q_d P_i^{\text{fs}} + Q_i^{\text{rv}} Q_d Q_i^{\text{fs}} (P_i^{\text{hw}})^2 \\ &\quad + Q_i^{\text{rv}} Q_d Q_i^{\text{fs}} \left(1 - (P_i^{\text{hw}})^2\right) P_i^{\text{fbv}}; \quad i = 1, \ldots, n \end{aligned} \tag{34}$$

$$P_i^{\text{rv}} = \sum_{t=1}^{p_i - 1} \sum_{t'=t+1}^{p_i} P_{i,t,t'}^{\text{rv}} y_{it} y_{it'} + \sum_{t=1}^{p_i} P_{i,\text{ihd},t}^{\text{rv}} y_{it} v_i + \sum_{t=1}^{p_i} P_{i,\text{fab},t}^{\text{rv}} y_{it} w_i + P_{i,\text{ihd},\text{fab}}^{\text{rv}} v_i w_i; \forall\, i \tag{35}$$

$$P_i^{\text{hw}} = \sum_{j=1}^{m_i} \frac{\left(1 - \overline{R}_{<ij}^{\text{hw}}\right) x_{ij}}{(z_i + 1)}; \quad i = 1, \ldots, n \tag{36}$$

$$P_i^{\text{fbv}} = \sum_{t=1}^{p_i-1}\sum_{t'=t+1}^{p_i}\left[\left(1-\overline{R}_{it}^{\text{cots}}\right)y_{it}\right]\cdot\left[\left(1-\overline{R}_{it'}^{\text{cots}}\right)y_{it'}\right]+\sum_{t=1}^{p_i}\left[\left(1-\overline{R}_{it}^{\text{cots}}\right)y_{it}\right]\cdot\left[\left(1-R_{<i}^{\text{ihd}}\right)v_i\right]$$
$$\left.+\sum_{t=1}^{p_i}\left[\left(1-\overline{R}_{it}^{\text{cots}}\right)y_{it}\right]\cdot\left[\left(1-R_{<i}^{\text{fab}}\right)w_i\right]+\left[\left(1-R_{<i}^{\text{ihd}}\right)v_i\right]\cdot\left[\left(1-R_{<i}^{\text{fab}}\right)w_i\right];\ i=1,\ldots,n\right\} \quad (37)$$

In order to solve this problem, fuzzy interactive approach was used [26]. The membership function for reliability and cost objective functions can be given as

$$\mu_{R(x)} = \begin{cases} 1, & \text{if } R(x) \geq R_u \\ \frac{R(x)-R_l}{R_u-R_l}, & \text{if } R_l < R(x) < R_u \\ 0, & \text{if } R(x) \leq R_l \end{cases} \quad \mu_{C(x)} = \begin{cases} 1, & \text{if } C(x) \leq C_l \\ \frac{C_u-C(x)}{C_u-C_l}, & \text{if } C_l < C(x) < C_u \\ 0, & \text{if } C(x) \geq C_u \end{cases}$$

where, R_l, C_l are the worst lower bounds and R_u, C_u are the best upper bounds of reliability and Cost objective resp. Following Bellman-Zadeh's maximization principle [18], the fuzzy biobjective optimization model is formulated as

Problem (P3)

$$\begin{aligned} &\text{Max } \alpha \\ &\text{subject to} \\ &\qquad \alpha \leq \mu_{R(x)} \\ &\qquad \alpha \leq \mu_{C(x)} \\ &\qquad 0 \leq \alpha \leq 1 \\ &\qquad X \in \overline{S} \end{aligned}$$

This problem is then solved following the steps of fuzzy interactive approach mentioned in [26]. The optimal value of α represents the best compromise solution between the two objectives. Hence the solution provides optimal mix of hardware and software components under reuse-build-or-buy scheme, giving trade-off between reliability and cost of the fault tolerant-embedded system.

4 Illustrative Example

An illustration was considered to formulate the optimization model for selection of hardware and software components under reuse-build-buy decision. Four subsystems-based embedded systems connected in series were considered. For development of each subsystem, one or more hardware components can be procured from vendor, while for software components a choice from COTS, in-house developed and/or reusable component exists. Table 1 depicts data set for cost, reliability, and delivery time of available hardware instances. Table 2 depicts data set for cost,

Table 1 Data for H/w components

SS	H/w instance	Cost	Rel	DT
1	1	11	0.95	9
	2	6	0.915	13
2	1	19	0.92	9
	2	18	0.94	12
	3	20	0.925	9
3	1	22	0.81	11
	2	15	0.82	20
	3	19	0.81	14
4	1	18	0.83	8

Table 2 Data COTS components

SS	COTS instance	Cost	SLOC	Rel	DT	ET
1	1	82	887	0.945	15	0.36
	2	74	866	0.89	12	0.32
	3	66	915	0.87	11	0.34
2	1	56	588	0.925	9	0.27
	2	65	620	0.94	9	0.25
3	1	48	410	0.91	7	0.71
	2	52	428	0.935	8	0.65
4	1	76	597	0.905	9	0.52
	2	61	615	0.92	12	0.53
	3	68	610	0.9	11	0.54

SLOC, reliability, execution time, and delivery time of COTS components. Table 3 presents data for unitary development (or fabrication) cost, estimated development time, execution time, and SLOC for built components and data for estimated fabrication time, execution time, and SLOC of fabricated reusable components.

For simplicity, it is assumed that in RB/1/1 architecture, $P^{rv}_i = 0.002\ \forall\ i$, $P_d = 0.002$, $P^{fs}_i = 0.003\ \forall\ i$, and the values of $\tau^{hd}{}_i$, $\tau^{fab}{}_i$ $\pi^{hd}{}_i$ and $\pi^{fab}{}_i$ are 0.05, 0.05, 0.002, and 0.002, respectively, for any subsystem. Limit on total SLOC of the system is set as $L = 4000$, while on the delivery time of system, limit is set as

Table 3 Dataset for in-house built and fabricated software components

Subsystem	Dev/fab cost	In-house developed components			Fabricated reusable components		
		Est. DT	SLOC	ET	Est. fab time	SLOC	ET
1	52	17	840	0.35	–	–	–
2	36	8	506	0.26	6	635	0.30
3	32	9	415	0.72	–	–	–
4	40	–	–	–	4	590	0.51

Table 4 Solution depicting reliability, cost and selected components

SS	z_1	Components selected	SS rel	Syst rel	Syst cost	A
1	0	$x_{11} = 1,\ y_{12} = 1$	0.845	0.768	872.2	0.79
2	1	$x_{23} = 2,\ y_{20} = 1,\ y_{21} = 1$	0.987			
3	1	$x_{32} = 2,\ y_{31} = 1,\ y_{32} = 1$	0.955			
4	1	$x_{41} = 2,\ y_{43} = 1,\ w_4 = 1$	0.964			

$D = 25$. Bounds set on execution time of ith subsystem from E_1 to E_4 are 0.4, 0.6, 1.4, and 1.1.

Solution: The problem is solved using LINGO. Upper and lower bounds for reliability and cost objectives were derived as $R_l = 0.387$, $R_u = 0.8644$, $C_l = 294$, $C_u = 3301$. Following solution is obtained (Table 4).

5 Conclusion

Biobjective optimization was formulated in this paper, with the objective of maximizing reliability simultaneously minimizing the cost under reuse-build-buy approach using RB/1/1 architecture. Considering the imprecision in parameters due to human judgments, a defuzzification approach was used to obtain crisp data.

Appendix 1: TFN for Reliability and DT of Hardware

SS	Hardware instance	Reliability			Delivery time		
		a_1	a_2	a_3	a_1	a_2	a_3
1	1	0.91	0.96	0.97	7	9	11
	2	0.9	0.915	0.93	11	12	17
2	1	0.895	0.915	0.955	7	8	13
	2	0.93	0.935	0.96	10	12	14
	3	0.88	0.93	0.96	8	9	10
3	1	0.79	0.8	0.85	10	11	12
	2	0.8	0.81	0.86	18	20	22
	3	0.79	0.81	0.83	12	13	18
4	1	0.81	0.835	0.84	6	7	8

Appendix 2: TFN for Reliability, Delivery Time and Execution Time of COTS Components

SS	COTS instance	Reliability			Delivery time			Execution time		
		a_1	a_2	a_3	a_1	a_2	a_3	a_1	a_2	a_3
1	1	0.92	0.95	0.96	13	14	19	0.33	0.36	0.38
	2	0.83	0.90	0.92	11	12	13	0.29	0.31	0.35
	3	0.83	0.87	0.90	9	11	13	0.31	0.34	0.36
2	1	0.89	0.93	0.94	7	9	11	0.26	0.27	0.28
	2	0.93	0.93	0.96	8	9	10	0.23	0.24	0.28
3	1	0.90	0.91	0.92	6	7	8	0.68	0.71	0.73
	2	0.91	0.94	0.95	5	7	13	0.63	0.65	0.67
4	1	0.89	0.90	0.93	7	8	13	0.49	0.52	0.53
	2	0.87	0.92	0.95	10	12	14	0.53	0.54	0.59
	3	0.81	0.93	0.91	10	11	12	0.51	0.54	0.56

Appendix 3: TFN for ET of Build Components and ET of Reusable Components

SS	Execution time of in-house developed components			Execution time of fabricated reusable components		
	a_1	a_2	a_3	a_1	a_2	a_3
1	0.31	0.355	0.38	–	–	–
2	0.225	0.27	0.275	0.275	0.295	0.335
3	0.69	0.725	0.74	–	–	–
4	–	–	–	0.48	0.515	0.53

References

1. Afonso, F., Silva, C., Brito, N., Montenegro, S., Tavares, A.: Aspect-Oriented Fault Tolerance for Real-Time Embedded Systems. ACM (2008).1-978-60558-142-2
2. Avizienis, A., Laprie, J.-C., Randell, B.: Fundamental concepts of dependability. Technical Report 739, Department of Computer Science, University of Newcastle upon Tyne (2001)
3. Narayanan, V., Xie, Y.: Reliability concerns in embedded system designs. Computer **39**(1), 118–120 (2006)
4. Kaur, R., Arora, S., Madan, S., Jha, P.C.: Bi-objective Optimization Model for Fault-Tolerant Embedded Systems under Build-or-Buy Strategy incorporating Recovery Block Scheme. World Scientific Publishing Co. (2015), Communicated

5. Pradhan, D.K.: Fault-Tolerant Computer System Design. Prentice-Hall, Inc. (1996)
6. Randell, B., Lee, P., Treleaven, P.C.: Reliability issues in computing system design. ACM Comput. Surv. **10**(2), 123–165 (1978)
7. Scott, R.K., Gault, J.W., McAllister, D.F.: Modeling fault tolerant software reliability. In: Proceedings of the Third Syrup. Reliability in Distributed Software and Database Systems, pp. 15–27 (1983)
8. Scott, R.K., Gault, J.W., McAllister, D.F., Wiggs, J.: Experimental validation of six fault-tolerant software reliability models. IEEE Fault Tolerant Comput. Syst. **14**, 02–107 (1984)
9. Scott, R.K., Gault, J.W., McAllister, D.F.: Fault tolerant software reliability modelling. IEEE Trans. Soft. Eng. **13**(5), 582–592 (1987)
10. Levitin, G.: Optimal structure of fault-tolerant software systems. Reliab. Eng. Syst. Saf. **89**(3), 286–295 (2005)
11. Laprie, J.-C., Arlat, J., Beounes, C., Kanoun, K.: Definition and analysis of hardware- and software-fault-tolerant architectures. IEEE Comput. 39–51 (1990)
12. Lyu, M.R.: Handbook of Software Reliability Engineering. IEEE Computer Society Press, Mc-Graw-Hill (1996)
13. Coit, D.W., Jin, T., Wattanapongsakorn, N.: System optimization with component reliability estimation uncertainty: a multi-criteria approach. IEEE Trans. Reliab. **53**(3), 369–380 (2004)
14. Tavakkoli-Moghaddam, R., Safari, J., Sassani, F.: Reliability optimization of series-parallel systems with a choice of redundancy strategies using a genetic algorithm. Reliab. Eng. Syst. Saf. **93**, 550–556 (2008)
15. Wattanapongsakorn, N., Levitan, S.: Reliability optimization models for fault-tolerant distributed systems. In: Proceedings of Annual Reliability & Maintainability Symposium, IEEE, pp. 193–199 (2001)
16. Wattanapongsakorn, N., Levitan, S.P.: Reliability optimization models for embedded systems with multiple applications. IEEE Trans. Reliab. **53**(3), 406–416 (2004)
17. Wattanapongskorn, N., Coit, D.W.: Fault-tolerant embedded system design and optimization considering reliability estimation uncertainty. Reliab. Eng. Syst. Saf. **92**(4), 395–407 (2007)
18. Bellman, R.E., Zadeh, L.A.: Decision making in a fuzzy environment. Manag. Sci. **17**, 141–164 (1970)
19. Kwong, C.K., Mu, L., Tang, J.F., Luo, X.G.: Optimization of software components selection for component-based software system development. Comput. Ind. Eng. **58**(4), 618–624 (2010)
20. Bryce, M., Bryce, T.: Make or Buy Software? J. Syst. Manag. **38**(8), 6–11 (1987)
21. Haines, C., Carney, D., Foreman, J.: Component-Based Software Development/ COTS Integration. http://www.sei.cmu.edu/str/descriptions/cbsd_body.html
22. Ivica, C.: Component-based software engineering for embedded systems. In: Proceedings of the 27th International Conference on Software Engineering, ACM, pp. 712–713 (2005)
23. Wu, Z.Q., Tang, J.F., Kwong, C.K.: Chan. C.Y.: A model and its algorithm for software reuse optimization problem with simultaneous reliability and cost consideration. Int. J. Innovative Comput. Inf. Control **7**(5), 2611–2622 (2011)
24. Sametinger, J.: Software Engineering with Reusable Components. Springer Science & Business Media (1997)
25. Kaur, R., Arora, S., Jha, P.C., Madan, S.: Fuzzy multi-criteria approach for modular software system incorporating reuse-build-buy decision under recovery block scheme., In: Proceedings of 2nd International Conference on Computing for Sustainable Global Development, pp. 655–660 (2015)
26. Gupta, P., Mehlawat, M.K., Verma, S.: COTS selection using fuzzy interactive approach. Optim. Lett. **6**, 273–289 (2012). doi:10.1007/s11590-010-0243-5

Hybrid Multi-criteria Decision-Making Approach for Supplier Evaluation in Fuzzy Environment

Nidhi Bhayana, Kanika Gandhi and P.C. Jha

Abstract The competitive business environment has enforced companies to revisit their strategies of purchasing and evaluation of suppliers. The supplier evaluation is considered multi decisive factor problem because it includes intangible and tangible attributes. The current paper introduces an approach for optimization of cost and carbon emission by evaluating the best performing suppliers, incorporating multiple techniques, viz., factor analysis, technique for order preference by similarity to ideal solution (TOPSIS), and fuzzy analytic hierarchy process (FAHP). A questionnaire is built and 160 manufacturers are surveyed. Factor analysis is then performed to identify possible factors considering different variables (criteria) chosen from literature. These factors are considered as criteria in FAHP where triangular fuzzy numbers are used to compute criteria weights followed by TOPSIS to get alternative weights. Then these alternative weights are used in bi-objective optimization model to minimize the cost incurred during procurement, transportation and holding, and minimizing the carbon emissions while transporting products to the direct retailers. For solution process, fuzzy goal programming approach has been used. Model is validated on a case data set.

Keywords Supplier evaluation · Multi-criteria decision-making (MCDM) · Factor analysis · Fuzzy set theory · AHP · TOPSIS · Carbon emission

Nidhi Bhayana (✉) · P.C. Jha
Department of Operational Research, University of Delhi, Delhi, India
e-mail: bhayananidhi@yahoo.com

P.C. Jha
e-mail: jhapc@yahoo.com

Kanika Gandhi
Bhavan's Usha & Lakshmi Mittal Institute of Management,
K.G. Marg, Delhi, India
e-mail: gandhi.kanika@gmail.com

M. Pant et al. (eds.), *Proceedings of Fifth International Conference on Soft Computing for Problem Solving*, Advances in Intelligent Systems and Computing 437, DOI 10.1007/978-981-10-0451-3_86

1 Introduction

Supply chain (SC) is defined as an integration of system consists of entities like manufacturers, suppliers, customers, storehouses, and channels of distribution. The mentioned entities are systematized to procure material, transform them into final shape of a product, locating the depots for distribution and selecting appropriate conduit to transport. An optimally designed SC should reveal the "best" configuration and operation of all these partners. Moreover, uncertainties in the supply chain would lead to complexities, such as supply disruption, global price changes of commodity goods, etc. Thus, there is a need to optimally formulate and coordinate all activities of the supply chain partners to achieve seamless function for large-scale and complex supply chains under uncertainty. One of the important concerns in SC is supplier management that enables organization to remain competitive. Thus, competitive advantage is gained by the organization by choosing right suppliers to provide products or services. As a result, in supply chain field, the concerns of supplier selection have captivated much attention. Supplier selection decision processes are considered complicated decisions and are based on multi-criteria that makes it a MCDM problem. The multi-criteria in the process should be reduced and grouped to understand the variable explained. Thus, factor analysis is used to reduce criteria/data in many variables, where association among interconnected variables are studied and formed them in terms of small number of essential factors. In general, factor analysis methodology is utilized to form some important factors, and to elucidate the associations among the variables and further identifies a fresh set of unassociated variables to reinstate the original set in subsequent multivariate analysis [11]. Wong [17] shows the association between SC integration and operational performance and the effects of uncertainties of environment on the association. Cao [3] explains the SC partnership and investigates the effect on performance of the company. Merschmann [12] shows the association between uncertainties in environment, SC flexibility, and performance of the company during a survey of manufacturing firms.

To identify the best performing suppliers, the evaluation by each expert is to be integrated and identifying relevant factors. Many researchers have proposed various MCDM techniques to provide an effective solution to supplier selection problem: Data envelopment analysis (DEA), AHP, analytic network process (ANP), artificial neural network (ANN), etc. Among these methods, Saaty [15] first developed AHP which helps in constructing a complex multi-attribute model, and presents a goal to make a decision from a set of strategies [5]. In spite of its high fame and simplicity, AHP is criticized for its lack of ability to effectively manage the intrinsic uncertainties and indistinctness related with the perception of decision-maker to crisp values. To deal with this, one of the most frequently used methods is Chang's Extent Analysis [4]. Next Hwang and Yoon [8] develops an approach named TOPSIS, for the final ranking of alternatives, which selects alternatives with simultaneous shortest distant from the positive ideal solution and the farthest distant from the negative ideal solution. Gumus [7] uses AHP and TOPSIS methods to evaluate harmful waste transportation firms.

Supply chain optimization under uncertainty involves the incorporation of decision-making processes to manage flow of goods, information and service from the source to the destinations. Nowadays, buyers are not only concerned with the selection of best performing supplier but they are also take into consideration the optimum order and type of vehicle to transport in order to be economical and environmentally conscious. Esfandiari [6] presents a multi-objective programming problem to minimize cost of purchase, unit rejection, and late deliveries and maximizes the total scores from the supplier evaluation process. Narasimhan [14] develops a multi-objective programming model to solve supplier selection problem keeping five criteria of performance. Wadhwa [16] proposes a multi-objective programming model to evaluate and select supplier with minimization functions: price, lead time, and rejections.

The current paper proposes a bi-objective optimization model that elucidates the integration of procurement and distribution to coordinate among many suppliers and retailers. According to various criteria (variables) that exist in the literature, finding the best performing supplier among several alternatives is a key problem in the supply chain. To identify possible factors considering different variables, factor analysis is performed. Subsequently, these factors are considered as criteria in FAHP where we use scale of triangular fuzzy numbers, provided by decision-maker, to get weights of criteria in pairwise comparison matrices, followed by TOPSIS to get alternative weights. The closeness coefficient are used as an input value in the bi-objective optimization model to minimize procurement, transportation and inventory cost keeping right amount of inventory and minimizing the carbon emissions when transporting products to the direct retailers.

2 Problem Statement

Over the past many years, there has been pressure on industry to reduce greenhouse gas emissions called “carbon emission” resulting from products or activities and to identify strategies to reduce their climate change. The electronics manufacturing industry is no exception. Therefore, minimizing carbon emission which is an environmental concern is being discussed by many organizations. Same concern is being discussed in this paper. A microwave oven manufacturing company shows concern to reduce carbon emission during its supply chain. The company is interested in procurement—distribution of two types of models because of high demand. As an illustrative case data set, the models are supplied by three suppliers to ten retail stores. Their concern is to optimize the cost incurred during inventory carrying, procurement and transportation cost and minimizing carbon emission. Company’s first concern is to evaluate and select suppliers on the basis of multi-criteria (variables). From the literature and discussion with company, 30 criteria have been extracted. Since some variables are interrelated with each other, we use factor analysis to reduce those 30 variables into factors. Five factors are extracted from these criteria. Afterwards, we have applied fuzzy AHP and TOPSIS to evaluate suppliers. Their next concern is about procurement and distribution process. Company transports the demanded quantity to each retail store which costs them higher because of transporting less quantity in a bigger truck which also leads to higher carbon emission. For this issue, model is developed to minimize the procurement, transportation, and inventory costs with carbon emission of transporting the products to direct retailers. Moreover, they want to know how much amount of inventory is needed and how much to procure to satisfy the demand. In this section, the methodology demonstration is done to exemplify the evaluation of suppliers.

2.1 Factor Analysis Process [11]

Based on the following steps given below, various variables are converted into different factors using factor analysis:

Step 1 *Formulate the problem and Collect the data from respondents by interviewing using a scale (here 7-point scale is used) for all the variables*: Here, our objective is to find various factors out of 30 qualitative variables. For this, a sample of 160 manufacturers was interviewed through questionnaire. They have been asked to indicate the degree of agreement with the following statements using a 7-point scale, (1-Strongly disagree, 2-moderately disagree, 3-disagree, 4-neither disagree nor agree, 5-agree, 6-moderately agree, 7-strongly agree). The factors and attributes are mentioned in Table 1.

Step 2 *Construction of Correlation matrix*: First, a correlation matrix is constructed using IBM SPSS 20.0 in which variables must be correlated if the factor analysis is appropriate. Bartlett's test of sphericity [1], tests the null hypotheses that the variables are uncorrelated in the population. High value of test favors the rejection of null hypothesis. Kaiser-Meyer-Olkin (KMO) [10] checks the adequacy of sampling process. The desirable value of KMO is 0.5.

The null hypothesis, that the population correlation matrix is an identity matrix, is rejected by the Bartlett's test of sphericity and the value of the KMO statistic is 0.662 shown in Table 2.

Step 3 *Principal Component Analysis*: The method considers total variance in the data. The variance value of each factor tells us the priority of each factor. Priority of first factor is higher than the other four factors. Similarly, second factor is more important than the third factor and so on. Hence, these priorities are used in the pairwise comparison of all the criteria (Due to page limitation, variance table is not shown).

Step 4 *Rotate Factors*: Afterwards, factors can be rotated to get the factor loadings values in which correlations between factors and variables are represented.

Table 1 Factors and their attributes (qualitative variables)

Management	Quality management system (QMS)	Environmental management system (EMS)	Information system (IS)	Innovation capability (IC)	Honesty (H)	Procedural compliment (PC)	Technical problem solving (TPS)	Infrastructure (I)	Service performance (SP)
Cost	Product price (PP)	Logistics cost (LC)	Value added cost (VAC)	Discount for bulk order (DBO)	Discount for early payment (DEP)	Ordering cost (OC)			
Delivery	Order lead time (OLT)	On time delivery (OTM)	Trade restrictions (TR)	Reliable delivery method (RDM)	Error free product type and quantity (EFPQ)				
Flexibility	Inventory availability (IA)	Customization (C)	Negotiability (N)	Variety flexibility (VF)	Service flexibility (SF)				
Quality	Product reliability (PR)	ISO certifications (ISO)	Warranty (W)	Return rate (RR)	Long durability (LD)				

Table 2 KMO and Bartlett's test

Kaiser-Meyer-Olkin measure of sampling Adequacy		0.662
Bartlett's test of sphericity	Approx. chi-square	7267.441
	Df	435
	Significant	0.000

Large absolute value of a coefficient signifies the close relationship between variable and factor. Using factor loadings, number of factors has been determined. (Due to page limitation, rotated component matrix is not shown).

Step 5 *Interpretation*: The factor can be interpreted by recognizing that variable which has high loadings under same factor. After getting rotated factor matrix (not shown here), variables with high coefficients for factor 1 are QMS, EMS, IS, IC, H, PC, TPS, I, SP, thus this factor is labeled as Management. Factor 2 is highly related with PP, LC, VAC, DBO, DEP, OC and is named as Cost. Variables OLT, OTM, TR, RDM, and EFPQ correlate with factor 3 after rotation, thus called as Delivery criteria. There are relatively high correlations among variables IA, C, N, VF and SF, so they are expected to correlate with the same factor which is being labeled as Flexibility. The remaining variables PR, ISO, W, RR and LD, correlate highly with factor 5 and hence are labeled as Quality. But negative coefficient for QMS variable leads to positive elucidation that QMS is important.

2.2 *Fuzzy Analytical Hierarchy Process [4]*

Factors which are determined using factor analysis will be considered here as criteria. Then fuzzy AHP (FAHP) is applied to get the weights of attributes. FAHP methodology is designed for managerial problems by integrating fuzzy set theory and analytic hierarchy process (AHP). Certain characteristics of fuzzy methodology and AHP empower the decision-maker to incorporate both their knowledge, which is mainly qualitative, and quantitative information into the decision model [9]. Based on factor analysis done above, five factors identified are used as qualitative criteria in FAHP problem: Management (F_1), Cost (F_2), Delivery (F_3), Flexibility (F_4), and Quality (F_5). As it is impossible to do mathematical operations directly on linguistic standards, thus linguistic scale must be transformed into fuzzy scale. In this paper, we are considering Likert scale in triangular fuzzy scale. The triangular fuzzy scale is given in Table 3 which will be used for the assessment model.

In current paper, Chang's extent analysis [4] method is used. After getting pairwise comparisons between criteria and alternatives, we apply Chang's extent

Table 3 Triangular fuzzy scale

Linguistic scale	Fuzzy scale	Fuzzy reciprocal scale
Equal	(1,1,1)	(1,1,1)
Important	(0.8,2,2.99)	(1/2.99,1/2,1/0.8)
More important	(1.8,3,3.99)	(1/3.99,1/3,1/1.8)
Very important	(2.8,4,4.99)	(1/4.99,1/4,1/2.8)

analysis. Using pairwise comparison of all attributes with respect to main objective, a fuzzy comparison matrix of attributes is formed as shown in Table 4.

The stepwise explanation of the Chang's extent analysis [4] method for solving FAHP is given as

Let $X = (x_1, x_2 \ldots x_n)$ be an object set, and $U = (u_1, u_2, \ldots, u_m)$ be a goal set. Using extent analysis method, each object is considered and Chang's extent analysis for each objective, g_i, is performed, respectively. Thus, m extent analysis values for each object can be attained, with the following notations:

$$M_{g_i}^1, M_{g_i}^2, \ldots, M_{g_i}^m,\ i = 1,\ 2, \ldots, n \tag{1}$$

where $M_{g_i}^j\,(j = 1,\ 2\ldots m)$ are triangular fuzzy numbers in which l, m, and u are the parameters which represents least possible value, most possible value, and u largest possible value, respectively. A triangular fuzzy number is represented as (l, m, u).

Step 1: Fuzzy synthetic extent value with respect to the ith object is defined as

$$\mathrm{SE}_i = \sum_{j=1}^{m} M_{g_i}^j \otimes \left[\sum_{i=i}^{n}\sum_{j=1}^{m} M_{g_i}^j\right]^{-1} \tag{2}$$

The fuzzy synthetic extent values for five attributes are represented by SE_1, SE_2, SE_3, SE_4, and SE_5, respectively. Using Eq. (2), we get
$\mathrm{SE}_1 = (0.16, 0.37, 0.78)$; $\mathrm{SE}_2 = (0.1, 0.24, 0.59)$; $\mathrm{SE}_3 = (0.08, 0.19, 0.45)$; $\mathrm{SE}_4 = (0.05, 0.12, 0.32)$; $\mathrm{SE}_5 = (0.04, 0.07, 0.17)$;
Step 2: The possibility degree of $M_1 = (l_1, m_1, u_1) \geq M_2 = (l_2, m_2, u_2)$ defined as

$$V(M_1 \geq M_2) = \sup_{x \geq y}\left[\min(\mu_{M_1}(x), \mu_{M_2}(y))\right] \tag{3}$$

For a pair (x, y) *with* $x \geq y$ and $\mu_{M_1}(x) = \mu_{M_2}(y) = 1$, we have $V(M_1 \geq M_2) = 1$. For convex fuzzy numbers M_1 and M_2, $V(M_1 \geq M_2) = 1$ iff $m_1 \geq m_2$;

$$V(M_2 \geq M_2) = \mathrm{hgt}(M_1 \cap M_1 = \mu_{M_1}(d)) = (l_1 - u_2)/(m_2 - u_2 - m_1 + l_1) \tag{4}$$

where d is the ordinate of the highest intersection point D between μ_{M_1} and μ_{M_2}. We need both the values of $V(M_1 \geq M_2)$ and $V(M_2 \geq M_1)$ to compare M_1 and M_2.

Table 4 Fuzzy comparison matrix of criteria with respect to the overall goal

Obj	F_1	F_2	F_3	F_4	F_5	Weight
F_1	(1,1,1)	(0.8,2,2.99)	(1.8,3,3.99)	(1.8,3,3.99)	(2.8,4,4.99)	0.36
F_2	(1/2.99,1/2,1/.8)	(1,1,1)	(0.8,2,2.99)	(0.8,2,2.99)	(1.8,3,3.99)	0.27
F_3	(1/3.99,1/3,1/1.8)	(1/2.99,1/2,1/.8)	(1,1,1)	(0.8,2,2.99)	(1.8,3,3.99)	0.22
F_4	(1/3.99,1/3,1/1.8)	(1/2.99,1/2,1/.8)	(1/2.99,1/2,1/.8)	(1,1,1)	(0.8,2,2.99)	0.14
F_5	(1/4.99,1/4,1/2.8)	(1/3.99,1/3,1/1.8)	(1/3.99,1/3,1/1.8)	(1/2.99,1/2,1/.8)	(1,1,1)	0.01

The possibility degree of SE_i over SE_j $(i \neq j)$ is calculated by Eq. (4).

$$V(SE_1 \geq SE_2) = 1, V(SE_1 \geq SE_3) = 1, V(SE_1 \geq SE_4) = 1, V(SE_1 \geq SE_5) = 1, V(SE_2 \geq SE_1) = 0.75$$
$$V(SE_2 \geq SE_3) = 1, V(SE_2 \geq SE_4) = 1, V(SE_2 \geq SE_5) = 1, V(SE_3 \geq SE_1) = 0.62, V(SE_3 \geq SE_2) = 0.88$$
$$V(SE_3 \geq SE_4) = 1, V(SE_3 \geq SE_5) = 1, V(SE_4 \geq SE_1) = 0.39, V(SE_4 \geq SE_2) = 0.66, V(SE_4 \geq SE_3) = 0.77$$
$$V(SE_4 \geq SE_5) = 1, V(SE_5 \geq SE_1) = 0.017, V(SE_5 \geq SE_2) = 0.30, V(SE_5 \geq SE_3) = 0.41, V(SE_5 \geq SE_4) = 0.68$$

If value of $(l_1 - u_2) > 0$, normalized the matrix elements and then follow the same process to get attribute weight vector.
Step 3: The possibility degree for a convex fuzzy number to be larger than k convex fuzzy numbers M_i $(i = 1, 2, \ldots, k)$ can be defined by

$$V(M \geq M_1, M_2, \ldots, M_k) = V[(M \geq M_1) \text{ and } (M \geq M_2) \text{ and} \ldots \text{and } (M \geq M_k)] = \min V(M \geq M_i) \text{ where } i = 1,\ 2,\ \ldots,\ k \quad (5)$$

$$\text{Assume that } d'(F_1) = \min V(SE_i \geq\ SE_k) \text{ for } k = 1,\ 2, \ldots,\ n;\ k \neq i. \quad (6)$$

$$\text{Then the weight vector is given by } W = \left(d'(F_1), d'(F_2), \ldots, d'(F_l)\right)^{T} \quad (7)$$

where, F_l $(l = 1, 2, \ldots, L)$ are L elements.
The minimum possibility degree of the superiority of each attribute over another is decided with the help of Eq. (6), and is given as
$d'(F_1) = \min V(SE_1 \geq SE_2, SE_3, SE_4, SE_5) = \min (1, 1, 1, 1) = 1$
Thus, weight vector are obtained as $W' = (1, 0.75, 0.62, 0.39, 0.02)^T$
Step 4: Using normalization, the weight vectors in normalized form are

$$W = (d(F_1), d(F_2), \ldots, d(F_l))^{T}, \text{ where } W \text{ is a non-fuzzy number.} \quad (8)$$

Final weights of the decision attributes with respect to the main goal is determined using normalization which are given as $W = (0.36, 0.27, 0.22, 0.14, 0.01)^T$. Fuzzy comparison matrices of decision alternatives and corresponding weight vectors of each alternative with respect to attributes F_1, F_2, F_3, F_4, and F_5 have been discussed. Comparison matrix for F_1 is given in Table 5 and remaining matrices are same.
Step 5: After getting weights of all criteria and alternatives, priority weights (closeness coefficient) of alternatives is obtained using TOPSIS discuss in Sect. 2.3.

Table 5 Fuzzy comparison matrix of decision alternatives with respect to attribute F_1

F_1	S_1	S_2	S_3	Weight
S_1	(1,1,1)	(1/2.99,1/2,1/.8)	(0.8,2,2.99)	0.33
S_2	(0.8,2,2.99)	(1,1,1)	(0.8,2,2.99)	0.46
S_3	(1/2.99,1/2,1/.8)	(1/2.99,1/2,1/.8)	(1,1,1)	0.20

Table 6 Summary combination of weights of all the criteria and alternatives

Criteria	F_1	F_2	F_3	F_4	F_5
Weights	0.3602	0.2717	0.2220	0.1401	0.0060
S_1	0.33	0.25	0.19	0.14	0.05
S_2	0.46	0.25	0.39	0.46	0.37
S_3	0.20	0.50	0.42	0.41	0.58

Table 7 Normalized decision matrix

Criteria	F_1	F_2	F_3	F_4	F_5
Weights	0.3602	0.2717	0.2220	0.1401	0.0060
S_1	0.55028	0.41	0.31	0.22	0.07
S_2	0.76427	0.41	0.65	0.73	0.53
S_3	0.33628	0.81	0.69	0.65	0.84

2.3 *Technique for Order Preference by Similarity to Ideal Solution*

Step 1: A decision matrix of each alternative with respect to each attribute is established. The decision matrix we obtained using AHP calculations is as follows (Table 6):

Step 2: The decision matrix $[H_{sl}]$ in normalized form is calculated as follows (Table 7):

$$H_{sl} = O_{sl} \Big/ \sqrt{\sum_{l=1}^{L} O_{sl}^2}, \quad s = 1, 2, \ldots, S; \; l = 1, 2, \ldots, L \tag{9}$$

For criteria F_1, the normalized values can be obtained using Eq. (9) is as follows:

$$\sqrt{\sum_{l=1}^{L} O_{sl}^2} = \sqrt{(0.33)^2 + (0.46)^2 + (0.20)^2} = \sqrt{0.3669} = 0.6058; \; H_{11} = \frac{0.33}{0.6058} = 0.55028$$

Step 3: The weighted normalized decision matrix is calculated by multiplying the normalized decision matrix by its associated weights. The weighted normalized value M_{ij} is calculated using the formula (Table 8):

Table 8 Weighted normalized decision matrix

	F_1	F_2	F_3	F_4	F_5
S_1	0.1982	0.1121	0.0680	0.0304	0.0004
S_2	0.2753	0.1121	0.1448	0.1020	0.0032
S_3	0.1211	0.2206	0.1540	0.0910	0.0050

Table 9 PIS and NIS of all criteria

	F_1	F_2	F_3	F_4	F_5
S_1	0.1982	0.1121	0.0680	0.0304	0.0004
S_2	0.2753	0.1121	0.1448	0.1020	0.0032
S_3	0.1211	0.2206	0.1540	0.0910	0.0050
N^+	0.2753	0.1121	0.1540	0.1020	0.0050
N^-	0.1211	0.2206	0.0680	0.0304	0.0004

$$N_{sl} = W_1 H_{sl},\ s = 1,2,\ldots,S;\ l = 1,2,\ldots,L \tag{10}$$

where, W_l is weight of the lth criterion.

For criteria F_1 and alternative S_1, we have $N_{11} = W_1{*}H_{11} = 0.3602{*}0.5503 = 0.1982$

Step 4: Determine the positive ideal solution (PIS) and negative ideal solution (NIS) respectively as

$$\begin{aligned} N^+ &= \{n_1^+, n_2^+, \ldots, n_l^+\} = \{(\text{Max}\, N_{sl}/l \in L), (\text{Min}\, N_{sl}/l \in L')\} \text{ and } N^- \\ &= \{n_1^-, n_2^-, \ldots, n_l^-\} = \{(\text{Min}\, N_{sl}/l \in L), (\text{Max}\, N_{sl}/l \in L')\} \end{aligned}$$

where L is related with positive criteria and L' is related with the negative criteria. Since Management (F_1) is a benefit criteria so we find N^+ and N^- as: N^+= max {0.1982, 0.2753, 0.1211} = 0.2753 and N^- = min {0.1982, 0.2753, 0.1211} = 0.1211. PIS and NIS are given in Table 9.

Step 5: The separation measures using the l-dimensional Euclidean distance is calculated. The separation measure E_s^+ of each alternative is given as

$$E_s^+ = \sqrt{\sum_{l=1}^{L} (N_{sl} - n_l^+)^2},\ s = 1,2,\ldots,S \tag{11}$$

Similarly, the separation measure E_s^- of each alternative is given as

$$E_s^- = \sqrt{\sum_{l=1}^{L} (N_{sl} - n_l^-)^2},\ s = 1,2,\ldots,S \tag{12}$$

Table 10 Separation measure of all alternatives

	Distance (E_s^+)	Distance (E_s^-)
S_1	0.1360	0.1331
S_2	0.0094	0.2158
S_3	0.1889	0.1053

Table 11 Closeness coefficient of all alternatives

Alternatives	S_1	S_2	S_3
CC_s	0.4947	0.9581	0.358

The separation measure for each alternative is calculated by above two equations mentioned in Table 10.

Step 6: Closeness coefficient (CC_s) for each alternative is calculated and rank them in descending order. CC_s can be expressed as

$$CC_s = (E_i^-)/(E_i^- + E_i^+) \tag{13}$$

where, the index value of CC_s lies between 0 and 1. The alternative's better performance depends upon the index value, i.e., higher the index value, higher is the performance. Closeness coefficient of all alternatives is given in Table 11.

Step 7: After getting closeness coefficients of each alternative, mathematical model has been applied in the next section to minimize cost and carbon emission.

2.4 Mathematical Model

In this section, we have developed the mathematical model to minimize inventory carrying cost, procurement cost, transportation cost with carbon emission by tactically connecting suppliers to the retail stores.

2.4.1 Assumptions, Sets, Parameters and Decision Variables

Assumptions: Initial inventory at all the retail stores is zero; shortages are not allowed; each supplier can supply both the models; supply is instantaneous

Sets: p = Product, $p = 1,\ldots,P$; s = supplier, $s = 1, \ldots, S$; r = retail store, $r = 1, \ldots, R$; t = time period, $t = 1, \ldots, T$; c = carbon emission break, $c = 1, \ldots, C$.

Parameters: H_{prt}: per unit inventory holding cost of pth product at rth retail store in time period t; P_{psrt}: Procurement cost of pth product supplied by sth supplier at rth retail store in time period t; TC_{psrt}: transportation cost per weight of pth product transporting products from sth supplier to rth retail store in time period t; Cap_{prt}: capacity of the pth product at the rth retail store during time period t; D_{prt}: demand of the pth product at the rth retail store during time period t; CO_{srct}: carbon emission value of cth vehicle load level from sth supplier to rth retail store in time period t; A_{srct}: load threshold beyond which a particular vehicle type is used; DG_s: closeness coefficient scores assigned to sth supplier; Sw: acceptable supplier weight given by company; IN: initial inventory at the rth retail store; w_p: weight of pth product; CapS_{pst}: capacity of the pth product at the sth supplier in time period t, CapT_{srt}:

capacity of the truck delivering products from sth supplier to rth retail store in time period t, MSL_s: minimum stock level of sth supplier; $\tilde{C}$ is fuzzy total cost; C_o and C^* are the aspiration and tolerance level of fuzzy total cost; $\widetilde{\mathrm{CB}}$ is fuzzy total carbon emission; CB_0 and CB^* are the aspiration and tolerance level of fuzzy carbon emission.

Decision variables: I_{prt}: ending inentory of the pth product at rth retail store in time period t; X_{psrt}: quantity flowing of pth product from sth supplier to rth retail store in time period t; V_{st}: 1, if sth supplier is selected for delivering pth product in time period t or 0 otherwise; J_{srct}: 1, if cth vehicle as per load gets activated for transporting weighted quantity from sth supplier to rth retail store in time period t or 0 otherwise; L_{srt}: weighted quantity flowing of all products from sth supplier to rth retail store in time period t.

Complexities in the relationship among different factors imply a significant degree of uncertainty in supply chain planning decisions. In the process, the relationship effectiveness is influenced by uncertainties and supply chains coordination affecting its performance considerably. Thus, fuzzy optimization model for vague aspiration levels is formulated, decided by decision-maker on the basis of knowledge and past experience on fuzzy total cost and fuzzy carbon emission. First objective function includes the cost of holding inventory at retail store, cost of procurement, and transportation cost.

$$\text{Min } \tilde{C} = \sum_{p=1}^{P}\sum_{r=1}^{R}\sum_{t=1}^{T} H_{\mathrm{prt}} I_{\mathrm{prt}} + \sum_{p=1}^{P}\sum_{s=1}^{S}\sum_{r=1}^{R}\sum_{t=1}^{T} P_{\mathrm{psrt}} X_{\mathrm{psrt}} V_{st} + \sum_{p=1}^{P}\sum_{s=1}^{S}\sum_{r=1}^{R}\sum_{c=1}^{C}\sum_{t=1}^{T} \mathrm{TC}_{\mathrm{psrt}} L_{\mathrm{srt}} V_{\mathrm{st}} J_{\mathrm{srct}} \tag{14}$$

Second objective function, on the basis of vehicle load and carbon emission value, minimizes the total carbon emission.

$$\text{Min } \widetilde{\mathrm{CB}} = \sum_{s=1}^{S}\sum_{r=1}^{R}\sum_{c=1}^{C}\sum_{t=1}^{T} \mathrm{CO}_{\mathrm{srct}} J_{\mathrm{srct}} \tag{15}$$

The three equations given below are inventory balancing equations at retail stores with minimum stock level of inventory which are complemented by the capacity constraint of supplier and at retail store:

$$I_{\mathrm{prt}} = IN + \sum_{s=1}^{S} X_{\mathrm{psrt}} V_{\mathrm{st}} - D_{\mathrm{prt}} \quad \forall p, r, t = 1; \tag{16}$$

$$I_{\mathrm{prt}} = I_{\mathrm{prt}-1} + \sum_{s=1}^{S} X_{\mathrm{psrt}} V_{\mathrm{st}} - D_{\mathrm{prt}} \quad \forall p, r, t > 1 \tag{17}$$

$$\sum_{t=1}^{T} I_{\text{prt}} + \sum_{s=1}^{S}\sum_{t=1}^{T} X_{\text{psrt}} V_{\text{st}} \geq \sum_{r=1}^{R}\sum_{t=1}^{T} D_{\text{prt}} \quad \forall p, r; \tag{18}$$

$$I_{\text{prt}} \geq \text{MSL}_s * V_{\text{st}} \, \forall p, r, t, s \tag{19}$$

$$I_{\text{prt}} + \sum_{s=1}^{S} X_{\text{psrt}} V_{\text{st}} \leq \text{Cap}_{\text{prt}} \, \forall p, r, t; \tag{20}$$

$$\sum_{p=1}^{P} X_{\text{psrt}} V_{\text{st}} \leq \sum_{p=1}^{P} \text{CapS}_{\text{pst}} \quad \forall s, r, t \tag{21}$$

Next equation is a connector between procurement and distribution followed by an equation for determining number of trucks required to transport products from supplier to retail stores.

$$L_{\text{srt}} = \sum_{p=1}^{P} w_p X_{\text{psrt}} \quad \forall s, r, t; \tag{22}$$

$$\begin{aligned} Z_{\text{srt}} &= L_{\text{srt}} / \text{CapT}_{\text{srt}} \, \forall s, r, t; \\ V_{\text{st}}, J_{\text{srct}} &= 1 \text{ or } 0; X_{\text{psrt}}, I_{\text{prt}} > 0 \end{aligned} \tag{23}$$

The mathematical model given above is a nonlinear fuzzy bi-objective with linear constraints which cannot be solved by simple mathematical programming model. Hence, fuzzy goal programming method is employed with fixed priorities for objectives.

2.4.2 Fuzzy Solution Algorithm [18]

Step 1. Using ranking technique as defuzzification function, crisp corresponding to the fuzzy parameters is calculated as $R(A) = (a_l + 2a_m + a_u)/4$ where a_l, a_m, a_u are the triangular fuzzy numbers (TFN).

Step 2. Since business is extremely unpredictable and demands of the customers changes in every small period, an accurate assessment of cost and carbon emission aspirations is a major area to discuss about. Hence, incorporating tolerance and aspiration level with the main goal is a better way to come out of such situation. So the model discussed in Sect. 2.4 can be rewritten as

$$\text{Find } X : \;\; C(X) \widetilde{\leq} C_0; \;\; \text{CB}(X) \widetilde{\leq} \text{CB}_0; \;\; \text{AX} \leq B, \, X \in S, X \geq 0$$

Step 3. Define membership functions for each of the fuzzy objective function.

$$\mu_C(X) = \begin{cases} 1; & C(X) \leq C_0 \\ \frac{C* - C(X)}{C* - C_0}; & C_0 \leq C(X) \leq C*; \\ 0; & C(X) \geq C* \end{cases}$$

$$\mu_{CB}(X) = \begin{cases} 1; & \mathrm{CB}(X) \leq \mathrm{CB}_0 \\ \frac{\mathrm{CB}_* - \mathrm{CB}(X)}{\mathrm{CB}_* - \mathrm{CB}_0}; & \mathrm{CB}_0 \leq \mathrm{CB}(X) \leq \mathrm{CB}* \\ 0; & \mathrm{CB}(X) \geq \mathrm{CB}* \end{cases}$$

Step 4. Extension principles are employed for the identification of the fuzzy decision, which results in a crisp mathematical programming problem given by

$$\text{Max } \alpha \text{ s.t. } w_1\alpha \leq \mu_C(X);\ w_2\alpha \leq \mu_{CB}(X);\ \alpha \in [0,1],\ X \geq 0,\ w_1, w_2 \geq 0, w_1 + w_2 = 1$$

where, α corresponds to the degree to which the decision-maker's aspiration is satisfied. Using the standard mathematical programming algorithms, above problem can be easily solved.

Step 5. While solving the problem using steps 1–4 given above, the objective of the problem is also treated as a constraint [2]. For the decision-maker, each constraint is considered to be an objective and the mathematical programming problem is a fuzzy bi-objective mathematical model. Moreover, each objective can have a different importance level and accordingly weights can be allocated to measure relative importance. Weighted min–max approach is used to solve the resulting problem. On substituting values of $\mu_C(X)$ and $\mu_{CB}(X)$ for the problem

$$\begin{aligned} &\text{Max } \alpha \\ &\text{s.t } C(X) \leq C_0 + (1 - w_1\alpha)(C* - C_0);\ \mathrm{CB}(X) \leq \mathrm{CB}_0 + (1 - w_2\alpha)(\mathrm{CB}* - \mathrm{CB}_0) \\ &\alpha \geq 0,\ X \geq 0, \alpha \in [0,1];\ w_1, w_2 \geq 0, w_1 + w_2 = 1 \end{aligned}$$

Step 6. If a feasible solution is not found in Step 5, then fuzzy goal programming approach is used to attain a compromised solution as given by [13].

2.4.3 Fuzzy Goal Programming Approach [13]

After solving the above problem, we found that the problem (P1) has no feasible results, i.e., the target cannot be attained for a feasible value of $\alpha \in [0, 1]$. Thus, fuzzy goal programming technique (FGP) is applied to achieve a compromised solution. FGP is used to solve the crisp goal programming problem. Any membership function can have 1 as a maximum value; maximization of $\alpha \in [0, 1]$ which can be attained by minimizing the negative deviational variables of goal programming (i.e., η) from 1. Fuzzy goal programming formulation for the given problem (P1) with negative and positive deviational variables ηj, ρj is given as

$$\text{Min } u \text{ s.t. } \frac{C_0 - C(X)}{C_0 - C*} + \eta_{1k} - \rho_{1k} = 1; \quad \frac{\text{CB}_0 - \text{CB}(X)}{\text{CB}_0 - \text{CB}*} + \eta_{2k} - \rho_{2k} = 1$$

$$u \geq w_j * \eta_j; \; j = 1, 2; \; \eta_j * \rho_j = 0; \; j = 1, 2; \; \alpha = 1 - u; \; \eta_j, \rho_j \geq 0, X \geq 0$$

The above mentioned model is solved by coded model in Lingo 11.0 to find the solution.

3 Case Study Data Set and Solution

The complete data provided by the company is difficult to present in the paper due to the limited space. A small set of data and their ranges are given: Weights of model I and model II of the product is 18.5 and 17.5 kg, respectively. Holding cost at the retail store varies from Rs. 92 to Rs. 116 for model I and 108–128 for model II of the product. Demand of model I at each retail store varies from 125 to 200 units and for model II, it lies between 125 and 150 units. Capacity range at each retail store for model I is 138–220 units and for model II, it is 138–165 units whereas, at each supplier it lies between 179 and 286 for model I and for model II it varies between 222 and 286 units. The major portion in the total cost is procurement and transportation cost. Purchase cost for model I vary from Rs. 11,500 to Rs. 14,500 and for model II, it varies from Rs. 13,500 to Rs. 16,000. Transportation cost varies from Rs. 4.6 to Rs. 6.4 per weight between supplier and retail store. The second important objective for the company is to minimize carbon emission which varies as per vehicle type. The model checks which truck type to choose as per load to be transported from suppliers to retail stores with different levels of carbon emission ranges from 300 to 476 gm/km. The truck with capacities 900, 1100, and 800 are available from suppliers S_1, S_2, and S_3, respectively.

3.1 *Result Analysis*

On the basis of five factors extracted from 30 variables, FAHP and TOPSIS have helped in identifying the two best out of three suppliers. This supplier has enough capacity to fulfill the demand of all retailers; therefore quantity is procured from the above best performing suppliers discussed in Table 12.

Table 12 Procured quantity of model I from second supplier

Period	R1	R2	R3	R4	R5	R6	R7	R8	R9	R10
T1	210	153	158	195	195	168	163	163	200	168
T2	179	164	159	195	179	159	196	175	200	164
T3	151	178	162	203	182	173	182	178	203	168
T4	116	145	171	178	149	171	149	175	143	166

Table 13 Weighted quantity transported & vehicle type selected from 2nd supplier to retail store in all periods

	R1/VT	R2/VT	R3/VT	R4/VT	R5/VT	R6/VT	R7/VT	R8/VT	R9/VT	R10/VT
T1	6186.25/C3	5577.5/C2	5486.25/C2	6083.25/C3	6267/C3	5593/C2	5588/C2	5771.75/C2	6359.5/C3	5776.75/C2
T2	5560.25/C2	5182.375/C2	5448.625/C2	6105.375/C3	6006.75/C3	5269/C2	5944.25/C2	5463.875/C2	6307.75/C3	5540.875/C2
T3	5313/C2	5640.312/C2	5624.312/C2	6378.188/C3	5447.438/C2	5553.75/C2	5734.125/C2	5384/C2	6264.188/C3	5577.188/C2
T4	4595.5/C2	4802.312/C2	5353.312/C2	5828.188/C2	4946.312/C2	5341.75/C2	5298.625/C2	5857.875/C2	5176.062/C2	5557.688/C2

Table 14 Number of trucks use to transport products from supplier to retailer store

	R1	R2	R3	R4	R5	R6	R7	R8	R9	R10
T1	6	5	5	6	6	5	5	5	6	5
T2	5	5	5	6	5	5	5	5	6	5
T3	5	5	5	6	5	5	5	5	6	5
T4	4	4	5	5	4	5	5	5	5	5

Above tables explains quantity procured as per the cost, available inventory, and better transportation by minimizing carbon emission from best performing selected suppliers. It can be observed from the above tables that each retailer did not consider all the suppliers to procure. Generally, inventory varies from 5 to 17 and in the last period of the tenure, inventory level restricted to 5 only. High product cost leads to high procurement cost. Thus, from Step 1 of the fuzzy goal programming approach, which helps to generate a trade-off between the two objectives, resulting cost aspiration level is Rs. 169,845,800 and carbon emission aspiration level is 15,599. From Step 2, we found that the cost values have been compromised in order to reduce the carbon emission values. The cost is Rs. 173,733,700 and carbon emission value is 15,564.

Company has variant capacity vehicles. The vehicle's carbon emission is also dependent upon its size. The model helped company to choose the best vehicle type (VT) and determine the number of trucks as per the weighted load to be transported from supplier to retail stores given in Tables 13 and 14.

4 Conclusion

Generally, multi-attribute problems hold uncertain and vague data, and to overcome from these situations fuzzy set theory is sufficient to deal with it. Different variables have been determined which are based on the current business scenario and experience of the experts in their respective fields. These variables are grouped together based upon their characteristics named as factors using factor analysis followed by fuzzy AHP to get the weights of all alternatives as well as criteria. TOPSIS was then applied to evaluate suppliers. This paper analyzed a hybrid approach for evaluating suppliers and then developed mathematical model that minimizes cost of procurement, holding, and transportation with minimum carbon emission. These days, minimizing carbon emission is an important aspect. Thus, there is a need of selecting right vehicle with distinct capacities which generates different carbon emission. The proposed model was fuzzy bi-objective model with constraints such as selecting efficient supplier, inventory balancing, supplier's capacity, retail store capacity, load transportation threshold for choosing vehicle type, and number of trucks to be determined for transporting products from suppliers to retail store. For solution process, fuzzy goal programming approach was used to create a trade-off between objectives with respect to constraints.

References

1. Bartlett, M.S.: Tests of significance in factor analysis. Br. J. Psychol. **3**(Part II), 77–85 (1950)
2. Bellman, R.E., Zadeh, L.A.: Decision-making in a fuzzy environment. Manag. Sci. **17**, 141–164 (1970)
3. Cao, M., Zhang, Q.: Supply chain collaboration: impact on collaborative advantage and firm performance. J. Oper. Manag. **29**(3), 163–180 (2011)
4. Chang, D.Y.: Applications of the extent analysis method on fuzzy AHP. Eur. J. Oper. Res. **95**, 649–55 (1996)
5. Cho, D.W., Lee, Y.H., Ahn, S.H., Hwang, M.K.: A framework for measuring the performance of service supply chain management. Comput. Ind. Eng. **62**, 801–818 (2012)
6. Esfandiari, N., Seifbarghy, M.: Modeling a stochastic multi-objective supplier quota allocation problem with price-dependent ordering. Appl. Math. Model. **37**, 5790–5800 (2013)
7. Gumus, A.-T.: Evaluation of hazardous waste transportation firms by using a two step fuzzy-AHP and TOPSIS methodology. Expert System. Appl. **36**(2), 4067–4074 (2009)
8. Hwang, C.L., Yoon, K.: Multiple Attribute Decision Making: Methods and Applications. Springer, Heidelberg (1981)
9. Isaai, M.T., Kanani, A., Tootoonchi, M., Afzali, H.R.: Intelligent timetable evaluation using fuzzy AHP. Expert Syst. Appl. **38**(4), 3718–3723 (2011)
10. Kaiser, H.F.: A second generation Little Jiffy. Psychometrika **35**, 401–415 (1970)
11. Malhotra, N.K. (ed.): Marketing Research: An applied orientation, 5th edn. PHI Publishers, Delhi (2007)
12. Merschmann, U., Thonemann, U.W.: Supply chain flexibility, uncertainty and firm performance: an empirical analysis of German manufacturing firms. Int. Prod. Econ. **130**(1), 43–53 (2011)
13. Mohamed, R.H.: The relationship between goal programming and fuzzy programming. Fuzzy Sets Syst. **89**, 215–222 (1997)
14. Narasimhan, R., Talluri, S., Mahapatra, S.K.: Multiproduct, multi-criteria model for supplier selection with product life-cycle considerations. Decis. Sci. **37**(4), 577–603 (2006)
15. Saaty, T.L.: The Analytic Hierarchy Process. McGraw-Hill (1980)
16. Wadhwa, V., Ravindran, A.R.: Vendor selection in outsourcing. Comput. Oper. Res. **34**(12), 3725–3737 (2007)
17. Wong, C.Y., Boon-Itt, S., Wong, C.W.Y.: The contingency effects of environmental uncertainty on the relationship between supply chain integration and operational performance. J. Oper. Manag. **29**(6), 604–615 (2011)
18. Zimmermann, H.J.: Description and optimization of fuzzy systems. Int. J. Gen. Syst. **2**, 209–215 (1976)

Mathematical Modeling of Multistage Evaporator System in Kraft Recovery Process

Om Prakash Verma, Toufiq Haji Mohammed, Shubham Mangal and Gaurav Manik

Abstract In order to reduce the energy consumption of the concentrating the black liquor in Kraft recovery process in paper mill, a mathematical models is developed and proposed for the analysis of multi stage evaporator (heptad's effect evaporator system). These multistage evaporator is modeled based on the flow of black liquor and steam operation option. In the present study, live steam split, liquor feed split, pre-heaters and hybrid of these energy saving scheme are coupled with backward, forward and mixed feed multi stage evaporator system. The systematic material balance and heat balance equations are described as the form of matrix equation to realize the generality of the mathematical models. The advantage of these models are its simplicity, representation of equations in matrix form and linearity. The proposed mathematical models can be easily solved by numerical methods combining with the iteration method, which has one more advantages of less sensitive to its initial values, high convergence speed and stability. The developed models can also be easily simulate under different operating strategies once knowing the liquor, steam and effects parameters.

Keywords Multi stage evaporator · Heptad's effect evaporator · Steam split · Feed split · Pre-heater · Steam economy

O.P. Verma · T.H. Mohammed · Shubham Mangal · Gaurav Manik (✉)
Department of Polymer and Process Engineering, Indian Institute of Technology Roorkee, Saharanpur Campus, Saharanpur, India
e-mail: manikfpt@iitr.ac.in

O.P. Verma
e-mail: opiitroorkee@gmail.com

T.H. Mohammed
e-mail: toufiq2t@gmail.com

Shubham Mangal
e-mail: mangalshubham161@gmail.com

M. Pant et al. (eds.), *Proceedings of Fifth International Conference on Soft Computing for Problem Solving*, Advances in Intelligent Systems and Computing 437, DOI 10.1007/978-981-10-0451-3_87

1 Introduction

Heat demands in pulp and paper mills occur in the following stage: Kraft pulping, Mechanical pulping, and paper machine etc. Although all the stages mentioned are crucial in the production process, but Kraft pulping process consist of multi stage evaporator (MSE) house consumes around 24–30 % of energy of its total energy and therefore, makes it as largest energy intensive section. MSE is not only used in pulp and paper mills but even used in sugar, pharmaceuticals, desalination, juice production, dairy production industries. In the present work, the pulp and paper industry has been considered for the prime focus of the investigation. Predominantly, pulp and paper mills uses the Kraft process in which black liquor is generated as spent liquor and MSE is used to concentrate the weak black liquor which consumes a large amount of thermal energy in the form of steam. In current investigation, an effort has been carried out to improve energy efficiency in term of steam economy (SE), either by improving its design or by operation and this work is done only by developing mathematical modeling of MSE system with different configurations. In the present work, the mathematical models are proposed for heptad's effect type multi stage evaporator system. The mathematical model analysis is based on different configurations including backward feed, forward feed, mixed feed, steam splitting, condensate flashing, pre-heaters and hybrid of these combinations.

The mathematical models of MSE system have been analyzed since last many decades and a few literatures reported in this area are: Kern [6], Holland [4], Nishitani and Kunugita [12], Aly and Marwan [1], Mirinda and Simpson [11]. Lambert et al. [10] developed a nonlinear mathematical model for MEE system and reduced this model to linear form and solved these equations by the Gauss elimination numerical technique and hence, the results of linear and nonlinear models solutions were compared. Zain and Kumar [15] reported nonlinear models have certain stability and convergence problem, whereas, solution of linear models having faster, stable and desirable convergence. The thermal integration of multiple effect evaporator system (MEE) was studied by Higa et al. [3] for sugar plant based on grand composite curve. Khademi et al. [7] studied steady state simulation and optimization of a six effect evaporator system in a desalination process. The present mathematical models development is based on Kaya and Sarac [5] work. A simplified mathematical model was developed for MEE system. Models developed by Kaya and Sarac [5] were a provision of concurrent, countercurrent and parallel flow operation. Further, these models were applied on sugar factory's data for case study and compare the results for no pre-heating case with preheating case. Bhargava et al. [2] investigate a nonlinear model represented in single cubical polynomial equation and solved by generalized cascade algorithm iteratively for septuple effect flat falling film evaporator system. The model proposed were capable of simulating process of evaporation considering variations in boiling point rise, overall heat transfer coefficient, heat loss, flow sequences, liquor/steam splitting, feed product, condensate flashing, vapor bleeding and physico-thermal properties of black liquor. Khanam and Mohanty [8] developed the models based on concepts of stream analysis and temperature path for MEE system with induction of condensate flashing. Sagharichiha et al. [14] simulate forward feed vertical tube multi effect evaporators in desalination industries. Ruan et al. [13] proposed the mathematical models for the countercurrent multiple effect evaporation system contains various energy saving schemes such as steam jet heat pump technology, solution flash and condensed water flashed. Ruan et al. [13] found heat pump is an effective energy saving measure in countercurrent multi evaporator system. Model developed were arrangement of prediction and prevention of scale formation. Under the above background the focus of this study is to develop simplified mathematical models for MSE, specifically, Heptads' effects type evaporator system with different configurations including backward feed, forward feed, mixed feed, steam splitting, condensate flashing, pre-heaters and hybrid of these combinations used in pulp and paper industries.

2 Generalized Modeling of a Heptads' Effect Evaporator System

In modeling of MEE system, a pressure and temperature value is known for each effect of the MEE system. The essential enthalpies for these temperature values are found from empirical correlations with the help of industrial data for the entire operating process temperature range (52–148 °C), and these are illustrated in Figs. 1, 2 and 3.

Latent heat of vaporization (λ), enthalpy of vapor (H) and enthalpy of condensate (h_c) are found to fit a second order of polynomial. The best fit correlations for λ_i, H_i and h_ci are represented by Eqs. (1)–(3) for the *i*th effect.

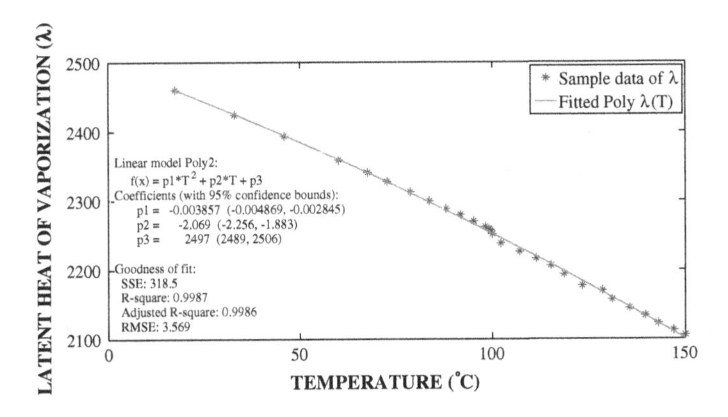

Fig. 1 Curve fitting of latent heat of vaporization (λ) (kJ/kg) data

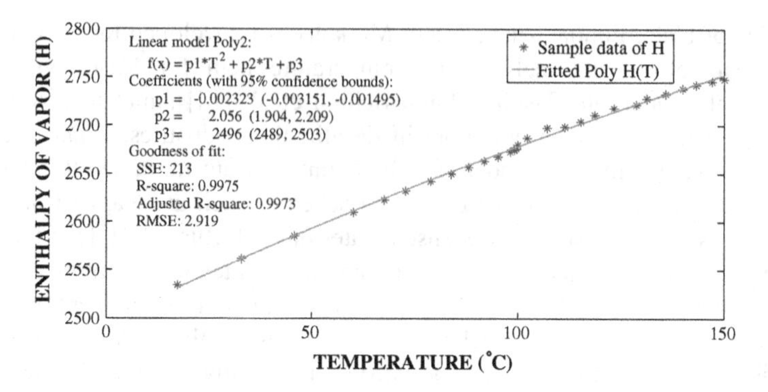

Fig. 2 Curve fitting of enthalpy of vapor (H) (kJ/kg) data

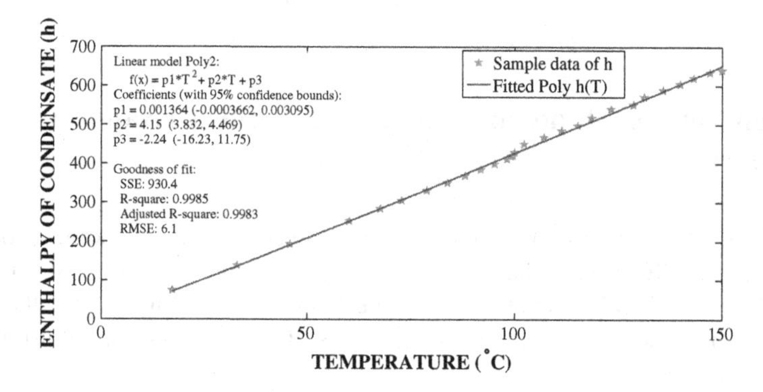

Fig. 3 Curve fitting of enthalpy of condensate (h_c) in (kJ/kg) data

$$\lambda_i = -0.003857T_i^2 - 2.069T_i + 2497 \tag{1}$$

$$H_i = -0.0002045T_i^2 + 1.677T_i + 2507 \tag{2}$$

$$h_{ci} = 0.001364T_i^2 + 4.15\,T - 2.24 \tag{3}$$

Bhargava et al. [2] developed correlations for the enthalpy of black liquor based on paper mill data is presented in Eq. (4).

$$h_{Li} = 4.187(1 - 0.54x_i)T_i \tag{4}$$

Once when empirical correlations are found then next is mathematical models development. Mathematical models emanate from the basic idea of conservation laws including mass, component and energy balances for each effect of the evaporator system. These models are then transformed to linearly independent equations. The developed empirical correlations from Eqs. (1)–(4) are reposed in models. In MSE, for N number of effect, N + 1 number of linear equations is obtained and hence, in current work, for heptads' effect evaporator system for each configuration *eight* number of equations has been obtained. The obtained equations are usually a function of the amount of steam feed and the amount of solution vapor generate at each effect in MSE system. The advantage of the developed models is its linear property, so the convergence and stability problem can't be reached. The study has been done for different feasible permutation configurations for heptads' effect evaporator system are listed in Table 1.

2.1 Backward Feed Configuration: Model-A

If the black liquor solution is fed to the last effect (i.e. 7th effect) of evaporator system and steam is fed to the 1st effect of evaporator system, this configuration type is called backward feed configuration. The concentrated black liquor flow from 7th effect to 1st effect and finally, the concentrated black liquor as a product is obtained from 1st effect of the MSE system. The solution vapor obtained at each effect is further used as a heat source for next effect of MSE. This type of configuration and mathematical model is derived from Eqs. (5)–(19) and represented in matrix form by Eq. (21) and illustrated in Fig. 4.

For the 1st effect, the energy balance equation is given by

$$V_1\lambda_1 + L_2h_{L2} = V_2H_2 + L_1h_{L1} \tag{5}$$

$$V_1\lambda_1 - V_2H_2 = L_1h_{L1} - L_2h_{L2} \tag{6}$$

$$V_1\lambda_1 - V_2H_2 = L_1h_{L1} - (V_2 + L_1)h_{L2} \tag{7}$$

$$V_1\lambda_1 + V_2(h_{L2} - H_2) = (h_{L1} - h_{L2})\left(\frac{L_fX_f}{X_1}\right) \tag{8}$$

For the 2nd effect, the energy balance is given by

$$V_2\lambda_2 + L_3h_{L3} = V_3H_3 + L_2h_{L2} \tag{9}$$

$$V_2\lambda_2 - V_3H_3 = L_2h_{L2} - L_3h_{L3} \tag{10}$$

$$V_2\lambda_2 - V_3H_3 = L_2h_{L2} - (V_3 + L_2)h_{L3} \tag{11}$$

$$V_2\lambda_2 + V_3(h_{L3} - H_3) = L_2(h_{L2} - h_{L3}) \tag{12}$$

$$V_2\lambda_2 + V_3(h_{L3} - H_3) = (V_2 + L_1)(h_{L2} - h_{L3}) \tag{13}$$

Table 1 Description of used configurations

Model	Configurations property	Remark
Model-A	Backward feed	Liquor feed to 7th effect and steam to 1st effect
Model-B	Forward feed	Liquor feed to 1st effect and steam to 1st effect
Model-C	Mixed feed	Liquor feed to 6th effect and steam to 1st effect
Model-D	Backward feed with steam splitting	Liquor feed to 7th effect and steam split to 1st and 2nd effect with split ratio Y
Model-E	Forward feed with steam splitting	Liquor feed to 1st effect and steam split to 1st and 2nd effect with split ratio Y
Model-F	Mixed feed with steam splitting	Liquor feed to 6th effect and steam split to 1st and 2nd effect with split ratio Y
Model-G	Backward feed with feed split	Liquor feed to 7th and 6th effect simultaneously with feed ratio K and steam to 1st effect
Model-H	Forward feed with feed split	Liquor feed to 1st and 2nd effect simultaneously with feed ratio K and steam to 1st effect
Model-I	Mixed feed with feed split	Liquor feed to 6th and 5th effect simultaneously with feed ratio K and steam to 1st effect
Model-J	Backward feed coupled with Feed and Steam split	Liquor feed to 7th and 6th effect simultaneously with feed ratio K and split to 1st and 2nd effect with split ratio Y
Model-K	Forward feed with coupled with Feed and Steam split	Liquor feed to 1st and 2nd effect simultaneously with feed ratio K and split to 1st and 2nd effect with split ratio Y
Model-L	Mixed feed with coupled with Feed and Steam split	Liquor feed to 6th and 5th effect simultaneously with feed ratio K and split to 1st and 2nd effect with split ratio Y
Model-M	Backward feed with coupled with Feed split, Steam split and Pre-heater	Liquor feed to 7th and 6th effect simultaneously with feed ratio K and split to 1st and 2nd effect with split ratio Y. Pre-heater used to concentrate the feed liquor
Model-N	Forward feed coupled with Feed, Steam split and Pre-heater	Liquor feed to 1st and 2nd effect simultaneously with feed ratio K and split to 1st and 2nd effect with split ratio Y. Pre-heater used to concentrate the feed liquor
Model-O	Mixed feed with coupled Feed split, Steam split and Pre-heater	Liquor feed to 6th and 5th effect simultaneously with feed ratio K and split to 1st and 2nd effect with split ratio Y. Pre-heater used to concentrate the feed liquor

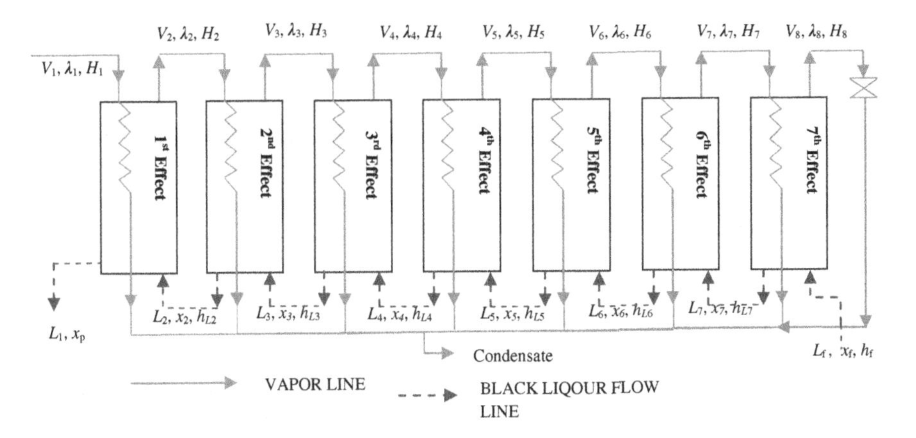

Fig. 4 Backward feed heptad's effect evaporator system

$$V_2(\lambda_2 + h_{L3} - h_{L2}) + V_3(h_{L3} - H_3) = (h_{L2} - h_{L3})\left(\frac{L_f X_f}{X_1}\right) \tag{14}$$

Similarly, for the 3rd–7th effect, the energy balances equations are given by

$$V_2(h_{L4} - h_{L3}) + V_3(\lambda_3 + h_{L4} - h_{L3}) + V_4(h_{L4} - H_4) = (h_{L3} - h_{L4})\left(\frac{L_f X_f}{X_1}\right) \tag{15}$$

$$\begin{aligned} &V_2(h_{L5} - h_{L4}) + V_3(h_{L5} - h_{L4}) + V_4(\lambda_4 + h_{L5} - h_{L4}) + V_5(h_{L5} - H_5) \\ &= (h_{L4} - h_{L5})\left(\frac{L_f X_f}{X_1}\right) \end{aligned} \tag{16}$$

$$\begin{aligned} &V_2(h_{L6} - h_{L5}) + V_3(h_{L6} - h_{L5}) + V_4(h_{L6} - h_{L5}) + V_5(\lambda_5 + h_{L6} - h_{L5}) \\ &+ V_6(h_{L6} - H_6) = (h_{L5} - h_{L6})\left(\frac{L_f X_f}{X_1}\right) \end{aligned} \tag{17}$$

$$\begin{aligned} &V_2(h_{L7} - h_{L6}) + V_3(h_{L7} - h_{L6}) + V_4(h_{L7} - h_{L6}) + V_5(h_{L7} - h_{L6}) \\ &+ V_6(\lambda_6 + h_{L7} - h_{L6}) + V_7(h_{L7} - H_7) = (h_{L6} - h_{L7})\left(\frac{L_f X_f}{X_1}\right) \end{aligned} \tag{18}$$

$$\begin{aligned} &V_2(h_{Lf} - h_{L7}) + V_3(h_{Lf} - h_{L7}) + V_4(h_{Lf} - h_{L7}) + V_5(h_{Lf} - h_{L7}) \\ &+ V_6(h_{Lf} - h_{L7}) + V_7(\lambda_7 + h_{Lf} - h_{L7}) + V_8(h_f - H_8) \\ &= (h_{L7} - h_{Lf})\left(\frac{L_f X_f}{X_1}\right) \end{aligned} \tag{19}$$

$$V_2 + V_3 + V_4 + V_5 + V_6 + V_7 + V_8 = L_f\left(1 - \frac{X_f}{X_1}\right) \tag{20}$$

In matrix form the above set of equations are represented as by Eq. (21)

$$\begin{bmatrix} \lambda_1 & (h_{L2}-H_2) & 0 & 0 & 0 & 0 & 0 & 0 \\ 0 & (\lambda_2+h_{L3}-h_{L2}) & (h_{L3}-H_3) & 0 & 0 & 0 & 0 & 0 \\ 0 & (h_{L4}-h_{L3}) & (\lambda_3+h_{L4}-h_{L3}) & (h_{L4}-H_4) & 0 & 0 & 0 & 0 \\ 0 & (h_{L5}-h_{L4}) & (h_{L5}-h_{L4}) & (\lambda_4+h_{L5}-h_{L4}) & (h_{L5}-H_5) & 0 & 0 & 0 \\ 0 & (h_{L6}-h_{L5}) & (h_{L6}-h_{L5}) & (h_{L6}-h_{L5}) & (\lambda_5+h_{L6}-h_5) & (h_{L6}-H_6) & 0 & 0 \\ 0 & (h_{L7}-h_{L6}) & (h_{L7}-h_{L6}) & (h_{L7}-h_{L6}) & (h_{L7}-h_{L6}) & (\lambda_6+h_{L7}-h_{L6}) & (h_{L7}-H_7) & 0 \\ 0 & (h_{Lf}-h_{L7}) & (h_{Lf}-h_{L7}) & (h_{Lf}-h_{L7}) & (h_{Lf}-h_{L7}) & (h_{Lf}-h_{L7}) & (\lambda_7+h_{Lf}-h_{L7}) & (h_{Lf}-H_8) \\ 0 & 1 & 1 & 1 & 1 & 1 & 1 & 1 \end{bmatrix} \begin{bmatrix} V_1 \\ V_2 \\ V_3 \\ V_4 \\ V_5 \\ V_6 \\ V_7 \\ V_8 \end{bmatrix} = \begin{bmatrix} (h_{L1}-h_{L2})L_f(x_f/x_1) \\ (h_{L2}-h_{L3})L_f(x_f/x_1) \\ (h_{L3}-h_{L4})L_f(x_f/x_1) \\ (h_{L4}-h_{L5})L_f(x_f/x_1) \\ (h_{L5}-h_{L6})L_f(x_f/x_1) \\ (h_{L6}-h_{L7})L_f(x_f/x_1) \\ (h_{L7}-h_{Lf})L_f(x_f/x_1) \\ L_f(1-x_f/x_1) \end{bmatrix} \tag{21}$$

2.2 *Forward Feed Configuration: Model-B*

If the weak black liquor and steam flow with vapor generated at each effect used as heat source for the next effect of MSE system flow in same direction, then such configuration type is called forward feed flow. In such a case, un-concentrated black liquor solution is fed to 1st effect of MSE system and final concentrated product is obtained from the last 7th effect by moving through 1st–7th effect and is illustrated in Fig. 5.

The energy balances for 1st is derived from Eqs. (22)–(25)

$$V_1\lambda_1 + L_f h_{Lf} = V_2 H_2 + L_1 h_{L1} \tag{22}$$

$$V_1\lambda_1 - V_2 H_2 = L_1 h_{L1} - (V_2 + L_1) h_{Lf} \tag{23}$$

$$V_1\lambda_1 + V_2(h_{Lf} - H_2) = L_1(h_{L1} - h_{Lf}) \tag{24}$$

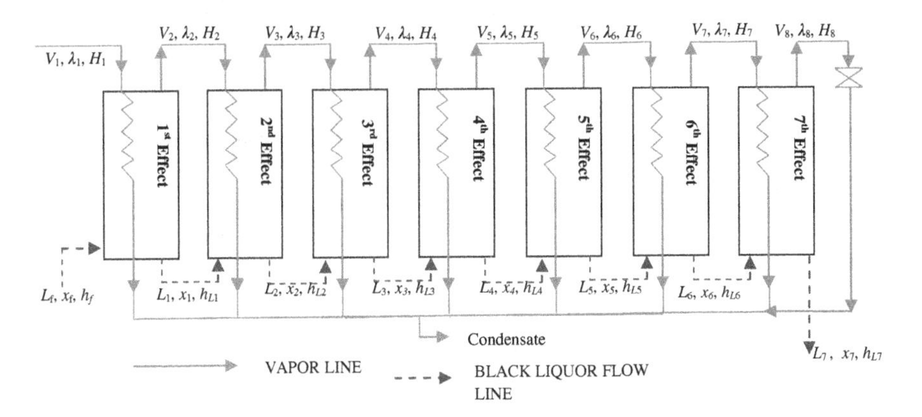

Fig. 5 Forward feed heptad's effect evaporator system

$$V_1\lambda_1 + V_2(h_{Lf} - H_2) + V_3(h_{Lf} - h_{L1}) + V_4(h_{Lf} - h_{L1}) + V_5(h_{Lf} - h_{L1}) + V_6(h_{Lf} - h_{L1}) + V_7(h_{Lf} - h_{L1}) + V_8(h_{Lf} - h_{L1}) = (h_{L1} - h_{Lf})\left(\frac{L_f X_f}{X_7}\right) \quad (25)$$

Similarly, the mathematical equations for 2nd–7th effect are written from Eqs. (26)–(31)

$$V_2\lambda_2 + V_3(h_{L1} - H_3) + V_4(h_{L1} - h_{L2}) + V_5(h_{L1} - h_{L2}) + V_6(h_{L1} - h_{L2}) + V_7(h_{L1} - h_{L2}) + V_8(h_{L1} - h_{L2}) = (h_{L2} - h_{L1})\left(\frac{L_f X_f}{X_7}\right) \quad (26)$$

$$V_3\lambda_3 + V_4(h_{L2} - H_4) + V_5(h_{L2} - h_{L3}) + V_6(h_{L2} - h_{L3}) + V_7(h_{L2} - h_{L3}) + V_8(h_{L2} - h_{L3}) = (h_{L3} - h_{L2})\left(\frac{L_f X_f}{X_7}\right) \quad (27)$$

$$V_4\lambda_4 + V_5(h_{L3} - H_5) + V_6(h_{L3} - h_{L4}) + V_7(h_{L3} - h_{L4}) + V_8(h_{L3} - h_{L4}) = (h_{L4} - h_{L3})\left(\frac{L_f X_f}{X_7}\right) \quad (28)$$

$$V_5\lambda_5 + V_6(h_{L4} - H_6) + V_7(h_{L4} - h_{L5}) + V_8(h_{L4} - h_{L5}) = (h_{L5} - h_{L4})\left(\frac{L_f X_f}{X_7}\right) \quad (29)$$

$$V_6\lambda_6 + V_7(h_{L5} - H_7) + V_8(h_{L5} - h_{L6}) = (h_{L6} - h_{L5})\left(\frac{L_f X_f}{X_7}\right) \quad (30)$$

$$V_7\lambda_7 + V_8(h_{L6} - H_8) = (h_{L7} - h_{L6})\left(\frac{L_f X_f}{X_7}\right) \tag{31}$$

The final representation of the models in the matrix form is given as Eq. (32)

$$\begin{bmatrix} \lambda_1 & (h_{Lf} - H_2) & (h_{Lf} - h_{L1}) & (h_{Lf} - h_{L1}) & (h_{Lf} - h_{L1}) & (h_{Lf} - h_{L1}) & (h_{Lf} - h_{L1}) & (h_{Lf} - h_{L1}) \\ 0 & \lambda_2 & (h_{L1} - H_3) & (h_{L1} - h_{L2}) & (h_{L1} - h_{L2}) & (h_{L1} - h_{L2}) & (h_{L1} - h_{L2}) & (h_{L1} - h_{L2}) \\ 0 & 0 & \lambda_3 & (h_{L2} - H_4) & (h_{L2} - h_{L3}) & (h_{L2} - h_{L3}) & (h_{L2} - h_{L3}) & (h_{L2} - h_{L3}) \\ 0 & 0 & 0 & \lambda_4 & (h_{L3} - H_5) & (h_{L3} - h_{L4}) & (h_{L3} - h_{L4}) & (h_{L3} - h_{L4}) \\ 0 & 0 & 0 & 0 & \lambda_5 & (h_{L4} - H_6) & (h_{L4} - h_{L5}) & (h_{L4} - h_{L5}) \\ 0 & 0 & 0 & 0 & 0 & \lambda_6 & (h_{L5} - H_7) & (h_{L5} - h_{L6}) \\ 0 & 0 & 0 & 0 & 0 & 0 & \lambda_7 & (h_{L6} - H_8) \\ 0 & 1 & 1 & 1 & 1 & 1 & 1 & 1 \end{bmatrix} \begin{bmatrix} V_1 \\ V_2 \\ V_3 \\ V_4 \\ V_5 \\ V_6 \\ V_7 \\ V_8 \end{bmatrix} = \begin{bmatrix} (h_{L1} - h_{Lf})L_f(x_f/x_7) \\ (h_{L2} - h_{L1})L_f(x_f/x_7) \\ (h_{L3} - h_{L2})L_f(x_f/x_7) \\ (h_{L4} - h_{L3})L_f(x_f/x_7) \\ (h_{L5} - h_{L4})L_f(x_f/x_7) \\ (h_{L6} - h_{L5})L_f(x_f/x_7) \\ (h_{L7} - h_{L6})L_f(x_f/x_7) \\ L_f(1 - x_f/x_7) \end{bmatrix} \tag{32}$$

2.3 Mixed Feed Configuration: Model-C

In the mixed feed configuration type, the un-concentrated black liquor is fed in any of the effect between 1st and 7th effect of the system and final concentrated product is obtained from any of the effect. In the present investigations, un-concentrated black liquor is fed to 6th effect and output is fed to 7th effect and then it flow to 5th, then 4th–1st effect. As shown in Fig. 6, the steam is provided to 1st effect of MSE. The liquor vapor flows in the same direction as explained earlier. The model equations remain same as in backward feed from 1st effect to 4th effect given in Eqs. (33)–(36).

$$V_1\lambda_1 + V_2(h_{L2} - H_2) = (h_{L1} - h_{L2})\left(\frac{L_f X_f}{X_1}\right) \tag{33}$$

$$V_2(\lambda_2 + h_{L3} - h_{L2}) + V_3(h_{L3} - H_3) = (h_{L2} - h_{L3})\left(\frac{L_f X_f}{X_1}\right) \tag{34}$$

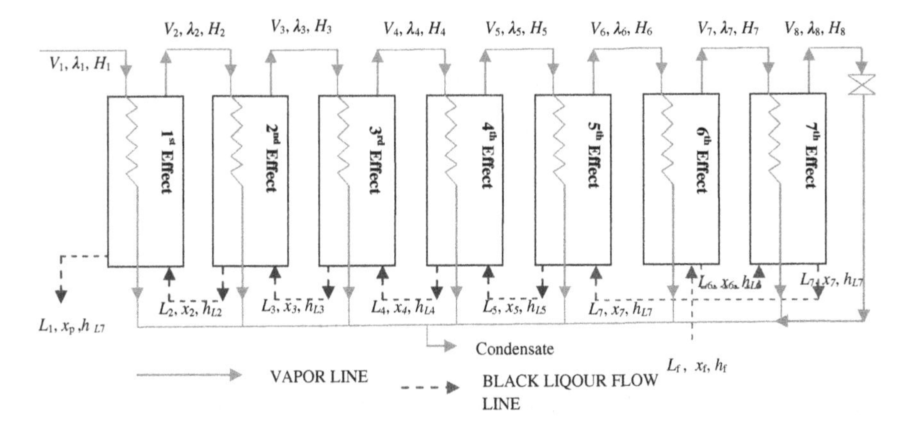

Fig. 6 Mixed feed heptad's effect evaporator system

$$V_2(h_{L4} - h_{L3}) + V_3(\lambda_3 + h_{L4} - h_{L3}) + V_4(h_{L4} - H_4) = (h_{L3} - h_{L4})\left(\frac{L_f X_f}{X_1}\right) \tag{35}$$

$$\begin{aligned} &V_2(h_{L5} - h_{L4}) + V_3(h_{L5} - h_{L4}) + V_4(\lambda_4 + h_{L5} - h_{L4}) \\ &+ V_5(h_{L5} - H_5) = (h_{L4} - h_{L5})\left(\frac{L_f X_f}{X_1}\right) \end{aligned} \tag{36}$$

For the 4th effect, the energy balances are different from the backward feed and is derived as

$$V_5\lambda_5 + L_7 h_{L7} = V_6 H_6 + L_5 h_{L5} \tag{37}$$

$$V_5\lambda_5 - V_6 H_6 = L_5 h_{L5} - L_7 h_{L7} \tag{38}$$

$$V_5\lambda_5 - V_6 H_6 = L_5 h_{L5} - (V_6 + L_5) h_{L7} \tag{39}$$

$$V_5\lambda_5 + V_6(h_{L7} - H_6) = (V_5 + L_4)(h_{L5} - h_{L7}) \tag{40}$$

$$V_5(\lambda_5 + h_{L7} - h_{L5}) + V_6(h_{L7} - H_6) = (V_4 + L_3)(h_{L5} - h_{L7}) \tag{41}$$

$$\begin{aligned} &V_4(h_7 - h_5) + V_5(\lambda_5 + h_{L7} - h_{L5}) + V_6(h_{L7} - H_6) \\ &= (V_3 + L_2)(h_{L5} - h_{L7}) \end{aligned} \tag{42}$$

$$\begin{aligned} &V_2(h_{L7} - h_{L5}) + V_3(h_{L7} - h_{L5}) + V_4(h_{L7} - h_{L5}) + V_5(\lambda_5 + h_{L7} - h_{L5}) \\ &+ V_6(h_{L7} - H_6) = (h_{L5} - h_{L7})\left(\frac{L_f X_f}{X_1}\right) \end{aligned} \tag{43}$$

In similar ways, Eqs. (44)–(45) are modeled for 6th and 7th effects

$$\begin{aligned} & V_2(h_{Lf} - h_{L6}) + V_3(h_{Lf} - h_{L6}) + V_4(h_{Lf} - h_{L6}) + V_5(h_{Lf} - h_{L6}) \\ & \quad + V_6(\lambda_6 + h_{Lf} - h_{L6}) + V_7(h_{Lf} - H_7) + V_8(h_{Lf} - h_{L6}) \\ & \quad = (h_{L6} - h_{Lf})\left(\frac{L_f X_f}{X_1}\right) \end{aligned} \tag{44}$$

$$\begin{aligned} & V_2(h_{L6} - h_{L7}) + V_3(h_{L6} - h_{L7}) + V_4(h_{L6} - h_{L7}) + V_5(h_{L6} - h_{L7}) \\ & \quad + V_6(h_{L6} - h_{L7}) + V_7\lambda_7 + V_8(h_{L6} - H_{L8}) = (h_{L7} - h_{L6})\left(\frac{L_f X_f}{X_1}\right) \end{aligned} \tag{45}$$

In similar ways, the final representation of Model-C in matrices is given by Eq. (46)

$$\begin{bmatrix} \lambda_1 & (h_{L2} - H_2) & 0 & 0 & 0 & 0 & 0 & 0 \\ 0 & (\lambda_2 + h_{L3} - h_{L2}) & (h_{L3} - H_3) & 0 & 0 & 0 & 0 & 0 \\ 0 & (h_{L4} - h_{L3}) & (\lambda_3 + h_{L4} - h_{L3}) & (h_{L4} - H_4) & 0 & 0 & 0 & 0 \\ 0 & (h_{L5} - h_{L4}) & (h_{L5} - h_{L4}) & (\lambda_4 + h_{L5} - h_{L4}) & (h_{L5} - H_5) & 0 & 0 & 0 \\ 0 & (h_{L7} - h_{L5}) & (h_{L7} - h_{L5}) & (h_{L7} - h_{L5}) & (\lambda_5 + h_{L7} - h_{L5}) & (h_{L6} - H_6) & 0 & 0 \\ 0 & (h_{Lf} - h_{L6}) & (h_{Lf} - h_{L6}) & (h_{Lf} - h_{L6}) & (h_{Lf} - h_{L6}) & (\lambda_6 + h_{Lf} - h_{L6}) & (h_{Lf} - H_7) & (h_{Lf} - h_{L6}) \\ 0 & (h_{L6} - h_{L7}) & (h_{L6} - h_{L7}) & (h_{L6} - h_{L7}) & (h_{L6} - h_{L7}) & (h_{L6} - h_{L7}) & \lambda_7 & (h_{L6} - H_8) \\ 0 & 1 & 1 & 1 & 1 & 1 & 1 & 1 \end{bmatrix} \begin{bmatrix} V_1 \\ V_2 \\ V_3 \\ V_4 \\ V_5 \\ V_6 \\ V_7 \\ V_8 \end{bmatrix} = \begin{bmatrix} (h_{L1} - h_{L2})L_f(x_f/x_1) \\ (h_{L2} - h_{L3})L_f(x_f/x_1) \\ (h_{L3} - h_{L4})L_f(x_f/x_1) \\ (h_{L4} - h_{L5})L_f(x_f/x_1) \\ (h_{L5} - h_{L7})L_f(x_f/x_1) \\ (h_{L6} - h_{Lf})L_f(x_f/x_1) \\ (h_{L7} - h_{L6})L_f(x_f/x_1) \\ L_f(1 - x_f/x_1) \end{bmatrix} \tag{46}$$

2.4 Configurations with Steam Split

In above Models, live steam is fed only to 1st effect. By introducing the concepts of steam split, now live steam is split among 1st and 2nd effect with steam split fraction Y and $(1 - Y)$, and hence, the vapor generated from these two effects are combined together and provided to 3rd effect of the MSE system. Further, in similar ways, the vapor generated from the 3rd effect is utilized as heating source for 4th effect and so on. Model-D to Model-F are derived with the same concepts discussed here.

2.4.1 Backward Feed Steam Split Configuration: Model-D

In this configuration, as similar way like backward feed Model-A, the black liquor is fed to last 7th effect and concentrated liquor as a product is taken from 1st effect. Steam is split among 1st and 2nd effect with steam split fraction Y and $(1-Y)$ as shown in Fig. 7. Furthermore, the vapor produced from these effects are combined together and sent to 3rd effect of the MSE system.

For the 1st effect, the energy balances equation is given by Eqs. (47) and (49) represents model for the solution vapor obtained at 3rd effect is further used as a heat source for next effect of MSE. Hence, for Model-D, the mathematical model is derived below.

1st effect.

$$YV_1\lambda_1 + L_2h_{L2} = V_2H_2 + L_1h_{L2} \tag{47}$$

$$YV_1\lambda_1 - V_2H_2 = L_1h_{L2} - L_2h_{L2} \tag{48}$$

$$YV_1\lambda_1 + V_2(h_{L2} - H_2) = L_f(x_f/x_1)(h_{L1} - h_{L2}) \tag{49}$$

In similar way, around 2nd effect the model is given by

$$(1-Y)V_1\lambda_1 + V_2(h_{L3} - h_{L2}) + V_3(h_{L3} - H_3) = L_f(x_f/x_1)(h_{L2} - h_{L3}) \tag{50}$$

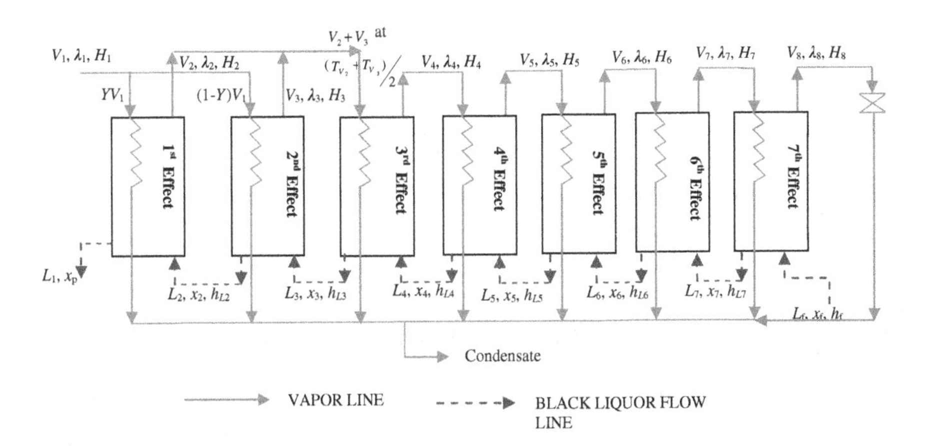

Fig. 7 Backward feed coupled with steam split heptads effect evaporator system

But, for the 3rd effect, as discussed earlier, the vapor produced from 1st and 2nd effects are combined together and sent to 3rd effect of the MSE system hence, the model will change slightly. The latent heat of vaporization λ is calculated at average temperature of 1st and 2nd effect, mathematically, at $\frac{T_{V2}+T_{V3}}{2}$. Defining, $\frac{T_{V2}+T_{V3}}{2} = \lambda_{avg}$. Hence, the energy balance equation is written as by Eq. (51)

$$(V_2 + V_3)\lambda_{avg} + L_4 h_{L4} = V_4 H_4 + L_3 h_{L3} \tag{51}$$

After solving Eq. (51),

$$\begin{aligned} &V_2(\lambda_{avg} + h_{L4} - h_{L3}) + V_3(\lambda_{avg} + h_{L4} - h_{L3}) + V_4(h_{L4} - H_4) \\ &= L_f(x_f/x_1)(h_{L3} - h_{L4}) \end{aligned} \tag{52}$$

From 4th to 7th effect, the mathematical equation will remain same as in Model-A. Hence, the final representation of mathematical model for Model-D is represented in matrix form is give by Eq. (53)

$$\begin{bmatrix} Y\lambda_1 & (h_{L2}-H_2) & 0 & 0 & 0 & 0 & 0 & 0 \\ (1-Y)\lambda_1 & (h_{L3}-h_{L2}) & (h_{L3}-H_3) & 0 & 0 & 0 & 0 & 0 \\ 0 & (\lambda_{avg}+h_{L4}-h_{L3}) & (\lambda_{avg}+h_{L4}-h_{L3}) & (h_{L4}-H_4) & 0 & 0 & 0 & 0 \\ 0 & (h_{L5}-h_{L4}) & (h_{L5}-h_{L4}) & (\lambda_4+h_{L5}-h_{L4}) & (h_{L5}-H_5) & 0 & 0 & 0 \\ 0 & (h_{L6}-h_{L5}) & (h_{L6}-h_{L5}) & (h_{L6}-h_{L5}) & (\lambda_5+h_{L6}-h_{L5}) & (h_{L6}-H_6) & 0 & 0 \\ 0 & (h_{L7}-h_{L6}) & (h_{L7}-h_{L6}) & (h_{L7}-h_{L6}) & (h_{L7}-h_{L6}) & (\lambda_6+h_{L7}-h_{L6}) & (h_{L7}-H_7) & 0 \\ 0 & (h_{Lf}-h_{L7}) & (h_{Lf}-h_{L7}) & (h_{Lf}-h_{L7}) & (h_{Lf}-h_{L7}) & (h_{Lf}-h_{L7}) & (\lambda_7+h_{Lf}-h_{L7}) & (h_{Lf}-H_8) \\ 0 & 1 & 1 & 1 & 1 & 1 & 1 & 1 \end{bmatrix} \begin{bmatrix} V_1 \\ V_2 \\ V_3 \\ V_4 \\ V_5 \\ V_6 \\ V_7 \\ V_8 \end{bmatrix} = \begin{bmatrix} (h_{L1}-h_{L2})L_f(x_f/x_1) \\ (h_{L2}-h_3)L_f(x_f/x_1) \\ (h_{L3}-h_{L4})L_f(x_f/x_1) \\ (h_{L4}-h_{L5})L_f(x_f/x_1) \\ (h_{L5}-h_{L6})L_f(x_f/x_1) \\ (h_{L6}-h_{L7})L_f(x_f/x_1) \\ (h_{L7}-h_{Lf})L_f(x_f/x_1) \\ L_f\left(1-\frac{x_f}{x_1}\right) \end{bmatrix} \tag{53}$$

2.4.2 Forward Feed with Steam Split Configuration: Model-E

In this case, the black liquor and steam with vapor is fed in same direction but only the change is live steam is split in first two effects as in Model-D. The vapor generated from 1st and 2nd effect are combined together and sent to 3rd effect and then vapor produced at 3rd effect is further used as heat source for the next effect.

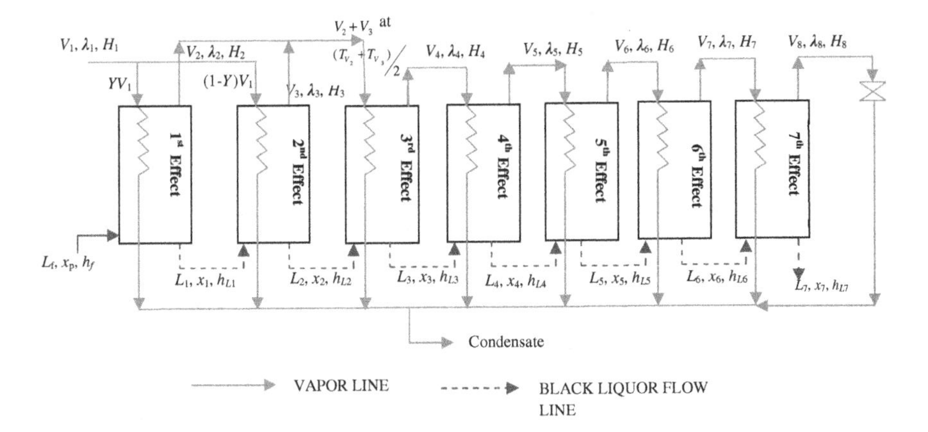

Fig. 8 Forward feed coupled with steam split heptads effect evaporator system

The un-concentrated liquor is fed to 1st effect and final concentrated product is obtained from last 7th effect as shown in Fig. 8. The mathematical model will remain same from 4th to last 7th effect and for 1st–3rd effect, the mathematical model are derived as per steam split concept in which the steam is split with fraction Y and $(1 - Y)$ at 1st and 2nd effect.

The energy balance at 1st effect is given by Eq. (54) and the final model at 1st effect is given by Eq. (57)

$$Y\lambda_1 V_1 + L_f h_{Lf} = V_2 H_2 + L_1 h_{L1} \tag{54}$$

$$Y\lambda_1 V_1 - V_2 H_2 = L_1 h_{L1} - L_f h_{Lf} \tag{55}$$

$$Y\lambda_1 V_1 + V_2(h_f - H_2) = (V_3 + L_2)(h_{L1} - h_{Lf}) \tag{56}$$

$$\begin{aligned} &Y\lambda_1 V_1 + V_2(h_f - H_2) + V_3(h_f - h_1) + V_4(h_f - h_1) + V_5(h_f - h_1) \\ &\quad + V_6(h_f - h_1) + V_7(h_f - h_1) + V_8(h_f - h_1) = (h_{L1} - h_{Lf})L_f(x_f/x_7) \end{aligned} \tag{57}$$

At 2nd effect, the same V_1 is fed but with steam fraction $(1 - Y)$, therefore, the derived model is given by Eq. (58)

$$\begin{aligned} &(1 - Y)\lambda_1 V_1 + V_3(h_{L1} - H_3) + V_4(h_{L1} - h_{L2}) + V_5(h_{L1} - h_{L2}) + V_6(h_{L1} - h_{L2}) \\ &\quad + V_7(h_{L1} - h_{L2}) + V_8(h_{L1} - h_{L2}) = (h_{L2} - h_{L1})L_f(x_f/x_7) \end{aligned} \tag{58}$$

At 3rd effect, the vapor produced from 1st and 2nd effect are combined, it means $(V_2 + V_3)$ is now fed to 3rd effect but now latent heat of vaporization is calculated at temperature $\frac{T_{V2} + T_{V3}}{2} = \lambda_{\text{avg}}$.

Therefore, using the energy balances derived model at 3rd effects is given by Eq. (59).

$$\begin{aligned} & V_2\lambda_{avg} + V_3\lambda_{avg} + V_4(h_{L2} - H_4) + V_5(h_{L2} - h_{L3}) + V_6(h_{L2} - h_{L3}) \\ & \quad + V_7(h_{L2} - h_{L3}) + V_8(h_{L2} - h_{L3}) = (h_{L3} - h_{L2})L_f(x_f/x_7) \end{aligned} \tag{59}$$

Hence, the final representation of Model-E is represented in matrix form is given by Eq. (60)

$$\begin{bmatrix} Y\lambda_1 & (h_{Lf} - H_2) & (h_{Lf} - h_{L1}) & (h_{Lf} - h_{L1}) & (h_{Lf} - h_{L1}) & (h_{Lf} - h_{L1}) & (h_{Lf} - h_{L1}) & (h_{Lf} - h_{L1}) \\ (1-Y)\lambda_1 & 0 & (h_{L1} - H_3) & (h_{L1} - h_{L2}) & (h_{L1} - h_{L2}) & (h_{L1} - h_{L2}) & (h_{L1} - h_{L2}) & (h_{L1} - h_{L2}) \\ 0 & \lambda_{avg} & \lambda_{avg} & (h_{L2} - H_4) & (h_{L2} - h_{L3}) & (h_{L2} - h_{L3}) & (h_{L2} - h_{L3}) & (h_{L2} - h_{L3}) \\ 0 & 0 & 0 & \lambda_4 & (h_{L3} - H_5) & (h_{L3} - h_{L4}) & (h_{L3} - h_{L4}) & (h_{L3} - h_{L4}) \\ 0 & 0 & 0 & 0 & \lambda_5 & (h_{L4} - H_6) & (h_{L4} - h_{L5}) & (h_{L4} - h_{L5}) \\ 0 & 0 & 0 & 0 & 0 & \lambda_6 & (h_{L5} - H_7) & (h_{L5} - h_{L6}) \\ 0 & 0 & 0 & 0 & 0 & 0 & \lambda_7 & (h_{L6} - H_8) \\ 0 & 1 & 1 & 1 & 1 & 1 & 1 & 1 \end{bmatrix} \begin{bmatrix} V_1 \\ V_2 \\ V_3 \\ V_4 \\ V_5 \\ V_6 \\ V_7 \\ V_8 \end{bmatrix} = \begin{bmatrix} (h_{L1} - h_{Lf})L_f(x_f/x_7) \\ (h_{L2} - h_{L1})L_f(x_f/x_7) \\ (h_{L3} - h_{L2})L_f(x_f/x_7) \\ (h_{L4} - h_{L3})L_f(x_f/x_7) \\ (h_{L5} - h_{L4})L_f(x_f/x_7) \\ (h_{L6} - h_{L5})L_f(x_f/x_7) \\ (h_{L7} - h_{L6})L_f(x_f/x_7) \\ L_f\left(1 - \frac{x_f}{x_1}\right) \end{bmatrix} \tag{60}$$

2.4.3 Mixed Feed with Steam Split Configuration: Model-F

Mathematical model for Model-F is derived using the models of Model-C and concepts of steam split as discussed above and the model is illustrated in Fig. 9 and the final representation of model is represented in matrix form and given by Eq. (61).

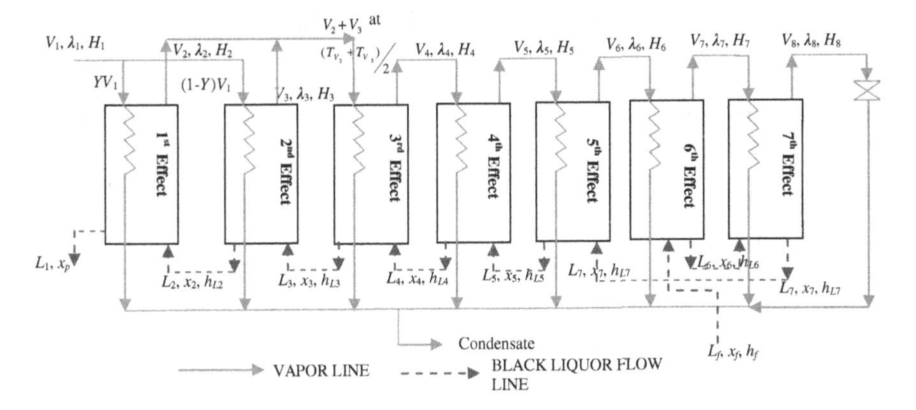

Fig. 9 Mixed feed coupled with steam split heptads effect evaporator system

$$
\begin{bmatrix}
Y\lambda_1 & (h_{L2}-H_2) & 0 & 0 & 0 & 0 & 0 & 0 \\
(1-Y)\lambda_1 & (h_{L3}-h_{L2}) & (h_{L3}-H_3) & 0 & 0 & 0 & 0 & 0 \\
0 & (\lambda_{avg}+h_{L4}-h_{L3}) & (\lambda_{avg}+h_{L4}-h_{L3}) & (h_{L4}-H_4) & 0 & 0 & 0 & 0 \\
0 & (h_{L5}-h_{L4}) & (h_{L5}-h_{L4}) & (\lambda_4+h_{L5}-h_{L4}) & (h_{L5}-H_5) & 0 & 0 & 0 \\
0 & (h_{L7}-h_{L5}) & (h_{L7}-h_{L5}) & (h_{L7}-h_{L5}) & (\lambda_5+h_{L7}-h_{L5}) & (h_{L7}-H_6) & 0 & 0 \\
0 & (h_{Lf}-h_{L6}) & (h_{Lf}-h_{L6}) & (h_{Lf}-h_{L6}) & (h_{Lf}-h_{L6}) & (\lambda_6+h_{Lf}-h_{L6}) & (h_{Lf}-H_7) & (h_{Lf}-h_{L6}) \\
0 & (h_{L6}-h_{L7}) & (h_{L6}-h_{L7}) & (h_{L6}-h_{L7}) & (h_{L6}-h_{L7}) & (h_{L6}-h_{L7}) & \lambda_7 & (h_{L6}-H_8) \\
0 & 1 & 1 & 1 & 1 & 1 & 1 & 1
\end{bmatrix}
\begin{bmatrix} V_1 \\ V_2 \\ V_3 \\ V_4 \\ V_5 \\ V_6 \\ V_7 \\ V_8 \end{bmatrix}
=
\begin{bmatrix}
(h_{L1}-h_{L2})L_f(x_f/x_1) \\
(h_{L2}-h_{L3})L_f(x_f/x_1) \\
(h_{L3}-h_{L4})L_f(x_f/x_1) \\
(h_{L4}-h_{L5})L_f(x_f/x_1) \\
(h_{L5}-h_{L7})L_f(x_f/x_1) \\
(h_{L6}-h_{Lf})L_f(x_f/x_1) \\
(h_{L7}-h_{L6})L_f(x_f/x_1) \\
L_f(1-x_f/x_1)
\end{bmatrix}
\tag{61}
$$

2.5 Configurations with Feed Split

Previous literature Kumar et al. [9] used this configuration in developing the dynamic model of MEE in pulp and paper industry and observes the dynamic behavior for the temperature, concentration of each effect of the system. In this configuration, the un-concentrated black liquor is split in fraction K and $(1-K)$ and simultaneously fed to two effect.

2.5.1 Backward Feed with Feed Split Configuration: Model-G

The un-concentrated black liquor is fed in last two effect 7th and 6th effects simultaneously with fraction K and $(1-K)$ respectively as shown in Fig. 10. The live steam is fed as same in Model-A.

The equations for the 1st–5th effect will remain same as in Model-A and for 6th and 7th effect due to flow of liquor fraction $(1-K)$, the equation will change and shown in Eqs. (62)–(68)

$$V_6\lambda_6 + L_7h_{L7} + (1-K)L_fh_{Lf} = V_7H_7 + L_6h_{L6} \tag{62}$$

$$V_6\lambda_6 - V_7H_7 = +L_6h_{L6} - L_7h_{L7} - (1-K)(V_8+L_7)h_{Lf} \tag{63}$$

$$V_6\lambda_6 - V_7H_7 + V_8(1-K)h_{Lf} = +L_6h_{L6} - L_7h_{L7} - (1-K)(V_7+L_7)h_{Lf} \tag{64}$$

$$V_6\lambda_6 + V_7\{(1-K)h_{Lf} - H_7\} + V_8(1-K)h_{Lf} = (V_6+L_5)\{h_{L6} - (1-K)h_{Lf}\} \tag{65}$$

$$\begin{aligned} &V_6\{\lambda_6 - h_{L6} + (1-K)h_{Lf}\} + V_7\{(1-K)h_{Lf} - H_7\} + V_8(1-K)h_{Lf} \\ &\quad = (V_5+L_4)\{h_{L6} - (1-K)h_{Lf}\} \end{aligned} \tag{66}$$

$$\begin{aligned} &V_5\{(1-K)h_{Lf} - h_{L6}\} + V_6\{\lambda_6 - h_{L6} + (1-K)h_{Lf}\} + V_7\{(1-K)h_{Lf} - H_7\} \\ &\quad + V_8(1-K)h_{Lf} = (V_4+L_3)\{h_{L6} - (1-K)h_{Lf}\} \end{aligned} \tag{67}$$

$$\begin{aligned} &V_2\{(1-K)h_{Lf} - h_{L6}\} + V_3\{(1-K)h_{Lf} - h_{L6}\} + V_4\{(1-K)h_{Lf} - h_{L6}\} \\ &\quad + V_5\{(1-K)h_{Lf} - h_{L6}\} + V_6\{\lambda_6 - h_{L6} + (1-K)h_{Lf}\} \\ &\quad + V_7\{(1-K)h_{Lf} - H_7\} + V_8(1-K)h_{Lf} = \{h_{L6} - (1-K)h_{Lf}\}L_f(x_f/x_1) \end{aligned} \tag{68}$$

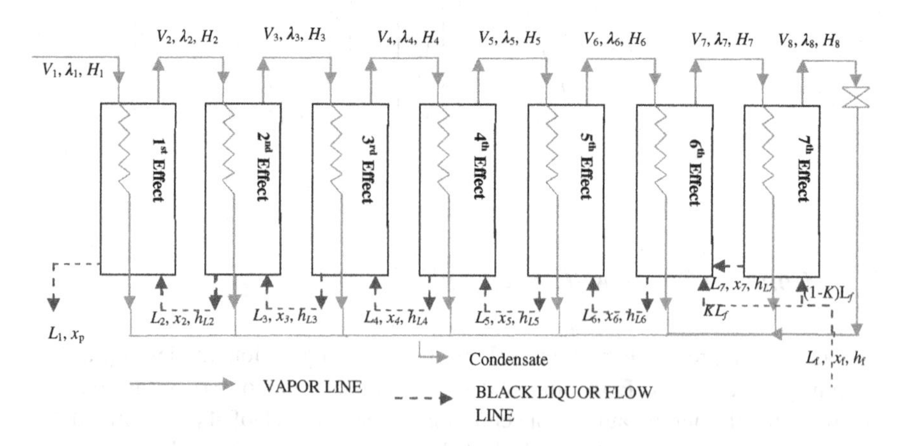

Fig. 10 Backward feed coupled with steam split heptads effect evaporator system

For 7th effect, the equation turns out into Eq. (69)

$$\begin{aligned} &V_2\{Kh_{Lf} - h_{L7}\} + V_3\{Kh_{Lf} - h_{L7}\} + V_4\{Kh_{Lf} - h_{L7}\} \\ &\quad + V_5\{Kh_{Lf} - h_{L7}\} + V_6\{Kh_{Lf} - h_{L7}\} + V_7\{\lambda_7 + Kh_{Lf} - h_{L7}\} \\ &\quad + V_8(Kh_{Lf} - H_8) = \{h_{L7} - Kh_{Lf}\}L_f(x_f/x_1) \end{aligned} \tag{69}$$

The final representation of the Model-G in matrix form is represented by Eq. (70)

$$\begin{bmatrix} \lambda_1 & (h_{L2}-H_2) & 0 & 0 & 0 & 0 & 0 & 0 \\ 0 & (\lambda_2+h_{L3}-h_{L2}) & (h_{L3}-H_3) & 0 & 0 & 0 & 0 & 0 \\ 0 & (h_{L4}-h_{L3}) & (\lambda_3+h_{L4}-h_{L3}) & (h_{L4}-H_4) & 0 & 0 & 0 & 0 \\ 0 & (h_{L5}-h_{L4}) & (h_{L5}-h_{L4}) & (\lambda_4+h_{L5}-h_{L4}) & (h_{L5}-H_5) & 0 & 0 & 0 \\ 0 & (h_{L6}-h_{L5}) & (h_{L6}-h_{L5}) & (h_{L6}-h_{L5}) & (\lambda_5+h_{L6}-h_{L5}) & (h_{L6}-H_6) & 0 & 0 \\ 0 & \{(1-K)h_{Lf}-h_6\} & \{(1-K)h_{Lf}-h_6\} & \{(1-K)h_{Lf}-h_6\} & \{(1-K)h_{Lf}-h_6\} & \{\lambda_6-h_{L6}+(1-K)h_{Lf}\} & \{(1-K)h_{Lf}-H_7\} & \{(1-K)h_{Lf}\} \\ 0 & (Kh_{Lf}-h_7) & (Kh_{Lf}-h_7) & (Kh_{Lf}-h_7) & (Kh_{Lf}-h_7) & (Kh_{Lf}-h_7) & \{\lambda_7+Kh_{Lf}-h_{L7}\} & (Kh_{Lf}-H_8) \\ 0 & 1 & 1 & 1 & 1 & 1 & 1 & 1 \end{bmatrix} \begin{bmatrix} V_1 \\ V_2 \\ V_3 \\ V_4 \\ V_5 \\ V_6 \\ V_7 \\ V_8 \end{bmatrix} = \begin{bmatrix} (h_{L1}-h_{L2})L_f(x_f/x_1) \\ (h_{L2}-h_{L3})L_f(x_f/x_1) \\ (h_{L3}-h_{L4})L_f(x_f/x_1) \\ (h_{L4}-h_{L5})L_f(x_f/x_1) \\ (h_{L5}-h_{L6})L_f(x_f/x_1) \\ \{h_{L6}-(1-K)h_{Lf}\}L_f(x_f/x_1) \\ (h_{L7}-Kh_{Lf})L_f(x_f/x_1) \\ L_f(1-x_f/x_1) \end{bmatrix} \tag{70}$$

2.5.2 Forward Feed with Feed Split Configuration: Model-H

In this configuration, as shown in Fig. 11, the live vapor is fed in similar ways as in forward feed configuration but the feed black liquor is split in first two effects, i.e.,

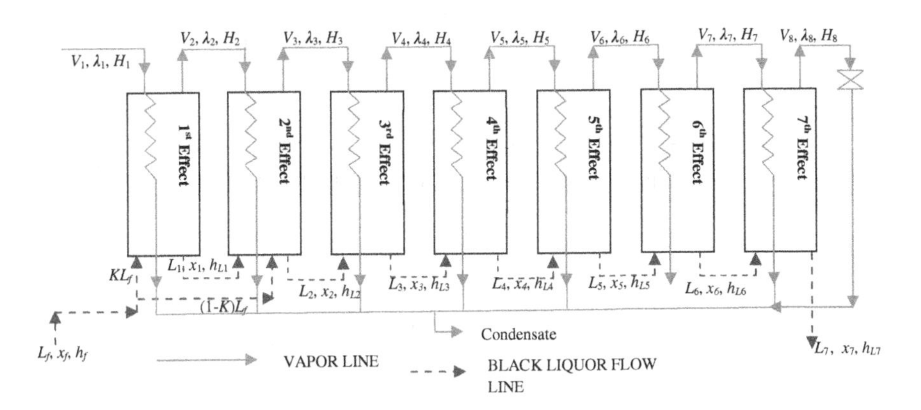

Fig. 11 Forward feed coupled with feed split heptads effect evaporator system

1st and 2nd effects with split fraction K and $(1 - K)$. Therefore, the mathematical model will change for first two effects and will remain same from 3rd to 7th effects. Hence, for first two effects, mathematical models are derived from Eqs. (71)–(84).

For 1st effect,

$$V_1\lambda_1 + KL_f h_{Lf} = V_2H_2 + L_1h_{L1} \tag{71}$$

$$V_1\lambda_1 - V_2H_2 = L_1h_{L1} - (V_2 + L_1)h_f \tag{72}$$

$$V_1\lambda_1 + V_2(h_f - H_2) = L_1(h_{L1} - h_f) \tag{73}$$

$$V_1\lambda_1 + V_2(h_f - H_2) = \left\{V_3 + L_2 - (1 - K)L_f\right\}(h_{L1} - h_f) \tag{74}$$

$$V_1\lambda_1 + V_2(h_f - H_2) + V_3(h_f - h_1) = \left\{V_4 + L_3 - (1 - K)L_f\right\}(h_{L1} - h_f) \tag{75}$$

$$\begin{aligned} &V_1\lambda_1 + V_2(h_f - H_2) + V_3(h_f - h_1) + V_4(h_f - h_1) \\ &= \left\{V_5 + L_4 - (1 - K)L_f\right\}(h_{L1} - h_f) \end{aligned} \tag{76}$$

$$\begin{aligned} &V_1\lambda_1 + V_2(h_f - H_2) + V_3(h_f - h_{L1}) + V_4(h_f - h_{L1}) + V_5(h_f - h_{L1}) \\ &\quad + V_6(h_f - h_{L1}) + V_7(h_f - h_{L1}) + V_8(h_f - h_{L1}) \\ &\quad = \{L_7 - (1 - K)L_f\}(h_{L1} - h_f) \end{aligned} \tag{77}$$

$$\begin{aligned} &V_1\lambda_1 + V_2(h_f - H_2) + V_3(h_f - h_{L1}) + V_4(h_f - h_{L1}) + V_5(h_f - h_{L1}) \\ &\quad + V_6(h_f - h_{L1}) + V_7(h_f - h_{L1}) + V_8(h_f - h_{L1}) \\ &\quad = L_f\{(x_f/x_7) - (1 - K)L_f\}(h_{L1} - h_f) \end{aligned} \tag{78}$$

For the 2nd effect,

$$V_2\lambda_2 + L_1h_{L1} + (1 - K)L_f h_{Lf} = V_3H_3 + L_2h_{L2} \tag{79}$$

$$V_2\lambda_2 + V_3(h_1 - H_3) = L_2h_{L2} - L_2h_{L1} + (1 - K)L_f h_{L1} - (1 - K)L_f h_f \tag{80}$$

$$\begin{aligned} V_2\lambda_2 + V_3(h_1 - H_3) + V_4(h_{L1} - h_{L2}) &= (V_5 + L_4)(h_{L2} - h_{L1}) \\ &\quad + (1 - K)L_f(h_{L1} - h_f) \end{aligned} \tag{81}$$

$$\begin{aligned} &V_2\lambda_2 + V_3(h_1 - H_3) + V_4(h_{L1} - h_{L2}) + V_5(h_{L1} - h_{L2}) \\ &= (V_6 + L_5)(h_{L2} - h_{L1}) + (1 - K)L_f(h_{L1} - h_f) \end{aligned} \tag{82}$$

$$\begin{aligned} &V_2\lambda_2 + V_3(h_1 - H_3) + V_4(h_{L1} - h_{L2}) + V_5(h_{L1} - h_{L2}) \\ &\quad + V_6(h_{L1} - h_{L2}) + V_7(h_{L1} - h_{L2}) + V_8(h_{L1} - h_{L2}) \\ &= L_7(h_{L2} - h_{L1}) + (1 - K)L_f(h_{L1} - h_f) \end{aligned} \tag{83}$$

$$\begin{aligned} & V_2\lambda_2 + V_3(h_1 - H_3) + V_4(h_{L1} - h_{L2}) + V_5(h_{L1} - h_{L2}) \\ & \quad + V_6(h_{L1} - h_{L2}) + V_7(h_{L1} - h_{L2}) + V_8(h_{L1} - h_{L2}) \\ & \quad = L_f\{(x_f/x_1)(h_{L2} - h_{L1}) + (1-K)L_f(h_{L1} - h_f)\} \end{aligned} \tag{84}$$

The final representation for Model-H is given by Eq. (85)

$$\begin{bmatrix} \lambda_1 & (h_f - H_2) & (h_{Lf} - h_{L1}) & (h_f - h_{L1}) & (h_f - h_{L1}) & (h_f - h_{L1}) & (h_f - h_{L1}) & (h_f - h_{L1}) \\ 0 & \lambda_2 & (h_{L1} - H_3) & (h_{L1} - h_{L2}) & (h_{L1} - h_{L2}) & (h_{L1} - h_{L2}) & (h_{L1} - h_{L2}) & (h_{L1} - h_{L2}) \\ 0 & 0 & \lambda_3 & (h_{L2} - H_4) & (h_{L2} - h_{L3}) & (h_{L2} - h_{L3}) & (h_{L2} - h_{L3}) & (h_{L2} - h_{L3}) \\ 0 & 0 & 0 & \lambda_4 & (h_{L3} - H_5) & (h_{L3} - h_{L4}) & (h_{L3} - h_{L4}) & (h_{L3} - h_{L4}) \\ 0 & 0 & 0 & 0 & \lambda_5 & (h_{L4} - H_6) & (h_{L4} - h_{L5}) & (h_{L4} - h_{L5}) \\ 0 & 0 & 0 & 0 & 0 & \lambda_6 & (h_{L5} - H_7) & (h_{L5} - h_{L6}) \\ 0 & 0 & 0 & 0 & 0 & 0 & \lambda_7 & (h_{L6} - H_8) \\ 0 & 1 & 1 & 1 & 1 & 1 & 1 & 1 \end{bmatrix} \begin{bmatrix} V_1 \\ V_2 \\ V_3 \\ V_4 \\ V_5 \\ V_6 \\ V_7 \\ V_8 \end{bmatrix} = \begin{bmatrix} L_f\{(x_f/x_7) - (1-K)\}(h_{L1} - h_{Lf}) \\ L_f\{(x_f/x_7)(h_{L2} - h_{L1}) + (1-K)(h_{L1} - h_{Lf})\} \\ L_f(x_f/x_7)(h_{L3} - h_{L2}) \\ L_f(x_f/x_7)(h_{L4} - h_{L3}) \\ L_f(x_f/x_7)(h_{L5} - h_{L4}) \\ L_f(x_f/x_7)(h_{L6} - h_{L5}) \\ L_f(x_f/x_7)(h_{L7} - h_{L6}) \\ L_f(1 - x_f/x_7) \end{bmatrix} \tag{85}$$

2.5.3 Mixed Feed with Feed Split Configuration: Model-I

This configuration emanates from mixed feed configuration as in Model-C and liquor feed split concept. The black liquor is to be fed is split among 5th and 6th effects with split fraction $(1 - K)$ and K, respectively, and the liquor produced at last effect is fed to 4th effect. Live fed vapor propagates in forward direction as shown in Fig. 12. Therefore, the mathematical models will remain same for 1st–3th effects but mathematical model changes for 4th and 7th effects.

The developed model for 4th–7th are shown from Eqs. (86)–(). For 5th effect,

$$V_4\lambda_4 + L_7h_{L7} + L_5h_{L5} = V_5H_5 + L_4h_{L4} \tag{86}$$

$$V_4\lambda_4 + V_5(h_{L7} - H_5) = L_4(h_{L4} - h_{L7}) + L_5(h_{L7} - h_{L5}) \tag{87}$$

$$\begin{aligned} & V_2(h_{L7} - h_{L4}) + V_3(h_{L7} - h_{L4}) + V_4(\lambda_4 + h_{L7} - h_{L4}) + V_5(h_{L7} - H_5) \\ & \quad + V_6(h_{L7} - h_{L5}) = L_f\{(x_f/x_1)(h_{L4} - h_{L7}) + (1-K)(h_{L7} - h_{L5})\} \end{aligned} \tag{88}$$

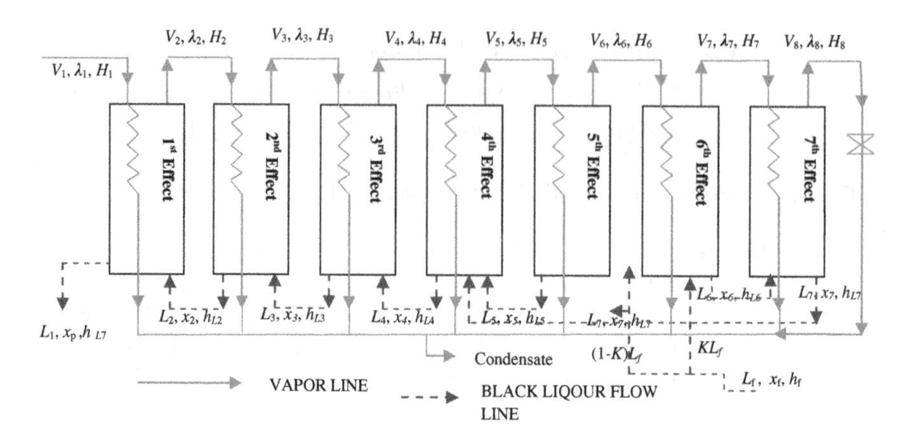

Fig. 12 Mixed feed coupled with feed split heptads effect evaporator system

For 5th effect,

$$V_5\lambda_5 + V_6(h_{L5} - H_6) = (1-K)L_f(h_5 - h_f) \tag{89}$$

For 6th effect,

$$V_6\lambda_6 + V_7(h_{L6} - H_7) = KL_f(h_{L6} - h_f) \tag{90}$$

For the last 7th effect,

$$V_7(\lambda_7 + h_{L6} - h_{L7}) + V_8(h_{L7} - H_8) = KL_f(h_{L7} - h_{L6}) \tag{91}$$

Hence, the final mathematical model for Model-I that will turnout in matrix form is given by Eq. (92)

$$\begin{bmatrix} \lambda_1 & (h_{L2}-H_2) & 0 & 0 & 0 & 0 & 0 & 0 \\ 0 & (\lambda_2+h_{L3}-h_{L2}) & (h_{L3}-H_3) & 0 & 0 & 0 & 0 & 0 \\ 0 & (h_{L4}-h_{L3}) & (\lambda_3+h_{L4}-h_{L3}) & (h_{L4}-H_4) & 0 & 0 & 0 & 0 \\ 0 & (h_{L7}-h_{L4}) & (h_{L7}-h_{L4}) & (\lambda_4+h_{L7}-h_{L4}) & (h_{L7}-H_5) & (h_{L7}-h_{L5}) & 0 & 0 \\ 0 & 0 & 0 & 0 & \lambda_5 & (h_{L5}-H_6) & 0 & 0 \\ 0 & 0 & 0 & 0 & 0 & \lambda_6 & (h_{L6}-H_7) & 0 \\ 0 & 0 & 0 & 0 & 0 & 0 & (\lambda_7+h_{L6}-h_{L7}) & (h_{L7}-H_8) \\ 0 & 1 & 1 & 1 & 1 & 1 & 1 & 1 \end{bmatrix}$$
$$\begin{bmatrix} V_1 \\ V_2 \\ V_3 \\ V_4 \\ V_5 \\ V_6 \\ V_7 \\ V_8 \end{bmatrix} = \begin{bmatrix} L_f(x_f/x_1)(h_{L1}-h_{L2}) \\ L_f(x_f/x_1)(h_{L2}-h_{L3}) \\ L_f(x_f/x_1)(h_{L3}-h_{L4}) \\ L_f\{(x_f/x_1)(h_{L4}-h_{L7})+(1-K)(h_{L7}-h_{L5})\} \\ L_f(1-K)(h_{L5}-h_{Lf}) \\ L_fK(h_{L6}-h_{Lf}) \\ L_fK(h_{L7}-h_{L6}) \\ L_f(1-x_f/x_1) \end{bmatrix} \tag{92}$$

2.6 Configuration with Steam Split and Feed Split

2.6.1 Backward Feed Configuration with Steam Split and Feed Split: Model-J

Using the concepts of steam split and feed split the new model is developed in which steam is split among first two effects 1st and 2nd effects and un-concentrated black liquor is fed among the last two effects 6th and 7th effects. The steam is split with fraction y and $(1 - Y)$ in 1st and 2nd effects, respectively, and black liquor is split with split fraction K and $(1 - K)$ in 7th and 6th effects, respectively, as shown in Fig. 13.

The mathematical model combines the Model-D and Model-G. The final model in matrix form is given by Eq. (93).

$$\begin{bmatrix} Y\lambda_1 & (h_{L2}-H_2) & 0 & 0 & 0 & 0 & 0 & 0 \\ (1-Y)\lambda_1 & (h_{L3}-h_{L2}) & (h_{L3}-H_3) & 0 & 0 & 0 & 0 & 0 \\ 0 & (\lambda_{\text{avg}}+h_{L4}-h_{L3}) & (\lambda_{\text{avg}}+h_{L4}-h_{L3}) & (h_{L4}-H_4) & 0 & 0 & 0 & 0 \\ 0 & (h_{L5}-h_{L4}) & (h_{L5}-h_{L4}) & (\lambda_4+h_{L5}-h_{L4}) & (h_{L5}-H_5) & 0 & 0 & 0 \\ 0 & (h_{L6}-h_{L5}) & (h_{L6}-h_{L5}) & (h_{L6}-h_{L5}) & (\lambda_5+h_{L6}-h_{L5}) & (h_{L6}-H_6) & 0 & 0 \\ 0 & (h_{L7}-h_{L6}) & (h_{L7}-h_{L6}) & (h_{L7}-h_{L6}) & (h_{L7}-h_{L6}) & (\lambda_6+h_{L7}-h_{L6}) & (h_{L7}-H_7) & 0 \\ 0 & 0 & 0 & 0 & 0 & 0 & \lambda_7 & (h_{L7}-H_8) \\ 0 & 1 & 1 & 1 & 1 & 1 & 1 & 1 \end{bmatrix} \begin{bmatrix} V_1 \\ V_2 \\ V_3 \\ V_4 \\ V_5 \\ V_6 \\ V_7 \\ V_8 \end{bmatrix} = \begin{bmatrix} (h_{L1}-h_{L2})(L_f x_f/x_1) \\ (h_{L2}-h_{L3})(L_f x_f/x_1) \\ (h_{L3}-h_{L4})(L_f x_f/x_1) \\ (h_{L4}-h_{L5})(L_f x_f/x_1) \\ (h_{L5}-h_{L6})(L_f x_f/x_1) \\ (h_{L6}-h_{L7})(L_f x_f/x_1)+L_f(1-K)(h_{L7}-h_{Lf}) \\ KL_f(h_{L7}-h_{Lf}) \\ L_f(1-x_f/x_1) \end{bmatrix} \tag{93}$$

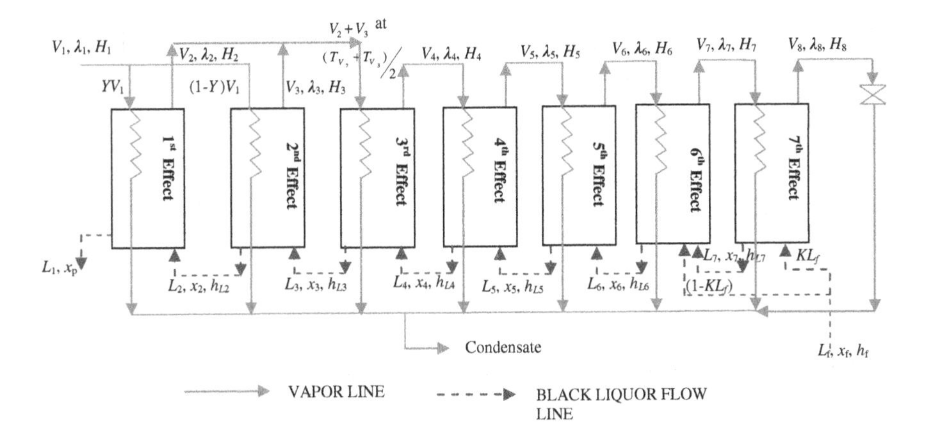

Fig. 13 Backward feed coupled with steam split and feed split heptads effect evaporator system

2.6.2 Forward Feed Configuration with Steam Split and Feed Split: Model-K

In this configuration, combining the concepts of Model-E and Model-H and obtained the improved Model-K and shown in Fig. 14. The mathematical model is represented in matrix form and is given by Eq. (94)

$$\begin{bmatrix} Y\lambda_1 & (h_{Lf}-H_2) & (h_{Lf}-h_{L1}) & (h_{Lf}-h_{L1}) & (h_{Lf}-h_{L1}) & (h_{Lf}-h_{L1}) & (h_{Lf}-h_{L1}) & (h_{Lf}-h_{L1}) \\ (1-Y)\lambda_1 & 0 & (h_{L1}-H_3) & (h_{L1}-h_{L2}) & (h_{L1}-h_{L2}) & (h_{L1}-h_{L2}) & (h_{L1}-h_{L2}) & (h_{L1}-h_{L2}) \\ 0 & \lambda_{\text{avg}} & \lambda_{\text{avg}} & (h_{L2}-H_4) & (h_{L2}-h_{L3}) & (h_{L2}-h_{L3}) & (h_{L2}-h_{L3}) & (h_{L2}-h_{L3}) \\ 0 & 0 & 0 & \lambda_4 & (h_{L3}-H_5) & (h_{L3}-h_{L4}) & (h_{L3}-h_{L4}) & (h_{L3}-h_{L4}) \\ 0 & 0 & 0 & 0 & \lambda_5 & (h_{L4}-H_6) & (h_{L4}-h_{L5}) & (h_{L4}-h_{L5}) \\ 0 & 0 & 0 & 0 & 0 & \lambda_6 & (h_{L5}-H_7) & (h_{L5}-h_{L6}) \\ 0 & 0 & 0 & 0 & 0 & 0 & \lambda_7 & (h_{L6}-H_8) \\ 0 & 1 & 1 & 1 & 1 & 1 & 1 & 1 \end{bmatrix} \begin{bmatrix} V_1 \\ V_2 \\ V_3 \\ V_4 \\ V_5 \\ V_6 \\ V_7 \\ V_8 \end{bmatrix} = \begin{bmatrix} L_f\{(x_f/x_7)-(1-K)\}(h_{L1}-h_{Lf}) \\ L_f(x_f/x_7)(h_{L2}-h_{L1})+(1-K)L_f(h_{L1}-h_{Lf}) \\ L_f(x_f/x_7)(h_{L3}-h_{L2}) \\ L_f(x_f/x_7)(h_{L4}-h_{L3}) \\ L_f(x_f/x_7)(h_{L5}-h_{L4}) \\ L_f(x_f/x_7)(h_{L6}-h_{L5}) \\ L_f(x_f/x_7)(h_{L7}-h_{L6}) \\ L_f\{1-(x_f/x_7)\} \end{bmatrix} \tag{94}$$

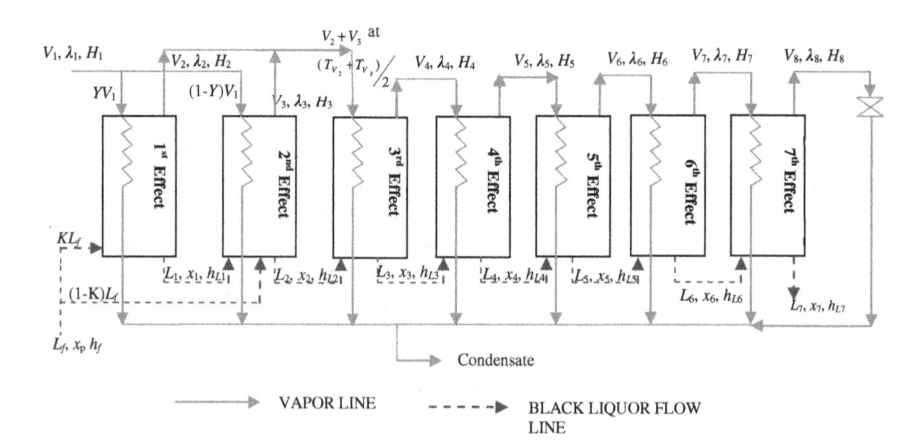

Fig. 14 Forward feed coupled with steam split and feed split heptads effect evaporator system

2.6.3 Mixed Feed Configuration with Steam Split and Feed Split: Model-L

Mixed feed configuration coupled with steam split and feed split is shown in Fig. 15. In similar way, the final equation for Model-L is represented by Eq. (95)

$$\begin{bmatrix} Y\lambda_1 & (h_{L2}-H_2) & 0 & 0 & 0 & 0 & 0 & 0 \\ (1-Y)\lambda_1 & (h_{L3}-h_{L2}) & (h_{L3}-H_2) & 0 & 0 & 0 & 0 & 0 \\ 0 & (\lambda_{avg}+h_{L4}-h_{L3}) & (\lambda_{avg}+h_{L4}-h_{L3}) & (h_{L4}-H_4) & 0 & 0 & 0 & 0 \\ 0 & (h_{L7}-h_{L4}) & (h_{L7}-h_{L4}) & (\lambda_4+h_{L7}-h_{L4}) & (h_{L7}-H_5) & (h_{L7}-h_{L5}) & 0 & 0 \\ 0 & 0 & 0 & 0 & \lambda_5 & (h_{L5}-H_6) & 0 & 0 \\ 0 & 0 & 0 & 0 & 0 & \lambda_6 & (h_{L6}-H_7) & 0 \\ 0 & 0 & 0 & 0 & 0 & 0 & (\lambda_7+h_{L7}-h_{L6}) & (h_{L7}-H_8) \\ 0 & 1 & 1 & 1 & 1 & 1 & 1 & 1 \end{bmatrix} \begin{bmatrix} V_1 \\ V_2 \\ V_3 \\ V_4 \\ V_5 \\ V_6 \\ V_7 \\ V_8 \end{bmatrix} = \begin{bmatrix} (h_{L1}-h_{L2})(L_f x_f/x_1) \\ (h_{L2}-h_{L3})(L_f x_f/x_1) \\ (h_{L3}-h_{L4})(L_f x_f/x_1) \\ (h_{L4}-h_{L7})(L_f x_f/x_1)+\{(1-K)L_f(h_{L7}-h_{L5})\} \\ (1-K)L_f(h_{L5}-h_{Lf}) \\ KL_f(h_{L6}-h_{Lf}) \\ KL_f(h_{L7}-h_{L6}) \\ L_f(1-x_f/x_1) \end{bmatrix} \quad (95)$$

2.7 Configuration Coupled with Pre-heater in MSE System

Pre-heater is used to heat the black liquor which is going to feed in MSE system and some part of vapor produced at several effects (Example: In backward feed

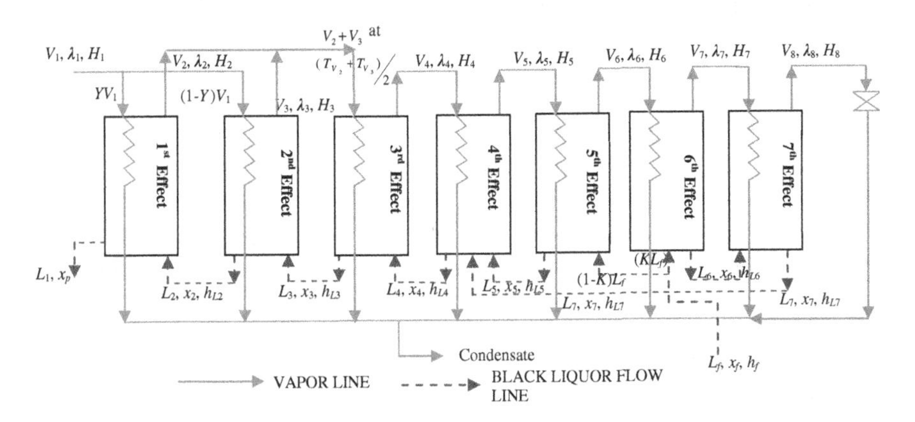

Fig. 15 Mixed feed coupled with steam split and feed split heptads effect evaporator system

configuration pre-heater is provided at 6th and 7th effect as shown in Fig. 16) worked as a heat source for the Pre-heater. Using this concept, new models Model-M, N and O combined with feed and steam split configurations are derived.

2.7.1 Backward Feed Configuration Coupled with Steam Split, Feed Split and Pre-heater: Model-M

In this operation, live steam is split among first two effects 1st and 2nd effects with split fraction Y and $(1 - Y)$, un-concentrated black liquor is taking from pre-heater and split to last two effects 6th and 7th effects and pre-heater is installed as shown in Fig. 16.

Some part of solution vapor obtained from 6th effect is sent to last effect, i.e., 7th effect and the remains are sent to the second pre-heater, while solution vapor of last effect, i.e., 7th effect is completely sent to pre-heater as heat source. The un-concentrated black liquor solutions temperature increases as $\Delta T_1 = T_0 - T$ in the first pre-heater and its temperature increase $\Delta T_2 = T - T_1$ in the second pre-heater. The un-concentrated solution is split and fed to last two effects. The concentrated black liquor solution as a product is obtained from 1st effects.

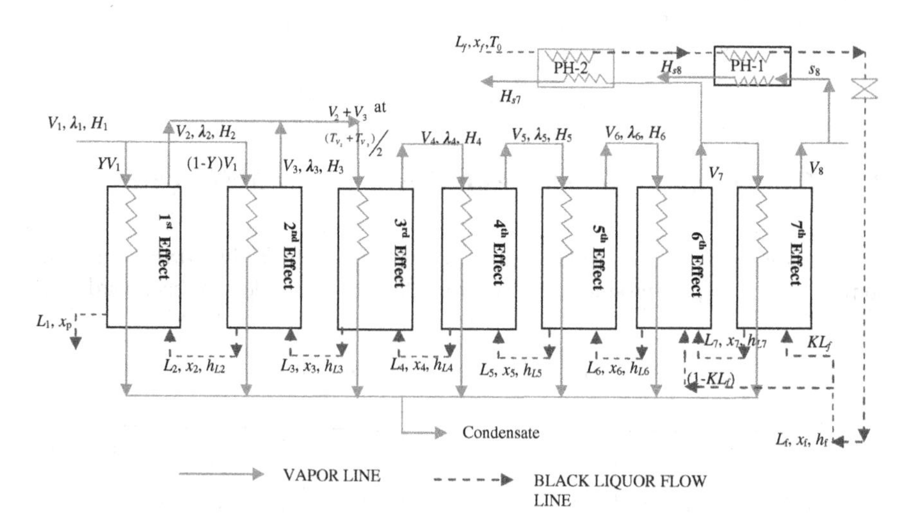

Fig. 16 Backward feed coupled with steam split, feed split and pre-heater heptads effect evaporator system

For the 1st–5th effects,

$$YV_1\lambda_1 + V_2(h_{L2} - H_2) = (h_{L1} - h_{L2})(L_f x_f / x_1) \tag{96}$$

$$(1 - Y)\lambda_1 V_1 + (h_{L3} - h_{L2})V_2 + (h_{L3} - H_3)V_3 = (h_{L2} - h_{L3})(L_f x_f / x_1) \tag{97}$$

$$\begin{aligned}(\lambda_{\text{avg}} + h_{L4} - h_{L3})V_2 + (\lambda_{\text{avg}} + h_{L4} - h_{L3})V_3 + (h_{L4} - H_4) \\ = (h_{L3} - h_{L4})(L_f x_f / x_1)\end{aligned} \tag{98}$$

$$\begin{aligned}(h_{L5} - h_{L4})V_2 + (h_{L5} - h_{L4})V_3 + (\lambda_4 + h_{L5} - h_{L4})V_4 \\ + (h_{L5} - H_5)V_5 = (h_{L4} - h_{L5})(L_f x_f / x_1)\end{aligned} \tag{99}$$

$$\begin{aligned}(h_{L6} - h_{L5})V_2 + (h_{L6} - h_{L5})V_3 + (h_{L6} - h_{L5})V_4 + (\lambda_5 + h_{L6} - h_{L5})V_5 \\ + (h_{L5} - H_5)V_6 = (h_{L5} - h_{L6})(L_f x_f / x_1)\end{aligned} \tag{100}$$

For the 6th effects,

$$V_6\lambda_6 + L_7 h_{L7} + (1 - K)L_f h_f = V_7 H_7 + L_6 h_6 \tag{101}$$

$$V_6\lambda_6 + V_7(h_{L7} - H_7) = L_6(h_{L6} - h_{L7}) + L_f(1 - K)(h_{L7} - h_{Lf}) \tag{102}$$

$$\begin{aligned}V_6(\lambda_6 + h_{L7} - h_{L6}) + V_7(h_{L7} - H_7) = (V_5 + L_4)(h_{L6} - h_{L7}) \\ + L_f(1 - K)(h_{L7} - h_{Lf})\end{aligned} \tag{103}$$

$$\begin{aligned}V_2(h_{L7} - h_{L6}) + V_3(h_{L7} - h_{L6}) + V_4(h_{L7} - h_{L6}) + V_5(h_{L7} - h_{L6}) \\ + V_6(\lambda_6 + h_{L7} - h_{L6}) + V_7(h_{L7} - H_7) \\ = L_f(x_f / x_1)(h_{L6} - h_{L7}) + L_f(1 - K)(h_{L7} - h_{Lf})\end{aligned} \tag{104}$$

For the 7th effects,

$$(V_7 - S_7)\lambda_7 + KL_f h_{Lf} = V_8 H_8 + L_7 h_{L7} \tag{105}$$

$$V_7\lambda_7 + KL_f h_{Lf} = V_8 H_8 + (KL_f - V_8)h_{L7} + L_f C_p \Delta T_1 \tag{106}$$

Hence, the final linear equation for the Model-M in matrix form is given by Eq. (107)

$$\begin{bmatrix} Y\lambda_1 & (h_{L2}-H_2) & 0 & 0 & 0 & 0 & 0 & 0 \\ (1-Y)\lambda_1 & (h_{L3}-h_{L2}) & (h_{L3}-H_3) & 0 & 0 & 0 & 0 & 0 \\ 0 & (\lambda_{\text{avg}}+h_{L4}-h_{L3}) & (\lambda_{\text{avg}}+h_{L4}-h_{L3}) & (h_{L4}-H_4) & 0 & 0 & 0 & 0 \\ 0 & (h_{L5}-h_{L4}) & (h_{L5}-h_{L4}) & (\lambda_4+h_{L5}-h_{L4}) & (h_{L5}-H_5) & 0 & 0 & 0 \\ 0 & (h_{L6}-h_{L5}) & (h_{L6}-h_{L5}) & (h_{L6}-h_{L5}) & (\lambda_5+h_{L6}-h_{L5}) & (h_{L6}-H_6) & 0 & 0 \\ 0 & (h_{L7}-h_{L6}) & (h_{L7}-h_{L6}) & (h_{L7}-h_{L6}) & (h_{L7}-h_{L6}) & (\lambda_6+h_{L7}-h_{L6}) & (h_{L7}-H_7) & 0 \\ 0 & 0 & 0 & 0 & 0 & 0 & \lambda_7 & (h_{L7}-H_8) \\ 0 & 1 & 1 & 1 & 1 & 1 & 1 & 1 \end{bmatrix} \begin{bmatrix} V_1 \\ V_2 \\ V_3 \\ V_4 \\ V_5 \\ V_6 \\ V_7 \\ V_8 \end{bmatrix} = \begin{bmatrix} (h_{L1}-h_{L2})(L_f x_f/x_1) \\ (h_{L2}-h_{L3})(L_f x_f/x_1) \\ (h_{L3}-h_{L4})(L_f x_f/x_1) \\ (h_{L4}-h_{L5})(L_f x_f/x_1) \\ (h_{L5}-h_{L6})(L_f x_f/x_1) \\ (h_{L6}-h_{L7})(L_f x_f/x_1)+L_f(1-K)(h_{L7}-h_{Lf}) \\ KL_f(h_{L7}-h_{Lf})+L_f C_p \Delta T_1 \\ L_f(1-x_f/x_1) \end{bmatrix} \tag{107}$$

2.7.2 Forward Feed Configuration Coupled with Feed Split and Pre-heater: Model-N

In this case, live steam V_1 is fed to 1st effect and some part of solution vapor V_2 obtained in this steps is sent to 2nd effects evaporator, i.e., $(V_2 - S_2)$ and S_2 remains are sent to the 2nd pre-heater, while some part of solution vapor V_3 of 2nd evaporator is sent to 3rd effect evaporator $(V_3 - S_3)$ and remaining S_3 is given to 1st pre-heater as a heat source. The solution obtained at 3rd effects is used further to 4th and so on. The un-concentrated solution black liquor temperature increases as $\Delta T_1 = (T_0 - T)$ in 1st pre-heater and its temperature increases $\Delta T_2 = (T - T_1)$ in 2nd pre-heater. Finally, the un-concentrated solution coming from 2nd pre-heater is split among 1st and 2nd effects with fraction K and $(1 - K)$, the concentrated product of each effects after 2nd effect is fed to next one. The product is obtained in the last effects, i.e., 7th effect and shown in Fig. 17.

For the 1st effects, the energy balances is

$$V_1\lambda_1 + KL_f h_f = L_1 h_{L1} + V_2 H_2 \tag{108}$$

$$V_1\lambda_1 - V_2 H_2 = (L_f - V_2)h_{L1} - KL_f h_f \tag{109}$$

$$V_1\lambda_1 + V_2(h_{L1} - H_2) = L_f(h_{L1} - Kh_f) \tag{110}$$

For the 2nd effects, the energy balances is

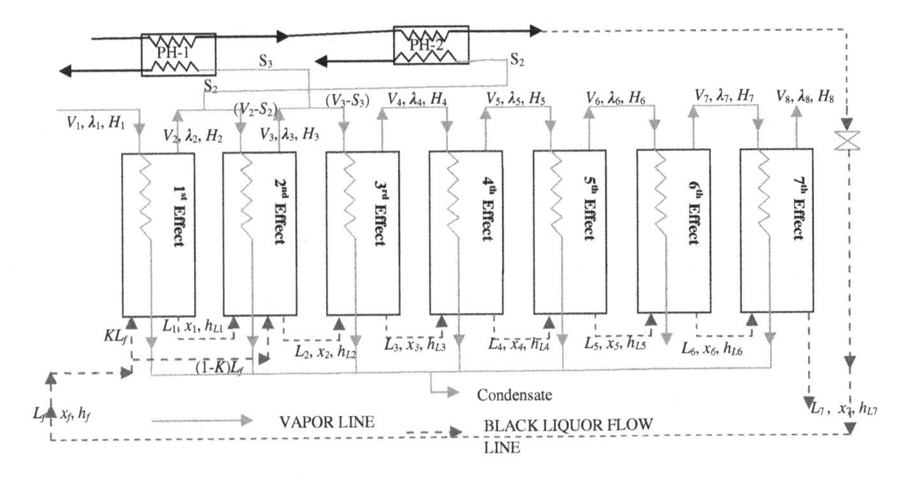

Fig. 17 Forward feed coupled with feed split and pre-heater heptads effect evaporator system

$$(V_2 - S_2)\lambda_2 + L_1 h_{L1} + (1 - K)L_f h_{Lf} = V_3 H_3 + L_2 h_{L2} \tag{111}$$

$$\begin{aligned} V_2\lambda_2 + V_3(h_{L2} - H_3) = L_1(h_{L2} - h_{L1}) \\ + (1 - K)L_f(h_{L2} - h_f) + L_f C_p(T - T_1) \end{aligned} \tag{112}$$

$$\begin{aligned} V_2(h_{L2} - h_{L1} + \lambda_2) + V_3(h_{L2} - H_3) = L_f\{(h_{L2} - h_{L1}) + (1 - K)(h_{L2} - h_f)\} \\ + L_f C_p(T - T_1) \end{aligned} \tag{113}$$

In similar way, for 3rd to 7th effect the mathematical model are derived and represented in matrix form which is given by Eq. (114)

$$\begin{bmatrix} \lambda_1 & (h_{L1} - H_2) & 0 & 0 & 0 & 0 & 0 & 0 \\ 0 & (\lambda_2 + h_{L2} - h_{L1}) & (h_{L2} - H_3) & 0 & 0 & 0 & 0 & 0 \\ 0 & 0 & \lambda_3 & (h_{L2} - H_4) & (h_{L2} - h_{L3}) & (h_{L2} - h_{L3}) & (h_{L2} - h_{L3}) & (h_{L2} - h_{L3}) \\ 0 & 0 & 0 & \lambda_4 & (h_{L3} - H_5) & (h_{L3} - h_{L4}) & (h_{L3} - h_{L4}) & (h_{L3} - h_{L4}) \\ 0 & 0 & 0 & 0 & \lambda_5 & (h_{L4} - H_6) & (h_{L4} - h_{L5}) & (h_{L4} - h_{L5}) \\ 0 & 0 & 0 & 0 & 0 & \lambda_6 & (h_{L5} - H_7) & (h_{L5} - h_{L6}) \\ 0 & 0 & 0 & 0 & 0 & 0 & \lambda_7 & (h_{L6} - H_8) \\ 0 & 1 & 1 & 1 & 1 & 1 & 1 & 1 \end{bmatrix}$$

$$\begin{bmatrix} V_1 \\ V_2 \\ V_3 \\ V_4 \\ V_5 \\ V_6 \\ V_7 \\ V_8 \end{bmatrix} = \begin{bmatrix} L_f(h_{L1} - Kh_f) \\ L_f\{(h_{L2} - h_{L1}) + (1 - K)(h_{L2} - h_f)\} + L_f C_p \Delta T_2 \\ L_f(x_f/x_7)(h_{L3} - h_{L2}) + L_f C_p \Delta T_1 \\ L_f(x_f/x_7)(h_{L4} - h_{L3}) \\ L_f(x_f/x_7)(h_{L5} - h_{L4}) \\ L_f(x_f/x_7)(h_{L6} - h_{L5}) \\ L_f(x_f/x_7)(h_{L7} - h_{L6}) \\ L_f\{1 - (x_f/x_7)\} \end{bmatrix} \tag{114}$$

2.7.3 Mixed Feed Configuration Coupled with Steam Split, Feed Split and Pre-heater: Model-O

In this configuration, live steam V_1 is split among 1st and 2nd effects and the vapor produced at these two effects, i.e., V_2 and V_3 are combined together and sent to 3rd effect at average temperature of $T_{avg} = \frac{T_{V_2} + T_{V_3}}{2}$ and the vapor solution produced at 3rd effect is used further in 4th effect and so on but at 6th effect, some part of solution vapor, V_7 obtained is sent to 7th effect, i.e., $(V_7 - S_7)$ and S_7 remains is sent to the 2nd pre-heater, while of solution vapor produced at last effect, i.e., 7th effect, V_8 is sent to 1st pre-heater as a heat source. The un-concentrated solution black liquor temperature increases as $\Delta T_1 = (T_0 - T)$ in 1st pre-heater and its temperature increases $\Delta T_2 = (T - T_1)$ in 2nd pre-heater. Finally, the un-concentrated solution coming from 2nd pre-heater is split among 5th and 6th effects with fraction K and $(1 - K)$, as illustrate in Fig. 18.

Writing the energy balance at each effect and rearranging these equation and finally, express the mathematical model in matrix form as is given by Eq. (115).

$$\begin{bmatrix} Y\lambda_1 & (h_{L2} - H_2) & & 0 & 0 & 0 & 0 & 0 \\ (1-Y)\lambda_1 & (h_{L3} - h_{L2}) & (h_{L3} - H_3) & 0 & 0 & 0 & 0 & 0 \\ 0 & (\lambda_{avg} + h_{L4} - h_{L3}) & (\lambda_{avg} + h_{L4} - h_{L3}) & (h_{L4} - H_4) & 0 & 0 & 0 & 0 \\ 0 & (h_{L7} - h_{L4}) & (h_{L7} - h_{L4}) & (\lambda_4 + h_{L7} - h_{L4}) & (h_{L7} - H_5) & (h_{L7} - h_{L5}) & 0 & 0 \\ 0 & 0 & 0 & 0 & \lambda_5 & (h_{L5} - H_6) & 0 & 0 \\ 0 & 0 & 0 & 0 & 0 & \lambda_6 & (h_{L6} - H_7) & 0 \\ 0 & 0 & 0 & 0 & 0 & 0 & (\lambda_7 + h_{L7} - h_{L6}) & (h_{L7} - H_8) \\ 0 & 1 & 1 & 1 & 1 & 1 & 1 & 1 \end{bmatrix} \begin{bmatrix} V_1 \\ V_2 \\ V_3 \\ V_4 \\ V_5 \\ V_6 \\ V_7 \\ V_8 \end{bmatrix} = \begin{bmatrix} \left(\frac{L_f x_f}{x_1}\right)(h_{L1} - h_{L2}) \\ \left(\frac{L_f x_f}{x_1}\right)(h_{L2} - h_{L3}) \\ \left(\frac{L_f x_f}{x_1}\right)(h_{L3} - h_{L4}) \\ \left(\frac{L_f x_f}{x_1}\right)(h_{L4} - h_{L7}) + \{(1-K)L_f(h_{L7} - h_{L5})\} \\ (1-K)L_f(h_{L5} - h_f) \\ KL_f(h_{L6} - h_f) \\ KL_f(h_{L7} - h_{L6}) + L_f C_p \Delta T_1 \\ L_f\left(1 - \frac{x_f}{x_1}\right) \end{bmatrix} \quad (115)$$

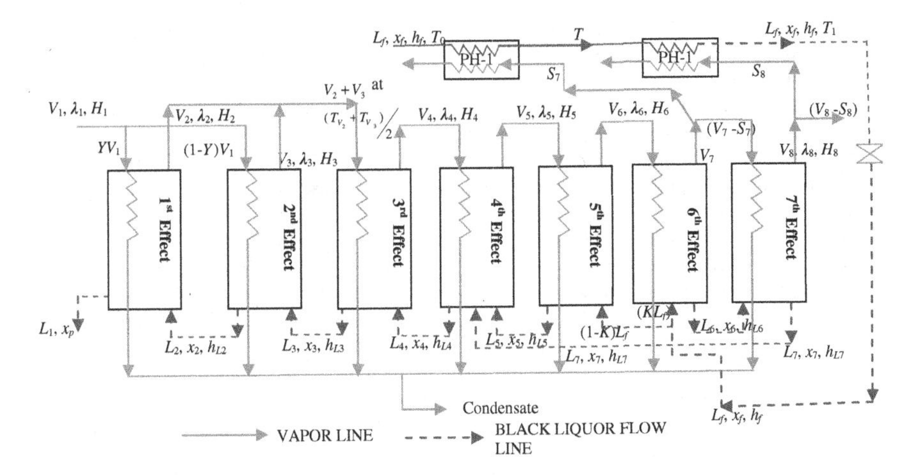

Fig. 18 Mixed feed coupled with steam split, feed split and pre-heater heptads effect evaporator system

3 Result and Discussions

In the present study, different types of mathematical models Model-A to O has been developed and proposed for the analysis of steam economy and consumption in order to reduce energy in Kraft recovery process used in paper industries. The proposed model helps to find the numerical solution easily for optimization of the energy economically and efficiently. There are three important parameters defined to measure the performance of the Pulp and Paper industries are (i) *Steam Economy* (SE) (ii) *Vaporization Capacity* (VC) and (iii) *Steam Consumption* (SC). The models derived above are simple, linear and most important, representation of equations in matrix form. Due to representation of equations in matrix form, a number of optimization technique or the solution technique to solve these developed models can be applied with the iteration methods. It makes the model less sensitive toward its initial values. Since, the models are linear and less sensitive to initial values, the convergence rate will be very high and solution will be stable.

References

1. Aly, N.H., Marwan, M.A.: Dynamic response of multi-effect evaporators. Desalination **114**(2), 189–196 (1997)
2. Bhargava, R., Khanam, S., Mohanty, B., Ray, A.K.: Simulation of flat falling film evaporator system for concentration of black liquor. Comput. Chem. Eng. **32**(12), 3213–3223 (2008)
3. Higa, M., Freitas, A.J., Bannwart, A.C., Zemp, R.J.: Thermal integration of multiple effect evaporator in sugar plant. Appl. Therm. Eng. **29**(2), 515–522 (2009)

4. Holland, C.D.: Fundamentals and Modelling of Separation Processes. Prentice Hall Inc., Englewood cliffs, New Jersey (1975)
5. Kaya, D., Sarac, H.I.: Mathematical modeling of multiple-effect evaporators and energy economy. Energy **32**(8), 1536–1542 (2007)
6. Kern, D.Q.: Process heat transfer. Tata McGraw-Hill Education (1950)
7. Khademi, M.H., Rahimpour, M.R., Jahanmiri, A.: Simulation and optimization of a six-effect evaporator in a desalination process. Chem. Eng. Process. **48**(1), 339–347 (2009)
8. Khanam, S., Mohanty, B.: Energy reduction schemes for multiple effect evaporator systems. Appl. Energy **87**(4), 1102–1111 (2010)
9. Kumar, D., Kumar, V., Singh, V.P.: Modeling and dynamic simulation of mixed feed multi-effect evaporators in paper industry. Appl. Math. Model. **37**(1), 384–397 (2013)
10. Lambert, R.N., Joye, D.D., Koko, F.W.: Design calculations for multiple-effect evaporators. 1. Linear method. Ind. Eng. Chem. Res. **26**(1), 100–104 (1987)
11. Miranda, V., Simpson, R.: Modelling and simulation of an industrial multiple effect evaporator: tomato concentrate. J. Food Eng. **66**(2), 203–210 (2005)
12. Nishitani, H., Kunugita, E.: The optimal flow-pattern of multiple effect evaporator systems. Comput. Chem. Eng. **3**(1), 261–268 (1979)
13. Ruan, Q., Jiang, H., Nian, M., Yan, Z.: Mathematical modeling and simulation of countercurrent multiple effect evaporation for fruit juice concentration. J. Food Eng. **146**, 243–251 (2015)
14. Sagharichiha, M., Jafarian, A., Asgari, M., Kouhikamali, R.: Simulation of a forward feed multiple effect desalination plant with vertical tube evaporators. Chem. Eng. Process. **75**, 110–118 (2014)
15. Zain, O.S., Kumar, S.: Simulation of a multiple effect evaporator for concentrating caustic soda solution-computational aspects. J. Chem. Eng. Jpn. **29**(5), 889–893 (1996)

Author Index

M. Pant et al. (eds.), *Proceedings of Fifth International Conference on Soft Computing for Problem Solving*, Advances in Intelligent Systems and Computing 437, DOI 10.1007/978-981-10-0451-3

Printed in the United States
By Bookmasters